STATICS

AND

MECHANICS OF MATERIALS

An Integrated Treatment

JAMES W. DALLY

ROBERT J. BONENBERGER, JR.

University of Maryland, College Park

College House Enterprises, LLC
Knoxville, Tennessee

The manuscript was prepared with 11 point Times New Roman font. The Word files were then converted to pdf files using Adobe Acrobat DC Pro.

College House Enterprises, LLC.
5713 Glen Cove Drive
Knoxville, TN 37919, U. S. A.
Phone (865) 558 6111
FAX (865) 558 6111
email jwd@www.collegehousebooks.com
Visit our web site http://www.collegehousebooks.com

ISBN 978-1-935673-35–4

This book is for Anne
who has shared Jim's dreams for so many years

and to the memory of Ann and Bob
who had encouraged and supported Bob until they were called.

ABOUT THE AUTHORS

James W. Dally obtained a Bachelor of Science and a Master of Science degree, both in Mechanical Engineering from the Carnegie Institute of Technology. He obtained a Doctoral degree in mechanics from the Illinois Institute of Technology. He has taught at Cornell University, Illinois Institute of Technology, the U. S. Air Force Academy and served as Dean of Engineering at the University of Rhode Island. He is currently a Glenn L. Martin Institute Professor of Engineering (Emeritus) at the University of Maryland, College Park.

Dr. Dally has also held positions at the Mesta Machine Co., IIT Research Institute and IBM. He is a fellow of the American Society for Mechanical Engineers, Society for Experimental Mechanics, and the American Academy of Mechanics. He was appointed as an honorary member of the Society for Experimental Mechanics in 1983 and elected to the National Academy of Engineering in 1984. Professor Dally was selected by his peers to receive the Senior Faculty Outstanding Teaching Award in the College of Engineering and the Distinguish Scholar Teacher Award from the University. He was also a member of the University of Maryland team receiving the 1996 Outstanding Educator Award sponsored by the Boeing Co.

Professor Dally has co-authored several other books: *Experimental Stress Analysis, Experimental Solid Mechanics, Photoelastic Coatings, Instrumentation for Engineering Measurements, Packaging of Electronic Systems, Production Engineering and Manufacturing, and Introduction to Engineering Design, Books 1, 2, 3, 4, 5, 6, 7, 8, 9, 10 and 11.* He has authored or coauthored about 200 scientific papers and holds five patents.

Robert J. Bonenberger, Jr. obtained B.S.E., M.S. and Ph.D. degrees in Mechanical Engineering from the University of Maryland, Baltimore County and College Park campuses. He has taught undergraduate and graduate students in mechanics, strength of materials, and experimental stress analysis. Currently, he is an Assistant Research Scientist and Lecturer in the Clark School of Engineering and a member of the Keystone Program at the University of Maryland, College Park.

Previously, Dr. Bonenberger worked in the Fracture Mechanics Section at the Naval Research Laboratory, both as a postdoctoral fellow and as a contract employee. He is a member of the American Society for Mechanical Engineers, Society for Experimental Mechanics, American Society for Materials, and American Society for Engineering Education. His research interests include material behavior at high strain rates, experimental stress analysis, and fracture mechanics. He has authored or co-authored 20 scientific papers.

PREFACE

This textbook has been prepared to support an integrated course offering for Statics and Mechanics of Materials. Hopefully the integration of the topics of these two closely related courses will make the learning experience for the students easier and more meaningful. Statics provides the first exposure of engineering students to the study of mechanics. While Statics is a relatively simple subject, many students find it difficult, and they often perform far below our expectations. In an effort to improve the curriculum, several members of the faculty at the University of Maryland have been working to enhance the student's learning experience when studying the first two courses in mechanics. This textbook indicates some of the changes in the philosophy adopted by the faculty when presenting the subject matter traditionally offered in introductory mechanics courses.

The changes in the philosophy were based on five premises:

1. Present the fundamental concepts in a more interesting manner.
    - Change the approach to make it more realistic and less abstract.
    - Couple the mechanics content tightly to design.
2. Provide a smooth transition from Introduction to Engineering Design to an integrated treatment that encompasses both Statics and Mechanics of Materials.
    - Provide the analysis methods and the scaling relations for verifying the safety of the design of this model.
3. Emphasize modeling structural components by stressing throughout the text the importance of preparing a complete free body diagram (FBD).
    - Show the method for constructing complete FBDs.
    - Integrate the FBD with the application of the equilibrium equations.
    - Approach the solution of equilibrium problems with equilibrium relations based on force and moment components.
    - Introduce vectors in conjunction with forces and moments, but only employ vector analysis after both FBDs and equilibrium concepts have been firmly established.
4. Integrate the content of Statics with that of Mechanics of Materials.
    - The concepts of normal and shearing stresses are introduced early in Statics.
    - Topics associated with Statics and with Mechanics of Materials are merged in Chapters 4 through 9.
    - Chapters 10 though 15 describe topics typically found in Mechanics of Materials but with a heavy emphasis on design.
    - Fracture mechanics, a topic new to Mechanics of Materials courses, is introduced in Chapter 16.
5. Emphasize the design of structural components for safety.
    - Stress is compared to strength to give safety factors for components.
    - Behavior of engineering materials is introduced and physical properties such as strength and modulus of elasticity are described.
    - Sizing of structural components for safety and cost are demonstrated.

This textbook includes sufficient material for students to study both Statics and Mechanics of Materials. We began developing notes for this book with a pilot offering in the spring semester of 1999. Many revisions were made before a limited first edition of the textbook was published in the summer of 1999. The second edition was published in 2000 and was used by about a thousand students. The third edition, written in 2001, included six new chapters to expand the coverage necessary for a complete Mechanics of Materials course. This new edition includes homework problems following each of its 16 chapters, with a sufficient number of problems for student assignments for a two semester course. Also the errors discovered during the extensive usage have been corrected; however, errors always occur even with careful proof reading by many diligent people. We would greatly appreciate students and instructors calling any errors they find to our attention. The e-mail address of one of the authors is given on the copyright page.

ACKNOWLEDGEMENTS

As is always the case when major changes are made to the curriculum, the College administrators must lead the way. We are indeed fortunate to have several administrators who not only supported this effort, but also insisted that the mechanics offerings be markedly improved. They seek not only a much more favorable educational experience for the students, but also much better understanding and retention of the course content by the students. We thank Dr. William L. Fourney, Chair of the Aerospace and Aeronautics Department, Dr. Avram Bar-Cohen, Chair of the Mechanical Engineering Department, and Dr. Gregory Baecher, Chair of the Civil Engineering Department for their support.

Special thanks are due to several individuals for their significant contributions to the development of this textbook. Dr. Bill Fourney led our efforts to change the curriculum to provide a more effective approach for presenting Statics and Mechanics of Materials so students would better understand the material and retain this knowledge. Bill has stayed with the project, taught sections of students every semester, and provided the leadership needed to keep others involved and interested. Dr. Hugh Bruck also made major contributions. He also taught an early pilot section and made many excellent suggestions for changes in the sequence of the content. Dr. Gary Pertmer, Assistant Dean, has assumed the leadership role in organizing the many sections of this course taught each year. William Rudd, former Dean of the School of Architecture at the University of Tennessee, was extremely helpful in providing several excellent photographs of bridges that were used as illustrations in Appendix F.

Many instructors teaching the course made valuable suggestions for improvements to the textbook. These include: Dr. Mary Bowden, Dr. Hugh Bruck, Dr. James Duncan, Dr. Bongtae Han, Dr. Kwan-Nan Yeh, Mr. Christopher Baldwin, and Mr. Thomas Beigel of the University of Maryland, College Park; Dr. Abhijit Nagchaudhuri of the University of Maryland, Eastern Shore; and Dr. Asif Shakur, of Salisbury State University.

James W. Dally
Robert J. Bonenberger
College Park, MD

April, 2017

CONTENTS

CHAPTER 9 FRICTION

CHAPTER 10 STRESSES IN BEAMS

CHAPTER 11 TORSION OF STRUCTURAL ELEMENTS

CHAPTER 12 STRESS STATES AND TRANSFORMATIONS

CHAPTER 13 DEFLECTION OF BEAMS

CHAPTER 14 BUCKLING OF COLUMNS

CHAPTER 15 ENERGY METHODS

CHAPTER 16 FRACTURE MECHANICS

APPENDICES

LIST OF SYMBOLS

A	area
%A	percent reduction in area
a	acceleration vector
a, b, c, ….	dimensions
a_{CR}	critical crack length
C	constant, center dimension
D	diameter
d	diameter or distance
%e	percent elongation
E	elastic modulus or Young's modulus
E	strain energy
E	modulus scale factor
e	eccentricity
e	strain energy density
F	force magnitude
F	force as a vector
F_f	friction force or failure force
FBD	free body diagram
g	gravitational constant
G	universal gravitational constant
G	shear modulus
GF	gage factor
H	horizontal shear force
HP	horsepower
h	height
i, j, k	unit vectors
I	moment of inertia
J	polar moment of inertia
K	stress concentration factor
K_I	stress intensity factor
K_{IC}	plane strain fracture toughness
K_c	fracture toughness
k	number, spring rate
L	length dimension
L	load scale factor
ln	natural logarithm
M	moment magnitude
M_Y	yield moment
M_L	limit moment
M	moment as a vector
M	multiplier
MA	mechanical advantage
MOS	margin of safety
m	mass, subscript for model
N	normal force or number
n	number
P	internal force
P_{CR}	critical buckling force
p	pressure, subscript for prototype
Q	first moment of the area
q	distributed loading
r	radius, radius of gyration, distance

r	position vector
R	reaction force, radius, resistance
S	vector sum
S	geometric scale factor
s	distance or dimension
S_{design}	design strength
S_f	fatigue strength
S_y	yield strength
S_{ys}	yield strength in shear
S_u	ultimate tensile strength
S_{us}	ultimate tensile strength in shear
SF	safety factor
SE	structural effectiveness
SR	slenderness ratio
T	temperature, torque
t	time
u	unit vector
V	shear force
v	velocity vector
V	volume
W	weight, watt
W	work
w	width dimension
x, y z	Cartesian coordinates
Z	section modulus
α	coefficient of thermal expansion
α, β, γ	direction cosines
Δ	delta
δ	deflection or displacement
ε	strain
ε_T	true strain
ϕ	angle of friction, angle of twist
γ	shear strain
κ	curvature
π	3.1416 radians
μ	coefficient of friction
ν	Poisson's ratio
ρ	radius of curvature
Σ	Summation sign
σ	stress
σ_a	alternating cyclic stress
σ_{design}	design stress
σ_f	failure stress
σ_m	mean cyclic stress
σ_T	true stress
σ_{fail}	stress to produce failure
σ_{yield}	stress to produce yielding
θ	angle
θ_s	angle of repose
τ	shear stress
ω	angular velocity

CHAPTER 1

BASIC CONCEPTS IN MECHANICS

1.1 INTRODUCTION

The subject of mechanics is usually divided into four different courses, which include:

1. Statics
2. Dynamics
3. Mechanics of Materials
4. Fluid Mechanics

Statics and dynamics both deal with rigid bodies that are subjected to a system of forces. In the traditional study of statics, we are concerned with determining either internal and/or external forces acting on a structural element that is in a state of equilibrium (usually at rest). In dynamics, the forces acting on the body produce motion and the body accelerates or decelerates. The analysis in dynamics deals with determining position, velocity (angular or linear) and acceleration as some function of time. Newton's laws guide our study of mechanics[1]. Consider Newton's second law:

$$\sum \mathbf{F} = \frac{d}{dt}(m\mathbf{v}) \tag{1.1}$$

where $\Sigma\mathbf{F}$ is the sum of all of the forces acting on the rigid body
m is the mass of the body
v is the velocity
d/dt is the derivative operator

In dividing the study of mechanics into its four subjects, scholars have considered the special situation where the velocity is constant (often zero) and developed **statics** based on this simplification of Newton's second law. In this special situation:

$$\sum \mathbf{F} = 0 \tag{1.2}$$

Of course, we study motion in dynamics—the velocity of the rigid body is changing ($\mathbf{v} \neq 0$) and the general form of Eq. (1.1) applies. However, in most situations the mass of the rigid body is constant (dm/dt = 0) and if this is the case, Eq. (1.1) reduces to:

$$\sum \mathbf{F} = m\frac{\mathbf{dv}}{\mathbf{dt}} = m\mathbf{a} \tag{1.3}$$

where $\mathbf{a} = d\mathbf{v}/dt$ is the acceleration of the rigid body.

In the study of both statics and dynamics, the material from which the body is manufactured is of no concern providing the body remains essentially rigid under the action of the imposed forces.

---

[1] Sir Isaac Newton (1642-1727) formulated three laws of motion and the law of universal gravitational attraction.

However, in the course mechanics of materials[2], the deformation of the body is an essential consideration in the analysis. In this course, small deformations of the body are assumed (plane sections remain plane). This assumption enables us to determine the distribution of internal forces and stresses in the body. The material from which the body is fabricated is of critical importance in mechanics of materials for two reasons. First, the deformations of the body due to the forces are markedly affected by the rigidity of the material (its elastic modulus). Second, the behavior of the body—whether it fails or not—depends on its strength. Of course, strength is a physical property of the material.

In fluid mechanics the situation is entirely different. The body is either a gas or a liquid. The deformations are so large as to be considered as flows. The flow can be compressible (gasses under higher pressures) or incompressible (liquids or gasses at low pressures). The flow can occur in closed channels or open channels. It may be internal to a conduit or external to some surface. The flow may be stable (laminar) or unstable (turbulent). The phase of the material may change during a process and the resulting flow consists of two-phases (some liquid and some gas). Because of the complexities inherent in fluid mechanics, it is studied after a student has established a thorough understanding of the other three branches of mechanics.

Before beginning the design analysis of any component, vehicle, structure or even a model of a structure, it is essential to thoroughly understand several basic concepts in mechanics and the behavior of materials. The first two principal subject areas in mechanics of materials are **Statics** and **Mechanics of Materials**. In this chapter, we introduce several of the basic concepts and physical laws included in these two subjects. These concepts and/or laws provide a foundation upon which the design analysis of machine components and structures is based. We also briefly describe the history of mechanics to give you a sense of the age of the principles used in analyzing modern machine elements and structural components.

1.2 STATICS AND MECHANICS OF MATERIALS

This textbook integrates two closely related subjects in mechanics, namely **statics** and **mechanics of materials**. In studying **statics**, we assume the body under consideration is perfectly rigid. As such it does not deform under the action of applied forces. You will solve many different types of problems determining **external forces** acting on some structures and the **internal forces** developed in others by using only **free body diagrams** and the **equilibrium equations**. The equilibrium equations are developed from the laws of motion proposed by Sir Isaac Newton to describe the balance of forces acting on a stationary body or on a body moving with a constant velocity. Solutions for forces acting on bodies in equilibrium employ only three basic steps:

- Construct a complete set of free body diagrams.
- Apply the appropriate equations of equilibrium.
- Execute the mathematics required to solve one or more equations of equilibrium.

Statics is a relatively easy subject that you can quickly master. You will find it easy to solve many different problems that are approached with the procedure outlined above.

Mechanics of materials is a relatively simple extension of statics, which accounts for the effect that **material deformations**[3] have on the **internal stresses** generated in a body by applied forces. To solve the problems that arise in mechanics of materials, we begin with the same three steps, and then add two more. The additional steps are to accommodate the effect of the deformations when the body is loaded:

- Construct a complete set of free body diagrams.
- Apply the appropriate equations of equilibrium.
- Assume the geometry of the deformations (usually plane sections remain plane).
- Employ the appropriate relations between stress and strain.
- Execute the mathematics required in solving the equations.

---

[2] The course mechanics of materials is also known as mechanics of deformable bodies or strength of materials.

[3] In mechanics of materials, we assume the deformations are so small that they do not significantly affect the magnitude or the direction of the internal and external forces acting on the body.

By comparing these lists, it is evident that statics and mechanics of materials are closely related. Indeed, we must use the equilibrium equations in statics before we can begin to solve typical problems in mechanics of materials. However, the solutions to statics problems are obtained on a global scale, whereas the solutions to mechanics of materials problems are obtained on a local scale.

In studying mechanics of materials, we quickly encounter the concepts of stress and strain. Mathematically these concepts are somewhat complex because they are tensor quantities, but physically they are simple and easily understood. Stress is a concept based on the equilibrium of a portion of a body. Consider a part of a body produced by sectioning. The internal force acting on the section cut is developed by some distribution of stresses over the area exposed by the cut.

Strain, on the other hand, is a geometric concept. We determine strain by the change in geometry that occurs when a body deforms under load. When considered individually, both stress and strain are independent of the material from which the body is made. It is only when we write a relation for strain in terms of stress, or vice versa, that we must consider the properties of the material used to fabricate the body.

Materials used in constructing various structures are largely ignored in the study of statics. The equilibrium equations are the same for all materials, and the internal and external forces for statically determinant structures[4] do not depend upon the materials employed to fabricate the structure. Materials are much more important in studies of the mechanics of materials. We usually relate stress and strain in our solutions, and must employ the elastic constants that describe the rigidity of the materials. In addition, we may be required to predict the margin of safety for a structure subjected to specific loading. In the solution to problems dealing with structural failure, we employ the appropriate **"strength"** of the material.

In this textbook, we stress the physical aspects of both statics and mechanics of materials. We consider it essential that you construct complete free body diagrams to model the structure and to define the unknown forces. We are more interested in your understanding of the equations of equilibrium than your use of **vector algebra** to solve them. Of course, it is important that you be able to correctly manipulate the equations resulting from the application of the principle of equilibrium. Whether these equations are written using **trigonometric** functions or in **vector** format is of lesser concern providing they are correct.

The general subject of mechanics and materials is both interesting and important. Let's try to enjoy the experience of learning this fundamental subject. In writing this textbook, we have used many examples to:

1. Illustrate drawing free body diagrams.
2. Show the correct procedures for applying the equilibrium equations.
3. Show the geometric changes due to deformations of the body under load.
4. Illustrate distributions of stress over areas exposed by section cuts.
5. Demonstrate the procedures and techniques involved in solving different types of problems.

1.3 HISTORY OF MECHANICS

As a student of the first course in mechanics, you are probably in your late teens or early twenties. It might be hard for you to imagine that mechanics has been under development for more than 2000 years. Early pioneers in mechanics, about two or three hundred years ago, understood most of the material included in this book. Although research is still conducted in mechanics, this first course covers the classical content that is considered the foundation for mechanics. To illustrate that mechanics is a classical subject, let's note the contributions of a few of the pioneers who established the foundations of mechanics. For a much more extended treatment of the history of mechanics see references [1, 2].

One of the earliest contributors to our knowledge of mechanics was Archimedes (287 – 212 BC) who discovered that we could use a lever to increase the weight that a person could lift. The concept of the lever with its

---

[4] Structures are classified as statically determinant if the reactions at their supports may be determined using only the equations of equilibrium.

fulcrum used in lifting a weight W with a force W/4 is illustrated in Fig. 1.1. Archimedes also described the use of pulleys and inclined planes for moving materials at construction sites. On a different application of engineering in ancient times, Archimedes is credited with discovering the law of buoyancy where the upward force on a submerged body is equal to the weight of the water displaced by that body.

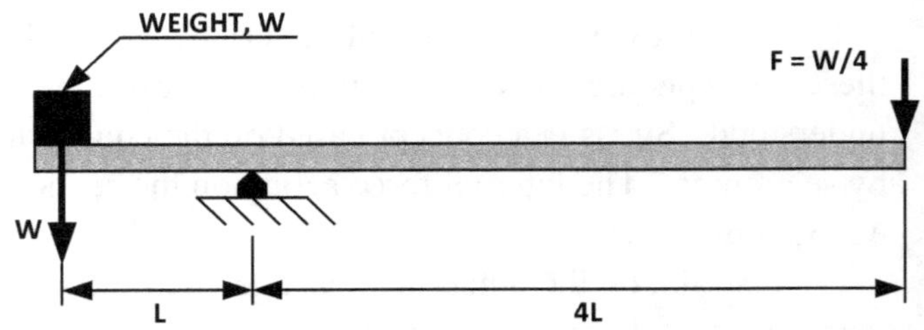

Fig. 1.1 The lever with a fulcrum permits an applied force to be increased with the lever ratio.

Leonardo da Vinci (1452 – 1519) was a man with incredible talent and one of the great Renaissance masters. He was a painter, sculptor, musician, architect and engineer. At the age of about 30, he served as the principal engineer for the Duke of Milan supervising the construction of bridges and war machines for the Duke's military ventures [3]. Leonardo is not known for his mathematical discoveries, but rather for his engineering innovations that implied his thorough understanding of mechanics.

The 17th century was the height of the Renaissance period for science and probably the most productive for developing the foundations of mechanics. During this century, Robert Hooke (1635 – 1703) introduced the concept of elasticity by noting that a body deformed in proportion to the applied forces. This concept leads to the well-known Hooke's law:

$$\sigma = E\,\varepsilon \tag{1.4}$$

where σ is the stress, ε is the strain, and E is the material property known as the modulus of elasticity.

The linearity of Hooke's law is illustrated in Fig. 1.2. Robert Hooke also developed springs and used them in watches and clocks to improve their accuracy and to store energy. He discussed planetary motion, but did not develop the mathematics necessary to describe the observed motion of the planets. It was Sir Isaac Newton who extended Hooke's early ideas about planetary motion and wrote the gravitational law that governs the motion of the planets.

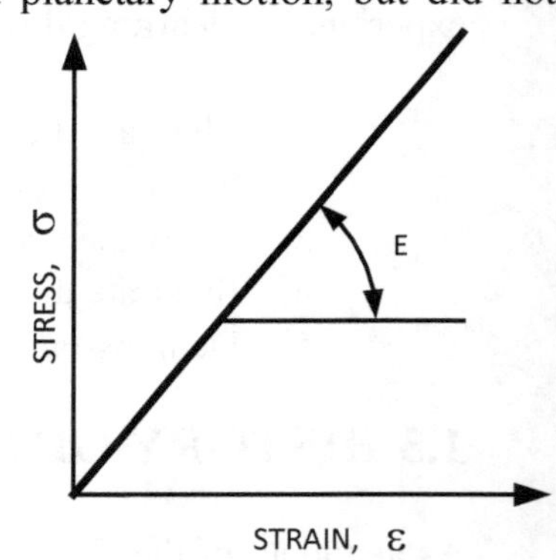

Fig. 1.2 The linear elastic relation between stress and strain.

Sir Isaac Newton (1642 – 1727) was a giant in both mathematics and mechanics. He was only 24 years old when he developed the mathematics known today as differential calculus. Later Newton turned his attention to planetary motion, and formulated his famous laws of motion. We will cover these three laws in more detail in the next section. However, these laws are the foundation of much of the content contained in mechanics courses on statics and dynamics. He is best known for his universal law of gravity [4] that explains the attractive forces between two bodies. The gravitational force is an internal force because it acts on each elemental volume of mass within a body. The attractive forces **F** between two masses, illustrated in Fig. 1.3, are given by:

$$F = G\, m_a\, m_b / r^2 \tag{1.5}$$

where F is the magnitude of the gravitational force.

$G = 6.673 \times 10^{-11}$ $m^3/(kg - s^2)$ is the universal gravitational constant.

m_a and m_b are the masses of bodies A and B.

r is the distance between the centers of the two bodies.

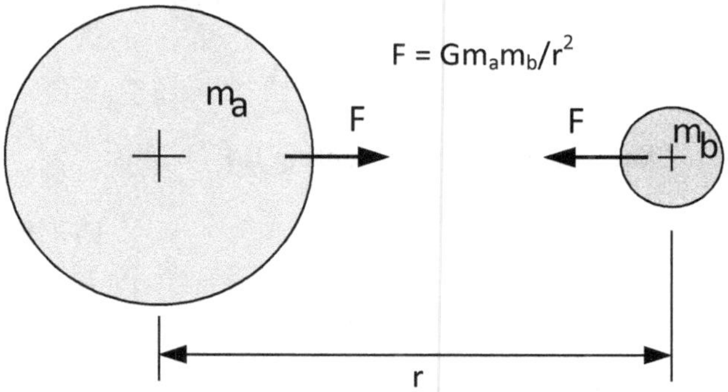

Fig. 1.3 Illustration of the attractive forces on both bodies developed by gravity.

Because Newton's law of gravity is extremely important, we will describe it in more detail later in this Chapter.

Jacob Bernoulli[5] (1654 – 1705) made two contributions to mechanics of materials; one was correct and the other was in error. We cite his contributions to indicate that even the great and famous make mistakes from time to time. Jacob was interested in determining the elastic curve representing the deflection of a beam. He correctly showed that the curvature of the beam was proportional to the bending moment at each point along the length of the beam. He then made the error of assuming that the elastic curve was positioned along the bottom edge of the beam. We will establish in a later chapter that the elastic curve must coincide with the neutral axis of the beam.

The history of mechanics is rich with accomplishments of many mathematicians. As the subject evolved, engineers designing safe efficient structures and machines such as airplanes, automobiles, bridges, machines, skyscrapers, and space satellites reduced the mathematical formulations to practice.

1.4 NEWTON'S LAWS OF MOTION

Sir Isaac Newton wrote three laws of motion that form the foundation for both statics and dynamics. They are:

(1) If the sum of all of the forces acting on a body is zero (i. e. $\Sigma \mathbf{F} = 0$), the body will:

   a. Remain at rest.
   b. Move at a constant velocity.

(2) If the sum of all of the forces **F** acting on a body is not zero, the body will undergo a time rate of change of the linear momentum (m**v**) given by:

$$\sum \mathbf{F} = \frac{d}{dt}(m\mathbf{v}) \qquad (1.1)$$

When the mass m of the body remains constant with respect to time dm/dt = 0, and Eq. (1.1) reduces to:

$$\sum \mathbf{F} = m\frac{\mathbf{dv}}{\mathbf{dt}} = m\mathbf{a} \qquad (1.3)$$

[5] There were three famous Bernoulli's, namely Jacob, John and Daniel. John, the younger brother of Jacob, was a splendid mathematician. Daniel, the son of John, was famous because of his work in fluid mechanics, his outstanding book Hydrodynamica, and for his suggestion to Euler for deriving the equations for elastic curves.

(3) The force exerted by body A on body B is equal in magnitude but opposite in direction to the force that body B exerts on body A. This third law is sometimes called the law of action and reaction.

1.4.1 Newton's First Law

The first of Newton's laws is written as:

$$\Sigma \mathbf{F} = 0 \qquad (1.2)$$

Another way of representing Eq. (1.2) is:

$$\mathbf{F}_1 + \mathbf{F}_2 + \ldots\ldots + \mathbf{F}_n = 0 \qquad (1.2a)$$

where n forces are acting on the body.

Equation (1.2), in one form or the other, is used extensively in both statics and mechanics of materials. It is used every time that we write the **equilibrium** equations. To illustrate the meaning of the mathematical symbol **ΣF**, let's examine the drawing of the potato like body presented in Fig. 1.4. We have four forces acting on this body. They are pointed in four different directions and have four different magnitudes. The forces are **vector** quantities that must be characterized by specifying both a magnitude and a direction[6]. When the **vector sum** of these forces is zero, the body is in equilibrium.

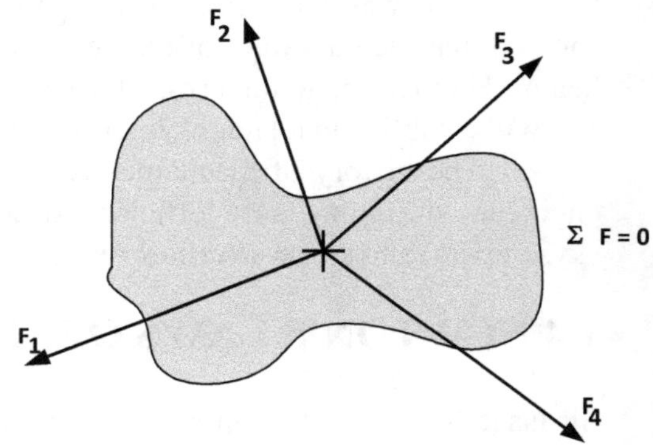

Fig. 1.4 Four forces acting on a body that is in equilibrium if $\Sigma \mathbf{F} = 0$.

We may recast the vector representation of Eq. (1.2) by writing the equivalent **scalar equations** as:

$$\Sigma F_x = 0 \quad \text{and} \quad \Sigma F_y = 0 \qquad (1.2b)$$

$$F_{1x} + F_{2x} + \ldots\ldots + F_{nx} = 0 \quad \text{and} \quad F_{1y} + F_{2y} + \ldots\ldots + F_{ny} = 0 \qquad (1.2c)$$

We have accounted for the direction of the forces in this equation by considering only those forces in either the x or y directions. When the forces are constrained to the x and y directions, we employ the **scalar form of the equilibrium equations**. However, when the forces act in directions other than the x and y, the vectors must first be decomposed into their **components** acting in the x and y direction before they can be used in the scalar equilibrium equations.

Let's consider an example to demonstrate the application of the equilibrium equation. We will first consider only forces acting in either the x or y directions.

EXAMPLE 1.1

Consider the forces acting on the body shown in Fig. E1.1, and determine if it is in equilibrium.

---

[6] We use the bold font to represent forces as vector quantities when both magnitude and direction are to be specified. Sometimes we consider only the magnitude of the force, and in this case normal fonts are used to represent this scalar magnitude.

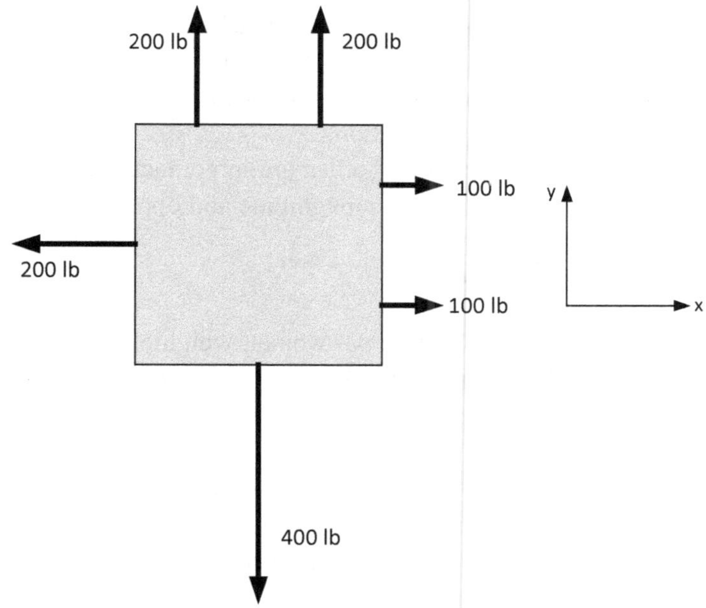

Fig. E1.1

Solution: We have inserted a x-y coordinate system in Fig. E1.1 as a reference to employ in summing the forces acting on the square block. We apply Eq. 1.2 to determine if this block is in equilibrium. When we write Σ **F** = 0, it is important to note that **F** is a **vector** quantity and the summation of the forces is the **vector form of the equilibrium equation**. Let's take this fact into account by considering first all of the forces in the x direction, and then all of the forces in the y direction. With this approach, we are accounting for both the directions and the magnitude of each of the six forces acting on the block. We write:

$$\Sigma F_x = 0 \quad \text{and} \quad \Sigma F_y = 0$$

$$+100 + 100 - 200 = 0 \quad \text{and} \quad +200 + 200 - 400 = 0$$

These results show that the equilibrium conditions are satisfied in both the x and y directions. The forces in the positive directions of both x and y cancel with those in the negative directions. Hence, the body is in equilibrium.

1.4.2 Newton's Second and Third Laws

Newton's second law Σ **F** = d/dt(m**v**) = m**a** is the equation used most frequently in the study of dynamics. When the sum of the forces is not zero, the body of mass m is subjected to an acceleration **a**. Depending on your choice of disciplines, you may pursue Newton's second law later in the curriculum.

The third law is often called the law of action and reaction. We illustrate the concept of active and reactive forces in Fig. 1.5.

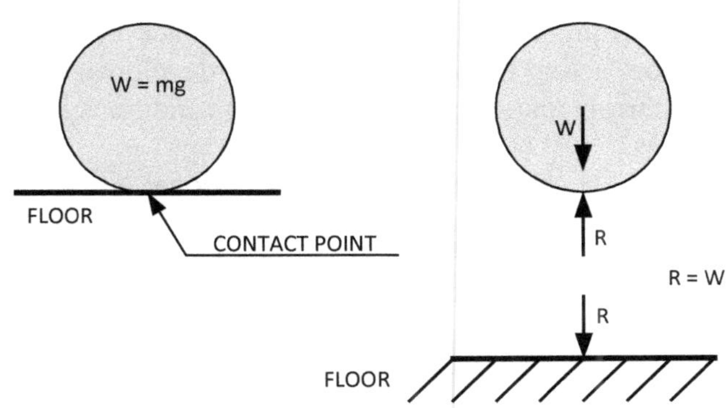

Fig. 1.5 The active force W due to gravity produces a reaction force R at the contact point.

In this illustration, a spherical shaped mass m with a weight W rests on the floor at a contact point. The sphere is in equilibrium under the action of two forces. The first force is the weight W due to gravity that acts downward. The second is the reaction force developed at the contact point. The reaction force R is equal in magnitude to the weight W, but opposite in direction.

When any two bodies are in contact (i.e. the sphere and the floor), two forces develop at the contact point. These forces are equal in magnitude and opposite in direction.

EXAMPLE 1.2

Suppose that a professional woman weighing 120 lb is approaching an elevator. Let's determine the reaction forces that develop between the floor and her feet.

Solution: We start by drawing a free body diagram of the woman. In the free body diagram shown in Fig. E 1.2, the woman is separated from the floor. She is suspended in space and maintained in equilibrium by vertical (y direction) forces applied to her feet.

At the two contact points with the floor, we have drawn arrows downward representing the forces applied to the floor (W/2). The reaction forces R = W/2 are drawn upward on each
foot. Clearly the applied and reaction forces are equal in magnitude and opposite in direction. Is the woman in equilibrium?

Fig. E 1.2 Free body diagram of a woman standing on a floor.

1.5 FORCES

Statics involves a study of forces that act on and within members of a structure. We seek to determine the external forces acting on bodies, and forces developed by stresses within structural members (internal forces). Some forces occurring under static (steady state) conditions include:

- Gravitational
- Pressure acting over a defined area
- Friction
- Magnetic
- Electrostatic

In addition, forces developed under dynamic conditions are referred to as inertial forces and include:

- Centrifugal
- Centripetal
- Coriolis

First, let's examine the forces due to gravity because they are by far the most important. We continuously work and expend huge amounts of energy to overcome gravitational forces. Gravitational forces are the primary concern when we design building and bridge structures against failure by collapse or rupture. Even in vehicle design where other dynamic forces are significant, gravitational forces are critical in the design of both the structure and the power train.

Weight is a force produced by the Earth's gravitational pull on the mass of our body as illustrated in Fig. 1.6. Suppose we examine the force on a body due to gravity by modifying Eq. 1.5, and letting:

$m_a = m_e$ the mass of the Earth.
$m_b = m_b$ the mass of our body.
$r = R_e$ the radius of the Earth.

Then we rewrite Eq. 1.5 to give:

$$F = G\, m_e\, m_b / R_e^2 \qquad (1.6)$$

In setting $r = R_e$, we assumed that Earth bound bodies, either those of people or objects, are very small compared to the radius of the Earth which is 3960 mi. or 6.37×10^6 m.

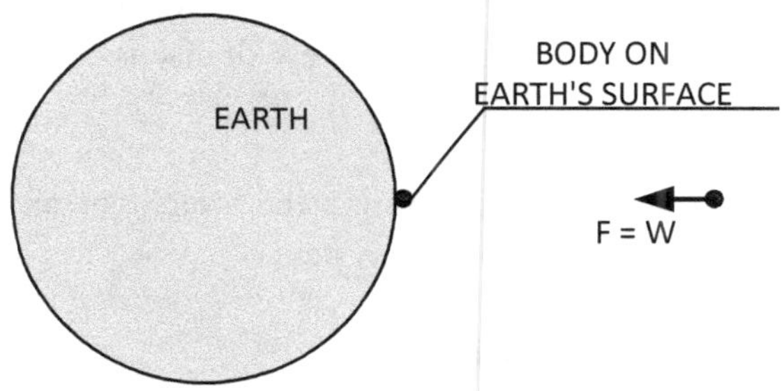

Fig. 1.6 Bodies on Earth's surface are small relative to the Earth's radius.

Next, we collect together all of the quantities in Eq. 1.6 that have anything to do with the Earth, and set them equal to g_e.

$$g_e = Gm_e / R_e^2 \qquad (1.7)$$

Note that g_e is the **gravitational constant** equal to 32.17 ft/s^2 or 9.807 m/s^2. We will drop the subscript in subsequent discussion of the gravitational constant with the understanding that g is to be applied to Earth bound bodies.

Strictly speaking g is not constant since it varies a bit as we move from one location to another. The Earth is not a perfect sphere, and R does not remain constant as we move from Pikes' Peak to Death Valley. However, the variations are so small that we neglect them without introducing significant error in our design analyses.

Combining Eqs. 1.6 and 1.7 gives:

$$F = m_b\, g = W \qquad (1.8)$$

The force F in Eq. 1.8 is the weight W of a body on Earth having a mass m_b.

From our definition, it clear that the units of g are ft/s^2 or m/s^2; hence, g is an acceleration. Indeed, if we jump off of a ladder, our body accelerates with a = g until we hit the ground. Clearly, this relationship is consistent with Newton's second law.

If we travel from Washington, D. C. to Denver, CO, we observe that our weight remains essentially the same. So we get confused and begin to think that the constant in Eq. 1.8 is our weight. It is a reasonable thought, but erroneous. When measuring weight, it is essentially constant if we remain Earth bound. However, the constant quantity in Eq. 1.8 is not the weight W, but the mass m_b. To prove it, go to the moon, and measure your weight. It is known that we weigh much less on the moon—about one sixth as much as here on Earth. Since our mass m_b is constant, we weigh less because the gravitational constant for the moon is only about g/6. The smaller gravitational constant for the moon is due to its much smaller mass when compared to the mass of the Earth. The mass of our body is the same whether we are on the moon, Mars the Earth or anywhere in space.

EXAMPLE 1.3

(a) Determine the weight of an object with a mass of 22 kg (a kilogram is the unit for mass in the International System).

Solution:

We use Eq. 1.8 to calculate the weight as:

$$F = W = mg = (22\ \text{kg})(9.807\ \text{m/s}^2) = 215.8\ \text{N}$$

where N is the symbol for **newton**, which is the unit for force in the SI system. One newton is also equivalent to 1kg-m/s^2. We will discuss units used in the SI system in more detail in the next section.

(b) Determine the weight of an object with a mass of 14 slugs (a slug is the unit for mass in the U.S. Customary system).

Solution:

Using Eq. 1.8 again gives:

$$F = W = mg = (14\ \text{slug})(32.17\ \text{ft/s}^2) = 450.4\ \text{lb}$$

The slug has units of (lb-s^2)/ft. An object weighing 32.17 lb on Earth has a mass of 1 slug.

Did you drive your car to the university today? If so, the pressure developed by the combustion of gasoline within the cylinders of your car's engine provided the force to propel it along the roads. Pressure, p acts over an area, A of some surface to create a force.

$$F = p\,A \qquad (1.9)$$

where A is the cross sectional area of the piston.

We illustrate the force F produced by the action of the pressure p on the piston shown in Fig 1.7. The magnitude of the force is determined by using Eq. 1.9; its direction is normal to the surface of the piston.

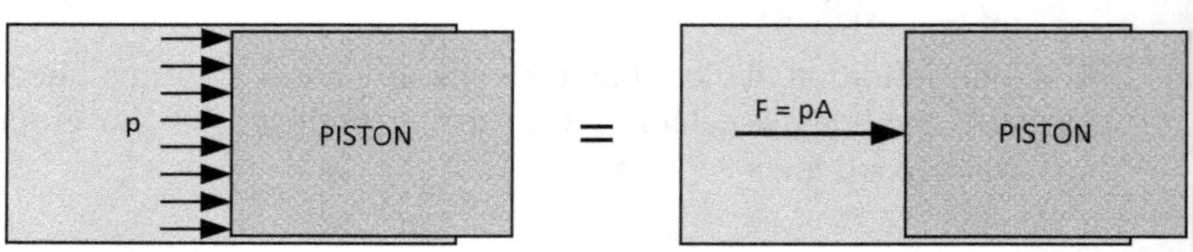

Fig. 1.7 The pressure acting on the piston produces a force F = pA.

1.6 BASIC QUANTITIES AND UNITS

In the study of mechanics, we will encounter four basic quantities—length, time, force and mass. These quantities are shown with their respective units for the SI and the U. S. Customary systems of units in Table 1.1.

Table 1.1
Basic Quantities and Units

System of Units	Length	Time	Mass	Force	g
International System of Units (SI)	meter (m)	second (s)	kilogram (kg)	newton (N), (kg-m)/s^2	9.807 m/s^2
U. S. Customary (FPS)	foot (ft)	second (s)	slug (lb-s^2)/ft	pound (lb)	32.17 ft/s^2

The basic units are not independent because Eq. 1.3 requires that the units be **dimensionally homogenous**. To maintain the dimension homogeneity of Eq. 1.3, we define the units for length, time and mass in the **SI** system, and then derive the remaining basic unit for the force, the **newton**, in terms of those units. For the **U. S. Customary** system, the basic unit for mass, the **slug**, is derived in terms of the units for length, time, and force.

In the International System of Units (SI), the length is given in meters (m), the time in seconds (s), and the mass in kilograms (kg). The unit for force is called a newton (N) in honor of Sir Isaac. The newton is derived from Eq. 1.3 so that a force of 1N will impart an acceleration of 1 m/s^2 to a mass of 1 kg [i. e. 1 N = (1 kg) (1 m/s^2)]. For dimension homogeneity, it is clear that N is equivalent to (kg-m)/s^2. In the SI system the gravitation constant g = 9.807 m/s^2. With this value of the acceleration due to gravity on Earth, the weight of a mass of 1 kg is:

$$W = mg = (1\text{kg})(9.807\ \text{m/s}^2) = 9.807\ \text{N}$$

In the U. S. Customary System, the length is given in feet (ft), the force in pounds (lb), and the time in seconds (s). The unit for mass is called a slug, which is derived from Eq. 1.3 so that a force of 1 lb will impart an acceleration of 1 ft/s^2 to a mass of 1 slug [i. e. 1 lb = (1 slug)(1 ft/s^2)]. For dimension homogeneity, it is clear that a slug is equivalent to (lb-s^2)/ft. In the U. S. Customary System the gravitation constant g = 32.17 ft/s^2. With this value of the acceleration due to gravity on Earth, the mass of a body weighing 32.17 lb is:

$$m = W/g = 32.17\ \text{lb}/(32.17\ \text{ft/s}^2) = 1\ \text{slug}$$

In addition to the quantities presented in Table 1.1, we deal with several other quantities in our studies of statics and mechanics of materials. These quantities are listed in Table 1.2

Table 1.2
Units of Other Frequently Used Quantities

System of Units	Moment M	Stress σ	Strain ε
SI	newton-meter, (N-m)	pascal, (Pa) = (N/m^2) mega pascal, (MPa) = (N/mm^2)	dimensionless
U. S. Customary, (FPS)	foot-pound, (ft-lb)	pound/square foot, (lb/ft^2) pound/square in., (lb/in.^{2}) = psi	dimensionless

We have shown the quantities in Table 1.2 in terms of the basic units for length, time and force. However, in practice other units are often employed. For instance, a moment may be expressed as in-lb instead of ft-lb, and a stress expressed as MPa instead of Pa or psi (lb/in^2) instead of (lb/ft^2). We are fortunate that strain is a dimensionless quantity; we will not have to convert from one system of units to another, or to convert from one unit to another within the same system. A technique to convert from one system of units to another is described in the next section. Often the metric prefixes are employed in expressing numerical results. Indeed, we used a metric prefix M with the unit for stress (Pa) in the previous paragraph. The metric prefixes are useful in dealing with either very large or very small numbers. Accordingly, it is common practice to employ them together with the symbol for the units (e.g. MPa).

Table 1.3
SI Prefixes

Multiplication Factor	Prefix Name	Prefix Symbol
10^{18}	exa	E
10^{15}	peta	P
10^{12}	tera	T
10^{9}	giga	G
10^{6}	mega	M
10^{3}	kilo	k
10^{2}	hecto*	h
10^{1}	deka*	da
10^{-1}	deci*	d
10^{-2}	centi*	c
10^{-3}	milli	m
10^{-6}	micro	μ
10^{-9}	nano	n
10^{-12}	pico	p
10^{-15}	femto	f
10^{-18}	atto	a

*To be avoided when possible.

1.7 CONVERSION OF UNITS

We show two different techniques for the conversion of units from one system to another. The first technique utilizes the conversion factors listed in Table 1.4.

Table 1.4
Unit Conversion Factors

Quantity	**U. S. Customary**	**SI Equivalent**
Acceleration	ft/s^2 in/s^2	0.3048 m/s^2 0.0254 m/s^2
Area	ft^2 in^2	0.0929 m^2 645.2 mm^2
Distributed Load	lb/ft lb/in.	14.59 N/m 0.1751 N/mm
Energy	ft-lb	1.356J
Force	kip = 1000 lb lb	4.448 kN 4.448N
Impulse	lb-s	4.448 N-s
Length	ft in mi	0.3048 m 25.40 mm 1.609 km
Mass	lb mass slug ton mass	0.4536 kg 14.59 kg 907.2 kg
Moments or Torque	ft-lb in-lb	1.356 N-m 0.1130 N-m

Quantity	U. S. Customary	SI Equivalent
Area Moment of Inertia	in^4	0.4162 x 10^6 mm^4
Power	ft-lb/s hp	1.356 W 745.7 W
Stress and Pressure	lb/ft^2 lb/in^2 (psi) ksi = 1000 psi	47.88 Pa 6.895 kPa 6.895 MPa
Velocity	ft/s in/s mi/h (mph)	0.3048 m/s 0.0254 m/s 0.4470 m/s
Volume	ft^3 in^3 gal	0.02832 m^3 16.39 cm^3 3.785 L
Work	ft-lb	1.356 J

Consider the following examples to illustrate the use of this conversion table.

EXAMPLE 1.4

(a) Determine the SI equivalent of a force of 200 kip.

Solution: Using the data listed in Table 1.4 found in Section 1.7, we write:

F = (200 kip) (4.448 kN/kip) = 889.6 kN

(b) Determine the SI equivalent of the mass of 4 slugs.

Solution: Using the data listed in Table 1.4 in the row for mass, we write:

m = (4 slug) (14.59 kg/slug) =58.36 kg

EXAMPLE 1.5

(a) Determine the U. S. Customary equivalent of a force of 350 N.

Solution: Using the data listed in Table 1.3 in the row for force, we write:

F = (350 N) (lb/4.448 N) = 78.69 lb

(b) Determine the U. S. Customary equivalent of a mass of 84 kg.

Solution: Using the data listed in Table 1.3 in the row for mass, we write:

m = (84 kg)/(slug/14.59 kg) = 5.757 slug

In solving these exercises, we demonstrated the technique employing data from Table 1.4.
A second technique, for converting units from one system to another, is based on multiplying the term to be converted by a combination of units, which equals one. Let's consider a simple conversion from say 16 MPa to its equivalent value expressed in terms of psi.

$$\underset{1^{st}}{16\ \text{MPa}} \times \underset{2^{nd}}{\frac{10^6\ \text{Pa}}{\text{MPa}}} \times \underset{3^{rd}}{\frac{\text{N}/\text{m}^2}{\text{Pa}}} \times \underset{4^{th}}{\frac{\text{m}^2}{(39.37)^2\,\text{in}^2}} \times \underset{5^{th}}{\frac{\text{lb}}{4.448\ \text{N}}} \times \underset{6^{th}}{\frac{\text{psi}}{\text{lb}/\text{in}^2}} = 2321\ \text{psi}$$

Let's examine this relation term by term:

1. The first term in the expression is a stress of 16 MPa. Recall that MPa is the unit for stress in the SI system.
2. The next unit term converts the metric prefix M to 10^6. Examine Table 1.3 for definitions of the metric prefixes.
3. The next unit term converts Pa to N/m^2.
4. The next unit term converts m to in.—note both the conversion factor and units are squared.
5. The next unit term converts N to lb.
6. The last term before the equal sign recognizes the symbol psi for lb/in^2.

If you find that a conversion table is not convenient and you are able to remember a few basic conversions, multiplication by a number of unit terms is a good method for converting units from one system to another.

EXAMPLE 1.6

Using the unit multiplication method convert the stress of 20 ksi to its equivalent in the SI system of units.

Solution:

$$20\text{ ksi}\times\frac{10^3\text{ psi}}{\text{ksi}}\times\frac{\text{lb}/\text{in}^2}{\text{psi}}\times\frac{4.448\text{ N}}{\text{lb}}\times\frac{(39.37)^2\text{in}^2}{\text{m}^2}\times\frac{\text{Pa}}{\text{N}/\text{m}^2}\times\frac{\text{M}}{10^6}=137.9\text{MPa}$$

This approach is longer than the more direct conversion method. However, if you do not remember that a stress of 1 ksi is equivalent to 6.895 MPa, it provides a means of performing the conversion using only the most fundamental of the conversion factors.

1.8 SIGNIFICANT FIGURES

With the advent of the hand held calculator, we usually obtain solutions with ten or more figures. These results are misleading because the data used in the formulas to calculate these ten figure results are rarely accurate to more than 0.2% or one part in 500. To avoid the implication of fictitious accuracies, we recommend that the results from your calculator be written with four significant figures. This practice yields a computational accuracy of 1/1000 or 0.1% that is consistent with the accuracy of the physical data and the analytical model upon which the formula is based.

Let's consider several numerical results and convert them to results acceptable in engineering practice.

EXAMPLE 1.7

Convert the following ten digit numerical results shown in the table below to a format with four significant figures.

Solution:

Table E1.7
Examples for Writing Numbers with Four Significant Figures

Ten Digit Number	Four Significant Figures	Ten Digit Number	Four Significant Figures
(a) 6.142857143	6.143	(f) 0.182926829	0.1829
(b) 0.000304136	3.041×10^{-4}	(g) 1226308544	1.226×10^9
(c) 0.026941846	0.02694	(h) 1628720.675	1.629×10^6
(d) 308.5225000	308.5	(i) 0.0000005984	5.984×10^{-7}
(e) 22.76923077	22.77	(j) 88.62508918	88.63

(a) In converting we rounded the fourth digit up since the fifth digit was larger than 5.
(b) We retained the first four numbers after the leading zeros. The zeros were accommodated by using 10^{-4}.
(c) We retained the first four numbers after the leading zero. Since only one zero was involved, we retained it in the representation.
(d) We maintained the fourth digit since the fifth digit was less than 5.
(e) We rounded the fourth digit up since the fifth digit was larger than 5.
(f) We maintained the fourth digit since the fifth digit was less than 5.
(g) We retained the first four numbers since the fifth digit was less than 5. The very large number was represented by using 10^9.
(h) We rounded the fourth digit up since the fifth digit was larger than 5. The very large number was represented by using 10^6.
(i) We retained the first four numbers after the leading zeros. The zeros were accommodated by using 10^{-7}.
(j) We rounded the fourth digit up since the fifth digit was equal to 5.

1.9 SCALARS, VECTORS AND TENSORS

In mechanics we deal with three types of quantities, namely **scalars, vectors** and **tensors**. In normal every day life it is not essential to recognize the difference among them. If we have true friends, food, shelter and some money, all is well. Incidentally, true friends, food, shelter and money are all scalar quantities. We need only to count these items in describing them. For example, we might have $74 and be content after eating two big Macs. Other quantities encountered in engineering that are scalar and listed in Table 1.4 include:

- Area
- Energy
- Length
- Mass
- Moment of inertia
- Power
- Volume
- Work

Scalar quantities are the easiest of the three with which to work. We can use simple arithmetic to add, subtract, multiply and divide when we use them. No additional operations are required to manipulate scalar quantities. This not the case with vectors and tensors. Additional mathematical operations and descriptions must be introduced to manipulate equations containing vectors or tensors.

Unfortunately we must deal with both vector and tensor quantities in mechanics. They are not **that** difficult, but we will need to recognize they are not scalar quantities. We will also need mathematics higher than arithmetic to deal with them. Let's start with quantities that must be described with vectors. Vector quantities require two descriptors—magnitude and direction. The magnitude indicates the size of the quantity, and the direction gives its orientation. Quantities that we cited in Table 1.4 which require vector representation for their complete description include:

- Acceleration
- Force
- Impulse
- Moments
- Velocity

We have already discussed gravitational forces. Our weight is the magnitude of the force due to the pull of gravity on the mass of our body. The direction of the gravitational force relative to an Earth bound coordinate system is toward the center of the Earth.

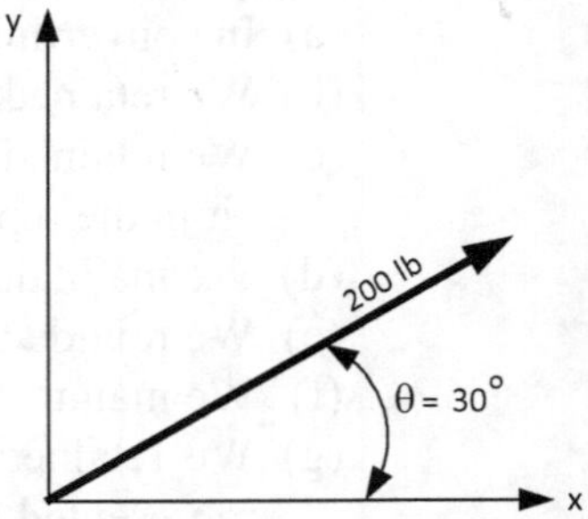

Fig. 1.8 A force vector with a magnitude—200 lb and direction—30° relative to the x-axis.

We represent a typical vector quantity, such as a force, with an arrow as shown in Fig. 1.8. The length of the arrow is proportional to the magnitude of the force, and its inclination relative to the x-y coordinate system gives its direction. We will use vectors extensively in our studies of statics, learning to accommodate both their magnitude and directions in the solution of many different types of equilibrium problems. Vectors are independent of the orientation chosen for the x-y coordinate system. However, the scalar value of the vector's direction and the components of the vector in the x and y directions are dependent on the orientation of the reference coordinate system. This property of vectors will be of importance when coordinate systems are chosen to solve equilibrium problems, because choices can be made that simplify the solution to the problem without changing the physical description of the problem.

Tensor quantities are even more difficult to describe than vector quantities because their complete description requires information about three characteristics. Like vectors, we must specify the magnitude and direction. In addition we must specify the orientation of the plane upon which a tensor quantity acts. Stresses, σ are tensor quantities. We demonstrate the three characteristics—magnitude, direction, and the orientation of the plane upon which the stresses act in Fig. 1.9.

The round bar is loaded with forces in the axial direction. A section cut exposes an internal surface (plane) of the bar normal to its axis. Stresses with a magnitude of 140 psi act in the axial direction.

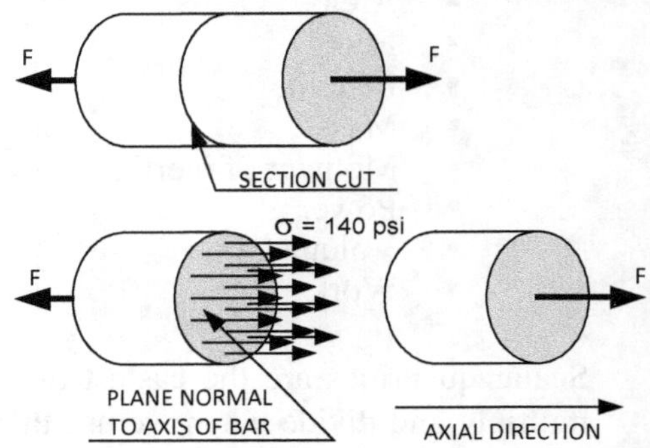

Fig. 1.9 Stresses of 140 psi act in the direction of the axis of the bar. The stresses are acting on the plane normal to the bar's axis.

1.10 SUMMARY

In statics we assume that the body is perfectly rigid—it does not deform under load. With this assumption, a free body diagram, and the equations of equilibrium, we are able to solve the many different types of problems arising in statics. Mechanics of materials is an extension of statics where very small changes in the geometry of the body due to deformations are used to determine the stresses and strains. Simple procedures are outlined for the solution of typical problems arising in both subjects.

A very brief review of the history of mechanics is given to show that both subjects have been well understood for many years. The contributions of Archimedes, Da Vinci, Hooke, Newton, and Jacob Bernoulli are mentioned.

Newton's three laws of motion that form the foundation for mechanics have been described. Newton's laws lead to Eqs. 1.2 and 1.3 which are so important that they are repeated here:

$$\Sigma \mathbf{F} = 0 \quad (1.2)$$

$$\Sigma \mathbf{F} = m\mathbf{a} \quad (1.3)$$

A simple example is provided to illustrate the use of the equilibrium equations.

Forces are described in considerable detail. Since forces due to gravity and pressure are so common, we provided the equations used to determine their magnitudes. The force or weight due to gravity is:

$$F = m_b\, g = W \qquad (1.8)$$

The force due to pressure is:

$$F = p\, A \qquad (1.9)$$

The basic quantities and their units that are employed in statics and mechanics of materials have been given. A table listing the conversion factors for the SI and U. S. Customary systems has been included. Examples showing different techniques for converting units from one system to the other are presented.

REFERENCES

1 Timoshenko, S. P., History of Strength of Materials, Dover edition, Dover Publications, New York, NY 1983.
2 Todhunter, I. and K. Pearson, History of the Theory of Elasticity and Strength of Materials, Cambridge University Press, Cambridge, U. K., 1886.
3 Uccelli, A., Leonardo da Vinci, Reynal and Co., New York, NY 1956.
4 Newton, Sir Isaac, Philosophiae Naturalis Principia Mathematica, 1687.

PROBLEMS

1.1 Design a lever that will permit you to lift a weight of 1,000 N a distance of 100 mm. In the design, prepare a dimensioned sketch of the lever, state the force that you apply to the lever, and show the analysis proving that you will be able to lift the weight and move it upward by the specified distance.

1.2 The *Sojourner*, a mechanical rover that explored Mars during a mission in 1997, had a mass of 10.5 kg. Determine the weight of *Sojourner* (in newtons) on (a) Earth, (b) Mars. Assume Mars has a mass of 0.64×10^{24} kg and a radius of 3390 km.

1.3 Weather satellites are often placed in geosynchronous orbit around Earth, so that they appear stationary in the sky when viewed from a point on the equator. The orbital altitude for such a satellite is 35,800 km. If a satellite weighs 500 N on the Earth's surface, calculate its weight in orbit. Assume Earth has a mass of 5.976×10^{24} kg and a radius of 6370 km.

1.4 An astronaut weighs 150 lb on the Earth's surface. When orbiting Earth in the space shuttle *Atlantis*, however, the astronaut's weight is measured as 130 lb. Determine the altitude of the shuttle's orbit. Assume Earth has a mass of 5.976×10^{24} kg and a radius of 6370 km.

1.5 Suppose that you are in a space ship traveling in a circular orbit about the Earth, as shown in the figure to the right. Using a spreadsheet, prepare a graph showing your weight as a function of your position relative to the surface of the Earth. Consider radii from $1.0(R_e)$, on the Earth's surface, to $15(R_e)$.

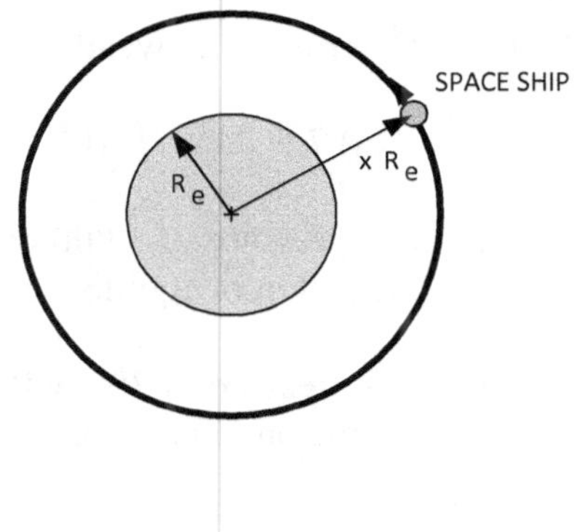

1.6 Determine the gravitational force exerted on the Earth by (a) the sun, (b) the moon. Also find the ratio of the forces (F_{sun}/F_{moon}). Assume masses of 5.976×10^{24} kg for the Earth, 1.990×10^{30} kg for the sun and 7.350×10^{22} kg for the moon. The mean distance (center-to-center) between the sun and the Earth is 149.6×10^{6} km and between the moon and the Earth is 384×10^{3} km.

1.7 Prove whether or not the blocks, shown in the figure (a) and (b) below, are in equilibrium.

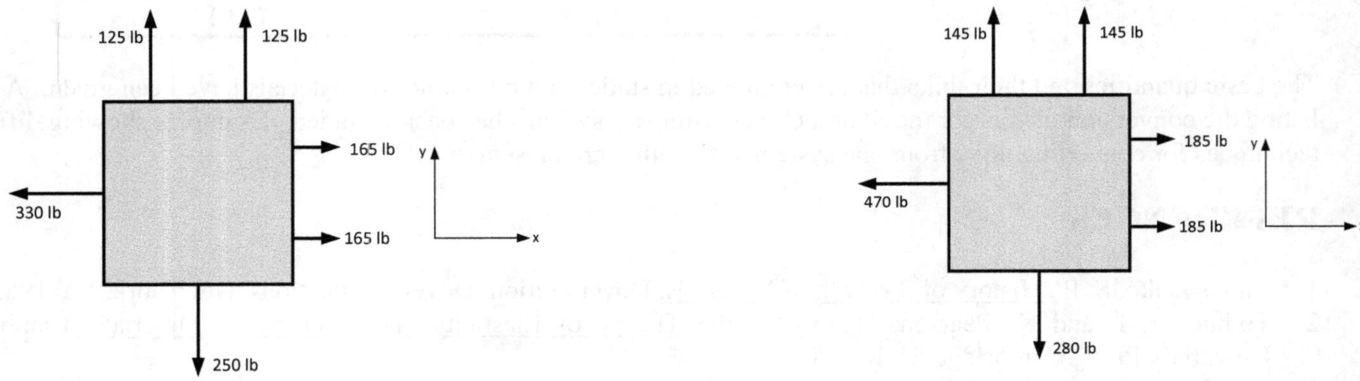

(a) (b)

1.8 Prove whether or not the blocks, shown in the figures (a) and (b) below, are in equilibrium.

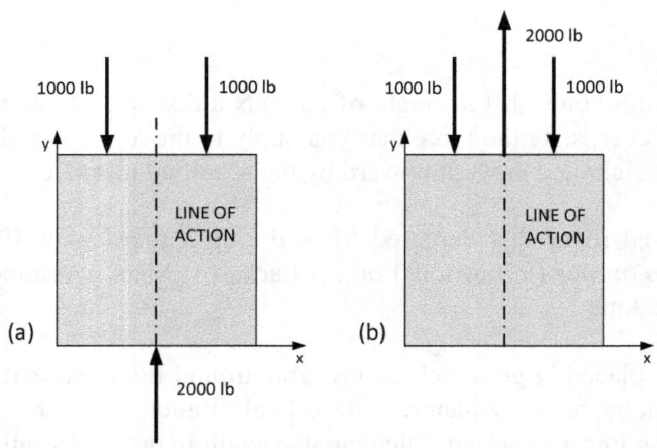

1.9 Draw a free body diagram of an automobile parked on a level street showing the reaction forces on the tires. Also show the active forces on the pavement. Assume that the auto weighs 3,200 lb. State the assumption made in distributing the weight of the auto among the four tires.

1.10 Determine the weight of a body with the following masses in SI units:

(a) 65 kg (b) 19 kg (c) 58 kg (d) 17.3 kg (e) 34.7 kg (f) 81.2 kg

1.11 Determine the weight of a body with the following masses in U. S. Customary units:

(a) 7.5 slug (b) 12 slug (c) 65 slug (d) 4.2 slug (e) 33.7 slug (f) 49.4 slug

1.12 If a pressure of 3,000 psi acts on a piston with a diameter of 1.75 in., determine the force required to maintain the piston in equilibrium.

1.13 If a pressure of 300 MPa acts on a piston with a diameter of 150 mm, determine the force required to maintain the piston in equilibrium.

1.14 If a force of 3,000 lb acts on a piston with a diameter of 5.0 in., determine the pressure required to maintain the piston in equilibrium.

1.15 If a force of 5,000 N acts on a piston with a diameter of 100 mm, determine the pressure required to maintain the piston in equilibrium.

1.16 If a piston is in equilibrium under a pressure of 1.5 ksi and a force of 10.5 kip, determine the diameter of the piston.

1.17 If a piston is in equilibrium under a pressure of 55 MPa and a force of 675 kN, determine the diameter of the piston.

1.18 List the four basic quantities and give their units in both the SI and U. S. Customary systems.

1.19 Determine the SI equivalent of the masses listed below.

(a) 7.5 slug (b) 12 slug (c) 65 slug (d) 4.2 slug (e) 33.7 slug (f) 49.4 slug

1.20 Determine the U. S. Customary equivalent of the masses listed below.

(a) 65 kg (b) 19 kg (c) 58 kg (d) 17.3 kg (e) 34.7 kg (f) 81.2 kg

1.21 Determine the SI equivalent of the following forces:

(a) 2,400 lb (b) 425 lb (c) 688 lb (d) 18 tons (e) 52 kip (f) 95 kip

1.22 Determine the U. S. Customary equivalents for the following forces:

(a) 2,234 N (b) 89.8 N (c) 13.2 kN (d) 96.9 mN (e) 142.3 μN (f) 44.8 MN

1.23 Determine the SI equivalent of the following lengths:

(a) 13 ft
(b) 153 in.
(c) 41.3 yards
(d) 16.4 miles
(e) 6 ft - 2 in.
(f) 1.8 miles - 662 yards

1.24 Determine the U. S. Customary equivalents for the following lengths:

(a) 7.4 m
(b) 500 mm
(c) 12.9 cm
(d) 13.2 km
(e) 18.8 mm
(f) 64.3 cm

1.25 Convert the following speeds into SI equivalent units:

(a) 60 mph
(b) 45 mph
(c) 30 mph
(d) 22 ft/s
(e) 90 ft/s
(f) 580 mph
(g) 16 ft/s
(h) 224 ft/s
(i) 187 mph

1.26 Convert the following speeds into U. S. Customary equivalent units:

(a) 120 km/h
(b) 61 km/h
(c) 88 km/h
(d) 123 m/s
(e) 510 m/s
(f) 17.5 km/s
(g) 312 m/h
(h) 16.7 m/s
(i) 43.5 km/s

1.27 Convert the following stresses into SI equivalent units:

(a) 12,000 psi
(b) 2,320 psi
(c) 1,980 psi
(d) 132.5 ksi
(e) 30×10^6 psi
(f) 10.5×10^6 ksi
(g) 194,200 psi
(h) 32 psi
(i) 96,230 psi

1.28 Convert the following stresses into U. S. Customary equivalent units:

(a) 1,000 MPa
(b) 121 MPa
(c) 86.4 MPa
(d) 1,400 kPa
(e) 144 kPa
(f) 9,642 kPa
(g) 207 GPa
(h) 16,430 MPa
(i) 42.3 kPa

1.29 Prepare a short written description of the differences among scalars, vectors, and tensors. Also prepare a sketch illustrating a quantity of each type.

1.30 Consider each quantity listed in Table 1.4 and classify it as a scalar, vector or tensor.

Table 1.4
Unit Conversion Factors

Quantity	U. S. Customary	SI Equivalent
Acceleration	ft/s^2	0.3048 m/s^2
	in/s^2	0.0254 m/s^2
Area	ft^2	0.0929 m^2
	in^2	645.2 mm^2
Distributed Load	lb/ft	14.59 N/m
	lb/in.	0.1751 N/mm
Energy	ft-lb	1.356J
Force	kip = 1000 lb	4.448 kN
	lb	4.448N
Impulse	lb-s	4.448 N-s
Length	ft	0.3048 m
	in	25.40 mm
	mi	1.609 km
Mass	lb mass	0.4536 kg
	slug	14.59 kg
	ton mass	907.2 kg
Moments or Torque	ft-lb	1.356 N-m
	in-lb	0.1130 N-m
Area Moment of Inertia	in^4	0.4162 x 10^6 mm^4
Power	ft-lb/s	1.356 W
	hp	745.7 W
Stress and Pressure	lb/ft^2	47.88 Pa
	lb/in^2 (psi)	6.895 kPa
	ksi = 1000 psi	6.895 MPa
Velocity	ft/s	0.3048 m/s
	in/s	0.0254 m/s
	mi/h (mph)	0.4470 m/s
Volume	ft^3	0.02832 m^3
	in^3	16.39 cm^3
	gal	3.785 L
Work	ft-lb	1.356 J

CHAPTER 2

FORCES AND MOMENTS

2.1 INTRODUCTION

Complex structures, vehicles and machines are constructed from many different structural elements such as rods, columns, struts, beams, cables, etc. To properly size these elements so they will perform their function safely for the duration of their design life, we must determine the forces and the moments acting at each point on each element. These forces and moments are vector quantities that are defined by specifying both their magnitude and direction. In this chapter, two different approaches for dealing with vector quantities are described:

1. A relatively simple approach based on trigonometry, which is easy to use when analyzing two-dimensional (plane) structures.
2. A more elegant technique utilizing vector algebra, which is very useful in analyzing three-dimensional structures. Unfortunately, this technique requires an understanding of a mathematical topic that is new to many engineering students.

We will solve two-dimensional examples employing the trigonometric techniques. We will also introduce vector algebra and demonstrate the vector dot and cross products to show their application in determining the unknown forces and moments in structural elements.

2.2 INTERNAL AND EXTERNAL FORCES

In dealing with forces, we distinguish between those that are applied to the structure **(external)**, and those that develop within a structural element **(internal)**. The external forces include the active loads applied to the structure such as those shown in Fig. 2.1a-c, and the reaction forces, shown in Fig. 2.1 d, that develop at the supports to maintain the structure in equilibrium.

In Fig. 2.1a, the simply supported beam is loaded with a **concentrated force** at a local point near its center. The concentrated force, applied at a point, is an idealization. Forces are always distributed over some area; however, with concentrated forces, we assume that the area is so small that it approaches a point. The symbol F will be used to designate the magnitude of concentrated forces.

In Fig. 2.1b, the beam is loaded with a **uniformly distributed load** that is applied over most of its length. Uniformly distributed loads along beams are specified in terms of force/unit length (i. e. lb/ft or N/m). The symbol q is used to designate the magnitude of the distributed forces applied to a beam.

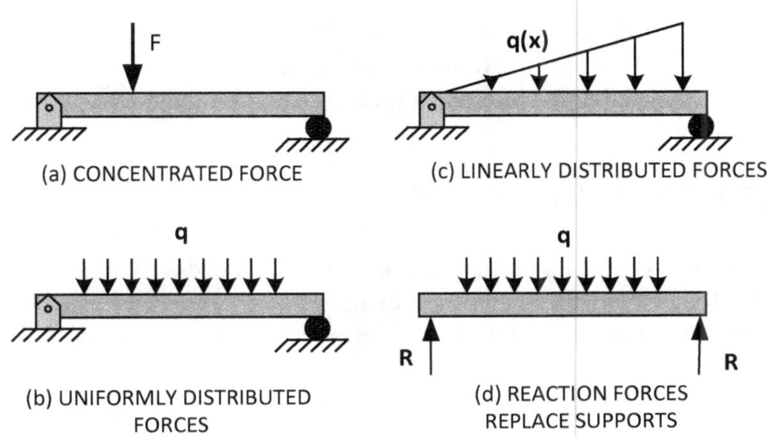

Fig. 2.1 Examples of different types of external forces applied to the structure (beam).

A distributed load that is increasing as we move from the left end of the beam to the right is illustrated in Fig. 2.1c. Again, the symbol q is used to designate the magnitude of the distributed forces; however, in this case we must recognize that q is a function of position x along the length of the beam, designated by q(x).

The last example of external forces is shown in Fig. 2.1d. The beam and its loading are identical with that shown in Fig. 2.1b, but the supports have been removed. The reaction forces developed by the supports to maintain the beam in equilibrium are shown as concentrated forces. The symbol R will be used to designate the magnitude of the reaction forces.

Internal forces develop within a structural member due to the action of the applied external forces. These internal forces are not visible—although we try to visualize them by making imaginary cuts through a structural member. Let's examine a bar subjected to two external forces F applied at each end as shown in Fig. 2.2. We make a section cut perpendicular to the axis in the central region of the bar. This is an imaginary cut, not a real one, but it permits us to visualize either segment of the bar. We examine the segment on the left, and find a **normal stress**, σ, which is uniformly distributed over the area exposed by the section cut. When this stress is integrated over the area of the bar, an internal force P_{int} is generated that acts along the axis of the bar. The magnitude of P_{int} is given by:

$$P_{int} = \int \sigma \, dA \qquad (2.1)$$

Fig. 2.2 (a) A tension bar with external loads and a section cut dividing the bar into two parts. (b) Two different representations of the left side of the bar.

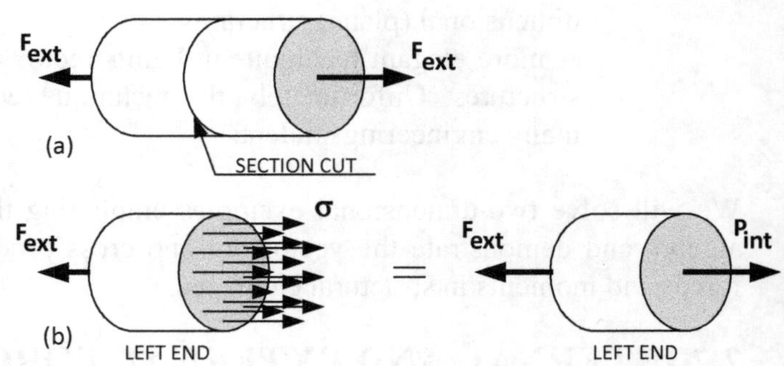

The left end of the bar must be in equilibrium, which implies that:

$$\Sigma F_x = 0 \qquad (1.2 \text{ bis})$$

Summing the forces in the x direction gives:

$$P_{int} = F_{ext} \qquad (2.2)$$

In this elementary example of a rod in tension, we have found the relation between the internal force P within the bar and the external force F applied to its ends. We will use the same three-step approach throughout the book in solving much more complex problems. In your solutions to the exercises remember to:

1. Make an appropriate section cut.
2. Use the equations of equilibrium.
3. Solve for internal forces in structural members.

2.3 FORCE VECTORS

Forces are vector quantities, and as such it is necessary to specify both their magnitude and direction. Thus far, we have focused on the magnitude to illustrate the physical aspects of different types of forces. Let's now move to a more mathematical discussion of forces and vectors. A typical force vector is presented with an arrow as shown in Fig. 2.2.

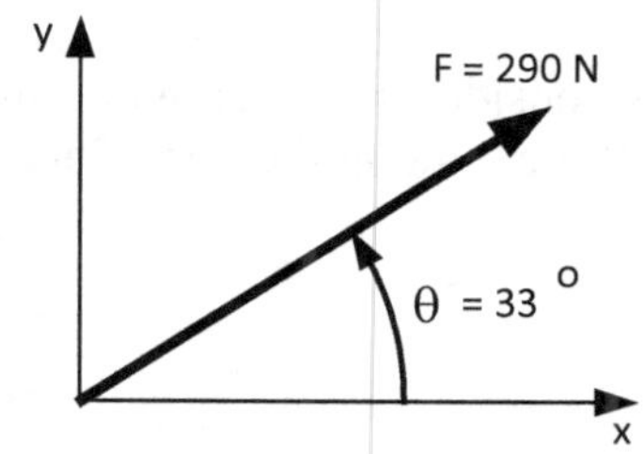

Fig. 2.3 A force vector is represented by an arrow with a length proportional to its magnitude (290 N). The direction of the vector (θ = 33°) is specified relative to the x-axis.

The length of the arrow indicates the magnitude, and the orientation of the arrow relative to the x-axis gives the direction. The tail end of the arrow is positioned at the origin of the coordinate system, and the arrowhead indicates the tip.

EXAMPLE 2.1

Construct a graph showing a force vector with a magnitude of 3700 N and a direction of 150° relative to an x-y coordinate system.

Solution:

First, define an x-y coordinate system making certain to include the negative side of the x-axis. Next, layout a line from the origin inclined by 180° – 150° = 30° relative to the negative x-axis. Select a scale for the magnitude; we suggest 1 mm = 100 N. Measure a length of 37 mm along this line, and place an arrowhead at this point. Place arrowheads on the coordinate system, and include the measure of the angle and the magnitude of the force vector. The result obtained is illustrated in Fig. E2.1:

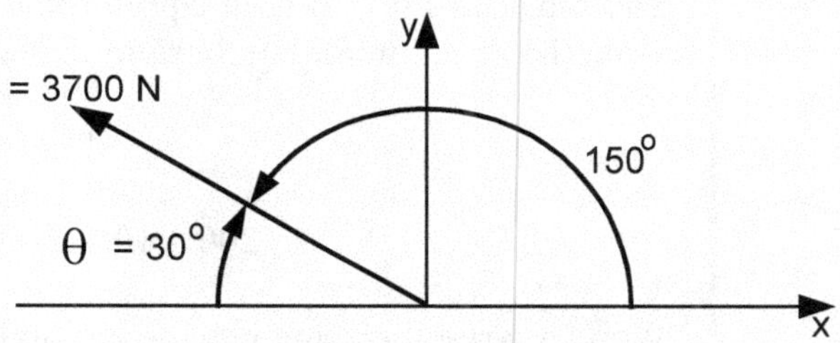

Fig. E2.1

Force vectors have a line of action, as indicated in Fig. 2.4, which is collinear with the direction of the vector.

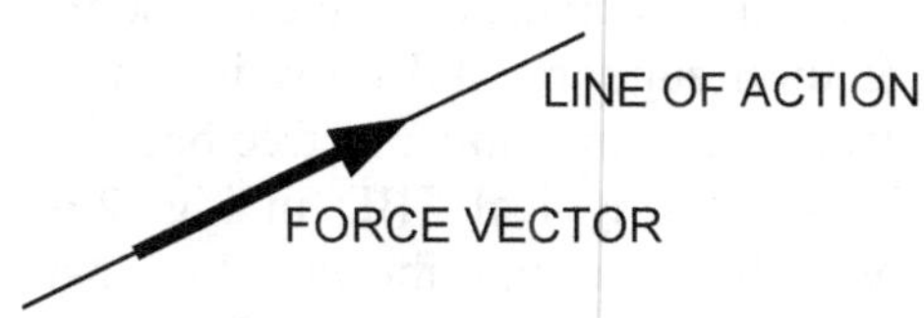

Fig. 2.4 The line of action is collinear with the force vector.

We can move the force vector along the line of action if it is convenient to do so. Shifting the force along the line of action does not change the state of equilibrium of the body. To illustrate this concept of the **transmissibility** of forces, suppose the body shown in Fig. 2.5a is in equilibrium with the force F_1 applied at point A. The body remains in equilibrium as we move the force F_1 along the line of action from point A to point B as indicated in Fig. 2.5b. In the first instance, we are pushing on the body with F_1, and in the second, we are pulling on the body. In both cases the force F_1 acts on the body with the same magnitude and direction; hence the body remains in equilibrium.

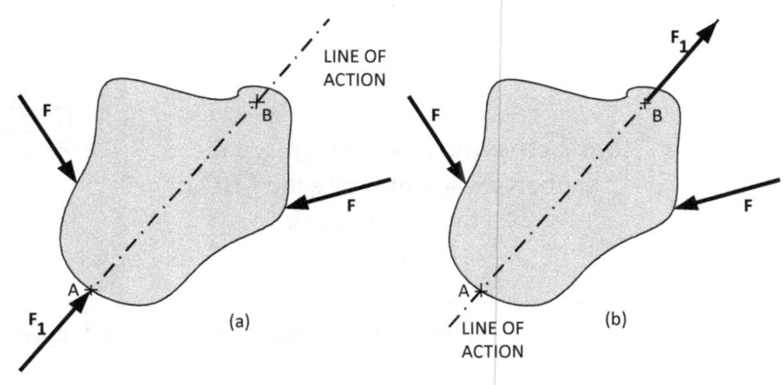

Fig. 2.5 The body remains in equilibrium as we shift $\mathbf{F_1}$ along the line of action from A to B.

EXAMPLE 2.2

Let's consider the block with three applied forces as shown below. Are both blocks in equilibrium after we shift the 200 N force along the line of action from the bottom edge to the top edge?

Fig. E2.3

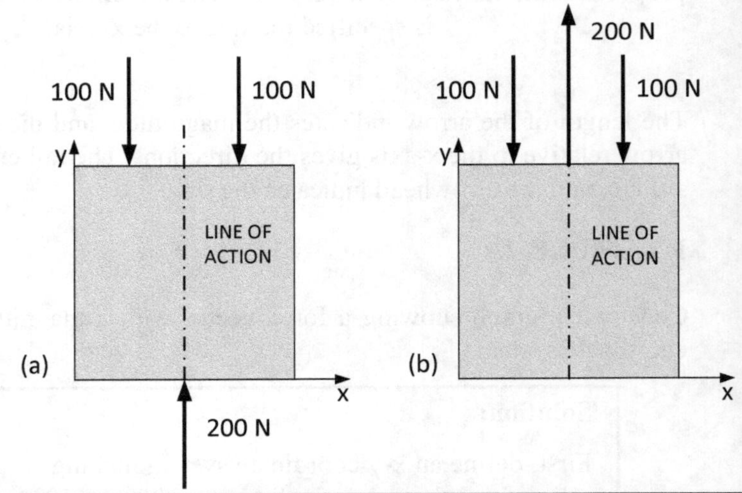

Solution:

The block in Fig. (a) is in equilibrium with a 200 N force applied to the bottom edge. From Eq. 2.4b, we write the equation for equilibrium in the y direction to obtain:

$$\Sigma F_y = 0$$

$$200 - 100 - 100 = 0$$

We can write the same equations for the block shown in Fig. (b). The fact that the 200 N force has been shifted along the line of action does not change the equilibrium relation.

We have already made a distinction between applied (active) forces F and reactive forces R. Reactive forces develop at structural supports and grow in magnitude as the active loads are applied to maintain the body in equilibrium. Reactive forces are shown in free body diagrams (FBDs)[1] when the supports are removed from a structural member. We illustrate a typical FBD in Fig. 2.6 showing three reactive forces that maintain a uniformly loaded, simply supported beam in equilibrium. In constructing the FBD, we remove the supports and apply the reactive forces at their location. The magnitude of R is determined from the equilibrium relations as will be discussed later in Chapter 3. The directions of both R_{yL} and R_{yR} are perpendicular to the surface of the beam. The direction of R_x is taken in the positive x direction. The two reactive forces R_{yL} and R_x at the left support replace the effect of the pinned joint. It is emphasized that you must specify directions for the reactive forces, which are consistent with the type of supports that are removed in constructing the FBD.

Fig. 2.6 Reactive forces in (b) replace the supports when drawing the FBD of the beam in (a).

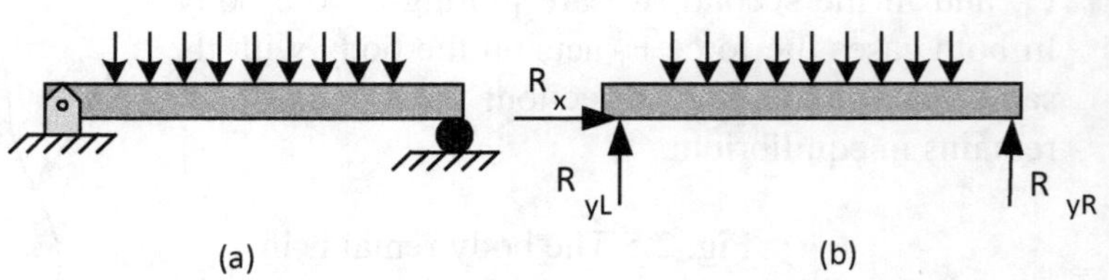

[1] We will use the abbreviation FBD to represent free body diagram throughout the remainder of this textbook.

EXAMPLE 2.3

To illustrate the direction of reactive forces, let's consider a circular pipe supported in a right-angled rack as shown below. Draw a FBD of the pipe and show the angle that the reaction forces make with a horizontal line.

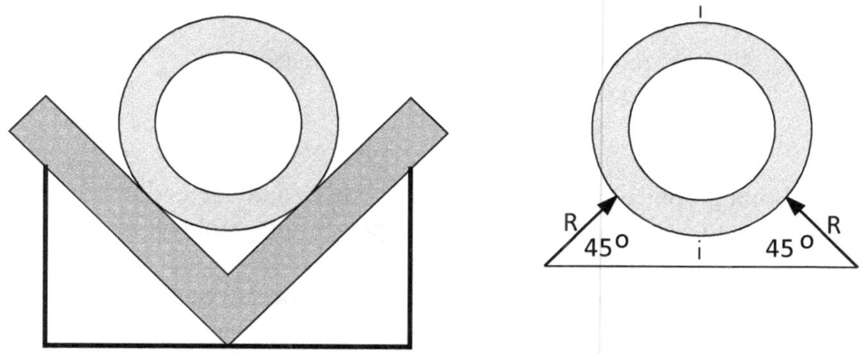

Fig. E2.3

Solution:

We remove the rack and support the pipe with a pair of reaction forces R. These forces are perpendicular to the surface of the pipe at the contact points. The lines of action of the reaction forces are radial and pass through the center of the pipe[2]. Since the rack is symmetric with a 90° included angle, it is easy to visualize the reaction forces at a 45° angle relative to a horizontal line. The third force W represents the weight of the pipe, which acts downward along the pipe's centerline.

2.4 ADDING AND SUBTRACTING VECTORS

Suppose that we have two force vectors **A** and **B**, shown in Fig. 2.7a, that are to be added to give the vector sum $\mathbf{S_v} = \mathbf{A} + \mathbf{B}$. There are several approaches that we can follow to determine this vector sum; however, to begin let's visualize the process of adding vectors by examining Fig. 2.7b. In this illustration, we have shifted vector **B** while retaining its direction and magnitude. Its new position places its tail at the tip of vector **A**. We then construct a line from the tail of vector **A** to the tip of vector **B** to obtain both the magnitude and direction of the vector sum $\mathbf{S_v}$.

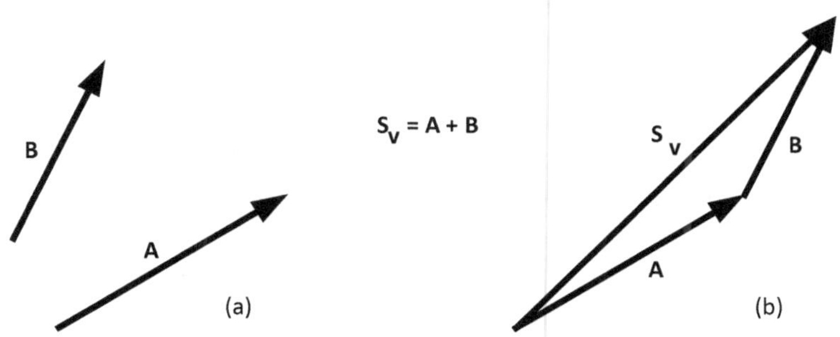

Fig. 2.7 An illustration of the physical interpretation of vector addition $\mathbf{S_v} = \mathbf{A} + \mathbf{B}$.

Now that we can visualize adding **A** + **B**, let's add **B** + **A**. Will reversing the order of adding vectors **A** and **B** give the same result? Let's repeat the process of moving vectors to obtain the results shown in Fig. 2.8. In this case, we have shifted vector **A**, maintaining its magnitude and direction, so that its tail coincides with the tip of vector **B**. The line from the tail of **B** to the tip of **A** is the vector $\mathbf{S_v}$. From these illustrations, it is evident that:

$$\mathbf{S_v} = \mathbf{A} + \mathbf{B} = \mathbf{B} + \mathbf{A} \tag{2.3}$$

This relation is known as the **commutative property of vector addition**.

---

[2] Friction forces at the contact points are assumed to be negligible.

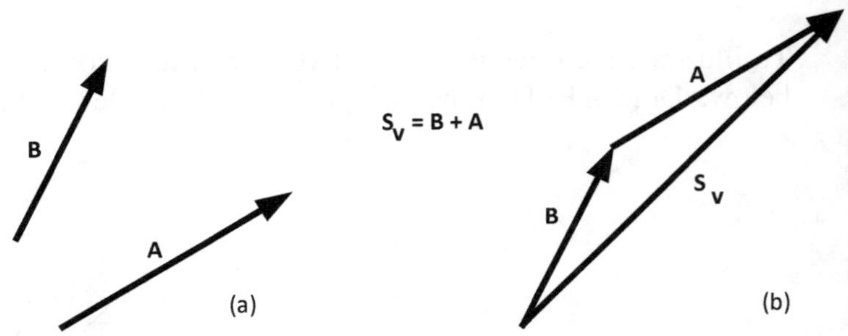

Fig. 2.8 Illustration of the physical interpretation of vector addition $\mathbf{S}_v = \mathbf{B} + \mathbf{A}$.

With a graphical approach, it is possible to add vectors and estimate the magnitude and direction of $\mathbf{S}_v$ using only a scale and a protractor. We will not pursue the graphical technique in this text because there is a much more accurate trigonometric method discussed later.

Let's consider vector subtraction where we seek $\mathbf{D}_v$: the vector formed by **A** – **B**. The subtraction process is illustrated in Fig. 2.9. In Fig. 2.9a, we show the same two vectors **A** and **B**. In Fig. 2.9b, we rotate **B** though 180° to reverse its direction and obtain a new vector – **B**. Next, we add **– B** to **A** following the same procedure used for vector addition to determine $\mathbf{D}_v$ as shown in Fig. 2.9c.

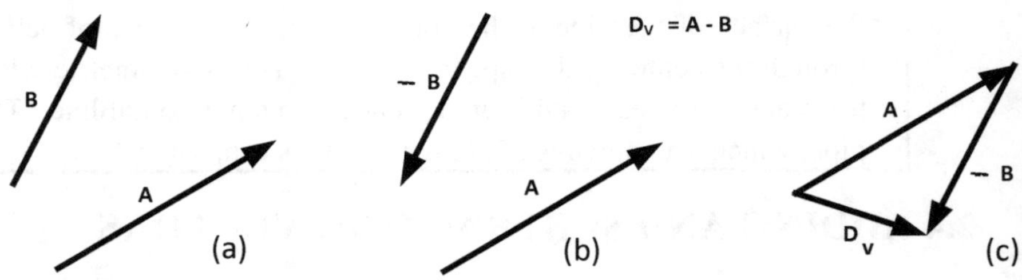

Fig. 2.9 An illustration of the process of vector subtraction.

From Figs. 2.7 – 2.9, which illustrate the addition or subtraction of two vectors, it is clear that both $\mathbf{S}_v$ and $\mathbf{D}_v$ are sides of a triangle. The trigonometric properties of a triangle will be valuable in determining the unknown magnitudes and directions of either $\mathbf{S}_v$ or $\mathbf{D}_v$ in vector addition or subtraction. Let's define the sides and the included angles of a triangle as indicated in Fig. 2.10.

In analyzing triangles, we make use of the **sine law:**

$$\frac{A}{\sin a} = \frac{B}{\sin b} = \frac{C}{\sin c} \qquad (2.4)$$

And the **cosine law:**

$$C = \sqrt{A^2 + B^2 - 2AB\cos(c)} \qquad (2.5)$$

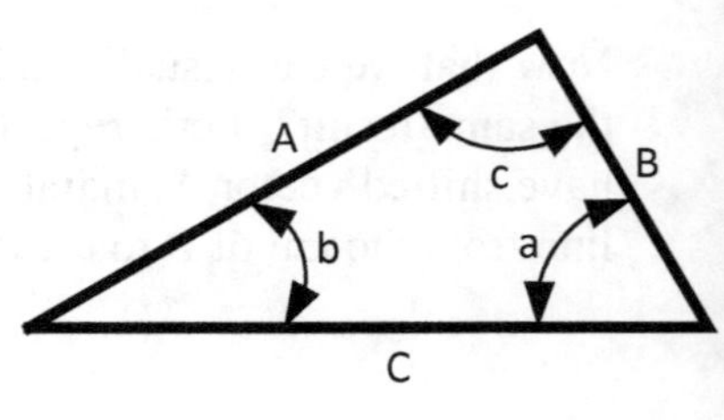

Fig. 2.10 Triangle with sides A, B, and C and included angles of a, b, and c.

The examples shown below illustrate the use of the sine and cosine laws in adding two vectors.

EXAMPLE 2.4

Add the vectors **A** and **B** defined in Fig. E2.4a to obtain the vector sum $\mathbf{S_v}$.

Fig. E2.4

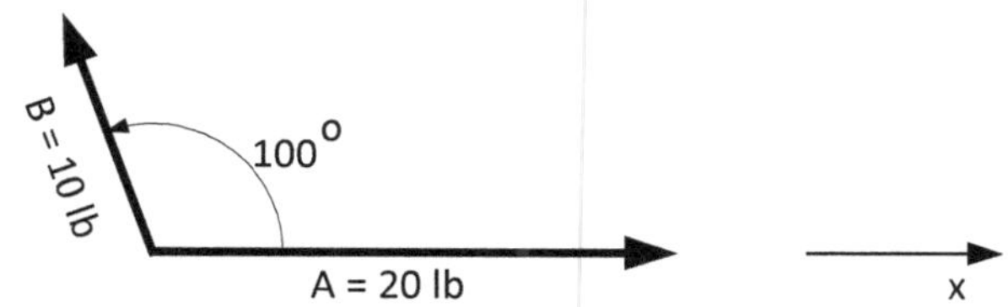

Solution:

We begin by constructing an A, B and C triangle where the magnitude of $\mathbf{S_v}$ is the same as the side C. The construction of the triangle is presented in Fig. E2.4b.

Fig. E2.4b

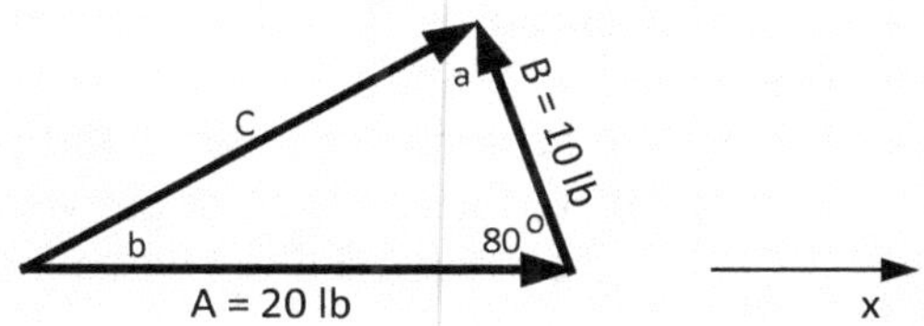

We solve for the magnitude of C using the cosine law, which is given in Eq. (2.5).

$$C = \left[A^2 + B^2 - 2AB\cos(c)\right]^{1/2}$$
$$C = \left[(20)^2 + (10)^2 - (2)(20)(10)\cos(80)\right]^{1/2}$$
$$C = \left[500 - 400(0.1736)\right]^{1/2} = \sqrt{430.5}$$
$$C = 20.75\ \text{lb}$$

Now that we have the magnitude of $\mathbf{S_v}$, the sine law given in Eq. (2.4) may be used to determine its direction. The direction of $\mathbf{S_v}$ relative to the x-axis is given by the angle b.

$$\frac{B}{\sin(b)} = \frac{C}{\sin(c)}$$
$$\sin(b) = \frac{B\sin(c)}{C} = \frac{(10)\sin(80)}{20.75} = 0.4746$$
$$b = 28.33^{\circ}$$

The direction of $\mathbf{S_v}$ is at angle of 28.33° relative to the x-axis. The solution is complete because we have determined both the magnitude and the direction of the vector sum $\mathbf{S_v}$. Note, the results are given with four significant figures, and the units for both the magnitude and the direction are stated.

EXAMPLE 2.5

Two forces **A** and **B** are applied to lift the weight shown in Fig. E2.5. If the weight W is to move vertically upward without swinging when it clears the floor, determine the angle θ to specify for the direction of the force **B**.

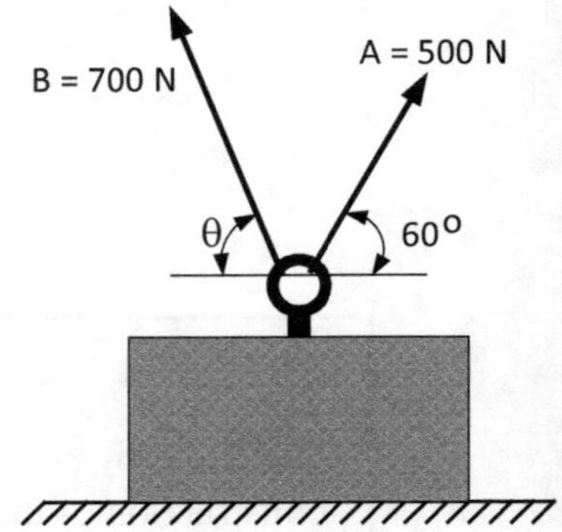

Fig. E2.5

Solution:

To begin the solution of this problem, draw the force triangle as shown in Fig. E2.5a. Note that the direction of $\mathbf{S_v}$, the resultant of the vector addition **A** + **B**, must be vertical if the weight is to be lifted without swinging.

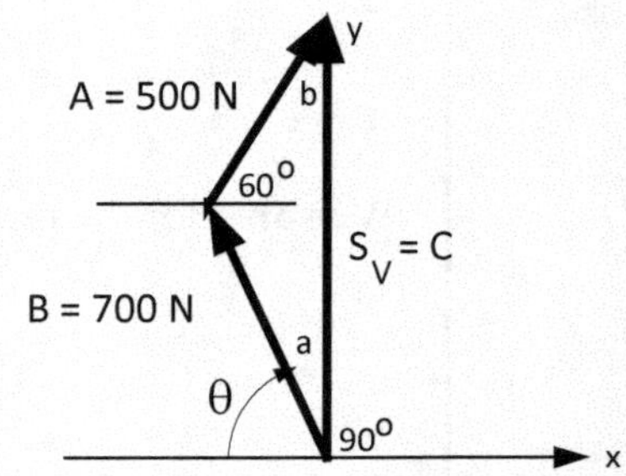

Fig. E2.5a

We will use the force triangle shown in Fig. E2.5b to solve for the angle θ. First, recall from your studies of plane geometry that the sum of the included angles within a triangle is 180°:

$$a + b + c = 180°$$

$$(90° - \theta) + b + (\theta + 60°) = 180°$$

$$b = 30°$$

Next use the law of sines given by Eq. (2.4) to solve for θ.

$$\frac{A}{\sin a} = \frac{B}{\sin b}$$

$$\frac{500}{\sin(90 - \theta)} = \frac{700}{\sin 30}$$

$$\sin(90 - \theta) = (5/7)(1/2) = 0.3571$$

$$90 - \theta = 20.92^{\circ} \Rightarrow \theta = 69.08^{\circ}$$

We must apply the force B at an angle θ = 69.08° if the weight is to lift vertically from the floor without swinging.

EXAMPLE 2.5 (Continued)

What weight can be lifted from the floor with the forces **A** and **B** applied as shown in Fig. E2.5a?

Solution:

The magnitude of the vector sum Sv or the side C of our force triangle gives the weight that can be lifted. Let's use the law of cosines given by Eq. (2.5) to determine C.

$$C^2 = A^2 + B^2 - 2AB\cos(c)$$

$$C^2 = (500)^2 + (700)^2 - 2(500)(700)\cos(129.1)$$

$$C^2 = 118.1 \times 10^4 \Rightarrow C = 1087\text{ N}$$

These results indicate that only 1087 N can be lifted even though forces of 500 N and 700 N (a total of 1200 N) were applied in the lifting process. Only the vertical components of the forces A and B contributed to lifting. The horizontal components cancelled each other, and did not contribute to lifting the weight. We will explore the components of a force in the next section.

2.5 COMPONENTS OF A FORCE VECTOR

In section 2.4, we discussed the techniques for combining two force vectors into a vector sum $\mathbf{S_v}$. Let's now reverse the process, and consider a technique for converting a single force vector into two equivalent vectors. Suppose we have a force vector **C** as shown in Fig. 2.11a. We can resolve this vector into a set of two vectors **A** and **B** that is equivalent to **C**. We show such a set of vectors in Fig. 2.11b. Clearly, an infinite set of selections of **A** and **B** can be made that are equivalent to **C**. The only requirement on the choice of **A** and **B** is that they satisfy the equation for vector addition:

$$\mathbf{A} + \mathbf{B} = \mathbf{C} \tag{2.6}$$

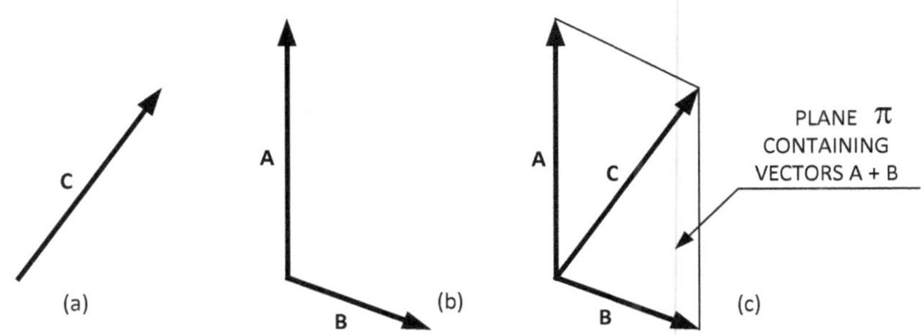

Fig. 2.11 Resolution of a vector **C** into an equivalent vector pair **A** and **B**.

We illustrate the vector addition involved in resolving a single vector into two equivalent vectors with the parallelogram presented in Fig. 2.11c. Note that the vector **C** and its equivalents **A** and **B** are coplanar—they all lie in the same plane π.

Let's next consider a special case of resolving a vector into an equivalent two vector pair oriented along the Cartesian axes. Suppose a force vector **F** lies in the x-y plane as illustrated in Fig. 2.12a.

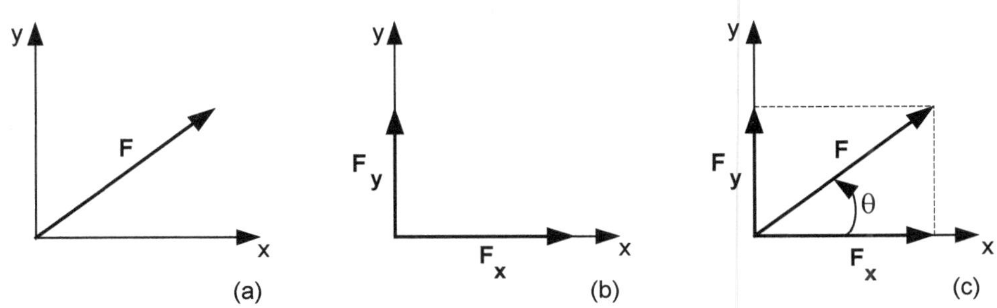

Fig. 2.12 Resolution of vector **F**, lying in the x-y plane, into two Cartesian force vectors $\mathbf{F_x}$ and $\mathbf{F_y}$.

The equivalent Cartesian force vectors $\mathbf{F_x}$ and $\mathbf{F_y}$ are shown in Fig. 2.12b where

$$\mathbf{F} = \mathbf{F_x} + \mathbf{F_y} \tag{2.7}$$

The magnitudes of the Cartesian force vectors are given by:

$$F_x = F \cos\theta \qquad \text{and} \qquad F_y = F \sin\theta \tag{2.8}$$

The relation between F and the Cartesian components F_x and F_y is given by:

$$F = [F_x^2 + F_y^2]^{1/2} \tag{2.9}$$

Resolution of a vector into its Cartesian components is extremely important, as this concept will be used repeatedly in applying the equations of equilibrium. Let's next consider two examples to demonstrate the procedures followed in determining the Cartesian components of a vector.

EXAMPLE 2.6

Determine the x and y components of the 600 lb force shown in the Fig. E2.6. Also check the solutions for F_x and F_y by verifying the value of 600 lb assigned to F.

Fig. E2.6

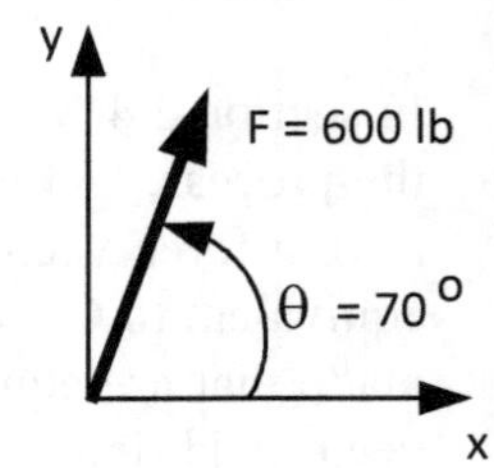

Solution:

From Eqs. (2.8), we write:

$$F_x = F\cos\theta = 600 \cos 70° = (600)(0.3420) = 205.2 \text{ lb}$$

$$F_y = F\sin\theta = 600 \sin 70° = (600)(0.9397) = 563.8 \text{ lb}$$

We use Eq. (2.9) to verify the value of 600 lb for the magnitude F.

$$F = [F_x^2 + F_y^2]^{1/2} = [(205.2)^2 + (563.8)^2]^{1/2} = 600 \text{ lb}$$

EXAMPLE 2.7

Determine the x and y components of the 50 kN force shown in Fig. E2.7. Also check the solutions for F_x and F_y by verifying the value of 50 kN assigned to F.

Fig. E2.7

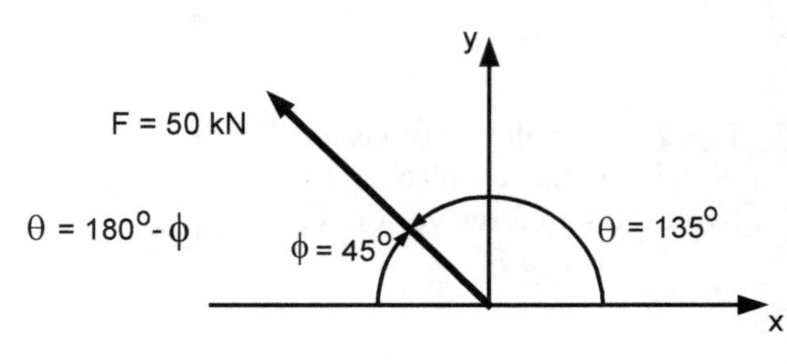

Solution:

From Eqs. (2.8), we write:

$$F_x = F \cos \theta = 50 \times 10^3 \cos 135° = (50 \times 10^3)(-0.7071) = -35.36 \text{ kN}$$

$$F_y = F \sin \theta = 50 \times 10^3 \sin 135° = (50 \text{ x } 10^3)(0.7071) = +35.36 \text{ kN}$$

We use Eq. (2.9) to verify the value of 50 kN for the magnitude F.

$$F = [F_x^2 + F_y^2]^{1/2} = [(-35.36)^2 + (35.36)^2]^{1/2} = \sqrt{2}\,(35.36) = 50 \text{ kN}$$

2.5.1 Adding Three or More Coplanar Vectors

In the previous discussion and examples, we consider ED the addition and subtraction of two vectors, **A** and **B**. A graphical method for performing the addition and subtraction was described and an analytical approach was discussed and demonstrated with examples. The analytical approach involved forming a triangle with sides A, B and C and employing the sine and cosine laws to solve for the unknown vector **C**. This approach is effective if you are required to add two vectors **A** and **B**, but the approach becomes tedious if it is necessary to add three or more vectors together. A much more efficient approach involves the use of adding the Cartesian components of each vector to form a vector sum.

Suppose we have the three vectors, **A**, **B** and **C** that lie in a common plane (coplanar) as shown in Fig. 2.13. Let's form the vector sum $\mathbf{S_v} = \mathbf{A} + \mathbf{B} + \mathbf{C}$.

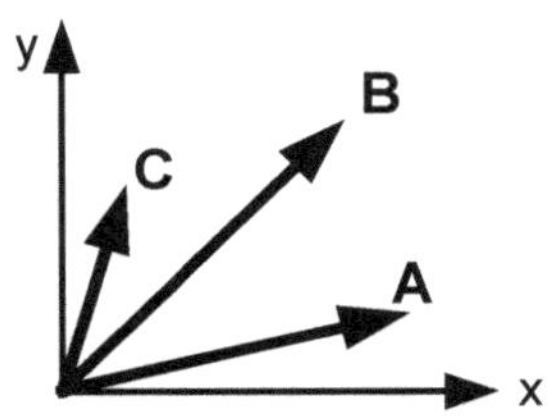

Fig. 2.13 Three vectors A, B and C.

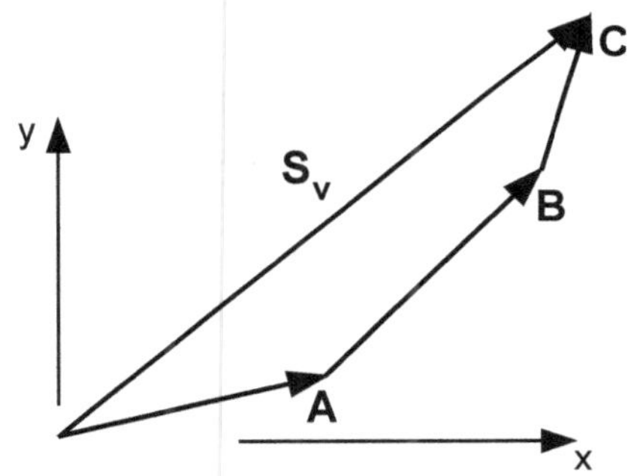

Fig. 2.14 Vector addition of three vectors A, B and C by a graphical technique.

Of course, we may add the three vectors using a graphical technique by connecting them together tip to tail as shown in Fig. 2.14.

Suppose we attempt to perform the addition of the three vectors by using the triangle approach described previously in Section 2.4. Clearly, the geometric shape in Fig. 2.14 is not a triangle because it has four sides (tetragon). To employ the laws of sines and cosines, it is necessary to form two triangles by dividing the tetragon as illustrated in Fig. 2.15.

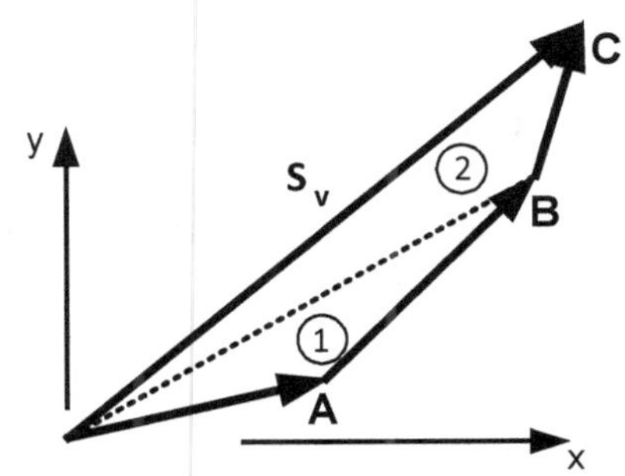

Fig. 2.15 Dividing the vector tetragon into two triangles.

It is possible to solve for the vector sum $\mathbf{S_v}$ if we employ the sine and cosine laws twice; however, this approach is time consuming and mathematical errors are common. It is recommended that you resolve the three (or more) vectors into their Cartesian components, and add them in the manner described in the following paragraphs and examples.

Suppose you resolve the vectors **A**, **B** and **C** and determine their Cartesian components using Eq. (2.8). These Cartesian components are illustrated in Fig. 2.16.

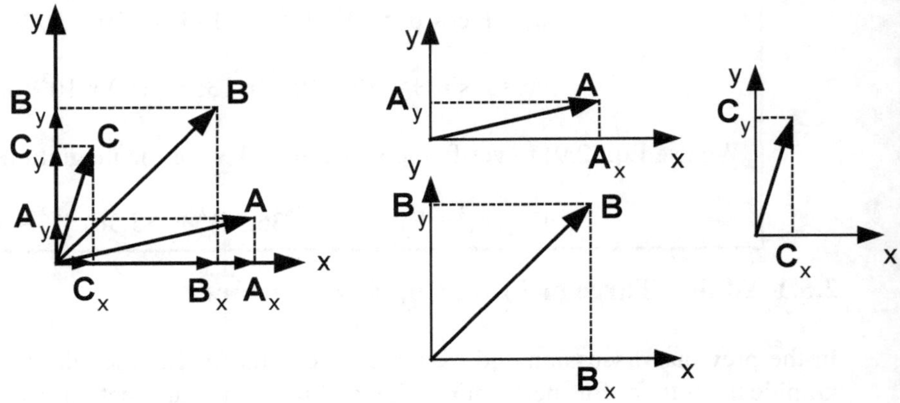

Fig. 2.16 Resolving vectors **A**, **B**, and **C** into Cartesian components A_x, A_y, B_x, B_y, C_x and C_y.

The next step is to add the Cartesian components together to obtain:

$$S_{vx} = A_x + B_x + C_x \qquad \text{and} \qquad S_{vy} = A_y + B_y + C_y \tag{2.10}$$

Substituting Eq. (2.10) into Eq. (2.9) yields the magnitude of the sum of the three vectors as:

$$S_v = [(A_x + B_x + C_x)^2 + (A_y + B_y + C_y)^2]^{1/2} \tag{2.11}$$

The vector $\mathbf{S_v}$ and its Cartesian components S_{vx} and S_{vy} are presented in Fig. 2.17. Also shown in this figure is the angle θ that defines the direction of the vector relative to the x-axis.

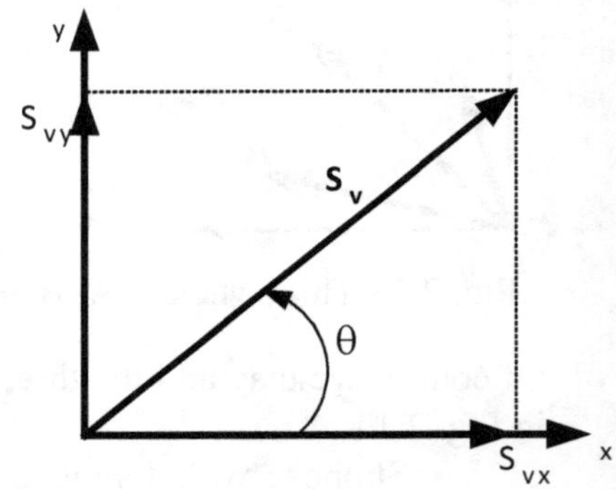

Fig. 2.17 The vector $\mathbf{S_v}$ and its Cartesian components S_{vx} and S_{vy}.

The direction of the vector $\mathbf{S_v}$ is given by θ, which is determined from:

$$\theta = \sin^{-1}(S_{vy}/S_v) = \cos^{-1}(S_{vx}/S_v) = \tan^{-1}(S_{vy}/S_{vx}) \tag{2.12}$$

EXAMPLE 2.8

Three vectors $\mathbf{F_1}$, $\mathbf{F_2}$, and $\mathbf{F_3}$ are illustrated in Fig. E2.8. Determine the sum of these three vectors and calculate the angle θ the vector $\mathbf{S_v}$ makes with the x-axis.

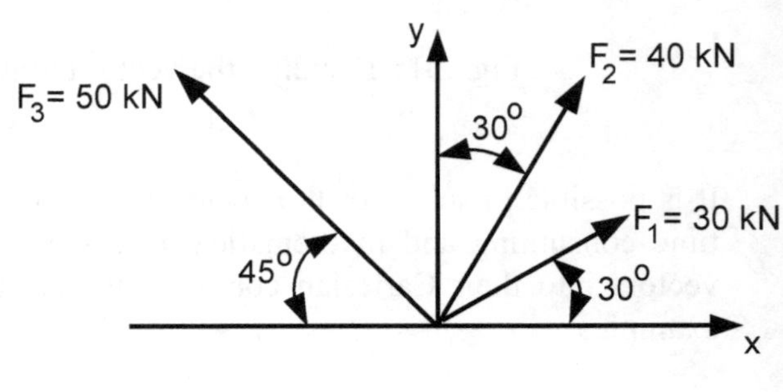

Fig. E2.8

Solution: Determine the Cartesian components of the three forces using Eq. (2.8) as:

$F_{1x} = 30 \cos(30°) = 25.98$ kN $F_{1y} = 30 \sin(30°) = 15.00$ kN
$F_{2x} = 40 \cos(90° - 30°) = 20.00$ kN $F_{2y} = 40 \sin(60°) = 34.64$ kN
$F_{3x} = 50 \cos(180° - 45°) = -35.36$ kN $F_{3y} = 50 \sin(135°) = 35.36$ kN

Using Eq. (2.10) determine the Cartesian components of the vector sum $\mathbf{S_v}$ as:

$$S_{vx} = F_{1x} + F_{2x} + F_{3x} = 25.98 + 20.00 - 35.36 = 10.62 \text{ kN}$$
$$S_{vy} = F_{1y} + F_{2y} + F_{3y} = 15.00 + 34.64 + 35.36 = 85.00 \text{ kN}$$

The magnitude S_v is determined from Eq. (2.11) as:

$$S_v = [S_{vx}^2 + S_{vy}^2]^{1/2} = [(10.62)^2 + (85.00)^2]^{1/2} = (7338)^{1/2} = 85.66 \text{ kN}$$

Finally, the direction of the vector $\mathbf{S_v}$ relative to the x-axis is given by Eq. (2.12) as:

$$\theta = \sin^{-1}(S_{vy}/S_v) = \sin^{-1}(85.00/85.66) = 82.88°$$

EXAMPLE 2.9

The gusset plate illustrated in Fig. E2.9 is subjected to four forces from the attached uniaxial structural members. If the vector sum of these four forces is zero, determine the forces F_3 and F_4.

Fig. E2.9

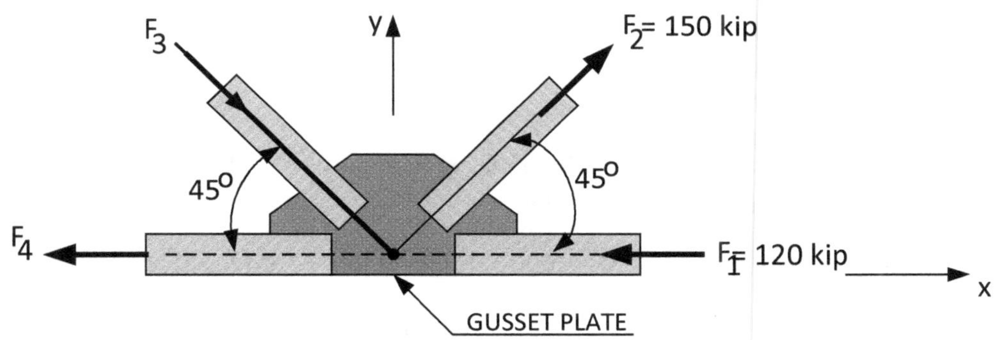

Solution:

The fact that the vector sum of the four forces $\mathbf{S_v} = 0$, implies that:

$$S_{vx} = S_{vy} = 0 \qquad \text{(a)}$$

Use Eq. (2.8), Eq. (2.10) and Eq. (a) to obtain:

$$S_{vy} = F_1 \sin(0°) + F_2 \sin(45°) - F_3 \sin(45°) + F_4 \sin(0°) = 0$$

$$F_2 = F_3 = 150 \text{ kip} \qquad \text{(b)}$$

Finally, Eq. (2.8), Eq. (2.10), Eq. (a) and Eq. (b) enable us to write:

$$S_{vx} = -F_1 \cos(0°) + F_2 \cos(45°) + F_3 \cos(45°) - F_4 \cos(0°) = 0$$

$$F_4 = \cos(45°)[F_2 + F_3] - F_1 = (0.7071)(300) - 120 = 92.13 \text{ kip}$$

EXAMPLE 2.10

Determine the magnitude and the direction of the force **F** in Fig. E2.10 if the box is lifted from the floor without swinging. The weight of the box is 5600 N.

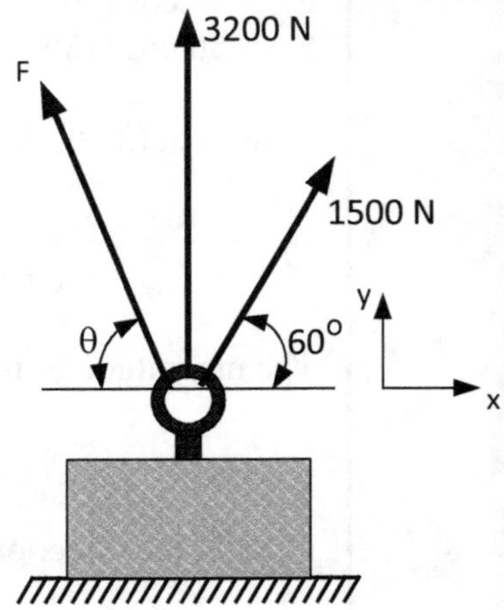

Fig. E2.10

Solution:

If the box is lifted without swinging the component of the vector sum of the three forces in the x direction must be zero, or $S_{vx} = 0$

From Eq. (2.8) and Eq. (2.10), we may write:

$$S_{vx} = 1500 \cos(60°) - F \cos\theta = 0 \qquad \text{(a)}$$

$$F \cos\theta = 750 \text{ N} \qquad \text{(b)}$$

For the box to lift from the floor the vertical sum of the three forces must equal the box's weight of 5600 N. This fact permits us to write:

$$S = W = 5600 = 1500 \sin(60°) + 3200 + F \sin\theta \qquad \text{(c)}$$

$$F \sin\theta = 1101 \text{ N} \qquad \text{(d)}$$

Dividing the results of Eq. (d) by Eq. (b) yields:

$$(\sin\theta/\cos\theta) = \tan\theta = 1101/750 = 1.4679 \qquad \text{(e)}$$

$$\theta = 55.74° \quad \text{and} \quad F = 1101/\sin(55.74°) = 1332 \text{ N} \qquad \text{(f)}$$

2.6 CONCURRENT AND COPLANAR FORCES

Concurrent and **coplanar** are descriptive words identifying certain vector systems. Let's start with a concurrent vector system. When two or more forces (or vectors) act at a point, as illustrated in Fig. 2.18, they are concurrent. When a body is subjected to a concurrent set of forces, it does not exhibit a tendency to rotate (i.e. the moments about the point Q are zero). Equilibrium of the body is determined by using only Eq. (1.2).

$$\Sigma \mathbf{F} = 0 \qquad \text{(1.2 bis)}$$

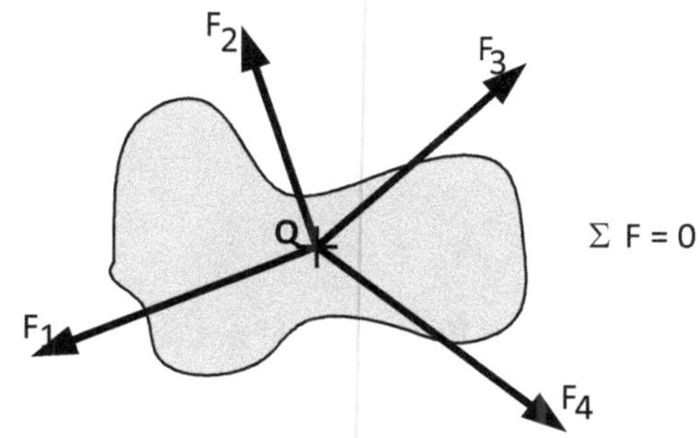

Fig. 2.18 Forces $\mathbf{F}_1$, $\mathbf{F}_2$, $\mathbf{F}_3$, and $\mathbf{F}_4$ are concurrent because they all are applied to point Q.

While the equilibrium of forces in Eq. (1.2) will prevent translational motion of a body, it does not guarantee that the body will not rotate. To understand what makes bodies rotate, the concept of **moments** must be introduced. Moment vectors, **M**, can be produced by force systems that are not concurrent. The equilibrium condition for moments is given by:

$$\Sigma \mathbf{M} = 0 \tag{2.13}$$

This relation is satisfied automatically for concurrent force systems. We will discuss moments produced by forces in much more detail in Sections 2.8 and 2.9.

Coplanar forces (or vectors) all lie in the same plane. If we have any two vectors, it is easy to show that they are coplanar regardless of their orientation. However, if we deal with three or more vectors, the vector system may or may not be coplanar. Examples of planar and non-planar vector systems are shown in Figs. 2.19 and 2.20 respectively.

Fig. 2.19 Examples of coplanar forces.

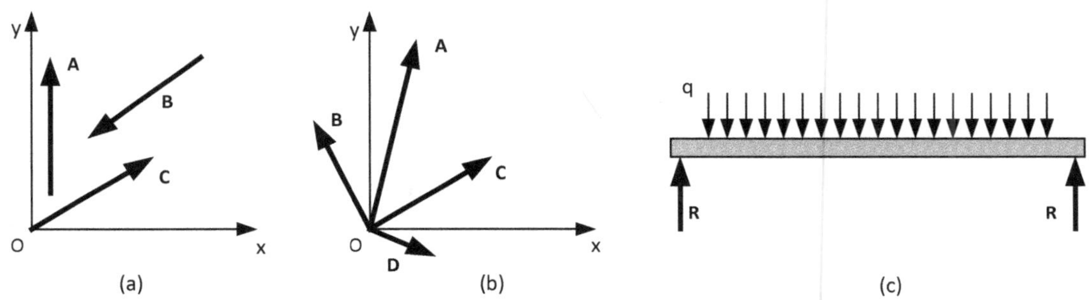

In Fig. 2.19a, the vectors **A**, **B**, and **C** all lie in the x-y plane, and the vector system is coplanar. However, this system is not concurrent. Vectors **A** and **C** both can be shifted along their lines of action to produce an intersection of their tail points, but it is impossible to shift vector **B** along its line of action to intersect at this point. Consequently, vector **B** is offset from this intersection point, and produces a moment about it.

In Fig. 2.19b, the vectors **A**, **B**, **C** and **D** all lie in the x-y plane, and the vector system is coplanar. Moreover, this system is concurrent because all four of the vectors intersect at the origin O.

The force vectors in Fig. 2.19c are coplanar because both of the reaction forces **R** and the individual forces constituting the distributed load q all act in the plane of the beam. This system is not concurrent because all of the force vectors are parallel, and cannot be shifted along their lines of action to intersect at a common point.

The three dimensional force systems illustrated in Figs. 2.20a and 2.20b are not coplanar. In Fig. 2.20a, we show a Cartesian force system with vectors $\mathbf{F}_x$, $\mathbf{F}_y$ and $\mathbf{F}_z$ that coincide with the x, y and z axes, respectively. Since these three forces pass though the origin, the system is concurrent. In Fig. 2.20b, the vector $\mathbf{F}_1$ lies in the x-z plane, $\mathbf{F}_2$ lies in the y-z plane and $\mathbf{F}_3$ lies in the x-y plane. Clearly, the system is not coplanar nor is it concurrent.

Classification of force vector systems is important because it indicates the number of equations of equilibrium that provide useful relations in solving for the forces and moments occurring in structures. We will present the equilibrium relations for the four possible classifications shown below in Chapter 3.

- Coplanar and concurrent.
- Coplanar and non-concurrent.
- Non-coplanar and concurrent
- Non-coplanar and non-concurrent.

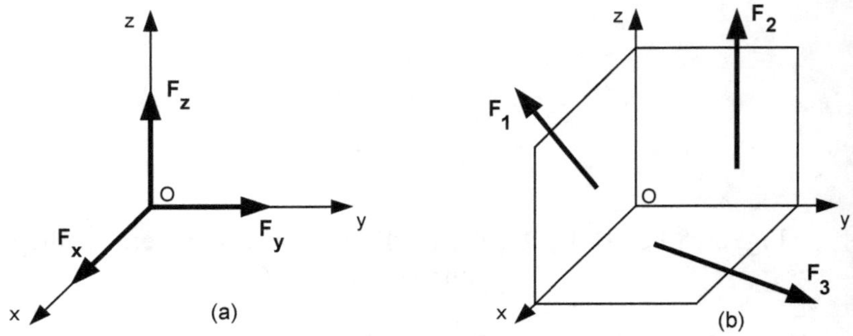

Fig. 2.20 Examples of non-coplanar vectors.

Three-dimensional force systems, like those illustrated in Fig. 2.20, are referred to as **space forces**. They are more complex because of the additional equations that must be satisfied to maintain a body in equilibrium when it is subjected to a three-dimensional system of forces.

2.7 SPACE FORCES

When forces are coplanar, we place them on a x-y plane and then apply the equilibrium equations to solve for the unknown forces acting on the structure under analysis. However, in some structures we must consider space forces, which cannot be completely characterized on the x–y plane. Consider the space force **F** shown in Fig. 2.21, which is represented with Cartesian vectors $\mathbf{F_x}$, $\mathbf{F_y}$ and $\mathbf{F_z}$.

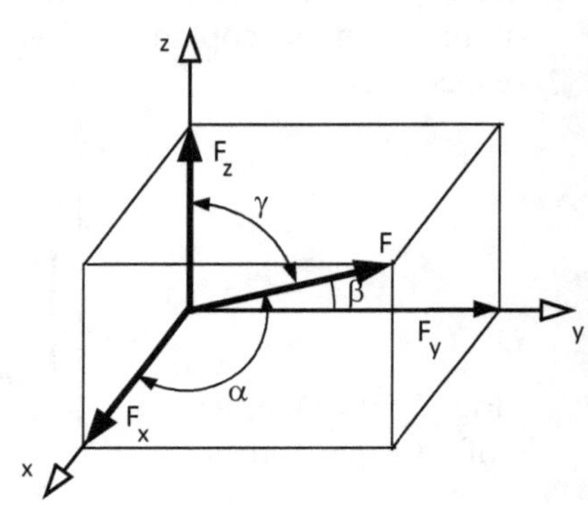

Fig. 2.21 A space force F is shown with its equivalent Cartesian vectors $\mathbf{F_x}$, $\mathbf{F_y}$ and $\mathbf{F_z}$.

The space force **F** has both magnitude and direction. Its magnitude is a scalar quantity F, and its direction relative to the three-dimensional Cartesian coordinate system is given by the angles α, β and γ. These quantities, known as the coordinate direction angles, are measured from the line of action of **F** to the positive x, y and z-axes.

The coordinate direction angles of the space vector F are given by:

$$\cos\alpha = \frac{F_x}{F} \qquad \cos\beta = \frac{F_y}{F} \qquad \cos\gamma = \frac{F_z}{F} \tag{2.14}$$

Also, recall the important identity for the coordinate direction angles, which is:

$$\cos^2\alpha + \cos^2\beta + \cos^2\gamma = 1 \tag{2.15}$$

From Eq. (2.14), it is evident that the magnitudes of the Cartesian components of the space force are given by:

$$F_x = F\cos\alpha \qquad F_y = F\cos\beta \qquad F_z = F\cos\gamma \tag{2.16}$$

The terms cos α, cos β and cos γ that appear in Eq. (2.14) and Eq. (2.16) are called direction cosines of the vector F.

Finally, the magnitude of F is given by the positive square root of the sum of the squares of its three components.

$$F = \sqrt{F_x^2 + F_y^2 + F_z^2} \tag{2.17}$$

EXAMPLE 2.11

A space force with a magnitude of 4500 N is positioned at the origin of a Cartesian coordinate system. The force is oriented so that the angles $\alpha = 45°$ and $\beta = 60°$. Determine the Cartesian components of the space force and the angle that it makes with the z-axis.

Solution:

From Eq. (2.15), we write:

$$\cos^2 \alpha + \cos^2 \beta + \cos^2 \gamma = \cos^2 (45°) + \cos^2 (60°) + \cos^2 \gamma = 1$$

This relation reduces to:

$$\cos^2 \gamma = 1 - 0.5 - 0.25 = 0.25$$

Solving for the angle γ yields:

$$\cos \gamma = 0.5 \qquad \Rightarrow \qquad \Rightarrow \qquad \gamma = 60°$$

Next, let's recall Eq. (2.16) and apply it to determine the Cartesian components of the space force.

$$F_x = F \cos \alpha \qquad F_y = F \cos \beta \qquad F_z = F \cos \gamma$$

$$F_x = 4500 \cos 45° \qquad F_y = 4500 \cos 60° \qquad F_z = 4500 \cos 60°$$

$$F_x = 3182 \text{ N} \qquad F_y = 2250 \text{ N} \qquad F_z = 2250 \text{ N}$$

We may check the accuracy of the result by employing Eq. (2.17).

$$F = \sqrt{F_x^2 + F_y^2 + F_z^2}$$

Substituting numerical values into this relation gives:

$$F = \sqrt{(3182)^2 + (2250)^2 + (2250)^2} = \sqrt{2025 \times 10^4} = 4500 \text{ N}$$

The results confirm the accuracy of the computations.

EXAMPLE 2.12

A space force is represented by its Cartesian components as:

$$F_x = 707 \text{ lb} \qquad F_y = 883 \text{ lb} \qquad F_z = 264 \text{ lb}$$

Determine the magnitude and the coordinate direction angles of the space force.

Solution: Let's first determine the magnitude of the space force by using Eq. (2.16)

$$F = \sqrt{F_x^2 + F_y^2 + F_z^2} = \sqrt{(707)^2 + (883)^2 + (264)^2} = 1162 \text{ lb}$$

The coordinate direction angles are calculated from Eq. (2.14) as:

$$\cos\alpha = \frac{F_x}{F} = \frac{707}{1162} = 0.6087$$

$$\cos\beta = \frac{F_y}{F} = \frac{883}{1162} = 0.7602$$

$$\cos\gamma = \frac{F_z}{F} = \frac{264}{1162} = 0.2273$$

Solving for the coordinate direction angles yields:

$\alpha = 52.51°$ $\beta = 40.52°$ $\gamma = 76.86°$

Again we may check the accuracy of our calculations by using Eq. (2.15). Substituting the results for the coordinate direction angles into this relation yields:

$$\cos^2\alpha + \cos^2\beta + \cos^2\gamma = (0.6087)^2 + (0.7602)^2 + (0.2273)^2 = 1.000$$

Because the identity for the coordinate direction angles is satisfied, we may conclude that the calculation was performed correctly.

2.7.1 Unit Vectors

Let's define a force vector **F** in terms of unit vectors as shown below:

$$\mathbf{F} = F_x\,\mathbf{i} + F_y\,\mathbf{j} + F_z\,\mathbf{k} \tag{2.18}$$

where F_x, F_y, F_z are the magnitudes of the Cartesian force components, and **i, j, k** are the Cartesian unit vectors in the x, y and z directions, respectively.

As the name implies, the unit vectors have a magnitude of unity. They provide directions along the x, y and z-axes in Eq. (2.18). The unit vectors **i, j** and **k** are illustrated in Fig. 2.22.

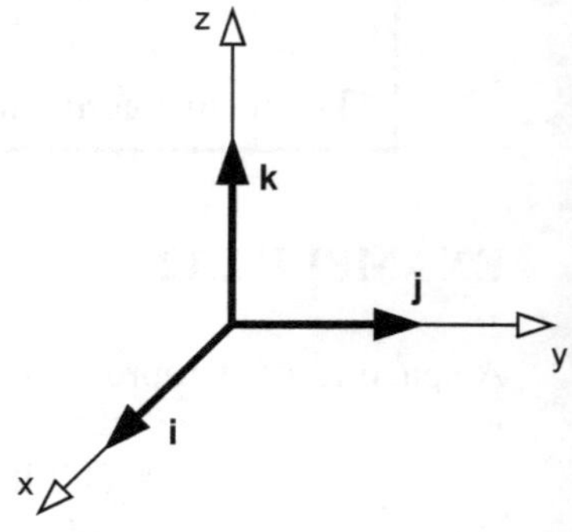

Fig. 2.22 Unit vectors **i, j** and **k** oriented along the x, y and z axes, respectively.

It is also useful to employ unit vectors that are oriented in an arbitrary direction. For example consider a vector **F** as shown in Fig. 2.23.

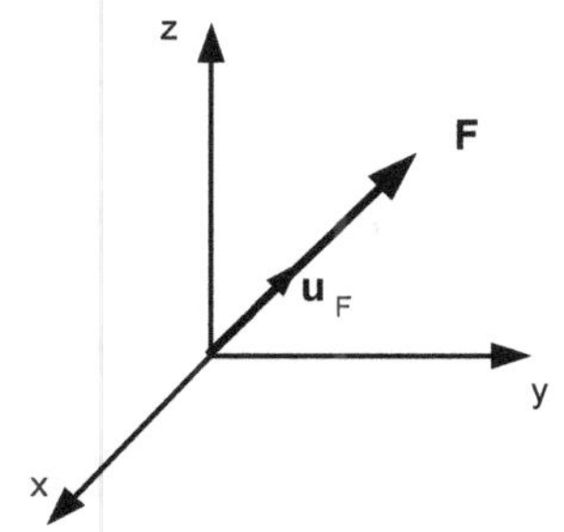

Fig. 2.23 A unit vector $\mathbf{u}_F$ gives the direction of the vector **F**.

The direction of this unit vector is given by:

$$\mathbf{u}_F = \mathbf{F}/F \quad (2.19)$$

Hence, the vector **F** may be written as:

$$\mathbf{F} = F\mathbf{u}_F \quad (2.20)$$

A comparison of Eqs. (2.18) and (2.20) indicates that we may write the equation for a force using either the Cartesian unit vectors, which are directed along the x, y and z-axes, or by a single unit vector along the line of action of the force. The direction of the force may also be expressed using either the unit vector $\mathbf{u}_F$ or the coordinate direction angles. This fact implies a relation between these quantities. To show this relation, we substitute Eq. (2.18) into Eq. (2.19) to obtain:

$$\mathbf{u}_F = \mathbf{F}/F = (F_x/F)\mathbf{i} + (F_y/F)\mathbf{j} + (F_z/F)\mathbf{k} \quad (2.21)$$

Substituting Eq. (2.14) into Eq. (2.21) yields:

$$\mathbf{u}_F = \cos\alpha\,\mathbf{i} + \cos\beta\,\mathbf{j} + \cos\gamma\,\mathbf{k} \quad (2.22)$$

EXAMPLE 2.13

A space force is represented by its Cartesian components as:

$F_x = 633$ N $\qquad F_y = 216$ N $\qquad F_z = 521$ N

Write an expression for the force vector **F** in terms of the Cartesian unit vectors and determine the magnitude of the force F.

Solution:

Using Eq. (2.18), we write:

$$F = (633\,\mathbf{i} + 216\,\mathbf{j} + 521\,\mathbf{k})\text{ N}$$

The magnitude F is calculated from Eq. (2.17) as:

$$F = \sqrt{F_x^2 + F_y^2 + F_z^2} = \sqrt{(633)^2 + (216)^2 + (521)^2} = 847.8\text{ N}$$

EXAMPLE 2.14

Determine the unit vector describing the direction of the space force specified in Example 2.13. Also, calculate the coordinate direction angles for this force vector.

Solution:

To determine the unit vector $\mathbf{u}_F$, we substitute the results from Example 2.13 into Eq. (2.21) to obtain:

$$\mathbf{u}_F = \mathbf{F}/F = (F_x/F)\mathbf{i} + (F_y/F)\mathbf{j} + (F_z/F)\mathbf{k} = (633/847.8)\,\mathbf{i} + (216/847.8)\,\mathbf{j} + (521/847.8)\,\mathbf{k}$$

$$\mathbf{u}_F = 0.7466\,\mathbf{i} + 0.2548\,\mathbf{j} + 0.6145\,\mathbf{k}$$

The coordinate direction angles for the space force are determined from Eq. (2.14) as:

$$\cos\alpha = \frac{F_x}{F} = \frac{633}{847.8} = 0.7466$$

$$\cos\beta = \frac{F_y}{F} = \frac{216}{847.8} = 0.2548$$

$$\cos\gamma = \frac{F_z}{F} = \frac{521}{847.8} = 0.6145$$

Solving for the coordinate direction angles gives:

$\alpha = 41.70°$ $\beta = 75.24°$ $\gamma = 52.08°$

EXAMPLE 2.15

Express the space force **F**, shown in Fig. E2.15, as a Cartesian vector.

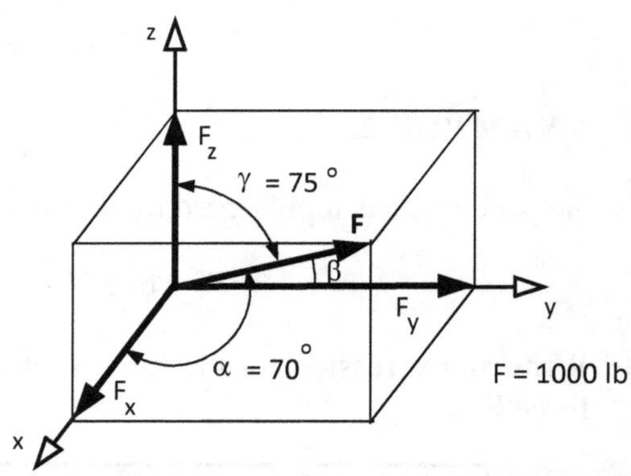

Fig. E2.15

Solution:

Because only two of the direction cosine angles have been specified, we begin by using Eq. (2.15) to determine the unknown angle γ.

$$\cos^2\beta = 1 - \cos^2\alpha - \cos^2\gamma = 1 - \cos^2(70°) - \cos^2(75°)$$

$$\cos^2\beta = 1 - 0.1170 - 0.06699 = 0.8160$$

$$\cos\beta = \pm\,0.9033 \quad \Rightarrow \quad \Rightarrow \quad \beta = 25.40° \text{ or } 154.6°$$

By inspection of Fig. E2.15, it is evident that β is an acute angle; therefore:

$$\beta = 25.40°$$

Next, employ Eq. (2.16) to determine the Cartesian components of the space force **F** as:

$$F_x = F \cos \alpha = 1000 \cos (70.0°) = 342.0 \text{ N}$$

$$F_y = F \cos \beta = 1000 \cos (25.4°) = 903.3 \text{ N}$$

$$F_z = F \cos \gamma = 1000 \cos (75.0°) = 258.8 \text{ N}$$

Finally, let's check the accuracy of these results by using Eq. (2.17).

$$F = \sqrt{F_x^2 + F_y^2 + F_z^2} = \sqrt{(342.0)^2 + (903.3)^2 + (258.8)^2} = 999.9 \text{ N}$$

The result of 999.9 N for the magnitude of F is within 1 part in 10,000 for the 1000 N force. The very small difference, of no consequence, is due to rounding errors.

2.7.2 Position Vectors

A position vector is a fixed vector in space. We use a position vector to locate a point in space relative to another point. A position vector is similar to a force vector in that it has both magnitude and direction. An example of a position vector is presented in Fig. 2.24. In this case, the position vector **r** locates a point A (x, y, z) relative to the origin O of the coordinate system.

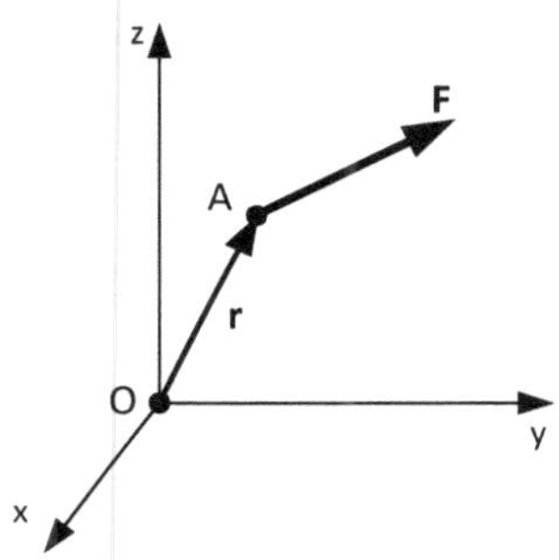

Fig. 2.24 This position vector **r** locates the point of application of **F** relative to the origin.

The position vector in Fig. (2.24) is expressed as:

$$\mathbf{r} = x\mathbf{i} + y\mathbf{j} + z\mathbf{k} \tag{2.23}$$

The coordinate direction angles of the position vector **r** are:

$$\cos \alpha = x/r$$

$$\cos \beta = y/r \tag{2.24}$$

$$\cos \gamma = z/r$$

Finally, the magnitude of the position vector r is given by:

$$r = [x^2 + y^2 + z^2]^{1/2} \tag{2.25}$$

EXAMPLE 2.16

A force vector **F** is applied to a rectangular space frame at point A as shown in Fig. E2.16. Write the equation for the position vector **r** locating the point of application relative to the origin of the Cartesian coordinates. Also determine its magnitude and direction.

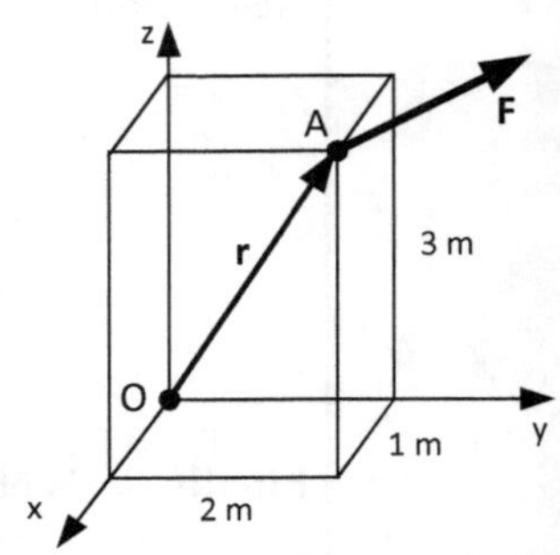

Fig E2.16

Solution:

Selecting the dimensions of the rectangular space frame from Fig. E2.16 and using Eq. (2.23) enables us to write:

$$\mathbf{r} = (1\,\mathbf{i} + 2\,\mathbf{j} + 3\,\mathbf{k})\ \text{m}$$

The magnitude of the position vector is given by substituting the coordinates into Eq. (2.25):

$$r = \sqrt{x^2 + y^2 + z^2} = \sqrt{(1)^2 + (2)^2 + (3)^2} = 3.742\ \text{m}$$

Next, we employ Eq. (2.24) to determine the coordinate direction angles for the vector **r**.

$$\cos\alpha = x/r = 1/3.742 = 0.2673 \quad \Rightarrow \quad \alpha = 74.50°$$

$$\cos\beta = y/r = 2/3.742 = 0.5345 \quad \Rightarrow \quad \beta = 57.69°$$

$$\cos\gamma = z/r = 3/3.742 = 0.8018 \quad \Rightarrow \quad \gamma = 36.70°$$

Let's check the accuracy of the calculations by using Eq. (2.15).

$$\cos^2\alpha + \cos^2\beta + \cos^2\gamma = (0.2673)^2 + (0.5345)^2 + (0.8018)^2 = 1.000$$

Okay! The results for the direction cosine angles check.

Let's consider a more general case for a position vector that locates point B relative to point A. In this case neither point A nor B are located at the origin of the coordinate system. The position vector for this situation is presented in Fig. 2.25.

The **i**, **j** and **k** components of the position vector **r** in Fig. 2.25 are written by taking the coordinates of the tip of the vector B (x_B, y_B, z_B) and subtracting them from the coordinates of the tail A(x_A, y_A, z_A).

$$\mathbf{r} = (x_B - x_A)\mathbf{i} + (y_B - y_A)\mathbf{j} + (z_B - z_A)\mathbf{k} \qquad (2.26)$$

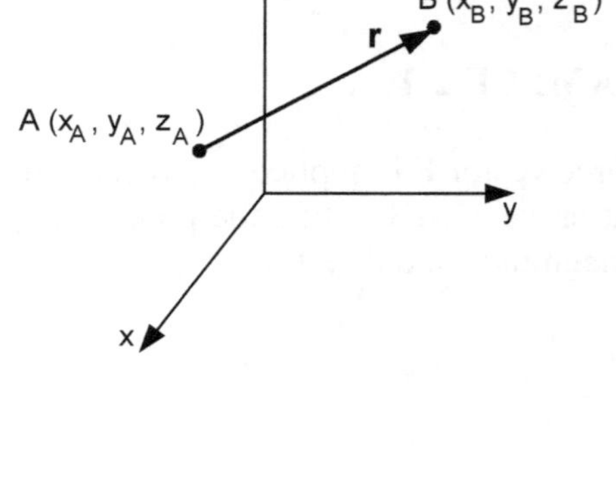

Fig. 2.25 A position vector r that locates point B relative to point A.

The coordinate direction angles of the position vector **r** are:

$$\cos\alpha = (x_B - x_A)/r$$

$$\cos\beta = (y_B - y_A)/r \qquad (2.27)$$

$$\cos\gamma = (z_B - z_A)/r$$

These coordinate direction angles are measured from a local coordinate system positioned at point A (the tail of the position vector **r**).

The magnitude of the position vector in Fig. 2.25 is given by:

$$\mathbf{r} = \sqrt{(x_B - x_A)^2 + (y_B - y_A)^2 + (z_B - z_A)^2} \qquad (2.28)$$

EXAMPLE 2.17

Determine the magnitude and direction of the position vector that extends from point A to B in Fig. 2.25 if the coordinates of points A and B are given by:

A(2, 0, −6) and B(−4, 4, 6)

The coordinates are expressed in m.

Solution:

Let's write the expression for r by using Eq. (2.26).

$$\mathbf{r} = (x_B - x_A)\mathbf{i} + (y_B - y_A)\mathbf{j} + (z_B - z_A)\mathbf{k} = (-4 - 2)\,\mathbf{i} + (4 - 0)\,\mathbf{j} + (6 + 6)\,\mathbf{k}$$

$$\mathbf{r} = (-6\,\mathbf{i} + 4\,\mathbf{j} + 12\,\mathbf{k})\text{ m}$$

The magnitude of the position vector is given by:

$$\mathbf{r} = \sqrt{(-6)^2 + (4)^2 + (12)^2} = 14\text{ m}$$

The coordinate direction angles are computed from Eq. 2.27.

$$\cos\alpha = (x_B - x_A)/r = -6/14 = -0.4286 \quad \Rightarrow \quad \alpha = 115.4°$$

$$\cos\beta = (y_B - y_A)/r = 4/14 = 0.2857 \quad \Rightarrow \quad \beta = 72.40°$$

$$\cos\gamma = (z_B - z_A)/r = 12/14 = 0.8571 \quad \Rightarrow \quad \gamma = 31.00°$$

We may also ascertain the direction of **r** by writing the expression for the unit vector $\mathbf{u}_r$. Let's adapt Eq. (2.21) and write:

$$\mathbf{u}_r = \mathbf{r}/r = (-6\,\mathbf{i} + 4\,\mathbf{j} + 12\,\mathbf{k})/14 = -0.4286\,\mathbf{i} + 0.2857\,\mathbf{j} + 0.8571\,\mathbf{k}$$

Note the coefficients of the unit vectors are identical to the results for the coordinate direction cosines.

EXAMPLE 2.18

A helicopter is attempting to tow a disabled supply truck along a road parallel to the y-axis in Fig. E2.18a. Unfortunately, the pilot is not properly aligned for the task with an initial position given by the coordinates x = – 10, y = – 15, and z = 20 feet relative to the hitch point on the truck. If a force component F_y of 1250 lb is required to tow the truck, determine the force that the helicopter must apply to the towline.

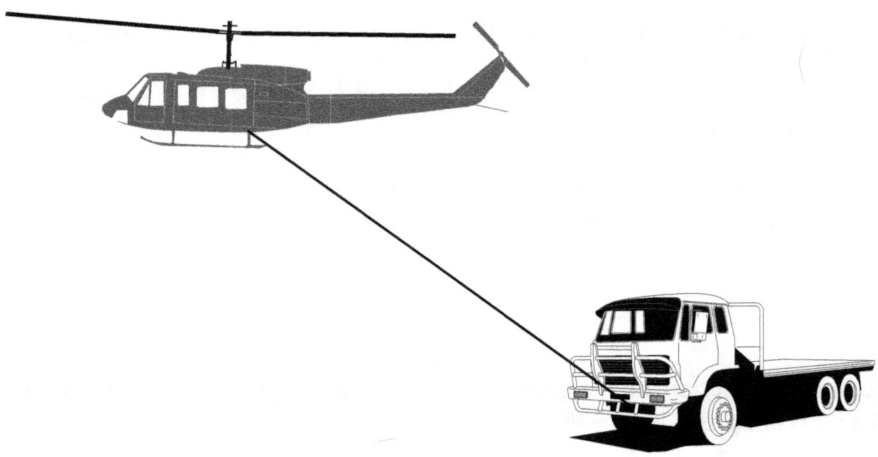

Fig. E2.18

Solution:

Prepare a drawing of the truck tow point and the helicopter hitch point relative to a three-dimensional Cartesian coordinate system as shown in Fig. E2.18a. We will use the tow point on the truck as the origin of the Cartesian coordinates.

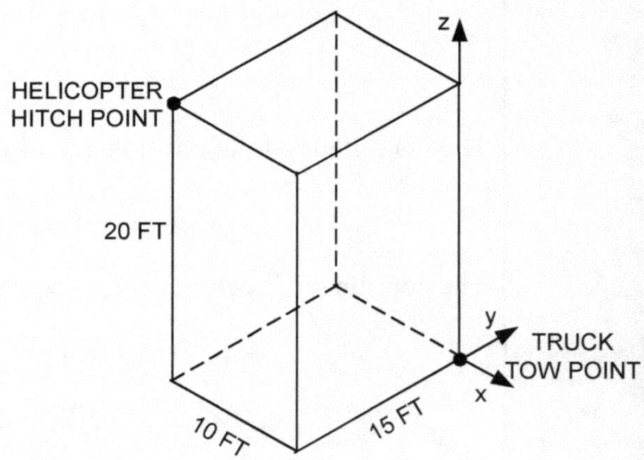

Fig. E2.18 a

Let's begin by determining the coordinate direction angles of the force that the helicopter will exert on the truck. The magnitude of position vector from the truck tow point to the helicopter hitch point is given by Eq. (2.25).

The coordinate direction angles are given by Eq. (2.24) as indicated below:

$$r = \sqrt{x^2 + y^2 + z^2} = \sqrt{(-10)^2 + (-15)^2 + (20)^2} = 26.926 \text{ ft}$$

$$\cos\alpha = \frac{x}{r} = \frac{-10}{26.926} = -0.3714 \quad \Rightarrow \alpha = 111.8^{\circ}$$

$$\cos\beta = \frac{y}{r} = \frac{-15}{26.926} = -0.5571 \quad \Rightarrow \beta = 123.9^{\circ}$$

$$\cos\gamma = \frac{z}{r} = \frac{20}{26.926} = 0.7428 \quad \Rightarrow \gamma = 42.03^{\circ}$$

Since the force and the rope are along the same line of action, the coordinate direction angles are identical for the position vector **r** and the force **F**. Recognizing this fact permits us to solve for the magnitude of the force necessary to tow the truck by using Eq. (2.14).

$$F = F_y/\cos\beta$$

$$F = -1250/(-0.5571) = +2244 \text{ lb}$$

The negative sign for the force component F_y is used to recognize that the force is being applied in the negative y direction.

EXAMPLE 2.19

A large pontoon is moored at a dock with a line attached to the cylinder at a mooring cleat as indicated in Fig. E2.19. If the cleat is at a position on the pontoon given by coordinates, x = 4 m, y = – 6 m and z = – 3 m, determine the tension in the mooring line if the wind and current are producing a drag force of 665 N in the x direction.

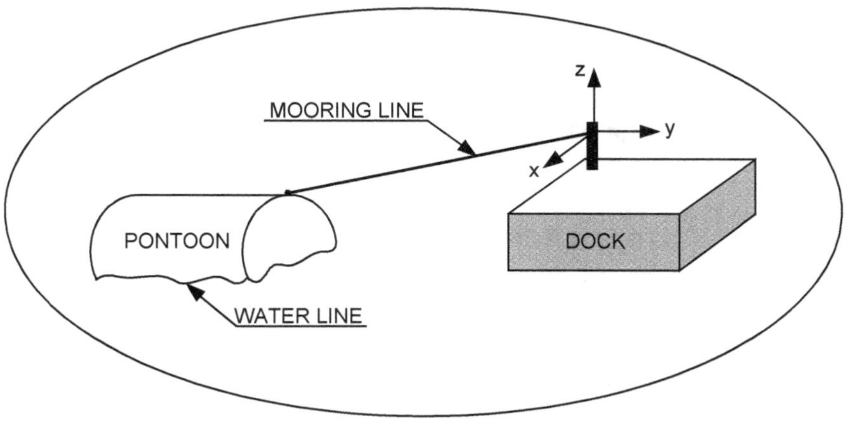

Fig. E2.19

Solution:

This situation is similar to the previous example where a rope or line is used to apply a force. In these situations, the position vector describing the orientation of the line and the force coincide. Since we know the coordinates of the rope, we can determine its length and orientation from Eqs. (2.24) and (2.25). The length of the position vector from the dock to the cleat is:

$$r = \sqrt{(4)^2 + (-6)^2 + (-3)^2} = 7.810 \text{ m}$$

The coordinate direction angles are given by modifying Eq. (2.24) as indicated below:

$$\cos\alpha = \frac{r_x}{r} = \frac{4}{7.810} = 0.5121 \quad \Rightarrow \alpha = 59.20^\circ$$

$$\cos\beta = \frac{r_y}{r} = \frac{-6}{7.810} = -0.7682 \quad \Rightarrow \beta = 140.2^\circ$$

$$\cos\gamma = \frac{r_z}{r} = \frac{-3}{7.810} = -0.3841 \quad \Rightarrow \gamma = 121.6^\circ$$

The magnitude of the force on the mooring line is determined from Eq. (2.16) as:

$$F = F_x/\cos\alpha = 665/0.5121 = 1299 \text{ N}$$

2.8 MOMENTS

If you have ever used a wrench or screwdriver to tighten a bolt or screw, you have generated a moment. A moment M_o is produced when a force F, as shown in Fig. 2.26, is applied in such a manner that it tends to cause a body to rotate about point O. The magnitude of a moment produced by a force is dependent on the location of the point O, and is given by:

$$M_o = F\,d \tag{2.29}$$

where d is the perpendicular distance from the point O to the line of action of F.

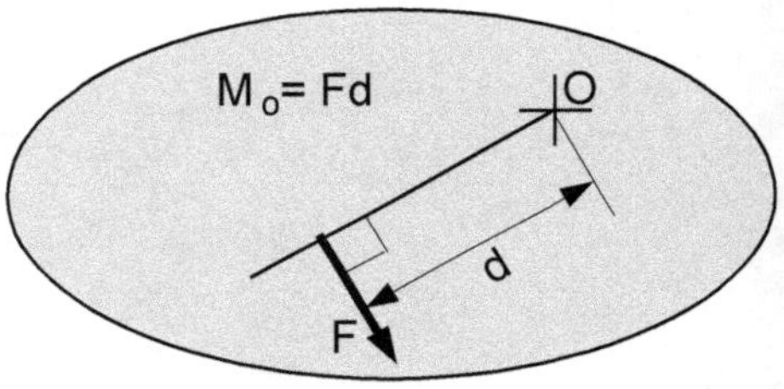

Fig. 2.26 A moment produced by a force depends on the position of the point O.

The units of a moment $\mathbf{M_o}$ are given as N-m or ft-lb as was stated in Table 1.2. $\mathbf{M_o}$ is a vector quantity that must be specified with both magnitude and direction. The magnitude is given by Eq. 2.29, and the direction of the vector is perpendicular to the plane in which both F and d lie. As shown in Fig. 2.27, the moment $\mathbf{M_o}$ has a sense of direction. In this illustration, the moment $\mathbf{M_o}$ tends to rotate the body in a **counterclockwise** direction and is **positive**.

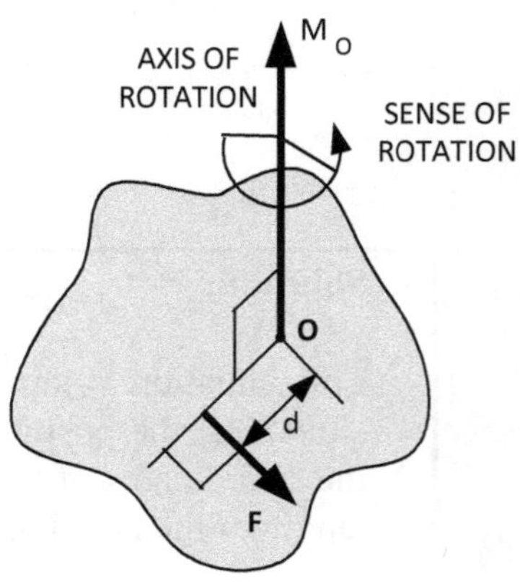

Fig. 2.27 A graphic illustration of the moment $\mathbf{M_o}$ as a vector quantity.

To determine the sign of the moment, we use the **right hand rule**. In applying this rule, place the palm of your right hand along the axis of rotation and point your fingers in the direction of the force and rotate your hand. If the direction of the force causes you to rotate **counterclockwise**, with your thumbs pointing up from the plane in which both F and d lie, then the moment is **positive**. However, if you must rotate your hand **clockwise**, with your thumb pointing downward, the moment is **negative**.

EXAMPLE 2.20

A hexagonal headed bolt is tightened with a wrench as shown in Fig. E2.20. A 35-lb force is applied to the handle of the wrench to produce a moment (torque). If the distance from the centerline of the bolt to the point of application of the force is d = 10 in., find the applied torque.

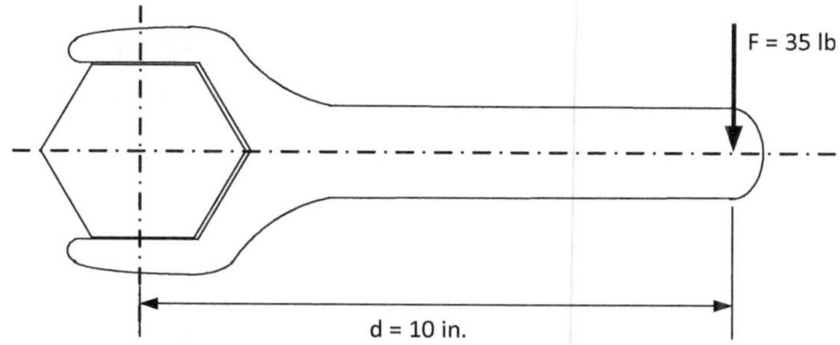

Fig. E2.20

Solution:

From Eq. 2.29, we write

$$M = F\,d = (35)(10) = 350 \text{ in-lb} = 29.17 \text{ ft-lb}$$

We consider this moment to be **negative** since it tends to produce a **clockwise** rotation about the head of the bolt.

EXAMPLE 2.21

Suppose a force of 125 N was applied to the wrench as shown in Fig. E2.21. Determine the torque (moment) applied to the hex headed bolt.

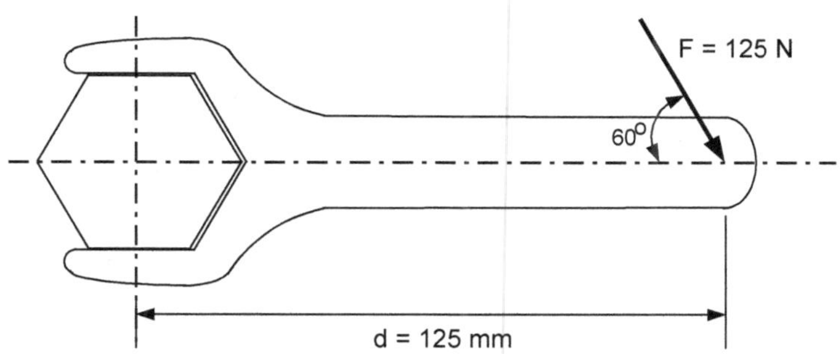

Fig. E2.21

Solution:

Let's resolve the 125 N force into two components—one perpendicular to the axis of the wrench and the other parallel.

The component of force parallel to the axis of the wrench is obtained by:

$$F_{=} = 125 \cos(60°) = 62.5 \text{ N}$$

Note that $F_{=}$ does not produce a moment (torque) on the bolt since its moment arm d = 0.

The component perpendicular to the axis of the wrench is given by:

$$F_\perp = 125 \sin(60°) = 108.3 \text{ N}$$

To determine the moment that this component of force produces on the head of the bolt, we use Eq. 2.29 and write:

$$M = F\, d = (108.3)(0.125) = 13.53 \text{ N-m}$$

Again, this moment is **negative** because it tends to produce a **clockwise** rotation about the head of the bolt.

2.9 VECTOR MECHANICS

2.9.1 Expressing Forces and Moments as Vectors

When we encounter three-dimensional structures, the analysis becomes more complex often involving six equations of equilibrium and several free body diagrams. Visualization of the structure and force components becomes more difficult. To alleviate the complexity and reduce visualization difficulties, we often use a vector mechanics approach. When we express each force **F** or moment **M** in a complete vector representation, the equilibrium relations reduce to two vector equations:

$$\mathbf{\Sigma F = 0} \qquad \textbf{and} \qquad \mathbf{\Sigma M = 0}$$

The **Σ** symbol implies that we sum all of the forces and all of the moments produced by the forces acting on the structure. Hence, we may write these relations as:

$$\mathbf{\Sigma F = F_1 + F_2 + \ldots\ldots\ldots + F_n = 0} \tag{2.30}$$

$$\mathbf{\Sigma M = M_1 + M_2 + \ldots\ldots\ldots + M_n = 0} \tag{2.31}$$

where forces 1, 2, 3 **.......** n act on the structure.

We set the sums to zero assuming the body is in equilibrium. Next let's use unit vectors in expressing the force vectors $\mathbf{F_1}$, $\mathbf{F_2}$, to $\mathbf{F_n}$ as:

$$\mathbf{F_1} = F_{1x}\,\mathbf{i} + F_{1y}\,\mathbf{j} + F_{1z}\,\mathbf{k}$$

$$\mathbf{F_2} = F_{2x}\,\mathbf{i} + F_{2y}\,\mathbf{j} + F_{2z}\,\mathbf{k}$$

$$\mathbf{F_n} = F_{nx}\,\mathbf{i} + F_{ny}\,\mathbf{j} + F_{nz}\,\mathbf{k} \tag{2.32}$$

where F_{1x}, F_{1y}, F_{1z}, etc. are the magnitudes of the Cartesian force components.

Substituting Eq. (2.32) into Eq. (2.30) yields:

$$\mathbf{\Sigma F} = (F_{1x} + F_{2x} + \ldots + F_{nx})\mathbf{i} + (F_{1y} + F_{2y} + \ldots + F_{ny})\mathbf{j} + (F_{1z} + F_{2z} + \ldots + F_{3z})\mathbf{k} = 0 \tag{2.33}$$

Inspection of this relation shows that:

$$F_{1x} + F_{2x} + \ldots. + F_{nx} = \mathbf{\Sigma} F_x = 0$$

$$F_{1y} + F_{2y} + \ldots. + F_{ny} = \mathbf{\Sigma} F_y = 0 \tag{2.34}$$

$$F_{1z} + F_{2z} + \ldots. + F_{nz} = \mathbf{\Sigma} F_z = 0$$

These three relations represent the equilibrium equations expressed in terms of scalar force components. We will employ these relations throughout most of this textbook. With the vector mechanics approach, they are combined into a single relation—Eq. (2.33).

EXAMPLE 2.22

A space force **F** with a magnitude of 800 lb has coordinate direction angles of $\alpha = 45^\circ$ and $\beta = 45^\circ$. Write a vector equation for **F** and prepare a drawing showing the force vector in a three-dimensional Cartesian coordinate system.

Solution: The specification of the force **F** in the problem statement is not complete because the coordinate direction angle γ is not given. We may determine angle γ by recalling Eq. (2.15).

$$\cos^2 \alpha + \cos^2 \beta + \cos^2 \gamma = \cos^2 (45^\circ) + \cos^2 (45^\circ) + \cos^2 \gamma = 1$$

$$\tfrac{1}{2} + \tfrac{1}{2} + \cos^2 \gamma = 1 \qquad \Rightarrow \qquad \cos^2 \gamma = 0 \qquad \Rightarrow \qquad \gamma = 90^\circ$$

The Cartesian components of the force vector are given by Eq. (2.16) as:

$$F_x = F \cos \alpha \qquad F_y = F \cos \beta \qquad F_z = F \cos \gamma$$

$$F_x = 800 \cos (45^\circ) = 565.7 \text{ lb} \qquad F_y = 800 \cos (45^\circ) = 565.7 \text{ lb} \qquad F_z = 800 \cos (90^\circ) = 0$$

Next, we may write the vector equation for the 800 lb force as:

$$\mathbf{F} = (565.7\,\mathbf{i} + 565.7\,\mathbf{j}) \text{ lb}$$

Note, the multiplier of the **k** unit vector is absent from this result because $F_z = 0$.

Finally, the force vector **F** is shown in the x – y plane in Fig. E2.22.

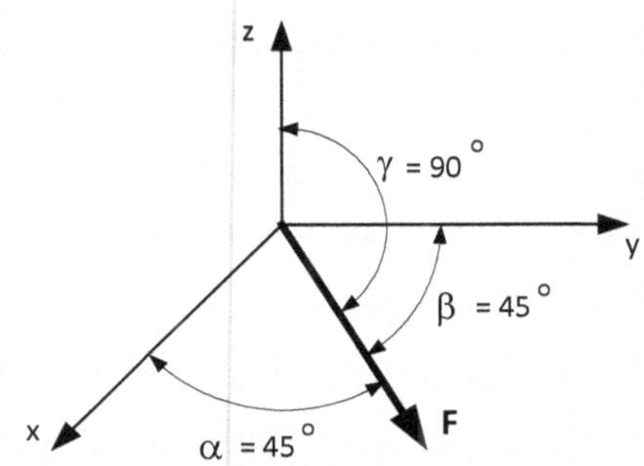

Fig. E2.22

EXAMPLE 2.23

A space force $\mathbf{F_1}$ with a magnitude of 20 kN has coordinate direction angles of $\alpha = 60^\circ$ and $\beta = 60^\circ$ and $\gamma = 135^\circ$. A second space force $\mathbf{F_2}$ with a magnitude of 12 kN has coordinate direction angles of $\alpha = 90^\circ$ and $\beta = 60^\circ$ and $\gamma = 30^\circ$. Write a vector equation for the summation of these two space forces.

Solution: The summation of $\mathbf{F_1}$ and $\mathbf{F_2}$ is given by Eq. (2.33) as:

$$\Sigma\mathbf{F} = (F_{1x} + F_{2x})\mathbf{i} + (F_{1y} + F_{2y})\mathbf{j} + (F_{1z} + F_{2z})\mathbf{k}$$

Let's first determine the Cartesian components of the space forces from Eq. (2.16).

$F_{x1} = F_1 \cos \alpha_1$ $\quad$ $F_{y1} = F_1 \cos \beta_1$ $\quad$ $F_{z1} = F_1 \cos \gamma_1$

$F_{x1} = 20 \cos (60°) = 10$ kN $\quad$ $F_{y1} = 20 \cos (60°) = 10$ kN $\quad$ $F_{z1} = 20 \cos (135°) = -14.14$ kN

$F_{x2} = F_2 \cos \alpha_2$ $\quad$ $F_{y2} = F_2 \cos \beta_2$ $\quad$ $F_z = F \cos \gamma_2$

$F_{x2} = 12 \cos (90°) = 0$ kN $\quad$ $F_{y2} = 12 \cos (60°) = 6$ kN $\quad$ $F_{z2} = 12 \cos (30°) = 10.39$ kN

We then sum the forces to obtain:

$$\Sigma \mathbf{F} = (10 + 0)\mathbf{i} + (10 + 6)\mathbf{j} + (-14.14 + 10.39)\mathbf{k}$$

$$\Sigma \mathbf{F} = \mathbf{F}_1 + \mathbf{F}_2 = (10\,\mathbf{i} + 16\,\mathbf{j} - 3.75\,\mathbf{k})\text{ kN}$$

While we have summed forces in Example 2.12, the sum has not been set to zero as shown in Eq. (2.33). The sum of the forces acting on a body is zero when the body is at rest or moving at a constant velocity. In this example problem, we have not considered equilibrium because our purpose was to demonstrate the technique for adding together two vectors.

2.9.2 The Vector Dot Product

In several examples, we have determined components of force vectors by using Eqs. (2.8) and (2.16). These relations are based on a trigonometric approach where the vector and its components are easy to visualize. However, in some three-dimensional problems, visualization is much more difficult. In these instances, vector algebra methods for determining components of force and moment vectors are useful. To introduce those aspects of vector algebra useful in design analysis two different methods for vector multiplication—the dot product and the cross product are described. The dot product is useful in determining the component of a force along some arbitrary line or the angle between two lines. The cross product is used to determine the moment due to a vector force about some arbitrary point.

Let's define the dot product of vectors **A** and **B** as shown below:

$$\mathbf{A} \bullet \mathbf{B} = AB \cos \theta \tag{2.35}$$

where the angle θ is defined in Fig. 2.28.

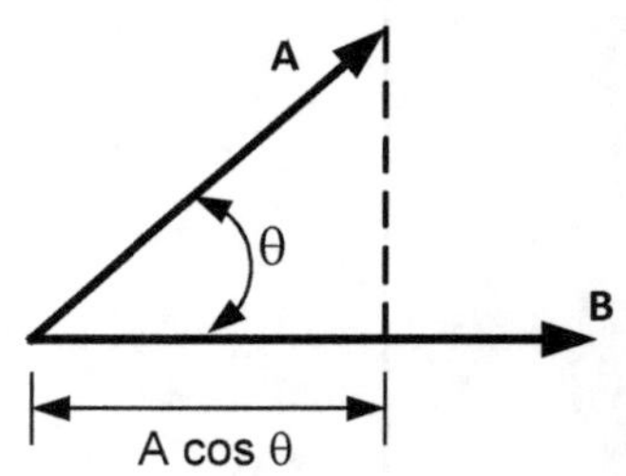

Fig. 2.28 Vectors A and B and their dot product.

In Fig. 2.28, we show the projection of the vector **A** as a dimension measured along vector **B**. The vector dot product **A • B** is represented by the projection (A cosθ) multiplied by the magnitude B. The dot product of two vectors is a scalar quantity with units dependent on the product of the units of A and B. As a scalar quantity **A • B** has a magnitude, but not a direction. We cannot draw an arrow to represent it, because it does not have an orientation.

Next let's consider the dot product of different pairs of unit vectors. It is easy to show that:

$$\mathbf{i} \bullet \mathbf{i} = (1)(1) \cos (0°) = 1 \qquad \text{and} \qquad \mathbf{i} \bullet \mathbf{j} = (1)(1) \cos (90°) = 0.$$

We list all possible combinations of the dot product of two unit vectors below:

$$\begin{array}{lll} \mathbf{i} \bullet \mathbf{i} = 1 & \mathbf{i} \bullet \mathbf{j} = 0 & \mathbf{i} \bullet \mathbf{k} = 0 \\ \mathbf{j} \bullet \mathbf{j} = 1 & \mathbf{j} \bullet \mathbf{k} = 0 & \mathbf{j} \bullet \mathbf{i} = 0 \\ \mathbf{k} \bullet \mathbf{k} = 1 & \mathbf{k} \bullet \mathbf{i} = 0 & \mathbf{k} \bullet \mathbf{j} = 0 \end{array} \tag{2.36}$$

Moments are a product of a force times a distance. In our previous discussion, we were always careful to define the distance as the perpendicular distance from the line of action of the force to the point about which the moments were determined. The definition of the moment does not change, but with the vector mechanics approach, we find it helpful to define the moment in terms of a vector cross product as:

$$\mathbf{M_O} = \mathbf{r} \times \mathbf{F} \tag{2.42}$$

where **r** is the position vector that locates point Q relative to the origin of a Cartesian coordinate system as shown in Fig. 2.30.

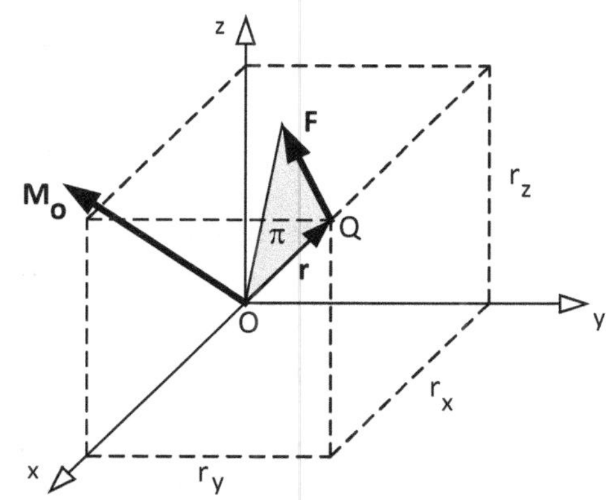

Fig. 2.30 A drawing showing the geometric representation of $\mathbf{M_O} = \mathbf{r} \times \mathbf{F}$.

The moment $\mathbf{M_O}$, relative to the origin O of the coordinate system, is due to the force **F** that is applied at point Q. The location of point Q relative to the origin O is established with a position vector **r**. Vectors **F** and **r** form a plane π that is shown in Fig. 2.30; the moment vector $\mathbf{M_O}$ is perpendicular to this plane.

The position vector is determined by the location of point Q as:

$$\mathbf{r} = r_x\,\mathbf{i} + r_y\,\mathbf{j} + r_z\,\mathbf{k} \tag{2.43}$$

where r_x, r_y and r_z are the coordinates of point Q.

Using Eq. (2.18) we write an expression for the force vector as:

$$\mathbf{F} = F_x\,\mathbf{i} + F_y\,\mathbf{j} + F_z\,\mathbf{k}$$

From Eq. (2.41) it is easy to show that the cross vector products of the unit vectors are given by:

$$\begin{array}{lll} \mathbf{i} \times \mathbf{i} = 0, & \mathbf{j} \times \mathbf{j} = 0, & \mathbf{k} \times \mathbf{k} = 0 \\ \mathbf{i} \times \mathbf{j} = \mathbf{k}, & \mathbf{j} \times \mathbf{k} = \mathbf{i}, & \mathbf{k} \times \mathbf{i} = \mathbf{j} \\ \mathbf{j} \times \mathbf{i} = -\mathbf{k} & \mathbf{k} \times \mathbf{j} = -\mathbf{i} & \mathbf{i} \times \mathbf{k} = -\mathbf{j} \end{array} \tag{2.44}$$

Combining Eqs. (2.42), (2.43), (2.18) and (2.45), we obtain:

$$\mathbf{M_O} = \mathbf{r} \times \mathbf{F} = (r_y\,F_z - r_z\,F_y)\mathbf{i} + (r_z\,F_x - r_x\,F_z)\mathbf{j} + (r_x\,F_y - r_y\,F_x)\mathbf{k} \tag{2.45}$$

We may also express the moment in terms of components relative to the coordinate axes as:

$$\mathbf{M_O} = \mathbf{r} \times \mathbf{F} = M_x\,\mathbf{i} + M_y\,\mathbf{j} + M_z\,\mathbf{k} \tag{2.46}$$

Comparing the results of Eqs. (2.45) with those of Eq. (2.46) yields the equations for the components of the moments about the coordinate axes as:

$$M_x = (r_y\,F_z - r_z\,F_y)$$

$$M_y = (r_z\,F_x - r_x\,F_z)$$

$$M_z = (r_x\,F_y - r_y\,F_x) \tag{2.47}$$

When employing the vector cross product **r × F** to determine the moment $\mathbf{M_O}$, we simultaneously obtain the relations for the moments M_x, M_y and M_z.

In writing Eq. (2.45), we often employ the determinant given by:

$$\mathbf{M}_O = \begin{vmatrix} \mathbf{i} & \mathbf{j} & \mathbf{k} \\ r_x & r_y & r_z \\ F_x & F_y & F_z \end{vmatrix} \qquad (2.48)$$

Let's apply these results to three example problems to demonstrate the use of the vector cross product in the determination of moments due to forces.

EXAMPLE 2.26

A force vector **F** with the following components (F_x = 3000 N, F_y = 5000 N, and F_z = 4500 N) is applied to a structure at point Q. If point Q is given by the coordinates x = 6 m, y = 5 m, and z = 12 m, determine the moment about the origin O of the coordinate system.

Solution:

The moment is given by Eq. (2.42) as **r × F**. To execute this simple cross product, we write both **r** and **F** in vector format as:

$$\mathbf{r} = r_x\,\mathbf{i} + r_y\,\mathbf{j} + r_z\,\mathbf{k} = (6\,\mathbf{i} + 5\,\mathbf{j} + 12\,\mathbf{k})\text{ m}$$

$$\mathbf{F} = (3000\,\mathbf{i} + 5000\,\mathbf{j} + 4500\,\mathbf{k})\text{ N}$$

To execute the cross product, let's employ Eq. (2.45) to obtain:

$$\mathbf{M_O} = \mathbf{r} \times \mathbf{F} = (r_y F_z - r_z F_y)\,\mathbf{i} + (r_z F_x - r_x F_z)\,\mathbf{j} + (r_x F_y - r_y F_x)\,\mathbf{k}$$

$$\mathbf{M_O} = [(5)(4500) - (12)(5000)]\,\mathbf{i} + [(12)(3000) - (6)(4500)]\,\mathbf{j}$$

$$+ [(6)(5000) - (5)(3000)]\,\mathbf{k}$$

$$\mathbf{M_O} = (-37{,}500\,\mathbf{i} + 9{,}000\,\mathbf{j} + 15{,}000\,\mathbf{k})\text{ N-m}$$

The magnitude of the moment is given by the square root of the sum of the squares of the moment components as:

$$M_O = [M_x^2 + M_y^2 + M_z^2]^{1/2} \qquad (2.49)$$

$$M_O = [(-37.5)^2 + (9)^2 + (15)^2]^{1/2} = 41.38\text{ kN-m}$$

EXAMPLE 2.27

If a structure is loaded at point Q with a force given by:

$$\mathbf{F} = (150\,\mathbf{i} + 240\,\mathbf{j} + 80\,\mathbf{k})\text{ lb}$$

Determine the moment $\mathbf{M_O}$ using the determinant format presented in Eq. (2.48). Note that point Q is located at the position x = 3, y = 6 and z = 5 ft.

Solution: Let's first write the relation for the position vector **r** as:

$$\mathbf{r} = r_x\,\mathbf{i} + r_y\,\mathbf{j} + r_z\,\mathbf{k} = (3\,\mathbf{i} + 6\,\mathbf{j} + 5\,\mathbf{k})\text{ ft} \qquad \text{(a)}$$

Recall Eq. (2.48) as:

$$\mathbf{M_O} = \begin{vmatrix} \mathbf{i} & \mathbf{j} & \mathbf{k} \\ r_x & r_y & r_z \\ F_x & F_y & F_z \end{vmatrix}$$

Substituting into the matrix, the coefficients from the relations for the position and force vectors gives:

$$\mathbf{M_O} = \begin{vmatrix} \mathbf{i} & \mathbf{j} & \mathbf{k} \\ 3 & 6 & 5 \\ 150 & 240 & 80 \end{vmatrix}$$

Solving the determinant gives:

$$\mathbf{M_O} = [(48 - 120)\mathbf{i} + (75 - 24)\mathbf{j} + (72 - 90)\mathbf{k}](10)$$

$$\mathbf{M_O} = (-720\,\mathbf{i} + 510\,\mathbf{j} - 180\,\mathbf{k})\text{ ft-lb}$$

EXAMPLE 2.28

The boom of a crane extends from its base at point O to its tip at point Q as indicated in Fig. E2.28. A three-dimensional coordinate system has been established in this illustration with the point O at the origin and the point Q defined with coordinates (1,1,5). A force with a magnitude of 12 kN is applied by the boom onto a cable that extends from the tip of the boom to point P. Point P is located on the x – y plane with coordinates (3,6, 0). If the coordinates are expressed in meters, determine the following quantities:

(a) The vector representation of the force **F**.
(b) The components F_x, F_y and F_z.
(c) The moment $\mathbf{M_O}$.
(d) The unit vector describing the direction of $\mathbf{M_O}$.
(e) The moments about the coordinate axes—M_x, M_y and M_z.
(f) The magnitude of the moment.
(g) The angle between the force and the boom OQ.
(h) The projection of F along the boom OQ.

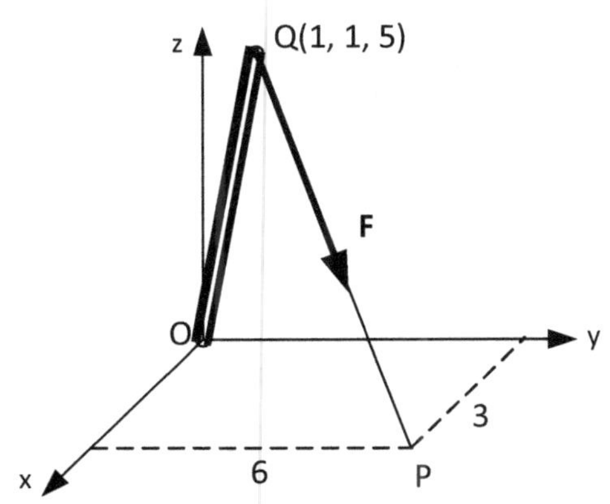

Fig. E2.28

Solution:

Let's begin by determining the unit vector that describes the direction of the applied force. Since the force is directed along the cable that originates at point Q and extends through point P, we know its orientation from the coordinates of these two points. The unit vector $\mathbf{u_F}$ is given by:

$$\mathbf{u_F} = (1/u_F)[(x_P - x_Q)\mathbf{i} + (y_P - y_Q)\mathbf{j} + (z_P - z_Q)\mathbf{k}] \quad (2.50)$$

where u is the magnitude of the length of the line from Q to P given by:

$$u_F = [(x_P - x_Q)^2 + (y_P - y_Q)^2 + (z_P - z_Q)^2]^{1/2} \quad (2.51)$$

Substituting the coordinates of points P and Q into Eqs. (2.50) and (2.51) gives:

$$\mathbf{u_F} = \frac{(3-1)\mathbf{i}+(6-1)\mathbf{j}+(0-5)\mathbf{k}}{\left[(3-1)^2+(6-1)^2+(0-5)^2\right]^{1/2}}$$

$$\mathbf{u_F} = 0.2722\mathbf{i}+0.6804\mathbf{j}-0.6804\mathbf{k} \quad \text{(a)}$$

The vector representation of the force F is then given by:

$$\mathbf{F} = F\,\mathbf{u_F} = (3.266\,\mathbf{i} + 8.165\,\mathbf{j} - 8.165\,\mathbf{k})\text{ kN} \quad \text{(b)}$$

The components of the force vector are:

$$F_x = 3.266\text{ kN}; \qquad F_y = 8.165\text{ kN}; \qquad F_z = -\,8.165\text{ kN} \quad \text{(c)}$$

From the coordinates of point Q, it is clear that the position vector **r** is given by:

$$\mathbf{r} = \mathbf{i} + \mathbf{j} + 5\,\mathbf{k} \quad \text{(d)}$$

The moment is determined from Eq. (2.48) as:

$$\mathbf{M}_O = \mathbf{r} \times \mathbf{F} = \begin{vmatrix} \mathbf{i} & \mathbf{j} & \mathbf{k} \\ 1 & 1 & 5 \\ 3.266 & 8.165 & -8.165 \end{vmatrix}$$

Evaluating this determinant yields:

$$\mathbf{M_O} = [(-8.165)(1) - (8.165)(5)]\mathbf{i} + [(5)(3.266) + (8.165)(1)]\mathbf{j} + [(8.165)(1) - (3.266)(1)]\mathbf{k}$$

$$\mathbf{M_O} = (-\,49.0\,\mathbf{i} + 24.5\,\mathbf{j} + 4.90\,\mathbf{k})\text{ kN-m} \quad \text{(e)}$$

The components of this vector relative to the Cartesian coordinates are:

$$M_x = -\,49.0\text{ kN-m}; \qquad M_y = 24.5\text{ kN-m} \qquad M_z = 4.90\text{ kN-m} \quad \text{(f)}$$

The magnitude of the moment M_O is determined from Eq. (2.49) as:

$$M_O = [(-49.0)^2 + (24.5)^2 + (4.90)^2]^{1/2} = 55.0 \text{ kN-m} \tag{g}$$

The unit vector describing the direction of the moment vector M_O is:

$$\mathbf{u}_{Mo} = (1/M_O)[\mathbf{M_O}] = (1/55.0)[-49.0\,\mathbf{i} + 24.5\,\mathbf{j} + 4.90\,\mathbf{k}]$$

$$\mathbf{u}_{Mo} = -0.891\,\mathbf{i} + 0.445\,\mathbf{j} + 0.089\,\mathbf{k} \tag{h}$$

Let's use Eq. (2.38) to determine the angle θ between the force vector **F** and a vector **QO** that extends from point Q to the origin O.

$$\theta = \cos^{-1}[\mathbf{F} \bullet \mathbf{QO}/(F)(QO)] \tag{i}$$

Note that **QO** = – **r**; hence, we use Eqs. (b) and (d) to express Eq. (i) as:

$$\theta = \cos^{-1}\left[\frac{(3.266\mathbf{i} + 8.165\mathbf{j} - 8.165\mathbf{k}) \bullet (-\mathbf{i} - \mathbf{j} - 5\mathbf{k})}{(12)(5.196)}\right] \tag{j}$$

$$\theta = \cos^{-1}[0.4714] = 61.87^{\text{O}}$$

The projection of the force F along the boom is given by:

$$F_{QO} = F \cos\theta = 12 \cos(61.87°) = 5.657 \text{ kN} \tag{k}$$

Another approach to determine F_{QO} involves employing Eq. (2.39). Then we write:

$$F_{QO} = \mathbf{F} \bullet \mathbf{u}_{QO} = \mathbf{F} \bullet (-\mathbf{r}/r) \tag{l}$$

You may wish to verify that Eq. (j) gives the same result as Eq. (k).

2.10 SUMMARY

Both external and internal forces have been described. External forces are applied to the structure either by loading or by reactions at the supports. Internal forces develop within structural members due to the action of the external forces. Section cuts on the structural member are made to visualize the stresses that generate the internal forces.

Forces are vector quantities that require the specification of both magnitude and direction for a complete description. We represent forces with arrows where its length is proportional to the magnitude and its orientation relative to a suitable coordinate system gives its direction. The line of action of a force is collinear with the force vector. It is possible to slide a force vector along the line of action to a new position without affecting equilibrium of a body.

The process of vector addition and subtraction has been described. We showed that the vector sum $\mathbf{S}_v = \mathbf{A} + \mathbf{B} = \mathbf{B} + \mathbf{A}$ could be represented by a triangle with sides of length A, B and Sv. The sine and cosine laws are employed to determine the magnitude and direction in vector addition or subtraction.

Vector decomposition, the resolution of a single force into two components, was described. Resolution of a force into its two Cartesian components is of particular significance. The equations involved are summarized below.

Classifications of vector systems were discussed with definitions and examples of concurrent and coplanar systems. For concurrent systems, where all of the forces are applied at a common point, it is important to recognize that the equilibrium equation $\Sigma\mathbf{M} = 0$ is satisfied automatically.

A brief discussion was included on moments. Moments, like forces, are vector quantities with both a magnitude and direction. Moments tend to cause bodies upon which they act to rotate. The direction of a moment is perpendicular to the plane in which the force and moment arm lie. Moments are considered **positive** when they tend to produce a **counterclockwise** rotation of the body upon which they act.

Space forces were described with force components in three directions (x, y and z). Coordinate direction angles were defined to provide the orientation of the force vectors in the three-dimensional coordinate system. Unit vectors **i**, **j** and **k** directed along the Cartesian axes were defined. It was shown that the direction of space forces could be expressed in terms of a unit vector $\mathbf{u}_F$. A position vector **r** was introduced to locate the point of application of a force in space.

Finally, vector mechanics was introduced and the three dimensional vector representations of forces, moments and unit vectors were given. These representations permit us to show separately the direction and magnitude of a vector quantity. The vector dot product was described together with two useful applications in analysis of forces acting on structures. The vector cross product was also defined. It is used extensively in determining the moments produced by forces acting on three-dimensional structures. Examples demonstrating the application of the vector dot and cross products were provided.

Key relations you will use on many occasions in the analysis of structures that were introduced in this Chapter are summarized below.

$$P_{int} = \int \sigma \, dA \qquad (2.1)$$

$$\frac{A}{\sin a} = \frac{B}{\sin b} = \frac{C}{\sin c} \qquad (2.4)$$

$$C = \sqrt{A^2 + B^2 - 2AB\cos(c)} \qquad (2.5)$$

For two dimensional forces:

$$\mathbf{F} = \mathbf{F}_x + \mathbf{F}_y \qquad (2.7)$$

$$F_x = F\cos\theta$$
$$F_y = F\sin\theta \qquad (2.8)$$

$$F = [F_x^{\,2} + F_y^{\,2}]^{1/2} \qquad (2.9)$$

$$M_o = (F)(d) \qquad (2.29)$$

We indicate the direction of the space forces with coordinate direction angles defined by:

$$\cos\alpha = \frac{F_x}{F} \quad \cos\beta = \frac{F_y}{F} \quad \cos\gamma = \frac{F_z}{F} \qquad (2.14)$$

Alternatively the direction may be specified by unit vectors directed along the line of action of the space force.

$$\mathbf{u}_F = \mathbf{F}/F \qquad (2.19)$$

For space forces (three-dimensional) and moments it is often easier to visualize the vectors and more efficient to perform the analysis by employing vector mechanics. The relations frequently used in for three-dimensional analyses include vector representation of forces, moments and position. Also, extensive utilization of the dot and cross products facilitates the computation required.

$$\mathbf{F} = F_x\,\mathbf{i} + F_y\,\mathbf{j} + F_z\,\mathbf{k} \tag{2.18}$$

$$\mathbf{A} \bullet \mathbf{B} = (A)(B)\cos\theta \tag{2.35}$$

$$\mathbf{A} \bullet \mathbf{B} = A_x B_x + A_y B_y + A_z B_z \tag{2.37}$$

$$\theta = \cos^{-1}[(\mathbf{A} \bullet \mathbf{B})/(A)(B)] \tag{2.38}$$

$$A_u = \mathbf{A} \bullet \mathbf{u} \tag{2.39}$$

$$\mathbf{C} = \mathbf{A} \times \mathbf{B} = [(A)(B)\sin\theta]\mathbf{u} \tag{2.41}$$

$$\mathbf{M_O} = \mathbf{r} \times \mathbf{F} \tag{2.42}$$

$$\mathbf{r} = r_x\,\mathbf{i} + r_y\,\mathbf{j} + r_z\,\mathbf{k} \tag{2.43}$$

$$\mathbf{M_O} = \mathbf{r} \times \mathbf{F} = (r_y F_z - r_z F_y)\mathbf{i} + (r_z F_x - r_x F_z)\mathbf{j} + (r_x F_y - r_y F_x)\mathbf{k} \tag{2.45}$$

$$\mathbf{M_O} = \mathbf{r} \times \mathbf{F} = M_x\,\mathbf{i} + M_y\,\mathbf{j} + M_z\,\mathbf{k} \tag{2.46}$$

$$\mathbf{M}_O = \begin{vmatrix} \mathbf{i} & \mathbf{j} & \mathbf{k} \\ r_x & r_y & r_z \\ F_x & F_y & F_z \end{vmatrix} \tag{2.48}$$

PROBLEMS

2.1 Prepare a free body diagram (FBD) showing the following loads applied to a beam:

(a) concentrated load
(b) uniformly distributed load
(c) non-uniform distributed load
(d) two reaction loads

2.2 Suppose you are to analyze the rope used in a tug of war. Make a section cut of the rope at a location between the two teams and draw a diagram representing the stresses acting on the area exposed by the section cut. Also draw a diagram representing the internal force acting on the area exposed by the section cut.

2.3 Show a graphic representation of a force vector with a magnitude (F) and an orientation (θ) relative to the positive x-axis for:

(a) F = 7,500 lb and θ = 105°
(b) F = 1,400 N and θ = 210°
(c) F = 11.2 kN and θ = 48°
(d) F = 1.620 kip and θ = 155°

2.4 Show that the body, illustrated in figures (a) and (b) below, is in equilibrium before and after sliding the 4,000 lb force along its line of action.

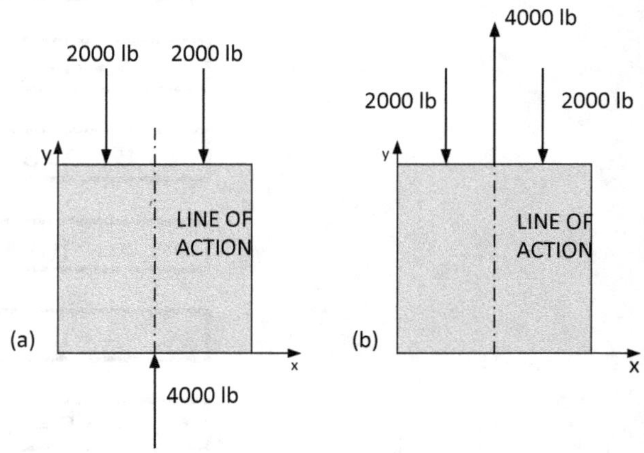

2.5 Draw a disk shaped body and apply a vertical force directed downward at its center. Also show a reactive force at its point of contact with a horizontal support. (a) Describe the condition for equilibrium of the disk. (b) Slide the applied force along its line of action to another location and describe the "new" condition for equilibrium.

2.6 Prepare a drawing showing the reactive forces on the short beam illustrated in the figure to the right.

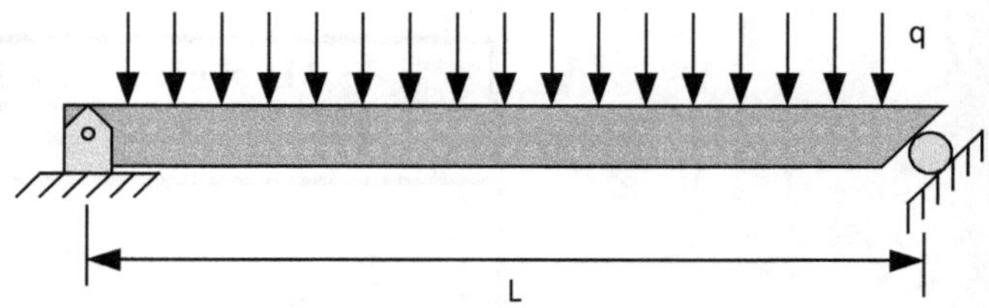

]

2.7 Prepare three drawings showing the reactive forces on each of the three cylinders in the stack shown in the figure to the right. Note that the cylinders are of the same diameter, and they are bonded together at the three contact points.

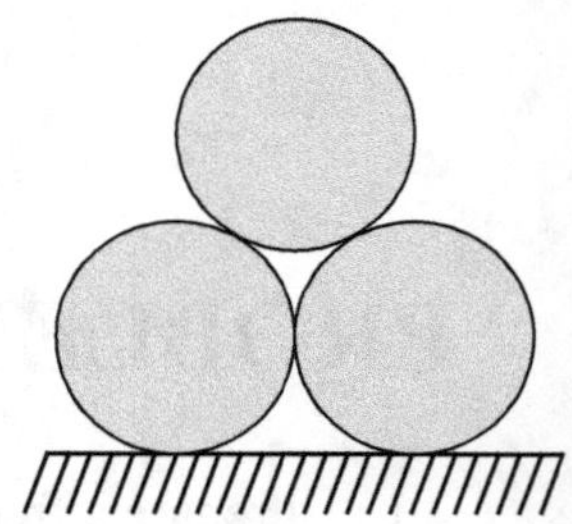

2.8 For the vectors **A** and **B,** illustrated in the figure shown to the left, determine the magnitude and direction of the following resultant vectors:

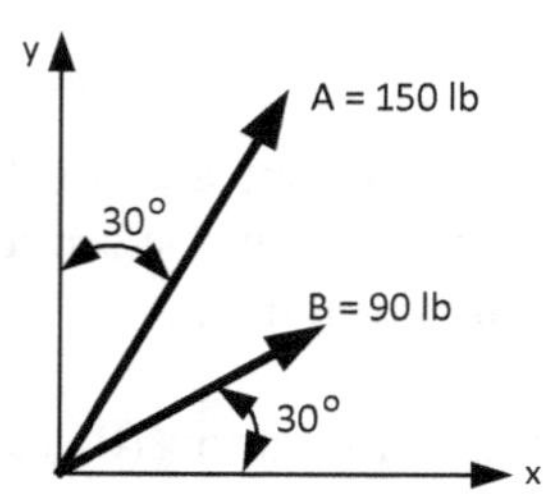

(a) $\mathbf{S_v} = \mathbf{A} + \mathbf{B}$
(b) $\mathbf{D_v} = \mathbf{A} - \mathbf{B}$
(c) $\mathbf{D_v} = \mathbf{B} - \mathbf{A}$.

2.9 For the vectors **A** and **B,** illustrated in the figure shown to the right, determine the magnitude and direction of the following resultant vectors:

(a) $\mathbf{S_v} = \mathbf{A} + \mathbf{B}$
(b) $\mathbf{D_v} = \mathbf{A} - \mathbf{B}$
(c) $\mathbf{D_v} = \mathbf{B} - \mathbf{A}$.

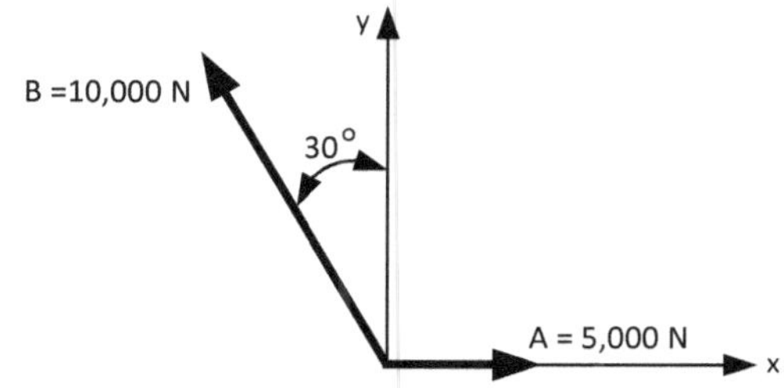

2.10 For the vectors **A** and **B,** illustrated in the figure shown to the left, determine the magnitude and direction of the following resultant vectors:

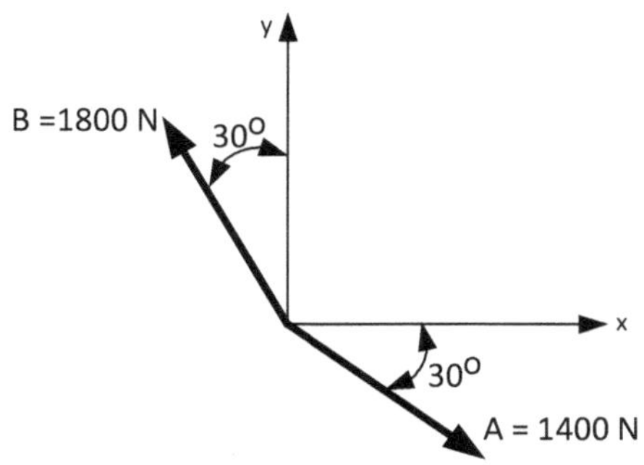

(a) $\mathbf{S_v} = \mathbf{A} + \mathbf{B}$
(b) $\mathbf{D_v} = \mathbf{A} - \mathbf{B}$
(c) $\mathbf{D_v} = \mathbf{B} - \mathbf{A}$.

2.11 Prepare a drawing similar to the one shown below illustrating the vector subtraction $\mathbf{D_v} = \mathbf{B} - \mathbf{A}$.

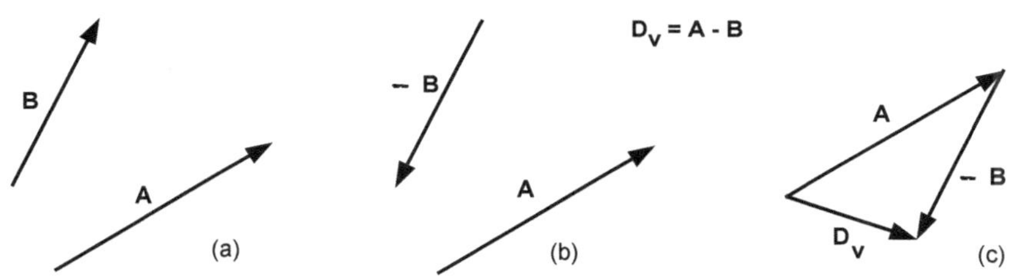

2.12 From the figure shown to the right, determine the Cartesian components of the force vectors: (a) **A** and (b) **B**.

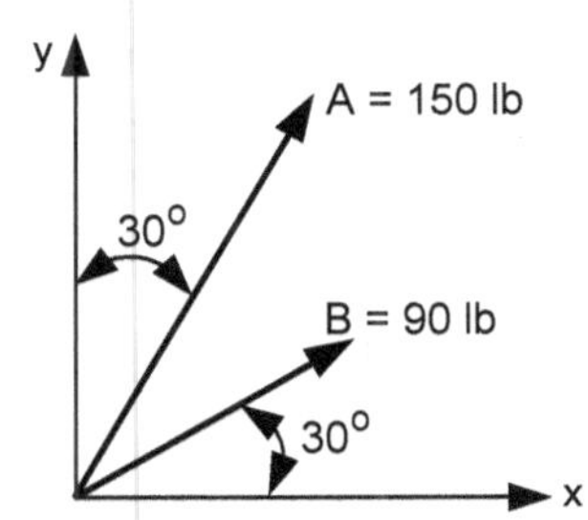

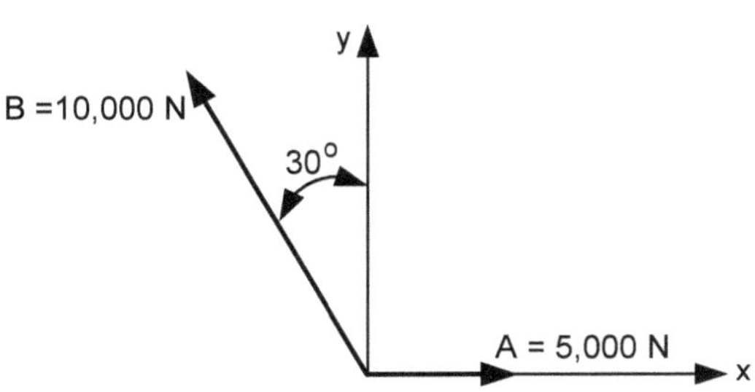

2.13 From the figure shown to the left, determine the Cartesian components of the force vectors: (a) **A** and (b) **B**.

2.14 From the figure shown to the right, determine the Cartesian components of the force vectors: (a) **A** and (b) **B**.

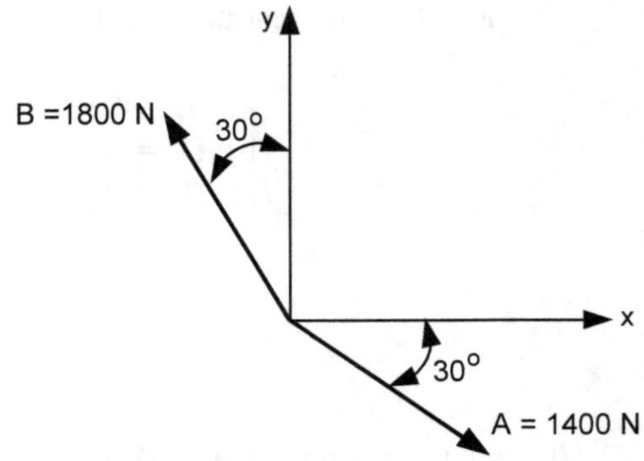

2.15 Using the Cartesian components of vectors **A** and **B**, in the figure shown to the left, determine the magnitude and direction of:

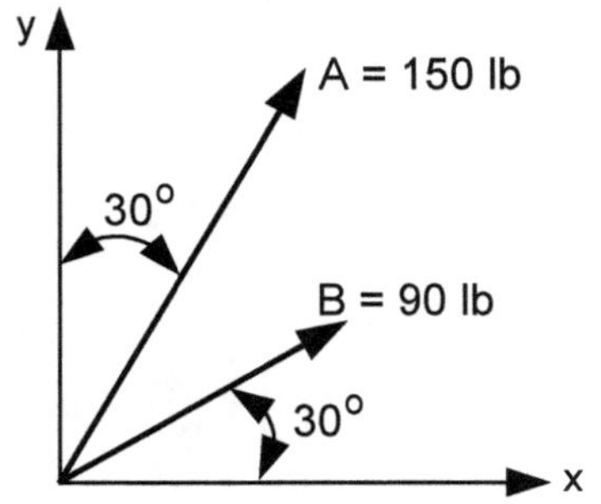

(a) $\mathbf{S_v} = \mathbf{A} + \mathbf{B}$
(b) $\mathbf{D_v} = \mathbf{A} - \mathbf{B}$
(c) $\mathbf{D_v} = \mathbf{B} - \mathbf{A}$.

2.16 Using the Cartesian components of vectors **A** and **B**, in the figure shown to the right, determine the magnitude and direction of:

(a) $\mathbf{S_v} = \mathbf{A} + \mathbf{B}$
(b) $\mathbf{D_v} = \mathbf{A} - \mathbf{B}$
(c) $\mathbf{D_v} = \mathbf{B} - \mathbf{A}$.

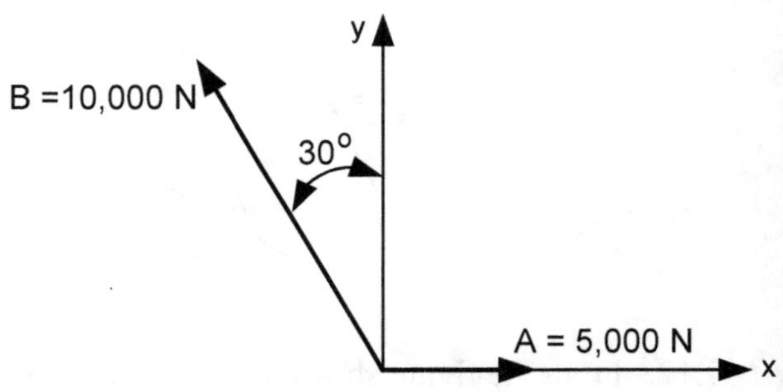

2.17 Using the Cartesian components of vectors **A** and **B**, in the figure shown to the left, determine the magnitude and direction of:

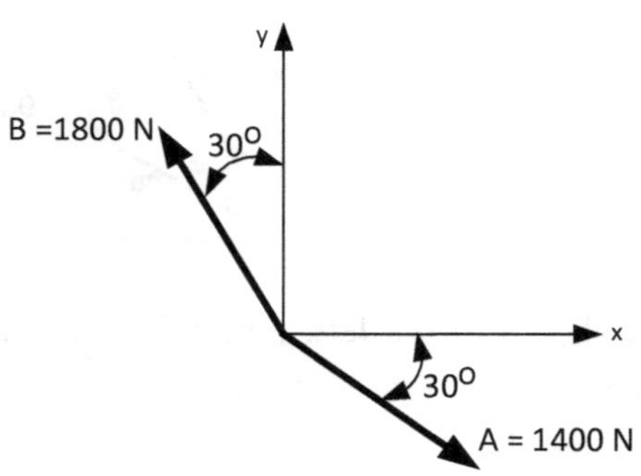

(a) $\mathbf{S_v} = \mathbf{A} + \mathbf{B}$
(b) $\mathbf{D_v} = \mathbf{A} - \mathbf{B}$
(c) $\mathbf{D_v} = \mathbf{B} - \mathbf{A}$.

2.18 Determine the force components along the x and y' axes due to the force of 18,440 lbs that is pulling on the eyebolt shown in the figure to the right.

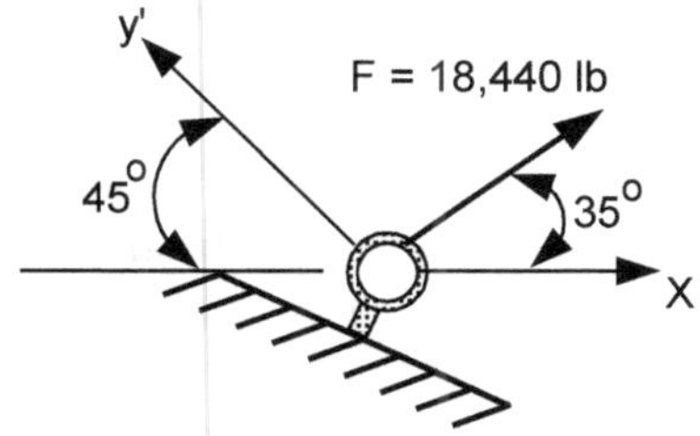

2.19 A linkage, shown in the figure to the left, is subjected to a force of 2,150 N. Determine the components of the force along the axes of links AB and AC.

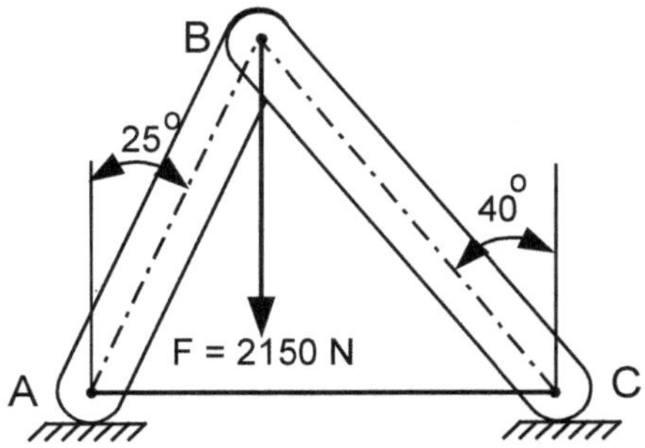

2.20 Determine the angle θ so that the link AB, shown in the figure to the right, has a component of force of 1,250 lb along its axis. The angle ϕ is fixed at 6.4 degrees.

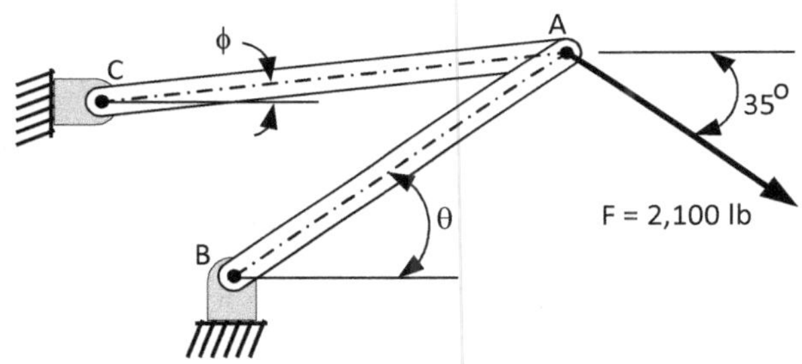

2.21 Determine the vector sum $\mathbf{S}_v$ of the three vectors shown in the figure to the right. Specify its direction relative to the x-axis.

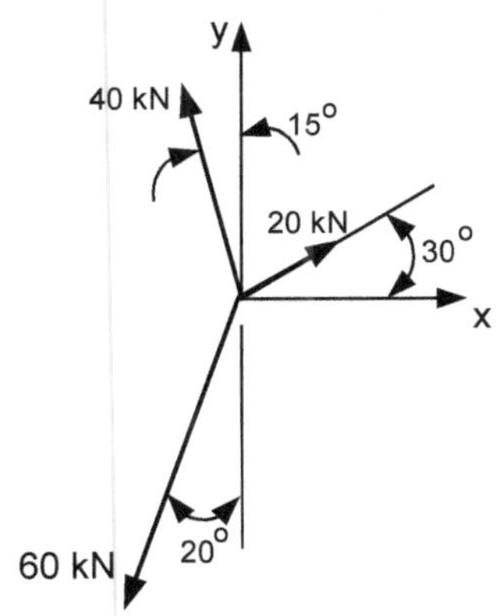

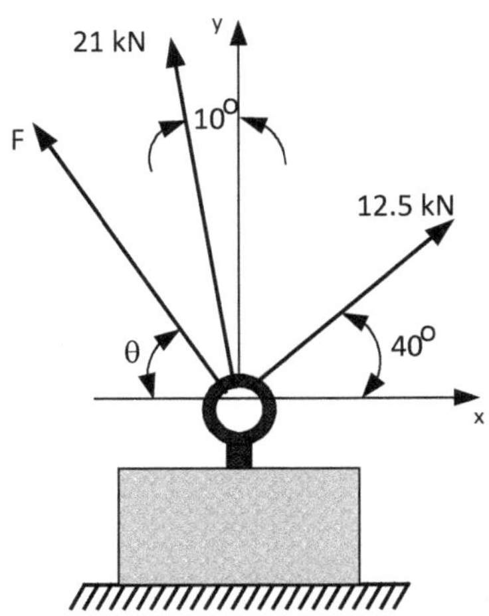

2.22 Determine the magnitude and the direction of the force **F**, shown in the figure to the left if the box is to be lifted from the floor without swinging. The weight of the box is 35 kN.

2.23 If θ is fixed at 55°, in the figure shown to the right, calculate the maximum weight of the box that can be lifted without it swinging when it clears the floor.

2.24 The gusset plate, illustrated in the figure below, is subjected to four forces due to the attached uniaxial structural members. If the vector sum of the four forces is zero, determine the forces F_1 and F_2.

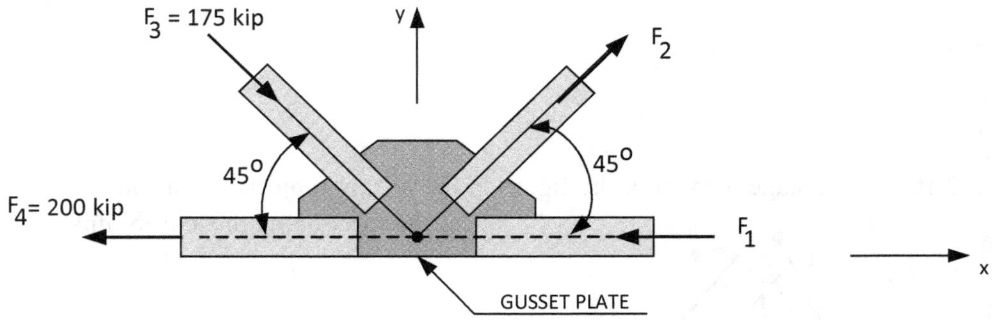

2.25 Three forces act on the bracket, illustrated in the figure to the right. Determine F_1 and θ if the magnitude of one of the Cartesian components of the vector sum S_{vx} = 16,200 lb. Note, S_{vy} = 0.

2.26 If the vector sum $\mathbf{S_v}$ of the three forces is 17,500 lb, and is oriented at 10° relative to the x-axis, determine F_1 and θ. Reference the figure to the right.

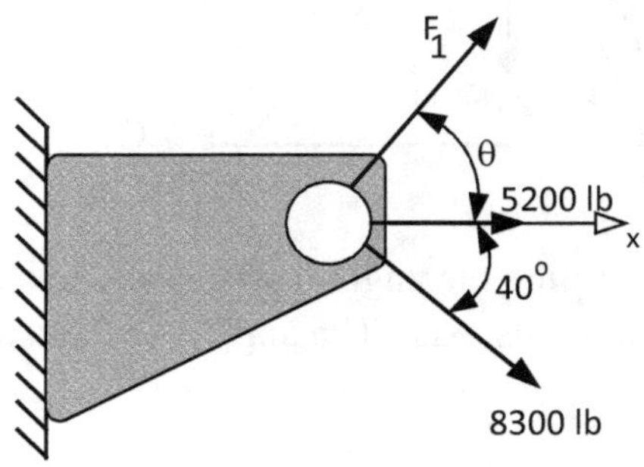

2.27 Define coplanar forces, and show a sketch illustrating four coplanar force vectors.

2.28 Define concurrent forces, and show a sketch illustrating four concurrent force vectors.

2.29 Prepare a sketch of three forces that are:

(a) coplanar, concurrent (b) non-coplanar, concurrent (c) coplanar, non-concurrent

2.30 Suppose you exert a force of 125 N on a single ended lug wrench with a 320-mm long handle to loosen a wheel bolt. What moment (torque) are you applying to the bolt?

2.31 You exert a force of 150 N with each hand on the handles of a double-ended lug wrench to loosen a wheel bolt. If each handle is 225 mm long, what moment (torque) are you applying to the bolt?

2.32 Determine the moment about point O due to the force shown in the figure to the right.

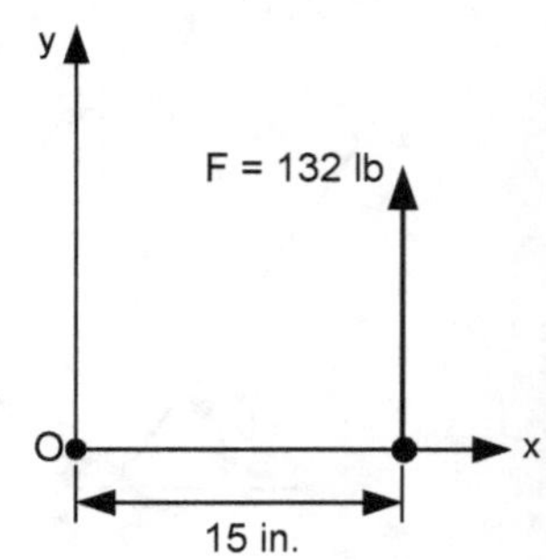

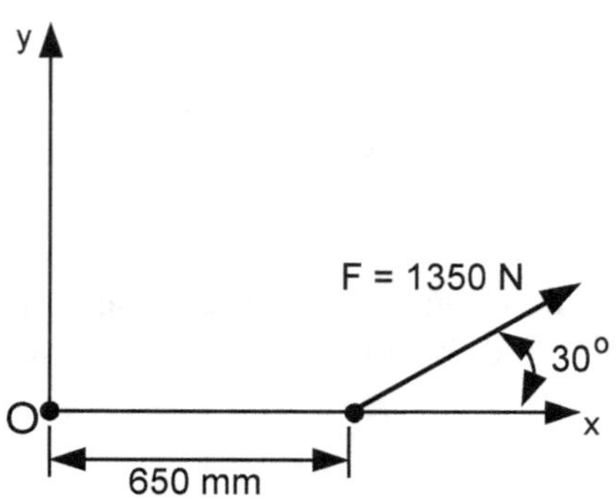

2.33 Determine the moment about point O due to the force shown in the figure to the left.

2.34 Determine the total moment about point O due to the two forces shown in the figure to the right.

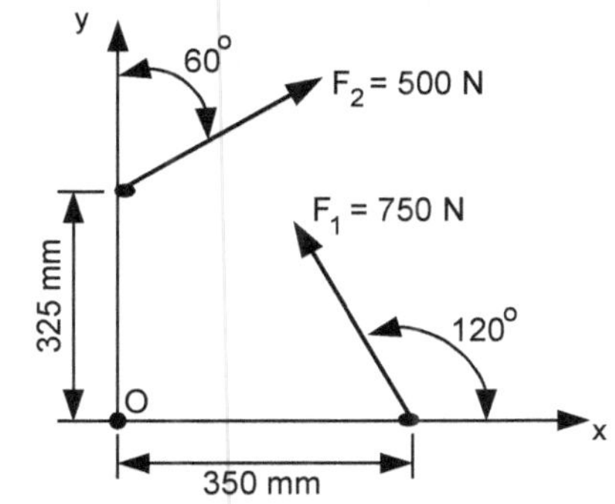

2.35 Express the force **F** or moment **M** in vector format if its Cartesian components are:

(a) $F_x = 2{,}100$ N, $F_y = -2{,}300$ N, and $F_z = 1{,}250$ N
(b) $F_x = 78$ lb, $F_y = -185$ lb, and $F_z = 124$ lb
(c) $M_x = 300$ N-m, $M_y = 400$ N-m, and $M_z = 500$ N-m
(d) $M_x = 740$ in-lb, $M_y = 425$ in-lb, and $M_z = -520$ in-lb

2.36 Determine the magnitude and direction, relative to the positive x-axis, for each of the vectors listed:

(a) $F_x = 1{,}400$ N, $F_y = 900$ N and $F_z = 0$
(b) $F_x = 355$ lb, $F_y = 520$ lb and $F_z = 0$
(c) $M_x = 1{,}800$ N-m, $M_y = 1{,}350$ N-m and $M_z = 0$
(d) $M_x = 655$ ft-lb, $M_y = -335$ ft-lb and $M_z = 0$

2.37 A position vector **r** is constructed in space between the origin (0, 0, 0) and point P. Write **r** as a Cartesian vector and determine the magnitude (r) and coordinate direction angles (α, β, γ) if:

(a) P = (7, 4, 6) in.
(b) P = (110, 65, 70) mm
(c) P = (0.120, 0.155, 0.170) m
(d) P = (–10, 3, –8) ft

2.38 Determine the coordinate direction angles (α, β, γ) for the following forces:

(a) $\mathbf{F} = (4\,\mathbf{i} + 5\,\mathbf{j} - 6\,\mathbf{k})$ N
(b) $\mathbf{F} = (-3\,\mathbf{i} + 2\,\mathbf{j} - 6\,\mathbf{k})$ lb
(c) $\mathbf{F} = (-3\,\mathbf{i} - 3\,\mathbf{j} + 7\,\mathbf{k})$ kN
(d) $\mathbf{F} = (5\,\mathbf{i} - 2\,\mathbf{j} - 8\,\mathbf{k})$ kip

2.39 A space force **F** has a known magnitude (F) and two known coordinate direction angles (α, β). Write a vector equation for **F**, specify the three components (F_x, F_y, F_z) and prepare a drawing showing the force vector in a three-dimensional Cartesian coordinate system for:

(a) F = 1,200 N, $\alpha = 60°$ and $\beta = 30°$
(b) F = 1,500 lb, $\alpha = 75°$ and $\beta = 50°$
(c) F = 17 kN, $\alpha = 135°$ and $\beta = 60°$
(d) F = 25 kip, $\alpha = 50°$ and $\beta = 140°$

2.40 Write a vector equation for the summation of two space forces $\mathbf{F}_1$ and $\mathbf{F}_2$. The force F_1 is 22.5 kN in magnitude with coordinate direction angles of $\alpha_1 = 60°$, $\beta_1 = 60°$ and $\gamma_1 = 135°$. The force F_2 is 30.0 kN in magnitude with coordinate direction angles of $\alpha_2 = 60°$, $\beta_2 = 30°$ and $\gamma_2 = 90°$. Prepare a drawing showing this resultant force vector in a three-dimensional Cartesian coordinate system.

2.41 Determine the angle between the following pairs of space forces:

(a) $\mathbf{F}_1 = (3\,\mathbf{i} + 6\,\mathbf{j} - 3\,\mathbf{k})$ N and $\mathbf{F}_2 = (-2\,\mathbf{i} + 5\,\mathbf{j} + 7\,\mathbf{k})$ N
(b) $\mathbf{F}_1 = (5\,\mathbf{i} - 1\,\mathbf{j} + 4\,\mathbf{k})$ lb and $\mathbf{F}_2 = (8\,\mathbf{i} - 2\,\mathbf{j} + 3\,\mathbf{k})$ lb
(c) $\mathbf{F}_1 = (-2\,\mathbf{i} + 6\,\mathbf{j} - 3\,\mathbf{k})$ kN and $\mathbf{F}_2 = (-3\,\mathbf{i} + 4\,\mathbf{j} - 4\,\mathbf{k})$ kN

2.42 Determine the magnitude of a single force component produced by two space forces ($\mathbf{F}_1$, $\mathbf{F}_2$) that is directed along a line of action which lies in the x–y plane and makes an angle of θ with the positive x-axis:

(a) $\mathbf{F}_1 = (3\,\mathbf{i} + 6\,\mathbf{j} - 3\,\mathbf{k})$ lb; $\mathbf{F}_2 = (-2\,\mathbf{i} + 3\,\mathbf{j} + 5\,\mathbf{k}.)$ lb; and $\theta = 30°$
(b) $\mathbf{F}_1 = (4\,\mathbf{i} + 3\,\mathbf{j} - 2\,\mathbf{k})$ kN; $\mathbf{F}_2 = (3\,\mathbf{i} - 5\,\mathbf{j} + 6\,\mathbf{k})$ kN; and $\theta = 40°$
(c) $\mathbf{F}_1 = (2\,\mathbf{i} - 4\,\mathbf{j} - 2\,\mathbf{k})$ kip; $\mathbf{F}_2 = (-4\,\mathbf{i} + 3\,\mathbf{j} - 5\,\mathbf{k})$ kip; and $\theta = 120°$

2.43 A force vector **F**, with components F_x, F_y and F_z is applied to a structure at point Q. Determine the moment of **F** about the origin O of the coordinate system when:

(a) $F_x = 4{,}000$ N, $F_y = 2{,}500$ N, $F_z = 1{,}500$ N and Q = (2.3, 3.4, 4.6) m
(b) $F_x = -2{,}000$ lb, $F_y = 2{,}500$ lb, $F_z = 1{,}200$ lb and Q = (3.4, –2.5, 4.6) ft
(c) $F_x = 2.3$ kN, $F_y = 5.5$ kN, $F_z = 3.3$ kN and Q = (1.4, 3.9, 2.6) m

2.44 If a structure is loaded at point Q with a force **F**, determine the moment $\mathbf{M_O}$ about the origin O. Prepare a drawing showing the moment vector $\mathbf{M_O}$ in a three-dimensional Cartesian coordinate system. Values for **F** and Q are:

(a) $\mathbf{F} = (125\,\mathbf{i} + 112\,\mathbf{j} + 45\,\mathbf{k})$ lb and Q = (2, –3, 15) ft
(b) $\mathbf{F} = (3.8\,\mathbf{i} + 3.3\,\mathbf{j} - 4.5\,\mathbf{k})$ kN and Q = (–2.2, 2.6, –1.8) m
(c) $\mathbf{F} = (55\,\mathbf{i} - 89\,\mathbf{j} + 42\,\mathbf{k})$ lb and Q = (–3, 4, 8) ft

2.45 The cell phone tower illustrated in the figure to the right is supported by three cables that are maintained with tension forces $F_A = F_B = F_C = 800$ lb. The cables are anchored into the ground plane at locations A, B and C. Because the structural strength and rigidity of the tower is along its axis, we design the cable support system to exhibit a vector sum $\mathbf{S_v}$ that is directed along the axis of the tower from point D to the tower support point O. If the tower height H = 120 ft and the anchor locations are given in the figure, determine the position for the anchor at point B to achieve the design objective.

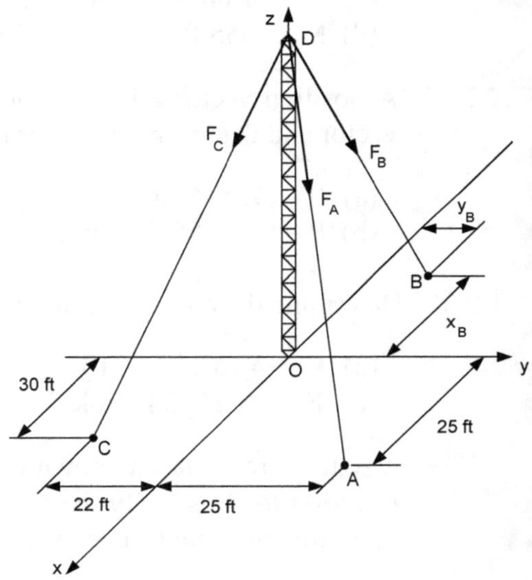

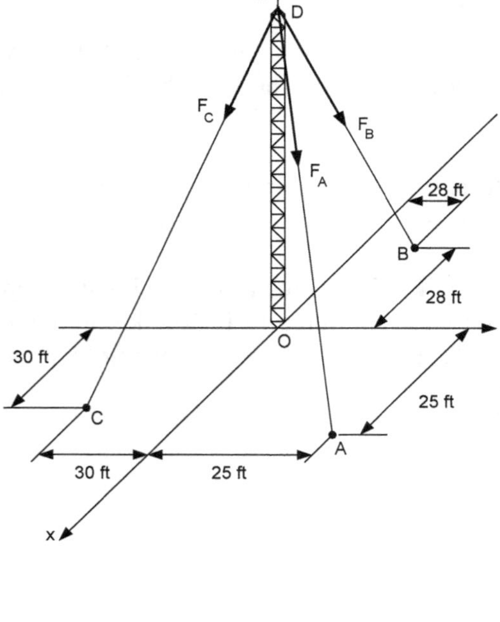

2.46 The cell phone tower, illustrated in the figure to the left, is supported by three cables that are maintained with tension forces $F_A = 800$ lb and $F_C = 1{,}000$ lb. The cables are anchored into the ground plane at locations A, B and C. Because the structural strength and rigidity of the tower is along its axis, we design the cable support system to exhibit a vector sum $\mathbf{S_v}$ that is directed along the axis of the tower from point D to the tower support point O. If the tower height H = 120 ft and the anchor locations are given in the figure, determine the force F_B that must be maintained in cable BD to achieve the design objective.

2.47 The boom of a crane extends from its base at point O to its tip at point Q as indicated in the figure below. A three-dimensional coordinate system has been established in this illustration with the point O at the origin and the point Q defined with coordinates (– 1, 3, 8). A force with a magnitude of 21 kN is applied by the boom onto a cable that extends from the tip of the boom to point P. Point P is located on the x–y plane with coordinates shown in the figure below. If the coordinates are expressed in meters, determine the following quantities:

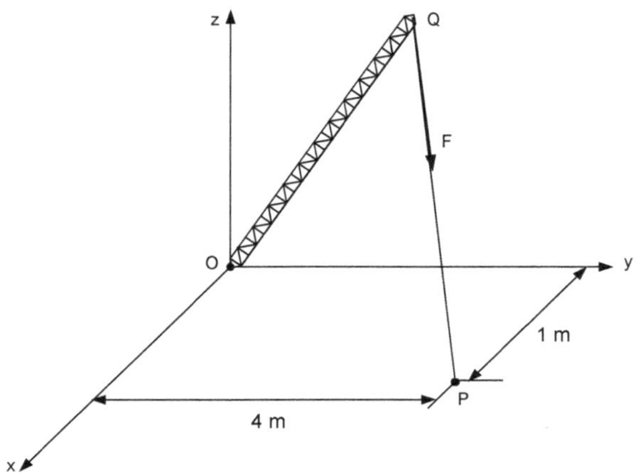

1. The force **F** (in vector form).
2. The force components F_x, F_y, and F_z.
3. The moment $\mathbf{M_O}$ (in vector form).
4. The moment components M_x, M_y, and M_z.
5. The magnitude of the moment (M_O).
6. The unit vector giving the direction of $\mathbf{M_O}$.
7. The angle between **F** and boom OQ.
8. The projection of **F** along boom OQ.

2.48 The boom of a crane extends from its base at point O to its tip at point Q as indicated in the figure below. A three-dimensional coordinate system has been established in this illustration with the point O at the origin and the point Q defined with coordinates (3, 2, 18). A force with a magnitude of 7,500 lb is applied by the boom onto a cable that extends from the tip of the boom to point P. Point P is located on the x–y plane with coordinates shown in the figure below. If the coordinates are expressed in feet, determine the following quantities:

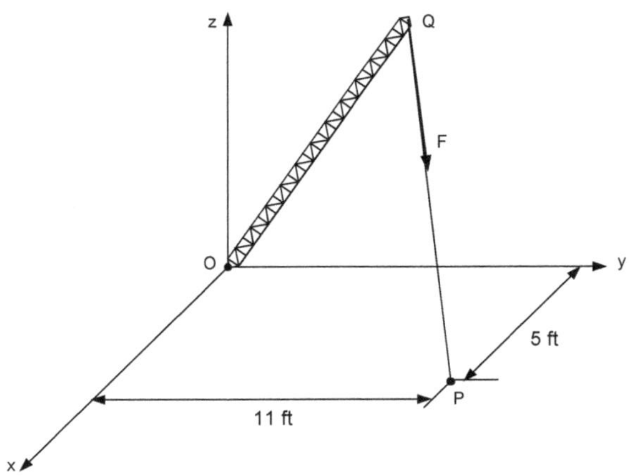

1. The force **F** (in vector form).
2. The force components F_x, F_y, and F_z.
3. The moment $\mathbf{M_O}$ (in vector form).
4. The moment components M_x, M_y, and M_z.
5. The magnitude of the moment (M_O).
6. The unit vector giving the direction of $\mathbf{M_O}$.
7. The angle between **F** and boom OQ.
8. The projection of **F** along boom OQ.

Notes:

CHAPTER 3

FREE BODY DIAGRAMS AND EQUILIBRIUM

3.1 INTRODUCTION

Equilibrium is an extremely important concept because it provides us with an approach for determining unknown forces that act on a body. We understand that a body is in equilibrium if it is at rest or traveling at some constant velocity. For example, the beam shown in Fig. 3.1 is at rest because the loads acting downward are resisted by the two reactive forces produced by the supports. The entire assembly is fixed to bedrock.

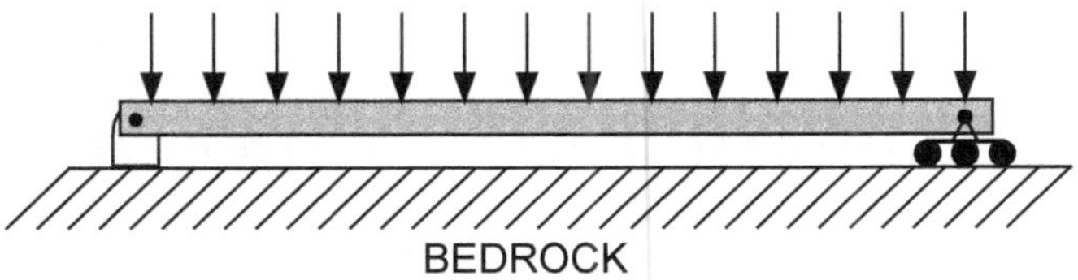

Fig. 3.1 A simply supported beam is at rest and in equilibrium.

Bodies that move may or may not be in equilibrium. If a body is in equilibrium, it moves with a constant velocity and a constant **momentum**[1]. The effect of unbalanced forces ($\Sigma F \neq 0$) acting on a body is to change its momentum by altering its velocity. The rate of that change is proportional to the unbalanced force. Therefore, when a body moves at constant velocity (zero acceleration) it possesses constant momentum and is in equilibrium.

The automobile in Fig. 3.2 is in equilibrium because the forward thrust due to the action of the wheels on the pavement is exactly equal to the rolling friction and drag forces that tend to impede the forward motion of the auto. The sum of the forces in the direction of motion is zero, the momentum is constant, the acceleration is zero, and the auto is in equilibrium although it may be moving at a velocity of 65 MPH.

Fig. 3.2 An automobile traveling at constant velocity is in equilibrium.

3.2 EQUATIONS OF EQUILIBRIUM

In Chapters 1 and 2, two equations of equilibrium were described:

$$\Sigma \mathbf{F} = 0 \qquad \text{and} \qquad \Sigma \mathbf{M}_O = 0$$

These relations are represented in a vector format; however, they may also be represented as Cartesian components of the forces and the moments:

$$\Sigma F_x = 0; \qquad \Sigma F_y = 0; \qquad \Sigma F_z = 0 \qquad (3.1)$$

$$\Sigma M_x = 0; \qquad \Sigma M_y = 0; \qquad \Sigma M_z = 0 \qquad (3.2)$$

[1] Momentum, M a vector quantity, is defined as M = m**v**; where **v** is the velocity of the body.

The two vector equations of equilibrium are equivalent to the six scalar equations of equilibrium. We will develop many of the solutions to problems arising in both statics and mechanics of materials by using Eqs. (3.1) and (3.2). The decision regarding which form of the equations to use will be left to your discretion, although in many cases you will find the scalar equations much easier to apply.

Although there are six scalar equations of equilibrium, it is not always necessary to use all six of them to solve an equilibrium problem. By classifying different systems of forces that act on a body, only those equations, which provide relevant information for solving equilibrium problems, may be identified. This classification greatly simplifies the associated equilibrium problems. Force systems are classified as follows:

- Non-coplanar and non-concurrent.
- Non-coplanar and concurrent.
- Coplanar and non-concurrent.
- Coplanar and concurrent.

Let's now examine the relevant equations of equilibrium for each of these force system.

3.2.1 Non-coplanar, Non-concurrent Force Systems

The three-dimensional force system, illustrated in Fig. 3.3, is **non-coplanar** because forces with components oriented in the x, y, and z directions act on the body. It is **non-concurrent** because the lines of action of the forces $\mathbf{F_1}$, $\mathbf{F_2}$ and $\mathbf{F_3}$ do not intersect at a common point. Moments are produced by forces $\mathbf{F_1}$, $\mathbf{F_2}$ and $\mathbf{F_3}$, and clearly there are components of the forces in all three directions. When the body in Fig. 3.3 is in equilibrium, the directions and magnitudes of $\mathbf{F_1}$, $\mathbf{F_2}$ and $\mathbf{F_3}$ must satisfy the **six** Cartesian component equations of equilibrium:

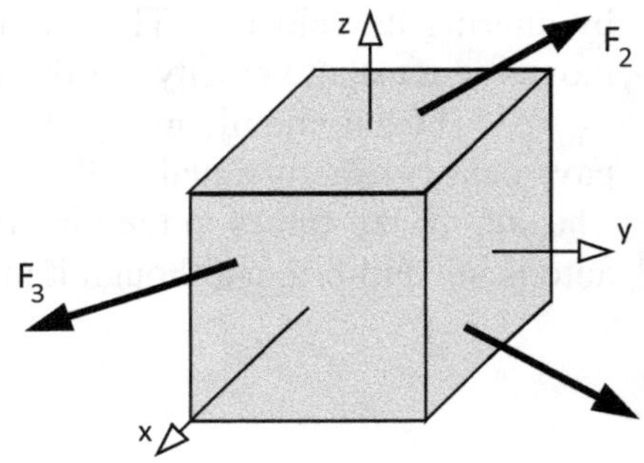

Fig. 3.3 An illustration of a non-coplanar, non-concurrent force system.

The **six** Cartesian component equations of equilibrium are:

$$\Sigma F_x = 0 \qquad \Sigma F_y = 0 \qquad \Sigma F_z = 0$$

$$\Sigma M_x = 0 \qquad \Sigma M_y = 0 \qquad \Sigma M_z = 0$$

3.2.2 Non-coplanar, Concurrent Force Systems

The three-dimensional force system, illustrated in Fig. 3.4, is **non-coplanar** because forces $\mathbf{F_1}$, $\mathbf{F_2}$ and $\mathbf{F_3}$ acting on the body exhibit components in the x, y and z directions.

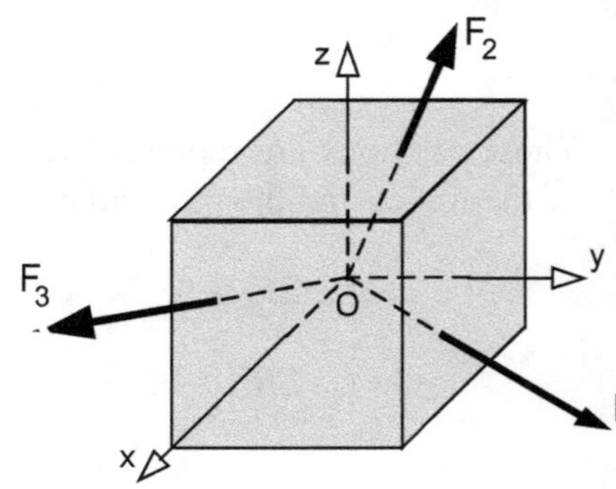

Fig 3.4 Example of a **non-coplanar, concurrent** force system.

This force system is **concurrent** because all of the forces pass through a common point (the origin of the Cartesian coordinate system in this case). The force system does not produce moments about the origin; hence, the equilibrium relation $\Sigma\mathbf{M} = 0$ is independent of the magnitude and direction of the forces $\mathbf{F_1}$, $\mathbf{F_2}$ and $\mathbf{F_3}$. However, these forces must satisfy the three remaining Cartesian equations of equilibrium:

$$\Sigma F_x = 0 \qquad \Sigma F_y = 0 \qquad \Sigma F_z = 0$$

3.2.3 Coplanar, Non-concurrent Force Systems

A coplanar non-concurrent force system is illustrated in Fig 3.5.

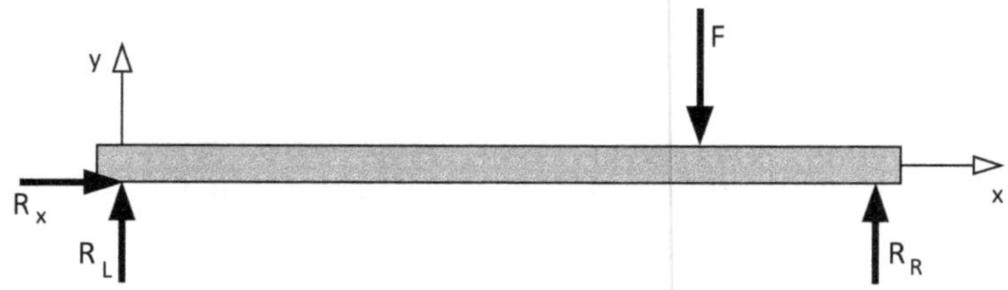

Fig. 3.5 A coplanar, non-concurrent force system.

Clearly, all of the forces acting on the beam lie in the x-y plane making this system **coplanar**. Three of the forces are parallel and their lines of action cannot intersect; hence, the system is **non-concurrent**. In this case, the forces acting on the body, F, R_x, R_L and R_R, must satisfy three of the six equations of equilibrium, namely:

$$\Sigma F_x = R_x = 0, \qquad \Sigma F_y = 0, \qquad \text{and } \Sigma M_z = 0$$

The remaining three equations of equilibrium are satisfied independently of the solution for F, R_x, R_L and R_R.

3.2.4 Coplanar, Concurrent Force Systems

A **coplanar, concurrent force system** is shown in Fig. 3.6.

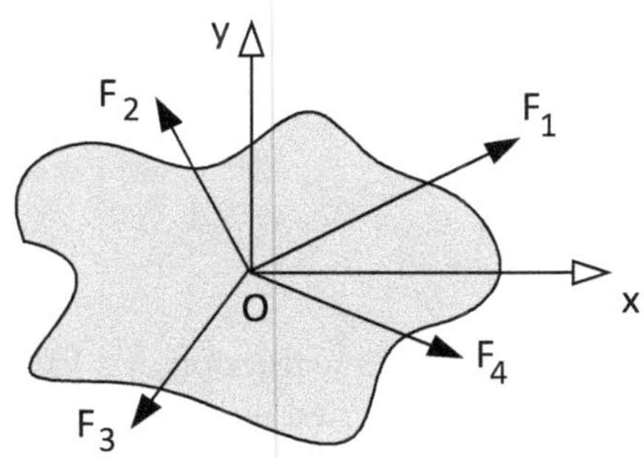

Fig. 3.6 A coplanar and concurrent system of forces.

This force system is **coplanar** because the body and all of the forces acting on it lie in the x-y plane. It is **concurrent** because the lines of action of all the forces pass through point O. This is the simplest of the four force systems because equilibrium is satisfied when:

$$\Sigma F_x = 0 \qquad \text{and} \qquad \Sigma F_y = 0$$

The other four of the six equilibrium relations are satisfied automatically and provide no useful information. Due to concurrency, we recognize that no moments occur about point O and the equations $\Sigma M_x = \Sigma M_y = \Sigma M_y = 0$ are satisfied regardless of the values assigned to the forces. Also, because the forces are coplanar, they all lie in the x-y plane and $\Sigma F_z = 0$.

3.3 FREE BODY DIAGRAMS (FBDs)

The equations of equilibrium that were described in the previous section apply to:

- A single body or member.
- A structure made of several members.
- A portion of a multi-member structure formed by a section cut.

- A part of a body or structure that has been formed by section cuts.
- An element removed from a body.

Before utilizing the equations of equilibrium, we must first construct a **free body diagram (FBD)** of the member being analyzed. A FBD is a model representation of the body being analyzed. The purpose of the model is to simplify the physical representation of the structure by omitting fine details that are not necessary for solving the equilibrium problem.

An example for the construction of a FBD is illustrated in Fig. 3.7. The beam is subjected to a uniformly distributed load of magnitude q (N/m), and supported near each end with simple supports. In Fig. 3.7a, we show the uniformly distributed load with shaded rectangle placed over the span of the beam. The uniformly distributed load may also be represented with a series of arrows as illustrated in Fig. 3.7b. We will use both of these techniques to model uniformly distributed loads. The simple supports are modeled with a pin and clevis on the left side and a roller on the right side. The pin and clevis holds the beam in a fixed position at its left end while the roller permits the beam to expand and/or contract with changes in temperature.

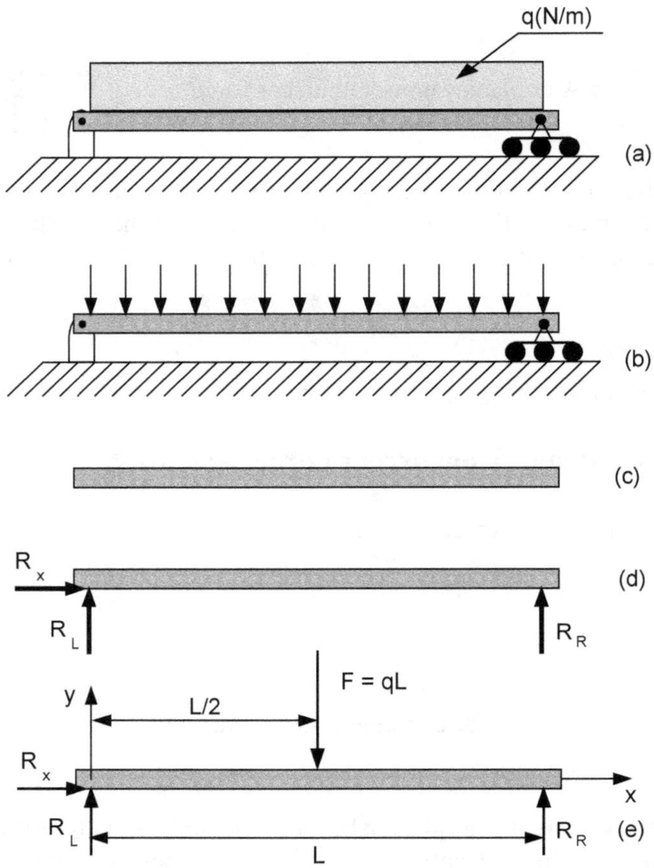

Fig. 3.7 Construction of a FBD for a simply supported uniformly loaded beam.

A four-step procedure is employed in drawing the free body of the beam.

1. Isolate the body (a beam in this case) by removing the supports and the uniformly distributed load. We show the isolated beam in Fig. 3.7c.
2. The supports are replaced with the **reaction** loads R_L, R_R and R_x. The pin and clevis at the left support produces reactions in both the x and y directions as shown in Fig. 3.7d. However, the roller, which is free to rotate, produces a reactive force normal to the surface of the beam (the y direction). At this stage of the analysis, the magnitude of these forces is not known.
3. The uniformly distributed load q applied over the length L of the beam is replaced with a concentrated force F. The magnitude of the force is F = (qL). This force is due to gravity and acts downward as shown in Fig. 3.7e.
4. Finally, we dimension the FBD and establish a coordinate system that will facilitate the equilibrium analysis. In dimensioning the FBD, we place the concentrated force at the location of the centroid (center) of the shaded rectangular area representing the uniformly distributed load (e.g. at L/2 from the left support).

3.3.1 Modeling Loads

The FBD is a model of a structure or some part of a structure. To prepare the FBD, we model both the loads that act on the structure and its supports. Let's first consider modeling loads due to gravity, which are the most commonly encountered forces. Suppose we have a block with a mass m as shown in Fig. 3.8a. We modeled the effect of the block with a concentrated force F = mg. This concentrated force is applied at the center of the block (its centroid). Because the force is due to gravity, the direction of the force is downward (in the negative y direction).

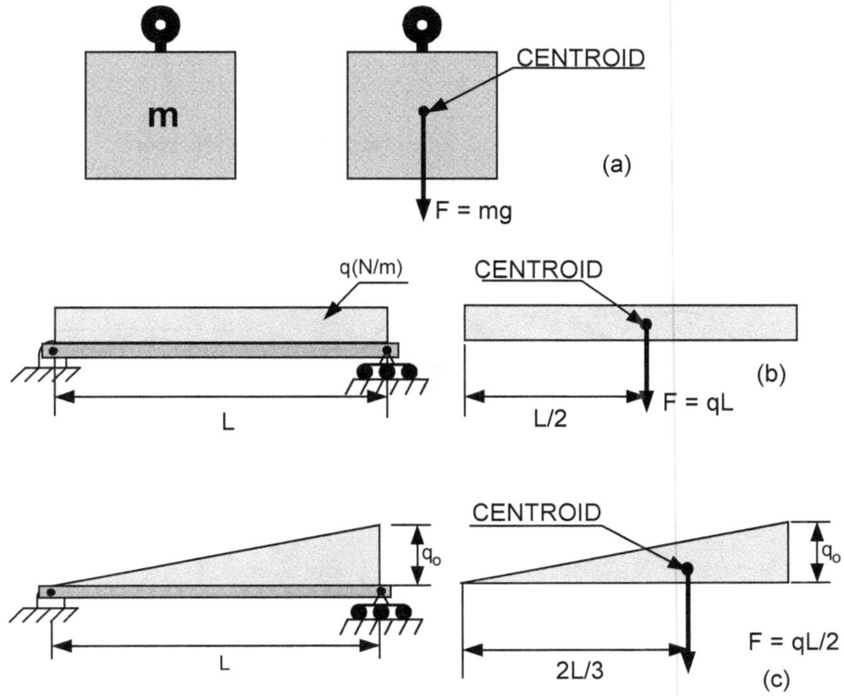

Fig. 3.8 Modeling loads due to gravity.

In Fig. 3.8 b, we encountered loads distributed over the span of a beam. In this illustration, the load is uniformly distributed with a magnitude q expressed in terms of force per unit length (N/m). We represented this uniformly distributed load with a rectangular area of height q and length L. A concentrated force F, the static equivalent of the uniformly distributed load, is given by the area of the rectangular (F = qL). The concentrated force is applied at the centroid of the rectangular area (x = L/2); because the load is due to gravity, the force acts downward.

In Fig. 3.8 c, the load is again distributed over the length of the beam, but it increases as a linear function of x (e.g. $q(x) = q_ox/L$). We represented this load with a triangular area with an altitude q_o and a base of L. A concentrated force, the static equivalent of the distributed load, is given by the area of the triangle as $F = q_oL/2$. It is applied at the centroid of the triangle (x = 2L/3) and acts downward.

When modeling to solve for reaction forces at structural supports, we replace distributed loads with statically equivalent concentrated forces. However, later when we are concerned with internal moments and shear forces, it is not possible to use this simplified modeling technique.

3.3.2 Modeling Supports

The pin and clevis and roller are often used to support beams. The pin and clevis is stationary and does not permit the beam to move in either the x or y directions. On the other hand, the roller permits the beam to expand or contract with changes in temperature. When the pin and the roller are removed in the construction of a FBD, they are replaced with reaction forces shown in Fig. 3.9. For the pivot, which restrains motion in the x and y directions, reaction forces R_L (perpendicular to the bottom surface of the beam) and R_x (parallel to the bottom surface of the beam) are required to represent this support. For the roller (on a frictionless surface), a single reaction force perpendicular to the bottom surface of the beam is sufficient for the representation on a FBD.

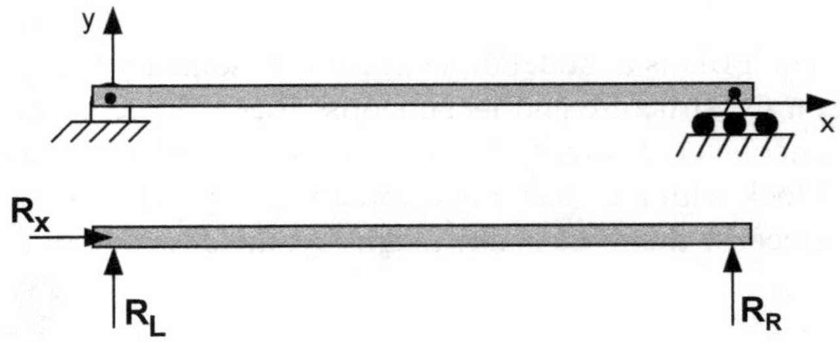

Fig. 3.9 Modeling the pivot and roller type supports.

There are many support conditions and connections to structures. In constructing FBDs, we model the structure by removing these supports and replace them with one or more reactive forces or reactive moments. We will list several different types of supports or connections and their reactive forces and moments in the illustrations presented in Fig. 3.10.

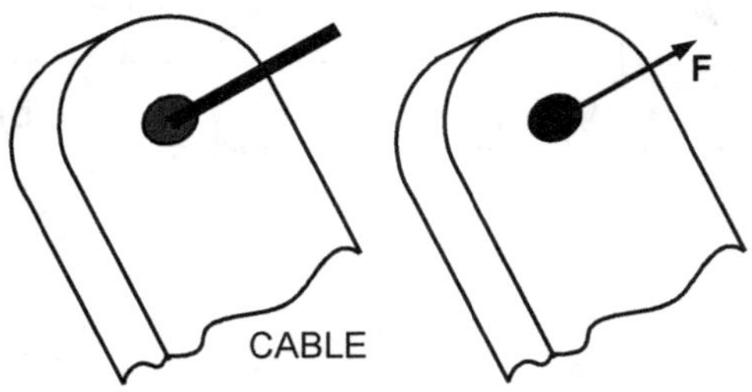

Fig. 3.10a A cable connection is represented with a single tension force acting along the cable in the direction away from the structural element.

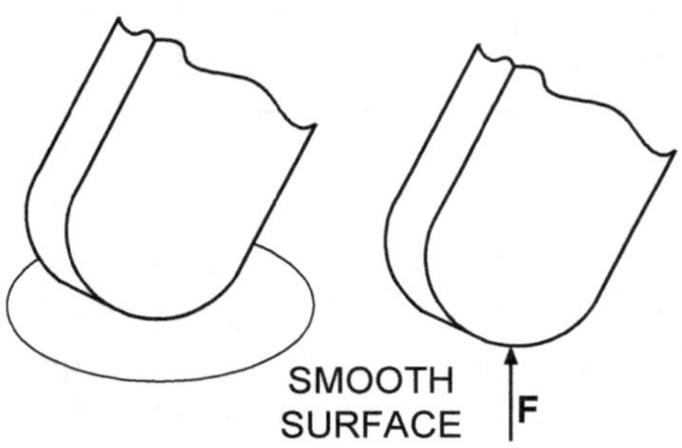

Fig. 3.10b A structural element contacting a smooth (frictionless) surface is modeled with a concentrated force that acts perpendicular to the surface at the point of contact.

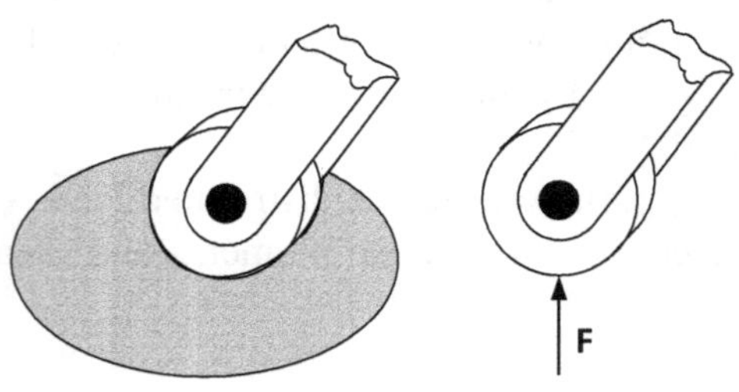

Fig. 3.10c A roller or rocker on a flat surface is represented with a concentrated force that acts perpendicular to the surface at the point of contact.

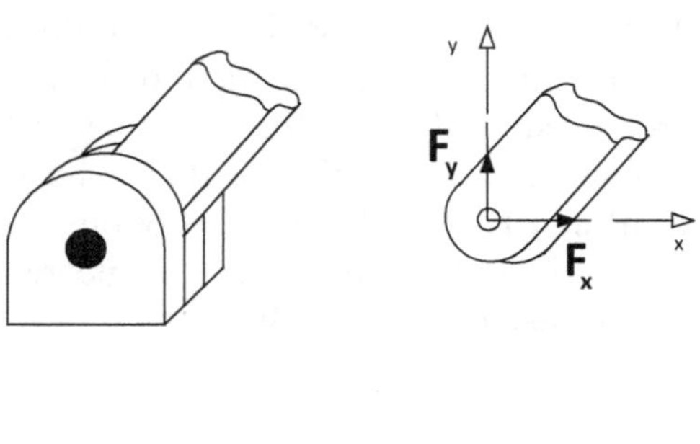

Fig. 3.10d A pin and clevis connection is represented with two Cartesian forces perpendicular to the pin.

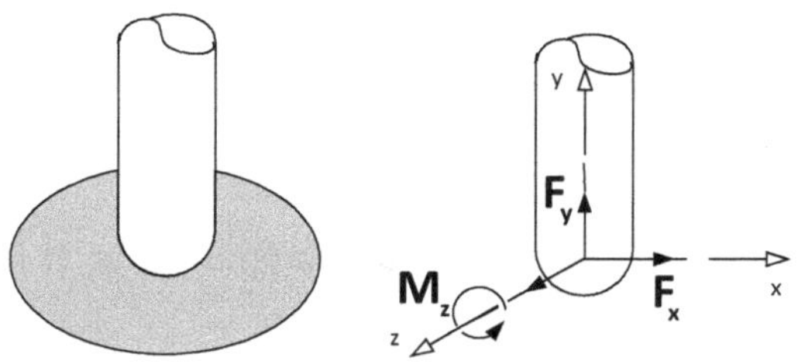

Fig. 3.10e A fixed support is represented with three possible reactions (two Cartesian forces and one Cartesian moment).

EXAMPLE 3.1

Draw a FBD of the ring that is used to attach the cables to the weight shown in Fig. E3.1.

Fig. E3.1

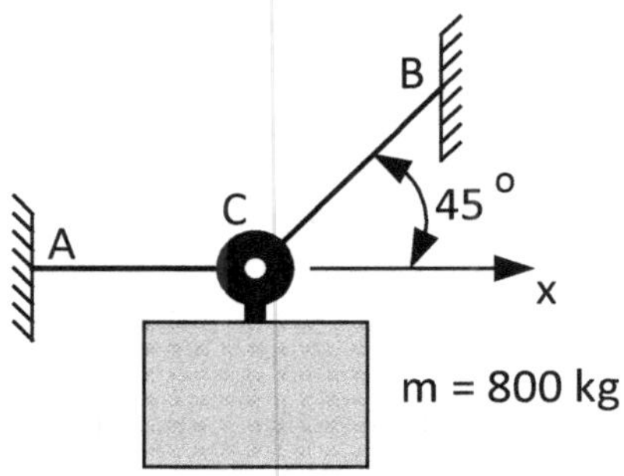

Solution:

The tension developed in the cables CA and CB maintains the 800-kg mass in equilibrium. We will determine the tension in the cables later in Section 3.5, but the first step in solving for the cable tensions is to construct the appropriate FBD. In this example, the small ring located at point C is the appropriate free body.

Step 1. Isolate the body. The isolated ring is shown in Fig. E3.1a.

Fig. E3.1a to c

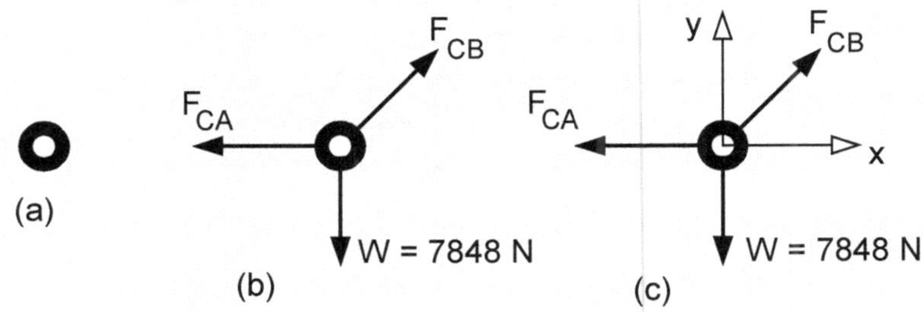

Step 2. Replace the cables with force vectors as shown in Fig. E3.1b. The forces F_{CA} and F_{CB} are of unknown magnitude; however, we know they are oriented in the direction of cables CA and CB respectively.

Step 3. Replace the mass with a force vector acting vertically downward through the center of gravity of the block. The weight of the mass is given by W = 800 kg x 9.81 N/kg = 7848 N.

Step 4. Dimension the FBD and establish a coordinate system. Since the force system is concurrent linear dimensions are not relevant. We show the 45° angle that F_{CB} makes with the x-axis. A x-y coordinate system is placed on the FBD with the origin located at the center of the ring.

Let's consider another example showing the procedure for drawing FBDs. In this instance, we will draw a FBD for a coplanar, non-concurrent force system.

EXAMPLE 3.2

The beam in Fig. E3.2 is subjected to a distributed load that increases from zero at the left support to q_o at the right support. Draw a FBD of the beam.

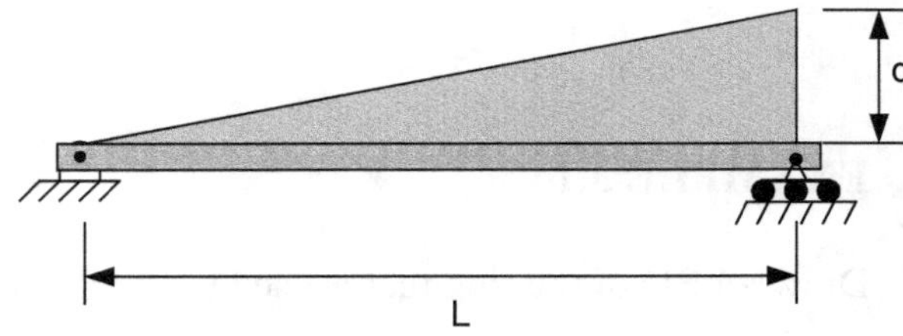

Fig. E3.2

Solution:

Again we model the distributed loading with arrows as shown in Fig. E3.2a. Then we utilize the four-step procedure to prepare the FBD.

Step 1 Isolate the body. The isolated beam is shown below in Fig. E3.2b.

Step 2 The supports are replaced with the reaction loads R_L, R_R and R_x. At this stage of the analysis, the magnitude of these forces is not known. However, at the roller, we know that the direction of the reaction forces is normal to the surface of the beam as shown in Fig. E3.2c. At the pivot, reaction forces R_x and R_L are shown in the x and y directions.

Step 3 Replace the linearly distributed load with a concentrated force F as shown in Fig. E3.2d. The area under the triangle representing the linearly increasing distributed load gives the magnitude of F. Hence;

$$F = (1/2)\, q_o\, L$$

The force F is positioned at (2/3) L, which corresponds to the location of the centroid of the area of a triangle. Of course, the force is directed downward to coincide with the direction of the gravitational field.

Step 4. Dimension the FBD by adding the coordinate system shown in Fig. E3.2d.

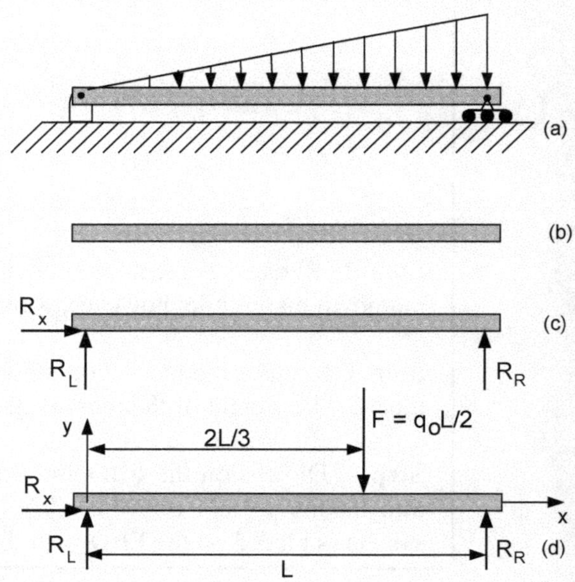

Fig. E3.2 a-d

The FBD is complete because:

1. All the forces acting on the beam are shown
2. The forces are located in the proper positions
3. All the required dimensions are given
4. A coordinate system is provided for reference.

You should use this checklist to determine if the free bodies that you prepare in executing the assigned exercise are complete.

EXAMPLE 3.3

Three smooth (frictionless) cylinders each with a diameter D and a mass of 4 slugs are constrained by a channel with a width W as illustrated in Fig. E3.3. Draw individual FBDs for cylinders A and C if the ratio W/D = 2.5.

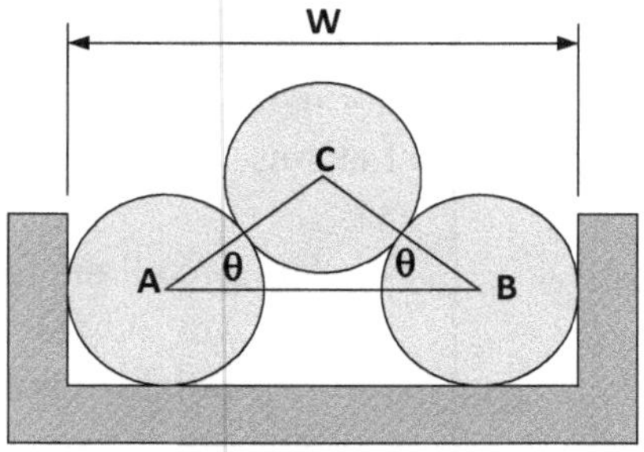

Fig. E3.3

Solution:

Before we construct the FBDs it is necessary to determine the stacking angle θ of the cylinders. From the geometry of the cylinders in the channel, we write:

$$W = D + 2D\cos\theta \qquad \Rightarrow\Rightarrow \qquad \theta = \cos^{-1}[(W/D) - 1]/2 \qquad \text{(a)}$$

$$\theta = \cos^{-1}(2.5 - 1)/2 = 41.41° \qquad \text{(b)}$$

Next let's determine the weight W of one of the cylinders:

$$F = mg = (4\ \text{slug})(32.17\ \text{lb/slug}) = 128.7\ \text{lb} \qquad \text{(c)}$$

An examination of cylinder A indicates that four forces must be placed on the FBD—three contact forces and one gravitational force. We show the FBD for cylinder A in Fig. E3.3a.

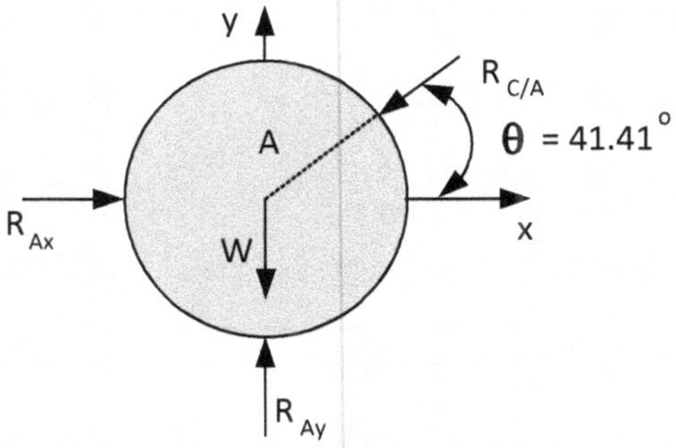

Fig. E3.3a

The reaction forces R_{Ax} and R_{Ay} are due to the surfaces of the channel contacting cylinder A. Because the cylinder is smooth (frictionless), the reaction forces are normal to the contact surface. We have used a double subscript to identify them—the first subscript identifies the cylinder in question and the second gives the direction of the force. Another reaction force $R_{C/A}$ is due to the action of cylinder C on cylinder A. It is normal to the surface of the cylinder at the point of contact. The subscript C/A indicates that cylinder C is acting on cylinder A. The angle $\theta = 41.41°$ is specified on the FBD. Finally the weight W is shown acting downward (in the negative y direction) from the center of the cylinder.

Following similar procedures, we draw the FBD for cylinder C in Fig. E3.3b.

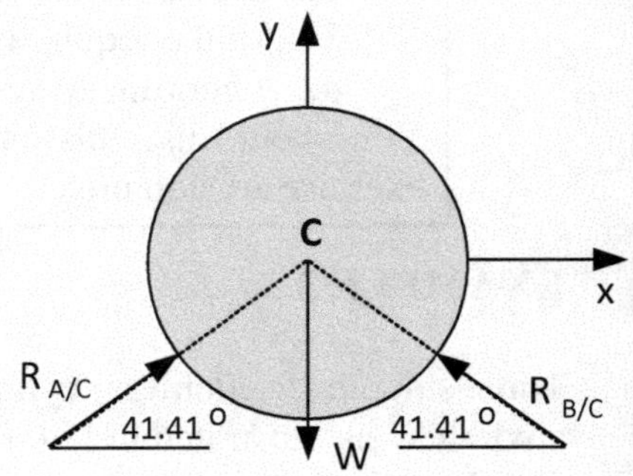

Fig. E3.3b

In this instance, we have the two reaction forces $R_{A/C}$ and $R_{B/C}$ both normal to the surfaces in contact. The weight W acts downward from the center of gravity of the cylinder. Let's use the checklist to ascertain if the FBDs are complete.

- ❑ Are all the forces acting on each cylinder? Yes, we have counted them.
- ❑ Are the forces located in the proper positions? Yes, we have positioned the reaction forces at the points of contact and the gravitational forces at the center of gravity of the cylinders.
- ❑ Are the required dimensions given? In this case the forces are concurrent and dimensions are not necessary. However, the directions of the reaction forces are given by the angle θ.
- ❑ Is a coordinate system provided for reference? Yes, we have placed coordinates on both FBDs with the origins at the center of each cylinder.

EXAMPLE 3.4

A hex nut is being tightened with the wrench shown in Fig. E3.4. The dimension of the flat on the hex nut is 40 mm. Prepare FBD diagrams for the nut and the wrench.

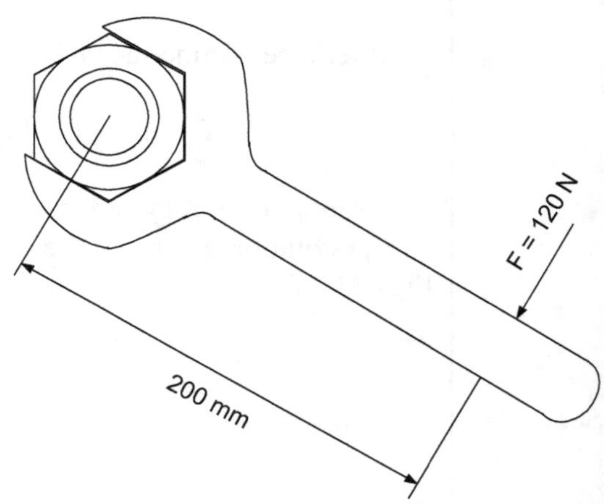

Fig. E3.4

Solution:

The free body diagrams of the nut and the wrench are shown in Fig. E3.4a and E3.4b. When the force is applied to the wrench, it contacts the hex nut at two corners producing reaction forces R_1 and R_2. These forces are perpendicular to the surfaces of the hex nut. The threads on the nut react with those on the bolt to produce a resistive moment (torque) M_z and a reactive force R_3. The angle defining the direction of the reaction forces acting on the flats is 60° because of the geometry of the hex configuration. The distance between the forces is the dimension of the flats on the hex nut. The FBD of the wrench is presented in Fig. E3.4b.

The reaction forces R_1 and R_2 are applied at locations corresponding to the corners of the hex nut. Of course, they are equal in magnitude and opposite in direction of those shown on the FBD of the hex nut. The force applied to the wrench handle is shown. Dimensions locating the point of application of the force F and the reaction forces are provided. Angles are not shown because they are implied by the configuration of the hex nut.

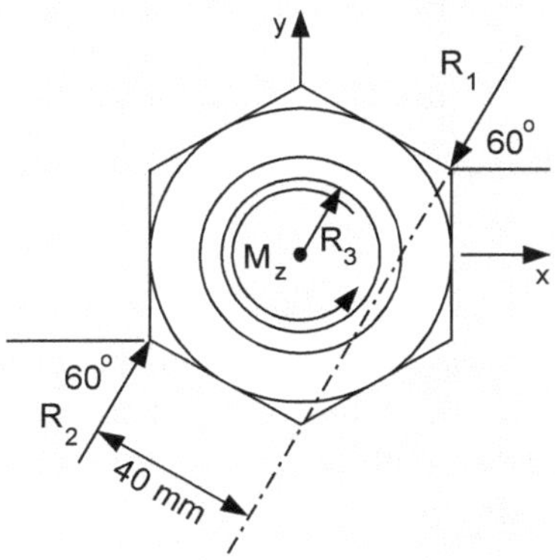

Fig. E3.4a

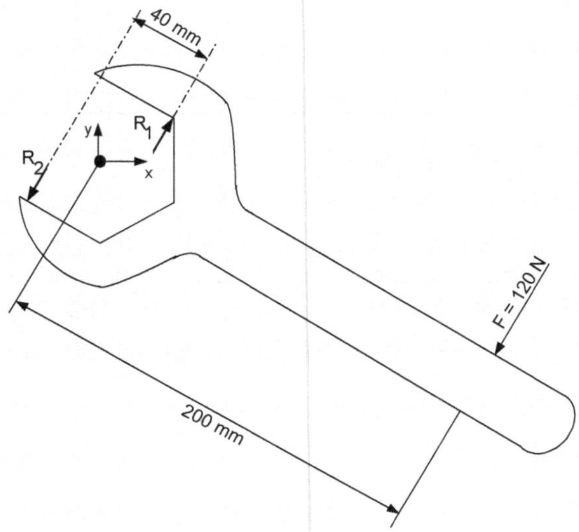

Fig. E3.4b

EXAMPLE 3.5

Prepare a FBD of the wheeled crane lifting a weight W.

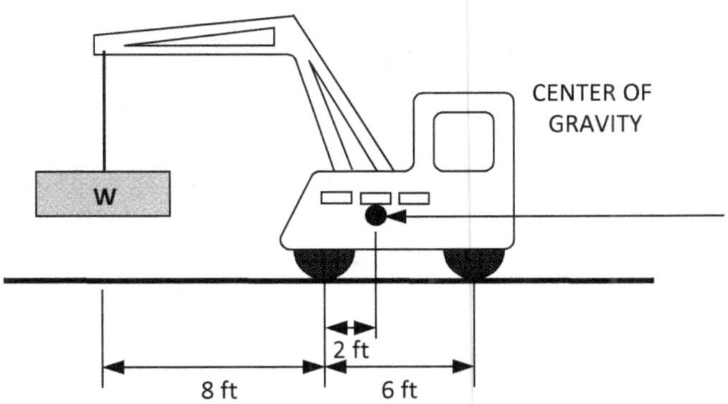

Fig. E3.5

Solution:

Let's isolate the body and apply the reaction loads R_L and R_R at the wheels as shown in Fig. E3.5a. Next, the forces W_L and W_C due to gravity are applied at the center of gravity of the weight and the crane, respectively. The dimensions locating the points of application of the forces are specified. Finally, a Cartesian coordinate system with its origin at the center of mass of the crane is established.

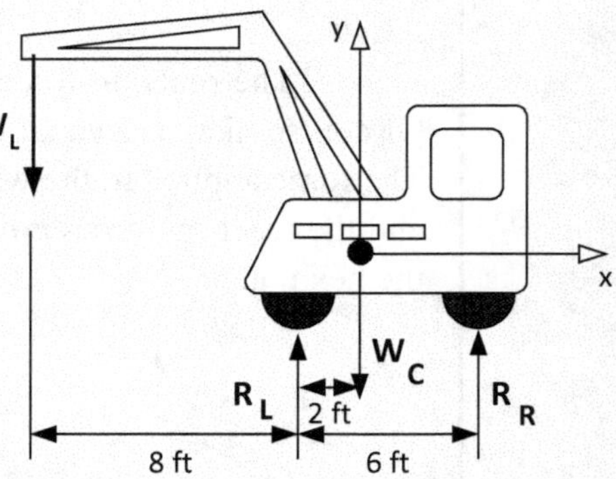

Fig. E3.5a

EXAMPLE 3.6

A small stadium for football games and other sporting events is constructed with a partial roof covering the seating arrangements. One of the many supports for the roof is illustrated in Fig. E3.6. A uniformly distributed load of q is applied to the inclined member that supports the roof. Prepare a FBD of this structural element. Note that the support is modeled as a two-dimensional structure.

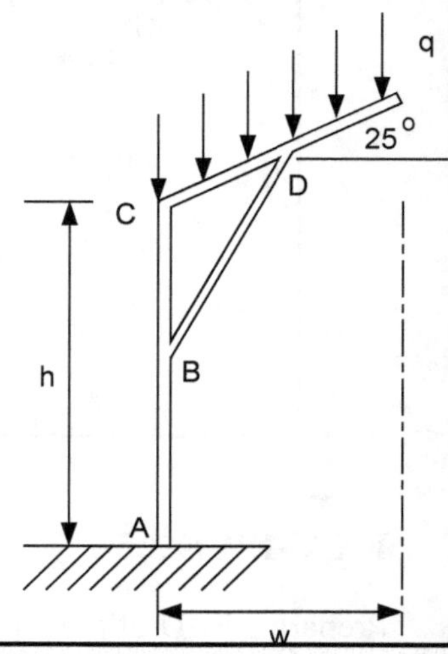

Fig. E3.6

Solution:

We first isolate the stadium component by removing the uniformly distributed load and the support. The uniformly distributed load is replaced with a statically equivalent force F = qw applied at a distance w/2 from the vertical member. The fixed end support of the vertical member is then replaced with reaction forces R_x and R_y and a reaction moment M_z. We show the moment M_z as positive (counterclockwise). Since the structure is coplanar, the other reactions at point A vanish ($R_z = M_x = M_y = 0$). A Cartesian coordinate system is placed on the FBD with its origin at point A. The dimensions of the applied force F are specified. The complete FBD is presented in Fig. E3.6a.

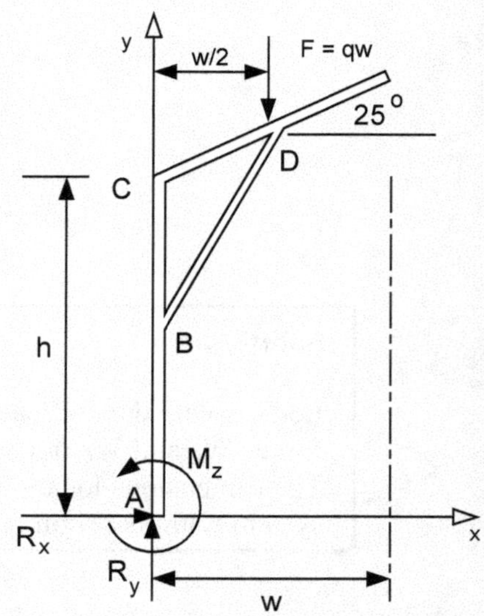

Fig. E3.6a

3.4 FBDs OF PARTIAL BODIES

We are very interested in determining internal forces in structural members because these internal forces produce the stresses and deflections that might cause failure of the structure. Consequently, determining these internal forces will be very important to you when designing your structures. The approach followed in studying internal forces is to:

1. Make an imaginary cut through the structural member being studied.
2. Draw a FBD of one part of this member.
3. Account for the effect of the cut away portion of the member by applying internal forces.
4. Solve for the internal forces or moments using the appropriate equations of equilibrium.

This procedure for constructing FBDs of partial bodies is demonstrated in Example 3.7.

EXAMPLE 3.7

Consider the stepped tension bar loaded with external forces shown in Fig. E3.7. Draw a FBD of the portion of the bar to the left side of section cut A-A.

Fig. E3.7

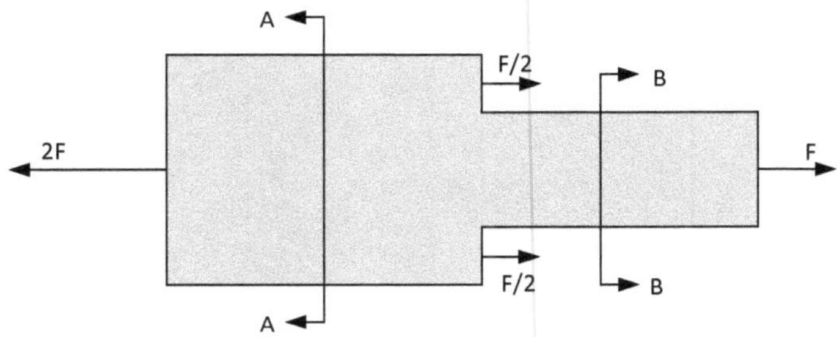

Solution:

Step 1. Isolate the partial body, which is the part of the tension bar to the left of the section cut A-A as shown in Fig. E3.7a.

Step 2. Replace the force 2F on the left part of the partial body as shown in Fig. E3.7b.

Step 3. Place an internal force P on the right face of the partial body to account for the right side of the tension bar that has been cut away as shown in Fig. E3.7c.

Step 4. Establish a coordinate system. In this case, dimensions are not necessary because the forces are **collinear** and **concurrent**. The complete FBD of the left portion of the tension member is shown below in

Fig. E3.7a

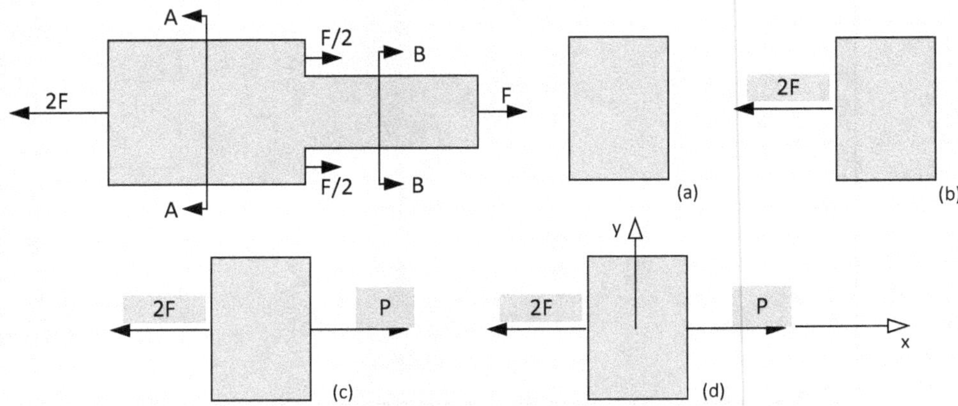

EXAMPLE 3.8

Prepare a FBD of the right portion of the tension bar shown in Fig. E3.7 from section A-A to the bar's right hand end.

Solution:

Step 1. Isolate the right portion of the bar as shown in Fig. E3.8a.

Step 2. Replace the force F and the two forces F/2 on the right part of the body as shown in Fig. E3.8b.

Step 3. Place an internal force P on the left face of the body to account for the left end of the tension bar that has been cut away as shown in Fig. E3.8c.

Step 4. Establish a coordinate system. Because the forces are symmetric about the x-axis, dimensions are not necessary. The complete FBD of the right hand portion of the tension member is shown below in Fig. E3.8c.

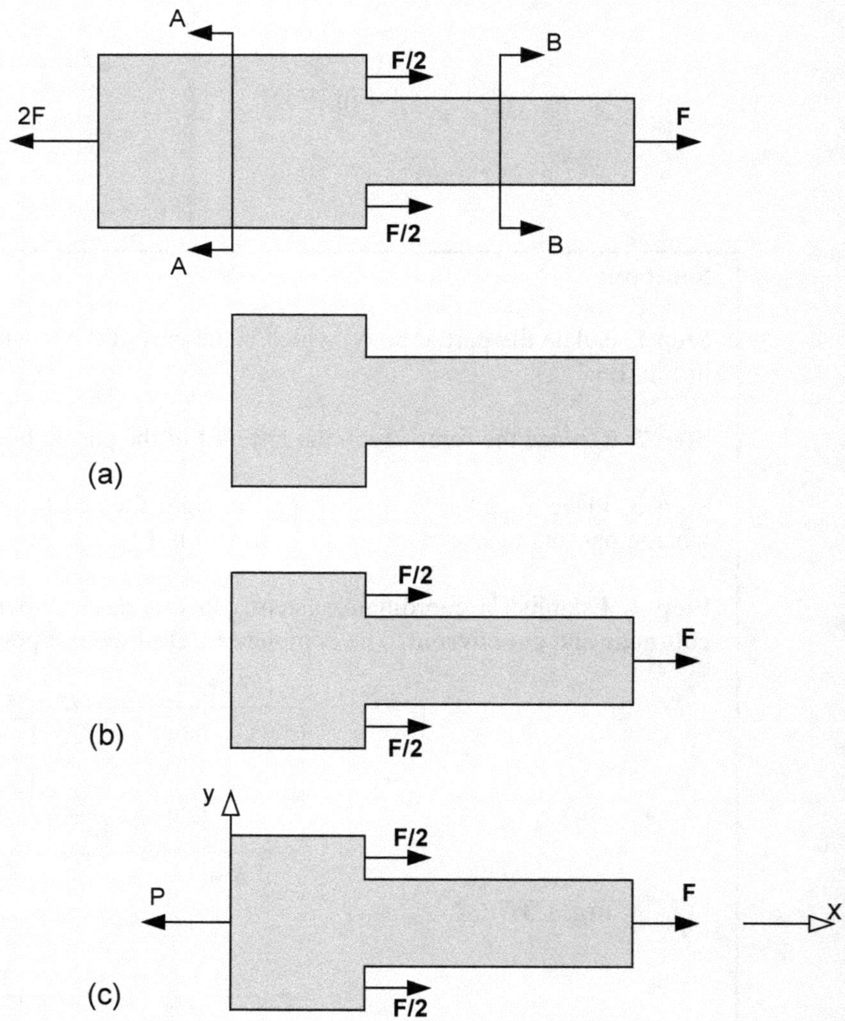

Fig. E3.8 a-c

EXAMPLE 3.9

Prepare a FBD of the left side of a simply supported beam subjected to a uniformly distributed load as shown in Fig. E3.9.

Fig. E3.9

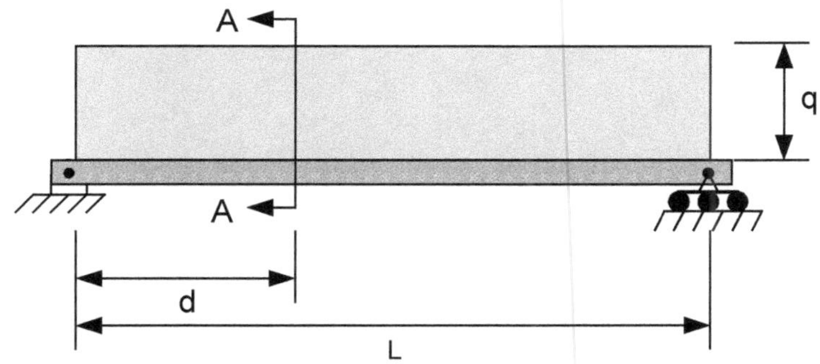

Solution:

Step 1. Isolate the segment of the beam to the left of section A-A as shown in Fig. E3.9a.

Step 2. Replace the uniformly distributed load q with a concentrated force F = qd positioned at d/2 from the left end of the beam.

Step 3. Place reaction forces R_x and R_y to account for the effect of the pivot support. Place internal forces P and V and an internal moment M_z on the exposed face to account for the effect of the right end of the beam that has been cut away. We apply a positive force P (in the positive x direction) and a positive moment M_z (counterclockwise). The force V is applied in the negative y direction. The reason for this choice will be explained later in the text.

Step 4. Dimension the FBD and establish a coordinate system. In analyzing beams, we usually place the origin of the coordinate system at the left end. The complete FBD of the left hand portion of the beam is shown in Fig. E3.9a.

Fig. E3.9a

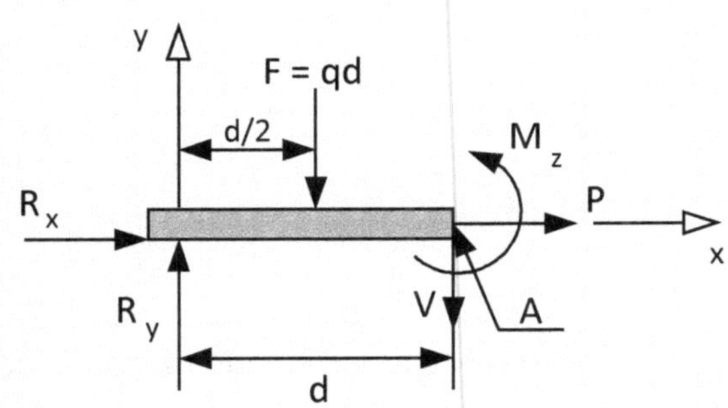

EXAMPLE 3.10

A new factory for heavy machinery has an assembly line with component parts stored above floor level. The components used in assembly are in elevated bins (#1, #2, and #3) as shown in Fig. E3.10. The elevated bins are suspended from a frame ABCD. Prepare a FBD of the left portion of the frame from Section A-A to the roller support at point A.

Fig. E3.10

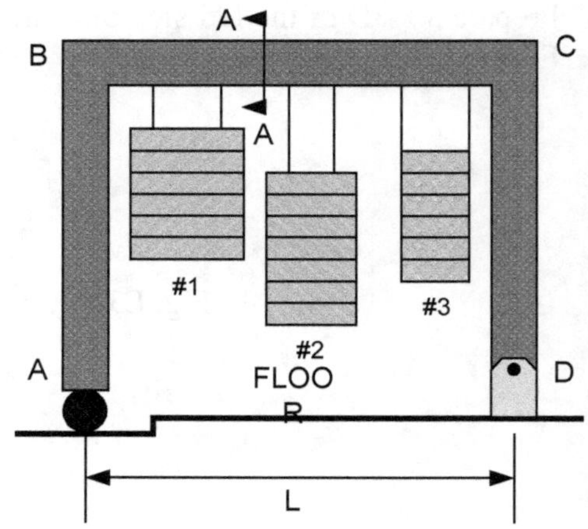

Solution:

Step 1. Isolate the segment of the frame to the left of section A-A as shown in Fig. E3.10a.

Step 2. Replace the loads due to the component parts on elevated bin #1 with concentrated forces $F_1/2$ and $F_1/2$ positioned at the location of the cables supporting this bin.

Fig. E3.10a

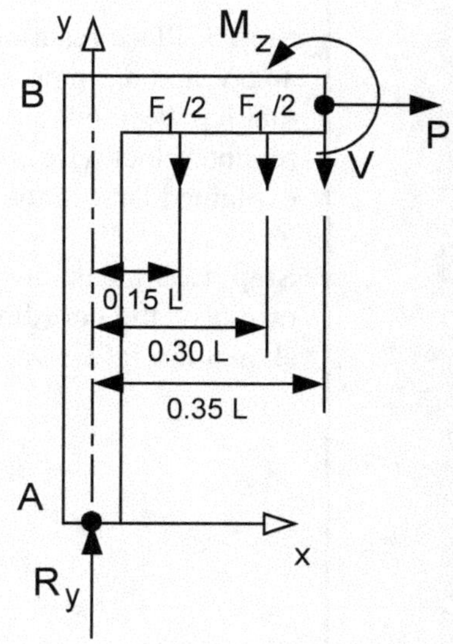

Step 3. Place a reaction force R_y at point A to account for the effect of the roller support. Place an internal forces P and V and an internal moment M_z on the exposed face to account for the right hand portion of the frame that has been cut away. We apply a positive force P (in the positive x direction) and a positive moment M_z (counterclockwise). The force V is applied in the negative y direction. The reason for this choice will be explained later in the text.

Step 4. Dimension the FBD and establish a coordinate system. We have placed the origin of the coordinate system at point A and have scaled the dimensions locating the cables for bin 1. The complete FBD of the left hand portion of the frame is shown in Fig. E3.10a.

3.5 SOLVING FOR REACTIONS

The FBD provides a model to assist us in writing equilibrium equations that are used in the solution for unknown forces and reactions. The diagram indicates the known and unknown forces and their directions. It also provides the dimensions needed to compute moments. Let's consider the FBDs, which we developed in Examples 3.1 to 3.10, and solve for the unknown forces by using the appropriate equations of equilibrium.

EXAMPLE 3.11

Recall the 800 kg mass that was supported by cables AC and BC as shown in the illustration in Example 3.1. In Section 3.3, we constructed a FBD for the ring that ties the weight and the cables together. For your convenience it is repeated in Fig. E3.11. Solve for the unknown forces acting on the ring.

Fig. E3.11

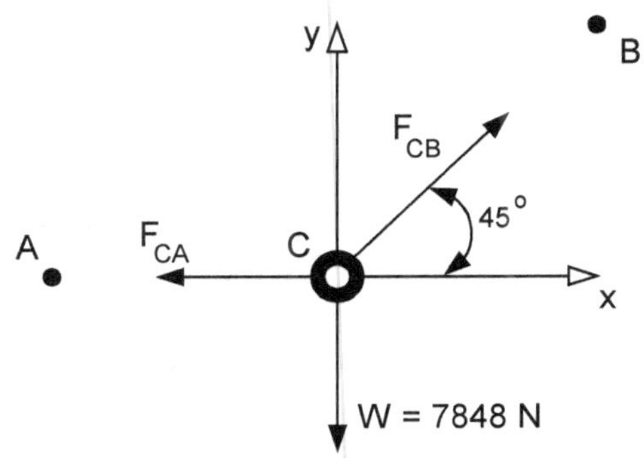

Solution:

The solution for the unknown forces and reactions is accomplished using the equations of equilibrium with a simple six-step procedure. We demonstrate this step-by-step approach for the cable-mass arrangement represented by the FBD.

Step 1. Prepare a complete FBD of the ring as shown above.

Step 2. Classify the force system. In this example, the force system is coplanar and concurrent.

Step 3. Write the relevant equations of equilibrium. For coplanar and concurrent force systems, only two of the six equations of equilibrium provide relevant information.

$$\Sigma F_x = 0 \qquad \Sigma F_y = 0$$

Step 4. Substitute the unknowns from the FBD into the relevant equations of equilibrium.

$$\Sigma F_y = 0 \qquad \Sigma F_x = 0$$

$$\Sigma F_y = F_{CB} \sin (45°) - 7{,}848 \text{ N} = 0 \qquad \Sigma F_x = F_{CB} \cos (45°) - F_{CA} = 0$$

Step 5. Execute the solution.

$$\Sigma F_y = F_{CB} \sin (45°) - 7{,}848 \text{ N} = 0 \qquad \Sigma F_x = F_{CB} \cos (45°) - F_{CA} = 0$$

$$F_{CB} = 7{,}848 \text{ N}/0.7071 \qquad F_{CA} = (11{,}100 \text{ N})(0.7071)$$

$$F_{CB} = 11{,}099 \text{ N} \Rightarrow 11{,}100 \text{ N} \qquad F_{CA} = 7{,}848 \text{ N}$$

Step 6. Check the solution and interpret the results. In this example, the equations are simple, and we may easily determine if the numerical results are realistic by checking the signs in each of the equilibrium equations. Note that **forces directed in the negative coordinate directions are considered negative quantities** in writing the ΣF terms. You should always make certain that the forces are expressed in the appropriate units. In this problem, the mass was given in kg (SI units), and this fact dictates that the force be expressed in newtons (N). Finally, we make a sketch of the force vectors shown in Fig. E3.11a verifying the results.

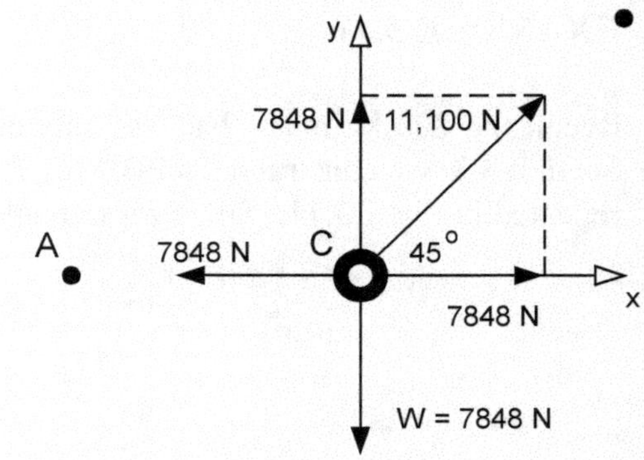

The 11,100 N force produces 7,848 N components in the positive x and y directions, which cancel out the 7,848 N forces, directed in the negative x and y directions. Clearly, the forces in the x and y directions are balanced, and the ring is in equilibrium.

EXAMPLE 3.12

Next consider the beam that was illustrated in Example 3.2. Recall that we constructed the FBD to model this beam as indicated below:

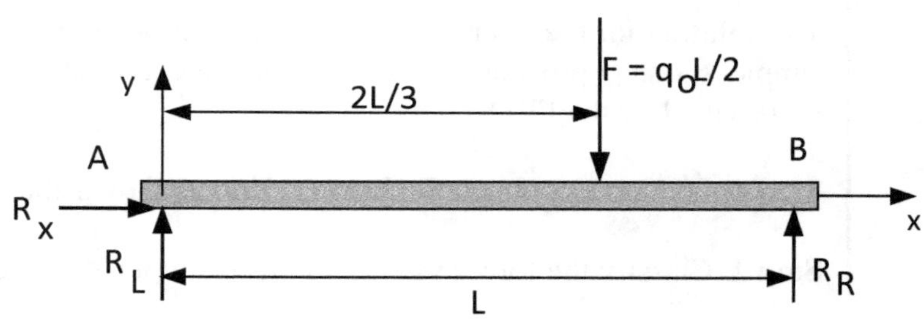

Fig. E3.12

Let's execute the six-step procedure to solve for the unknown magnitudes of the reaction forces R_x, R_L and R_R.

Solution:

Step 1. Prepare a complete FBD of the beam as shown above. Note the force F was determined from the area of the triangle representing the linear distributed load as:

$$F = (1/2)\, q_o\, L$$

Step 2. Classify the force system. In this problem, the force system is coplanar and non-concurrent.

Step 3. Write the relevant equations of equilibrium. For coplanar and non-concurrent force systems, only three of the six equations of equilibrium are relevant.

$$\Sigma F_x = 0 \qquad \Sigma F_y = 0 \qquad \Sigma M_A = 0$$

To determine the moments to substitute into the equation $\Sigma M = 0$, it is necessary to select some arbitrary point on the FBD to use as a reference for determining the moments. We have selected point A because it eliminates the moment due to the force R_L and simplifies the resulting equations.

Step 4. Substitute the unknowns from the FBD into the relevant equations of equilibrium.

$$\Sigma F_x = 0 \quad \Rightarrow \quad R_x = 0$$

$$\Sigma F_y = 0 \quad \Rightarrow \quad \Sigma F_y = R_L + R_R - (1/2)\, q_o\, L = 0 \qquad \text{(a)}$$

$$\Sigma M_A = 0$$

$$\Sigma M_A = R_R\, L - (1/2)\, q_o\, L\, (2/3)L = 0 \qquad \text{(b)}$$

Step 5. Execute the solution.

$$\Sigma M_A = R_R\, L - (1/2)\, q_o\, L\, (2/3)L = 0 \qquad \text{(c)}$$

$$\Sigma F_y = R_L + R_R - (1/2)\, q_o\, L = 0 \qquad \text{(d)}$$

Solving Eq. (c) for R_R yields:

$$R_R = q_o\, L/3 \qquad \text{(e)}$$

and Eq. (d) gives:

$$R_L = q_o\, L/6 \qquad \text{(f)}$$

Step 6. Check the solution and interpret the results. In this example, the solution involves solving two linear algebraic equations. We check the signs in each of the equilibrium equations. Note that **forces directed in the negative coordinate directions are considered negative quantities** in writing the ΣF terms. The **moments tending to rotate the beam counterclockwise are treated as positive quantities, and those tending to produce clockwise motion are treated as negative quantities**. You should make certain that the forces and the moments are expressed in the appropriate units. In this problem, no units were specified since the applied loading and the beam's length were given as symbols q_o and L respectively. However, let's suppose that L = 14 ft and q_o = 1200 lb/ft. With L and q_o specified we may determine numerical values and units for R_L and R_R as:

$$R_R = q_o\, L/3 = (1200 \text{ lb/ft})(14 \text{ ft})/3 = 5600 \text{ lb}$$

$$R_L = q_o\, L/6 = (1200 \text{ lb/ft})(14 \text{ ft})/6 = 2800 \text{ lb}$$

$$F = (1/2)\, q_o\, L = (1200 \text{ lb/ft})(14 \text{ ft})/2 = 8400 \text{ lb} = R_R + R_L$$

The result for the force F provides a check on the solution, and the units for all of the results are specified.

Let us continue to develop skills for solving equilibrium problems. For the next example, consider the cantilever beam that is shown in Fig. 3.11. A cantilever beam is a long slender structural member. It is built-in (supported so that the beam cannot rotate) on the left-hand end, and free (not supported in any manner) on the right-hand end. A beam is usually loaded with transverse forces distributed along its length. The cantilever beam in Fig. 3.11 is loaded with a transverse concentrated force F at its free end.

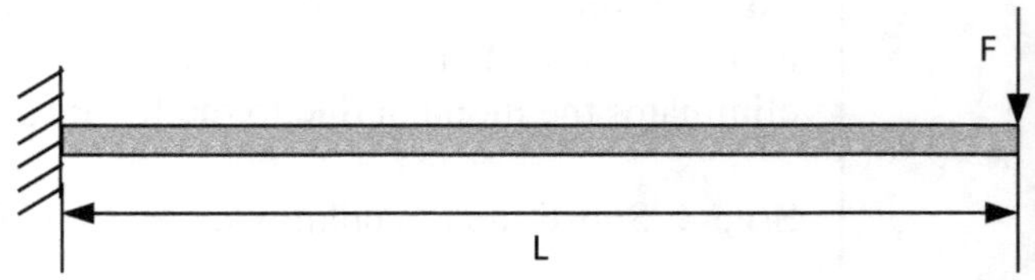

Fig. 3.11 A cantilever beam loaded at its free end with a concentrated force F.

EXAMPLE 3.13

Determine the reactions (forces and moments) at the built-in end of the cantilever beam illustrated in Fig. 3.11.

Solution:

Let's execute the six-step procedure to determine the unknown magnitudes of the reaction forces R_x and R_L and the reaction moment M_W due to the built-in support.

Step 1. Prepare a FBD to provide a model of the beam as shown in Fig. E3.13. Note the built-in end support is removed, and replaced with the reaction forces R_L and R_x and the reaction moment M_W. The moment M_W is taken as a positive quantity and is shown in the counter clockwise direction. A two-dimensional Cartesian coordinate system was added for reference.

Fig. E 3.13

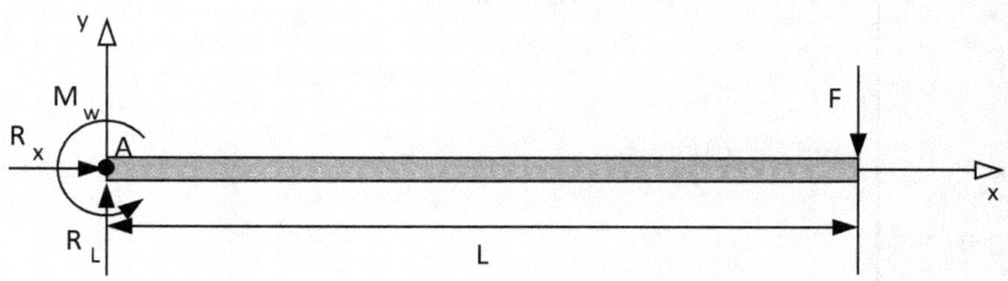

Step 2. Classify the force system. In this problem, the force system is coplanar and non-concurrent.

Step 3. Write the relevant equations of equilibrium. For coplanar and non-concurrent force systems, only three of the six equations of equilibrium are relevant.

$$\Sigma F_x = 0 \qquad \Sigma F_y = 0 \qquad \Sigma M_A = 0$$

Step 4. Substitute the unknowns from the FBD into the relevant equations of equilibrium.

$$\Sigma F_x = 0$$

$$R_x = 0 \qquad \text{(a)}$$

With the applied forces all in the transverse direction, the reaction force R_x at the built-in support will always be zero as shown in Eq. (a).

$$\Sigma F_y = R_L - F = 0 \qquad \text{(b)}$$

$$\Sigma M_A = M_W - F\,L = 0 \qquad \text{(c)}$$

Step 5. Execute the solution. Solving Eq. (a) for R_L yields:

$$R_L = F \qquad \text{(d)}$$

and Eq. (c) gives:

$$M_W = FL \qquad \text{(e)}$$

Step 6. Check the solution and interpret the results. In this example, the mathematics involves solving two very simple linear algebraic equations. We check the signs in each of the equilibrium equations. The forces directed in the negative coordinate directions are considered negative quantities in writing the ΣF terms. The moments tending to rotate the beam counterclockwise are treated as positive quantities, and those tending to produce clockwise motion are treated as negative quantities in writing the ΣM terms. The reaction moment M_W was assumed to be in the counterclockwise direction. The solution $M_W = FL$ is a positive quantity indicating this assumption was correct. **If we had assumed the incorrect direction, the solution of the equilibrium relation would have resulted in a negative quantity.** You should make certain that the forces and the moments are expressed in the appropriate units. In this problem, no units were specified since the applied loading and the length of the beam were given as symbols F and L, respectively. However, let's suppose that L = 6 m and F = 20 kN. Determining numerical values and units for R_L and M_W, we find:

$$R_L = F = 20 \text{ kN}$$

$$M_W = F\,L = (20 \text{ kN})(6 \text{ m}) = 120 \text{ kN-m}$$

For another example, let's determine the reaction forces at the supports for the truss bridge structure shown in Fig. 3.12. The truss bridge is supported at one end with a pin and clevis and at the other with a roller. **The roller is often used to support one end of a long structure because it rotates to accommodate the changes in the structure's length due to temperature changes.** With the roller support, the truss in Fig. 3.12 is free to expand and contract with temperature, and thermal stresses due to constraint do not develop.

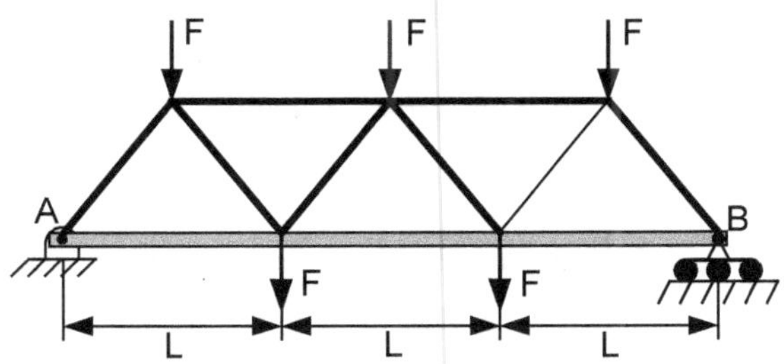

Fig. 3.12 The truss is loaded at the intersection of the members with a load of F.

EXAMPLE 3.14

Determine the reaction forces at the supports for the truss presented in Fig. 3.12.

Solution:

Step 1. Prepare a FBD to provide a model of the truss as shown in Fig. E3.14. Note the supports were removed and replaced with the reaction forces R_L and R_R. A two-dimensional Cartesian coordinate system was added for reference.

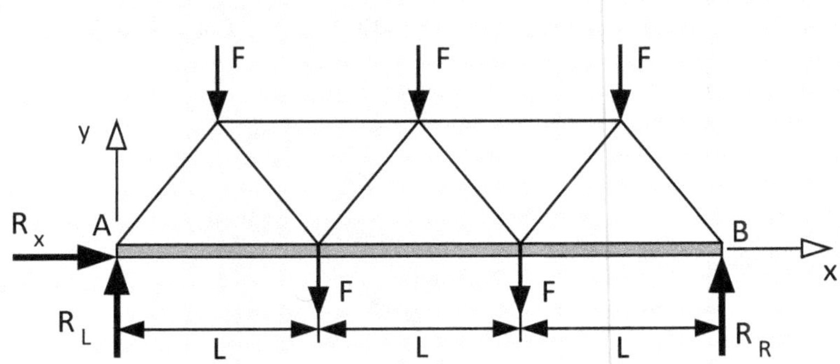

Fig. E3.14

Step 2. Classify the force system. In this instance, the force system is coplanar and non-concurrent.

Step 3. Write the relevant equations of equilibrium. For a coplanar and non-concurrent force system, only three of the six equations of equilibrium are relevant.

$$\Sigma F_x = 0 \qquad \Sigma F_y = 0 \qquad \Sigma M_A = 0$$

Step 4. Substitute the unknowns from the FBD into the relevant equations of equilibrium.

$$\Sigma F_x = 0 \quad \Rightarrow \quad R_x = 0 \qquad \text{(a)}$$

$$\Sigma F_y = R_L + R_R - 5F = 0 \qquad \text{(b)}$$

$$\Sigma M_A = R_R\,(3L) - F(L/2) - FL - F(3L/2) - F(2L) - F(5L/2)\ 0 \qquad \text{(c)}$$

Step 5. Execute the solution. Solving Eq. (c) for R_R yields:

$$R_R = (2.5)\,F \qquad \text{(d)}$$

and Eq. (b) gives:

$$R_L = (2.5)\,F \qquad \text{(e)}$$

Step 6. Check the solution and interpret the results. Again, the mathematics involves solving two simple linear algebraic equations. We check the signs in each of the equilibrium equations. The moments tending to rotate the truss counterclockwise are treated as positive quantities, and those tending to produce clockwise motion are treated as negative quantities in writing the ΣM terms. Once again, if we had assumed the incorrect direction for the reaction forces, the solution of the equilibrium relation would have been a negative quantity. You should make certain that the forces and the moments are expressed in the appropriate units. In this problem, no units were specified since the applied loading and the lengths of the truss segments are given as symbols F and L, respectively. However, let's suppose that L = 10 m and F = 12 kN. Determining numerical values and units for R_L and R_R, we find:

$$R_R = R_L = (2.5)\,F = 30 \text{ kN}$$

Notice that the reaction forces are equal. We expect this result because the geometry of the truss and its loading are symmetric. Also recognize that the reaction forces are independent of the length of the truss. Can you explain the reason for the reaction forces to be independent of the length of the truss?

EXAMPLE 3.15

Using the FBDs presented in Fig. E3.15a and Fig. E3.15b, determine the contact forces between the cylinders described in Example 3.3.

Solution:

Step 1. Prepare a complete FBD of the cylinders A and B.

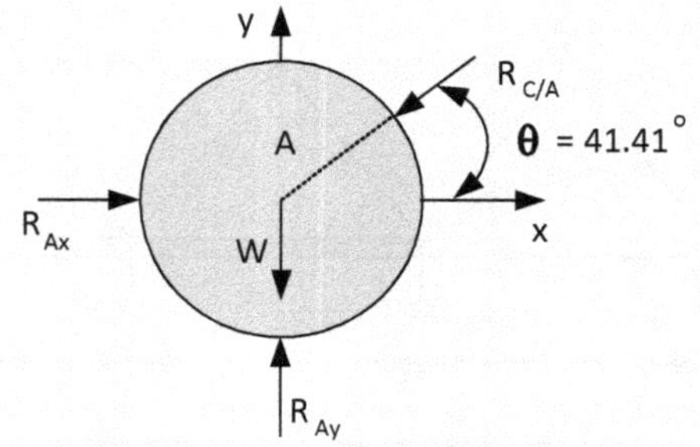

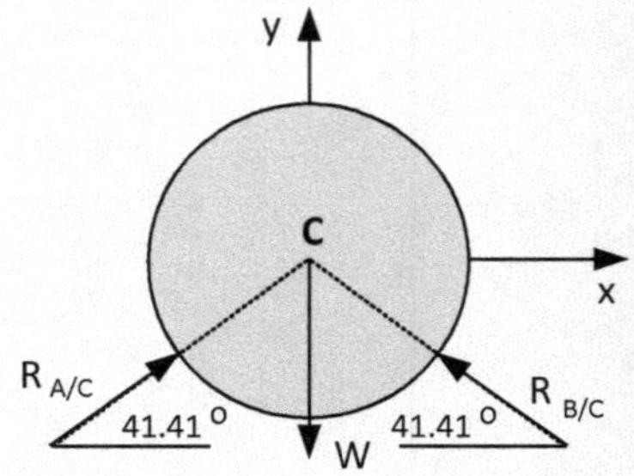

Step 2. Classify the force system. The force system acting on each cylinder is coplanar and concurrent.

Step 3. Write the relevant equations of equilibrium. For coplanar and concurrent force systems, only two of the six equations of equilibrium provide relevant information.

$$\Sigma F_x = 0 \qquad \Sigma F_y = 0$$

Step 4. Consider cylinder C first, substitute the unknowns from the FBD, presented in Fig. E3.3b, into the relevant equations of equilibrium, and execute the solution.

$$\Sigma F_x = R_{A/C} \cos(41.41°) - R_{B/C} \cos(41.41°) = 0$$

$$R_{A/C} = R_{B/C} \qquad \text{(a)}$$

$$\Sigma F_y = R_{A/C} \sin(41.41°) + R_{B/C} \sin(41.41°) - 128.7 = 0$$

$$R_{A/C} = R_{B/C} = 128.7/[2 \sin(41.41°)] = 97.29 \text{ lb} \qquad \text{(b)}$$

Step 5. Consider cylinder A, substitute the unknowns from the FBD, shown in Fig. E3.3a, into the relevant equilibrium equations and execute the solution.

$$\Sigma F_x = R_{Ax} - R_{C/A} \cos(41.41°) = 0$$

$$R_{Ax} = (97.29)(0.7500) = 72.97 \text{ lb} \qquad \text{(c)}$$

$$\Sigma F_y = R_{Ay} - W - R_{C/A} \sin(41.41°) = 0$$

$$R_{Ay} = 128.7 + (97.29)(0.6614) = 193.1 \text{ lb} \qquad \text{(d)}$$

Step 6. Check the solution and interpret the results. In this example, the equations are simple, and we may easily determine if the numerical results are realistic by checking the signs in each of the equilibrium equations and by repeating the calculations. Note that **forces directed in the negative coordinate directions are considered negative quantities** in writing the ΣF terms. You should always make certain that the forces are expressed in the appropriate units.

EXAMPLE 3.16

For the hex nut and wrench described in Example 3.4, determine the torque acting at the threads of the nut and the reaction forces R_1 and R_2.

Solution:

Let's refer to the FBD for the wrench presented in Fig. E3.4b. An examination of the FBD indicates that the force system is coplanar and non-concurrent. Writing the equilibrium relations yields:

$$\Sigma F_{\perp} = R_1 - R_2 - 120 = 0 \qquad \Sigma F_{\parallel} = 0 \qquad \text{(a)}$$

where the subscripts ⊥ and || refer to directions perpendicular and parallel to the wrench axis, respectively.

$$R_1 = R_2 + 120 \qquad \text{(b)}$$

$$\Sigma M_O = 20(R_1 + R_2) - (200)(120) = 0 \qquad \text{(c)}$$

From Eq. (a) and Eq. (b), we solve for R_1 and R_2 to obtain:

$$R_1 = 660 \text{ N} \quad \Rightarrow \quad R_2 = 540 \text{ N} \qquad \text{(d)}$$

To determine the torque T applied to the nut, refer to the FBD in Fig. E3.4a.

$$\Sigma M_O = M_z - 20\, R_1 - 20\, R_2 = 0$$

$$T = M_z = 20(660 + 540) = 24{,}000 \text{ N-mm} = 24 \text{ N-m} \qquad \text{(e)}$$

The positive sign for the moment M_z, which acts on the threads of the nut, indicates that the direction assigned to M_z in Fig. E3.4a was correct.

EXAMPLE 3.17

Determine the reactions at the wheels of the crane using the FBD presented in Fig. E3.17. The weight lifted is 1,200 lb and the wheeled crane weighs 6,400 lb.

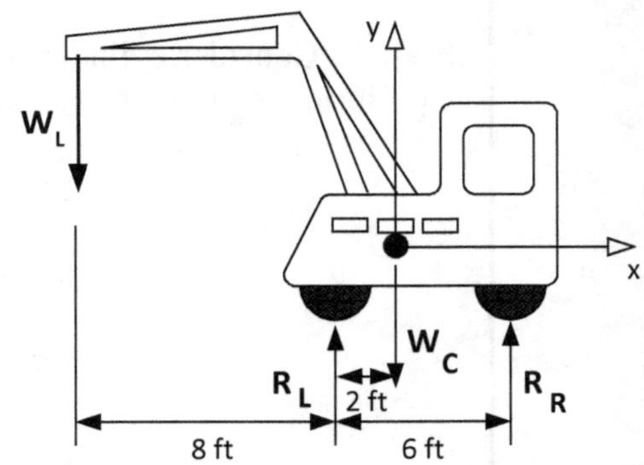

Fig. E3.17

Solution:

Let's refer to the FBD for the crane presented in Fig. E3.17. An examination of the FBD indicates that the force system is coplanar and non-concurrent. Writing the equilibrium relations yields:

$\Sigma F_x = 0$ is satisfied regardless of the values determined for the reaction forces.

$$\Sigma F_y = R_L + R_R - W_C - W_L = 0 \qquad \text{(a)}$$

$$R_L = W_C + W_L - R_R \qquad \text{(b)}$$

$$\Sigma M_O = (6 - 2)R_R - 2R_L + 10W_L = 0 \qquad \text{(c)}$$

Substituting Eq. (b) into Eq. (c) and solving for R_R yields:

$$R_R = (1/3)(W_C - 4W_L) \qquad \text{(d)}$$

$$R_L = (1/3)(2W_C + 7W_L) \qquad \text{(e)}$$

Substituting $W_C = 6{,}400$ lb and $W_L = 1{,}200$ lb into Eqs. (d) and (e) gives:

$$R_R = 533.3 \text{ lb} \quad \Rightarrow\Rightarrow \quad R_L = 7{,}067 \text{ lb} \qquad \text{(f)}$$

The results obtained may be checked by substituting the numerical values for R_L, R_R W_C and W_L into Eq. (a) and verifying the equality.

EXAMPLE 3.18

Using the FBD in Fig. E3.18, determine the reactions at the base of the vertical member that supports the roof of the football stadium.

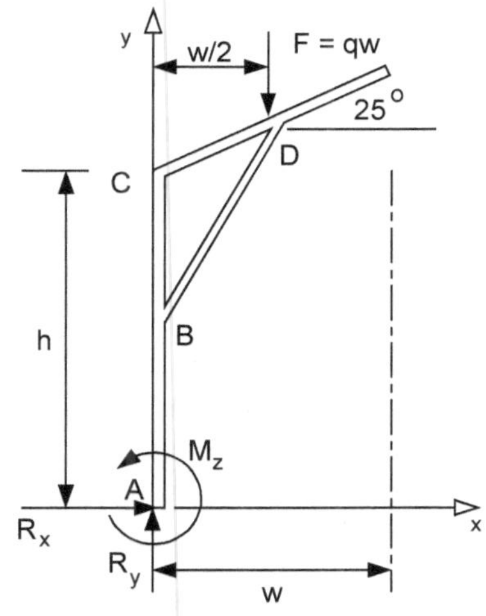

Fig. E3.18

Solution:

The force system is classified as coplanar and non-concurrent. Accordingly we may write three equilibrium relations.

$$\Sigma F_x = 0 \quad \Rightarrow \quad R_x = 0 \qquad \text{(a)}$$

$$\Sigma F_y = R_y - qw = 0 \quad \Rightarrow \quad R_y = qw \qquad \text{(b)}$$

$$\Sigma M_A = M_z - (qw)(w/2) \quad \Rightarrow \quad M_z = qw^2/2 \qquad \text{(c)}$$

A numerical value for the uniformly distributed load q has not been provided, nor has the value for the dimension w. How are we to check the solution? Consideration of the units of the quantities in Eq. (b) and Eq. (c) provides useful information. In Eq. (b), we know the reaction R must have a force unit F, the distributed load q a unit of F/L, and the distance w a unit of L. We then write the equation to check the units as:

$$F = (F/L)L \quad \Rightarrow \quad F = F \qquad \text{(d)}$$

Similarly for the moment M in Eq. (c), it must have a unit FL. We check the homogeneity of the units in Eq. (c) by:

$$FL = (F/L)(L)^2 \quad \Rightarrow \quad FL = FL \qquad \text{(e)}$$

The results from Eqs. (d) and (e) confirm that the units in our solution are homogenous as required.

3.6 FORCES IN CABLE AND PULLEY ARRANGEMENTS

Pulleys and cables[2] are often used in lifting devices such as cranes, derricks, and elevators. The pulley is like a wheel because it rotates about a shaft with very low friction. In fact, we will assume that the pulley is frictionless. At the perimeter of the pulley, a groove is cut to contain the cable and to prevent the cable from slipping off. The essential details of a pulley are illustrated in Fig. 3.13.

[2] A cable is like a rope except that wire is used for the cable's fibers instead of hemp or some other organic material.

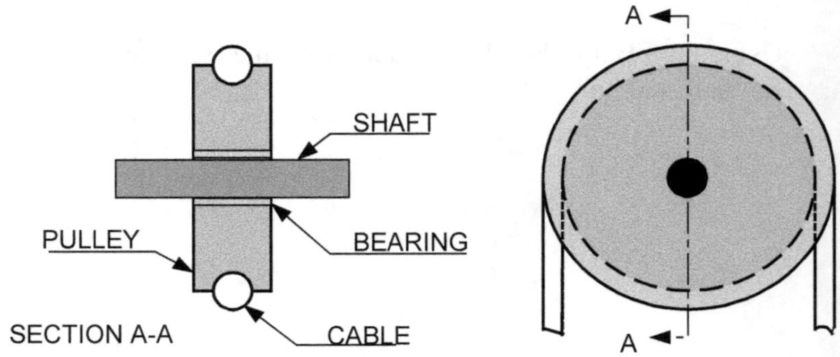

Fig. 3.13 Details of a cable and pulley assembly.

The pulley/cable system has two primary advantages. First, it can change the direction of a force. For instance, as we pull down on the cable at the left side of the pulley in Fig. 3.13, the cable on the right side will move up. Using this property, we may transport masses to significant heights. Second, pulleys and a cable properly arranged act as a machine and amplify the force capability of a person or a mechanical device. Let's consider two different examples that illustrate the advantages of a cable/pulley system either to change the direction or to amplify a force.

EXAMPLE 3.19

The cable pulley system presented in Fig. E3.19 is being used to lift a mass of 15 slugs from the floor. Determine the force required to lift this mass and the reaction force acting on the pin in the pulley.

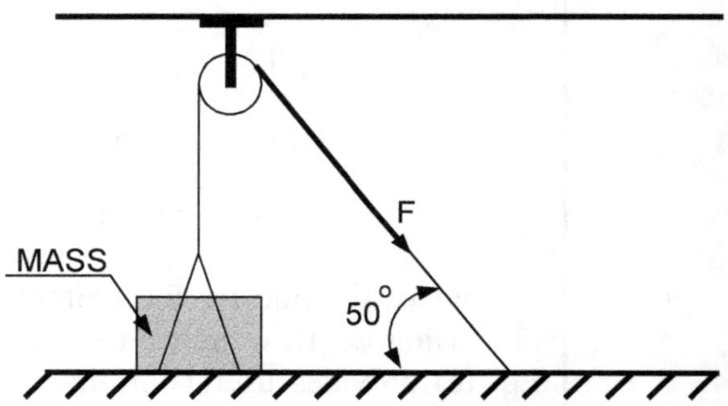

Fig. E3.19

Solution:

Step 1. Prepare two FBDs of the pulley/cable system (one of the pulley and the other of the mass) as shown in Fig. E3.19a. After lifting the mass from the floor, we cut the cable from the mass and replace it with the force F_1. A force W, acting downward, equal to the weight of the mass is applied. The weight of the mass is determined from Eq. (1.8) as:

$$W = mg = (15)(32.17) = 482.6 \text{ lb}$$

Then we remove the support attachment from the pulley and replace it with force components F_{2x} and F_{2y}. Since we have cut the cable to the mass, it is necessary to apply a force F_1 to replace the cable completing the FBD of the pulley.

Step 2. Classify the force systems. For the mass, the force system is collinear, coplanar and concurrent. For the pulley, the force system is coplanar and non-concurrent.

Step 3. Write the relevant equations of equilibrium. For the mass, only one of the six equations of equilibrium is relevant.

$$\Sigma F_y = 0$$

For the pulley, only three of the six equations of equilibrium are relevant.

$$\Sigma F_x = 0 \qquad \Sigma F_y = 0 \qquad \Sigma M_O = 0$$

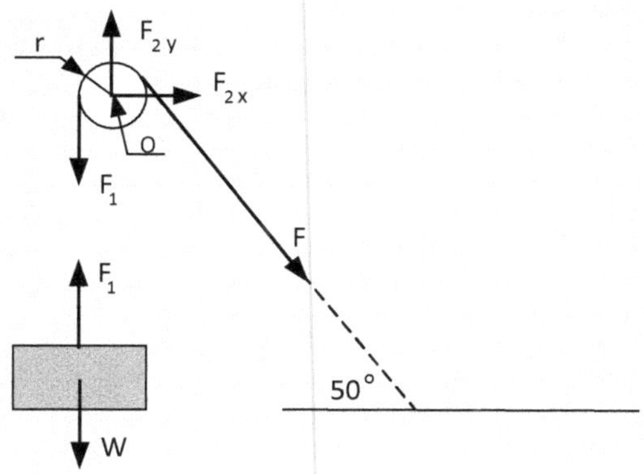

Step 4 and 5. Substitute the unknowns from the FBD into the relevant equations of equilibrium for both the mass and the pulley and then execute the solutions. For the mass, we write:

$$\Sigma F_y = F_1 - W = 0$$

$$F_1 = W = 482.6 \text{ lb} \qquad \text{(a)}$$

The pulley, with a radius r, is assumed to be frictionless. We write moments about the center O of the pulley to obtain:

$$\Sigma M_O = (F_1)(r) - (F)(r) = 0$$

$$F = F_1 = 482.6 \text{ lb} \qquad \text{(b)}$$

It is clear from Eq. (b) that **the tension in the cable remains constant across the pulley** even though the direction of the forces F and F_1 are different.

Next, consider the sum of the forces in the x and y directions for the pulley.

$$\Sigma F_x = F_{2x} + F \cos 50° = 0$$
$$F_{2x} = -(482.6)\cos 50° = -310.2 \text{ lb} \qquad \text{(c)}$$

The minus sign indicates that the direction for F_{2x} in Fig. E3.19a is not correct. The force component is actually in the negative x direction.

Next, consider the sum of the forces in the y direction.

$$\Sigma F_y = F_{2y} - F \sin 50° - F_1 = 0$$

$$F_{2y} = 482.6\,(1 + 0.7660) = 852.3 \text{ lb} \qquad \text{(d)}$$

$$F_2 = [F_{2x}^2 + F_{2y}^2]^{1/2} = [(-310.2)^2 + (852.3)^2]^{1/2} = 907.0 \text{ lb}$$

Step 6. Check the solution and interpret the results. Again, the mathematics involves solving simple linear algebraic equations. We check the signs in each of the equilibrium equations and repeat the calculation. We check to ascertain that the correct units for each of the unknown forces have been assigned. We compare the forces F, F_1 and F_2 and find that their magnitudes are reasonable.

EXAMPLE 3.20

For the pulley and cable arrangement shown in Fig. E3.20, determine the force F required to maintain a weight W of 10 kN in equilibrium.

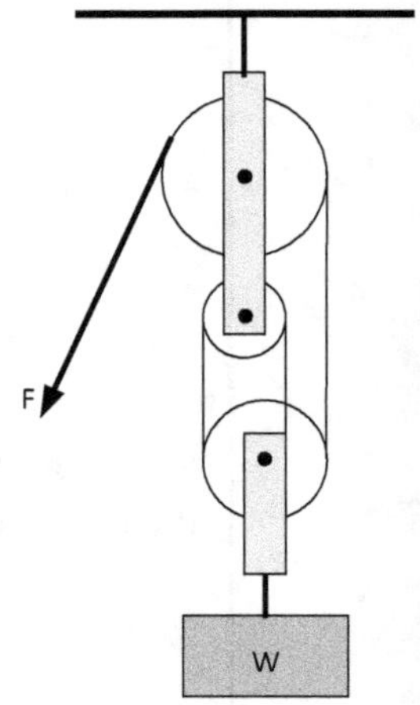

Fig. E3.20

Solution:

Step 1. Prepare a FBD of the pulley/cable system as shown in Fig. E3.20a. We cut the three cables above the lower pulley and apply a force F to each cable. Recall that the tension in the cable is a constant across a pulley when the friction at the shaft is zero. A force W = 10 kN, acting downward replaces the weight.

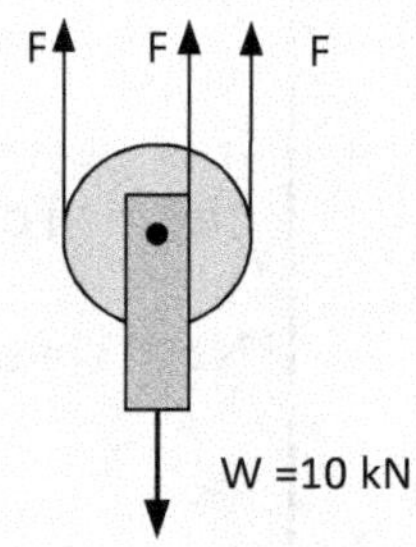

Step 2. Classify the force systems. For the lower pulley, the force system is coplanar and non-concurrent.

Step 3. Write the relevant equations of equilibrium. For the pulley, only one of the six equations of equilibrium is relevant.

$$\Sigma F_y = 0$$

Step 4 and 5. Substitute the unknowns from the FBD into the relevant equations of equilibrium for the pulley and then execute the solutions.

$$\Sigma F_y = 3\ F - W = 0$$

$$F = W/3 = 3.333 \text{ kN} \quad \text{(a)}$$

Step 6. Check the solution and interpret the results. Again, the mathematics involved is trivial. What is important is the realization that this pulley/cable arrangement permits one to lift a weight of 10 kN with a force of only 3.33 kN. The pulley/cable arrangement of Fig. E3.20 is a machine enabling one to amplify applied forces. While the forces are amplified, the work to lift the weight remains constant. The forces are decreased by a factor of three but the distance through which the force is moved to lift the weight is increased by that same factor. No work is gained by a pulley/cable arrangement.

EXAMPLE 3.21

Consider the pulley arrangement that supports blocks A, B and C as shown in Fig. E3.21. Blocks A and B each have a mass of 80 kg. The angle $\beta = 30°$. Determine the angle θ and the mass of block C if the system is in equilibrium.

Fig. E3.21

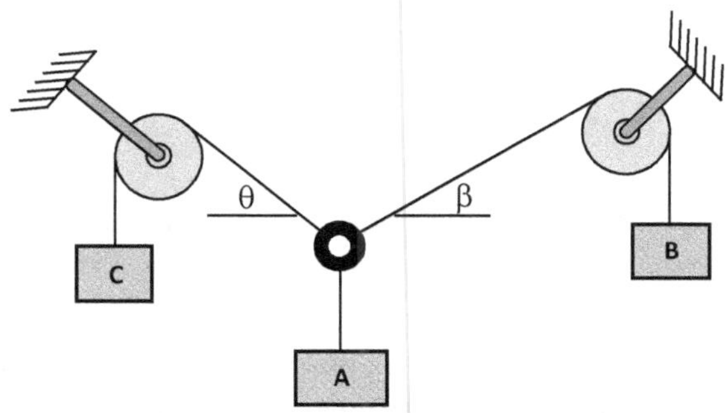

Solution:

Prepare a FBD of the ring as shown in Fig. E3.21a.

Fig. E3.21a

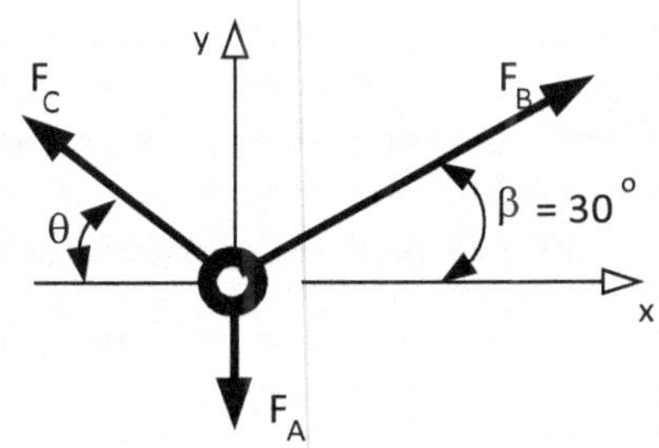

An examination of the FBD indicates that the force system is coplanar and concurrent. Accordingly, we may write:

$$\Sigma F_x = 0 \Rightarrow \quad F_B \cos(30°) - F_C \cos\theta = 0 \quad \text{(a)}$$

$$\Sigma F_y = 0 \Rightarrow \quad F_B \sin(30°) + F_C \sin\theta - F_A = 0 \quad \text{(b)}$$

Since we know the mass $m_A = m_B = 80$ kg for blocks A and B, we may write:

$$F_A = F_B = m_A\, g \quad \text{(c)}$$

Substitute Eq. (c) into Eq. (b) and solve for F_C:

$$F_C = [1/(\sin\theta)][F_A - F_B/2] = m_A g/2\sin\theta \quad \text{(d)}$$

Substitute Eq. (d) into Eq. (a) and solve for the angle θ:

$$m_A g \cos(30°) - [m_A g/(2\sin\theta)]\cos\theta = 0 \Rightarrow \quad \tan\theta = 1/[(2)(0.8660)] = 0.5774$$

$$\theta = 30° \quad \text{(e)}$$

Finally, to determine the mass of block C, substitute Eq. (e) into Eq(d):

$$F_C = m_A\, g/(2\sin\theta) = m_C\, g \quad \Rightarrow \quad m_C = 80/[(2)(0.500)] = 80 \text{ kg} \quad \text{(f)}$$

3.7 FORCES IN SPRINGS

Helical springs are often used to develop forces and to store energy in mechanical systems. These springs are usually wound from wire in the form of a helix as shown in Fig. 3.14. The spring shown in Fig. 3.14 develops a tension force because the hooks at both ends enable the spring to be stretched.

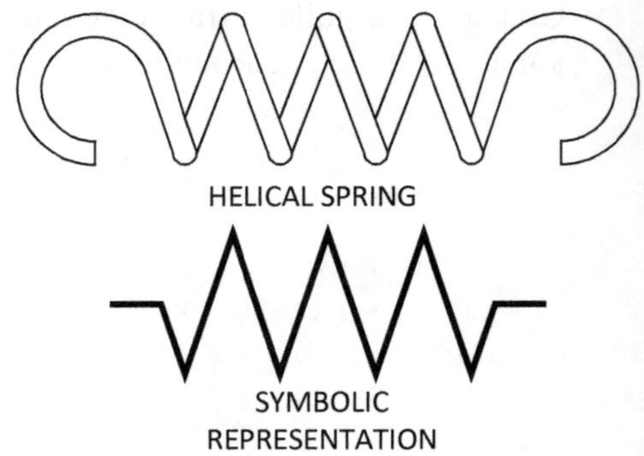

Fig. 3.14 A tension type helical spring.

If the spring is manufactured with flat ends, it is employed to support compression forces that squeeze the coils together. The stiffness or rigidity of the spring is controlled by the size of the wire used in winding the spring, the number of coils and the inside and outside diameters of the helix. The stiffness of the spring is known as the **spring rate** k. The force F required to elongate a tension spring or to squeeze a compression spring is given by:

$$F = k\delta \quad (3.3)$$

where δ is the amount of extension or compaction of the helical spring.

The extension (or compression) δ of the spring is established by:

$$\delta = L_d - L_o \quad (3.4)$$

where L_d and L_o are deformed and original length of the spring , respectively.

Because helical springs are difficult to draw, we represent them with a zigzag line drawing as illustrated in Fig. 3.14.

EXAMPLE 3.22

Suppose a tension spring with an original length of 9 inches is extended to a deformed length of 10.6 inches. Determine the tension force required to stretch the spring to its deformed length if the spring rate is 28 lb/in.

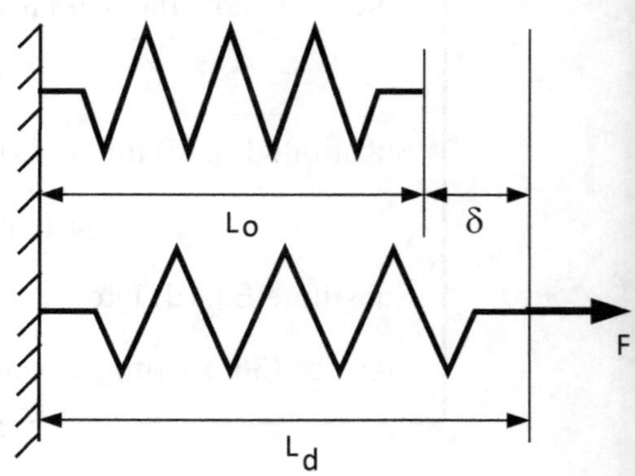

Solution:

From Eq. (3.4), we may write:

$$\delta = L_d - L_o = 10.6 - 9 = 1.6 \text{ in.} \qquad \text{(a)}$$

Substitute the results from Eq. (a) into Eq. (3.3) to obtain:

$$F = k\delta = (28)(1.6) = 44.8 \text{ lb} \qquad \text{(b)}$$

EXAMPLE 3.23

Suppose a compression spring with an original length of 350 mm is compacted to a deformed length of 273 mm. Determine the compression force required to compress the spring to its deformed length if the spring rate is 85 N/mm.

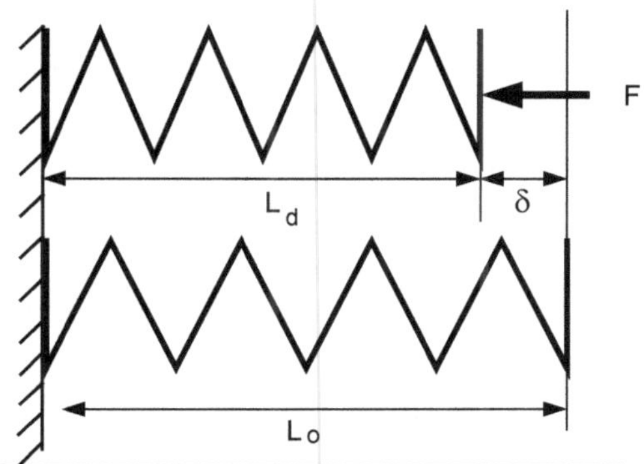

Solution:

From Eq. (3.4), we may write:

$$\delta = L_d - L_o = 273 - 350 = -77 \text{ mm} \qquad \text{(a)}$$

Substitute the results from Eq. (a) into Eq. (3.3) to obtain:

$$F = k\delta = (85)(-77) = -6.545 \text{ kN} \qquad \text{(b)}$$

Note the negative sign indicates that the force acting on the spring is compressive.

3.8 SOLVING FOR INTERNAL FORCES

We now have learned to use FBDs and the equations of equilibrium to solve for cable tension and reactions at the supports of beams and other structures. Let's now consider using the equilibrium relations to solve for the internal forces that develop within a structural member due to external loading.

EXAMPLE 3.24

Consider the stepped tension bar loaded with external forces illustrated in Fig. E3.24.

(a) Determine the internal force P that occurs in the thick portion of the stepped bar.

(b) Determine the internal force P that occurs in the thin portion of the stepped bar.

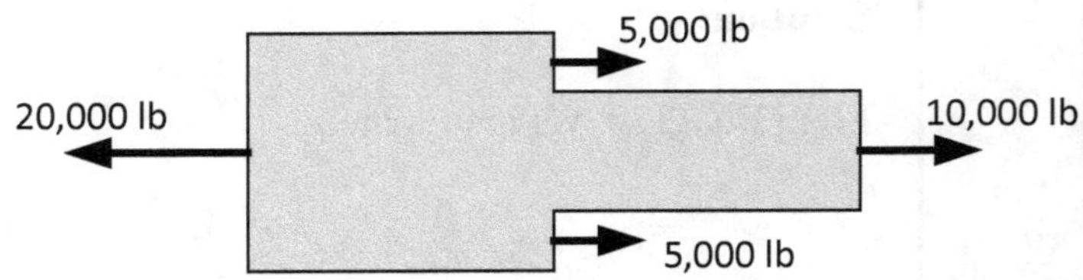

Fig. E3.24

Solution:

First, let's execute the six-step procedure to solve for the unknown magnitude of the internal force P in the thicker portion of the stepped tension bar.

Step 1. Prepare an appropriate FBD of the bar as shown in Fig. E3.24a. We begin this FBD by making a section cut A-A, which frees the left end of the bar. We will use the left end of the bar with the equilibrium relations to determine the internal force P. The FBD of the left end of the bar is also presented in Fig. E3.24a.

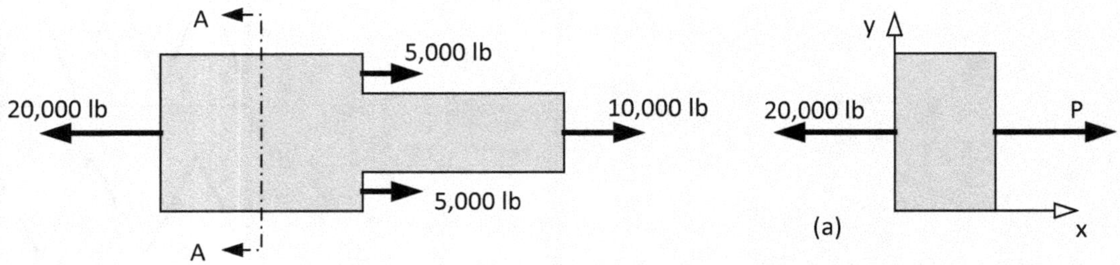

Fig. 3.24a

Step 2. Classify the force system. In this problem, the force system is coplanar and parallel.

Step 3. Write the relevant equations of equilibrium. In this case, only one of the six equations of equilibrium is relevant.

$$\Sigma F_x = 0$$

It is evident from inspection that the relations $\Sigma F_y = 0$ and $\Sigma M = 0$ are satisfied regardless of the value determined for the unknown force P.

Step 3. Substitute the unknowns from the FBD into the relevant equations of equilibrium.

$$\Sigma F_x = 0$$

$$\Sigma F_x = P - 20{,}000 \text{ lb} = 0$$

Step 5. Execute the solution, which is obvious in this example.

$$P = 20{,}000 \text{ lb}$$

Step 6. Check the solution. In this problem the equations were so simple that they were solved by inspection. We have included the units in the result. Since the bar is in tension, we understand that the internal force must be positive as indicated in our solution.

Making a section cut across the thinner end of the bar and repeating the six-step process described above solves part (b) of this example. The result obtained is P = 10,000 lb over the length of the thinner portion of the stepped tension bar. A force diagram showing the value of P over the length of the stepped tension bar is shown in Fig. E3.24b.

We have now explored the **solution space** for this problem, and have plotted all of the possible values of P occurring in the tension bar in the force-position diagram shown above. By preparing a graph of the solution for each position along the length of the bar, it is possible to determine the locations where the critical loads will occur. It will be these locations that are of concern when designing structures to resist failure.

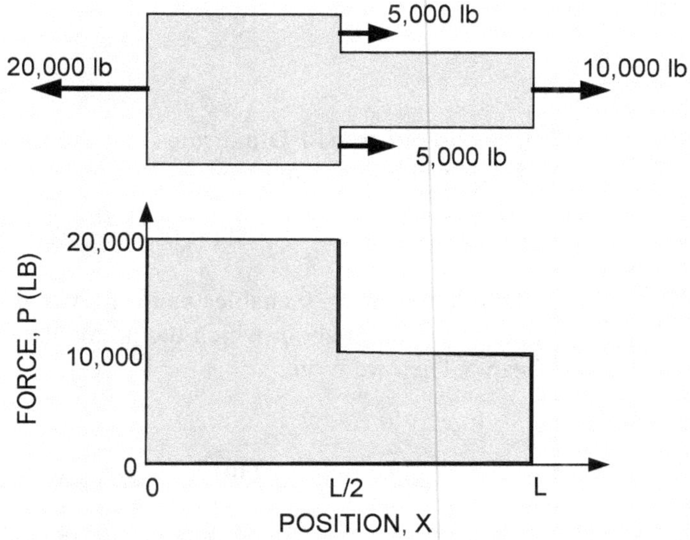

Fig. E3.24b

This example has shown that it is very easy to determine the internal forces in a tension bar. Let's consider a more complex internal force problem—a simply supported beam with a transverse concentrated force that is shown in Example 3.25.

EXAMPLE 3.25

Determine the internal force and internal moment developed in a simply supported beam with a transverse concentrated load at the locations defined by x = 2 m, x = 3.5 m and x = 5.0 m, where x is defined from the left end of the simply supported beam shown in Fig. E3.25.

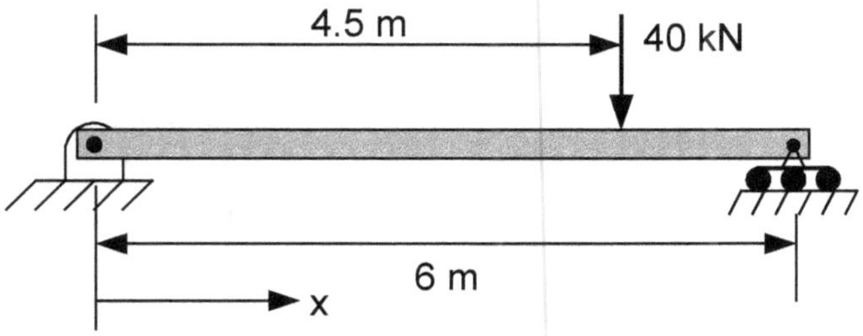

Fig. E3.25

Solution:

Let us begin the solution by solving for the reactions R_x, R_L and R_R due to the simple supports. We construct the FBD for the entire beam following the procedure described in previous examples.

- Remove the supports.
- Add the reaction forces R_x, R_L and R_R.
- Define a coordinate system.

The FBD for the beam is shown in Fig. E3.25a.

Fig. E3.25a

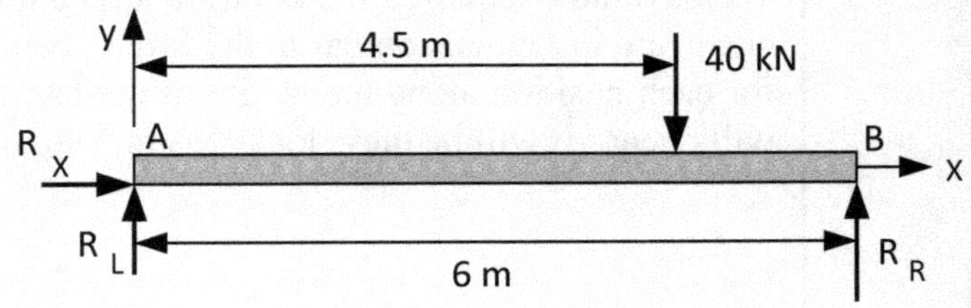

Inspection of this FBD indicates that the force system is coplanar and non-concurrent. Hence the equations of equilibrium that apply are:

$$\Sigma F_x = 0 \qquad \Sigma F_y = 0 \qquad \Sigma M = 0$$

The relation $\Sigma F_x = 0$ enables us to show that $R_x = 0$. The relation $\Sigma M = 0$ is very useful since we are free to choose the point about which the moments are to be determined. Let's select point B at the right end of the beam. Then we write:

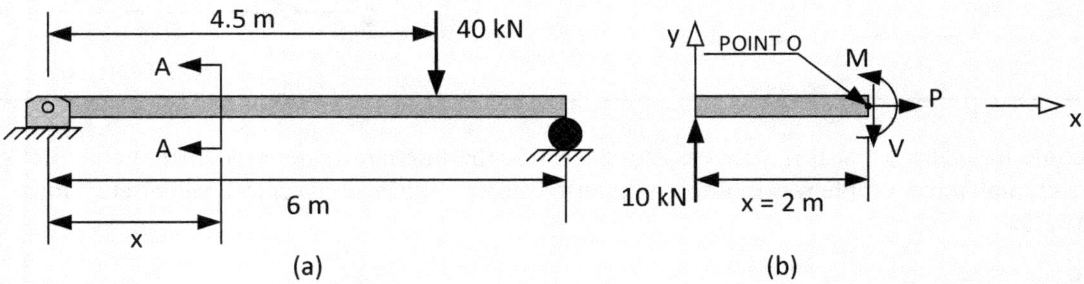

$$\Sigma M_B = (40 \text{ kN})(1.5 \text{ m}) - (6 \text{ m})R_L = 0$$

$$R_L = (40 \text{ kN}/6 \text{ m})(1.5 \text{ m}) = 10 \text{ kN}$$

To find the reaction R_R, we use the remaining equation of equilibrium.

$$\Sigma F_y = R_L + R_R - 40 \text{ kN} = 0$$

$$R_R = 40 \text{ kN} - 10 \text{ kN} = 30 \text{ kN}$$

Now that the external reaction forces acting at the beam supports have been determined, we proceed to solve for the internal forces at x = 2 m by making a section cut A-A through the beam at this location (see Fig. E3.25b). We then remove the left portion of the beam, freed by the section cut, as a free body. The FBD, which is the model for the equilibrium analysis, is shown in inset (b) of this figure.

The FBD shows an internal force P (tensile) in the positive x direction, and another internal shear force V, which we will consider positive when it acts in the negative y direction. An internal moment M_z about the point O is shown indicating rotation in the counterclockwise direction. We have drawn the arrows representing P, V and M assuming they are all positive quantities.

An examination indicates the force system is coplanar and non-concurrent. The appropriate equations of equilibrium are:

$$\Sigma F_x = 0 \qquad \Sigma F_y = 0 \qquad \Sigma M_o = 0$$

From the relation $\Sigma F_x = 0$, we find:

$$\Sigma F_x = P = 0$$

It is clear that the internal axial force P in a transversely loaded beam vanishes.

From the relation $\Sigma F_y = 0$, we find:

$$\Sigma F_y = 10\text{ kN} - V = 0$$

$$V = +\ 10\text{ kN}$$

The positive sign for V indicates our assumption was correct—this vertical force is in the negative y direction. The internal force V is a shear force, which we will discuss in much more detail in Chapter 10.

The relation $\Sigma M_O = 0$ permits us to solve for the internal moment $(M_z)_A$.

$$\Sigma M_O = (M_z)_A - (10\text{ kN})(2\text{ m}) = 0$$

$$(M_z)_A = 20\text{ kN-m}$$

In reality, this internal moment is not produced by a rotating force vector as symbolized in the FBD. Instead, it is produced by a linear distribution of bending stresses that act normal to the surface exposed in section A-A. The rotating force vector is used to simplify the representation of the moment in the FBD while capturing the rotational effect it has on the body.

Let's continue our analysis of the internal forces in the beam by considering the location x = 4.5 m. Note that we freed the left portion of the beam with the section cut B-B. The cut was at x = 4.5 m, but incrementally to the left side of the 40 kN concentrated force as shown in insert (a) of Fig. E3.25c. Accordingly, the 40 kN force is not included in the FBD shown in inset (b) of Fig. E3.25c.

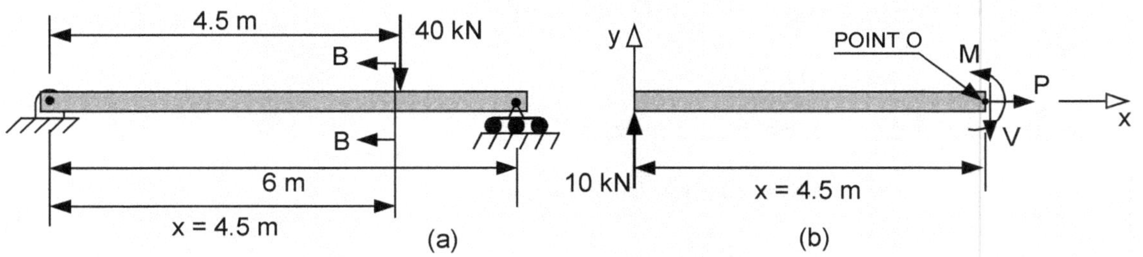

Fig. E3.25c

An examination indicates the force system is coplanar and non-concurrent. The appropriate equations of equilibrium are:

$$\Sigma F_x = 0 \qquad \Sigma F_y = 0 \qquad \Sigma M_o = 0$$

From the relation $\Sigma F_x = 0$ we again find:

$$\Sigma F_x = P = 0$$

From the relation $\Sigma F_y = 0$ we find:

$$\Sigma F_y = 10\text{ kN} - V = 0$$

$$V = +\ 10\text{ kN}$$

This result is identical to that determined for V at x = 2 m. Indeed, the shear force V is a constant for the region of the beam $0 < x < 4.5$ m.

The relation $\Sigma M_O = 0$ permits us to solve for the internal moment $(M_z)_B$.

$$\Sigma M_O = (M_z)_B - (10\ kN)(4.5\ m) = 0$$

$$(M_z)_B = 45\ kN\text{-}m$$

If we examine the results for M_z at x = 2 and 4.5 m, it is clear that the moment increases linearly with x. Indeed, we can write an equation showing the variation of the moment with location x as:

$$M_z = 10\ x \qquad \text{for } 0 < x < 4.5$$

The moment M_z is in units of kN-m when x is measured in meters. We must constrain this relation for M_z to the range of x covered by the FBDs that have been considered in the analysis. At this stage in the analysis, it is clear that the moment increases from zero at the left end of the beam to a maximum at x = 4.5 m. To determine the internal forces and moments in the beam for values of x > 4.5 m, it is necessary to consider a third FBD of the beam with a section cut C-C located at x = 5 m as presented in Fig. 3.25d.

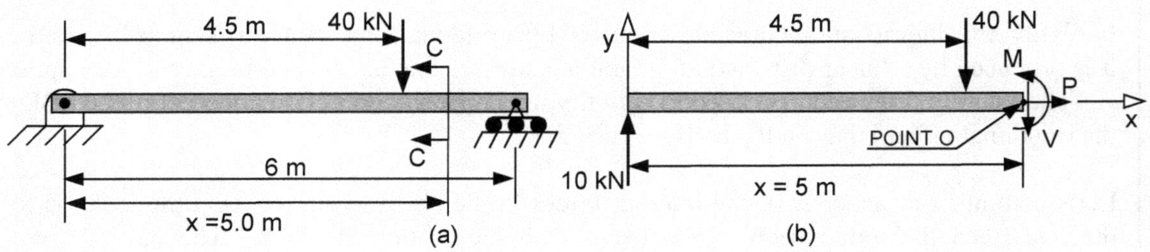

Fig. E32.5d

Following the procedure established previously, we again find:

$$\Sigma F_x = P = 0$$

From the relation $\Sigma F_y = 0$, we determine V as:

$$\Sigma F_y = 10\ kN - V - 40\ kN = 0$$

$$V = -30\ kN$$

Observe that the sign for the shear force is negative. This fact indicates that our assumption for V oriented in the negative y direction was not correct. For x > 4.5 m, the shear force is directed upwards in the positive y direction. The shear force V undergoes a step change when x locates a point incrementally to the right of the concentrated load of 40 kN.

The relation $\Sigma M_O = 0$ permits us to solve for the internal moment $(M_z)_C$.

$$\Sigma M_O = (M_z)_C - (10\ kN)(5\ m) + (40\ kN)(0.5\ m) = 0$$

$$(M_z)_C = 30\ kN\text{-}m$$

It is interesting to observe that the value of M_z at x = 5 m is less than its value at x = 4.5 m. If we let x = 6 m, it is easy to show that $M_z = 0$. Indeed, at both supports M = 0 because the beam is free to rotate at its ends. The simple supports cannot provide constraint against rotation and the moments must vanish at their locations.

In Example 3.25, we examined the internal forces and moments in the transversely loaded beam at three different positions along its length. We also observed that the internal moments vanish at the ends of the beam. Let's use this information to draw the diagrams presented in Fig. 3.15 showing the shear force V and the internal moment M as a function of position along the length of the beam.

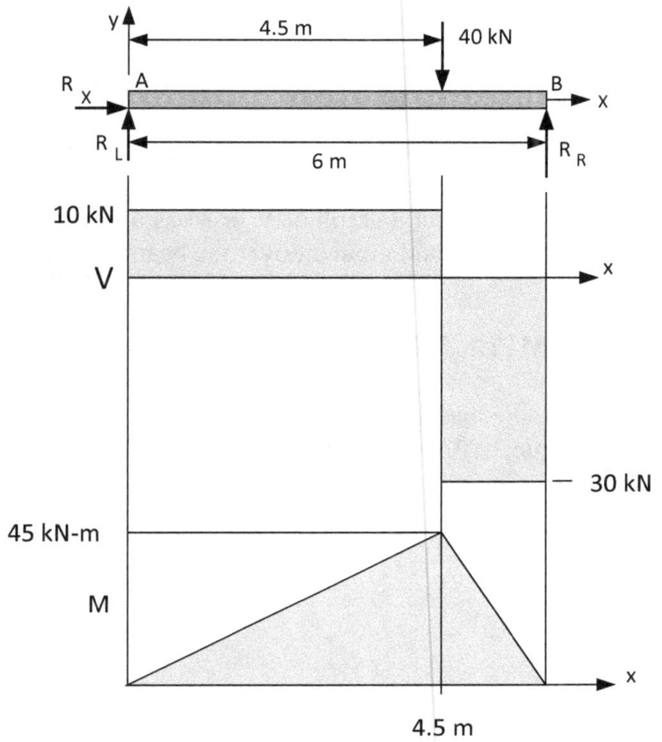

Fig 3.15 Shear and bending moment diagrams for a simply supported beam with a single transverse concentrated force at x = 4.5 m.

We will describe shear and bending moment diagrams in much more detail in Chapter 10 when considering bending and shear stresses produced by transverse loads applied to beams.

EXAMPLE 3.26

Determine the internal forces and moment in the uniformly loaded beam presented in Fig. E3.26. Solving for the reactions at the pivot support at the left end of the beam gives $R_x = 0$ and $R_y = qL/2$.

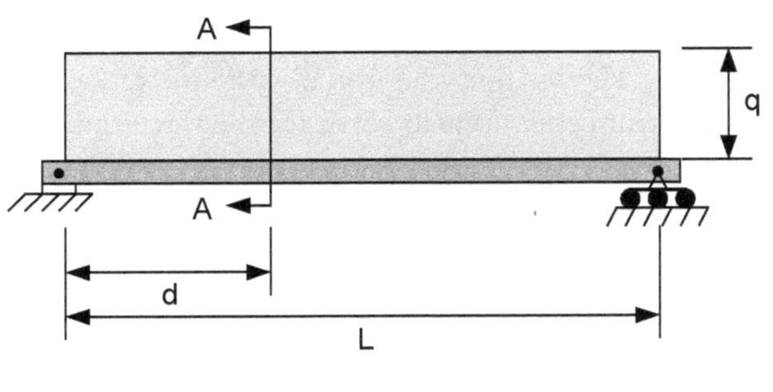

Fig. E3.26

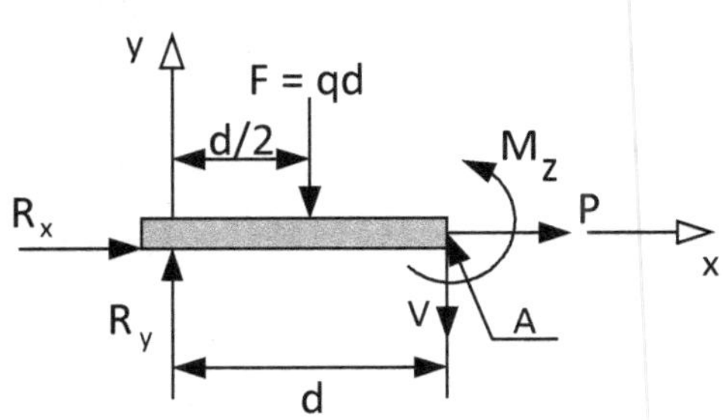

Fig. E3.26a

Solution:

Examination of Fig. E3.26a, shows that the force system is coplanar and non-concurrent. Hence, we write three equilibrium equations to solve for the three unknowns P, V and M_z.

$$\Sigma F_x = R_x + P = 0 \qquad \Rightarrow\Rightarrow \qquad P = -R_x = 0 \qquad \text{(a)}$$

$$\Sigma F_y = R_y - qd - V = 0 \quad \Rightarrow\Rightarrow \quad V = qL/2 - qd = q(L/2 - d) \quad \text{(b)}$$

$$\Sigma M_A = M_z - R_y\, d + (Fd)/2 = 0 \quad \Rightarrow\Rightarrow \quad M_z = (qL/2)d - (qd)(d/2)$$

$$M_z = (qL/2)d - (qd)(d/2) = (qd/2)(L - d) \quad \text{(c)}$$

The results show that the internal force V and the internal moment M_z depend on the dimension d that locates the section cut. V is a maximum when d = 0, and M_z is a maximum when d = L/2. You may wish to construct a graph of V or M as a function of d as it ranges from 0 to L, to show the variation of the internal forces and moment over the beam's length.

EXAMPLE 3.27

Determine the internal forces and moment at section A-A in the frame of the factory building described previously in Example 3.10. Note that L = 120 ft, R_y = 60 tons, and F_1 = 25 tons.

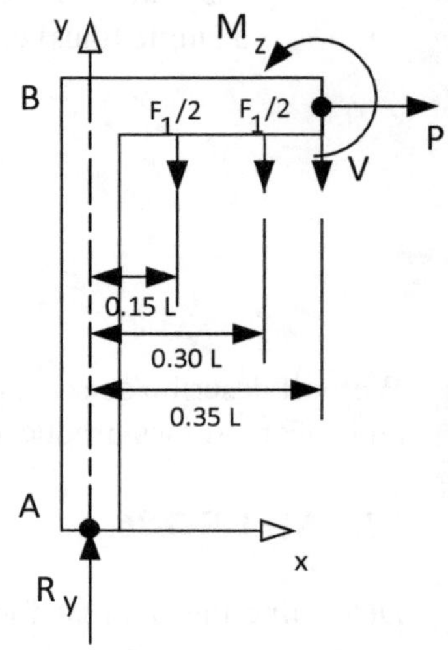

Fig. E3.27

Solution:

Examination of the FBD presented in Fig. E3.27, shows that the force system is coplanar and non-concurrent. Accordingly, we may write three equilibrium equations to solve for the three unknowns P, V and M_z.

$$\Sigma F_x = P = 0 \quad \text{(a)}$$

$$\Sigma F_y = R_y - F_1/2 - F_1/2 - V = 0$$

$$V = (60)(2000) - (25)(2000) = 70{,}000 \text{ lb} \quad \text{(b)}$$

$$\Sigma M_O = M_z - R_y\,(0.35L) + (F_1/2)(0.05L + 0.20L) = 0$$

$$M_z = (2000)(120)[(60)(0.35) - (25/2)(0.25)] = 4.290 \times 10^6 \text{ ft-lb} \quad \text{(c)}$$

The results show that the internal force V and the internal moment M_z depend on the position of the section cut.

EXAMPLE 3.28

The C-clamp shown in Fig. E3.28 is tightened onto a block. If the screw applies a force F = 600 lb to the block, determine the internal forces at the section cut A-A along the straight portion of its back.

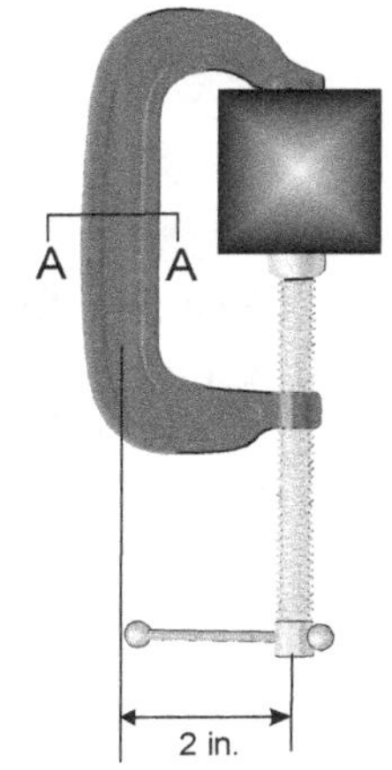

Fig. E3.28

Solution:

Next classify the force system as coplanar, non-concurrent. Accordingly, we have three relevant equations of equilibrium. Writing these relations yields:

$$\Sigma F_x = V = 0 \qquad \text{(a)}$$

$$\Sigma F_y = F - P = 0 \Rightarrow\Rightarrow \qquad P = F = 600 \text{ lb} \qquad \text{(b)}$$

$$\Sigma M_O = 2F + M_z = 0 \qquad \Rightarrow\Rightarrow \quad M_z = -2\,F = -1{,}200 \text{ in.-lb} \qquad \text{(c)}$$

The negative sign for M_z indicates that the direction of the rotating vector shown in Fig. E3.28a is not correct. This rotating vector should be in the clockwise direction on the FBD to maintain equilibrium of the top portion of the C-clamp.

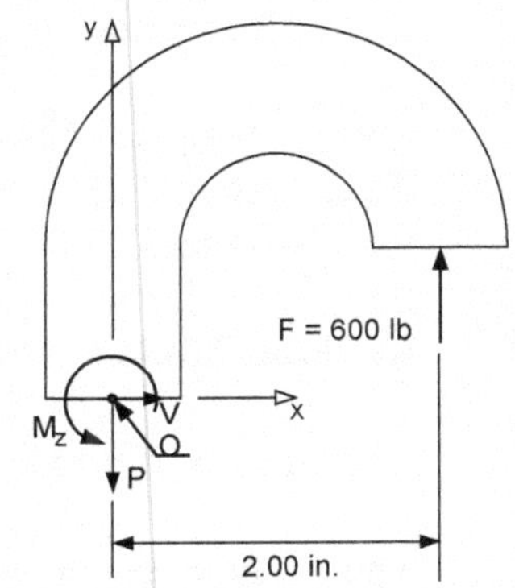

Fig. E3.28a

3.9 SUMMARY

The two equilibrium equations in vector form, $\Sigma\mathbf{F} = 0$ and $\Sigma\mathbf{M} = 0$, may be represented by the six equations involving Cartesian components of the forces and moments as:

$$\Sigma F_x = 0, \qquad \Sigma F_y = 0, \qquad \text{and } \Sigma F_z = 0 \qquad (3.1)$$

$$\Sigma M_x = 0, \qquad \Sigma M_y = 0, \qquad \text{and } \Sigma M_z = 0 \qquad (3.2)$$

It is useful in selecting the appropriate equilibrium relations to classify the force system acting on a structure into one of the following categories.

- Coplanar and concurrent.
- Coplanar and non-concurrent.
- Non-coplanar and concurrent.
- Non-coplanar and non-concurrent.

When the force system is classified it is apparent which of the six Cartesian equilibrium relations provide meaningful results.

It is difficult to overemphasize the importance of preparing complete and correct FBDs. The FBD represents a model of the structure under consideration. It provides a guide in writing the equilibrium equations, information on pertinent dimensions, and a record of the assumptions regarding the direction of the forces and moments. A systematic method for constructing FBDs includes the following four-step procedure:

1. Isolating the body.
2. Replacing supports with reaction forces and/or moments.
3. Replacing distributed loads with concentrated loads positioned at the centroid of the area representing the distributed load.
4. Dimensioning the diagram and defining a coordinate system.

Methods for solving equilibrium problems in statics are covered in detail. We suggest a systematic six step technique that includes:

1. Prepare a complete FBD of the component under consideration.
2. Classify the force system.
3. Write the relevant equations of equilibrium.
4. Substitute the known and unknown quantities shown in the FBD into the relevant equations of equilibrium.
5. Execute the solution.
6. Check the solution and interpret the results.

Many examples are provided to demonstrate techniques for solving for reaction forces in structures with coplanar force systems. The following relevant equilibrium relations were employed depending on the classification of the force systems.

Coplanar and concurrent: $\Sigma F_x = 0 \quad \Rightarrow \quad \Sigma F_y = 0$

Coplanar and non-concurrent: $\Sigma F_x = 0 \quad \Rightarrow \quad \Sigma F_y = 0 \quad \Rightarrow \quad \Sigma M_z = 0$

The methods employed in the analysis of non-coplanar force systems will be described in Chapter 7.

An analysis of cable/pulley arrangements was made to illustrate the ability of pulleys to change the direction of an applied force and with certain arrangements to amplify an applied force. Also the helical spring was introduced as a device that develops a force in proportion to its deformation δ. The relation governing this force is given by:

$$F = k\delta \qquad (3.3)$$

A method of sectioning structural members provides a means to isolate portions of the structure. FBDs of these partial members yield models that are analyzed to solve for the internal forces and moments. Examples are provided to show solution techniques for determining internal forces in tension members and internal shear forces and bending moments in beams, as well as in other components and structures.

An extended example was covered to introduce the concept of shear force and bending moment diagrams that will be used extensively in Chapter 10.

PROBLEMS

3.1 An auto weighing 17,500 N is traveling on a straight highway down a hill with a grade of 3% at a constant velocity of 85 km/h. Construct a FBD of the auto showing all of the forces acting on it. A highway with a grade of 3% elevates 3 m for every 100 m in the horizontal direction.

3.2 A sports utility vehicle weighing 21,500 N is traveling on a straight highway up a hill with a grade of 2% at a constant velocity of 95 km/h. Construct a FBD of the vehicle showing all of the forces acting on it. A highway with a grade of 2% elevates 2 m for every 100 m in the horizontal direction.

3.3 An auto weighing 3,400 lb is traveling on a straight highway up a hill with a 2% grade at a constant velocity of 65 MPH. Construct a FBD of the auto showing all of the forces acting on it. A highway with a grade of 2% elevates 2 ft for every 100 ft in the horizontal direction.

3.4 A sports utility vehicle weighing 4,500 lb is traveling on a straight highway down a hill with a 3.5% grade at a constant velocity of 70 MPH. Construct a FBD of the auto showing all of the forces acting on it.

3.5 Prepare a sketch illustrating the force system described below and write its relevant equilibrium relations.

(a) Coplanar and concurrent
(b) Non-coplanar and concurrent
(c) Coplanar and non-concurrent
(d) Non-coplanar and non-concurrent

3.6 Construct a FBD for the mass and two-cable assembly shown in the figure to the right. Also determine the tension forces in cables AC and BC.

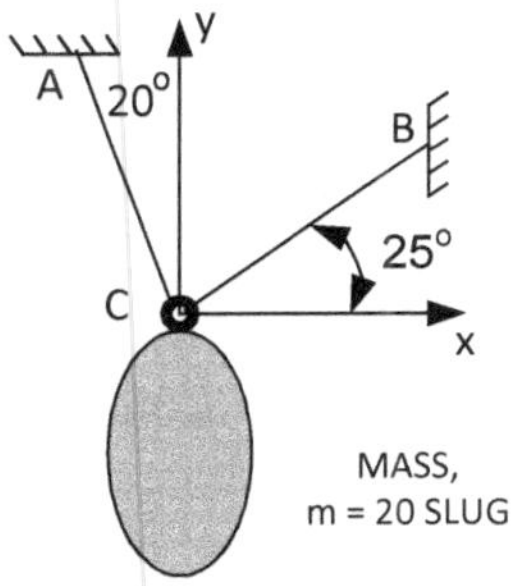

3.7 Construct a FBD for the mass and two-cable assembly shown in the figure to the left. Also determine the tension forces in cables AC and BC.

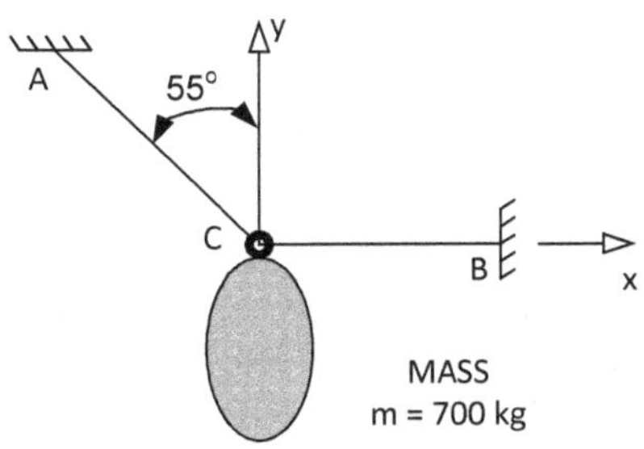

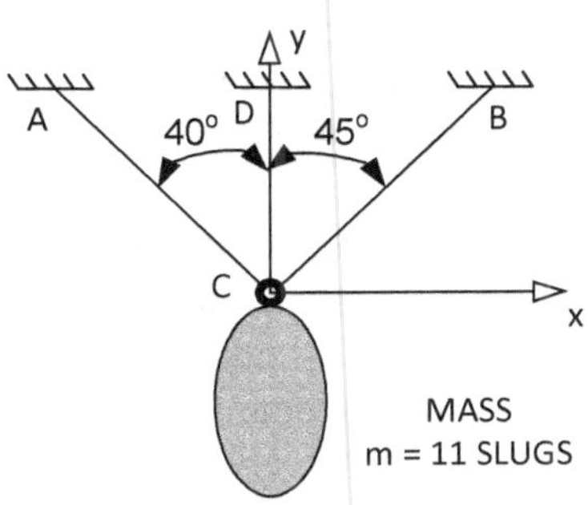

3.8 Construct a FBD for the mass and three-cable assembly shown in the figure to the right. Determine the tension forces in cables AC, BC and CD. Assume cable CD supports 30% of the mass.

3.9 Determine the forces in cables AC and BC that support the signal lights at a traffic intersection as illustrated in the figure to the left. The signal lights have a mass m = 40 kg. The dimensions for the cable arrangement are given in meters.

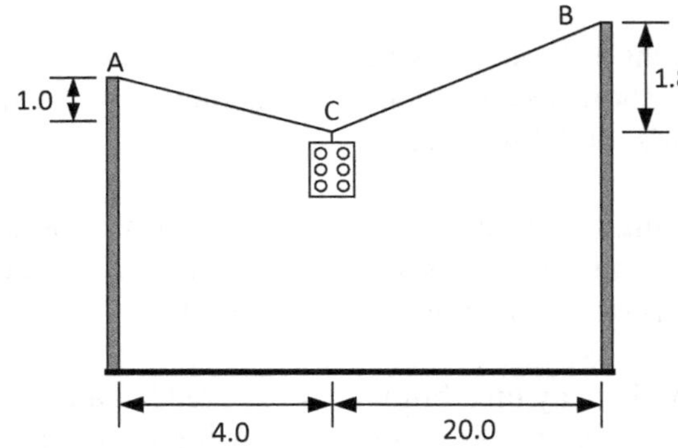

3.10 A container A that holds a 1.0 ton weight ton is lifted from the floor with a cable and pulley system shown in the figure to the right. Determine the equation for the force F required to lift the container as a function of α. Evaluate this equation to determine the force F if α = 20°. Comment on the effectiveness of this cable pulley system.

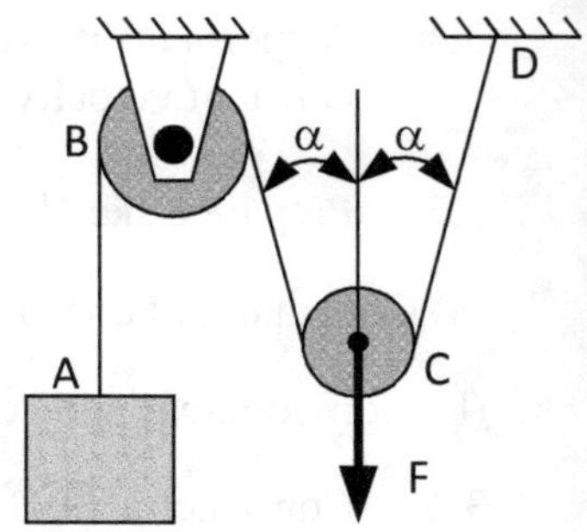

3.11 A container holding a mass m = 850 kg is lifted from the floor with a cable and pulley system shown in the figure to the right. Determine the force F required to lift the container as a function of α. Evaluate this equation to determine the force F if α = 17°. Comment on the effectiveness of this cable pulley system.

3.12 Using the pulley and cable arrangement shown in the figure to the right, design a lifting arrangement that requires only a force of 1,000 lb to lift a container weighing 1,800 lb. Assume that the pulleys are frictionless.

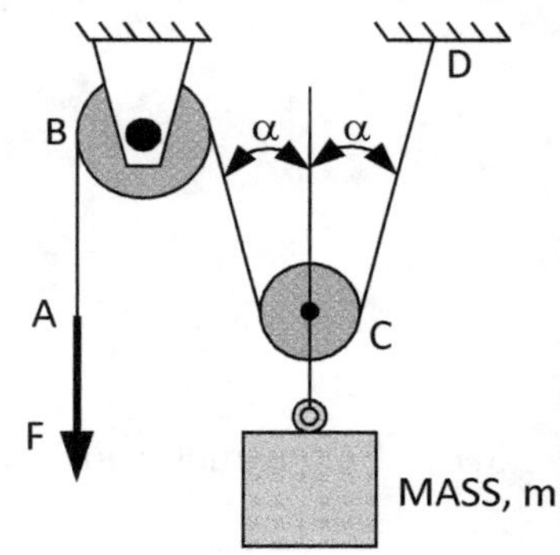

3.13 Using pulleys and cables, design a lifting arrangement that requires only a force of 500 lb to lift a one-ton container. Assume that the pulleys are frictionless.

3.14 For the pulleys and cable arrangement shown in Problem 3.11, determine the force required to lift a 4.5 kN weight as a function of the angle α the cables make with pulley C. Let the angle α vary from zero to 90°. It is suggested that you use a spreadsheet for these calculations and prepare a graph showing the results.

3.15 A differential hoist is illustrated in the figure below and to the right. At the top, two pulleys with radii r_1 and r_2 are fastened together so they turn as a single unit. A continuous cable passes around the smaller pulley, with a radius r_2, and then around the lower pulley with a radius r_3, and finally around the larger pulley with a radius r_1. We assume that the pulleys are frictionless relative to the shafts passing through their hubs. We also assume that the cables do not slip in the grooves of the pulleys. If the diameter of the lower pulley is $D = 2r_3 = r_1 + r_2$, determine the equation for the applied force F in terms of the load W, which is maintained in equilibrium.

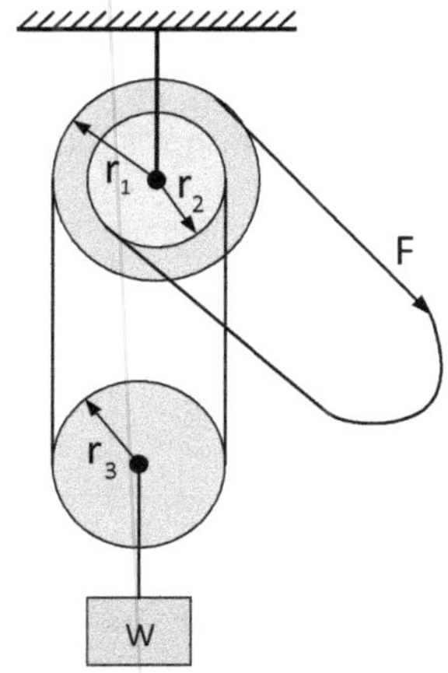

3.16 An artist has designed a bell using a long piece of pipe as the vibrating sound source. The pipe is to be supported using the wire ring arrangement shown in the figure to the left. Determine the force in each of the wires. The pipe has a mass of 43 kg.

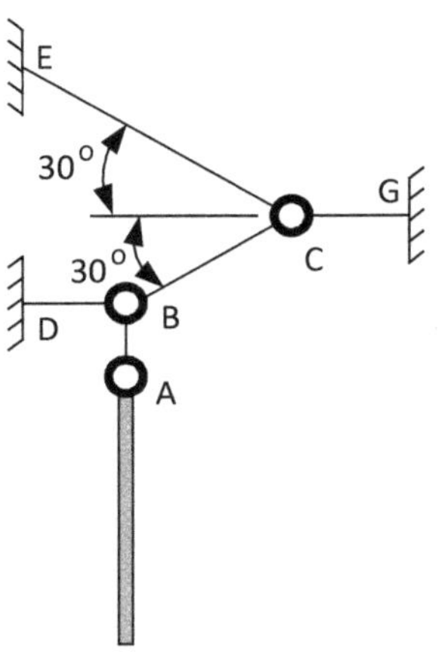

3.17 The pulley and cable, shown in the figure to the right, support a mass m = 12 slug suspended at point A. The cable B-A-C is 65 ft long and sags prior to the addition of the pulley and the container.

(a) If the pulley diameter is small relative to the span of 50 ft between the walls at B and C, determine the height h associated with the equilibrium position of the pulley and container.
(b) If the pulley and container are released from point C, describe the motion, which you would observe as the system reaches equilibrium.
(c) If the pulley and container are released from point B, describe the motion, which you would observe as the system reaches equilibrium.

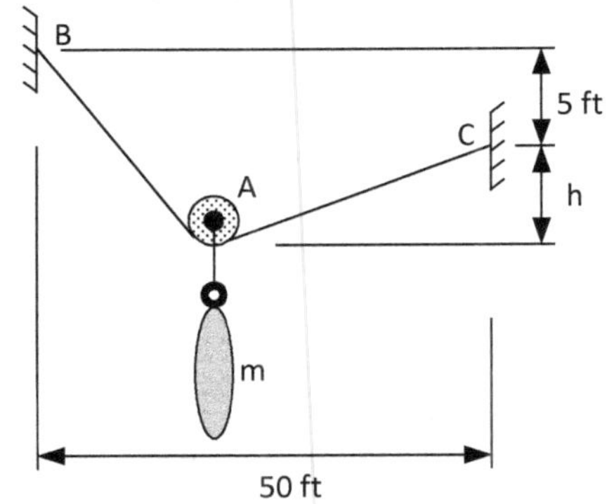

3.18 Two masses m_B and m_C are supported by the cable arrangement shown in the figure to the right. Determine the mass m_C and the forces in the cables AB, BC and CD if the mass m_B = 16.0 kg.

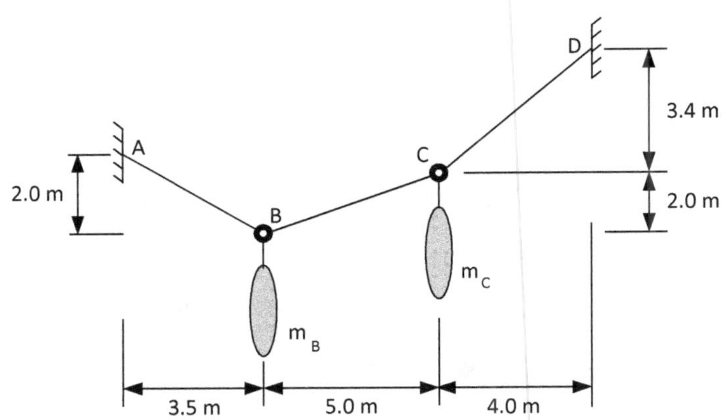

3.19 A bungee cord is stretched across a river as shown in the figure below. A stuntman weighing 140 lb is to cross the river by tightrope walking on the bungee cord. If the cord has a linear spring constant k = 60 lb/in., determine if the stuntman will remain dry.

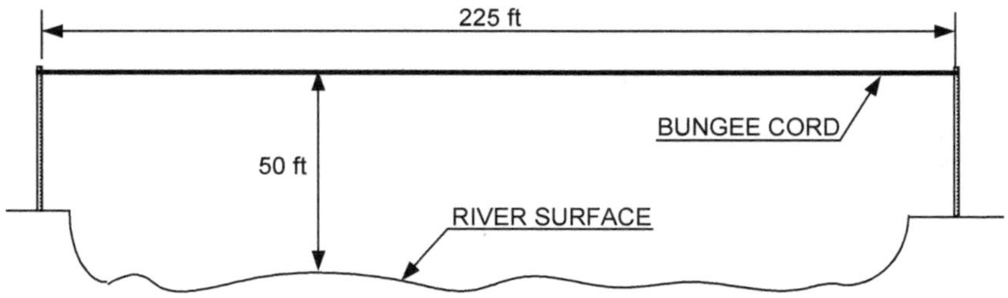

3.20 For the conditions described in Problem 3.19, determine the spring constant of the bungee cord if the stuntman is to barely wet the soles of his shoes.

3.21 Three springs are connected in series as shown in the figure to the right. Determine the extension of the entire arrangement if a force of 23.4 lb is applied to the last spring. Also determine the effective spring rate for the three springs in this series arrangement.

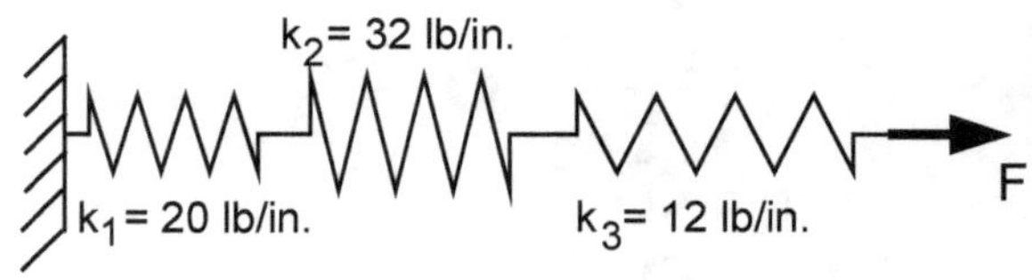

3.22 Three springs are connected in series as shown in the figure to the left. Determine the extension of the entire arrangement if a force of 70 N is applied to the last spring. Also determine the effective spring rate for the three springs in the series arrangement.

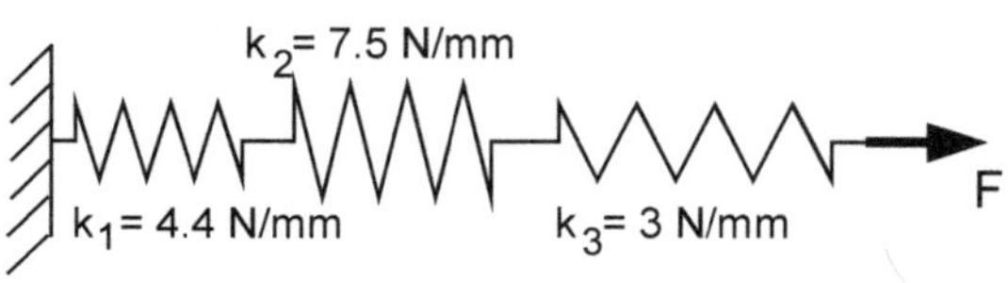

3.23 For the parallel spring arrangement determine the deflection of the lower bar shown in the figure to the right. The spring rates k_A = 12 N/mm. and k_B = 23 N/mm. The force F = 1145 N. Also determine the effective spring rate for the three springs in parallel.

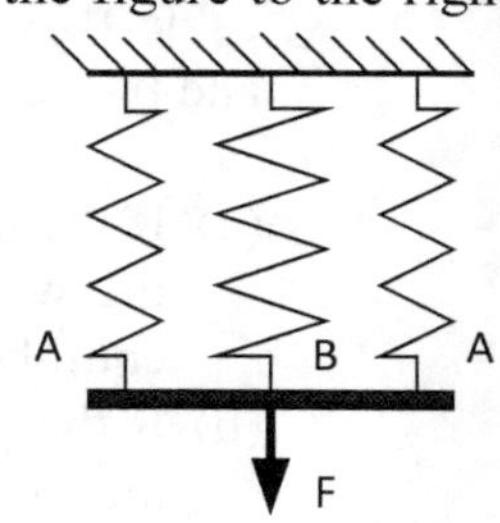

3.24 Three springs are used to support the rigid bar with a mass m as illustrated in the figure to the right. If k_1 = 20 lb/in., k_2 = 28 lb/in., k_3 = 36 lb/in. and m = 1.3 slug, determine the amount each spring elongates. The bar is constrained and remains parallel to the top surface to which the springs are attached.

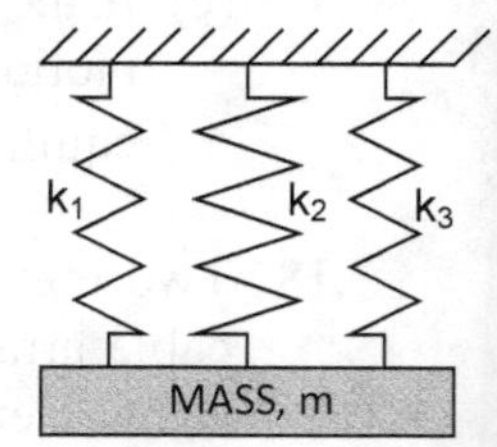

3.25 A cable and a spring are connected together at point C as shown in the figure to the left. Determine the equilibrium position of point C if a force F = 400 N is applied. The undeformed length L of both the cable and the spring before the application of the load is shown in the figure together with the spring rate k.

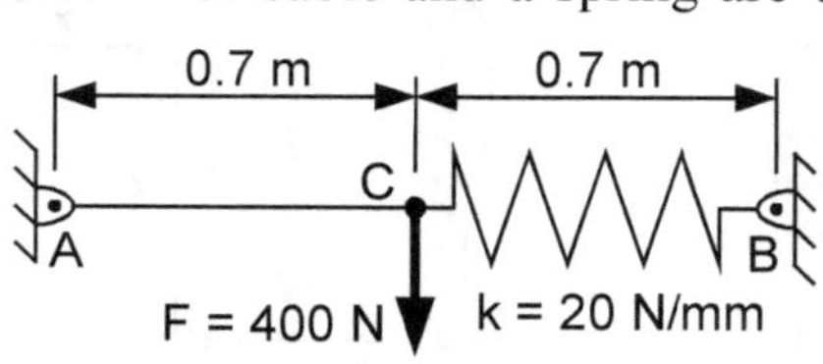

3.26 For the beam, shown in the figure to the right, solve for the reactions at the pin and clevis and the roller.

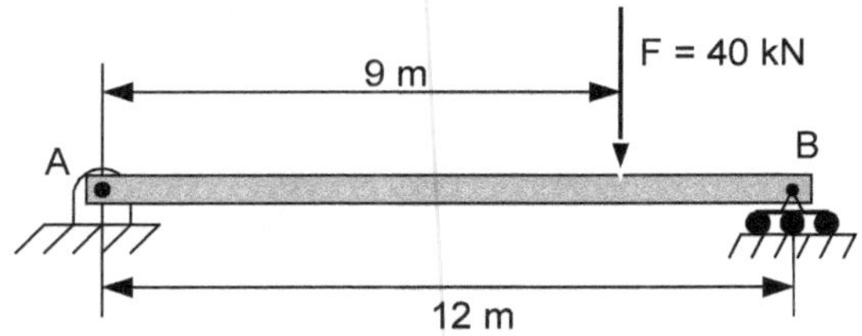

3.27 For the beam, shown in the figure to the left, solve for the reactions at the pin and clevis and the roller.

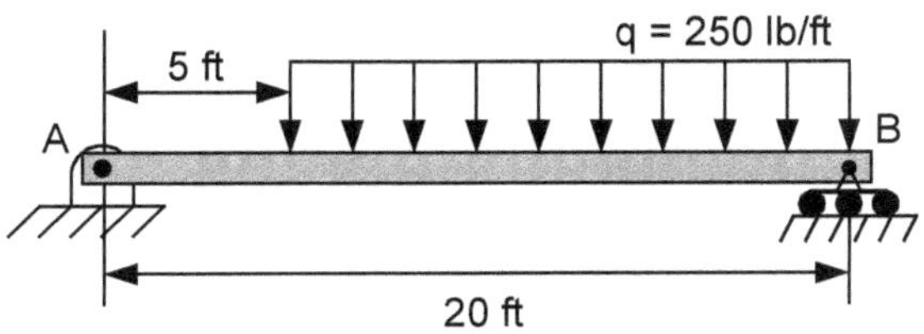

3.28 For the beam, shown in the figure to the right, solve for the reactions at the pin and clevis and the roller.

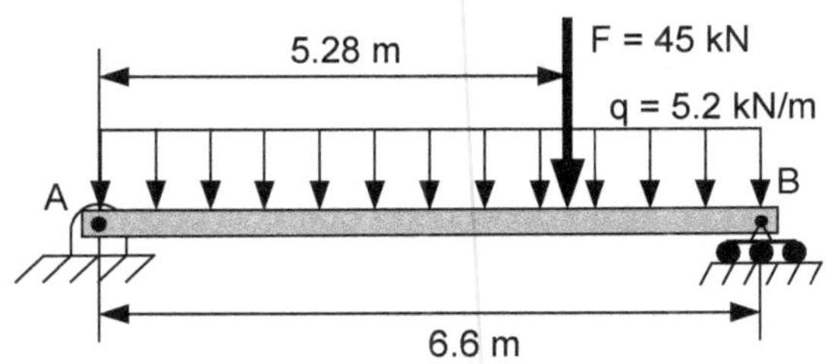

3.29 For the beam, shown in the figure to the right, solve for the reactions at the built-in end of the cantilever beam.

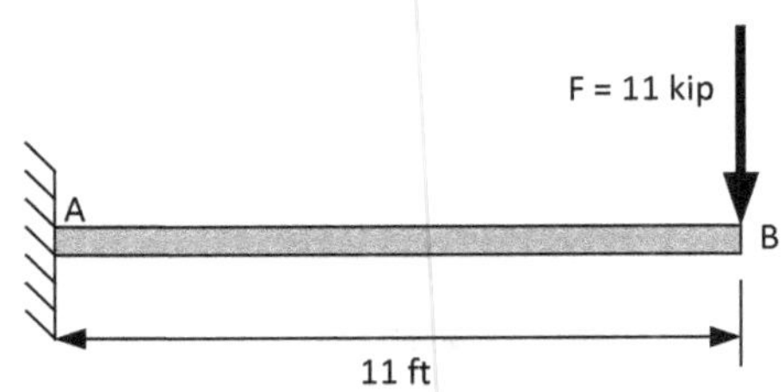

3.30 For the beam, shown in the figure to the left, solve for the reactions at the built-in end of the cantilever beam.

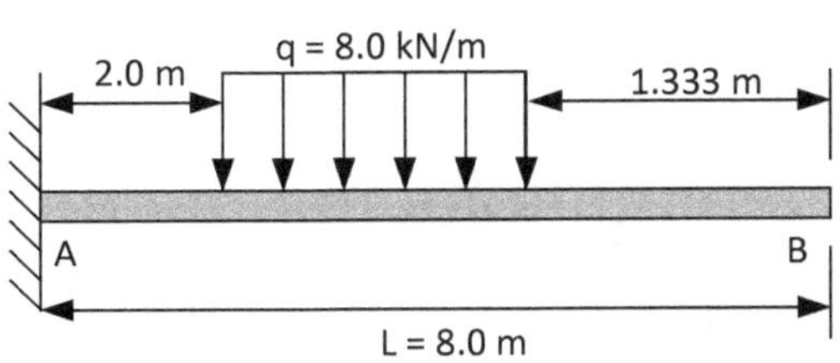

3.31 For the beam, shown in the figure to the right, solve for the reactions at the two supports.

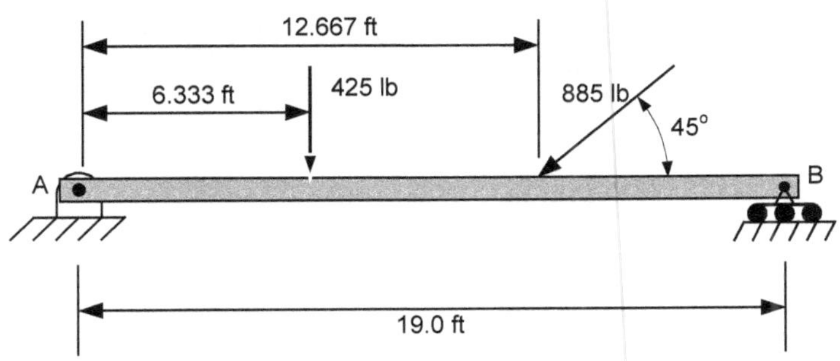

3.32 For the beam, shown in the figure below, determine the reactions at the two supports. The dimensions of the beam and the mass are given in the figure. Note that the arm CD is rigidly attached to the beam. Also, the radii of the two pulleys are the same and equal to 0.25 m.

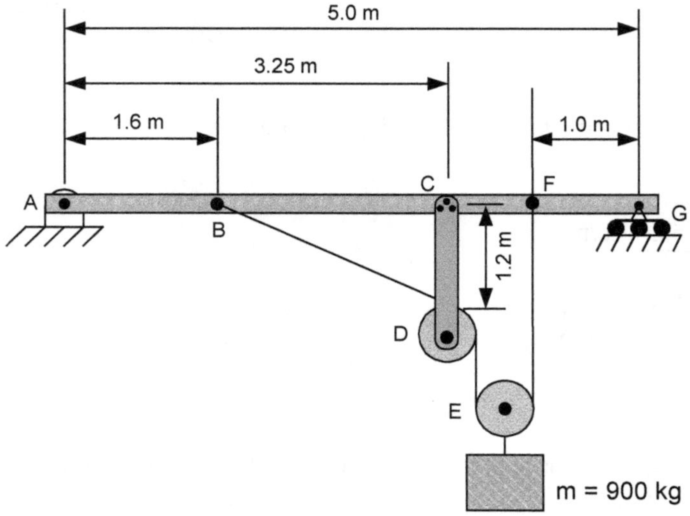

3.33 For the truss, shown in the figure to the right, determine the reactions at the two supports.

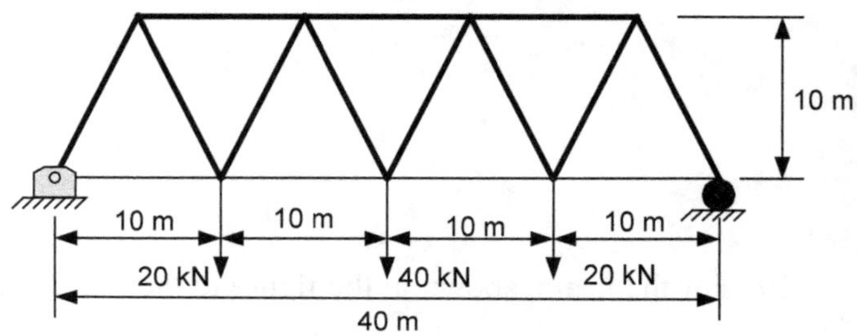

3.34 A long slender pole is to be erected by a winch and cable arrangement that is mounted on a truck, as illustrated in the figure to the left. If the pole weighs 2,000 lb and is 40 ft long, determine the reactions at point O and the force in the cable. Consider angles θ varying from 20° to 75°. A pivot is attached to the base of the pole so that it may be rotated without a resisting moment (i.e. acts like a pin).

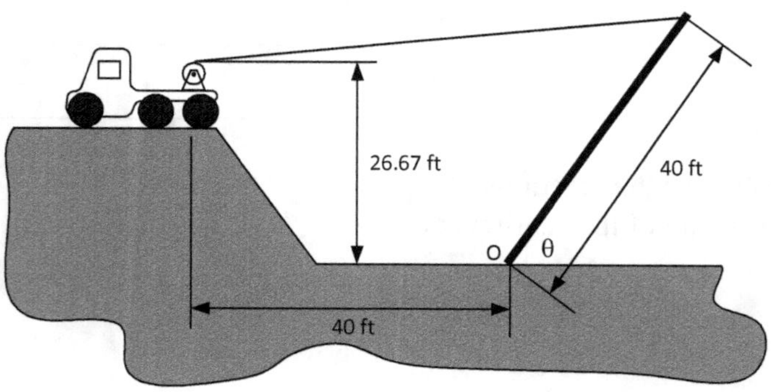

3.35 For the crane, described in the figure to the right, determine the maximum load W_L that can be lifted before the crane tips. Note, the crane weights 30 kN.

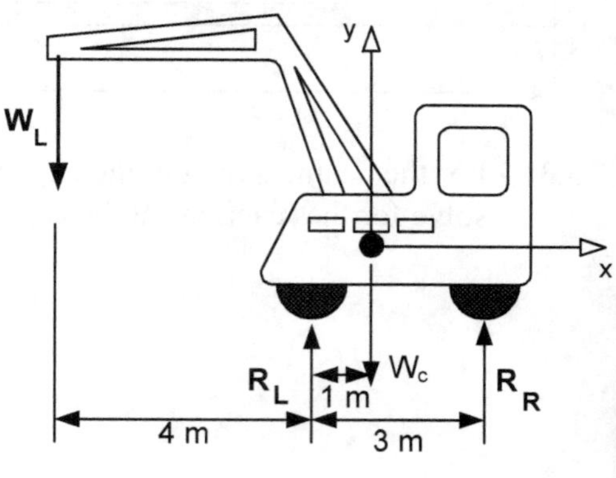

3.36 For the crane, illustrated in the figure to the left, determine the maximum load that can be lifted before the crane tips. Note, the crane weights 3.5 ton.

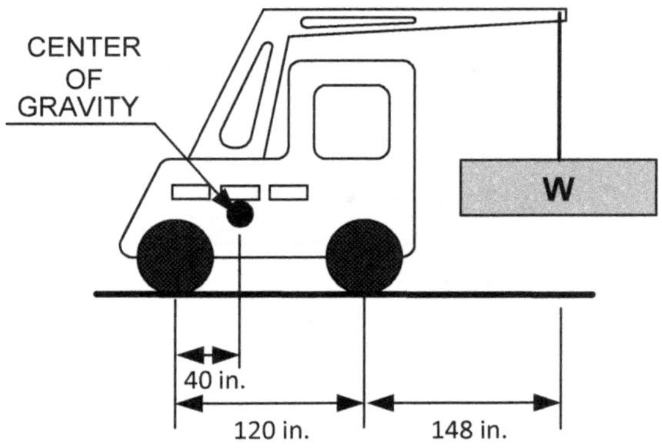

3.37 Determine the reaction forces and moments at the support for the stadium structure shown in the figure to the right.

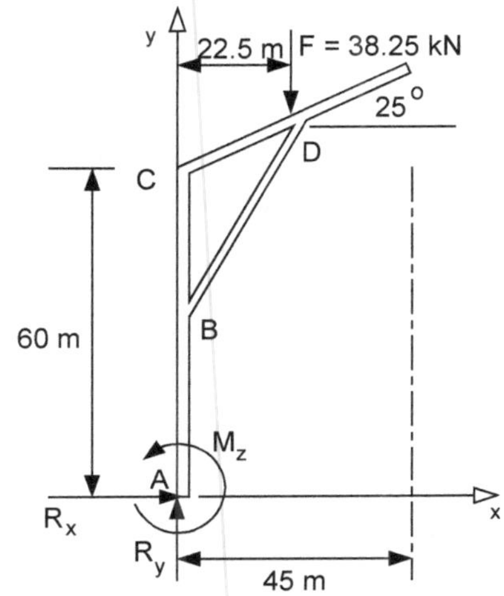

3.38 Determine the reaction forces at pins A and B that support the gusset plate shown in the figure to the left. Assume that the vertical components of the reaction forces at the pins are equal (i.e. $R_{Ay} = R_{By}$)

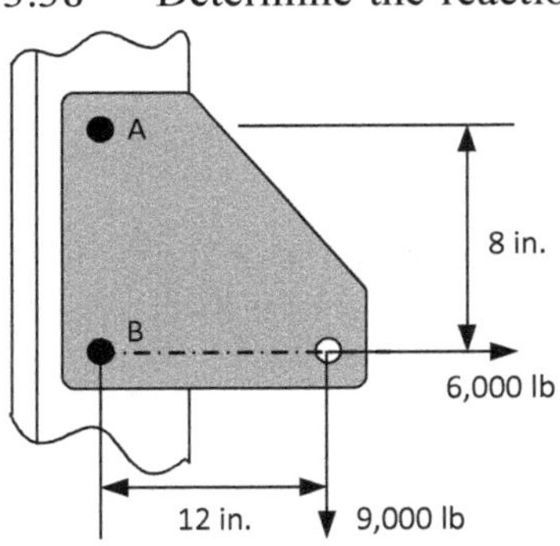

3.39 For the tension bar, shown in the figure to the right, determine the internal tension force P at sections A-A and B-B.

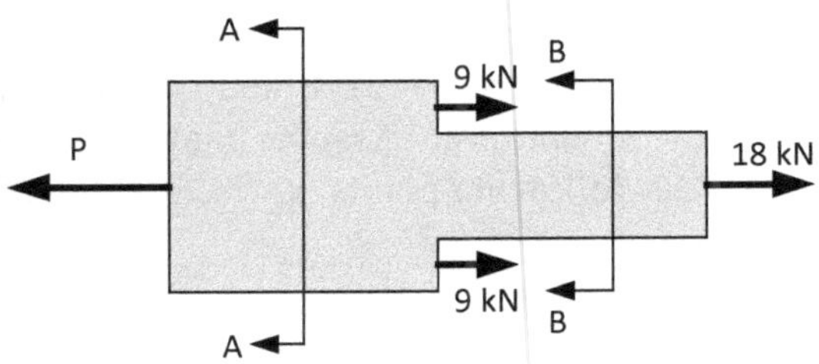

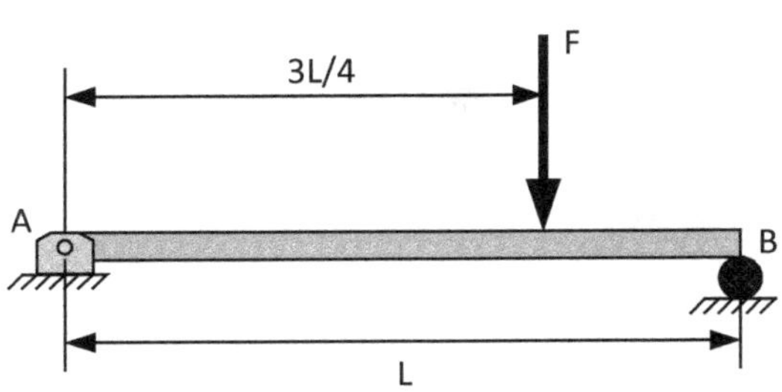

3.40 For the beam, shown in the figure to the left, determine the internal shear force V and the internal moment M at the position x = L/2.

3.41 For the beam, shown in the figure above and to the left, determine the internal shear force V and the internal moment M as a function of position for $0 < x < L$. Prepare a graph of M and V as a function of x.

3.42 For the beam, shown in the figure to the right, determine the internal shear force V and the internal moment M at position $x = L/3$. Present your results in terms of q and L, which are known quantities.

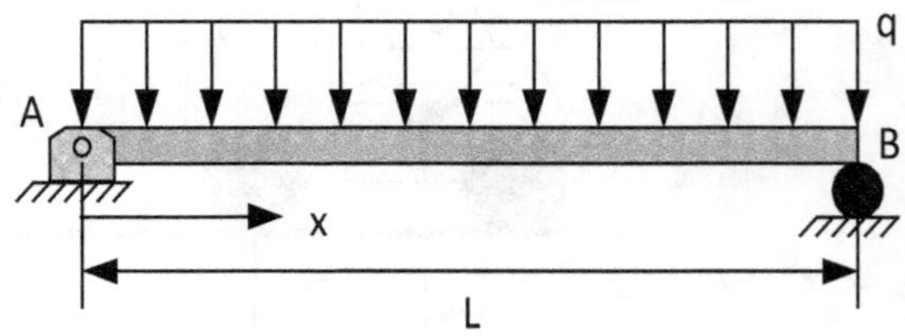

3.43 For the beam, shown, in the figure to the right, determine the internal shear force V and the internal moment M as a function of position x. Prepare a graph of M and V as a function of x.

3.44 For the beam, shown in the figure to the right, determine the internal shear force V and the internal moment M at position $x = 3L/4$. Present your results in terms of F and L, which are known quantities.

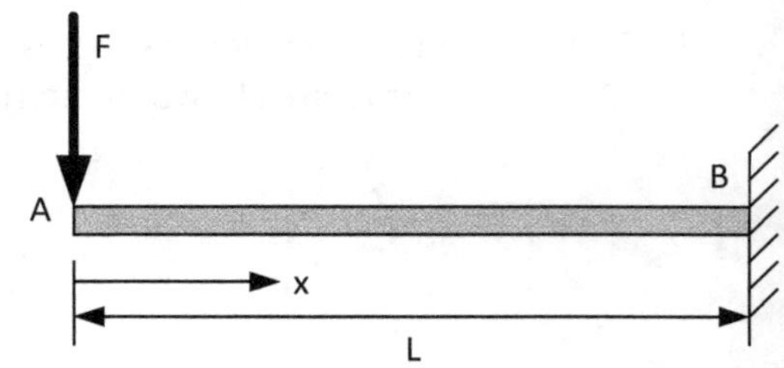

3.45 For the beam, shown in the figure to the right, determine the internal shear force V and the internal moment M as a function of position x. Prepare a graph of M and V as a function of x.

3.46 For the beam, shown in the figure to the lower right, determine the internal shear force V and the internal moment M at the position x = 12 ft.

3.47 For the beam, shown in the figure to the right, determine the internal shear force V and the internal moment M as a function of position x. Prepare a graph of M and V as a function of x.

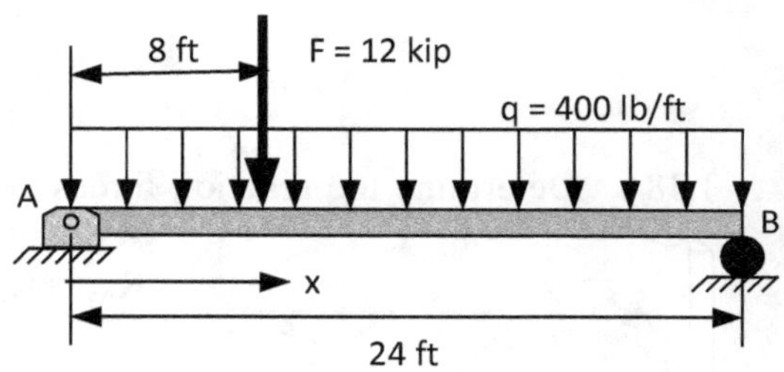

3.49 A new factory for heavy machinery has an assembly line with component parts stored above floor level. The components used in assembly are in elevated bins (#1, #2, and #3) as shown in the figure to the right. The elevated bins are suspended from a frame ABCD. Prepare a FBD of the left portion of the frame from Section A-A to the roller support at point A. Determine the internal forces and moment at section A-A as a function of the loads W_1, W_2 and W_3. Locate Section A-A at position $x = 3L/8$ and consider the loads W_1, W_2 and W_3 as known quantities. Assume that the centers of each bin are located at the following points: $x_1 = L/4$, $x_2 = L/2$ and $x_3 = 3L/4$.

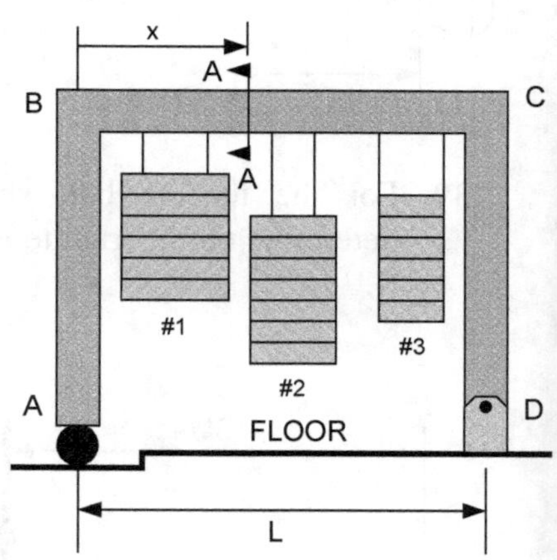

CHAPTER 4

AXIALLY LOADED STRUCTURAL MEMBERS

4.1 INTRODUCTION

Large structures like bridges, sports stadiums, skyscrapers, etc. are fabricated from many small structural elements. There are several reasons for constructing big structures from much smaller components. First, manufacturing facilities limit the size (width and thickness) of structural elements that are produced. The rolls used in steel and aluminum mills to produce bars, plates and sheets usually are only six to eight feet long. Even more limiting is the requirement for transport from the manufacturing facility to the construction site. If highways are used for transport, state law limits both the width and length of the structural members.

In this chapter we will consider two types of long thin structural members—flexible and stiff. Both types of these structural members will be subjected to axial loading. Wire rope or cables, which are flexible, are capable of supporting only tensile loading that tends to stretch the cable. However, rods or bars, which are stiff[1], are capable of supporting both tensile and compressive loading.

As structural elements, wire rope, cable, rods and bars subjected to axial loading have a significant advantage. They are subjected to a uniform state of stress over their cross sectional area. Also, if they are two-force members loaded at their two ends, the stresses are uniform over their length. Thus, the entire volume of the rod or bar is subjected to the same stress, which is the optimum condition for minimum weight design.

Examples of structural applications of long thin members in bridge construction are numerous. For instance, very high-strength wire rope is used in the construction of suspension bridges and in cable stayed bridges. Also, many bridges are designed with trusses deployed on the two sides of the structure. These trusses are fabricated from many bars or rods connected together at bolted or welded joints. We analyze these rods and bars using the same equations established for wire rope and cables. The difference is that the rods and bars are sufficiently stiff to support compressive stresses.

4.2 CHARACTERISTICS OF CABLES, RODS AND BARS

What characteristics do cables, wire ropes, rods and bars have in common and how do they differ? Cables, rods and bars exhibit lengths that are very long when compared to the dimensions of the cross section. A cable differs from a rod or a bar because it is flexible whereas the rod and bar are stiff. **Cables** have a circular cross section that is uniform along its length. A cable is made from many strands of wire that are twisted together to form a larger diameter very high-strength structural member. Even though a typical cable is larger in diameter than a wire, it is treated as a flexible member capable of supporting only tensile forces. We will use wire, wire rope and cable interchangeably in this text. The cross sectional area of **bars and rods**[2] may be of arbitrary shape.

Because cables are flexible, they can be wrapped about pulleys, formed into loops, and sometimes knotted. Indeed, they are so flexible that we cannot push on them because they buckle. It is possible to pull, but not push. This fact means that internal tensile forces develop within a thin flexible member but not compression forces. However, rods and bars are sufficiently stiff to support compressive forces and stresses. They are used as tie rods when loaded in tension, and as columns when loaded in compression. We will assume in this chapter that the cross section of the column is adequate and that failure by buckling is not an issue.

---

[1]Rods or bars actually stretch or compress under the action of axial loads, but this deformation is small and is neglected in determining the loads and stresses. However, this very small deformation is considered when determining the strain and deflection of the long thin structural members.

[2]In most applications, the cross section of a rod is circular and that of a bar rectangular; however, this distinction is not always maintained and one can expect to find cross sections of arbitrary shape for both rods and bars.

We are considering the simplest structural application—the use of long thin members under axial loading. The geometry of a structural member (long length with small cross sectional dimensions) leads us to constrain the direction of the loading and to make an important assumption about the deformation of the member that significantly simplifies the analysis.

1. The internal and external forces supported by the member act along its length.
2. Plane sections before loading remain plane after loading and subsequent deformation of the member.

Constraining the loads in the uniaxial direction is consistent with the title of the chapter. We are considering only the effects of axial loading on cables, rods and bars. Later in Chapter 10 on beams, we will consider bending induced by transverse loading.

The assumption that plane sections before loading remain plane after loading has also been confirmed by experiment. Suppose we draw a straight scribe line on a long thin wire or bar as shown in Fig. 4.1 a. This line represents the edge of a plane through the cross section of the member. Now apply an axial load F to the wire as indicated in Fig. 4.1b. The wire will stretch a small amount δ, but the line remains straight and the plane through the cross section remains plane (i.e. flat). This is a very important observation because it implies that the normal stresses σ are uniformly distributed across this plane.

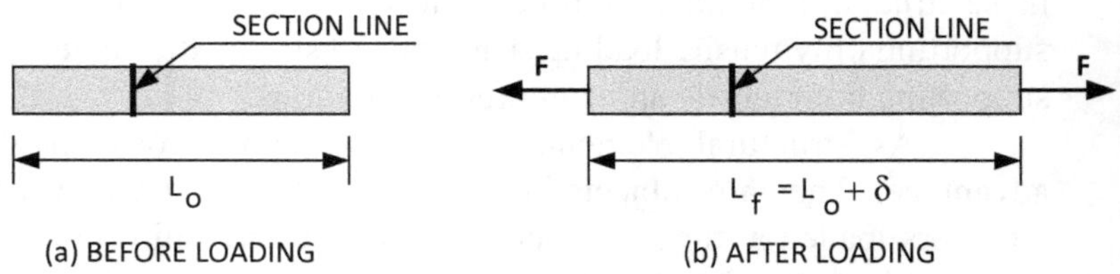

Fig. 4.1 A section line drawn across the member remains straight after loading.

4.3 DESIGN ANALYSIS METHODS

In the design analysis, we compute the axial forces acting on the uniaxial member, the amount it stretches under load, and the stresses developed. Next, we compare the stresses acting on the structural member with its strength to determine a safety factor. The size of the safety factor is evaluated to determine if this member is safe to be employed in a structure used by the public. This chapter describes methods that will enable you to size cables, rods and bars to provide structures with adequate safety margins.

We will divide this presentation into two parts—the first dealing with cables and the second involving rods and bars. The reason for the separation is the difference in the structural applications. The flexible cables are often used with pulleys for lifting and for high-strength tension members. Rods and bars are employed as tie rods or columns in structural applications. While the applications differ, the equations governing the behavior of the flexible and the stiff members are the same.

4.3.1 Design Analysis of Cables and Wire Rope

What happens if we pull on a wire and continue to pull with increasing force? The wire stretches, and stretches still more until it fails by breaking. Let's explore the stretch of the wire, and the consequences of different types of failure of the wire in subsequent sections.

Stretch of Wire under Load

When we pull on a wire it stretches as shown in Fig. 4.1b. We define δ, the **stretch** or **extension** of the wire, as:

$$\delta = L_f - L_o \tag{4.1}$$

where L_f is the length of the wire under load and L_o is the original length.

A FBD of a segment of wire is shown in Fig. 4.2a. The equilibrium relation ($\Sigma F_x = 0$) indicates that the internal force P is equal to the external force F for these uniaxial members.

Fig. 4.2 FBD of a long thin member subject to axial load.

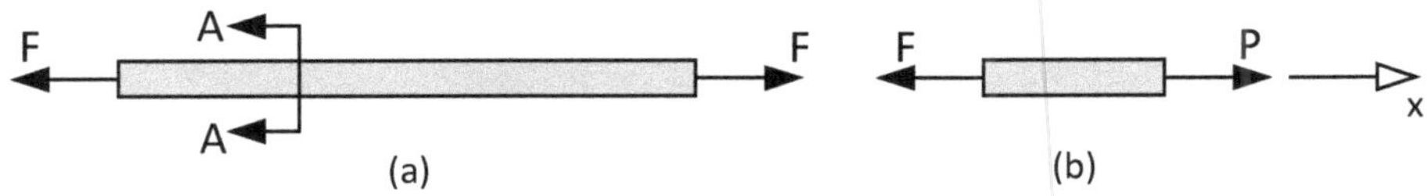

We determine the stretch δ from:

$$\delta = (PL)/(AE) \qquad (4.2)$$

where P is the internal force in the wire, in N or lb, L is the length of the wire, in m or in, $A = \pi r^2$ is the cross sectional area of the wire, in m^2 or in^2, r is the radius of the wire, in m or in, and E is the **modulus of elasticity**, in GPa or psi.

We will derive Eq. (4.2) later in this section. It is important at this time to understand that the extension of the wire δ is a very small quantity if the wire is fabricated from a metal such as steel or aluminum. To demonstrate the magnitude of the extension, consider the following example.

EXAMPLE 4.1

A No. 30 gage[3] steel music wire (0.080 in. in diameter) is employed to lift a weight of 500 lb. If the wire is 6 ft. in length, determine the amount that the wire stretches under load.

Solution: In Appendix B, we note that the modulus of elasticity for steel is listed as $E = 30 \times 10^6$ psi. Next, let's substitute the values for P, L, $A = \pi r^2$ and E into Eq. (4.2) to obtain:

$$\delta = (PL)/(AE) = (500 \text{ lb})(6 \text{ ft})(12 \text{ in/ft})/[\pi\,(0.040 \text{ in.})^2\,(30 \times 10^6 \text{ lb/in}^2)]$$

$$\delta = 0.2387 \text{ in.}$$

Is 0.2387 in. a small extension of the wire? Small is a relative term and must be compared to another value to be judged. Let's compare the amount of this extension to the original length of the wire by computing the ratio:

$$\delta/L_o = 0.2387/(6)(12) = 0.003316 \text{ or } 0.3316\%$$

The 500 lb load stretches the wire by less than a one third of one percent. That is a very small extension when compared to the original length of the wire.

Let's conduct a simple experiment and observe the behavior of a wire under the action of a monotonically increasing tensile load. If we measure the load F applied to the wire, the extension δ of the wire, and equate F_{EXT} to P_{INT}, we can construct the graph shown in Fig. 4.3 a.

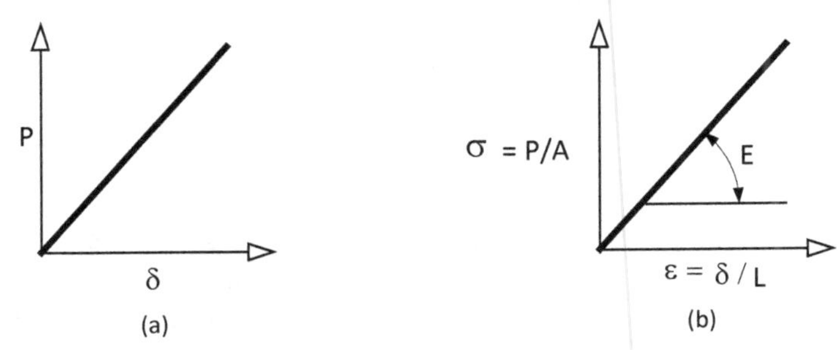

Fig. 4.3 Graph of load versus stretch in (a) and stress versus strain in (b).

[3]There are several standards for gage sizes that define the diameter of wire or the thickness of sheet metal. Four of the commonly referenced standards are presented in Appendix A.

Note a linear relationship exists between P and δ as indicated by Eq. (4.2). This P-δ curve is a straight line until the wire begins to fail by **yielding**. The linear portion of the P-δ relation is the **elastic region** of the load-deflection response. Equation (4.2) is valid only in this elastic region.

Let's modify the axes in Fig. 4.3a by dividing the force P by the area A and the extension δ by the length L as indicated below:

$$\varepsilon = \delta/L \qquad (4.3)$$

$$\sigma = P/A \qquad (4.4)$$

where ε is the strain and σ is the stress in the wire, respectively.

These relations show that the **normal strain**, ε, is the change in length divided by the original length; strain is a dimensionless quantity. The normal stress σ is the internal force P divided by the area A, over which it acts. The stress is expressed in terms of Pa (Pascal) in SI units or psi (lb/in^2) in U. S. Customary units. We have shown the graph of stress versus strain in Fig. 4.3b. As expected there is a linear relation between stress and strain until the stress is sufficient to cause the wire to yield. The **slope** of the σ - ε line is the **modulus of elasticity** E of the material from which the wire is fabricated. The **modulus of elasticity** is a **material property**, which is independent of the shape of the body.

It is evident from the linear response in the stress-strain diagram that:

$$\sigma = E\varepsilon \qquad (4.5)$$

This stress-strain relation is known as **Hooke's law**. It is named after Robert Hooke who is credited with the discovery of elasticity in the 17th century. A word of caution—Hooke's law is valid only for uniaxial states of stress that arise in long thin structural members. We will introduce another more complex form of the stress-strain relation to accommodate multi-axial stress fields in a later chapter.

Let's combine Eqs. 4.3, 4.4 and 4.5 in the manner shown below:

$$\sigma = P/A = E\varepsilon = (E\delta/L)$$

and solve this expression for δ to derive Eq. (4.2):

$$\delta = (PL)/(AE) \qquad (4.2\ \text{bis})$$

EXAMPLE 4.2

Determine the strain that develops in a 7 m long wire, which is stretched by 12 mm.

Solution:

Using Eq. (4.3), we write:

$$\varepsilon = \delta/L = 12 \times 10^{-3}\ \text{m} / 7\ \text{m} = 0.001714\ \text{dimensionless}$$

Note that strain is dimensionless because we divided a change in length by the original length. Also, observe that we have determined the strain without knowledge of the material used in fabricating the wire. Strain is a geometric concept. The strain imposed on a wire or rod may be determined without knowledge of the load or the modulus of elasticity if the amount of deformation is known.

EXAMPLE 4.3

Determine the strain that develops in a 10 ft long wire that is subjected to a force of 160 lb. The wire size is Gage 10 (0.10189 in. in diameter) and it is fabricated from an aluminum alloy 2024-T4.

Solution:

Using Eq. (4.3), we write:

$$\varepsilon = \delta/L$$

Recall Eq. (4.2):

$$\delta = (PL)/(AE)$$

Combining Eqs. (4.2) and (4.3) yields:

$$\varepsilon = P/(AE) \qquad (4.6)$$

From Appendix B, we find the modulus of elasticity E for aluminum alloy 2024-T4 is 10.4×10^6 psi. Next, we compute the cross sectional area of the wire as:

$$A = \pi d^2/4 = \pi(0.1019)^2/4 = 0.008155 \text{ in.}$$

Substituting values for P, A and E into Eq. (4.6) gives:

$$\varepsilon = 160 \text{ lb}/(0.008155 \text{ in.}^2 \times 10.4 \times 10^6 \text{ lb/in.}^2) = 0.001886$$

In this example, it was necessary to use the value of the modulus of elasticity in determining the strain. Why? The modulus of elasticity was needed because the deformation (stretch) of the wire was not specified in the problem statement. It was necessary to determine the deformation of the wire from the load using Eq. (4.2). This step in the analysis required us to introduce the modulus of elasticity.

Internal Forces and Stresses

Internal forces that develop within a structural member when it is subjected to external loads generate stresses. In Chapter 2, we introduced the concept of internal forces and showed that they are determined from the appropriate equations of equilibrium in Chapter 4. We also indicated the necessity for making an imaginary section cut to expose the cross sectional area of the member over which the stresses act. In this section, we emphasize the connection between the external forces, internal forces, and the stresses that develop at some point in the member.

To illustrate these connections, examine Fig 4.4, which shows an axial member subjected to external axial forces F.

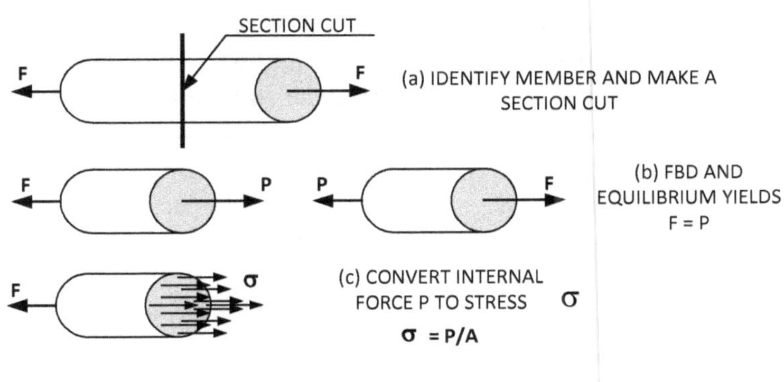

Fig. 4.4 An illustration showing the transition from external forces to internal forces to stresses.

1. We identify the member of interest and the forces acting on the body in Fig. 4.4a.
2. A section cut is made at a location of interest along the length of the member. We then remove one end or the other to construct the FBD depicted in Fig. 4.4b.
3. Next, the equilibrium relation $\Sigma F_x = 0$ is written to prove that $F_{EXT} = P_{INT}$.
4. Finally, from Eq. (4.4) ($\sigma = P/A$), we convert the internal force P into the normal stress σ as indicated in Fig. 4.4c.

In this illustration, the internal force P is a constant over the length of the uniaxial member and the location of the section cut is not important. However, you should be aware that in many applications the internal forces often vary from one location to another. In these situations the location of the section cut is vitally important. Also observe in Fig. 4.4c that the distribution of the stress σ over the cross sectional area exposed by the section cut is uniform. The uniform distribution is a result of two conditions:

- Plane sections remain plane for long thin members subjected to axial loading.
- Internal moments do not develop under the applied external loading.

In the most general case of loading, we relate the internal force and the normal stress over an area A by:

$$P_{INT} = \int \sigma \, dA \qquad (4.7)$$

However, when the stress is uniformly distributed as it is in this illustration Eq. (4.7) reduces to:

$$P_{INT} = \sigma \int dA = \sigma A \qquad (4.8)$$

EXAMPLE 4.4

Determine the stress in a 1.5 mm diameter wire subjected to an axial load of 250 N.

Solution: From Eq. (4.4), we write

$$\sigma = P/A$$

From a FBD identical to the one shown in Fig. 4.4b and the equilibrium relation $\Sigma F_x = 0$, we understand that:

$$F = P = 250 \text{ N}$$

The area $A = \pi d^2/4 = \pi(1.5 \text{ mm})^2/4 = 1.767 \text{ mm}^2$. Substituting these values into Eq. (4.4) gives:

$$\sigma = 250\text{N}/ 1.767 \text{ mm}^2 = 141.5 \text{ N/mm}^2 = 141.5 \text{ MPa}$$

Note that one N/mm^2 is equivalent to one MPa. Later in this Chapter, we will present an interpretation of this result after we describe in detail the strength of structural materials.

EXAMPLE 4.5

A large diameter wire rope fabricated from steel with a cross sectional area of 6.4 in^2 is to support a portion of the roadway on a suspension bridge. The highway engineers have specified that the maximum load imposed on the cable is not to exceed 150 ton. Determine the stress and strain in the wire rope when subjected to the specified load.

Solution:

Let's compute the stress from Eq. (4.4) as:

$$\sigma = P/A = (150 \text{ ton})(2000 \text{ lb/ton})/6.4 \text{ in.}^2 = 46{,}880 \text{ psi} = 46.88 \text{ ksi}$$

Next, calculate the strain ε from Eq. (4.6) to obtain:

$$\varepsilon = P/(AE) = 300{,}000\ \text{lb}/[(6.4\ \text{in}^2)(30 \times 10^6\ \text{lb/in}^2)] = 1.563 \times 10^{-3}$$

In determining the strain, we obtained the value of E = 30 × 10[6] psi for steel from Appendix B. We also introduced ksi, a unit in the U. S. Customary system. The conversion factor between psi and ksi is given by ⇒ 1 ksi = 1,000 psi.

Failure

Suppose we take a length of wire, grip it in a universal-testing machine, and pull until it fails. The behavior of the wire under increasing load depends on the material from which the wire was drawn[4]. Nearly all materials exhibit a **linear elastic** response like that shown in Fig. 4.3 at lower stress levels; however, at higher stresses significantly different behavior is observed for different materials. We have classified these behaviors as:

- **Brittle** with abrupt failure and small, elastic deformations.
- Yielding with **strain hardening** and plastic deformation prior to rupture.
- Yielding with **strain softening** and plastic deformation prior to rupture.

These three behaviors are illustrated graphically in Figs. 4.5 and 4.6.

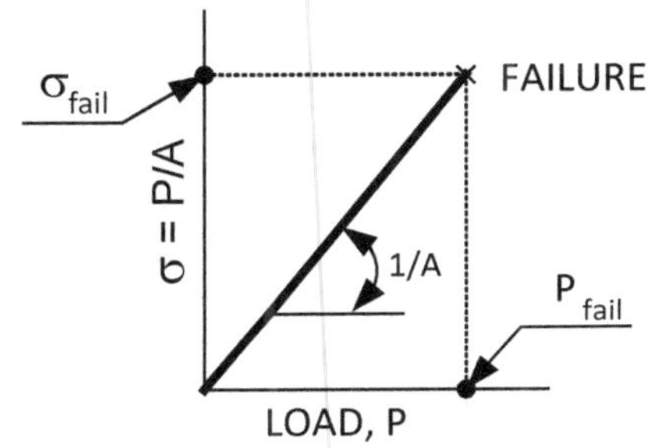

Fig. 4.5 Linear-elastic response until brittle failure.

The graph in Fig. 4.5 indicates the linear elastic response of a brittle material. The stress increases linearly with both axial load and strain. The slope of the stress-load line is the reciprocal of the area (1/A), and the slope of the stress-strain line is the modulus of elasticity E as indicated in Fig. 4.3. When we increase the load to a critical value P_{fail}, the uniaxial member breaks. The failure occurs at a specific value of stress, σ_{fail}. We define this failure stress as the **ultimate tensile strength,** S_u of the material of the wire.

$$S_u = \sigma_{fail} = P_{fail}/A \qquad (4.9)$$

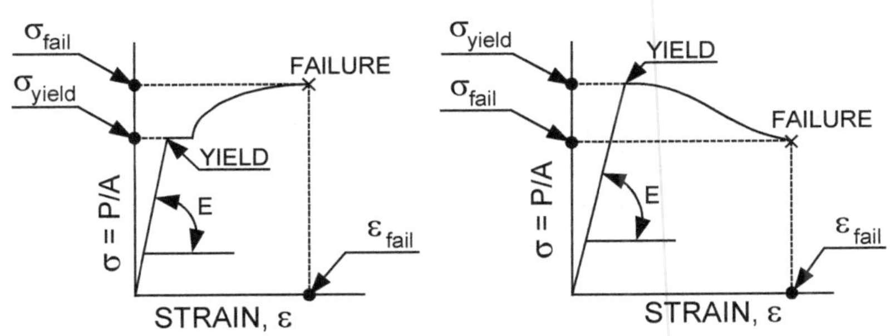

Fig. 4.6 Materials yield with strain hardening (left) and strain softening (right).

The two graphs in Fig. 4.6 illustrate material behavior when yielding occurs. The graph to the left in Fig. 4.6 is typical of a material that yields and then strain hardens. As the axial load on the wire increases, the stress increases linearly until the material begins to yield at σ_{yield}. At that point, the uniaxial member continues to stretch, but the load and stress remain essentially constant. After some degree of post-yield stretch, the material stiffens and the load and stress begin to increase. The stress increases until the wire ruptures at σ_{fail}.

[4]Drawing a circular rod through a smaller diameter die produces wire. The drawing process improves the surface finish and enhances the strength of the wire.

We establish two strengths for the wire from this graph. The **yield strength** S_y given by:

$$S_y = \sigma_{yield} = P_{yield} / A \tag{4.10}$$

and the **ultimate tensile strength,** S_u given by Eq. (4.9).

The graph to the right side in Fig. 4.6 is typical of a material that yields and then strain softens. As the load increases, the stress increases linearly until the material begins to yield at σ_{yield}. At that point, the uniaxial member continues to stretch, but the load and stress remain essentially constant. After some degree of post-yield stretch, the material softens and the load and stress begin to decrease with a continued increase in the deformation. The strain increases until the wire ruptures at σ_{fail}.

Strength and Safety Factor

We have defined two different strengths (i.e. yield and ultimate tensile) in the previous section. Since a structural member may fail by excessive deformation, we may choose to limit the stress applied to the structure so that it is less than the yield strength S_y. On the other hand in some structures, we can tolerate plastic deformation in one or more members without compromising the structure's function. In these cases, we can tolerate stresses exceeding the yield strength, S_y but they must be less than the ultimate strength, S_u. For example, when designing a bridge, it is important that its shape remain fixed under normal service loads. In this case, the yield strength is specified as the maximum limit for the design stress in the bridge to prevent post-yield deformations. However, in designing a bridge to remain standing during a collision with a freighter or during an earthquake, the ultimate strength of the structural members is the important criterion.

In designing a structural member to carry a specified load, we always size the member so that the **design stress,** σ_{design}, is less than the strength based on either the yield or failure criteria. It would not be prudent to permit the design stress to equal the strength of the member. To size our structural members so they are safe, we usually employ a **factor of safety, SF** in the analysis. We define the safety factor as the ratio of a strength divided by the design stress. This definition leads to the two relations given below:

$$\mathbf{SF_u} = S_u / \sigma_{design} \tag{4.11}$$

$$\mathbf{SF_y} = S_y / \sigma_{design} \tag{4.12}$$

where S_u the ultimate strength of the material, in units of MPa or psi, S_y the yield strength of the material, in units of MPa or psi, σ_{design} the design stress given by P_{design}/A, in units of MPa or psi ,P_{design} the load that is specified in designing the structural member, in N or lb.

Once again, the value for the factor of safety depends upon the application. Ideally, you would like the factor of safety to be as large as possible. However, this desire for excessive safety factors must be balanced by practical considerations such as economics, aesthetics, functionality, and ease of assembly. In designing, you may be concerned with the cost and weight of its components. The factor of safety you decide to specify should reflect these concerns.

EXAMPLE 4.6

A hoisting cable with a diameter of 7/16 in. is fabricated from many strands of an improved plow steel wire. The manufacturer of the wire certifies its breaking load as 16,500 lb. Determine the strength of the wire used in the manufacture of the cable. Also discuss the assumption pertaining to the wire rope made in the analysis and its implication on the strength of the strands of wire.

Solution: Recall Eq. (4.9), which gives:

$$S_u = \sigma_{fail} = P_{fail} / A$$

The cross sectional area of the cable will be less than $A = \pi d^2/4 = 0.1503 \text{ in}^2$. Substituting this value for the area into Eq. (4.9) yields:

$$S_u = 16{,}500 \text{ lb}/ 0.1503 \text{ in.}^2 = 109.8 \text{ ksi}$$

We have assumed the cross sectional area of the wire rope to be equivalent to that of a solid wire 7/16 in. in diameter. Wire rope is made of many very small diameter wires that are twisted together to form strands. The strands in turn are formed in a helix about a fiber core. The wire rope in this example has a designation of 6 × 19 (6 strands with 19 small wires in each strand). The cross section of this 6 by 19-wire rope is illustrated below.

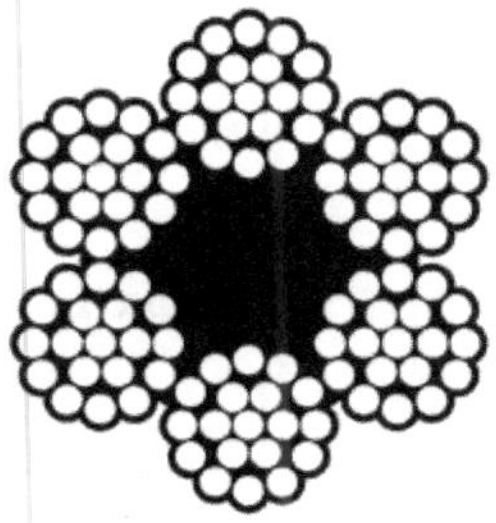

Fig. 4.7 A wire rope is fabricated from many strands of small-diameter high-strength wire.

The assumption of a solid cross sectional area overestimates the cross sectional area, A by a factor of more than two. The ultimate tensile strength of the small diameter wire used to form the strands of wire for a typical cable is in excess of 200 ksi.

EXAMPLE 4.7

A solid hard drawn copper wire exhibits an ultimate tensile strength of 390 MPa. If the diameter of the wire is 1.6 mm, determine the load required to break the wire.

Solution: Recall Eq. (4.9), which gives:

$$S_u = \sigma_{fail} = P_{fail} / A$$

Rearrange this relation and substitute known quantities to give:

$$P_{fail} = S_u A = (390 \times 10^6 \text{ N/m}^2)(\pi/4)(1.6 \text{ mm})^2 \times 10^{-6} \text{ m}^2/\text{mm}^2 = 784.1 \text{ N}$$

An axial load of approximately 800 N is required to rupture the hard drawn copper wire.

EXAMPLE 4.8

A monofilament, nylon fishing line is rated at 10-lb test. If the line is 0.012 inch in diameter, determine the strength of the nylon in the form of a small diameter line. Comment on the effect that the filament geometry has on the strength of a polymer like nylon-6/6.

Solution: Recall Eq. (4.9), which gives:

$$S_u = \sigma_{fail} = P_{fail} / A$$

$$S_u = (4)(10)/ [\pi (0.012)^2] = 88{,}420 \text{ psi}$$

Monofilaments of polymers like nylon-6/6 are drawn from a melt into thin fibers or lines. In this process, the long molecules of the polymer are aligned with the axis of the filament thus enhancing its strength.

EXAMPLE 4.9

A No. 14 gage (0.080 in. diameter) black annealed steel wire is listed in a material handbook with a yield strength of 220 MPa and an ultimate tensile strength of 340 MPa. Determine the axial load that will cause the wire to yield.

Solution: Recall Eq. (4.10), and rearrange it to give:

$$S_y = \sigma_{yield} = P_{yield} / A$$

$$P_{yield} = S_y A \qquad \text{(a)}$$

Since we have mixed units in the problem statement, let's convert the units for the diameter of the wire to the SI system and determine the cross sectional area A in mm^2.

$$A = \pi d^2/4 = \pi\,(0.080 \text{ in.})^2\,(25.4 \text{ mm/in.})^2\,/4 = 3.243 \text{ mm}^2$$

Next, substitute into Eq. (a) to obtain:

$$P_{yield} = S_y A = (220 \text{ MPa})[(\text{N}/(\text{mm}^2)/\text{MPa})](3.243 \text{ mm}^2) = 713.4 \text{ N}$$

EXAMPLE 4.10

A 5/8 in. diameter stainless steel (type 304) wire rope with a 6 × 19 configuration is listed in a catalog with a breaking load of 35,000 lb. If a safety factor of 2.4 is to be employed in the design of a structure using this type of cable, specify the design load.

Solution: Recall Eqs. (4.9) and (4.11) and combine them to obtain:

$$\mathbf{SF_u} = S_u / \sigma_{design} = AS_u / P_{design} = P_{fail} / P_{design} \qquad \text{(a)}$$

Solving Eq. (a) for the design load P_{design} gives:

$$P_{design} = P_{fail} / \mathbf{SF_u} = 35{,}000/\ 2.4 = 14{,}580 \text{ lb}$$

EXAMPLE 4.11

A single #20 gage steel wire with a solid cross section is 0.0348 in. in diameter. It is to carry a load of 100 N with a safety factor of 2.8. Specify the steel alloy from which the wire should be manufactured to avoid failure by rupture.

Solution: Recall Eq. (4.11) and solve it to obtain an expression for S_u:

$$\mathbf{SF_u} = AS_u / P_{design}$$

$$S_u = (P_{design})\mathbf{SF_u}/A = 4(100 \text{ N})(2.8)/[\pi\,(0.0348 \text{ in.})^2\,(25.4 \text{ mm/in.})^2] = 456.3 \text{ MPa}$$

It is necessary to refer to a materials handbook to select a suitable material for this application. Reference to Appendix B-2 indicates three steel alloys and four stainless steel alloys with ultimate tensile strengths exceeding the requirement for S_u in this example.

EXAMPLE 4.12

A cold drawn alloy steel wire, with a yield strength of 125,000 psi, is to carry a load of 600 lb while incorporating a safety factor of 2.25. You are to inform the purchasing representative regarding the size of the wire to be ordered.

Solution: To determine the size of the wire in this application, recall Eq. (4.12).

$$\mathbf{SF}_y = S_y / \sigma_{design} = AS_y / P_{design}$$

Solve this relation for the area to obtain:

$$A = \mathbf{SF}_y\ (P_{design})/S_y = \pi d^2/4$$

$$A = (2.25)(600)/125{,}000 = 0.0108\ in^2 = \pi d^2/4$$

$$d = 0.1173\ in.$$

Do you call the purchasing representative and order a cold drawn alloy steel wire with a diameter of 0.1173 in? **No!!!** If you make this mistake, you will soon learn that steel wire is available only with a standard diameter. Economic laws dictate that the most cost effective way to mass-produce materials is by limiting inventories to only those sizes and geometries that satisfy most customers. In Appendix A standard wire sizes and the corresponding gage numbers that are used to order them are listed. For this example, four entries from the listings in Appendix A for standard steel wire are presented below:

Gage No.[5]	Diameter (in.)	Area (in^2)
10	0.1350	0.01431
11	0.1205	0.01140
12	0.1055	0.00874
13	0.0915	0.00658

The response to the purchasing representative is to procure Gage No. 11 wire. It is important for you to specify a standard size in ordering structural members. Standard sizes are available with minimum delay at the lowest cost. Analyses rarely give dimensions for structural members that correspond to the available standard sizes. The usual practice is to increase the size determined in the analysis to the next larger dimension available as a standard. This approach enhances safety while minimizing cost and delivery time.

4.3.2 Design Analysis of Rods and Bars

All of the equations [(4.1) to (4.12)], derived in Section 4.3.1 for cables, are also valid for rods and bars. If the external axial forces are tensile (tend to pull the bar apart), the internal forces and stresses are tensile and denoted with a positive sign. On the other hand, if the external axial forces are compressive (tend to squash the bar), the internal forces and stresses are compressive and denoted with a minus sign.

We must qualify the capability of a rod or bar to carry compressive loads. If the rod is very long and slender and the compressive force too high, the rod may buckle. **Buckling** is an unstable condition, and if the critical load is exceeded, the rod fails suddenly and catastrophically. In this chapter, we will assume that the rods or bars loaded in compression are sized to resist buckling; consequently, they fail due to excessive compressive stresses. However, the tendency for these structures to buckle cannot always be ignored. Theories describing the failure of **columns** due to buckling will be introduced later in the Chapter 14.

Centroids

The relations derived for the stresses and deformations of cable assumed that the internal and external forces acted through the **centroid** of its cross sectional area. For cables or wire, with circular cross sections, the location of the centroid is obvious—it is at the center of the circle. However, for cross sections of more complex shapes, the location of the centroid is not obvious.

[5] There are several different standards that refer to gage numbers for wire and steel sheet. In this example we list the American Steel and Wire Co. standard that is commonly used for steel wire.

A centroid is the point coinciding with the center of gravity of a two-dimensional area. For the circular cross section shown in Fig. 4.8, the center of the circle clearly locates the centroid. The center also serves as the origin for the centroidal axes x_c and y_c. For cross sectional shapes such as ellipses, circles, squares and rectangles, the center may be located by inspection because these shapes are all symmetric about both horizontal and vertical axes. The centroid is located at the intersection of the two axes of symmetry. However, for non-symmetric figures, such as a triangle, a portion of a circular area, a parabolic area etc., locating the center of the gravity of the area by inspection is not possible.

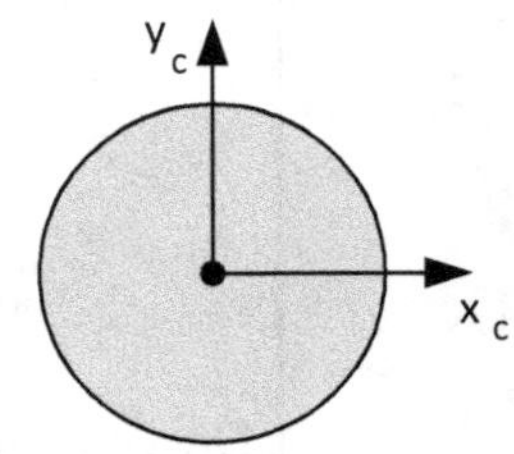

Fig. 4.8 A circular cross section with the center as the centroid and the centroidal axes x_c -y_c.

One approach for determining the location of a centroid is to consider the area as a flat plate positioned in the x-y plane with its outer normal vector in the z direction as illustrated in Fig. 4.9. Let's examine Fig. 4.9a and observe that the weight W of the plate acts downward through the **center of gravity** defined by the point G_c. The center of gravity and the centroid will be located at the same point for a plane body with uniform thickness and density. While the gravitational forces are distributed over the area of the plate, we represent the total weight of the plate as a concentrated force of magnitude W that acts downward though the center of gravity.

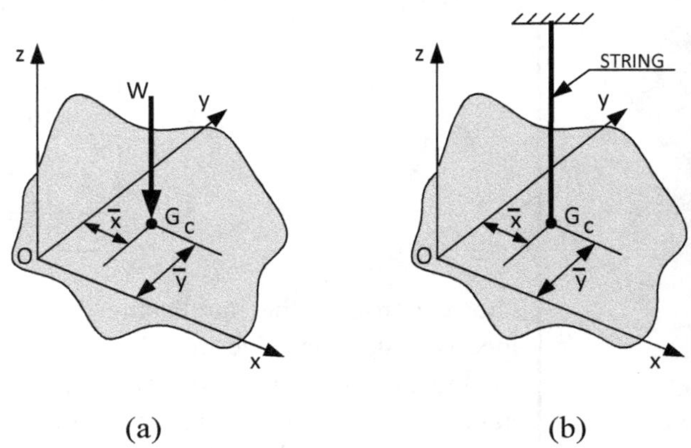

Fig. 4.9 A flat plate of arbitrary area with its weight acting though its center of gravity G_c.

The center of gravity may be established experimentally or by analysis. The experimental approach is illustrated in Fig. 4.9b. Suppose we represent the plate with a sheet of cardboard cut to the shape of the arbitrary cross section. We scribe a Cartesian coordinate system on the cardboard to provide a reference for measuring the location of the center of gravity G_c. A string is fixed to the cardboard model at some point P near its center, and the model is suspended in space hanging from this string. In most cases, we cannot precisely estimate the location of the centroid of the model, and the model flops over onto its side. However, when the points P and G_c coincide, the model will hang from the string with its plane horizontal and with its outer normal (the z axis) parallel to the string.

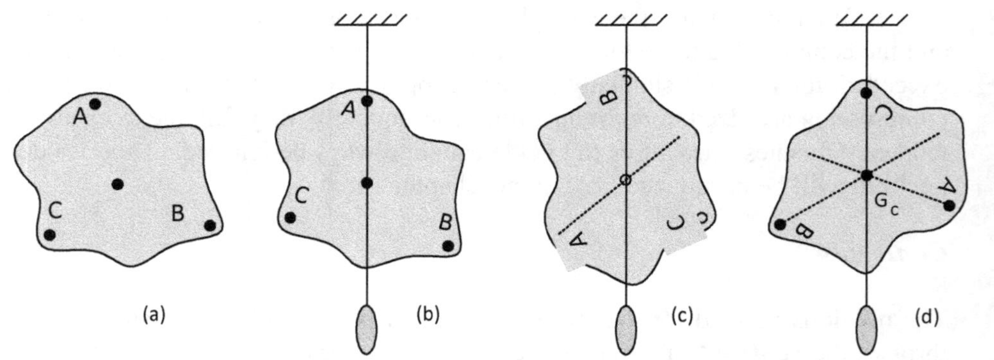

Fig. 4.10 Experimental determination of the centroid of an arbitrary area.

Another experimental method for locating the center of gravity (also the centroid) is depicted in Fig. 4.10. Again we begin with a cardboard model of the area in question. With this approach we punch three small holes at points (A, B and C) located near the boundary of the model as illustrated in Fig. 4.10 a. We hang the model from a pin inserted through the hole at point A, and draw a straight line vertically downward from point A as shown in Fig. 4.10b. A plumb line or a level is used to aid in drawing a perfectly vertical line. We repeat this process with the model suspended from points B and then C as indicated in Figs. 4.10c and 4.10d. The three straight lines intersect; thus, locating the center of gravity or the centroid at a point G_c.

These experimental methods are effective in locating the position of the centroid and give an insight to the meaning of the centroid, center of gravity and the **first moment of an area**. There is one qualification that must be imposed on the fabrication of the model—the sheet of cardboard must be homogenous (of uniform thickness and density).

Analytical methods for locating the centroid of an arbitrary shaped area are covered in Appendix C.

Stresses in a Uniform Bar or Rod

The procedure for determining the normal stresses σ is the same as described previously for wires and cables except for the fact that the internal forces may be compressive or tensile. We begin by constructing a FBD to show the point of application and the directions of the internal and external forces. From the equilibrium relations, we determine the internal forces and their sign. Finally, Eq. (4.4) is employed to determine the normal stresses σ. We demonstrate this procedure with the example problems presented below.

EXAMPLE 4.13

Determine the stress in the bar shown in Fig. E4.13 when subjected to a compressive axial force of F that acts through the centroid of the bar. The following numerical parameters define the bar and the applied load: F = 800 kN, L = 0.500 m, w = 100 mm and h = 50 mm.

Fig. E4.13

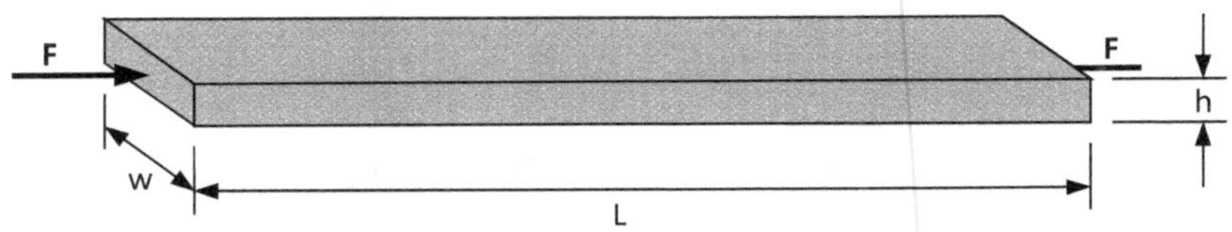

Solution: We make a section cut near the center of the bar, and construct the free body diagram shown below:

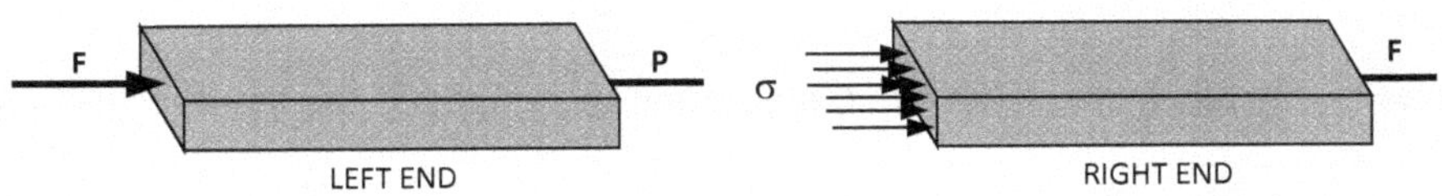

The stresses σ visible on the face exposed by the section cut are due to the internal force P. It is clear from $\Sigma F_x = 0$ that P = F; then from Eq. (4.4), we write:

$$\sigma = P/A = -\ 800 \times 10^3\ \text{N}/(100)(50)\text{mm}^2$$

$$\sigma = -\ 800/5 = -\ 160\ \text{N/mm}^2 = -\ 160\ \text{MPa}$$

The axial stress is a negative number (– 160 MPa) indicating that a compressive stress develops in the bar due to the compressive axial loading. Also observe that the value of the stress is independent of the material from which the bar is fabricated.

EXAMPLE 4.14

Determine the stress in a two-foot long rod that has a diameter of 4 in. The bar is fabricated from mild steel with a yield strength S_y = 30,000 psi, and is subjected to an axial compressive load of 100,000 lb. Also, compute the safety factor of the bar against failure by yielding.

Solution: Again, we begin with a drawing of the rod and a free body diagram in Fig. E4.14 showing the forces F and P and the normal stresses σ.

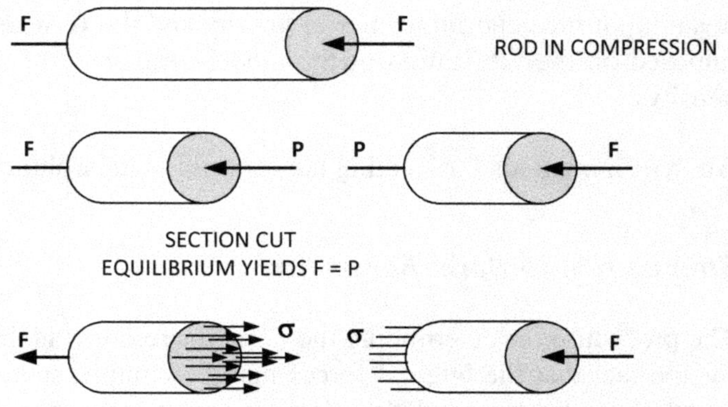

Fig. E4.14

From Eq. (4.4), we write:

$$\sigma = P/A = -\ 100{,}000\ \text{lb}/[\pi \times (2)^2] = -\ 7{,}958\ \text{psi}$$

We determine the safety factor against yielding from Eq. (4.12).

$$\mathbf{SF}_y = S_y/S_{design} = -\ 30{,}000/-\ 7{,}958 = 3.77$$

Note, the safety factor is always a positive quantity. In this problem, we compared a compressive strength of – 30,000 psi (taken as a negative quantity) with the design stress of –7,958 psi.

Deflection of Axially Loaded Bars and Rods

The deformation of rods and bars may be described using Eqs. (4.1) and (4.2), as illustrated in the following examples.

EXAMPLE 4.15

If the bar described in Example 4.13 is fabricated from steel, determine the length of the bar after the application of the compressive force.

Solution: An inspection of the free body diagram shown in Fig E4.13a indicates that the internal force P is constant from one end of the bar to the other and equal to the external force F. Using Eqs. (4.1) and (4.2), we may write:

$$\delta = L_f - L_0 = PL_0/(AE) \qquad \text{(a)}$$

Solving Eq. (a) for L_f, gives:

$$L_f = L_0\ [1 + P/(AE)] \qquad \text{(b)}$$

Substituting numerical values for the known parameters in Eq. (b) yields:

$$L_f = 0.5\ \{1 - [(800 \times 10^3)/(0.05)(0.1)(207 \times 10^9)]\}$$

$$L_f = 0.5[1 - 0.7729 \times 10^{-3}]\ \text{m} = (500 - 0.3864)\ \text{mm} = 499.6\ \text{mm} \qquad \text{(c)}$$

Since the new length of the rod is 499.6 mm, it is apparent that it contracted by 0.4 mm under the action of the compressive load. The value of P in Eq. (4.2) is treated as a negative number when the axial load on the bar is compressive. Also, the value of the modulus of elasticity was taken as 207 GPa from Appendix B.

EXAMPLE 4.16

Determine the axial deflection of the rod described in Example 4.14. Repeat the solution for the rod if it is fabricated from an aluminum alloy.

Solution: An inspection of the free body diagram in Fig. E4.14 indicates that the internal force P is constant over the entire length of the rod. From Eq. (4.2), we write:

$$\delta = PL/(AE) \qquad \text{(a)}$$

Substituting the values for the known quantities for the steel rod in Eq. (a) gives:

$$\delta = -\ [(100{,}000)(2)(12)]/\ [(\pi)(2)^2\ (30 \times 10^6)] = -\ 0.006366 \text{ in.}$$

Substituting the values for the known quantities for the aluminum rod in Eq. (a) gives:

$$\delta = -\ [(100{,}000)(2)(12)]/\ [(\pi)(2)^2\ (10.4 \times 10^6)] = -\ 0.01836 \text{ in.}$$

Both solutions carry a negative sign indicating that the rod is compressed (shortened) by the action of the compressive force. It is interesting to observe that the rod fabricated from an aluminum alloy exhibited nearly three times the deflection of the steel rod. The reason for this difference is in the lower modulus of elasticity of aluminum relative to steel. The modulus of elasticity for aluminum and steel used in this calculation was 10.4×10^6 and 30×10^6 psi, respectively as cited in Appendix B.

4.4 SHEAR STRESSES

In the preceding discussions, the stresses created by internal forces were normal to the area exposed by the section cut. To emphasize this fact, we called these **normal stresses**. A second type of stress exists—a **shear stress**. As the name implies, the shear stress lies in the plane of the area exposed by the section cut. To show shear stresses in a more graphical manner, consider the stubby beam-like member loaded with the force F in Fig. 4.11.

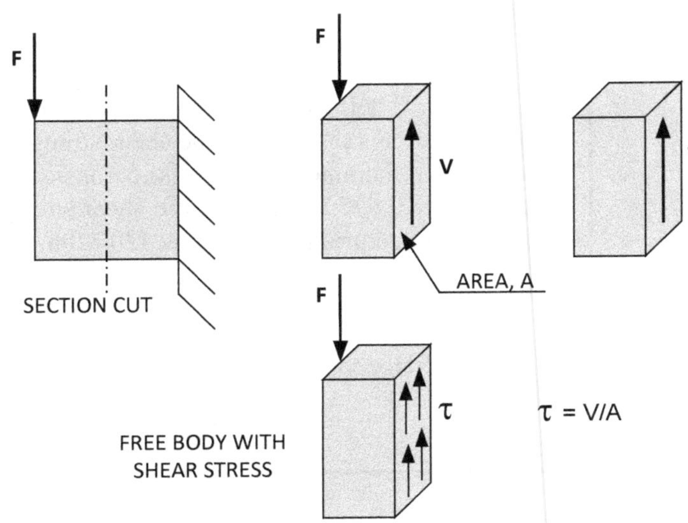

Fig. 4.11 A section cut made in a stubby beam-like member produces a free body.

First, cut the stubby member to create a free body of the left end, and then apply an internal shear force V in the plane of the area exposed by the section cut[6]. From $\sum F_y = 0$, we determine that F = V. The shear force V is produced by a shear stress τ. The relation between shear stress τ and the shear force V is:

$$\tau = V/A \tag{4.13}$$

We have assumed that the shear stress τ is uniformly distributed over the area of the stubby member. Later, in the discussion of beam theory found in Chapter 10, we will note that the shear stress is not uniformly distributed over the cross sectional area of a beam. However, for many block-like members, the assumption of a uniform distribution of shear stresses is a reasonable approximation.

EXAMPLE 4.17

A key is employed to keep a gear from slipping on a shaft when transmitting power. Forces F of 50 kN are created on the key at the locations shown in Fig. E4.17. Determine the shear stresses in the key if it is 8mm wide, 9 mm high and 50 mm long.

Solution: We construct a FBD of the key, presented in Fig. E4.17, showing the equal and opposite forces F and the shear plane between the gear and the shaft. We section the key along the shear plane, and draw another free body diagram of its lower portion. On this second free body diagram, we illustrate the shear stresses τ that occur on the shear plane.
The shear stresses on the key are given by Eq. (4.13) as:

$$\tau = V/A = (50 \times 10^3)/(0.008)(0.050) = 125 \text{ MPa}$$

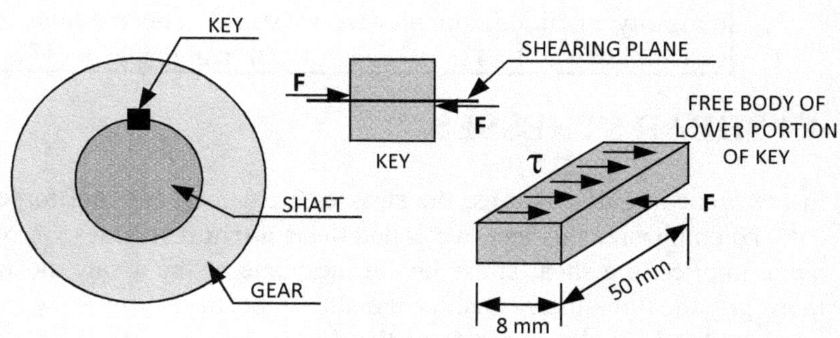

Fig. E4.17 A section cut showing shear forces acting on a key, which prevents the gear from slipping on the shaft

The average shear stress acting on the key is 125 MPa. To analyze the impact of this solution on the design of a structure, it is necessary to compare the imposed stress with the shear strength of the material from which the key is fabricated. The shear strength is usually lower than the yield or ultimate tensile strength of a material. A common practice is to consider either the yield or tensile strength of a material and multiply that value by 0.577 to estimate the shear strength. For example, if the tensile yield strength, S_y of the steel used in manufacturing the key was 320 MPa, the yield strength in shear S_{ys} is estimated as:

$$S_{ys} = 0.577\ S_y = (0.557)(320) = 184.6 \text{ MPa}$$

The safety factor for the key is determined from:

$$SF = S_{sy}/\tau = 184.6/125 = 1.477$$

[6] To maintain the focus of the discussion on shear forces, we have not included the internal moment acting at the section cut on the FBD in Fig. 4.11.

In interpreting the results for normal stresses, we recognize the difference between tensile (+) and compressive (–) values for the results. The reason for the careful distinction is the fact that the strength of material subjected to tensile or compressive stresses is often different. However, shear strength is not sensitive to the sign of the shear stress τ. The strength of a material to an imposed shear stress is not dependent on the direction of the imposed shear force. For this reason, we neglect the sign of the shear force in our analysis of stress on machine components.

4.5 STRESSES ON OBLIQUE PLANES

In making section cuts on bars and rods, we have limited our choices to sections perpendicular to the axis of the bar. This restriction was helpful because it simplified the state of stress that was examined. With these perpendicular section cuts, we considered only the normal stresses acting on the area exposed by the cut.

We showed that shear stresses exist in Section 4.4 and computed their magnitude in Example 4.17. In this case, we restricted the section cut to one parallel to the imposed system of equal and opposite forces. With this restriction on the section cut, only shear forces are developed. Clearly, by restricting the section cut we develop special cases where normal stresses exist in the absence of shear stresses and vice versa.

Let's treat the more general case with the bar cut at an arbitrary angle as shown in Fig 4.12.

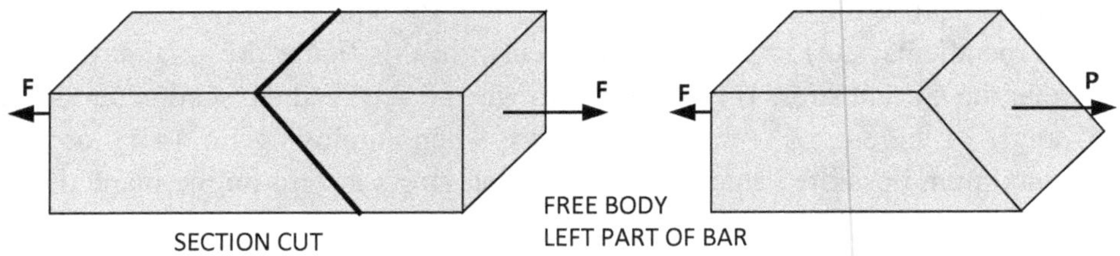

Fig. 4.12 Bar with axial loading with a section cut at an arbitrary angle.

Let's cut the free body of Fig. 4.12 with still another section to produce a triangular shape as shown in Fig. 4.13a. The triangular shape in Fig. 4.13b is defined with the angle θ. The left side of this triangle has an area A; therefore, the hypotenuse of the triangle has an area = A/cos θ. Note, the x-axis is coincident with the axis of the bar; the n axis is coincident with the outer normal to the inclined cut; the t axis is tangential to the inclined cut.

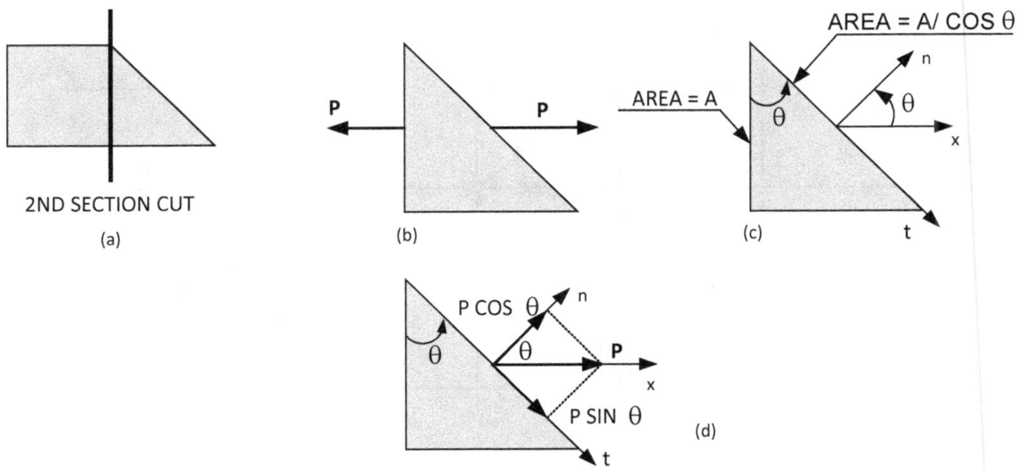

Fig. 4.13 Free body diagrams of the bar with an inclined cut: (a) The section cut is perpendicular to the axis of the bar to produce a triangle; (b) Internal forces acting on both faces of the element; (c) The area of the left side and the hypotenuse is illustrated; (d) The forces on the surface of the inclined cut are resolved into components.

As shown in Fig. 4.13c, the internal force P has been resolved into components along the n and t axes to yield:

$$P_n = P \cos \theta \qquad \text{(a)}$$

$$V_t = P \sin \theta \qquad \text{(b)}$$

To ascertain the stresses on the inclined surface, we simply divide either P_n or V_t by the area formed with the inclined section cut in accordance with Eq. (4.4) or Eq. (4.13). The stress σ_θ in the normal direction (n) is determined from Eq. (4.4) as:

$$\sigma_\theta = (P \cos \theta)/(A/\cos \theta) = (P/A) \cos^2 \theta \qquad \text{(c)}$$

If we define the axial stress $\sigma_x = P/A$, then we may write:

$$\sigma_\theta = (\sigma_x)\cos^2 \theta \qquad (4.14)$$

The shear stress τ_θ that acts along the face of the inclined surface is given by Eq. (4.13) as:

$$\tau_\theta = P \sin \theta/(A/\cos \theta) = (P/A) \sin\theta\cos \theta \qquad \text{(d)}$$

This relation is rewritten as:

$$\tau_\theta = (\sigma_x/2) \sin 2\theta \qquad (4.15)$$

Numerical results are presented in Fig. 4.14 for both σ_θ and τ_θ as the angle of the inclined section varies from zero (a perpendicular cut) to 90° (a parallel cut). In this figure the axial stress σ_x = 100 units. From Fig. 4.14 it is evident that the normal stress σ is a maximum when θ = 0° and the section cut is perpendicular to the axis of the bar. For this angle of the cut, $\sigma_\theta = \sigma_x = P/A$. Thus, when employing Eq. (4.4) to compute the normal stress, we determine the maximum possible value of σ_θ. The shear stress is zero on the plane defined by θ = 0° where the normal stress is a maximum.

Both the normal and shear stress vanish when the section cut is made at 90°. The shear stress is a maximum when the section cut is defined by θ = 45°. When θ = 45°, the normal and shear stresses are equal to each other $\sigma_{45} = \tau_{45} = \sigma_x/2$.

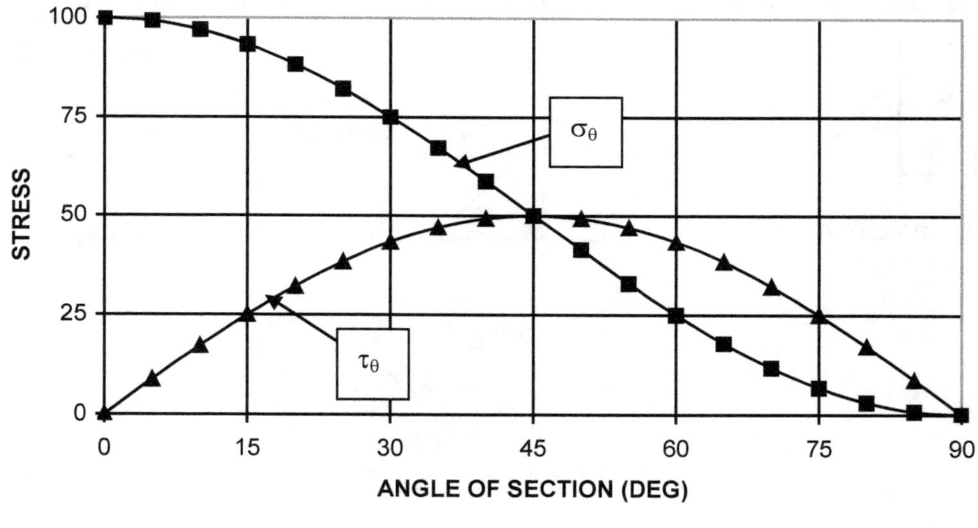

Fig. 4.14 Variation in σ_θ and τ_θ with the angle of the inclined section.

The example of the axially loaded bar, with an inclined section cut, illustrates why stresses must be treated as tensor quantities. As we vary the angle of inclination of the section cut, two parameters are changing relative to either the normal or the shear stresses. First, the magnitude of the force components in the n and t direction changes with the angle θ because forces are vector quantities. Second, the area of the inclined surface exposed by the section cut increases with θ. Both of these parameters affect the magnitude of the normal and shear stresses; therefore stresses must be treated as tensor quantities—not vectors.

EXAMPLE 4.18

A circular rod 30 mm in diameter and 2 m long is subjected to an axial load of 120 kN. Determine:

- The maximum normal stress and the plane upon which it acts.
- The maximum shear stress and the plane upon which it acts.
- Draw a FBD of the section for the maximum normal stress.
- Draw a FBD of the section for the maximum shear stress.

Solution: The maximum normal stress occurs on a plane perpendicular to the axis of the rod where $\theta = 0°$. Equation (4.14) applies.

$$\sigma_{\theta=0} = \sigma_x \cos^2\theta = (P/A)\cos^2\theta = (120 \times 10^3)/[\pi\,(0.015)^2]\cos^2 0° = 169.8 \text{ MPa}$$

The maximum shear stress occurs when $\theta = 45°$ as indicated by the results depicted in Fig. E4.14. From Eq. (4.15), we write:

$$\tau_{\theta=45} = (\sigma_x/2)\sin 2\theta = (169.8/2)\sin 90° = 84.90 \text{ MPa}$$

The FBDs for the right portion of the bar showing the maximum normal and shear stresses are shown below:

Fig. E4.18 Section cuts for (a) maximum normal stress and (b) maximum shear stress.

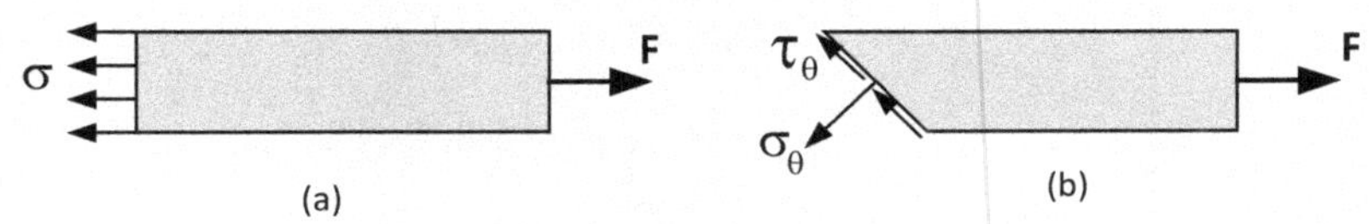

The FBD in Fig. E4.18a shows the normal stresses σ acting on a plane area perpendicular to the axis of the rod. The shear stresses are shown in Fig. E4.18b where they act on a surface inclined at a 45° angle to the axis of the rod.

4.6 AXIAL LOADING OF A TAPERED BAR

The discussion of stress and deflection of a rod or bar has been limited to those members with a uniform cross sectional area. Of course, uniform bars and rods are most commonly employed in building structures because they are easy to design with and less expensive to manufacture; however, in some instances it may be more desirable to use members that are not uniform. In these cases, we must accommodate the effect of the changing cross sectional area over the length of the bar on both the stresses and displacement.

4.6.1 Normal Stresses in a Tapered Bar

The normal stresses that occur in a tapered bar or rod are determined using Eq. (4.4). The only consideration made to account for the taper in the bar is to adjust the area A to correspond with the position along the length of the bar. Let's consider the tapered bar presented in Fig. 4.15. The thickness of the bar, b, is constant along its entire length. The height, h_x, of the bar varies with position x according to the relation:

$$h_x = h_1 + (h_2 - h_1)(x/L)$$

The area, A_x, at any position x along the length of the bar is then:

$$A_x = bh_x = b[h_1 + (h_2 - h_1)(x/L)] \qquad (4.16)$$

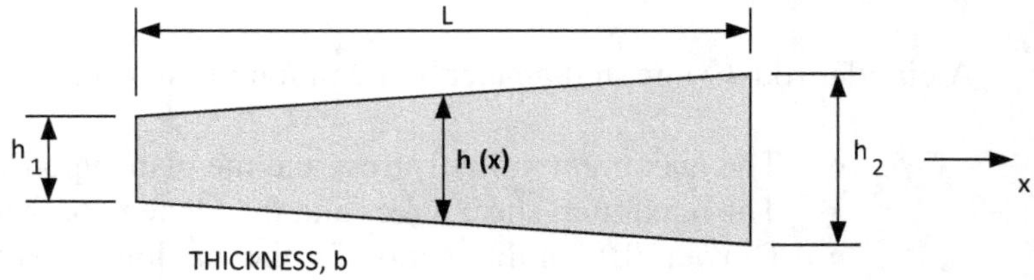

Fig. 4.15 Geometry of a tapered bar with a uniform thickness.

The normal stresses σ_x due to an axial force are determined by substituting Eq. (4.16) into Eq. (4.4) to obtain:

$$\sigma_x = P/A_x = P/\{b[h_1 + (h_2 - h_1)(x/L)]\} \qquad (4.17)$$

where L is the length of the tapered bar.

EXAMPLE 4.19

For a tapered bar similar to the one shown in Fig. 4.15, determine the normal stress as a function of position, x along the length of the bar. The geometry of the bar is given by $h_1 = 2$ in., $h_2 = 6$ in., $b = 3$ in. and $L = 36$ in. The axial load imposed on the tapered bar is 5,000 lb.

Solution: We employ Eq. (4.17) and write:

$$\sigma_x = P/\{b[h_1 + (h_2 - h_1)(x/L)]\} = 5{,}000/\{3[2 + (6 - 2)(x/36)]\} \qquad \text{(a)}$$

$$\sigma_x = 15{,}000/(18 + x)$$

The stress $\sigma_x = 833.3$ psi is a maximum at $x = 0$, and decreases to a minimum value of $\sigma_x = 277.8$ psi when $x = 36$ in.

4.6.2 Deflection of Tapered Bars

The deflection of a tapered rod may be determined from Eq. (4.2) although we must again modify the approach to accommodate for the changing cross sectional area over the length of the bar. To determine the axial deflection, this accommodation is more difficult since the total deformation of the bar is the sum of the incremental deflections at each position x along the entire length of the bar. We begin by considering an incremental length dx at some position x as shown in Fig. 4.16.

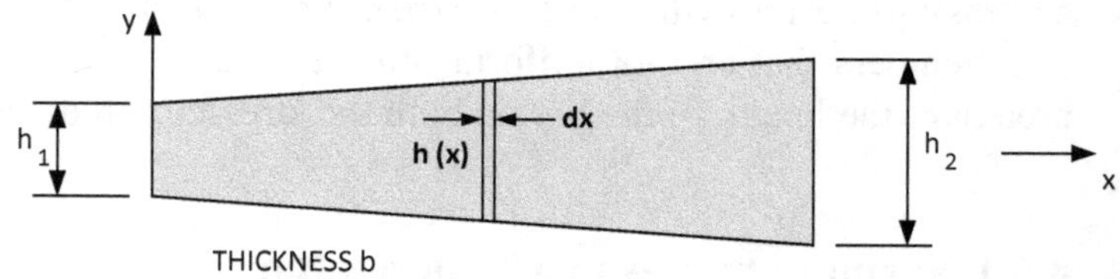

Fig. 4.16 An incremental length (slice) dx at position x along the length of the tapered bar.

To accommodate for the taper in the bar, we determine the incremental deflection dδ of a bar of length dx. Since dx approaches zero, we treat the cross sectional area as a constant over the incremental length. Accordingly, we modify Eq. (4.2) to read:

$$d\delta = P\,dx/(A_x E) \qquad (4.18)$$

where A_x is the cross sectional area of the bar that is a function of x given in Eq. (4.16).

Substitute Eq. (4.16) into Eq. (4.18), simplify and integrate to obtain:

$$\delta = \frac{PL}{Eb}\int_0^L \frac{dx}{h_1 L + (h_2 - h_1)x} \qquad \text{(a)}$$

$$\delta = \frac{PL}{Eb}\left(\frac{1}{h_2 - h_1}\right)\ln\left(\frac{h_2}{h_1}\right) \tag{4.19}$$

EXAMPLE 4.20

Determine the deflection of the tapered bar described in Example 4.19 if the bar is fabricated from an aluminum alloy with a modulus of elasticity E = 10.4 × 10^6 psi.

Solution: Let's solve this example problem by recalling Eq. (4.19).

$$\delta = \frac{PL}{Eb}\left(\frac{1}{h_2 - h_1}\right)\ln\left(\frac{h_2}{h_1}\right)$$

Substituting the parameters describing the geometry and the material constant for the bar into this relation yields:

$$\delta = [(5000)(36)/(10.4 \times 10^6)(3)][1/(6 - 2)]\ln(6/2)] = 1.585 \times 10^{-3} \text{ in.}$$

Examine the magnitude of the deflection of the tapered bar. Is the deflection large or small? What reference do you use to judge? Clearly, the axial extension of the bar is small. For small quantities, we sometimes use the human hair as a reference. The diameter of a single strand of hair is slightly more than 2 × 10^{-3} in.; hence, the extension is about 75% of a hair diameter.

4.7 AXIAL LOADING OF A STEPPED BAR

In some structures, bars of different cross sectional areas are employed where the area changes abruptly at some position along the length of the bar. An illustration of a stepped bar where the cross section undergoes an abrupt change is presented in Fig. 4.17.

The procedure for determining the normal stresses and deflection remains the same as for a bar with a uniform cross sectional area:

1. Draw a free body diagram of each section of the bar.
2. Use the equilibrium relations to establish the internal axial forces acting on each section.
3. Determine the stresses from Eq. (4.4) using the area of the section of interest.
4. Solve for the deflection of each segment of the bar from Eq. (4.2) and then add them to obtain the total deflection.

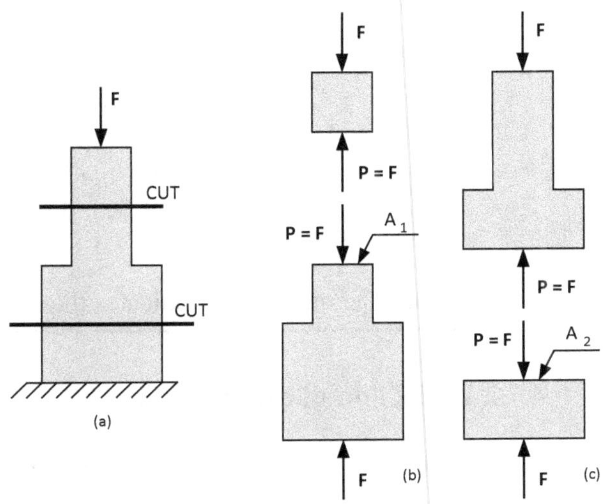

Fig. 4.17 A stepped bar with free body diagrams for each section of the bar.

4.7.1 Normal Stresses in Stepped Bars

The normal stress varies with the cross sectional area of the bar. For example, in Fig 4.17 the normal stress in the upper section of the bar is given by $\sigma = P/A_1$, and the stress in the lower section of the bar is $\sigma = P/A_2$.

EXAMPLE 4.21

For the stepped bar, shown in Fig. E4.21, determine the normal stress in each of the two sections of the bar if $F_1 = 40$ kN; $F_2 = 80$ kN; $w_1 = 50$ mm; $b_1 = 60$ mm; $w_2 = 90$ mm; $b_2 = 60$ mm; $L_1 = 200$ mm and $L_2 = 300$ mm

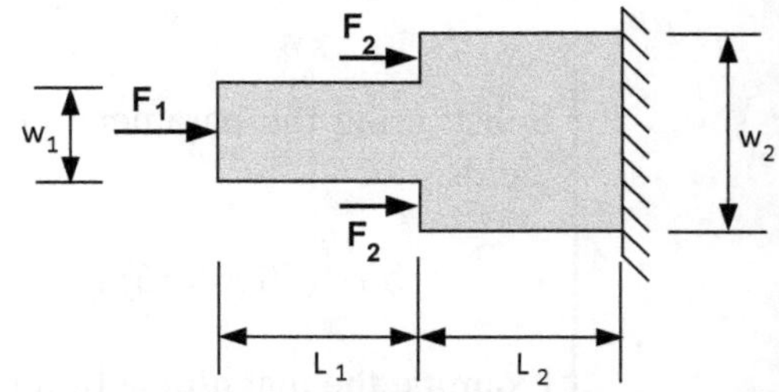

Fig. E4.21 The thickness of the bar is given by b_1 and b_2.

Solution:

Let's begin the solution by drawing free bodies of both sections of the bar. We make two section cuts A and B as illustrated in Fig. E4.21a. Then we draw free bodies associated with the portion of the bar to the left side of the section cut as shown below:

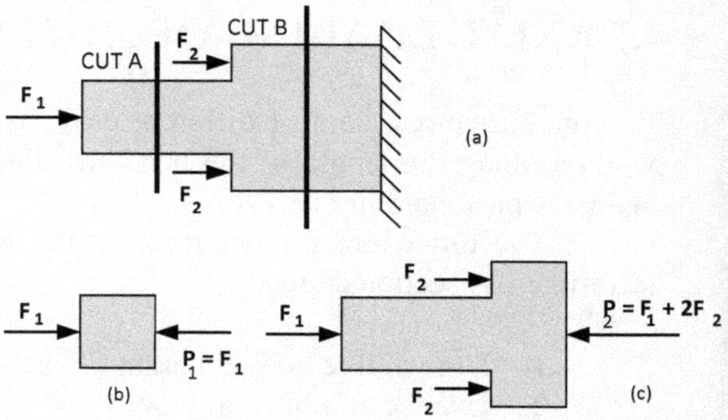

Fig. E4.21a FBDs of stepped sections.

From the equilibrium relations, it is clear that the internal force $P_1 = F_1$ in the smaller section of the stepped bar. Also, for the larger section of the bar, the internal force $P_2 = F_1 + 2F_2$.

Finally, from Eq. (4.4) for the smaller section of the bar, we write:

$$\sigma_1 = P_1/A_1 = -(40 \times 10^3)/[(50)(60)] = -13.33 \text{ N/mm}^2 = -13.33 \text{ MPa}$$

For the larger section of the bar, we write:

$$\sigma_2 = P_2/A_2 = -\{[40 + (2)(80)] \times 10^3\}/[(90)(60)] = -37.03 \text{ N/mm}^2 = -37.03 \text{ MPa}$$

Note, the minus sign indicates the stresses in the bar are compressive.

4.7.2 Deflection of Stepped Bars

To compute the axial extension or compression of the stepped bar, we consider each uniform section of the bar separately. Equation (4.2) is valid for each section because each has a uniform cross sectional area. For a stepped bar comprised of n uniform sections, we superimpose the individual deflections δ to obtain:

$$\delta_{total} = \delta_1 + \delta_2 + \delta_3 + \ldots.. + \delta_n \qquad (4.20)$$

EXAMPLE 4.22

If the stepped bar defined in Example 4.21 is fabricated from a titanium alloy with a modulus of elasticity E = 114 GPa, determine the total deflection of the bar.

Solution: We recognize that the deflection of the stepped bar is determined from Eq. (4.20) and Eq. (4.2) as:

$$\delta_{total} = \delta_1 + \delta_2 = (P_1 L_1)/(A_1 E) + (P_2 L_2)/(A_2 E) \qquad (a)$$

Since the modulus of elasticity is constant for both sections of the bar, Eq. (a) reduces to:

$$\delta_{total} = \delta_1 + \delta_2 = \frac{1}{E}\left[\frac{P_1 L_1}{A_1} + \frac{P_2 L_2}{A_2}\right] \qquad (b)$$

Substituting the numerical parameters for the unknown quantities in this relation yields:

$$\delta_{total} = \delta_1 + \delta_2 = \frac{1}{114 \times 10^9}\left[\frac{(-40 \times 10^3)(0.2)}{(0.05)(0.06)} + \frac{(-200 \times 10^3)(0.3)}{(0.09)(0.06)}\right] = -0.1209 \text{ mm}$$

Let's interpret this solution. The negative sign indicates the bar was compressed and the deformations reduced its length. The original length of the bar was $L_1 + L_2 = 500$ mm. When the total deformation of the bar is compared to this length, we find the deformation is very small—only 0.024%. This example again emphasizes that deformations of metallic members are usually very small. For this reason, we neglect these deformations and use the original lengths of structural members when substituting into the equilibrium equations.

4.8 STRESS CONCENTRATIONS

In our discussion of stresses in uniform, tapered or stepped bars, we have assumed that the stresses are uniformly distributed over the cross sectional area of the bar. For the uniform and the tapered bars, this is a valid assumption except near the ends of the bar where the external forces are applied. However, for the stepped bar, the stresses are not uniformly distributed across the section in the vicinity of the step. Indeed, whenever we encounter a discontinuity such as a step or a hole in a bar, the stresses tend to concentrate at that discontinuity. As a consequence, the use of Eq. (4.4) to compute the stresses seriously underestimates their actual value. We must account for the effect of the structural discontinuities by determining a suitable stress concentration factor. The topic of stress concentration factors will be introduced in the subsection below. However, since the topic is extremely important, the reader is referred to an excellent book by R. E. Peterson for more details [1].

4.8.1 Stress Concentration Factors Due to Circular Holes in Bars

Let's consider the bar with a centrally located circular hole subjected to an axial tension force as shown in Fig. 4.18. The stress distribution in a section removed three or more diameters from the hole is uniform with a magnitude given by Eq. (4.4) as $\sigma_o = P/(bw)$. However, on the section through the center of the hole the stress distribution shows significant variation. The stresses increase sharply adjacent to the discontinuity (the hole) and concentrate at this location. The maximum value of the normal stress occurs adjacent to the hole as indicated in Fig 4.19

We are interested in determining the maximum stress, σ_{MAX} adjacent to the hole. It is convenient to express the maximum stresses in terms of a stress concentration factor by employing:

$$\sigma_{MAX} = K\, \sigma_{NOM} \tag{4.21}$$

where K is the stress concentration factor and σ_{NOM} is the nominal stress.

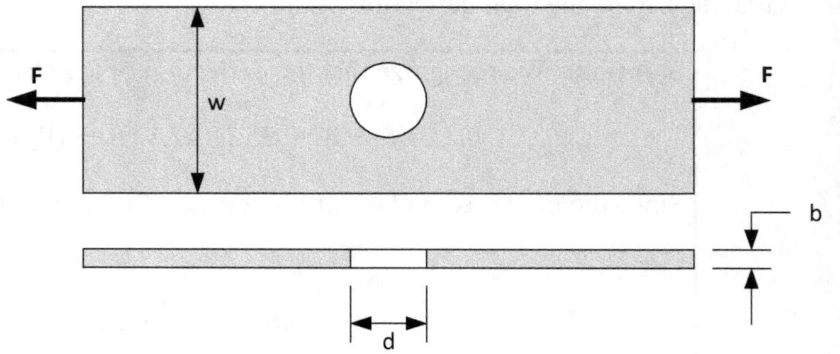

Fig. 4.18 A centrally located circular hole in an axially loaded bar.

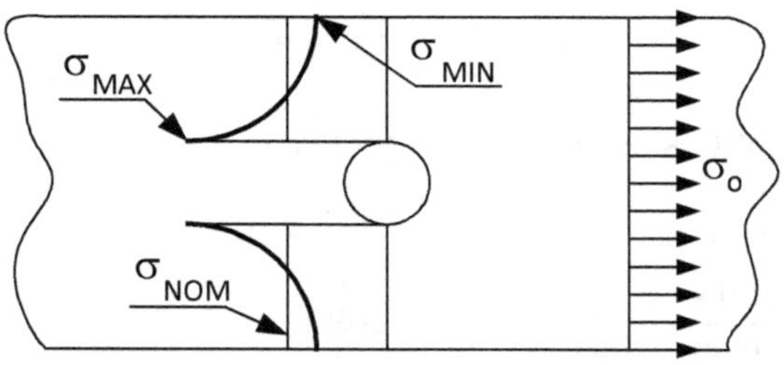

Fig. 4.19 Distribution of stress across a section through the center of the hole shows the concentration of stresses adjacent to the boundary of the hole.

The **stress concentration factor** K for a uniform thickness bar with a central circular hole subjected to axial loading is a function of the geometry depending on the ratio of d/w as shown in Fig. 4.20.

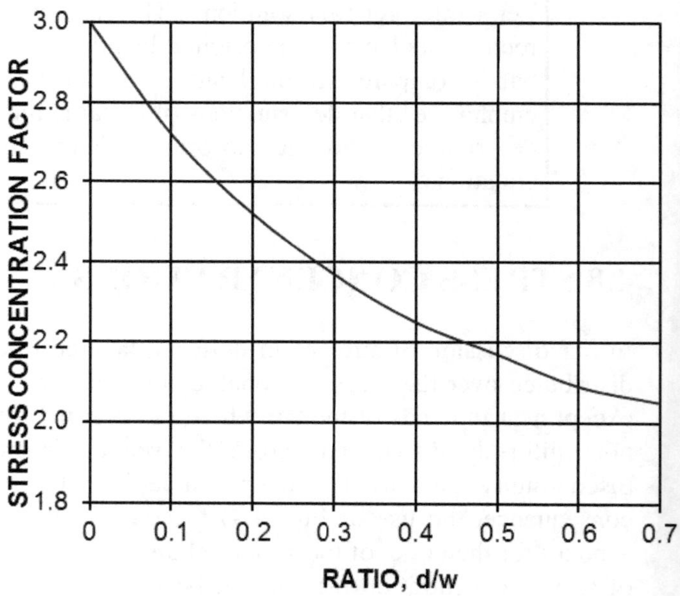

Fig. 4.20 Stress concentration factor for a central circular hole in an axially loaded bar.

The **nominal stress** is the average stress across the net section containing the hole, and is given by:

$$\sigma_{NOM} = P/A_{NOM} = P/[(w - d)b] \tag{4.22}$$

where b is the thickness of the bar, w is the bar width and d is the hole diameter.

The uniform stress σ_o and the nominal stress σ_{NOM} are related by:

$$P = \sigma_{NOM}\, A_{NOM} = \sigma_o\, A_{UNF} \tag{a}$$

Substituting for the areas in Eq. (a) and simplifying yields:

$$\sigma_{NOM} = [w/(w - d)]\sigma_o \qquad (4.23)$$

The nominal stress is always greater than the uniform stress since the factor w/(w – d) is always greater than one.

EXAMPLE 4.23

A centrally located hole of diameter d = 0.5 in. is drilled in a long thin bar that is subjected to an axial load of 8,000lb. If the bar is defined by w = 1.5 in., and b = 0.50 in., determine the nominal stress, the stress concentration, and the maximum stress.

Solution: The nominal stress is given by Eq. (4.22) as:

$$\sigma_{NOM} = P/[b(w - d)] = 8{,}000/[(1.5 - 0.5)(0.5)] = 16{,}000 \text{ psi}$$

The stress concentration K is determined from Fig. 4.20, by locating the intercept of a vertical line originating at d/w = 0.5/1.5 = 0.3333 with the curve. This intercept gives K = 2.33.

Finally, the maximum stress is given by Eq. (4.21) as:

$$\sigma_{MAX} = K\,\sigma_{NOM} = (2.33)(16{,}000) = 37.28 \text{ ksi}$$

By drilling a hole in the bar, we increase the maximum stress significantly. The procedure using the stress concentration factor provides a simple yet effective approach for solving a very difficult stress analysis problem. Also, you should be aware that the uniform stress in the bar prior to drilling the hole was σ_o = 10.67 ksi; therefore, the presence of the hole increased the stresses by a factor of 3.50.

4.8.2 Stress Concentrations in Shouldered Bars

The stepped bar is another configuration with an abrupt change in the section of the bar. This geometric discontinuity produces a non-uniform distribution of stresses and a concentration of stress at the fillet used in transitioning from one section of the bar to the other. The geometric parameters involved in characterizing the stress concentration factor are w_2/w_1 and r/w_1 as defined in Fig. 4.21.

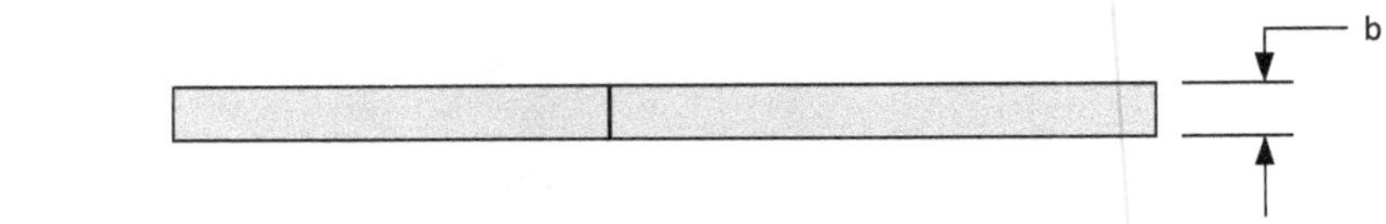

Fig. 4.21 Geometric parameters controlling the stress concentration in a stepped bar.

The maximum stress occurs at the fillet in the transition from the narrow section of the bar (w_1) to the wider section (w_2). The maximum stress is determined from Eq. (4.21). However, for the case of the stepped bar the nominal stress is defined as:

$$\sigma_{NOM} = P/(w_1 b) \qquad (4.24)$$

The stress concentration factor K for the stepped bar depends on two different ratios r/w_1 and w_2/w_1. Curves showing the stress concentration factor K as a function of these ratios are presented in Fig. 4.22.

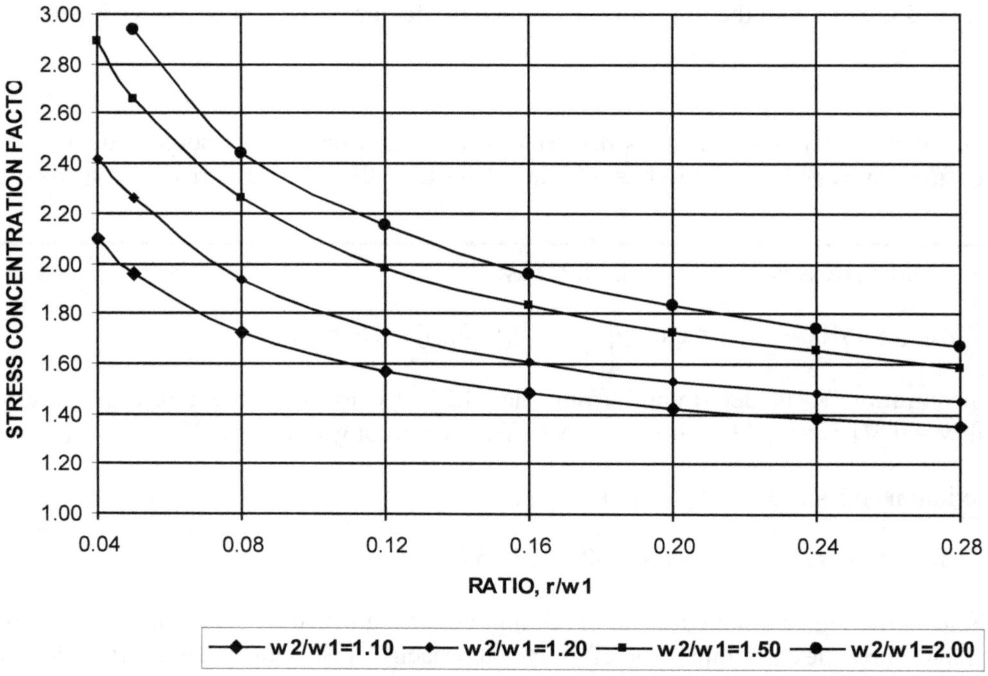

Fig. 4.22 Stress concentration factor for a stepped bar subjected to axial forces.

EXAMPLE 4.24

Consider the axially loaded stepped bar shown in Fig. E4.24, and determine the maximum stress at the fillet in the transition region of the bar.

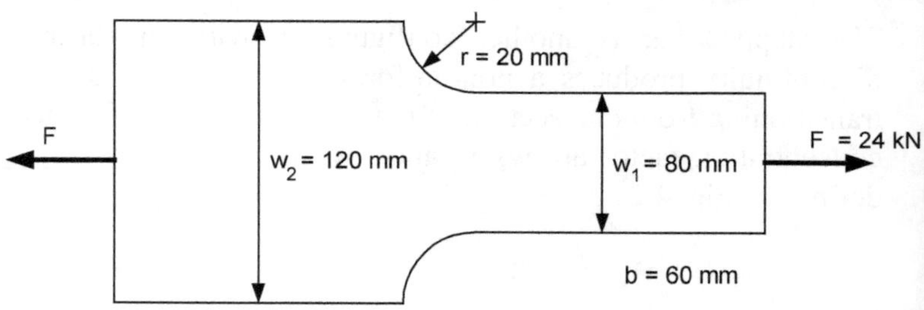

Fig. E4.24

Solution: From Eq. (4.21), we write $\sigma_{MAX} = K\,\sigma_{NOM}$.

From Eq. (4.24), write:

$$\sigma_{NOM} = P/w_1b = (24 \times 10^3)/[(80)(60)] = 5 \text{ MPa}$$

To determine the stress concentration factor K, we note that $r/w_1 = 20/80 = 0.25$, and $w_2/w_1 = 120/80 = 1.5$. Reading the chart for the stress concentration factor in Fig. 4.22, we find that K is:

$$K = 1.65 - (1/4)(1.65 - 1.58) = 1.63.$$

Finally, Eq. (4.21) yields:

$$\sigma_{MAX} = K\,\sigma_{NOM} = 1.63 \times 5 = 8.15 \text{ MPa}$$

4.9 SCALE MODELS

We have described methods for determining the stresses, strains and deflection of bars and rods subjected to axial tension or compression forces. The analysis required to determine these quantities is relatively simple when the bar can be isolated and the forces acting on it are known. However, if the bar is part of a complex structure, determining the forces acting on the bar may prove to be more difficult. In some cases, scale models of the structure under consideration are constructed and then subjected to various loads to verify structural integrity.

Geometric Scale Factor

Scale models of structures are usually much smaller than the real thing. Let's call the real structure—a bridge, building or stadium—the prototype, and consider a scale model that is 100 times smaller than the prototype. In this case, the geometric scale factor $\mathbf{S} = 1/100$. If the same scale factor S, is used for all three dimensions (i.e., length, width and thickness), we may write:

$$L_m = \mathbf{S}\, L_p$$

$$w_m = \mathbf{S}\, w_p \qquad (4.25)$$

$$b_m = \mathbf{S}\, b_p$$

where L, w and b represent length, width and thickness and subscripts m and p refer to the model and the prototype respectively.

We note from Eq. (4.25), that the scale model is exactly the same shape as the prototype except for its size. It is geometrically proportional.

Scaling Factor for Stresses

Let's continue with the concept of scaling and consider the stresses occurring in both the model and the prototype. The stress σ in a bar subjected to axial loading is given by Eq. (4.4) as:

$$\sigma = P/A = P/(w\, b) \qquad (4.26)$$

where P is the axial load applied to the bar, and $A = w\, b$ is the cross sectional area of a bar of width w and depth b.

If the left side of Eq. (4.26) is divided by its right side, a unit dimensionless ratio involving the stress and its controlling parameters is formed:

$$\sigma(w\, b)/P = 1 \qquad (4.27)$$

Let's designate a model and a prototype in Eq. (4.27) by writing:

$$\sigma_m\, (w_m\, b_m)/P_m = \sigma_p\, (w_p\, b_p\,)/P_p \qquad (a)$$

Rearranging the terms in Eq. (a) gives:

$$\sigma_m = (P_m/\, P_p)(w_p\, /w_m\,)(b_p\, /\, b_m\,)\sigma_p \qquad (b)$$

Substituting Eq. (4.25) into Eq. (b) yields:

$$\sigma_m = (P_m/\, P_p)(1/\mathbf{S}^2)\sigma_p \qquad (c)$$

Examining Eq. (c) shows that the stresses in the model and the prototype are related by the geometric scale factor $\mathbf{S}$ and load scale factor $\mathbf{L}$, which is defined by:

$$\mathbf{L} = (P_m / P_p) \qquad (4.28)$$

Finally, we substitute Eq. (4.28) into Eq. (c) and obtain:

$$\sigma_m = (\mathbf{L} / \mathbf{S}^2)\, \sigma_p \qquad (4.29)$$

Clearly, the stresses produced in the model are related to the scale factors for the load and the geometry. It is also important to recognize that the geometric scale factor usually differs significantly from the load scale factor.

EXAMPLE 4.25

Suppose we construct a scale model of the Golden Gate Bridge using a scale factor $\mathbf{S}$ of 1/500. Determine the size of the model if:

- Total bridge length⇒ 8981 feet
- Length of suspended structure⇒6450
- Length of main span⇒4200 feet
- Length of each side span⇒1125 feet
- Width of bridge⇒90 feet
- Diameter of main cable⇒36-3/8 inch
- Width of roadway between curbs⇒60 feet
- Lanes of vehicular traffic⇒6
- Weight of main span per lineal foot⇒21,300 lb
- Live load capacity per lineal foot⇒4,000 lb

Solution: The model bridge is constructed in three parts including the main span and the two side spans. The length of the main span of the model is given by Eq. (4.25) as:

$$L_m = \mathbf{S}\, L_p = (1/500)(4200) = 8.4 \text{ ft} = 100.8 \text{ in.}$$

The width of the model bridge is determined in the same manner as:

$$w_m = \mathbf{S}\, w_p = (1/500)(90) = 0.18 \text{ ft} = 2.16 \text{ in.}$$

The diameter, D of the main cable on the model is:

$$D_m = \mathbf{S}\, D_p = (1/500)(36.375) = 0.073 \text{ in.}$$

All geometric features of the model of the bridge are determined in this manner.

EXAMPLE 4.26

Let's suppose a very small strain gage is installed on a steel bar of a model used to simulate a critical structural member in a prototype. The gage provides a measurement of the strain equal to 4000×10^{-6} when the model bridge was fully loaded. If the scaling factor for the load $\mathbf{L} = 1/100{,}000$, determine the stress in the main cable of the prototype.

Solution: The stress acting on the bar in the model is given by Hooke's law as:

$$\sigma = E\varepsilon \qquad (4.5\ \text{bis})$$

Substituting numerical values for the modulus of elasticity and the strain into Eq. (4.5) gives the stress in the main cable of the model as:

$$\sigma_m = E_m\, \varepsilon_m = (30 \times 10^6)(4000 \times 10^{-6}) = 120{,}000 \text{ psi}$$

where the modulus of elasticity $E = 30 \times 10^6$ psi for steel.

Solving Eq. (4.29) for the stress on the prototype gives:

$$\sigma_p = (\mathbb{S}^2/\ \mathbb{L})\, \sigma_m \qquad (4.30)$$

Substituting the scale factors for $\mathbb{S}$ and $\mathbb{L}$ into Eq. (4.30) yields:

$$\sigma_p = (\mathbb{S}^2/\ \mathbb{L})\, \sigma_m = [(1/500)^2/(1/100{,}000)][120{,}000] = (10/25)(120{,}000) = 48{,}000 \text{ psi}$$

This example illustrates that the scale factors for the load and the geometry of the structure should be selected to limit the stresses induced in the model. In most instances, the scale factor for the load is much smaller than the scale factor for the geometry.

Scaling Factor for Displacements

A model of the structure may also provide displacement measurements that may be used to predict displacements in the prototype. To develop the displacement relation between the model and the prototype, we again seek a unit dimensionless quantity that includes the variables controlling the displacement. Recalling Eq. (4.2) it may be shown that the displacement δ of a rod of length L subjected to an axial load P is given by:

$$\delta = PL/AE = PL/[(w\ b)E] \qquad (4.31)$$

If the left side of Eq. (4.31) is divided by its right side, a unit dimensionless ratio is formed:

$$[\delta(w\ b)E]/(PL) = 1 \qquad (4.32)$$

Let's identify the model and the prototype with appropriate subscripts in Eq. (4.32) and write:

$$\delta_m\, (w_m\, b_m)\, E_m /(P_m\, L_m) = \delta_p\, (w_p\, b_p)\, E_p /(P_p\, L_p) \qquad (a)$$

Solving Eq. (a) for δ_p and substituting Eq. (4.25) and (4.28) into the result, yields:

$$\delta_p = [(\mathbb{S}\ \mathbb{E})/\ \mathbb{L}]\, \delta_m \qquad (4.33)$$

where $\mathbb{E} = E_m/E_p$ is the modulus scale factor between the model and the prototype.

If the same materials are employed in the manufacture of both the model and the prototype the modulus scale factor is one. However, selection of the model materials is not restricted. We may use a wide variety of materials to fabricate the model providing they respond in an elastic and linear manner.

EXAMPLE 4.27

Suppose that a model of a truss type bridge structure is fabricated from members formed from sheet aluminum. The prototype structure is to be fabricated from steel with a span of 600 feet. The model is geometrically scaled so that its span is six feet. The capacity of the live load on the prototype is 15,000 lb/ft, and the model is loaded with 25 lb/ft. If the model deflects a distance of two inches under full load at the center of the span, determine the deflection of the prototype under the design load.

Solution: From Eq. (4.33), we write:

$$\delta_p = (\mathbf{S}\,\mathbf{E}/\mathbf{L})\delta_m$$

Note the scale factors are given by:

$$\mathbf{S} = 6/600 = 1/100$$

$$\mathbf{L} = 25/15{,}000 = 1.667 \times 10^{-3}$$

$$\mathbf{E} = 10 \times 10^6 / 30 \times 10^6 = 1/3$$

Substituting the scale factors into Eq. (4.33) yields:

$$\delta_p = (\mathbf{S}\,\mathbf{E}/\mathbf{L})\delta_m = [(1/100)(1/3)/(1.667 \times 10^{-3})](2) = 4.0 \text{ in.}$$

In this instance, the deflection of the prototype is twice as large as the deflection of the model. The choice of materials for the model and the scaling factors for both the load and geometry influence the differences in the stress, and deflection between the model and the prototype.

4.10 STATICALLY INDETERMINATE AXIAL MEMBERS

Structures are considered to be statically determinate when the unknown forces, either internal forces or external reactions, can be determined using only the equilibrium relations. However, in some structures the number of unknown forces exceed the number of applicable equilibrium equations, and a solution is not possible without the introduction of additional equations. When the number of unknown forces exceeds the number of applicable equilibrium equations, the structure is **statically indeterminate.**

The approach for solving for the unknown forces in statically indeterminate structures is to use the equilibrium relations and then to introduce additional equations based on the deformation of the structure. The solution involves two separate but compatible parts:

1. Prepare the traditional FBD together with all of the applicable equations of equilibrium.
2. Prepare displacement diagrams together with deformation equations that are consistent with the forces shown on the FBD.

Let's consider an example to demonstrate this approach in solving for the unknown forces in a statically indeterminate structure.

EXAMPLE 4.28

A 14 ft. long column supporting the corner of a very large building is fabricated from a steel pipe with an outside diameter of 8 in. and an inside diameter of 7 in. The pipe is filled with medium strength concrete to form a composite column that supports a 100-ton axial load as illustrated in Fig. E4.28. Determine the forces and stresses in the steel pipe and concrete plug.

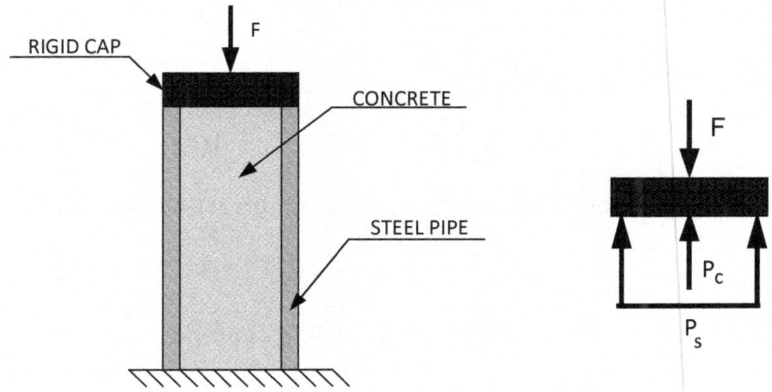

Fig. E4.28a

Solution:

Step 1: Prepare FBDs of the rigid cap, steel pipe and concrete plug as shown in Fig. E4.28a, and write the only applicable equilibrium equation, which is $\Sigma F_y = 0$.

$$\Sigma F_y = P_c + P_s - F = 0$$

$$P_c + P_s = 100 \text{ ton} \qquad \text{(a)}$$

where P_c and P_s are the internal forces in the concrete plug and steel pipe, respectively.

Because we have two unknown forces P_c and P_s and only a single applicable equation of equilibrium, the unknown forces are statically indeterminate.

Step 2: Prepare a drawing shown in Fig. E4.28b that depicts an exaggerated view of the deformations occurring in the concrete plug and the steel pipe when subjected to the internal forces P_c and P_s. Because the two structural elements deform together under the action of the axial loads, we write the deformation equation.

$$\delta_c = \delta_s \qquad \text{(b)}$$

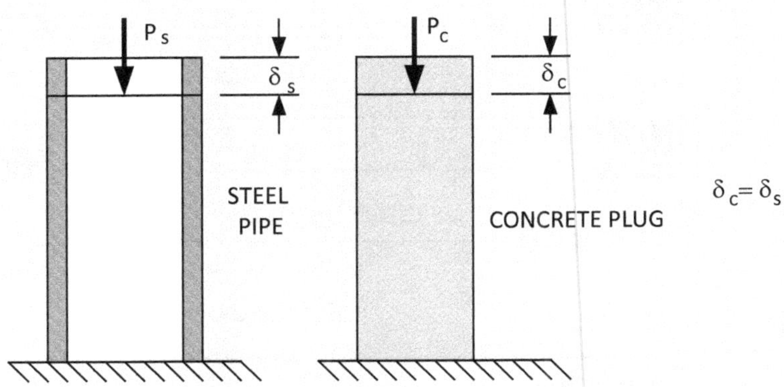

Fig. E4.28b

Recall Eq. (4.2) and write:

$$\delta_c = P_cL_c / A_cE_c = \delta_s = P_sL_s / A_sE_s \qquad \text{(c)}$$

Solve Eq. (c) for P_c noting that $L_c = L_s$, to obtain:

$$P_c = (A_c/A_s)(E_c/E_s)P_s \qquad \text{(d)}$$

Equation (d) provides the second relation needed to solve for the unknown internal forces.

Step 3: Solve for the internal force P_s by substituting Eq. (c) into Eq. (a) to obtain:

$$P_s = 100/[1 + (A_c/A_s)(E_c/E_s)] \quad \text{(e)}$$

The area ratio is given by:

$$A_c/A_s = [d_i^2/(d_o^2 - d_i^2)] = (7.0)^2/[(8.0)^2 - (7.0)^2] = 3.267 \quad \text{(f)}$$

Using values from Appendix B-1, the modulus ratio is given by:

$$E_c/E_s = 3.6/30 = 0.12 \quad \text{(g)}$$

Substituting the results from Eqs. (f) and (g) into Eq. (e) gives:

$$P_s = 71.84 \text{ ton} \quad \text{(h)}$$

Then from Eqs. (a) and Eq. (h), it is clear that:

$$P_c = 28.16 \text{ ton} \quad \text{(i)}$$

Step 4: Solve for the stresses in the concrete plug and steel pipe as:

$$\sigma_c = P_c/A_c = [4(28.16)(2000)]/[\pi(7.0)^2] = 1{,}463 \text{ psi}$$

$$\sigma_s = P_s/A_s = [4(71.84)(2000)]/\{\pi[(8.0)^2 - (7.0)^2]\} = 12{,}200 \text{ psi}$$

Note that both of these stresses are compressive.

EXAMPLE 4.29

A grade 8 steel bolt ¾ in. in diameter (10 threads/in.) is employed to clamp an aluminum bushing between two rigid platens as illustrated in Fig. E4.29. The aluminum bushing has a 2.0 in. outside diameter and 1.0 in. inside diameter and is 10.00 in. long. After the unit is assembled with a snug fit, the nut is tightened by ½ of a turn. Determine the axial stresses in the bolt and the bushing. Also determine the deflection of the aluminum bushing.

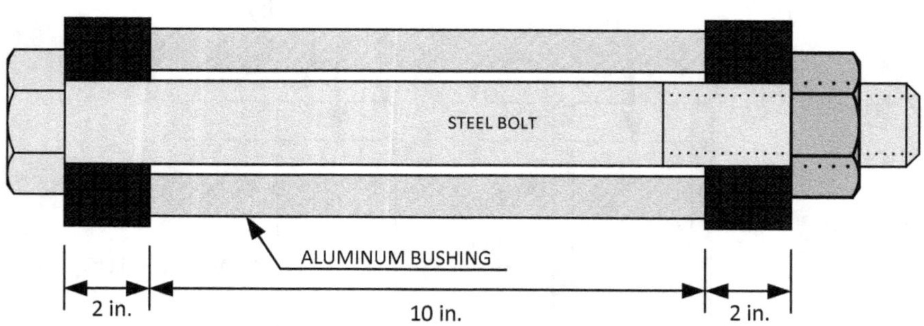

Fig. E4.29

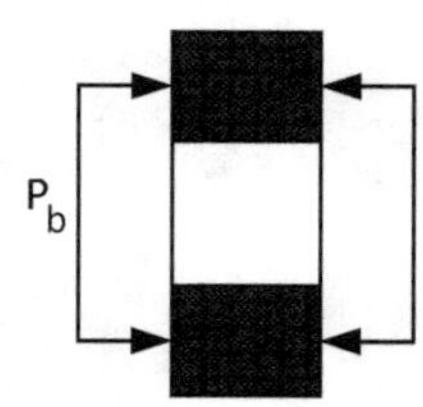

Fig. E4.29a

Solution:

Step 1: Prepare FBDs of one of the rigid platens as shown in Fig. E4.29a, and write the only applicable equilibrium equation, which is $\Sigma F_x = 0$.

$$\Sigma F_x = -P_a + P_b = 0$$

$$P_a = P_b \quad \text{(a)}$$

where P_a and P_b are the internal forces in the aluminum bushing and the steel bolt, respectively. They are equal in magnitude and opposite in sign. Because we have two unknown forces P_a and P_b and only a single applicable equation of equilibrium, the unknown forces are statically indeterminate.

Step 2: Prepare a drawing shown in Fig. E4.29b that depicts an exaggerated view of the deformations occurring in the bolt and the aluminum bushing when subjected to the internal forces P_a and P_b. When the nut is tightened the steel bolt tends to stretch and the aluminum bushing is compressed (shortened). The total of the two deformations must equal the amount of displacement induced by rotating the nut by ½ of a turn, which is (1/2)(1/10) = 0.050 in.

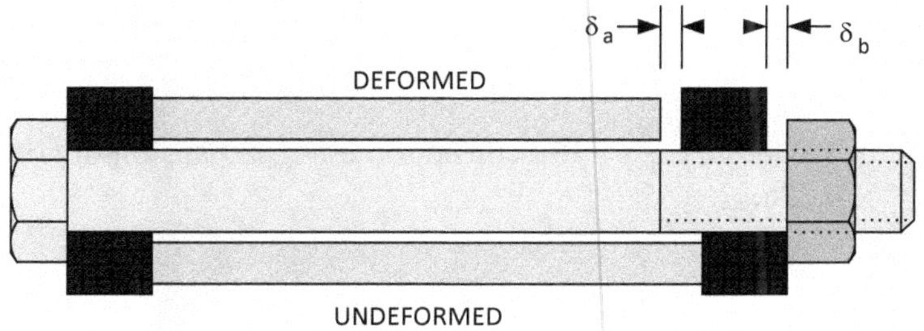

Fig. E4.29b

$$\delta_a + \delta_b = .050 \text{ in.} \qquad \text{(b)}$$

Step 3: Recall Eq. (4.2) and Eq. (a) and write:

$$\delta_a = P_aL_a/A_aE_a = P_b\, L_a/A_aE_a = [(4)(10)/\{\pi[(2)^2 - (1)^2](10.4 \times 10^6)\}]\, P_b = 0.4081 \times 10^{-6}\, P_b$$

$$\delta_b = P_bL_b/A_bE_b = \{(4)(14)/[\pi(0.75)^2\,(30 \times 10^6)]\}P_b = 1.0563 \times 10^{-6}\, P_b \qquad \text{(c)}$$

Substitute Eq. (c) into Eq. (b) and solve for P_b.

$$P_b\,[0.4081 + 1.0563] = 0.050 \times 10^6$$

$$P_a = P_b = 34{,}140 \text{ lb} \qquad \text{(d)}$$

Step 4: The stresses are given by Eq. (4.4) as:

$$\sigma_b = P_b/A_b = (4)(34{,}140)/[\pi(0.75)^2] = 77.290 \text{ ksi (tension)}$$

$$\sigma_a = P_a/A_a = (4)(34{,}140)/[3\pi] = 14.490 \text{ ksi (compression)}$$

Step 5: The amount of the compression of the aluminum bushing is given by Eq. (4.2) as:

$$\delta_a = P_aL_a/A_aE_a = [(34{,}140)(10)(4)]/[(3\pi)(10.4 \times 10^6)] = 0.01393 \text{ in. (shorter)}$$

These two examples demonstrate the approach employed to solve statically indeterminate problems. This approach involves two important steps—writing the applicable equations of equilibrium based on an accurate FBD and writing a deformation equation based on a drawing showing the deformation of the structural elements involved.

4.10.1 Thermal Stresses

Thermal stresses are produced in structural elements by the constraint of the free expansion of a material subjected to a temperature change. To illustrate this fact, consider a long thin rod of length L_o at a temperature T_o. When the temperature of the rod increases to say T_1, the rod undergoes a free expansion with an attendant increase in length δ_T given by:

$$\delta_T = L_o\,\alpha(T_1 - T_o) = L_o\,\alpha\Delta T \qquad (4.34)$$

where α is the thermal coefficient of expansion.

It is clear from the definition of strain and Eq. (4.3) that a thermal strain ε_T accompanies a free expansion.

$$\varepsilon_T = \delta_T / L_o = \alpha \Delta T \tag{4.35}$$

Thermal stresses in a free expansion process are zero because the temperature change does not generate the internal forces P required for thermal stresses. Thermal strains induce thermal stresses if, and only if, the structural element is constrained in some manner. When the structural element is totally constrained and free expansion is not permitted, the thermal stresses in the uniaxial rod are given by:

$$\sigma_T = E \alpha \Delta T \tag{4.36}$$

To derive Eq. (4.36), consider the free expansion of a long thin rod due to a temperature change ΔT as shown in Fig. 4.23a.

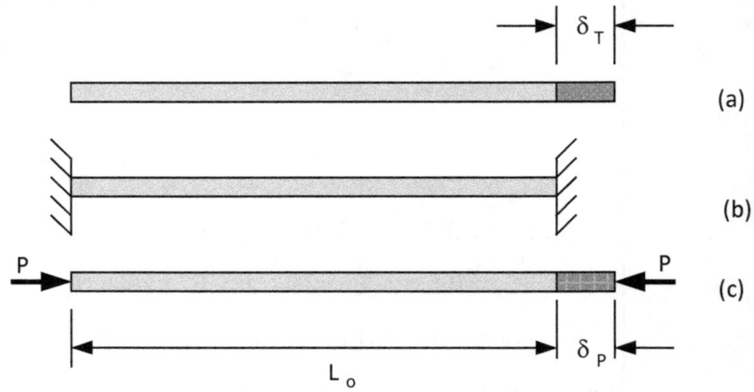

Fig. 4.23 Rod with free expansion, constraint and internal forces P.

If the rod is constrained within rigid walls as illustrated in Fig. 4.23b, the length remains at L_o after the temperature change ΔT. As the rod attempts to expand against the constraint offered by the rigid walls, an internal force P is produced as shown in Fig. 4.23c. The internal force P is sufficiently large to produce an axial displacement (contraction) equal to the thermal expansion. Hence, we can write:

$$\delta_T = \delta_P \tag{4.37}$$

Substituting Eqs. (4.2) and (4.34) into Eq. (4.37) and simplifying yields:

$$\sigma_T = P/A = E \alpha \Delta T \tag{4.36bis}$$

The results obtained by applying Eq. (4.36) should be considered as an upper bound on the thermal stresses generated by a temperature change ΔT. The derivation of this relation assumed total constraint of the free expansion of the bar. In practice, total constraint is difficult to achieve and because the displacements are small, even slight relief in the constraint markedly reduces the magnitude of the thermal stresses.

We have included a table showing the coefficient of thermal expansion, modulus of elasticity and thermal stresses for a ΔT = 100° F or 100° C in Table 4.1.

Table 4.1
Coefficient of thermal expansion, modulus of elasticity and thermal stress induced by ΔT = 100° F or 100 ° C in various engineering materials.

Material	$E \times 10^6$ psi	$\alpha \times 10^{-6}$/°F	σ_T (ksi)	$E \times 10^9$ (Pa)	$\alpha \times 10^{-6}$/°C	σ_T (MPa)
Steel	30	6.3	18.90	207	11.3	233.9
Aluminum	10.4	12.9	13.42	72	23.2	167.0
Brass	16	11.1	17.76	110	20.0	220.0
Stainless Steel	27.5	9.6	26.40	190	17.3	328.7
Titanium	16.5	4.9	8.09	114	8.8	100.3

EXAMPLE 4.30

Two heavy blocks of steel are permanently fastened together by a steel shrink link that is fitted into pockets machined into the two blocks as illustrated in Fig. E4.30. If a shrink link has a body length of 100.0 mm at a temperature of 275° C, determine the clamping force provided by the shrink link at an ambient temperature of 20° C. Also determine the stress in the link after it has cooled to ambient temperature. The body of the link has a square cross section with a side dimension of 50 mm.

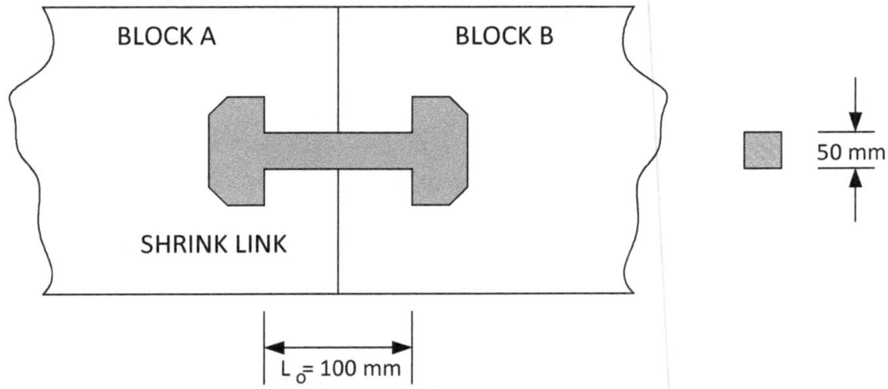

Fig. E4.30

Solution:

Step 1: Upon cooling to ambient temperature the link attempts to shrink an amount δ_T given by Eq. (4.34) as:

$$\delta_T = L_o\, \alpha\Delta T = 100(11.3 \times 10^{-6})(275 - 20) = 0.2882 \text{ mm} \qquad \text{(a)}$$

Step 2: We assume that the free contraction of the link is totally constrained by the two steel blocks. This constraint condition yields:

$$\delta_T = \delta_P \qquad \text{(b)}$$

Substituting Eq. (4.2) into Eq. (b) gives the internal force P developed in the shrink link as:

$$P = (\delta_P AE)/L = [(0.2882)(2500)(207 \times 10^3)]/100 = 1491 \text{ kN} \qquad \text{(c)}$$

Step 3: The thermal stress induced in the link by the $\Delta T = 255°$ C is given by Eq. (4.36) as:

$$\sigma_T = P/A = E\alpha\Delta T = (1491 \times 10^3)/(0.050)^2 = 596.4 \text{ MPa} \qquad \text{d)}$$

EXAMPLE 4.31

A thin retaining ring, fabricated from stainless steel is fitted over the polished steel journal of a solid shaft 7.000 inches in diameter. The outside diameter of the retaining ring is 7.375 in. and its length is 0.50 in. To fit the ring over the journal it is heated to 350 °F and slid on the shaft to its designated position. The ring becomes snug at a temperature of 295 °F. Determine the stress in the retaining ring and the clamping pressure between the ring and the journal.

Solution:

Step 1: Upon cooling from the snug temperature of 295° F to the ambient temperature of 75° F, the ring attempts to shrink diametrically by an amount δ_T given by Eq. (4.34) as:

$$\delta_T = d_o\, \alpha\Delta T = (7.000)(9.6 \times 10^{-6})(295 - 75) = 14.78 \times 10^{-3} \text{ in.} \qquad \text{(a)}$$

Step 2: We assume that the free contraction of the retaining ring is totally constrained by the polished journal of the solid steel shaft. This constraint condition yields:

$$\delta_T = \delta_P \quad \text{(b)}$$

Consider the circumference $C = \pi d_o$ of the retaining ring. Then the free contraction ΔC of the circumference of the ring is given by:

$$\Delta C = \pi \Delta d = \pi \delta_T = \pi \delta_P \quad \text{(c)}$$

The increase in the circumference of the ring due to the generation of an internal force P is obtained by substituting into Eq. (4.2) as:

$$\Delta C = (P\pi d)/(AE) = \pi \delta_T \quad \text{(d)}$$

Solving Eq. (d) for the internal force P in the ring gives:

$$P = (\delta_T AE)/d_o = (14.78 \times 10^{-3})(0.375/2)(0.5)(27.5 \times 10^6)/7.000 = 5.444 \text{ kip} \quad \text{(e)}$$

Step 3: The thermal stress induced in the retaining ring due to $\Delta T = 220°$ F is given by Eq. (4.36) as:

$$\sigma_T = P/A = E\alpha\Delta T = (27.5 \times 10^6)(9.6 \times 10^{-6})(220) = 58.08 \text{ ksi} \quad \text{(f)}$$

Step 4: To determine the interfacial pressure between the ring and the journal, consider the FBD of the ring shown in Fig. E4.31.

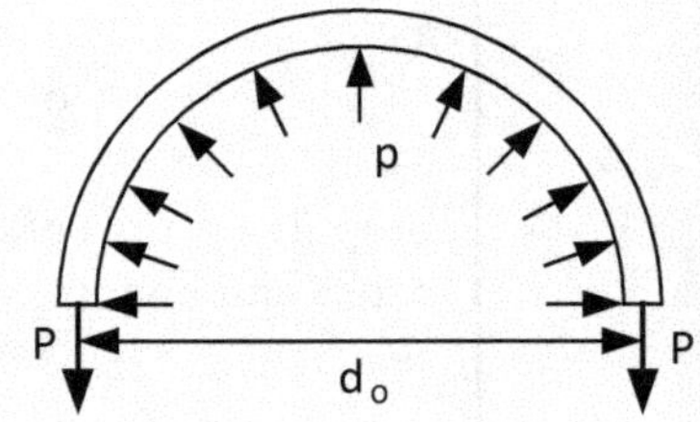

Fig. E4.31

Writing $\Sigma F_y = 0$ gives an expression containing the interfacial pressure p as:

$$\Sigma F = p(db) - 2P = 0 \qquad p = 2P/(db) \quad \text{(g)}$$

where b = 0.5 in. is the width of the ring

The interfacial pressure is then given by:

$$p = 2(5.444 \times 10^3)/(7.0)(0.5) = 3{,}111 \text{ psi} \quad \text{(h)}$$

4.11 SUMMARY

This chapter treats structural members that are long and thin. Flexible members, such as wire and cable, only support axial tensile loads. They cannot support compressive forces because they buckle under very low loads. In addition, they cannot support transverse loads because their transverse stiffness is also negligible. Under increasing tension forces applied along the axis of these long, thin, and flexible members, they stretch, yield and finally rupture.

Rods and bars are sufficiently stiff to carry compressive forces. In this chapter, we consider only axial loading of the bars and rods. The case of transverse loading is deferred until Chapter 10. The equations derived for stresses and deflections in wire and cable are also applicable to bars and rods. When dealing with compressive forces and stresses, a minus sign is used to indicate the direction of the loading. We also assume compressively loaded bars and rods are sized so they will not fail by buckling. Examples are presented for determining stresses and deflections of uniform axial bars.

Relations have been derived to determine the stretch (axial deformation). Stress and strain have been defined and the stress-strain relation (Hooke's law) has been discussed. Several examples have been provided to guide you in determining, stress, strain, and axial extension.

We have shown that the internal force P is equal to the external force F for these uniaxial members. The internal force is produced by normal stresses that are uniformly distributed over the cross section of the uniaxial member.

We describe failure by yielding and rupture and introduce the stress-strain curve for different types of materials in the process. Ultimate tensile strength and yield strength are defined. Safe design philosophy is discussed together with the use of factors of safety in design. Examples are provided to demonstrate problem-solving techniques dealing with strength and safety factors. The importance of specifying standard sizes of structural members in design is illustrated in one of the examples.

The concept of models that are geometrically similar to actual structures is introduced. The stresses and displacements produced by loading or deforming a model are related to those developed in the prototype (structure). Relations involving scaling factors are employed to determine the stresses and displacements in the structure based on measurements made on a geometrically scaled model.

The equations developed for uniform members subjected to axial forces are summarized below:

$$\delta = L_f - L_o \qquad (4.1)$$

$$\delta = (PL)/(AE) \qquad (4.2)$$

$$\varepsilon = \delta/L \qquad (4.3)$$

$$\sigma = P/A \qquad (4.4)$$

$$\sigma = E\varepsilon \qquad (4.5)$$

$$\varepsilon = P/(AE) \qquad (4.6)$$

$$P_{INT} = \int \sigma \, dA \qquad (4.7)$$

$$P_{INT} = \sigma \int dA = \sigma A \qquad (4.8)$$

$$S_u = \sigma_{fail} = P_{fail} / A \qquad (4.9)$$

$$S_y = \sigma_{yield} = P_{yield} / A \qquad (4.10)$$

$$\mathbf{SF_u} = S_u / \sigma_{design} \qquad (4.11)$$

$$\mathbf{SF_y} = S_y / \sigma_{design} \qquad (4.12)$$

$$\tau = V/A \qquad (4.13)$$

Stresses on a section with an inclined cut at an angle θ were determined using the equilibrium relations and the simple definition for normal and shear stresses as:

$$\sigma_\theta = (\sigma_x)\cos^2\theta \qquad (4.14)$$

$$\tau_\theta = (\sigma_x/2)\sin 2\theta \qquad (4.15)$$

In Fig. 4.14, we note that both the normal stress σ_θ and the shear stress τ_θ vary with the angle θ. The normal stress is a maximum when $\theta = 0°$, and the section cut is perpendicular to the axis of the bar. The shear stress vanishes on this section. The shear stress is a maximum when $\theta = 45°$. On this plane, $\sigma_{45} = \tau_{45} = \sigma_x /2$.

Tapered bars subjected to axial loading were described. For the stresses, Eq. (4.4) was modified to account for the variable cross sectional area as a function of position x along the length of the tapered bar.

$$\sigma_x = P/A_x \qquad (4.17)$$

For deflection of the tapered bar, we considered an incremental length dx along the length of the bar and modified Eq. (4.2) to read as:

$$d\delta = P\,dx/(A_x\,E) \qquad (4.18)$$

To determine the deflection δ of the tapered bar, we write an expression for A_x and integrate Eq. (4.18) from zero to L. Examples are provided to demonstrate the computation technique for tapered bars.

We have also described techniques for determining the stresses in stepped bars. The stresses in individual sections of the bar are computed from Eq. (4.4) using the appropriate cross sectional area for the section under consideration. Displacements of each section of the bar are determined from Eq. (4.2) and then superimposed to give the total deflection as shown in Eq. (4.20).

$$\delta_{total} = \delta_1 + \delta_2 + \delta_3 + \ldots\ldots + \delta_n \qquad (4.20)$$

Stress concentrations that develop at discontinuities in bars and rods have been discussed. These stress concentrations occur for all types of loading: axial, transverse and torsion. The stress concentration increases the maximum stresses by a significant amount. We determine the maximum stresses from:

$$\sigma_{MAX} = K\,\sigma_{NOM} \qquad (4.21)$$

The stress concentration factor K is defined for a bar with a central circular hole in Fig. 4.20 and for a stepped bar in Fig. 4.22. The nominal stresses for these two discontinuous bars are given by:

$$\sigma_{NOM} = P/(w - d)b \qquad (4.22)$$

$$\sigma_{NOM} = P/w_1 b \qquad (4.24)$$

Examples demonstrating the method for determining the stress concentration factor, and the nominal and maximum stresses are provided.

Modeling, an approach to experimentally determine stresses and displacements in large complex structures, was introduced. The relations for the stresses and displacement between the model and the structure (prototype) are given by:

$$\sigma_m = (\mathbf{L} / \mathbf{S}^2)\,\sigma_p \qquad (4.29)$$

$$\delta_p = [\,(\mathbf{S}\,\mathbf{E})/\,\mathbf{L}]\,\delta_m \qquad (4.33)$$

Structures are considered to be statically determinate when the unknown forces, either internal forces or external reactions, can be determined using only the equilibrium relations. However, in some structures the number of unknown forces exceed the number of applicable equilibrium equations, and a solution is not possible without the introduction of additional equations. When the number of unknown forces exceeds the number of applicable equilibrium equations, the structure is **statically indeterminate.** The approach for solving for the unknown forces in statically indeterminate structures is to use the equilibrium relations and then to introduce additional equations based on the deformation of the structure. The solution involves two separate but compatible parts:

1. Prepare the traditional FBD together with all of the applicable equations of equilibrium.
2. Prepare displacement diagrams together with deformation equations that are consistent with the forces shown on the FBD.

Thermal stresses are produced in structural elements by the constraint of the free expansion of a material subjected to a temperature change. When the free expansion or contraction is totally constrained the thermal stresses are given by:

$$\sigma_T = E\alpha\Delta T \tag{4.36}$$

REFERENCES

1. Peterson, R. E., Stress Concentration Design Factors, 2nd Ed., Wiley & Sons, New York, 1990.

PROBLEMS

4.1 A steel music wire (Gage No. 10 with a diameter of 0.024 in.) is employed to support an axial load P of 8.64 lb. If the wire was initially 22 ft in length, determine its final length and the stress and strain in the wire.

4.2 A steel music wire is tightened on a guitar by rotating the wire supporting post through an angle of 11°. If the post has a diameter d = 0.25 in. and the length of the wire is 32 in., determine the strain induced in the guitar string.

4.3 Referring to Problem 4.2, determine the additional stress induced in the guitar string as it is tightened.

4.4 A hemp rope 15 m long and 45 mm in diameter is supported from one of the roof beams in a gymnasium. Three gymnasts climb this rope together. The lead gymnast has a mass of 55 kg, the next 72-kg and the trailing gymnast 52 kg.

(a) Determine the maximum stress in this rope.
(b) Determine the stress in the rope at a location between the lead and intermediate gymnast.
(c) Determine the stress in the rope at a location between the intermediate and trailing gymnast.

4.5 A suspension bridge is to carry a roadway and traffic that may weigh up to 5,000 ton. If the safe load that can be imposed on the wire rope to be used in the construction is 175 ton, specify the number of cables to be employed. Justify your answer.

4.6 Describe the constraints on the relation $\sigma = E\varepsilon$.

4.7 Determine the design load that can be specified for Gage No. 00 steel wire with a strength of 530 MPa, if the design specification calls for the safety factor of 2.5.

4.8 Discuss the factors to consider in establishing the safety factor in a design specification.

4.9 For the cable-pulley-mass arrangement shown in the figure to the right, determine the stresses in the cable. The effective cable diameter d is 0.500 in.

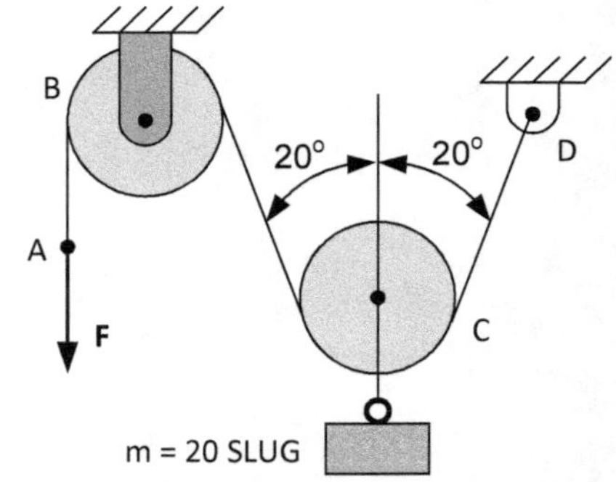

4.10 Cold drawn steel alloy wire exhibits an ultimate tensile strength of 600 MPa and a yield strength of 500 MPa. If this alloy is employed, specify the gage of the wire if it is to support a design load of 7.5 kN with a safety factor of 3.2. Specify the gage number required based on both yielding and rupturing as modes of failure.

4.11 Determine the safety factor for the wire CD in the cable-weight arrangement shown in the figure to the right. The steel cables are each 10 m long, exhibit a yield strength of 500 MPa and have an effective cross sectional area of 8 mm^2. Assume that the forces in the horizontal wires are equal in magnitude and opposite in direction ($F_{CB} = -F_{CA}$). Also assume $F_{CB} = (0.18)F_{CD}$.

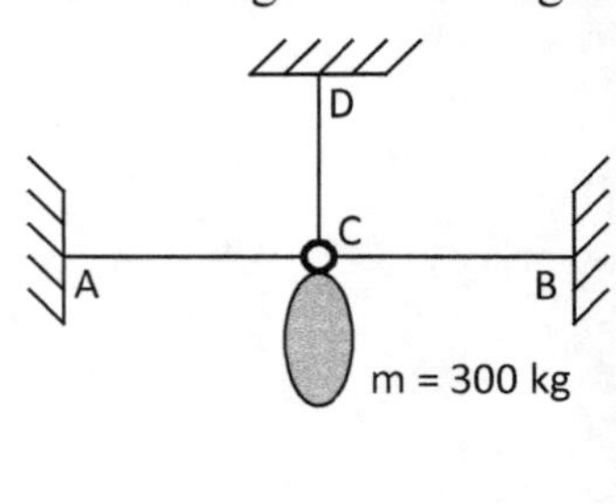

4.12 Determine the mass m, shown in the diagram to the right, that is required to yield the cable CD if the cables are each 22 ft long, exhibit a yield strength of 80ksi and have an effective cross sectional area of 0.015 in.^{2} Assume that the forces in the horizontal wires are equal in magnitude and opposite in direction ($F_{CB} = - F_{CA}$). Also assume $F_{CB} = (0.25)F_{CD}$. Is the cable-weight arrangement stable after yield? Why?

4.13 The wire-mass system, shown in the figure to the left, is constructed using steel wire. Specify a suitable alloy and the gage number for the wires if the criterion for failure is based on yielding. Use the same diameter for both wires. The safety factor imposed on the design is 2.9.

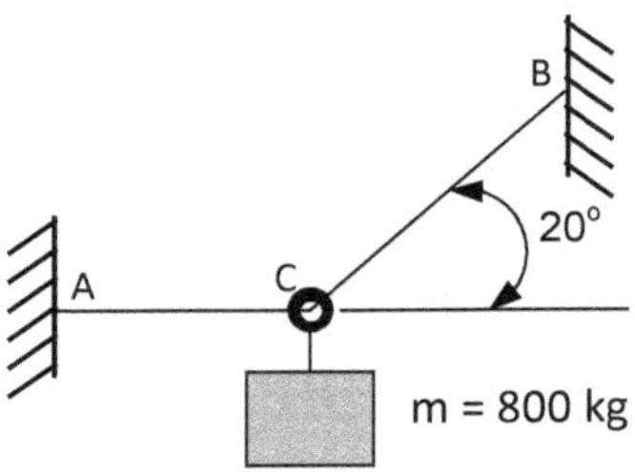

4.14 The wire-mass system, shown in the figure to the right, is constructed using aluminum wire. Specify a suitable alloy and the gage number for the wires if the criterion for failure is based on ultimate tensile strength. Use the same diameter for both wires. The safety factor imposed on the design is 1.8.

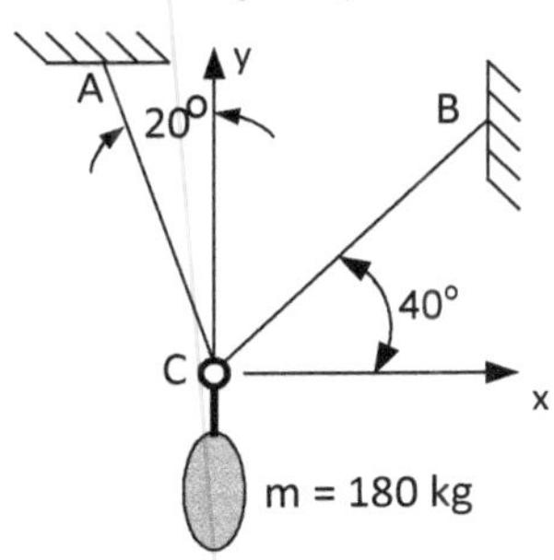

4.15 The wire-mass system ,shown in the figure to the left, is constructed using steel wire fabricated from 1010 A. A safety factor of 2.9, based on the ultimate tensile strength, is to be employed for both wires. Find the maximum weight that can be lifted (in pounds) and specify the required gage numbers for each wire.

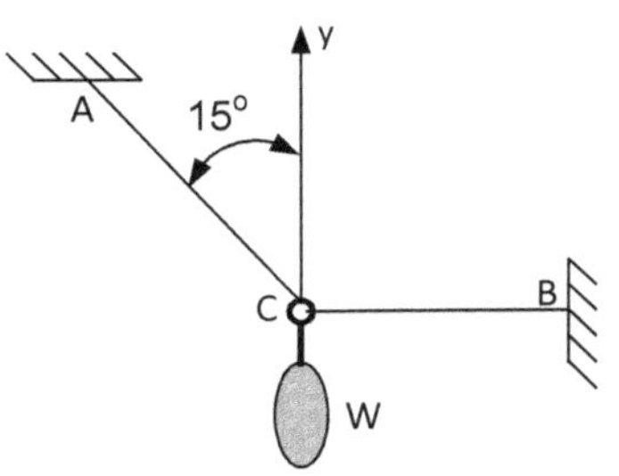

4.16 Describe in an engineering brief the differences between a wire rope and a rod.

4.17 Describe in an engineering brief the similarities between a wire rope and a rod.

4.18 A long thin steel bar with a length of 1.8 m, a width of 50 mm, and a thickness of 20 mm is subjected to an axial force of 22 kN. Determine the tensile stress and the axial deformation of the bar.

4.19 A bar fabricated from steel with a tensile strength of 54.0 ksi is subjected to an axial tensile force of 20 kip. The bar is designed with a safety factor of 3.2. Determine the design stress for the bar and the required cross sectional area.

4.20 A bar fabricated from an aluminum alloy with a tensile strength of 400 MPa is subjected to an axial tensile force of 75 kN. The bar is designed with a safety factor of 2.5. Determine the design stress for the bar and the required cross sectional area.

4.21 A bar fabricated from brass with a tensile strength of 247 MPa is subjected to an axial compressive force of 76 kN. The bar is designed with a safety factor of 2.8. Determine the design stress for the bar and the required cross sectional area. Assume the tensile and compressive stresses are equal.

4.22 A key used to lock a gear onto a shaft is subjected to a shear force of 8.0 kip. If the key is 0.25 in. wide by 0.375 in. high and 1.5 in. long, determine the shear stress acting on the key. Draw a free body diagram showing this shear stress.

4.23 If the key in Problem 4.22 is machined from a steel alloy with a tensile yield strength of 48,000 psi, determine the safety factor for the key.

4.24 A rectangular bar, shown in the figure to the right, is subjected to an axial tensile force F = 30 kN. Determine the normal stress and the shear stress on an inclined plane if the angle of the section cut is $\theta = 30°$.

4.25 A rectangular bar, similar to the one shown in the figure to the right, is adhesively bonded along a 20° section cut. The normal stress in the bond line is limited to 2.5 ksi and the shear stress to 1.5 ksi. If the bar has a cross sectional area A = 1.75 in.^{2} and a safety factor of 3.0 is specified, determine the largest axial force that can be applied to the bar.

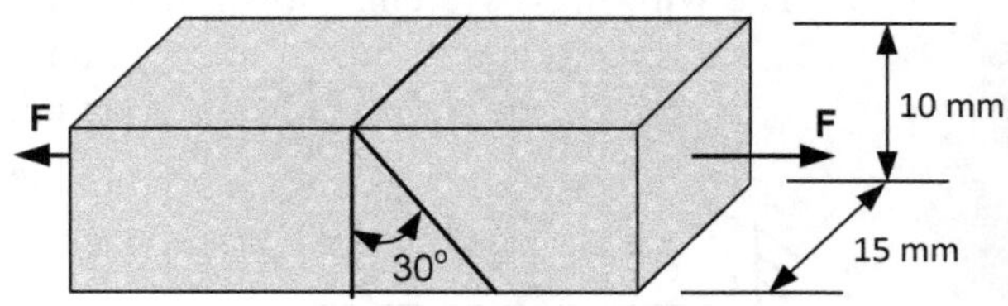

4.26 A rod with a circular cross section with a diameter d is fabricated from two pieces as illustrated in the figure to the right. If a force F is applied to the bar, derive the expression for the normal and shear stress acting in the plane of the adhesive joint as a function of angle ϕ.

4.27 For the bar illustrated in the figure to the right, determine the shear stresses in the adhesive joint if the angle ϕ is varied from 0° to 90°. The axial force applied to the bar is 2.0 kN and its diameter is 20 mm. Also compute the normal stresses acting on the adhesive joint. Hint: Use a spreadsheet to determine the shear and tensile stresses in the adhesive joint for the range of ϕ specified.

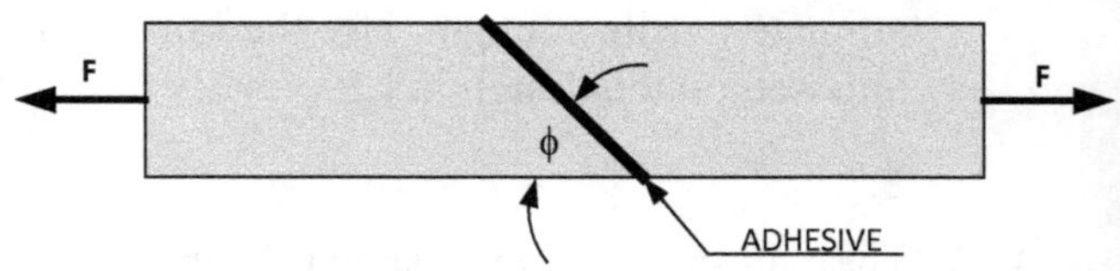

4.28 A tensile bar, defined in the figure to the right, has a cross sectional area of 175 mm^2 and is subjected to a tensile force F. The stresses on the inclined plane A—B are $\sigma_\theta = 81$ MPa and $\tau_\theta = -27$ MPa. Determine the stress σ_x, the angle θ and the force F.

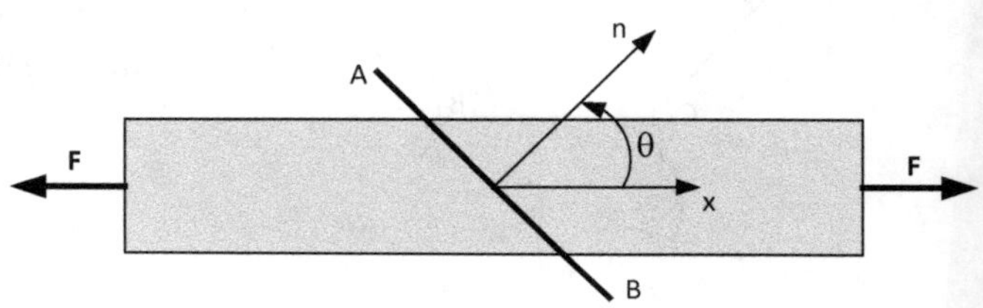

4.29 The tensile bar, defined above and in the figure to the right, has a cross sectional area of 175 mm^2 and is subjected to a tensile force F. The stresses on the inclined plane A—B are $\sigma_\theta = 81$ MPa and $\tau_\theta = -27$ MPa. Determine the shear and normal stresses acting on an inclined plane with $\theta = 40°$.

4.30 For the tensile bar in Problem 4.29, prepare a graph that shows the stresses σ_θ and τ_θ as a function of the angle of the inclined cut as θ varies from 0 to 90°.

4.31 Consider the tapered bar presented in the figure below and to the right . The tapered bar is subjected to an axial force of 500 kN. Prepare a graph of the axial stress σ_x as a function of x as it varies from zero to 4 m.

4.32 Consider the tapered bar presented in Problem 4.31. If the bar is fabricated from an aluminum alloy, determine the extension of the bar when it is subjected to an axial tensile force of 500 kN.

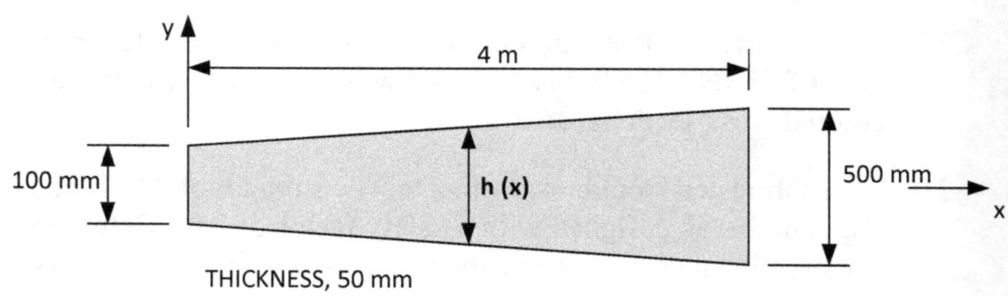

4.33 Consider the tapered bar presented in the figure to the right. If the bar is fabricated from an aluminum alloy, determine the extension of the bar when it is subjected to a tensile force of 120 kN. In this solution, do not employ Eq. (4.19). Instead, use Eq. (4.18) and perform a numerical integration on a spreadsheet.

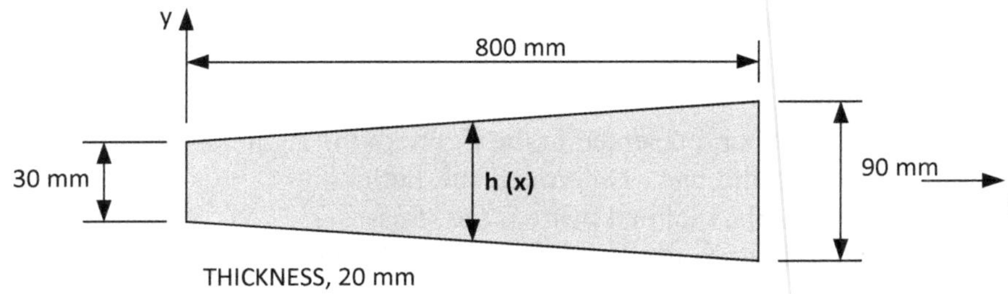

4.34 Describe the procedure employed to solve for the stresses and deflection in a stepped bar subjected to axial loading.

4.35 For the stepped bar, illustrated in the figure to the right, determine the stresses in both portions of the bar and its total deflection. The bar is fabricated from steel.

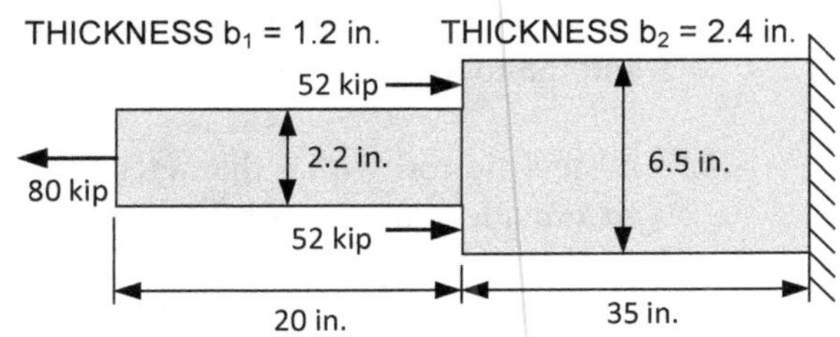

4.36 A tensile bar 40 in. long, shown in the figure below, is subjected to an axial force of 10 kip. Determine: (a) the nominal stress, (b) the stress concentration factor K, and (c) the maximum stress.

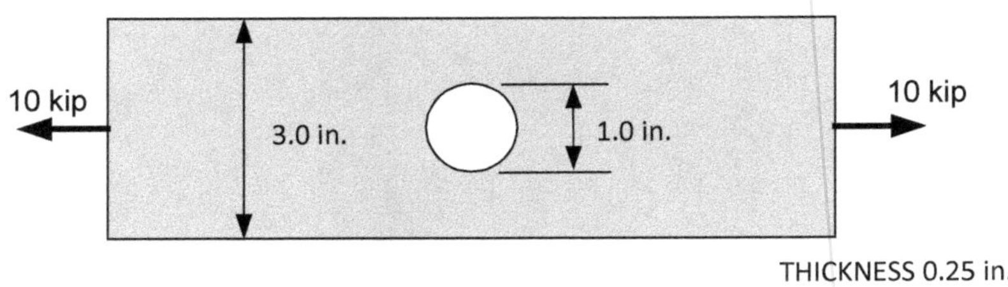

4.37 A stepped tensile bar, illustrated in the figure below, with fillets at the transition between the small and the large section is subjected to an axial tensile force of 75 kN. Determine the maximum stress at the fillets.

4.38 A scale model of a large structure has been fabricated from steel and tested. A strain gage on one member of the structure indicated an axial strain of $\varepsilon = 1{,}450$ μm/m. Determine the stress in the corresponding member of the prototype. The numerical parameters defining the scaling factors for the loads and the size of the structural member are $w_m = 1.6$ mm, $w_p = 80$ mm, $b_m = 2.0$ mm, $b_p = 100$ mm, $L_m = 14$ mm and $L_p = 7.0$ m. The scaling factor L = 1/10,000.

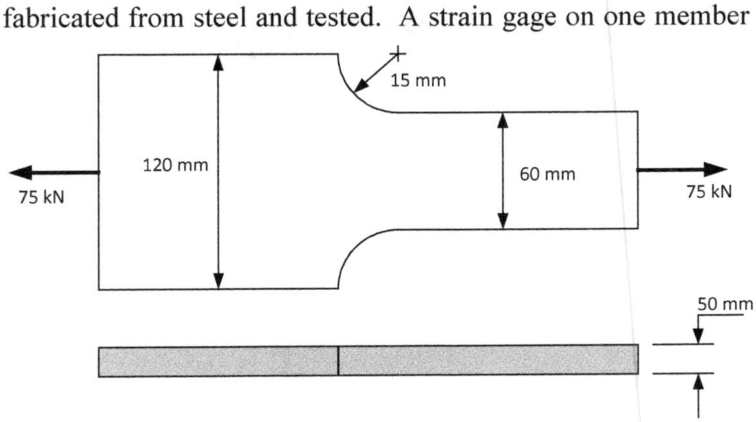

4.39 Suppose that a model of a structure is fabricated from members formed from sheet aluminum. The prototype structure is to be fabricated from steel with an open span of 200 feet. The model is geometrically scaled so that its span is four feet. The capacity of the live load on the prototype is 150 lb/ft, and the model is loaded with 2.5 lb/ft. If the model deflects a distance of 0.220 in. under full load at the center of the span, determine the deflection of the prototype under the design load.

4.40 The tensile bar, presented in the figure to the right, is fabricated with an adhesive joint inclined at an angle ϕ to the axis of the bar. Determine the optimum angle ϕ for the inclined plane if the stresses in the adhesive are not to exceed either 3.5 ksi in tension or 2.8 ksi in shear. Also determine the maximum force that can be applied to the member without exceeding the stresses in the adhesive joint.

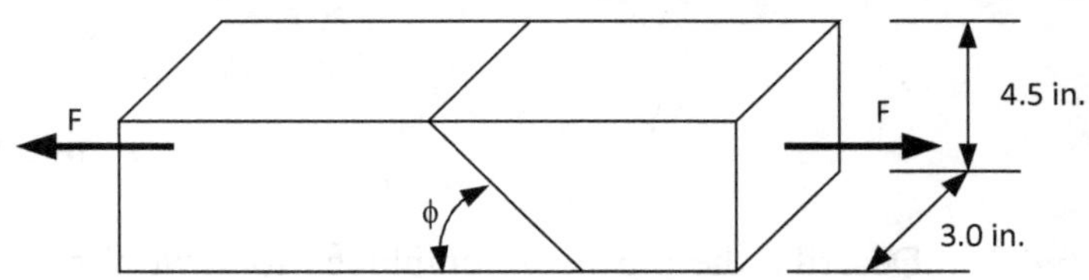

4.41 A structural member is fabricated from a solid round bar of steel with a diameter d = 50 mm. If the member is 6.0 m long, determine the maximum axial force that can be applied if the axial stress is not to exceed 175 MPa and the total elongation is not to exceed 0.14% of its length.

4.42 A steel tie rod with a diameter of 1.0 in. and length of 34.0 in. is employed to compress a brass bushing with an outside diameter of 3.0 in. and a length of 24.0 in. as shown in the figure to the right. Determine the minimum wall thickness of the bushing if the deflection of the tie rod is limited to $\delta = 0.020$ in. Note that F = 10 kip.

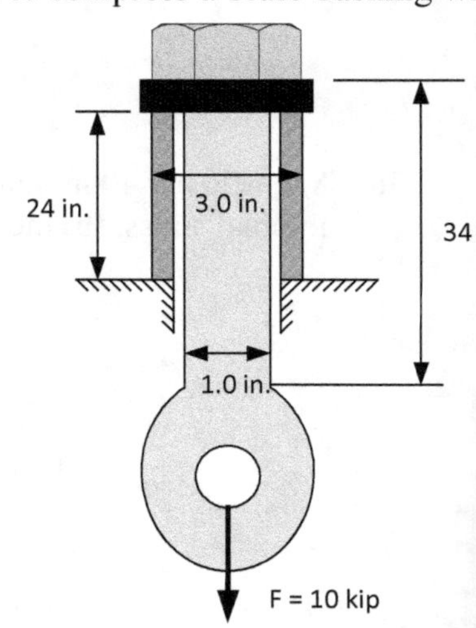

4.43 A short column with a height of 2.0 m is fabricated by adhesively bonding aluminum faceplates to a core of plastic foam as shown in the figure to the left. The foam plastic core has a square cross section with an elastic modulus of 1200 psi. Determine the stresses in the aluminum plates and the plastic foam. Also, determine the displacement of the column under the action of the applied force F = 35 kN.

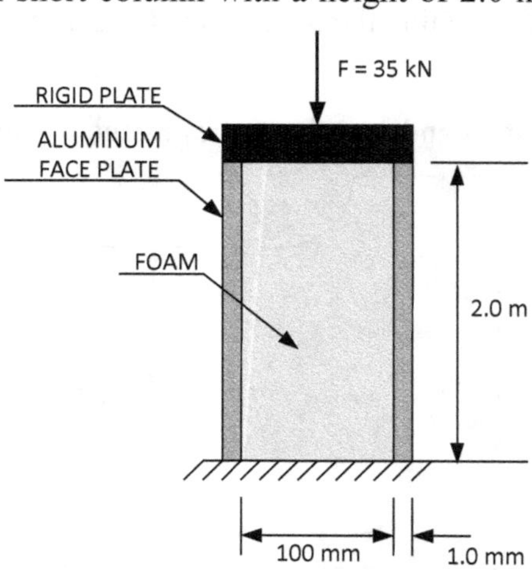

4.44 A grade 8 steel bolt, 1.0 in. in diameter (14 threads/in.) is employed to clamp a brass bushing between two rigid platens as illustrated in the figure below. The brass bushing has a 2.5 in. outside diameter and 1.5 in. inside diameter and is 12.00 in. long. After the unit is assembled with a snug fit, the nut is tightened by 1/3 of a turn. Determine the axial stresses in the bolt and the bushing. Also determine the deflection of the brass bushing.

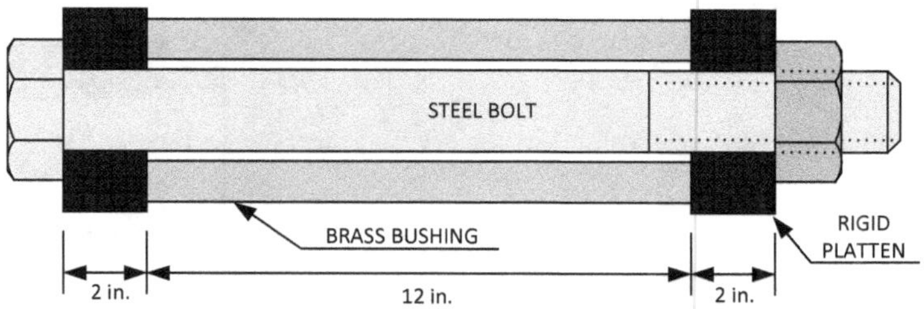

4.45 Twenty-five steel reinforcing rods with a ¼ in. diameter are placed within a high strength concrete column with a square cross section that supports an applied force of 50 kip. Determine the stresses in the steel reinforcing bars and the concrete. Also determine the amount of deflection of the column. Reference the figure to the right.

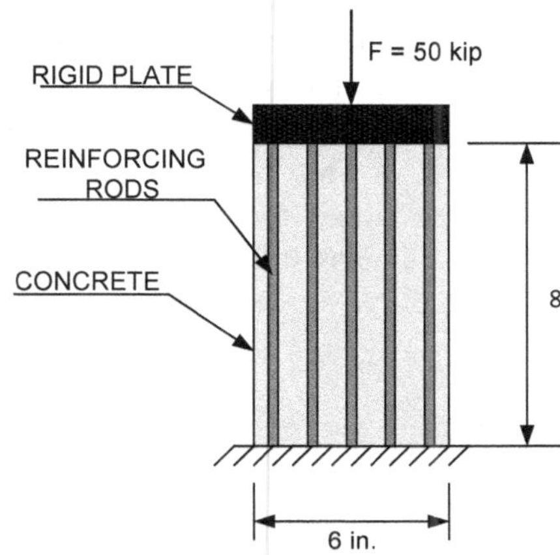

4.46 A long aluminum rod is connected to a shorter brass cylinder with a bolted flange as shown in the figure to the left. Prior to assembly a gap of δ = 0.20 mm occurred between the flange plates. Bolts were inserted and tightened bringing the flange plates together. Determine the stresses induced in both the rod and the cylinder by the assembly operation. Also determine the displacement of the face of each flange.

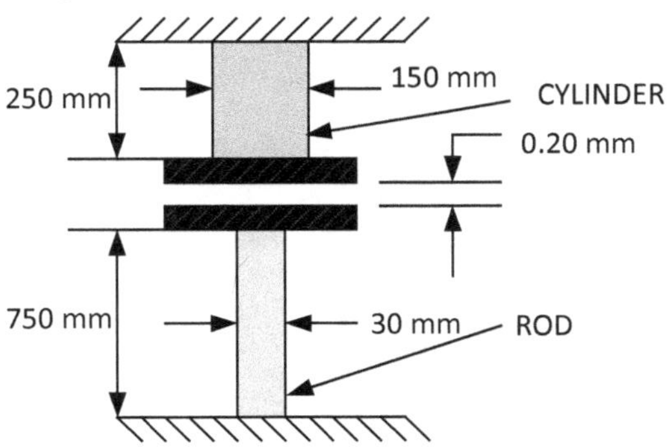

4.47 A copper and stainless steel rod are assembled between two rigid walls at an ambient temperature of 20° C as shown in the figure to the right. If the temperature is increased by an amount ΔT = 120 °C, determine the thermal stresses induced in each rod. Also determine the change in length of each member.

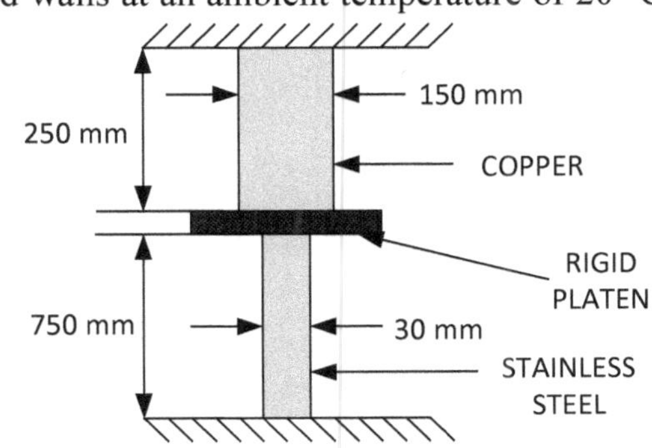

4.48 A high strength steel bolt with a diameter of 1.0 in. passes through an aluminum bushing with a cross sectional area of 3.0 in.^{2} as shown in the figure below. The unit is assembled with a snug fit at an ambient temperature 70 °F. If the temperature of the assembly increases by $\Delta T = 60$ °F, determine the thermal stresses induced in the bolt and the bushing.

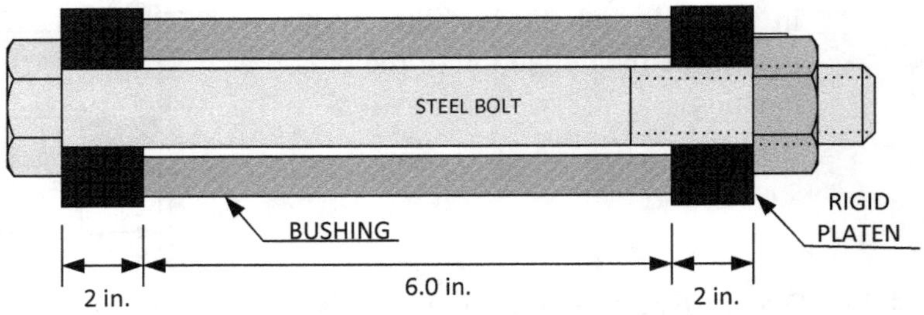

4.49 A three bar suspension system, shown in the figure to the right, is assembled at an ambient temperature of 20° C. Determine the axial stresses in each bar after a force F of 100 kN is applied to the rigid platen and the temperature is increased by $\Delta T = 80$ °C. The bars labeled A are fabricated from aluminum with a cross sectional area of 750 mm^2 and the bar labeled B is fabricated from stainless steel with a cross sectional area of 1,250 mm^2.

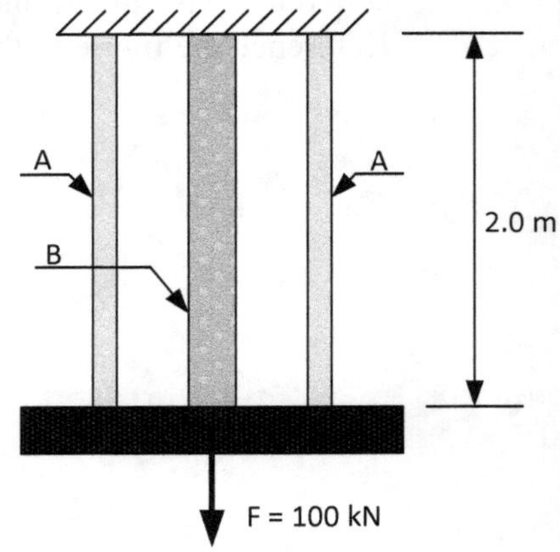

CHAPTER 5

MATERIAL PROPERTIES

5.1 INTRODUCTION

Today structures are usually fabricated with steel or steel reinforced concrete. These materials are subjected to stresses due to gravitational loads and due to live loads. The gravitational loads (weight of the structure itself) are constant with respect to time. The live loads are due to vehicular traffic for bridges, placement of furniture or equipment, or storage of inventory, which varies with time. To ensure the safety of the structure, the strength of the materials (steel, wood, concrete, etc.) must be adequate to support the applied stresses with a comfortable, but not excessive margin of safety. This premise seems simple, but on careful study, we find several strengths are needed to characterize the behavior of structural materials—yield, tensile or ultimate, and fatigue. To perform design analysis, it is also necessary to describe elastic constants that define the stiffness of these materials. To judge the adequacy of these materials for large structures, we examine their ductility (i.e. elongation and percent reduction in area). In this chapter, we define these different strengths, and demonstrate methods for accounting for the strength of construction materials in the design of structural components.

5.2 TYPES OF STRUCTURAL FAILURES

Structures fail while in service for three different reasons. First, by fracturing where the structure breaks into two or more pieces. This is the most catastrophic mode of failure since the structure collapses and usually falls. Often life and property is endangered or lost, and careers are ruined.

The second type of failure is due to yielding. When a structural member is stressed beyond its capabilities, the material, from which the member is fabricated, yields and redistributes the excessive stresses. In yielding, the structure undergoes excessive deformation, but usually it does not break or become unstable. The structure fails because of excessive plastic deformations that alter the geometry of one or more of the structural members. No one is satisfied with a beam that is permanently bowed or bent. Usually these deformed structures have to be straightened or replaced in the repair of the structure.

The third mode of failure is due to excessive elastic deformation. In this case, the structure remains intact, and after the load is removed, it returns to its original shape. The structure fails because small elastic deformations change its shape and alter its performance. For example, suppose the structure is a tool holder for a lathe that is used in a cutting operation removing material from a shaft. In order to maintain the tolerances on the diameter of the shaft, it is necessary to limit the deformations of the tool holder to the specified tolerances. If the tool holder deflects by a significant amount, it fails because it is too flexible and cannot maintain the tool in a relatively fixed location.

In analyzing a structure to ensure that it does not fracture or yield, we compare the stress developed in the structure with the strength of the material used in its fabrication. In making the comparison, we employ one or more strength criteria, which characterize the capability of the material. These include:

- Yield strength
- Ultimate tensile strength
- Fatigue strength
- Compressive strength

A table giving some of the properties characterizing commonly employed engineering materials is shown in Appendix B.

5.3 THE TENSILE TEST

We determine several material properties, as shown in Appendix B, by conducting standardized tensile tests[1]. For example, the yield and ultimate tensile strength are determined in a tensile test. To begin, we prepare **tensile specimens** of the materials under investigation. The standard size of the specimen is shown in Fig. 5.1.

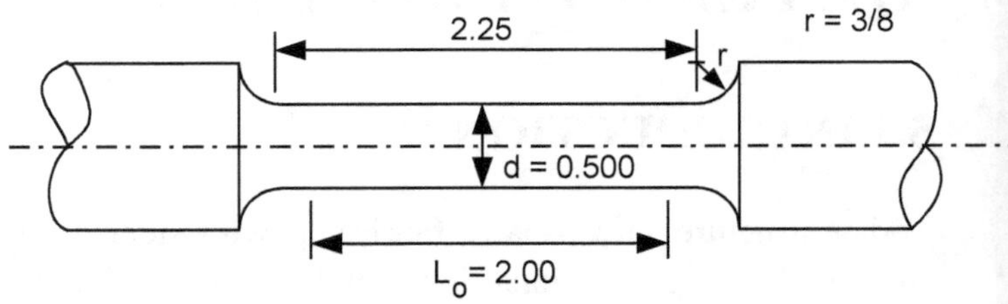

Fig. 5.1 Tensile test specimen (dimensions in inches).

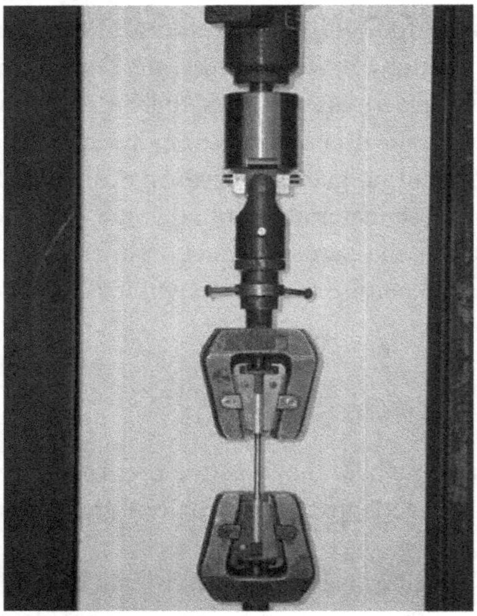

Fig. 5.2 Tensile specimen, wedge-grips, flex-joints and load cell in a testing machine.

The tensile specimen is mounted in a **universal testing machine** similar to the one illustrated in Fig. 5.2. The machine may be a mechanical type with one head fixed and the other driven by screws, or it may be a hydraulic machine where the movable head is driven by a hydraulic cylinder. In either instance, the testing machine applies a load F along the axis of the tensile specimen. The load is applied slowly until the specimen yields and/or fails by fracturing.

During the tension test, we measure the applied load F and the stretch δ over the gage length L_o of the specimen. A **load cell** on the universal testing machine indicates the applied load F and an **extensometer** mounted on the specimen, shown in Fig. 5.3, measures the stretch δ. The electrical signals from the load cell and the extensometer are recorded together on a x-y chart to provide a load-deformation curve that is proportional to the stress-strain curve. The load-deflection (stretch) curves recorded during the tension tests are converted into stress-strain curves that characterize the tensile behavior of the metallic material by utilizing:

$$\sigma = F/A_o \qquad (4.4)$$

$$\varepsilon = \delta/L_o \qquad (4.3)$$

where L_o and A_o are the initial gage length and cross sectional area of the tensile specimen.

[1] The American Society for Testing Materials (ASTM) publishes standards that define the test specimen and procedures for measuring the material properties described in this chapter. The standard for tension testing of metallic materials is E8 – 99, (see the *Annual Book of ASTM Standards* volume 03.01).

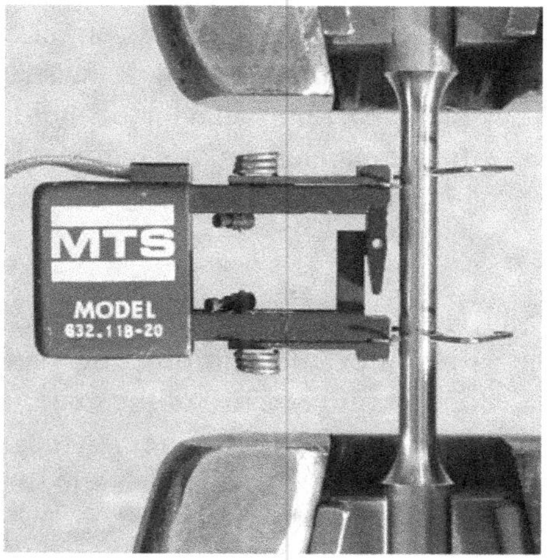

Fig. 5.3 A tension specimen with an extensometer for measuring the stretch δ.

This procedure gives the **engineering stress** and the **engineering strain,** which differ from the **true stress** and the **true strain**. The distinction between the "engineering" and "true" quantities is described in Section 5.6.

For brittle materials, which do not exhibit significant plastic deformation before fracture, the stress-strain curve is nearly linear until failure, as indicated in Fig. 5.4. The stress σ_f producing the failure of the brittle specimen is recorded during the tensile test. It is this particular value of the stress that defines the material property known as the ultimate tensile strength S_u for a brittle material.

$$S_u = \sigma_f \tag{5.1}$$

A photograph of a tension specimen that exhibited brittle failure is depicted in Fig. 5.5. Note the absence of any necking or extensive plastic deformation in the uniform section of the specimen. Brittle failure is dangerous because it is sudden and catastrophic. The fracture initiates without warning and the cracks propagate across the specimen (structure) in microseconds. Of course, in the selection of materials for our designs, we avoid the use of brittle materials in structures to preclude the possibility of a catastrophic failure. Brittle materials such as gray cast iron are sometimes used because of its excellent casting properties and for its ability to damp vibrations. In the usage of brittle materials, we are always careful to maintain a state of compressive stress in the structure.

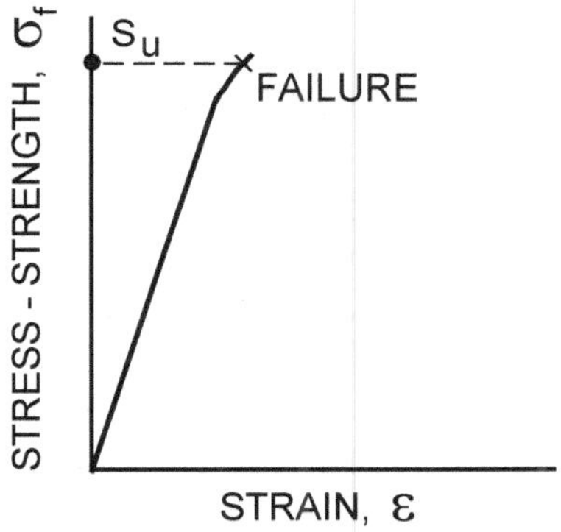

Fig. 5.4 Stress-strain curve for a brittle metallic material.

Fig. 5.5 Failure of a brittle material occurs suddenly with little plastic deformation.

Structures are designed with ductile materials, which yield and undergo extensive plastic deformation prior to rupture. A typical stress-strain curve for a ductile material (mild steel) is presented in Fig. 5.6.

An inspection of the stress-strain diagram shows that there are four different regions with each corresponding to a different behavior of the ductile steel. The first is called the **elastic region** where the material responds in a linear manner. In this region, Hooke's law applies and we write:

$$\sigma = E\,\varepsilon \qquad (4.5)$$

Hooke's law is a mathematical model of material behavior, but it is valid only in the elastic region. The elastic region extends until the low carbon steel (or some other ductile material) begins to yield. When the material yields, the linear response of the material ceases and Eq. (4.5) is no longer valid.

The second region on the stress-strain diagram depicts the initiation of yielding where the stress in the specimen exceeds the **elastic limit**. Slip planes have developed in the specimen, and deformation by slip along these planes is occurring with small or negligible increases in the stress. The stress-strain relationship is non-linear in this region. In some ductile materials, a small decrease in the applied stress is observed as the tensile specimen yields and continues to elongate by slip.

The third region describes a material behavior known as **strain hardening**. The easy slip that occurred during the initial yielding phase becomes more difficult to induce. As a consequence, higher stresses are required to continue deforming the tensile specimen. The stresses increase with increases in the strain; however, the relationship is not linear and Eq. (4.5) is not valid in this region.

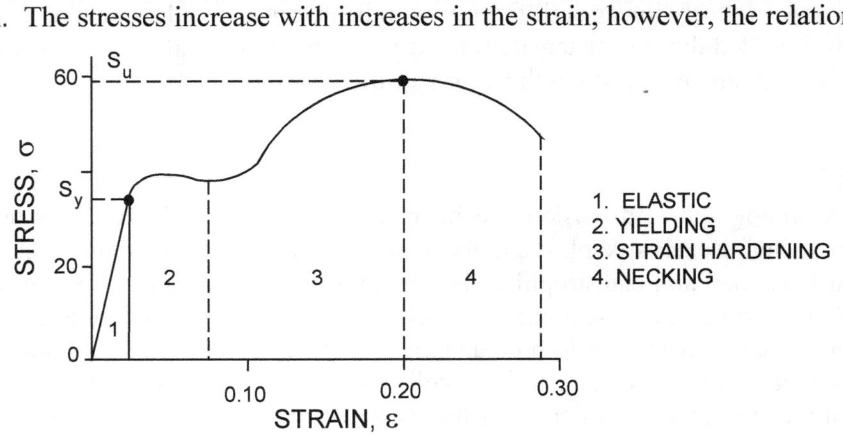

Fig. 5.6 Stress-strain diagram for low carbon steel.

In the final region, the tensile specimen undergoes a dramatic change in appearance: it begins to **neck**. The deformation becomes localized to a small area along the length of the bar. During this phase of the deformation, the specimen resembles an hourglass. The neck decreases in diameter with increasing deformation until the specimen fails by rupturing. The axial deformation, which occurs as the tensile specimen necks, requires no increase in the applied loads. Indeed, the load may actually decrease significantly during the necking phase of the deformation processes. The appearance of a ruptured tension specimen fabricated from low carbon steel is illustrated in Fig. 5.7.

Regions 2, 3 and 4 are often combined and called the **plastic regime** for a ductile material.

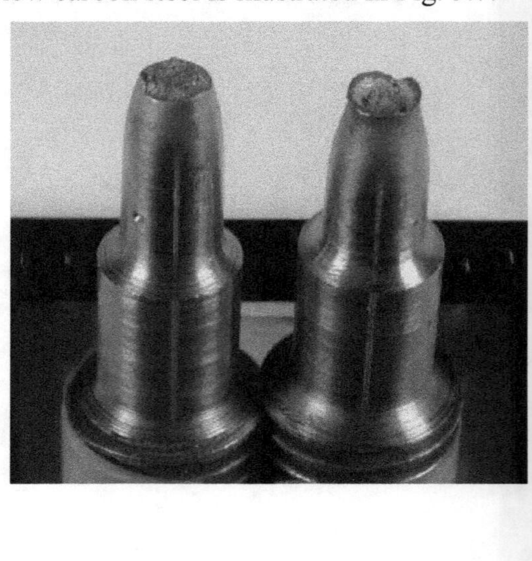

Fig. 5.7 Ductile failure of a tensile specimen.

5.4 MATERIAL PROPERTIES

The tensile test provides several material properties that are important in the analysis of engineering components and structures. These properties include two measures of strength, two measures of ductility, and two elastic constants. Let's first discuss strength.

5.4.1 Measures of Strength

The two measures of strength determined in a tensile test of a ductile material are the yield strength and the ultimate tensile strength. The yield strength, as the name implies, is the stress required to induce yielding:

$$S_y = \sigma_y \tag{5.2}$$

To establish the yield stress σ_y, we examine the stress-strain diagram and attempt to identify the stress when yielding initiated. For some ductile materials with stress-strain diagrams similar to that shown in Fig. 5.6, the precise identification of σ_y is clear. However, the yield behavior of other materials is much less well defined. For example, suppose a material exhibits the stress-strain behavior as indicated in Fig. 5.8 a. Where is the yield point?

It is evident in Fig. 5.8a that the stress-strain curve is non-linear, but we might differ in defining the point where slip and yield initiated. To eliminate the ambiguity in the definition of the yield point from a stress-strain diagram, the offset method is employed. We construct a line parallel to the linear portion of the σ-ε curve in the elastic region. This line is offset along the strain axis by 0.2% or $\varepsilon = 0.002$. The intersection of the offset line with the σ-ε curve defines the yield point as illustrated in Fig. 5.8 b. With the **0.2% yield stress** σ_y defined, we use Eq. (5.2) to establish the yield strength S_y for the material.

The ultimate tensile strength is established from the maximum stress on the specimen that occurs at the onset of necking. We have defined this point on Fig. 5.8 b, and have established the ultimate stress σ_u. The ultimate tensile strength S_u for a ductile material is given by:

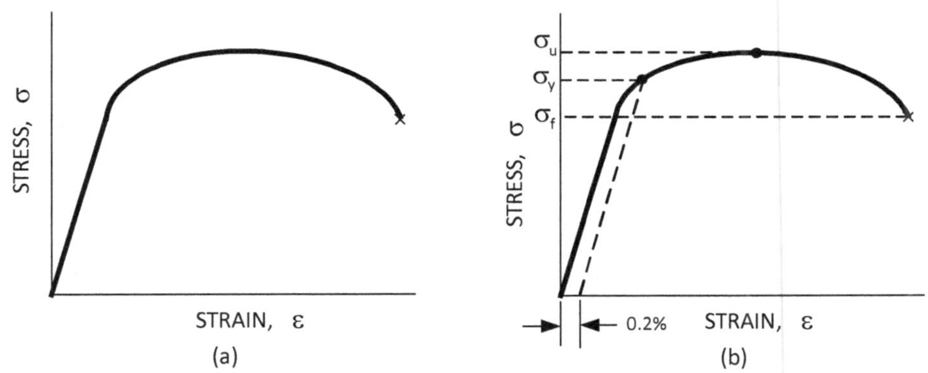

Fig. 5.8 The offset method for determining the stress σ_y for yielding.

$$S_u = \sigma_u \tag{5.3}$$

The stress at failure, σ_f for a ductile material, shown in Fig. 5.8 b, is only of academic interest. If the structure or machine component is properly designed, the deformation state will probably be limited to the elastic region. In some instances, plastic deformations are tolerated in design, but these exceptions are limited to yielding in small local regions under conditions of constant loading. Global yielding is not tolerated, and yielding under the action of cyclic stresses is a recipe for disaster. For this reason, the stresses imposed on a structure are usually less than σ_y and certainly less than σ_u.

5.4.2 Measures of Ductility

There are two common measures of ductility for metallic materials. The first is the **percent elongation** given by:

$$\%\,e = \frac{L_f - L_0}{L_0} \times 100 \tag{5.4}$$

where L_o is the gage length of the tensile specimen — usually 2.0 in. or 50 mm and L_f is the final deformed length of the specimen between the marks for the gage length.

For low carbon steels, the percent elongation is in the range from 20 to 35%; however, the percent elongation decreases with increasing carbon content in the steel and increasing strength.

The second measure of ductility is the **percent reduction in area** that is given by:

$$\% A = \frac{A_0 - A_f}{A_0} \times 100 = \left[\left(\frac{d_0^2 - d_f^2}{d_0^2}\right)\right] \times 100 \quad (5.5)$$

where A_o is the initial cross sectional area of the tensile specimen given by $\pi d_o^2/4$ and $A_f = \pi d_f^2/4$ is the final cross sectional area of one of the ruptured ends of the specimen.

The percent reduction in area for low carbon steel is typically in the range of 60 to 70%. As was the case with percent elongation, the percent reduction in area decreases as the carbon content of steel is increased to enhance its strength.

There is a trade-off between ductility and strength for metallic alloys. To illustrate the loss in ductility with increasing strength, let's examine typical stress-strain curves for three different types of steels that are shown in Fig. 5.9. As the strength increases the strain to failure decreases with an accompanying decrease in the ductility. For low carbon steels, the strain to failure usually exceeds 60 to 70%. This value decreases to 30 to 40% for higher carbon steels. The very high strength steel alloys fail at strains ranging from 10 to 20%.

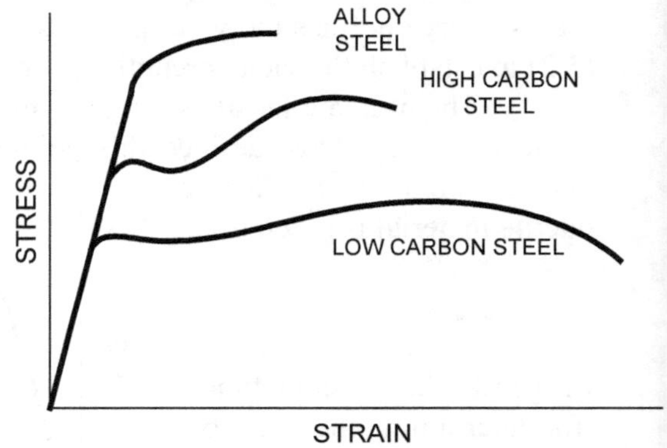

Fig. 5.9 Stress-strain curves for three different types of steels.

EXAMPLE 5.1

A steel supplier provides you with data from a recent series of tensile tests of two different steels that they sell to your corporation. Your manager questions the ductility of both materials, and asks you to determine it. An examination of the supplier's data indicates:

Lower cost steel at \$9.00/100lb	Higher cost steel at \$9.74/100lb
L_f = 66 mm and d_f = 9.1 mm	L_f = 61 mm and d_f = 10.8 mm

The gage length L_o and diameter d_o for specimens from both types of steel was 50 mm and 12.5 mm respectively.

Solution: For the measure of ductility known as the percent elongation, Eq. (5.4) gives:

$\%e = 100(66 - 50)/50 = 32\%$ for the lower cost steel.

$\%e = 100(61 - 50)/50 = 22\%$ for the higher cost steel.

For the measure of ductility known as the percent reduction in area, Eq. (5.5) gives:

$\%A = 100(d_o^2 - d_f^2)/d_o^2 = 100[(12.5)^2 - (9.1)^2]/(12.5)^2 = 47.0\% \Rightarrow$ lower cost steel.
$\%A = 100(d_o^2 - d_f^2)/d_o^2 = 100[(12.5)^2 - (10.8)^2]/(12.5)^2 = 25.3\% \Rightarrow$ higher cost steel.

When your manager asks which steel has the higher ductility, what is your response?

5.4.3 Elastic Constants

Modulus of Elasticity

In this discussion of the elastic constants, we limit the application of these constants to characterize deformations only in the linear elastic region of the elastic-plastic regime. In this region, the material is elastic and recovers completely when the load or stress is removed from the specimen. The slope of the stress-strain curve is defined as the **modulus of elasticity** or Young's modulus. The slope illustrated in Fig. 5.10 is given by:

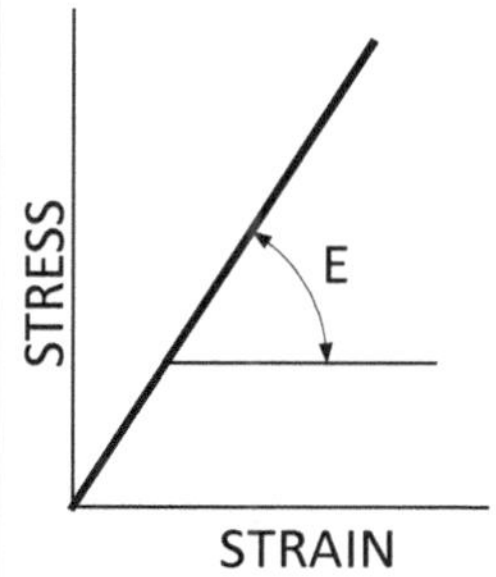

Fig. 5.10 The slope of the stress-strain curve determines the modulus of elasticity.

$$E = \sigma/\varepsilon \qquad (4.5)$$

EXAMPLE 5.2

After conducting a tensile test with a mild steel specimen, you measure the slope $\Delta P/\Delta\delta$ of the trace on the load-deflection curve (P, δ) and find:

$$\Delta P/\Delta\delta = 7850 \text{ lb}/ 2.68 \times 10^{-3} \text{ in.}$$

If the specimen diameter is 0.504 in. and the gage length of the extensometer is 2.00 in., determine the modulus of elasticity E.

Solution: From Eq. 5.5, we write:

$$E = \sigma/\varepsilon = (P/A) / (\delta/L) = PL / \delta A = \Delta PL/\Delta\delta A$$

$$E = [(7850\text{lb})(2 \text{ in.})]/[(2.68 \times 10^{-3} \text{ in.})(\pi\ (0.252)^2 \text{ in.}^2)] = 29.36 \times 10^6 \text{ psi}$$

Poisson's Ratio

Another elastic constant, **Poisson's ratio**, may be measured in the tensile test; however, strain gages must be attached to the specimen to measure the strain in both the axial and the transverse directions. Poisson's ratio is defined as:

$$\nu = -\varepsilon_t/\varepsilon_a \qquad (5.6)$$

where ε_t is the strain in the transverse direction, and ε_a is the strain in the axial direction.

When a specimen is pulled in tension, its length extends and its diameter contracts. This contraction is usually 1/10 to 1/2 of the amount of the specimen's axial stretch. Extension of the specimen in the elastic region produces a contraction that is usually too small to be observed during the tensile test.

Poisson's ratio is a measure of the relative contraction occurring in the diameter of the tensile specimen as it is loaded. Although too small to be observed, the Poisson effect is a very important phenomenon because it significantly affects the analysis of structures subjected to multiaxial states of stress, where stresses are imposed in more than one direction.

To describe the Poisson contraction in more detail, consider the drawing of a rectangular bar of an elastic material with dimensions L, W, and D as shown in Fig. 5.11a. Next, apply an axial strain ε_a and write:

$$\varepsilon_a = \Delta L/L \qquad (a)$$

In the deformed state, the dimensions of the bar change to L + ΔL, W + ΔW, and D + ΔD as shown in Fig. 5.11b. The strains ε_t in the x and y directions (both x and y are transverse directions) are both due to the Poisson effect and are equal. They are given by:

$$\varepsilon_{tx} = \Delta W/W = \varepsilon_{ty} = \Delta D/D \qquad \text{(b)}$$

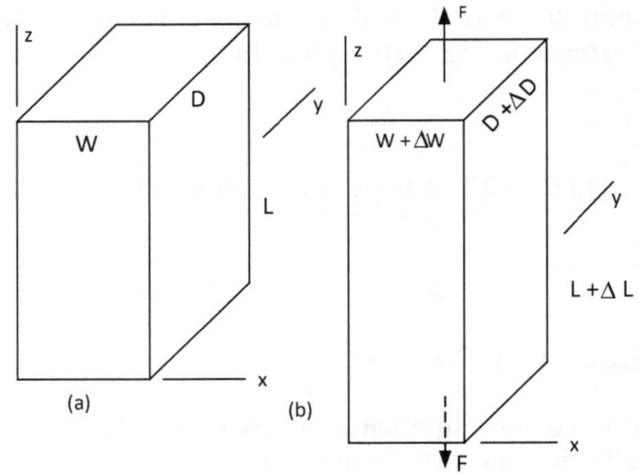

Fig. 5.11 A rectangular bar before and after imposing an axial deformation.

If we substitute Eq. (5.6) into Eq. (b), we obtain:

$$\varepsilon_t = -\nu\varepsilon_a = -\nu\Delta L/L = \Delta W/W = \Delta D/D \qquad \text{(c)}$$

From Eq. (c) it is clear that the changes in the transverse dimensions of the bar due to the Poisson effect are given by:

$$\Delta W = -\nu\Delta L(W/L); \qquad \Delta D = -\nu\Delta L(D/L) \qquad \text{(d)}$$

From this elementary analysis of the deformed geometry of a rectangular bar, we note that the material property **Poisson's ratio, ν** gives the transverse deformation in terms of the axial deformation when a structural member is subjected to a uniaxial loading. The quantities ΔW and ΔD are negative when ΔL is positive because the Poisson effect produces a contraction in the transverse (x and y) directions when the body is extended in the axial (z) direction.

EXAMPLE 5.3

You cut a circular hole with a 100-mm diameter in a large sheet of dental dam (very thin rubber sheet). If the sheet, originally 1,000 mm long by 500 mm wide, is stretched until it is 1,300 mm long, determine the new width of the sheet and the dimensions of the deformed hole. Assume that rubber is perfectly elastic with a Poisson's ratio, $\nu = ½$.

Solution: Recall Eq. (5.6) and write:

$$\varepsilon_t = -\nu\varepsilon_a = -\nu\,\Delta L/L_o = -(1/2)\,(1300 - 1000)/1000 = -0.150$$

$$\varepsilon_a = \Delta L/L_o = (1300 - 1000)/1000 = 0.300$$

Since these strains are imposed over the entire sheet, determine the deformed width W_{NEW} from:

$$W_{NEW} = W_o + \Delta W = W_o + \varepsilon_t\,W_o = (1 + \varepsilon_t)W_o = (1 - 0.150)500 = 425 \text{ mm}$$

The dimensions of the deformed hole are given by:

$$D_a = D_o + \Delta D_a = D_o + \varepsilon_a\,D_o = (1 + \varepsilon_a)D_o = (1 + 0.300)100 = 130 \text{ mm}$$

$$D_t = D_o + \Delta D_t = D_o + \varepsilon_t D_o = (1 + \varepsilon_t)D_o = (1 - 0.150)100 = 85 \text{ mm}$$

where D_a and D_t are the axial and transverse diameters of the hole after deformation.

What is the geometric form of the deformed hole in the rubber sheet?

EXAMPLE 5.4

Determine the change in volume of a rectangular bar subjected to an axial strain of $\varepsilon_a = 2 \times 10^{-3}$. The dimensions of the bar before deformation were L = 4W = 3D = 15 in. The material from which the bar is fabricated has a Poisson's ratio of $\nu = ¼$.

Solution: The original volume, **V** of the bar is given by:

$$\mathbf{V} = L \times W \times D = L^3/12 = (15)^3/12 = 281.25 \text{ in}^3 \qquad \text{(a)}$$

From Eq. (5.6), we determine the new dimensions of the bar as:

$$W_{NEW} = (1 - \varepsilon_a/4)W; \quad D_{NEW} = (1 - \varepsilon_a/4)D,; \quad L_{NEW} = (1 + \varepsilon_a)L \qquad \text{(b)}$$

We rewrite Eq. (b) as:

$$W_{NEW} = (1 - \varepsilon_a/4)L/4; \quad D_{NEW} = (1 - \varepsilon_a/4)L/3; \quad L_{NEW} = (1 + \varepsilon_a)L \qquad \text{(c)}$$

The new volume is given by:

$$V_{NEW} = L_{NEW} \times W_{NEW} \times D_{NEW} = (1 + \varepsilon_a)(1 - \varepsilon_a/4)(1 - \varepsilon_a/4)(L^3/12) \qquad \text{(d)}$$

From Eq. (a) and (d), it is evident that:

$$\Delta V = [(1 + \varepsilon_a)(1 - \varepsilon_a/4)(1 - \varepsilon_a/4) - 1](L^3/12) \qquad \text{(e)}$$

Substituting $\varepsilon = 2 \times 10^{-3}$ into Eq. (e) yields:

$$\Delta V = [(1 + 2 \times 10^{-3})(1 - 0.5 \times 10^{-3})(1 - 0.5 \times 10^{-3}) - 1]\,V \qquad \text{(f)}$$

Performing the calculation gives:

$$\Delta V = 0.00099825\,V = (0.00099825)(281.25) = 0.2808 \text{ in}^3$$

The percentage change in the volume is given by:

$$\Delta V/V = [(1 + \varepsilon_a)(1 - \nu\varepsilon_a)(1 - \nu\varepsilon_a) - 1] = 0.000998 = 0.0998\%$$

From these results it is evident that the change in the volume is extremely small for strains in the elastic region.

Shear Modulus

Another elastic constant that we often use in the analysis of structures subjected to shear stress is the **shear modulus**, G. The shear modulus relates the shear stress to the shear strain by:

$$\tau = G \gamma \qquad (5.7)$$

where τ is the shear stress and γ is the shear strain.

The shear strain γ is defined as the change in angle of two perpendicular lines when a body is deformed. For instance, suppose we have a Cartesian coordinate system scribed on a body with the x and y axes serving as the two perpendicular lines. When the body deforms under the action of a shear stress τ, these two lines rotate, and the angle at their intersection changes from 90° to some other angle 90° ± γ as illustrated in Fig. 5.12.

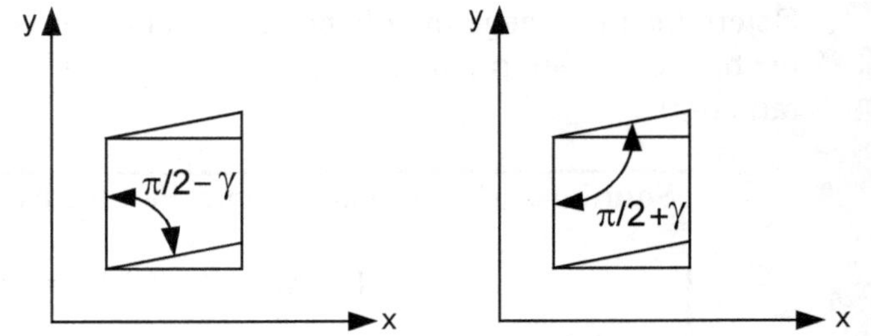

Fig. 5.12 Shear strain is the change in a right angle when a body deforms.

The shear modulus G is related to the modulus of elasticity and Poisson's ratio by:

$$G = \frac{E}{2(1 + \nu)} \qquad (5.8)$$

EXAMPLE 5.5

Determine the shear modulus G for:

(a) Steel with the modulus of elasticity $E = 30 \times 10^6$ psi and Poisson's ratio $\nu = 0.30$.
(b) An aluminum alloy with E = 73 GPa and Poisson's ratio $\nu = 0.33$.

Solution: From Eq. (5.8) we write for steel the following relation:

$$G = (30 \times 10^6) / [2(1 + 0.30)] = 11.54 \times 10^6 \text{ psi for steel.}$$

For the aluminum alloy, the shear modulus is given by:

$$G = (73 \times 10^9) / [2(1 + 0.33)] = 27.44 \text{ GPa for an aluminum alloy.}$$

5.5 HOOKE'S LAW

In the elastic region, there is a linear relationship between stress and strain. In the simple case of uniaxial loading, which produces a uniaxial state of stress in long thin structural members such as bars and beams, Hooke's law gives the relationship:

$$\sigma = E \varepsilon \qquad (4.5)$$

However, some structural members are not long and thin. For example, tubes or cylindrical pressure vessels are subjected to biaxial stresses. Stresses, due to an applied pressure p, develop in both the axial (x) and the circumferential (y) directions as shown in Fig. 5.13.

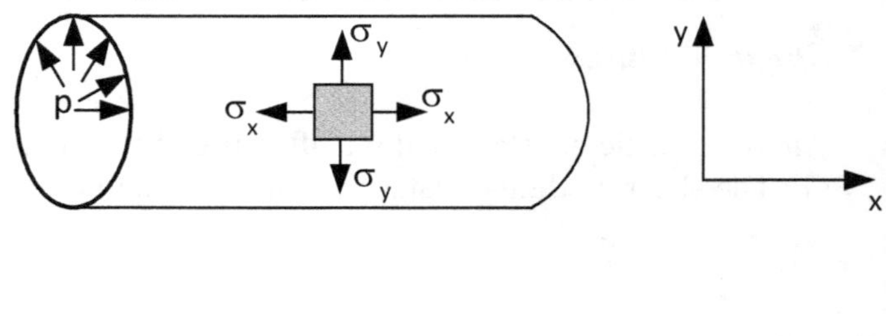

Fig. 5.13 Biaxial stresses induced by pressure loading a cylindrical vessel.

When a biaxial state of stress occurs in a structural member, the simple form of Hooke's law given in Eq. (4.5) is not valid. The Poisson effect produces an interaction between the stresses and strains in the two orthogonal directions. To accommodate the Poisson effect, we express the strains in terms of the stresses as:

$$\varepsilon_x = \frac{1}{E}\left(\sigma_x - \nu\sigma_y\right)$$
$$\varepsilon_y = \frac{1}{E}\left(\sigma_y - \nu\sigma_x\right) \qquad (5.9)$$

The stresses in terms of the strains are:

$$\sigma_x = \frac{E}{\left(1-\nu^2\right)}\left(\varepsilon_x + \nu\varepsilon_y\right)$$
$$\sigma_y = \frac{E}{\left(1-\nu^2\right)}\left(\varepsilon_y + \nu\varepsilon_x\right) \qquad (5.10)$$

We employ Eqs. (5.9) and (5.10) with two-dimensional structural members such as tubes, plates and shells.

EXAMPLE 5.6

The stresses in the wall of a cylindrical pressure vessel are:

$\sigma_x = 7{,}500$ psi — axial direction
$\sigma_y = 15{,}000$ psi — hoop direction

If the pressure vessel is fabricated from steel plates with $E = 30 \times 10^6$ psi and $\nu = 0.30$, determine the strain in the hoop and axial directions.

Solution: Recall Eq. (5.9) and write:

$$\varepsilon_x = \frac{1}{E}\left(\sigma_x - \nu\sigma_y\right) = \frac{1}{30}\left[7{,}500 - (0.30)(15{,}000)\right] \times 10^{-6} = 100 \times 10^{-6}$$
$$\varepsilon_y = \frac{1}{E}\left(\sigma_y - \nu\sigma_x\right) = \frac{1}{30}\left[15{,}000 - (0.30)(7{,}500)\right] \times 10^{-6} = 425 \times 10^{-6}$$

It is interesting to observe that the pressure vessel develops biaxial tensile stresses with a ratio of 2 hoop to 1 axial; however, the ratio of strain is 4.25 hoop to 1 axial. The change in the ratio of strains is due to the Poisson effect.

EXAMPLE 5.7

Electrical resistance strain gages are employed to measure the strain in the hoop and the axial directions on a high-pressure pipeline that runs close to a residential neighborhood. The strains are:

$$\varepsilon_a = \varepsilon_x = 600 \times 10^{-6} \quad \text{and} \quad \varepsilon_h = \varepsilon_y = 2650 \times 10^{-6}$$

If the pipe in the line is fabricated from a high strength alloy steel with E = 200 GPa and Poisson's ratio $\nu = 0.29$, determine the axial and hoop stresses and comment on the safety of the pipeline.

Solution: Recall Eqs. (5.10) and write:

$$\sigma_a = \sigma_x = \frac{E}{(1-\nu^2)}(\varepsilon_x + \nu\varepsilon_y) = \frac{200}{1-(0.29)^2}[600 + (0.29)(2650)] \times 10^{9-6} = 298.7 \text{ MPa}$$

$$\sigma_h = \sigma_y = \frac{E}{(1-\nu^2)}(\varepsilon_y + \nu\varepsilon_x) = \frac{200}{1-(0.29)^2}[2650 + (0.29)(600)] \times 10^{9-6} = 616.6 \text{ MPa}$$

Comment on the safety of the pipeline:

Although the specific type of steel has not been defined, an examination of the strength of the steels listed in Appendix B indicates that the hoop stress is very high. Material samples should be obtained and tensile tests conducted to determine the yield and tensile strength of the actual steel used in constructing the pipeline. Preliminary analysis indicates that the safety of the pipeline is suspect.

5.6 TRUE STRESS AND TRUE STRAIN

In the standard tensile test, we define the (engineering) stress as:

$$\sigma = P/A_o \tag{4.4}$$

where A_o is the cross sectional area of the tensile specimen.

As we load the tensile specimen, the diameter of the bar decreases due to Poisson contraction. Clearly, the cross sectional area of the specimen is changing during the standard tensile test. How do we handle this situation?

There are two approaches. First, with the engineering approach, we ignore the changes in the cross sectional area and compute the stress (engineering) using Eq. (4.4). We also determine the strain (engineering) by:

$$\varepsilon = \delta/L_o \tag{4.3}$$

where L_o is the initial gage length.

Obviously the gage length of the tensile specimen is increasing as we load the tensile bar. However, if we treat the strain as an "engineering" quantity, it is computed using Eq. (4.3). The errors resulting by neglecting the changes in the area A and the gage length L are small when the analysis is limited to stresses and strains in the elastic region

The second approach is to accommodate both the changing cross sectional area A and the gage length L in the analysis of the results from a tensile test. With this approach the "true" stress σ_T is given by:

$$\sigma_T = F/A_F \qquad (5.11)^{2}$$

where A_F is the cross sectional area of the tensile specimen at any applied force F.

The true strain[3] is also defined by considering an incremental change in length dL over a gage length L, which is increasing with the applied load. With this definition the true strain is given by:

$$d\varepsilon_T = dL/L \tag{5.12}$$

[2] We interchange the external force F and the internal force P in the equation for stress in the specimen because clearly P = F in a tensile specimen.

[3] True strain is sometimes referred to as "natural" or "logarithmic" strain.

We integrate Eq. (5.12) as the gage length increases from L_o to L_F and obtain:

$$\varepsilon_T = \int_{L_0}^{L_F} \frac{dL}{L} = \ln \frac{L_F}{L_0} \qquad (5.13)$$

The engineering strain ε is related to the true strain by:

$$\varepsilon_T = \ln (L_F/L_o) = \ln [(L_o + \Delta L)/L_o] = \ln (1 + \varepsilon) \qquad (5.14)$$

If the results of a tensile test are interpreted using the definitions of true stress and true strain, the stress-strain curve in the plastic regime is changed markedly. A typical true stress-true strain diagram is illustrated in Fig. 5.14. Inspection of this figure shows that the true stress-strain curve differs dramatically from the engineering stress-strain curve. For the elastic region where the strains are very small, the two curves are nearly identical. However, for larger strains significant differences are evident. The true stress is larger than the engineering stress for a given load and the true strain is smaller than the engineering strain. The true stress does not decrease with the onset of necking because the area at the neck is decreasing more rapidly than the load. The engineering stress, which is based on the initial area A_0 of the specimen, decreases because the load decreases as the neck develops.

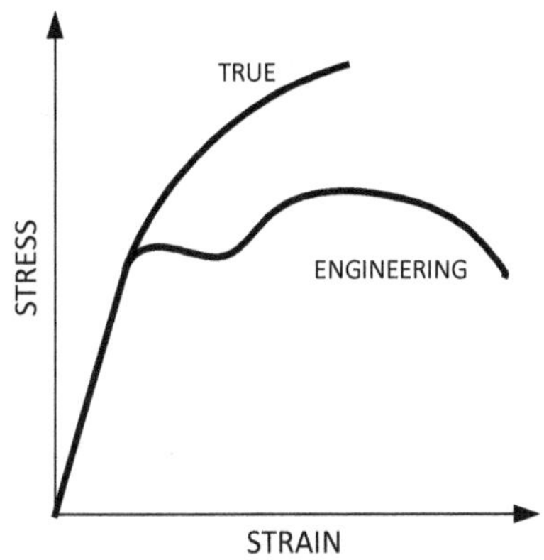

Fig. 5.14 Illustration of a true-stress—true-strain diagram.

EXAMPLE 5.8

The rupture load on a standard tensile specimen with an initial diameter d_o = 0.500 in. and initial gage length L_o = 2.00 in. was 12,000 lb. If the diameter d_F at the neck of the ruptured specimen was 0.415 in. and the gage length after failure was L_F = 2.650 in., determine:

a. The true stress at failure.
b. The true strain at failure.
c. The engineering stress at failure.
d. The engineering strain at failure.

Solution: The true stress at failure is given by Eq. (5.11) as:

$$\sigma_T = F/A = 4F/\pi d_F^2 = [(4)(12000)]/[\pi(0.415)^2] = 88,710 \text{ psi}$$

The true strain at failure is given by Eq. (5.14) as:

$$\varepsilon_T = \ln (L_F/L_o) = \ln (2.650/2.000) = 0.281 = 28.1\%$$

The engineering stress at failure is given by Eq. (4.4) as:

$$\sigma = F/A_o = 4F/\pi d_o^2 = [(4)(12000)]/[\pi(0.500)^2] = 61,120 \text{ psi}$$

The engineering strain at failure is given by Eq. (4.3) as:

$$\varepsilon = (L_F - L_o)/L_o = (2.650 - 2.000)/2.000 = 0.325 = 32.5\%$$

5.7 FATIGUE STRENGTH

Many structures are loaded only once. For example, the structural members in the Hancock tower in Chicago were loaded by the dead weight of all of the structural components as the tower was constructed. The people, furnishings and equipment going into the tower added some weight, but this additional load was small when compared to the weight of the basic structure. For structures and machine components loaded only once, either the yield strength or the ultimate tensile strength is adequate to predict the safety of the structure.

However, many structures and machine components are subjected to repeated loading. Very heavy tractor-trailer trucks, for example, traverse a bridge. With the passage of each truck every bridge member is subjected to another cycle of loading. Consider the engine of your automobile. If you are driving along a highway with the tachometer registering 3600 RPM, you are subjecting the crankshaft in your engine to 3600 cycles of load for each minute of operation.

There are two detrimental effects due to the cyclic loading of structures and machine components. First, the design strength of the material from which the structure is fabricated is lowered. We design to a fatigue strength S_f that is a function of the cyclic stresses and the number of cycles of load imposed onto the structure.

Second, fatigue failures in **ductile** materials are of a **brittle** nature (i.e., structural failure and collapse occur catastrophically). The failure mechanism in fatigue is markedly different from that observed in yielding or rupture. In fatigue, microscopic cracks are initiated due to accumulated irreversible slip in a very thin layer of material adjacent to the surface. These cracks grow larger and extend into the material until reaching critical size. At this point, the cracks become unstable and extend at very high speed across the structural member producing sudden and catastrophic collapse. Examination of a surface of a fatigue failure shows the brittle nature of the phenomena. Although the material may be classified as ductile based on the results of a tensile test, there is no visible sign of coarse slip or necking associated with failure in fatigue.

5.7.1 The S_f - N Diagram

To accommodate for the degrading effects of cyclic loading, we compare the applied stresses with the **fatigue strength, S_f** of the material from which the structure is fabricated. A typical example of the fatigue strength for low carbon steel is presented in the S_f-N diagram of Fig. 5.15.

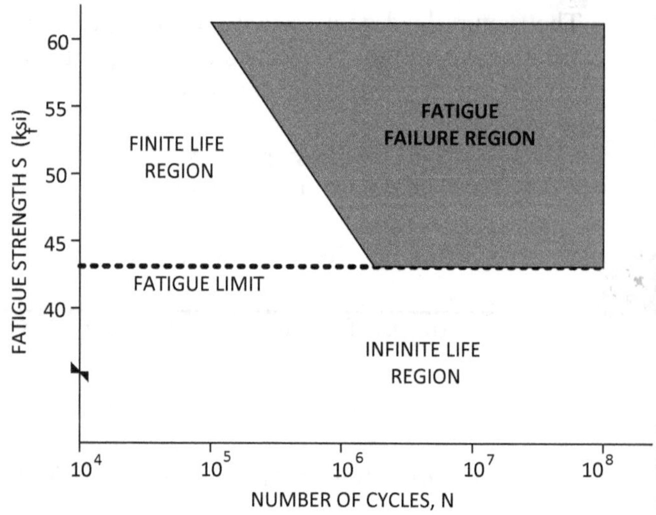

Fig. 5.15 S_f – N curve for high strength low alloy steel.

The S_f - N diagram is a graphical representation of both safe and critical states of cyclic stresses. The S_f - N diagram is divided into three different regions:

1. A failure region (shaded) where many cycles of high stresses produce a fatigue crack leading to fracture.
2. Safe regions where component stresses lower than the fatigue (endurance) limit ensure infinite cyclic life.
3. A finite-life region where a specified number of cycles can be endured at a specified stress level that is greater than the fatigue limit.

For infinite life, the fatigue strength is often called the endurance limit S_e where:

$$S_e = S_f \qquad \text{for } N > 10^6 \text{ cycles} \qquad (5.15)$$

For finite life, with the number of cycles less than 10^6, the fatigue strength S_f is larger than S_e. The value of S_f used in a design analysis is determined from the S_f-N diagram for the material used to fabricate the structural elements. Example 5.10 demonstrates the techniques used to determine the strength S_f associated with finite life.

In comparing the cyclic stresses with the strength as defined in Fig. 5.15, we use the alternating portion of the applied stresses. The alternating stress, σ_a, and the mean stress, σ_m, for different types of cyclic loading are defined in the stress-time diagrams presented in Fig. 5.16. We compute the mean stress for cyclic loading by identifying the maximum and minimum stress in a given cycle of applied loading. The mean stress is the average of the maximum and minimum stress. The relation used to determine the mean cyclic stress is given by:

$$\sigma_m = (\sigma_{max} + \sigma_{min})/2 \qquad (5.16)$$

The alternating stress is determined from the difference in the maximum and minimum stress during a typical load cycle. The relation used to determine the alternating cyclic stress is given by:

$$\sigma_a = (\sigma_{max} - \sigma_{min})/2 \qquad (5.17)$$

The method used to determine the safety of a structural member when subjected to cyclic loading is described in Example 5.9.

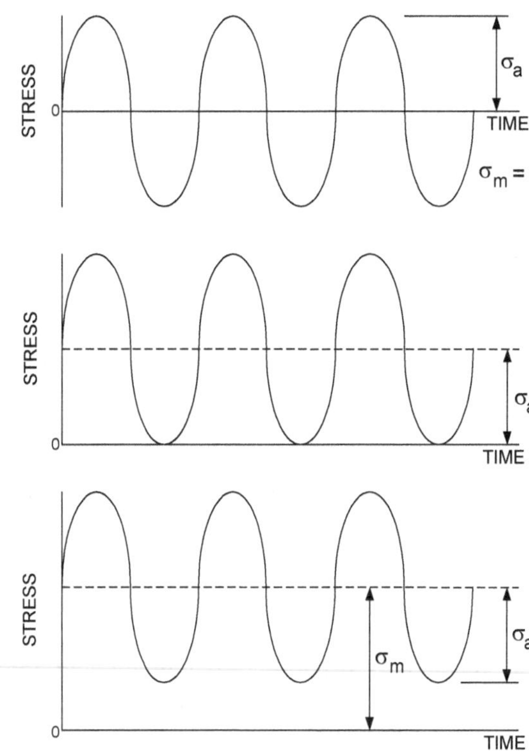

Fig. 5.16 Different types of cyclic stresses imposed on a structure.

EXAMPLE 5.9

Consider a cross beam used in constructing a bridge. The dead weight of the bridge structure produces a stress of 22,000 psi in this beam. However, when a fully loaded tractor-trailer crosses over the beam the stress increases to 65,000 psi. The beam is fabricated from a material with the fatigue properties described in Fig. 5.15. The strengths, yield and ultimate tensile, are 54 ksi and 88 ksi respectively. Will the beam fail in fatigue? If failure occurs, predict the cyclic life of the structure. Neglect the effect of the mean stress on fatigue strength in this example.

Solution: Determine the alternating stresses imposed on the bridge beam from Eq. (5.17).

$$\sigma_a = (\sigma_{max} - \sigma_{min})/2 = (65{,}000 - 22{,}000)/2 = 21.5 \text{ ksi}$$

Next, compare this value to the fatigue limit of 43 ksi that is obtained by reading the S_f - N curve in Fig. 5.15. Clearly S_f = 43 ksi > σ_a = 21.5 ksi; the alternating stresses are less than the endurance limit and the bridge beam is safe. Indeed, the safety factor is determined from a modified form of Eq. (4.11) as:

$$\mathbf{SF_f} = S_f/\sigma_a \qquad (5.18)$$

$$\mathbf{SF_f} = 43/21.5 = 2.00$$

The mean stress imposed on the bridge beam during the cyclic loading is determined from Eq. (5.16) as:

$$\sigma_m = (\sigma_{max} + \sigma_{min})/2 = (65{,}000 + 22{,}000)/2 = 43.5 \text{ ksi}$$

This is a relatively large mean stress to impose on a beam. As a consequence, the fatigue strength will be lower than that obtained from the S_f–N curve to account for the effects of the large tensile mean stress. This solution is modified in Example 5.11 to account for the effects of mean stresses in reducing the allowable alternating stresses.

EXAMPLE 5.10

Consider the crankshaft of a high performance racecar which is subjected to maximum and minimum stresses of + 345 MPa and – 345 MPa. If the crankshaft is fabricated from the high strength low alloy steel described in Fig. 5.15, determine its fatigue life.

Solution: Let's first determine the alternating stress σ_a for the crankshaft from Eq. (5.17).

$$\sigma_a = (\sigma_{max} - \sigma_{min})/2 = [345 - (-345)]/2 = 345 \text{ MPa}$$

Next convert this result from MPa units to ksi units so that we may employ a graphical approach using the fatigue properties of the high strength low alloy steel shown in Fig. 5.15.

$$\sigma_a = 345 \text{ MPa} \times 1 \text{ ksi}/6.895 \text{ MPa} = 50.0 \text{ ksi}$$

Plot σ_a = 50.0 ksi at point A on the ordinate of the S_f-N diagram shown in Fig. E5.10. Extend a line parallel to the abscissa until it intersects the shaded (fatigue failure) region and plot point B. Drop a vertical line from point B until it intersects the axis defining the cyclic life to establish point C. Point C gives the anticipated cyclic life, which in this example is less than 10^6 cycles. For a crankshaft operating at 6000 RPM in a high-performance racing car, the anticipated life is less than three hours. Clearly, this life is too short by several orders of magnitude and the design of the crankshaft is inadequate. Either the crankshaft size must be increased to lower the stress level or a different material with a higher endurance limit S_e must be selected.

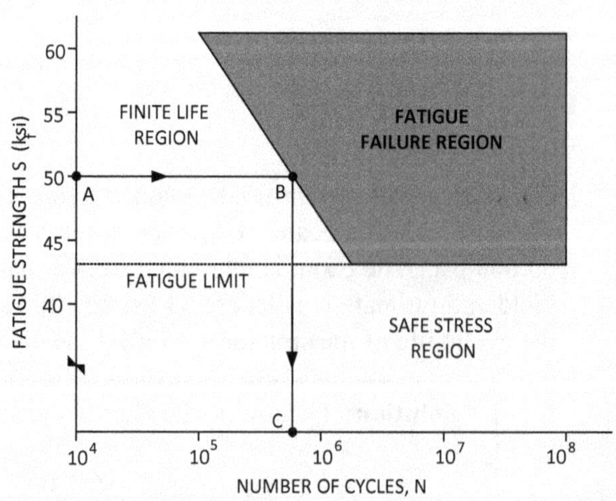

Fig. E5.10

5.7.2 The Goodman Diagram

In some instances, the cyclic loading on a structural member produces a combination of alternating and mean stresses. When mean stresses are superimposed on the alternating stresses, the effect is to decrease the fatigue strength of the material. Goodman and Gerber have developed empirical methods to determine a modified fatigue strength that accounts for the detrimental effects of combined alternating and mean stresses. The modified fatigue strength S_a is presented as a function of the cyclic mean stress in Fig. 5.17. Examination of Fig. 5.17 reveals that the modified fatigue strength S_a is equal to S_e when $\sigma_m = 0$ for all the empirical relations. The decrease in the modified fatigue strength is linear with the increase in σ_m, with the Goodman method. The fatigue strength S_a becomes zero when $\sigma_m = S_u$, for the Goodman method and Gerber methods.

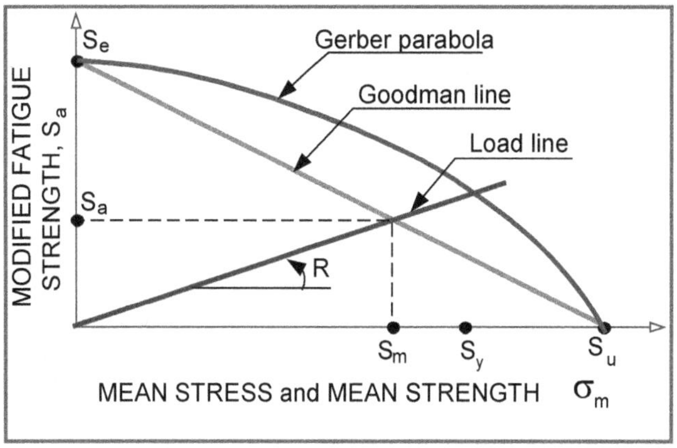

Fig. 5.17 Modified fatigue strength S_a decreases as the cyclic mean stresses increase.

Let's consider the Goodman method, because it is the most commonly used technique and it is conservative compared to the Gerber method. We show a load line with a slope R in Fig. 5.17. The slope R is given by the ratio:

$$R = \sigma_a/\sigma_m = S_a/S_m \quad (5.19)$$

The load line extends from the origin and intersects the Goodman line. At this point, vertical and horizontal lines are drawn to intersect the axes, defining the allowable alternating strength S_a and the allowable alternating mean strength S_m. We may also characterize the modified fatigue strength S_a in equation format for the Goodman and Gerber methods as:

$$S_a = S_e\,[1 - (S_m/S_u)] \qquad \text{(Goodman)} \qquad (5.20a)$$

$$S_a = S_e\,[1 - (S_m/S_u)^2] \qquad \text{(Gerber)} \qquad (5.20b)$$

where S_e is the endurance limit, S_u is the ultimate tensile strength, and S_m is the mean strength.

A graph of S_a as a function of S_m for Eqs. (5.20) is shown in Fig. 5.17. The result shown as the blue curve is a parabola and the red curve is linear. Fatigue test data for both steels and aluminum alloys are usually located in the region between the red and the blue curves. The red line due to Goodman's method is more conservative than the Gerber method.

EXAMPLE 5.11

Let's reconsider the bridge described in Example 5.9. In the previous solution, we neglected the influence of the mean stress on the fatigue strength. In this solution, let's account for the degrading effect of the cyclic mean stress on the fatigue strength by using the Goodman method for adjusting the fatigue strength. Determine the safety factor using the values determined for the allowable alternating and mean fatigue strengths. Recall from Example 5.9 that the fatigue strength was 43 ksi.

Solution:

Recall the results for the alternating and mean stresses from Example 5.9.

$$\sigma_a = (\sigma_{Max} - \sigma_{Min})/(2) = (65{,}000 - 22{,}000)/(2) = 21.5 \text{ ksi} \quad \text{(a)}$$

$$\sigma_m = (\sigma_{Max} + \sigma_{Min})/(2) = (65{,}000 + 22{,}000)/(2) = 43.5 \text{ ksi} \quad \text{(b)}$$

Determine the slope R of the load line from Eq. (5.19) as:

$$R = \sigma_a/\sigma_m = 21.5/43.5 = 0.4943 \quad \text{(c)}$$

Next, let's determine the allowable alternating and mean strengths. Substituting Eq. (5.10) into Eq. (5.20a) yields:

$$S_a = R\, S_e\, S_u/(R\, S_u + S_e) \quad (5.21a)$$

and

$$S_m = S_a/R \quad (5.21b)$$

Solving for S_a and S_m gives:

$$S_a = \frac{(0.4943)(43)(88)}{(0.4943)(88)+43} = 21.62 \text{ ksi}$$

and (d)

$$S_m = S_a/R = 21.62/0.4943 = 43.74 \text{ ksi}$$

The safety factor is given by setting $S_a = S_f$ in Eq. (5.18) as:

$$SF_f = S_a/\sigma_a = 21.62/21.5 = 1.006 \quad \text{(e)}$$

When accounting for the mean stresses on fatigue strength, the safety factor was reduced slightly more than 1.0. In the previous solution for Example 5.9, the safety factor $\mathbf{SF_f} = 2.0$; whereas, in this analysis the safety factor is reduced to 1.006. Clearly, the effect of the mean stress is to reduce the allowable fatigue strengths S_a and S_m. With a safety factor of only 1.006, the beam size must be increased or a substitute material employed with higher fatigue strength.

5.8 SUMMARY

Structures and machine components fail because of fracturing, yielding or excessive elastic deformation. In predicting either fracture or yielding, we compare the stress imposed on the structure with the strength of the material from which it is fabricated. In some analyses, we employ a handbook to provide the strength of the materials, and in more critical designs we conduct standardized tensile tests to establish the strength of the specific materials to be employed in construction.

The standard ASTM tensile test is introduced in Section 5.3. Using a standard tensile specimen and standard test procedures a force-deflection curve is generated. We convert this force-deflection data to a stress-strain curve, and then interpret the curve to give the strength, ductility and the elastic constants. Care is exercised in the interpretation to distinguish between "brittle" and "ductile" materials because the yield strength cannot be measured for brittle materials. For ductile materials, procedures are described for measuring both the yield strength and the ultimate tensile strength.

The ductility of a material is indicated by two measurements — the percent elongation and the percent reduction in area.

$$\%\, e = \frac{L_f - L_0}{L_0} \times 100 \qquad (5.4)$$

$$\%\, A = \frac{A_0 - A_f}{A_0} \times 100 \qquad (5.5)$$

Methods for determining both of these quantities have been given. We noted that for most types of steel the ductility decreased with increasing carbon content used for strength enhancement.

Two elastic constants, the modulus of elasticity E and Poisson's ratio ν, are measured in a tensile test. The modulus of elasticity is determined from the slope of the stress-strain curve in the elastic region. Poisson's ratio is determined from the ratio of strain in both the axial and transverse directions measured during a uniaxial tension test.

$$\nu = -\,\varepsilon_t/\varepsilon_a \qquad (5.6)$$

The shear modulus, another elastic constant often used in analysis of shear stresses and torsion loading of circular shafts, is introduced.

$$\tau = G\,\gamma \qquad (5.7)$$

$$G = \frac{E}{2(1 + \nu)} \qquad (5.8)$$

Clearly, the shear modulus G is not independent as it is a function of E and ν.

Two versions of Hooke's law were discussed. The historical version, due to Robert Hooke, where $\sigma = E\varepsilon$, is valid only when the state of stress is uniaxial. However, when the structure is subjected to biaxial stresses, the uniaxial form of Hooke's law is not valid. Instead we must represent the relations between stress and strain with the biaxial equations given by:

$$\varepsilon_x = \frac{1}{E}\left(\sigma_x - \nu\sigma_y\right); \quad \varepsilon_y = \frac{1}{E}\left(\sigma_y - \nu\sigma_x\right) \qquad (5.9)$$

$$\sigma_x = \frac{E}{\left(1-\nu^2\right)}\left(\varepsilon_x + \nu\varepsilon_y\right); \quad \sigma_y = \frac{E}{\left(1-\nu^2\right)}\left(\varepsilon_y + \nu\varepsilon_x\right) \qquad (5.10)$$

The difference between engineering and true stress and strain is described. Definitions of true stress and true strain are given, and a relation between true strain and engineering strain is derived.

$$\varepsilon_T = \int_{L_0}^{L_F} \frac{dL}{L} = \ln \frac{L_F}{L_0} \qquad (5.13)$$

$$\varepsilon_T = \ln (L_F/L_o) = \ln [(L_o + \Delta L)/L_o] = \ln (1 + \varepsilon) \quad (5.14)$$

The difference between engineering and true strains is not significant if the strains are small, but the difference is significant for large values. For this reason, we usually employ engineering stress and strain if we are performing analyses of structures subjected to stresses and strains in the elastic region. Structures subjected to stresses beyond the yield conditions deform significantly, and we usually perform plastic analysis using the definitions of true stress and true strain.

The concept of failure by cyclic loading is introduced. The repeated loading induces fatigue damage in structural members, which produces a brittle failure at stresses that are often lower than the measured yield stress. The $S_f - N$ curve that describes the relation between the number of cycles and the fatigue strength of a material is presented. Examples are provided that illustrate an approach for determining the safety of a structure or a machine component subjected to fatigue loading.

REFERENCES

1. Hertzberg, R. W., Deformation and Fracture Mechanics of Engineering Materials, 4th Edition, John Wiley, New York, NY 1996.
2. Madayag, A. F. Metal Fatigue, John Wiley, New York, NY, 1969.
3. Forrest, P. G., Fatigue of Metals, Pergamon Press, Oxford, UK, 1962.

PROBLEMS

5.1 Sketch a stress-strain diagram for a relatively brittle material and identify the ultimate tensile strength on the diagram.

5.2 Sketch a stress-strain diagram for a relatively ductile material and identify the yield strength and ultimate tensile strength on the diagram.

5.3 Sketch a stress-strain diagram for a ductile material that does not exhibit a well-defined yield point. Identify on the diagram the yield strength and the ultimate tensile strength.

5.4 Write a test plan for conducting a standard tensile test.

5.5 Determine the yield strength and the ultimate tensile strength for the material represented by the stress-strain diagram shown in the figure to the right.

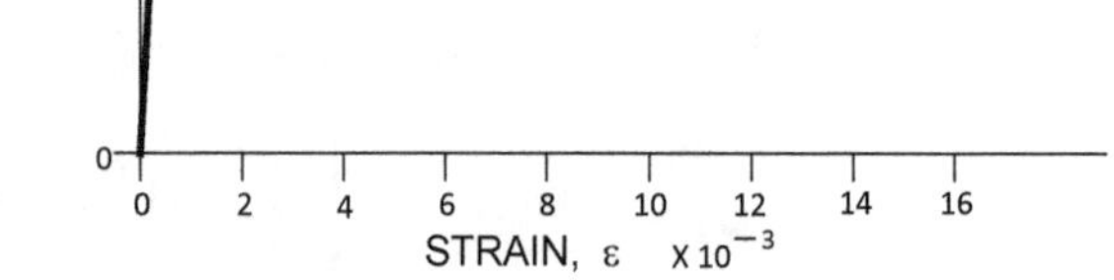

5.6 Determine the yield strength and the ultimate tensile strength for the material represented by the stress-strain diagram shown in the figure to the left.

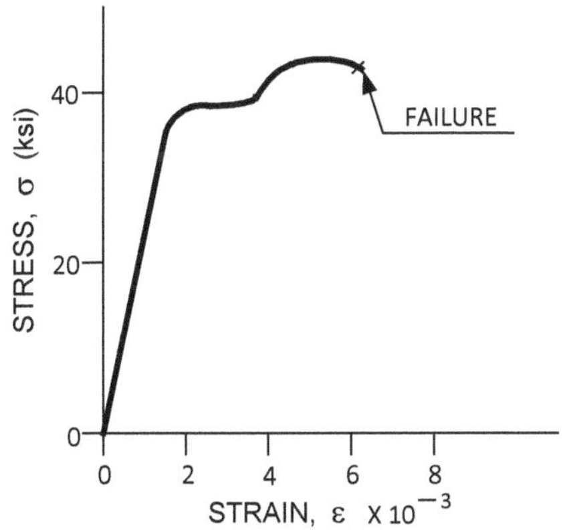

5.7 Determine the percent elongation and the percent reduction in area if measurements of L_f = 66.4 mm and d_f = 10.52 mm were made during a standard tensile test. The initial length and diameter of the standard tensile specimen is L_o = 50 mm and d_o = 12.7 mm.

5.8 In conducting a tensile test of an aluminum alloy, you adjust the signals from the load cell and the extensometer to zero and then load the specimen. You dwell at several loads in the elastic region of the stress-strain response, and record readings of the load F and the stretch δ. The values of F and δ are plotted on an F-δ diagram to obtain the slope $\Delta F/\Delta\delta$ = (16,000 N)/(100 × 10^{-3} mm). Determine the elastic modulus of the aluminum alloy if the initial length and diameter of the tensile specimen is L_o = 50 mm and d_o = 12.7 mm.

5.9 If a steel alloy exhibits a Poisson's ratio of $\nu = 0.30$, determine the diameter of a tensile specimen when it is subjected to an elastic stress of $\sigma = 52.6$ ksi. The dimensions L_o and d_o for the specimen are 2.00 in. and 0.500 in., respectively.

5.10 A long (3 m) sheet of an aluminum alloy is stretched in a manufacturing process until it is 3.8 m in length. If the sheet was initially 0.6 m wide and 2 mm thick, determine its new width and thickness. In the plastic regime, Poisson's ratio is 0.5 for all metallic materials because volume is conserved in the plastic deformation process.

5.11 A long (3.8 m) sheet of an aluminum alloy is stretched in a manufacturing process until it is 4.6 m in length. If a central hole initially 75 mm in diameter was drilled in the sheet prior to stretching, determine its new dimensions. In the plastic regime, Poisson's ratio is 0.5 for all metallic materials because volume is conserved in the plastic deformation process.

5.12 Prove that volume of an object is conserved in a uniaxial stress state if $\nu = ½$.

5.13 If a steel alloy is linearly elastic until the applied axial stress equals the yield strength, determine the strain at yield for a structural steel with a yield strength of 38 ksi and a modulus of elasticity of 30 × 10^6 psi.

5.14 A long, thin, aluminum-alloy bar is subjected to an axial stress of 227 MPa. Determine the strain in the axial and transverse directions of the bar.

5.15 A spherical pressure vessel, fabricated from a steel alloy, is subjected to a biaxial stress field with $\sigma_x = \sigma_y$ = 20.6 ksi. Determine the strains ε_x and ε_y.

5.16 An electrical strain gage is placed with an arbitrary orientation on a spherical pressure vessel fabricated from a titanium alloy. The gage provides a strain measurement ε = 1,140 × 10^{-6}. Determine the stresses σ_x and σ_y in terms of MPa.

5.17 Prove that the ratio of hoop strain to axial strain for a cylindrical pressure vessel is given by:

$$\varepsilon_h/\varepsilon_a = (2 - \nu)/(1 - 2\nu).$$

5.18 An orthogonal pair of electrical strain gages is placed on a cylindrical pressure vessel fabricated from an aluminum alloy. The strain gages provide measurements of the axial strain ε_a = 264 × 10^{-6} and hoop strain ε_h = 1,232 × 10^{-6}. Determine the axial stress σ_a and the hoop stress σ_h in terms of MPa.

5.19 A standard tensile specimen ruptured with an applied force F = 11.3 kip. Your measurement of the diameter at the neck region after failure gives d_f = 0.366 in. Determine the engineering stress and the true stress at failure.

5.20 Determine the engineering strain and the true strain for a tensile specimen with L_o = 2.00 in. if the final length, L_f is given in the table shown below:

Final Length L_f (in.)	Engineering Strain	True Strain
2.1		
2.2		
2.3		
2.4		
2.5		
2.6		
2.7		
2.8		
2.9		

5.21 Determine the engineering strain and the true strain for a tensile specimen with L_o = 50 mm if the final length L_f is given in the table below:

Final Length L_f (mm)	Engineering Strain	True Strain
53		
56		
59		
62		
65		
68		
71		
74		
77		

5.22 Complete the conversion table that relates engineering strain and true strain.

Engineering Strain	True Strain
0.001	
0.002	
0.005	
0.010	
0.020	
0.050	
0.100	
0.200	
0.500	
1.000	

5.23 The maximum stress imposed at the root of a tooth on a spur gear is σ_{max} = 78.9 ksi and the minimum stress σ_{min} = 0. Neglecting the influence of the cyclic mean stress on the fatigue life, determine the anticipated life of the gear, if the $S_f - N$ curve presented to the right characterizes its fatigue properties.

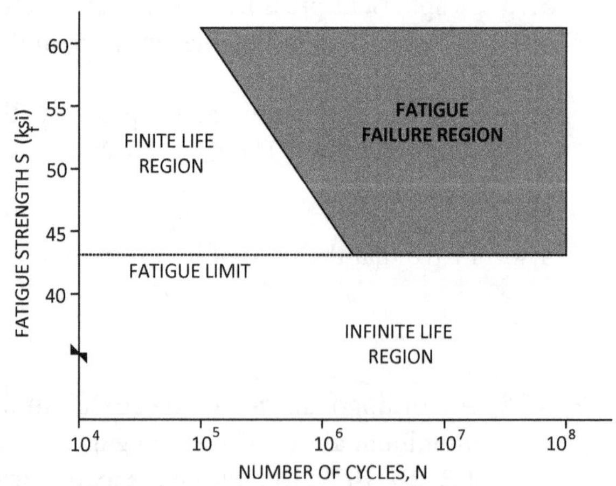

5.24 A rule for approximating the fatigue strength is $S_f = S_u/2$. Using this rule, determine the maximum stress that can be imposed on a structural member with a dynamic load that is twice the static load as illustrated in the figure below. The structural member is fabricated from steel with $S_u = 340$ MPa. Neglect the influence of the cyclic mean stresses on the fatigue behavior.

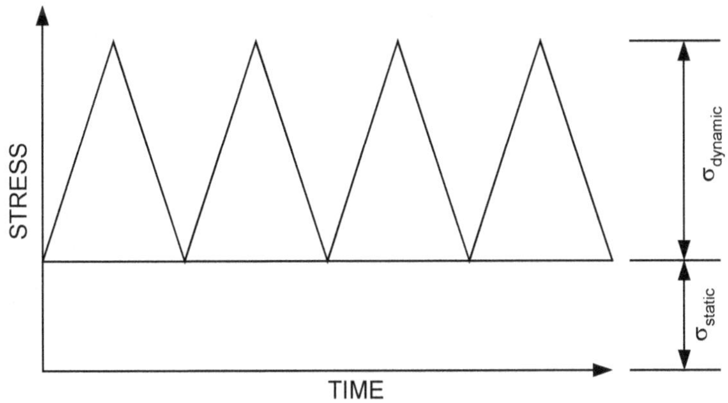

5.25 Reconsider Problem 5.23 taking into account the effect of the cyclic mean stress.

5.26 Reconsider Problem 5.24 taking into account the effect of the cyclic mean stress.

5.27 Repeat Problem 5.24 with the dynamic load equal to the static load.

5.28 Repeat Problem 5.24 with the dynamic load equal to one half of the static load.

5.29 Beginning with Eq. (5.9) derive Eq. (5.10).

5.30 Convert the stress-strain curve shown in the figure to the right to a true stress-true strain curve.

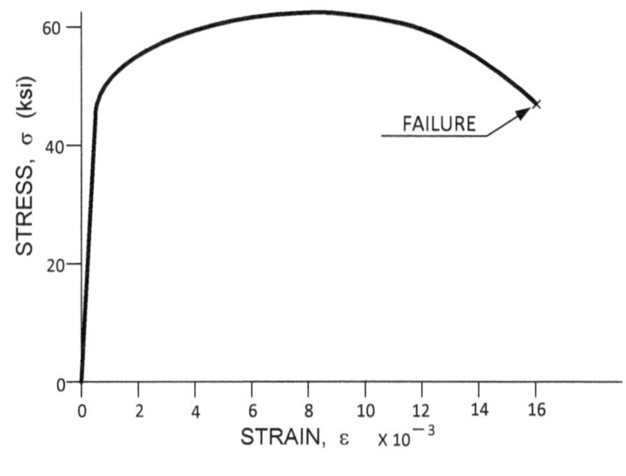

5.31 Convert the stress-strain curve shown in the figure to the left to a true stress-true strain curve.

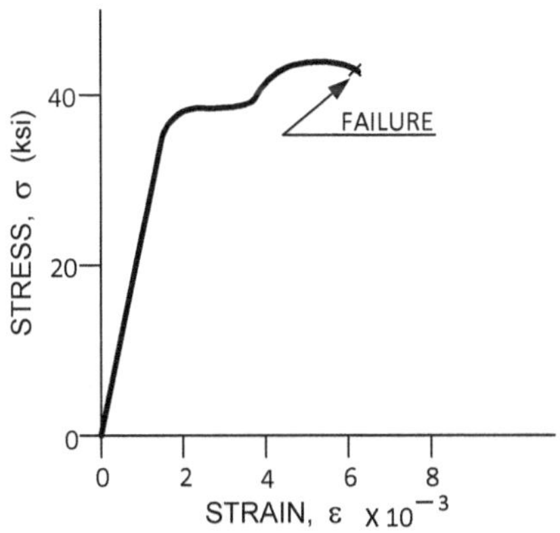

5.32 At the proportional limit in a tensile test, a 2 in. gage length had elongated by 0.003 in. and the diameter of the standard specimen was smaller by 0.00026 in. The load measured by a load cell fitted to the tensile machine at the proportional limit was 5,000 lb. Determine the modulus of elasticity, Poisson's ratio and the proportional limit for this material.

5.33 A rod with a diameter of 1.00 in. and length of 8.0 ft undergoes an extension of 0.220 in. when subjected to an axial force of 54.0 kip. The diameter of the rod decreases by 0.0007 in. at this load. Determine the modulus of elasticity, Poisson's ratio and the shear modulus for the rod's material.

5.34 A standard tensile specimen was fitted with an extensometer with a 2.00-inch gage length and tested until failure. The force and extension measured during the test is presented in the table below. The diameter of the specimen at the fracture neck was 0.413 in. Analyze the data using a spreadsheet and determine the following quantities.

a. Modulus of elasticity
b. Yield strength (0.2% offset)
c. Ultimate strength
d. Fracture stress
e. Strain at fracture
f. Percent elongation
g. Percent reduction in area.

Load (kip)	Extension (in.)	Load (kip)	Extension (in.)
0	0	10.8	0.251
2.0	0.0021	11.2	0.282
4.0	0.0039	11.6	0.303
6.0	0.0058	11.6	0.318
7.0	0.0072	11.5	0.327
7.8	0.0084	11.2	0.351
8.3	0.0280	10.8	0.367
8.7	0.0530	10.1	0.394
9.2	0.0860	9.7	0.402
9.8	0.1450	9.6	0.420
10.4	0.2220		

5.35 A standard tensile specimen 12.7 mm in diameter was fitted with an extensometer with a 50 mm gage length and tested until failure. The force and extension measured during the test is presented in the table below. The diameter of the specimen at the fracture neck was 9.0 mm. Analyze the data using a spreadsheet and determine the following quantities.

a. Modulus of elasticity
b. Yield strength (0.2% offset)
c. Ultimate strength
d. Fracture stress
e. Strain at fracture
f. Percent elongation
g. Percent reduction in area.

Load (kN)	Extension (mm.)	Load (kN)	Extension (mm)
0	0	35.0	3.54
5.0	0.010	37.0	4.52
10.0	0.019	38.0	5.53
15.0	0.029	38.0	6.18
20.0	0.042	36.0	6.66
25.0	0.048	33.0	7.35
26.0	0.50	30.0	7.86
27.0	0.95	27.0	8.42
29.0	1.50	23.0	8.78
31.5	2.21	19.0	9.32

CHAPTER 6

TRUSSES

6.1 INTRODUCTION

Trusses are structures commonly used to efficiently span relatively long distances. As such, they are employed in the design of bridges or the support for roofs of stadiums and both large and small buildings. A **truss** is defined as a large structure made of many smaller uniaxial members (bars, rods or cables) that are connected together to form a strong and stiff arrangement. The term truss is derived from the Middle English word *trusse*, which means, "bundle". Several of the geometries used in truss design are presented in Fig. 6.1.

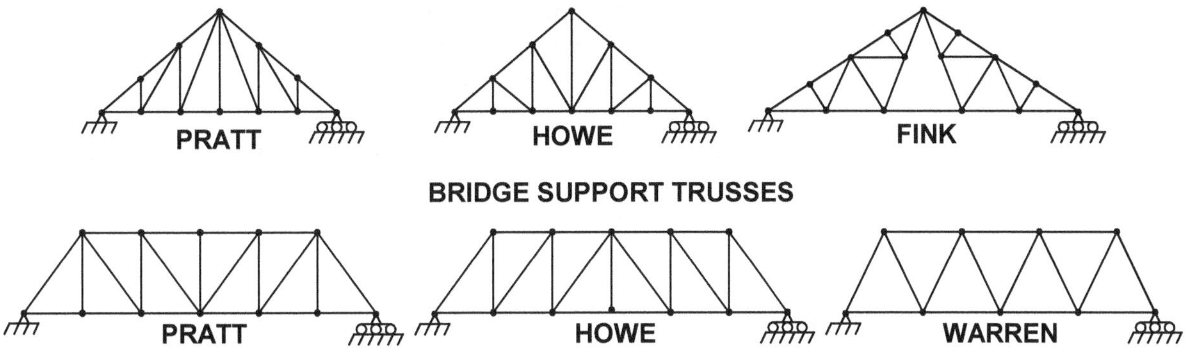

Fig. 6.1 Different types of trusses used in the design of bridges and roof supports.

6.1.1 Stability

If we examine the geometry of the trusses presented in Fig. 6.1, it is clear that they are all made from straight members arranged to form an array of triangles. The triangular form is very important because it insures stability of the truss regardless of the direction of the applied forces. Let's examine the stability of a four-member frame with a rectangular arrangement of the members as defined in Fig. 6.2. The rectangular frame in Fig. 6.2a is fabricated with joints at the four corners A, B, C and D. If a horizontal force F is applied to this frame at point C, the joints tend to act as hinges[1], and the frame rotates as shown in Fig. 6.2b. If the force F is maintained as the frame rotates, the structure collapses.

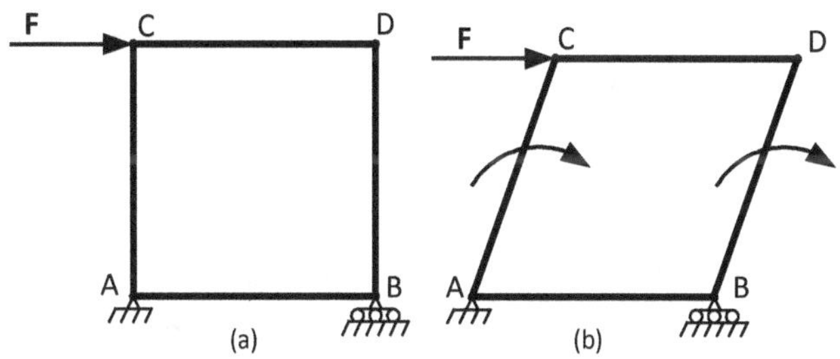

Fig. 6.2 A four-bar frame with a rectangular arrangement is unstable.

To stabilize the rectangular frame of Fig. 6.2, we add a fifth bar between points A and D as shown in Fig. 6.3. With the application of the horizontal force F, the joints will again tend to rotate. However, significant rotations of the

[1] Joints can be designed to resist rotation; however, it is usually more efficient and less costly to insure stability by adding an additional member to the structure that prevents rotation.

structure require member AD to elongate by yielding. If member AD is designed with sufficient strength, the truss cannot deform and its stability is ensured.

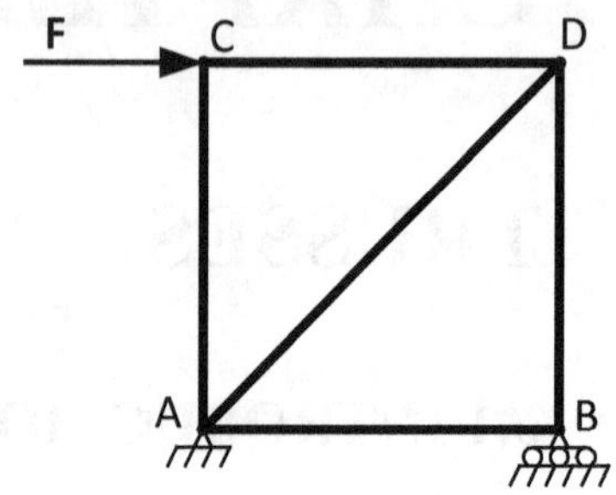

Fig. 6.3 Converting the rectangular structure into a truss with triangular elements assures its stability.

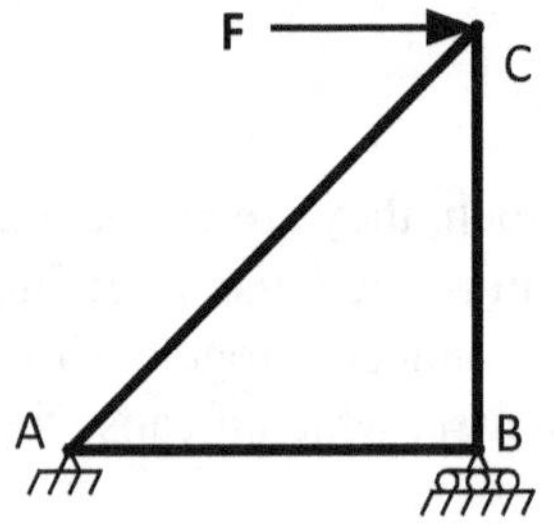

Fig. 6.4 Triangular structural elements are stable.

Note the addition of the member AD has converted the rectangular truss arrangement into one consisting of two triangles. The triangular element is the essence of stability of a structural arrangement. A simple three-member triangular truss is illustrated in Fig. 6.4 with a horizontal force F applied at point C. Even if the joints are made with frictionless pins (free to rotate), the structure remains stable. Stability is insured by the presence of member AC that must experience large deformations before rotation of member BC can occur.

Examination of the truss geometries presented in Fig. 1.6 shows the repeated use of the triangular elements to increase the size of the truss. With the addition of the new members, the number of joints used in fabricating the truss is also increased. For trusses fabricated from simple triangular elements, the number of joints and the number of members are related by:

$$n = 2k - 3 \tag{6.1}$$

where n is the number of members and k is the number of joints.

6.1.2 The Truss Members

The members of a truss are long and thin. As such, they cannot support transverse (lateral) forces; consequently loads applied to the structure must be placed at the joints. The weight of most truss members is usually negligible in comparison to the applied forces. However, if the weight of a specific member is significant, because of an exceptionally high density or extremely long length, the weight of the member is divided by two and applied to its joints—thus accounting for the effect of the member's weight in the analysis.

The joints are usually constructed using a gusset plate to effect the connection of the various members that meet at a specified point. We illustrate a gusset plate used in fabricating a joint between four members in Fig. 6.5. The members are fastened to the gusset plate by riveting, bolting or welding.

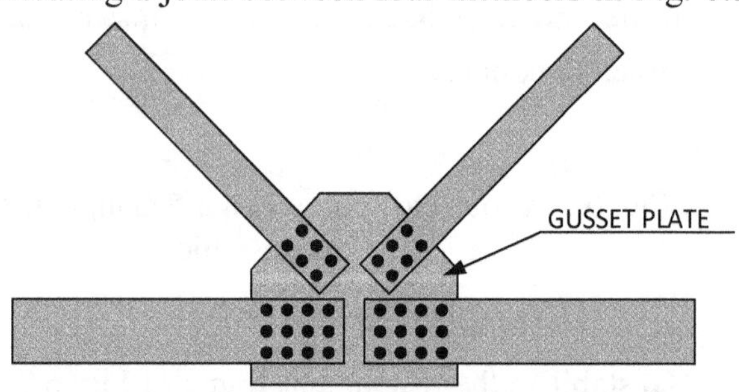

Fig. 6.5 A joint formed with a gusset plate at the junction of four truss members.

Even though the joint between two or more members may resist rotation and support moments, we assume that the joint is free to rotate. This assumption does not produce significant errors in the analysis if the individual truss members are long and flexible. Also this assumption permits us to treat each truss member as a **two-force member**—free of moments and transverse forces. For a two-force member, the force acting on the member must be transmitted through the axis of the member. Consequently, the direction of the force is established by the orientation of the member. The fact that the individual truss members are two-force members permits us to represent members, joints, and sections with the three different FBDs shown in Fig. 6.6.

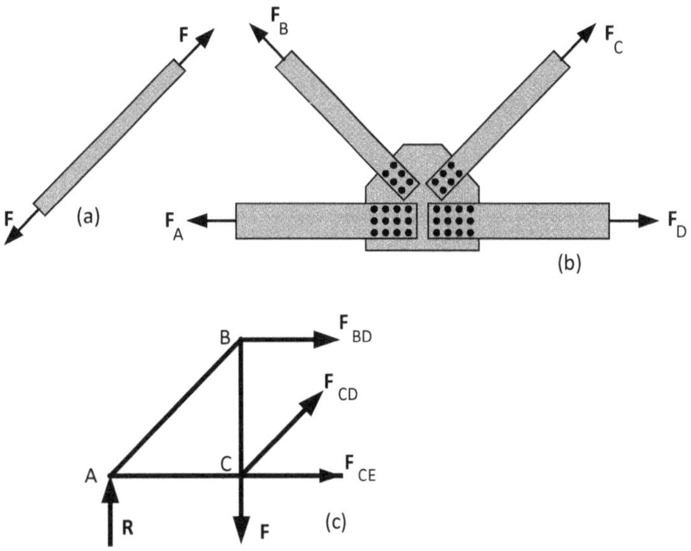

Fig. 6.6 FBDs for a truss: (a) an individual member from a truss, (b) a joint from a truss, and (c) a section of a truss.

6.2 METHOD OF JOINTS

In conducting a design analysis of a truss, we employ either the method of joints or the method of sections. Both methods employ a FBD of a portion of the truss as illustrated in Fig. 6.6. With the method of joints, we "remove" a specified joint from the truss and draw a free diagram of that joint including both internal and external forces that act on it. With the method of sections, we make a section cut to remove either the left or right side of the truss. We then prepare a FBD of one side or the other of the sectioned truss.

The method of joints is a simple extension of the material already covered in Chapters 3, 4 and 5. The analysis incorporates six steps:

1. Draw a FBD of the entire truss structure.
2. Apply the equations of equilibrium to solve for the reactions at the supports.
3. Select a joint and construct a FBD that includes both the internal and external forces.
4. Apply the equations of equilibrium to solve for the forces in the members that are connected at the subject joint. Since we have assumed that the joint acts like a particle, the joint is represented as a point and equilibrium is satisfied if $\mathbf{\Sigma F} = 0$. It is not necessary to consider $\sum \mathbf{M} = 0$.
5. Determine the stresses in the individual members by using $\sigma = P/A$, and note if the stresses are tensile or compressive.
6. Compare the stress in each member with the strength of the material used to fabricate the truss and then establish the margin of safety for each member.

To illustrate the procedure for a design analysis of a simple truss, consider the following example of a Howe truss.

EXAMPLE 6.1

Consider the Howe truss shown in Fig. E6.1. The total span of the truss is $S_p = 4s = 4 \times 12 = 48$ ft, and the height h = 16 ft. External forces F_C, F_E and F_G are applied at joints C, E and G. Determine the margin of safety for truss members AB and AC. The following parameters are necessary for a numerical solution for the margin of safety.

The forces are $F_C = 100{,}000$ lb, $F_E = 150{,}000$ lb, and $F_G = 100{,}000$ lb

The cross sectional areas of the bars employed in the construction of the truss are $A_{AC} = 4$ in^2. and $A_{AB} = 6$ in^2. These bars are fabricated from hot rolled mild steel with a yield strength of 38,000 psi.

Fig. E6.1

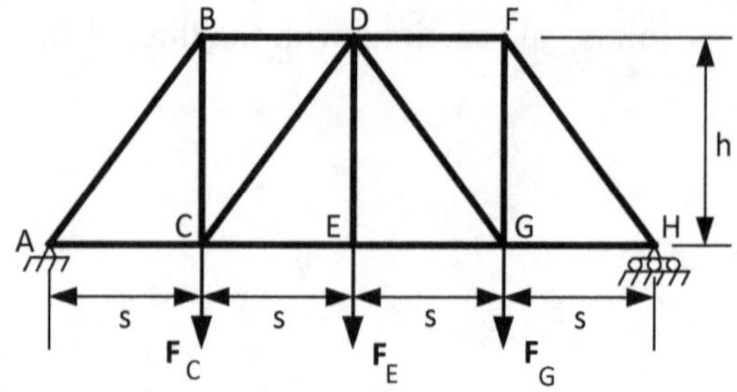

Solution:

Step 1: Draw a FBD of the entire truss structure.

Since it is necessary to determine the reactions at the left and right hand supports of the truss, the supports are removed and replaced with reaction forces R_L, R_R and R_x in the FBD shown In Fig. E6.1a:

Fig. E6.1a

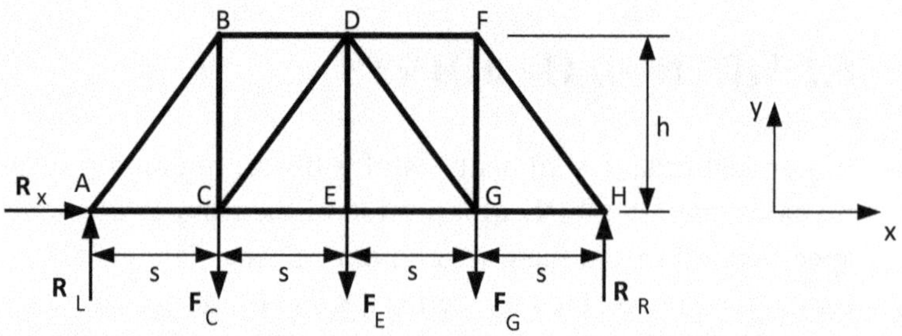

Step 2: Apply the equations of equilibrium to solve for the reactions at the supports. It is immediately evident from $\Sigma F_x = 0$ that the reaction force at the left support $R_x = 0$.

Consider the sum of the moments about point A and write:

$$\Sigma M_A = 4sR_R - 3sF_G - 2sF_E - sF_C = 0 \qquad \text{(a)}$$

Solving Eq. (a) for R_R yields:

$$R_R = (1/4)(3F_G + 2F_E + F_C) = (1/4)(300 + 300 + 100) = 175 \text{ kip} \qquad \text{(b)}$$

From equilibrium we write:

$$\Sigma F_y = R_L + R_R - F_C - F_E - F_G = 0$$

$$R_L = 100 + 150 + 100 - 175 = 175 \text{ kip} \qquad \text{(c)}$$

Both R_L and R_R are equal to 175 kip. The equality of the two reactions is expected because the geometry and the loading of the truss are both symmetrical.

Step 3: Select a joint and construct a FBD that includes both the internal and external forces acting at that joint. We choose the joint at point A because members AB and AC, the subject of the analysis, meet at this joint. The FBD is presented in Fig. E6.1b:

Fig. E6.1b

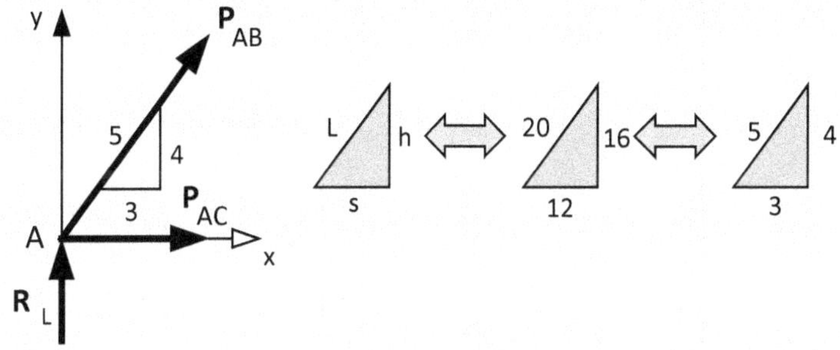

Note, the dimensions of the span and height of the primary triangular element in the truss provides the orientation of the internal force P_{AB} in member AB. In this FBD we have used:

- The symbol P to represent internal forces.
- The symbol R to represent reaction (external) forces.
- Subscripts to identify the truss members.
- The internal forces P_{AB} and P_{AC} as both positive (tensile)[2].

Step 4: Apply the equations of equilibrium to solve for the forces in the members that connect at the subject joint. Since we have assumed that the joint acts like a pin, the joint is represented as a point and equilibrium is satisfied if $\mathbf{\Sigma F} = 0$, or $\Sigma F_x = 0$ and $\Sigma F_y = 0$.

Let's begin with:

$$\Sigma F_y = R_L + (4/5)\, P_{AB} = 0$$

Solve this relation for P_{AB} by using the results of Eq. (c) to obtain:

$$P_{AB} = -\,(5/4)\, R_L = -\,(5/4)\, 175 = -\,218.8 \text{ kip} \qquad \text{(d)}$$

The negative sign for P_{AB} indicates that it is a compressive force. Our initial assumption indicating this force was tensile (+) was in error. This error does not cause a problem—we note the sign, declare that member AB is in compression, and treat P_{AB} as a negative quantity in subsequent steps in the analysis.

Next, write the equilibrium relation for the force components in the x direction.

$$\Sigma F_x = (3/5)P_{AB} + P_{AC} = 0$$

$$P_{AC} = (-\,3/5)P_{AB} = -\,(3/5)(-\,218.8) = +\,131.3 \text{ kip} \qquad \text{(e)}$$

The plus sign for P_{AC} indicates that the internal force in truss member AC is tensile. Our assumption about the direction of the force P_{AC} in constructing the FBD for joint A was correct.

Step 5: Determine the stresses in the individual members of the truss from $\sigma = P/A$, and note if the stresses are tensile or compressive.

For member AB and AC, we write:

$$\sigma_{AB} = P_{AB} / A_{AB} = -\,218.8 \text{ kip} / 6 \text{ in}^2 = -\,36.46 \text{ ksi} \qquad \text{(f)}$$

$$\sigma_{AC} = P_{AC} / A_{AC} = +\,131.3 \text{ kip} / 4 \text{ in}^2 = +\,32.82 \text{ ksi} \qquad \text{(g)}$$

The minus sign for σ_{AB} indicates that the stress in member AB is compressive[3]. The stress in member AC is tensile as indicated by the positive sign.

[2]It is usual practice to assume that the unknown forces in truss members are tensile.

Step 6: Compare the stress in each member to the strength of the material used to fabricate the truss and then establish the margin of safety for the structure.

From Eq. 4.12, we determine the safety factor for members AB and AC as:

$$SF_{AB} = S_y/\sigma_{AB} = 38.0/36.46 = 1.042 \quad \text{(h)}$$

$$SF_{AC} = S_y/\sigma_{AC} = 38.0/32.82 = 1.158 \quad \text{(i)}$$

Finally, let's determine the margin of safety MOS according to its definition:

$$MOS = SF - 1 \quad (6.2)$$

Then for members AB and AC, we determine:

$$MOS_{AB} = 1.042 - 1 = 0.042 = 4.2\% \quad \text{(j)}$$

$$MOS_{AC} = 1.158 - 1 = 0.158 = 15.8\% \quad \text{(k)}$$

Okay! We have completed our solution, and it is mandatory to interpret the results for the safety of both members AB and AC. **Clearly, these margins of safety are too small.** If the external loads are increased by only five percent, member AB will fail by yielding and begin to deform. This deformation will change the shape of the structure and redistribute the internal forces. The structure is in significant danger—collapse is pending.

When designing structures, it is common to incorporate safety factors of two or three which provide margins of safety of 100 to 200%. These safety factors are to accommodate for uncertainties in the construction and operation over the life of the structure. Uncertainties include:

- A possible increase in loading, deliberate or accidental, any time during the operation of the structure.
- Variations in the strength of the materials used in construction.
- Effect of fatigue (cyclic loading) in generating flaws (cracks) in the structure.
- Effects of corrosion on reducing section size or on inducing crack initiation in one or more members of the structure.

In this example problem, it is necessary to increase either the area of the cross section of members AB and AC or to increase the yield strength of the material from which the members are fabricated.

EXAMPLE 6.2

Determine the internal forces in members AB, AC, BC, BD, BE, DE, and DF of the truss defined in Fig. E6.2. Also determine the margin of safety for the highest stressed member. The external forces applied at the joints C, E, G, K, and I are all equal to 100 kip. The cross sectional area of all of the members used to construct the truss is equal to 10 in^2. The material used in fabricating the truss is hot rolled steel with a yield strength of 54 ksi. Note, s and h are 10 and 20 ft respectively.

[3]Since the stress in member AB is compressive, there is a possibility that the member may fail by buckling. Another analysis must be conducted to determine if the critical buckling load on this member has been exceeded. Methods for determining the critical buckling load on bars loaded in axial compression will be provided in a later chapter.

Fig. E6.2

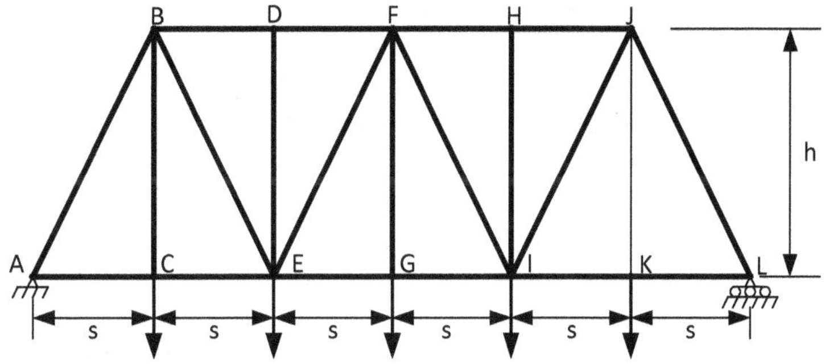

Solution:

Step 1: Construct a FBD of the entire truss:

Fig. E6.2a

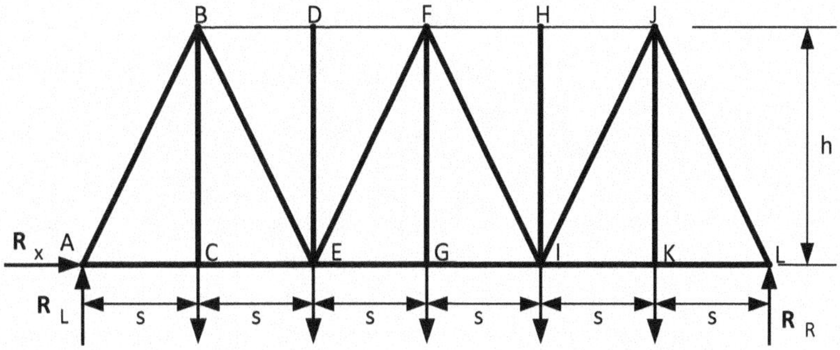

Step 2: Solve for the reactions at the supports.

Again it is clear from $\Sigma F_x = 0$ that $R_x = 0$.

Inspection of the structure shows that it is symmetric with respect to both geometry and loading. For this reason, it is evident that $R_L = R_R = R$.

$$\Sigma F_y = 2R - 5F = 0$$

$$R = (5/2)100 = 250 \text{ kip} \qquad \text{(a)}$$

Step 3: Select the joints necessary for solution of the unknown forces and prepare FBDs as shown in Fig. E6.2b.

Fig. E6.2b

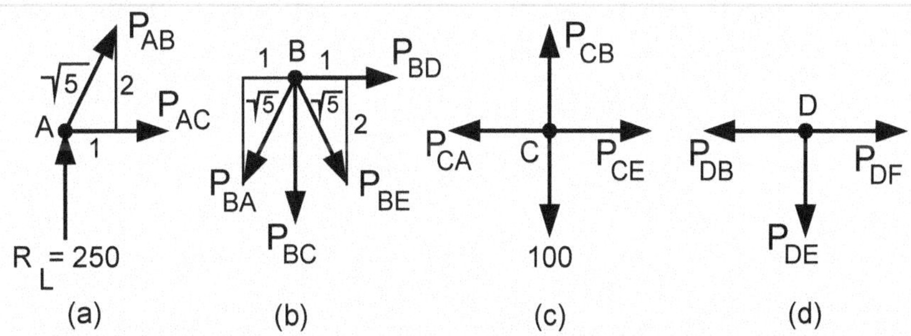

We have selected joints A, B, C and D because the forces occurring at these joints in the FBDs correspond to the unknown forces listed in the problem statement. Note, we have assumed all of the internal forces to be tension as is the usual practice. Also, the first subscript gives the letter of the joint described in the FBD. The second subscript gives the direction of the force. Since the truss members are two-force members in equilibrium, $P_{AB} = P_{BA}$, etc.

Step 4: Use equilibrium relations as necessary to solve for the unknown forces.

We have four FBDs—one for joint A, B, C and D. It is possible to write a total of eight relevant equilibrium relations if they are required. Let's write equilibrium equations beginning with the FBD of joint A, then B, C and D.

For joint A:

$$\Sigma F_y = R_L + (2/\sqrt{5})\, P_{AB} = 0$$

$$P_{AB} = -(\sqrt{5}/2)R_L = -(\sqrt{5}/2)(250) = -\,279.5 \text{ kip} \qquad \text{(b)}$$

Note, the minus sign for P_{AB} indicates the internal force in member AB is compressive.

$$\Sigma F_x = P_{AC} + (1/\sqrt{5})\, P_{AB} = 0$$

$$P_{AC} = -\,(1/\sqrt{5})P_{AB} = -\,(1/\sqrt{5})(-\,279.5) = 125 \text{ kip} \qquad \text{(c)}$$

For joint B:

Examining the FBD for joint B shows that four forces act at this point. We know only one of these forces, namely P_{BA}. The remaining three forces are unknown quantities. Since only two meaningful equilibrium relations may be written for a concurrent force system, the analysis of joint B is statically indeterminate. The analysis of joint B is possible only if we reduce the number of unknown forces to two. With the reduction of forces at joint B as a motivation, let's move onto joint C.

For joint C:

$$\Sigma F_x = P_{CE} - P_{CA} = 0$$

$$P_{CE} = P_{CA} = P_{AC} = 125 \text{ kip} \qquad \text{(d)}$$

$$\Sigma F_y = P_{CB} - 100 = 0$$

$$P_{CB} = 100 \text{ kip} \qquad \text{(e)}$$

Consider joint B again:

From the analysis of Joint C, we determined that P_{CB} = 100 kip. Since P_{CB} and P_{AB} are known, only two unknown forces (P_{BE} and P_{BD}) act at joint B; therefore, we may proceed with the solution by writing equilibrium relations:

$$\Sigma F_y = -\,P_{BC} - (2/\sqrt{5})(P_{BA} + P_{BE}) = 0$$

$$P_{BE} = -\,(\sqrt{5}/2)P_{BC} - P_{BA} = -\,(\sqrt{5}/2)100 - (-\,279.5) = +\,167.7 \text{ kip} \qquad \text{(f)}$$

$$\Sigma F_x = P_{BD} - (1/\sqrt{5})(P_{BA} - P_{BE}) = 0$$

$$P_{BD} = (1/\sqrt{5})(P_{BA} - P_{BE}) = (1/\sqrt{5})[\,-\,279.5 - 167.7] = -\,200.0 \text{ kip} \qquad \text{(g)}$$

Finally, consider joint D:

$$\Sigma F_y = P_{DE} = 0 \qquad \text{(h)}$$

$$\Sigma F_x = -P_{DB} + P_{DF} = 0$$

$$P_{DB} = P_{DF} = -200.0 \text{ kip} \qquad \text{(i)}$$

Since $P_{DE} = 0$, member DE of the truss structure is a zero-force member. We will discuss the arrangement of members at a joint in a truss that results in zero-force members later in this chapter.

Let's summarize the results of the analysis in Table 6.1.

Table 6.1
Summary of forces and stresses in Example 6.2

MEMBER	FORCE kip	AREA in^2	STRESS ksi	SAFETY FACTOR	MARGIN OF SAFETY
AB	279.5 C	10	27.95 C	1.93	0.93
AC	125.0 T	10	12.50 T	4.32	3.32
BC	100.0 T	10	10.00 T	5.40	4.40
BD	200.0 C	10	20.00 C	2.70	1.70
BE	167.7 T	10	16.77 T	3.22	2.22
DE	0	10	0	∞	∞
DF	200.0 C	10	20.00 C	2.70	1.70

Step 5: Solve for the stresses.

Let's determine the stresses in member AB of the truss.

$$\sigma_{AB} = P_{AB}/A_{AB} = -279.5/10 = -27.95 \text{ ksi}$$

The results for the stresses in the other members in the truss are shown in Table 6.1.

Step 6: Determine the safety factor and the margin of safety for the most highly stressed member.

Inspection of Table 6.1 indicates that member AB is the most highly stressed member in the truss. We determine its safety factor as:

$$SF = S_y/\sigma$$

For member AB:

$$SF = S_y/\sigma = 54/27.95 = 1.93$$

$$MOS = SF - 1 = 1.93 - 1 = 0.93 = 93\%$$

Finally, let's interpret the results that are presented in Table 6.1 from a designer's viewpoint. First, we note a significant difference in the stresses from one member to another. It is clear that the choice of 10 in^2 for the cross sectional areas of all of the members was not appropriate. A more uniformly safe design would have been possible if the areas had been adjusted for the different forces imposed on the different members. Area adjustment is possible, but it is preferable to use structural members with standard section sizes. It is not usually economically feasible to specify sectional areas that differ from the areas available with standard size members.

Second, the safety factor and the margin of safety for members AB, BD and DF are probably too low. We normally seek a safety factor of 2 to 3 to account for contingencies that have been described previously in Example 6.1.

Third, two **zero-force members** have been incorporated in the design: members DE and IH carry no force. The designer may have added them to the truss structure for aesthetic reasons, to provide backup if certain diagonal members failed by yielding, or in anticipation of loads not considered in this example. However, if a good reason for using these members does not exist, they should be eliminated. They add no strength to the structure and increase its cost, weight, and construction time.

EXAMPLE 6.3

A scissors truss illustrated in Fig. E6.3 is loaded with forces F_1, F_2 and F_3 at the joints located at points B, C and D, respectively. If the truss is fabricated from HR 1020 steel with a yield strength of 290 MPa, determine the size of members AB, AF, BC and BF. The safety factor is specified as 2.6. The forces and dimensions are:

$F_1 = 10$ kN, $F_2 = 12$ kN, $F_3 = 15$ kN ⇒⇒ $h = 5$ m, $s = 8$ m

Fig. E6.3

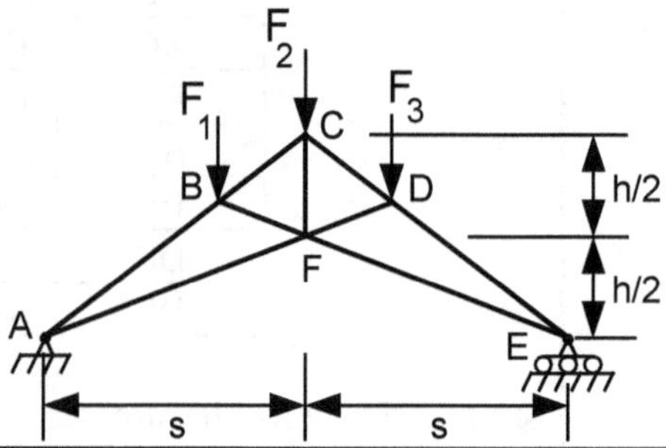

Solution:

Step 1: Construct a FBD of the entire truss as indicated in Fig. E6.3a. We remove the supports and replace them with the reactions for a pivot and roller. However, we encounter a problem in dimensioning the FBD. The dimensions locating points B and D (the points of application of forces F_1 and F_3) are not given explicitly in Fig. E6.3. It will be necessary to consider the geometry of the truss and compute the x coordinate of these points. Let's begin by forming the triangle ADE as shown in Fig. E6.3b. Note that the member EF has been removed to make the triangle more apparent.

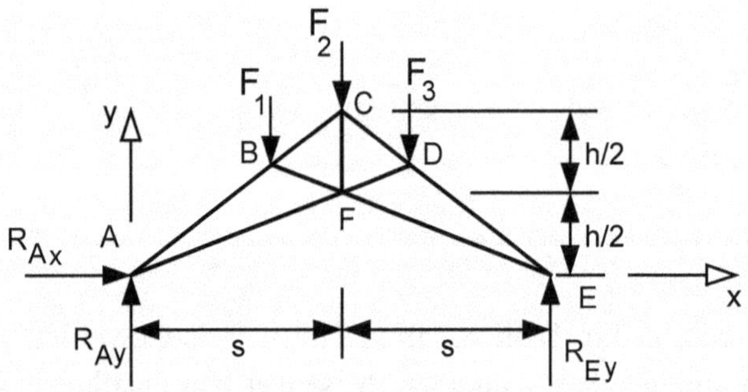

Fig. E6.3a

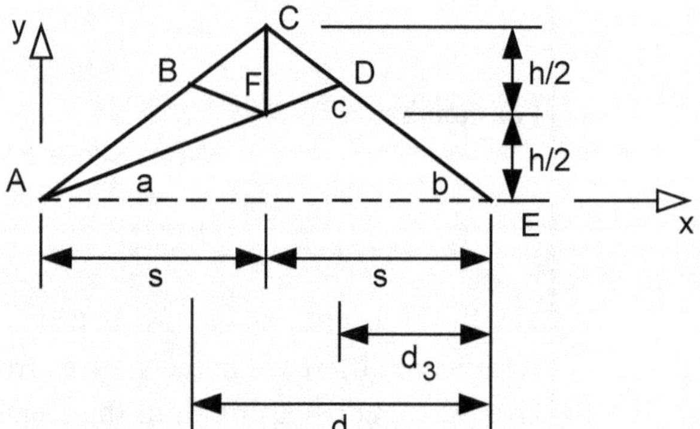

Solving for the angles a, b and c in the triangle ADE gives:

Fig. E6.3b

$$a = \tan^{-1}(h/2s) = \tan^{-1}(5/16) = 17.35°$$
$$b = \tan^{-1}(h/s) = \tan^{-1}(5/8) = 32.00°$$
$$c = 180° - a - b = 180° - 17.35° - 32.00° = 130.65°$$

The law of sines is employed to write:

$$\sin (c)/2s = \sin (a)/DE$$

Solving this relation for DE gives:

$$DE = (2s)[\sin (a)/\sin (c)] = 16[\sin (17.35°)/\sin (130.65)] = 6.289 \text{ m}$$

The distances d_3 and d_1 locating the points of application of forces F_3 and F_1 relative to point E are given by:

$$d_3 = DE \cos (b) = 6.289 \cos(32.00°) = 5.333 \text{ m}$$

Because of geometric symmetry, we may write:

$$d_1 = 2s - d_3 = 16 - 5.333 = 10.67 \text{ m}$$

Step 2: Solve for the reactions at the supports.

Again it is clear from $\Sigma F_x = 0$ that $R_{Ax} = 0$.

Inspection of the structure shows that it is not symmetric with respect to the loading. For this reason, $R_{Ay} \neq R_{Ey}$. Let's consider moments about point E and write:

$$\Sigma M_E = -2sR_{Ay} + d_1 F_1 + s F_2 + d_3 F_3 = 0$$

Solving for R_{Ay} gives:

$$R_{Ay} = (d_1 F_1 + s F_2 + d_3 F_3)/(2s) = [(10.67)(10) + (8)(12) + (5.333)(15)]/16 = 17.67 \text{ kN}$$

Let's solve for R_{Ey} by writing:

$$\Sigma F_y = R_{Ay} + R_{Ey} - F_1 - F_2 - F_3 = 0$$

$$R_{Ey} = 10 + 12 + 15 - 17.67 = 19.33 \text{ kN}$$

Step 3: Select the joints necessary for solution of the unknown forces and prepare FBDs as shown in Fig. E6.3c.

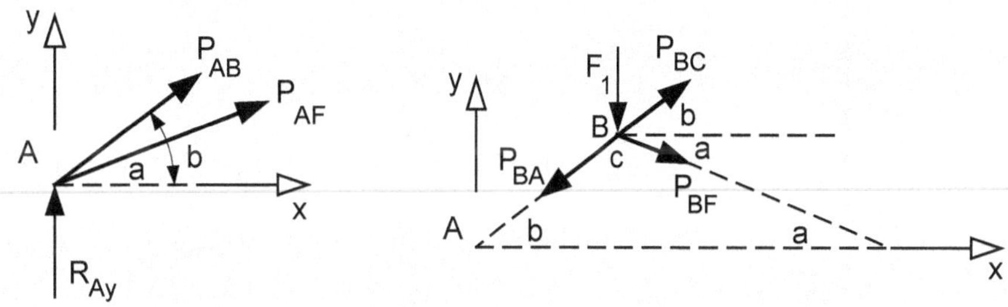

Fig. E6.3c

We have selected joints A, and B because the forces occurring at these joints in the FBDs correspond to the unknown forces listed in the problem statement. Note, we have assumed all of the internal forces to be tension as is the usual practice. Also, the first subscript gives the letter of the joint described in the FBD. The second subscript gives the direction of the force. Since the truss members are two-force members in equilibrium, $P_{AB} = P_{BA}$, etc.

Step 4: Use equilibrium relations as necessary to solve for the unknown forces.

We have two FBDs—one for joint A and another for joint B. It is possible to write a total of four relevant equilibrium relations if they are required. Let's begin to write equilibrium equations beginning with the FBD of joint A.

For joint A:

$$\Sigma F_y = R_{Ay} + P_{AB} \sin (b) + P_{AF} \sin (a) = 0$$

$$P_{AB} = -[R_{ay} + P_{AF} \sin (a)] / \sin (b) = -[17.67 + P_{AF} \sin (17.35°)]/\sin (32.00°)$$

$$P_{AB} = -33.34 \text{ kN} - 0.5627 P_{AF} \qquad \text{(a)}$$

$$\Sigma F_x = P_{AB} \cos (b) + P_{AF} \cos (a) = 0$$

$$P_{AF} = -P_{AB} \cos (b)/ \cos (a) = -P_{AB} \cos (32.00°)/ \cos (17.35°) = -0.8884 P_{AB} \qquad \text{(b)}$$

Combining the results from Eqs. (a) and (b) yields:

$$P_{AB} = -66.67 \text{ kN} \qquad \Rightarrow\Rightarrow \qquad P_{AF} = 59.23 \text{ kN} \qquad \text{(c)}$$

Note, the minus sign for P_{AB} indicates the internal force in member AB is compressive.

For joint B:

Examining the FBD for joint B shows that four forces act at this point. We know two of these forces, namely P_{BA} and F_1. The remaining two forces are unknown quantities that are determined from the two equilibrium relations that apply for a concurrent force system.

$$\Sigma F_x = P_{BC} \cos (b) + P_{BF} \cos (a) - P_{BA} \cos (b) = 0$$

$$P_{BC} \cos (32.00°) + P_{BF} \cos (17.35°) - P_{BA} \cos (32.00°) = 0$$

$$(0.8480)\, P_{BC} + (0.9545) P_{BF} - (-66.67)(0.8480) = 0$$

$$(0.8480)\, P_{BC} + (0.9545) P_{BF} + 56.54 = 0 \qquad \text{(d)}$$

$$\Sigma F_y = P_{BC} \sin (b) - P_{BF} \sin (a) - P_{BA} \sin (b) - F_1 = 0$$

$$P_{BC} \sin (32.00°) - P_{BF} \sin (17.35°) - P_{BA} \sin (32.00°) - 10.0 = 0$$

$$(0.5299)\, P_{BC} - (0.2982) P_{BF} - (-66.67)(0.5299) - 10.0 = 0 \qquad \text{(e)}$$

Solving Eqs. (d) and (e) for the two unknowns gives:

$$P_{BF} = -11.20 \text{ kN} \qquad \Rightarrow\Rightarrow \qquad P_{BC} = -54.10 \text{ kN} \qquad \text{(f)}$$

The minus signs in these results indicate that both members BC and BF are subjected to internal compression forces.

Step 5: Solve for the allowable design stresses in the members.

Let's determine the allowable design stresses from Eq. (4.12) as:

$$\sigma_{design} = S_y / \mathbf{SF_y} = 290/2.6 = 111.5 \text{ MPa} \qquad \text{(g)}$$

Step 6: Determine the areas required to limit the design stresses to the allowable level determined in Eq. (g).

$$A = P/\sigma_{dessign} = P/111.5 \quad \text{(h)}$$

The results from Eq. (h) are presented in Table 6.2.

Table 6.2
Results showing required dimensions for scissors truss members

Member	P (kN)	A (mm)	d_1 (mm)	d_2 (mm)
AB	–66.67	597.9	27.59	30
AF	59.23	531.2	26.01	30
BC	–54.10	485.2	24.86	30
BF	– 11.20	100.4	11.31	15

Inspection of Table 6.2 indicates the areas for each member. We have assumed that rods with a diameter d are to be employed in fabricating the scissors truss. In Table 6.2, the diameter d_1 is the computed value. Since rods of these sizes are not commercially available, we have increased the diameter to d_2 that corresponds to standard sizes that are available without special order.

6.3 ZERO-FORCE MEMBERS

Zero-force members are usually easy to identify in a structure if we examine all of the joints in the structure that do not carry externally applied loads. Let's consider two examples of joints with a zero-force member as shown in Fig. 6.7.

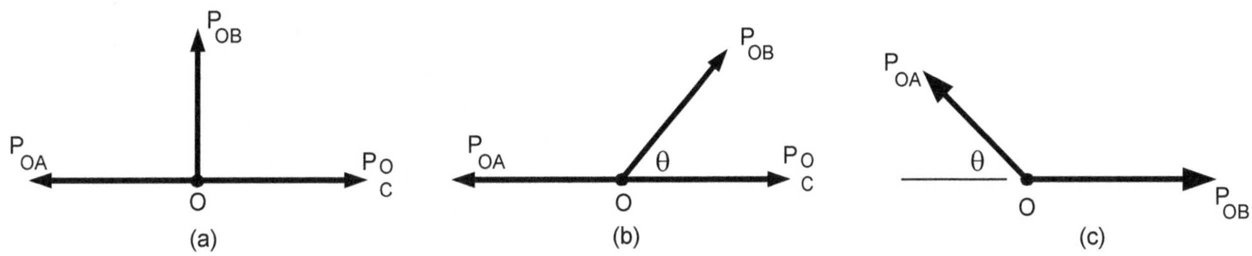

Fig. 6.7 Joint O with a zero-force member OB.

Let's begin by considering the equilibrium of joint O in Fig. 6.7a. Write $\Sigma F_y = 0$ and it is clear that:

$$\Sigma F_y = P_{OB} = 0$$

because the forces P_{OA} and P_{OC} do not have force components in the y direction. Also, from $\Sigma F_x = 0$, we may write that:

$$\Sigma F_x = -P_{OA} + P_{OC} = 0$$

$$P_{OA} = P_{OC}$$

Next, examine joint O in Fig. 6.7b. While force P_{OB} is inclined at some angle θ with respect to the x-axis, its component in the y direction is zero because neither P_{OA} nor P_{OC} have components in the y direction. Since P_{OB} is transmitted through a two-force member, the component of P_{OB} in the x direction is also zero. Thus, OB is a zero-force member.

Finally, consider the joint in Fig. 6.7c. Summing forces in the y direction yields $P_{OA} = 0$. Then summing forces in the x direction gives $P_{OB} = 0$. In this case both forces at joint O are zero. The three cases illustrated in Fig. 6.7 provide the only three configurations where zero force members can be identified by inspection.

EXAMPLE 6.4

For the truss defined in Fig. E6.4, determine the forces, stresses, and safety of members GI, HI, JI and HJ. The dimensions h = s = 5 m. The yield strength of the structural steel from which the truss is fabricated is 270 MPa. The following cross sectional areas have been specified for these members:

$A_{GI} = 2000\ mm^2$, $A_{HI} = 2500\ mm^2$, $A_{JI} = 1200\ mm^2$, and $A_{HJ} = 1200\ mm^2$

Fig. E6.4

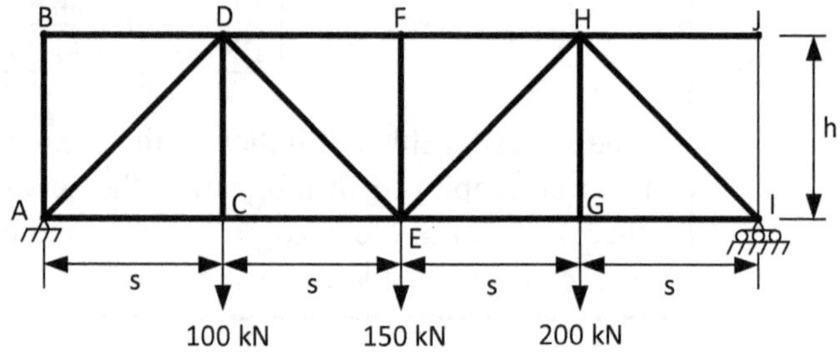

Solution:

Step 1: Draw the FBD to permit solution for the unknown reaction forces as illustrated in Fig. E6.4a.

Fig. E6.4a

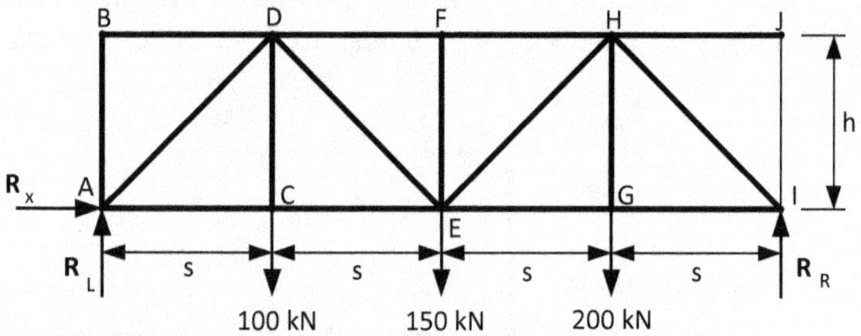

Step 2: Use the equilibrium relations to solve for the reaction forces R_x, R_R and R_L.

$$\Sigma M_A = 4sR_R - 3s(200) - 2s(150) - s(100) = 0$$

$$R_R = \tfrac{1}{4}(600 + 300 + 100) = 250\ kN \quad \text{(a)}$$

$$\Sigma F_y = R_L + R_R - 100 - 150 - 200 = 0$$

$$R_L = 450 - 250 = 200\ kN \quad \text{(b)}$$

$$\Sigma F_x = R_x = 0$$

Step 3: Select joints J and I and construct FBDs of each showing the internal and external forces acting at these joints as shown in Fig. 6.4b.

Fig. 6.4b

Step 4: Use equilibrium relations as necessary to solve for the unknown forces.

Consider joint J.

It is evident by inspection of the FBD for the joint at point J that:

$$P_{JH} = P_{JI} = 0 \qquad \text{(c)}$$

Both JI and JH are zero-force members that do not contribute to the strength or rigidity of the structure.

For joint I, we write:

$$\Sigma F_y = 250 + (1/\sqrt{2})P_{IH} = 0$$

$$P_{IH} = -(\sqrt{2})(250) = -353.6 \text{ kN} \qquad \text{(d)}$$

$$\Sigma F_x = -P_{IG} - (1/\sqrt{2})P_{IH} = 0$$

$$P_{IG} = -(1/\sqrt{2})P_{IH} = -(1/\sqrt{2})(-353.6) = 250 \text{ kN} \qquad \text{(e)}$$

The analysis of the forces in the four members defined in the problem statement is complete.

Step 5: Determine the stresses in member IH and IG. Note, the stresses in the zero-force members are obviously zero.

$$\sigma_{IH} = P_{IH}/A_{IH} = -353.6 \times 10^3/2500 = -141.4 \text{ N/mm}^2 = -141.4 \text{ MPa} \qquad \text{(f)}$$

$$\sigma_{IG} = P_{IG}/A_{IG} = 250 \times 10^3/2000 = 125 \text{ N/mm}^2 = 125 \text{ MPa} \qquad \text{(g)}$$

Step 6: Compute the safety factor and the margin of safety for these two members.

For member IH:

$$SF = S_y/\sigma = 270/141.4 = 1.91$$

$$MOS = SF - 1 = 1.91 - 1 = 0.91 = 91\%$$

For member IG

$$SF = S_y/\sigma = 270/125 = 2.16$$

$$MOS = SF - 1 = 2.16 - 1 = 1.16 = 116\%$$

An examination of the results indicates that the zero-force members contribute nothing to the strength and rigidity of the structure. They should be removed to reduce the weight and cost of the structure unless there is another compelling reason for their inclusion in the design. The other two members of the truss (IG and IH) appear to be properly sized with safety factors of about two. One member IH is loaded in compression and its resistance to buckling must be ascertained in a subsequent analysis.

6.4 METHOD OF SECTIONS

The method of sections, like the method of joints is based on construction of a FBD and the subsequent application of the equations of equilibrium. The difference is in the selection of the portion of the truss used for the FBD. With the method of sections, we cut the truss and remove a complete section of it for the FBD. A typical example of a section cut through a truss and the two FBDs that are produced is shown in Fig. 6.8. The section cut passes through three members of the truss—BD, CD and CE. We have the option of examining either the left or right side of the truss. FBDs of both sides are shown in Fig. 6.8. At the location of the section cut, the internal forces in the members are displayed in the FBDs.

Fig. 6.8 FBDs resulting from the section cut of the truss.

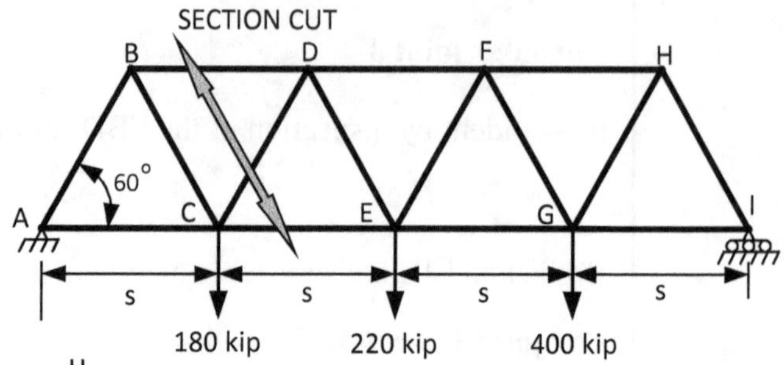

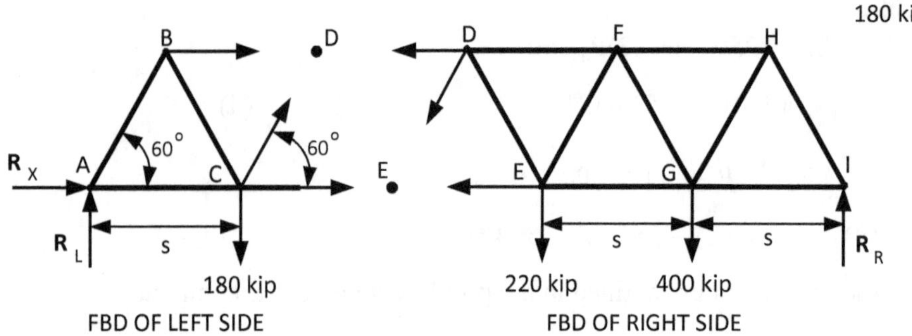

The FBDs provide a guide in writing the appropriate equilibrium relations. For the sections of the truss, either the right side or the left, we recognize the system of forces is planar but not concurrent. Accordingly, we write the following three equilibrium equations:

$$\Sigma F_x = 0; \qquad \Sigma F_y = 0; \qquad \text{and} \qquad \Sigma M_O = 0$$

and solve for three unknown internal forces in members of the truss that have been exposed by the section cut. Let's consider an example to demonstrate the procedure for the method of sections. You will note that we follow a similar six-step process as described for the method of joints.

EXAMPLE 6.5

Consider the equilateral triangular section truss defined in Fig. 6.8, and determine the margin of safety for members BD, CD and CE. The span s of each triangular section is 20 ft and the yield strength of the structural steel used in fabrication is 42 ksi. The cross sectional areas of the three members are $A_{BD} = 20$ in^2, $A_{CD} = 12$ in^2 and $A_{CE} = 24$ in^2.

Solution:

Step 1: Construct a FBD of the entire truss structure as shown in Fig. E6.5.

Fig. E6.5

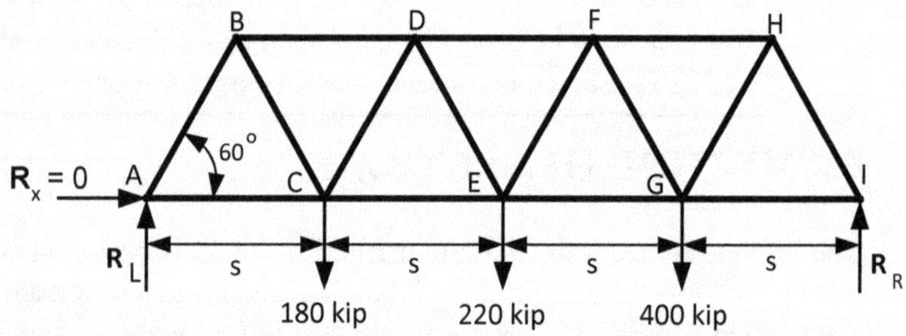

Step 2: Apply the equilibrium relations to determine the reactions at the supports.

$$\Sigma F_x = R_x = 0$$

$$\Sigma M_A = (4s)R_R - (3s)(400) - (2s)(220) - (s)(180) = 0$$

$$R_R = ¼(1200 + 440 + 180) = 455 \text{ kip} \quad (a)$$

$$\Sigma F_y = R_L - 180 - 220 - 400 + 455 = 0$$

$$R_L = 345 \text{ kip} \quad (b)$$

Step 3: Select a portion of the truss and make a section cut through the three members BD, CD and CE. Construct a FBD of that portion as shown in Fig. E6.5a. In this example, the left side is used for the FBD because it is easier to analyze than the right side.

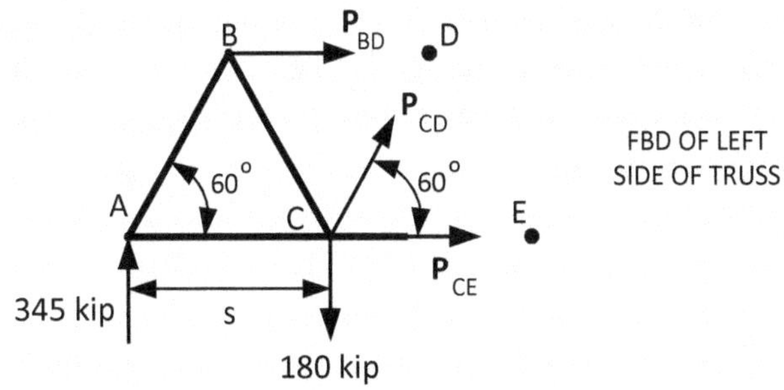

Fig. E6.5a

Step 4: Apply the equilibrium relations to solve for the forces P_{BD}, P_{CD} and P_{CE}.

$$\Sigma M_C = -345(s) - P_{BD}(s) \sin(60°) = 0$$

$$P_{BD} = -345/0.8660 = -398.4 \text{ kip} \quad (c)$$

$$\Sigma F_y = 345 - 180 + P_{CD} \sin(60°) = 0$$

$$P_{CD} = (180 - 345)/0.8660 = -190.5 \text{ kip} \quad (d)$$

$$\Sigma F_x = P_{BD} + P_{CD} \cos(60°) + P_{CE} = 0$$

$$P_{CE} = -(-398.4) - (-190.5)(0.5) = 493.6 \text{ kip} \quad (e)$$

Step 5*:* Determine the stresses in the members.

$$\sigma_{BD} = P_{BD}/A_{BD} = -398.4/20 = -19.92 \text{ ksi} \quad (f)$$

$$\sigma_{CD} = P_{CD}/A_{CD} = -190.5/12 = -15.88 \text{ ksi} \quad (g)$$

$$\sigma_{CE} = P_{CE}/A_{CE} = 493.6/24 = 20.57 \text{ ksi} \quad (h)$$

Step 6*:* Determine the safety factors and margins of safety for the three members.

For member BD:

$$SF = S_y/\sigma = 42/19.92 = 2.11; \qquad MOS = SF - 1 = 2.11 - 1 = 111\% \quad (i)$$

For member CD:

$$SF = S_y/\sigma = 42/15.88 = 2.64; \qquad MOS = SF - 1 = 2.64 - 1 = 164\% \quad (j)$$

For member CE:

$$SF = S_y/\sigma = 42/20.57 = 2.04; \qquad MOS = SF - 1 = 2.04 - 1 = 104\% \quad (k)$$

From an examination of the results for these three truss members, it is apparent that the design is reasonable with safety factors exceeding two in all cases. The design is not perfectly balanced because the margin of safety for member CD is somewhat higher than the other two members. If standard section sizes are commercially available, cost and weight can be reduced by reducing the cross sectional area of this member from 12 in^2 to 9 or 10 in^2. Of course we note that members BD and CD are loaded in compression and their critical buckling load must be determined to complete the design analysis.

EXAMPLE 6.6

For the truss shown in Fig. E6.6, determine the cross sectional area required for members DF, EF and EG if the safety factor for the design is to exceed 2.8. The span s for each section of the truss is 8 m. The yield strength of the hot rolled steel members used in fabricating the truss is 320 MPa.

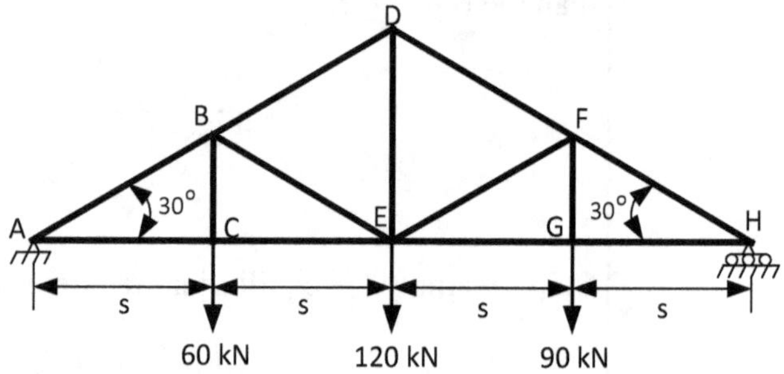

Solution:

Step 1: We construct a FBD of the entire structure as shown in Fig. E6.6a.

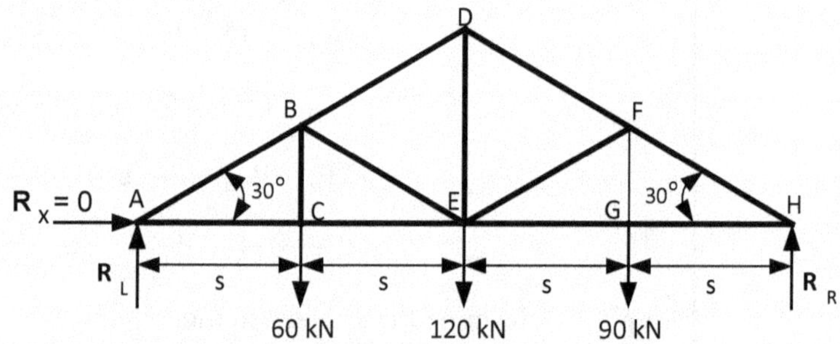

Fig. 6.6a

Step 2: Apply the equilibrium relations to solve for the reaction forces R_L and R_R.

$$\Sigma F_x = R_x = 0$$

$$\Sigma M_A = -60(s) - (120)(2s) - (90)(3s) + (4s)R_R = 0$$

$$R_R = ¼(60 + 240 + 270) = 142.5 \text{ kN} \qquad \text{(a)}$$

$$\Sigma F_y = R_L - 60 - 120 - 90 + R_R = 0$$

$$R_L = -270 + 142.5 = 127.5 \text{ kN} \qquad \text{(b)}$$

Step 3: Make a section cut through the truss to expose the internal forces in members DF, EF and EG. Construct a FBD of the right hand side of the truss showing the unknown internal forces P_{DF}, P_{EF}, and P_{EG} as presented in Fig. E6.6b.

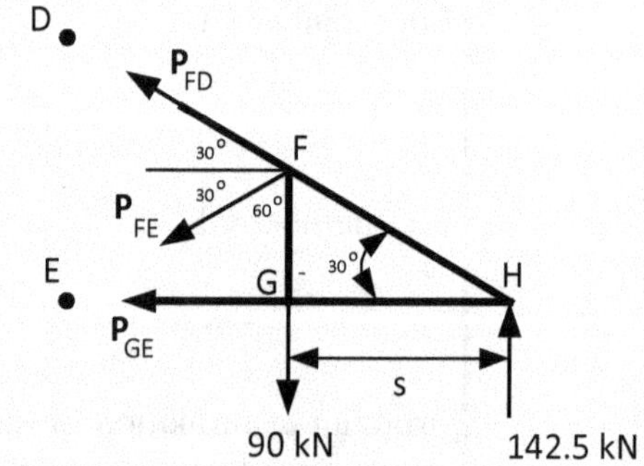

Fig. E6.6b

Step 4: Apply the equilibrium relations to solve for the unknown internal forces.

$$\Sigma M_F = (142.5)(s) - (P_{GE})(s)\tan(30°) = 0$$

$$P_{GE} = 142.5/\tan(30°) = 246.8 \text{ kN} \quad \text{(c)}$$

$$\Sigma F_y = 142.5 - 90 - P_{FE}\sin(30°) + P_{FD}\sin(30°) = 0$$

$$P_{FE} = P_{FD} + 105 \quad \text{(d)}$$

$$\Sigma F_x = -P_{GE} - P_{FD}\cos(30°) - P_{FE}\cos(30°) = 0$$

$$P_{FD} = -[P_{FE} + P_{GE}/\cos(30°)] \quad \text{(e)}$$

Substituting Eqs. (c) and (d) into Eq. (e) yields:

$$P_{FD} = -[P_{FE} + 246.8/0.8660] = -[P_{FD} + 105 + 285.0]$$

$$P_{FD} = -195.0 \text{ kN} \quad \text{(f)}$$

Next, substitute Eq. (f) into Eq. (d) to obtain:

$$P_{FE} = P_{FD} + 105 = -195 + 105 = -90.0 \text{ kN} \quad \text{(g)}$$

Step 5: Determine the stresses in the three members.

$$\sigma_{GE} = P_{GE}/A_{GE}$$

$$\sigma_{FD} = P_{FD}/A_{FD}$$

$$\sigma_{FE} = P_{FE}/A_{FE} \quad \text{(h)}$$

It is impossible to solve for the stresses implicitly because the cross sectional areas of the three members are unknown. We will use Eqs. (h) after developing a relation for the stresses in terms of the safety factor. The result will give us a relationship between the allowable stress and the internal force on each member.

Step 6: Determine the allowable stresses and then the cross sectional areas by using the information regarding the safety factor and the yield strength required for the truss members.

$$SF = S_y/\sigma \qquad \sigma = S_y/SF \quad \text{(i)}$$

Substitute Eq. (i) into Eqs. (h) and solve for the cross sectional area of each member.

$$\sigma_{GE} = P_{GE}/A_{GE} = S_y/SF = 320/2.8 = 114.3 \text{ MPa}$$

$$\sigma_{FD} = P_{FD}/A_{FD} = S_y/SF = 320/2.8 = 114.3 \text{ MPa}$$

$$\sigma_{FE} = P_{FE}/A_{FE} = S_y/SF = 320/2.8 = 114.3 \text{ MPa} \quad \text{(j)}$$

From Eqs. (j), (c), (f) and (g), it is evident that:

$$A_{GE} = P_{GE}/114.3 = 246.8 \times 10^3/114.3 = 2159 \text{ mm}^2$$

$$A_{FD} = P_{FD}/114.3 = 195 \times 10^3/114.3 = 1706 \text{ mm}^2$$

$$A_{FE} = P_{FE}/114.3 = 90 \times 10^3/114.3 = 787.4 \text{ mm}^2 \quad \text{(k)}$$

The results of Eq. (k) give the cross sectional areas required to achieve the specified safety factors; however, sizes such as these are not normally used in design. Instead, we seek a standard size section slightly larger than that determined in the analysis. Hot rolled steel sections are available in a wide range of sizes and shapes from a large number of suppliers of metal products. We use standard sizes because it reduces costs and delivery time.

EXAMPLE 6.7

Determine the forces and the stresses in members CE, DE and DF, of the bowstring truss shown in Fig. E6.7. Note that the numerical parameters that control the forces and stresses are $F_1 = 6.0$ kip, $F_2 = 5.0$ kip and $F_3 = 4.0$ kip, s = 9 ft, $A_{CE} = 1.2$ in.^{2}, $A_{DE} = 1.0$ in.^{2}, and $A_{DF} = 1.4$ in.^{2}

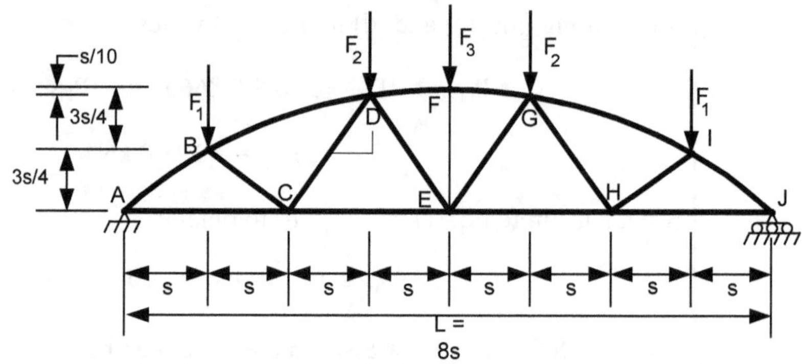

Fig. E6.7

Step 1: We construct a FBD of the entire structure as shown in Fig. E6.7a.

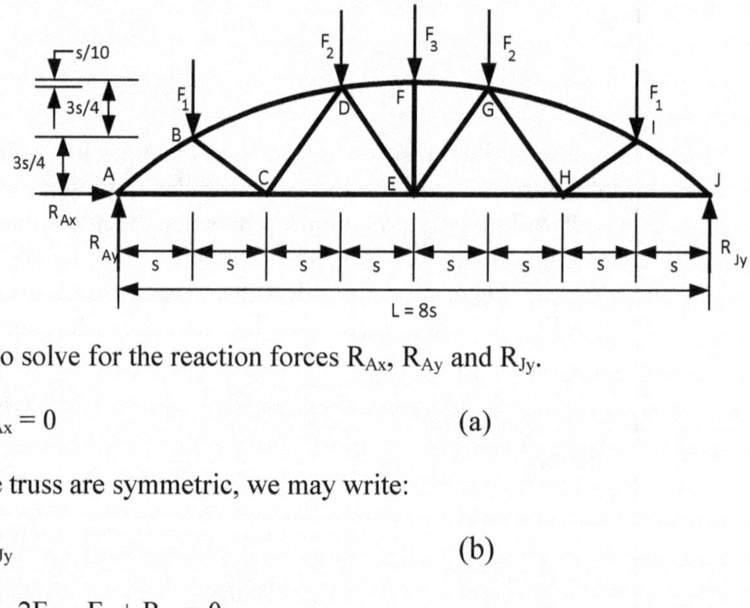

Fig. E6.7a

Step 2: Apply the equilibrium relations to solve for the reaction forces R_{Ax}, R_{Ay} and R_{Jy}.

$$\Sigma F_x = R_{Ax} = 0 \tag{a}$$

Since the loading and the geometry of the truss are symmetric, we may write:

$$R_{Ay} = R_{Jy} \tag{b}$$

$$\Sigma F_y = R_{Ay} - 2F_1 - 2F_2 - F_3 + R_{Jy} = 0$$

$$R_{Ay} = R_{Jy} = (1/2)[2(6.0) + 2(5.0) + 4.0] = 13.0 \text{ kip} \tag{c}$$

Step 3: Make a section cut through the truss to expose the internal forces in members CE, DE, and DF. Construct a FBD of the left hand side of the truss showing the unknown internal forces P_{CE}, P_{DE}, and P_{DF} as presented in Fig. E6.7b.

Fig. E7.7b

$$\alpha = \tan^{-1}(1/10) = 5.710°$$
$$\beta = \tan^{-1}[(6/4) - (1/10)]$$
$$\beta = \tan^{-1}(1.4) = 54.46°$$

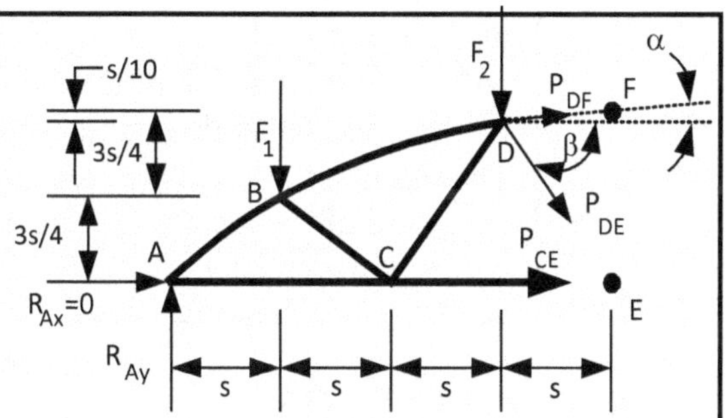

Step 4: Apply the equilibrium relations to solve for the unknown internal forces. Select point E and write $\Sigma M_E = 0$ to obtain:

$$\Sigma M_E = (s)F_2 + (3s)F_1 - (4s)R_{Ay} - (P_{DF})[(s)\sin\alpha + (1.4s)\cos\alpha)] = 0 \quad \text{(d)}$$

Substituting numerical parameters into Eq. (d) and solving for P_{DF} yields:

$$P_{DF} = -\ 19.43 \text{ kip} \quad \text{(e)}$$

Consider equilibrium of the sectioned structure in the y direction:

$$\Sigma F_y = R_{Ay} - F_1 - F_2 + P_{DF}\sin\alpha - P_{DE}\sin\beta = 0 \quad \text{(f)}$$

Substituting numerical parameters into Eq. (f) and solving for P_{DE} yields:

$$P_{DE} = 0.08215 \text{ kip} \quad \text{(g)}$$

Consider equilibrium of the sectioned structure in the x direction:

$$\Sigma F_x = R_{Ax} + P_{CE} + P_{DF}\cos\alpha + P_{DE}\cos\beta = 0 \quad \text{(h)}$$

Substituting numerical parameters into Eq. (h) and solving for P_{CE} yields:

$$P_{CE} = 19.29 \text{ kip} \quad \text{(i)}$$

The minus sign for the numerical result for member DF indicates that it is subjected to a compressive force. However, the other two members of the bowstring truss are in tension.

Step 5: Determine the stresses in the three members. We may solve for the stresses in members DE and CE using Eq. (5.4) as:

$$\sigma_{DE} = P_{DE}/A_{DE} = 0.08215 \text{ kip}/1 \text{ in.}^2 = 0.08215 \text{ ksi}$$

$$\sigma_{CE} = P_{CE}/A_{CE} = 19.29 \text{ kip}/1.2 \text{ in.}^2 = 16.08 \text{ ksi}$$

It is not possible for us to determine the stresses in member DF because it is curved. The curvature of the member does not affect the determination of the force P_{DF} since DF is a two-force member. For this reason, the internal forces acts along a line of action passing through points D and F on the bowstring truss as shown in Fig. E6.7c.

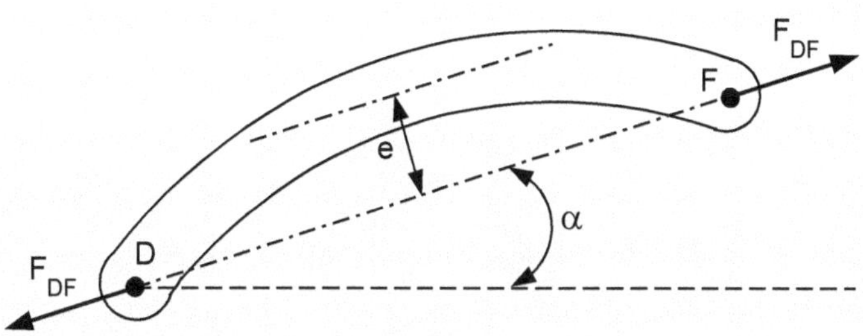

Fig. E6.7c

The force external F_{DF} is equal to the internal force P_{DF} both of which act along the line of action defined with respect to the x-axis by the angle α. However, the centerline of the curved member does not coincide with the line of action of the two-forces. Indeed, an eccentricity e is evident in Fig. E6.7c. This eccentricity is important when determining the stresses in the curved member because it causes bending of the member. The relation for stresses given by Eq. (4.4) is not valid in this case because it does not account for either the curvature of the member or the eccentricity. Both of these topics are beyond the scope of this textbook and will be covered later in a more advanced course.

6.5 SUMMARY

Trusses are fabricated from bars and cables to provide a structure capable of spanning long distances. As such they are often used in the design of bridges and roofs to cover large buildings. Stability of the truss is achieved by constructing it from a series of triangular arrangements.

The bars or cables in a truss are two force members connected together by means of joints. We assume that the joints are capable of rotation and cannot support a moment. Since the bars and cables are long and flexible, it is necessary to restrict the point of load application to only the joints of the truss.

Two methods for analysis of trusses are described. The first, the method of joints, entails removing a specified joint from the truss and constructing a FBD showing both the internal and external forces acting on it. The joint is represented as a particle acted upon by a planar concurrent force system. As such two equilibrium relations apply:

$$\Sigma F_x = 0 \qquad \text{and} \qquad \Sigma F_y = 0$$

These relations are sufficient to solve for any two unknown forces that act on the joint.

The method of sections involves making a section cut through the truss to divide it into two or more parts. A FBD is prepared for one section of the truss showing both the internal and external forces. The forces acting on a particular section are planar but not concurrent; hence, the equilibrium relations that apply are:

$$\Sigma F_x = 0; \qquad \Sigma F_y = 0; \qquad \text{and} \qquad \Sigma M_O = 0$$

These relations are sufficient to solve for any three unknown forces that act on the portion of the truss under consideration. Examples illustrating the procedure for using the method of joints and the method of sections were provided.

A procedure was described for identifying zero-force members in a truss by examining the joints not subjected to externally applied loads. When a joint is acted upon by three internal forces and two of these forces are collinear, then the remaining force must be zero. Also, when two non-collinear members meet at a joint both are zero force members. Zero force members do not contribute to the strength or rigidity of a structure and should be eliminated unless there is a compelling reason for their inclusion in the design.

PROBLEMS

6.1 Why is a truss used in constructing a bridge or a roof covering for a large structure?

6.2 What is the inherent geometric form that is used repeatedly in all truss designs? Why?

6.3 If a truss is fabricated from 37 members, determine the number of joints that will be required in constructing the truss.

6.4 Why is it necessary to employ an odd number of members in constructing a truss?

6.5 Prepare a sketch of a joint using a gusset plate for connecting four uniaxial bars in a truss structure. Identify all the components included in the joint that you have designed.

6.6 Why is it important to restrict the application of external loads on a truss to only the joints?

6.7 Consider the four-unit Howe truss shown in the figure to the right. If the external forces are as specified in this figure, use the method of joints to determine the forces in members AC, BC, CD and CE.

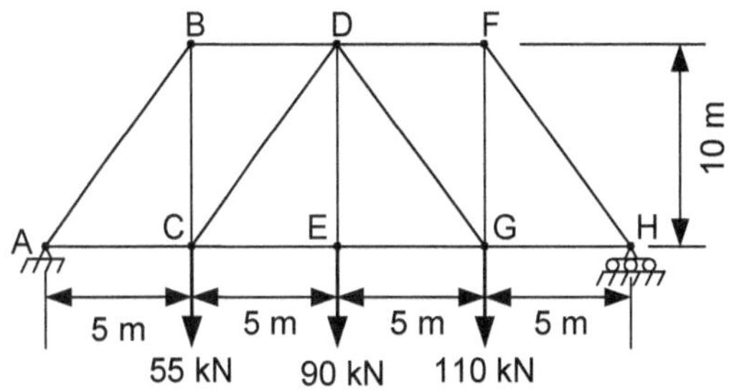

6.8 Consider the four-unit Howe truss shown in the figure to the left. If the external forces are as specified in this figure, use the method of joints to determine the forces in members AC, BC, CD and CE.

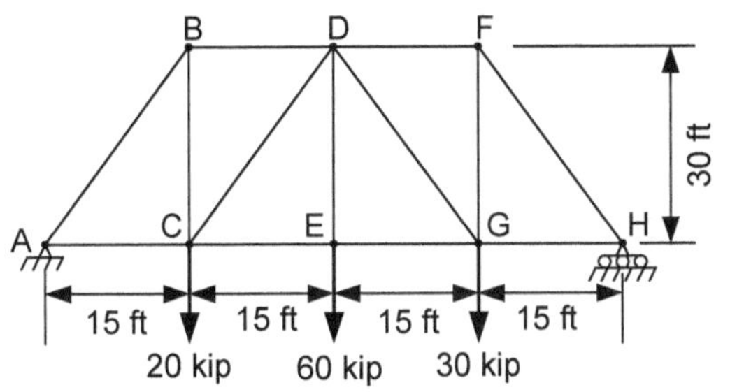

6.9 Consider the four-unit Howe truss shown in the figure to the right. If the external forces are as specified in this figure, use the method of joints to determine the forces in members GE, GD, GF and GH.

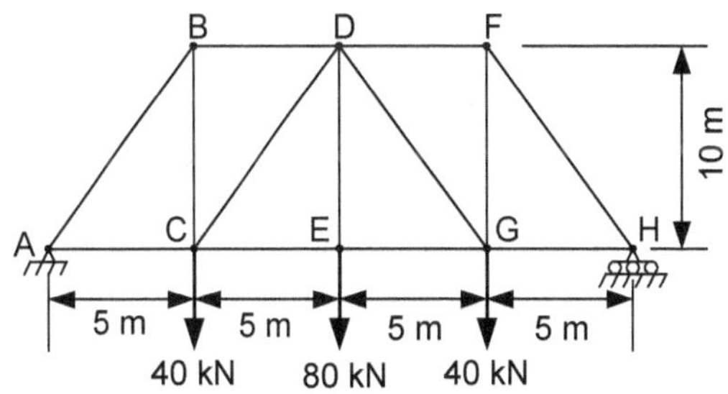

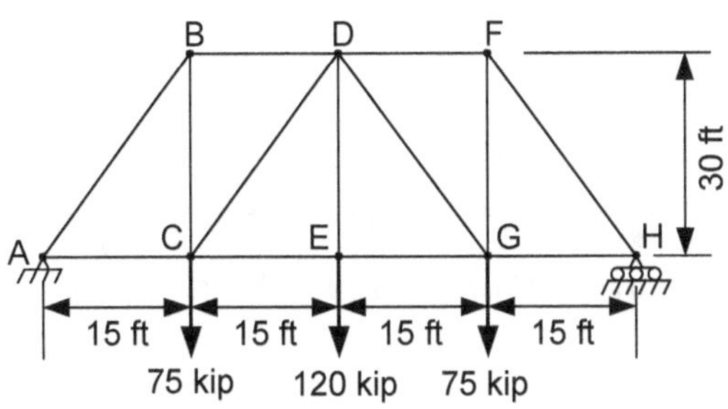

6.10 Consider the four-unit Howe truss shown in the figure to the left. If the external forces are as specified in this figure, use the method of joints to determine the forces in members AC, BC, CD and CE.

6.11 If the safety factor SF for each of the members in the truss of Problem 6.7 is specified as 3.0, determine the minimum required cross sectional area for members AC, BC, CD and CE. Note the yield strength S_y of the hot rolled structural steel used in fabricating the truss is 200 MPa.

6.12 If the safety factor SF for each of the members in the truss of Problem 6.8 is specified as 3.2, determine the minimum required cross sectional area for members AB, AC, BD and BC. Note the yield strength S_y of the hot rolled structural steel used in fabricating the truss is 30 ksi.

6.13 A scissors truss, illustrated in the figure to the right, is loaded with three forces at the joints B, C and D. If the truss is fabricated from 1018 A steel, determine the size of members AB, AF, BC and BF. The safety factor is specified as 3.0.

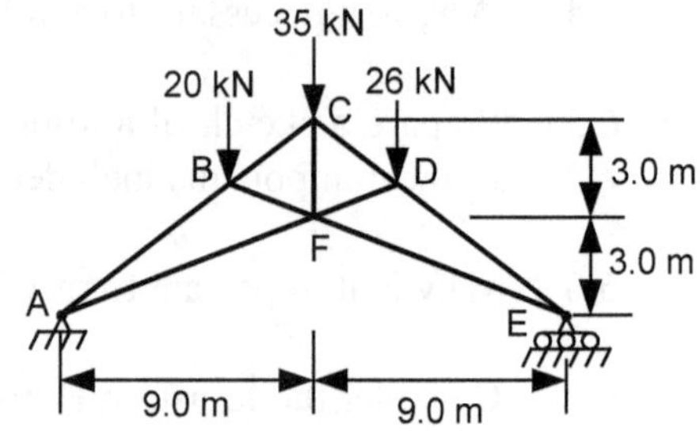

6.14 A scissors truss, illustrated in the figure to the left, is loaded with three forces at the joints B, C and D. If the truss is fabricated from 1020 HR steel, determine the size of members AB, AF, BC and BF. The safety factor is specified as 2.9.

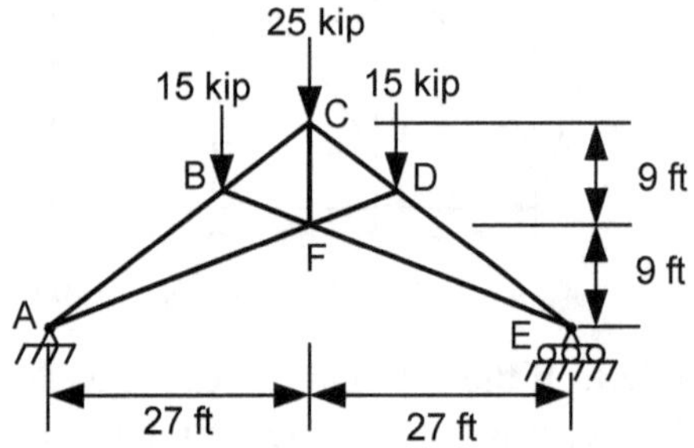

6.15 For the Howe truss, shown in the figure to the right, use the method of sections to determine the forces in the members BD, CD and CE.

6.16 For the Howe truss, shown in the figure to the right, use the method of sections to determine the forces in the members DF, DG and EG.

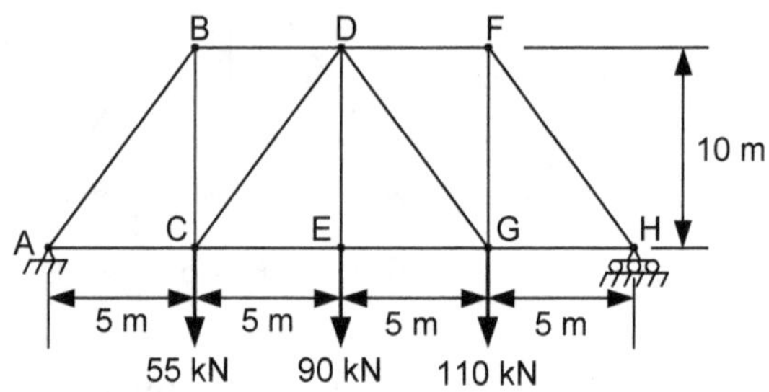

6.17 For the truss, defined in the figure to the right, use the method of sections to determine the forces in members DF, DB and DE. The external forces applied at joints C, E, G, I, and K are each equal to 15 kip.

6.18 If the cross sectional areas of all of the members in the truss defined in Problem 6.17 are equal to A = 2.3 in.^{2}, determine the safety factor for structural members DF, DB and DE. The yield strength of the steel used in the truss members is 35 ksi.

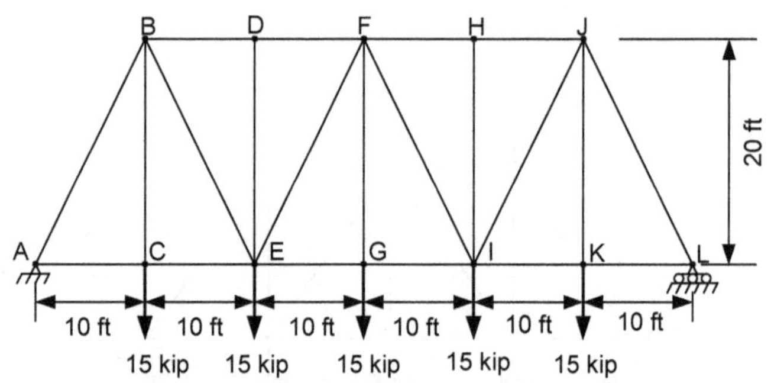

6.19 By inspection identify the zero-force members in the truss defined in Problem 6.17.

6.20 For the truss, defined in the figure to the right, use the method of joints to determine the forces in members DC, DE and DF.

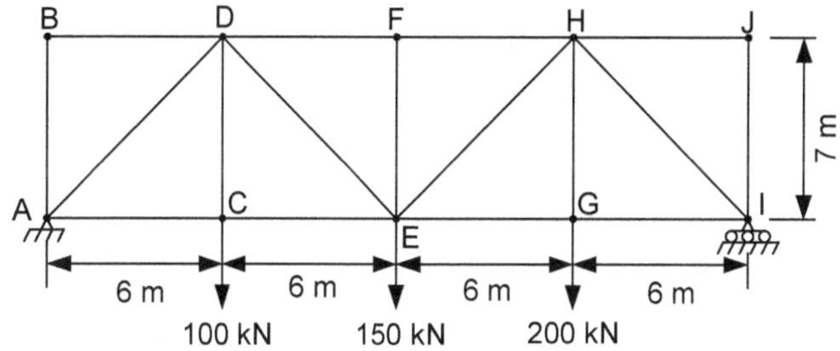

6.21 If the cross sectional areas of all of the members in the truss defined in Problem 6.20 are equal to 3,000 mm^2, determine the safety factor for the members DC, DE and DF. Note the yield strength of the steel used in the truss is 310 MPa.

6.22 Use the method of sections with the truss, shown in the figure for Problem 6.20, to determine the forces in members FH, EH, and EG.

6.23 Let s/h be a variable in the truss structure defined in the figure to the left. Determine the forces in the members AC and AD as a function of the s/h ratio. We suggest you use a spreadsheet to perform the calculations. Consider the ratio s/h over the range from 0.5 to 2.0 varying in steps of 0.1. Prepare a graph of the results for the forces in members AC and AD as a function of s/h.

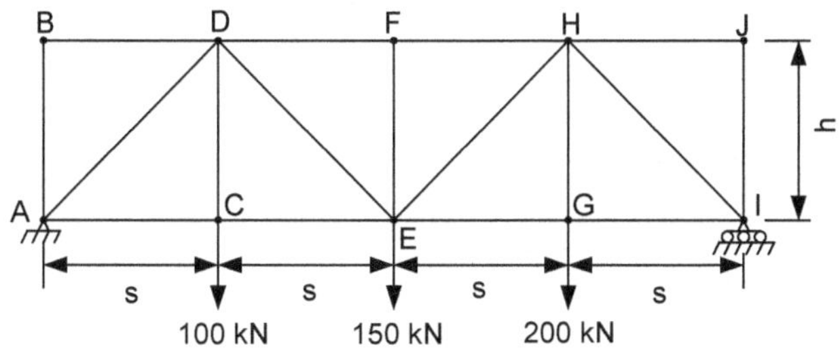

6.24 Determine the forces in members CE, DE and DF of the bowstring truss shown in the figure to the right. The span s = 12 ft.

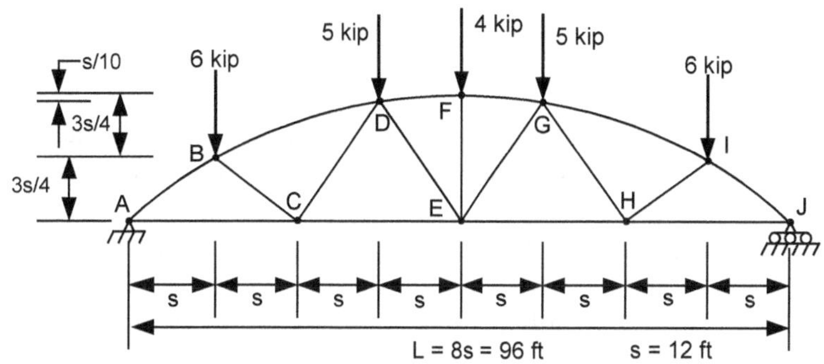

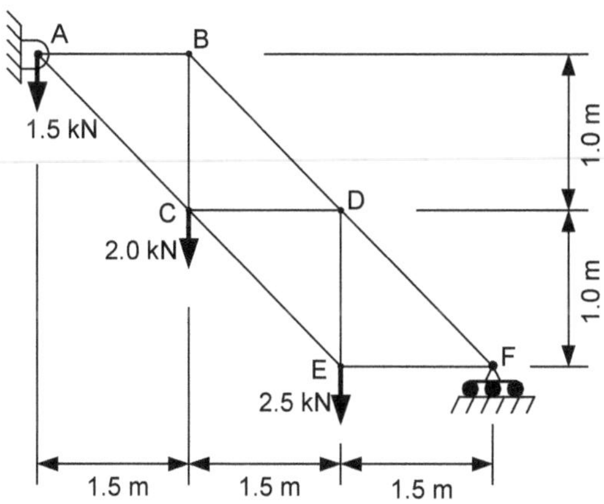

6.25 Determine the forces in members AC, CD, CE and DF of the inclined truss shown in the figure to the left.

6.26 Determine the forces in members CD, CJ, and KJ of the truss shown in the figure below and to the right.

6.27 Determine the forces in members DE, EJ, and JI of the truss shown in the figure to the right.

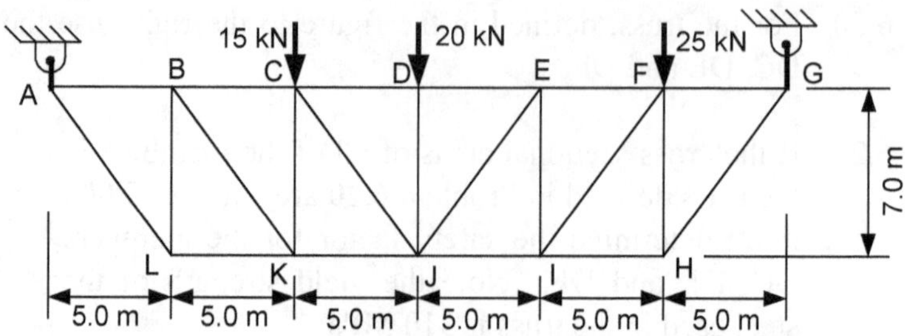

6.28 Determine the forces in members EF, IF, and IH, of the truss shown in the figure below and to the right.

6.29 Determine the forces in members BC, BK, and LK, of the truss shown in the figure to the right.

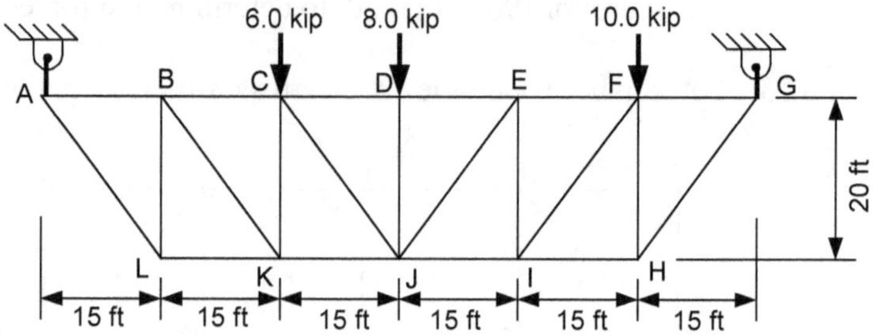

6.30 For the cantilever truss, shown in the figure to the right, determine the stress in member BC if its cross sectional area is 2.5 in^2. Also determine the safety factor for this truss member if it is fabricated from 1020 HR steel.

6.31 For the cantilever truss, shown in the figure to the right, determine the stress in member AB if its cross sectional area is 3.5 in^2. Also determine the safety factor for this truss member if it is fabricated from 1020 HR steel.

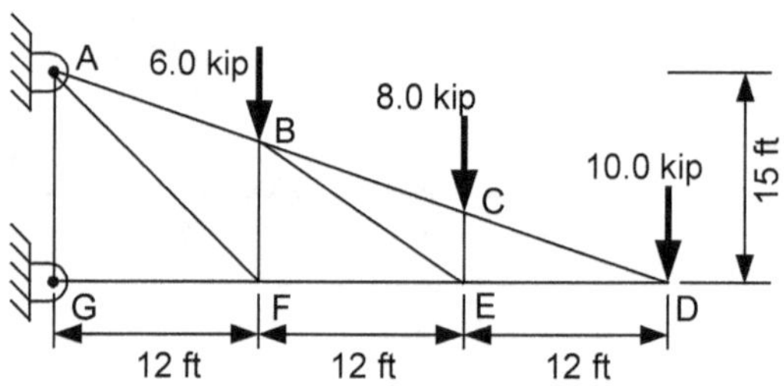

6.32 For the cantilever truss, shown in the figure to the right, determine the force in member GF. Specify the minimum cross sectional area if the member is to exhibit a safety factor of 3.0 relative to the yield strength of its material. This truss member is fabricated from 1020 HR steel.

6.33 For the cantilever truss, shown in the figure to the right, determine the stress in member CD if its cross sectional area is 900 mm^2. Also determine the safety factor for this truss member if it is fabricated from 1020 HR steel.

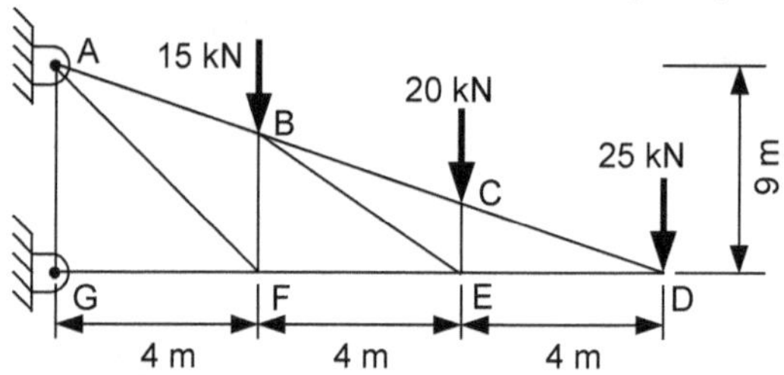

6.34 For the cantilever truss shown in Problem 6.33, determine the force in member ED. Specify the minimum cross sectional area if this member is to exhibit a safety factor of 3.0 relative to the yield strength of its material. This truss member is fabricated from 1212 HR steel.

6.35 For the truss, presented in the figure below, determine the cross sectional area required for members AB, BD, DE, EF and CF if they are fabricated from 1020 HR steel and a safety factor of 3.2 is specified.

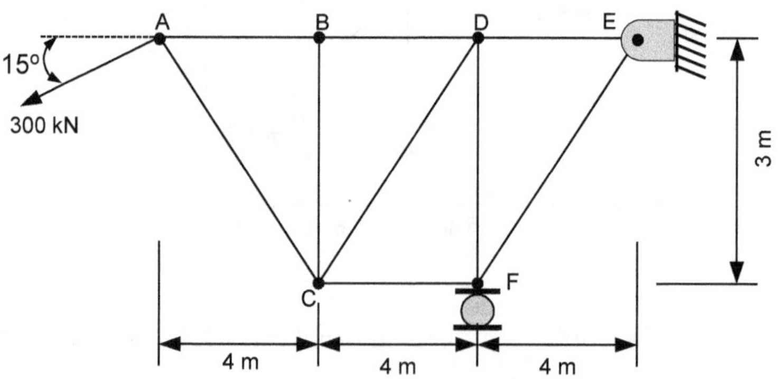

CHAPTER 7

SPACE STRUCTURES AND 3-D EQUILIBRIUM

7.1 INTRODUCTION

A space structure is a three-dimensional system, fabricated from uniaxial members, that spans in two-directions. The space structure incorporates members that extend into space with force components in the x-y, x-z and y-z planes. Space structures may be fabricated with rigid joints, which support moments, or pinned joints, which do not. In this chapter, we will first consider space structures with pinned joints that cannot support moments. Then we will study more complex three-dimensional structures. Trusses, described in Chapter 6, are two-dimensional structures because all of the members involved in their construction lie in the x-y plane. Trusses span in only one direction. We can perform a design analysis of a two-dimensional truss by using only two or possibly three of the six equilibrium relations. However, the three-dimensional nature of space structures is more complex requiring the repeated application of the six equations of equilibrium to determine the forces in its uniaxial members.

The geometry of space structures is quite diverse with domes, tetrahedrons and half-octahedrons. An example of a space structure constructed with the half octahedrons is presented in Fig. 7.1. Space structures are often employed to support horizontal, flat roofs that cover large buildings. However, they may be employed in other applications including walls and sloped roofs. Because space structures are efficient, they cost less than a system of trusses spanning the primary direction with beams (purlins) spanning in the opposite direction.

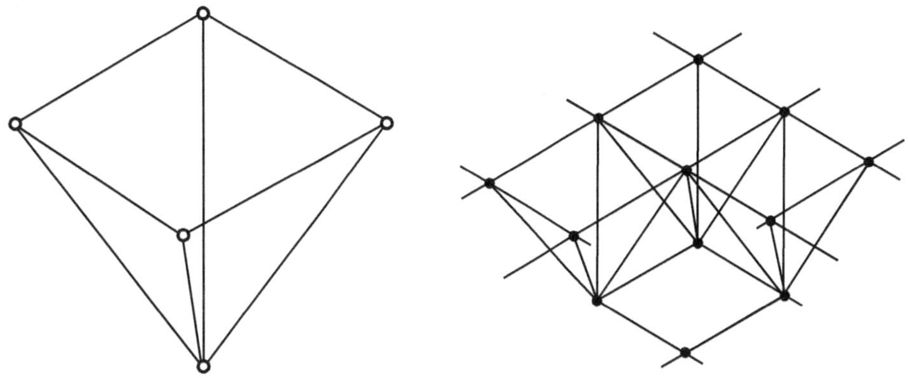

Fig. 7.1 Half-octahedrons arranged to form a space structure.

Another illustration of a space structure is a communication tower that is an extremely tall structure usually built at a high elevation to facilitate the distance of unobstructed propagation of microwave, radio or television signals. The structure has a very high aspect ratio and is inherently unstable. Stability is achieved with cables mounted near the top of the transmission tower that extend downward and outward. These cables are anchored into the ground to provide a large footprint for the tower. Since stability is achieved with cables, the base of the tower is usually pinned. In three dimensions the equivalent for a pinned joint is a ball and socket joint. A ball and socket joint provides reaction forces R_x, R_y and R_Z but moments are not supported. We show a schematic illustration of a communication tower with four cables anchored to the ground plane in Fig. 7.2.

The forces acting on a communications tower are due to:

- The dead weight of the tower and cables.
- The pretension, if any, applied to the cables
- The weight of any communication equipment installed on the tower.
- Snow and/or ice accumulations.
- Transverse loads due to winds
- In some parts of the country earthquake loading should be considered.

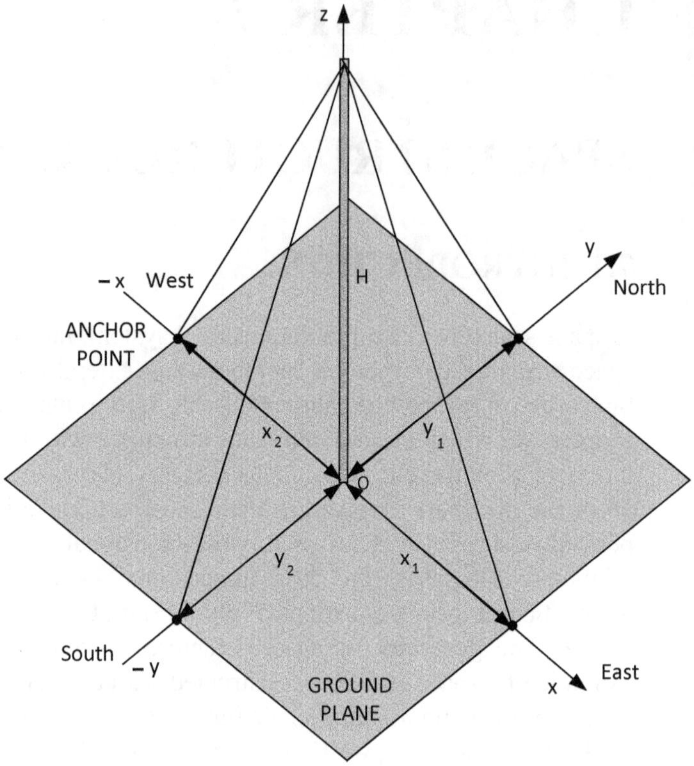

Fig. 7.2 Schematic illustration of a cable stabilized communications tower.

To perform a design analysis of three-dimensional structures, such as a communications tower, we follow the same six-step procedure as outlined previously for trusses. The difference in the analysis involves the added complexity due to the inclusion of the third dimension. The additional dimension requires additional FBDs, and the application of all six of the equilibrium relations.

The structure is always constrained by supports at several locations. These supports are modeled in FBDs with some combination of concentrated forces and moments that effectively provides the constraint the supports provide. The forces and the moments produced by common supports are listed in Table 7.1.

Table 7.1
Different types of supports or connections and their reactive forces and moments.

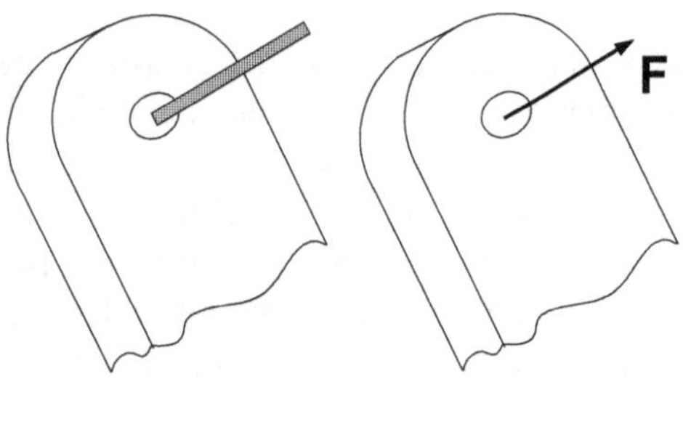

A. A cable connection is modeled with a single force acting in the direction of the cable and away from the structure.

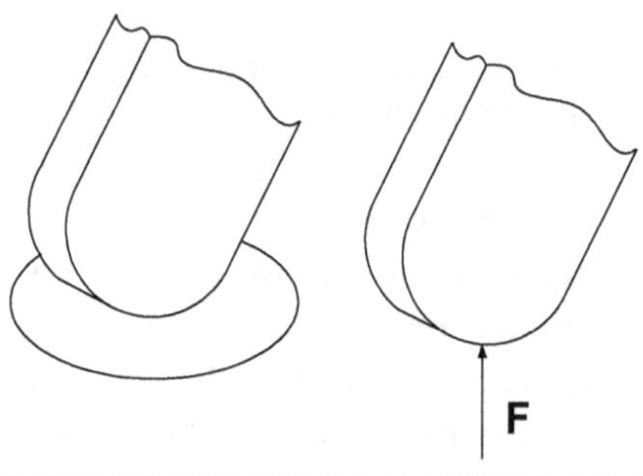

B. A rocker on a frictionless surface is modeled with a concentrated force perpendicular to the surface at the point of contact.

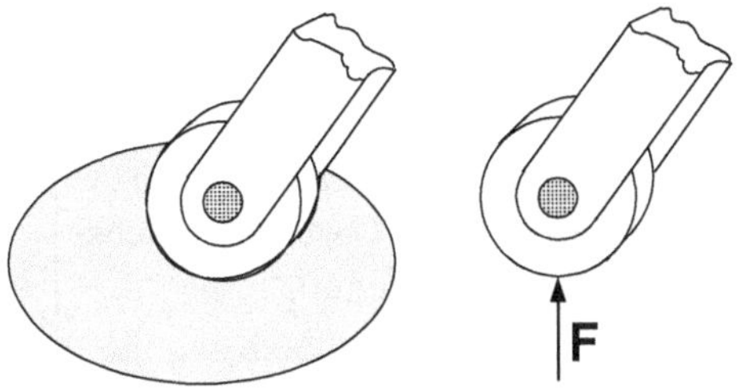

C. A roller on a flat surface is represented with a concentrated force perpendicular to the surface at the point of contact.

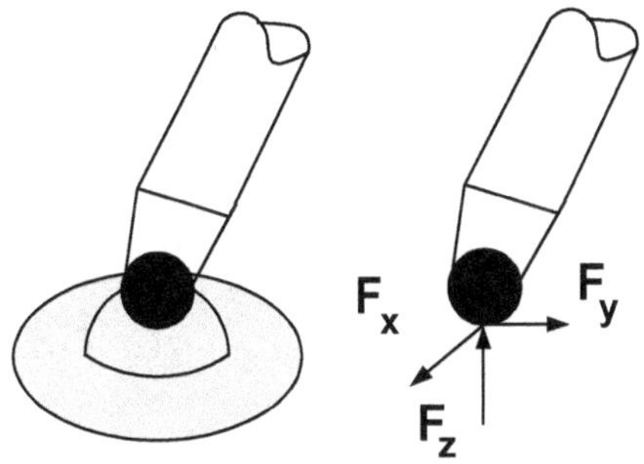

D. A ball and socket joint is modeled with three Cartesian force components.

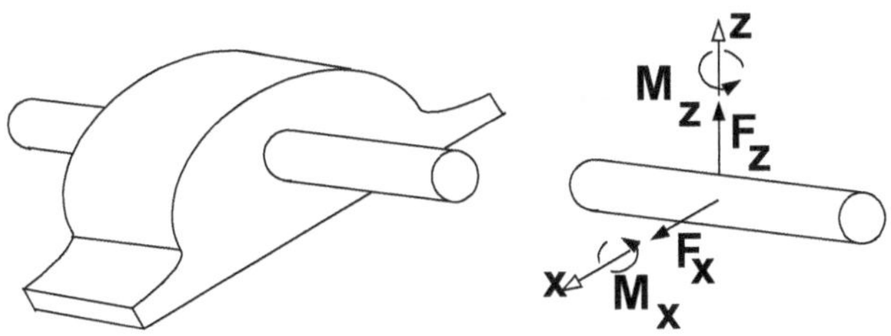

E. A journal bearing is represented with two forces perpendicular to the shaft and two moments about axes perpendicular to the shaft.

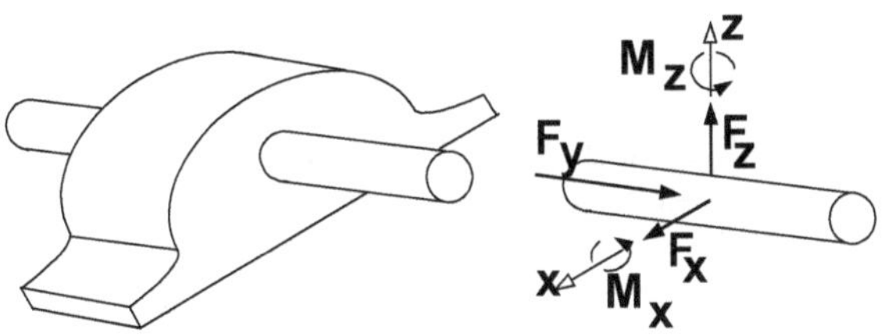

F. A thrust bearing is represented with three Cartesian forces and two moments about axes perpendicular to the shaft.

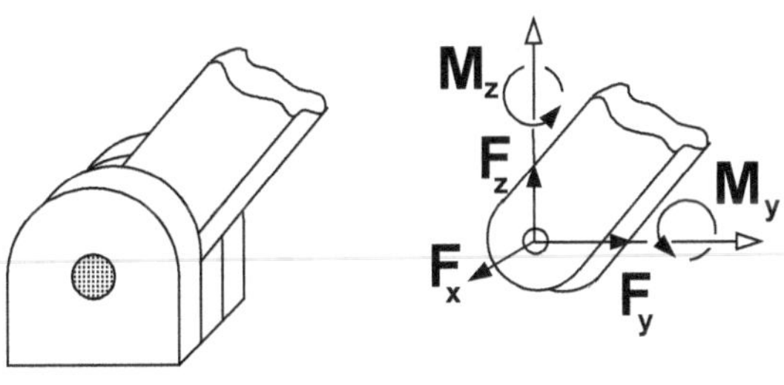

G. A pin and clevis connection is represented with three Cartesian forces and two moments about axes perpendicular to the pin.

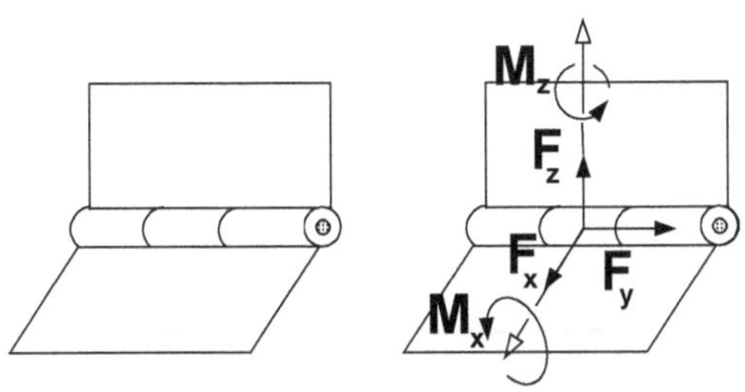

H. A hinge connection is represented with three Cartesian forces and two moments about axes perpendicular to the hinge pin.

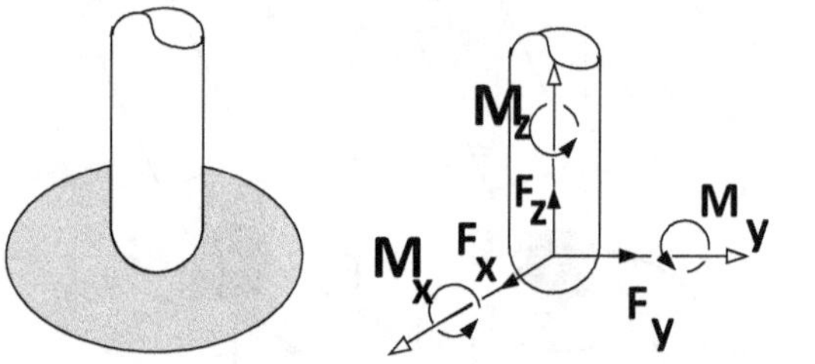

I. A fixed support is modeled with six possible reactions (three Cartesian forces and three Cartesian moments).

7.2 SOLUTIONS USING TRIGONOMETRIC ANALYSIS

In the discussion of three-dimensional structures, we will present two different mathematical approaches. The first approach, consistent with previous treatments, employs trigonometric methods for determining force and moment components in each of the three directions. The second approach employs vector analysis. The first approach is intuitively obvious, while the second is more systematic and mathematically more elegant.

Let's begin with the trigonometric approach to determine the force and moment components by performing a design analysis of the communications tower.

EXAMPLE 7.1

For the communications tower described in Fig. 7.2, determine the safety factors[1] for the cable and the tower structure if:

- The dead weight of the tower is 20,000 lb.
- The receivers and transmitters mounted on the tower weigh 2,000 lb.
- The tower height H is 500 ft and the material used in fabrication has a yield strength of 35,000 psi.
- Each of the four cables is 5/8 in. in diameter with a rated breaking strength of 41,200 lb. The length of the cables is adjusted so that they are snug but they are not pretensioned.
- The anchor points for the tower are $x_1 = x_2 = y_1 = y_2 = 200$ ft.
- The maximum transverse load anticipated from a southwest wind is 1600 lb. We assume that this load is applied at a distance $z = 2H/3$ from the ground plane.
- The anchor points are oriented so that the x-axis is due east and the y-axis is due north.
- The cross sectional area at the base of the tower is 11 in^2.
- A ball and socket joint supports the tower at point O.

Solution:

Step 1: Draw a FBD of the structure as shown in Fig. E7.1.

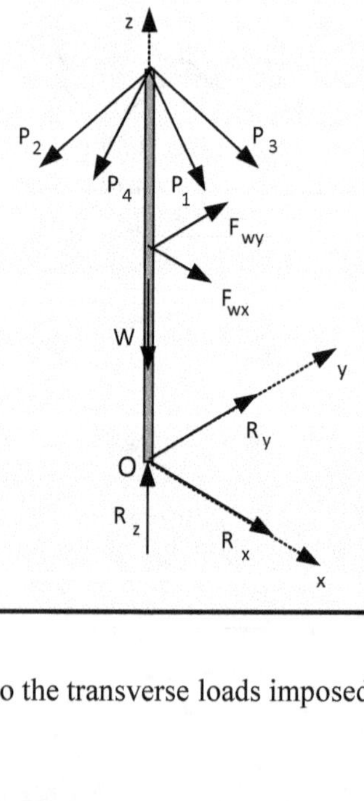

Fig. E7.1

[1] We will neglect bending in this analysis; however the tower is subjected to bending stresses due to the transverse loads imposed by the wind.

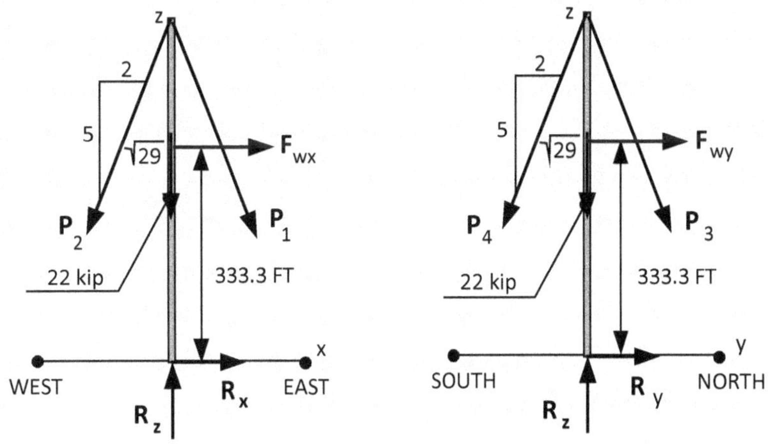

Fig. E7.1a

The view on the left is observed by standing to the south of the tower and looking north. The view on the right is observed by standing in the east and looking west. By drawing two different views of the FBDs, we are able to show the components of the wind loading and the forces in all four of the cables. The FBDs are not complete because we have not shown the magnitude of the wind forces F_{wx} and F_{wy}. Since the wind is from the southwest, we represent the wind force with the vector diagram shown in Fig. E7.1b, and solve for the force components in the x and y directions.

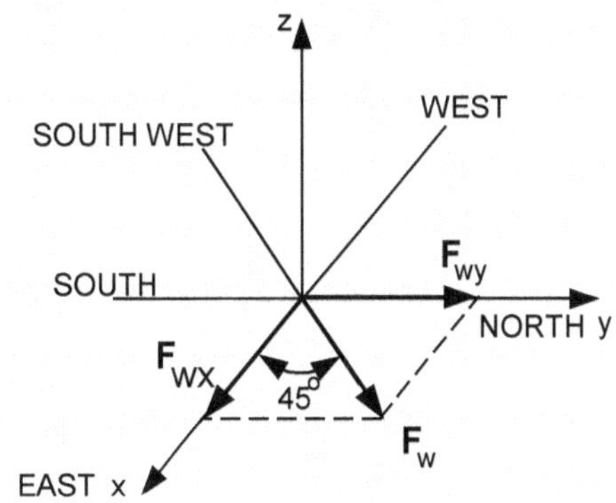

Fig. E7.1b

It is evident that:

$$F_{wy} = F_w \sin(45°) = (1600)(0.7071) = 1131 \text{ lb}$$

$$F_{wx} = F_w \cos(45°) = 1131 \text{ lb} \qquad \text{(a)}$$

Step 2: Consider equilibrium of the structure. From the FBDs in Fig. E7.0a, we will write several moment equations.

First, consider moments about the y-axis and write:

$$\Sigma M_y = (2/\sqrt{29})P_2\,(500) - F_{wx}\,(333.3) = 0$$

$$P_2 = [(1131)(333.3)(\sqrt{29})]/(1000) = 2030 \text{ lb} \qquad \text{(b)}$$

Since the cables supporting the tower are not pretensioned, $P_1 = P_3 = 0$ when the wind has components from the west and south. Next, consider the moments about the x-axis and write:

$$\Sigma M_x = (2/\sqrt{29})\,P_4(500) - F_{wy}\,(333.3) = 0$$

$$P_4 = [(1131)(333.3)(\sqrt{29})]/(1000) = 2030 \text{ lb} \qquad \text{(c)}$$

Finally, note that the relation $\Sigma M_z = 0$ is satisfied regardless of the values of the applied forces. There are no forces that tend to twist the structure, because all the forces intersect the z-axis.

Consider equilibrium of forces in the three directions and write:

$$\Sigma F_x = -(2/\sqrt{29})P_2 + F_{wx} + R_x = 0$$

$$R_x = (2/\sqrt{29})P_2 - F_{wx} = (2/\sqrt{29})(2030) - 1131 = -377 \text{ lb} \qquad \text{(d)}$$

$$\Sigma F_y = -(2/\sqrt{29})P_4 + F_{wy} + R_y = 0$$

$$R_y = (2/\sqrt{29})P_4 - F_{wy} = (2/\sqrt{29})(2030) - 1131 = -377 \text{ lb} \qquad \text{(e)}$$

The negative signs for R_x and R_y indicate that the directions for these reactions should be reversed in the FBDs presented in Fig. E7.1.

$$\Sigma F_z = -22{,}000 - (5/\sqrt{29})P_2 - (5/\sqrt{29})P_4 + R_z = 0$$

$$R_z = 22{,}000 + (10/\sqrt{29})(2030) = 25{,}770 \text{ lb} \qquad \text{(f)}$$

Step 3: Determine the stresses in the structure.

At the base of the tower, the normal stress σ_z is given by:

$$\sigma_z = R_z/A = -25{,}770/11 = -2343 \text{ psi}$$

The stress at the base of the tower is compressive.

It is not necessary to compute the stress in the cables because the breaking strength is given in terms of the force applied to the cable, and we have already made this determination.

Step 4: Determine the safety factor and the margin of safety for the tower and the cables.

For the tower:

$$SF = S/\sigma = 35{,}000/2343 = 14.9 \qquad MOS = SF - 1 = 13.9$$

For the cables:

$$F = P_b/P = 41{,}200/2030 = 20.3 \qquad MOS = SF - 1 = 19.3$$

Interpretation: The results indicate that the safety factors are very large. Develop arguments supporting the use of these large margins of safety in the design of a communication tower. Also, develop arguments for redesign with smaller diameter cables, smaller footprint and smaller cross sectional area for the tower structure.

EXAMPLE 7.2

A small hoist is constructed with two cables (AE and CE) and a strut (DE) as shown in Fig. E7.2. If the applied force F = 3 tons, determine the size of the cables and strut if a safety factor of 3.5 is specified. The yield strength of the cables is specified by the supplier as 150,000 psi, and the yield strength of the low carbon steel used for the strut is 40,000 psi. The dimensions for the crane assembly are AB = BC = 6 ft; BD = 8 ft; BE = 12 ft.

Fig. E7.2

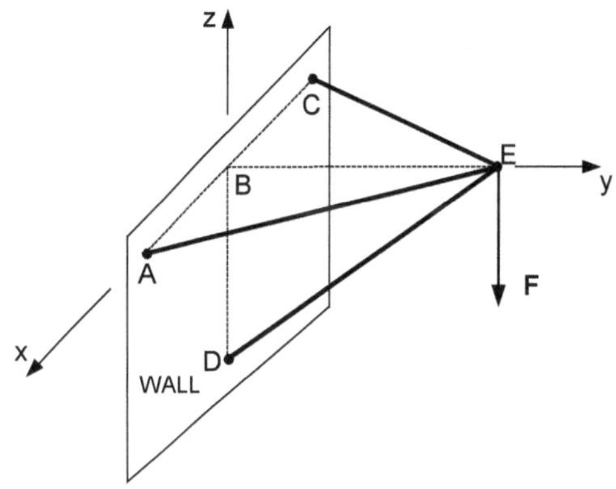

Solution:

Step 1: Draw a FBD of the structure as shown in Fig. E7.2a.

Fig. E7.2a

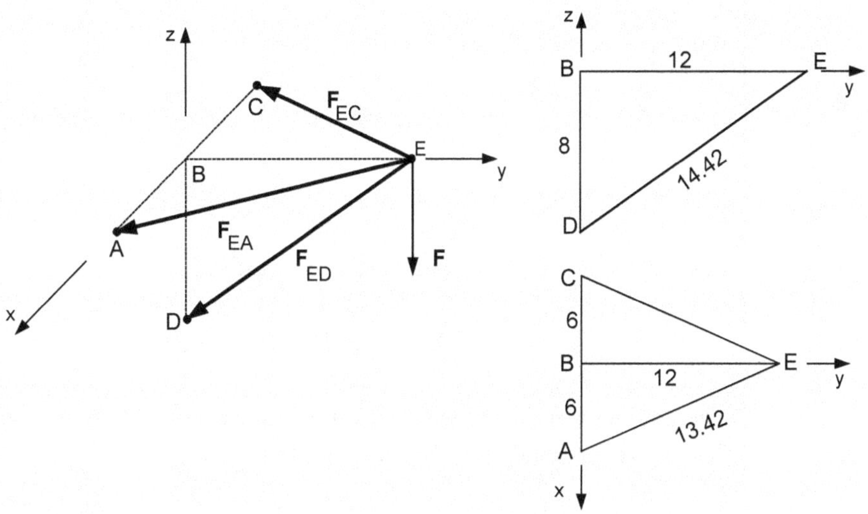

We have removed the wall and have drawn force vectors for the cables and strut. We have also presented two drawings—one for the y-z plane and the other for the x-y plane. These drawings, showing the dimensions of the arrangement, are useful in determining the components of the forces.

Step 2: Consider equilibrium of the structure. From the FBDs, it is clear that the system is non-coplanar and concurrent. Since the system is concurrent it is not necessary to write moment equations. The forces in the cables and strut are determined from equilibrium of the forces at point E as:

$$\Sigma F_x = F_{EA}\,(6/13.42) - F_{EC}\,(6/13.42) = 0 \qquad \text{(a)}$$

$$\Sigma F_y = -\,F_{EA}\,(12/13.42) - F_{EC}\,(12/13.42) - F_{ED}\,(12/14.42) = 0 \qquad \text{(b)}$$

$$\Sigma F_z = -\,F_{ED}\,(8/14.42) - F = 0 \qquad \text{(c)}$$

From Eq. (c), it is clear that F_{ED} is given by:

$$F_{ED} = -\,(14.42/8.00)\,F = -\,(1.8025)(3) = (5.408\text{ ton})(2000\text{ lb/ton}) = -\,10{,}815\text{ lb} \qquad \text{(d)}$$

From Eq. (a), it is evident that:

$$F_{EA} = F_{EC} \qquad \text{(e)}$$

Substituting Eqs. (d) and (e) into Eq. (b) yields:

$$-2\,F_{EA}\,(12/13.42) - (-10{,}815)\,(12/14.42) = 0$$

$$F_{EA} = F_{EC} = 5032 \text{ lb} \tag{f}$$

Step 3: Determine the stresses in the cables and the strut.

The stresses are determined from the safety factor and the strength of the materials used in fabricating the components.

$$\sigma = S_y/SF \tag{g}$$

For the cables, the allowable design stress is:

$$\sigma_{design} = S_y/SF = 150{,}000/3.5 = 42{,}860 \text{ psi} \tag{h}$$

For the strut, the allowable design stress is:

$$\sigma_{design} = S_y/SF = 40{,}000/3.5 = 11{,}430 \text{ psi} \tag{i}$$

Step 4: Determine the cross sectional area required for the cables and the strut.

For the cables, the cross sectional area required is:

$$A = P/\sigma = F_{EA}/\sigma = 5{,}032/\,42{,}860 = 0.1174 \text{ in}^2 \tag{j}$$

The diameter of the cable specified would be d = 7/16 in. providing an area $A = 0.1503$ in^2.
The cross sectional area required for the strut is:

$$A = P/\sigma = F_{EA}/\sigma = 10{,}815/\,11{,}430 = 0.9462 \text{ in}^2 \tag{k}$$

EXAMPLE 7.3

The Patcenter in Princeton, NJ is a large open structure with a cable-stayed roof. In Fig. E7.3, we show one of the nine tubular steel masts that are uniformly spaced at 9 m intervals along the depth of the building to support the roof structure. In this design, the roof hangs from cables[2] since the vertical side members are used only to prevent wind uplift. The tubular mast, 15 m high is supported with a 9 m wide by 6 m high rectangular steel frame. Determine the force in the primary rod stay and the tubular steel masts. Assume the roof is pinned to the steel mast at point C.

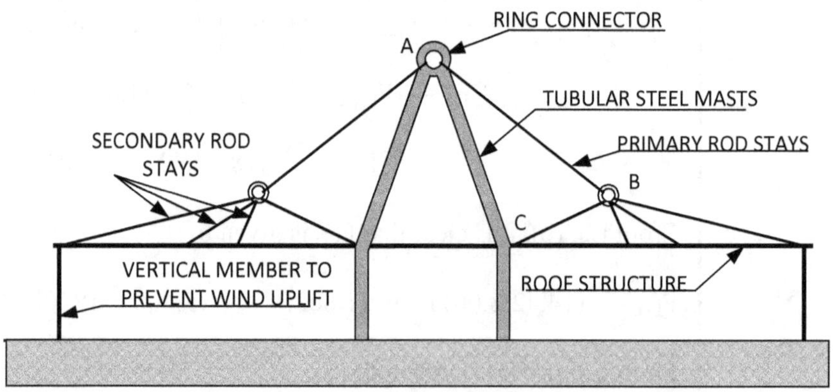

Fig. E7.3

[2] Thin steel tie rods were actually employed instead of cables.

Solution:

Step 1. Prepare FBDs for the structural members. Add dimensions, reactions and applied forces to the FBDs. The dimensions shown in Fig. 7.3a were obtained by scaling a drawing of the front view of the Patcenter [1].

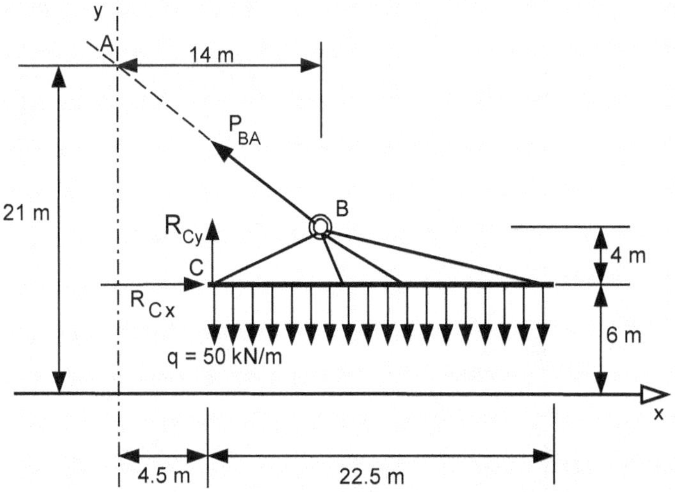

Fig. 7.3a

The uniformly distributed load of 50 kN/m was specified to account for the dead weight of the roof structure and snow/water accumulation[3]. To obtain this distributed load, we multiplied the roof area (9 m by 22.5 m) by a pressure of approximately 5.56 kN/m^2, and then divided by the length (22.5 m) over which the force was distributed.

The second FBD is a model of the tubular steel frame as shown in Fig. E7.3b. We have recognized the symmetry of the structure and have equated internal forces on the left and right sides of the structure. We have also considered that the internal forces in the tubular members act along their axes. The angles α and β in Fig. E7.3b are given by:

$$\alpha = \tan^{-1}(11/14) = 38.16° \qquad \beta = \tan^{-1}(4.5/15) = 16.70° \qquad \text{(a)}$$

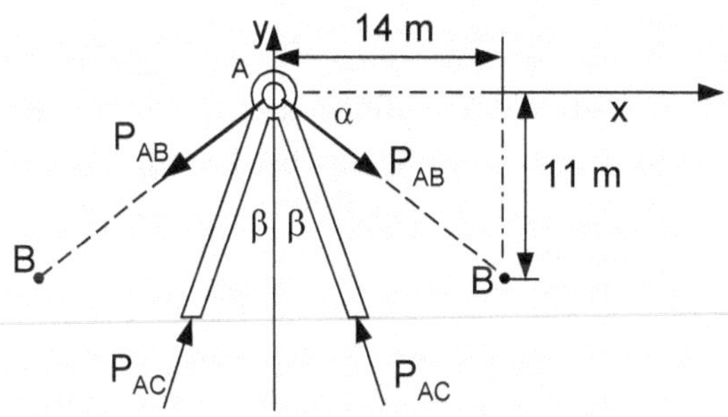

Fig. E7.3b

Step 2: Consider equilibrium of the structure. From the FBD in Fig. E7.3a, we write the moment equations about point C as:

$$\Sigma M_C = -(50)(22.5)(22.5/2) + P_{BA} \sin\alpha\,(14 - 4.5) + P_{BA} \cos\alpha\,(4) = 0 \qquad \text{(b)}$$

Solving Eq. (b) for P_{BA} yields:

$$P_{BA} = 1404 \text{ kN} \qquad \text{(c)}$$

[3] The snow loading assumes a water-saturated layer of snow 0.3 m deep.

Consider equilibrium in the x direction and write:

$$\Sigma F_x = R_{Cx} - P_{BA} \cos \alpha = 0 \tag{d}$$

$$R_{Cx} = 1404 \cos (38.16°) = 1104 \text{ kN} \tag{e}$$

$$\Sigma F_y = R_{Cy} + P_{BA} \sin \alpha - (50)(22.5) = 0 \tag{f}$$

$$R_{Cy} = 257.5 \text{ kN} \tag{g}$$

Next refer to the FBD in Fig. E7.3b, and determine the internal forces acting on the tubular A frame. Note that this FBD represents a coplanar concurrent force system.

Because the structure is symmetric the relation $\Sigma F_x = 0$ does not provide any additional information. However, considering the equilibrium equation in the y direction yields:

$$\Sigma F_y = 2P_{AC} \cos \beta - 2\, P_{AB} \sin \alpha = 0 \tag{h}$$

Solving Eq. (h) for P_{AC} gives:

$$P_{AC} = [(1404) \sin (38.16°)] / \cos (16.70°) = 905.7 \text{ kN} \tag{1}$$

The Patcenter is a modern structure with large open spaces that provide flexibility in the arrangement of laboratories and offices. Except for the vertical members supporting the A frame and the vertical tie bar to stabilize the roof against wind loading, it is free of columns. Modeling with FBDs and applying the simple equations of equilibrium enabled the analysis of the internal forces in selected members of this structure.

7.3 SOLUTIONS USING VECTOR MECHANICS

7.3.1 Expressing Forces and Moments as Vectors

When we encounter three-dimensional structures, the analysis becomes complex—often involving all six of the equations of equilibrium and several FBDs. Visualization of the structure and force components becomes more difficult. To alleviate the complexity and reduce visualization difficulties, we often use vector mechanics. When we express each force **F** or moment **M** in a complete vector representation, the equilibrium relations reduce to two vector equations:

$$\Sigma \mathbf{F} = 0 \qquad \text{and} \qquad \Sigma \mathbf{M} = 0 \tag{7.1}$$

Moments are a product of a force times a distance. In most of our previous solutions to example problems, we were careful to define the distance as the perpendicular distance from the line of action of the force to the point about which the moments were determined. The definition of the moment does not change, but with the vector mechanics approach described previously[4] in Chapter 2, we define the moment in terms of a vector cross product.

$$\mathbf{M} = \mathbf{r} \times \mathbf{F} \tag{7.2}$$

where r is a position vector given by:

$$\mathbf{r} = r_x\, \mathbf{i} + r_y\, \mathbf{j} + r_z\, \mathbf{k} \tag{7.3}$$

[4] For the convenience of the reader several of the more important equations from Chapter 2 are summarized in this section.

The force vector is written as:

$$\mathbf{F} = F_x\,\mathbf{i} + F_y\,\mathbf{j} + F_z\,\mathbf{k} \tag{7.4}$$

Recall, the properties of the cross vector product of the unit vectors:

$$\mathbf{i} \times \mathbf{i} = 0, \qquad \mathbf{j} \times \mathbf{j} = 0, \qquad \text{and} \qquad \mathbf{k} \times \mathbf{k} = 0$$

$$\mathbf{i} \times \mathbf{j} = -\,\mathbf{j} \times \mathbf{i} = \mathbf{k}, \qquad \mathbf{j} \times \mathbf{k} = -\,\mathbf{k} \times \mathbf{j} = \mathbf{i}, \qquad \mathbf{k} \times \mathbf{i} = -\,\mathbf{i} \times \mathbf{k} = \mathbf{j} \tag{7.5}$$

Combining Eqs. (7.2) to (7.5), we obtain:

$$\mathbf{M} = \mathbf{r} \times \mathbf{F} = (r_y\,F_z - r_z\,F_y)\mathbf{i} + (r_z\,F_x - r_x\,F_z)\mathbf{j} + (r_x\,F_y - r_y\,F_x)\mathbf{k} \tag{7.6}$$

or

$$\mathbf{M} = \mathbf{r} \times \mathbf{F} = M_x\,\mathbf{i} + M_y\,\mathbf{j} + M_z\,\mathbf{k} \tag{7.7}$$

Comparing the results of Eqs. (7.6) with those of Eq. (7.7) yields:

$$M_x = (r_y\,F_z - r_z\,F_y)$$

$$M_y = (r_z\,F_x - r_x\,F_z)$$

$$M_z = (r_x\,F_y - r_y\,F_x) \tag{7.8}$$

When employing the vector cross product **r × F** to determine the moment **M,** we simultaneously write the equations for the moments M_x, M_y and M_z about the three Cartesian axes. In writing Eq. (7.6), we often employ the determinant given by:

$$\mathbf{M} = \begin{vmatrix} \mathbf{i} & \mathbf{j} & \mathbf{k} \\ r_x & r_y & r_z \\ F_x & F_y & F_z \end{vmatrix} \tag{7.9}$$

Let's apply these results to a previous example and solve for the forces acting on the communications tower. As we repeat the solution to this example problem, a step-by-step procedure to follow in a design analysis of a three-dimensional structure is outlined.

EXAMPLE 7.4

Consider again the communications tower of Example 7.1. However, use the vector mechanics approach in the solution for the unknown forces acting on the tower.

Solution:

Step 1: Construct a FBD of the communications tower. We always begin a solution with a FBD regardless of the mathematical approach used in the solution of the problem. We will use the FBD shown previously in Example 7.1 to assist us in writing the necessary equations for the vectors.

Step 2: Count the forces acting on the structure and write the vector equation for each force. We have five forces acting on the tower including:

- Two-forces applied by the cables (two of the four cables are slack).

- The wind force.
- The reaction force at the base (origin) of the tower.
- The weight of the tower and transmission equipment.

The vector representing the force $\mathbf{P_2}$ on the cable anchored on the x-axis (west) is written as:

$$\mathbf{P_2} = P_2[(-2/\sqrt{29})\mathbf{i} + (-5/\sqrt{29})\mathbf{k}] \quad \text{(a)}$$

Similarly, the force $\mathbf{P_4}$ on the cable anchored on the y-axis (south) is:

$$\mathbf{P_4} = P_4[(-2/\sqrt{29})\mathbf{j} + (-5/\sqrt{29})\mathbf{k}] \quad \text{(b)}$$

Recall that $\mathbf{P_1} = \mathbf{P_3} = 0$ because these cables are slack under the specified wind conditions and do not support the tower. However, when the wind direction changes these cables become functional.

The wind force $\mathbf{F_3}$ is written as:

$$\mathbf{F_3} = F_3[\cos(45°)\,\mathbf{i} + \sin(45°)\,\mathbf{j}] = (\sqrt{2}/2)F_3[\mathbf{i} + \mathbf{j}] \quad \text{(c)}$$

Note, the vector diagram shown in Fig. E7.1b is helpful in writing Eq. (c).

The reaction force $\mathbf{R_4}$ is written as:

$$\mathbf{R_4} = R_{4x}\,\mathbf{i} + R_{4y}\,\mathbf{j} + R_{4z}\,\mathbf{k} \quad \text{(d)}$$

The dead weight of the tower structure and the equipment is expressed as:

$$\mathbf{F_5} = -22{,}000\,\mathbf{k} \quad \text{(e)}$$

Okay! We have described the five active forces acting on the communications tower in vector format.

Step 3: Write the equations for the position vectors for the five active forces. Let's consider the forces in order (i.e. 1, 2, …..5).

For the two active cables that are attached at the top of the tower, we write the relations for the position vectors as:

$$\mathbf{r_1} = H\,\mathbf{k} \quad \text{(f)}$$

$$\mathbf{r_2} = H\,\mathbf{k} \quad \text{(g)}$$

For the wind force, the position vector is given by:

$$\mathbf{r_3} = (2/3)H\,\mathbf{k} \quad \text{(h)}$$

Finally, for both the reaction and dead weight forces, we note that their lines of action pass through the origin and they do not produce a moment about the origin. Hence:

$$\mathbf{r_4} = \mathbf{r_5} = 0 \quad \text{(i)}$$

Step 4: Employ the equilibrium relations to solve for the unknown forces.

First, consider the moments:

$$\Sigma \mathbf{M} = \Sigma \mathbf{r} \times \mathbf{F} = \mathbf{r}_1 \times \mathbf{P}_2 + \mathbf{r}_2 \times \mathbf{P}_4 + \mathbf{r}_3 \times \mathbf{F}_3 \qquad \text{(j)}$$

Note, the forces due to the reaction and the dead weight forces do not appear in Eq. (j) because $\mathbf{r}_4 = \mathbf{r}_5 = 0$.

Substituting the results for the forces and the position vectors in Eq. (7.9) yields three determinants:

$$\mathbf{M} = P_2 \begin{vmatrix} \mathbf{i} & \mathbf{j} & \mathbf{k} \\ 0 & 0 & H \\ \frac{-2}{\sqrt{29}} & 0 & \frac{-5}{\sqrt{29}} \end{vmatrix} + P_4 \begin{vmatrix} \mathbf{i} & \mathbf{j} & \mathbf{k} \\ 0 & 0 & H \\ 0 & \frac{-2}{\sqrt{29}} & \frac{-5}{\sqrt{29}} \end{vmatrix} + \frac{\sqrt{2}}{2} F_3 \begin{vmatrix} \mathbf{i} & \mathbf{j} & \mathbf{k} \\ 0 & 0 & \frac{2H}{3} \\ 1 & 1 & 0 \end{vmatrix} = 0 \qquad \text{(k)}$$

Expanding these determinants gives:

$$\mathbf{M} = [(2/\surd 29)HP_4 - (\surd 2/2)F_3\,(2H/3)]\,\mathbf{i} + [-(2/\surd 29)HP_2 + (\surd 2/2)F_3\,(2H/3)]\,\mathbf{j} = 0 \qquad \text{(l)}$$

It is evident from Eq. (l) that:

$$M_x = [(2/\surd 29)HP_4 - (\surd 2/2)F_3\,(2H/3)] = 0$$
$$M_y = [-(2/\surd 29)HP_2 + (\surd 2/2)F_3\,(2H/3)] = 0 \qquad \text{(m)}$$

Recall that $F_3 = 1600$ lb and solve Eqs. (m) to obtain:

$$P_2 = P_4 = 2030 \text{ lb} \qquad \text{(n)}$$

The results for P_2 and P_4 as expected are identical with the solutions determined previously in Example 7.1.

Next, consider the forces:

$$\Sigma \mathbf{F} = \mathbf{P}_2 + \mathbf{P}_4 + \mathbf{F}_3 + \mathbf{R}_4 + \mathbf{F}_5 = 0 \qquad \text{(o)}$$

Substituting Eqs. (a) through (e) into Eq. (o) yields:

$$\Sigma \mathbf{F} = P_2[(-2/\surd 29)\mathbf{i} + (-5/\surd 29)\mathbf{k}] + P_4[(-2/\surd 29)\mathbf{j} + (-5/\surd 29)\mathbf{k}]$$
$$+ F_3[\cos(45°)\,\mathbf{i} + \sin(45°)\,\mathbf{j}] + R_{4x}\,\mathbf{i} + R_{4y}\,\mathbf{j} + R_{4z}\,\mathbf{k} + (-22{,}000)\,\mathbf{k} = 0 \qquad \text{(p)}$$

We collect the coefficients of the unit vectors **i**, **j**, and **k** and set each of these to zero. It is clear that the components of the sum of the forces must be zero in each of the three directions.

The coefficient of the unit vector **i** is:

$$\Sigma F_x = P_2(-2/\surd 29) + F_3 \cos(45°) + R_{4x} = 0 \qquad \text{(q)}$$

The coefficient of the unit vector **j** is:

$$\Sigma F_y = P_4(-2/\surd 29) + F_3 \sin(45°) + R_{4y} = 0 \qquad \text{(r)}$$

The coefficient of the unit vector **k** is:

$$\Sigma F_z = P_2(-5/\surd 29) + P_4(-5/\surd 29) + R_{4z} - 22{,}000 = 0 \qquad \text{(s)}$$

From Eqs (q) and (r), we show that:

$$R_{4x} = R_{4y} = -377.5 \text{ lb} \qquad \text{(t)}$$

The negative sign indicates that the direction of the forces R_{4x} and R_{4y} shown in the FBD should be reversed.

And from Eq. (s), we solve for R_{4z} as:

$$R_{4z} = (5/\sqrt{29})(P_2 + P_4) + 22{,}000$$

$$R_{4z} = (10/\sqrt{29})(2030) + 22{,}000 = 25{,}770 \text{ lb} \qquad \text{(u)}$$

Comparisons of the results for the forces acting on the communication tower are identical with those determined in Example 7.1. The approach in these two examples was the same—we employed FBDs and the equations of equilibrium in both cases. However, in Example 7.1, we employed the six equations of equilibrium involving the components of the forces and moments. In Example 7.4 we used the two vector equations of equilibrium, a complete vector representation for the moments and forces, and determined the moments by using the vector cross product.

EXAMPLE 7.5

For the derrick shown in Fig. E7.5, determine the margin of safety for the three supporting cables if the force F = 32 kN. The anchor points P and Q for two of the cables lay in the x-y plane with coordinates P(4, – 3, 0) and Q(– 4, –3, 0). The derrick pole is 5 m high and the boom is 6 m long. The derrick pole and boom are separate components. The boom is attached to the pole with a sleeve type bearing at point R that enables the boom to rotate about the pole. However, the boom is constrained from sliding up or down the pole. The cables are all 20 mm in diameter with a specified strength of 700 MPa. The pole is supported by the floor at point O with a ball and socket joint.

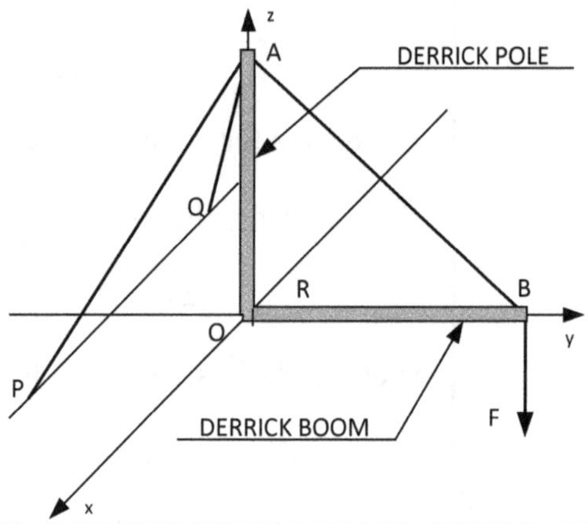

Fig. E7.5

Solution:

Step 1: Let's consider the boom and pole of the derrick separately. First, prepare a FBD of the boom showing all of the unknown forces as shown in Fig. E7.5a. Examination of this figure indicates that we have one cable force F_{AB}, the force F due to the load, and the reaction forces between the boom and the derrick pole. Note that no reactive moments are present at point R, only reactive forces.

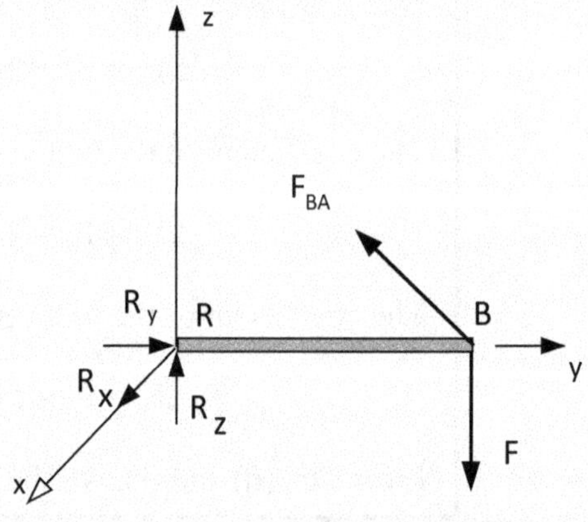

Fig. E7.5a

Step 2: Let's write the relations for the forces in vector format.

$$\mathbf{F_{BA}} = F_{BA}\,\mathbf{u_{BA}} \qquad \mathbf{R} = R_x\mathbf{i} + R_y\,\mathbf{j} + R_z\,\mathbf{k} \qquad \mathbf{F} = -\,F\,\mathbf{k} = -\,32\,\mathbf{k} \qquad \text{(a)}$$

Step 3: Next write the equation for the unit vector that provides the direction of the cable force F_{BA} and the position vector defining the location of point B relative to the origin:

$$\mathbf{u_{BA}} = (-\,6\,\mathbf{j} + 5\,\mathbf{k})/(36 + 25)^{1/2} = -\,0.7682\,\mathbf{j} + 0.6402\,\mathbf{k} \qquad \text{(b)}$$

$$\mathbf{r_B} = 6\,\mathbf{j} \qquad \text{(c)}$$

Step 4: Apply the equilibrium equations to the derrick boom.

$$\Sigma\mathbf{M_O} = \mathbf{r_B} \times \mathbf{F_{BA}} + \mathbf{r_B} \times \mathbf{F} = 0 \qquad \text{(d)}$$

$$\sum \mathbf{M}_O = \begin{vmatrix} \mathbf{i} & \mathbf{j} & \mathbf{k} \\ 0 & 6 & 0 \\ 0 & -0.7682 & 0.6402 \end{vmatrix} F_{BA} + \begin{vmatrix} \mathbf{i} & \mathbf{j} & \mathbf{k} \\ 0 & 6 & 0 \\ 0 & 0 & -32 \end{vmatrix} = 0$$

$$\Sigma\mathbf{M_O} = (6)(0.6402)F_{BA} - (6)(32)\,\mathbf{i} = 0$$

$$F_{BA} = 49.98 \text{ kN} \qquad \text{(e)}$$

$$\Sigma\mathbf{F} = \mathbf{F_{BA}} + \mathbf{R} + \mathbf{F} = 0 \qquad \text{(f)}$$

$$\Sigma\mathbf{F} = (R_x)\,\mathbf{i} + (-\,0.7682\,F_{BA} + R_y)\,\mathbf{j} + (0.6402\,F_{BA} + R_z - 32)\,\mathbf{k} = 0$$

Considering the coefficient of the **i**, **j** and **k** terms yields:

$$R_x = 0$$

$$R_z = 32 - 0.6402\,F_{BA} \qquad R_z = 32 - (0.6402)(49.98) = 0 \qquad \text{(g)}$$

$$R_y = 0.7682\,F_{BA} \qquad R_y = (0.7682)(49.98) = 38.4 \text{ kN} \qquad \text{(h)}$$

We observe that R_y acts along the axis of the boom tending to compress it. Having completed the analysis of the boom, we continue by considering the pole.

Step 5: Let's prepare a FBD of the pole showing all of the unknown forces as shown in Fig. E7.5b. Inspection of this FBD shows five unknown forces (F_{AP}, F_{AQ}, R_{xp}, R_{yp} and R_{zp}) and two known forces (F_{AB} and R_y). The reaction R_y is the boom pushing on the pole; whereas, the reactions R_{xp}, R_{yp} and R_{zp} are due to the interaction of the pole with the deck supporting the derrick.

Step 6: Let's write the vector equations for the forces acting on the pole:

$$\mathbf{F_{AB}} = 49.98\,\mathbf{u_{AB}} \qquad \mathbf{F_{AP}} = F_{AP}\,\mathbf{u_{AP}}$$

$$\mathbf{F_{AQ}} = F_{AQ}\,\mathbf{u_{AQ}} \qquad \mathbf{R_p} = R_{xp}\,\mathbf{i} + R_{yp}\,\mathbf{j} + R_{zp}\,\mathbf{k} \qquad \text{(i)}$$

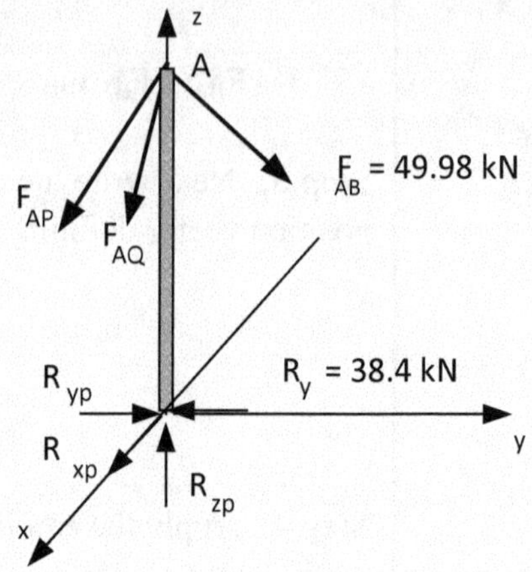

Fig. E7.5b

Step 7: Next write the equations for the unit vectors and position vector that provide the directions of the cable forces and their location relative to the origin:

$$\mathbf{u_{AB}} = (6\,\mathbf{j} - 5\,\mathbf{k})/(36 + 25)^{1/2} = 0.7682\,\mathbf{j} - 0.6402\,\mathbf{k}$$

$$\mathbf{u_{AP}} = (4\,\mathbf{i} - 3\,\mathbf{j} - 5\,\mathbf{k})/(16 + 9 + 25)^{1/2} = 0.5657\,\mathbf{i} - 0.4243\,\mathbf{j} - 0.7071\,\mathbf{k} \qquad \text{(j)}$$

$$\mathbf{u_{AQ}} = (-\,4\,\mathbf{i} - 3\,\mathbf{j} - 5\,\mathbf{k})/(16 + 9 + 25)^{1/2} = -\,0.5657\,\mathbf{i} - 0.4243\,\mathbf{j} - 0.7071\,\mathbf{k}$$

$$\mathbf{r_A} = 5\,\mathbf{k}$$

Step 8: Apply the equilibrium equations to the derrick pole.

$$\Sigma\mathbf{M_O} = \mathbf{r_A} \times \mathbf{F_{AB}} + \mathbf{r_A} \times \mathbf{F_{AP}} + \mathbf{r_A} \times \mathbf{F_{AQ}} = 0 \qquad \text{(k)}$$

$$\sum M_O = \begin{vmatrix} \mathbf{i} & \mathbf{j} & \mathbf{k} \\ \mathbf{0} & \mathbf{0} & \mathbf{5} \\ \mathbf{0} & \mathbf{0.7682} & \mathbf{-0.6402} \end{vmatrix} 49.98 + \begin{vmatrix} \mathbf{i} & \mathbf{j} & \mathbf{k} \\ \mathbf{0} & \mathbf{0} & \mathbf{5} \\ \mathbf{0.5657} & \mathbf{-0.4243} & \mathbf{-0.7071} \end{vmatrix} F_{AP} + \begin{vmatrix} \mathbf{i} & \mathbf{j} & \mathbf{k} \\ \mathbf{0} & \mathbf{0} & \mathbf{5} \\ \mathbf{-0.5657} & \mathbf{-0.4243} & \mathbf{-0.7071} \end{vmatrix} F_{AQ} = 0$$

Evaluation of this relation gives the coefficients of the unit vectors. For **i**, we obtain:

$$(-5)(0.7682)(49.98) + (5)(0.4243)\,F_{AP} + (5)(0.4243)F_{AQ} = 0 \qquad \text{(l)}$$

For **j**, we obtain:

$$(5)(0.5657)\,F_{AP} - (5)(0.5657)\,F_{AQ} = 0 \qquad \text{(m)}$$

Note that the coefficient of the unit vector **k** is zero. From Eq. (m), it is clear that:

$$F_{AP} = F_{AQ} \qquad \text{(n)}$$

Substituting Eq. (n) into Eq. (l) gives:

$$F_{AP} = F_{AQ} = [(5)(0.7682)(49.98)]/4.243 = 45.24 \text{ kN} \qquad \text{(o)}$$

This completes the solution for the forces in the cables. In a more complete analysis, we would employ $\mathbf{\Sigma F} = \mathbf{0}$ to determine the reactions at the base of the pole. It is suggested that you verify the results for $R_{px} = 0$, $R_{py} = 38.4$ kN and $R_{pz} = 95.99$ kN.

Step 9: Let's determine the stresses in each of the cables using the results for the forces given in Eqs. (e) and (o).

$$\sigma_{AB} = F_{AB}/A = (49.98 \times 10^3) /[\pi (10)^2] = 159 \text{ MPa}$$

$$\sigma_{AP} = \sigma_{AQ} = F_{AP}/A = (45.24 \times 10^3) /[\pi(10)^2] = 144 \text{ MPa}$$

Step 10: Finally, we determine the margin of safety for each of the cables.

For cable AB:

$$MOS = SF - 1 = S/\sigma - 1 = 700/159 - 1 = 3.40$$

For cables AP and AQ:

$$MOS = SF - 1 = S/\sigma - 1 = 700/144 - 1 = 3.86$$

Step 11: To complete the analysis, the results should be interpreted. We find, in this example, margins of safety of about 3.5 for all of the cables. The design is reasonably well balanced with stresses on all three cables nearly equal. The margin of safety is comfortable and certainly not excessive for a derrick application where large loads are lifted in the presence of workers.

EXAMPLE 7.6

A space truss shown in Fig. E7.6, is supported at point A with a ball and socket joint and with cables that are anchored at points B and D as shown in the illustration. A force **F** is applied to joint E. **F** is represented in vector format as:

$$\mathbf{F} = [10\,\mathbf{i} + 8\,\mathbf{j} - 15\,\mathbf{k}] \text{ kips}$$

Determine the internal forces in the three members that intersect at joint E. Also specify the diameter of the rods, if the truss is fabricated from 1020 HR steel. A safety factor of 3.4 based on yield strength is specified by the building code.

Fig. E 7.6

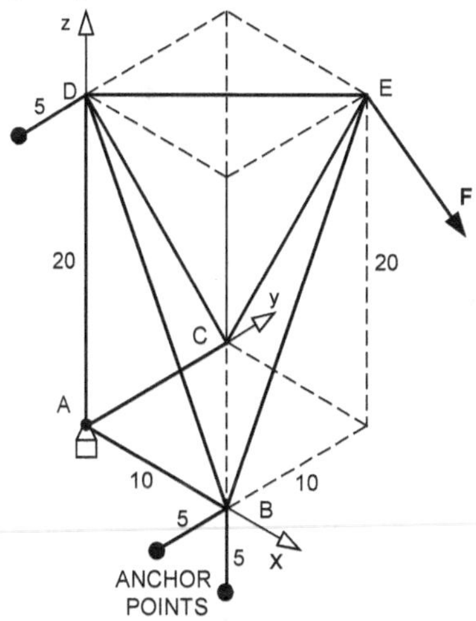

Solution:

Step 1: Let's consider point E of the space truss and prepare a FBD showing all of the unknown forces as shown in Fig. E7.6a. Examination of this figure indicates that we have a concurrent three-dimensional force system.

Step 2: Let's write the relations for the forces in vector format.

$$\mathbf{P_{EB}} = P_{EB}\,\mathbf{u_{EB}} \qquad \mathbf{P_{EC}} = P_{EC}\,\mathbf{u_{EC}} \qquad \mathbf{P_{ED}} = P_{ED}\,\mathbf{u_{ED}}$$

$$\mathbf{F} = 10\,\mathbf{i} + 8\,\mathbf{j} - 15\,\mathbf{k} \qquad \text{(a)}$$

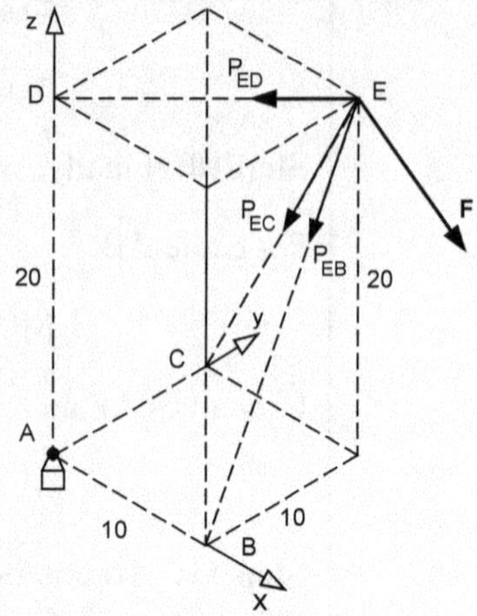

Fig. E7.6a

Step 3: Next write the equation for the unit vectors that gives the direction of the forces F_{EB}, F_{EC}, and F_{ED}.

$$\mathbf{u_{EB}} = (-\,10\,\mathbf{j} - 20\,\mathbf{k})/(100 + 400)^{1/2} = -\,0.4472\,\mathbf{j} - 0.8944\,\mathbf{k}$$

$$\mathbf{u_{EC}} = (-\,10\,\mathbf{i} - 20\,\mathbf{k})/(100 + 400)^{1/2} = -\,0.4472\,\mathbf{i} - 0.8944\,\mathbf{k}$$

$$\mathbf{u_{ED}} = (-\,10\,\mathbf{i} - 10\,\mathbf{j})/(100 + 100)^{1/2} = -\,0.7071\,\mathbf{i} - 0.7071\,\mathbf{j} \qquad \text{(b)}$$

Step 4: Apply the equilibrium equation to the force system acting at point E.

$$\Sigma\,\mathbf{F} = 0 \qquad \text{(c)}$$

Substituting Eqs. (a) and (b) into Eq. (c) yields:

$$\Sigma\,\mathbf{F} = \mathbf{i}\,(-0.4472P_{EC} - 0.7071\,P_{ED} + 10) + \mathbf{j}\,(-0.4472\,P_{EB} - 0.7071\,P_{ED} + 8)$$

$$+\,\mathbf{k}\,(-0.8944\,P_{EB} - 0.8944\,P_{EC} - 15) = 0 \qquad \text{(d)}$$

Setting each of the coefficients of the unit vectors to zero yields the three simultaneous equations given by:

$$-0.4472P_{EC} - 0.7071\,P_{ED} + 10 = 0$$

$$-0.4472\,P_{EB} - 0.7071\,P_{ED} + 8 = 0 \qquad \text{(e)}$$

$$-0.8944P_{EB} - 0.8944\,P_{EC} - 15 = 0$$

Solving Eq. (e) for the forces gives:

$$P_{EC} = -\,6.149\text{ kip} \qquad P_{EB} = -\,10.62\text{ kip} \qquad P_{ED} = 18.03\text{ kip} \qquad \text{(f)}$$

Step 5: Determine the allowable design stress from the yield strength of the 1020 HR steel and the specified safety factor.

$$\sigma_{design} = S_y\,/\mathbf{SF_y} = 42/3.4 = 12.35\text{ ksi} \qquad \text{(g)}$$

Step 6: Solve for the cross sectional areas and diameters of the three members:

$$A = P/\sigma_{design} \qquad d = (4A/\pi)^{1/2}$$

$$A_{EC} = 6.149/12.35 = 0.4979\text{ in.}^2 \qquad d = 0.7962\text{ in.} \;\Rightarrow d = 7/8\text{ in.}$$

$A_{EB} = 10.62/12.35 = 0.8599\ in.^2$ $\quad d = 1.046\ in.$ $\quad \Rightarrow d = 1\ ¼\ in.$

$A_{ED} = 18.03/12.35 = 1.460\ in.^2$ $\quad d = 1.363\ in.$ $\quad \Rightarrow d = 1\ ½\ in.$ (h)

Our steel supplier lists the standard sizes for 1020 HR steel rods from 1 to 2 in. in diameter as 1, 1 ¼, 1 ½, 1 ¾ and 2 in.

EXAMPLE 7.7

A tetrahedral space truss, shown in Fig. E7.7, supports a massive scoreboard and several sets of remotely controlled spotlights in an amphitheater. The base ABC of the space truss lies in the x-y plane (horizontal). The base triangle is connected to anchors in the roof beams by long cables. Determine the size of the solid round rods required to fabricate members AB, AC and AD of the space truss. The safety factor for public arenas is 10, based on yield strength, and the material employed for all of the members is 1018A steel. The scoreboard and spotlights weigh 250 kN.

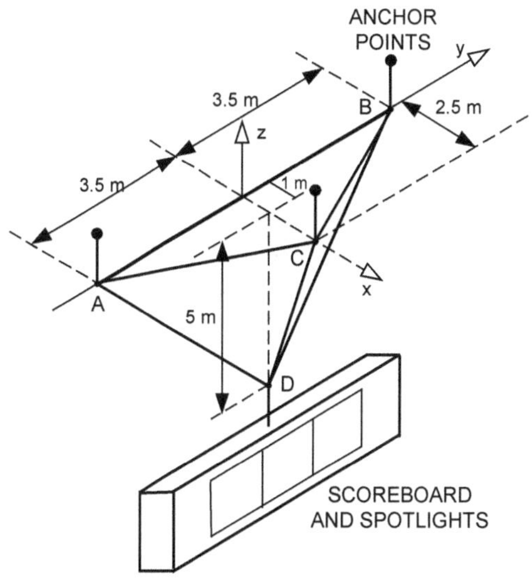

Fig. E7.7

Solution:

Step 1: Let's consider the space truss in its entirety and prepare a FBD showing the three unknown forces applied by the anchor cables as shown in Fig. E7.7a.

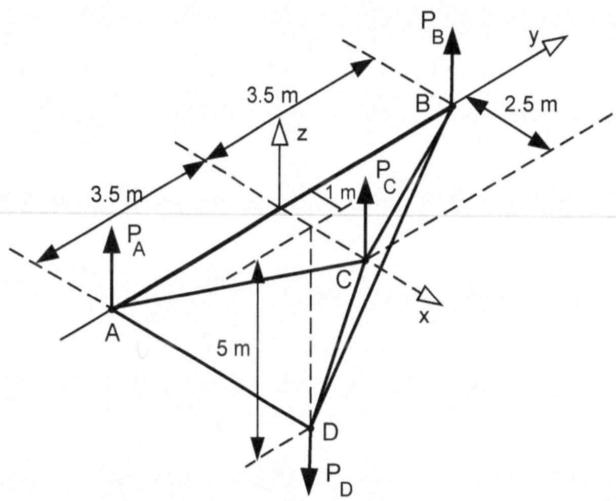

Fig. E7.7a

Step 2: Let's write the equations for the moments about point C in vector format and use the results to solve for P_A.

$$\Sigma \mathbf{M}_C = \mathbf{r}_{CA} \times \mathbf{P}_A + \mathbf{r}_{CB} \times \mathbf{P}_B + \mathbf{r}_{CD} \times \mathbf{P}_D = 0 \quad \text{(a)}$$

The force vectors and position vectors are written as:

$\mathbf{P_A} = P_A\,\mathbf{k} \qquad \mathbf{P_B} = P_B\,\mathbf{k} \qquad \mathbf{P_C} = P_C\,\mathbf{k} \qquad \mathbf{P_D} = -\,W\,\mathbf{k}$ (b)

$\mathbf{r_{CA}} = -\,2.5\,\mathbf{i} - 3.5\,\mathbf{j} \qquad \mathbf{r_{CB}} = -\,2.5\,\mathbf{i} + 3.5\,\mathbf{j} \qquad \mathbf{r_{CD}} = -\,1.5\,\mathbf{i} - 5\,\mathbf{k}$ (c)

Substituting the results of Eqs. (b) and (c) into Eq. (a) yields:

$$\mathbf{M_C} = \begin{vmatrix} \mathbf{i} & \mathbf{j} & \mathbf{k} \\ -2.5 & -3.5 & 0 \\ 0 & 0 & P_A \end{vmatrix} + \begin{vmatrix} \mathbf{i} & \mathbf{j} & \mathbf{k} \\ -2.5 & +3.5 & 0 \\ 0 & 0 & P_B \end{vmatrix} + \begin{vmatrix} \mathbf{i} & \mathbf{j} & \mathbf{k} \\ -1.5 & 0 & -5 \\ 0 & 0 & -W \end{vmatrix} = 0 \qquad \text{(d)}$$

Expanding the determinants gives two equations:

$\mathbf{i}(-3.5\,P_A + 3.5\,P_B) = 0 \qquad \mathbf{j}(2.5\,P_A + 2.5\,P_B - 1.5\,W) = 0$ (e)

Recall W = 250 kN and solve Eq. (e) to obtain:

$P_A = P_B \qquad P_A = 75\text{ kN}$ (f)

Step 3: Let's consider joint A from the space truss and prepare a FBD showing the three unknown forces P_{AB}, P_{AC}, and P_{AD} as shown in Fig. E7.7b.

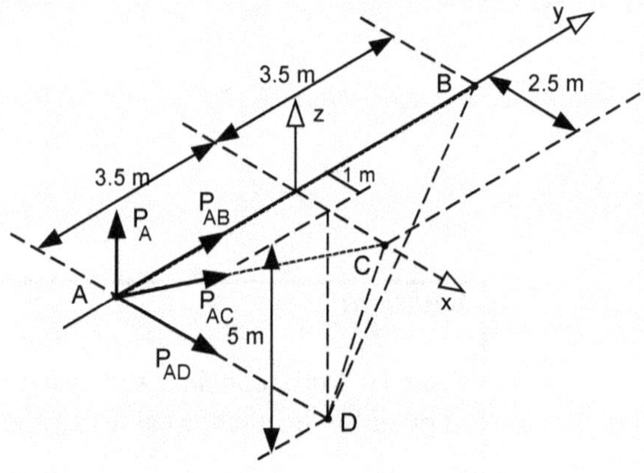

Fig. E7.7b

Step 4: Write the equilibrium equation for the concurrent force system acting on joint A as:

$$\Sigma\mathbf{F} = \mathbf{P_{AB}} + \mathbf{P_{AC}} + \mathbf{P_{AD}} + \mathbf{P_A} = 0 \qquad \text{(g)}$$

To employ this relation we write the forces in vector format as:

$\mathbf{P_{AB}} = P_{AB}\,\mathbf{u_{AB}} \qquad \mathbf{P_{AC}} = P_{AC}\,\mathbf{u_{AC}} \qquad \mathbf{P_{AD}} = P_{AD}\,\mathbf{u_{AD}}$ (h)

The unit vectors are given by:

$$\mathbf{u_{AB}} = \mathbf{r_{AB}}/r_{AB} = 7\mathbf{j}/7 = \mathbf{j}$$

$$\mathbf{u_{AC}} = \mathbf{r_{AC}}/r_{AC} = (2.5\,\mathbf{i} + 3.5\mathbf{j})/[(2.5)^2 + (3.5)^2]^{1/2} = 0.5812\,\mathbf{i} + 0.8137\,\mathbf{j}$$

$$\mathbf{u_{AD}} = \mathbf{r_{AD}}/r_{AD} = (1.0\,\mathbf{i} + 3.5\mathbf{j} - 5\mathbf{k})/[(1.0)^2 + (3.5)^2 + (5)^2]^{1/2}$$

$$\mathbf{u_{AD}} = 0.1617\,\mathbf{i} + 0.5659\,\mathbf{j} - 0.8084\,\mathbf{k} \qquad \text{(i)}$$

Substituting the results of Eqs. (h) and (i) into Eq. (g) yields:

$\mathbf{i}\,(0.5812\ P_{AC} + 0.1617\ P_{AD}) + \mathbf{j}\,(0.8137\ P_{AC} + 0.5659\ P_{AD} + P_{AB}) + \mathbf{k}\,(75 - 0.8084\ P_{AD}) = 0$

Solving this equation gives:

$$P_{AD} = 92.77\ \text{kN} \qquad P_{AC} = -25.81\ \text{kN} \qquad P_{AB} = -31.50\ \text{kN} \qquad \text{(j)}$$

Step 5: Determine the allowable design stress from the yield strength of the 1018A steel and the specified safety factor.

$$\sigma_{design} = S_y / \mathbf{SF}_y = 221/10 = 22.1\ \text{MPa} \qquad \text{(k)}$$

Step 6: Solve for the cross sectional areas and diameters of the three members:

$$A = P/\sigma_{design} \qquad d = (4A/\pi)^{1/2} \qquad \text{(l)}$$

$A_{AD} = 92.77 \times 10^3 /22.1 = 4197\ \text{mm}^2$ $\quad d = 73.10$ mm $\quad \Rightarrow d = 75$ mm

$A_{AC} = 25.81 \times 10^3/22.1 = 1168\ \text{mm}^2$ $\quad d = 38.56$ mm $\quad \Rightarrow d = 40$ mm

$A_{AB} = 31.50 \times 10^3/22.1 = 1425\ \text{mm}^2$ $\quad d = 42.60$ mm $\quad \Rightarrow d = 45$ mm $\quad$ (m)

The diameters of the bars have been converted to the standard metric sizes available. A second analysis will be necessary to insure that the long thin rods used for members AC and AB, that are subjected to compressive forces, do not buckle.

EXAMPLE 7.8

The long horizontal boom of a construction crane is fabricated from a lattice of steel bars. A four-cell segment of the boom, presented in Fig. E7.8, shows that the lattice is periodic from points A to E with identical cells at 8 ft intervals. The rectangle AEOK lies in the x-y plane and the equilateral triangle AKF lies in the x-z plane. The lengths of the members AF = FK = AK = 6 ft. The crane is lifting a load F = 12 kip at point P. Determine the internal forces in members CD, MN, HI, HN, HD, and MD using the method of sections.

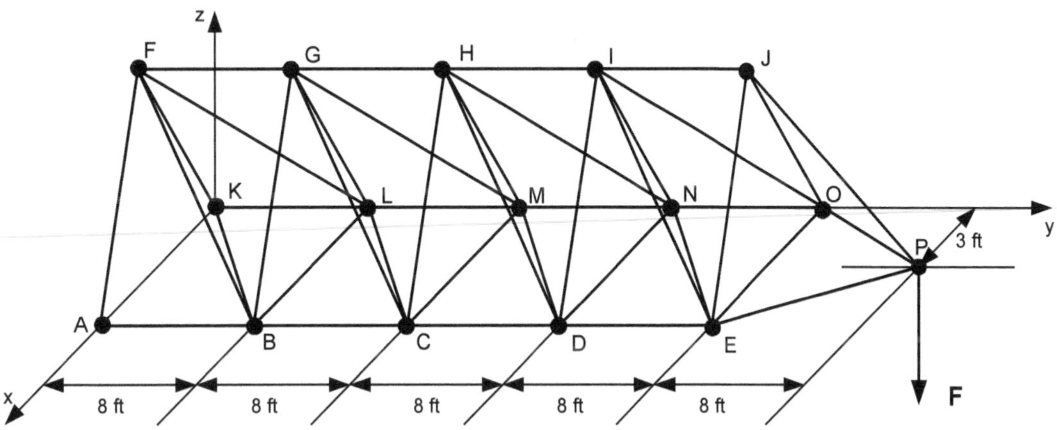

Solution:

Step 1: Let's make a section cut perpendicular to the y-axis between points C and D and prepare a FBD of the right side of the boom. This FBD, presented in Fig. E7.8a, shows six unknown internal forces acting on the members CD, MN, HI, HN, DH, and DM.

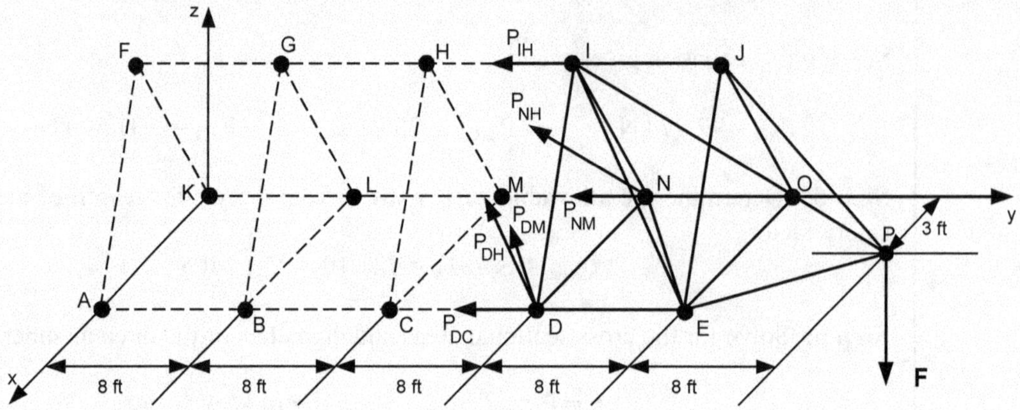

Fig. E7.8a

Step 2: Let's write the equations for the moments about point D in vector format and use the results to obtain three equations in terms of the unknown internal forces P_{NM}, P_{NH} and P_{IH}.

$$\Sigma \mathbf{M_D} = \mathbf{r_{DN}} \times \mathbf{P_{NM}} + \mathbf{r_{DN}} \times \mathbf{P_{NH}} + \mathbf{r_{DI}} \times \mathbf{P_{IH}} + \mathbf{r_{DP}} \times \mathbf{F} = 0 \qquad \text{(a)}$$

The force vectors, unit vectors and position vectors in Eq. (a) are written as:

$$\mathbf{P_{NM}} = P_{NM}\, \mathbf{u_{NM}} \qquad \mathbf{P_{NH}} = P_{NH}\, \mathbf{u_{NH}} \qquad \mathbf{P_{IH}} = P_{IH}\, \mathbf{u_{IH}} \qquad \mathbf{F} = -F\,\mathbf{k} \qquad \text{(b)}$$

$$\mathbf{u_{NM}} = \mathbf{r_{NM}}/\, r_{NM} = -8\,\mathbf{j}/8 = -\mathbf{j}$$

$$\mathbf{u_{NH}} = \mathbf{r_{NH}}/\, r_{NH} = (+3\mathbf{i} - 8\mathbf{j} + 5.196\,\mathbf{k})/[(3)^2 + (8)^2 + (5.196)^2]^{1/2} = 0.3\,\mathbf{i} - 0.8\,\mathbf{j} + 0.5196\,\mathbf{k} \qquad \text{(c)}$$

$$\mathbf{u_{IH}} = \mathbf{r_{IH}}/\, r_{IH} = -8\,\mathbf{j}/8 = -\mathbf{j}$$

Substituting Eq. (c) into Eq. (b) gives the internal forces in vector notation as:

$$\mathbf{P_{NM}} = -P_{NM}\,\mathbf{j} \qquad \mathbf{P_{NH}} = P_{NH}\,(0.3\,\mathbf{i} - 0.8\,\mathbf{j} + 0.5196\,\mathbf{k}) \qquad \mathbf{P_{IH}} = -P_{IH}\,\mathbf{j} \qquad \mathbf{F} = -F\,\mathbf{k} \qquad \text{(d)}$$

The position vectors are written as:

$$\mathbf{r_{DN}} = -6\,\mathbf{i} \qquad \mathbf{r_{DI}} = -3\,\mathbf{i} + 5.196\,\mathbf{k} \qquad \mathbf{r_{DP}} = -3\,\mathbf{i} + 16\,\mathbf{j} \qquad \text{(e)}$$

Substituting the results of Eqs. (d) and (e) into Eq. (a) yields:

$$\mathbf{M_D} = P_{NM}\begin{vmatrix} \mathbf{i} & \mathbf{j} & \mathbf{k} \\ -6 & 0 & 0 \\ 0 & -1 & 0 \end{vmatrix} + P_{NH}\begin{vmatrix} \mathbf{i} & \mathbf{j} & \mathbf{k} \\ -6 & 0 & 0 \\ 0.3 & -0.8 & 0.5196 \end{vmatrix} + P_{IH}\begin{vmatrix} \mathbf{i} & \mathbf{j} & \mathbf{k} \\ -3 & 0 & 5.196 \\ 0 & -1 & 0 \end{vmatrix} + \begin{vmatrix} \mathbf{i} & \mathbf{j} & \mathbf{k} \\ -3 & 16 & 0 \\ 0 & 0 & -F \end{vmatrix} = \mathbf{0} \qquad \text{(f)}$$

Expanding the determinants gives three equations:

$$\mathbf{i}(5.196\, P_{IH} - 16\, F) = 0$$

$$\mathbf{j}(3.118\, P_{NH} - 3F) = 0 \qquad \text{(g)}$$

$$\mathbf{k}(6\, P_{NM} + 4.8\, P_{NH} + 3\, P_{IH}) = 0$$

Recall F = 12 kip and solve Eqs. (g) to obtain:

$$P_{IH} = 36.95 \text{ kip} \qquad P_{NH} = 11.55 \text{ kip} \qquad P_{NM} = -27.71 \text{ kip} \qquad \text{(h)}$$

Step 3: Let's write the equation for equilibrium of the forces acting on the right side of the crane boom that is freed by the section cut:

$$\Sigma \mathbf{F} = \mathbf{P_{DC}} + \mathbf{P_{DH}} + \mathbf{P_{DM}} + \mathbf{P_{NM}} + \mathbf{P_{NH}} + \mathbf{P_{IH}} + \mathbf{F} = 0 \qquad \text{(i)}$$

Let's write each of these forces in vector notation as:

$$\mathbf{P_{DC}} = -P_{DC}\,\mathbf{j}$$

$$\mathbf{P_{DH}} = P_{DH}(-0.3\,\mathbf{i} - 0.8\,\mathbf{j} + 0.5196\,\mathbf{k})$$
$$\mathbf{P_{DM}} = P_{DM}(-0.6\,\mathbf{i} - 0.8\,\mathbf{j})$$

$$\mathbf{P_{NM}} = 27.71\,\mathbf{j} \qquad \text{(j)}$$

$$\mathbf{P_{NH}} = 3.465\,\mathbf{i} - 9.240\,\mathbf{j} + 6.000\,\mathbf{k}$$

$$\mathbf{P_{IH}} = -36.95\,\mathbf{j}$$

$$\mathbf{F} = -12.0\,\mathbf{k}$$

Substitute Eqs. (j) into Eq. (i) and collect all of the coefficients of **i, j** and **k** to obtain three equations containing the three unknown internal forces:

$$-0.3\,P_{DH} - 0.6\,P_{DM} + 3.465 = 0$$

$$-P_{DC} - 0.8\,P_{DH} - 0.8\,P_{DM} + 27.71 - 9.240 - 36.95 = 0 \qquad \text{(k)}$$

$$0.5196\,P_{DH} + 6.000 - 12.00 = 0$$

Solving Eqs. (k) yields:

$$P_{DH} = 11.55 \text{ kip} \qquad P_{DM} = 0 \qquad P_{DC} = -27.71 \text{ kip} \qquad \text{(l)}$$

We note that the forces $P_{DC} = P_{NM} = -27.71$ kip are compressive along the stringers in the x-y plane. The force $P_{IH} = 36.95$ kip along the top stringer is tensile as expected. Because the internal force in the diagonal member DM = 0, the space truss exhibits symmetric behavior with $P_{DC} = P_{NM}$ and $P_{DH} = P_{NH}$.

7.4 THREE DIMENSIONAL EQUILIBRIUM

The emphasis in the preceding sections of this chapter has been on the application of the equilibrium equations to solve for unknown forces in space structures. However, applications of the equilibrium relations to other three-dimensional bodies such as machine components are often required in engineering studies. While the geometry of the bodies differs between structures and machine components, the approach is the same. We prepare FBDs, write appropriate equilibrium equations for each FBD and solve the equations for the unknown forces. To demonstrate these non-structural applications, let's consider two examples.

EXAMPLE 7.9

A local firm has constructed a small crane consisting of a boom supported by two steel wires BC and DE as shown in Fig. E7.9. The boom is fixed to the supporting wall with a ball and socket joint at point A. The wires are anchored into the wall at points C and E. Each wire has a diameter of 0.0625 in. and an ultimate tensile strength of 200 ksi. Determine the maximum weight that can be supported by the boom, the support reactions at point A and the forces in both support wires.

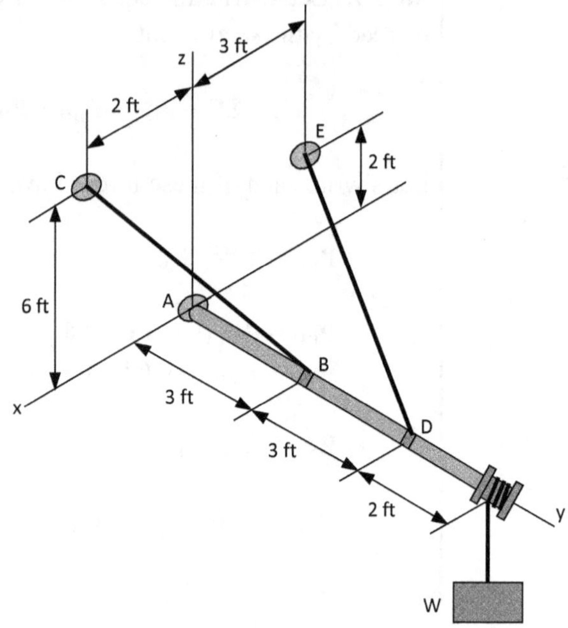

Fig. E7.9

Solution:

Step 1: Let's consider the crane in its entirety and prepare a FBD showing the three unknown forces at the ball and socket joint and the forces in the wires as shown in Fig. E7.9a.

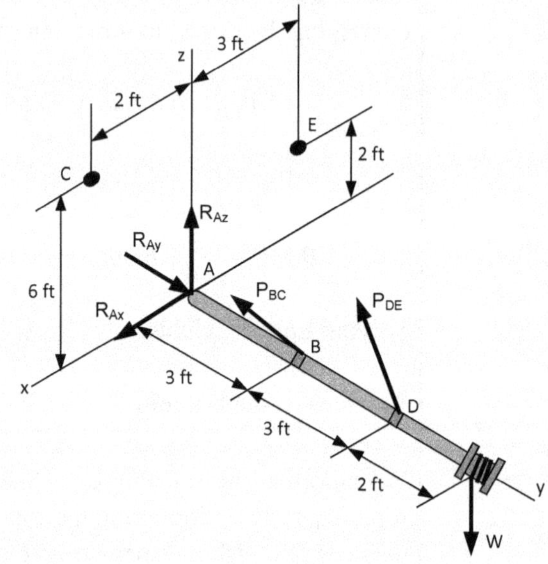

Fig. E7.9a

Step 2: Write the equations for the forces in the wires using a vector format.

$$\mathbf{P_{BC}} = P_{BC}\,\mathbf{u_{BC}} = P_{BC}\,(2\,\mathbf{i} - 3\,\mathbf{j} + 6\,\mathbf{k})/(7) \qquad \text{(a)}$$

$$\mathbf{P_{DE}} = P_{DE}\,\mathbf{u_{DE}} = P_{DE}\,(-3\,\mathbf{i} - 6\,\mathbf{j} + 2\,\mathbf{k})/(7) \qquad \text{(b)}$$

Step 3: Consider equilibrium of the structure. From the FBD in Fig. E7.9a, we write the six equations of equilibrium as:

$$\Sigma F_x = R_{Ax} + (2/7)P_{BC} - (3/7)\,P_{DE} = 0 \qquad \text{(c)}$$

$$\Sigma F_y = R_{Ay} - (3/7)P_{BC} - (6/7)\,P_{DE} = 0 \qquad \text{(c)}$$

$$\Sigma F_z = R_{Az} + (6/7)P_{BC} + (2/7)\,P_{DE} - W = 0 \qquad \text{(d)}$$

$$\Sigma M_x = (3)(6/7)P_{BC} + (6)(2/7)P_{DE} - (8)W = 0 \qquad \text{(e)}$$

$$\Sigma M_y = 0 \qquad \text{(f)}$$

$$\Sigma M_z = -\,(3)(2/7)P_{BC} + (6)(3/7)P_{DE} = 0 \qquad \text{(g)}$$

Step 4: Solve Eq. (g) and determine the maximum tension allowable in either wire to obtain:

$$P_{BC} = 3P_{DE} \qquad \text{(h)}$$

$$P_{Max} = SA = (200 \times 10^3)[(\pi/4)(0.0625)^2] = 613.6 \text{ lb} \qquad \text{(i)}$$

Step 5: Use Eqs. (h) and (i) to solve for the forces in the wires. Because $P_{BC} > P_{DE}$ it is evident that:

$$P_{BC} = P_{Max} = 613.6 \text{ lb} \quad \text{and} \quad P_{DE} = 204.5 \text{ lb} \qquad \text{(j)}$$

Step 6: Solve for the maximum weight that can be lifted by substituting Eq. (j) into Eq. (e):

$$W_{Max} = (1/8)[(18/7)(613.6) + (12/7)(204.5)] = 241.1 \text{ lb} \qquad \text{(k)}$$

Step 7: Solve for the reaction forces at point A by substituting Eqs. (j) and (k) into Eqs. (c), (d) and (e) to obtain:

$$R_{Ax} = -\,87.67 \text{ lb} \qquad R_{Ay} = 438.3 \text{ lb} \qquad R_{Az} = -\,343.3 \text{ lb} \qquad \text{(l)}$$

EXAMPLE 7.10

A hand operated lifting mechanism called a windless utilizes a crank to rotate a drum as shown in Fig. E7.10. The shaft of the mechanism is supported by a wall mounted ball and socket joint at point A and a smooth journal bearing at point B. The handle of the crank in the position shown is in the y-z plane. For this position of the crank handle determine the force P required to lift a weight W = 150 lb. Also determine the reactions at the ball and socket joint and the journal bearing.

Solution:

Step 1: Let's consider the crane in its entirety and prepare a FBD showing the three unknown forces at the ball and socket joint, the forces at the journal bearing and the applied forces as shown in Fig. E7.10a.

Step 2: Let's employ vector mechanics in this solution by writing the equation for the moment vector as:

$$\Sigma \mathbf{M} = \mathbf{r_W} \times \mathbf{F_W} + \mathbf{r_{AB}} \times \mathbf{F_B} + \mathbf{r_F} \times \mathbf{F} = 0 \qquad \text{(a)}$$

The forces are given by:

$$\mathbf{W} = -\,150\,\mathbf{k} \qquad \mathbf{F_B} = F_{Bx}\,\mathbf{i} + F_{Bz}\,\mathbf{k} \qquad \mathbf{F} = F\,\mathbf{i} \qquad \text{(b)}$$

The position vectors are given by:

$$\mathbf{r_W} = 0.5\,\mathbf{i} + 2\,\mathbf{j} \qquad \mathbf{r_{AB}} = 4\,\mathbf{j} \qquad \mathbf{r_F} = 6\,\mathbf{j} - 1\,\mathbf{k} \qquad \text{(c)}$$

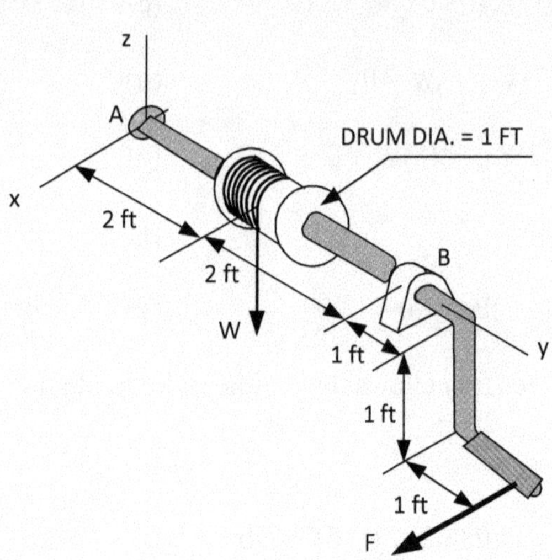

Fig. E7.10

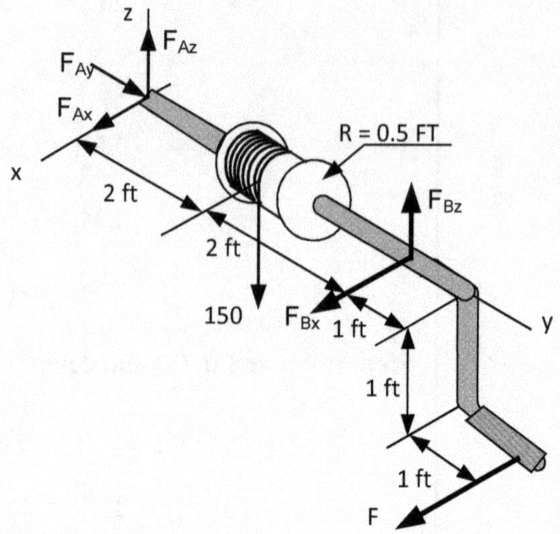

Fig. E7.10a

Step 3: Substitute the results from Eqs. (a) and (b) into Eq. (7.9) and solve the determinates to obtain:

$$\Sigma M = \begin{vmatrix} \mathbf{i} & \mathbf{j} & \mathbf{k} \\ 0.5 & 2 & 0 \\ 0 & 0 & 0 \end{vmatrix} + \begin{vmatrix} \mathbf{i} & \mathbf{j} & \mathbf{k} \\ 0 & 4 & 0 \\ F_{Bx} & F_{By} & F_{Bz} \end{vmatrix} + \begin{vmatrix} \mathbf{i} & \mathbf{j} & \mathbf{k} \\ 0 & 6 & -1 \\ F & 0 & 0 \end{vmatrix} = 0 \qquad \text{(d)}$$

$$(300\,\mathbf{i} + 75\,\mathbf{j}) + (4F_{Bz}\,\mathbf{i} - 4\,F_{Bx}\,\mathbf{k}) + F\,(-\,\mathbf{j} - 6\,\mathbf{k}) = 0 \qquad \text{(e)}$$

Rearrange the terms in Eq. (e) and write:

$$(-300 + 4F_{Bz})\mathbf{i} + (75 - F)\,\mathbf{j} + (-4F_{Bx} - 6F)\,\mathbf{k} = 0 \qquad \text{(f)}$$

Equate each coefficient in Eq. (f) to zero to obtain:

$$(-300 + 4F_{Bz}) = 0 \qquad (75 - F) = 0 \qquad (-4F_{Bx} - 6F) = 0 \qquad \text{(g)}$$

Solve Eqs. (g) for the unknowns F, F_{Bx} and F_{Bz} to obtain:

$$F = 75\text{ lb} \qquad F_{Bz} = 75\text{ lb} \qquad F_{Bx} = -\,112.5\text{ lb} \qquad \text{(h)}$$

Step 4: To determine the forces at the ball and socket joint, let's use $\Sigma\mathbf{F} = 0$ and write:

$$\Sigma F_x = F_{Ax} + F_{Bx} + F = 0 \qquad \Sigma F_y = F_{Ay} = 0 \qquad \Sigma F_z = F_{Az} + F_{Bz} - 150 = 0 \qquad \text{(i)}$$

Substituting the results from Eqs. (h) into Eqs (i) yields:

$$F_{Ax} = 37.5\text{ lb} \qquad F_{Ay} = 0 \qquad F_{Az} = 75\text{ lb} \qquad \text{(j)}$$

7.5 SUMMARY

In the discussion of three-dimensional structures, we presented two different mathematical approaches. The first, consistent with most of the previous discussion, employed trigonometric methods for determining force and moment components in the three directions. The second approach employs vector analysis. The first approach is intuitively obvious and the second is more mathematically elegant. In both approaches, we begin with FBDs of the structure. We suggest the use of two or three-view FBDs to facilitate the visualization process.

With the trigonometric approach for determining the force and moment components, we employ six equilibrium relations in the solution of the unknown internal and external forces:

$$\Sigma F_x = 0; \quad \Sigma F_y = 0; \quad \text{and} \quad \Sigma F_z = 0$$

$$\Sigma M_x = 0; \quad \Sigma M_y = 0; \quad \text{and} \quad \Sigma M_z = 0$$

With the vector analysis approach, we write the vector relations defining all of the forces acting on the structure.

$$\mathbf{F_1} = F_{1x}\,\mathbf{i} + F_{1y}\,\mathbf{j} + F_{1z}\,\mathbf{k}$$
$$\cdots\cdots\cdots\cdots$$
$$\cdots\cdots\cdots\cdots$$
$$\mathbf{F_n} = F_{nx}\,\mathbf{i} + F_{ny}\,\mathbf{j} + F_{nz}\,\mathbf{k}$$

Then the position vectors that define the point of application of the forces relative to the point about which the moments act are written.

$$\mathbf{r_1} = r_{1x}\,\mathbf{i} + r_{1y}\,\mathbf{j} + r_{1z}\,\mathbf{k}$$
$$\cdots\cdots\cdots\cdots$$
$$\cdots\cdots\cdots\cdots$$
$$\mathbf{r_n} = r_{nx}\,\mathbf{i} + r_{ny}\,\mathbf{j} + r_{nz}\,\mathbf{k}$$

The moments are determined from the force vectors and the position vectors by employing the vector cross product.

$$\mathbf{M} = \mathbf{r} \times \mathbf{F}$$

Finally, the vector forms of the equilibrium relations are used to solve for the unknown internal and external forces acting on the three-dimensional structure.

$$\Sigma\mathbf{F} = 0 \ \text{ and } \quad \Sigma\mathbf{M} = 0$$

Three-dimensional structures are difficult to analyze in comparison to two-dimensional structures. In both cases, the approach is the same; construct the appropriate FBDs and apply the equilibrium relations. The difficulty in analyzing three-dimensional structures arises for two reasons. First, visualization is often a problem. Second, many forces are usually involved and the analysis becomes long and tedious. The need for visualization is mitigated to some degree by employing a vector mechanics approach where the process leads one to understand the geometry and dimensions of the body under consideration. Dividing the solution into different steps and completing each step as an individual analysis addresses the length of the analysis. This stepwise approach was demonstrated in several examples.

REFERENCES

1. Zalewski, W. and E. Allen, Shaping Structures: Statics, John Wiley & Sons, New York, NY, 1998, p 167-169.

PROBLEMS

7.1 The Patcenter in Princeton, NJ is a large open structure with a cable-stayed roof. In the figure presented below, one of the nine tubular steel masts that are uniformly spaced at 9 m intervals to support the roof structure is shown. In this design, the roof hangs from cables because the vertical side members are used only to prevent wind uplift. The tubular mast, 15 m high, is supported with a 9 m wide by 6 m high rectangular steel frame. Determine the force in the primary rod stay and the tubular steel masts if the uniformly distributed load on the roof is 70 kN/m. Assume the roof is pinned to the steel mast at point C. Hint: Refer to the Example 7.3 in the text for additional dimensions.

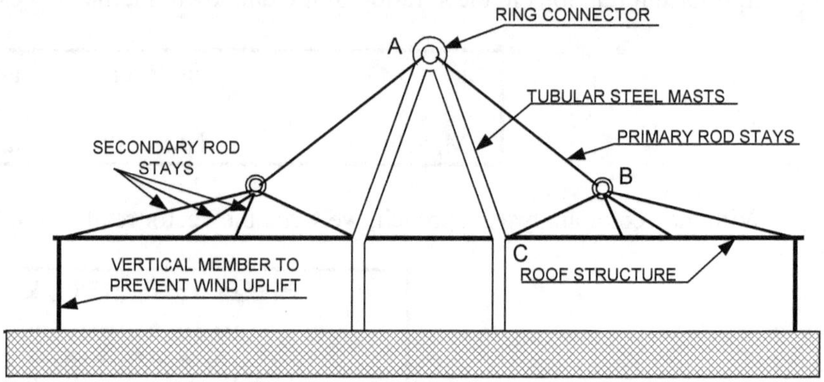

7.2 A new building similar to the Patcenter is under consideration by an architectural firm. They propose increasing the height of the mast from 15 m to 24 m to make the appearance of the structure more dramatic. If all of the other parameters are the same as given in Problem 7.1, prepare an analysis indicating the effect of this change.

7.3 A new building similar to the Patcenter is under consideration by an architectural firm. They propose a structure with the dimensions presented in the figure to the right. Determine the force in the primary rod stay AB and the tubular steel masts AC. Also find the reaction forces at point C.

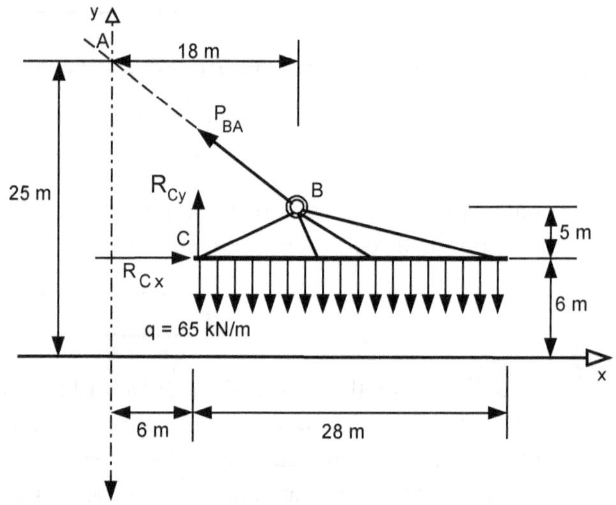

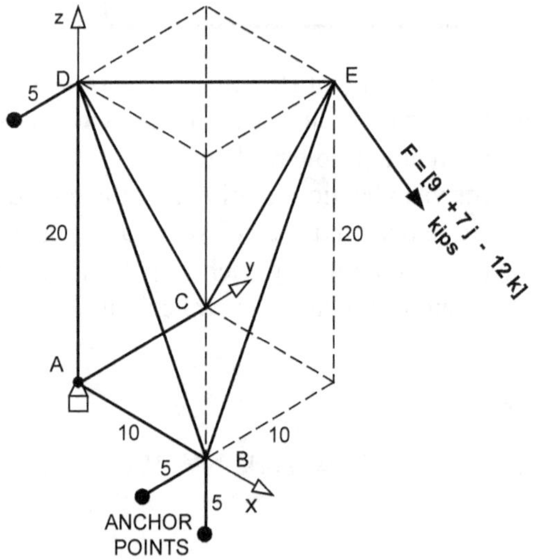

7.4 A space truss, shown in the figure to the left, is supported at point A with a ball and socket joint and with cables anchored at points B and D. A force F is applied to joint E. If F is represented in vector format as shown in the figure, determine the internal forces in the three members that intersect at joint E. Also specify the diameter of the rods, if the truss is fabricated from a 1018 A steel alloy. A safety factor of SF = 3.0 based on yield strength is specified by the building code. The truss dimensions are given in ft.

7.5 Determine the internal forces in members AB, DB and EB of the truss described in Problem 7.4.

7.6 Determine the internal forces in members AC, DC and EC of the truss described in Problem 7.4.

7.7 For the communications tower illustrated in the figure to the right, a wind force of 4,500 lb from the West is applied at a position z = 300 ft. The tower has a height H = 400 ft and a dead weight W = 50 kips. The anchor points on the ground plane are defined by: x_1 = 100ft, x_2 = 115 ft, y_1 = 130 ft and y_2 = 145 ft. The cables have a breaking strength of 26,600 lb and a ball and socket joint is used to support the tower at point O. Determine the loads in the four cables and the reaction forces at the base of the tower. Also determine the minimum safety factor.

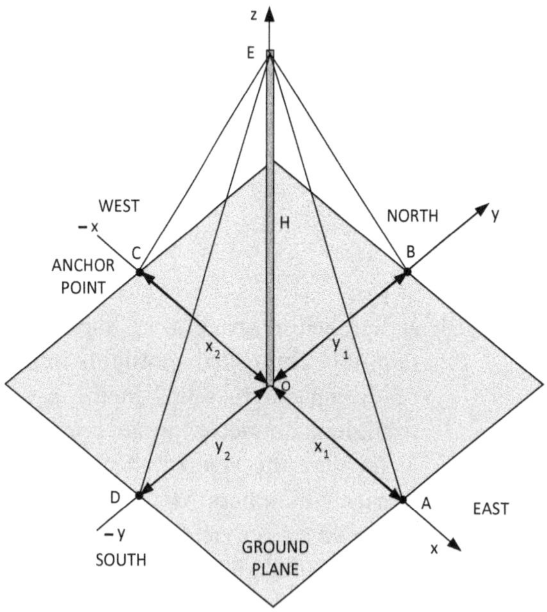

7.8 The anchor points on the ground plane for the communications tower shown in the figure to the right are changed to: x_1 = 60 m, x_2 = 70 m, y_1 = 80 m, y_2 = 90 m. The tower has a height of 120 m, a dead weight of 200kN and is subjected to a wind force of 12.6 kN from the East, that is applied at a position z = 80 m. The cables have a breaking strength of 130 kN, and the tower is supported by a ball and socket joint at point O. Determine the loads on the four cables and the reaction forces at the ball and socket joint. Also find the minimum safety factor.

7.9 If the steel used in fabricating the tower described in Problem 7.7 has a yield strength of 50 ksi, determine the cross sectional area required at the base of the tower. The safety factor for the tower is specified as 4.0. Does the wind loading condition affect the result?

7.10 If the steel used in fabricating the tower in Problem 7.8 has a yield strength of 360 MPa, determine the cross sectional area required at the base of the tower. The safety factor for the tower is specified as 5.0. Does the wind loading condition affect the result?

7.11 The safety factors for the communications tower described in Problems 7.7 and 7.8 are very large. Develop arguments supporting the use of relatively large safety factors in the design of communications towers. Also develop arguments for redesign with smaller diameter cable, smaller footprint and smaller cross sectional area at the base of the tower structure.

7.12 A severe ice storm strikes the communications tower coating all of the members with a thick layer of ice. The dead weight of the tower is increased from 200 kN to 320 kN. Determine the decrease in the safety factor for the conditions of Problem 7.8.

7.13 For the derrick shown in the figure below, determine the margin of safety for the three cables if the force F acting on the boom is 4.0 ton. The cable diameter is 0.5 in. for all three cables. The anchor points P and Q for two of the cables lie in the x-y plane with coordinates P= (12, –9, 0) ft and Q = (–12, – 9, 0) ft. The derrick pole is 15 ft high and the boom is 20 ft long. The derrick pole and the boom are separate components. The boom is attached to the pole at point R with a sleeve type bearing that enables the boom to rotate about the pole. However, the boom is constrained from sliding up or down the pole. The derrick pole is supported by a ball and socket joint at point O. Details of the bearing arrangement are also shown in the figure below. The cables have a specified breaking strength of 180 ksi.

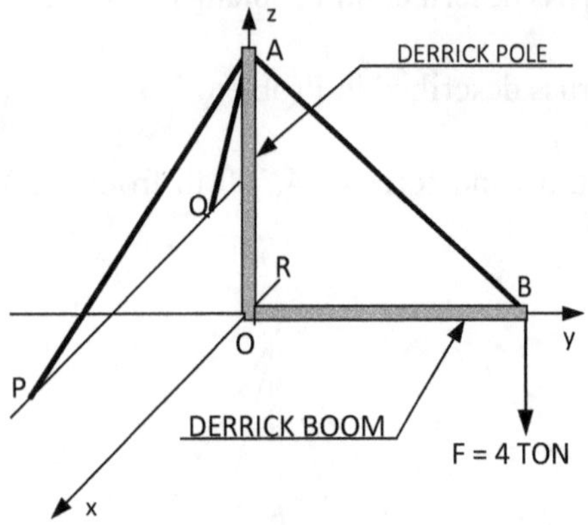

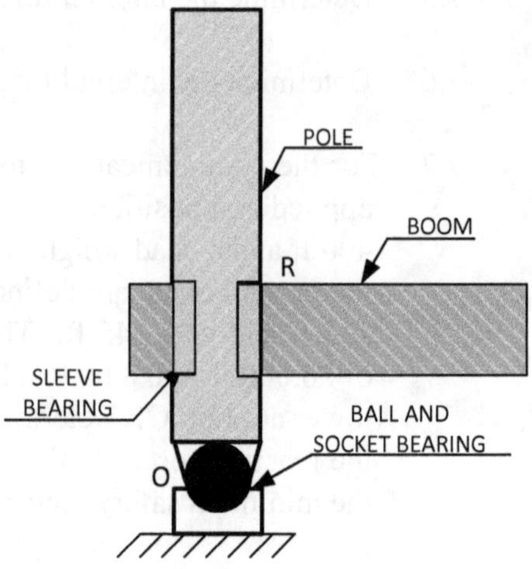

7.14 A tetrahedral space truss, shown in the figure to the right, supports a massive scoreboard and several sets of remotely controlled spotlights in an amphitheater. The base ABC of the space truss lies in the x-y plane (horizontal). The base triangle is connected to anchors in the roof beams by long cables. Determine the size of the solid round rods that are required to fabricate members AB, AC and AD of the space truss. The safety factor based on yield strength is specified as 6.0. 1020 HR steel is employed for all of the members and the weight of the scoreboard and spotlights is 20.0 kN.

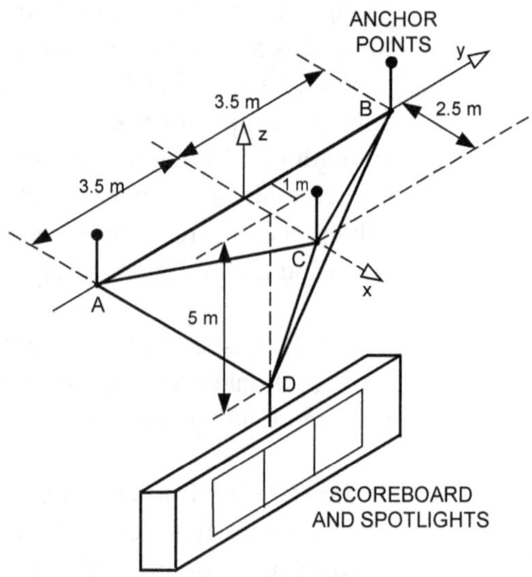

7.15 For the long horizontal boom of a construction crane, illustrated in the figure shown below, determine the internal forces in the members AB, BF, BK, FG, KL and FL. The load applied to the boom is F = 10 kip. Hint: See Example 7.8 in the text for additional details.

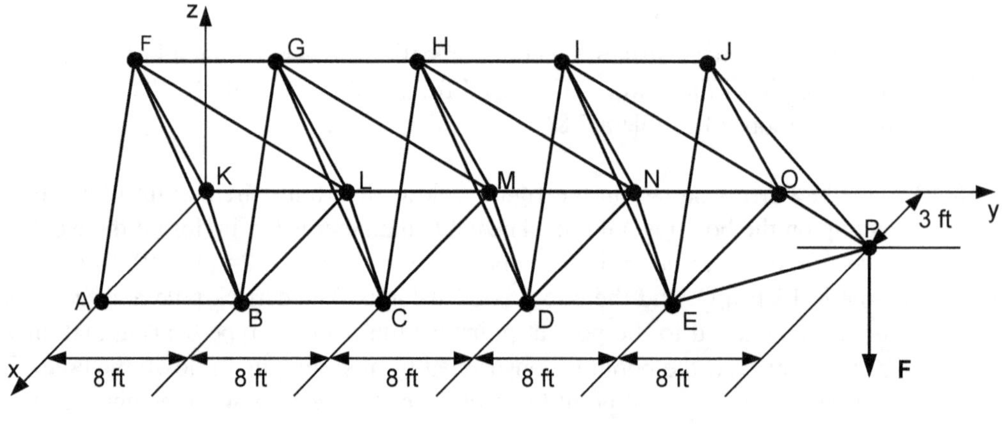

7.16 Repeat Problem 7.15 if the loading on the long horizontal boom is increased from 10 kip to 14 kip.

7.17 A hanging light assembly is positioned near the corner of a gymnasium as shown in the figure to the right. Determine the gage of the stainless steel wire required for the support of a light assembly weighing 1,100 N if a safety factor of 5.0 is specified. The wire is fabricated from 302A stainless steel. Points B and C are anchors that lie in the x-y plane and point D is an anchor that lies along the z-axis. Point A is not anchored; however it lies in the x-y plane.

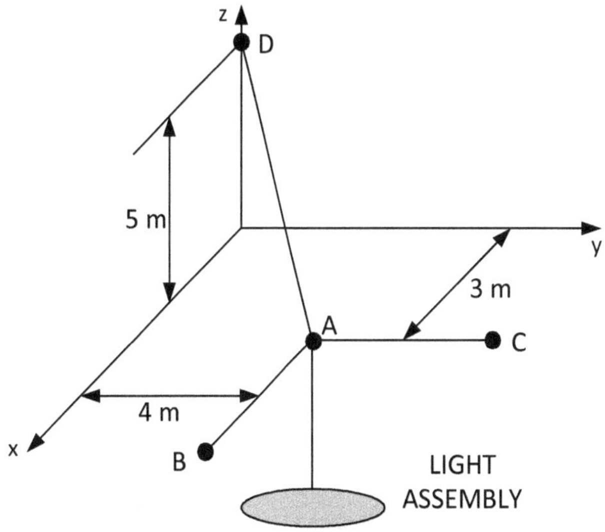

7.18 A hot air balloon, shown in the figure to the right, is moored to the ground with three cables that are anchored at points A, B and C. The coordinates (x, y, z) of the anchor points on the ground and on the basket are given in the figure. Determine the force exerted by each of the cables if the upward lift of the balloon is 950 lb. Assume that wind forces are negligible.

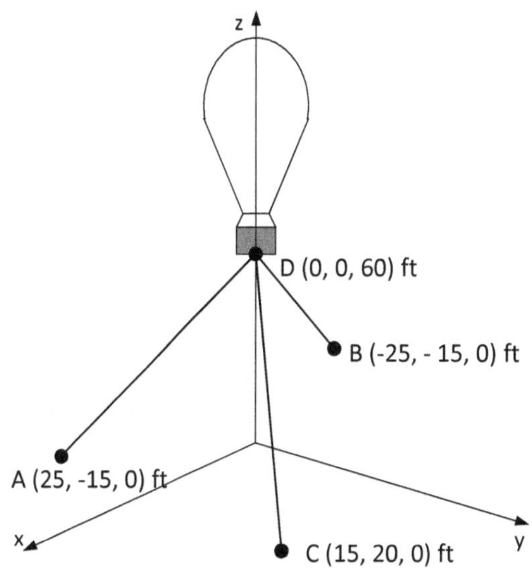

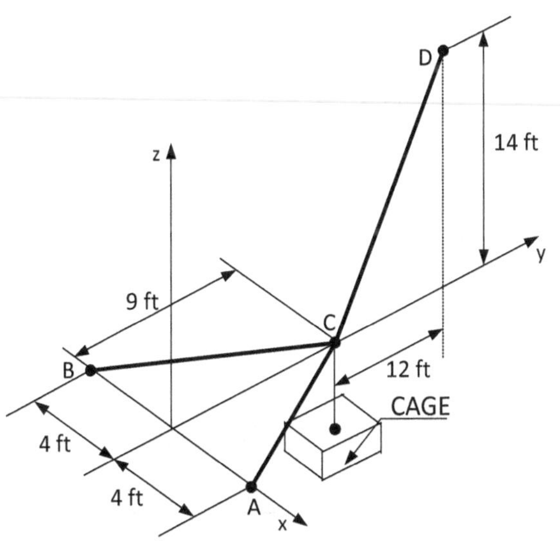

7.19 A circus cage, displayed in a large high ceiling auditorium, is supported above ground level by the three wires illustrated in the figure to the left. Determine the gage of the stainless steel wire required to support the cage weighing 2,500 lb if a safety factor of 3.5 is specified. The wire is fabricated from 4340 HR steel. The geometric parameters defining the assembly are listed in the table below. Points A, B and D locate the anchors for the cables.

7.20 A local firm has constructed a small crane consisting of a boom supported by two steel wires BC and DE as shown in the figure to the right. The boom is fixed to the supporting wall with a ball and socket joint at point A. The wires are anchored into the wall at points C and E. Each wire has a diameter of 0.125 in. and an ultimate tensile strength of 150 ksi. Determine the maximum weight that can be supported by the boom, the support reactions at point A and the forces in wires BC and DE.

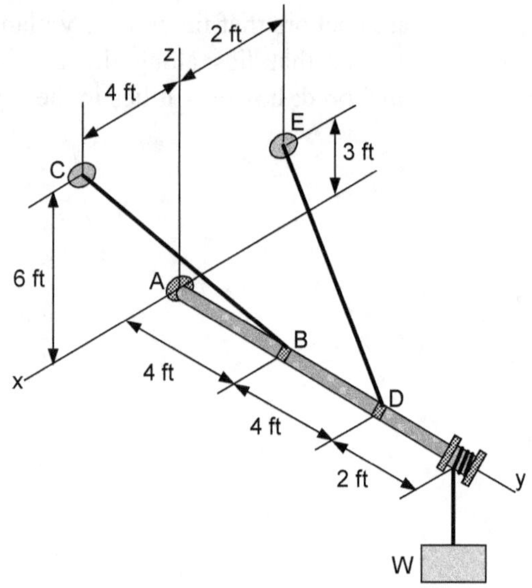

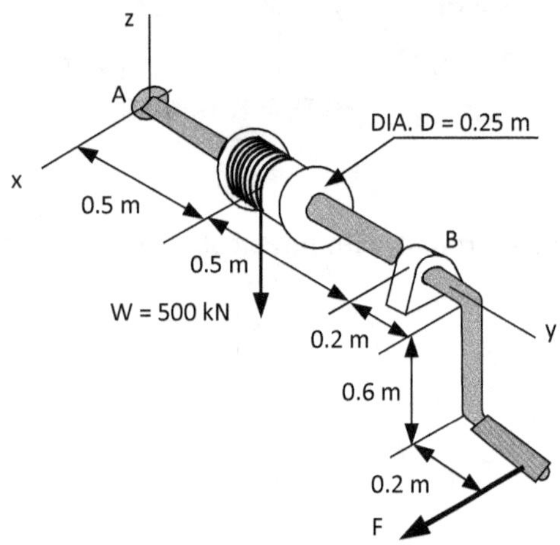

7.21 A hand operated lifting mechanism called a windless utilizes a crank to rotate a drum as shown in the figure to the left. The shaft of the mechanism is supported by a wall mounted ball and socket joint at point A and a smooth journal bearing at point B. The arm and handle of the crank in the position shown is in the y-z plane. For this position of the crank handle, determine the force F required to hold a weight of 500 kN in equilibrium. Also determine the reactions at the ball and socket joint and the journal bearing.

7.22 Repeat Problem 7.21 if the crank is rotated clockwise 90 degrees so that the arm and handle of the crank is in the x-y plane and the force F is acting in the positive z direction. For this position of the crank handle, determine the force F required to hold the weight of 500 kN in equilibrium. Also determine the reactions at the ball and socket joint and the journal bearing.

7.23 Repeat Problem 7.21 if the crank is rotated clockwise 180 degrees so that the handle of the crank is in the y-z plane and the force F is acting in the negative x direction. For this position of the crank handle determine, the force F required to hold the weight of 500 kN in equilibrium. Also determine the reactions at the ball and socket joint and the journal bearing.

7.24 A cross-like base and a short pole, as shown in the figure to the right, support a tabletop with a uniform thickness. Determine the largest weight W that can be applied to the table if it is placed at points A, B or C. The tabletop weighs 120 N/m^2. The dimensions of the tabletop and the base are given in the figure.

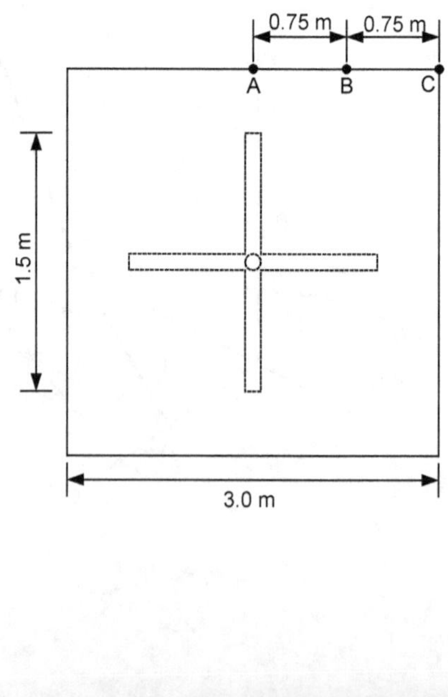

CHAPTER 8

FRAMES AND MACHINES

8.1 INTRODUCTION

Frames are similar to trusses in that they both are fabricated from long thin members. However, there are two important differences.

1. The joints in a truss are pinned and free to rotate; whereas, one or more joints in a frame may be rigid. Because rotation is constrained, the rigid joints are capable of providing reaction moments.
2. The forces applied to a truss act only at the joints; whereas, forces may be applied at any location on a frame. Thus, one or more members in a frame may be subjected to more than two forces.

All of the members in a truss are subjected to only two external forces; hence, the internal force must coincide with the line of action of these two equal and opposite forces. In a frame one or more members is subjected to more than two external forces. When we make a section cut across a multiforce member[1], it is necessary to assume the existence of an internal axial force P, an internal shear force V, and an internal moment M. Because of this fact (introducing three unknowns for each member cut), the method of joints and the method of sections are usually not effective when analyzing multiforce members.

To analyze a frame, we follow the same general approach as with trusses while carefully avoiding section cuts on any multiforce member:

1. Prepare a FBD of the entire frame.
2. Apply the appropriate equilibrium relations to solve for the reactions.
3. Dismember the frame and draw FBDs of each member.
4. Use Newton's law of action and reaction in preparing these FBDs.
5. Identify the two-force members, and show the two equal and opposite forces acting at the pinned joints.
6. Apply the appropriate equilibrium relations to solve for the internal forces and moments.

Machines are similar to frames in that they contain one or more multiforce members. However, machines differ from both frames and trusses because motion occurs as they act to convert an input force (or moment) to an output force (or moment). Machines may act to amplify forces, to attenuate forces or change the direction of forces. Simple machines include the lever, screw, inclined plane, and pulley. While machines may amplify or attenuate forces when they are transmitted from the input to the output, energy is not gained in the process. Energy is conserved in the process except when heat is generated due to frictional effects and this heat is dissipated into the atmosphere.

The approach for solving problems involving frames and machines will be demonstrated in the following examples. The importance of drawing complete and accurate FBDs is stressed in these examples.

[1] A multiforce member is subjected to three or more external forces, or alternatively two or more external forces and one or more external moments.

8.2 FRAMES—EXAMPLES AND SOLUTIONS

EXAMPLE 8.1

To demonstrate the method for analyzing frames, consider the simple hoist presented in Fig. E8.1. First, identify the multiforce and two-force members. Then solve for the reactions at points A and E. Determine all of the external forces acting on member CFG. Determine the internal forces and the internal moment acting on this member. The load F on the hoist is 480 N. Point E is a pinned support and point A is a roller support.

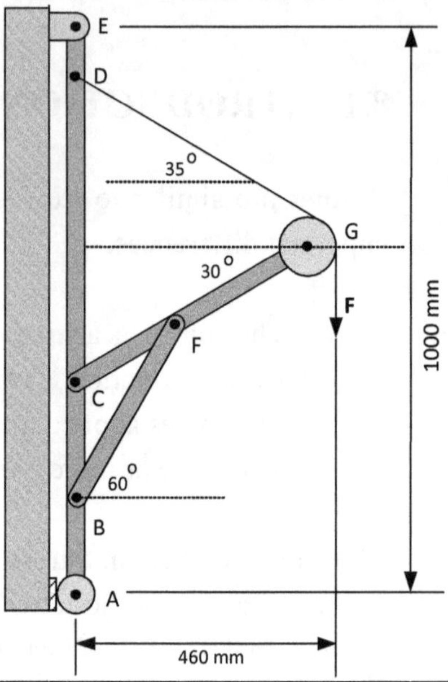

Fig. E8.1

Solution:

Step 1: Identify the multiforce members and two force members.

The vertical bar ABCDE and the inclined bar CFG are multiforce members. The cable DG and the strut BF are two-force members.

Step 2: Draw a FBD of the entire frame as shown in Fig E8.1a.

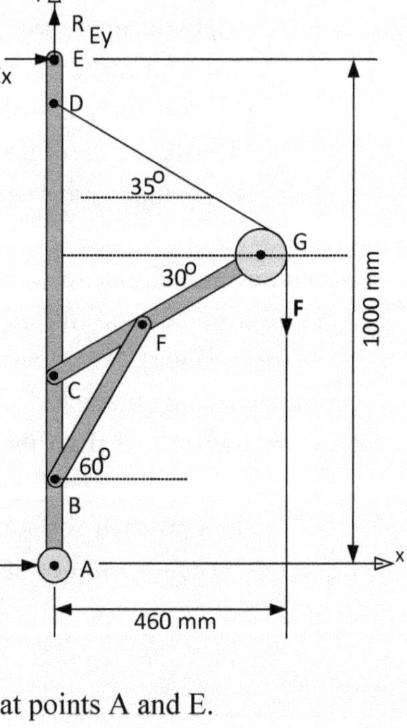

Fig. E8.1

Step 3: Let's employ the equilibrium relations to determine the reactions at points A and E.

$$\Sigma M_E = 1000\ R_{Ax} - 460\ F = 0 \quad \Rightarrow R_{Ax} = 0.46\ (480) = 220.8\ \text{N} \qquad \text{(a)}$$

$$\Sigma F_x = R_{Ax} + R_{Ex} = 0 \qquad \Rightarrow R_{Ex} = -R_{Ax} = -220.8\text{ N} \qquad \text{(b)}$$

$$\Sigma F_y = R_{Ey} - F = 0 \qquad \Rightarrow R_{Ey} = F = 480\text{ N} \qquad \text{(c)}$$

Step 4: Next, prepare the FBDs of the individual members as shown in Fig. E8.1b. Four FBDs are presented in this illustration to show the forces acting on members ABCDE, BF, CFG and the pulley at G. Note that we have used the law of action and reaction in preparing these FBDs. Let's examine member CFG to ascertain if it is possible to determine the forces applied to it. We have five forces acting on this member— R_{Cx}, R_{Cy}, R_B, R_{Gx} and R_{Gy}. The force system is coplanar and non-concurrent; hence, only three equilibrium equations are applicable. Since we have five unknowns and only three equations, it is necessary to write two additional equations in terms of two of the unknown forces. Clearly, we can write these additional equations by considering the equilibrium of the pulley at point G.

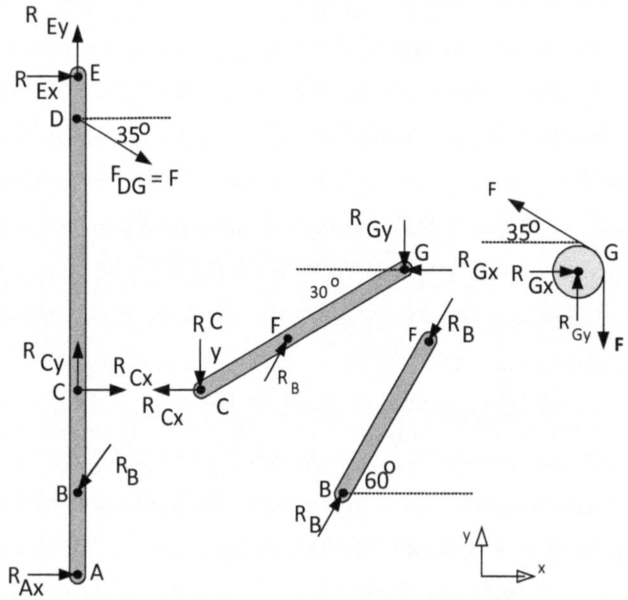

Fig. E8.1a

Step 5: Writing the equilibrium equations for the pulley gives:

$$\Sigma F_x = R_{Gx} - F\cos(35°) = 0 \qquad \Rightarrow R_{Gx} = 480\cos(35°) = 393.2\text{ N} \qquad \text{(d)}$$

$$\Sigma F_y = R_{Gy} + F\sin(35°) - F = 0 \Rightarrow R_{Gy} = F\,[1 - \sin(35°)] = 204.7\text{ N} \qquad \text{(e)}$$

Step 6: Now it is possible to solve for the forces acting on member CFG from the equilibrium relations. Dimensions were not included in the FBDs presented in Fig E8.16b to avoid confusing the already crowded drawing. Note for member CFG that distances CF = 200 mm, FG = 320 mm and CG = 520 mm. Therefore, the pulley G has a radius of 10 mm.

$$\Sigma M_C = [R_B\sin(30°)](200) + R_{Gx}\,[(520)\sin(30°)] - R_{Gy}\,[(520)\cos(30°)] = 0 \qquad \text{(f)}$$

Substituting the results from Eqs. (d) and (e) into Eq. (f) yields:

$$R_B = -100.5\text{ N}$$

The negative sign indicates that the direction of the force R_B in the FBDs is not correct. We showed R_B as a compressive force in the FBD of members BF and CFG. In reality it is a tensile force.

Continuing with the equilibrium analysis leads to:

$$\Sigma F_x = R_B\cos(60°) - R_{Cx} - R_{Gx} = 0 \qquad \Rightarrow \qquad R_{Cx} = -443.5\text{ N}$$

$$\Sigma F_y = R_B\sin(60°) - R_{Cy} - R_{Gy} = 0 \qquad \Rightarrow \qquad R_{Cy} = -291.7\text{ N}$$

We have completed the analysis of member CFG. It is possible to continue the application of the equilibrium relations to the member ABCDE to obtain results for R_B and R_C that will enable you to check the results from the analysis of member CFG.

The Effect of Rigid Joints in Frames

Let's explore the effect of converting the pinned joints to rigid ones on stability of a structure. A rectangular frame with pinned joints is illustrated in Fig. 8.1a. Clearly, the application of a horizontal force at the top left corner of the frame causes it to rotate about its pinned joints. The slightest horizontal force component will cause the rectangular frame to collapse. However, if the pins at the upper corners are replaced with gusset plates, that are sufficiently rigid to constrain rotation, as shown in Fig 8.1b, the rectangular frame is stable.

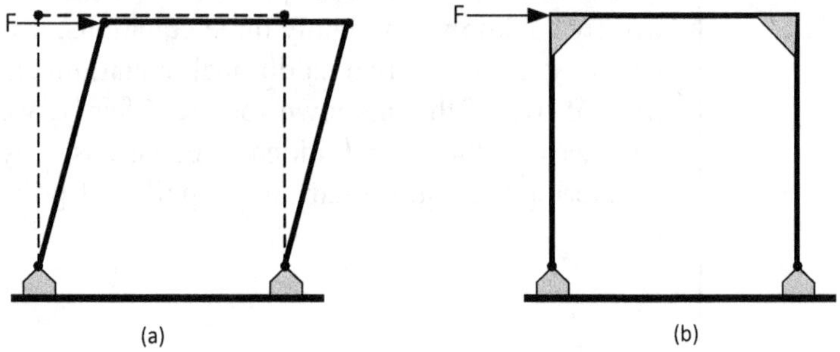

Fig. 8.1 Stability of a rectangular frame is achieved by making the joints rigid.

The forces applied to a frame need not be restricted to the location of the pins, and can be applied at any point along the length of any member. Of course the application of an additional force along its length produces a multiforce member as described in the previous section. A rectangular frame with a uniformly distributed force system is presented in Fig. 8.2.

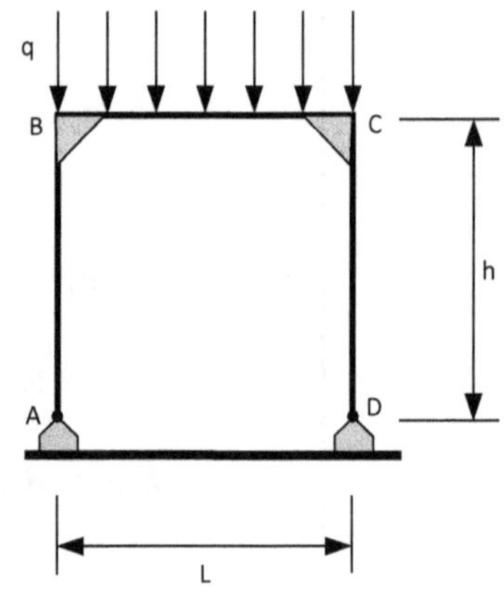

Fig. 8.2 Rectangular frame supporting a uniformly distributed force q.

Let's analyze this rectangular frame to determine the internal forces acting within its members.

EXAMPLE 8.2

Determine the internal forces and moments for the three members in the rectangular frame presented in Fig. 8.2.

Solution:

Step 1: The procedure for determining the internal forces and moments is identical to that previously established when analyzing trusses. However, it is critical to recognize that some of the members in the frame will be multiforce members. We begin with a FBD of the entire frame, presented in Fig. E8.2, and then write the appropriate equations of equilibrium.

Step 2: The force system in Fig. E8.2 is coplanar and non-concurrent. Accordingly, we write:

$$\Sigma F_x = R_{Ax} - R_{Dx} = 0$$

$$\Sigma F_y = R_{Ay} + R_{Dy} - qL = 0 \qquad \text{(a)}$$

$$\Sigma M_A = R_{Dy} L - qL(L/2) = 0$$

Solving Eqs. (a) yields:

$$R_{Ay} = R_{Dy} = qL/2 \qquad \Rightarrow \qquad R_{Ax} = R_{Dx} \qquad \text{(b)}$$

Fig. E8.2

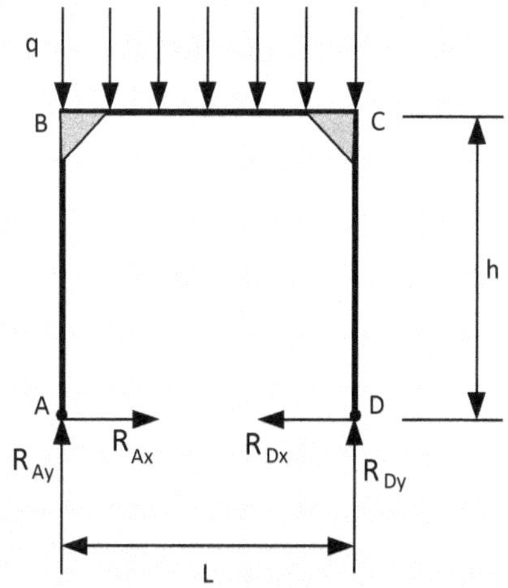

Step 3: Let's draw FBDs for each of the members in the frame, as shown in Fig. E8.2a. In preparing each of these FBDs, we make use of the law of action and reaction. Examination of the FBDs again indicates that the force system is coplanar and non-concurrent for each of its three members.

Fig. E8.2a

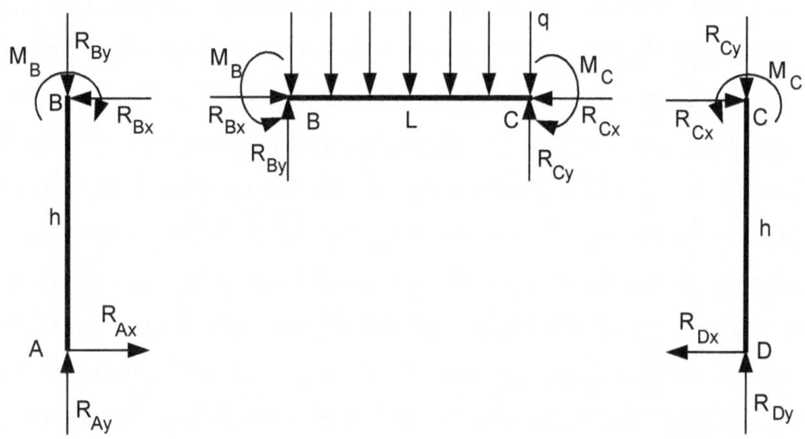

Step 4: Writing the equations of equilibrium and recognizing the symmetry of both the loading and the geometry yields:

$$M_B = M_C \qquad \Rightarrow R_{Bx} = R_{Cx} = R_{Ax} = R_{Dx} \qquad \Rightarrow R_{By} = R_{Cy} = R_{Ay} = R_{Dy} = qL/2 \qquad \text{(c)}$$

Consider the vertical member and write moments about point B to give R_{Ax} as:

$$R_{Ax} = M_B/h \qquad \text{(d)}$$

We have employed all of the equations of equilibrium for each of the three members; however, we have not determined the magnitude of the reactions in the x direction or the magnitude of the moment produced by the rigid constraint at points B and C. We have established the relation given in Eq. (d), but not the magnitude. The problem is statically indeterminate. To resolve the indeterminacy, it is necessary to consider the deformed shape of the frame, which is beyond the scope of this textbook.

EXAMPLE 8.3

A manual hydraulic pump with a 2.00 in. diameter piston is illustrated in Fig. E8.3. The fluid in the cylinder is pressurized when a force F is applied to the lever arm ADB. If the cylinder pressure is 500 psi, determine the force F applied to the lever arm and the force in the link CD.
Note that CD is a two-force member, so the internal force F_{CD} is directed along the axis of the link.

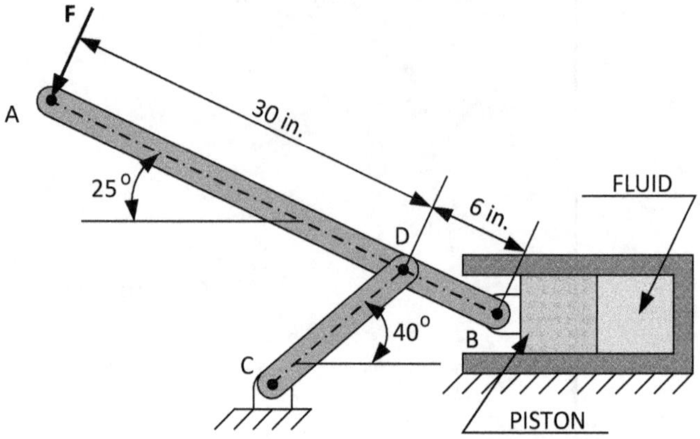

Fig. E8.3

Solution:

Step 1: We begin with a FBD of the entire pump, presented in Fig. E8.3a, and then write the appropriate equations of equilibrium.

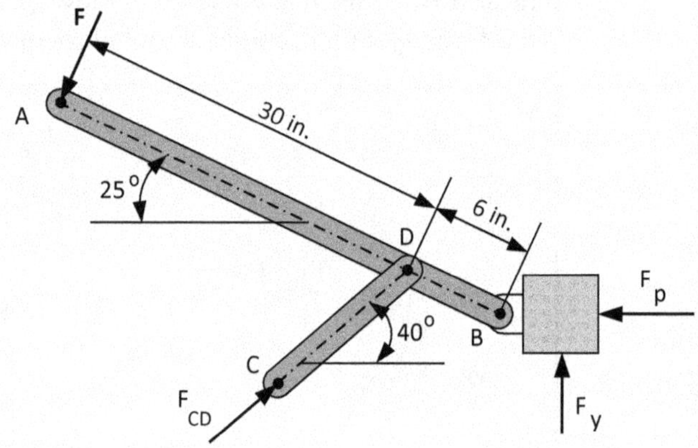

Fig. E8.3a

Step 2: The force F_p acting on the piston is given by:

$$F_p = pA = 500\ (\pi d^2/4) = 500\ [\pi\ (2)^2/4] = 550\pi = 1571 \text{ lb} \qquad \text{(a)}$$

Step 3: The force system in Fig. E8.3a is coplanar and non-concurrent. Accordingly, we write:

$$\Sigma\ F_x = F_{CD} \cos(40°) - F \sin(25°) - F_p = 0 \qquad \text{(b)}$$

$$\Sigma\ F_y = F_y + F_{CD} \sin(40°) - F \cos(25°) = 0 \qquad \text{(c)}$$

We cannot write a meaningful moment equation for the FBD in Fig. 8.3a because the location of the force F_y along the piston is not known.

Substituting Eq. (a) into Eq. (b) yields:

$$(0.7660)\ F_{CD} - (0.4226)\ F - 1571 = 0 \qquad \text{(d)}$$

Step 4: Let's draw FBDs for the two members in the pump's lever system, as shown in Fig. E8.3b. In preparing each of these FBDs, the law of action and reaction is used. Examination of the FBD for the lever arm indicates that the force system is coplanar and non-concurrent.

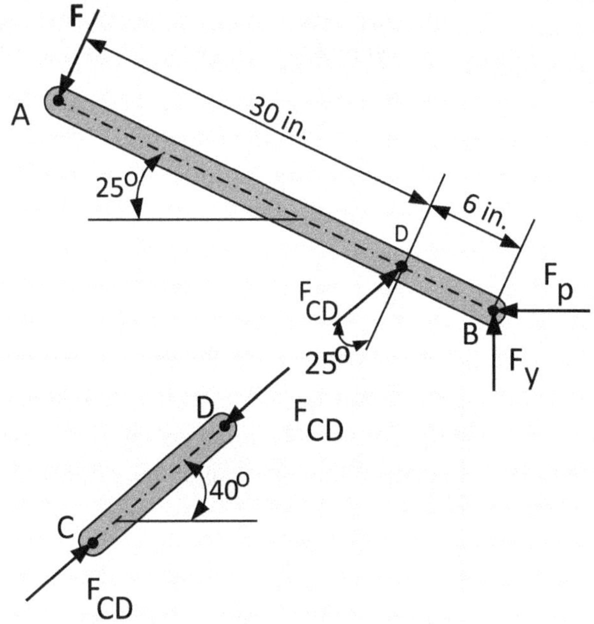

Fig. 8.3b

Step 5: Writing the moment equation of equilibrium for the lever arm about point B yields:

$$\Sigma M_B = (36.00)\ F - (6.00)[F_{CD} \cos(25°)] = 0$$

$$F = (1/6)\ F_{CD} \cos(25°) = 0.1511\ F_{CD} \qquad \text{(e)}$$

Combining Eq. (d) and Eq. (e) gives:

$$F_{CD} = \frac{1571}{[0.7660 - (0.4226)(0.1511)]} = 2237 \text{ lb} \qquad \text{(f)}$$

$$F = (0.1511)(2237) = 338.0 \text{ lb}$$

The force required on the lever to develop the pressure of 500 psi is very large for a manual pump. Clearly, most people would not be capable of generating a pump pressure 500 psi. It would be advisable to suggest a redesign of the lever to decrease the force required.

8.3 MACHINES—EXAMPLES AND SOLUTIONS

Machines are similar to frames in that they usually contain one or two multiforce members. The difference is that one or more of the members comprising the machine moves. The movement is necessary to amplify or attenuate a force or to change its direction. The equations employed in the analysis of machines are identical to those used in solving problems related to trusses and frames—the equilibrium relations. The motion that occurs in these machines is considered to be without acceleration; hence, the static equations of equilibrium apply. As with frames, the analysis usually requires a number of carefully prepared and accurate FBDs.

EXAMPLE 8.4

A pair of pliers clamps a small diameter rubber cylinder as shown in Fig. E8.4. If opposing forces F, equal to 30 lb, are applied to the handles of the pliers determine the force acting on the rubber cylinder. Also determine the work performed if each handle moves though a distance of 0.3 in. Finally, determine the diametrical compression of the rubber cylinder by the application of the forces.

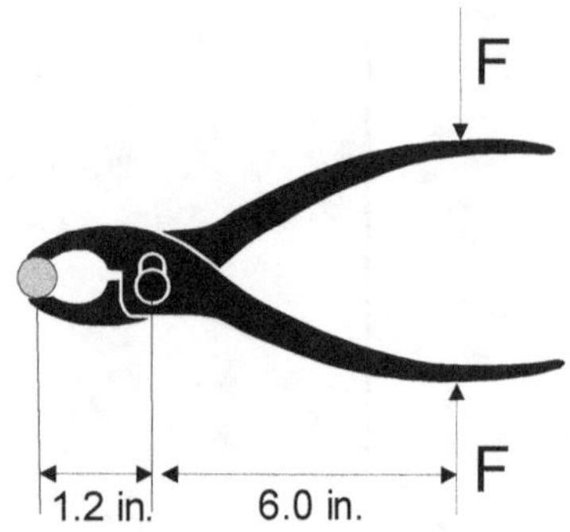

Fig. E8.4

Solution: Let's consider each member of the pliers individually and prepare a FBD for each as illustrated in Fig. E8.4a. The force acting on the rubber cylinder is R and the force acting at the pivot pin on the pliers is P.

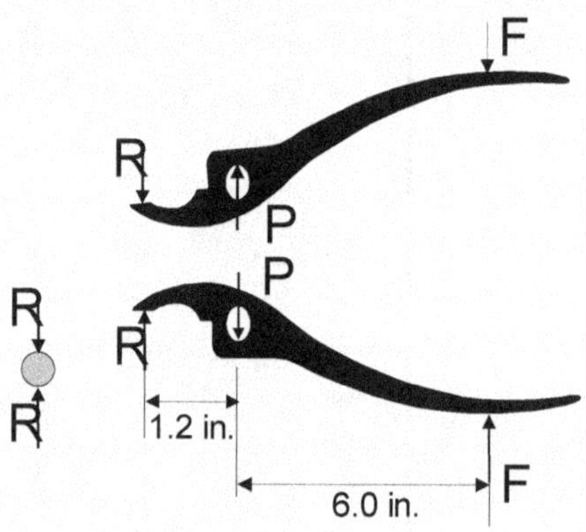

Fig. E8.4a

Three forces are applied to each handle of the pliers; hence the handles are considered as multiforce members. Writing the moment equation about point P, we obtain:

$$\Sigma M_P = (1.2)\,R - (6.0)\,F = 0 \quad \Rightarrow \quad R = (6.0/1.2)\,F = (5)(30) = 150 \text{ lb} \qquad \text{(a)}$$

The pliers exhibit a mechanical advantage of 5 since the output force R is five times the input force F. The mechanical advantage **MA** is given by:

$$\mathbf{MA} = R/F \qquad (8.1)$$

The work **W** performed is given by:

$$\mathbf{W} = F\,d \qquad (8.2)$$

where d is the distance that the force F is moved along its line of action by a machine.

In this case, each of the two forces is moved a distance 0.3 in. along the line of action of F; hence:

$$\mathbf{W} = 2Fd = 2\,(30)(0.3) = 18 \text{ in.-lb} \qquad \text{(b)}$$

Each handle of the pliers is a lever (curved to better fit the hand but a lever nevertheless). The pin holding the two handles together is the fulcrum. The diametrical compression of the rubber cylinder Δd is given by:

$$\Delta d = (1/MA)s_h \qquad (8.3)$$

$$\Delta d = (1/5)(2)(0.3) = 0.12 \text{ in.} \quad \text{(c)}$$

where s_h is the distance through which each handle moves.

EXAMPLE 8.5

Determine the reaction force at point G and the mechanical advantage of the toggle mechanism illustrated in Fig. E8.5. The dimensions are a = 310 mm, b = 220 mm, c = 160 mm, d = 100 mm, e = 140 mm and f = 120 mm. The input force F = 200 N. The contact at point G is made with a roller and pins are inserted at joints B, C, D and E.

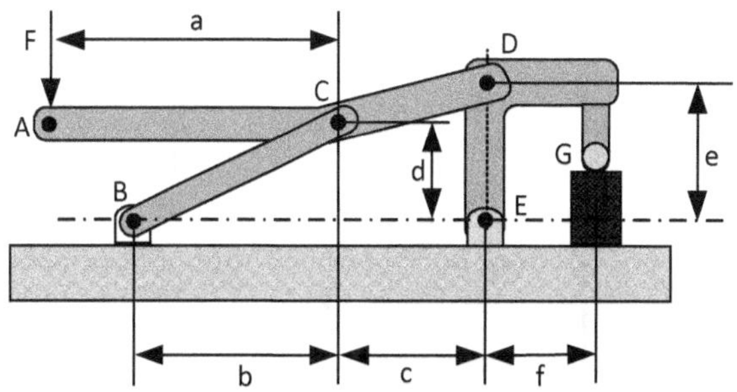

Fig. E8.5

Solution:

Step 1: Identify the link BC as a two-force member and ACD and EDG as multiforce members.

Step 2: Prepare FBDs of the three members as presented in Fig. E8.5a. Note that we have used the fact that the link BC is a two-force member. The force R_B acts along its axis. Also the reaction at pin C is equal and opposite to the reaction at pin B. We have used the laws of action and reaction in assigning forces and directions at all of the pins. Select member ACD for analysis because only three unknown forces are acting on it.

Step 3: Determine the angle θ and then apply the equations of equilibrium to member ACD.

$$\theta = \tan^{-1}(d/b) = \tan^{-1}(100/220) = 24.44° \quad \text{(a)}$$

$$\Sigma F_x = R_B \cos\theta - R_{Dx} = 0 \quad \Rightarrow \quad R_{Dx} = 0.9104\, R_B \quad \text{(b)}$$

$$\Sigma F_y = R_B \sin\theta - R_{Dy} - F = 0 \quad \Rightarrow \quad R_{Dy} = (0.4138)\, R_B - 200 \quad \text{(c)}$$

$$\Sigma M_C = F\, a - R_{Dy}\, c + R_{Dx}\,(e - d) = 0 \quad \text{(d)}$$

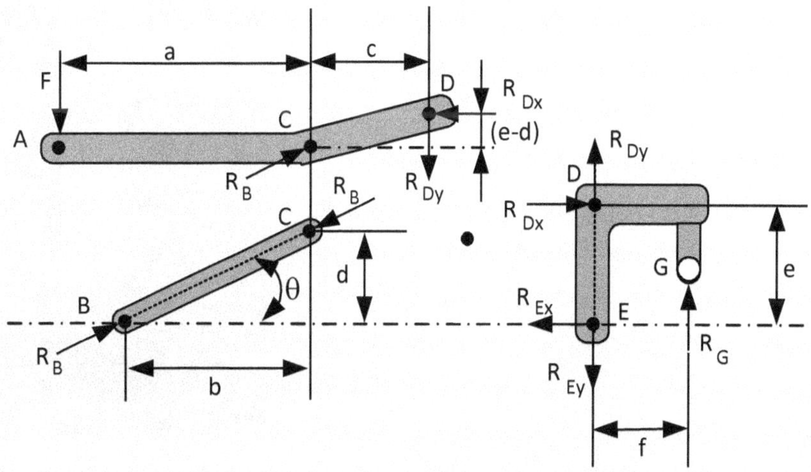

Substituting the results from Eqs. (b) and (c) into Eq. (d) yields:

$$\Sigma M_C = (200)(310) - [(-200) + (0.4138)\,R_B](160) + (0.9104)R_B\,(40) = 0 \qquad \text{(e)}$$

Solving Eq. (e) for R_B and substituting the result into Eqs. (b) and (c) yields:

$$R_B = 3{,}155\text{ N} \quad \Rightarrow \quad R_{Dx} = 2{,}873\text{ N} \;\Rightarrow\; R_{Dy} = 1{,}106\text{ N} \qquad \text{(f)}$$

Step 4: Apply one of the equations of equilibrium to member EDG.

$$\Sigma M_E = -\,R_{Dx}\,e + R_G\,f = 0$$

$$R_G = (e/f)\,R_{Dx} = (140/120)(2873) = 3{,}352\text{ N} \qquad \text{(g)}$$

Step 5: Determine the mechanical advantage from Eq. (8.1).

$$\mathbf{MA} = R/F = 3352/200 = 16.76 \qquad \text{(h)}$$

The toggle mechanism has amplified the input force of 200 N to an output force of 3.352 kN. The mechanical advantage of a toggle mechanism is significant. For this reason toggles are often used when input forces are limited and high output forces are necessary.

8.4 CONSTRUCTION EQUIPMENT

Construction equipment provides several real examples of machines that are used to modify forces. The backhoe, shown in Fig. 8.3, involves two large arms that support a bucket. Two hydraulic cylinders are employed to rotate the arms and another hydraulic cylinder is used to rotate the bucket. Both ends of each hydraulic cylinder are pinned, and the end of each of the large arms arm is pinned. The bucket rotation involves a linkage arrangement.

Fig. 8.3 Hydraulic cylinders actuate the arms and bucket on a backhoe.

EXAMPLE 8.6

A backhoe in action is illustrated in Fig. E8.6. Prepare a drawing showing the bucket, its linkage, the large arm supporting the bucket, and the hydraulic cylinder, which actuates the bucket. Draw FBDs of this machine showing the forces at the pins. If the bucket contains 800N of sand, determine the force exerted by the hydraulic cylinder. Dimensions of the components are presented in the FBDs given in the solution.

Fig. E8.6 A backhoe bucket containing 800 N of sand.

Solution:

Step 1: Prepare a drawing showing the arm and linkages supporting the bucket as presented in Fig. E8.6a.

Fig. E8.6a Note the bucket arm and the bucket are welded together.

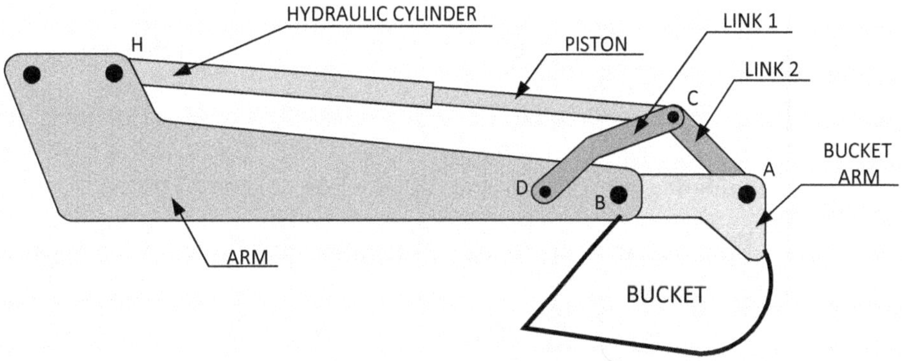

An examination of this drawing shows that Link 1 and 2 are both two force members and that the hydraulic actuator is also a two-force member. This fact will be instrumental in solving for the force generated by the hydraulic cylinder in supporting the loaded bucket.

Step 2: Prepare a FBD of the backhoe bucket as shown in Fig. E8.6b.

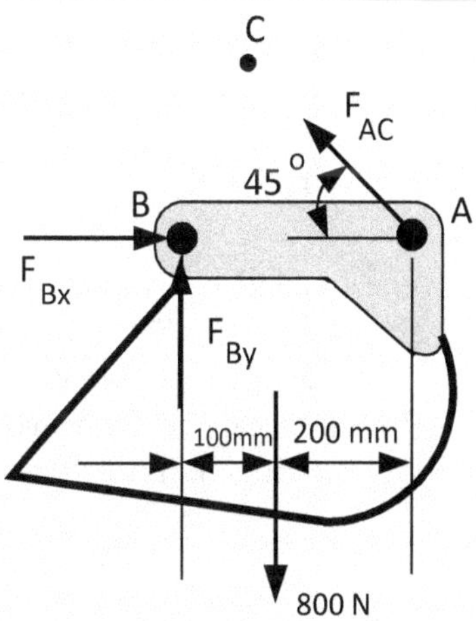

Step 3: Apply the equilibrium relations and solve for the three unknown forces.

$$\Sigma M_B = (300)\, F_{AC} \sin(45°) - (100)(800) = 0 \quad \text{(a)}$$

$$\Sigma F_y = F_{By} + F_{AC} \sin(45°) - 800 \quad \text{(b)}$$

$$\Sigma F_x = F_{Bx} - F_{AC} \cos(45°) = 0 \quad \text{(c)}$$

Solving Eqs. (a), (b) and (c) for the three unknown forces yields:

$$F_{AC} = 377.1 \text{ N} \qquad F_{By} = 533.4 \text{ N} \qquad F_{Bx} = 266.7 \text{ N} \quad \text{(d)}$$

Step 4: Prepare a FBD for the pin at point C where links 1 and 2 join the piston end of the hydraulic actuator as indicated in Fig. E8.6c.

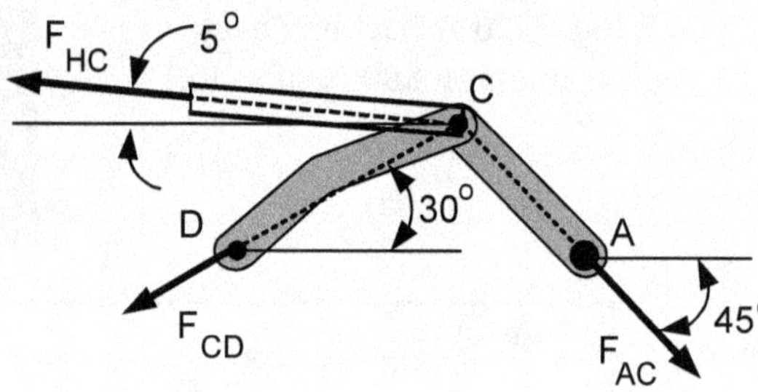

Fig. E8.6c

Step 5: Apply the equilibrium relations and solve for the two unknown forces.

$$\Sigma F_y = F_{HC} \sin(5°) - F_{AC} \sin(45°) - F_{CD} \sin(30°) = 0 \quad \text{(e)}$$

$$\Sigma F_x = - F_{HC} \cos(5°) + F_{AC} \cos(45°) - F_{CD} \cos(30°) = 0 \quad \text{(f)}$$

Combining the results in Eq. (d) with Eqs. (e) and (f) gives:

$$F_{HC}(0.08716) - F_{CD}(0.5000) = (377.1)(0.7071) = 266.65 \quad \text{(g)}$$

$$F_{HC}(0.9962) + F_{CD}(0.8660) = (377.1)(0.7071) = 266.65 \quad \text{(h)}$$

Solving these two equations for the two unknowns yields:

$$F_{CD} = -\,422.5 \text{ N} \qquad F_{HC} = 635.0 \text{ N}$$

Of course, the negative sign for F_{CD} indicates the force is compressive.

EXAMPLE 8.7

The Loadall is a versatile piece of construction equipment that can be used for several different purposes. The unit shown in Fig. E8.7 is fitted with a pair of forks for lifting pallets. These forks are fixed to the end of a telescoping boom, and are positioned over a dumpster in the photograph of Fig E8.7. If a pallet weighing 1200 lb were lifted from a truck bed with these forks, determine the actuator force necessary to rotate the boom and the forces at the bracket hinge pin.

Fig. E8.7 A Loadall unit equipped with a pair of forks for lifting pallets.

Solution:

Step 1: Prepare a drawing showing the telescoping arm and system supporting the forks as presented in Fig. E8.7a.

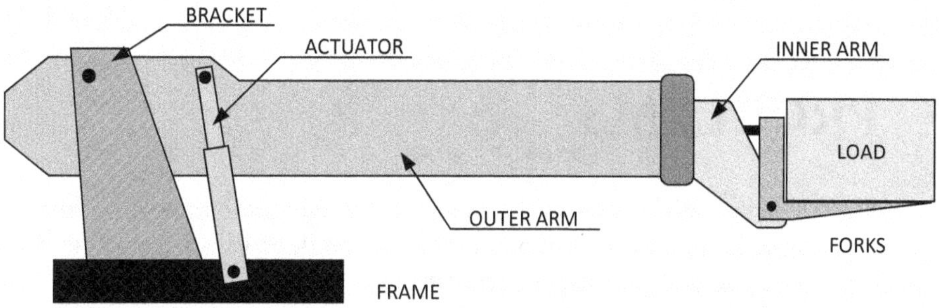

Fig. E8.7a

Step 2: Prepare a FBD of the Loadall mechanism as shown in Fig. E8.7b. Dimensions of the points where forces are applied are shown in the illustration. The weight of the arms acts through the center of gravity as shown in Fig. E8.7b.

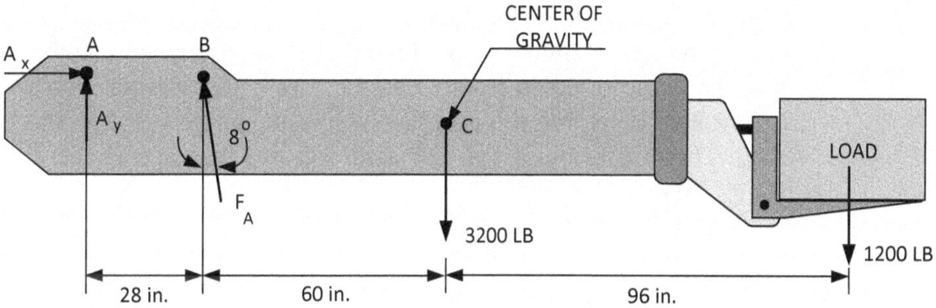

Fig. E8.7b

Step 3: Apply the equilibrium relations and solve for the three unknown forces—A_x, A_y and F_A.

$$\Sigma M_A = [F_A \cos(8°)](28) - (3{,}200)(28 + 60) - (1{,}200)(28 + 60 + 96) = 0$$

$$F_A = 18{,}120 \text{ lb} \qquad \text{(a)}$$

$$\Sigma F_x = A_x - F_A \sin(8°) = 0 \qquad A_x = (18{,}120)(0.1392) = 2{,}522 \text{ lb} \qquad \text{(b)}$$

$$\Sigma F_y = A_y + F_A \cos(8°) - 3{,}200 - 1{,}200 = A_y + (18{,}120)(0.9903) - 4{,}400 = 0$$

$$A_y = -\,13{,}540 \text{ lb} \qquad \text{(c)}$$

While the Loadall appears to be a complex machine capable of lifting and moving loads, it is relatively easy to analyze. The arm is a multiforce member, but if the FBD is prepared correctly, the solution for the actuator force F_A and the forces A_x and A_y at the hinge pin is straightforward.

8.5 SUMMARY

The method of analysis of frames and machines was introduced. Frames and machines contain one or more members that have three or more externally applied forces. For these multiforce members, the internal force does not coincide with the axis of the member. This fact is extremely important since it implies that two internal forces (P and V) and an internal moment M must be applied when making a section cut through a multiforce member. Because we introduce so many unknowns, section cuts through a multiforce member are usually avoided when analyzing frames and machines. The approach is to construct FBDs of the individual members of the structure and to apply the law of action and reaction as well as the equations of equilibrium to each member. Examples for both frames and machines were presented to demonstrate the method of analysis. Examples of construction equipment were described to show that complex machines could be analyzed using these relatively straightforward techniques.

PROBLEMS

8.1 For the hoist shown in the figure to the right, determine all of the external forces on member CFG as point D is moved along member AE so that the angle θ varies from 0 to 45°. We suggest that you use a spreadsheet in preparing this solution. Note the angle β = 60° and F = 580 N. The dimensions are: CF = 200 mm, FG = 320 mm and the pulley radius r = 10 mm.

8.2 For the hoist shown in the figure to the right, determine all of the external forces on member CFG as point B is moved along member AE so that the angle β varies from 35 to 75°. We suggest that you use a spreadsheet in preparing this solution. Note the angle θ = 30° and F = 240 N. The dimensions CF = 200 mm, FG = 320 mm and the pulley radius r = 10 mm.

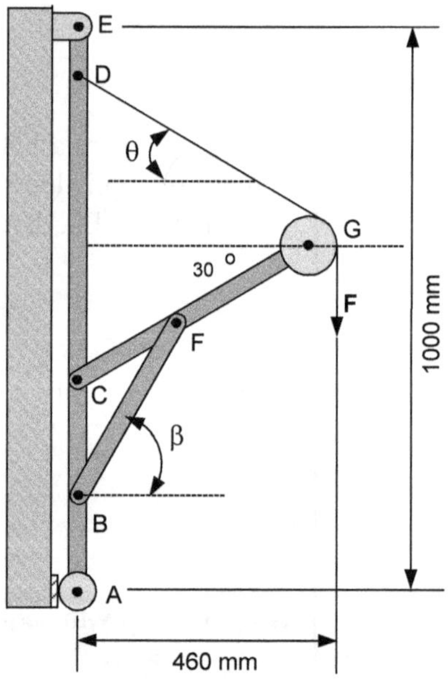

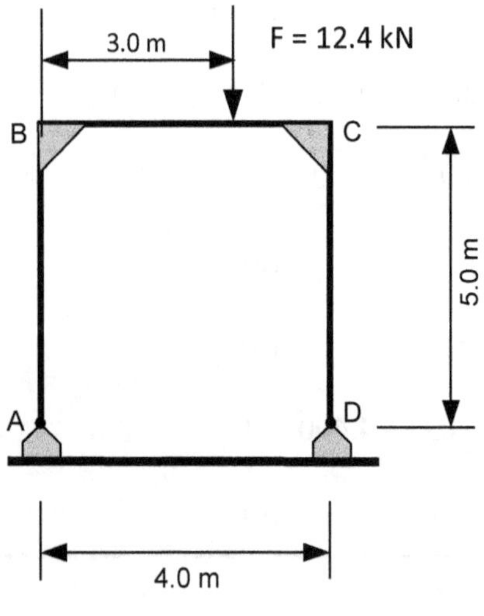

8.3 Prepare a FBD for the entire rectangular frame, shown in the figure to the left, if it is subjected to a concentrated force of 12.4 kN located a distance of 3.0 m from the left end of the horizontal member.

8.4 Determine the force F required to develop a pressure p = 2.1 MPa in the cylinder of the pump shown in the figure to the right. The piston area for the pump is 300 mm^2.

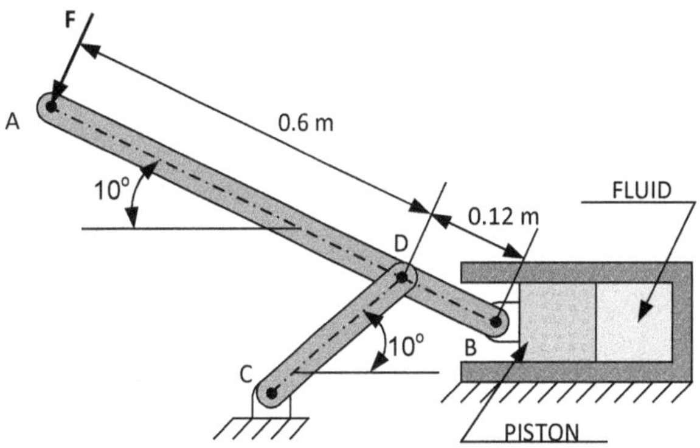

8.5 Determine the reaction force at point G and the mechanical advantage of the toggle mechanism illustrated in the figure to the right. A force F of 300 N is applied to the lever at point A. The contact at point G is made with a roller, and the links are connected with pins inserted at points B, C, D and E.

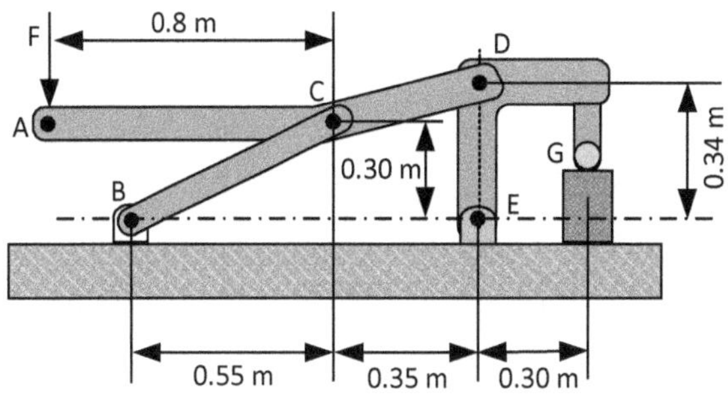

8.6 Determine the reaction force at point G and the mechanical advantage of the toggle mechanism illustrated in the figure to the right. The input force F is now 500 N. The contact at point G is made with a roller and the links are connected with pins inserted at points B, C, D and E.

8.7 Your manager believes that the dimension (e – d) for the toggle mechanism shown in the figure to the left is a critical design parameter. She asks you to determine the mechanical advantage of this mechanism if the dimension d is fixed and the dimension e is modified so that (e – d) varies from 5 mm to 50 mm. You may consider using a spreadsheet for this analysis.

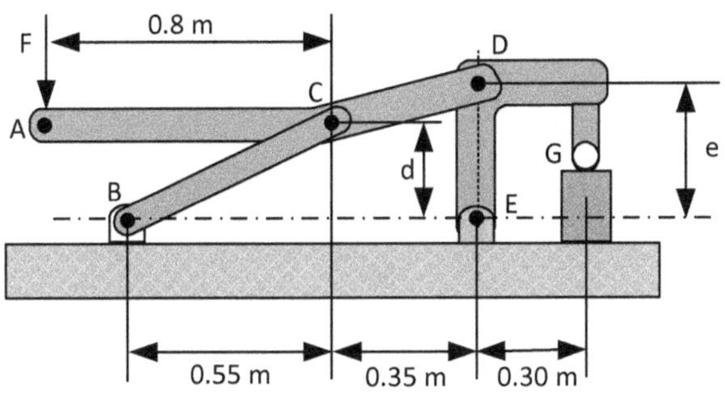

8.8 If the spring constant of the block under point G is 2.0 kN/mm, determine the vertical displacement of point G in Problem 8.5.

8.9 If the spring constant of the block under point G is 2.0 kN/mm, determine the work output of the toggle mechanism for the conditions described in Problem 8.5.

8.10 If the spring constant of the block under point G is 3.8 kN/mm, determine the vertical displacement of point G in Problem 8.6.

8.11 If the spring constant of the block under point G is 3.8 kN/mm, determine the work output of the toggle mechanism for the conditions described in Problem 8.6.

8.12 For the compaction press, shown in the figure to the right, determine the compressive force developed by the sliding platen if a force of 5 ton is applied to the toggle mechanism at point B. Also determine the mechanical advantage.

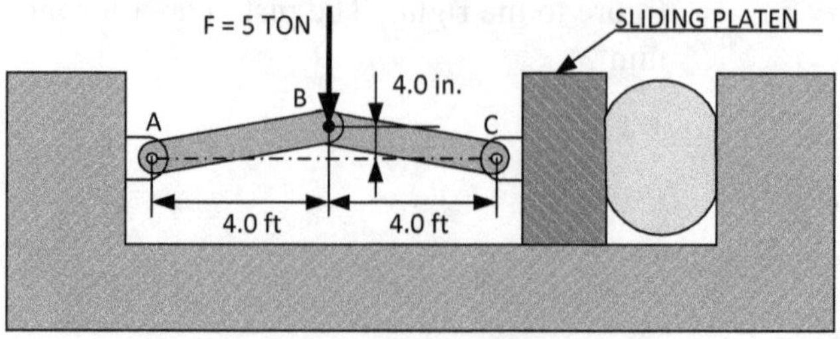

COMPACTION PRESS

8.13 A pair of pliers clamps a small diameter rubber cylinder as shown in the figure to the right. If opposing forces of 30 lb are applied to the handles of the pliers, determine the reaction forces acting on the cylinder. Also determine the work performed if each handle moves though a distance d = 0.14 in. Finally, determine the amount that the cylinder is squeezed by the application of the forces. The dimensions of the pliers are a = 1.0 in. and b = 5.5 in.

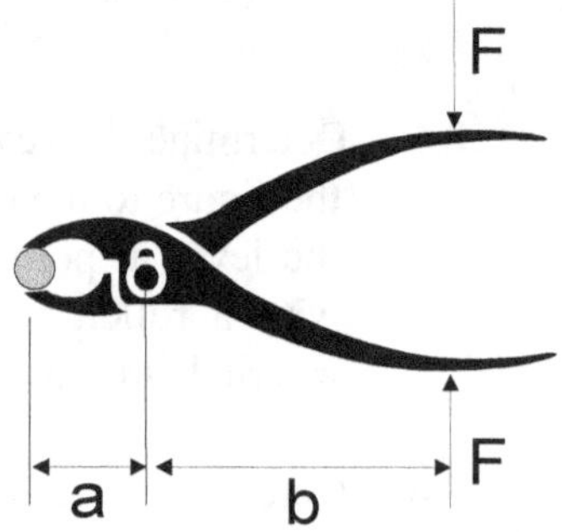

8.14 For the frame, shown in the figure to the left, determine the forces and moment at the fixed support at A, the forces at pin C and the force in link BD.

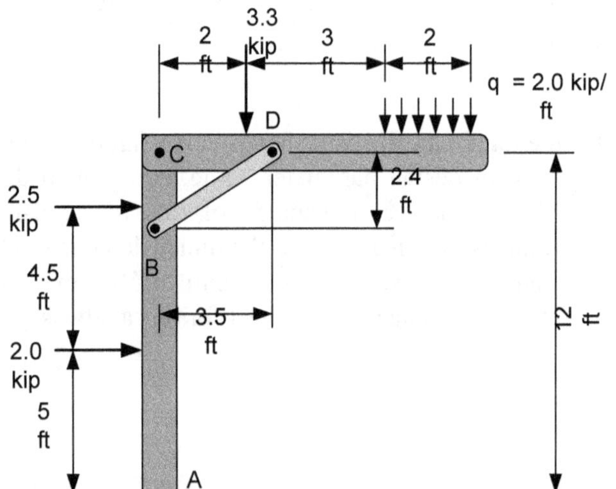

8.15 For the frame, illustrated in the figure to the right, determine all the forces acting on member ABCDE. Also determine the internal force in member AF. The attached weight W is 4.8 kN and the pulley radius is 150 mm. The dimensions in the figure are given in mm.

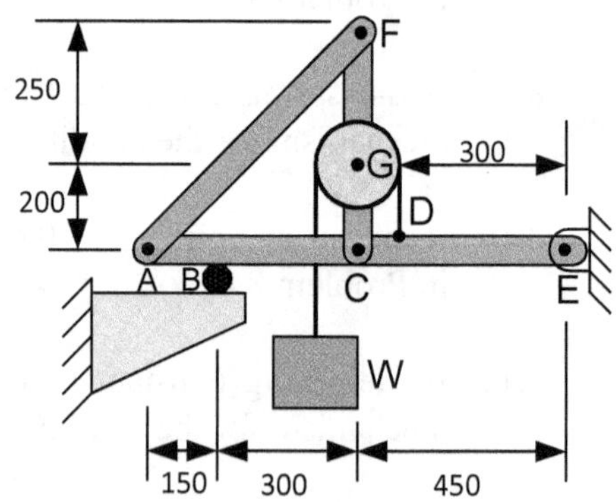

8.16 For the frame illustrated in the figure to the right, determine all the forces acting on member ABCDE. Also determine the internal force in member AF. The attached weight W is 4.8 kN and the pulley radius is 100 mm. The dimensions in the figure are given in mm.

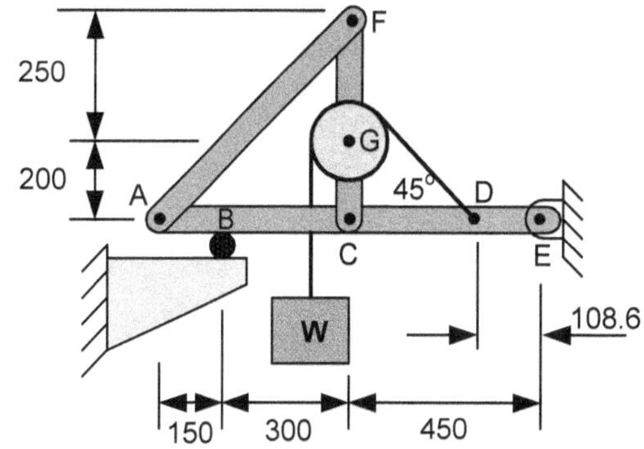

8.17 For the arm of the backhoe, shown in the figure below, determine the force that must be exerted by the hydraulic actuator AC to maintain the bucket loaded with a weight of 1.0 kN in equilibrium. Also find the forces in links BC and CE and the force acting on pin D. The bucket and the crosshatched appendage are welded together.

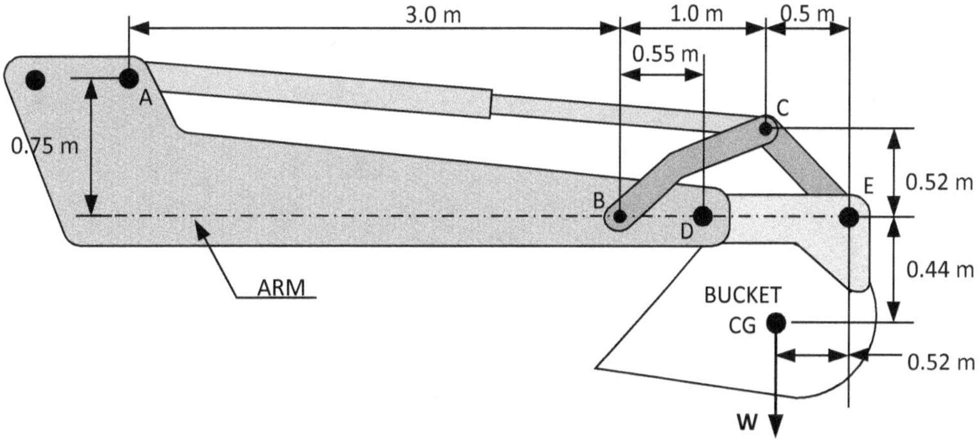

8.18 For the "Loadall" arm, presented in the figure below, determine the forces exerted by the hidden actuator and the visible actuator BC to maintain equilibrium. Also determine the forces acting at pins A and D. The arm weighs 1.5 kN and its center of gravity is located 1.8 m to the right of point A. All of the dimensions are given in meters.

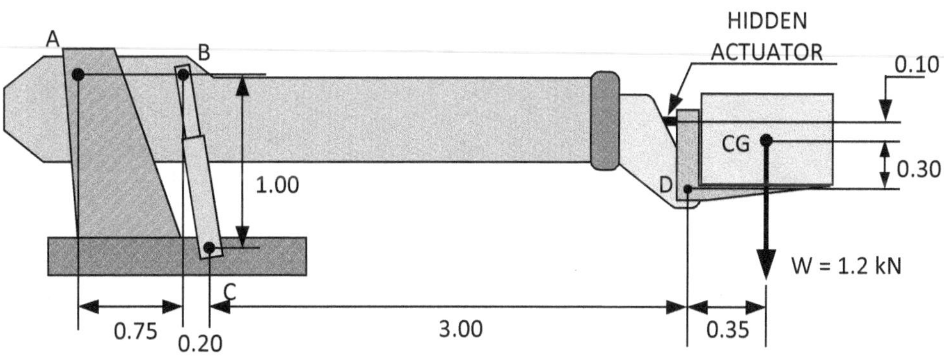

8.19 Levers are well known machines used for either amplifying or attenuating forces. When used in scales to weigh heavy objects, the levers are often arranged to compound the attenuation. Such an arrangement is illustrated in the figure below. If the pins are frictionless, show that the relation between the known scale weight w and the unknown scale weight W is:

$$W = w\left[\frac{(a+b+c+d-x)(a+b+c)b}{aec}\right]$$

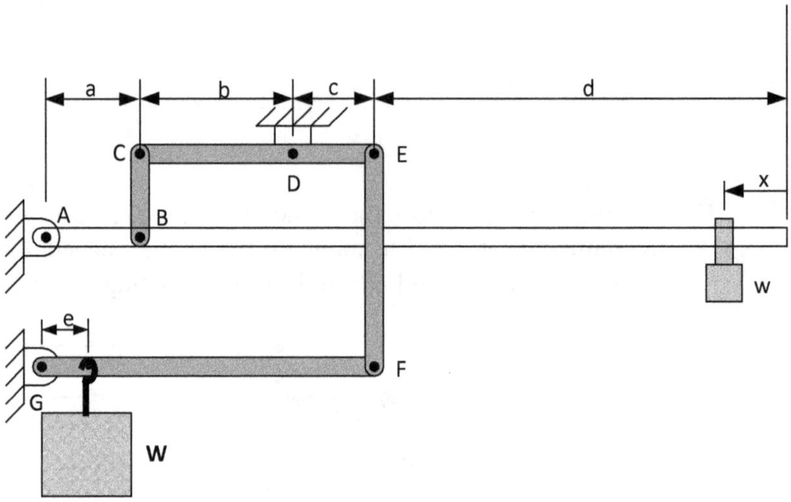

8.20 Select dimensions a through e if the scale, illustrated in the figure above, is to measure a weight W = 5.0 kN with a small sliding weight w = 10 N. The value of x = 200 mm is fixed in this design analysis.

8.21 The horizontal boom of a construction crane is counter balanced with a 15 kip weight that is centered at point B. A cable BFE anchored at points B and E supports the horizontal boom. The cable is maintained at a constant tension over its length with a small pulley located at point F. The horizontal boom is attached to the tower with a pin located at point C. The crane is lifting a load W = 10 kip from a hoist located at point D. Determine all of the forces acting on the horizontal boom, the cable tension, all of the reactions at point A and the force acting on the pin at point F. The weight of the boom is 5.0 kip and the weight of the tower member is 4.0 kip.

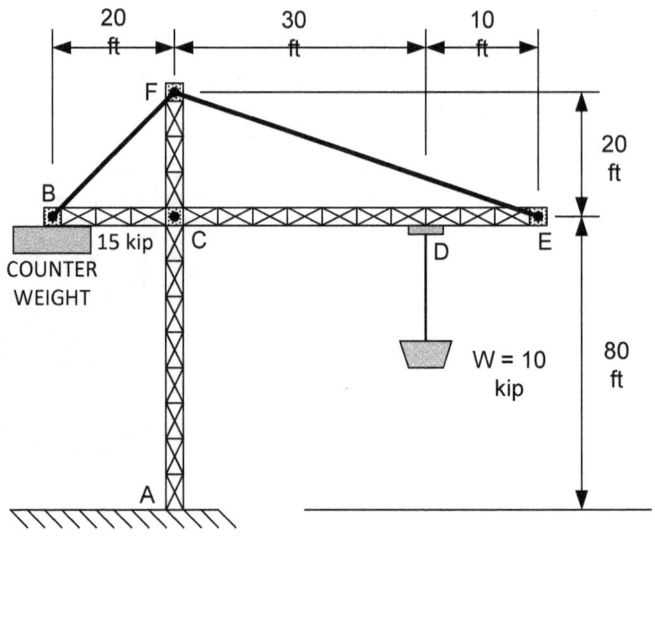

CHAPTER 9

FRICTION

9.1 INTRODUCTION

Friction occurs when two bodies are in contact and forces are applied to one or both bodies. Frictional forces develop at the contacting surfaces that tend to inhibit the sliding of one body relative to the other. For example, it is friction between the tires and road that keeps a car from sliding off when it rounds a turn. This ability of a tire to maintain contact with the road is often referred to as **traction**. We depict this situation with the illustration given in Fig. 9.1.

Fig. 9.1 Two bodies in contact with a force applied to the upper block.

If we slowly increase the lateral force applied to the upper block, we observe that it remains stationary. The block remains in equilibrium even with an applied force of significant magnitude. Is this behavior consistent with Newton's second law, which states $\Sigma \mathbf{F} = m\mathbf{a}$? Let's look more closely at the upper block in Fig. 9.1 by preparing a FBD showing **all** the forces that act on the body. The FBD for both the upper and lower blocks is shown in Fig. 9.2. Since the upper block is in equilibrium, it is clear that the frictional force must act in the opposite direction to the applied force. The frictional force on the fixed (lower) block is the same magnitude, but it is in the opposite direction because of the law of action and reaction. Normal forces also develop between the two contacting surfaces. The normal forces are perpendicular to the contacting surfaces, equal in magnitude and opposite in direction.

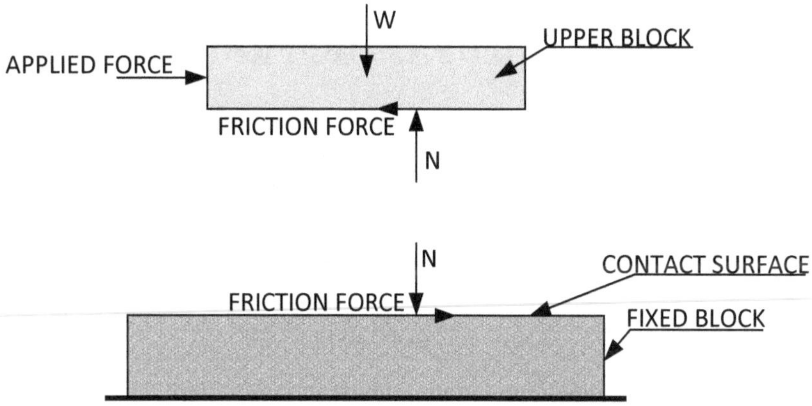

Fig. 9.2 Free body diagram showing the friction forces developed on both bodies at the contact surface.

If we continue to increase the applied force, the upper block will eventually begin to slide over the surface of the lower block, indicating some upper limit on the frictional forces. Before we explore this upper limit, let's consider several factors that influence friction. We know that friction only occurs with two bodies in contact so it is reasonable to examine in detail the two contact surfaces. Consider a magnified view of the surfaces presented in Fig. 9.3.

We note in Fig. 9.3 that smooth surfaces are not really smooth when magnified sufficiently. Instead, the surfaces are made up of valleys and peaks, which we call **asperities**. Contact occurs only when two opposing asperities are aligned. Since the alignment of two opposing peaks is rare, the actual area of contact A_c is small compared to the area A of the entire surface. For reasonably smooth surfaces, the ratio of A_c /A is:

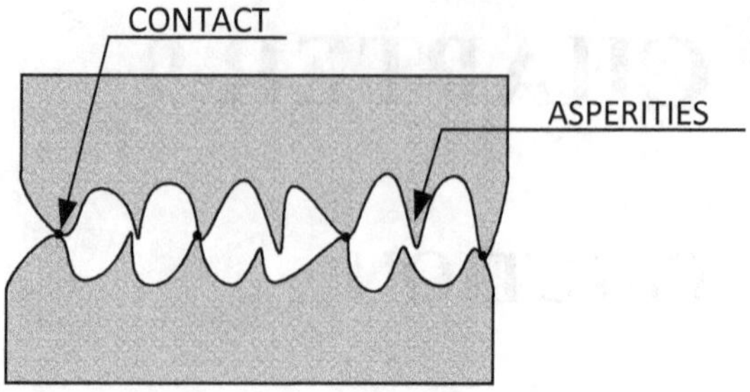

Fig. 9.3 Schematic illustration of contacting surfaces.

$$A_c /A \leq 0.001 \quad (9.1)$$

When a lateral force is applied to one of the bodies a resistive force (friction) develops because some of the asperities are coupled together. In some cases, these asperities mechanically interfere with each other, and must be sheared off before motion between the surfaces may occur. In other cases, the contacting asperities bond together, and this bond must be fractured before sliding may occur. The shearing of asperties and the fracture of the contact bonds produce wear of sliding surfaces.

In the study of mechanics, we ignore the microscopic characteristics of the surfaces and treat friction from a more pragmatic view point. We simply distinguish between dry and fluid friction. In **dry** or **Coulomb**[1] **friction** the asperities are in contact, and appreciable friction forces develop to resist sliding motion between the contacting surfaces. In **fluid friction**, we introduce a lubricant that separates the two surfaces as illustrated in Fig. 9.4.

When a lubricant is present the frictional forces are reduced significantly. This is the reason for using oil to lubricate the engines on our automobiles. The resistance is no longer due to breaking asperities free to permit sliding. Instead, much smaller forces are required to shear the lubricant. In this chapter we will be concerned with only dry or Coulomb friction. Lubrication theory is a topic introduced later in the Mechanical Engineering curriculum.

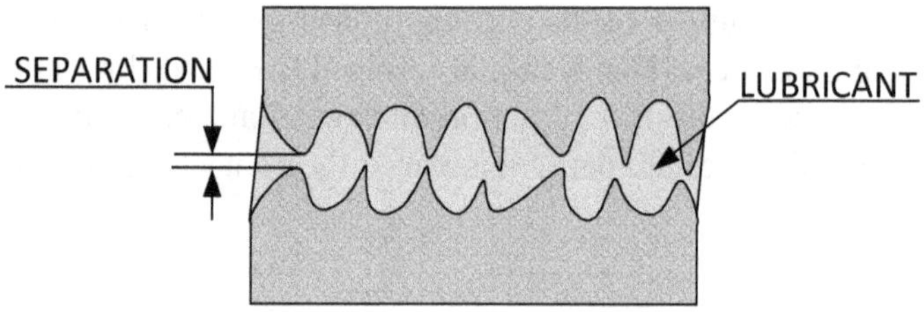

Fig. 9.4 Lubricant separates the surfaces preventing contact of the asperities.

9.2 STATIC AND DYNAMIC FRICTION

Let's return to the model depicted in Fig. 9.1 with two blocks in contact, and examine the conditions required to maintain the upper block in equilibrium. The body is two-dimensional and three equations of equilibrium must be satisfied, namely $\Sigma F_x = 0$, $\Sigma F_y = 0$ and $\Sigma M_O = 0$. The FBD of the upper block presented in Fig. 9.5, shows the applied lateral force F, the weight of the block W, the vertical reaction of the supporting block N, and the friction force F_f. The equilibrium relations yield:

$$\Sigma F_y = 0 \Rightarrow N = W \quad (a)$$

$$\Sigma F_x = 0 \Rightarrow F = F_f \quad (b)$$

The normal force N is located along the contact surface at position s, to satisfy the moment equation of equilibrium.

From the relation $\Sigma M_O = 0$, we may determine the location s for the normal force N as:

$$s = (Ww - Fh)/2W \quad (9.2)$$

[1] Charles Augustine Coulomb, a military engineer, developed the general theory of friction and published his work in a lengthy memoir. Théorie des Machines Simple, En ayant égard au frottement de leurs parties, et a la roideur des Cordages, Mémoires de Mathématique et de Physique, vol. X. 1785.

We note that s = w/2 in the absence of a lateral force F. When F increases the value of s decreases as the normal force shifts to the right. When F = (wW)/h, s = 0, and the block is on the verge of tipping. We will consider whether the block slides or tips later in this chapter.

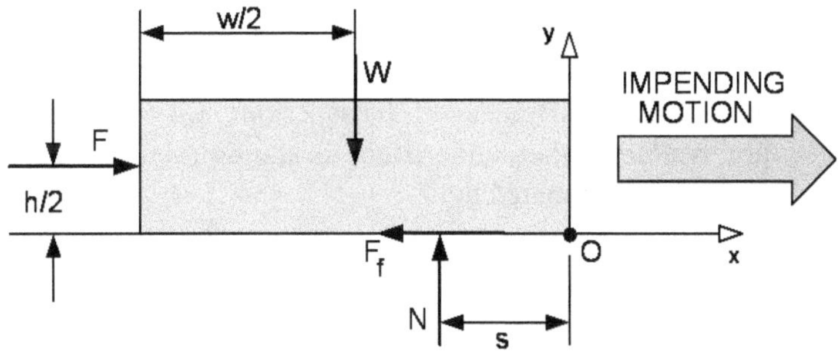

Fig. 9.5 FBD of a block prior to the initiation of sliding.

The magnitude of the frictional force is variable. The frictional force becomes as large as necessary to prevent sliding up to some maximum force when the asperities break free. The maximum frictional force is approximated by:

$$(F_f)_{max} = \mu N \tag{9.3}$$

where μ is the coefficient of friction.

The coefficient of friction depends on the materials in contact and to a significant degree on the surface finish and cleanliness of the contact surfaces. A listing of the coefficient of friction for a number of different materials is given in Table 9.1; however, it is important to note the very wide variations in μ for all of the material combinations. The variation is so large that it is advisable to conduct simple experiments to measure the coefficient of friction of the materials involved in a design of a machine component.

The maximum frictional force increases linearly with the normal force N and is considered to be independent of the area of the surfaces in contact providing that A_c/A remains very small, as indicated in Eq. (9.1). When the normal force becomes extremely large the asperities undergo plastic deformation and the ratio A_c/A increases markedly. Another situation leading to significant increases in A_c/A is when one of the two bodies is fabricated from soft materials such as rubber or plastics. In these two cases, the area of the surfaces is important, and Eq. (9.3) is no longer valid.

Table 9.1
Approximate coefficient of friction for contact between different materials.

Surface Materials	Coefficient of Friction (static)
Metal on metal	0.15 – 0.50
Metal on stone or concrete	0.30 – 0.70
Metal on leather	0.30 – 0.60
Metal on wood	0.20 – 0.60
Wood on wood	0.30 – 0.70
Wood on stone or concrete	0.30 – 0.70
Rubber on concrete	0.60 – 0.90
Metal on brake pad	0.20 – 0.40
Metal on Ice	0.03 – 0.05

For normal conditions Eq. (9.3) is valid, and the friction force increases linearly with the applied lateral force F as shown in Fig. 9.6 until motion (sliding or tipping) occurs.

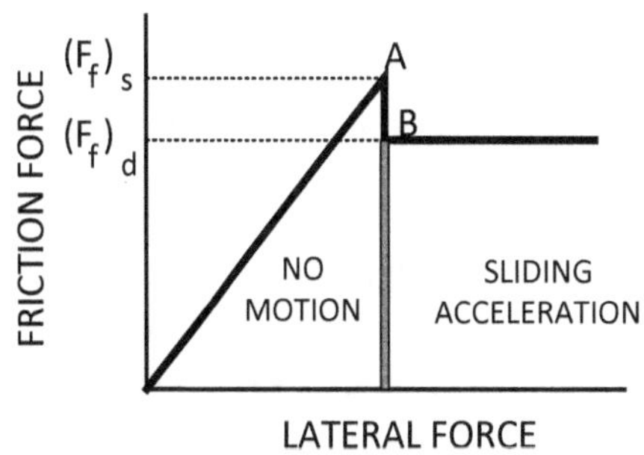

Fig. 9.6 Graph of frictional force with the applied lateral force.

The frictional force is exactly equal to the applied lateral force until the sliding initiates due to the failure of asperities. At this instant, the frictional force decreases abruptly (from points A to B in Fig. 9.6). After the onset of sliding, the frictional force remains constant at a value of $(F_f)_d$ with further increases in the applied lateral force. The friction force that develops under sliding conditions is smaller than the friction under static conditions. The mechanical interference of the asperities still occurs, but the time required for adhesion (bonding) of the asperities is not available when one surface is sliding relative to the other. Under sliding conditions, the kinetic friction force, $(F_f)_d$ is approximated by:

$$(F_f)_d = \mu_d N \tag{9.4}$$

where μ_d is the dynamic or kinetic coefficient of friction, which is usually approximated by multiplying the static value by 0.75 to 0.80.

9.3 MEASURING THE COEFFICIENT OF FRICTION

The coefficient of friction μ exhibits significant variation due to affinity of different surfaces, surface roughness, and the cleanliness of the two mating surfaces. For these reasons it is advisable to measure the coefficient of friction of the materials to be employed in the design of machine components and structures. The measurement of μ is relatively easy to accomplish using the simple experimental arrangement shown in Fig. 9.7.

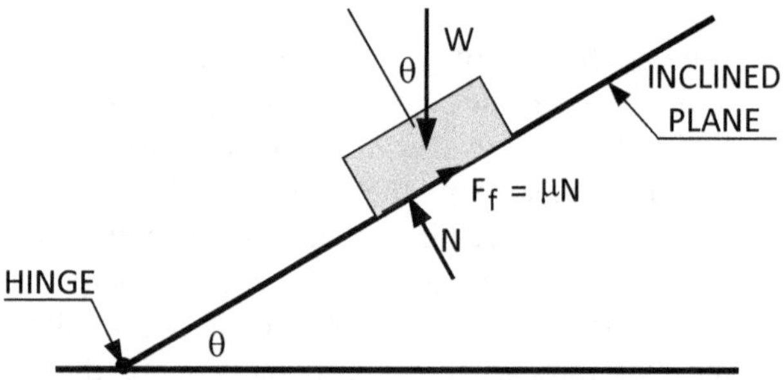

Fig. 9.7 Illustration of a method for measuring the static coefficient of friction.

A block of one material is placed on an inclined plane with a surface fabricated with the other material. The angle of inclination of the plane θ is slowly increased until the block begins to slide. The angle required to initiate movement (sliding) is known as the angle of impending motion θ_s. It is measured with a protractor. Let's perform an equilibrium analysis to show the relation between μ and θ_s.

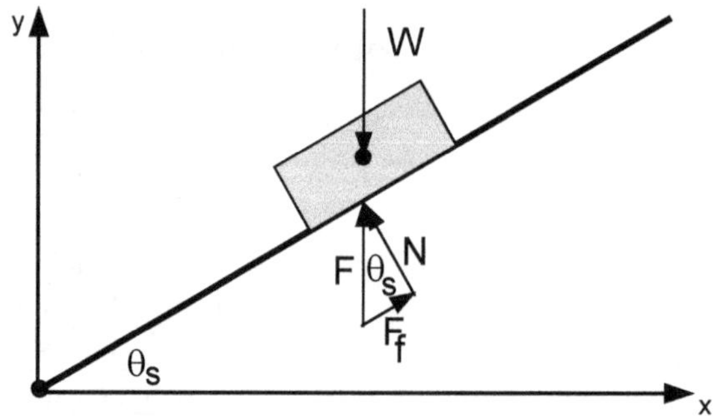

Fig. 9.8 FBD of the block on the incline plane at the position of impending motion.

We begin the equilibrium analysis by drawing a FBD of the block where the inclined plane is at the position of impending motion as shown in Fig. 9.8. We have rearranged the forces acting on the block to show the resultant force **F,** which is given by:

$$\mathbf{F} = \mathbf{N} + \mathbf{F_f} \tag{a}$$

Next, consider the equilibrium relation $\Sigma F_y = 0$, and write:

$$\Sigma F_y = F - W = 0 \quad \Rightarrow \quad F = W \tag{b}$$

We also note that:

$$N = F \cos \theta_s = W \cos \theta_s \tag{c}$$

$$F_f = F \sin \theta_s = W \sin \theta_s \tag{d}$$

Recall Eq. (9.3) and use Eqs. (c) and (d) to write:

$$\mu = F_f / N = \sin \theta_s / \cos \theta_s = \tan \theta_s \qquad (9.5)$$

Clearly, we can measure the coefficient of friction quickly and efficiently with this simple experiment. The angle θ_s is related to the coefficient of friction through the tangent function. Also, the angle θ_s is sometimes referred to as the angle of repose, which we will show is equal to the angle of friction ϕ.

Let's consider a block resting on the horizontal surface. Apply a lateral force F to the block until is just ready to slide (**impending motion**). We illustrate the FBD for the block under these loading conditions in Fig. 9.9.

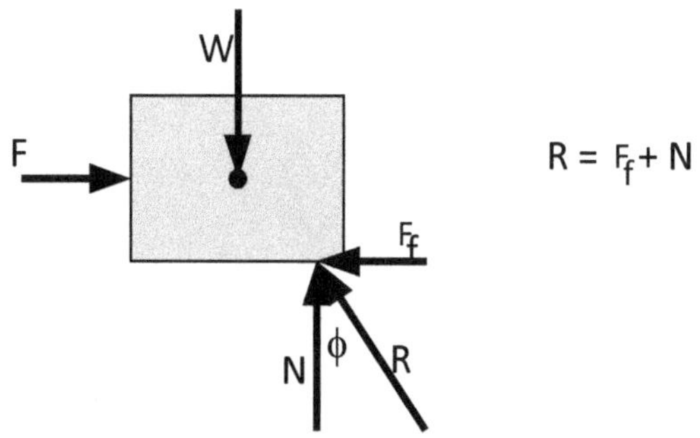

Fig. 9.9 FBD of a block subjected to applied forces associated with impending motion.

To analyze this FBD note that the angle of friction is defined as:

$$\tan \phi = F_f / N \qquad (a)$$

But with the block subject to **impending motion**, we may write $F_f = \mu N$. Substituting this relation into Eq. (a) leads to:

$$\tan \phi = \mu \qquad (9.6)$$

Combine Eqs. (9.5) and (9.6) to yield:

$$\theta_s = \phi \qquad (9.7)$$

From these simple derivations, we have proven that the angle of repose θ_s is equal to the angle of friction ϕ.

EXAMPLE 9.1

A sales representative states that the coefficient of friction μ for a flat neoprene-fabric belt is 0.6 when the belt is operated against a smooth metal pulley. You obtain a sample of the belt, and plan to conduct an inclined plane experiment to measure μ. As a part of your plan determine the expected angle of repose.

Solution:

Recall Eq. (9.5) and write:

$$\theta_s = \tan^{-1} \mu = \tan^{-1} (0.6)$$

$$\theta_s = 31.0°$$

Your plan for the friction experiment should permit the inclined plane to tilt to at least 40°, and the protractor for measuring the angle of repose should be accurate to 1/2°.

EXAMPLE 9.2

In conducting the experiments with the sample of flat neoprene/fabric belt, you record five different values of the angle θ_s including 29.5, 31.5, 31.0, 32.5, and 28.0. Determine the average value of μ and its range.

Solution:

Recall Eq. (9.5) and write:

$$\mu = \tan \theta_s \qquad \text{(a)}$$

The results from Eq. (a) are given in the table below:

θ_s	μ
29.5	0.566
31.5	0.613
31.0	0.601
32.5	0.637
28.0	0.532

The average value of μ is $\Sigma\mu/n = 2.949/5 = 0.590$; and the range in the measurements of μ is $\mu_{max} - \mu_{min} = 0.637 - 0.532 = 0.105$

EXAMPLE 9.3

You are attending a county fair and decide to watch an exhibition involving a harnessed team of mules that are pulling a sled with a load of weights. The load is pulled only short distances (about two meters). You are amazed at the size of the cast iron blocks that are placed on the sled, and decide to conduct an analysis to determine the force F with which the team pulls. Draw the FBD necessary to conduct the analysis, and list the information necessary to determine the force F.

Solution:

Let's represent the sled, its weight, and the weight of the cast iron blocks with a single rectangular block as shown in Fig. E9.3.

Fig. E9.3

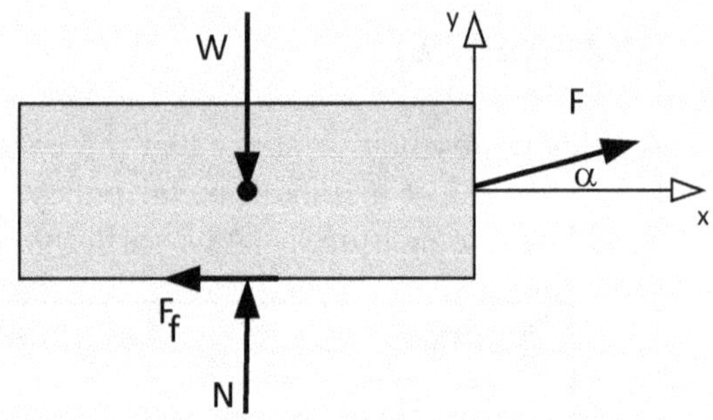

This FBD represents a model of the mule team, sled and the load of weights. The mule team pulls with a force F that is oriented at an angle α relative to the ground. The friction force F_f provides the resistance in the x direction.

To determine the force F, we must gather information defining the angle α, the friction coefficient, and the weight W. Suppose the sled together with the added weights total 5200 lb. The harness on the mules is shortened as much as possible to maximize the angle α at a value of 15°. The steel runners on the sled interact with a flat, bare, dry, clay field. If we assume that the coefficient of friction of metal (steel) on dry clay is 0.45, we have collected all of the information needed to determine F. Let's solve for the pull F provided by the team of mules.

Begin by writing the equilibrium relations $\Sigma F_x = 0$ and $\Sigma F_y = 0$, which yield:

$$\Sigma F_x = F \cos \alpha - F_f = 0 \qquad \text{(a)}$$

$$\Sigma F_y = F \sin \alpha + N - W = 0 \qquad \text{(b)}$$

Next recall Eq. (9.3) and note the maximum load pulled by the mule team occurs when motion of the sled is impending. In this instance Eq. (9.3) is valid and we may write:

$$(F_f)_{max} = \mu N$$

Substituting Eq. (9.3) into Eq. (a) yields:

$$F = \mu N/\cos \alpha \qquad \text{(c)}$$

Substituting Eq. (c) into Eq. (b), and solving for the unknown force N leads to:

$$\mu N \tan \alpha + N - W = 0$$

$$N = W/(1 + \mu \tan \alpha) \qquad (9.8)$$

Finally, substitute Eq. (9.8) into Eq. (c) to obtain:

$$F = \mu W/[\cos \alpha(1 + \mu \tan \alpha)] \qquad (9.9)$$

We have numbered the relations for the two unknown forces N and F in the sled problem as Eqs. (9.8) and (9.9). These relations may be used in the solution of N and F for any block that is moved by sliding providing μ, W, and α are known. Let's complete the solution and determine the capability of our mule team. Substituting into Eq. (9.9) gives:

$$F = [(0.45)(5200\text{ lb})]/[\cos 15°(1 + (0.45)\tan 15°)] = 2162\text{ lb}$$

EXAMPLE 9.4

Consider a wooden crate that is resting on a level concrete floor of a warehouse. A worker attempts to slide the crate a short distance as shown in Fig. E9.4. If the crate weighs 400 N and the coefficient of friction between the wooden crate and the concrete floor is $\mu = 0.5$, determine the force that must be applied to slide the crate. Note that the crate is short and the worker must lean over to apply a force at an angle of 30° relative to the floor.

Fig. E9.4

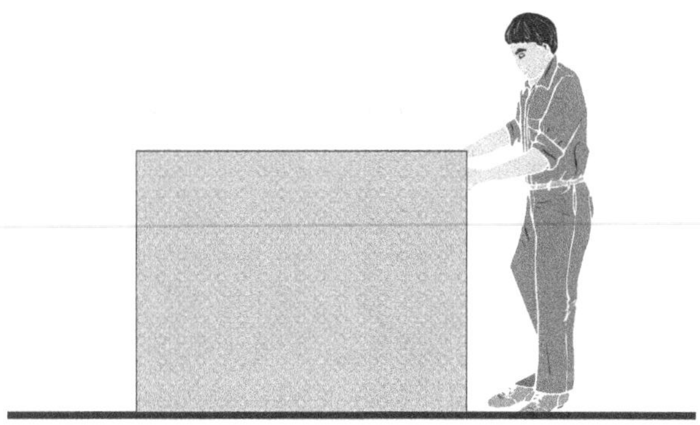

Solution:

Begin the solution by constructing a FBD to model the physical situation. We recognize this problem is identical to the problem in Example 9.3, except the worker is pushing with a force that has components in the negative x and y directions. We account for this fact by deriving Eqs. (9.8) and (9.9) again. Following the same procedure, we obtain:

Fig. E9.4a

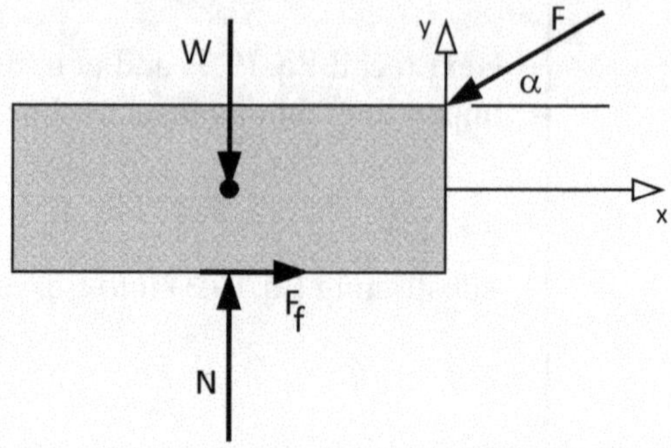

$$N = W/(1 - \mu \tan \alpha) \qquad (9.8a)$$

$$N = 400/[1 - (0.5) \tan (30°)] = 562.3 \text{ N}$$

$$F = \mu W/[\cos \alpha(1 - \mu \tan \alpha)] \qquad (9.9a)$$

$$F = (0.5)(400)/ \{\cos (30°)[1 - (0.5) \tan(30°)]\} = 325 \text{ N}$$

The result of 325 N is much greater than a force of 200 N, which would have been required if the worker had applied the force F in the direction of motion. There are two reasons for the significant increase in the required force. First, the worker is pushing with a component of force downward on the crate, which increases the normal and friction forces. Second, only the horizontal component of the applied force is useful in sliding the crate.

EXAMPLE 9.5

Suppose you are driving in a storm that is depositing a mixture of snow and ice on the highways. As a prudent driver, you reduce your speed significantly, and are making slow but steady progress toward your destination. However, you encounter a long hill with a grade of 4% as shown in Fig. E9.5. Unfortunately part way up the hill your front wheel drive car loses traction and stops. Your attempts to move the car forward fail because the tires spin as you depress the accelerator to increase the power applied to the wheels. In fact, your efforts cause the car to slip a short distance down the hill.

Determine the coefficient of friction that exists between the surface of the tire and the roadbed. Also explain why your attempts to move the car forward (up the hill) are futile.

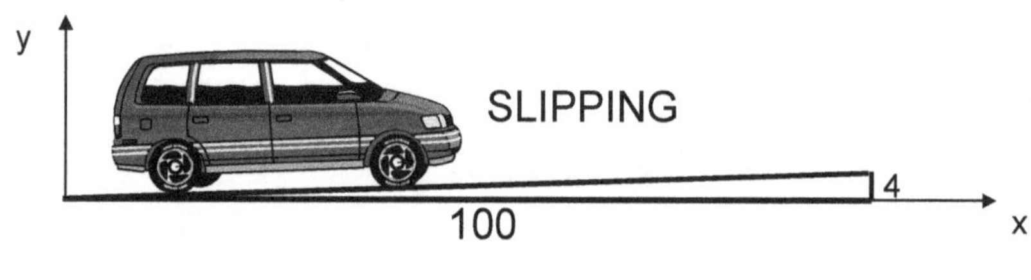

Solution:

Let's begin our analysis by drawing a FBD of one of the front wheels as shown in Fig. E9.5a.

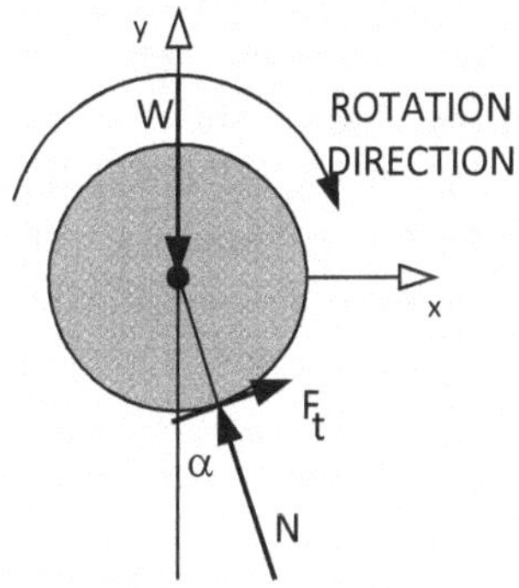

Fig. E9.5a

Clearly, we select a front wheel for the FBD because the car applies the traction force F_t to the pavement through the front wheels. Begin the analysis by writing the equilibrium relations.

$$\Sigma F_x = F_t \cos\alpha - N \sin\alpha = 0 \qquad \text{(a)}$$

$$\Sigma F_y = F_t \sin\alpha + N \cos\alpha - W = 0 \qquad \text{(b)}$$

Since the car slid down the hill, a condition of impending motion existed prior to the slide. Hence, we may write:

$$F_t = F_f = \mu N \qquad \text{(c)}$$

We have equated the traction force F_t with the friction force F_f in this equation.

Next, let's substitute Eq. (c) into Eq. (a) to give:

$$\mu N \cos\alpha - N \sin\alpha = 0 \qquad \text{(d)}$$

$$\mu = \tan\alpha = 0.04 \qquad \text{(e)}$$

The result in Eq. (e) shows that the friction coefficient is independent of the normal force N, and is equal to the grade of the hill. This fact is clear because N cancels out of Eq. (d). We will return to this point in the discussion following this example.

To continue the analysis, substitute Eq. (9.3) into Eq. (b) and write:

$$\mu N \sin\alpha + N \cos\alpha - W = 0 \qquad \text{(f)}$$

$$N = W/(\mu \sin\alpha + \cos\alpha) \qquad \text{(g)}$$

We have determined the relation for N to present a complete solution. The coefficient of friction is independent of both N and the weight W acting on the wheel.

Attempts to get the car moving are almost always unsuccessful because the wheels begin to spin as power is applied to the drive train. As the wheels spin, heat is generated melting the snow and ice mixture. Water is formed that better lubricates the surface of the ice and snow coating the highway, which reduces still further the coefficient of friction.

If you have ever driven much in snow and ice and have mastered the art of successfully negotiating slippery hills when all about you get stuck, the solution in Example 9.5 should certainly raise two very important questions.

1. Why do front wheel drive cars perform better in the snow and ice than rear drive cars?
2. Why do we spend significant amounts of money and put up with excessive road noise to equip our cars with snow tires?

Front wheel drive cars perform better in the snow than rear wheel drive cars because more weight is placed over the drive wheels. The analysis showed that the friction coefficient was independent of the weight on the wheel only because we did not consider the possibility of the tire tread being forced into the snow and ice mixture covering the pavement. Higher wheel forces may in many cases produce enough penetration of the tread into the coating of snow to give some degree of interlocking at the tire-snow interface. The analysis in Example 9.5 considered the surfaces of the wheel and the pavement to be perfectly rigid, which is not true.

Snow tires are also beneficial because the tread is designed to penetrate snow and gain traction by interlocking with a coating of even relatively hard packed snow. However, the benefit gained by a tread that penetrates a coating of snow largely vanishes if the pavement is coated with a rigid layer of ice.

9.4 FRICTION AND STABILITY

In many engineering applications friction is a culprit. It results in a loss of power, a loss of efficiency, and the generation of undesirable heat. However, in other applications friction is a benefit that enables us to move about with control and to maintain stability. The next series of examples demonstrate the role of friction in stability, in preventing slipping, in inducing tipping and in promoting rolling.

9.4.1 Stability

When you climb a ladder, you want it to remain stable or otherwise you will fall. The stability is maintained because friction keeps the ladder from slipping and sliding as you climb it. Often ladders are designed with rotating footpads with a rubber surface that improves frictional contact, increases the coefficient of friction, and enables these ladders to remain stable at smaller angles of inclination.

EXAMPLE 9.6

Consider the ladder that rests on a level concrete surface and leans against a vinyl-clad wall as shown in Fig. 9.10. A man with a mass of 81.5 kg climbs ¾ of the way up the ladder. We question the stability of the ladder. Is it stable or will it slip and fall?

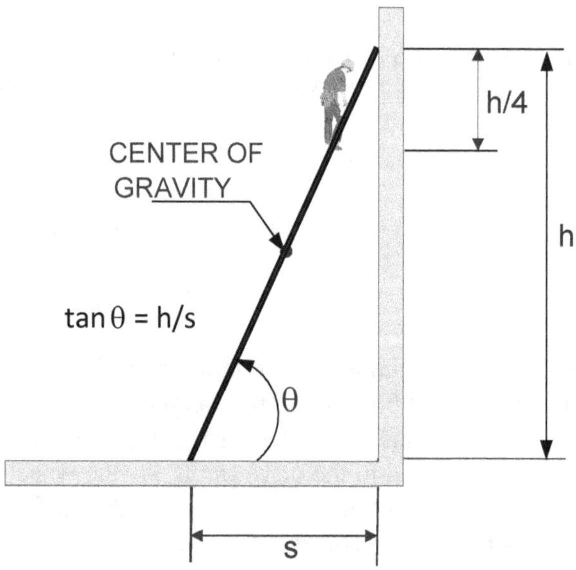

Fig. 9.10 Illustration of climbing a ladder.

Solution:

Before considering the stability of the ladder, we need to gather much more information about the ladder, its interactions with the floor and the wall, and its weight. A FBD of the ladder shown in Fig. E9.6 provides guidance before we seek the additional information necessary to determine if the ladder is stable.

Fig. E9.6a

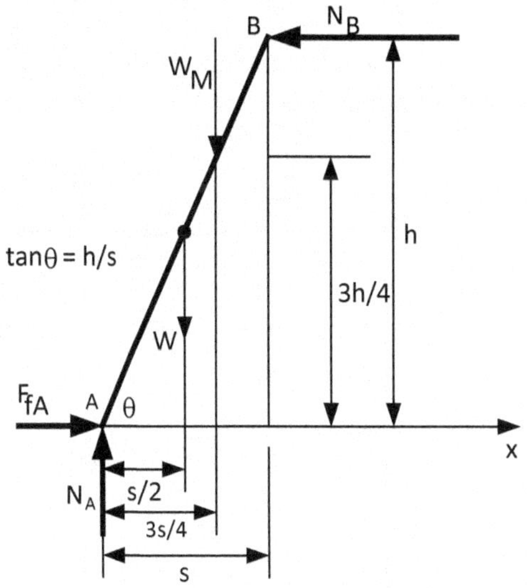

Let's introduce the parameters that control the stability of the ladder.

$$s = 3m;\ h = 7\ m;\ \theta = \tan^{-1}(h/s) = \tan^{-1}(7/3) = 66.80°;\ W = 250\ N;$$

$$W_M = (81.5)(9.81) = 800\ N;\ \mu_A = 0.6 \text{ and } \mu_B = 0.$$

Note that we have assumed that $\mu_B = 0$ because a wet vinyl clad surface exhibits very little frictional restraint.

In performing a stability analysis, we assume that the ladder is about to slip (implying impending motion). This assumption is important because it permits us to determine the maximum friction force that can be developed to resist motion (instability).

Let's apply the three equilibrium relations that must be satisfied if the ladder is stable.

$$\Sigma F_y = N_A - W - W_M = 0$$

$$N_A = W + W_M = 250 + 800 = 1050\ N \quad \text{(a)}$$

$$\Sigma M_A = N_B\, h - Ws/2 - 3W_M\, s/4 \quad \text{(b)}$$

$$N_B = (s)(2W + 3W_M)/4h = (3)(500 + 2400)/(4)(7) = 310.7\ N \quad \text{(c)}$$

$$\Sigma F_x = F_{fA} - N_B = 0 \quad \text{(d)}$$

$$F_{fA} = N_B = 310.7\ N \quad \text{(e)}$$

This result indicates that the friction force at the base of the ladder F_{fA} must be 310.7 N to maintain the ladder in a state of stable equilibrium. The maximum friction force that can develop at Point A is given by Eq. (9.3) as:

$$(F_{fA})_{max} = \mu_A\, N_A = (0.6)(1050) = 630\ N \quad \text{(f)}$$

Since $(F_{fA})_{max} = 630\ N \geq F_{fA} = 310.7$ N, the ladder will be stable under the conditions described above. The magnitude of the friction force that may develop at the base of the ladder is larger than that necessary to maintain equilibrium.

EXAMPLE 9.7

Let's consider the ladder shown in Fig. 9.10, and determine its stability if all the conditions are the same as described in Example 9.6 except for the following parameters:

$$h = 6\text{ m};\ s = 5\text{ m};\ \mu_A = 0.4,\ \text{and}\ \mu_B = 0$$

Solution:

Utilizing the FBD presented in Example 9.6, we explore the stability question by writing the equilibrium relations:

$$\Sigma F_y = N_A - W - W_M = 0 \qquad \text{(a)}$$

$$N_A = W + W_M = 250 + 800 = 1050\text{ N}$$

$$\Sigma M_A = N_B\,h - Ws/2 - 3W_M\,s/4$$

$$N = (s)(2W + 3W_M)/(4h) = (5)(500 + 2400)/(4)(6) = 604.2\text{ N} \qquad \text{(b)}$$

$$\Sigma F_x = F_{fA} - N_B = 0 \qquad \text{(c)}$$

Substituting the results from Eq. (b) into Eq. (c) gives:

$$F_{fA} = N_B = 604.2\text{ N} \qquad \text{(d)}$$

Next, we must determine if the friction force $F_{fA} = 604.2$ N at the base of the ladder can be developed. From Eq. (9.3), $(F_{fA})_{max}$ is given by:

$$(F_{fA})_{max} = \mu_A\,N_A = (0.4)(1050) = 420\text{ N} \qquad \text{(e)}$$

Clearly, the ladder is unstable and the base will slide away from the wall since:

$$F_{fA} = 604.2\text{ N} > (F_{fA})_{max} = 420\text{ N} \qquad \text{(f)}$$

EXAMPLE 9.8

Let's consider a slightly more complex question pertaining to the stability of the ladder shown in Fig. 9.10. The new parameters describing the condition of the ladder are:

$$h = 7\text{m};\ s = 4\text{m};\ \mu_A = 0.4;\ \text{and}\ \mu_B = 0.3$$

All of the other parameters are the same as those described in Example 9.6.

Solution:

Since the coefficient of friction at the point of contact of the ladder with the wall is not zero, we must modify the FBD by adding a friction force F_{fB} as shown in Fig. E9.8. Examination of the FBD indicates that we encounter four unknown forces, namely N_A, N_B, F_{fA}, and F_{fB}. However, we only have three equations of equilibrium to apply in our stability analysis. To resolve this difficulty, we assume that the ladder is in a condition of impending motion at the wall (point B), which enables us to use Eq. (9.3) and express F_{fB} in terms of N_B. This assumption reduces the number of unknown forces to three, which is equal to the number of relevant equilibrium relations.

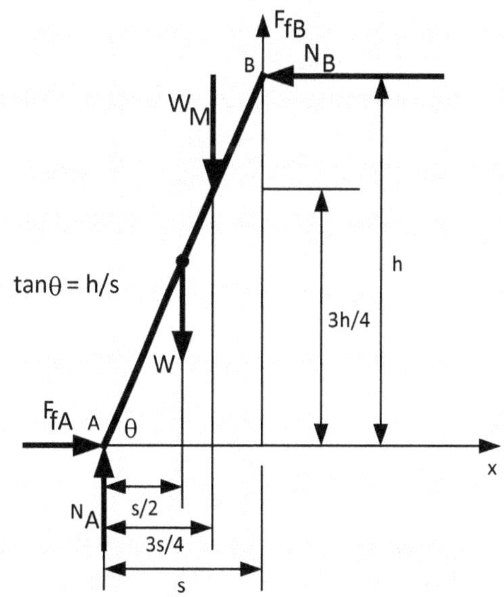

Fig. E9.8

Next we write the equation for the friction force at point B as:

$$F_{fB} = (F_{fB})_{max} = \mu_B N_B \qquad \text{(a)}$$

Applying the equilibrium relations yields:

$$\Sigma M_A = - Ws/2 - 3W_M s/4 + F_{fB} s + N_B h = 0 \qquad \text{(b)}$$

Substituting Eq. (a) into Eq. (b), and solving for N_B gives:

$$N_B = (s/4)(2W + 3W_M)/(\mu_B s + h) \qquad \text{(c)}$$

$$N_B = (4/4)(500 + 2400)/[(0.3)(4) + 7] = 354 \text{ N}$$

$$\Sigma F_y = N_A - W - W_M + \mu_B N_B = 0 \qquad \text{(d)}$$

Solving Eq. (d) for N_A yields:

$$N_A = W + W_M - \mu_B N_B = 250 + 800 - (0.3)(354) = 943.8 \text{ N} \qquad \text{(e)}$$

Next, we employ the final equilibrium relation to determine the friction force at the base of the ladder required to insure stability.

$$\Sigma F_x = F_{fA} - N_B = 0 \qquad \text{(f)}$$

$$F_{fA} = N_B = 354 \text{ N} \qquad \text{(g)}$$

From Eq. (9.3) and the condition of impending motion at point A, we write:

$$(F_{fA})_{max} = \mu_A N_A = (0.4)(943.8) = 377.5 \text{ N} \qquad \text{(h)}$$

Then we compare the maximum friction force possible with the required friction force as indicated below:

$$(F_{fA})_{max} = 377.5 \text{ N} > F_{fA} = 354 \text{ N} \qquad \text{(i)}$$

We conclude that the ladder is stable, but with only a modest margin of safety. If we had assumed impending motion at the floor rather than the wall, the solution would give a negative value for the force F_{fB}. Clearly the wall cannot pull on the ladder the assumption of impending motion at the floor was correct.

9.4.2 Tipping

Let's consider a different type of stability problem that involves sliding objects along some surface. When we apply a force to the object to overcome the frictional forces, the object may slide or it may tip. We seek to determine if the object will slide or tip for a given set of conditions describing the object. We will introduce this topic by considering a rectangular box resting on a horizontal surface as the object as illustrated in Fig. 9.11.

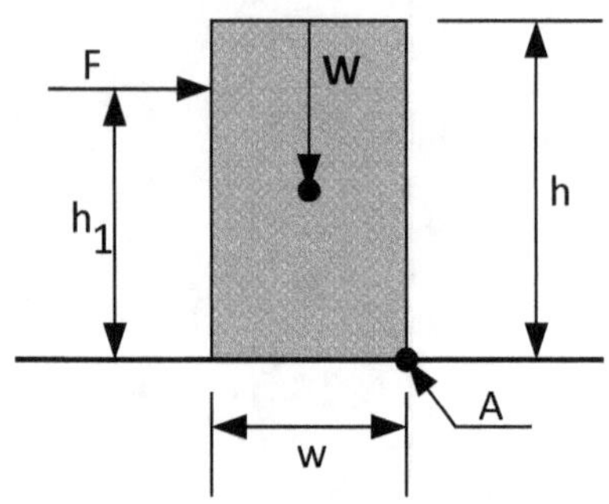

Fig. 9.11 A rectangular box being moved by sliding along a horizontal surface.

EXAMPLE 9.9

Consider a crate that weighs 400 lb resting on a level concrete floor. The coefficient of friction $\mu = 0.4$ between the floor and the box. The dimensions of the quantities defined in Fig. 9.11 are h = 4 ft, h_1 = 3 ft, and w = 1 ft. Determine the force required moving the box, and establish if it will slide or tip when the force F is applied.

Solution:

Begin with a FBD of the crate as shown in Fig. E9.9. Next, apply the equations of equilibrium.

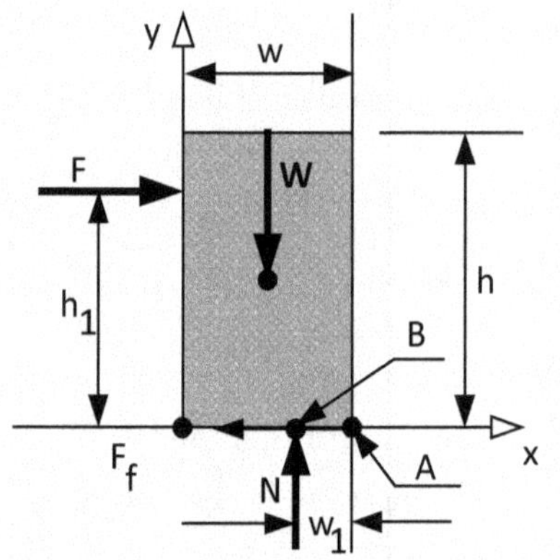

Fig. E9.9

$$\Sigma F_y = N - W = 0 \quad \text{(a)}$$

$$N = W = 400 \text{ lb}$$

$$\Sigma F_x = F - F_f = 0 \quad \text{(b)}$$

$$\Sigma M_B = W[(w/2) - w_1] - Fh_1 = 0 \quad \text{(c)}$$

At this stage, we have solved for one unknown (N); however, three unknowns remain (F, F and w) with only two applicable equations remaining. We must consider each case of impending motion separately to determine if the crate will tip or slide.

First, assume that the crate slides with $F_{Slide} = F_f$. Then the following relation holds:

$$F_{Slide} = F_f = \mu N = (0.4)(400) = 160 \text{ lb} \quad \text{(d)}$$

Now, let's consider the argument for tipping. When the crate begins to tip, the point of application of N (point B) shifts to the right hand edge of the crate (point A). For this situation, $w_1 = 0$ and Eq. (c) becomes:

$$F_{Tip} = (Ww)/(2h) = [(400)(1)]/[(2)(3)] = 66.67 \text{ lb} \quad \text{(e)}$$

Clearly, the crate will tip before it slides because $F_{Tip} < F_{Slide}$.

EXAMPLE 9.10

The crate in Example 9.9 tipped because we applied the force F near the top of the crate. For the same conditions, determine the largest value of h_l for applying the force F, which will result in the crate sliding without tipping.

Solution:

To solve this problem, recognize that tipping becomes possible when the point of application of the normal force N shifts to point A in Fig. 9.11, or when $w_l = 0$. Recall Eq. (c) from Example 9.9, let $w_l = 0$, and then solve for h_l.

$$h_l = (W/F)[(w/2 - w_l) \quad (a)$$

$$h_l = (400/160)[1/2 - 0] = 1.25 \text{ ft} \quad (b)$$

For h < 1.25 ft, the crate will slide.

9.4.3 Rolling

Let's next consider the role of friction in determining if a cylindrical object rolls or slides. To illustrate this situation, refer to the roller positioned adjacent to a step as shown in Fig. 9.12.

As we apply a force F to the cable wound about the smaller cylinder, the larger roller contacts the step. If the friction force developed at point A is sufficiently large, the cylinders will roll up the vertical face, and the assembly will negotiate the step. On the other hand, if the friction force is insufficient, the cylinders will slip and remain at the bottom of the step.

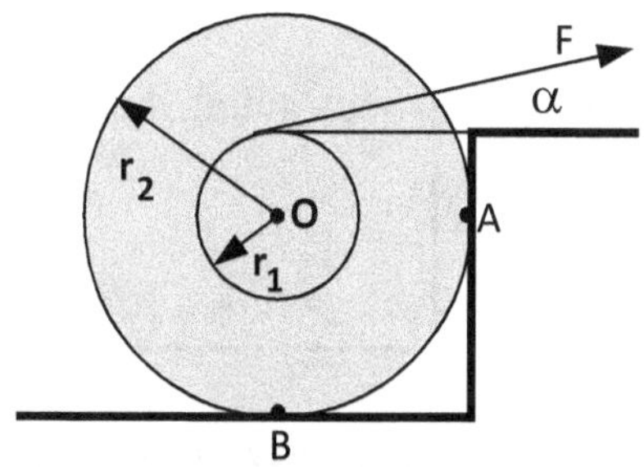

Fig. 9.12 Rolling a cylinder over a step.

Rolling friction can be both beneficial and detrimental in the same application. For example, when designing wheels for a vehicle it is necessary to maintain enough rolling friction to provide the thrust forces that are generated during acceleration of the vehicle. At the same time, the rolling friction also reduces the fuel efficiency of the car by resisting the forward motion of the vehicle when it travels at a constant velocity.

EXAMPLE 9.11

Consider a cylindrical arrangement similar to that shown in Fig. 9.12 with a force F applied at an angle $\alpha = 30°$. If the coefficient of friction at the contact points A and B is $\mu_A = \mu_B = 0.4$, determine if the cylinder will roll up and over the step or slip and remain in the corner as the force is increased. Also determine the force required to roll a cylindrical arrangement that weighs 600 N up and over the step. Note that $r_1 = 100$ mm, and $r_2 = 300$ mm.

Solution: In constructing the FBD shown in Fig. E9.11, we have assumed that the cylindrical assembly has been lifted from the horizontal surface ($N_B = 0$) and is beginning to roll up the step. To resolve the roll or slip question, write the equilibrium relations to obtain:

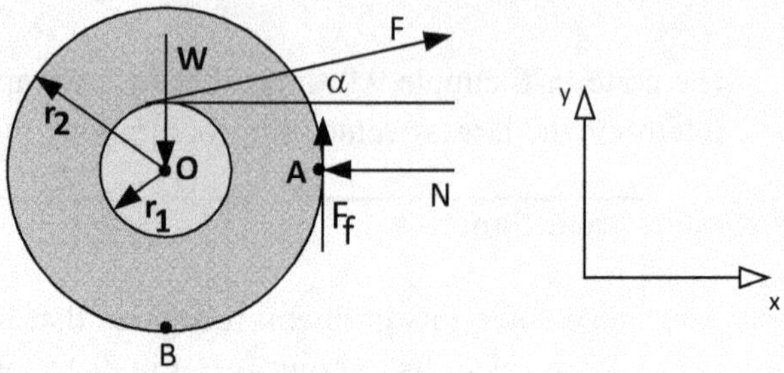

Fig. E9.11

$$\Sigma M_O = F_f\, r_2 - F\, r_1 \cos\alpha = 0 \quad \text{(a)}$$

$$F_f = (r_1/r_2)F \cos\alpha = (1/3)F(0.8660) = 0.2886\ F \quad \text{(b)}$$

$$\Sigma F_x = F \cos\alpha - N = 0; \;\Rightarrow\quad N = 0.866\ F \quad \text{(c)}$$

The maximum possible value of the friction force at point A is given by Eq. (9.3) as:

$$(F_f)_{max} = \mu N = (0.4)(0.8660)F = 0.346\ F \quad \text{(d)}$$

A comparison of the results of Eq. (b) and Eq. (d) shows that:

$$(F_f)_{max} > F_f \quad \text{(e)}$$

Clearly, the cylinder will roll up and over the step.

Finally, let's determine the force F required.

$$\Sigma F_y = F \sin\alpha - W + F_f = 0 \quad \text{(f)}$$

Substituting the results of Eq. (b) into Eq. (f) yields:

$$(0.5)F + 0.2886\ F = 600 \;\Rightarrow\quad F = 760.8\ N$$

Since $(F_f)_{max} > (F_f)$, the cylindrical assembly will roll up the step without slipping.

9.5 FRICTION EFFECTS ON WEDGES

A wedge is classified as a simple machine because it can be used to magnify a force if the effects of friction are not too large. The lifting of a block with a weight W by a tapered wedge is illustrated in Fig. 9.13. For the wedge to act as a machine, the force F used to drive the wedge under the block must be less that the weight W.

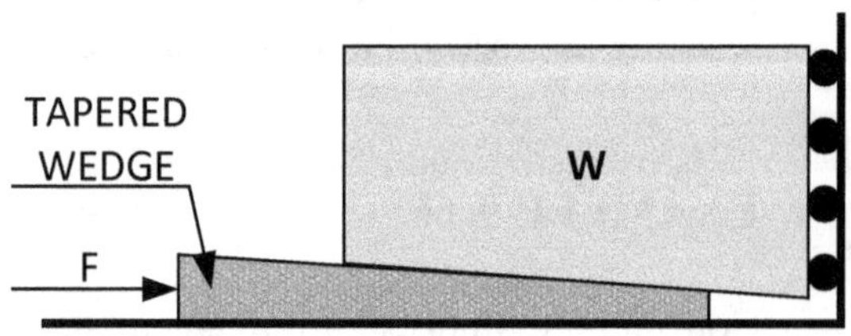

Fig. 9.13 A tapered wedge acts as a machine to lift a weight.

The use of a wedge as a simple machine has three modes of operation:

- Lifting
- Holding
- Lowering

Friction plays an important role in all three of these operations.

9.5.1 Lifting with a Wedge

Let's first consider the lifting operation and explore the role of friction. As usual, we begin with the FBD diagram of the block and the wedge shown in Fig. 9.14. Note that the force F, in Fig. 9.14, acts to the right in lifting the block. Also observe that we have assumed that the rollers eliminated the friction force F_{f3} on the vertical surface of the block.

Let's apply the equations of equilibrium to the block and then to the wedge to determine the relation between the applied force F and the weight W.

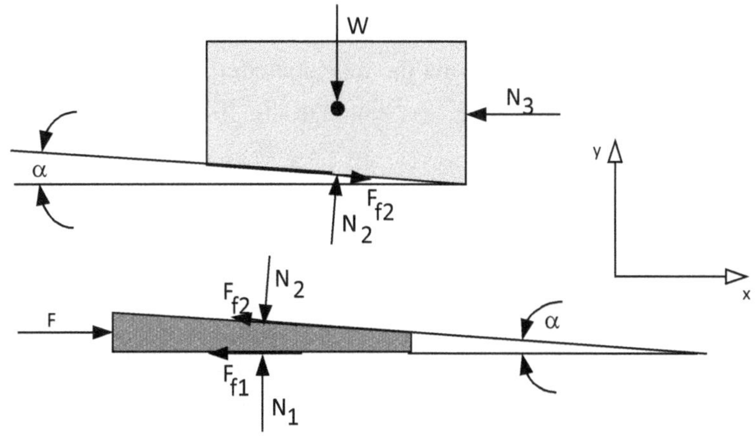

Fig. 9.14 FBD of a block and wedge in a lifting operation.

For the block, we write:

$$\Sigma F_y = N_2 \cos\alpha - F_{f2} \sin\alpha - W = 0 \quad \text{(a)}$$

Since motion occurs to move the wedge, we may write:

$$F_{f1} = \mu N_1 \quad \text{and} \quad F_{f2} = \mu N_2 \quad \text{(b)}$$

Combining Eq. (a) and Eq. (b) and solving for N_2 yields:

$$N_2 = W/(\cos\alpha - \mu \sin\alpha) \quad \text{(c)}$$

For the wedge one of the equilibrium equations gives:

$$\Sigma F_y = N_1 + F_{f2} \sin\alpha - N_2 \cos\alpha = 0 \quad \text{(d)}$$

Combining Eqs. (b), (c) and (d) yields:

$$N_1 = W \quad \text{(e)}$$

Applying the final relevant equilibrium relation for the wedge gives:

$$\Sigma F_x = F - F_{f1} - F_{f2} \cos\alpha - N_2 \sin\alpha = 0$$

$$F = \mu W + \mu N_2 \cos\alpha + N_2 \sin\alpha \quad \text{(f)}$$

$$F = W\left[\frac{2\mu + (1 - \mu^2)\tan\alpha}{1 - \mu\tan\alpha}\right] = CW \quad (9.10)$$

where C is the coefficient for the wedge acting as a simple machine. It is given by:

$$C = \frac{2\mu + (1 - \mu^2) \tan \alpha}{1 - \mu \tan \alpha} \tag{9.11}$$

For the wedge to be effective as a machine and lift a heavy weight W with a small force F, we seek to minimize the coefficient C. This minimization is accomplished by reducing the friction coefficient and by using small wedge angles. A graph showing the machine coefficient C as a function of the wedge angle for several different values of the friction coefficient is presented in Fig. 9.15.

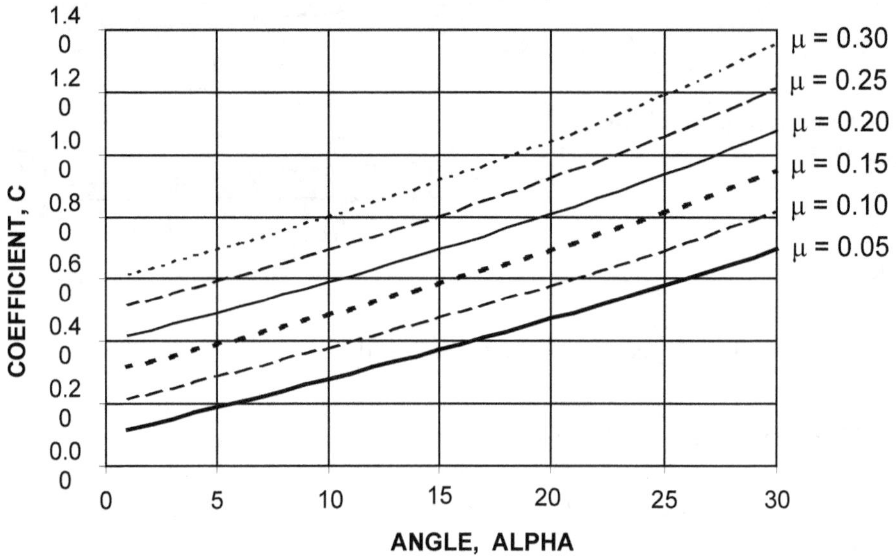

Fig. 9.15 Small forces lift heavy weights with wedges when α is small and μ is low.

EXAMPLE 9.12

You are attempting to level a heavy machine bed with a mass of 20,000 kg. To lift one end of the bed, you insert a wedge with an angle $\alpha = 10°$. If the coefficient of friction between the wedge and the machine bed is $\mu = 0.2$, determine the force necessary to lift the machine bed by 120 mm.

Solution:

The weight of the machine bed is given by:

$$W = (9.81)(20{,}000) = 196.2 \text{ kN}$$

Recall Eq. (9.10)

$$F = C\,W = (0.590)(196.2) = 115.8 \text{ kN}$$

where C = 0.590 is determined from the graph in Fig. 9.15 or by evaluating Eq. (9.11).

Suppose that the hydraulic jack that is available only has a capacity of 10 tons (89.0 kN). Clearly this jack cannot drive the wedge with the conditions listed above. What can you do to resolve this situation?

There are three possible solutions.

- Purchase a larger capacity jack. This option is expensive and will delay the project.
- Find a wedge with a smaller angle. This is a valid solution if a smaller angle wedge exists. If not ordering a new wedge will cost money and delay the project.
- Lubricate the wedge with a heavy grease which will reduce the friction coefficient to $\mu = 0.11$. This is the best solution because grease is readily available at a low cost.

If we apply a coating of grease to both sides of the wedge, the friction coefficient is reduced to $\mu = 0.11$. Then for a wedge angle of $\alpha = 10°$, the machine coefficient is C = 0.401. Then:

$$F = C\,W = (0.401)(196.2\ \text{kN}) = 78.68\ \text{kN}$$

This result for F is equivalent to 8.84 tons so the 10-ton capacity jack, which is available, provides sufficient force to drive the wedge.

9.5.2 Holding with a Wedge

Suppose the wedge is driven to raise the weight as illustrated in Fig. 9.13. The question is then whether or not the wedge will stay in place after the force F is removed. The weight acting downward is resisted by a normal force that has a component tending to push the wedge outward. Frictional forces must be sufficiently large to prevent the wedge from being forced out. If the friction forces are insufficient, the weight will fall, and the situation could be dangerous.

Let's explore the conditions under which the wedge will hold the load in the absence of the driving force F. As usual we begin with a free body diagram that is shown in Fig. 9.16.

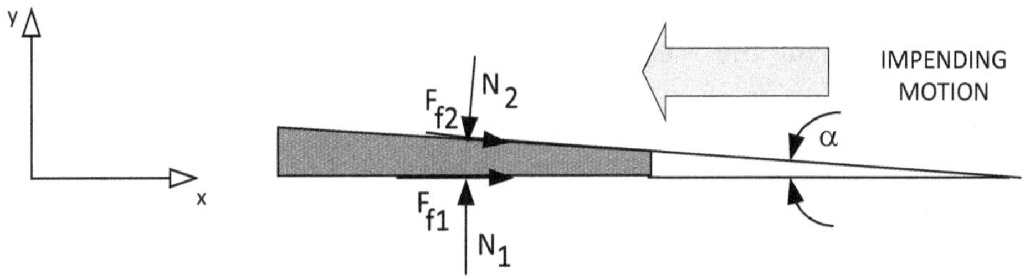

Fig. 9.16 FBD under holding conditions with no force F applied to the wedge.

If we consider the wedge and the weight coupled together as a single body, the equilibrium relation $\Sigma F_y = 0$ yields:

$$\Sigma F_y = 0 \quad \Rightarrow \quad W = N_1 \qquad \text{(a)}$$

For the wedge, we may write:

$$\Sigma F_y = W - N_2 \cos\alpha - F_{f2} \sin\alpha = 0 \qquad \text{(b)}$$

Since the wedge slipping implies impending motion, we may use Eq. (9.3) and write:

$$F_{f1} = \mu\, N_1 \quad \Rightarrow \quad F_{f2} = \mu\, N_2 \qquad \text{(c)}$$

$$\Sigma F_x = N_2 \sin\alpha - F_{f2} \cos\alpha - F_{f1} = 0 \qquad \text{(d)}$$

Combining Eqs. (a) through (d) leads to:

$$\tan\alpha = 2\mu/(1 - \mu^2) \qquad (9.12)$$

This relation obviously needs some interpretation because the quantities α and μ are usually specified conditions and are not physically related. We have developed a relation in Eq. (9.12) that describes the condition for the wedge to begin to slip from under the weight. For stability or instability of the weight we explore the inequalities on either side of Eq. (9.12), and write:

$$\tan \alpha < 2\mu/(1 - \mu^2) \qquad (9.13)$$

This relation defines the locking angle for a wedge; the wedge remains fixed after it is driven if this equation is satisfied.

On the other hand the wedge will slip, and the weight will fall (unstable) if:

$$\tan \alpha > 2\mu/(1 - \mu^2) \qquad (9.14)$$

EXAMPLE 9.13

A wedge with an angle α = 20° is used to lift a 1200 lb weight. The wedge is lubricated to facilitate the lifting operation, and its coefficient of friction with the mating surfaces is only 0.12. Determine if the wedge will slip when the driving force is removed.

Solution: Check for stability by employing Eq. (9.13).

$$\tan \alpha < 2\mu/(1 - \mu^2) \quad \Rightarrow \quad \text{stability condition}$$

$$\tan 20° = 0.3640 \text{ and } 2\mu/(1 - \mu^2) = (2)(0.12)/[1 - (0.12)^2] = 0.2435$$

Since tan 20° = 0.3640 > $[2\mu/(1 - \mu^{2)}]$ = 0.2436, Eq. (9.14) is satisfied and the instability condition exists. The wedge will slip, and the weight will fall.

9.5.3 Removing the Wedge

Suppose that the wedge is a tapered key that is used to fix a gear to a shaft. In some instances, the gear must be removed from the shaft to perform maintenance on the gear, shaft, or bearings. In these cases, we must apply a force to pull the wedge from the keyway. Let's determine the force F_R required to pull the key. As usual, we begin with the free body diagram presented in Fig. 9.17 to model the physical situation and to guide us in writing the appropriate relations.

Fig. 9.17 FBD of a wedge removed from a keyway.

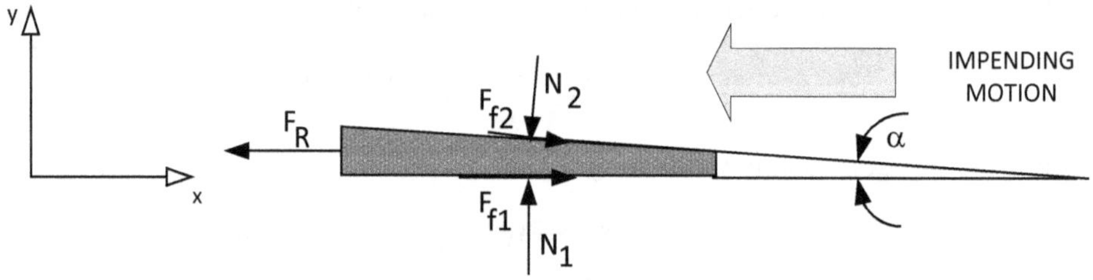

Let's assume the pressure of the gear acting on the wedge-like key produces a force W that is equal to force N_1 as defined in Fig. 9.17.

$$W = N_1 \qquad (a)$$

For the wedge, we may write:

$$\Sigma F_y = W - N_2 \cos \alpha - F_{f2} \sin \alpha \qquad (b)$$

Since pulling the wedge implies impending motion, we may use Eq. (9.3) and write:

$$F_{f1} = \mu N_1 \quad \Rightarrow \quad F_{f2} = \mu N_2 \tag{c}$$

$$\Sigma F_x = N_2 \sin \alpha - F_{f2} \cos \alpha - F_{f1} + F_R = 0 \tag{d}$$

Combining Eqs. (a) through (d) leads to:

$$F_R = W\,[2\mu - (1 - \mu^2)\tan \alpha] / [1 + \mu \tan \alpha] \tag{9.15}$$

EXAMPLE 9.14

Determine the force required to pull a key if the coefficient of friction $\mu = 0.8$ and the angle α of the key is 6°. The pressure between the gear and the key produces a force W of 2500 lbs.

Solution:

We employ Eq. (9.15) and write:

$$F_R = W\,[2\mu - (1 - \mu^2)\tan \alpha] / [1 + \mu \tan \alpha]$$

$$F_R = 2500\,[2(0.8) - (1 - 0.64)\tan(6°)] / [1 + (0.8)\tan(6°)]$$

$$F_R = 3602 \text{ lb}$$

The result may seem high, but the large coefficient of friction may imply that the key has corroded in the keyway.

9.6 FRICTION EFFECTS ON SCREWS

Screws are used as fasteners to hold two or more components together in an assembly. They are also used in jacks to lift bodies. In all probability, you have a screw jack in your automobile that is used in lifting one corner of a car in changing a flat tire. Screws are also used in positioning and measuring devices. For example, a screw is used to move the carriage that supports the cutting tool on a lathe, and a screw is used to control the position of the head on a micrometer.

A screw is very much like a wedge or an inclined plane that is wrapped around a cylinder as shown in Fig. 9.18. The inclined plane is wrapped around the cylinder in the from of a helix with an angle α to form the thread. A nut, not shown in Fig. 9.18, is fitted over the thread. As the thread is rotated, the nut is driven in one direction or the other depending on the direction of rotation. In one rotation, the nut is driven a distance of p, which is the pitch of the thread. In driving the carriage of a lathe, the thread rotates in a fixed position and the nut drives the carriage along the length of the lathe bed. In a typical jack, the nut is fixed to the support structure for the jack and the end (or both ends) of the screw moves to lift an auto.

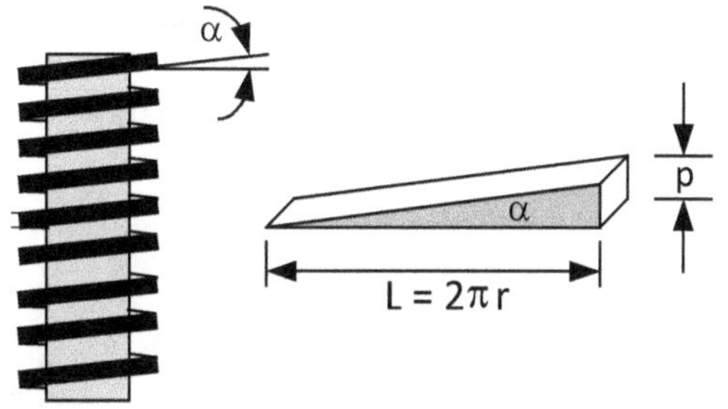

Fig. 9.18 Representing a screw thread with an inclined plane.

We apply a moment to turn the screw in either lifting or lowering a load. The frictional forces and the magnitude of the load determine the magnitude of the moment applied to the screw. Let's determine the magnitude of the moment required to lift a load.

9.6.1 Lifting a Load with a Screw

Suppose we are lifting a load of W with a screw jack. In designing the jack, we must determine the moment required to lift the load. Let's begin this analysis by modeling the thread on the cylinder with an inclined plane and the nut with a block as shown in the FBD of Fig. 9.19.

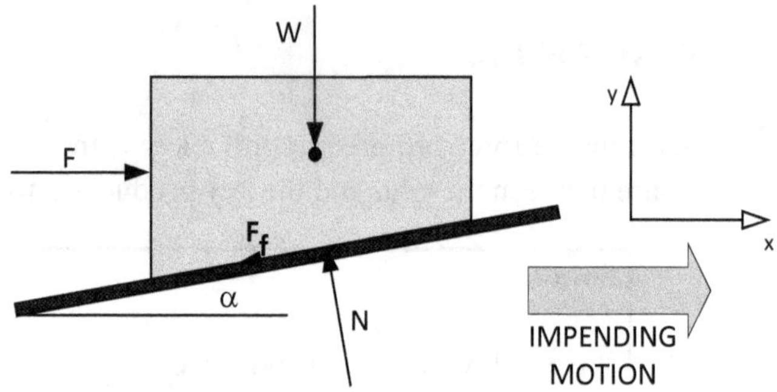

Fig. 9.19 FBD of a model of a screw and nut for lifting a load W.

We apply the equilibrium relations to determine the moment required to lift the load.

$$\Sigma F_y = N \cos\alpha - F_f \sin\alpha - W = 0 \qquad \text{(a)}$$

$$N = W/(\cos\alpha - \mu \sin\alpha)$$

$$\Sigma F_x = F - N \sin\alpha - F_f \cos\alpha = 0 \qquad \text{(b)}$$

$$F = N(\sin\alpha + \mu \cos\alpha)$$

$$F = W(\tan\alpha + \mu)/(1 - \mu \tan\alpha) \qquad \text{(c)}$$

However, the relation of the moment M to the force F is:

$$M = F\,r \qquad (9.16)$$

where r is the radius of the screw.

Then from Eqs (c) and (9.16), we write:

$$M = W\,r(\tan\alpha + \mu)/(1 - \mu \tan\alpha) \qquad (9.17)$$

9.6.2 Lowering a Load with a Screw

The analysis of the screw employed to lower a load is almost identical to that shown above. The only differences are the direction of the frictional force F_f and the applied force F as shown in Fig. 9.20.

We apply the equilibrium relations to determine the moment required to lower the load.

$$\Sigma F_y = N \cos\alpha + F_f \sin\alpha - W = 0 \qquad \text{(a)}$$

$$N = W/(\cos\alpha + \mu \sin\alpha)$$

$$\Sigma F_x = -F - N \sin\alpha + F_f \cos\alpha = 0 \qquad \text{(b)}$$

$$F = N(-\sin\alpha + \mu \cos\alpha)$$

$$F = W (\mu - \tan \alpha)/(1 + \mu \tan \alpha) \quad \text{(c)}$$

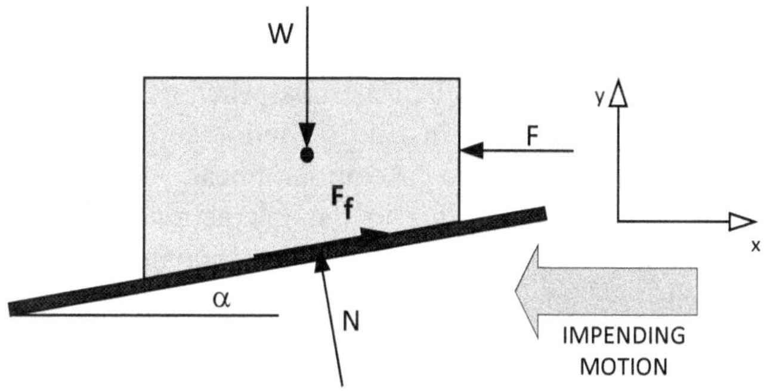

Fig. 9.20 The FBD of a screw and nut used in lowering the load W.

Then by substituting Eq. (c) into (9.16), we obtain:

$$M = W\, r (\mu - \tan \alpha)/(1 + \mu \tan \alpha) \quad (9.18)$$

9.6.3 Holding a Load with a Screw

The analysis of the screw employed to hold the load in a fixed position is also very similar to the two previous studies. The friction force F_f opposes the impending motion and the applied force F is absent. We show the FBD for the new model, which represents lowering the load in Fig. 9.21.

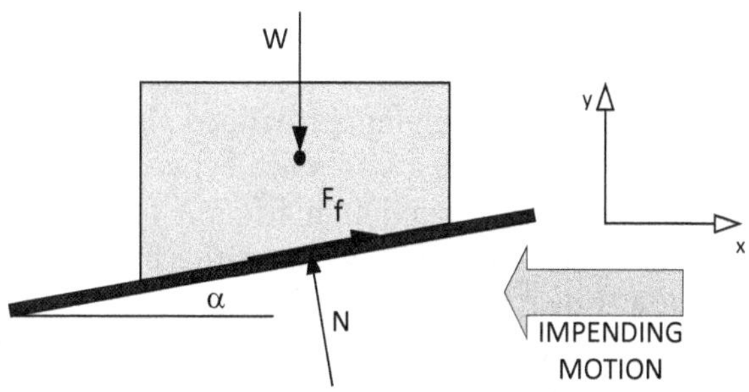

Fig. 9.21 The FBD for a screw holding a load without an applied moment (torque).

Apply the equilibrium relations to determine if the jack will be stable without a moment applied to the screw. This is an important consideration as self locking screws are essential for the safe operation of a lifting device. No one wants to have the load crashing down as soon as the moment is released from the screw.

Let's begin by writing the equations of equilibrium for the block shown in Fig. 9.21.

$$\Sigma F_x = - N \sin \alpha + F_f \cos \alpha = 0 \quad \text{(a)}$$

We again assume that motion is impeding and substitute Eq. (9.3) into Eq. (a) to obtain:

$$\mu = \sin \alpha / \cos \alpha = \tan \alpha \quad \text{(b)}$$

Equation (b) describes the equilibrium condition for the screw with the moment M = 0. However, to insure stability, we write the inequality condition as:

$$\tan \alpha < \mu \quad (9.19)$$

Substituting Eq. (9.6) into Eq. (9.19) gives:

$$\tan \alpha < \tan \phi \quad (9.20)$$

Clearly, Eq. (9.20) shows that the helix angle α of the screw must be less than the friction angle ϕ for the screw to be self-locking.

Another method for showing the self-locking condition for a screw is to set M = 0 in Eq. (9.18). Then it is clear that $(\mu - \tan \alpha) = 0$ with M = 0 and the conditions for stability given in Eq. (9.19) are obtained.

9.7 SUMMARY

Friction occurs when we have two bodies in contact and attempt to move one relative to another. Although motion may not occur, friction forces develop that must be considered in any equilibrium analysis of the individual bodies. Friction effects may be beneficial when friction forces are essential in maintaining stability and in providing traction. However, in many instances friction forces are detrimental because they result in losses of power, energy, and efficiency, and in the generation of heat.

Friction is due to the interactions of the asperities on both contacting surfaces. With dry friction, the frictional forces increase until they become sufficiently large to break the bonds between adjoining asperities and/or shear the interlocking asperities. Lubricating the surfaces markedly reduces the frictional forces because the lubricants physically separate the asperities. The small friction forces developed under fully lubricated conditions are due to shearing of the fluid lubricant.

The magnitude of the friction force is variable. It increases up to some maximum value to prevent relative motion between the two contacting surfaces. The maximum friction force is related to the shearing force necessary to destroy the bonds between the contacting asperities. It is approximated by:

$$(F_f)_{max} = \mu N \qquad (9.3)$$

A table providing approximate values for the coefficients of friction, μ for different material pairs is presented; however, a typical coefficient exhibits a large variation; therefore it is advisable to measure μ with a simple experiment involving an inclined plane as illustrated in Fig. 9.7.

There is a difference between the static and dynamic coefficients of friction. The static coefficient of friction describes the maximum friction force that may be developed before motion occurs. The dynamic coefficient describes a constant value of the friction force after motion begins. The dynamic coefficient is usually 20 to 25% lower than the static coefficient.

We often describe the coefficient of friction with either the friction angle or the angle of repose. These angles are equal. The angle of repose θ_s is measured with an inclined plane, and the friction angle ϕ is the angle between the resultant force and the friction force as shown in Fig. 9.9. The tangent of both of these angles is equal to the coefficient of friction.

$$\mu = \tan\theta_s = \tan\phi \qquad (9.5 - 9.7)$$

Many examples are presented to illustrate procedures useful in solving "friction" problems including:

- Construction of a FBD to model the physical situation and to show all of the forces acting on the body and their dimensions.
- Application of the appropriate equations of equilibrium.
- Execution of the mathematics required for the solution of these equations.

Friction dependent stability was demonstrated by considering a number of different examples involving the stability of a ladder. We assume a condition of impending motion and derive an inequality relation that indicates stability or the lack thereof:

$$F_f \leq (F_f)_{max} \quad \Rightarrow \quad \text{for stability.}$$

$$F_f \geq (F_f)_{max} \quad \Rightarrow \quad \text{for instability.}$$

Examples showing the role of friction in tipping versus sliding were considered. The determination involves locating the point of application of the normal reactive force. If this normal force is positioned to the right of point A in Fig. 9.11, the crate tips rather than slides.

One example was introduced to show the conditions for rolling or slipping. The determination of whether a cylinder rolls or slips is identical to the stability condition for the ladder.

$F_f \leq (F_f)_{max}$ ⇒ the rolling condition.

$F_f \geq (F_f)_{max}$ ⇒ the slipping condition.

We examined the role of friction relative to both the wedge and the screw, which are simple machines. In both components, three situations must be analyzed—lifting, holding and lowering. All of the equations necessary to determine the driving force and or moment (torque) required to perform these three functions for both the wedge and the screw are summarized below:

The force required to lift a weight W with a wedge is:

$$F = W\left[\frac{2\mu+(1-\mu^2)\tan\alpha}{1-\mu\tan\alpha}\right] = CW \qquad (9.10)$$

$$C = \frac{2\mu+(1-\mu^2)\tan\alpha}{1-\mu\tan\alpha} \qquad (9.11)$$

With no force applied a wedge will hold if the inequality below is satisfied:

$$\tan\alpha < 2\mu/(1-\mu^2) \qquad (9.13)$$

On the other hand the wedge will slip, and the weight will fall (unstable) if:

$$\tan\alpha > 2\mu/(1-\mu^2) \qquad (9.14)$$

The force F_R required to remove a wedge is given by:

$$F_R = W\,[2\mu-(1-\mu^2)\tan\alpha]\,/\,[1+\mu\tan\alpha] \qquad (9.15)$$

The moment that must be applied to a screw to lift a load W is given by:

$$M = W\,r\,(\tan\alpha+\mu)/(1-\mu\tan\alpha) \qquad (9.17)$$

The moment that must be applied to a screw to lower a load W is given by:

$$M = W\,r\,(\mu-\tan\alpha)/(1+\mu\tan\alpha) \qquad (9.18)$$

Finally the condition for stability of a screw in holding a load is given by:

$$\tan\alpha < \tan\phi = \mu \qquad (9.20)$$

The conditions for holding a wedge or screw in place when the load or moment is removed are particularly important, because if a screw or wedge does not hold when the load or moment is removed, the machine is unstable. The load will fall and injury to persons or property may result.

REFERENCES

1. Ling, F. F. and C. H. T. Pan, *Approaches to Modeling Friction and Wear*, Springer-Verlag, New York, 1986.

PROBLEMS

9.1 Write an engineering brief describing the factors affecting friction forces acting between two bodies.

9.2 Write an engineering brief explaining the reasons for the difference between static and dynamic friction coefficients.

9.3 Derive Eq. (9.2).

9.4 Explain the conditions that must exist before you apply the equation $(F_f)_{max} = \mu N$.

9.5 Suppose you develop a fixture consisting of an inclined plane and a block of some solid material to measure the material's coefficient of friction. You measure and record the angle of the inclined plane at the instant motion is initiated. Determine the average value of the friction coefficient from the five separate tests and its range using the data given in the table below.

Problem No.	Angle	Angle	Angle	Angle	Angle
9.5a	22.0°	22.2°	24.3°	21.8°	22.5°
9.5b	24.0°	23.2°	26.3°	24.8°	23.5°
9.5c	32.0°	31.2°	33.3°	32.8°	32.7°
9.5d	37.0°	38.2°	37.9°	39.0°	39.2°

9.6 Derive Eq. (9.7).

9.7 Determine the force F required to move the sled, illustrated in the figure to the right, if the coefficient of friction between the sled and the surface is $\mu = 0.28$.

9.8 Determine the force required to slide a box, illustrated in the figure to the left, across a level floor if the force is applied at an angle $\alpha = 10°$. The coefficient of friction between the box and the floor is $\mu = 0.25$.

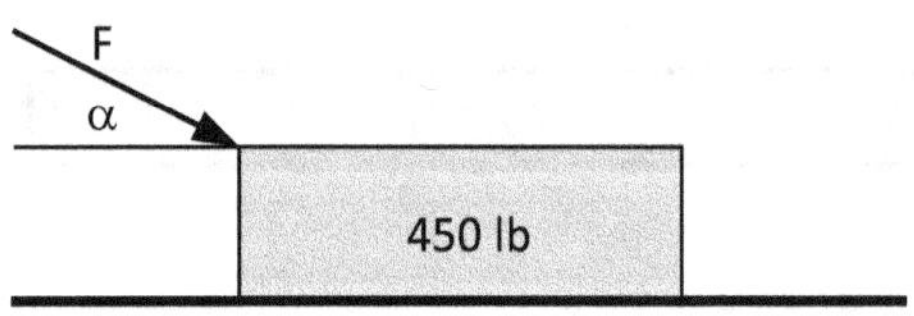

9.9 A worker attempts to slide a wooden crate along a level concrete floor in a warehouse as shown in the figure to the right. The weight of the crate is 80 lb and the worker weighs 120 lb. The coefficient of friction between the crate and the floor $\mu_c = 0.55$ and between the worker and the floor $\mu_w = 0.70$. Note that the crate is short and the worker must lean over to apply a force at an angle of 30° relative to the floor. Is it possible for the worker to move the crate? Justify your answer by showing all the relevant calculations.

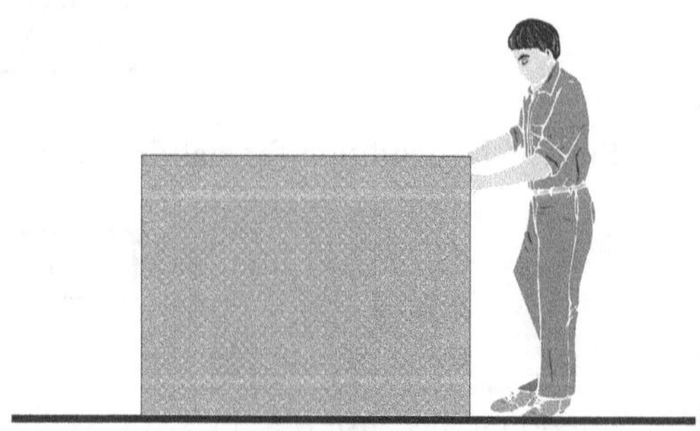

9.10 It snowed on the day of your final examination for this course, and you find the roads covered with a coating of ice and snow. In your attempt to drive to the University, you cannot negotiate a hill with a grade of 4°. Estimate the coefficient of friction between the road and the tire for your instructor, as you explain the reason for missing the final examination.

9.11 Two blocks rest on a surface as illustrated in the figure to the right. If $F_2 = 400$ lb, determine the maximum force F_1 that can be applied before one of the blocks moves. The coefficient of friction between the two blocks is 0.28 and between block B and the surface it is 0.22. Block A weighs 1,200 lb and Block B weighs 2,000 lb.

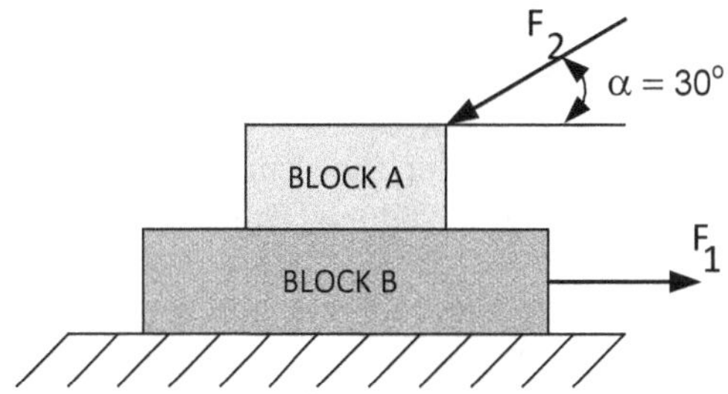

9.12 Determine if the ladder, shown in the figure to the left, is stable or not. The man on the ladder weighs 145 lb and the ladder weighs 52 lb. The coefficient of friction at the base of the ladder is 0.30 and the coefficient of friction between the ladder and the wall is 0.15.

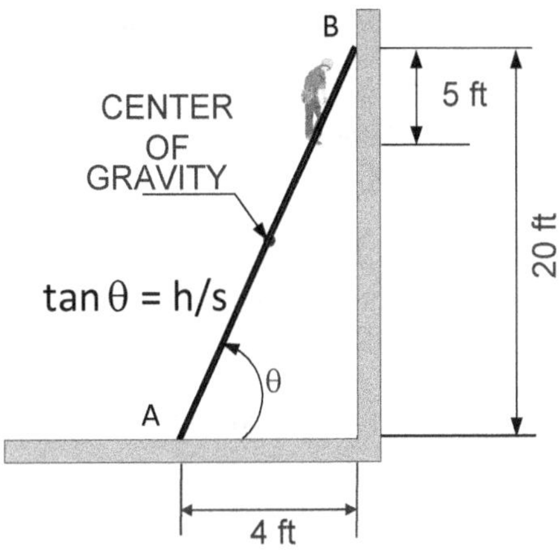

9.13 Determine the force required to move the crate, shown in the figure to the right, and indicate whether it will slide or tip. The crate weighs 720N and the coefficient of friction between it and the floor is $\mu = 0.42$.

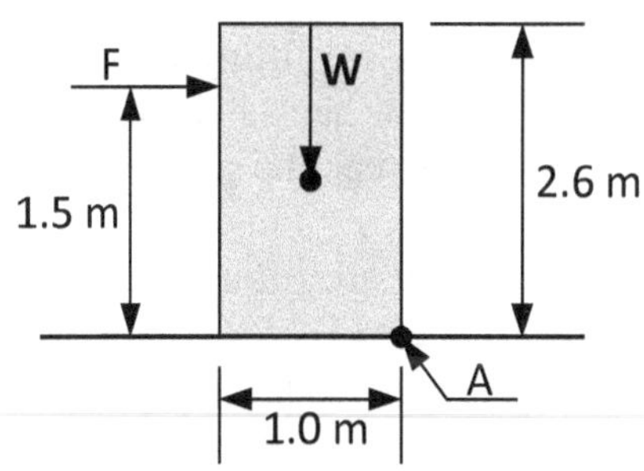

9.14 A cylindrical assembly is pulled by a force F to roll up and over a step as shown in the figure to the left. The cylinder weighs 420 N, and the coefficient of friction between the cylinder and the surfaces at points A and B is 0.43. The radii $r_1 = 0.5$ m and $r_2 = 1.0$ m. Determine if the cylinder will roll up and over the step or slip and remain in the corner. Show all of your work and justify your answer.

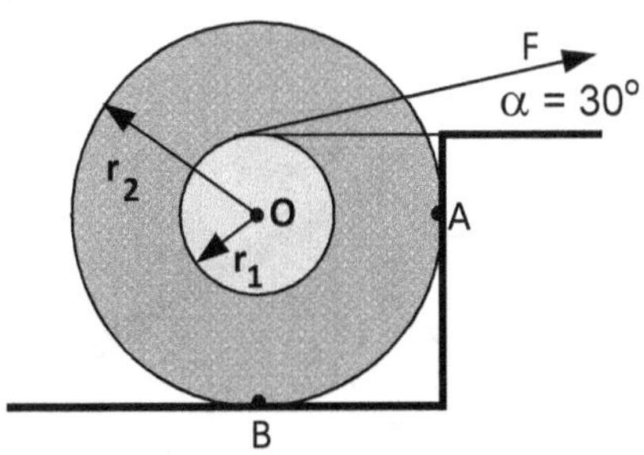

9.15 A wedge inserted at point A is used to lift one end of the machine bed shown in the figure to the right. If the wedge has an included angle $\alpha = 12°$, determine the force necessary to drive the wedge if the coefficient of friction with both surfaces $\mu = 0.35$, and the weight per unit area of the machine bed is 4 kN/m^2.

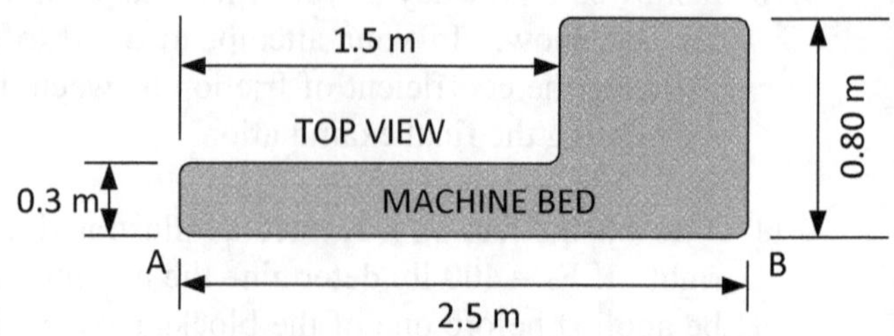

9.16 Reconsider Problem 9.15; however, the wedge is to lift the machine bed from point B and the angle of the wedge is reduced from 12° to 6°.

9.17 Determine the machine coefficient for the wedge, which measures its effectiveness as a simple lifting machine for the results obtained in Problem 9.15. Discuss the parameters leading to the best machine coefficient. Recall, work is defined as **F • d**.

9.18 If the machine bed in Problem 9.15 is lifted through a distance of 60 mm, determine the work expended in driving the wedge. Also determine the work accomplished in lifting the weight of the machine bed. Recall that work is defined as **F • d**.

9.19 If the machine bed in Problem 9.16 is lifted through a distance of 30 mm, determine the work expended in driving the wedge. Also determine the work accomplished in lifting the weight of the machine bed. Recall that work is defined as **F • d**.

9.20 For the conditions described in Problem 9.15, determine if the wedge will hold when the driving force is removed. Also determine the force F_R required to remove the wedge.

9.21 For the conditions described in Problem 9.16, determine if the wedge will hold when the driving force is removed. Also determine the force F_R required to remove the wedge.

9.22 You are lifting one corner of a car with a screw jack. The car weighs 3800lb, and the screw has a helix angle of 4°. The screw is lubricated with a thick, heavy grease and the coefficient of friction between the screw and the nut is $\mu = 0.11$. Determine the torque necessary to turn the screw.

9.23 You are lifting one corner of a heavy plate of steel with a screw jack. The plate weighs 10,000 N, and the screw has a helix angle of 3°. The screw is lubricated with a thick, heavy grease and the coefficient of friction between the screw and the nut is $\mu = 0.11$. Determine the torque necessary to turn the screw.

9.24 You are lifting one corner of a truck with a screw jack. The truck weighs 25 kN, and the screw has a helix angle of 4.5°. The screw is lubricated with a heavy grease and the coefficient of friction between the screw and the nut is $\mu = 0.11$. Determine the torque necessary to turn the screw.

9.25 Determine the locking angle α for helical threads if the coefficient of friction is 0.11.

9.26 Write an engineering brief describing the dangers associated with using a screw with a helical angle of more than 6° in the design of a jack to be used in lifting heavy weights. The screw is made from steel and the nut is made from brass.

9.27 A barrel with a diameter of 0.7 m and weighing 500 N is being rolled up an inclined ramp as shown in the figure to the right. The ramp weighs 250 N. If the coefficient of friction is 0.24 between all contacting surfaces, determine the distance x through which the barrel can be rolled before slippage occurs.

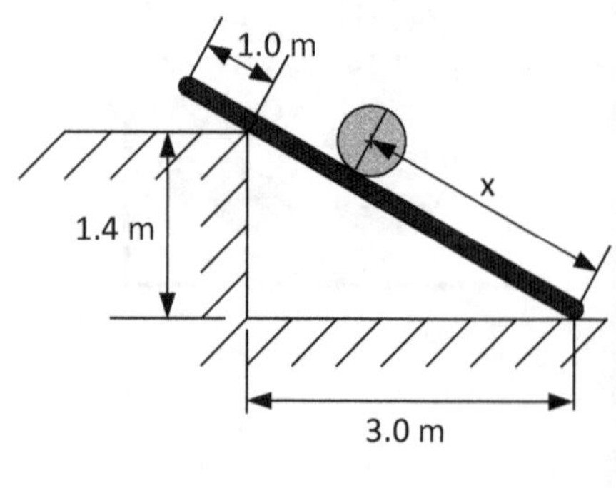

9.28 Block B with a mass m_B = 19 kg sits on top of a larger block A with a mass m_A = 32 kg. A horizontal force F is applied to block B as shown in the figure to the left. If the coefficient of friction between all of the surfaces is 0.28, determine the force F required to initiate motion. Describe the motion that occurs.

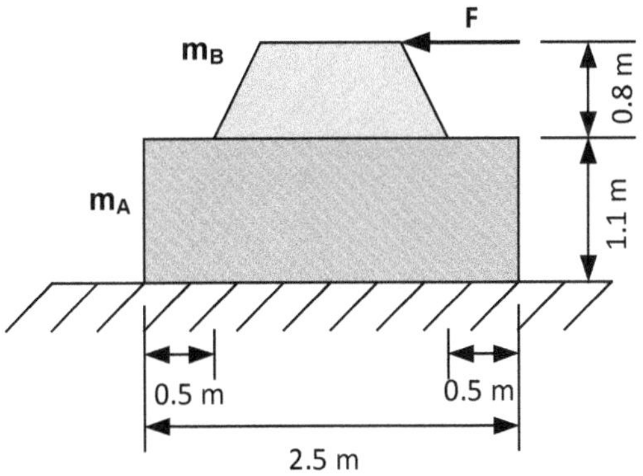

9.29 The links in the toggle mechanism, depicted in the figure to the right, are sufficiently light in weight to be neglected. The pins are frictionless. The coefficient of friction between the block and the surface is 0.34, and the block weighs 3.8 kN. Determine the angle θ that permits the application of the largest force F for which motion of the block does not occur.

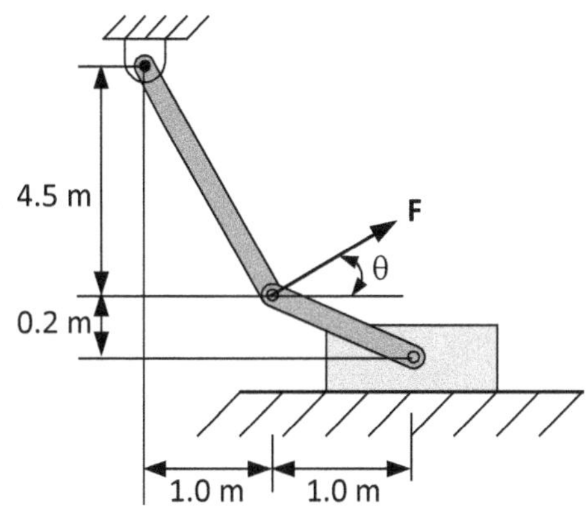

9.30 A carton weighing W_A = 85 lb has tipped and is resting against another carton weighing W_B = 125 as shown in the figure to the left. The coefficient of friction between the floor and the first (tipped) carton is μ_A = 0.45 and the coefficient of friction between the floor and the second (upright) carton is μ_B = 0.40. Determine if the boxes are in equilibrium in the position shown. Neglect the effect of friction between the cartons at the contact point.

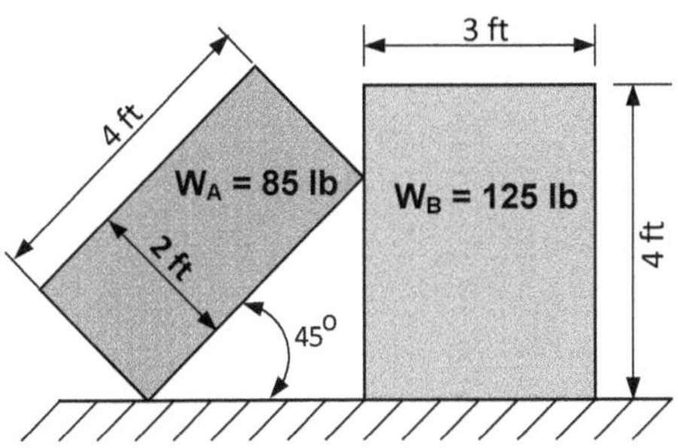

9.31 A crate with a weight W = 2.5 kN is connected to two opposing weights W_1 and W_2 by thin cables that pass over frictionless pulleys as shown in the figure to the right. Determine the minimum and maximum values for the weight W_1 so that motion does not occur. The coefficient of friction between the crate and the floor is 0.28.

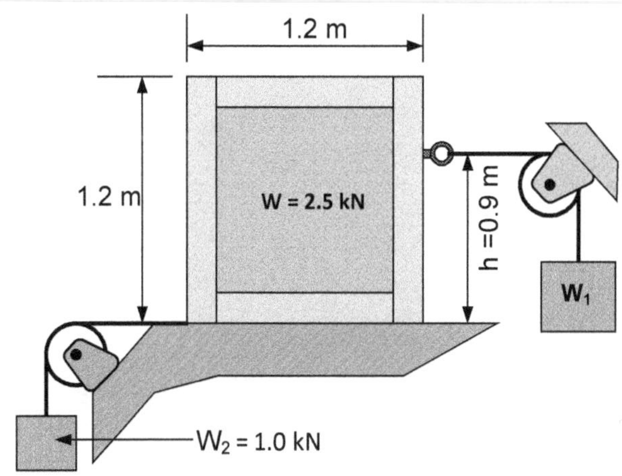

9.32 Consider the crate, shown in the figure to the right and above that is subjected to the conditions given in Problem 9.31 except for the height h. Determine the product of h times the weight W_1 that produces motion by slipping and tipping simultaneously.

CHAPTER 10

STRESSES IN BEAMS

10.1 PURE BENDING OF SYMMETRIC BEAMS

A beam is similar to a rod because both have the same geometry—long and slender. However, the beam is loaded differently. A rod is subjected to tensile or compressive forces applied in the axial direction. A beam is subjected to transverse forces applied in a perpendicular direction to the longitudinal axis of the long, thin member. The difference in the loading is depicted in Fig. 10.1.

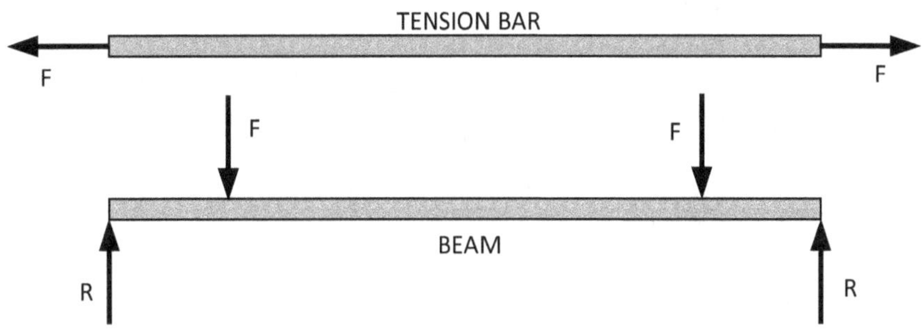

Fig. 10.1 Both the rod and beam are fabricated from long, thin members, but the direction of the applied external loads is different.

Now that we understand the geometry and the loading of a beam, let's consider what the expression "pure" bending implies. A beam is in "pure" bending if it is subjected to a constant moment over its length. When a straight beam is subject to a constant moment, as illustrated in Fig. 10.2, it deforms into a circular arc with a radius of curvature ρ.

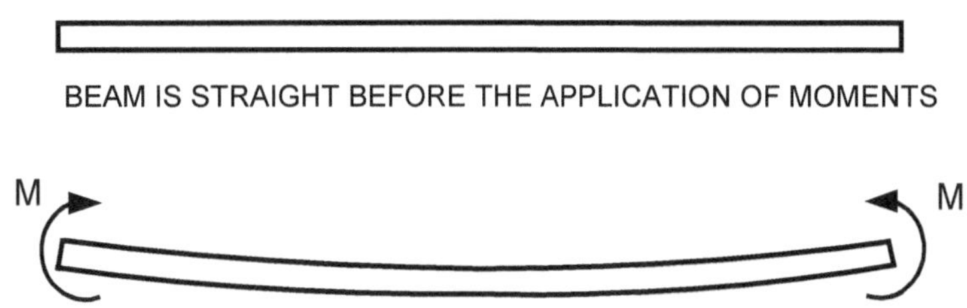

Fig. 10.2 A beam subjected to external moments M deforms into a circular arc with a radius ρ.

We limit our analysis to beams with symmetric cross sections. Several examples of symmetric cross sections are shown in Fig. 10.3. Clearly, the symmetry is only about the vertical, or y-axis. The theory for beams in bending accommodates the lack of symmetry with respect to the horizontal or z-axis.

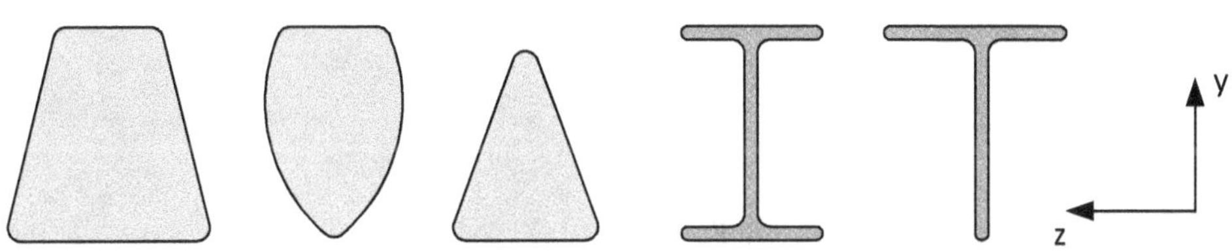

Fig. 10.3 Cross sectional areas of beams exhibiting symmetry about the vertical y-axis.

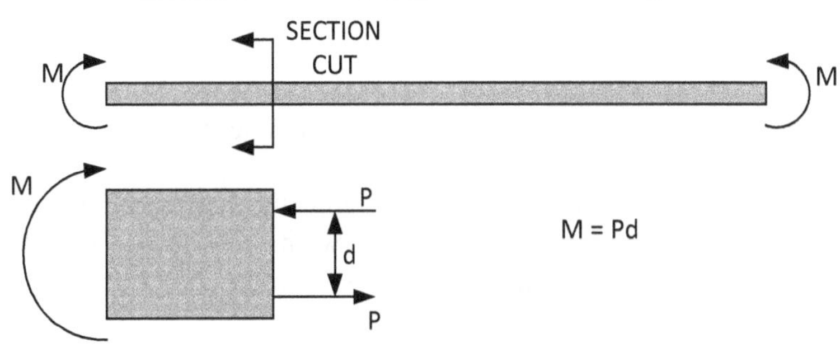

Fig. 10.4 The left-hand portion of the beam showing a couple (P)(d) acting on the area exposed by a section cut.

10.1.1 External Moments Produced by Internal Forces

Suppose we consider a beam with moments applied at its ends as shown in Fig. 10.2. Clearly, the entire beam is in equilibrium because the equilibrium equation $\Sigma M = 0$ is satisfied. The clockwise moment acting on the left end of the beam is equilibrated by the counterclockwise moment on the right end. Next, let's make a section cut at some position x along the length of the beam, as indicated in Fig. 10.4, and consider the left portion of the beam. In constructing a FBD, we must apply internal forces that will maintain equilibrium of this portion of the beam. Equilibrium is achieved by placing two equal and opposite internal forces P that act on the surface of the cross sectional area exposed by the section cut. Since the internal forces are equal and opposite, the equilibrium equation $\Sigma F_x = 0$ is satisfied. The equation of equilibrium for the moments is satisfied if:

$$\Sigma M = M - P\,d = 0$$

$$P\,d = M \qquad \text{(a)}$$

EXAMPLE 10.1

The concept of a pure moment acting on a beam may be difficult to comprehend. How can a pure moment be generated in a beam?

Solution: We produce moments by applying forces to a member or component in such a way that the forces create a moment. For instance, we pull with a force F on a wrench with a length d to create a moment (torque) on a bolt to screw it into a joint by overcoming resistance to the bolt's rotational motion. A method for producing a pure moment on a beam is to apply two forces to a simply supported beam in what is referred to as a **four-point bend configuration**, as illustrated in Fig. E10.1.

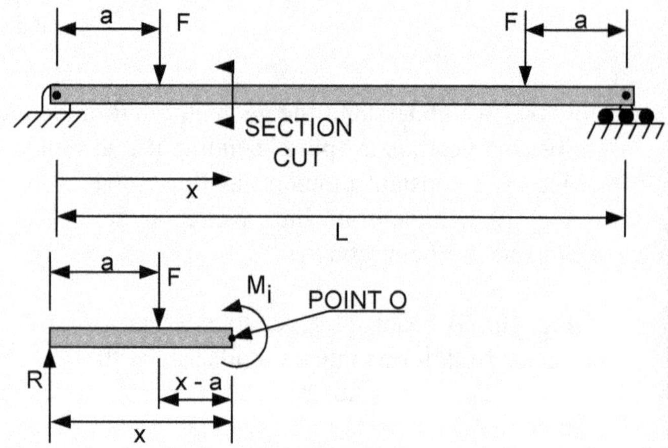

Fig. E10.1

To show that this arrangement of externally applied forces produce a pure moment, we make a section cut at position x, remove the left end of the beam, and construct a FBD. Let's apply the equilibrium equations to the FBD and write:

$$\Sigma F_y = R - F = 0, \qquad \Rightarrow \qquad R = F \qquad \text{(a)}$$

Next, consider moments about point O, and write:

$$\Sigma M_O = -Rx + F(x - a) + M_i = 0 \qquad \text{for } a < x < (L - a)$$

$$M_i = Fa \qquad \text{(b)}$$

where the subscript i refers to an internal moment produced by internal forces.

It is evident that only the moment M_i acts on the central segment of the beam. For the segment of the beam defined by $a < x < (L - a)$, we have a state of pure bending. No shear force V exists on the section and the net force in the x direction is zero.

10.1.2 Stress Distributions

We have used the concept of equilibrium to establish that equal and opposite internal forces P act on the surface exposed by the section cut. The forces produce an internal couple, $M_i = (P)(d)$ that counteracts the externally applied moment M. However, at this stage in the analysis, we cannot determine the individual values of either P or d, although we know the direction of the forces. We need the concept of **stress distributions** to continue the analysis.

The internal forces P are due to normal stresses σ_x, which act over the surface of the exposed cross sectional area. Clearly, the upper most force in Fig. 10.4 is compressive since it presses against the surface exposed by the section cut. Similarly, the lower force is tensile. We are concerned with the distribution of the normal stresses that produce these two forces. Several possible stress distributions are suggested in Fig. 10.5.

The distributions of normal stresses presented in Fig. 10.5 produce equal and opposite internal forces P separated by some distance d. Accordingly, each distribution satisfies the equilibrium equations. However, which, if any, of the three possibilities are correct? We have exhausted the equilibrium principle, and to continue it is necessary to consider the deformation of the beam. To begin this second phase of the analysis, let's make an assumption about the geometry of the beam in its deformed state.

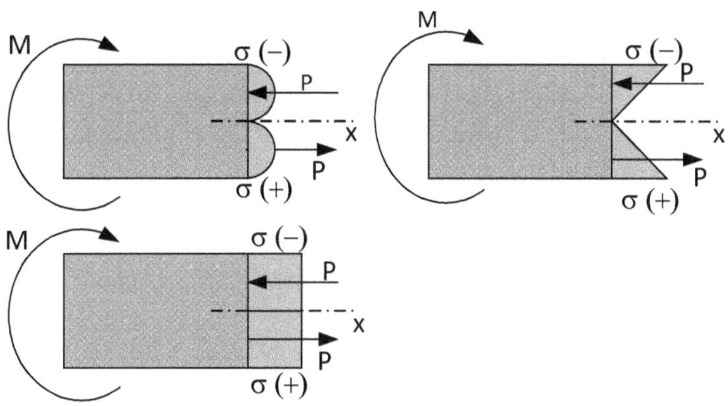

Fig. 10.5 Possible distributions of stress σ_x acting on an internal surface producing the couple M = P d.

10.2 DEFORMATION OF A BEAM IN BENDING

When a beam is subjected to a moment as shown in Fig. 10.2, it bends into a circular arc. Parallel vertical lines scribed on the sides of the beam remain straight after the beam undergoes deformation from a straight member to one with curvature. We show a segment of the beam before and after deformation into a circular arc in Fig. 10.6.

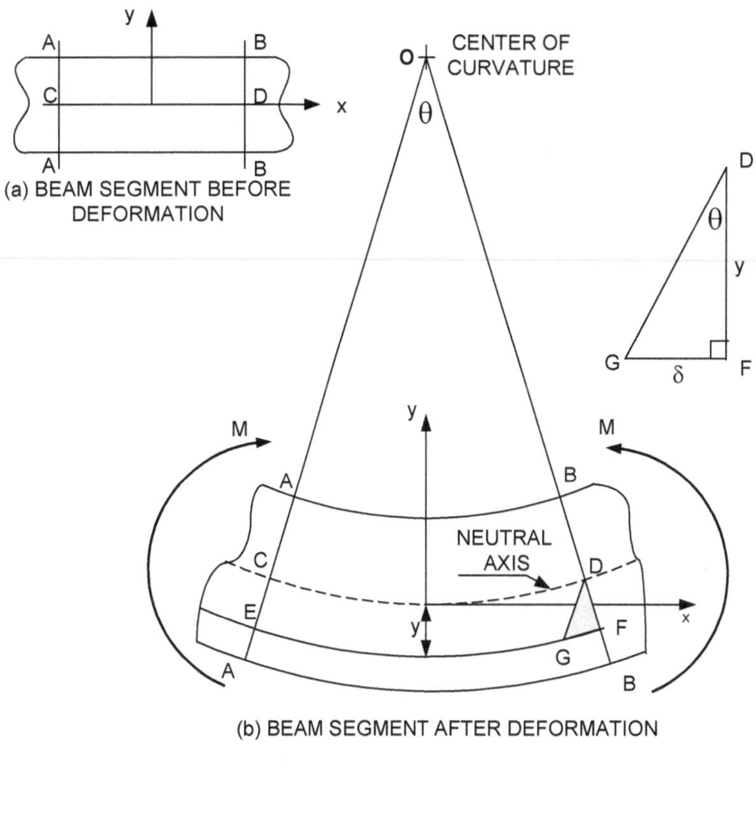

Fig. 10.6 Plane sections A – A and B – B of the beam remain plane after deformation.

The geometry of the deformed segment of the beam is extremely important. The circular arc is centered at point O. The lines A – A and B – B are straight and radial. They have each rotated about axes perpendicular to the plane of bending, and their extensions intersect at the center of curvature to form an angle θ. In addition to the lines A – A and B – B remaining straight, the transverse sections of the beam, originally plane, remain plane after bending.

The dotted line C – D is exactly the same length as it was before deformation. This line defines what is known as the **neutral axis**. The radius of curvature ρ of the circular arc is the distance from the center of curvature O to the neutral axis (the arc C – D). Lines (arcs) above the neutral axis, become shorter after deformation and lines (arcs) below the neutral axis, such as E – F, become longer. These changes in the length of lines (arcs) above and below the neutral axis result in longitudinal strain. The strains are compressive (negative) above the neutral axis when the beam has a positive curvature[1] as illustrated in Fig. 10.6. Similarly, the strains are tensile (positive) below the neutral axis when the beam has a positive curvature.

10.2.1 Strains in Beams with a Constant Radius of Curvature

Let's determine the magnitude of the strains by examining the changes in the length of line E - F. Clearly, the line E - F is lengthened in the deformation process. The distance GF in Fig. 10.6b shows the increase in the length, δ of line E - F. Note that we constructed line D - G parallel to line A – A to produce the shaded triangle in Fig. 10.6b. Recall Eq. 4.3 and write:

$$\varepsilon = \delta/L = GF/EG \qquad \text{(a)}$$

The similarity of triangle COD and the triangle GDF, shown in the inset of Fig. 10.6b, enables us to write:

$$CD = EG = \rho\theta; \; GF = -\,y\theta \qquad \text{(b)}$$

$$GF/EG = -\,y\theta/\rho\theta = -\,y/\rho \qquad \text{(c)}$$

Combining Eqs. (a) and (c) yields:

$$\varepsilon = -\,y/\rho \qquad (10.1)$$

where y is the distance from the neutral axis (CD) to the fiber in question (EF).

Note, the strain ε is a linear function of the position y. The strain is a minimum (zero) at the neutral axis and a maximum when y locates either the top or bottom fiber of the beam. If we consider a beam with a rectangular cross section and define its height as h, then $y_{max} = -\,h/2$ and from Eq. (10.1), it is evident that:

$$\varepsilon_{max} = h/(2\rho) \qquad (10.2)$$

The sign of the maximum strain is positive (tensile) for fibers along the bottom surface of the beam if the curvature is positive.

EXAMPLE 10.2

A thin strip of metal 1.00 mm thick and 10.0 mm wide is wrapped about a mandrel with a radius of 200 mm. Determine the maximum bending strain produced in the strip. Assume that the neutral axis is in the center of the cross sectional area as shown below:

---

[1] We define positive curvature of the beam in two ways. The curvature is positive if the beam bends so it will hold water (concave upward). The curvature is positive if the internal moment acting on the section under consideration is positive.

Fig. E10.2

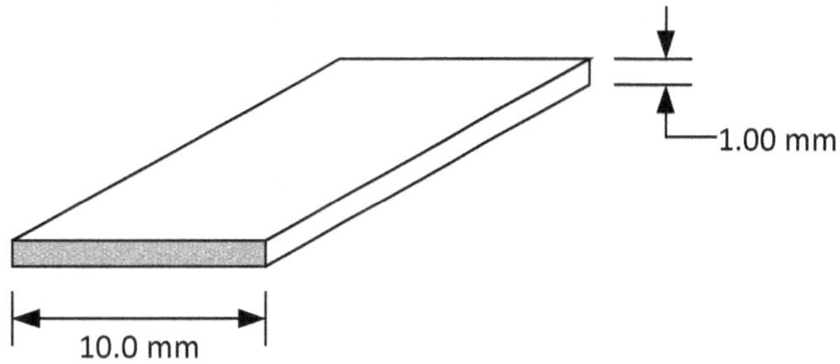

Solution:

From Eq. (10.2), we write:

$$\varepsilon_{max} = y_{max}/\rho = h/2\rho = 1.0/[(2)(200.5)] = 0.002494$$

While this is a very simple example, you should note that we have determined a strain for a beam in bending without employing the elastic constants for the material. Indeed, we did not even specify the metal employed in fabricating the strip in the problem statement. Strain is a geometric quantity (a change in length over original length). If we know the geometry before and after deformation, strain can be determined without knowledge of either the loads or the material properties.

10.2.2 Stresses in Beams

It is easy to convert the strains in the beam to stresses if we recall Hooke's Law—Eq. (4.5). The longitudinal fibers in the beam are subjected to a uniaxial state of stress with fibers on the convex side of the neutral axis in tension and those on the concave side in compression. By combining Eqs. (10.1) and (4.5), we obtain:

$$\sigma = E\varepsilon = -\,yE/\rho \qquad (10.3)$$

Examination of Eq. (10.3) indicates that the normal stress σ is linearly distributed with respect to y; a minimum of zero at the neutral axis where y = 0, and a maximum at the outer fibers in the beam where y is a maximum.

The fact that the stresses are not uniformly distributed has significant implications regarding the design of the cross section of the beam. We will discuss this important design consideration in much more detail in Section 10.3.

10.2.3 Locating the Neutral Axis

The results in Eq. (10.3) may be used to calculate the stress on a beam if we know the material from which it is fabricated, the location of the neutral axis, and the radius of curvature. However, in most cases we have knowledge of the applied external forces—not the radius of curvature. Also, we have not yet established the location of the neutral axis relative to the cross section of the beam.

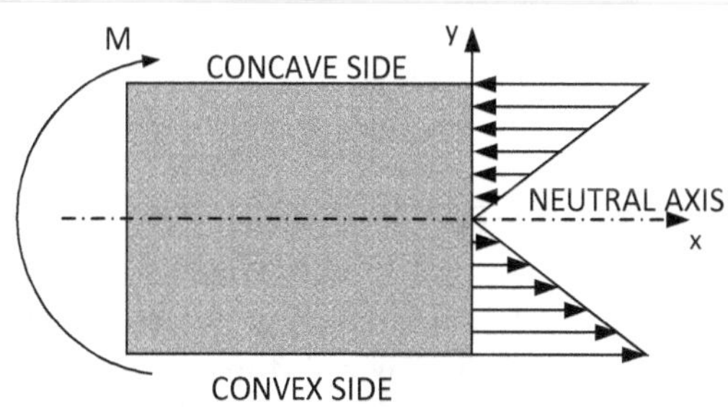

Fig. 10.7 The normal stresses σ are linear in y for a beam in bending.

To establish the position of the neutral axis of the beam, let's consider the portion of the beam shown in Fig. 10.7 as a free body. Writing the equilibrium relation $\Sigma F_x = 0$ and using Eq. (10.3) gives;

$$\Sigma F_x = \int \sigma\, dA = -\int (Ey/\rho)\, dA = -\,(E/\rho) \int y\, dA = 0$$

where A is the cross sectional area over which the stress σ is distributed.

Because (E/ρ) cannot be zero when the beam is deformed, it is evident that:

$$\int y\, dA = Q_x = 0 \qquad (10.4)$$

where Q_x is the first moment of the area A of the cross section of the beam.

For equilibrium of the beam in the x direction to be satisfied, it is necessary that the position of the neutral axis be adjusted so that $Q_x = 0$. If the neutral axis passes through the centroid of the area of the transverse section of the beam, Eq. (10.4) is satisfied. For common cross section shapes such as circular, rectangular, I, H, wide-flange, etc., the location of the centroid is at the intersection of the two axes of symmetry. Procedures for locating the centroid of areas without two axes of symmetry are presented in Appendix C.

10.2.4 The Relationship between Moment and Curvature

Now that a procedure has been presented for locating the neutral axis, it is necessary to develop an approach for determining the radius of curvature. The stresses σ are distributed so as to produce a moment M as shown in Fig. 10.7. We will use this fact in the following derivation. The moment of the force acting on an element dA, shown in Fig. 10.8, with respect to the neutral axis is:

$$dM = -\sigma\, dA\, (y) = -(-Ey/\rho)(dA)(y) = +(Ey/\rho)(dA)(y) \qquad (a)$$

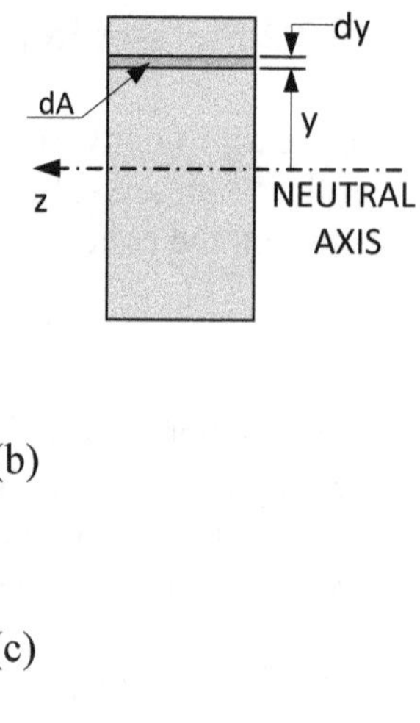

Fig. 10.8 Cross sectional area of a rectangular beam with elemental area dA.

Integrate Eq. (a) to obtain the internal moment M_i:

$$M_i = \int (Ey^2/\rho)\, dA = (E/\rho)\int y^2\, dA \qquad (b)$$

Recall that the second moment[2] of the area A relative to the neutral axis I_z is given by:

$$I_z = \int y^2\, dA \qquad (c)$$

Combining Eqs. (b) and (c), and using the equilibrium relation $\Sigma M = 0$ gives:

$$M_i = (EI_z/\rho) = M \qquad (10.5)$$

$$\kappa = 1/\rho = M/(EI_z) \qquad (10.6)$$

where $\kappa = 1/\rho$ is the curvature of the beam and M is the moment.

From Eq. (10.6), it is evident that the curvature κ is directly proportional to the moment M and inversely proportional with the flexural rigidity (EI_z) of the beam. We have dropped the subscripts on the moments since it is understood that the internal moment produces the stress σ.

Finally, by substituting Eq. (10.6) into Eq. (10.3) the relationship between the applied moment and the internal stress distribution can be obtained:

$$\sigma = -My/I_z \qquad (10.7)$$

---

[2] The second moment of the area is often called the moment of inertia. Methods for determining this quantity for different shaped cross sections are presented in Appendix C.

In Eq. (10.7), M is positive when the beam is deflected so that the convex side is on the bottom and y is positive in the upward direction. This relation is considered to be one of the most important of the many equations employed by engineers when designing structures that contain beams, and as such it will become essential to you in the design of the structure involved in your design project.

10.2.5 Stresses in Beams with Rectangular Cross Sections

Consider a beam with a rectangular cross section as defined in Fig. 10.9. For a rectangular cross section, the neutral axis coincides with the horizontal axis of symmetry:

$$y_{top} = h/2; \qquad y_{bottom} = -h/2 \qquad (a)$$

Substituting Eq. (a) into Eq. (10.7) yields:

$$\sigma_{max} = Mh/(2I_z) \quad \text{and} \quad \sigma_{min} = -Mh/(2I_z) \qquad (10.8)$$

The tensile stress is a maximum when y is equal to – h/2, and the compressive stress is a maximum when y is equal to h/2. The tensile and compressive stresses are equal because the neutral axis for the rectangular section divides the section into two equal parts.

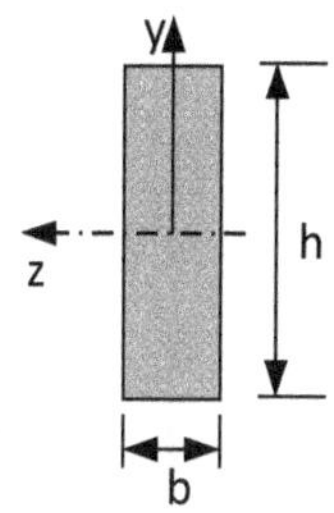

Fig. 10.9 Rectangular cross section of a beam with height h and width b.

Let's define the **section modulus**, Z, for a symmetric cross section of arbitrary geometry as:

$$Z = I_z / c \qquad (10.9a)$$

where c is the distance from the centroid to an outer fiber of the beam. For a rectangular cross section, c = h/2 and Eq. (10.9a) becomes:

$$Z = 2I_z/h \qquad (10.9b)$$

Then, we may express the maximum and minimum normal stress σ as:

$$\sigma_{max} = M/Z \quad \text{and} \quad \sigma_{min} = -M/Z \qquad (10.10)$$

For a rectangular cross section, the second moment of the area I_z and the section modulus are given by:

$$I_z = bh^3/12 \quad \text{and} \quad Z = bh^2/6 \qquad (10.11)$$

We now have the relations necessary to determine the stresses in a beam subjected to internal moments that are constant along its length. Let's consider two examples to illustrate the procedure for determining the stresses in beams with rectangular cross sections.

EXAMPLE 10.3

Consider a beam with a rectangular section with h = 4 in. and b = 0.75 in. Determine the margin of safety for the beam if it is loaded with a moment of 6,000 ft-lb and fabricated from steel with a yield strength of 42,000 psi.

Solution: Let's substitute Eq. (10.11) into Eq. (10.8) to determine the maximum stress.

$$\sigma_{max} = Mh/(2I_z) = 12\ Mh/(2bh^3) = 6M/(bh^2) \qquad (10.12)$$

$$\sigma_{max} = (6)(6000)(12)/[(0.75)(4)^2] = 36,000 \text{ psi}$$

Finally, the safety factor and the margin of safety are given by:

$$SF = S_y/\sigma = 42,000/36,000 = 1.167; \qquad MOS = SF - 1 = 16.7\%$$

The margin of safety for this beam is extremely small and careful consideration should be given to reducing either the applied moment or to increasing the size of the section.

EXAMPLE 10.4

Consider a beam with a rectangular section with h = 300 mm and b = 100 mm. Determine the largest moment that can be applied to the beam if it is fabricated from steel with a yield strength of 300 MPa. The beam is to have a safety factor of 3.0.

Solution: Determine the allowable design stress by considering the safety factor and the yield strength:

$$\sigma = S_y/SF = 300/3.0 = 100 \text{ MPa} \qquad \text{(a)}$$

Rearrange Eq. (10.12) to solve for the moment as:

$$M = \sigma(bh^2)/6 = (100)(100)(300)^2/6 = 150 \times 10^6 \text{ N-mm} = 150 \text{ kN-m} \qquad \text{(b)}$$

In this solution, we used the fact that 1MPa = 1 N/mm^2. Alternatively, we could have obtained the same result by expressing the stress in Pa and all lengths in m.

10.3 PROPERTIES OF CROSS SECTIONS

If we examine the stress distribution shown in Fig. 10.7, it is clear that the beam is not an efficient structural component. Only the top and the bottom fibers of the beam are stressed to the maximum. The interior region of the beam is under utilized. Indeed, near the neutral axis the stresses approach zero. A beam with a rectangular cross section is extremely inefficient because the average magnitude of stress over the section is only $\sigma_{max}/2$. The closer the average magnitude of stress is to the maximum stress, the more efficient will be the use of the cross sectional area of the beam.

To improve the effectiveness of beams, we removed the under-stressed, central-region of the cross section to produce I beams and WF (wide-flange) shapes. A typical example of a more efficient section is shown in Fig. 10.10.

For a beam section with a specified height h, we judge its structural effectiveness **SE** by the ratio of its section modulus to its cross sectional area.

$$\mathbf{SE} = Z/A \qquad (10.13)$$

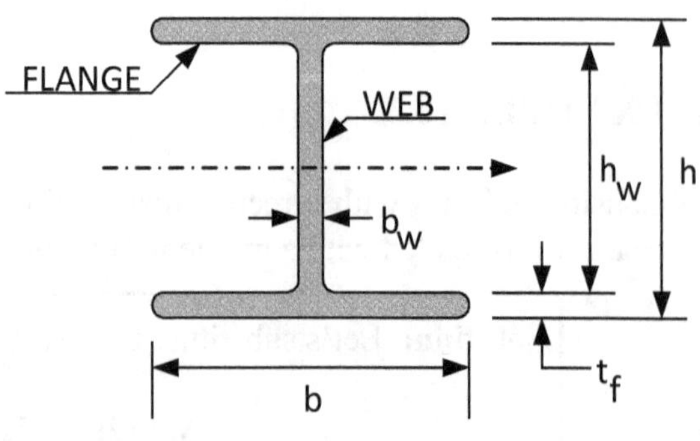

Fig. 10.10 Removing a large part of the under stressed material in a section of a beam increases its structural effectiveness.

The effectiveness of a beam with a rectangular cross section is given by Eq. (10.11) as:

$$\mathbf{SE} = bh^2/(6bh) = h/6 = 0.1667\ h \qquad (10.14)$$

For the wide flange section illustrated in Fig. 10.10, the relation for the effectiveness depends on several variables including b, h, b_w and h_w. To derive the relation for **SE**, we begin by determining I_z as:

$$I_z = I_{web} + I_{flange} \qquad (a)$$

$$I_{web} = b_w\, h_w^{\,3}/12 \qquad (b)$$

$$I_{flange} = 2\left[\frac{1}{12}bt_f^3 + bt_f\left(\frac{h_w + t_f}{2}\right)^2\right] \qquad (c)$$

Using Eqs. (a), (b) and (c) to solve for Z gives:

$$Z = \frac{2I_z}{h} = \frac{1}{h}\left[\frac{b_w h_w^3}{6} + \frac{bt_f^3}{3} + bt_f(h_w + t_f)^2\right] \qquad (d)$$

The area of the wide flange section is given by:

$$A = 2bt_f + b_w h_w$$

$$\mathbf{SE} = \frac{1}{h(2bt_f + b_w h_w)}\left\{\frac{b_w h_w^3}{6} + \frac{bt_f^3}{3} + bt_f(h_w + t_f)^2\right\} \qquad (10.15)$$

The result in Eq. (10.15) involves many terms and the improvement in the effectiveness of the section is not apparent simply by inspecting this relation. To ascertain the benefits of removing the under-stressed material, let's consider Example 10.5.

EXAMPLE 10.5

If the wide flanged section is fabricated with the following geometry, determine the effectiveness of the section and compare it with the effectiveness of a rectangular section.

$b_w = 0.1\ b$; $\quad h_w = 0.8\ h$; $\quad$ then $\quad t_f = 0.1\ h$

Solution: Substituting these quantities into Eq. (10.15) yields:

$$\mathbf{SE} = \frac{1}{h(0.2bh + 0.08bh)}\left\{\frac{0.1b(0.8h)^3}{6} + \frac{b(0.1h)^3}{3} + b(0.1h)[0.8h + 0.1h]^2\right\}$$

Reducing this expression gives:

$$\mathbf{SE} = 0.3209\ h$$

A comparison of this result with Eq. (10.14) indicates that the wide flanged section defined in this example with **SE** = 0.3209 h is nearly twice as efficient as the rectangular cross section with **SE** = 0.1667 h.

Determining I_z and Z for the various non-rectangular cross sections is a time consuming task. Fortunately, others have resolved this problem for us. The American Institute of Steel Construction (AISC) has prepared tables showing the properties of the commercially available sections. An abridged version of these tables is presented in Appendix D.

10.4 BENDING OF BEAMS WITH TRANSVERSE FORCES

Moments and stresses are induced in beams when they are subjected to transverse forces. In structures, the most significant loads are usually due to gravitational forces although in machine components the loading is due to a number of different effects and gravitational forces are often negligible. It is important, when modeling a beam, to consider both the various types of loading and supports for the beam. Typical examples of different loads and supports for the beams are illustrated in Fig. 10.11.

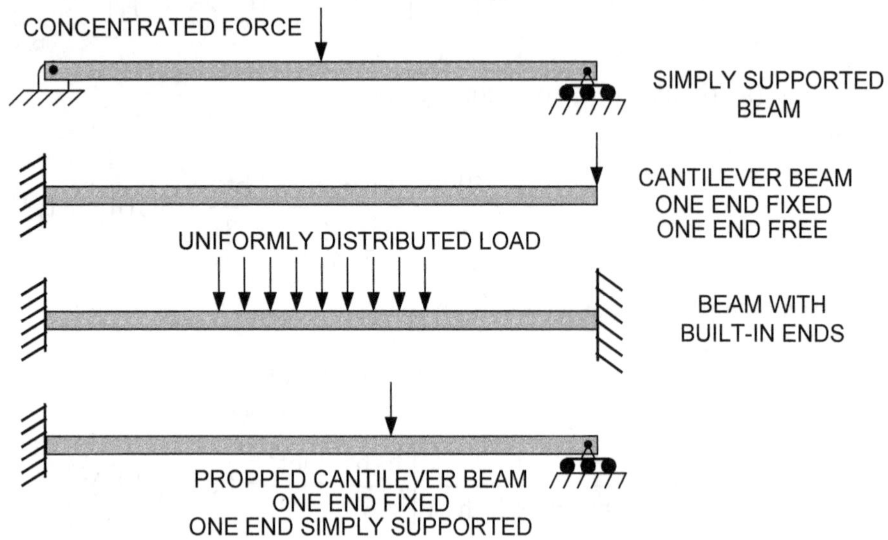

Fig. 10.11 Various types of transverse loading and supports for beams.

10.4.1 Shear and Bending Moments in Beams

Let's begin by considering the simply supported beam with a single concentrated force F as illustrated in Fig. 10.11. How does a concentrated force develop a moment? Where is the moment arm? We answer these questions by drawing several FBDs of the beam as shown in Fig. 10.12.

In Fig. 10.12a, we show a simply supported beam of length L with a concentrated force F applied a distance, a, from the left end. A free body of the entire beam, in Fig. 10.12b, shows the supports replaced with reaction forces R_L and R_R. When the beam deforms, its ends are free to rotate over the simple supports (the pin at A or the roller at B); consequently the moment M = 0 at both ends of the beam.

We use the equilibrium equations to solve for the reaction forces R_L and R_R.

$$\Sigma M_A = R_R\, L - F\, a = 0$$

$$R_R = F\, a/L \tag{a}$$

$$\Sigma F_y = R_L + R_R - F = 0$$

Substituting Eq. (a) into this relation gives:

$$R_L = F - R_R = F - F\, a/L = F[(L - a)/L] = F\, c/L \tag{b}$$

where c = L – a

Next, we make a section cut a distance x from the left end of the beam where $0 < x < a$ as shown in Fig. 10.12c. The left side of the beam is removed as a free body. We apply a positive[3] internal shear force V and a positive internal moment M to the surface exposed by the section cut as indicated in Fig. 10.12d.

From the equilibrium equations, we write:

$$\Sigma F_y = R_L - V = Fc/L - V = 0$$

$$V = F\,c/L \quad \Rightarrow \quad 0 < x < a \qquad \text{(c)}$$

$$\Sigma M_O = M - R_L\,x = M - F\,x\,c/L = 0$$

$$M = F\,x\,c/L \quad \Rightarrow \quad 0 < x < a \qquad \text{(d)}$$

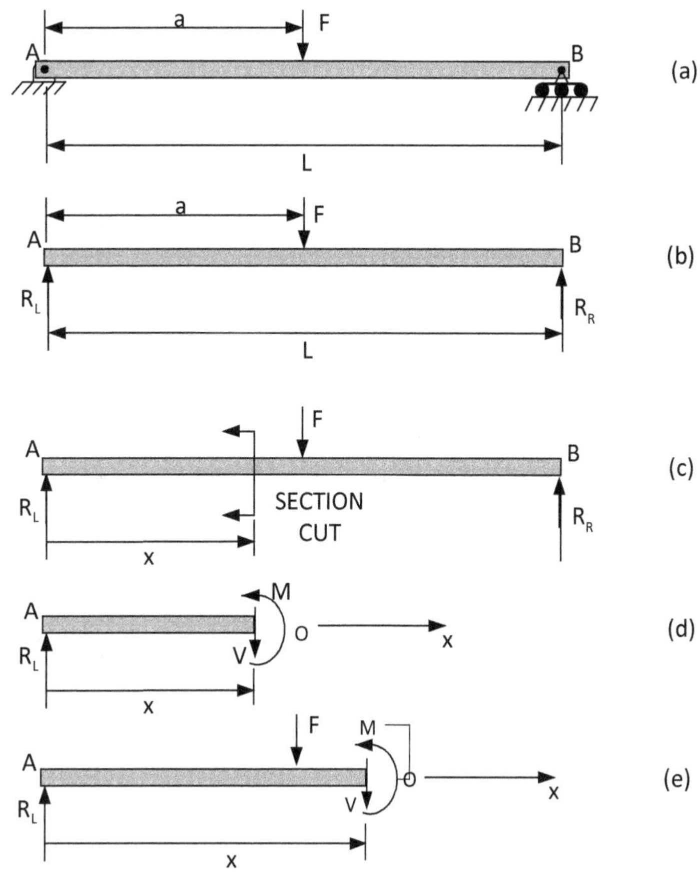

Fig. 10.12 FBDs of a simply supported beam with a concentrated load F.

To determine the relations for the shear force V and the internal moment M for the remaining part of the beam, we consider the FBD shown in Fig. 10.12e and repeat the equilibrium analysis.

$$\Sigma F_y = R_L - F - V = F(L - a)/L - F - V = 0$$

$$V = -F\,a/L \quad \Rightarrow \quad a < x < L \qquad \text{(e)}$$

$$\Sigma M_O = M - R_L\,x + F(x - a) = M - F\,x(L - a)/L + F(x - a) = 0$$

$$M = F\,a(L - x)/L \quad \Rightarrow \quad a < x < L \qquad \text{(f)}$$

Note, Eqs. (e) and (f) are only valid for $a < x < L$, and Eqs. (c) and (d) only apply for $0 < x < a$.

---

[3] We consider the internal shear force to be positive when it acts downward in the negative y direction on the FBD of the left end of the beam.

Equations (c) and (e) give the shear force V and Eqs. (d) and (f) give the internal moment M as a function of position x along the length of the beam. Examination of Eqs (d) and (f) shows that the maximum moment occurs at the load point x = a. Accordingly, we write:

$$M_{max} = F\,a\,c/L \qquad (10.16)$$

The maximum stress σ is given by:

$$\sigma_{max} = F\,a\,c/(L\,Z) \qquad (10.17)$$

A shear stress τ also develops due to the shear force, V; however, we will delay the discussion of shear stress until Section 10.4.3. For long thin beams, the shear stress τ is small when compared to the normal stress σ and is usually ignored without compromising the safety of the structure. However, for short beams with significant height, shear stresses must be considered. Shear stresses are also important when the beams are fabricated from fibrous materials like wood, because the shear strength in the direction of the fibers is very low.

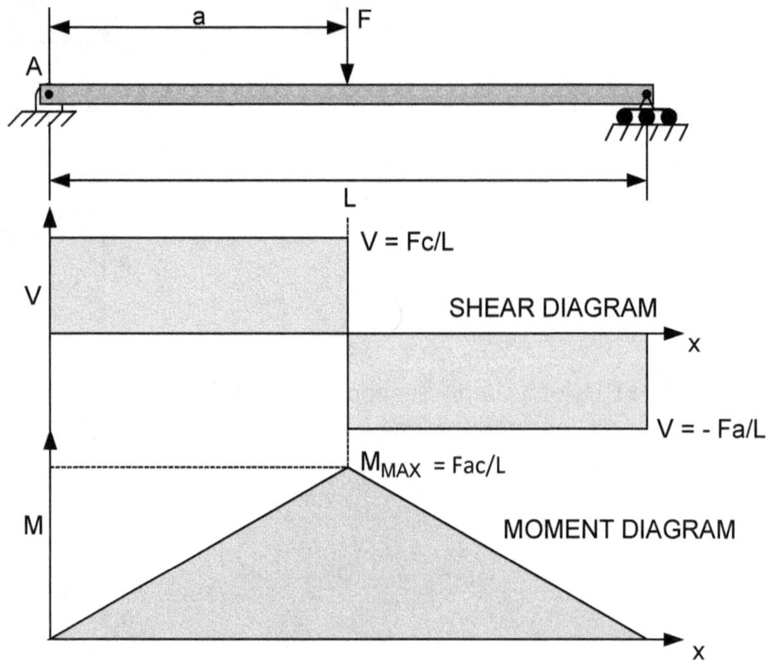

Fig. 10.13 Shear and bending moment diagrams show the distribution of V and M with x.

10.4.2 Shear and Bending Moment Diagrams

The equations for the shear force V and the bending moment M are often represented with diagrams that show the distribution of both of these quantities over the length of the beam. The diagrams are simply graphs of the equations developed by an analysis of the appropriate free bodies. An example of the shear and bending moment diagrams for the beam loaded with a single concentrated force F is presented Fig. 10.13.

Shear and bending moment diagrams are useful for two reasons. First, they aid in the visualization of the distribution of the shear force V and the bending moment M. Second, we use these diagrams to graphically determine the maximum moment and shear force. We have provided a table below to show the relationship among the loading condition and the lines or curves used in representing the shear and bending moment diagrams.

Table 10.1
Relationships among the loading conditions and the shear and bending moment diagrams

TYPE OF LOAD	SHEAR DIAGRAM	MOMENT DIAGRAM
NO LOAD	HORIZONTAL	OR RAMP
CONCENTRATED LOAD	OR VERTICAL JUMP	OR SLOPE CHANGE
UNIFORMLY DISTRIBUTED LOAD	OR RAMP	OR PARABOLA
LINEARLY DISTRIBUTED LOAD	OR 2nd DEGREE CURVE	OR 3rd DEGREE CURVE
CONCENTRATED MOMENT	HORIZONTAL	CCW OR CW VERTICAL JUMP

Let's consider an example to demonstrate the graphical method of determining the shear and moment diagrams.

EXAMPLE 10.6

For the simply supported beam loaded as indicated in Fig. E10.6, construct the shear and bending moment diagrams using a graphical approach.

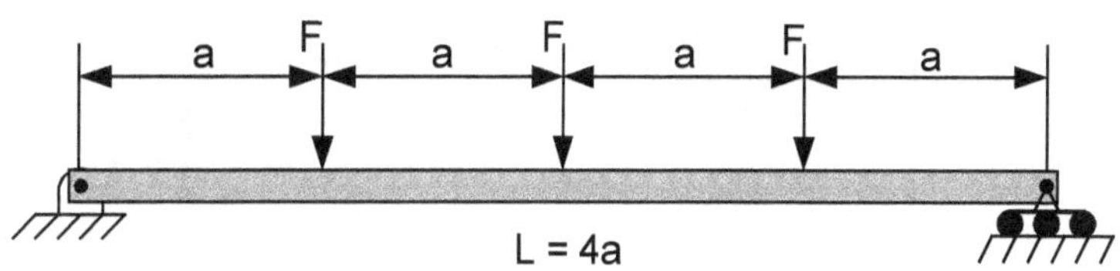

Fig. E10.6

Solution: The simply supported beam is symmetric; hence it is evident by inspection that:

$$R_L = R_R = 3F/2$$

We construct the FBD of the beam and then draw the shear diagram as shown below: To construct the shear diagram, begin at the left end of the beam at the origin of the V – x axes. Since the left hand reaction is upward and equal to 3F/2, we draw a line along the positive V axis (upward) scaled to a length proportional to 3F/2. Move to the right on the beam (the positive x direction) and observe the beam is free of loads until x = a. This observation means the shear force is constant from $0 < x < a$. At x = a, we encounter a concentrated force F which is directed downward. Since F is downward, we decrease the shear force by $\Delta V = -F$ so that $V = 3F/2 - F = F/2$ as shown in the shear diagram.

We proceed without change in the shear force until x = 2a where we again encounter another concentrated force of magnitude F. As before, we decrease V by this amount so that $V = F/2 - F = -F/2$.

We continue to increase x without change in the shear force until x = 3a where another concentrated force of magnitude F is encountered. As before, we decrease V by this amount so that V = – F/2 – F = – 3F/2.

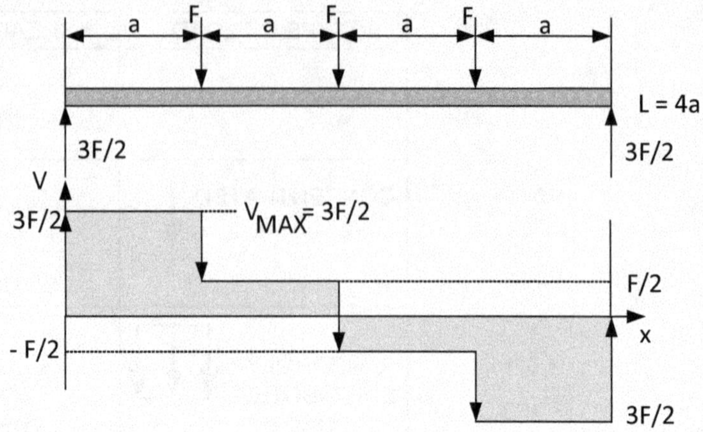

Fig. 10.6a

We continue to increase x without change in the shear force until x = 4a = L where we encounter the right hand reaction 3F/2 that is directed upward. Since the force is directed upward, we increase V by this amount so that V = – 3F/2 + 3F/2 = 0. It is evident by the construction demonstrated that concentrated forces produce step changes in the shear diagram. Construction of the shear diagram then consists of drawing straight-line segments and adding or subtracting the concentrated forces to the shear force.

To construct the moment diagram, we make use of a relation between the shear forces and bending moments[4], namely:

$$dM/dx = V \tag{10.18}$$

From Eq. (10.18), it is clear that the slope of the moment diagram at any position x is equal to the value of the shear force V at that same position. Integrating Eq. (10.18), we obtain:

$$M_{x_2} - M_{x_1} = \int_{x_1}^{x_2} V dx \tag{10.19}$$

For the shear diagram in this example, the areas under the V – x graph are simple rectangles; hence, the integral ∫Vdx between x_2 and x_1 is $V(x_2 - x_1)$. It is easy to draw the moment diagram shown in Fig. E10.6b using the shear diagram as a guide.

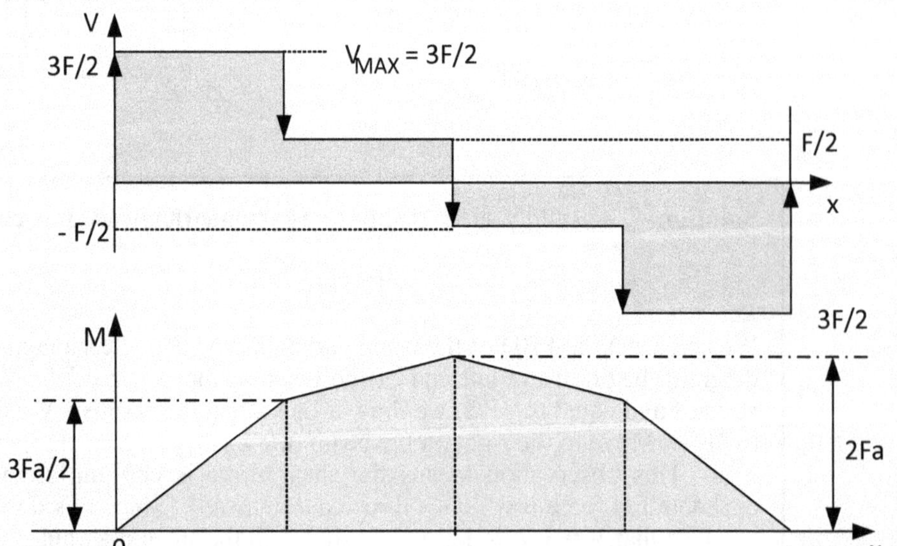

Fig. 10.6b

[4] We will defer the derivation of Eq. (10.18) until Section 10.4.3.

The area of the first rectangle between x = 0 and x = a is 3Fa/2. Since the area of the rectangle increases linearly with position x, we draw a straight line on the M – x diagram from the origin to the point with coordinates x = a, M = 3Fa/2. The area of the next rectangle from a < x < 2a is Fa/2. We add this area to the existing moment to obtain 3Fa/2 + Fa/2 = 2Fa. Again we draw a straight line on the M – x diagram from the point (a, 3Fa/2) to the next point with coordinates x = 2a, M = 2Fa. We may construct the remainder of the moment diagram by repeating the procedure. Note that the next two areas on the V-x diagram lie below the axis and are negative. These areas cause the M-x diagram to decrease and return to zero at x = 4a. The beam is symmetric; consequently, the moment diagram is also symmetric[5].

EXAMPLE 10.7

For the simply supported beam with a uniformly distributed load of magnitude q, write the equations for the internal shear force V and the moment M as functions of position x. Also construct the shear and bending moment diagrams.

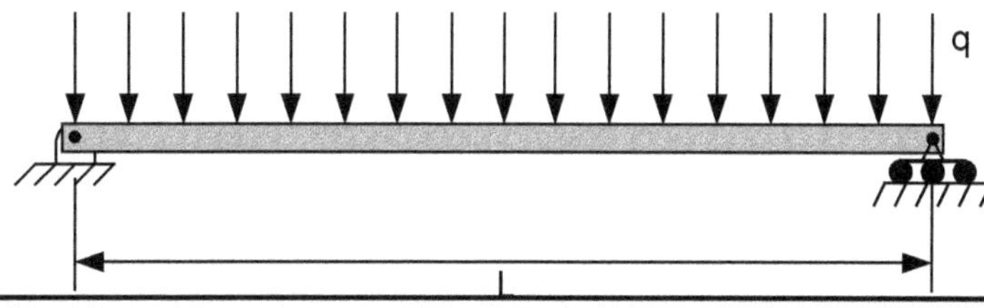

Solution:

The beam is symmetric so that the reactions R_L and R_R are equal. In addition, the total force in the negative y direction is F = qL; therefore, it is evident that

$$R_L = R_R = qL/2 \qquad \text{(a)}$$

Let's make a section cut at some arbitrary position x and construct a FBD of the left side of the beam as shown in Figs. E10.7a and inset (b).

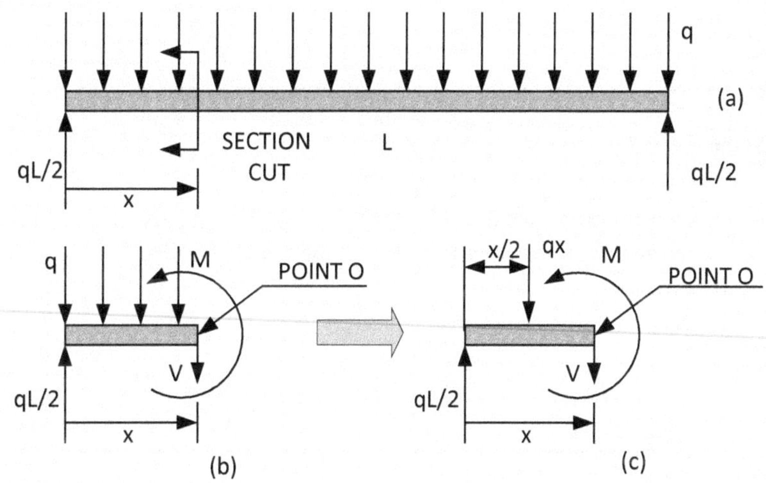

Fig. 10.7a

The FBD of the left side of the beam is shown with a uniformly distributed load acting over a length x of the beam segment. We replace this uniformly distributed load with an equivalent concentrated force qx located at the center of the segment as shown in Fig. E10.7a inset (c). Next, we apply the equilibrium equations and write:

[5] The shear diagram is also symmetric with respect to the magnitude of V, but the signs of the shear forces are not symmetric.

$$\Sigma F_y = qL/2 - qx - V = 0$$

$$V = (q/2)(L - 2x) \quad \Rightarrow \quad 0 < x < L \qquad \text{(b)}$$

$$\Sigma M_O = M + qx(x/2) - (qL/2)x = 0$$

$$M_x = (qx/2)(L - x) \quad \Rightarrow \quad 0 < x < L \qquad \text{(c)}$$

Inspection of Eq. (b) shows that V = 0 at x = L/2. This zero crossing of the shear force on the V – x diagram locates the position of the maximum moment. Solving for M_{max} at x = L/2 using Eq. (c) gives:

$$M_{max} = qL^2/8 \qquad \text{(d)}$$

We may construct graphs of Eqs. (b) and (c) showing V and M as a function of position along the length of the beam to produce the shear and moment diagrams. Alternatively we may construct them directly from the transverse forces applied to the beam as shown below.

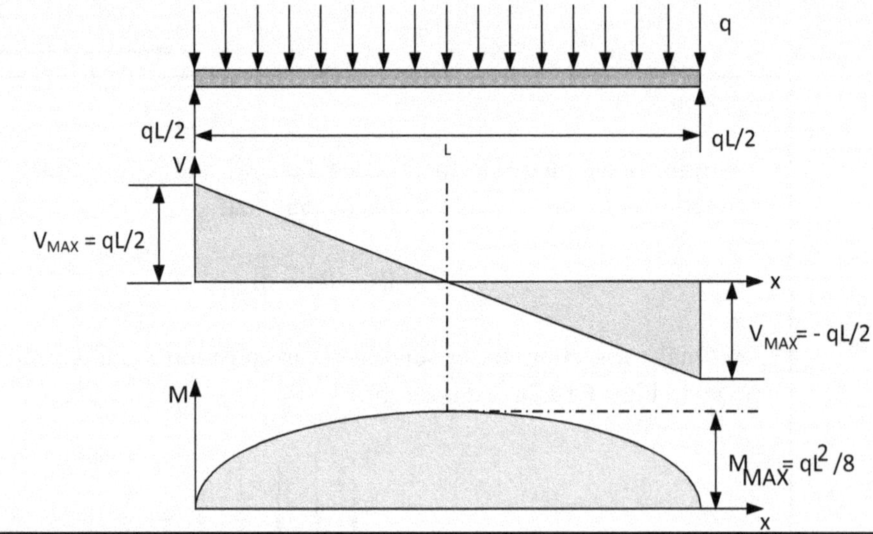

Fig. E10.7b

EXAMPLE 10.8

For the cantilever beam with a concentrated force F at its free end, write the equations for the internal shear force V and the moment M, and construct the shear and bending moment diagrams.

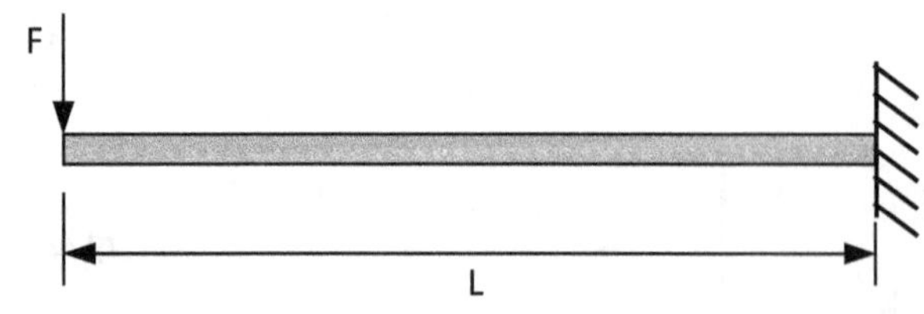

Solution: Let's begin by removing the built-in end and replacing it with a shear force V_{FE} and a moment M_{FE} when constructing the FBD of the cantilever beam as shown below:

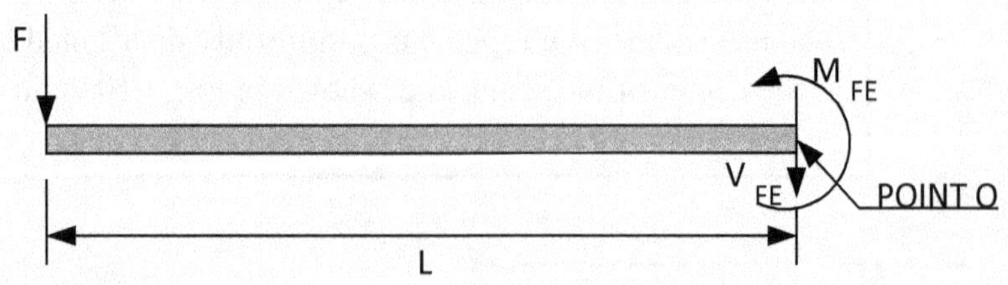

Fig. E10.8a

From the equilibrium equations, we write:

$$\Sigma F_y = -F - V_{FE} = 0$$

$$V_{FE} = -F \qquad (a)$$

$$\Sigma M_O = M_{FE} + FL = 0$$

$$M_{FE} = -FL \qquad (b)$$

The negative signs for V_{FE} and M_{FE} imply that the directions shown in the FBD are not correct. We change the directions in the FBD shown in Fig. E10.8 b.

Fig. E10.8b

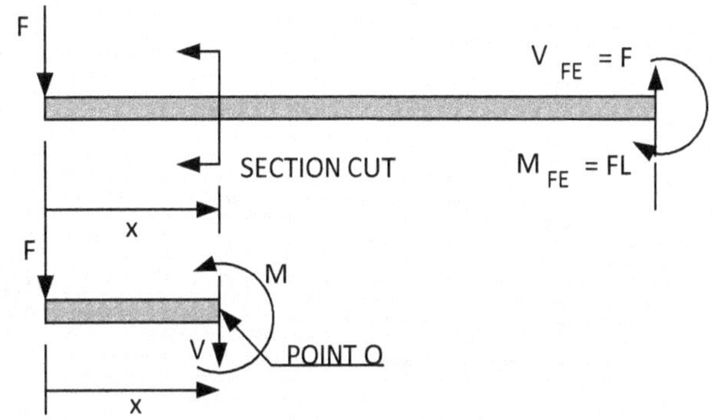

Next, we make a section cut at position x and remove the left side of the cantilever beam as a free body. Again, we write the equilibrium equations to obtain.

$$\Sigma F_y = -F - V = 0; \qquad V = -F \qquad (c)$$

$$\Sigma M = M + Fx = 0; \qquad M = -Fx \qquad (d)$$

The shear and moment diagrams are constructed by using either Eqs. (c) and (d) or by drawing the diagrams directly from the knowledge of the distribution of forces over the length of the beam as indicated in Fig. E10.8 c. The shear force, V is a constant along the entire length of the beam. The bending moment M increases linearly in x—starting at zero at the free end and becoming a maximum $M_{max} = -FL$ at the fixed end.

Fig. E10.8c

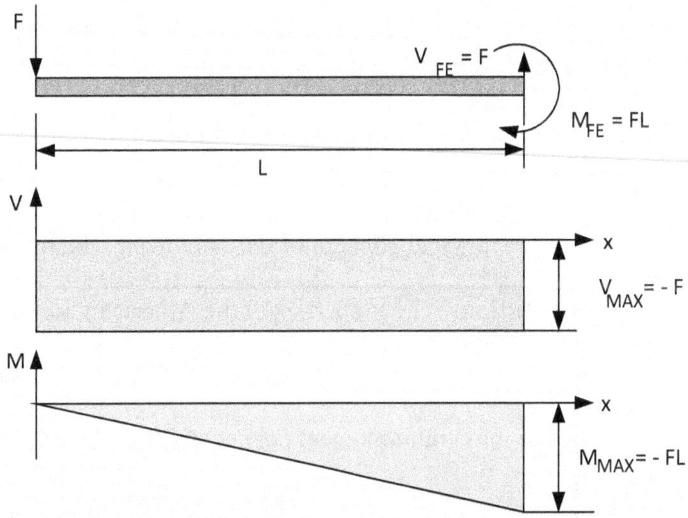

EXAMPLE 10.9

For the simply supported beam described in Fig. 10.13, determine the margin of safety if the beam and its loading are specified as:

- A wide flanged (WF) shape W 10 × 60
- F = 20 kip
- L = 24 ft and a = 10 ft.
- The yield strength of the hot rolled steel beam is 42 ksi.

Solution:

The properties of the WF section are given in Appendix D as:

$$I_z = 341 \text{ in.}^4 \qquad Z = 66.7 \text{ in.}^3 \qquad \text{(a)}$$

The maximum moment occurs at x = a = 10 ft and is given by Eq. (10.16) as:

$$M_{max} = F\,a\,c/L = (20 \times 10^3)(10)(24 - 10)/24 = 116.7 \times 10^3 \text{ ft-lb} \qquad \text{(b)}$$

The maximum stress is given by Eq. (10.10) as:

$$\sigma_{max} = M_{max}/Z = [(116.7 \times 10^3)(12)]/66.7 = 21.0 \text{ ksi} \qquad \text{(c)}$$

Finally, the margin of safety is given by:

$$MOS = SF - 1 = S_y/\sigma - 1 = 42.0/21.0 - 1 = 100\% \qquad \text{(d)}$$

The margin of safety is probably adequate at 100% unless there is considerable uncertainty about the loading. The beam is hot rolled to a standard section shape—its geometry is well-established and not subject to significant error in fabrication. The yield strength is typical of low alloy steel and it will not usually vary by more than ± 10% from one shipment of steel to another.

EXAMPLE 10.10

For the simply supported beam with a uniformly distributed load described in Example 10.7, determine the margin of safety if the beam and its loading are specified as:

- A S8 × 23 American Standard shape
- q = 1 kip/ft
- L = 16 ft
- The yield strength of the steel from which the beam is hot rolled is 38 ksi.

Solution: The properties of the American Standard section are given in Appendix D as:

$$I_z = 64.9 \text{ in.}^4 \qquad Z = 16.2 \text{ in.}^3 \qquad \text{(a)}$$

The maximum moment occurs at x = L/2 = 8 ft and is given by:

$$M_{max} = qL^2/8 = (1 \times 10^3)\,(16)^2/8 = 32{,}000 \text{ ft-lb} \qquad \text{(b)}$$

The maximum stress is given by Eq. (10.10) as:

$$\sigma_{max} = M_{max}/Z = (32{,}000)(12)/16.2 = 23.7 \text{ ksi} \quad \text{(c)}$$

Finally, the margin of safety is given by:

$$MOS = SF - 1 = S_y/\sigma - 1 = 38.0/23.7 - 1 = 60.3\% \quad \text{(d)}$$

The margin of safety is probably somewhat low at 60.3%. While the beam is hot rolled to a standard section shape (its geometry is well established and not subject to error in fabrication), any significant over loading can endanger the stability of the structure. Consideration should be given to increasing the section size to an American Standard S10 × 25.4.

EXAMPLE 10.11

For the cantilever beam with a concentrated force F applied at its free end described in Example 10.8, determine the margin of safety if the beam and its loading are specified as:

- A WT 5 × 30 structural tee
- F = 400 lb
- L = 15 ft
- The yield strength of the steel is 44 ksi.

Solution: The properties of the structural tee is given in Appendix D as:

$$I_z = 12.9 \text{ in.}^4 \qquad Z = 3.04 \text{ in.}^3 \quad \text{(a)}$$

The maximum moment occurs at x = L = 15 ft and is given by:

$$M_{max} = FL = (400)(15) = 6{,}000 \text{ ft-lb} \quad \text{(b)}$$

The maximum stress is given by Eq. (10.10) as:

$$\sigma_{max} = M_{max}/Z = (6{,}000)(12)/(3.04) = 23.68 \text{ ksi} \quad \text{(c)}$$

Finally, the margin of safety is given by:

$$MOS = SF - 1 = S_y/\sigma - 1 = 44.0/23.68 - 1 = 85.81\% \quad \text{(d)}$$

It appears that the margin of safety is marginal given the specification for the beam and the loading. Consideration should be given to selection of a larger or heavier beam with a larger section modulus, such as WT5 × 44 or WT6 × 25.

10.4.3 Shear Stresses in Beams

The transverse forces applied to beams produce internal shear forces V in both horizontal and vertical directions (H and V respectively). These shear forces give rise to shear stresses τ_H and τ_V. Let's first establish the relationship between the internal bending moment M and the internal shear force V. Consider the simply supported beam with a uniformly distributed load q as illustrated in Fig. 10.14a. We make two section cuts and remove the segment of the beam between the section cuts with a length Δx. The FBD of this beam segment is presented in Fig. 10.14b. Writing the equilibrium equations yields:

$$\Sigma F_y = V - q\Delta x - (V + \Delta V) = 0$$

$$\Delta V/\Delta x = dV/dx = -q \quad (10.20)$$

$$\Sigma M_O = M + \Delta M - M + q\Delta x^2/2 - V\Delta x = 0$$

$$\Delta M/\Delta x = dM/dx = V \qquad (10.18 \text{ bis})$$

Note, $q\Delta x^2/2$, a second order term, is neglected because it is small when compared to the other terms in the equation.

Next, let's sub-section this small segment of the beam with a horizontal cut some distance y above the neutral axis as shown in Fig. 10.14c. The FBD of the subsection of this beam segment is illustrated in Fig. 10.14d. We write the equilibrium equation $\Sigma F_x = 0$ for this free body to obtain:

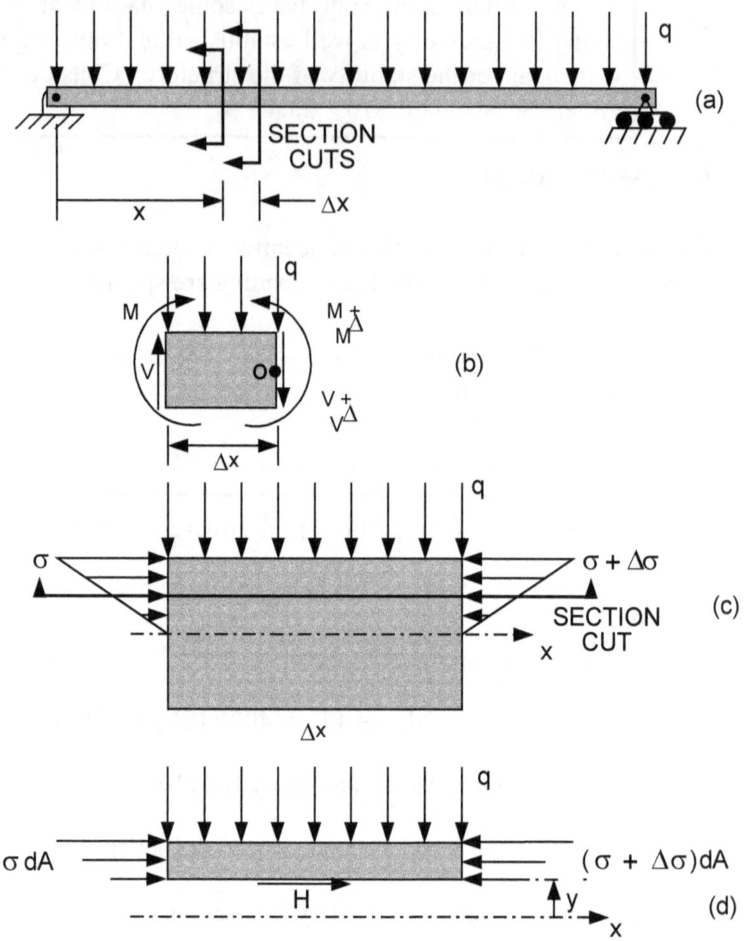

Fig. 10.14 Several FBDs that show the relation between M and V.

$$\Sigma F_x = -\int(\sigma + \Delta\sigma)dA + \int \sigma dA + H = 0$$

$$H = \int \Delta\sigma \, dA \qquad (a)$$

where H is a horizontal shear force.

From Eq. (10.7), we may write:

$$\Delta\sigma = -\Delta M y/I_z \qquad (b)$$

Substituting Eq. (10.18) into Eqs. (b) and (a) gives:

$$H = -(V\Delta x/I_z)\int y dA = -(V\Delta x/I_z)Q_1 \qquad (c)$$

where Q_1 is defined by:

$$Q_1 = \int_{y_1}^{c} y dA \qquad (10.21)$$

Q_1 is the first moment of the area from the section cut at position y_1 to the location of the outer fiber on the top surface of the beam at $y = c$.

The horizontal shear stress τ_H acting on the surface exposed by the section cut is given by:

$$\tau_H = H/A_H = -\ VQ_l/(bI_z) \qquad (10.22)$$

where $A_H = b\Delta x$ and b is the beam thickness.

We are not usually concerned with the sign in determining the shear stress. Failure of the material, from which structural components are fabricated, is independent of the direction of the shear stress. Finally, we show that the horizontal and vertical shear stresses[6] are equal.

$$\tau = \tau_V = \tau_H = VQ_l/(bI_z) \qquad (10.23)$$

The equality of the shear stresses ($\tau = \tau_V = \tau_H$) is established by taking moments about the point O for the element Δx by Δy shown in Fig. 10.15. This element may be removed from any plane body such as a beam. Also the location of point O is arbitrary.

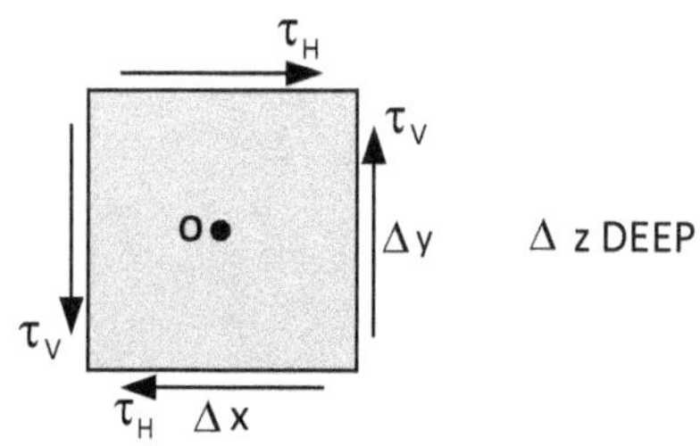

Fig. 10.15 An element Δx by Δy by Δz with shear stresses τ_V and τ_H acting on its faces.

$$\Sigma M_O = (\tau_V\ \Delta y\ \Delta z)\ \Delta x - (\tau_H\ \Delta x\ \Delta z)\ \Delta y = 0$$

$$\tau_V = \tau_H \qquad (10.24)$$

Examination of Eq. (10.23) indicates that the shear stress τ_V is a maximum when Q_l is a maximum. It is evident from Eq. (10.21) that Q_l is a maximum for any given section when $y_l = 0$. Therefore, the shear stress τ is a maximum at the neutral axis. As y_l approaches the outer fiber at $y_l = c$, Q_l approaches zero. For this reason, the shear stresses τ are zero on the top and bottom surfaces of the beam where the normal stresses σ are a maximum. We will demonstrate the method for solving for the shear stresses in a beam with a rectangular cross section in the following examples.

EXAMPLE 10.12

Consider a simply supported beam of length L with a concentrated force F of 10 kN applied at the center of the span. The beam cross section is rectangular with b = 50 mm and h = 125 mm and the beam length L = 5 m. Determine the maximum normal stress σ and the maximum shearing stress τ.

Solution: Since the simply supported beam is symmetric, it is clear that the reactions are equal to:

$$R_L = R_R = F/2 = 10/2 = 5.0 \text{ kN} \qquad (a)$$

Next, we construct the shear and moment diagrams as shown in Fig. E10.12.
From the shear and bending moment diagrams, we determine that:

$$V_{max} = 5 \text{ kN} \qquad \text{and} \qquad M_{max} = 12.5 \text{ kN-m} \qquad (b)$$

The maximum normal stress is determined from Eq. (10.12) as:

$$\sigma_{max} = 6M_{max}/(bh^2) = [(6)(12.5 \times 10^3 \text{ N–m})]/[0.05 \text{ m } (125)^2 \text{ mm}^2] = 96 \text{ MPa} \qquad (c)$$

The maximum shear stress is given by Eq. (10.23) as:

[6] The vertical shear stresses τ_V and vertical shear force V act on the surface exposed by a vertical section cut.

$$\tau_{max} = V(Q_1)_{max} /(bI_z) \qquad \text{(d)}$$

To evaluate Eq. (d), let's first determine $(Q_1)_{max}$ using Eq. (10.21)

$$Q_1 = \int_{y_1}^{c} ydA = b\int_{0}^{h/2} ydy = \frac{b}{2}\left[y^2\right]_0^{h/2} = \frac{bh^2}{8} \qquad \text{(e)}$$

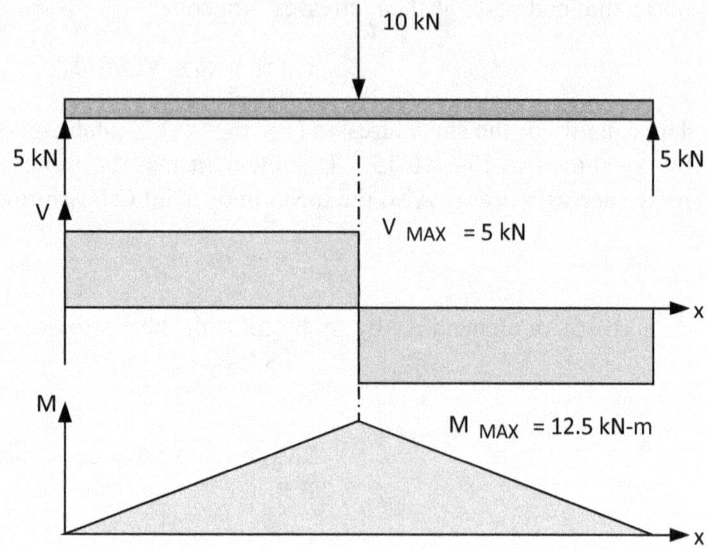

Fig. E10.12

We may also determine Q_1 without integrating by recalling that the first moment of the area A_1 shown in Fig. E 10.12 a is given by:

$$Q_1 = A_1\bar{y} \qquad \text{(f)}$$

where A_1 is the area of the cross section above the section cut and y is the distance from the neutral axis to the centroid of the area A_1.

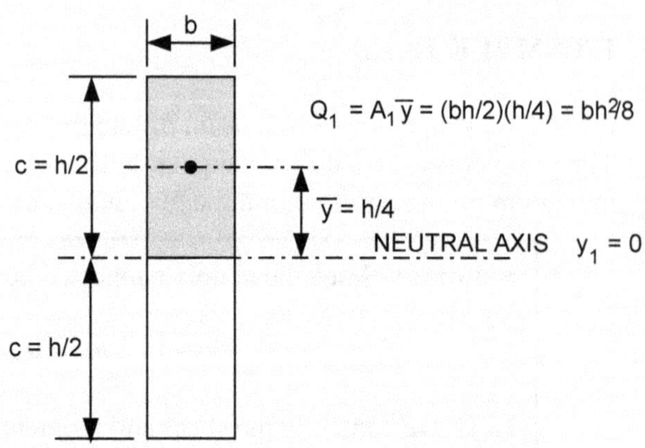

Fig. E10.12a

For a rectangular cross section the moment of inertia is:

$$I_z = bh^3/12 \qquad \text{(g)}$$

Substituting Eqs. (e) and (g) into Eq. (d) yields:

$$\tau_{max} = (3/2)(V/bh) = (3/2)(V/A) \qquad (10.25)$$

$$\tau_{max} = (1.5)(5 \times 10^3) /[(50 \times 125)] = 1.20 \text{ MPa} \qquad \text{(h)}$$

Comparison of the maximum normal stresses and the maximum shear stress for this example shows that the shear stresses are very small (1.25% of the maximum normal stresses). For long thin beams, the shear stresses are usually small enough to be neglected[7].

Fig. 10.12b

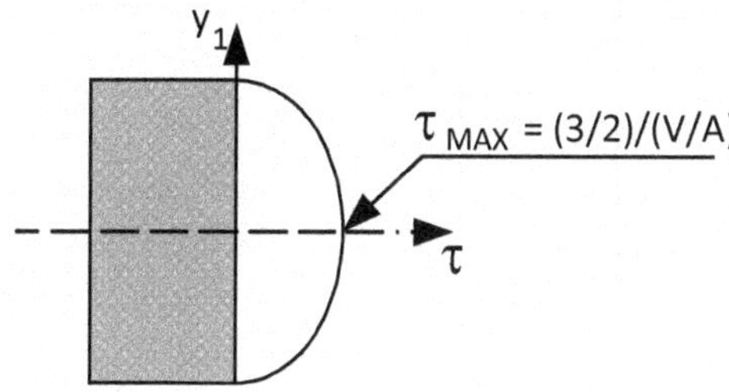

Equation (10.25) indicates that the maximum shear stress is 1.5 times the average shear stress. The shear stress is not distributed uniformly over the cross sectional area, but varies as a parabolic function; the maximum occurs at the neutral axis as shown in Fig. E10.12b.

EXAMPLE 10.13

Let's again consider a simply supported beam of length L = 5 m with a concentrated force F = 10 kN applied at the center of the span. The beam cross section is circular with a diameter d = 100 mm. Determine the maximum normal stress σ and the maximum shearing stress τ.

Solution:

The beam in this example is identical with that in the previous example except for its cross section. The reactions, shear and bending moment diagrams, maximum shear force and moment are the same:

$$R_L = R_R = F/2 = 10/2 = 5.0 \text{ kN} \quad \text{(a)}$$

$$V_{max} = 5 \text{ kN} \quad \text{and} \quad M_{max} = 12.5 \text{ kN-m} \quad \text{(b)}$$

Since the moment of inertia I_z for a circular area about its centerline is given by:

$$I_z = \pi d^4/64 = \pi r^4/4 \quad \text{(c)}$$

The maximum normal stress is given by:

$$\sigma_{max} = M_{max}d/(2I_z) = 64\ M_{max}d/(2\pi d^4) = 32\ M_{max}/(\pi d^3) \quad (10.26)$$

$$\sigma_{max} = (32)(12.5 \times 10^3)/[(\pi)(0.1)(100)^2] = 127.3 \text{ MPa} \quad \text{(d)}$$

The maximum shear stress is given by Eq. (10.23) as:

$$\tau_{max} = V(Q_1)_{max}/(bI_z) \quad \text{(e)}$$

We will determine Q_1 using the procedure described below. For the circular bar, the parameter b = 2r at the neutral axis, and $Q_1 = 2r^3/3$ as indicated in Fig. E10.13.

—

[7] The exception is when wood is used to fabricate the beam. The shear strength of wood is very low and even small shear stresses may be sufficient to split the wooden beam along its neutral axis.

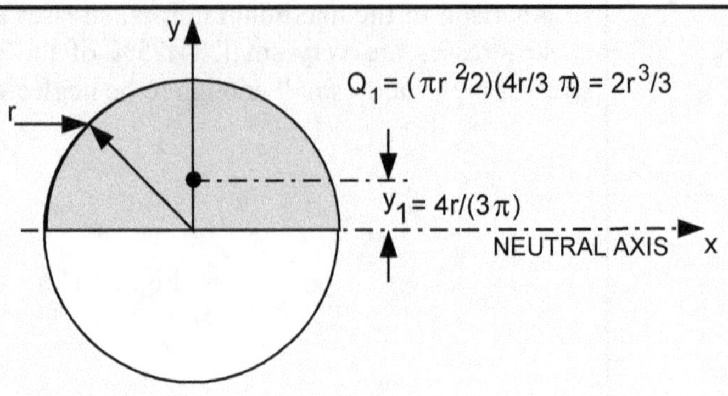

Substituting these results and Eq. (c) into Eq. (e) gives:

$$\tau_{max} = (4/3)(V/\pi r^2) = (4/3)(V/A) \qquad (10.27)$$

$$\tau_{max} = (4/3)(5 \times 10^3)/[\pi\,(50)^2] = 0.8488 \text{ MPa}$$

Again, it is evident that the maximum shear stress is very small (less than 1%) when compared to the normal stress.

10.5 BENDING OF COMPOSITE BEAMS

In our discussions of beams, we have considered only those fabricated from a single material such as steel, aluminum or wood. However, in engineering design there is an increasing trend to employ beams fabricated from two or more materials. These are called **composite beams**; they offer the opportunity of using each of the materials employed in their construction to advantage. You may want to consider using them in the design of your project structure.

10.5.1 Foam Core with Metal Cover Plates

Let's consider a composite beam fabricated with thin metal cover plates on the top and bottom with a plastic foam core as shown by the cross sectional area depicted in Fig. 10.16.

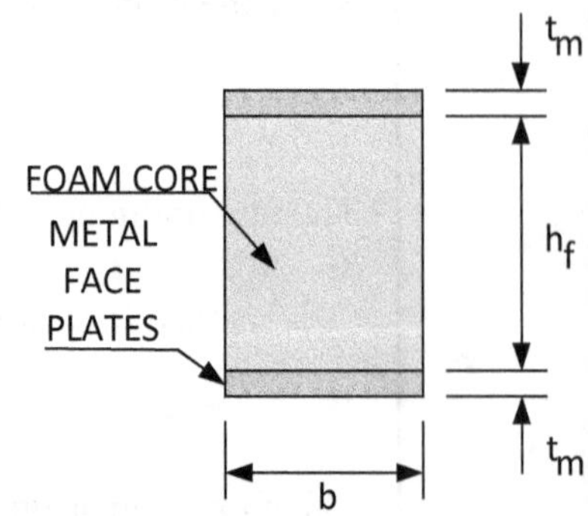

Fig. 10.16 Cross section of a composite beam with a foam core and metal cover plates.

The design concept for the foam core beam is to use light, low-strength foam to support the load-bearing metal plates positioned at the top and bottom, where they are the most efficient. To analyze the composite beam, we again assume that plane sections remain plane before and after loading so the strain is linearly distributed as shown in Fig. 10.17. Note that the strain is continuous across the interface between the foam and the cover plates.

The stress in the foam is given by:

$$\sigma_f = E_f\,\varepsilon \approx 0 \qquad (a)$$

We set the stress in the foam equal to zero because its modulus of elasticity E_f is small when compared to the modulus of elasticity of the metal. Also, the stress in the metal of the cover plates, σ_m is described by Hooke's law as:

$$\sigma_m = E_m\,\varepsilon = -\,E_m\,y/\rho \qquad \text{(b)}$$

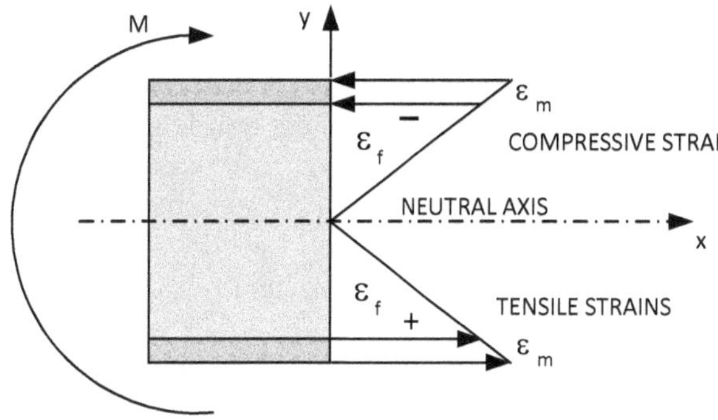

Fig. 10.17 The strain distribution is continuous across the foam-metal interface.

We substitute Eq. (10.6) into Eq. (b) to obtain:

$$\sigma_m = -\,M\,y/I_z \qquad \text{(10.7 bis)}$$

The relation for the stress is the same as that established previously in Section 10.2; however, the foam does not directly contribute to the load carrying capacity of the beam because its modulus of elasticity is negligible. For this reason, we do not consider the foam in determining I_z. Instead, we consider only the metal plates when writing the relation for I_z as:

$$I_z \cong 2Ad^2 = 2\left[bt_m\left(\frac{h_f + t_m}{2}\right)^2\right] = \frac{bt_m}{2}(h_f\ +\ t_m)^2 \qquad \text{(10.28)}$$

Combining Eq. (10.7) and (10.28) gives the maximum stress in the metal plates as:

$$\sigma_{max} = \frac{M(h_f + 2t_m)}{bt_m(h_f + t_m)^2} \qquad \text{(10.29)}$$

Let's utilize this result to determine the effectiveness of foam core-metal plate composite beams.

EXAMPLE 10.14

Consider a simply-supported, foam core, metal cover plate composite beam with a uniformly distributed load of magnitude q. Aluminum cover plates 0.063 in. thick, 10 in. wide and 10 ft long are adhesively bonded to a polystyrene foam core. The foam is 10 in. wide, 6 in. high and 10 ft long. If the yield strength of the aluminum cover plates is 32 ksi, determine the distributed load q_Y that produces yielding.

Solution: From Example 10.7, we note the maximum moment at x = L/2 as:

$$M_{max} = qL^2/8 = q(10 \times 12)^2/8 = 1800\ q \qquad \text{(a)}$$

where q is in units of lb/in.

When the composite beam yields, the stresses in the cover plates are:

$$\sigma_{max} = S_y = 32{,}000 \text{ psi.} \qquad \text{(b)}$$

Substituting Eqs.(a) and (b) into Eq. (10.29) gives:

$$\sigma_{max} = \frac{M(h_f + 2t_m)}{bt_m(h_f + t_m)^2} = S_y = 32{,}000 = \frac{1800q_Y(6 + 0.126)}{(10)(0.063)(6.063)^2}$$

Solving this relation for q_Y gives:

$$q_Y = 67.2 \text{ lb/in.} = 806 \text{ lb/ft} \quad \text{(c)}$$

The effectiveness of this beam may be judged by the ratio of its load carrying capacity to its weight per unit length. Its weight w per unit length is:

$$w = 2bt_m\gamma_m + bh_f\gamma_f \quad \text{(d)}$$

where $\gamma_m = 0.10$ lb/in.^{3} and $\gamma_f = 0.001$ lb/in.^{3} are the densities of the aluminum and foam respectively.

Substituting these quantities into Eq. (d) yields:

$$w = (2)(10)(0.063)(0.10) + (10)(6)(0.001) = 0.126 + 0.06 = 0.186 \text{ lb/in.}$$

The ratio of $q_Y/w = 67.2/0.186 = 361$

With the composite beam described, we can support a uniformly distributed load that is 360 times larger than the weight of the beam. That is a very good ratio. Composite structures with foam cores are used in aircraft structures where minimizing structural weight is a critical design criterion.

EXAMPLE 10.15

A cantilever beam 5 m long is subjected to a concentrated force of 1 kN applied at its free end. The beam has a rectangular cross section with b = 100 mm. Determine the height of the foam and the thickness of the aluminum cover plates if the maximum stress is limited to 200 MPa. In this determination recall that the standard thickness of stock aluminum sheet from which the cover plates are cut is 0.016, 0.020, 0.025, 0.032, 0.040, 0.050, 0.063, and 0.090 in.

Solution: For a cantilever beam, the maximum moment is given by:

$$M_{max} = FL = (1 \times 10^3)(5) = 5.0 \times 10^3 \text{ N-m} = 5.0 \times 10^6 \text{ N-mm} \quad \text{(a)}$$

Since the maximum stress is limited to 200 MPa, we may write:

$$\sigma_{max} = \frac{M(h_f + 2t_m)}{bt_m(h_f + t_m)^2} = 200 \text{N / mm}^2 \quad \text{(b)}$$

In composite beam construction, we often find that:

$$t_m << h_f; \qquad \text{hence,} \qquad h_f + 2t_m \approx h_f + t_m. \quad \text{(c)}$$

This approximation simplifies Eq. (b):

$$\sigma_{max} \approx \frac{M}{bt_m(h_f + t_m)} \quad \text{(d)}$$

Substituting known quantities into Eq. (d) and solving for h_f leads to:

$$h_f = \frac{M}{bt_m\sigma_{max}} - t_m = \frac{5.0 \times 10^6}{(100)t_m(200)} - t_m = \frac{250 \text{ mm}^2}{t_m} - t_m \quad \text{(e)}$$

In Eq. (e) we have expressed one unknown h_f in terms of another unknown t_m. However, the thickness of the metal cover plates is restricted to standard sizes. Let's evaluate Eq. (e) for all possible choices of t_m and then select the most suitable combination for the design of the beam. The computation was performed with the aid of a spreadsheet and the results are shown in Table E10.15.

Examination of this table shows that we have many different solutions to consider in designing a beam; however, not all of them are realistic. If we use very thin cover plates (0.406 mm), the height of the foam is excessive (615 mm). If we use thick cover plates (2.29 mm), the height of the foam core (107 mm) is about the same as the width of the beam (100 mm). There is no unique solution, but the cross section with $h_f = 307$ mm and $t_m = 0.8128$ mm gives a h/b ratio of about 3/1 that appears reasonable.

Table E 10.15
Height of Foam as a Function of Cover Plate Thickness

COVER PLATE THICKNESS	COVER PLATE THICKNESS	HEIGHT OF FOAM
t_m (in.)	t_m (mm)	h_f (mm)
0.016	0.4064	614.7
0.020	0.5080	491.6
0.025	0.6350	393.0
0.032	0.8128	306.7
0.040	1.0160	245.0
0.050	1.2700	195.5
0.063	1.6002	154.6
0.090	2.2860	107.0

10.5.2 Reinforced Concrete

Another method of producing composite beams combines steel and concrete that utilizes both materials to advantage. **Concrete** is a mixture of sand, aggregate and cement that is mixed with water, poured into a mold and cured into the shape of some structural member. Concrete is a durable material that has replaced stone in many structural applications. Like stone, it exhibits a satisfactory compressive strength, but it fails when subjected to relatively low tensile stresses. Its compressive strength ranges from 3,000 to 6,000 psi depending on the constituents used in the mix. However, its tensile strength is only 50 to 100 psi.
It is so weak in tension that it cannot be employed to fabricate beams without reinforcement.

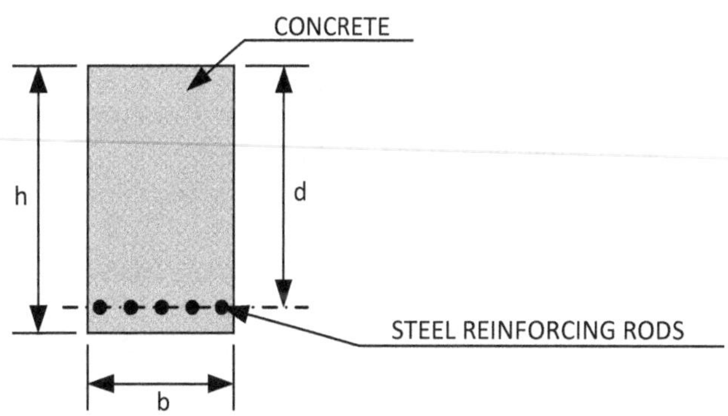

Fig. 10.18 Typical cross section of a reinforced concrete beam. Steel reinforcing rods are placed a short distance above the beam's bottom surface.

To reinforce concrete so that it may be used in beams, we embed steel rods in the concrete when it is cast into forms that define the shape of the beam. The concrete solidifies about the reinforcing steel rods, called **rebar**, producing a composite beam. A typical cross section of a concrete beam reinforced with steel rebar is illustrated in Fig. 10.18.

In preparing the drawing for Fig. 10.18, we assumed the beam was subjected to positive moments that produce tension in regions below the neutral axis. The positive bending moment produces tensile strains in both the concrete and the steel reinforcing rods because plane sections before deformation remain plane after loading for composite beams as well as homogenous beams. These tensile strains produce tensile stresses that are sufficient to fail the concrete. When the concrete fails, micro-cracks initiate and grow into the tension region of the cross section. However, these cracks arrest near the neutral axis. The steel does not fail and the beam remains intact and rigid. However, the concrete on the tension side of the section does not contribute to the strength and rigidity except to position the reinforcing rods some distance from the neutral axis of the beam. To accommodate for the lack of tensile strength of the concrete, we model the cross section of the beam as shown in Fig. 10.19.

The original cross section of the beam is shown in Fig. 10.19a. We remove all of the concrete located below the neutral axis from the model to account for the micro cracking in Fig. 10.19b. This leaves a rectangular section of concrete, s high by b wide, to resist the compressive forces acting above the neutral axis. We also indicate the total area A_s of the steel reinforcing rods that carry the tensile forces. The model in Fig. 10.19b is difficult to analyze because it involves two different materials, both of which contribute to the load carrying capacity of the beam.

To facilitate the analysis, we convert the composite (heterogeneous) model into a homogenous model by representing the steel area with an equivalent concrete area in Fig. 10.19c. The equivalent concrete area A_{eq} for the steel rods is:

$$A_{eq} = kA_s \tag{10.30}$$

where

$$k = E_s / E_c \tag{10.31}$$

Note that E_s and E_c are the elastic moduli of the steel and concrete, respectively.

Fig. 10.19 Modeling a steel-bar reinforced concrete beam.

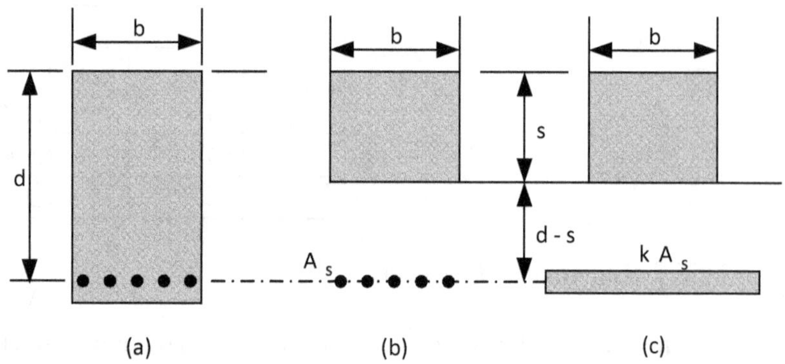

The analysis of the model shown in Fig. 10.19 requires that we first determine the position of the neutral axis. Recall that the neutral axis is coincident with the centroid of the section so that $\int y dA = 0$; hence, we may write:

$$(s)(b)(s/2) - kA_s(d - s) = 0$$

$$s^2b/2 + kA_s s - kA_s d = 0 \tag{10.32}$$

Solution of this quadratic equation gives the dimension s locating the neutral axis and establishing the dimensions of the concrete above the neutral axis.

The maximum stress σ_{con} in the concrete portion of the beam is given by:

$$\sigma_{con} = Ms/I_z \tag{10.33}$$

The stress in the steel reinforcing rods is given by:

$$\sigma_{steel} = kM(d - s)/I_z \tag{10.34}$$

Note in Eq. (10.34), we have incorporated the multiplier k in the relation to convert the results from an equivalent concrete beam to a real beam containing steel reinforcing rods.

Finally, the second moment of the area illustrated in Fig. 10.19c is given by:

$$I_z = bs^3/3 + kA_s(d - s)^2 \tag{10.35}$$

The first term refers to the concrete above the neutral axis, and the second term is from the parallel axis theorem (Ad^2) for the steel rods. We have neglected the moment of inertia of the rods about their centroidal axis (I_z^2). This term is usually very small (much less than 1%) for reinforced concrete beams so it can be omitted without impairing safety.

EXAMPLE 10.16

A concrete floor slab is reinforced with four ½ in. diameter steel bars placed on 8 in. centers. The slab is 34 in. wide and 6 in. thick, and the centerline of the steel rebar is 1 in. above the bottom surface of the slab. The slab is 20 ft. long and is simply supported when installed. If a uniformly distributed load q is applied to the floor, determine the largest magnitude of q if the stress in the steel is limited to 15,000 psi and in the concrete to 1,200 psi. The modulus of elasticity of the steel and the concrete is 30×10^6 and 2.5×10^6 psi respectively.

Solution:

Let's first determine the equivalent area of the model produced by the four steel rebars. From Eqs. (10.30) and (10.31), we write:

$$A_{eq} = kA_s = (E_s / E_c)(\pi d^2/4)N$$

where N is the number of rebars.

$$A_{eq} = (30/2.5)[\pi(0.5)^2/4]4 = (12)(\pi/16)(4) = 9.425 \text{ in}^2 \qquad \text{(a)}$$

Next, let's use Eq. (10.32) to locate the position of the neutral axis

$$(b/2)\, s^2 + kA_s s - kA_s d = (b/2)\, s^2 + A_{eq} s - A_{eq} d = 0$$

$$(34/2)s^2 + 9.425\, s - 9.425(6 - 1) = 0$$

Solving this quadratic equation gives:

$$s = 1.411 \text{ in.} \qquad \text{and } s = -1.965 \text{ in.} \qquad \text{(b)}$$

We select the positive value because the negative result has no physical meaning.

Having established the position of the neutral axis we may determine I_z from Eq. (10.35) as:

$$I_z = bs^3/3 + kA_s(d - s)^2 = 34(1.411)^3/3 + 9.425(5 - 1.411)^2 = 153.2 \text{ in}^4. \qquad \text{(c)}$$

The maximum stress σ_{con} in the concrete portion of the beam is given by Eq. (10.33) as:

$$\sigma_{con} = Ms/I_z = 1{,}200 = 1.411M/153.2 \qquad \text{(d)}$$

Solving this relation for M yields:

$$M_{max} = 130{,}300 \text{ in-lb}$$

For a uniformly loaded beam, the maximum moment is:

$$M_{max} = qL^2/8 = 130{,}300 \qquad \text{(e)}$$

$$q_{max} = (8)(130{,}300)/[(20)(12)^2] = 18.10 \text{ lb/in.} = 217.2 \text{ lb/ft} \qquad \text{(f)}$$

We reach the limit of the stress in the concrete when the uniformly distributed load q is 217.2 lb/ft. However, it is necessary to determine if the limiting stress in the steel rebar is achieved with a higher or lower value of q. The stress in the steel reinforcing rods is given by Eq. (10.34) as:

$$\sigma_{steel} = kM(d - s)/I_z = 15{,}000 = (30/2.5)(5 - 1.411)M/153.2 \qquad (g)$$

Solving this relation for M yields:

$$M_{max} = 53{,}360 \text{ in-lb}$$

And in a similar manner, we solve for the limiting value of q as:

$$q_{max} = 8\, M_{max}/L^2 = (8)(53{,}360)/[(20)(12)^2] = 7.412 \text{ lb/in.} = 88.9 \text{ lb/ft} \qquad (h)$$

Compare Eqs. (f) and (h) and select the smaller value to give q_{max} = 88.9 lb/in.

An examination of the results in Eqs. (f) and (h) indicates that the design of the reinforced slab is not balanced. The steel meets the limiting stress of 15, 000 psi at loads less than one half that required for the concrete to reach its limit stress. To improve the design of the slab the amount of steel should be increased to better balance the load carrying capacity of the steel and concrete. Problem 10.35 provides you with the opportunity to increase the size or number of the rebars to increase the amount of steel and better balance the design of the reinforced concrete beam.

EXAMPLE 10.17

A concrete beam is reinforced with three 1 in. diameter steel bars placed on 6 in. centers. The section of the beam is rectangular—14 in. wide and 24 in. high with the centerline of the steel rebar 1.5 in. above the bottom surface of the beam. The beam is 24 ft. long and simply supported. If it supports three concentrated loads F = 10 kip, as shown in Fig. E10.17, determine the margin of safety for the beam. The yield strength of the steel is 42 ksi and the compressive strength of the concrete is 4,500 psi. The modulus of elasticity of the steel and the concrete is 30×10^6 and 3.0×10^6 psi respectively.

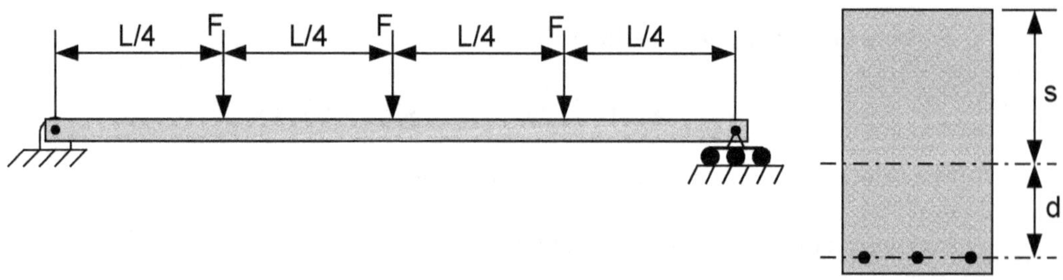

Fig. E 10.17

Solution: The reactions for the symmetrically loaded beam are:

$$R = R_L = R_R = 3F/2 = 3(10)/2 = 15 \text{ kip} \qquad (a)$$

The bending moment is a maximum at the center of the beam where x = L/2 and is:

$$M_{max} = R_L(L/2) - F(L/4) = FL/2 = (10{,}000)(24)(12)/2 = 1.44 \times 10^6 \text{ in-lb} \qquad (b)$$

After the point of failure and the maximum bending moment for the composite beam have been determined, the location of the neutral axis is ascertained from Eq. (10.32).

$$s^2b/2 + kA_s s - kA_s d = 0$$

$$7s^2 + 10(3\pi/4)s - 10(3\pi/4)(22.5) = 0$$

From the quadratic formula, we calculate that:

$$s = 7.18 \text{ in.} \qquad \text{and } s = -10.55 \text{ in.} \tag{c}$$

Again, ignore the negative root ands select s = 7.18 in.

From Eq. (10.35), we establish that I_z is:

$$I_z = bs^3/3 + kA_s(d - s)^2$$

$$I_z = 14(7.18)^3/3 + 10(3\pi/4)(22.5 - 7.18)^2 = 7257 \text{ in}^4. \tag{d}$$

The maximum stress σ_{con} in the concrete portion of the beam is given by Eq. (10.33) as:

$$\sigma_{con} = Ms/I_z = (1.44 \times 10^6)(7.18)/7257 = 1425 \text{ psi} \tag{e}$$

The maximum stress in the steel reinforcing rods is given by Eq. (10.34) as:

$$\sigma_{steel} = kM(d - s)/I_z = (10)(1.44 \times 10^6)(22.5 - 7.18)/7257 = 30.40 \text{ ksi} \tag{f}$$

The safety factors and the margins of safety are given by:

$SF_{con} = S_{con}/\sigma_{con} = 4500/1425 = 3.16$ $\qquad$ $MOS = SF_{con} - 1 = 3.16 - 1 = 216\%$

$SF_{steel} = S_{steel}/\sigma_{steel} = 42.0/30.4 = 1.38$ $\qquad$ $MOS = SF_{steel} - 1 = 1.38 - 1 = 38\%$

A review of these results indicates that the margin of safety for the steel is too low, and for the concrete, it is more than adequate. Clearly, the design of the composite beam is deficient. The fact that one safety factor is high while the other is low indicates that the amount of steel relative to the size of section is not properly balanced. In this example, the amount of steel is inadequate. Adding more steel (increasing A_s) increases s, decreases (d – s), and increases I_z. All of these changes decrease the stresses in the steel rebar. Problem 10.36 is provided to enable you to design a reinforced concrete beam of the size described in Example 10.17 with a margin of safety for the steel that is more comparable with the margin of safety for the concrete.

10.5.3 General Case of Beams with Two Materials

In Section 10.5.1, we studied the case of a composite beam fabricated with thin metal cover plates on the top and bottom and a plastic foam core in the center. We ignored the effect of the foam because its elastic modulus E_f was several orders of magnitude less than the elastic modulus of the metal cover plates E_m. This fact allowed us to assume that the cover plates carried the entire load in the beam. However, what would happen if the core were wood instead of foam? Wood is stronger than foam, and its stiffness is much closer to that of structural metals. It seems reasonable to expect that the wood would carry a portion of the load in the beam. In this section, we will study the case of composite beams fabricated from two materials that both share the load applied to the beam.

Let's begin with a composite "sandwich" beam composed of two materials having a cross section shown in Fig. 10.20. The core is made of one material (with E_1) and the cover plates are made of a second material (with E_2). To analyze the beam, we again assume that plane sections remain plane before and after loading so the strain is linearly distributed as shown in Fig. 10.21. Note that the strain is continuous across the material interface between the core and the cover plates.

The stresses in the core and cover plate materials are given by:

$$\sigma_1 = E_1 \varepsilon_1 \tag{a}$$

$$\sigma_2 = E_2 \varepsilon_2 \tag{b}$$

where the strains ε_1 and ε_2 depend on the distance from the neutral axis. At the interfaces between the core and the cover plates, the strains are equal in both materials ($\varepsilon_1 = \varepsilon_2$) because the distances from the neutral axis to the junctions are identical. However, Eqs. (a) and (b) indicate that the stresses are not equal ($\sigma_1 \neq \sigma_2$) because the elastic moduli E_1 and E_2 of the materials are different. The stress distribution across the cross section is shown in Fig. 10.22 and, unlike the strain distribution, is not continuous across the material interface.

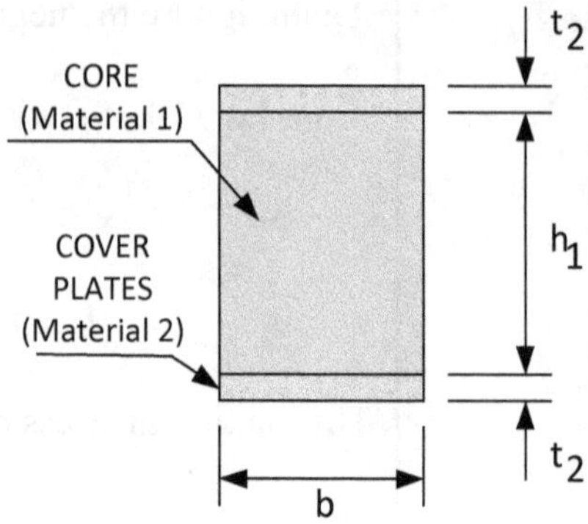

Fig. 10.20 Cross section of a composite "sandwich" beam with a core and cover plates.

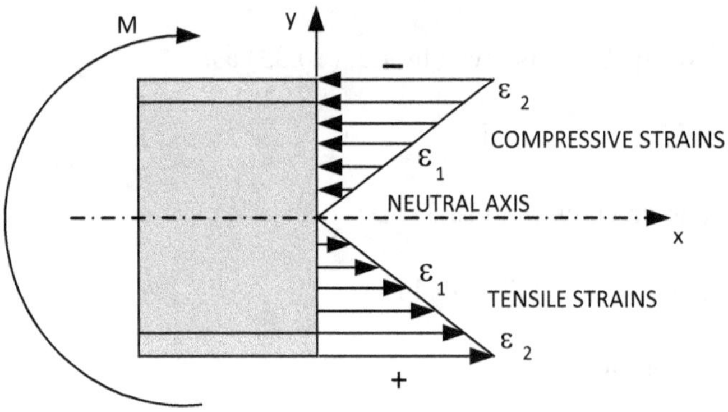

Fig. 10.21 The strain distribution is continuous across the material interface.

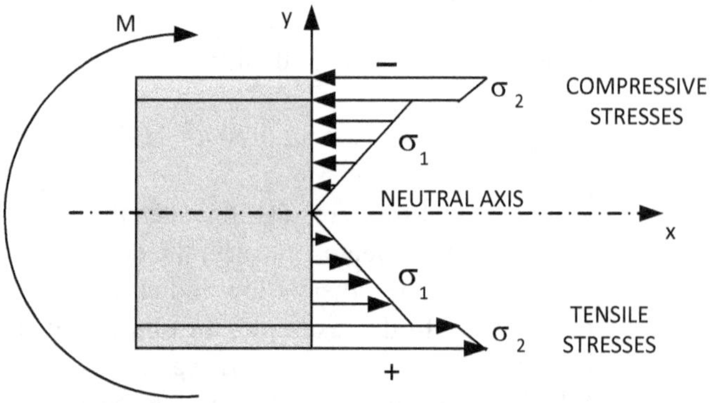

Fig. 10.22 The stress distribution is not continuous across the material interface.

To continue the analysis, we must convert the composite (heterogeneous) beam into a homogeneous beam by transforming the cross sectional area. First, we establish the transformation factor k as the ratio of the elastic moduli of the two materials:

$$k = E_1/E_2 \qquad (10.36)$$

The material in the core (E_1) will be changed into a modified material with a stiffness equivalent to the cover plate material (E_2). Let's assume that $E_2 > E_1$, so that the transformation factor k is less than 1. Next, we transform the cross section by adjusting the width of the modified material by the transformation factor:

$$b_{1T} = k\, b_1 \qquad (10.37)$$

The transformed cross section composed of only the cover plate material (2) is depicted in Fig. 10.23.

From the figure, we see that the width of the core material (1) has changed to the value given by Eq. (10.37). Because the transformation factor is less than 1, the new width is smaller than the original width. The transformed cross section appears as an I-beam composed entirely of the cover plate material (2). Notice that the cover plate material (2) remains unchanged after transformation. Also, the vertical dimensions in both materials are not altered, to preserve the linearity of strain across the section. In fact, the stress distribution also varies linearly across the transformed section. Indeed, the transformed section is completely equivalent to the original cross section in terms of beam stiffness.

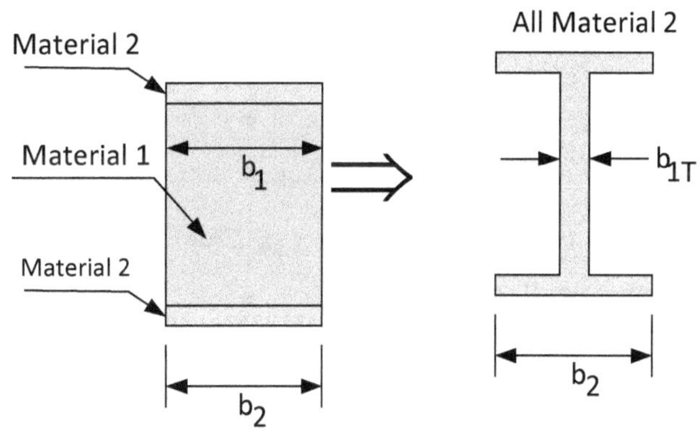

Fig. 10.23 Transformed cross section of a composite beam.

After transforming the cross section into a single material, we can now solve for stresses at any point along the new cross section by using the elastic flexure formula from Eq. (10.7):

$$\sigma_T = -M\,y/I_{Tz} \qquad (10.7\text{ bis})$$

In this equation, the moment of inertia I_{Tz} refers to the transformed section shown in Fig. (10.23). We can use concepts from Appendix C to determine the moment of inertia for this new cross section. Also, remember that stresses calculated from Eq. (10.7) are for the transformed section. We must use the following equations to determine stresses in the actual beam:

$$\sigma_1 = -k\,M\,y/I_{Tz} \qquad (10.38a)$$

$$\sigma_2 = -M\,y/I_{Tz} \qquad (10.38b)$$

Eqs. (10.38) show that the transformation factor is required to determine stresses in the modified core material (1) but not for the cover plate material (2), which was not transformed. The maximum stress in each material occurs at the farthest point from the neutral axis. This point occurs at the top and bottom edges of the beam for the cover plates, but is located at the interface between the core and cover plates for core material.

EXAMPLE 10.18

A composite "sandwich" beam is fabricated by adhesively bonding steel strips to the top and bottom faces of a western white pine beam. The cross section of the wood is 150 × 150 mm square, and the steel strips each have a 150 × 20 mm rectangular cross section. The beam is subjected to a maximum bending moment of 20 kN-m. Determine the maximum stress in the wood and in the steel. Assume the modulus of elasticity for the wood is 10 GPa and for the steel is 200 GPa.

Solution:

Start by transforming the beam into one equivalent material—steel. The appropriate transformation factor is given by Eq. (10.36) as:

$$k = E_w/E_s = (10)/(200) = 0.05 \qquad (a)$$

The thickness of the wood in the transformed section is determined from Eq. (10.37):

$$b_{1T} = k\,b_1 = (0.05)(150) = 7.5\text{ mm} \qquad (b)$$

A diagram of the original and transformed cross sections is shown in Fig. E10.18.

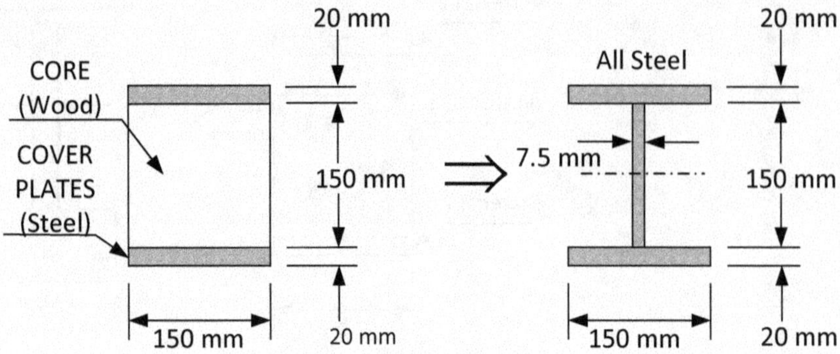

Fig. E10.18

The transformed cross section becomes an I-beam composed entirely of steel. Because this beam has vertical symmetry, the neutral axis is located at the centerline of the section, which is 95 mm from its bottom edge. Next, the transformed moment of inertia I_{Tz} must be determined. One method is to break the section into three rectangles, a vertical one for the web and two horizontal ones for the flanges, and find the moment of inertia of each rectangle about the neutral axis. The moment of inertia of the web is:

$$I_{wz} = (1/12)\, b_w\, (h_w)^3 = (1/12)(7.5)(150)^3 = 2.109 \times 10^6 \text{ mm}^4 \qquad \text{(c)}$$

The moment of inertia for each flange is more complex because the parallel axis theorem is involved:

$$I_{fz} = (1/12)\, b_f\, (h_f)^3 + A_f\, (d_f)^2 = (1/12)(150)(20)^3 + (150)(20)(95 - 10)^2 \qquad \text{(d)}$$

$$= 21.78 \times 10^6 \text{ mm}^4$$

Finally, the total moment of inertia for the cross section is determined from Eqs. (c) and (d) as:

$$I_{Tz} = I_{wz} + 2\, I_{fz} = (2.109 \times 10^6) + 2(21.78 \times 10^6) = 45.67 \times 10^6 \text{ mm}^4 \qquad \text{(e)}$$

The maximum stress in the steel occurs at the top and bottom surface of the beam where y = c = 95 mm. The maximum stress in the wood occurs at the junction between the core and the cover plates where y = 75 mm. Substituting these values, the bending moment, and I_{Tz} into Eqs. (10.38) gives the two stresses as:

$$\sigma_w = k\, M\, y/I_{Tz} = (0.05)(20 \times 10^6)(75)/(45.67 \times 10^6) = 1.642 \text{ MPa} \qquad \text{(f)}$$

$$\sigma_s = M\, y/I_{Tz} = (20 \times 10^6)(95)/(45.67 \times 10^6) = 41.60 \text{ MPa} \qquad \text{(g)}$$

Note that the moment terms appearing in (f) and (g) have been converted to N-mm to maintain consistent units. Also, the negative signs from Eqs. (10.38) have been dropped for convenience. The answers show that the maximum stress in the steel is over 25 times greater than the largest stress in the wood. Clearly the steel carries a much larger portion of the load applied to the beam than the wooden core. It is interesting to compare these results with the results for the beam with a foam core and metal cover plate described in Section 10.5.1. Recall in that earlier case, the contribution of the foam core was ignored and all of the beam's load carrying capacity was attributed to the cover plates. Using Eq. (10.29) from Section 10.5.1 in solving Example 10.18, gives a maximum stress in the steel of 43.83 MPa. Thus, by ignoring the wood completely, the stress in the steel is over-estimated by only about 5%.

10.6 PLASTIC BENDING

In our discussion of beams, we have considered only their elastic response and have limited our interest to situations where the stresses exhibited magnitudes lower than the yield strength. This approach insures an adequate margin of safety based on the elastic behavior of the beam's material. However, it is important to realize that beams do not suddenly collapse when the applied bending moment exceeds that required to produce stresses equal to the yield strength.

The behavior of a beam subjected to moments larger than the yield moment depends on the post-yield response of the beam material. In this discussion, we consider a well-known class of materials: **elastic perfectly-plastic**. The stress-strain diagram for an elastic perfectly-plastic material is shown in Fig. 10.24. The material responds elastically until it yields, and then additional strain is imposed at constant stress $\sigma = S_y$.

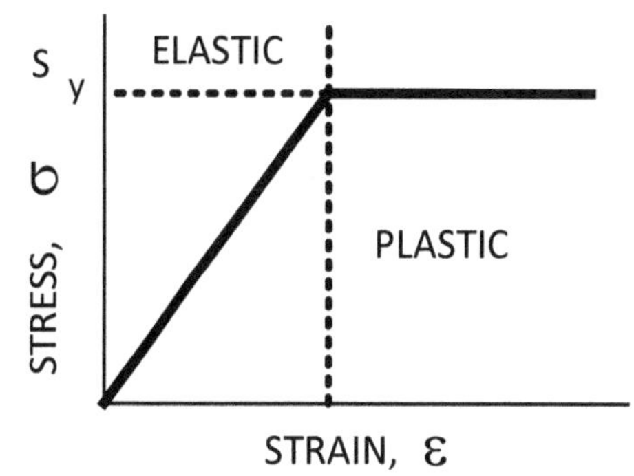

Fig. 10.24 The stress-strain diagram for an elastic perfectly-plastic material.

As we increase the moment applied to a beam, both the stress and the strain at the outer fibers of the beam increase linearly with the moment until $\sigma = S_y$ as illustrated in Fig. 10.25a. When the moment is increased by an additional increment, the strain continues to increase, and the distribution of strain from the neutral axis to the outside fibers remains linear with respect to the position y. As indicated in Fig. 10.25b, no discontinuity occurs at the elastic-plastic interface, as the curvature κ of the beam becomes larger.

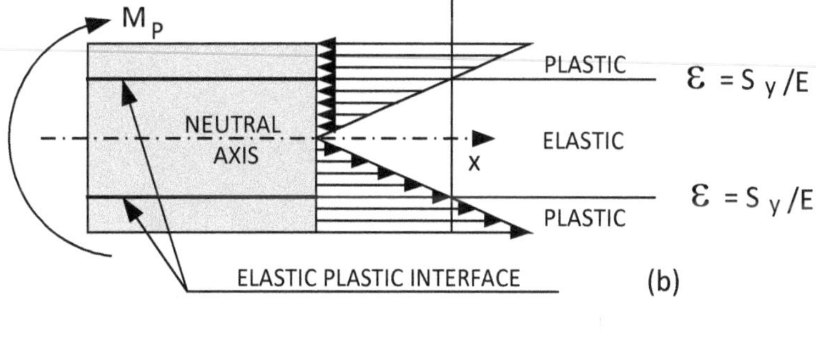

Fig. 10.25 Strain distribution due to pure bending: (a) at yield and (b) beyond yield.

The distribution of stress with position y across the section exhibits a discontinuity at the elastic-plastic interface as shown in Fig. 10.26. The maximum stress $\sigma_{MAX} = S_y$ is achieved at the outside fibers when M equals the yield moment M_Y. As the moment is increased beyond M_Y, the outer fibers yield, the stress in the plastic region remains constant at $\sigma = S_y$, and an elastic-plastic interface develops. This elastic-plastic interface separates the cross section into two regions—an elastic core near the neutral axis and plastic zones adjacent to the top and bottom surfaces.

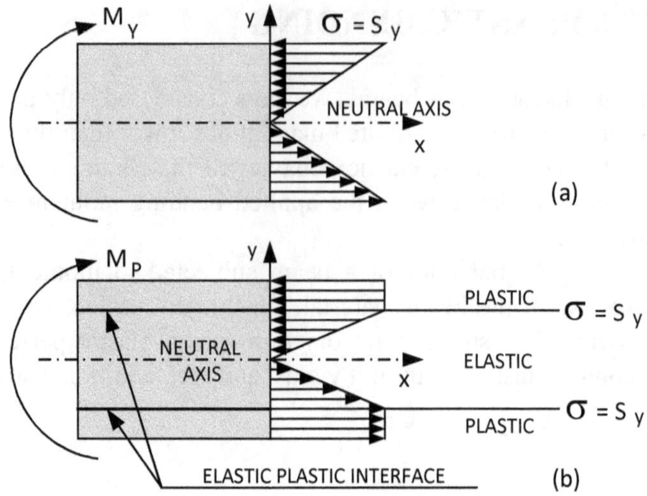

Fig. 10.26 Stress distributions (a) when $M = M_Y$; and (b) when $M_Y < M < M_L$.

The moment M_Y required to produce a stress $\sigma = S_y$ on the outside fibers of a beam with a rectangular cross section is given by:

$$M_Y = \frac{\sigma(2I_z)}{h} = \frac{S_y}{h}(2)\frac{1}{12}(bh^3) = S_y\frac{bh^2}{6} \tag{10.39}$$

The moment M_P that is required to drive the elastic-plastic interface to some position y_p is determined from Fig. 10.27. The equal and opposite forces F_e and F_p form two couples—one associated with the elastic region ($F_e\, d_e$) and the other with the plastic region ($F_p\, d_p$).

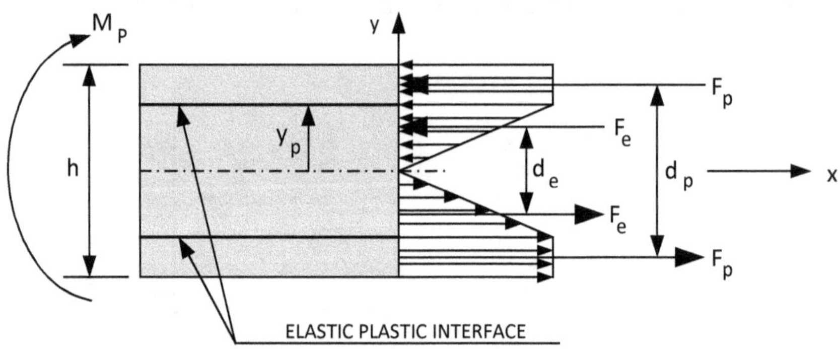

Fig. 10.27 Couples developed in the elastic and plastic regions of the cross section.

The forces, due to the stress σ, are given by:

$$F_p = (b/2)(h - 2y_p)S_y \quad \text{and} \quad F_e = (by_p/2)S_y \tag{a}$$

And the distances d_p and d_e are given by:

$$d_p = (1/2)(h + 2y_p) \quad \text{and} \quad d_e = (4/3)y_p \tag{b}$$

Summing moments about the origin of the x-y coordinate system gives:

$$\Sigma M_O = F_p\, d_p + F_e\, d_e - M_P = 0 \tag{c}$$

Substituting Eqs. (a) and (b) into Eq. (c) and solving for M_P yields:

$$M_P = S_y b\left(\frac{1}{2}(h - 2y_p)\frac{1}{2}(h + 2y_p) + \frac{y_p}{2}\frac{4y_p}{3}\right) = \frac{S_y b}{12}(3h^2 - 4y_p{}^2) \tag{10.40}$$

Equation (9.29) indicates that beams are capable of supporting moments larger than the yield moment M_Y. The beam will be damaged in the processes because it will be permanently deformed; however, it will not collapse. The question then arises as to how large of an overload we can place on the beam before it collapses.

To answer this question, consider the behavior of the beam when the elastic-plastic interface is driven into the neutral axis. The elastic core disappears and the entire cross section of the beam is in the plastic regime. At this point, the beam has become **fully plastic** reaching its **limit moment** where additional moments cannot be supported.

It is easy to determine the limit moment, M_L from Eq. (10.37) if we set $y_p = 0$, thereby moving the position of the elastic-plastic interface to the neutral axis. Then, Eq. (10.37) reduces to:

$$M_L = S_y\, bh^2/4 \qquad (10.41)$$

A comparison of the yield moment and the limit moment is possible from Eq. (10.39) and Eq. (10.41). The ratio M_L/M_Y is given by:

$$\frac{M_L}{M_Y} = \frac{S_y bh^2}{4}\frac{6}{S_y bh^2} = \frac{3}{2} \qquad (10.42)$$

Clearly for the rectangular cross section, the limit moment is 150% of the yield moment. This is a comfortable margin that adds to the safety factor if collapse is an important criterion for design. However, the ratio of M_L/M_Y is a function of the shape of the cross section. For WF and standard I beam shapes the ratio is usually about 1.10 to 1.15.

EXAMPLE 10.19

Consider the simply supported beam of length L = 8 m with a concentrated force **F** located 3 m from the left end as shown below. Determine the force required to produce the yield moment and the force required for the beam to collapse. The cross section of the beam is rectangular with a height h = 100 mm and a width b = 50mm. The beam is fabricated from steel with a yield strength of 300 MPa. We will assume that the steel responds like an elastic-perfectly-plastic material.

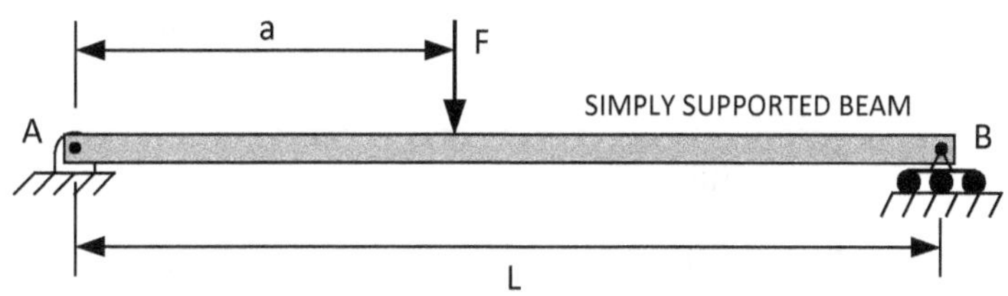

Fig. E10.19

Solution:

Recall from Section 10.4.1 that the reaction forces R_R and R_L are given by:

$$R_R = Fa/L \qquad \text{and} \qquad R_L = Fc/L \qquad (a)$$

where c = L – a.

In Eq. (10.16), we showed that the moment was a maximum at x = a (under the load point):

$$M_{max} = Fac/L \qquad (b)$$

Combining Eqs. (10.39) and (b) while setting $F = F_Y$ and $M_{MAX} = M_Y$ yields:

$$F_Y ac/L = S_y\, bh^2/6$$

$$F_Y = S_y\, bh^2L/(6ac) \qquad (c)$$

Substituting the parameters describing the beam and the material into Eq. (c) yields:

$$F_Y = \frac{S_y bh^2 L}{6ac} = \frac{(300 \times 10^6)(0.05)(0.1)^2(8)}{(6)(3)(8-3)} = 13.33\text{kN}$$

The force associated with the limit load is determined by substituting Eq. (10.41) into Eq. (b) or by recognizing that $M_L = 3M_Y/2$ as indicated in Eq. (10.42). Accordingly, we find that :

$$F_L = 3F_Y/2 = 20 \text{ kN}$$

When F_L produces the moment M_L, a **plastic hinge** forms under the load point and the beam collapses as indicated with the schematic illustration presented in Fig. E 10.19a.

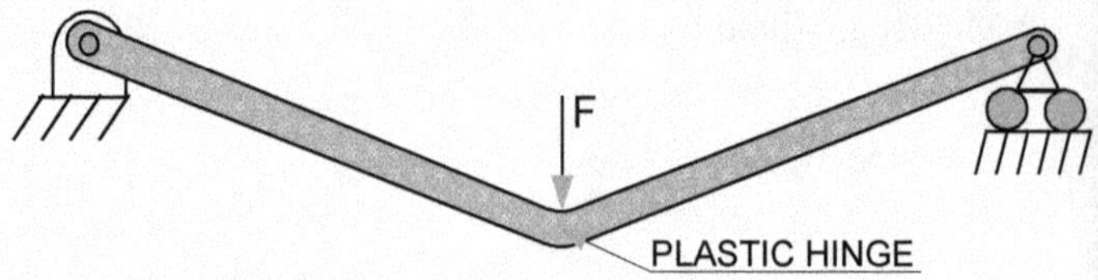

Fig. E 10.19a

A plastic hinge has formed under the load when $M = M_L$ with both sides of the beam rotating about this point. The hinge is not frictionless, because work ($M_L \times \theta$) is produced by the force F_L as it deflects the beam and rotates the plastic hinge.

EXAMPLE 10.20

Consider a simply supported beam of length L = 9 m with a uniformly distributed load q acting over its entire the length. Determine the magnitude of the uniformly distributed load required to produce the yield and limit moments. The cross section of the beam is rectangular with h = 120 mm and b = 40 mm. The beam is fabricated from steel with a yield strength of 360 MPa. We will assume that the steel acts like an elastic-perfectly-plastic material.

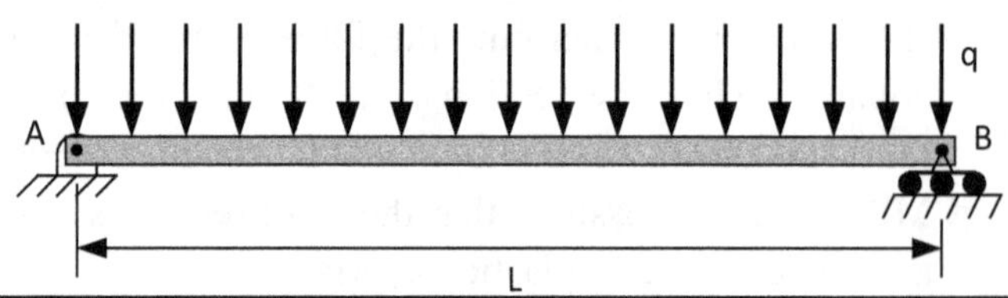

Fig. E10.20

Solution:

Recall from Example 10.7 that the reaction forces R_R and R_L are given by:

$$R_R = R_L = qL/2 \quad \text{(a)}$$

Also in Example 10.7, we showed that the moment was a maximum at x = L/2:

$$M_{MAX} = qL^2/8 \quad \text{(b)}$$

Combining Eqs. (10.39) and (b) yields:

$$M_{MAX} = q_Y L^2/8 = M_y = S_y\, bh^2/6$$

$$q_Y = (4/3)S_y b(h/L)^2 \quad \text{(c)}$$

Substituting the parameters describing the beam and its material into Eq. (c) yields:

$$q_Y = \frac{4S_y bh^2}{3L^2} = \frac{4(360 \times 10^6)(0.04)(0.12)^2}{3(9)^2} = 3.413\text{kN/m}$$

The distributed load associated with the limit load is determined by substituting Eq. (10.41) into Eq. (b).

$$q_L = \frac{2S_y bh^2}{L^2} = \frac{2(360\times10^6)(0.04)(0.12)^2}{(9)^2} = 5.120 \text{ kN/m}$$

When the distributed load $q = q_L$ a plastic hinge forms at the beam's center point and the beam collapses.

10.7 STRESS CONCENTRATIONS IN BEAMS

In our study of methods to analyze stresses in beams, we have assumed that the beams were continuous along their entire length. However, in design of structures it is common to cut notches in beams for one reason or another. In some instances, shouldered shafts or beams are employed and occasionally transverse holes are drilled in shafts to enable lubricant transfer. These geometric discontinuities produce stress concentrations in regions adjacent to the hole or notch. To determine the maximum stress, we account for these discontinuities by establishing the appropriate stress concentration factor. The stress concentration factor multiplies a nominal stress to give the maximum stress at the discontinuity. This procedure is identical to that described previously in Section 4.8.

10.7.1 Stress Concentration Factors

Due to Notches in Beams

Let's consider a beam with a rectangular cross section that is notched on its top and bottom surfaces as shown in Fig. 10.28.

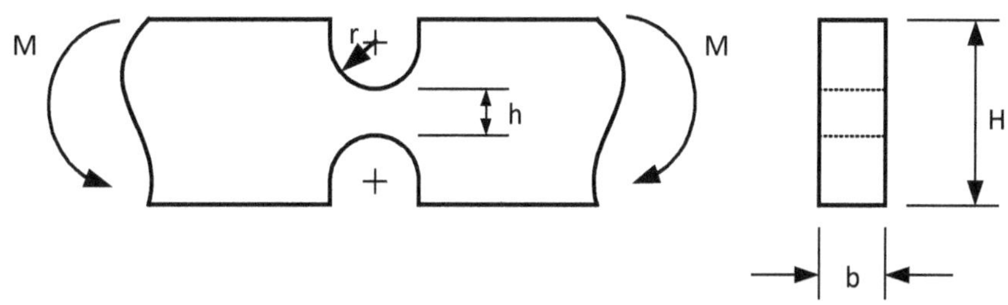

Fig. 10.28 A notched beam subjected to bending moments.

The stresses in the region adjacent to the notch are elevated due to the presence of the discontinuity. The maximum stress σ_{Max} occurs at the base of the notch, and is given by:

$$\sigma_{Max} = K\,\sigma_{Nom} \qquad (10.43)$$

where σ_{Nom} is the nominal stress and K is the stress concentration factor.

The nominal stress is the bending stress at the notched section that is given by:

$$\sigma_{Nom} = Mc/I_z = 6M/(bh^2) \qquad (10.44)$$

where b and h are defined in Fig. 10.28.

The stress concentration factor is a function of the geometry of the discontinuity and depends on the ratios H/h and r/h as shown in Fig. 10.29

EXAMPLE 10.21

A simply supported beam with a uniform load is shown in Fig. E10.21. If notches are cut on the top and bottom surfaces at the center span of the beam, determine the nominal stresses, the stress concentration factor and the maximum stresses at the notches. The beams dimensions are L= 20 ft, b = 2.5 in., H = 8 in. and h = 5 in. The notch radius r = 0.5 in. is centered on the top and bottom edges of the beam. The distributed load q = 0.8 kip/ft.

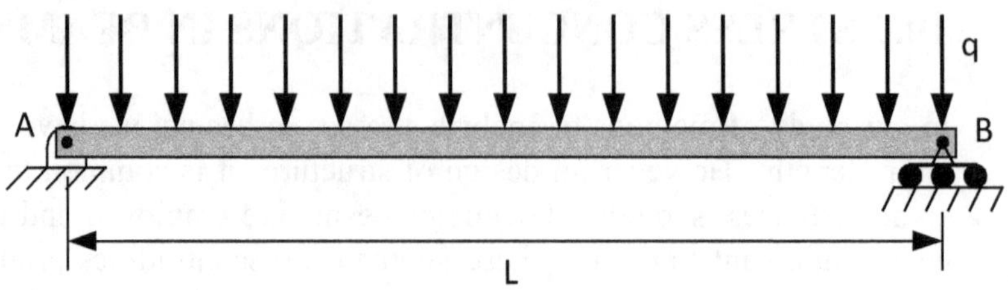

Fig. E10.21

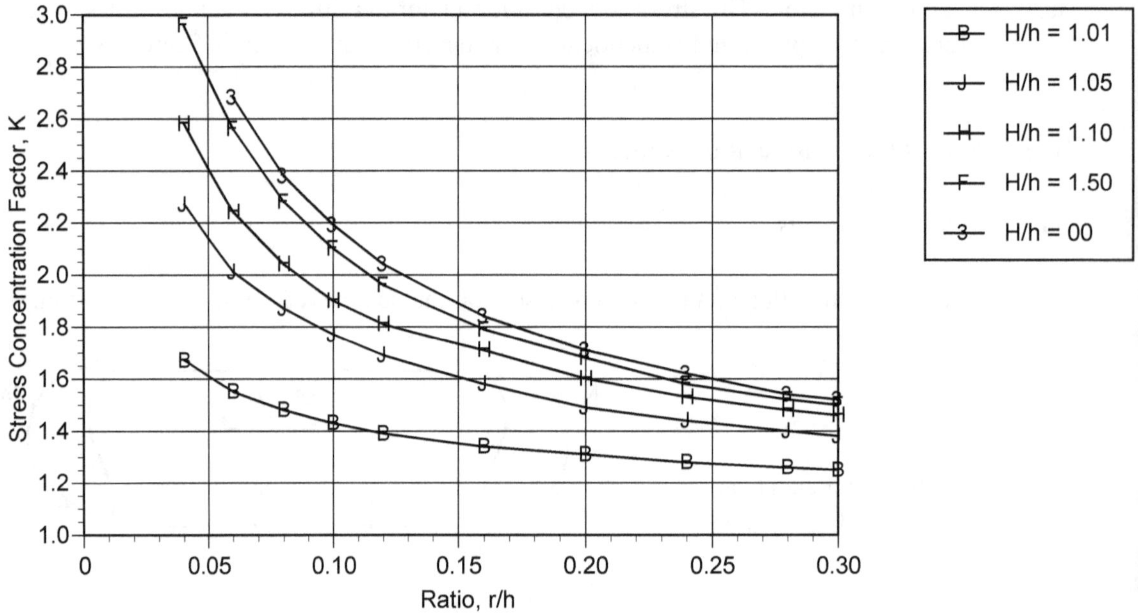

Fig. 10.29 Stress concentration factor for a notched beam in bending.

Solution: Let's begin by determining the moment at the midspan of the beam. In Example 10.7, we showed that the moment for a beam with a uniformly distributed load was a maximum at midspan given by:

$$M_{Max} = qL^2/8 = (800)(20)^2/8 = 40 \text{ kip-ft} = 480 \text{ kip-in.} \qquad (a)$$

Next calculate the nominal stress σ_{Nom} from Eq. (10.44) as:

$$\sigma_{Nom} = Mc/I_z = 6M/(bh^2) = 6(480 \times 10^3)/[(2.5)(5)^2] = 46.08 \text{ ksi} \qquad (b)$$

Reference Fig. 10.29 to find the stress concentration factor. The ratios giving the coordinates for K are r/h = 0.50/5.0 = 0.10 and H/h = 8.0/5 = 1.60. Then from Fig. 10.29, we establish that:

$$K = 2.11 \qquad (c)$$

Substituting Eqs. (b) and (c) into Eq. (10.43) yields:

$$\sigma_{Max} = K\,\sigma_{Nom} = (2.11)(46.08) = 97.23 \text{ ksi} \qquad (d)$$

Due to Grooves in Circular Shafts

Let's consider a shaft with a circular cross section that has a groove machined about its circumference as shown in Fig. 10.30. When the shaft is subjected to a bending moment M, the stresses in the region at the root of the groove are elevated due to the presence of the discontinuity. The procedure is the same as we described previously—we multiply a nominal value for the bending stresses by a suitable stress concentration factor to determine the maximum stresses in accordance with Eq. (10.43).

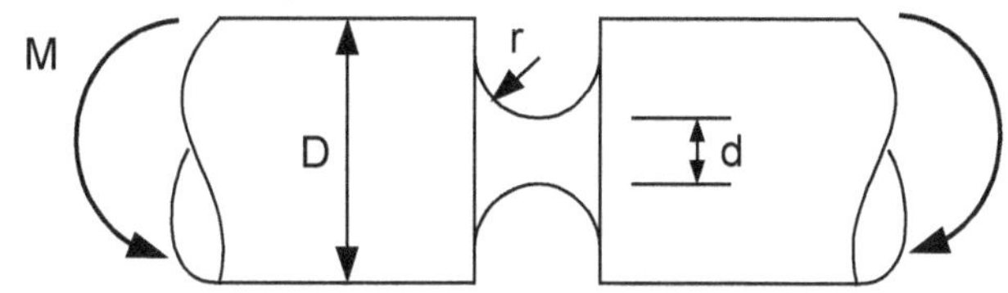

Fig. 10.30 A grooved shaft subjected to bending moments.

The nominal stress at the neck section where the groove is located is given by:

$$\sigma_{Nom} = 32M/\pi d^3 \quad (10.45)$$

The curves defining the stress concentration factors are presented in Fig. 10.31.

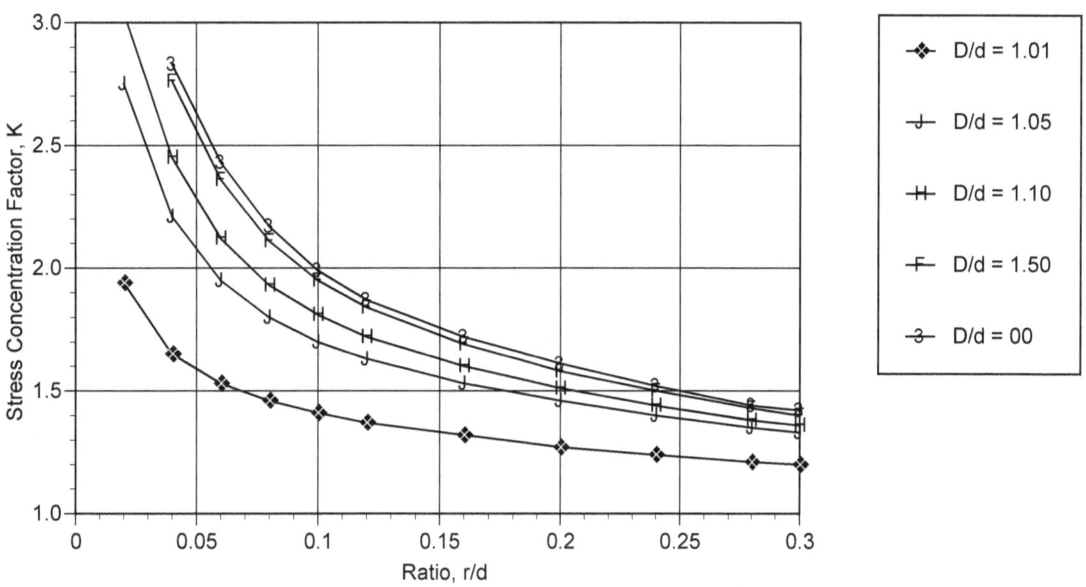

Fig. 10.31 Stress concentration factor for a grooved shaft.

EXAMPLE 10.22

A circular shaft is loaded in four-point bending producing a bending moment M = 5.0 kN-m over the central region of a circular shaft. If a groove is cut about the perimeter of the shaft at its midspan point, determine the nominal stresses, the stress concentration factor and the maximum stress at the base of the groove. The shaft's dimensions are D = 100 mm, d = 75 mm. The notch radius r = 15 mm is centered near the bottom of the grove.

Solution:

Let's begin by determining the calculating the nominal stress at the location of the groove from Eq. (10.45) as:

$$\sigma_{Nom} = 32M/\pi d^3 = (32)(5.0 \times 10^3)/[\pi(0.075)^3] = 120.7 \text{ MPa} \quad (a)$$

Reference Fig. 10.31 to find the stress concentration factor. The ratios giving the coordinates for K are r/d = 15/75 = 0.20 and D/d = 100/75 = 1.3333. Then from Fig. 10.31, we establish that K = 1.53.

Substituting the results for σ_{Nom} and K into Eq. (10.43) yields:

$$\sigma_{Max} = K\,\sigma_{Nom} = (1.53)(120.7) = 184.7 \text{ MPa}$$

Due to Shoulder Fillets in a Stepped Beam

Some beams are not continuous and are of two different heights—H and h. We show such a stepped beam in Fig. 10.32, which is acted upon by a bending moment M. A circular fillet is placed at the transition to mitigate the concentration of stresses at the reentrant corner. When the beam is subjected to bending, the stresses in the region adjacent to the fillet are elevated due to the presence of the step. The procedure to determine the stresses at the fillet is the same as we described previously.

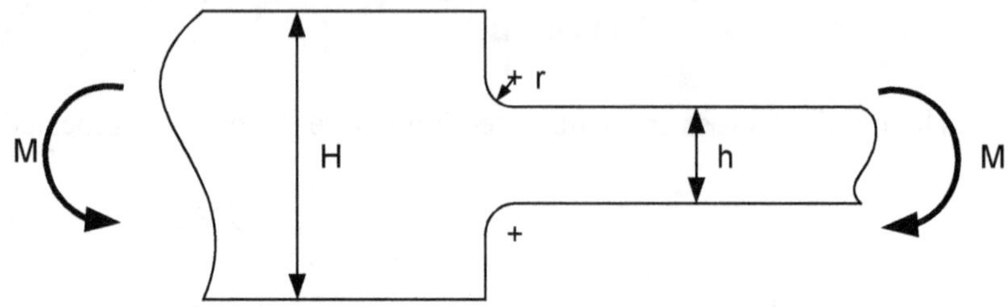

Fig. 10.32 A stepped beam subjected to bending.

The nominal stress σ_{Nom} for this case is determined from by Eq. (10.44), and the stress concentration factor is given in Fig. 10.33.

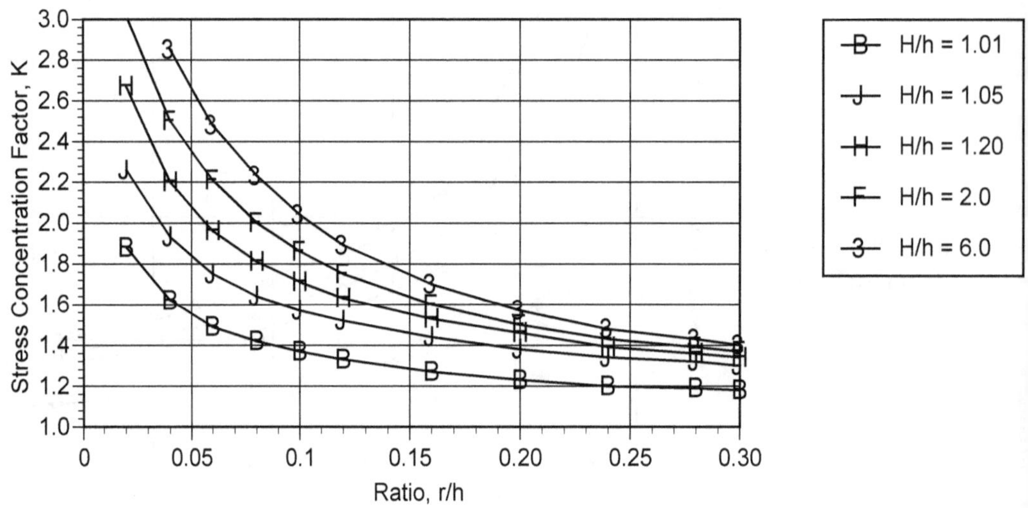

Fig. 10.33 Stress concentration factors for a stepped beam with shoulders.

Due to Shoulder Fillets in a Stepped Shaft

Many shafts are not continuous and are of two or more different diameters to accommodate the installation of gears, bearings, etc. We show such a stepped shaft in Fig. 10.34, which is acted upon by a bending moment M. A circular fillet is placed at the transition to mitigate the concentration of stresses at the reentrant corner. When the shaft is subjected to bending, the stresses in the region adjacent to the fillet are elevated due to the presence of the step. The procedure to determine the stresses at the fillet is the same as we described previously.

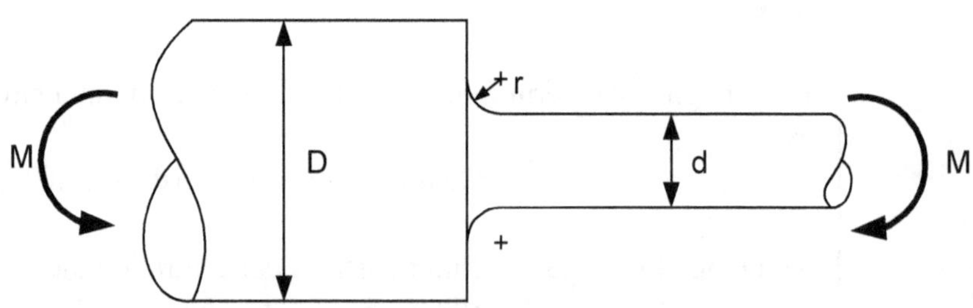

Fig. 10.34 Stepped shaft subjected to a bending moment M.

The nominal stress σ_{Nom} for this case is determined from by Eq. (10.45), and the stress concentration factor is given in Fig. 10.35.

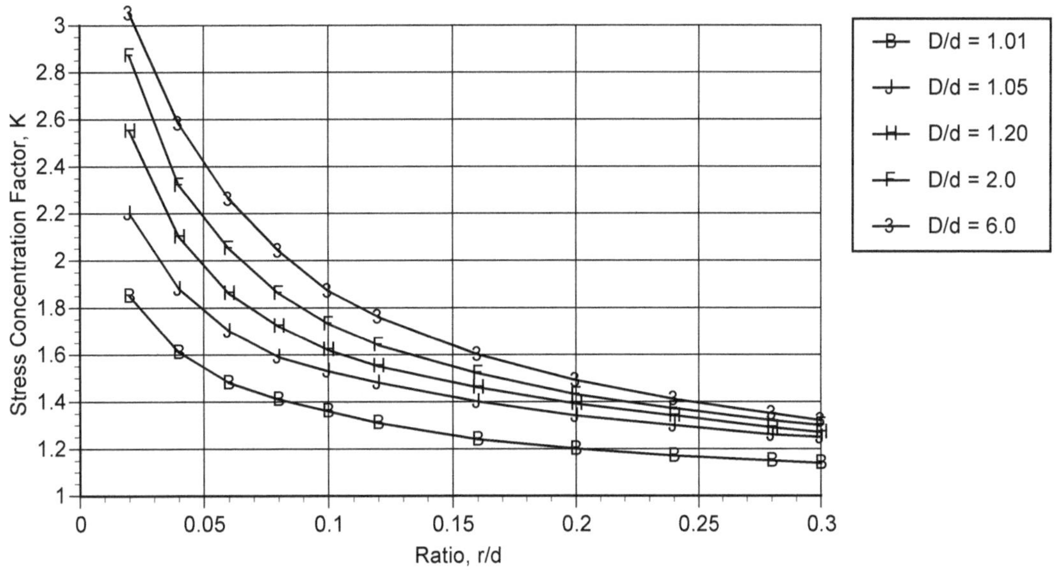

Fig. 10.35 Stress concentration factors for a stepped shaft.

EXAMPLE 10.23

A shaft with a circular cross section is designed with a step transition between two diameters—D = 2.0 in. and d = 1.0 in. The shaft is subjected to a bending moment of M = 150 ft-lb at the transition point. Determine the minimum radius of the fillet at the transition region if the maximum stress is not to exceed 40.0 ksi.

Solution:

Let's solve for the nominal stress at the transition region by using Eq. (10.45).

$$\sigma_{Nom} = 32M/(\pi d_3) = (32)(150)(12)/[\pi(1.0)^3] = 18.33 \text{ ksi} \qquad \text{(a)}$$

Next, determine the maximum allowable stress concentration factor from Eq. (10.43) as:

$$K_{Max} = \sigma_{Max}/\sigma_{Nom} = 40/18.33 = 2.182 \qquad \text{(b)}$$

Note that D/d = (2.0)/(1.0) = 2.0. Then select the dotted line curve for D/d = 2.0 from Fig. 10.35. Project a line horizontally from K = 2.182 and find the intercept with the dotted curve. Project a second line downward and read the intercept for r/d as 0.05. The radius of the fillet at the transition section must exceed r = (0.05)(1.0) = 0.05 in. so that $\sigma_{Max} \leq 40$ ksi.

Due to a Transverse Hole in a Circular Shaft

Many shafts are designed with transverse holes to enable the flow of lubricants or coolants to some other component in the system. We show a shaft with a small diameter transverse hole in Fig. 10.36. When the shaft is subjected to bending, the stresses in the region adjacent to the hole are elevated due to the presence of the discontinuity. The procedure to determine the maximum stresses at the hole is the same as we described previously.

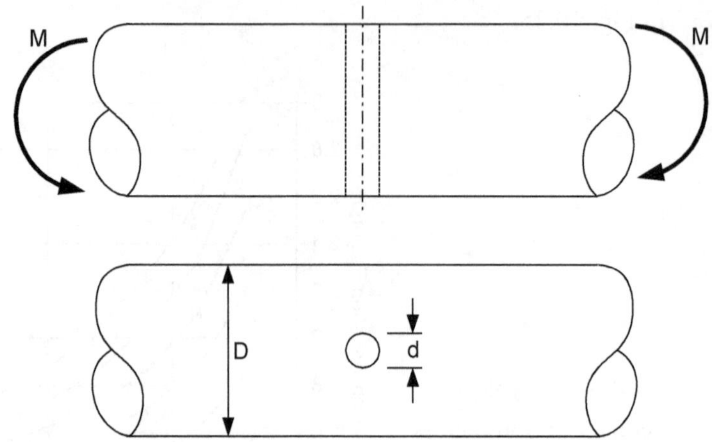

Fig. 10.36 A transverse hole in shaft with circular cross section that is subjected to a moment M.

The nominal stress σ_{Nom} at the location of the transverse hole is approximated by:

$$\sigma_{Nom} = \frac{M}{\frac{\pi D^3}{32} - \frac{D^2 d}{6}} \qquad (10.46)$$

The stress concentration factor, which is a function of the ratio of d/D, is presented in Fig. 10.37.

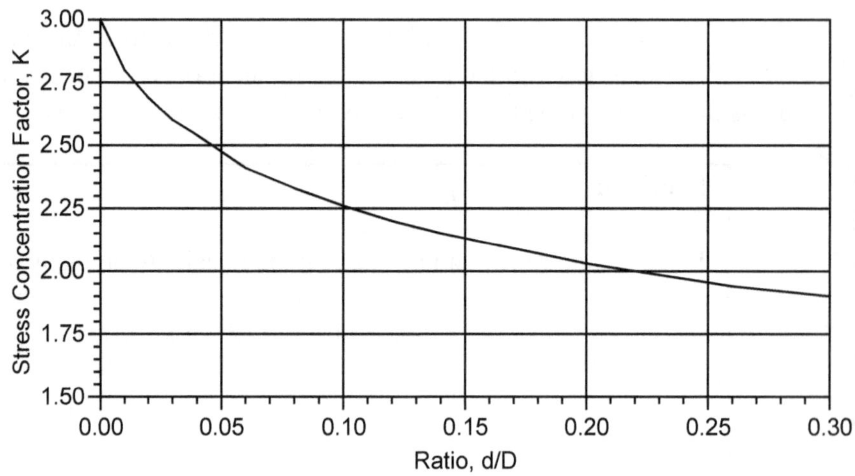

Fig. 10.37 Stress concentration factor for a shaft with a transverse hole.

EXAMPLE 10.24

A circular shaft with a diameter D = 125 mm is drilled with a transverse hole having a diameter d = 6 mm. If the shaft is subjected to a bending moment of 125 kN-m, determine the maximum stress adjacent the hole.

Solution: The nominal stress is determined from Eq. (10.46) as:

$$\sigma_{Nom} = \frac{M}{\frac{\pi D^3}{32} - \frac{D^2 d}{6}} = \frac{25 \times 10^6}{\frac{\pi (125)^3}{32} - \frac{(6)(125)^2}{6}} = 141.9 \text{ MPa} \qquad (a)$$

Using the ratio D/d = 6.0/125 = 0.048 and the curve in Fig. 10.37, we find that K = 2.49. The from Eq. (10.43) we write:

$$\sigma_{Max} = K\,\sigma_{Nom} = (2.49)(141.9) = 353.3 \text{ MPa}$$

10.8 SUMMARY

Beams are long thin members subjected to transverse forces—either reactionary or applied. These transverse forces produce internal moments in the beam, which bend it into a circular arc. The curvature of the arc depends on the internal moment and may vary along the length of the beam.

In bending, we assume that plane sections of the beam remain plane as the beam deforms. This assumption permits us to determine the longitudinal strain in the beam if we know its radius of curvature, ρ.

$$\varepsilon = -y/\rho \qquad (10.1)$$

The equations of equilibrium are employed to establish the position of the neutral axis of the beam at the centroid of its cross sectional area. We also showed the relation between the internal moment M and the radius of curvature as:

$$M = (EI_z/\rho) = \kappa EI_z \qquad (10.5)$$

With a knowledge of the relation between curvature and the moment, we were able to establish the important relationship between the internal moment and normal stress:

$$\sigma = -My/I_z \qquad (10.7)$$

Using Eq. (10.7), we consider beams with rectangular cross sections and showed that they were not efficient structural members because they have extensive regions near their neutral axis, which are subjected to relatively small stresses.

We introduced more efficient designs of cross sections by considering American standard I and WF shapes. An effectiveness factor SE = Z/A was introduced to judge the adequacy of the design of various cross sections for beams.

$$SE = \frac{1}{h(2bt_f + b_w h_w)}\left\{\frac{b_w h_w^3}{6} + \frac{bt_f^3}{3} + bt_f(h_w + t_f)^2\right\} \qquad (10.15)$$

In most cases, the moment varies along the length of the beam and it is necessary to determine the shear forces and bending moments as a function of position. We introduced methods for constructing free bodies of the beam to determine the reactions, shear forces and bending moments. Also, the method for constructing shear and bending moment diagrams was introduced. It is important to recall the relations between loads, shear force and bending moment as indicated below in constructing the shear and bending moment diagrams.

$$dV/dx = -q \qquad (10.20)$$

$$dM/dx = V \qquad (10.18)$$

$$M_{x_2} - M_{x_1} = \int_{x_1}^{x_2} V dx \qquad (10.19)$$

Several examples are presented to demonstrate methods for calculating stresses, safety factors, and margins of safety for beams with various types of supports subjected to both concentrated forces and uniformly distributed forces.

We have briefly introduced the topic of shear stress in beams, and showed that:

$$\tau = \tau_V = \tau_H = VQ_1/(bI_z) \qquad (10.23)$$

The shear stresses are not uniformly distributed over the cross section of the beam; they are a maximum at the neutral axis and zero at the top and bottom surfaces of the beam. In most cases, the shear stresses are small when compared to the normal stresses and they are often ignored. The exception is when the beams are fabricated from wood. The shear strength of wood is so small that even the low shear stresses may be sufficient to split wooden beams at the neutral axis.

The bending of composite beams fabricated from two materials was discussed. The first composite section described consisted of a foam core with adhesively bonded metal cover plates. The maximum stress in the metal cover plates is given by:

$$\sigma_{max} = \frac{M(h_f + 2t_m)}{bt_m(h_f + t_m)^2} \qquad (10.29)$$

We also developed methods for determining stresses in a concrete beam reinforced with steel rods and for a composite beam fabricated with a stiff core and cover plates. These methods involved the development of an equivalent section where one of the materials was transformed into an modified material with the same stiffness of the other material. The reinforced concrete beam was then analyzed using the equations listed below:

$$A_{eq} = kA_s \qquad (10.30)$$

$$k = E_s / E_c \qquad (10.31)$$

$$s^2 b/2 + kA_s s - kA_s d = 0 \qquad (10.32)$$

$$\sigma_{con} = Ms/I_z \qquad (10.33)$$

$$\sigma_{steel} = kM(d - s)/I_z \qquad (10.34)$$

$$I_z = bs^3/3 + kA_s(d - s)^2 \qquad (10.35)$$

The composite beam fabricated with a stiff core and cover plates was analyzed using the equations listed below:

$$k = E_1/E_2 \qquad (10.36)$$

$$b_{1T} = k\, b_1 \qquad (10.37)$$

$$\sigma_2 = - M\, y/I_{Tz} \qquad (10.38b)$$

$$\sigma_1 = - k\, M\, y/I_{Tz} \qquad (10.38a)$$

Examples were presented to demonstrate the analytical procedure and to introduce the concept of balanced design of composite beams.

The plastic response of beams was introduced for an elastic-perfectly-plastic material. We showed that the strain increased linearly with increasing moment even in the plastic regime. However, the stresses were limited to the yield strength of the beam's material. As a consequence, a discontinuity formed in the distribution of the stresses across the section. This discontinuity defines the elastic-plastic interface. As the moment increases the elastic-plastic interface is driven toward the neutral axis, a plastic hinge forms, and a simply supported beam collapses. Concepts of the yield moment, M_Y and the limit moment, M_L were introduced and equations were developed for computing these quantities for beams with rectangular cross sections.

$$M_Y = S_y \frac{bh^2}{6} \tag{10.39}$$

$$M_L = S_y\, bh^2/4 \tag{10.41}$$

Note that M is 1.5 times higher than M for beams with rectangular cross sections, but only 1.10 to 1.15 times higher for wide flange and standard I-beams.

Stress concentrations due to discontinuities in beams and circular shafts subjected to bending were presented. Stress concentration factors that account for the elevation of the stresses adjacent to the discontinuity were introduced for five different geometries. The maximum stresses are determined from Eq. (10.43).

$$\sigma_{Max} = K\, \sigma_{Nom} \tag{10.43}$$

PROBLEMS

10.1 Draw three different shapes of cross sectional areas that exhibit symmetry about the vertical or y axis.

10.2 For the beam shown in the figure to the right, construct the shear and bending moment diagrams and write the equations for the bending moment as a function of x as it varies from zero to L.

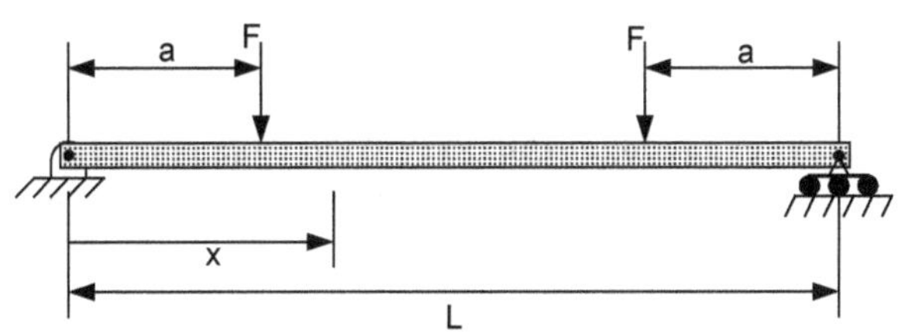

10.3 In a manufacturing process, an aluminum sheet with a thickness of 1.2 mm is passed around a roller with a 450 mm diameter. Determine the maximum strain in the sheet. Also determine the maximum stress.

10.4 A beam with a rectangular cross section has a depth b = 80 mm and a height h = 600 mm. If the yield strength of the beam's material is 200MPa, determine the maximum moment that the beam can support before it yields if the margin of safety is 100%.

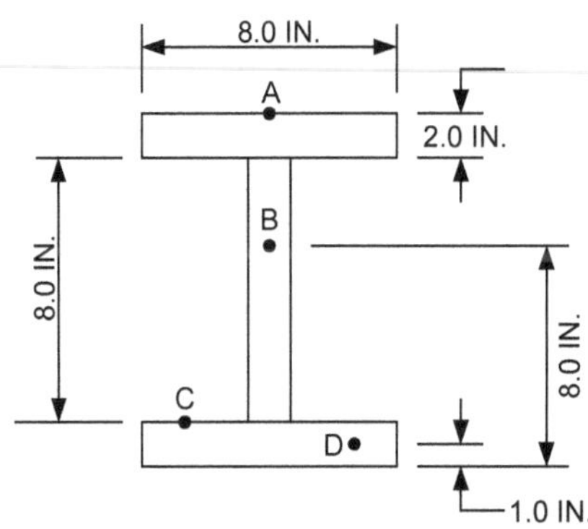

10.5 Three planks with dimensions of 2.0 by 8.0 inches are adhesively bonded together to form a timber I beam as shown in the figure to the right. Determine the stresses due to bending at points A, B, C and D. The bending moment applied to the beam is 27,000 ft-lb.

10.6 A WT6 × 60 structural T section is employed as a cantilever beam. The beam is fabricated from steel that has a yield strength of 40 ksi in tension and 85 ksi in compression. Bending occurs about the z-axis and a safety factor of at least 2.8 is required. Determine the largest positive bending moment that the beam can withstand if all of the requirements are satisfied. Also determine the maximum negative bending moment that the beam can withstand.

10.7 For the beam section—W406 × 100, calculate values of I_z and Z relative to the Z-Z axis from data given in Appendix D. Compare your values with those listed in the table.

10.8 Consider the flange and web section, defined in the figure to the left. If $b_w = xb$ and $h_w = yh$, determine the effectiveness SE as x varies from 0.05 to 0.3 and y varies from 0.6 to 0.95. It is suggested that you perform these calculations on a spreadsheet and prepare a graph displaying your results.

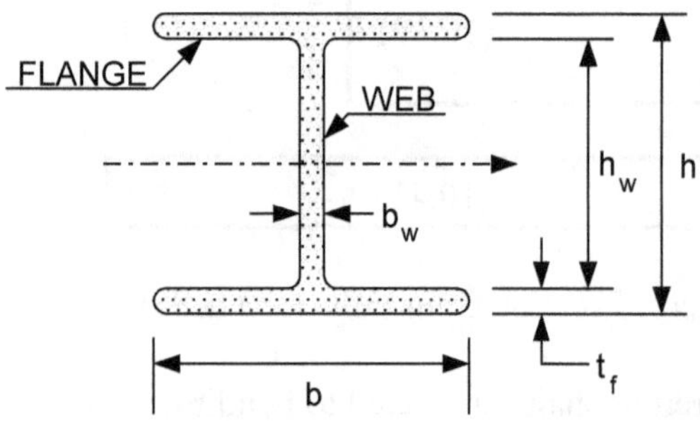

10.9 Two steel plates 9.0 in. by 0.75 in. are welded to the flanges of an S18 × 70 American Standard I-beam as shown in the figure to the right. Calculate the moment of inertia about each of the axes of symmetry.

10.10 If the thickness of the two plates were increased from 0.75 in. to 1.25 in. for the modified I-beam shown in the figure to the right, determine the moment of inertia about each of the axes of symmetry.

10.11 For the modified I-beam in the figure to the right with the 0.75 in. thick plates, determine the added capacity of the beam in supporting bending moments about the horizontal axis of symmetry.

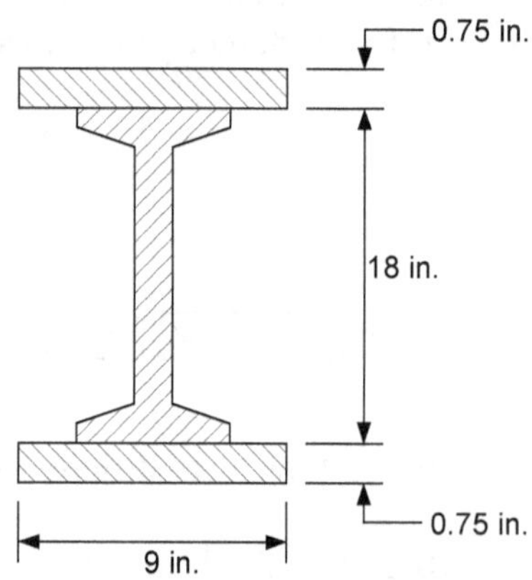

10.12 Four different beams with various end conditions are shown in the figure below. Construct FBDs of each beam showing the forces and/or moments at each support.

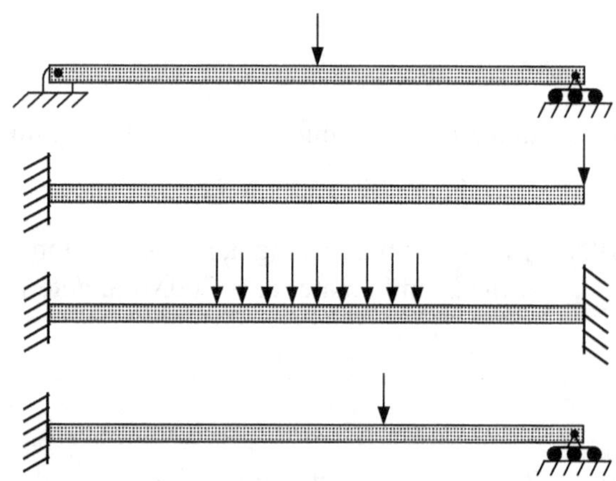

10.13 Write equations for the shear and bending moments as a function of position x across the length of the beam as shown in the figure to the right. Also construct the shear and bending moment diagrams.

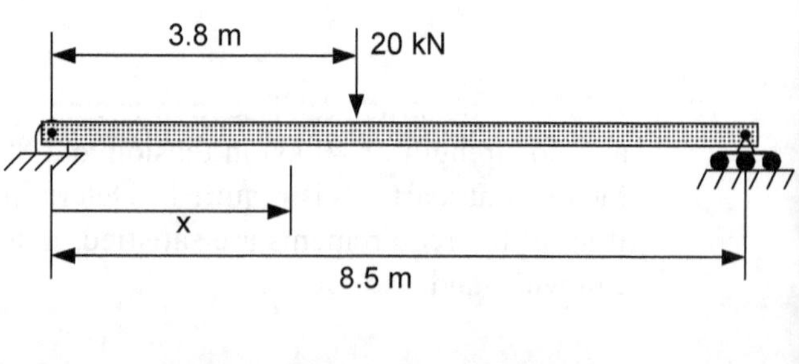

10.14 Write equations for the shear and bending moments as a function of position across the beam described in the figure to the left. Also construct the shear and bending moment diagrams.

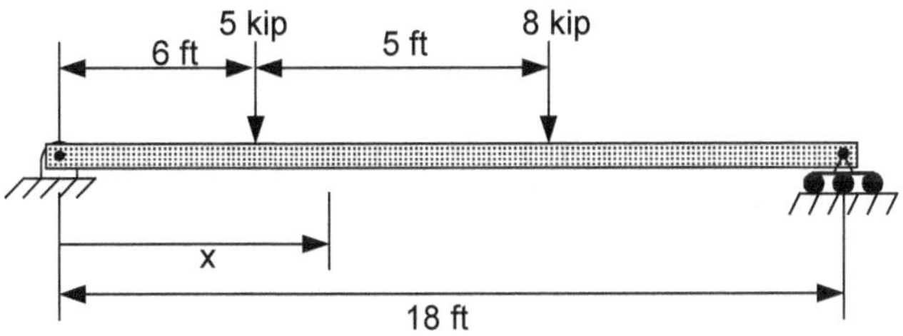

10.15 Write equations for the shear and bending moments as a function of position x across the beam described in the figure to the right. Also construct the shear and bending moment diagrams.

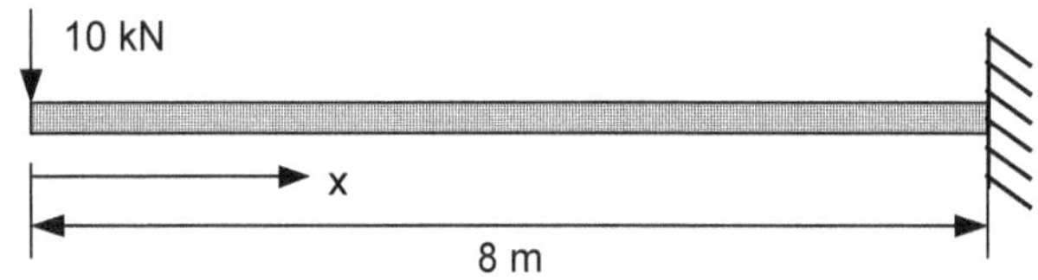

10.16 Write equations for the shear and bending moments as a function of position x across the beam described in the figure to the left. Also construct the shear and bending moment diagrams.

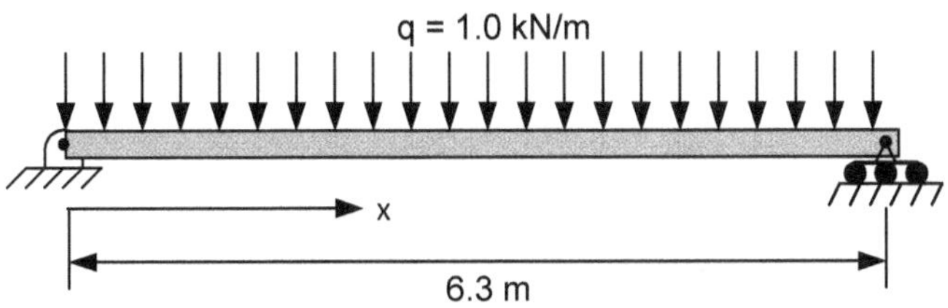

10.17 Write equations for the shear and bending moments as a function of position x across the beam described in the figure to the right. Also construct the shear and bending moment diagrams.

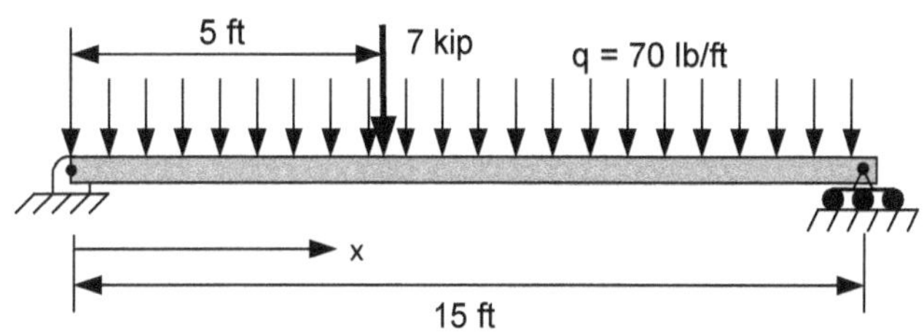

10.18 Write equations for the shear and bending moments as a function of position x across the beam described in the figure to the left. Also construct the shear and bending moment diagrams.

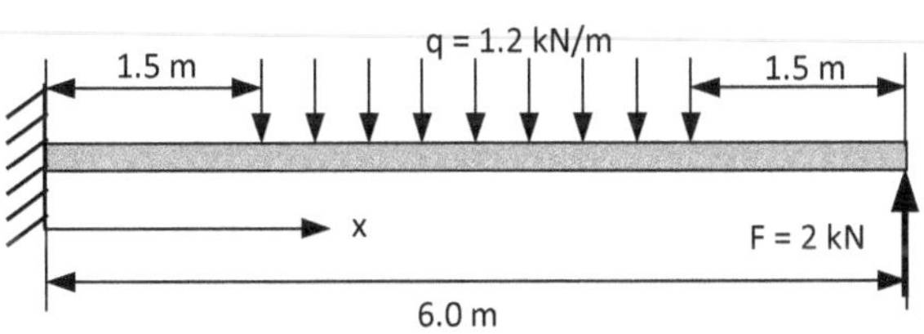

10.19 Select a WF section for the steel beam shown in the figure to the right. The margin of safety (MOS) is 200% and the yield strength of the material is 320 MPa.

10.20 Select an American Standard section for the steel beam shown in the figure to the right. The margin of safety (MOS) is 150% is and the yield strength of the material is 350 MPa.

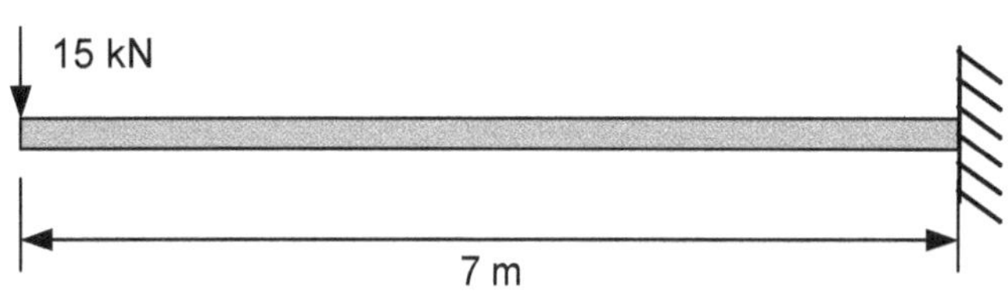

10.21 Select a structural tee section for the steel beam shown in the figure to the right. The margin of safety (MOS) is 200% and the yield strength of the material is 300MPa.

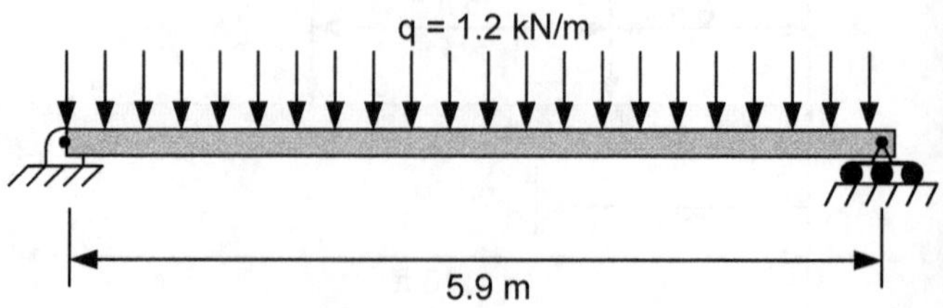

10.22 For the beam, described in the figure to the left, determine the maximum shear stress τ if the cross section of the beam is rectangular with a height h = 4.0 in. and a depth b = 1.0 in.

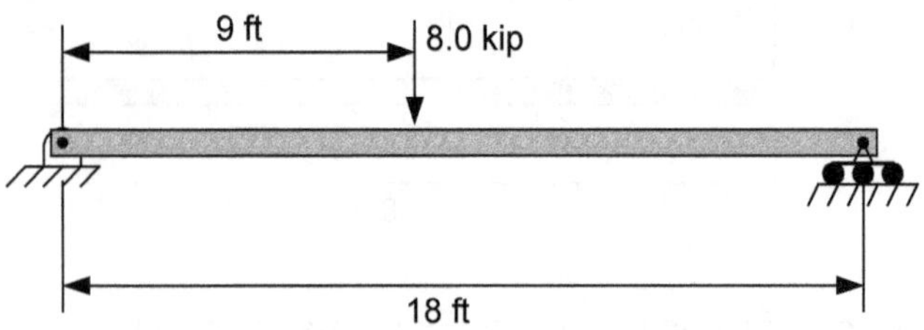

10.23 A circular shaft with a diameter of 50 mm is subjected to transverse loading from a gear set that produces a maximum shear force of 25 kN on the shaft. Determine the maximum shear stress τ_{Max} in the shaft due to this transverse load.

10.24 A timber company has designed a series of box beams with the geometry presented in the figure to the right. The beams are constructed by bonding the top and bottom boards to the edges of the vertical members at the four interfaces. Determine the maximum shear stress τ_{Max} and the shear stress at the adhesive joints. The vertical shear force V is 5.0 kN.

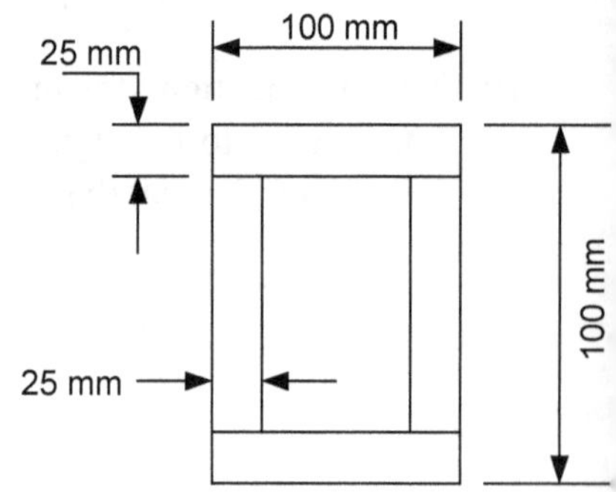

10.25 A timber company has designed a series of unsymmetrical T-sections to serve as beams as indicated in the figure to the left. The beams are constructed by bonding the top and bottom flanges to the edges of the vertical web at the two interfaces. Determine the maximum shear stress τ_{Max} and the shear stress at the adhesive joints. The vertical shear force V is 1,200 lb.

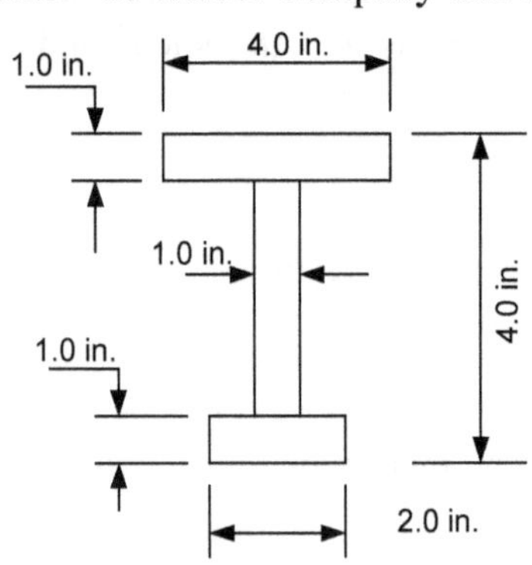

10.26 A timber company produces a series of I-beams fabricated from three pieces of timber as shown in the figure to the right. The flange of the I-beam is fastened to the web with spikes that are each capable of resisting a shear force of 960 lb. The beams are 12 ft long and are simply supported. They carry a concentrated force of 3,500 lb at midspan. Determine the maximum spacing of the spikes needed to transmit the shear from the web to the flange of the beam.

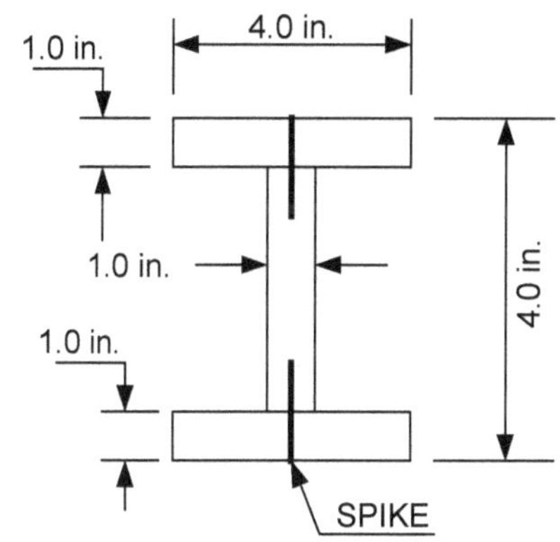

10.27 Derive Eq. (10.29).

10.28 Consider a simply-supported, foam core, metal cover plate composite beam as illustrated in the figure to the left. If the beam is loaded with a uniformly distributed load, determine the magnitude of q_Y that produces yielding. Aluminum cover plates are adhesively bonded to a polystyrene foam core. The beam is 2.5 m long and the yield strength of the aluminum is 280 MPa.

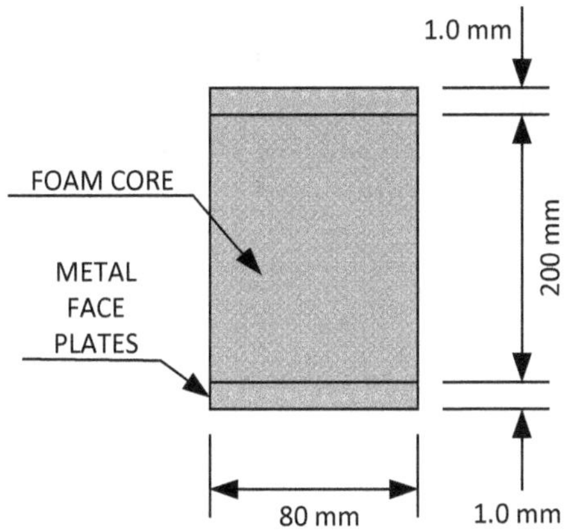

10.29 Design a portable, one-way, bridge beam with a single structural member consisting of metal cover plates and foam core. The bridge is to support the weight of four people uniformly distributed over the span of the bridge beam. Assume that the average weight of a typical person is 700 N. The bridge beam spans a ravine 5 m wide.

10.30 A cantilever beam 1.5 long is subjected to a concentrated force of 1,200 N applied at its free end. The beam has a rectangular cross section with a width of 90 mm. Determine the height of the foam h_f and the thickness of the cover plates t_m if the design stress in these plates is limited to 160 MPa. In this determination, recall that the standard thickness of aluminum sheet stock from which the cover plates are cut is 0.016, 0.020, 0.025, 0.032, 0.040, 0.050, 0.063, and 0.090 in.

10.31 A beam fabricated from western white pine, with dimensions shown in the figure to the right, has been reinforced by bonding steel strips on its top and bottom surfaces to form a composite beam. If the design stress is limited to 1.5 ksi in the wood and 30 ksi in the steel, determine the maximum bending moment if the beam bends about its horizontal axis of symmetry.

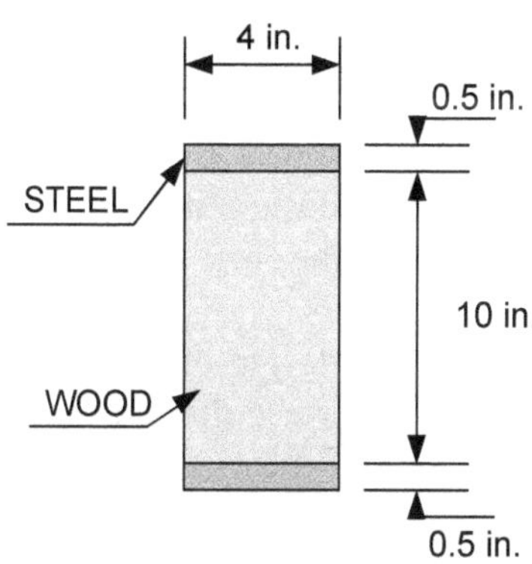

10.32 For the wood and steel composite beam described in Problem 10.31, determine the maximum bending moment if the beam is bent about its vertical axis of symmetry.

10.33 Derive Eqs. (10.32) and (10.35).

10.34 A concrete floor slab is reinforced with 12 steel rebars each with a diameter of ½ in. as illustrated in the figure below. The slab is 24 ft long and is simply supported when installed. If a uniformly distributed load q is applied to the floor, determine its maximum magnitude if the stress in the steel is limited to 20 ksi and in the concrete to 1,500 psi. Note the modulus of elasticity of the steel and the concrete is 30×10^6 and 2.8×10^6 psi, respectively. Comment on the balance of the design of this reinforced concrete slab.

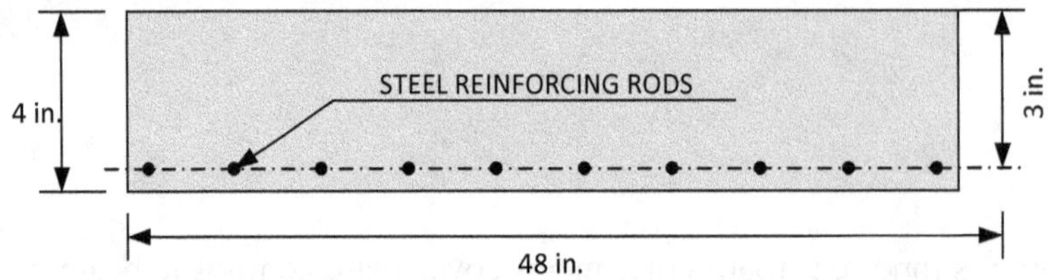

10.35 A concrete beam is reinforced with 8 steel rebars each with a ½ in. diameter as shown in the figure to the right. The beam is 20 ft long and simply supported. If it supports three concentrated forces each 6.0 kip positioned at L/4, L/2 and 3L/4, determine the margin of safety for the beam. The yield strength of the rebar is 36 ksi and the compressive strength of concrete is 4 ksi. The modulus of elasticity of the steel and the concrete is 30×10^6 and 2.8×10^6 psi, respectively.

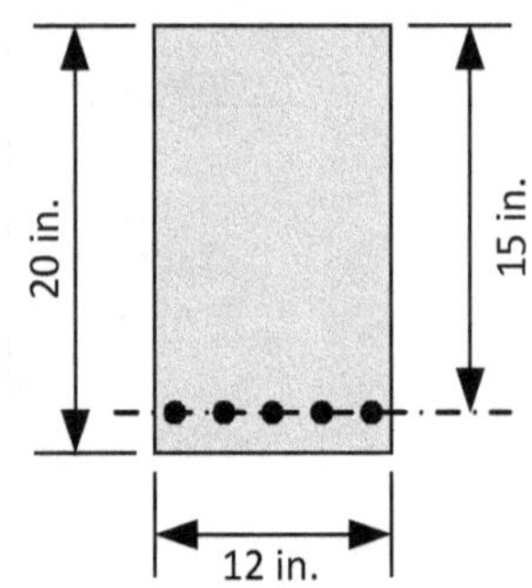

10.36 A high strength concrete beam is reinforced with 3 steel rods each with an area of 115 mm^2 located at positions shown in the figure to the left. For an applied bending moment of M = 5 kN-m, determine the maximum stress in the steel rebars and the concrete.

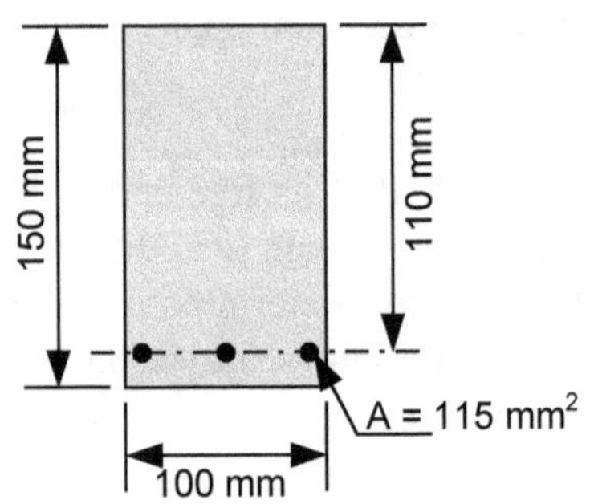

10.37 Discuss the advantage of using steel reinforced concrete beams as structural components in a building in the event of a severe fire. In your discussion, discuss the effect of prolonged exposure to flames on the strength of steel.

10.38 Redesign the steel reinforced concrete floor slab in Example 10.16 to improve the balance between the stresses in the steel and concrete.

10.39 Redesign the steel reinforced concrete beam in Example 10.17 to improve the balance in the margin of safety between the steel and concrete.

10.40 A beam with a rectangular cross section, defined in the figure to the right, is fabricated from steel with S_y = 320 MPa. Determine the yield moment M_Y.

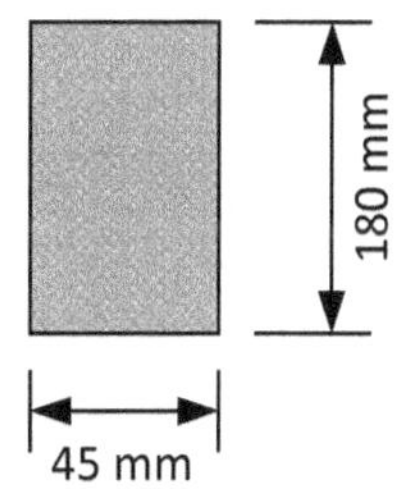

10.41 A beam with a rectangular cross section, defined in the Problem 10.40, is fabricated from steel with S_y = 320 MPa. Determine the moment M_p required to drive the elastic-plastic interface to the position $y_p = h/4$.

10.42 For a beam with a rectangular cross section, show that $M_P = M_Y$ when $y_p = h/2$. You may wish to examine Eq. (10.40) in developing this solution.

10.43 Determine the ratio of M_L/M_Y for the web and flange section, defined in the figure to the left, if $b_w = xb$ and $h_w = yh$. Let x vary from 0.05 to 0.3 and y vary from 0.6 to 0.95. It is suggested that you perform these calculations on a spreadsheet and prepare a graph displaying your results.

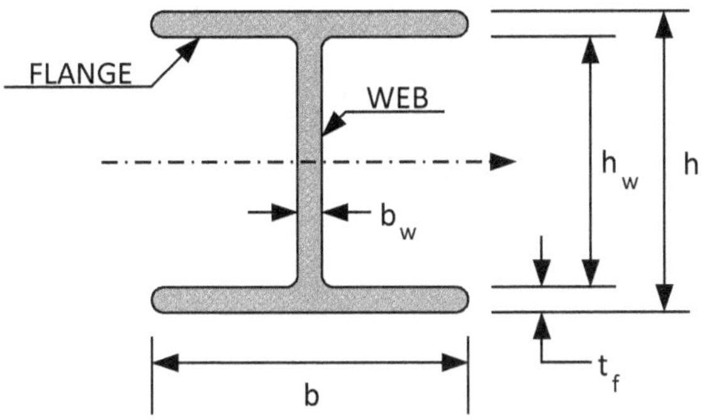

10.44 A notched cantilever beam, illustrated in the figures below, is notched at position x = 3.8 ft. Determine the maximum tensile stress on the beam.

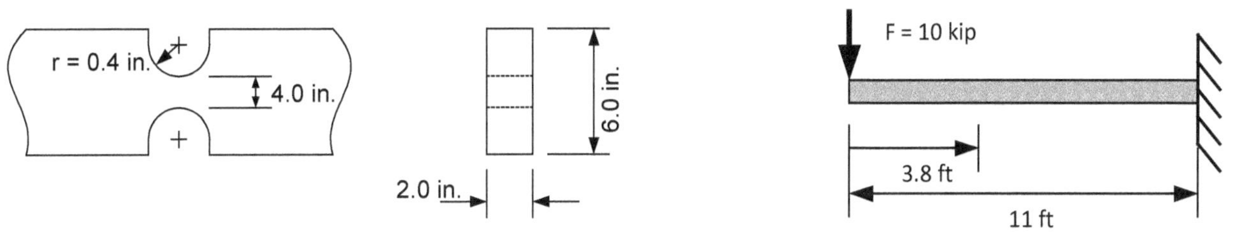

10.45 An architect has proposed a series of stepped roof beams to provide a visual impact for a public building with high pedestrian traffic. A representative beam with its loading is presented in the figures below. The beam's depth is 60 mm, and its yield strength is 240 MPa. As the city's structural engineer, you have the responsibility to approve or disapprove the design. Justify your decision.

10.46 A shaft in a drive train, illustrated in the figures below, is subjected to bending moments due to a transverse force of 3.5 kN imposed by a gear set. Determine the maximum stresses at the fillet due to the gear load.

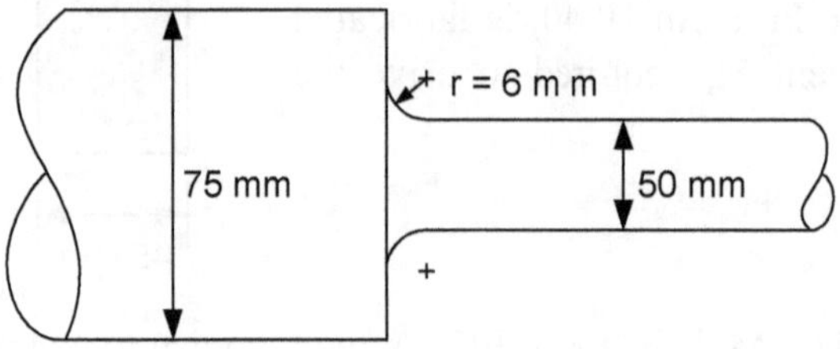

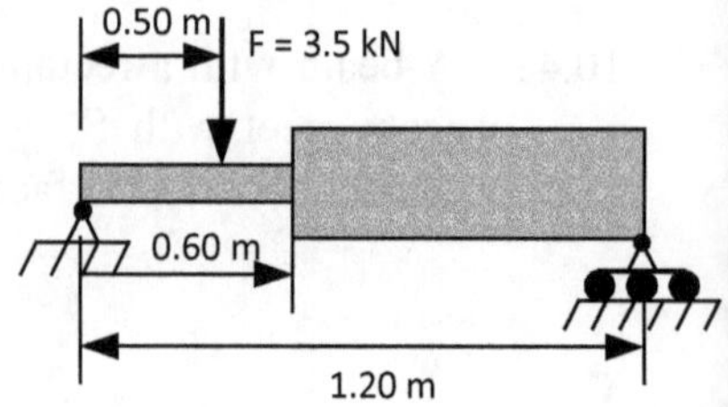

10.47 A shaft in a turbine engine is used as a conduit to transport coolant to the root of the turbine blades. The coolant enters and exits the shaft through small holes as shown in the figure below. Determine the maximum stress adjacent to the hole if the shaft supports a bending moment M = 400 N-m at the location of the hole.

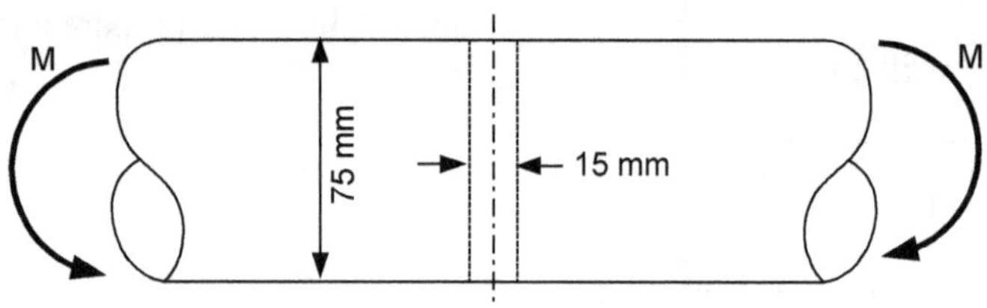

CHAPTER 11

TORSION OF STRUCTURAL ELEMENTS

11.1 TORSION LOADING

Circular shafts are commonly employed to transmit power from a motor or engine to some appliance. For example, when a power drill is used to drive a screw into a sheet of metal, a short circular shaft transmits the power from the electric motor to the chuck holding the screwdriver. When you operate a car, two shafts transmit power from the gearbox to the drive wheels. You may recall the relation between the power P transmitted and the torque T applied to the shaft as:

$$P = T\omega \tag{11.1}$$

where ω is the angular velocity of the shaft in radian/s, P is the power in W, where 1W = 1 N-m/s and T is the torque in N-m.

If Eq. (11.1) is employed, the power and torque are expressed in SI units. In U. S. Customary units, the power is expressed in horsepower HP and the torque in terms of ft-lb. The definition of horsepower[1] is given by:

$$1 \text{ HP} = 33{,}000 \text{ ft-lb/min} = 550 \text{ ft-lb/s} \tag{11.2}$$

We are particularly interested in the torsion load (torque) produced when a shaft transmits power. The torque produces stresses and angular deformations in the shaft. If either of these two quantities becomes excessive, the shaft will fail in service.

As a motor or engine turns, it acts on the shaft producing a torque T. We illustrate this action in Fig. 11.1 where two circular disks are connected by a circular shaft. A couple with forces F_1 act on the larger diameter disk producing a torque $T_1 = F_1 d_1$. Another couple with forces F_2 act on the smaller diameter disk producing a torque $T_2 = F_2 d_2$. If the shaft is either stationary or rotating at a constant angular velocity, it is in equilibrium and the torques T_1 and T_2 are equal in magnitude and opposite in sign. The shaft transmits the torque from one disk to the other.

In subsequent sections of this chapter we will show the relations for determining the stresses and strains induced in the shaft by this torque. We will also demonstrate the methods for calculating the stresses, strains and deformations as we design a shaft for an infinite service life.

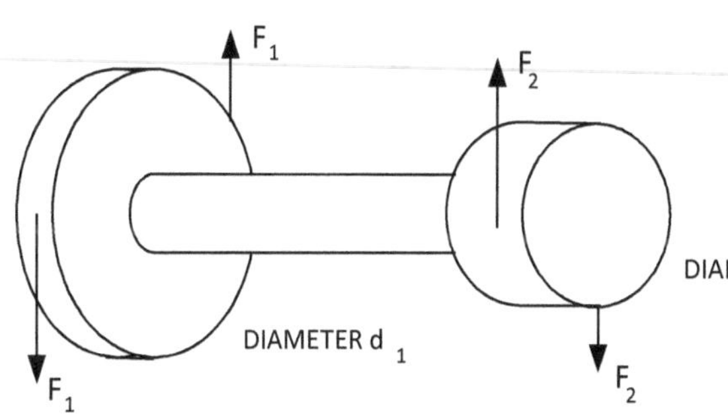

Fig. 11.1 A circular shaft transmitting torque $T_1 = F_1 d_1 = T_2 = F_2 d_2$.

[1] James Watt conducted experiments in 1782 to determine that a "brewery horse" was able to produce power equivalent to 32,400 ft-lb per minute. He standardized the conversion factor at 33,000 ft-lb per minute to classify the Boulton and Watt steam engine that his company was manufacturing at the time.

EXAMPLE 11.1

A small gasoline engine powers a string type trimmer used to cut grass and weeds. If the maximum power generated by the engine is 800 W, determine the torque imposed on the shaft from the engine to the string disk if it is rotating at 400 RPM.

Solution:

Using Eq. (11.1), we write:

$$T = P/\omega = 800/\,[2\pi\,(400/60)] = 19.10 \text{ N-m}$$

Note that the angular velocity of the shaft was converted from RPM to radians per second by employing the conversion shown below:

$$\omega \text{ (radian/s)} = 2\pi\, n/60$$

where n is the angular velocity in RPM.

We will return to this example later in this chapter and determine the stress in the shaft under these operating conditions.

EXAMPLE 11.2

The output shaft from an electric motor drives a compressor on a household refrigerator. If the motor operates at 1,780 RPM and is rated at 0.75 HP determine the torque on the shaft.

Solution:

Substituting conversion factors into Eq. (11.1) and solving for the torque on the motor— compressor shaft gives:

$$T = 33{,}000 \text{ HP}/(2\pi n) \qquad (11.3)$$

where HP is the power transmitted by the shaft in horsepower.

$$T = (33{,}000)(0.75)/[2\pi\,(1780)] = 2.213 \text{ ft-lb}$$

The torque applied to a shaft is a twisting moment. This moment (torque) acts about the centerline of the shaft and tends to twist it. Plane sections (disks) perpendicular to the axis of the circular shaft remain plane as the shaft is twisted. This fact is extremely important, because it enables us to determine the stresses and deformation from relatively elementary equations for shafting with circular cross sections. In the next section, we will show how the strains are produced by this twisting action on the shaft.

11.2 DEFORMATION OF A CIRCULAR SHAFT DUE TO TORSION

Let's consider a circular shaft of length L with one end built-in (fixed) and the other end free. A twisting moment (torque) is applied to its free end as illustrated in Fig. 11.2. A straight line A-B is scribed on the outside surface of the shaft prior to the application of the torque T. Upon the application of the torque, the shaft deforms, the scribe line rotates through an angle γ and point B moves to a new location given by point C.

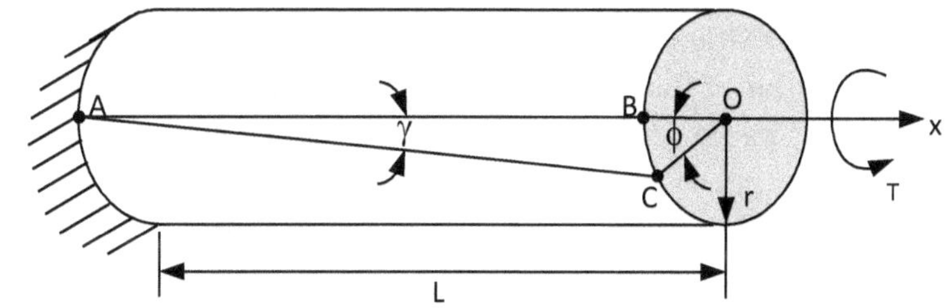

Fig. 11.2 Deformation of a circular shaft due to the application of a torque T at its free end.

The length of the arc B-C is given by:

$$BC = \gamma L \tag{11.4}$$

The angle ϕ is known as the angle of twist for the shaft. Let's examine the deformation of the circular shaft more closely by considering a square scribed on its surface at some arbitrary position along line A-B as shown on Fig. 11.3.

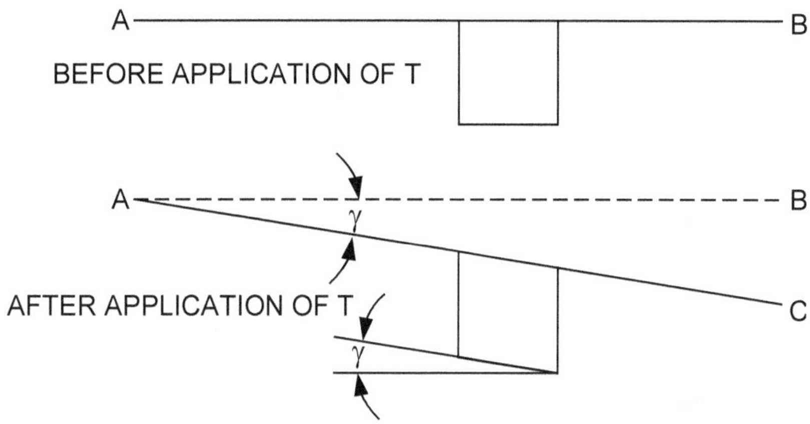

Fig. 11.3 Deformation of a small square scribed on the surface of the shaft before and after application of the torque.

When the shaft deforms under the action of the torque, the straight line A-B rotates, remains straight, and assumes the orientation of line A-C. The two sides of the inscribed square that are perpendicular to the axis of the shaft do not rotate. However, both of the sides parallel to the axis of the shaft rotate through the angle γϕ. Clearly, the small square is distorted as indicated in Fig. 11.3. The included angles of the square have changed 90° ± γ. Therefore, it is clear that a shearing strain γ occurs on the external surface of the circular shaft.

11.2.1 Shear Strain Due to Torque Applied to a Circular Shaft

To determine a relation for the magnitude of the shearing strain γ, let's consider the free end of a circular shaft subjected to an applied torque T as indicated in Fig. 11.4. The line OB rotates through an angle ϕ sweeping through an arc BC. The length of the arc B-C is given by:

$$BC = r\phi \tag{11.5}$$

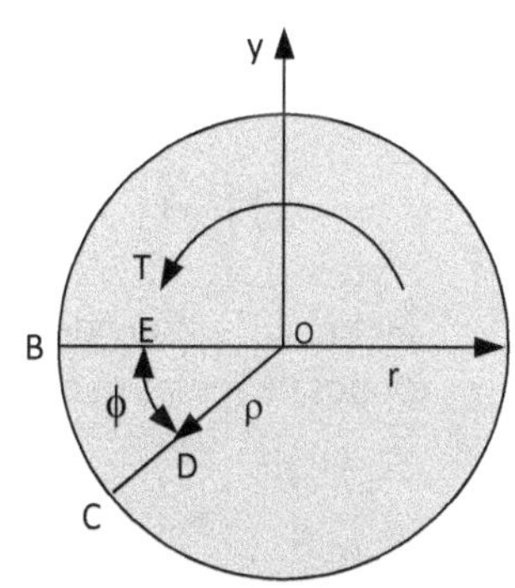

Fig. 11.4 Rotation of line OB on the free end of a circular shaft subjected to a torque T.

Substituting Eq. (11.4) into Eq. (11.5) yields:

$$\gamma_{\rho=r} = r\phi/L \tag{11.6}$$

The shear strain on the surface of the shaft where ρ = r is given by Eq. (11.6). The shear strain at some point D on the interior of the shaft is less because the arc length D-E is decreased. Following the same procedure, we write the shearing strain γ for an arbitrary point defined by the position radius ρ as:

$$\gamma_\rho = \rho\phi/L \qquad (11.7)$$

Examination of Eq. (11.7) shows that the shearing strain varies from zero at the centerline of the shaft to a maximum value at the surface that is given by:

$$\gamma_{Max} = r\phi/L \qquad (11.8)$$

Because the applied torque is constant along the length of the shaft, the strain is also constant along its length.

11.3 STRESSES PRODUCED BY TORSION

To determine the shear stresses acting on the circular shaft, we employ the relation between shear stress and shear strain.

$$\tau = G\gamma \qquad (11.9)$$

where G is the shear modulus of the shaft material given by:

$$G = \frac{E}{2(1+\nu)} \qquad (11.9a)$$

Substituting Eq. (11.7) into Eq. (11.9) yields:

$$\tau_\rho = G\phi\rho/L \qquad (11.10)$$

Again, we observe that the shear stress τ increases linearly with the position radius ρ from zero at the centerline to a maximum value at the surface of the shaft. This linear distribution of shear stress is illustrated in Fig. 11.5 a.

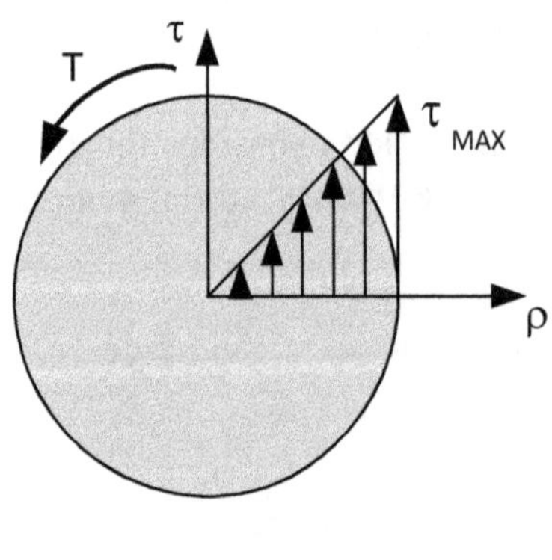

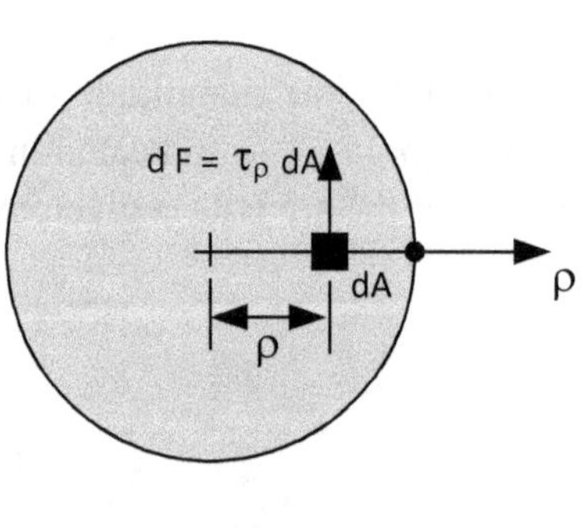

Fig. 11.5a Linear distribution of shear stress τ. Fig. 11.5b The incremental force dF = τdA.

Next, let's consider the equilibrium of the shaft with an applied torque. The shear stress produces an equal and opposite moment to the applied torque at the far end of the shaft. We write the equilibrium equation $\Sigma M_x = 0$ as:

$$\Sigma M_x = T - \int \rho\tau \, dA = 0 \qquad (11.11)$$

The sum of all the incremental forces ($dF = \tau_\rho \, dA$) produce the moment M_x about the centerline of the shaft as presented in Fig. 11.5b. Substituting Eq. (11.10) into Eq. (11.11) gives:

$$T = (G\phi \int \rho^2 \, dA)/L \qquad (11.12)$$

The term $\int \rho^2 \, dA$ is the polar moment of inertia J of a circular area about its center. Hence, we can rewrite Eq. (11.12) as:

$$T = G\phi J/L \qquad (11.13)$$

Finally, we note from Eq. (11.10) that $(G\phi/L) = \tau_\rho/\rho$. Accordingly, we can write:

$$\tau_\rho = T\rho/J \qquad (11.14)$$

Equation (11.14) is employed to determine the shear stresses in circular shafts when the applied torque is known. Clearly, the maximum shear stress occurs when $\rho = r$. In this instance:

$$\tau_{Max} = Tr/J = Td/(2J) \qquad (11.15)$$

Because $J = \pi r^4/2 = \pi d^4/32$ for solid circular shafts, we may express Eq. (11.15) as:

$$\tau_{Max} = 2\,T/(\pi r^3) = 16\,T/(\pi d^3) \qquad (11.16)$$

To demonstrate the method for determining stresses in circular shafts, let's consider two examples.

EXAMPLE 11.3

A solid circular shaft fabricated from 1020 HR steel is subjected to a torque T = 2,500 in-lb. Determine the shear stress τ as a function of the diameter of the shaft.

Solution:

To determine the shear stresses in the circular shaft, substitute the numerical parameters describing the shaft and the applied torsion into Eq. (11.16) to obtain:

$$\tau_{Max} = 16\,T/(\pi d^3) = 16\,(2500)/[\pi d^3] = 40 \times 10^3 \text{ in-lb}/\pi d^3$$

We now have an equation relating the shear stress on the surface to the shaft diameter. Let's study several cases by permitting the diameter of the shaft to vary from 0.25 to 2.00 in. Using a spreadsheet in performing the calculations, we obtain the results shown below:

d (in.)	d^3 (in.^{3})	Stress, τ (ksi)
0.250	0.0156	814.87
0.375	0.0527	241.44
0.500	0.1250	101.86
0.625	0.2441	52.15
0.750	0.4219	30.18
0.875	0.6699	19.01
1.000	1.0000	12.73
1.250	1.9531	6.52
1.500	3.3750	3.77
1.750	5.3594	2.38
2.000	8.0000	1.59

The numerical values for the shear stress vary from 814.87 ksi for a shaft with a diameter of 0.25 in. to only 1.59 ksi for a 2.0 in. diameter shaft. Obviously, a large range in the results is evident. The very large variations in τ, with small changes in d, are due to the fact that the diameter occurs as a cube in the denominator of Eq. (11.16).

To determine a suitable diameter of the shaft, we must compare the stresses to the strength of the 1020 HR steel. This task raises a question—is it possible to compare shear stress with yield strength or ultimate tensile strength? The answer is **no**. Yield and ultimate tensile strength for materials are measured in a tensile test and cannot be directly compared to shear strength. Let's consider another example to introduce an accepted technique for approximating the shear strengths of most metals.

EXAMPLE 11.4

A shaft diameter of 1.00 in. is selected for the shaft described in Example 11.3. Determine the safety factor for this shaft if the applied torque is 2,500 in.-lb.

Solution:

To account for the effects of shear stress on failure, we determine yield S_{ys} and ultimate S_{us} strengths in terms of shear by multiplying the yield and ultimate tensile strengths, S_y and S_u, by a constant = $\sqrt{3}/3 = 0.5774$:

$$S_{ys} = 0.5774\ S_y \tag{11.17a}$$

$$S_{us} = 0.5774\ S_u \tag{11.17b}$$

Substituting numerical values into Eq. (11.17) gives the yield and ultimate shear strength for 1020 HR steel as:

$$S_{ys} = 0.5774\ S_y = (0.5774)(42) = 24.25 \text{ ksi} \tag{a}$$

$$S_{us} = 0.5774\ S_u = (0.5774)(66) = 38.10 \text{ ksi} \tag{b}$$

The safety factor is determined by computing the ratio of strength to the stress according to:

$$SF = S_{ys} / \tau_{Max} \tag{11.18}$$

Substituting the results from Eq. (a) and the table for Example 11.3 into Eq. (11.18) gives:

$$SF = 24.25/12.73 = 1.905 \tag{c}$$

We have used the yield strength in shear for the 1020 HR steel in determining the safety factor. Reference to Eq. (c) shows a safety factor of nearly 2, which that should be adequate if the applied torque is specified accurately.

EXAMPLE 11.5

A shaft, fabricated from 302 A stainless steel, is to transmit 900 W of power while rotating at 1190 RPM. Specify the diameter of the shaft in millimeters if the safety factor is to be in the range of 2.50 to 3.00.

Solution:

Let's substitute Eq. (11.17 a) into Eq. (11.18) and rearrange terms to obtain:

$$\tau_{Max} = S_{ys} / SF = 0.5774\ S_y / SF \tag{a}$$

Reference to Appendix B-2 indicates that $S_y = 234$ MPa for 302 A stainless steel.

Consider the lowest safety factor SF = 2.50; for this choice, determine τ_{Max} from Eq. (a).

$$\tau_{Max} = S_{ys} / SF = 0.5774(234)/2.50 = 54.04 \text{ MPa} \qquad \text{(b)}$$

Then using Eq. (11.16) we write:

$$d^3 = (16T)/(\pi\tau_{Max}) \qquad \text{(c)}$$

To compute the required diameter d, it is necessary to determine the applied torque from Eq. (11.1).

$$T = P/\omega = [(900)(60)]/[(1190)(2\pi)] = 7.222 \text{ N-m} = 7222 \text{ N-mm} \qquad \text{(d)}$$

Substituting Eq. (d) into Eq. (c) gives:

$$d^3 = (16)(7222)/[\pi(54.08)] = 680.6 \text{ mm}^3 \qquad \text{(e)}$$

Consider d = 9 mm and note that d^3 = 729 mm^3. This choice is a standard size round bar in the SI system that is only slightly smaller in diameter than that necessary to achieve a value of d^3 = 680.6 mm^3. Let's check to determine if this 9 mm diameter bar is sufficient to provide a safety factor of 3.0 or less.

From Eq. (11.16), we write:

$$\tau_{Max} = (16T)/(\pi d^3) = (16)(7222)/[\pi\,(729)] = 50.45 \text{ MPa} \qquad \text{(f)}$$

Next, substitute the results from Eq. (f) into Eq. (a) to obtain:

$$SF = 0.5774 S_y/\tau_{Max} = (0.5774)(234)/50.45 = 2.678 \qquad \text{(g)}$$

This result indicates that the diameter of 9 mm is sufficient to provide a safety factor that lies within the range specified for this application.

11.3.1 Hollow Shafting

In many engineering applications weight is an important consideration because it often detrimentally affects performance and usually increases cost. It is possible to reduce the weight of shafting by using tubes instead of solid rounds. The weight penalty associated with solid round bars is due to the reduced stresses in central regions of the bar, which are much smaller than the maximum shear stress that occurs on the outer diameter of the shaft. With a tube, the central region of the shaft is removed and the resulting structure is more uniformly stressed and structurally more efficient.

Equation (11.14) derived for a solid circular shaft is also valid for a hollow circular shaft. The only difference that arises is in the expression for the polar moment of inertia J. For a solid shaft $J = \pi d^4/32$; however, for a hollow shaft the polar moment of inertia is given by:

$$J = (\pi/32)(d_o^4 - d_i^4) \qquad (11.19)$$

where d_i and d_o represent the inside and outside diameters, respectively.

Substituting the results for Eq. (11.19) into Eq. (11.14) yields:

$$\tau_\rho = 32T\rho/[\pi(d_o^4 - d_i^4)] \qquad (11.20)$$

The maximum shear stress occurs when $\rho = d_o/2$; hence Eq. (11.20) may be rewritten as:

$$\tau_{Max} = 16\,Td_o/[\pi(d_o^4 - d_i^4)] \qquad (11.21)$$

EXAMPLE 11.6

A hollow circular shaft is fabricated from a seamless tube of 1018 HR steel. It is subjected to a torque T of 2100 in-lb. If the outside diameter of the tube is fixed at 1.25 in., determine the shear stress τ_{Max} as a function of the inside diameter of the shaft.

Solution:

Let's employ Eq. (11.21) and substitute numerical values for the known quantities in this relation.

$$\tau_{Max} = (16)(2{,}100)(1.25)/\{\pi[(1.25)^4 - d_i^4)]\} = 13{,}370/[(2.441) - d_i^4)] \qquad \text{(a)}$$

Using a spreadsheet to evaluate Eq. (a) gives:

d_i (in.)	d_i^4 (in.)^4	Stress, τ (ksi)
0.000	0	5.477
0.125	0.0002	5.478
0.250	0.0039	5.486
0.375	0.0198	5.522
0.500	0.0625	5.621
0.625	0.1526	5.842
0.750	0.3164	6.293
0.875	0.5862	7.208
1.000	1.0000	9.278
1.125	1.6018	15.932

Examination of these results shows that the maximum shear stress increases as the inside diameter of the tube increases. Compared to the shear strength of the 1018 HR steel, which is S_{ys} = 0.5774(32 ksi) = 18.48 ksi, the maximum shear stresses are lower than this value for every inside diameter considered. However, the safety factor is lower that 2.0 when $d_i \geq 1.00$ in.

11.3.2 Section Modulus in Torsion

When considering beams in bending, we introduced a section modulus that defined the influence of the shape of the beam's cross-section on the flexural stress. Let's adapt the same idea to circular shafts in torsion by rewriting Eq. (11.21) as:

$$\tau_{Max} = 16\, Td_o/[\pi(d_o^4 - d_i^4)] = T/Z_T \qquad (11.22)$$

$$Z_T = \pi(d_o^4 - d_i^4)/(16d_o) \qquad (11.23)$$

Clearly, to reduce the maximum shear stress it is necessary to increase the section modulus Z_T. In a hollow shaft, Eq. (11.23) indicates that both the internal and external diameters of the shaft influence the section modulus. To limit the shear stress, we seek to increase Z_T, but without the penalty of a large increase in the weight of the shaft. To establish the most suitable approach to achieve large Z_T values without incurring weight penalties, let's first consider the weight W per unit length L of the shaft.

$$q = W/L = \gamma(\pi/4)(d_o^2 - d_i^2) \qquad (11.24)$$

where γ is the weight per unit volume of the shaft.

Because it is good design practice to increase the section modulus while limiting the increase in q, it is useful to examine the ratio Z_T/q. Dividing Eq. (11.23) by Eq. (11.24) yields:

$$Z_T/q = [\pi(d_o^4 - d_i^4)/(16d_o)]/\,[\gamma(\pi/4)(d_o^2 - d_i^2)] \quad \text{(a)}$$

$$Z_T/q = (d_o^2 + d_i^2)/(4\gamma d_o) \quad (11.25)$$

For a solid shaft, $d_i = 0$ and Eq. (11.25) reduces to:

$$Z_T/q = d_o/4\gamma \quad (11.26)$$

Comparing the ratio Z_T/q for the hollow and solid shafts provides a method for assessing the improvement in the design by employing hollow shafting. Accordingly, we form the ratio shown below:

$$\frac{\left[\frac{Z_T}{q}\right]_{Hollow}}{\left[\frac{Z_T}{q}\right]_{Solid}} = 1 + \left(\frac{d_i}{d_o}\right)^2 \quad (11.27)$$

As d_i becomes large and approaches d_o, this ratio in Eq. (11.27) approaches 2, indicating a hollow shaft with a very thin wall has nearly twice the section modulus per unit weight when compared to a solid shaft. Of course there is a limit to the diameter of the hole. When the wall thickness of a hollow shaft becomes two small, the shaft will fail by buckling under torsion.

The disadvantage of using hollow shafts in the design of power transmission shafts is cost and availability. It is more expensive to manufacture a tube than it is to manufacture a solid shaft. Sometimes the manufacturing costs are larger than the savings achieved by reducing the weight of the shaft. Also for equal values of Z_T, hollow shafts are larger in outside diameter than solid shafts; hence, larger diameter and more costly bearings are necessary to support hollow tubular shafts.

EXAMPLE 11.7

A 14 in. long shaft, fabricated from cold rolled alloy steel, exhibits a yield strength of 55 ksi and is to support a torque of 400 in.-lb. If a safety factor of at least 2.8 is employed, determine the weight of a solid shaft for this application. Also, determine the weight of a hollow shaft for this same application if $d_i = 0.8\ d_o$.

Solution:

The design stress τ_{Design} is given by Eq. (11.17a) and Eq. (11.18) as:

$$\tau_{Design} = (0.5774)S_y/SF \quad \text{(a)}$$

Substituting numerical parameters into Eq. (a) gives:

$$\tau_{Design} = (0.5774)(55)/(2.8) = 11.34 \text{ ksi} \quad \text{(b)}$$

The diameter is determined from Eq. (11.16) as:

$$d^3 = (16T)/(\pi\tau_{Design}) = (16)(400)/[\pi(11{,}340)] = 0.1796 \text{ in.}^3 \quad \text{(c)}$$

Solving Eq. (c) gives d = 0.5643 in. Selecting a commercially available standard size round bar gives:

$$d = 5/8 = 0.625 \text{ in.} \quad \text{(d)}$$

The weight W of the shaft is given by Eq. (11.24) as:

$$W = \gamma(\pi/4)d^2 L = 0.283(\pi/4)(0.625)^2 (14) = 1.216 \text{ lb} \quad \text{(e)}$$

where the weight density of steel is $\gamma = 0.283$ lb/in.^{3}

For the hollow shaft, the diameter is determined from Eq. (11.21) as:

$$[(d_o^4 - d_i^4)]/d_o = (16T)/\pi\tau_{Design} \quad \text{(f)}$$

Substitute $d_i = 0.8\ d_o$ into Eq. (f) and simplify to obtain:

$$d_o^3 = (16)(400)/[(0.5904)(\pi)(11{,}340)] = 0.3042 \text{ in.}^3 \quad \text{(g)}$$

Solving Eq. (g) for the outside diameter gives $d_o = 0.6725$ in. Selecting the next larger standard size tube gives:

$$d_o = 3/4 = 0.750 \text{ in. and } d_i = (0.8)(0.750) = 0.600 \text{ in.} \quad \text{(h)}$$

The weight is determined as:

$$W = \gamma(\pi/4)(d_o^2 - d_i^2) L = (0.283)(\pi/4)[(0.750)^2 - (0.600)^2](14) = 0.6301 \text{ lb} \quad \text{(i)}$$

The results show that the use of a hollow shaft in this application reduces the weight of the shaft from 1.216 lb to 0.6301 lb. This savings in shaft weight of 0.5859 lb (48.18%) reduces the overall weight of the product; however, the costs may increase because of the need to procure tubing instead of solid rod and to employ larger diameter bearings that are more expensive.

11.3.3 Structural Efficiency of Circular Shafts

The term structural efficiency **SE** was introduced in the discussion of beams in bending. It was defined as the section modulus of the beam divided by its cross sectional area. Adapting this definition to torsion of both solid and hollow circular shafts gives:

$$\mathbf{SE} = Z_T/A \quad (11.28)$$

Substituting Eq. (11.23) into Eq. (11.28) and simplifying yields:

$$\mathbf{SE} = \frac{d_o^2 + d_i^2}{4d_o} \quad (11.29)$$

As d_i increases from zero towards the value of d_o, the structural efficiency increases from 0.250 d_o to 0.500 d_o. Of course, d_i must always be less than d_o. In fact, the tubular shaft will buckle in torsion if the wall thickness becomes too small.

11.4 SHEAR STRESSES ON DIFFERENT PLANES

11.4.1 SHEAR STRESSES ON ORTHOGONAL PLANES

Let's consider an elemental area ΔA scribed on the surface of a circular shaft as shown in Fig. 11.6. The x-axis is parallel to the shaft's longitudinal axis and the y-axis is perpendicular to it. With the x and y axes defined in this manner, the shear stress $\tau_{xy} = \tau_{Max} = 16T/\pi d^3$. We remove this elemental area and draw a FBD of the element as shown in Fig. 11.7.

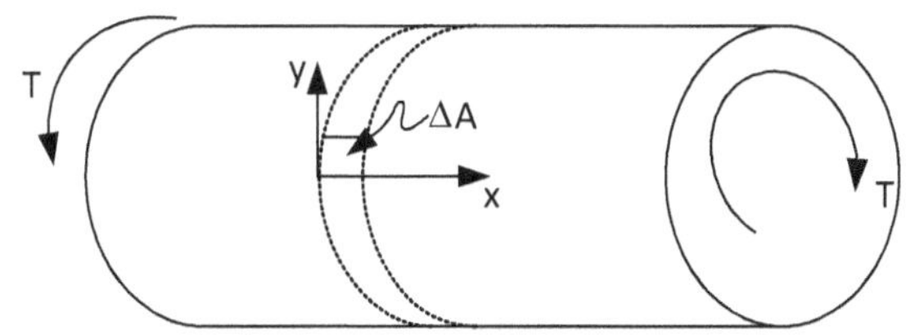

Fig. 11.6 An elemental area ΔA on the surface of a circular shaft subjected to torque T.

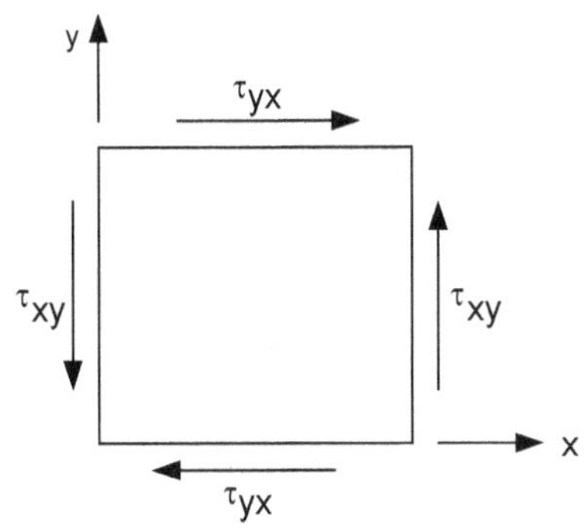

Fig. 11.7 A FBD of an elemental area ΔA from the surface of the shaft. All of the shear stresses shown are considered positive.

In drawing the FBD of the elemental area, we note that both of the edges of the element that are perpendicular to the x-axis are subjected to the shear stress $\tau_{xy} = \tau_{Max}$. The convention for the double subscript placed on τ is:

- The first subscript identifies the outer normal to the surface upon which the stress acts.
- The second subscript shows the direction in which the stress acts.
- The stress is considered positive if the outer normal is positive and the direction is positive.
- The stress is considered positive if the outer normal is negative and the direction is negative.

Let's convert the shear stresses acting on the elemental area to shear forces and then write the equilibrium equations. The element has dimensions Δx, Δy and Δz; hence, the shear forces acting on the sides that are perpendicular to the x-axis are given by:

$$V_y = \tau_{xy}\,\Delta y \Delta z \qquad (11.30\text{ a})$$

Similarly, the shear forces acting on the sides perpendicular to the y-axis are:

$$V_x = \tau_{yx}\,\Delta x \Delta z \qquad (11.30\text{ b})$$

Next, consider the equilibrium equation $\Sigma F_y = 0$. This relation shows that the shear force V_y acting on the right side of the element in the positive y direction must be balanced by an equal and opposite force acting on the left side of the element. Similarly the equation $\Sigma F_x = 0$ indicates that the shear force V_x acting on the top side of the element must be balanced by an equal and opposite force acting on the bottom side of the element. We have shown these shear stresses in Fig. 11.7 with the correct directions, and have satisfied these two of the three applicable equilibrium relations.

Finally, apply the equilibrium equation $\Sigma M_z = 0$, and write:

$$(\tau_{yx}\,\Delta x \Delta z)\Delta y = (\tau_{xy}\,\Delta y \Delta z)\Delta x$$

This relation reduces to:

$$\tau_{xy} = \tau_{yx} \qquad (11.31)$$

To satisfy the equilibrium equations, the shear stresses acting on orthogonal planes must be equal in magnitude.

11.4.2 SHEAR STRESSES ON OBLIQUE PLANES

To examine the shear stress acting on planes that are oblique relative to the shaft's axis, let's slice a corner from this elemental area as shown in Fig. 11.8. The inclined side of the triangular element has an area ΔA, and the vertical and horizontal sides have areas of ΔA cos θ and ΔA sin θ respectively. Two stress components act on this plane—one perpendicular to the surface (σ_n) and one along the surface (τ_{nt}). Applying the equations of equilibrium to the small triangular element gives:

$$\Sigma F_t = 0 \qquad \text{(a)}$$

$$\tau_{nt}\,\Delta A - [\tau_{xy}\,(\Delta A \cos\theta)]\cos\theta + [\tau_{yx}\,(\Delta A \sin\theta)]\sin\theta = 0 \qquad \text{(b)}$$

Solving Eq. (b) and recalling the results of Eq. (11.31) yields:

$$\tau_{nt} = \tau_{xy}\,(\cos^2\theta - \sin^2\theta) = \tau_{xy}\cos 2\theta \qquad (11.32)$$

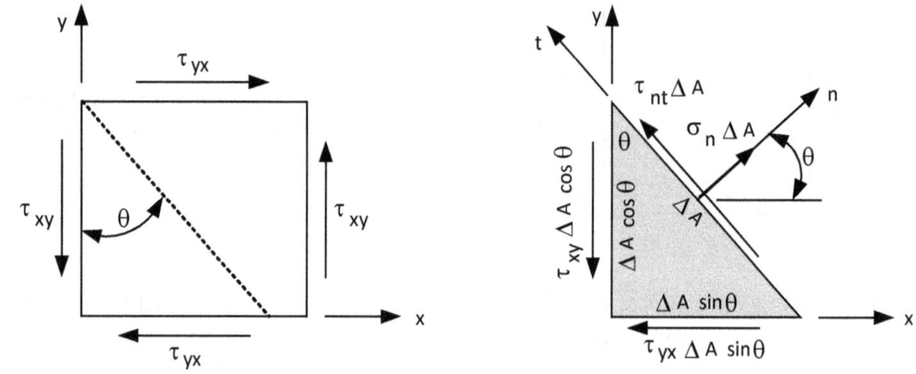

Fig. 11.8 Forces acting on a triangular element cut from an element.

$$\Sigma F_n = 0 \qquad \text{(c)}$$

$$\sigma_n\,\Delta A - [\tau_{xy}\,(\Delta A \cos\theta)]\sin\theta - [\tau_{yx}\,(\Delta A \sin\theta)]\cos\theta = 0 \qquad \text{(d)}$$

Solving Eq. (d) and using Eq. (11.13) yields:

$$\sigma_n = 2\,\tau_{xy}\cos\theta\sin\theta = \tau_{xy}\sin 2\theta \qquad (11.33)$$

Equation (11.33) shows that a normal stress σ_n occurs on all planes except those perpendicular or parallel to the axis of the shaft. Let's take τ_{xy} = 1 unit and evaluate Eq. (11.33) by permitting θ to vary from 0 to 360°. The results obtained for the normal stress σ_n, which acts on the oblique plane, are presented in Fig. 11.9.

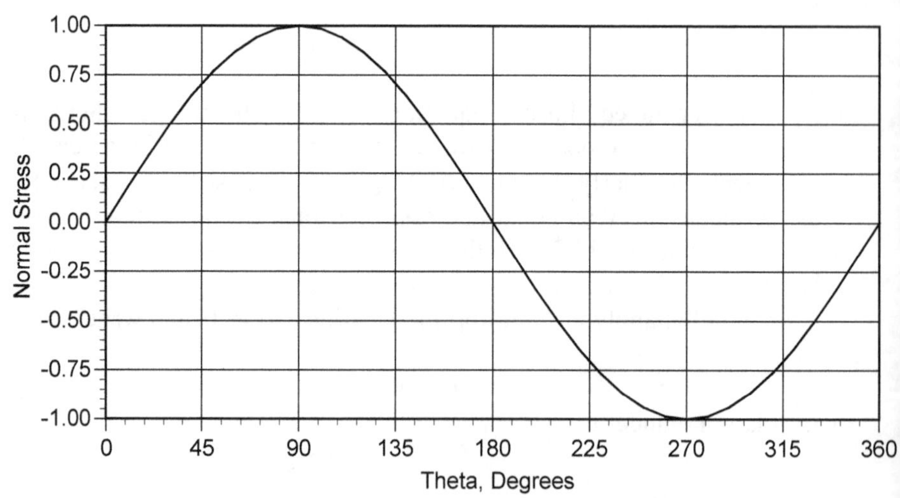

Fig. 11.9 The distribution of σ_n as a function of the orientation of the plane defined by θ.

EXAMPLE 11.8

If the shear stress $\tau_{xy} = 1$ unit, determine the shear stress τ_{nt} as a function of the orientation of the n-t plane defined by θ.

Solution:

Let's employ $\tau_{nt} = \tau_{xy} \cos 2\theta$ and solve for τ_{nt} as a function of θ. We will make use of a spreadsheet and vary θ from 0 to 360° in steps of 5° in this evaluation. A graph of the results is shown in Fig. E11.8. Also it is clear that $\tau_{Max} = \tau_{xy} = 1$ unit when $\theta = 0°$ and 180° and $\tau_{Min} = -\tau_{xy} = -1$ unit when $\theta = 90°$ and 270°. The shear stress vanishes on planes defined by $\theta = 45°, 135°, 225°, 315°$ etc.

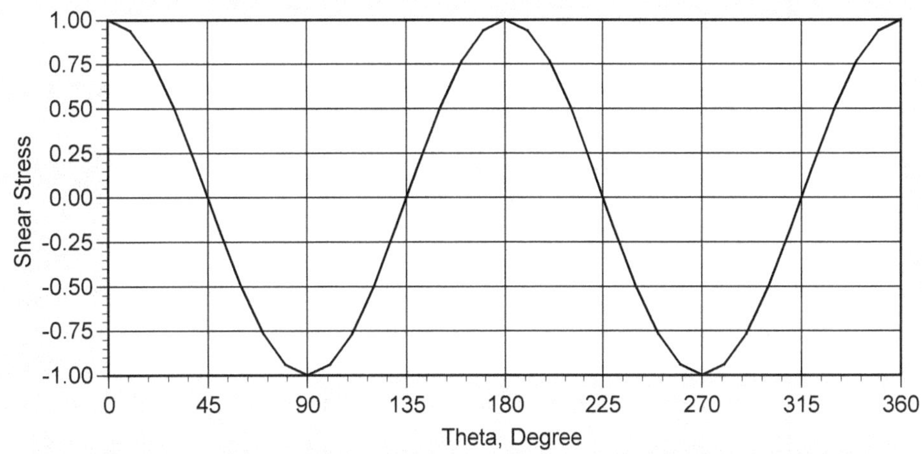

Fig. E11.8 Shear stress τ_{nt} as a function of the angle of inclination θ.

11.5 PRINCIPAL STRESSES IN SHAFTS

It is evident from Fig. 11.9 that the normal stress is a maximum ($\sigma_n = \tau_{xy} = \tau_{Max}$) when $\theta = 45°$ and 225° and a minimum ($\sigma_n = -\tau_{xy} = -\tau_{Max}$) when $\theta = 135°$ and 315°. Also the shear stress on these four planes is given by Eq. (11.32) as:

$$\tau_{nt} = \tau_{xy} \cos 2\theta = 0 \qquad \text{when } \theta = (1 + 2n)\pi/4 \text{ and } n = 0, 1, 2, \text{etc.}$$

When the normal stress is a maximum or a minimum acting on a plane with zero shear stress, a principal state of stress exists. The planes upon which the normal stress acts are called principal planes. We usually attempt to determine the principal planes and principal stresses because failure often occurs on these planes when the stresses become excessive.

For a circular bar subjected to a torque T, the principal plane makes an angle of 45° with the bar's axis as shown in Fig. 11.10. The principal plane spirals like a helix along the length of the bar. If this bar is made of a brittle material such as gray cast iron or chalk, its failure surface forms a helix as illustrated in Fig. 11.11.

Fig. 11.10 The principal planes for a circular shaft subjected to torque T are helical.

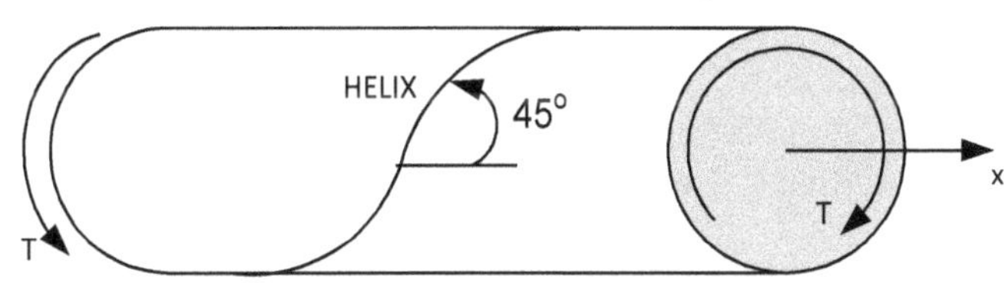

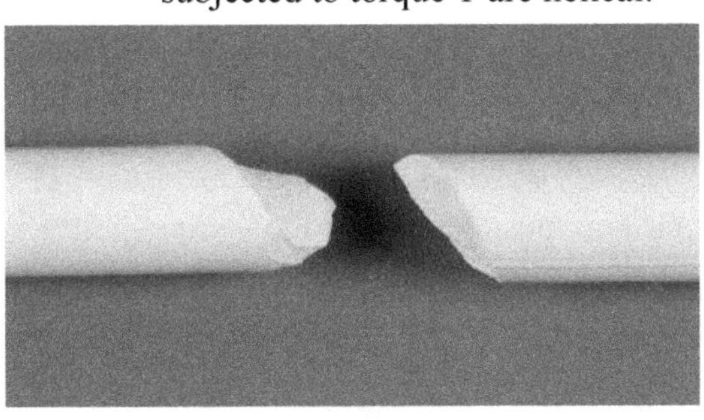

Fig. 11.11 Twisting a piece of chalk produces a helical failure making a 45° angle with its axis.

EXAMPLE 11.9

A circular shaft 10 mm in diameter is transmitting 4 kW from an electric motor to an air compressor. If the motor is operating at 1200 RPM, determine the maximum normal stress acting on the shaft. Also determine the plane upon which this maximum normal stress acts.

Solution

We solve for the torque acting on the shaft by using Eq. (11.11) to obtain:

$$T = P/\omega = (4000)(60)/(1200)(2\pi) = 31.83 \text{ N-m}$$

The shear stress may be determined from Eq. (11.16) as:

$$\tau_{Max} = \tau_{xy} = 16T/\pi d^3 = (16)(31.83)(1000)/[\pi(10)^3] = 162.1 \text{ N/mm}^2 = 162.1 \text{ MPa} \quad \text{(a)}$$

The relation between the shear stress and the normal stress is given by Eq. (11.33) as:

$$\sigma_n = \tau_{xy} \sin 2\theta$$

The normal stress is a maximum when $2\theta = \pi/2$ or $\theta = \pi/4$. This result for the principal angle is consistent with the helix presented in Figs. 11.10 and 11.11.

In this example, the maximum normal stress $(\sigma_n)_{Max}$ is the principal stress σ_1 given by:

$$(\sigma_n)_{Max} = \sigma_1 = \tau_{xy} = 162.1 \text{ MPa} \quad (11.34)$$

11.6 ANGLE OF TWIST

When torque is applied to a shaft, it twists through some angle ϕ about its longitudinal axis. This angle of twist of a circular shaft, illustrated in Fig. 11.2, is analogous to the stretch in a tension rod under the action of an axial force. To derive the expression for the angle of twist, rewrite Eq. (11.8) as:

$$\phi = \gamma_{Max} L/r \quad \text{(a)}$$

Substituting Eq. (11.9) into Eq. (a) yields:

$$\phi = (\tau_{Max} L)/(Gr) \quad \text{(b)}$$

Finally, substituting Eq. (11.15) into Eq. (b) gives:

$$\phi = (TL)/(GJ) \quad (11.35\text{ a})$$

Noting that $J = \pi d^4/32$, we may rewrite Eq. (11.35 a) as:

$$\phi = (32\, TL)/(\pi d^4\, G) \quad (11.35\text{ b})$$

where is the shear modulus defined in Eq. (11.9a).

If T, G or J are not constant along the length of the shaft, then Eq. (11.35a) must be modified to:

$$\phi_{Total} = \sum_{i=1}^{n} \frac{T_i L_i}{G_i J_i} \quad (11.35\text{c})$$

In this instance, the shaft must be divided into several sections and Eq. (11.35a) used to determine ϕ for each section. The total angle of twist, ϕ_{Total} is calculated by summing the ϕ from each section. Let's consider two examples to demonstrate the application of Eq. (11.35).

EXAMPLE 11.10

A solid steel shaft is 80 mm long and is subjected to a torque T of 200 N-m. If the shaft is 20 mm in diameter, determine the shear stress τ_{Max} and the angle of twist ϕ.

Solution

Let's employ Eq. (11.16) to solve for the maximum shear stress as:

$$\tau_{Max} = (16T)/(\pi d^3) = (16)(200)(1000)/[\pi(20)^3] = 127.3 \text{ MPa} \qquad \text{(a)}$$

The angle of twist is given by Eq. (11.35b) as:

$$\phi = (32\ TL)/(\pi d^4\ G) = (32)(200)(0.08)/[\pi(0.02)^4\ (79 \times 10^9)]$$

$$\phi = 0.01289 \text{ radian} = 0.7387° \qquad \text{(b)}$$

Inspection of Eq. (b) shows that the angle of twist is given in radians when Eq. (11.35) is employed. The result is converted to degrees by multiplying by the conversion factor 180°/π.

EXAMPLE 11.11

A line shaft, shown in Fig. E11.11, fabricated from high-strength steel incorporates five spur gears. Gear A provides the input torque T_A and gears B, C, D and E each remove torque from the shaft. Determine the angle of twist for the entire shaft AE and between gears at A and C. The shaft diameters and lengths are $d_{A\text{-}B} = 2.00$ in., $d_{B\text{-}C} = 1.75$ in., $d_{C\text{-}D} = 1.50$ in. and $d_{D\text{-}E} = 1.25$ in., $L_{A\text{-}B} = 30$ in., $L_{B\text{-}C} = 24$ in., $L_{C\text{-}D} = 36$ in. and $L_{D\text{-}E} = 48$ in. The applied torques are $T_A = 2400$ ft-lb, $T_B = 600$ ft-lb, $T_C = 800$ ft-lb, $T_D = 700$ ft-lb and $T_E = 300$ ft-lb. Neglect the effect of gear deformations.

Fig. E11.11

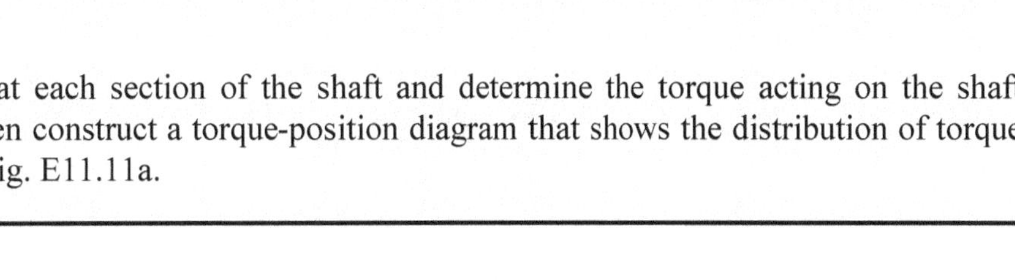

Solution:

First, prepare a series of FBDs at each section of the shaft and determine the torque acting on the shaft segments between the gears. Then construct a torque-position diagram that shows the distribution of torque along the line shaft as shown in Fig. E11.11a.

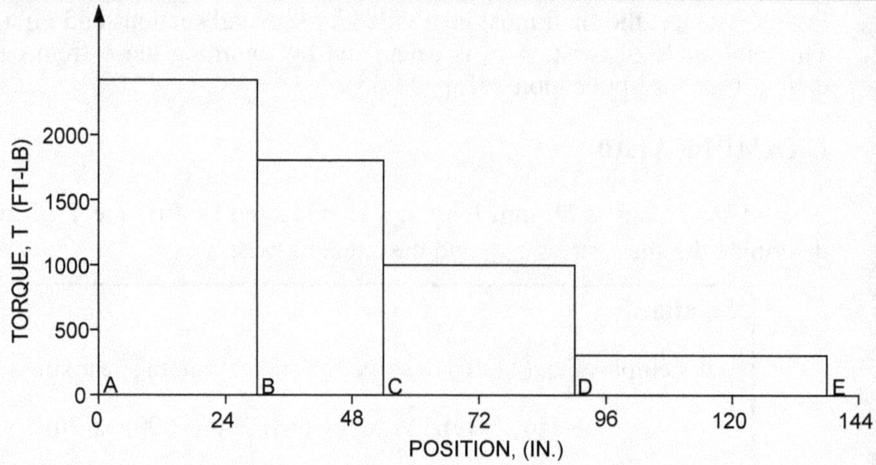

Fig. E11.11a

Next, write Eq. (11.35b) to determine the angle of twist for each segment of the line-shaft.

$$\phi_{A\text{-}B} = \frac{32TL}{\pi d^4 G} = \frac{32(-2400)(12)(30)}{\pi(2)^4\ (11.5\times10^6)} = -0.04783 \text{ radians} = -2.740^\circ$$

$$\phi_{B\text{-}C} = \frac{32TL}{\pi d^4 G} = \frac{32(-1800)(12)(24)}{\pi(1.75)^4\ (11.5\times10^6)} = -0.04896 \text{ radians} = -2.805^\circ$$

$$\phi_{C\text{-}D} = \frac{32TL}{\pi d^4 G} = \frac{32(-1000)(12)(36)}{\pi(1.5)^4\ (11.5\times10^6)} = 0.07558 \text{ radians} = -4.331^\circ$$

$$\phi_{D\text{-}E} = \frac{32TL}{\pi d^4 G} = \frac{32(-300)(12)(48)}{\pi(1.25)^4\ (11.5\times10^6)} = 0.06269 \text{ radians} = -3.592^\circ$$

The angle of twist for the entire shaft AE is given by Eq. (11.35c) as:

$\phi_{A\text{-}E} = \phi_{A\text{-}B} + \phi_{B\text{-}C} + \phi_{C\text{-}D} + \phi_{D\text{-}E} = -\ 2.740^\circ - 2.805^\circ - 4.331^\circ - 3.592^\circ = -\ 13.468^\circ$

The angle of twist between gears A and C can be determined from Eq. (11.35c) as:

$$\phi_{A\text{-}C} = \phi_{A\text{-}b} + \phi_{B\text{-}C} = -\ 2.740 - 2.805^\circ = -\ 5.545^\circ$$

The negative signs describe the sense of rotation (clockwise).

11.6.1 Torsion Bar Springs

When torque is applied to a shaft, it twists through an angle ϕ and strain energy is stored in the shaft. The strain energy **E** is equal to the work **W** input to the shaft as it twists. We depict this work as the triangular area in Fig. 11.12.

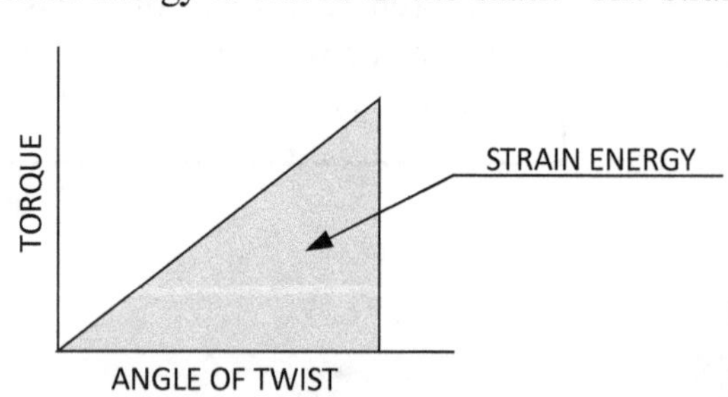

Fig. 11.12 The area under the torque-angle of twist curve is equal to the strain energy stored in a shaft.

The strain energy **E** is given by:

$$\mathbf{E} = (1/2)\,T\phi \qquad (11.36)$$

Substituting Eq. (11.35a) into this result gives:

$$\mathbb{E} = (T^2 L)/(2GJ) \qquad (11.37)$$

$$\mathbb{E} = (\phi^2 GJ)/(2L) \qquad (11.38)$$

The shaft acts as a torsion spring with the applied torque linearly proportional to the angle of twist.

$$T = k_T \, \phi \qquad (11.39)$$

where k_T is the spring rate given by:

$$k_T = GJ/L \qquad (11.40)$$

EXAMPLE 11.12

A steel torsion bar is incorporated into the chassis of an automobile to serve as a suspension spring. If the bar is fabricated from a steel rod 18 mm in diameter and 1.2 m long, determine its spring rate. Also determine its angle of twist if it stores strain energy $\mathbb{E}$ = 5.0 N-m as the automobile traverses a bump.

Solution:

We apply Eq. (11.40) to obtain:

$$k_T = GJ/L = [(79)(10^9)\ \pi\ (0.018)^4]/[(1.2)(32)] = 678.5 \text{ N-m}$$

From Eq. (11.38) and Eq. (11.40), it is evident that:

$$\phi = [2E/k_T\,]^{1/2} = [10/\ 678.5]^{1/2} = 0.1214 \text{ radians} = 6.956°$$

The result shows that the torsion spring undergoes a large angle of twist. Will it survive over the service life of the automobile?

11.7 STRESS CONCENTRATIONS IN CIRCULAR SHAFTS

In many applications, shafts are stepped with different diameters to accommodate gears and bearings. Stress concentrations occur at the fillets adjacent to a shaft's shoulder where one diameter transitions into another. The geometry leading to a stress concentration of this type is illustrated in Fig. 11.13. In applications where the loads are cyclic, the shafting is subjected to fatigue failure and the stress concentrations must be accounted for in the design analysis.

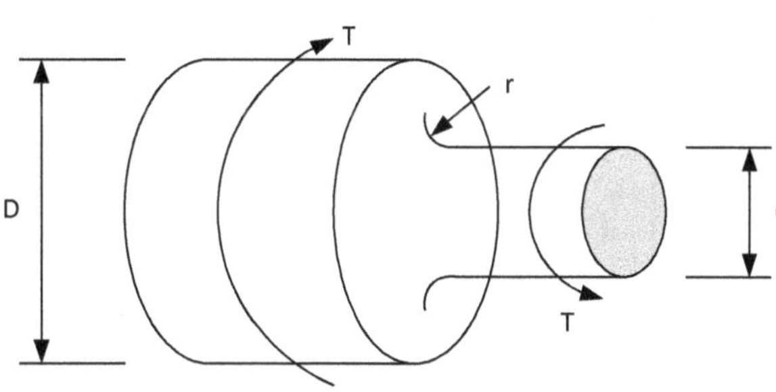

Fig. 11.13 Stress concentrations occur at the fillet located at the transition of the shaft diameters.

The effect of the stress concentration on elevating the shear stresses due to the applied torque is accommodated by introducing a torsion stress concentration factor K_{TS}. In this case, the maximum shear stress at the fillet is given by:

$$\tau_{MAX} = K_{TS}\,\tau_{NOM} = 16K_{TS}\,T/(\pi d^3) \qquad (11.41)$$

where the nominal shear stress $\tau_{NOM} = 16T/(\pi d^3)$ occurs in the smaller diameter shaft.

The stress concentration factor depends on the radius of the fillet and the mismatch in the diameters as shown in Fig. 11.14.

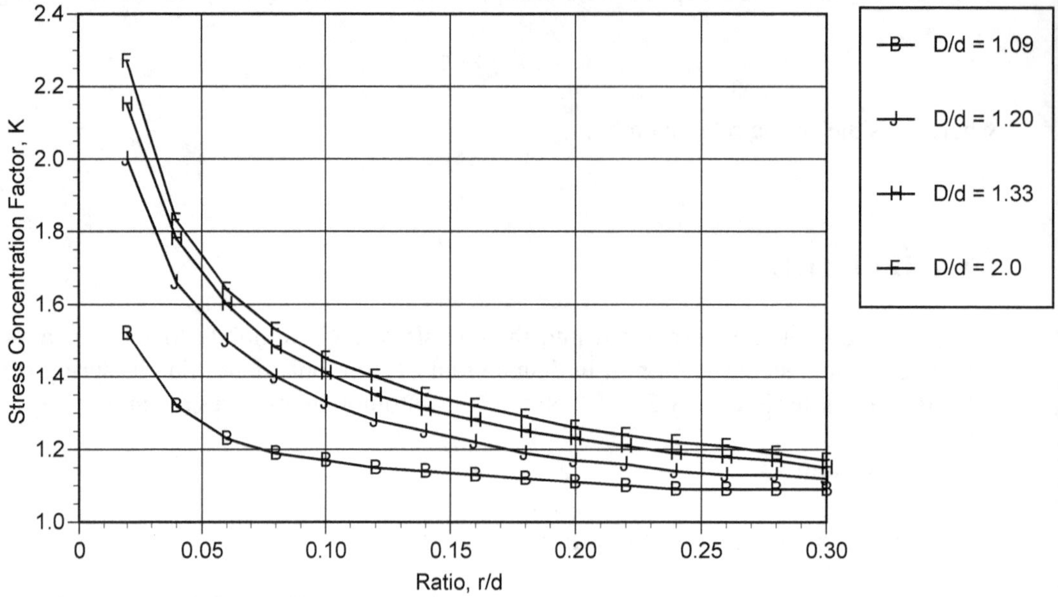

Fig. 11.14 Stress concentration factor for a shouldered shaft subject to torsion.

EXAMPLE 11.13

A shouldered shaft is subjected to a torque T = 1,500 N-m. The diameters are d = 120 mm and d = 60 mm. The radius of the transition fillet is 6.0 mm. Determine the maximum shear stress at the fillet.

Solution:

The nominal shear stress is given by:

$$\tau_{NOM} = 16T/(\pi d^3) = (16)(1{,}500)/[\pi\,(0.060)^3] = 35.37 \text{ MPa} \qquad (a)$$

To determine the stress concentration factor compute the ratios D/d and r/d as:

$$D/d = 120/60 = 2.00 \qquad \text{and} \qquad r/d = 6.0/60 = 0.10$$

Reference to the graph for the stress concentration factor K_{TS} shown in Fig. 11.14 gives:

$$K_{TS} = 1.45 \qquad (b)$$

Then from Eq. (11.41), we write:

$$\tau_{MAX} = K_{TS}\,\tau_{NOM} = 1.45(35.37) = 51.28 \text{ MPa} \qquad (c)$$

Examination of these results indicates that the presence of the shoulder on the shaft increased the shear stresses in the fillet from 35.37 MPa to 51.28 MPa. This increase is significant and it is clear that stress concentrations must be taken into account whenever we are dealing with cyclic loading or brittle materials.

11.8 DESIGN OF POWER TRANSMISION SHAFTING

Power transmission shafts connect a power source such as an electric motor or an internal combustion engine to some appliance such as a compressor or a machine tool. Shafts are circular in cross section because they rotate (often at very high speed) and are supported by bearings. Shafting is usually fabricated from alloy steel because of its superior strength. Because the shafting rotates and the stresses imposed are usually cyclical, the design is based on the fatigue strength of the steels. Also, stress concentrations are taken into account in a typical design analysis.

The relations for determining the shearing stress in shafting were introduced in a previous section of this chapter. The purpose of this section is to demonstrate the application of these relations in a design analysis of two challenging problems.

EXAMPLE 11.14

The electric motor shown in Fig. E11.14 is rated at 200 kW and operates at 1180 RPM. It powers a gear drive with a variable load which at times requires the motor to operate at its maximum rating. The shaft is supported by two bearings—one on each side of the gear drive. The shaft from the motor has a diameter d and the shaft at the gear has a diameter D. The radius at the transition is specified as r/d = 0.15 and the diameter mismatch ratio is specified as D/d = 1.33. Select a material for the shaft and determine the diameters d and D if the maximum shear stress in the material is not to exceed 1/3 the yield strength of the material in shear. Another constraint on the design is the angle of twist ϕ, which is not to exceed 2.5° measured from the motor to the gear drive.

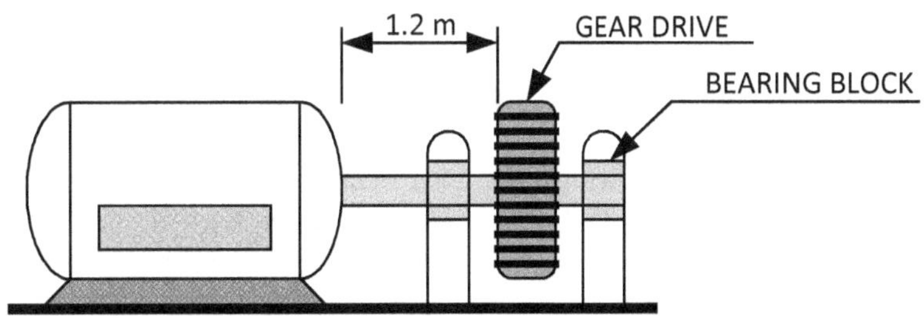

Fig. E11.14

Solution:

A step-by step approach will be used in the solution of this example.

Step 1: We begin by selecting the material for the shaft. Alloy steel, 4340 HR, is specified because it exhibits a very high strength. The high strength is important because it will reduce the diameter of the shaft and hence the diameter of the bearings and the armature in the electric motor. The cost savings in the bearings and armature will offset the added cost of the alloy steel.

From Appendix B-1 and B-2, we determine the material's strength and elastic modulus as:

$$G = 207 \text{ GPa} \qquad S_u = 1041 \text{ MPa} \qquad S_y = 910 \text{ MPa} \qquad \text{(a)}$$

Next we use Eq. (11.17a) to determine the yield strength of the 4340 HR alloy steel in shear.

$$S_{ys} = 0.5774S_y = (0.5774)(910) = 525.4 \text{ MPa} \qquad \text{(b)}$$

Step 2: Determine the maximum shear stress that can be imposed on the shaft from the problem statement and Eq. (b) as:

$$\tau_{MAX} = S_{ys}/3 = 525.4/3 = 175.1 \text{ MPa} \qquad \text{(c)}$$

Step 3: Employ Fig. 11.14 and the ratios r/d = 0.15 and D/d = 1.33 to establish the stress concentration factor at the fillet of the transition of the shaft diameters as:

$$K_{TS} = 1.31$$

Substitute this value of K_{TS} into Eq. (11.41) to obtain the nominal shear stress that acts on the smaller diameter shaft:

$$\tau_{NOM} = \tau_{MAX} / K_{TS} = 175.1/1.31 = 133.7 \text{ MPa} \qquad \text{(d)}$$

Step 4: Determine the torque applied to the shaft when the electric motor is operating at its rated capacity by using Eq. (11.1:

$$T = P/\omega = [(200 \times 10^3)(60)]/[(1180)(2\pi)] = 1{,}618.5 \text{ N-m} \qquad \text{(e)}$$

Step 5: If we employ a solid steel shaft, which is the most cost effective choice, Eq. (11.16) can be used together with Eq. (d) and Eq. (e) to determine the diameter d of the shaft leading from the motor to the gear drive as:

$$d^3 = (16\,T)/(\pi\tau_{NOM}) = (16)(1{,}618.5)/[\pi(133.7 \times 10^6)] = 61.65 \times 10^{-6} \text{ m}^3$$

$$d = 0.0395 \text{ m} = 39.5 \text{ mm} \qquad \text{(f)}$$

Select a standard size available in metric diameters, which is 40 mm. Then from the problem statement, D = 1.33 (40) = 53.2 mm. Because this is not a standard size, it is necessary to increase D to 55 mm. This change will not markedly affect the stress concentration; therefore it is not necessary to rework the analysis.

Step 6: Check to determine if the angle of twist ϕ is within the specified limit of 2.5° by employing Eq. (11.35b).

$$\phi = (32TL)/(\pi d^4 G) = [(32)(1{,}618.5)(1.2)]/[\pi(0.04)^4 (207 \times 10^9)] = 0.03773 \text{ radians} = 2.139°$$

This angle is within the specifications outlined in the problem statement. The final guidance for the drawing of the shaft is D = 55 mm, d = 40 mm and r = 6 mm.

EXAMPLE 11.15

The electric motor shown in Fig. E11.15 is rated at 125 HP and operates at 1780 RPM. It powers a drive shaft with three output gears. When the motor is operating at rated capacity gear A removes 1/2 the power developed by the motor, while gear B and gear C remove 1/3 and 1/6, respectively. The shaft is supported by four bearings as shown in Fig. E11.15. The shaft from the motor has a diameter d but it is shouldered with a diameter D at each of the three gear locations. The radius at the transition is specified as r/d = 0.20 and the diameter mismatch ratio is specified as D/d = 2.00. If an alloy steel with a yield strength in shear of 125 ksi is used to fabricate the shaft, determine the diameters d and D if the maximum shear stress in the material is not to exceed 1/4 the yield strength of the material in shear. Another constraint on the design is the angle of twist ϕ, which is not to exceed 3.0° measured from the motor to the gear C.

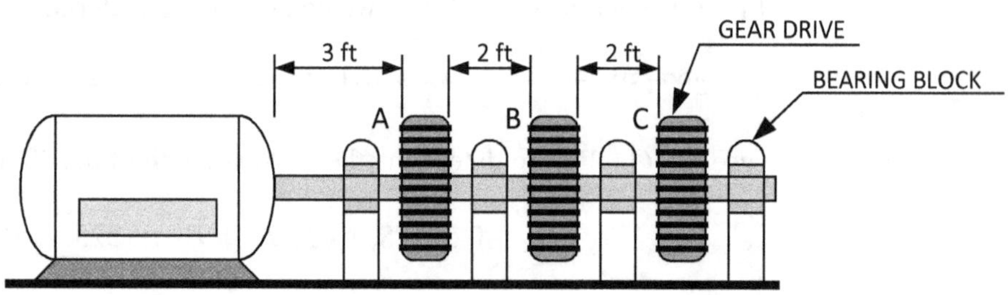

Fig. E11.15

Solution:

The shaft is of uniform diameter d from the motor to the free end except for the shoulder regions at the gears. A step-by-step solution is given below:

Step 1: Determine the torque exerted on the shaft when the electric motor is operated at its rated capacity from Eq. (11.3).

$$T = 33{,}000\ HP/2\pi n = (33{,}000)(125)/(2\pi)(1780) = 368.8 \text{ ft-lb} \qquad (a)$$

Step 2: Compute the torque removed at each gear by using the fractions listed in the problem statement and construct a torque position diagram as shown in Fig. E11.15a:
$T_A = 368.8\ (1/2) = 184.4$ ft-lb, $T_B = 368.8\ (1/3) = 122.9$ ft-lb, $T_C = 368.8\ (1/6) = 61.47$ ft-lb

Fig. 11.15a

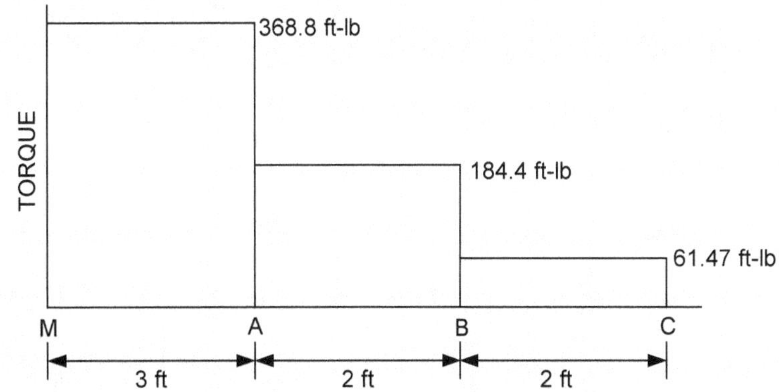

Step 3: From Fig. 11.14, determine the stress concentration factor K_{TS} for a shaft with r/d = 0.2 and D/d = 2.0:

$$K_{TS} = 1.26 \qquad (b)$$

Step 4: Determine the nominal shear stress that is allowable for the shaft by using Eq. (11.41) and the fact that $\tau_{MAX} = 125/4 = 31.25$ ksi:

$$\tau_{NOM} = \tau_{MAX}/K_{TS} = 31.25/1.26 = 24.80 \text{ ksi} \qquad (c)$$

Step 5: If a solid steel shaft is employed, Eq. (11.16) can be used together with Eq. (a) and Eq. (c) to determine the diameter d of the shaft leading from the motor to the gear drive as:

$$d^3 = (16\,T)/(\pi\tau_{NOM}) = (16)(368.8)(12)/[\pi(24.8 \times 10^3)] = 0.9088 \text{ in.}^3$$

$$d = 0.9686 \text{ in.} \qquad (d)$$

Select standard sizes as d = 1.00 in. and D = 2.00 in. for the shaft diameters.

Step 6: Determine the total angle of twist by summing the twist of the shaft between each of the gear drives. We will assume that the twist at the gears is sufficiently small to neglect. From Eqs. (11.35b) and (11.35c) we write:

$$\phi_{M\text{-}C} = \phi_{M\text{-}A} + \phi_{A\text{-}B} + \phi_{B\text{-}C} = [(32)/(\pi d^4 G)][T_{M\text{-}A}\, L_{M\text{-}A} + T_{A\text{-}B}\, L_{A\text{-}B} + T_{B\text{-}C}\, L_{B\text{-}C}]$$

$$\phi_{M\text{-}C} = \{32/[\pi(1)^4\,(11.5 \times 10^6)]\}\{(368.8)(12)(36) + (184.4)(12)(24) + (61.47)(12)(24)\}$$

$$\phi_{M\text{-}C} = 0.2038 \text{ radians} = 11.68° \qquad (e)$$

Clearly this angle of twist exceeds the allowable twist that is specified by a factor of nearly four. It is necessary to increase the shaft diameter to reduce ϕ. If we recognize that the angle of twist decreases as the fourth power of the shaft diameter, it is easy to compute the required diameter from:

$$(d_{2nd}/d_{1st})^4 = 4 \text{ or } d_{2nd} = (4)^{1/4}\, d_{1st} = (1.414)(1.00) = 1.414 \text{ in.} \qquad (f)$$

Selecting the next larger standard size gives d = 1.50 in., D = 3.0 in. and r = 0.30 in.

11.9 TORSION OF NON-CIRCULAR SHAFTS

The discussion of torsion thus far has focused on rotating shafts with circular cross sections because most applications involve transmitting power from a motor or engine to an appliance. However, structural engineers sometimes load members such as angle and channel sections with transverse forces that induce torque. The derivations of the equations for the maximum shear stress and the angle of twist in shafts with non-circular cross sections are beyond the scope of this text. For this reason, we limit this section to the definition of the cross section geometry and present equations for τ_{MAX} and ϕ.

Elliptical Cross Section

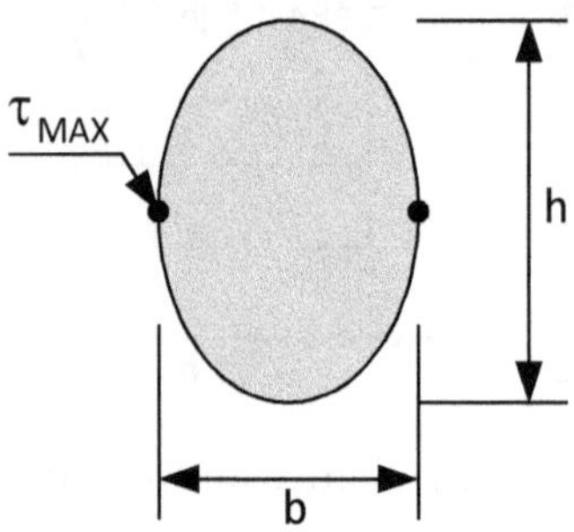

$$\tau_{MAX} = 16T/\pi b^2 h$$
$$\phi = (4\pi^2 TLJ)/(A^4 G) \qquad (11.42)$$
$$J = (\pi/64)(b h^3 + b^3 h)$$
$$A = \pi bh/4$$

Equilateral Triangle

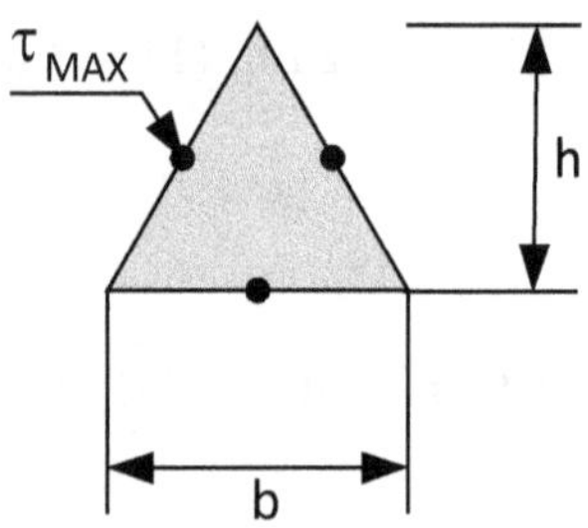

$$\tau_{MAX} = 20T/b^3$$
$$\phi = (TL)/(0.6\ GJ) = 46.2TL/b^4 G \qquad (11.43)$$

Hexagon

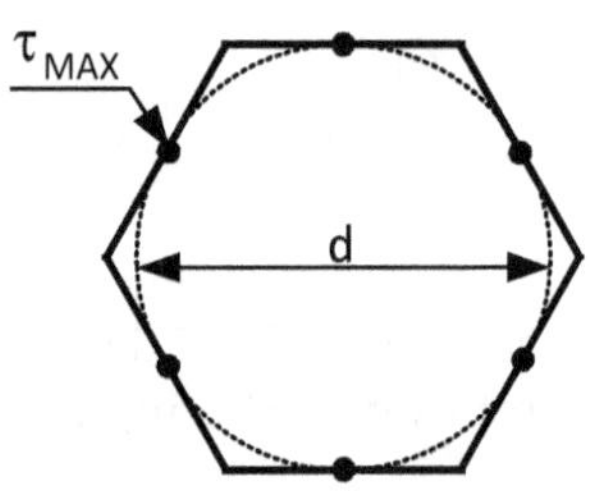

$$\tau_{MAX} = T/[0.217\ Ad]$$
$$\phi = (TL)/(0.133\ Ad^2 G) \qquad (11.44)$$
$$A\ 0.866d^2$$

Rectangle

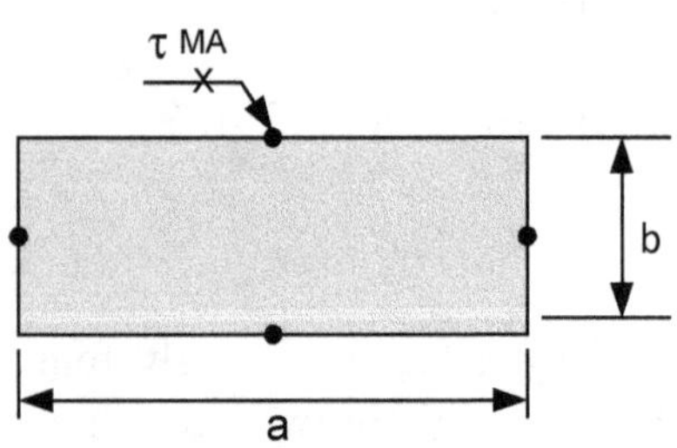

$$\tau_{MAX} = T/(\alpha ab^2) \qquad \text{for } a \geq b$$
$$\phi = (TL)/(\beta ab^3 G) \qquad (11.45)$$

$$\alpha = 0.3334 - 0.1904(b/a) - 0.2574(b/a)^2 + 1.0255(b/a)^3 - 1.0946(b/a)^4 + 0.3916(b/a)^5$$

$$\beta = 0.3333 - 0.1997(b/a) - 0.0790(b/a)^2 + 0.1892(b/a)^3 - 0.1766(b/a)^4 + 0.0739(b/a)^5$$

Thin Rectangle

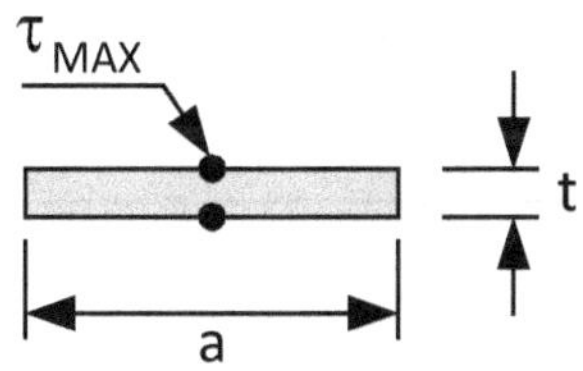

$$\tau_{MAX} = 3.0\ T/at^2$$
$$\phi = (3TL)/(a\ t^3\ G) \qquad (11.46)$$

Angle Section

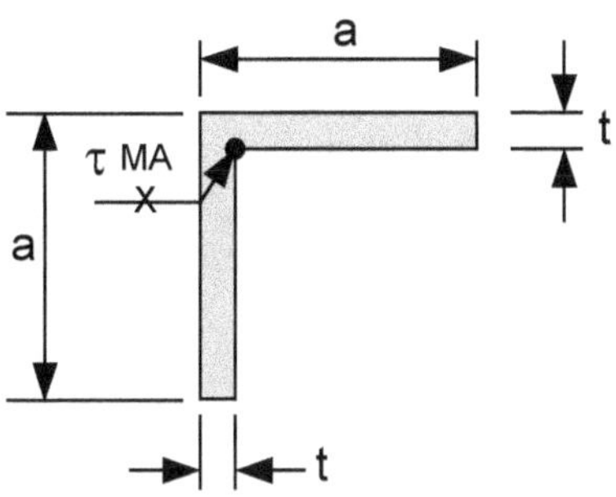

$$\tau_{NOM} = 3.0\ T/2at^2$$
$$\phi = (3TL)/(2a\ t^3\ G) \qquad (11.47)$$
for $t \ll a$

The maximum shear stress occurs at the reentrant corner of the angle, channel and I sections. The stress concentration associated with the reentrant corner is usually neglected if the section is fabricated from a ductile material and is not subjected to cyclic loading.

Channel Section

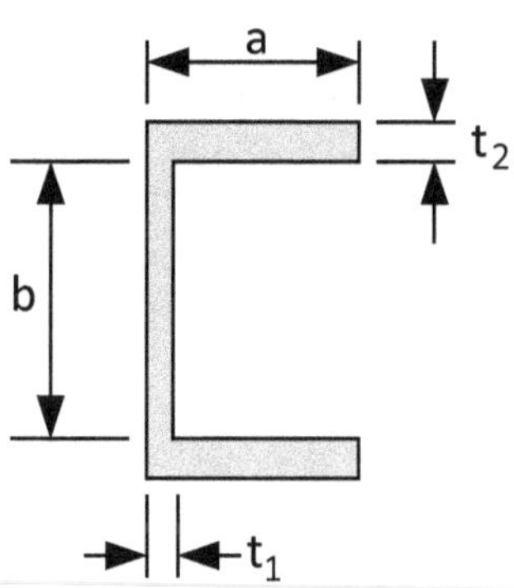

$$\tau_{NOM} = 3.0\ Tt_2/(bt_1^{\,3} + 2at_2^{\,3})$$
$$\phi = (3TL)/[(bt_1^{\,3} + 2at_2^{\,3})G] \qquad (11.48)$$

I Section

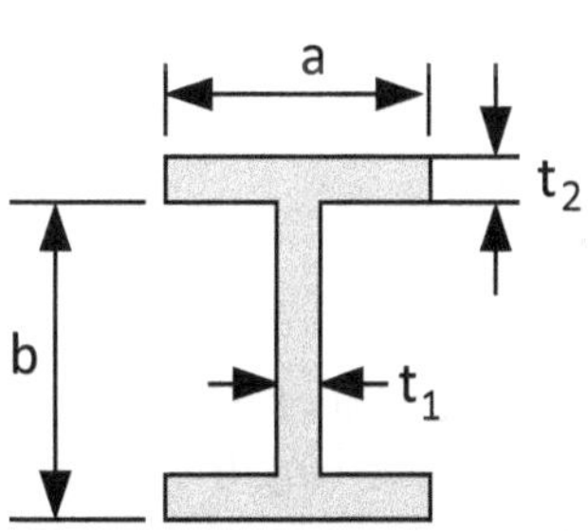

$$\tau_{NOM} = 3.0\ Tt_2/(bt_1^{\,3} + 2at_2^{\,3})$$
$$\phi = (3TL)/[(bt_1^{\,3} + 2at_2^{\,3})\ G] \qquad (11.49)$$

EXAMPLE 11.16

A torque T = 400 N-m is applied to a structural member with a square cross section. If the square section is 40 mm on a side and is 3 m long, determine the maximum stress and the angle of twist. Will the structural member yield if it is fabricated from 1010 A carbon steel?

Solution: First calculate the numerical factors α and β from Eq. (11.45):

For (b/a) = 1.0 ⇒⇒ $\alpha = 0.208$; $\beta = 0.141$

Reference Eq. (11.45) and write:

$$\tau_{MAX} = T/(\alpha ab^2) = (400)/[(0.208)(0.04)^3] = 30.05 \text{ MPa} \quad \text{(a)}$$

Next use Eq. (11.17a) to determine the yield strength of the 1010 A carbon steel in shear.

$$S_{ys} = 0.5774 S_y = (0.5774)(200) = 115.5 \text{ MPa} \quad \text{(b)}$$

It is clear that the structural member will not yield.

The angle of twist for the square cross section is given by:

$$\phi = (TL)/(\beta ab^3 G) = (400)(3)/[(0.141)(.04)^4 (79 \times 10^9)] = 0.04208 \text{ radians}$$

$$\phi = 2.411°$$

11.10 SHEAR CENTER

The topic of shear center combines the concept of shear stresses induced by transverse forces applied to beams and the angle of twist in torsion. Twisting of beams in bending did not arise in our previous discussion in Chapter 10 because we considered only beams with symmetric cross sections. If the cross section of the beam is symmetric about its vertical axis, the shear stresses do not produce a twisting moment. However, if the beam's cross-section is not symmetric, twisting will occur. To illustrate this point consider a cantilever beam loaded at its free end with a concentrated force F. The concentrated load is applied through the center of gravity of the channel section as shown in Fig. 11.15. When the load is applied, the beam will twist about its longitudinal axis through an angle ϕ.

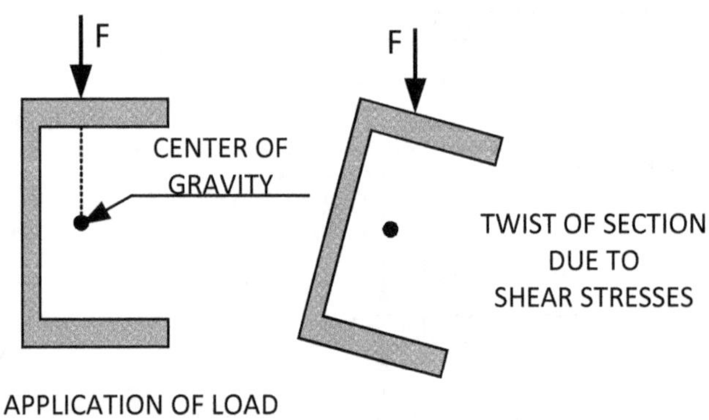

Fig. 11.15 Twisting of unsymmetrical beam sections induced by shear stresses.

Let's consider a cantilever beam fabricated from a channel section with the force F applied at its free end through its center of gravity as illustrated in Fig. 11.16a. A constant shear force V = F occurs over the length L of the beam. This shear force produces shear stresses in both the web and the flanges of the channel section that are given by Eq. (10.23) as:

$$\tau = \tau_V = \tau_H = VQ_1/(bI_z) \quad \text{(10.23 bis)}$$

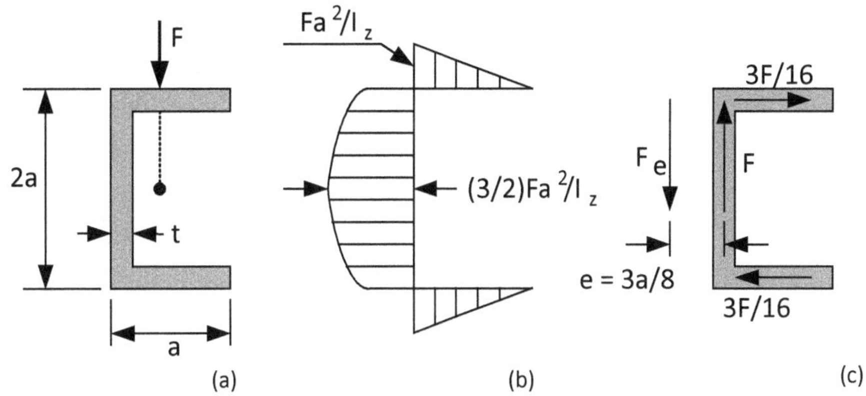

Fig. 11.16 Shear stress distribution in a channel section subjected to shearing forces V = F.

If the shear stresses in the channel section are determined from Eq. (10.23), the distribution of τ_V and τ_H are shown in Fig. 11.16b. In this solution of Eq. (10.23) higher powers of (t/a) have been disregarded because they are small in relation to the other terms. Integrating the shear stress distribution over the three rectangular areas comprising the channel section and substituting $I_z = 8a^3t/3$ gives the three shear forces shown in Fig. 11.16c. Examination of these forces indicates that the force F in the vertical portion of the channel is equal to the applied force F as expected. The two horizontal forces 3F/16 are equal and opposite and cancel so as to satisfy $\Sigma F_z = 0$. However, the two horizontal forces produce a couple, a torque T = (3/8)Fa, that tends to twist the channel section (in a clockwise direction) so as to produce the twisted shape presented in Fig. 11.15.

The shear center is established from this twisting moment. If the transverse force $F_e = F$ is applied at an eccentric position, e = (3/8)a, from the centerline of the vertical segment of the channel as shown in Fig. 11.16c, it produces a couple with T = (3/8)Fa in the counter clockwise direction. Thus, the couples cancel and the channel section does not twist when the transverse force is applied. When loading unsymmetrical sections, the transverse force should be applied through the shear center instead of the center of gravity.

Channel sections and other unsymmetrical sections are often employed in structural applications. In these cases, attachments are welded to the main beams to allow for the eccentric placement of the applied forces as indicated in Fig. 11.17.

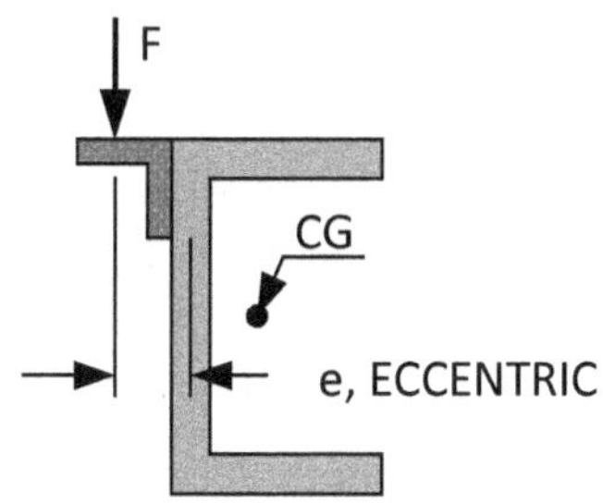

Fig. 11.17 Eccentric placement of the applied force to suppress twisting due to shear stresses.

EXAMPLE 11.17

The channel section presented in Fig. E11.17 is to be employed in fabricating a cantilever beam. Determine the eccentricity e of the transverse force F if the beam is to deflect without rotation.

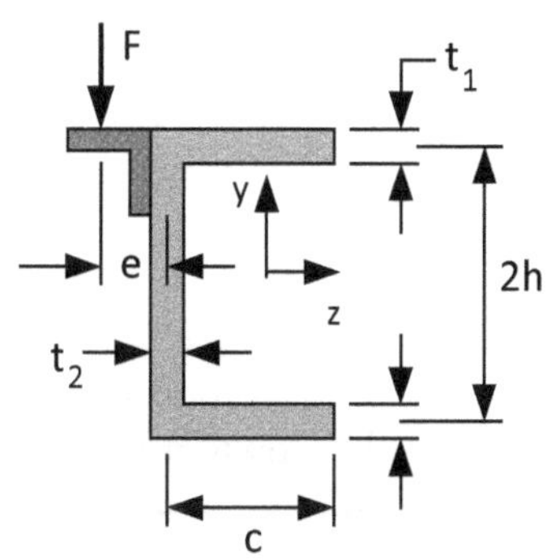

Solution:

Step 1: Prepare FBDs to show the shear stresses introduced by the shear force V = F. Section the top flange of the channel section a distance d from its edge as shown in Fig. E11.17a.

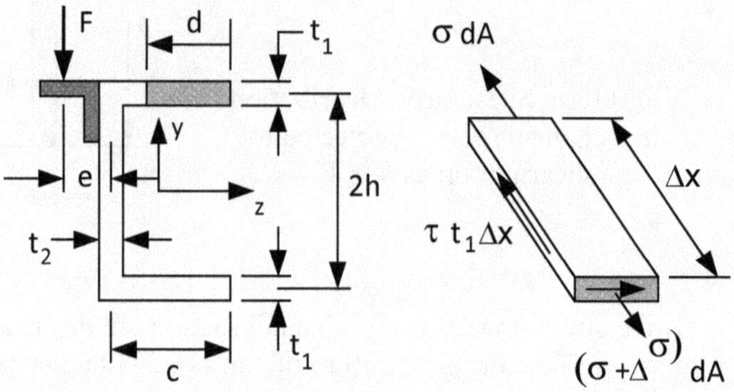

Fig. E11.17a

Step 2: The shear stress τ in the top flange is determined from Eq. (10.23) as:

$$\tau = VQ_1/I_z\, b = (Fhd/I_z) \qquad \text{(a)}$$

where $V = F$, $b = t_1$ and $Q_1 = hdt_1$

The shear stress in the flange is a linear function of d increasing from zero at the edge to a maximum value where the flange intersects the web of the channel. Clearly τ_{MAX} is given by:

$$(\tau_{MAX})_{Flange} = Fhc/I_z \qquad \text{(b)}$$

Step 3: The shear stress at the top of the web is also determined from Eq. (10.23) as:

$$\tau = VQ_1\,/I_z\, b = [(Fhct_1)\,/(t_2\, I_z)] \qquad \text{(c)}$$

where $V = F$, $b = t_2$ and $Q_1 = hct_1$

Step 4: The shear stress is a maximum at the center of the web where y = 0. From Eq. (10.23), we write:

$$\tau = VQ_1\,/I_z\, b = Fh[c(t_1/t_2) + (h/2)]/I_z \qquad \text{(d)}$$

The shear stress exhibits a parabolic distribution over the height of the web. If this shear stress is integrated over the area of the web, we find:

$$\int \tau dA_{Web} \approx F = V \qquad \text{(e)}$$

Step 5: The shear stress in the flanges varies linearly from zero to a maximum at the junction of the web and the flange. Integrating the shear stress over the area of the top flange gives the shear force H_1 as:

$$H_1 = \int \tau dA_{Flange} = (1/2)(ct_1)(Fhc/I_z) = (Fhc^2 t_1\,)/(2I_z) = -\,H_2 \qquad \text{(f)}$$

A second shear force H_2, equal in magnitude and opposite in direction, is induced in the lower flange as shown in Fig E11.17b.

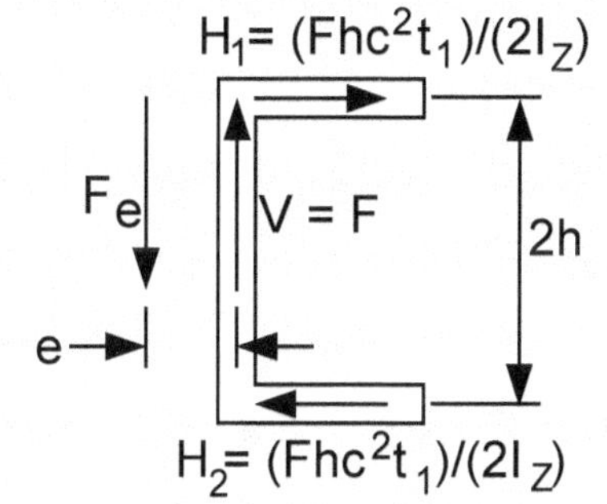

Fig. E11.17b

Step 6: The two equal and opposite shear forces, H_1 and H_2, produce a clockwise couple T equal to:

$$T = (2h)H_1 = (Fh^2 c^2 t_1)/(I_z) \qquad \text{(g)}$$

Step 7: An equal and opposite moment (torque) is produced by applying the force F with an eccentricity e as shown in Fig. E11.17b. The eccentricity is given by e = T/F as:

$$e = T/F = (h^2 c^2 t_1)/(I_z) \qquad \text{(h)}$$

The moment of inertia I_z is given by:

$$I_z = (2/3)t_2 h^3 + 2t_1 ch^2 \qquad \text{(i)}$$

Step 8: Substituting Eq. (i) into Eq. (h) gives the eccentricity necessary in applying the transverse force F to avoid twisting the channel section beam.

$$e = \left(\frac{3}{2}\right)\left[\frac{c^2 t_1}{t_2 h + 3t_1 c}\right] \qquad \text{(j)}$$

11.11 SUMMARY

The emphasis in this chapter was on the design of shafting fabricated from circular rods. Shafts are commonly employed to transmit power from an electric motor or an internal combustion engine to some appliance. In transmitting power, the shaft is subjected to a torque and it may fail due to shear stresses that produce yielding or due to an excessive angle of twist ϕ. The relation between power transmitted and the torque developed in the shaft was given as:

$$P = T\omega \qquad (11.1)$$

When a shaft is subjected to a torque T, it twisted through an angle ϕ. This deformation in turn produces a shearing strain γ given by:

$$\gamma_\rho = \rho\phi/L \qquad (11.7)$$

The shearing stress τ in the shaft is determined from the stress-strain relation for shearing deformation, which is given by:

$$\tau = G\gamma \qquad (11.9)$$

where the shear modulus is related to the elastic modulus by:

$$G = \frac{E}{2(1+\nu)} \qquad (11.9a)$$

The shear stress in terms of the torque and radial position ρ is:

$$\tau = T_\rho \rho / J \qquad (11.14)$$

The shear stress τ is a maximum on the outside surface of the shaft where ρ = r = d/2.

$$\tau_{Max} = 2\ T/(\pi r^3) = 16\ T/(\pi d^3) \qquad (11.16)$$

The yield strength in shear is much lower than the yield strength in either tension or compression. We approximate the yield and ultimate strength in shear as:

$$S_{ys} = 0.5774\ S_y \qquad (11.17a)$$

$$S_{us} = 0.5774\ S_u \qquad (11.17b)$$

We introduced the concept of section modulus in torsion Z_T for a hollow circular shaft by defining τ_{MAX} as:

$$\tau_{Max} = 16\ Td_o/[\pi(d_o^4 - d_i^4)] = T/Z_T \qquad (11.22)$$

$$\text{where } Z_T = \pi(d_o^4 - d_i^4)/(16 d_o) \qquad (11.23)$$

Using the section modulus in torsion, we introduced the concept of structural efficiency of a shaft in torsion SE following the same approach that we employed in the discussion of beams in bending.

$$SE = Z_T/A \qquad (11.28)$$

$$SE = \frac{d_o^2 + d_i^2}{4 d_o} \qquad (11.29)$$

We considered shear stress acting on orthogonal planes and established that they are equal in magnitude and opposite in direction.

$$\tau_{xy} = \tau_{yx} \qquad (11.31)$$

We examined the shear stresses on oblique planes and found that the shear and normal stresses on the oblique planes oscillate as sinusoids:

$$\tau_{nt} = \tau_{xy}(\cos^2\theta - \sin^2\theta) = \tau_{xy}\cos 2\theta \qquad (11.32)$$

$$\sigma_n = 2\ \tau_{xy}\cos\theta\sin\theta = \tau_{xy}\sin 2\theta \qquad (11.33)$$

We found that the normal stresses were a maximum on planes where the shear stress vanished. This fact enabled us to define principal planes and leads to the relation for the principal stress σ_1:

$$(\sigma_n)_{Max} = \sigma_1 = \tau_{xy} \qquad (11.34)$$

We considered the deformation of the shaft due to torsion and established the expression for the angle of twist as:

$\phi = (TL)/(GJ)$	(11.35 a)
$\phi = (32\ TL)/(\pi d^4\ G)$	(11.35 b)

Torsion rods are often employed as springs in suspension systems or to store strain energy. We derived the relation between the torsion and the angle of twist in terms of the spring rate k_T as:

$T = k_T\ \phi$	(11.39)
$k_T = GJ/L$	(11.40)

When shafts are designed with shoulders, a stress concentration occurs in the fillet at the transition from the smaller to the larger diameter. We accommodate the stress concentration by introducing an appropriate stress concentration factor that is dependent on the geometry of the transition region.

$$\tau_{MAX} = K_{TS}\ \tau_{NOM} = 16K_{TS}\ T/(\pi d^3) \qquad (11.41)$$

The stress concentration factor K_{TS} is given in Fig. 11.14.

Torsion of non-circular sections was introduced. When the section is non-circular, the plane sections do not remain plane when the bar is subjected to torsion. Instead the plane sections warp. This distortion complicates the derivation of the relations for the maximum shear stress and the angle of twist. We have provided these equations without derivation for the following cross sections—elliptical, equilateral triangle, hexagon, rectangle, thin rectangle and angle, channel, and I sections.

We reintroduced the concept of shear stresses produced as a beam bends under the action of transverse forces. If the cross section of the beam is not symmetric about a vertical axis, the beam will twist due to the effect of these shear stresses. The torsional couple produced by bending a cantilever beam fabricated from a channel section was discussed. The twisting of the beam can be avoided if the transverse force is applied through the shear center of the section. Example 11.17 demonstrated the analysis used to determine the location of the shear center.

PROBLEMS

11.1 A shaft is employed to transmit 700 W of power from an electric motor to an appliance. Determine the torque imposed on the shaft if the motor is operated at the angular velocity of 3600 RPM.

11.2 A shaft with a circular cross section is twisted through an angle $\phi = 3.1°$. If the diameter of the shaft is 15 mm and its length is 1.3 m, determine the maximum shearing strain γ.

11.3 Determine the maximum shearing stress τ_{max} for the conditions described in Problem 11.2 if the shaft is fabricated from steel.

11.4 A shaft with a circular cross section is subjected to a torque of 120 ft-lb. If the shaft's diameter is 0.750 in and its length is 15 in., determine the maximum shearing stress τ_{MAX}.

11.5 A shaft fabricated from 302 A stainless steel transmits 1.5 kW of power while rotating at an angular velocity of 1180 RPM. Specify a standard size round for the diameter of the shaft if a safety factor of 2.9 with respect to yielding is specified.

11.6 A hollow tube is employed in the design of a shaft to transmit 1.5 kW of power at 1180 RPM with a safety factor of 2.7 with respect to yielding. Determine the outside diameter d_o of the tube if its wall thickness is 0.05 d_o. The tube is fabricated from 1020 HR steel.

11.7 Clearly a hollow tube is structurally more efficient than a solid rod as a shaft for transmitting power. Cite several reasons for the limited usage of tubes to transmit power in actual practice.

11.8 Determine the section modulus of a circular tube with an outside diameter of 120 mm and an inside diameter of 90 mm.

11.9 Determine the structural efficiency SE of the shaft described in Problem 11.8.

11.10 For the inclined surface, defined in the figure to the right, determine the stresses τ_{nt} and σ_n on the inclined plane. The shear stress $\tau_{xy} = 6.5$ ksi and the angle $\theta = 22°$.

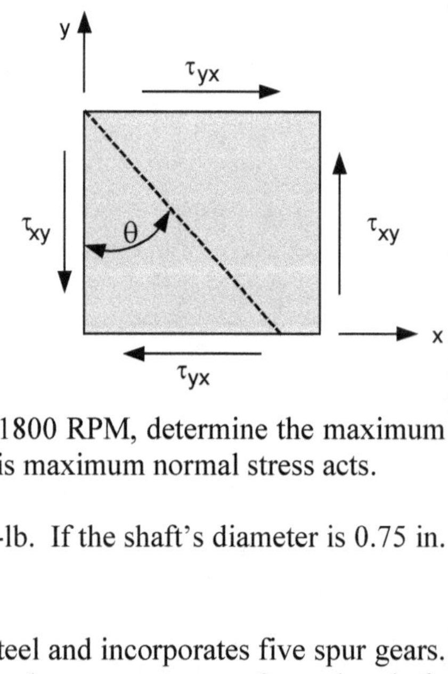

11.11 For the stress $\tau_{xy} = 75$ MPa specified in the figure to the right, prepare a graph of τ_{nt} and σ_n as θ varies from zero to 360°. Define the principal planes and the planes on which the shear stress is a maximum.

11.12 A solid circular shaft with a diameter of 1.0 in. is transmitting 7.0 HP from an electric motor to an air compressor. If the motor is operating at 1800 RPM, determine the maximum normal stress acting on the shaft. Also determine the plane upon which this maximum normal stress acts.

11.13 A steel shaft with a circular cross section is subjected to a torque of 120 ft-lb. If the shaft's diameter is 0.75 in. and its length is 15.0 in., determine its angle of twist ϕ.

11.14 A line shaft, shown in the figure below, is fabricated from high-strength steel and incorporates five spur gears. The gear A provides the input torque T_A and gears B, C, D and E each remove torque from the shaft. Determine the angle of twist for the entire shaft AE and between gears at A and C. The shaft diameters and lengths are given in the table below. The torque at each gear is also listed in this table. Neglect the effect of gear deformation on the angle of twist. Also determine the maximum shear stress in the shaft. Where does this shear stress occur?

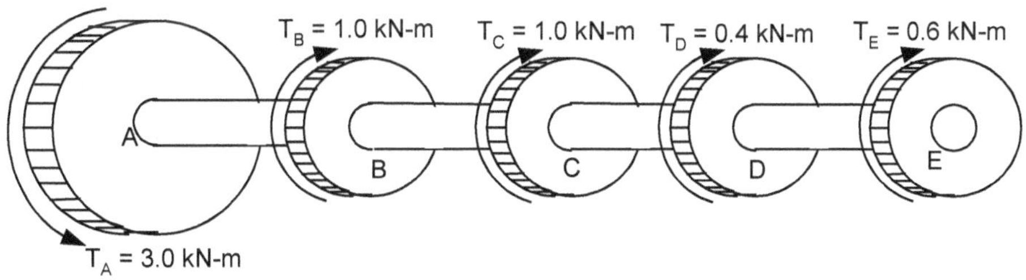

Prob. No.	d_{A-B}	d_{B-C}	d_{C-D}	d_{D-E}	L_{A-B}	L_{B-C}	L_{C-D}	L_{D-E}
11.14	40 mm	35 mm	30 mm	25 mm	0.4 m	0.4 m	0.5 m	0.7 m

11.15 A line shaft, shown in the figure below, is fabricated from high-strength steel and incorporates five spur gears. The gears A and E provide input torques T_A and T_E. Gears B, C, and D each remove torque from the shaft. Determine the angle of twist for the entire shaft AE and between gears at A and C. The shaft diameters and lengths are given in the table below. The torque at each gear is also listed in this table. Neglect the effect of gear deformation on the angle of twist. Also determine the maximum shear stress in the shaft. Where does this shear stress occur?

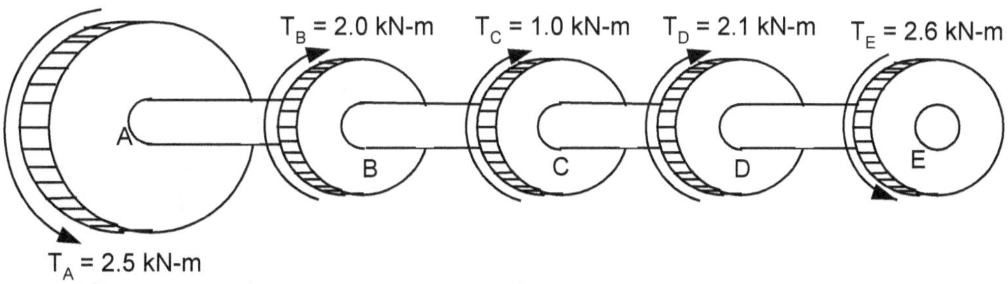

Prob. No.	d_{A-B}	d_{B-C}	d_{C-D}	d_{D-E}	L_{A-B}	L_{B-C}	L_{C-D}	L_{D-E}
11.15	40 mm	35 mm	30 mm	25 mm	0.4 m	0.4 m	0.5 m	0.7 m

11.16 A steel torsion bar is incorporated into the chassis of an automobile to serve as a suspension spring. If the bar is fabricated from steel rod with a diameter of 15 mm and a length of 1.1 m, determine its spring rate. Also determine its angle of twist if it stores a strain energy of $\mathbb{E}$ = 6.0 N-m as the automobile traverses a bump.

11.17 Determine the nominal and maximum shear stresses for a shouldered shaft shown below with a diameter D = 1.5 in. and a diameter d = 1.25 in. subjected to a torque of 500 ft-lb. The fillet radius at the shoulder of the shaft is 0.25 in.

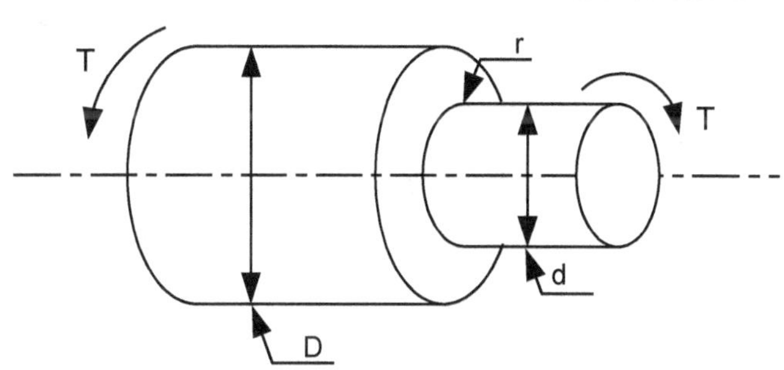

11.18 A shouldered shaft is designed such that the maximum shear stress τ_{Max} cannot exceed a critical value of 50 MPa. The shouldered shaft shown to the right has a diameter D = 40 mm and a diameter d = 30 mm. The fillet radius at the shoulder of the shaft is 6 mm. Determine the largest torque that the shaft can support.

11.19 The electric motor shown in the figure to the right has a power rating of 5.0 kW and operates at 1180 RPM. It powers a gear drive with a variable load, which sometimes requires the motor to operate at its maximum rating. The shaft is supported by two bearings—one on each side of the gear drive. The shaft from the motor has a diameter d and the shaft at the gear has a diameter D. The radius at the transition is specified as r/d = 0.15 and the diameter mismatch ratio is specified as D/d = 1.33. Select a material for the shaft and determine the diameters d and D if the maximum shear stress in the material is not to exceed 1/4 of the yield strength of the material in shear. Another constraint on the design is the angle of twist ϕ, which is not to exceed 2.0° measured from the motor to the gear drive.

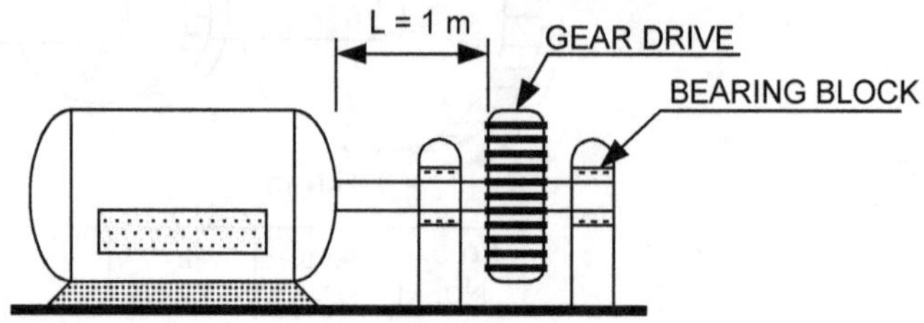

11.20 The electric motor, shown in the figure below, has a power rating of 20.0 HP and operates at 1780 RPM. It powers a drive shaft with three gears. When the motor is operating at rated capacity, gear A removes 1/2 of the power developed by the motor, while gear B and gear C remove 1/3 and 1/6 of the power, respectively. The shaft is supported by four bearings as shown in the figure. The shaft from the motor has a diameter d but it is shouldered with a diameter D at each of the three gear locations. The radius at the transition is specified as r/d = 0.20 and the diameter mismatch ratio is specified as D/d = 2.00. If an alloy steel with a yield strength in shear of 150 ksi is used to fabricate the shaft, determine the diameters d and D if the maximum shear stress in the material is not to exceed 1/3 of the yield strength of the material in shear. Another constraint on the design is the angle of twist ϕ, which is not to exceed 4.4° measured from the motor to the gear C.

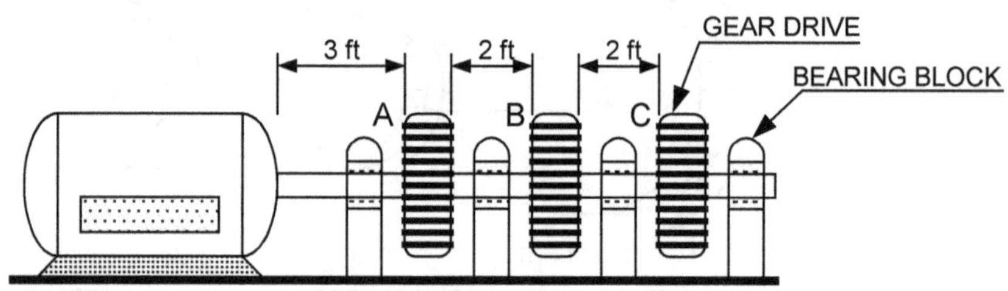

11.21 A torque of 20 ft-lb is applied to a structural member with an elliptical cross section that is fabricated from steel. The length of the structural member is 4.2 ft. The height h of the ellipse is 1.0 in. and its depth b is 0.35 in. Determine the maximum shear stress and the angle of twist for this structural member.

11.22 A torque of 25 N-m is applied to a steel structural member with an equilateral triangle for its cross section. The length of the structural member is 1.0 m. The equilateral triangle has a base b = 30 mm. Determine the maximum shear stress and the angle of twist for this structural member.

11.23 A torque of 30 ft-lb is applied to a steel structural member with a hexagon for its cross section. The length of the structural member is 6 ft. The inscribed diameter within the hexagon has a diameter of 0.75 in. Determine the maximum shear stress and the angle of twist for this structural member.

11.24 A torque of 800 N-m is applied to a steel structural member with a square cross section. The length of the structural member is 2.0 m. The side of the square section measures 40 mm. Determine the maximum shear stress and the angle of twist for this structural member.

11.25 Prove that the shear stress τ is zero at all four corners of a shaft with a square cross section that is subjected to a torque T.

11.26 Determine the eccentricity e locating the shear center for a beam fabricated from the channel section as shown in the figure to the right.

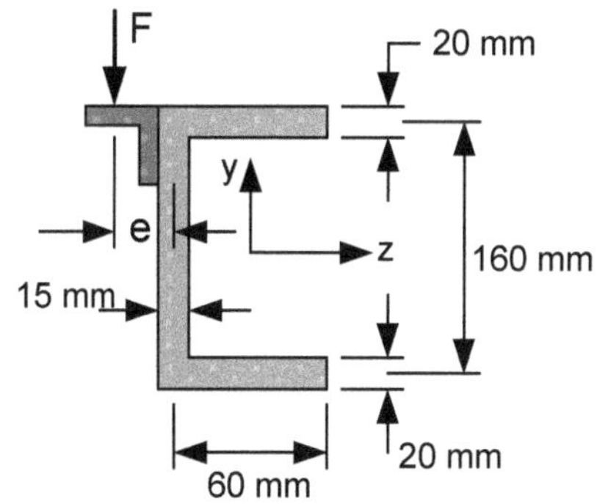

11.27 For the channel section, illustrated in the figure to the right, show that integrating the parabolic distribution of shear stresses over the area of the web yields a shear force V equal to:

$$V = F\left[\frac{2ct_1 + (3/5)ht_2}{2ct_1 + (2/3)ht_2}\right]$$

Explain why V is slightly less than F. Hint: the shear stress at the top of the web is given by:

$$\tau = F\left[\frac{ct_1 h}{t_2 I_z}\right]$$

Notes

CHAPTER 12

STRESS STATES AND TRANSFORMATIONS

12.1 INTRODUCTION

In our discussions we have with few exceptions, concentrated on two-dimensional force systems and uniaxial stress systems. Long thin rods used for tension members, beams and torsion bars lend themselves to a simple analysis, which generate uniaxial stresses. In a few instances, we have examined oblique planes and discovered that shear and normal stresses can occur together. However, we have not explored the implication of shear and normal stresses occurring on oblique planes for multi-dimensional states of stress.

If we consider a general three-dimensional body subjected to forces acting in all three dimensions, the stress system developed contains nine Cartesian components, which include the normal stresses σ_{xx}, σ_{yy} and σ_{zz} and the shear stresses τ_{xy}, τ_{xz}, τ_{yz}, τ_{yx}, τ_{zx}, andτ_{zy} as illustrated in Fig. 12.1.

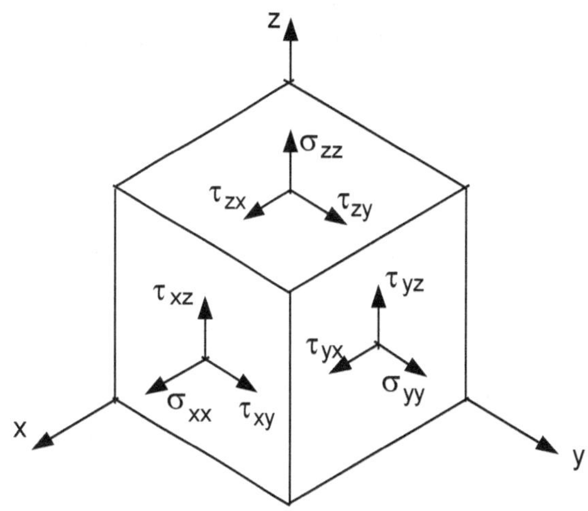

Fig. 12.1 Cartesian components of stress representing a three-dimensional state of stress.

The six Cartesian components of shearing stress may be reduced to three by recognizing that:

$$\tau_{xy} = \tau_{yx} \qquad \tau_{yz} = \tau_{zy} \qquad \tau_{xz} = \tau_{zx} \tag{12.1}$$

We have used a double subscript in Fig. 12.1 in designating the normal stresses. The first subscript is to denote the direction of the normal vector of the plane on which the stress acts and the second is to give the direction of the stress. The subscript convention also applies for the shear stress components.

In a general three-dimensional state of stress, it is necessary to specify six Cartesian components of stress (σ_{xx}, σ_{yy}, σ_{zz}, τ_{xy}, τ_{yz}, and τ_{zx}) to completely define the stress state. These general three-dimensional solutions are beyond the scope of this book. However, we will develop a special case of plane stress, which is an important state of stress with many applications.

For the general three-dimensional state, the strains are related to the stress by:

$$\varepsilon_{xx} = (1/E)[\sigma_{xx} - \nu(\sigma_{yy} + \sigma_{zz})]$$

$$\varepsilon_{yy} = (1/E)[\sigma_{yy} - \nu(\sigma_{xx} + \sigma_{zz})]$$

$$\varepsilon_{zz} = (1/E)[\sigma_{zz} - \nu(\sigma_{xx} + \sigma_{yy})] \tag{12.2}$$

$$\gamma_{xy} = \tau_{xy}/G$$

$$\gamma_{yz} = \tau_{yz}/G$$

$$\gamma_{xz} = \tau_{xz}/G$$

Solving Eq. (12.2) for the stresses in terms of the strains gives:

$$\sigma_{xx} = \frac{E}{(1+\nu)(1-2\nu)}\left[(1-\nu)\varepsilon_{xx} + \nu(\varepsilon_{yy} + \varepsilon_{zz})\right]$$

$$\sigma_{yy} = \frac{E}{(1+\nu)(1-2\nu)}\left[(1-\nu)\varepsilon_{yy} + \nu(\varepsilon_{zz} + \varepsilon_{xx})\right] \qquad (12.3)$$

$$\sigma_{zz} = \frac{E}{(1+\nu)(1-2\nu)}\left[(1-\nu)\varepsilon_{zz} + \nu(\varepsilon_{xx} + \varepsilon_{yy})\right]$$

$$\tau_{xy} = G\gamma_{xy}; \qquad \tau_{yz} = G\gamma_{yz}; \qquad \tau_{zx} = G\gamma_{zx}$$

An inspection of Eq. (12.3) shows that Hooke's law ($\sigma_{xx} = E\varepsilon_{xx}$) is not generally valid for three-dimensional states of stress. Only for the case of uniaxial tension (i.e. $\sigma_{yy} = \sigma_{zz} = 0$; $\varepsilon_{yy} = \varepsilon_{zz} = -\nu\varepsilon_{xx}$) does Eq. (12.3) take the simplified form of Hooke's law.

EXAMPLE 12.1

A three-dimensional stress state is defined by six Cartesian components of stress listed below.

$\sigma_{xx} = 10$ ksi, $\sigma_{yy} = 14$ ksi, $\sigma_{zz} = -9.5$ ksi, $\tau_{xy} = 12$ ksi, $\tau_{yz} = 8.4$ ksi, and $\tau_{zx} = -7.5$ ksi

Determine the state of strain, if the body is made from an aluminum alloy.

Solution:

From Appendix B-1, we obtain $E = 10.4 \times 10^6$ psi, $G = 3.9 \times 10^6$ psi and $\nu = 0.32$. Substituting the stress components and these values into Eq. (12.2) yields:

$$\varepsilon_{xx} = (1/E)[\sigma_{xx} - \nu(\sigma_{yy} + \sigma_{zz})] = [1/(10.4 \times 10^6)][10 - (0.32)(14 - 9.5)] \times 10^3 = 0.8231 \times 10^{-3}$$

$$\varepsilon_{yy} = (1/E)[\sigma_{yy} - \nu(\sigma_{xx} + \sigma_{zz})] = [1/(10.4\times10^6)][14 - (0.32)(10 - 9.5)]\times10^3 = 1.331 \times 10^{-3}$$

$$\varepsilon_{zz} = (1/E)[\sigma_{zz} - \nu(\sigma_{xx} + \sigma_{yy})] = [1/(10.4\times10^6)][-9.5 - (0.32)(10 + 14)]\times10^3 = -1.652 \times 10^{-3}$$

$$\gamma_{xy} = \tau_{xy}/G = (12 \times 10^3)/(3.9 \times 10^6) = 3.077 \times 10^{-3}$$

$$\gamma_{yz} = \tau_{yz}/G = (8.4 \times 10^3)/(3.9 \times 10^6) = 2.154 \times 10^{-3}$$

$$\gamma_{xz} = \tau_{xz}/G = (-7.5 \times 10^3)/(3.9 \times 10^6) = -1.923 \times 10^{-3}$$

Examination of the results indicates that the strains are small quantities (of the order of 10^{-3}) even when the stress state is of the order of 10^3.

EXAMPLE 12.2

A three-dimensional strain state is defined by six Cartesian components of strain listed below.

$\varepsilon_{xx} = 1.0 \times 10^{-3}$, $\varepsilon_{yy} = 1.4 \times 10^{-3}$, $\varepsilon_{zz} = -0.95 \times 10^{-3}$, $\gamma_{xy} = 1.2 \times 10^{-3}$, $\gamma_{yz} = 0.84 \times 10^{-3}$, and $\gamma_{zx} = -0.75 \times 10^{-3}$

Determine the state of stress, if the body is made from steel.

Solution:

From Appendix B-1, we obtain E = 207 GPa, G = 79 GPa and ν = 0.30. Substituting the strain components and these values into Eq. (12.3) yields:

$$\sigma_{xx} = \frac{E}{(1+\nu)(1-2\nu)}\left[(1-\nu)\varepsilon_{xx} + \nu(\varepsilon_{yy} + \varepsilon_{zz})\right]$$

$$\sigma_{xx} = \frac{207 \times 10^9}{(1.3)(0.4)}[(0.7)(1.0) + (0.3)(1.4 - 0.95)] \times 10^{-3} = 332.4 \text{ MPa}$$

$$\sigma_{yy} = \frac{E}{(1+\nu)(1-2\nu)}\left[(1-\nu)\varepsilon_{yy} + \nu(\varepsilon_{zz} + \varepsilon_{xx})\right]$$

$$\sigma_{yy} = \frac{207 \times 10^9}{(1.3)(0.4)}[(0.7)(1.4) + (0.3)(1.0 - 0.95)] \times 10^{-3} = 396.1 \text{ MPa}$$

$$\sigma_{zz} = \frac{E}{(1+\nu)(1-2\nu)}\left[(1-\nu)\varepsilon_{zz} + \nu(\varepsilon_{xx} + \varepsilon_{yy})\right]$$

$$\sigma_{zz} = \frac{207 \times 10^9}{(1.3)(0.4)}[(0.7)(-0.95) + (0.3)(1.0 + 1.4)] \times 10^{-3} = 21.90 \text{ MPa}$$

$$\tau_{xy} = G\gamma_{xy} = (79 \times 10^9)(1.2 \times 10^{-3}) = 94.80 \text{ MPa}$$

$$\tau_{yz} = G\gamma_{yz} = (79 \times 10^9)(0.84 \times 10^{-3}) = 66.36 \text{ MPa}$$

$$\tau_{zx} = G\gamma_{zx} = (79 \times 10^9)(-0.75 \times 10^{-3}) = -59.25 \text{ MPa}$$

12.1.1 Plane Stress

When a body is thin relative to its lateral dimensions such as that shown in Fig. 12.2, the stresses acting on all of the planes perpendicular to the normal vector z of the thin body are sufficiently small to be disregarded. Also on a free surface of any body, the normal and shear stresses acting on that surface vanish. In these situations, the stress state is called plane stress. For a plane state of stress, we set the stresses acting in the z direction equal to zero to give.

$$\sigma_{zz} = \tau_{xz} = \tau_{yz} = 0 \qquad (12.4)$$

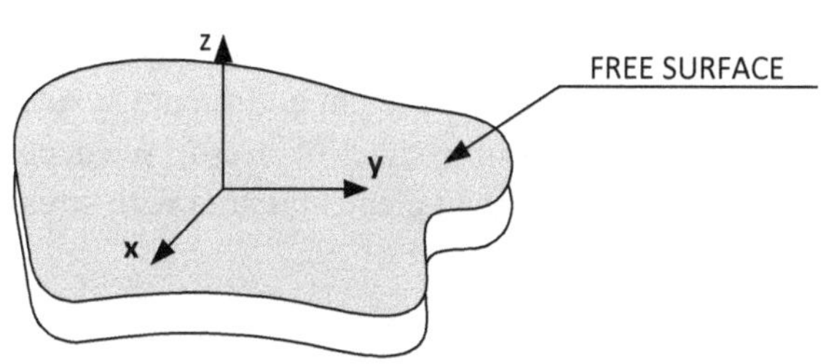

Fig. 12.2 Thin bodies exhibit plane stress conditions.

The remaining stresses σ_{xx}, σ_{yy} and τ_{xy} are specified when defining the stresses associated with a state of plane stress. For a state of plane stress, with $\sigma_{zz} = \tau_{xz} = \tau_{yz} = 0$, Eq. (12.2) reduces to:

$$\varepsilon_{xx} = (1/E)[\sigma_{xx} - \nu(\sigma_{yy})]$$

$$\varepsilon_{yy} = (1/E)[\sigma_{yy} - \nu(\sigma_{xx})] \qquad (12.5)$$

$$\varepsilon_{zz} = (1/E)[- \nu(\sigma_{xx} + \sigma_{yy})]$$

$$\gamma_{xy} = \tau_{xy}/G$$

Similarly, the stress-strain relations given in Eq. (12.3) reduce to:

$$\sigma_{xx} = \frac{E}{(1-\nu^2)}\left(\varepsilon_{xx} + \nu\varepsilon_{yy}\right)$$

$$\sigma_{yy} = \frac{E}{(1-\nu^2)}\left(\varepsilon_{yy} + \nu\varepsilon_{xx}\right) \qquad (12.6)$$

$$\tau_{xy} = G\gamma_{xy}$$

While the equations for the shear stresses do not change with the dimension of the state of stress, the expressions for the normal stresses differ markedly. Hooke's law ($\sigma_{xx} = E\varepsilon_{xx}$) is valid only for a uniaxial state of stress where $\sigma_{yy} = \sigma_{zz} = 0$.

EXAMPLE 12.3

The stresses on the external surface of a cylindrical pressure vessel in the hoop and axial direction are predicted from a simple analysis to be $\sigma_h = 28$ ksi and $\sigma_a = 14$ ksi when the vessel is pressurized to 125 psi. If a pair of strain gages is placed on the pressure vessel to measure the strains in the hoop and axial directions, predict the strains that will occur when the pressure of 125 psi is applied. The pressure vessel is fabricated from steel.

Solution:

From Appendix B-1, we note $E = 30 \times 10^6$ psi, $G = 11.5 \times 10^6$ psi and Poisson's ratio $\nu = 0.30$. Substituting the hoop and axial stresses in Eq. (12.5) gives:

$$\varepsilon_{xx} = (1/E)[\sigma_{xx} - \nu(\sigma_{yy})] = [1/(30 \times 10^6)][28 - (0.3)(14)] \times 10^{+3} = 0.7933 \times 10^{-3}$$

$$\varepsilon_{yy} = (1/E)[\sigma_{yy} - \nu(\sigma_{xx})] = [1/(30 \times 10^6)][14 - (0.3)(28)] \times 10^{+3} = 0.1867 \times 10^{-3}$$

12.2 STRESS EQUATIONS OF TRANSFORMATION

In developing the stress equations of transformation, we will treat the case of plane stress where $\sigma_{zz} = \tau_{xz} = \tau_{yz} = 0$. Suppose we know the state of stress (σ_{xx}, σ_{yy} and τ_{xy}) in a body relative to a set of coordinates say Oxy. Is this state of stress the most important combination of stresses that will determine the failure loads? Or is some other state of stress relative to a different coordinate system the critical state? To answer these questions it is necessary to explore all of the admissible states of stress. We accomplish this task by using the stress equations of transformation.

Consider an elemental area with stresses σ_{xx}, σ_{yy} and τ_{xy} as shown in Fig. 12.3a.

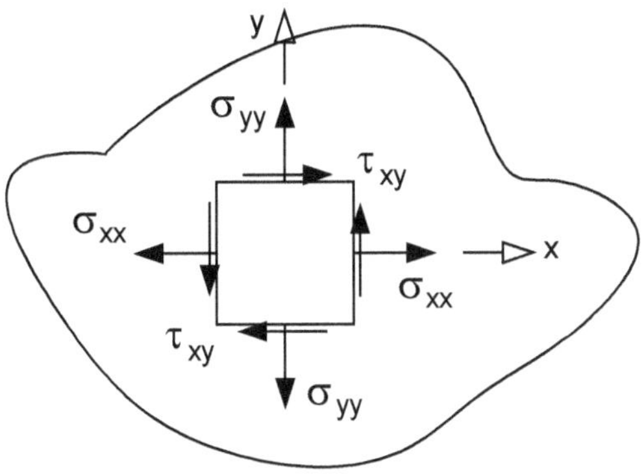

Fig. 12.3a An element with x-y orientation.

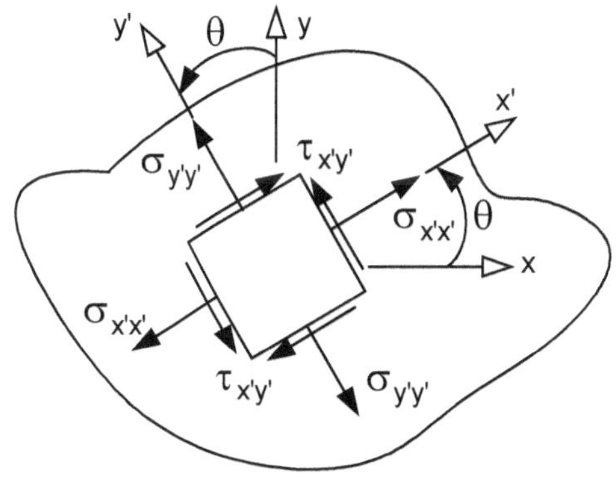

Fig. 12.3b An element with x'-y' orientation.

As we rotate the element about its center through some angle θ, the stresses on the element faces change from σ_{xx}, σ_{yy} and τ_{xy} to $\sigma_{x'x'}$, $\sigma_{y'y'}$ and $\tau_{x'y'}$. To explore stress space, it is necessary to establish relations for $\sigma_{x'x'}$, $\sigma_{y'y'}$ and $\tau_{x'y'}$ in terms of σ_{xx}, σ_{yy} and τ_{xy} and the angle θ.

Let's follow the same procedure as we used in Chapter 11 when considering stresses on oblique planes. Cut the left corner from the element shown in Fig. 12.3a at an angle θ so as to produce a wedge like element with an inclined surface that is parallel to the y' axis and perpendicular to the x' axis. Consider the element thickness in the z direction to be unity, and define the area of the inclined surface as ΔA as indicated in Fig. 12.4. The area of the short side of the wedge is given by ΔA sin θ and the area of the longer side of the wedge is ΔA cos θ. These areas are important because they factor into the equilibrium equations that we will employ later in this derivation.

Stresses $\tau_{x'y'}$ and $\sigma_{x'x'}$ act on the inclined surface of the wedge and τ_{xy}, σ_{xx} and σ_{yy} act on the other sides of the wedge. We multiply these stresses with the areas of the three sides of the wedge to obtain the forces shown in Fig. 12.4b. No forces are shown on the face of the wedge because the element was removed from a body in a plane state of stress where $\sigma_{zz} = \tau_{xz} = \tau_{yz} = 0$.

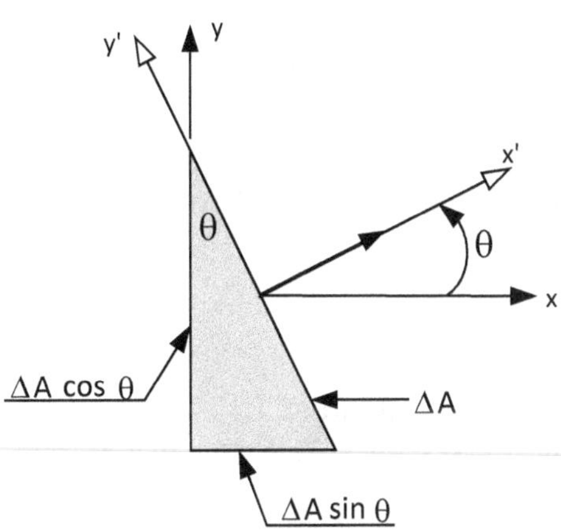

Fig. 12.4a A wedge like element cut from the x-y element at an angle θ.

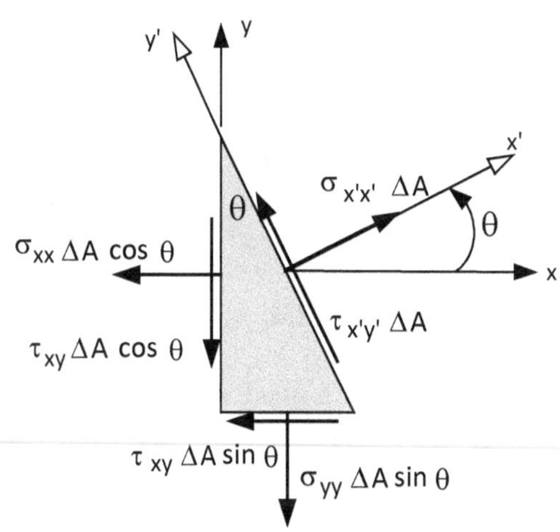

Fig. 12.4b Forces acting on the three sides of the wedge like element.

Next we write two equilibrium relations summing forces in the x' and y' directions:

$$\Sigma F_{x'} = 0 \qquad \begin{aligned} &\sigma_{x'x'}\Delta A - \sigma_{xx}(\Delta A \cos\theta)\cos\theta - \tau_{xy}(\Delta A \cos\theta)\sin\theta \\ &- \sigma_{yy}(\Delta A \sin\theta)\sin\theta - \tau_{xy}(\Delta A \sin\theta)\cos\theta = 0 \end{aligned}$$

$$\Sigma F_{y'} = 0 \qquad \begin{aligned} &\tau_{x'y'}\Delta A + \sigma_{xx}(\Delta A \cos\theta)\sin\theta - \tau_{xy}(\Delta A \cos\theta)\cos\theta \\ &- \sigma_{yy}(\Delta A \sin\theta)\cos\theta + \tau_{xy}(\Delta A \sin\theta)\sin\theta = 0 \end{aligned} \qquad (12.7)$$

Simplifying these two equations and solving for $\sigma_{x'x'}$ and $\tau_{x'y'}$ gives:

$$\sigma_{x'x'} = \sigma_{xx} \cos^2 \theta + \sigma_{yy} \sin^2 \theta + 2\tau_{xy} \sin \theta \cos\theta \qquad (12.8)$$

$$\tau_{x'y'} = -(\sigma_{xx} - \sigma_{yy}) \sin \theta \cos \theta + \tau_{xy} (\cos^2 \theta - \sin^2 \theta) \qquad (12.9)$$

These relations enable you to search solution space to determine the angle θ, which produces maximum and minimum values of both the normal and shear stresses.

By introducing four trigonometric identities, it is possible to convert Eqs. (12.8) and (12.9) into a double angle (2θ) format as shown below.

$$\sigma_{x'x'} = \frac{\sigma_{xx} + \sigma_{yy}}{2} + \frac{\sigma_{xx} - \sigma_{yy}}{2} \cos 2\theta + \tau_{xy} \sin 2\theta$$

$$\tau_{x'y'} = -\frac{\sigma_{xx} - \sigma_{yy}}{2} \sin 2\theta + \tau_{xy} \cos 2\theta \qquad (12.10)$$

A similar relation for $\sigma_{y'y'}$ may be derived by increasing the angle of the cut for the wedge from θ to 90° + θ. Alternatively, you may convert Eq. (12.8) or the first of Eq. (12.10) to an expression for $\sigma_{y'y'}$ by letting θ = θ + 90° to obtain:

$$\sigma_{y'y'} = \frac{\sigma_{xx} + \sigma_{yy}}{2} - \frac{\sigma_{xx} - \sigma_{yy}}{2} \cos 2\theta - \tau_{xy} \sin 2\theta \qquad (12.11)$$

The form of the relations for $\sigma_{x'x'}$ and $\sigma_{y'y'}$ is almost identical except for the signs of the last two terms. It is evident from inspection that we can add Eq. (12.10) and Eq. (12.11) to obtain:

$$\sigma_{x'x'} + \sigma_{y'y'} = \sigma_{xx} + \sigma_{yy} \qquad (12.12)$$

Equation (12.12) indicates that the sum of the normal stress components is independent of the angle of θ. The sum of the normal stresses is an invariant that does not depend on the orientation of the planes upon which the stresses act. This is one of three stress invariants.

EXAMPLE 12.4

We represent a plane state of stress at a point by a small elemental area ΔA shown in Fig. E12.4. If the stresses are σ_{xx} = 35 MPa, σ_{yy} = – 22 MPa and τ_{xy} = 16 MPa, determine the stresses on an element that is rotated 30° counterclockwise relative to the element shown in Fig. E12.4.

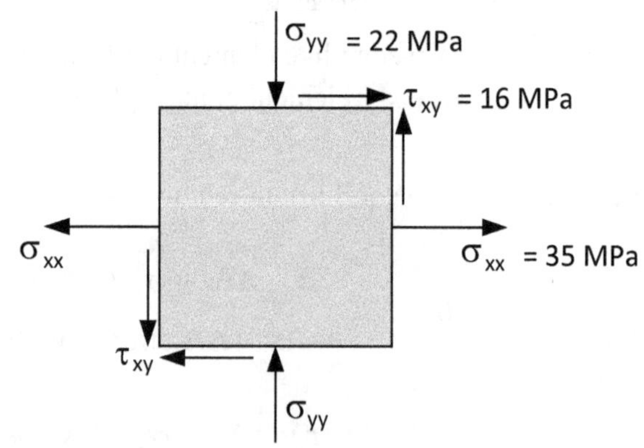

Fig. E12.4

Solution:

Using Eqs. (12.10) and (12.11) and setting θ = 30°, we write:

$$\sigma_{x'x'} = \frac{\sigma_{xx}+\sigma_{yy}}{2} + \frac{\sigma_{xx}-\sigma_{yy}}{2}\cos 2\theta + \tau_{xy}\sin 2\theta$$

$$\sigma_{x'x'} = \frac{35-22}{2} + \frac{35+22}{2}\cos 60 + 16\sin 60 = 34.61 \text{ MPa}$$

$$\tau_{x'y'} = -\frac{\sigma_{xx}-\sigma_{yy}}{2}\sin 2\theta + \tau_{xy}\cos 2\theta$$

$$\tau_{x'y'} = -\frac{35+22}{2}\sin 60 + 16\cos 60 = -16.68 \text{ MPa}$$

$$\sigma_{y'y'} = \frac{\sigma_{xx}+\sigma_{yy}}{2} - \frac{\sigma_{xx}-\sigma_{yy}}{2}\cos 2\theta - \tau_{xy}\sin 2\theta$$

$$\sigma_{y'y'} = \frac{35-22}{2} - \frac{35+22}{2}\cos 60 - 16\sin 60 = -21.61 \text{ MPa}$$

EXAMPLE 12.5

We represent a plane state of stress at a point by a small elemental area ΔA shown in Fig. E12.5. If the stresses are σ_{xx} = 15 ksi, σ_{yy} = 12 ksi and τ_{xy} = – 10 ksi, determine the stresses on an element that is rotated 40° clockwise relative to the element shown in Fig. E12.5.

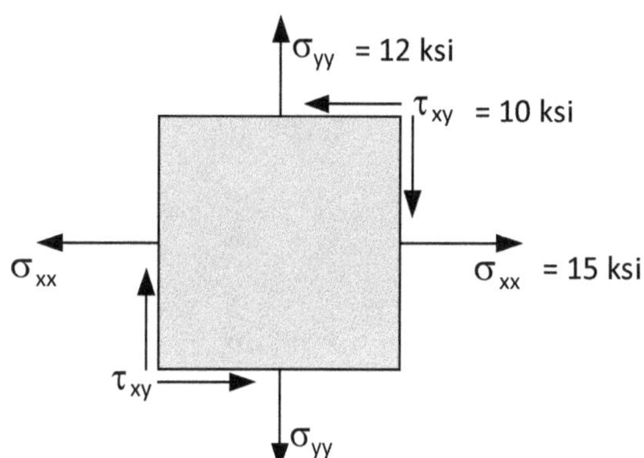

Fig. E12.5

Solution:

Using Eqs. (12.10) and (12.11) and setting θ = – 40°, we write:

$$\sigma_{x'x'} = \frac{\sigma_{xx}+\sigma_{yy}}{2} + \frac{\sigma_{xx}-\sigma_{yy}}{2}\cos 2\theta + \tau_{xy}\sin 2\theta$$

$$\sigma_{x'x'} = \frac{15+12}{2} + \frac{15-12}{2}\cos(-80) - 10\sin(-80) = 23.608 \text{ ksi}$$

$$\tau_{x'y'} = -\frac{\sigma_{xx}-\sigma_{yy}}{2}\sin 2\theta + \tau_{xy}\cos 2\theta$$

$$\tau_{x'y'} = -\frac{15-12}{2}\sin(-80) - 10\cos(-80) = -0.2593 \text{ ksi}$$

$$\sigma_{y'y'} = \frac{\sigma_{xx}+\sigma_{yy}}{2} - \frac{\sigma_{xx}-\sigma_{yy}}{2}\cos 2\theta \; - \; \tau_{xy}\sin 2\theta$$

$$\sigma_{y'y'} = \frac{15+12}{2} - \frac{15-12}{2}\cos(-80) \; + \; 10\sin(-80) \; = \; 3.391 \text{ ksi}$$

These two examples show the application of the stress equations of transformation for determining the Cartesian components of stress on an arbitrary element rotated an angle θ with respect to the x-y coordinate system. We will next use the stress equations of transformation to locate the planes where the normal and shear stresses are a maximum or minimum.

12.3 PRINCIPAL STRESSES

When considering all of the possible planes upon which normal stresses act, we seek those planes where the normal stresses are either a maximum or a minimum. We will define these planes by determining the angle θ_p, which designates the principal plane. To determine this angle, let's employ the standard calculus approach by differentiating the normal stress $\sigma_{x'x'}$ with respect to θ and set the result equal to zero.

$$\frac{d}{d\theta}\sigma_{x'x'} = -(\sigma_{xx}-\sigma_{yy})\sin 2\theta_p + 2\tau_{xy}\cos 2\theta_p = 0$$

$$\tan 2\theta_p = \frac{2\tau_{xy}}{\sigma_{xx}-\sigma_{yy}} \qquad (12.13)$$

From Eq. (12.13), it is easy to show that:

$$\cos 2\theta_p = \frac{(\sigma_{xx}-\sigma_{yy})/2}{\sqrt{\left(\left(\frac{\sigma_{xx}-\sigma_{yy}}{2}\right)^2 + \tau_{xy}^2\right)}}$$

$$\sin 2\theta_p = \frac{\tau_{xy}}{\sqrt{\left(\left(\frac{\sigma_{xx}-\sigma_{yy}}{2}\right)^2 + \tau_{xy}^2\right)}} \qquad (12.14)$$

Substitute Eq. (12.14) into Eq. (12.10) and simplifying yields:

$$\sigma_{\text{Max, Min}} = \frac{\sigma_{xx}+\sigma_{yy}}{2} + \frac{\sigma_{xx}-\sigma_{yy}}{2}\cos 2\theta_p \; + \; \tau_{xy}\sin 2\theta_p$$

$$\sigma_{\text{Max, Min}} = \frac{\sigma_{xx}+\sigma_{yy}}{2} \pm \sqrt{\left(\left(\frac{\sigma_{xx}-\sigma_{yy}}{2}\right)^2 + \tau_{xy}^2\right)} \qquad (12.15)$$

$$\tau_{pp} = -\frac{\sigma_{xx}-\sigma_{yy}}{2}\sin 2\theta_p \; + \; \tau_{xy}\cos 2\theta_p = 0$$

The maximum normal stress $\sigma_{Max} = \sigma_1$ is the largest algebraic value of the two numbers obtained in the solution of Eq. (12.15). The minimum normal stress $\sigma_{Min} = \sigma_2$ is the smaller of the two values. Both σ_1 and σ_2 are principal stresses, which act on orthogonal planes defined by Eq. (12.13). The shear stress on the principal planes $\tau_{pp} = 0$. We illustrate a typical principal state of stress in Fig. 12.5. In this illustration, the principal element is inscribed in a traditional Cartesian element although the Cartesian components of stress are not shown to avoid clutter. On the principal element, we have shown σ_1 and σ_2 as well as the shear stress τ_{pp} vanishing on its edges.

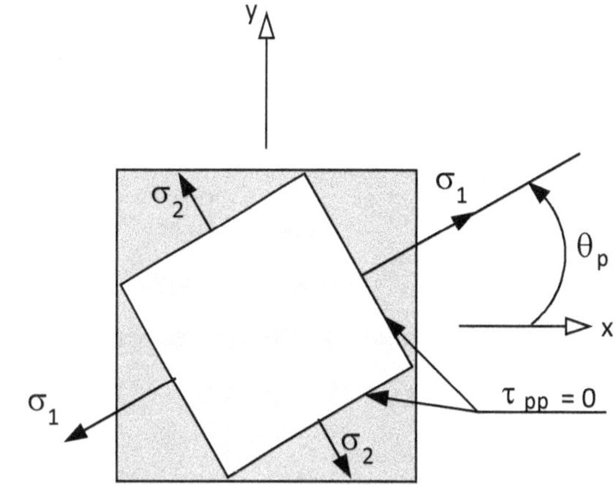

Fig. 12.5 A principal element inscribed in a traditional Cartesian element.

EXAMPLE 12.6

Consider a point, shown here as an elemental area, with Cartesian components of stress defined in Fig. E12.6. Determine the orientation of the principal planes and the stresses that act on the principal planes.

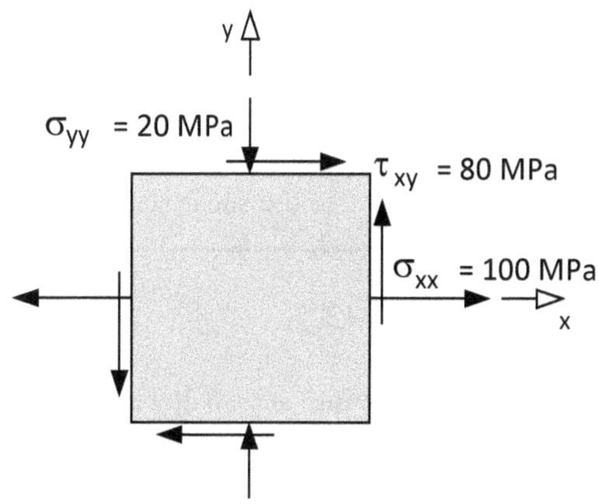

Fig. E12.6

Solution:

The principal angle θ_p is determined from Eq. (12.13) as:

$$\tan 2\theta_p = \frac{2\tau_{xy}}{\sigma_{xx} - \sigma_{yy}} = \frac{2(80)}{100+20} = 1.333$$

$$2\theta_p = \tan^{-1}(1.333) = 53.13^\circ \qquad \text{(a)}$$

$$2\theta_p = 180^\circ + 53.13^\circ = 233.13^\circ$$

$$\theta_p = 26.57^\circ \text{ and } 116.57^\circ \qquad \text{(b)}$$

The first principal angle (26.57°) defines the plane upon which the maximum principal stress σ_1 acts and the second principal angle (116.57°) defines the plane upon which the minimum principal stress σ_2 acts.

Let's determine the magnitude of the principal stresses by using Eq. (12.15) as:

$$\sigma_{\text{Max, Min}} = \frac{\sigma_{xx} + \sigma_{yy}}{2} \pm \sqrt{\left(\left(\frac{\sigma_{xx} - \sigma_{yy}}{2}\right)^2 + \tau_{xy}^2\right)}$$

$$\sigma_{\text{Max, Min}} = \frac{100 - 20}{2} \pm \sqrt{\left(\left(\frac{100 + 20}{2}\right)^2 + (80)^2\right)} \quad \text{(c)}$$

$$\sigma_{\text{Max, Min}} = 40 \pm 100 \text{ MPa}$$

$$\sigma_1 = 140 \text{ MPa and } \sigma_2 = -60 \text{ MPa} \quad \text{(d)}$$

We show these stresses acting on the principal element in Fig. E12.6a.

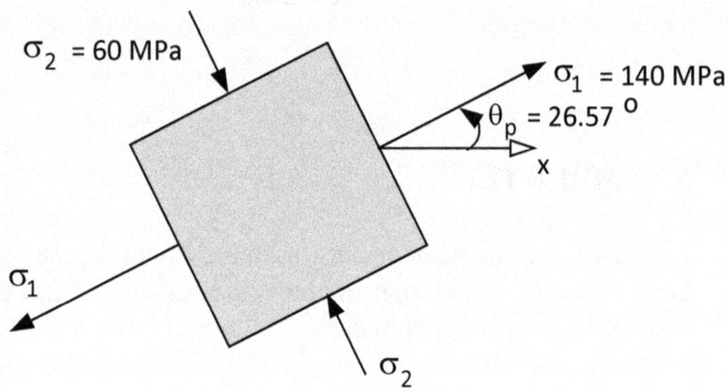

Fig. E12.6a

Of course the shear stresses on the principal element vanish in accordance with Eq. (12.15).

EXAMPLE 12.7

Consider a point, shown here as an elemental area, with Cartesian components of stress defined in Fig. E12.7. Determine the orientation of the principal planes and the stresses that act on the principal planes.

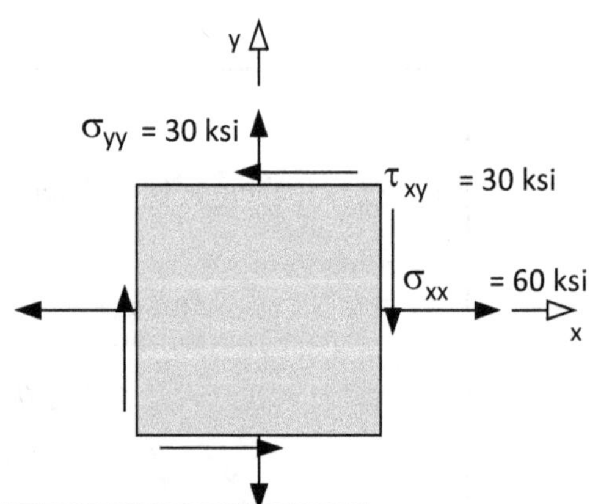

Fig. E12.7

Solution:

The principal angles θ_p are determined from Eq. (12.13) as:

$$\tan 2\theta_p = \frac{2\tau_{xy}}{\sigma_{xx} - \sigma_{yy}} = \frac{2(-30)}{60 - 30} = -2.00$$

$$2\theta_p = \tan^{-1}(-2.00) = -63.43^\circ \quad \text{(a)}$$

$$2\theta_p = 180^\circ - 63.43^\circ = 116.57^\circ$$

$$\theta_p = -31.72° \text{ and } 58.29° \qquad \text{(b)}$$

The first principal angle (31.72°) defines the plane upon which the maximum principal stress σ_1 acts and the second principal angle (121.72°) defines the plane upon which the minimum principal stress σ_2 acts.

Let's determine the magnitude of the principal stresses by using Eq. (12.15) as:

$$\sigma_{\text{Max, Min}} = \frac{\sigma_{xx} + \sigma_{yy}}{2} \pm \sqrt{\left(\left(\frac{\sigma_{xx} - \sigma_{yy}}{2}\right)^2 + \tau_{xy}^2\right)}$$

$$\sigma_{\text{Max, Min}} = \frac{60+30}{2} \pm \sqrt{\left(\left(\frac{60-30}{2}\right)^2 + (-30)^2\right)} \qquad \text{(c)}$$

$$\sigma_{\text{Max, Min}} = 45 \pm 33.54 \text{ ksi}$$

$$\sigma_1 = 78.54 \text{ ksi and } \sigma_2 = 11.46 \text{ ksi} \qquad \text{(d)}$$

We show these stresses acting on the principal element in Fig. E12.7a.

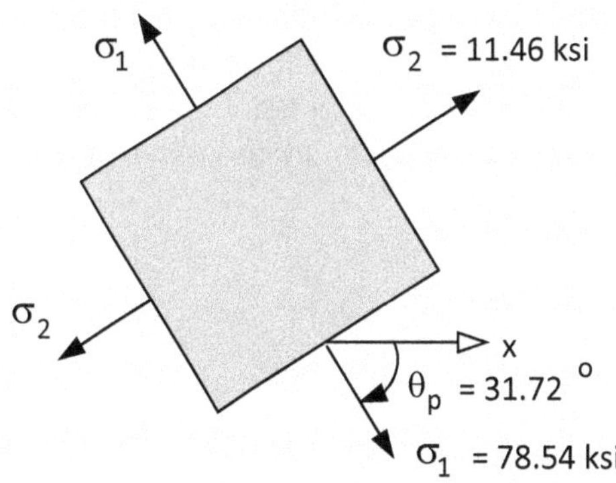

Fig. E12.7a

12.3.1 Maximum Shear Stresses

We used differential calculus to ascertain the principal planes defined by θ_p, which enabled us to determine the maximum and minimum normal stresses. Let's follow the same approach to determine the planes upon which the maximum shear stress occurs. Differentiate the second of Eq. (12.10) with respect to θ, to obtain:

$$\frac{d}{d\theta}\tau_{x'y'} = -\frac{\sigma_{xx} - \sigma_{yy}}{2}(2\cos 2\theta) - 2\,\tau_{xy}\sin 2\theta = 0 \qquad \text{(a)}$$

Solving Eq. (a) for θ_s gives:

$$\tan 2\theta_s = -\frac{\sigma_{xx} - \sigma_{yy}}{2\tau_{xy}} \qquad (12.16)$$

where the angle θ_s defines the planes upon which the maximum shear stresses occur.

Comparing Eq. (12.16) for $2\theta_s$ with Eq. (12.13) for $2\theta_p$, shows that they are negative reciprocals. This fact implies that the angles $2\theta_s$ and $2\theta_p$ are orthogonal. Hence, the angles θ_s and θ_p are separated by 45°. Clearly, the planes of maximum shear and the principal planes are oriented at 45° relative to one another.

Note from Eq. (12.16) that $\cos 2\theta_s$ and $\sin 2\theta_s$ are given by:

$$\sin 2\theta_s = \frac{-(\sigma_{xx} - \sigma_{yy})/2}{\sqrt{\left(\left(\frac{\sigma_{xx} - \sigma_{yy}}{2}\right)^2 + \tau_{xy}^2\right)}}$$

$$\cos 2\theta_s = \frac{\tau_{xy}}{\sqrt{\left(\left(\frac{\sigma_{xx} - \sigma_{yy}}{2}\right)^2 + \tau_{xy}^2\right)}} \tag{12.17}$$

Substituting Eq. (12.17) into Eq. (12.10) and simplifying yields:

$$\tau_{Max} = \sqrt{\left(\left(\frac{\sigma_{xx} - \sigma_{yy}}{2}\right)^2 + \tau_{xy}^2\right)} \tag{12.18}$$

EXAMPLE 12.8

Consider a point, shown here as an elemental area, with Cartesian components of stress defined in Fig. E12.8. Determine the orientation of the planes upon which the maximum shear stresses act and also determine the normal stresses that act on these planes.

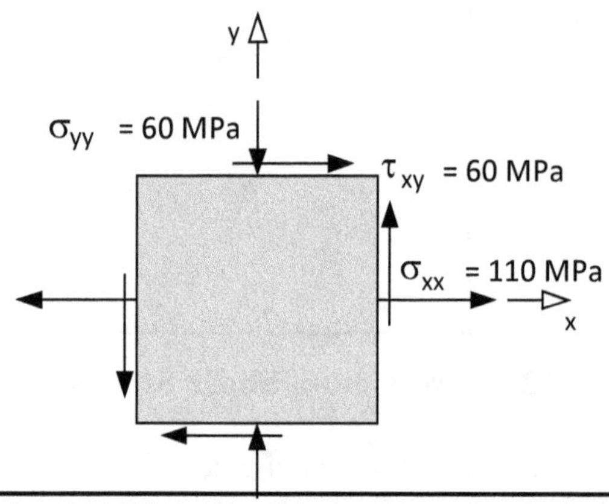

Solution:

The plane upon which the maximum shear stress acts is given by Eq. (12.16) as:

$$\tan 2\theta_s = -\frac{\sigma_{xx} - \sigma_{yy}}{2\tau_{xy}} = -\frac{(110+60)}{2(60)} = -1.417 \tag{a}$$

$$2\theta_s = \tan^{-1}(-1.417) = -54.79°$$

$$\theta_s = -27.39° \tag{b}$$

The maximum shear stress is calculated from Eq. (12.18) as:

$$\tau_{Max} = \sqrt{\left(\left(\frac{\sigma_{xx} - \sigma_{yy}}{2}\right)^2 + \tau_{xy}^2\right)} = \sqrt{\left(\left(\frac{110+60}{2}\right)^2 + (60)^2\right)} = 104.0 \text{ MPa} \tag{c}$$

To determine the normal stress acting on this plane, substitute the value of $2\theta_s = -54.79°$ into Eq. (12.10) to obtain:

$$\sigma_{x'x'} = \frac{\sigma_{xx}+\sigma_{yy}}{2} + \frac{\sigma_{xx}-\sigma_{yy}}{2}\cos 2\theta + \tau_{xy}\sin 2\theta$$

$$\sigma_n = \frac{110-60}{2} + \frac{110+60}{2}\cos(-54.79) + (60)\sin(-54.79) \qquad \text{(d)}$$

$$\sigma_n = 25 + 85(0.5766) + 60(-0.8170) = 25 \text{ MPa}$$

Using Eq. (12.10) and the angle θ_s is a cumbersome method to determine σ_n. We will show in the next section that on planes of maximum shear stress, σ_n is given by:

$$\sigma_n = (\sigma_{xx} + \sigma_{yy})/2 \qquad (12.19)$$

12.4 MOHR'S CIRCLE

A graphical technique, known as Mohr's circle, was employed for solving for $\sigma_{x'x'}$, $\sigma_{y'y'}$ and $\tau_{x'y'}$ prior to the development of efficient electronic computers. Today the equations of stress transformation are easy to solve; however, Mohr's circle remains useful because it enables one to visualize how the Cartesian components of stress vary with the angle θ that defines the plane upon which the stresses act. Also the magnitude and orientation of σ_1, σ_2 and τ_{Max} are apparent from a cursory inspection of Mohr's circle.

To establish the relation between a circle and the stress state, consider the equation of a circle with a radius R centered at x = C as shown in Fig. 12.6.

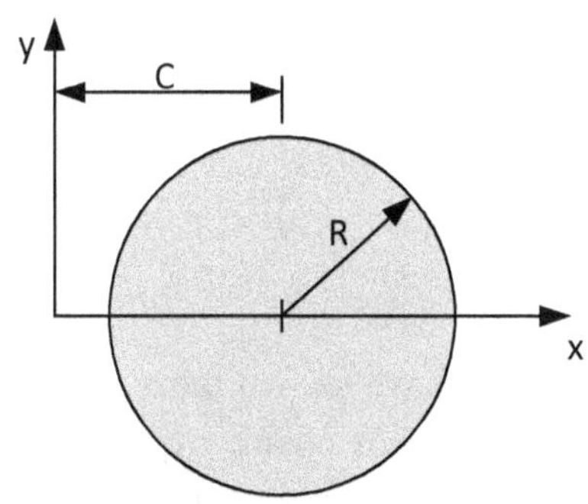

Fig. 12.6 Circle with radius R centered at x = C.

The equation for the circle is:

$$(x - C)^2 + y^2 = R^2 \qquad (12.20)$$

To show that the stress equations of transformation map into a similar circle, rewrite Eq. (12.10) to obtain:

$$\sigma_{x'x'} - \frac{\sigma_{xx}+\sigma_{yy}}{2} = \frac{\sigma_{xx}-\sigma_{yy}}{2}\cos 2\theta + \tau_{xy}\sin 2\theta$$

$$\tau_{x'y'} = -\frac{\sigma_{xx}-\sigma_{yy}}{2}\sin 2\theta + \tau_{xy}\cos 2\theta \qquad (12.21)$$

Square both sides of these two equations and add the two to obtain:

$$\left(\sigma_{x'x'} - \frac{\sigma_{xx}+\sigma_{yy}}{2}\right)^2 + \tau_{x'y'}^{\,2} = \left(\frac{\sigma_{xx}-\sigma_{yy}}{2}\right)^2 + \tau_{xy}^{\,2} \qquad (12.22)$$

Comparing the form of Eq. (12.20) with that of Eq. (12.22), permits us to establish the several facts about Mohr's circle.

1. The correspondence between x and $\sigma_{x'x'}$ indicates σ will be found along the abscissa.
2. Mohr's circle will be centered at C where:

$$C = (\sigma_{xx} + \sigma_{yy})/2 \qquad (12.23)$$

3. The correspondence between y and $\tau_{x'y'}$ indicates τ will be found along the ordinate.
4. The radius of Mohr's circle R is given by:

$$R = \sqrt{\left(\frac{\sigma_{xx} - \sigma_{yy}}{2}\right)^2 + \tau_{xy}^{\;2}} \qquad (12.24)$$

We present an example of Mohr's circle in Fig 12.7, showing the σ and τ axes, the center point located by C, the radius R, and the principal stresses σ_1 and σ_2 and the maximum shear stress τ_{Max}.

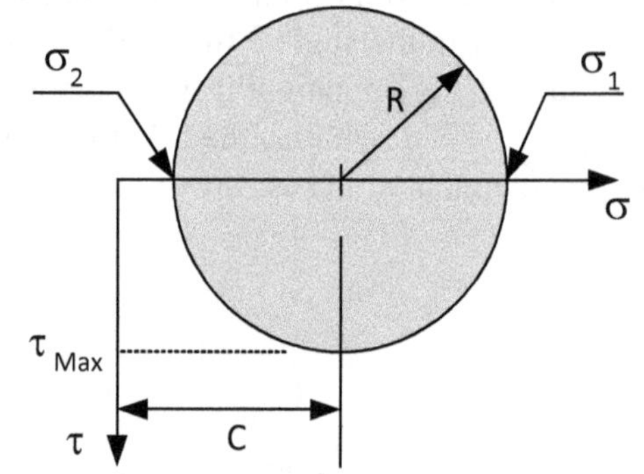

Fig. 12.7 Mohr's circle showing the principal stresses, and maximum shear stress.

We have shown the τ axis as positive downward in this representation. This choice is consistent with the definition of θ as a positive quantity for counterclockwise rotation and for the sign convention for τ_{xy}. Let's consider two examples to demonstrate the usefulness of Mohr's circle in visualizing a compete state of stress at a point.

EXAMPLE 12.9

Consider a point, shown here as an elemental area, with Cartesian components of stress defined in Fig. E12.9. Determine the principal stresses and the orientation of the principal planes. Also determine the maximum shear stresses and their orientation.

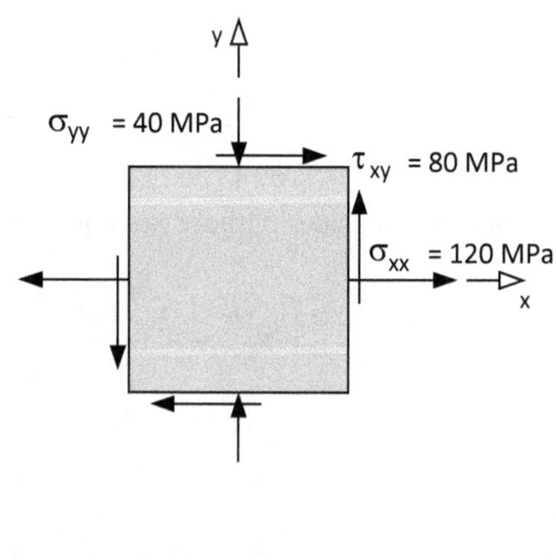

Fig. E12.9

Solution

Step 1: Use Eq. (12.23) and Eq. (12.24) to determine the position of the center of the circle and its radius R as:

$$C = (\sigma_{xx} + \sigma_{yy})/2 \qquad = (120 - 40)/2 = 40 \text{ MPa} \qquad \text{(a)}$$

$$R = \sqrt{\left(\frac{\sigma_{xx} - \sigma_{yy}}{2}\right)^2 + \tau_{xy}^{\ 2}} = \sqrt{\left(\frac{120+40}{2}\right)^2 + (80)^2} = 113.1 \text{ MPa} \qquad \text{(b)}$$

Step 2: Draw the σ and τ axes, locate the center of the circle on the σ axis at C = 40 MPa and construct a circle with the radius R = 113.1 MPa as shown in Fig. E12.9a. The principal stresses σ_1 and σ_2 are located where the circle intersects the σ axis and given by:

$$\sigma_1 = C + R \qquad \sigma_2 = C - R \qquad (12.25)$$

The maximum shear stress is given by:

$$\tau_{Max} = R \qquad (12.26a)$$

The normal stress acting on the plane associated with τ_{Max} is given by:

$$\sigma_n = C \qquad (12.26b)$$

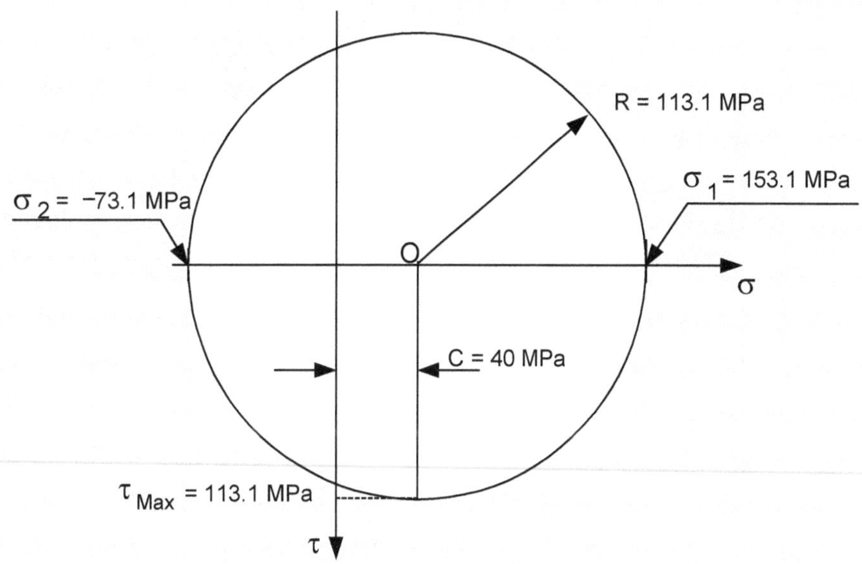

Fig. E12.9a

Step 3: To determine the orientation of the planes upon which the principal stresses and the maximum shear stress acts, let's establish the location of the x axis and y axes on Mohr's circle. Locate σ_{xx} = 120 MPa on the σ axis and draw a perpendicular line through this point. Measure down the perpendicular a distance equal to the value of τ_{xy} = 80 MPa. We measure downward because τ_{xy} is a positive quantity (counterclockwise) and the positive τ axis is downward. Point A, shown in Fig. E12.9b, represents the point with coordinates σ = 120 MPa and τ = 80 MPa. This point also falls on the circle that was drawn in Step 2. Clearly the line O – A represents the radius of Mohr's circle.

We have extended line O – A beyond the circle and labeled it as the x –axis. We have also extended this line upward and to the left until it intersect the circle at point B. If we drop a perpendicular line from point B to its intersection with the σ axis, the intersection occurs at σ_{yy} = – 40 MPa verifying the accuracy of the construction of the circle. The line O – B represents the y-axis. Note on this axis that τ_{xy} is negative (clockwise) and the negative τ axis is upward. Having established the positions of the x and y axes relative to the circle, the next step is to measure the angles defining the principal planes and the plane upon which the maximum shear stress acts.

Fig. E12.9b

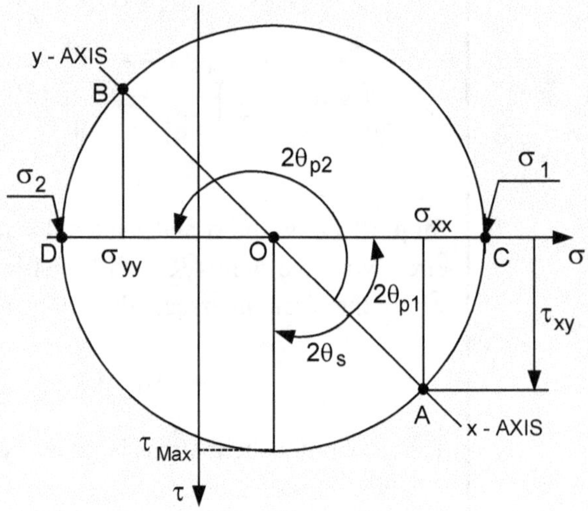

Step 4: From line O-A, we measure the angle AOC as illustrated in Fig. E12.9b, to obtain $2\theta_{p1} = 45°$. This observation indicates that the plane upon which σ_1 acts is located at $\theta_{p1} = 22.5°$. We also note that the angle $2\theta_{p2} = 2\theta_{p1} + 180°$. Thus, σ_2 acts on the plane located by $\theta_{p2} = 22.5° + 90° = 112.5°$. The angle $2\theta_{p1}$ can be verified by substituting the values for σ_{xx}, σ_{yy} and τ_{xy} into Eq. (12.13) to obtain:

$$\tan 2\theta_{p1} = \frac{2\tau_{xy}}{\sigma_{xx} - \sigma_{yy}} = \frac{2(80)}{120+40} = 1 \qquad \text{(d)}$$

$$2\theta_{p1} = \tan^{-1}(1) = 45^\circ$$

The angle $2\theta_s$ is measured in the clockwise (negative) direction. The measurement gives $2\theta_s = -45°$ as expected because the expression for $2\theta_s$ is the negative reciprocal of Eq. (12.13).

Step 5: Finally, let's locate the principal planes relative to the x and y axes as shown in Fig. E12.9c.

Fig E12.9c

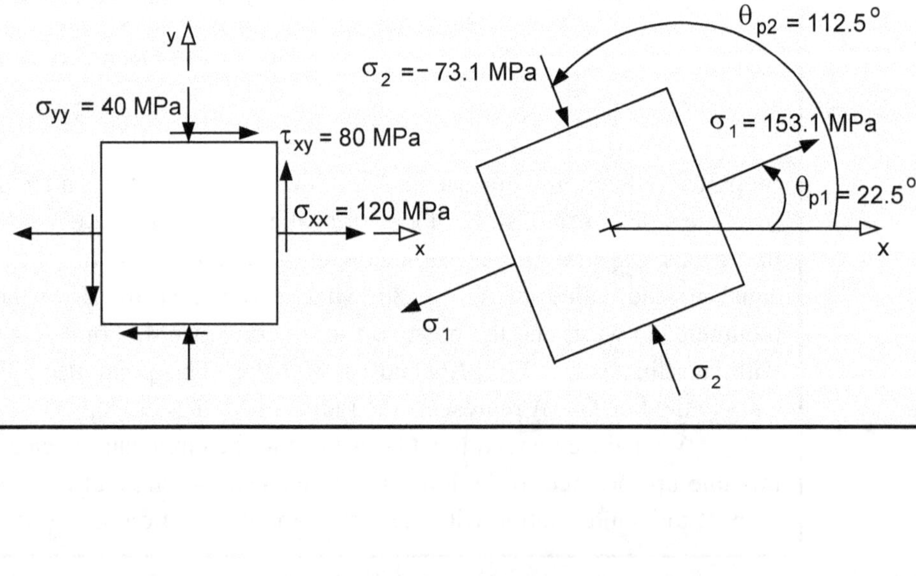

EXAMPLE 12.10

Consider a point, shown here as an elemental area, with Cartesian components of stress defined in Fig. E12.10. Using Mohr's circle, determine the principal stresses and the orientation of the principal planes. Also determine the maximum shear stresses and their orientation.

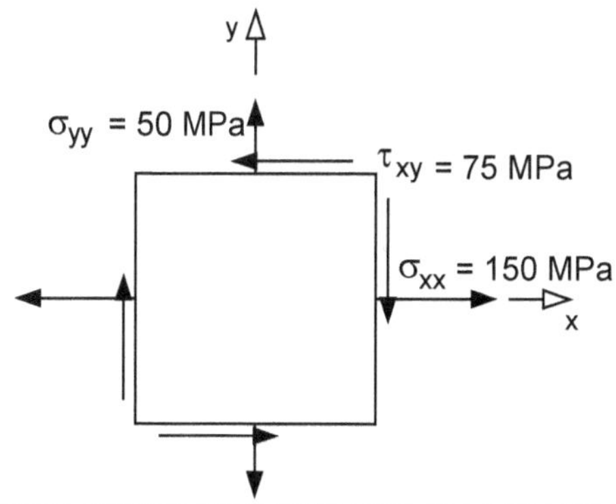

Fig. E12.10

Solution:

Step 1: Use Eq. (12.23) to determine the position of the center of the circle as:

$$C = (\sigma_{xx} + \sigma_{yy})/2 \qquad = (150 + 50)/2 = 100 \text{ MPa} \qquad \text{(a)}$$

Step 2: Layout the axes for Mohr's circle and locate the center point O at $\sigma = C$. Then locate $\sigma_{xx} = 150$ MPa along the σ axis and construct a perpendicular line upward from this intersection. Scale the length of the perpendicular line to be proportional to $\tau_{xy} = 75$ MPa as shown in Fig E12.10a. Because τ_{xy} is a negative quantity it is plotted in the negative τ direction. Draw the line O – A, which is the radius of Mohr's circle. This line also locates the x – axis relative to Mohr's circle. The radius can be determined from Eq. (12.24) as:

$$R = \sqrt{\left(\frac{\sigma_{xx} - \sigma_{yy}}{2}\right)^2 + \tau_{xy}{}^2} = \sqrt{\left(\frac{150-50}{2}\right)^2 + (75)^2} = 90.14 \text{ MPa} \qquad \text{(b)}$$

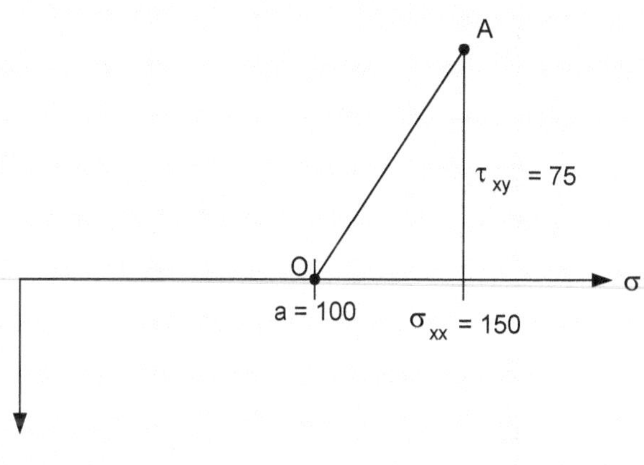

Fig. E12.10a

Step 3: Using the length of the line O - A as the radius R and the center of the circle at $\sigma = C$, we construct Mohr's circle as shown in Fig. E12.10b. The axes are scaled and the locations of σ_1, σ_2 and τ_{Max} are marked. From the scale imposed on the σ and τ axes we determine the approximate values of the principal and maximum shear stresses as:

$\sigma_1 = 190$ MPa $\qquad \sigma_2 = 10$ MPa $\qquad \tau_{Max} = 90$ MPa $\qquad \sigma_n = 100$ MPa

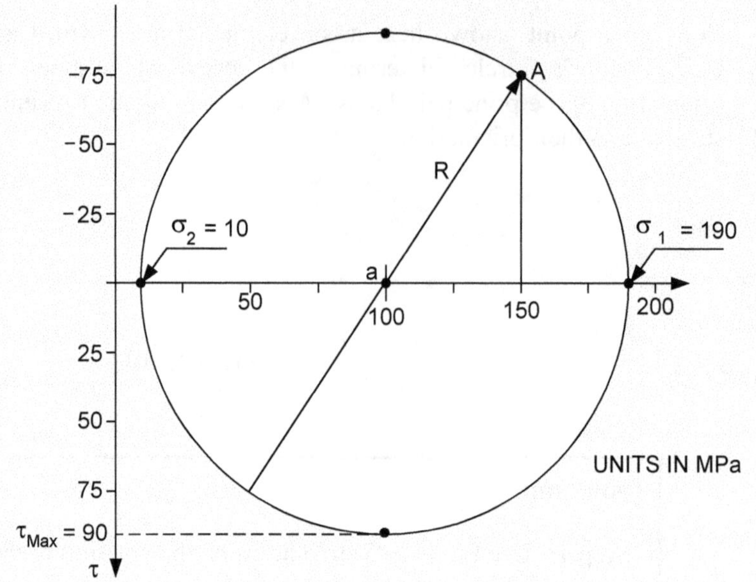

Fig. E12.10b

You can verify the accuracy of these results by employing Eq. (12.15) and Eq. (12.18).

Step 4: The location of the principal planes and the planes of maximum shear stress is determined by measuring the angles $2\theta_p$ and $2\theta_s$ as illustrated in Fig. E12.10c. Measurements of the angles AOC and AOD, both in the clockwise (negative) direction, shows that $2\theta_p = -56°$ and that $\theta_s = 146°$.

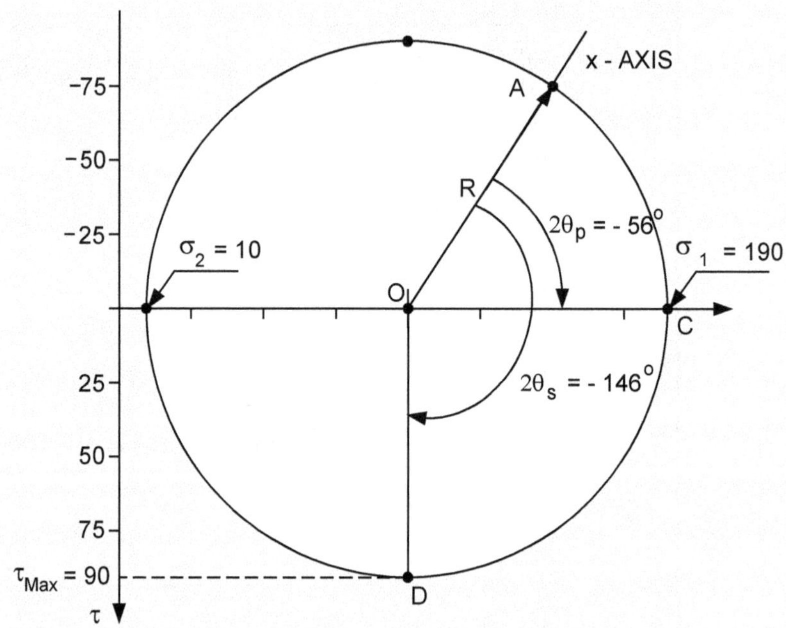

Fig. E12.10c

Step 5: Construct elements showing the planes upon which σ_1 and σ_2 act. We rotate a line through an angle θ_p measured in the clockwise direction from the x-axis to obtain the element shown in Fig. E 12.10d. Construct a similar element showing the plane upon which τ_{MAX} acts. Again we rotate a line an angle θ_s measured in the clockwise direction from the x-axis to obtain the element shown in Fig. E 12.10e.

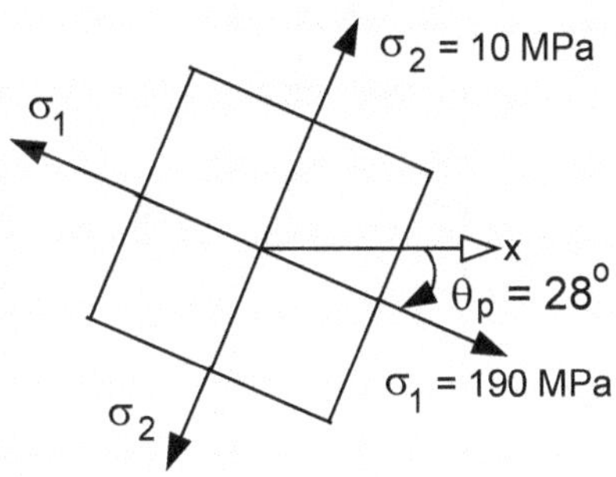

Fig. E12.10d

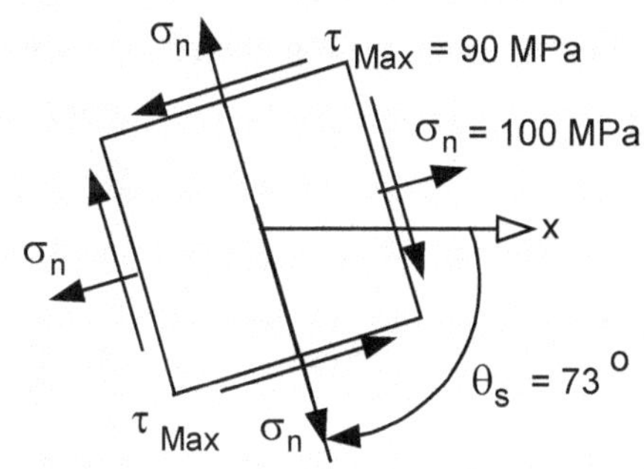

Fig. E12.10e

12.5 FAILURE THEORIES

In previous discussions pertaining to the safety of uniaxial members—beams, truss members, tie rods, etc., we indicated that failure would occur when the stress σ imposed on the member in question exceeded the strength of the material from which the member was fabricated. The equation for failure for these uniaxial members is given by:

$$\sigma_{xx} = \sigma_1 \geq S_y \qquad \sigma_{xx} = \sigma_1 \geq S_u \qquad (12.27)$$

where the choice of yield or ultimate tensile strength, S_y or S_u, depends on the admissible mode of failure.

For uniaxial members the stress state is also uniaxial with $\sigma_{yy} = \sigma_{zz} = 0$ and $\sigma_2 = \sigma_3 = 0$; hence, the conditions for structural failure given in Eq. (12.27) are valid. However, for more complex structural components such as plates and shells or machine components, the state of stress is multi-dimensional and more involved theories of failure are required. We will consider a two-dimensional state of stress—plane stress— in this section where the principal stresses are ordered so that:

$$\sigma_1 \geq \sigma_2 \qquad \sigma_3 = 0 \qquad (12.28)$$

Several different theories of failure are used today, but in the interest of brevity, we will describe only the three most popular ones.

1. The maximum principal stress theory.
2. The maximum shear stress theory.
3. The maximum distortional energy theory.

We will also assume that the failure occurs upon yielding and that the yield strength in tension is the same as the yield strength in compression.

12.5.1 The Maximum Principal Stress Theory

The maximum principal stress theory states that a structural component fails by yielding when the maximum principal stress σ exceeds the yield strength of the material from which it is fabricated.

$$|\sigma_1| \geq S_y \qquad |\sigma_2| \geq S_y \qquad (12.29)$$

If we consider all possible combinations of σ_1 and σ_2, it is possible to construct the yielding diagram presented in Fig. 12.8. If we plot a point with coordinates σ_1/S_y and σ_2/S_y and it falls within the square in Fig. 12.8, the structural component will not yield. However, if the point falls on the boundary or outside the boundary of the box, yielding will occur. This theory is employed in the analysis of uniaxial structural members where the influence of the second principal stress σ_2 is small. Unfortunately, failure of ductile materials subjected to biaxial states of stress occurs due to shear on planes of maximum shear stress. In these cases, the maximum principal stress theory is not conservative. Yielding by shear occurs at stress states lower than predicted by the maximum principal stress theory.

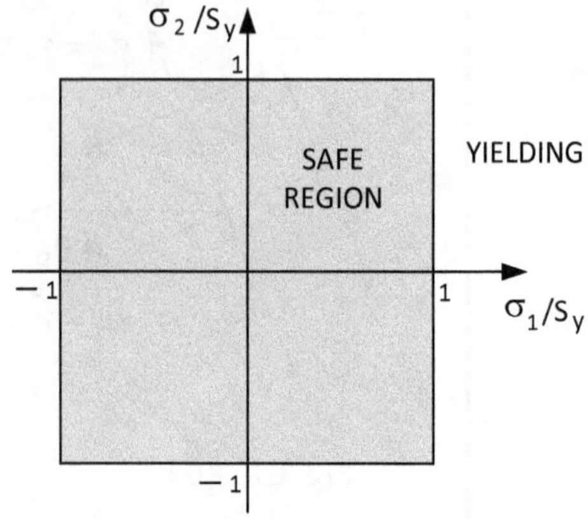

Fig. 12.8 Failure diagram for the maximum principal stress theory for yielding.

12.5.2 The Maximum Shear Stress Theory

The maximum shear stress theory for yielding, sometimes called the Tresca theory, is often employed to predict the onset of yielding in machine components fabricated from ductile materials. These ductile materials fail on maximum shear planes that are oriented at 45° to the principal planes. This theory indicates that yielding will occur when τ_{Max} in a machine component equals the maximum shear stress achieved at yield in a simple tensile test of the material.

Consider first the simple tension test where:

$$\sigma_1 = S_y \qquad \sigma_2 = \sigma_3 = 0 \qquad \tau_{Max} = (\sigma_1 - \sigma_2)/2 = S_y/2 \tag{12.30}$$

Next, consider the maximum shear stress in the structural component. Two possibilities must be considered:

1. When σ_1 and σ_2 are of opposite signs τ_{Max} is given by:

$$\tau_{Max} = (\sigma_1 - \sigma_2)/2 \tag{12.31}$$

Substituting Eq. (12.30) into Eq. (12.31) gives the equation governing yielding as:

$$|\sigma_1 - \sigma_2| \geq S_y \tag{12.32}$$

2. When σ_1 and σ_2 are the same sign τ_{Max} is given by:

$$\tau_{Max} = (\sigma_1 - \sigma_3)/2 = \sigma_1/2 \tag{12.33}$$

Substituting Eq. (12.30) into Eq. (12.33) gives the equation governing yielding as:

$$|\sigma_1| \geq S_y \tag{12.34}$$

If we consider all possible combinations of σ_1 and σ_2, it is possible to construct the yielding diagram presented in Fig. 12.9.

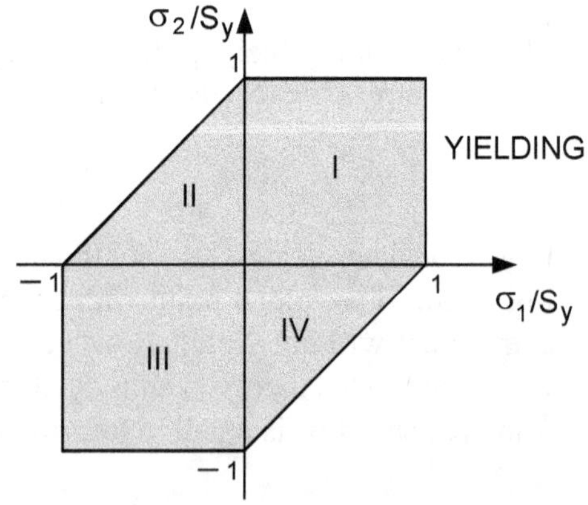

Fig. 12.9 Failure diagram for the Maximum shear stress theory of yielding.

12.5.3 Maximum Distortion Energy Theory

The maximum distortion energy theory, sometimes called the von Mises theory, predicts yielding in a machine component when the distortion energy per unit volume equal the distortion energy per unit volume achieved at yielding of the material in a simple tensile test.

For a two-dimensional stress state with $\sigma_3 = 0$, the distortion energy per unit volume U_d is given by:

$$U_d = (1/6G)[\sigma_1^2 - \sigma_1\sigma_2 + \sigma_2^2] \qquad (12.35)$$

In a tensile test with $\sigma_1 = S_y$ and $\sigma_2 = \sigma_3 = 0$, the distortion energy per unit volume is:

$$U_d = (1/6G)S_y^2 \qquad (12.36)$$

By substituting Eq. (12.36) into Eq. (12.35), we obtain the von Mises equation for yielding.

$$(\sigma_1/S_y)^2 - (\sigma_1/S_y)(\sigma_2/S_y) + (\sigma_2/S_y)^2 = 1 \qquad (12.37)$$

If we consider all possible combinations of σ_1 and σ_2, it is possible to construct the yielding diagram presented in Fig. 12.10. The boundary of the failure diagram is an ellipse that is slightly less conservative than the maximum principal stress theory for yielding in quadrants I and III. In quadrants II and IV where the principal stresses are of opposite sign the von Mises theory is less conservative than the maximum shear theory but more conservative that the maximum principal stress theory. The distortional energy theory is widely accepted as the most accurate theory for predicting yielding in ductile materials commonly used to fabricate structural elements and machine components.

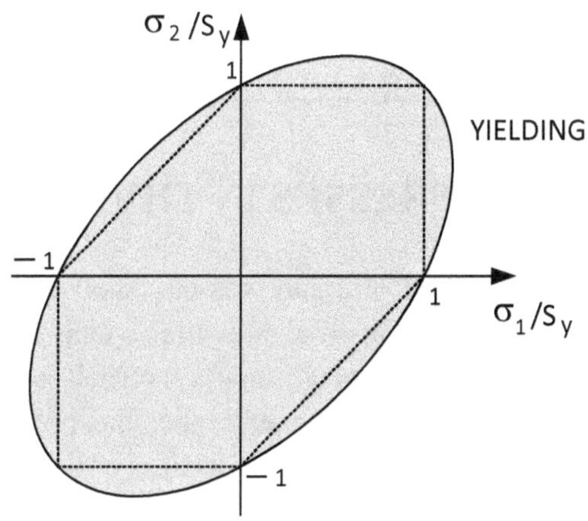

Fig 12.10 Failure diagram for the maximum distortion energy theory for yielding.

EXAMPLE 12.11

Your analysis of the principal stresses in a machine component indicates that $\sigma_1 = 38$ ksi and $\sigma_2 = -20$ ksi. If the component is fabricated from 1020 HR steel, will it yield in service?

Solution:

Because $S_y = 42$ ksi for 1020 HR steel, the prediction of yielding will depend on the theory of yielding employed in the analysis. Let's use the three different theories described in the previous subsections. For the maximum principal stress theory for yielding and Eq. (12.29), it is clear that yielding will not occur because:

$$\sigma_1 = 38 \text{ ksi} < S_y = 42 \text{ ksi} \qquad |\sigma_2| = |20| \text{ ksi} < S_y = 42 \text{ ksi} \qquad \text{(a)}$$

Because σ_1 and σ_2 are of opposite signs, the maximum shear stress theory for yielding and Eq. (12.32) gives:

$$(\sigma_1 - \sigma_2) = (38 + 20) = 58 \text{ ksi} > S_y = 42 \text{ ksi} \qquad \text{(b)}$$

From the results of Eq. (b), we predict that yielding will occur.

Using the maximum distortion energy theory and applying Eq. (12.37), we write:

$$(\sigma_1/S_y)^2 - (\sigma_1/S_y)(\sigma_2/S_y) + (\sigma_2/S_y)^2 \geq 1$$

$$(38/42)^2 - (38/52)(-20/42) + (-20/42)^2 = 1.476 > 1 \quad \text{(c)}$$

The maximum distortional energy theory also predicts yielding. Because the two conservative theories of failure that are widely accepted to predict yielding in ductile materials provide the same result, we indicate that the machine component will fail by yielding.

We indicate the same result on the graphic shown in Fig. E12.11, where the operating point σ_1/S_y = 0.905 and $\sigma_2/S_y = -0.476$ is plotted.

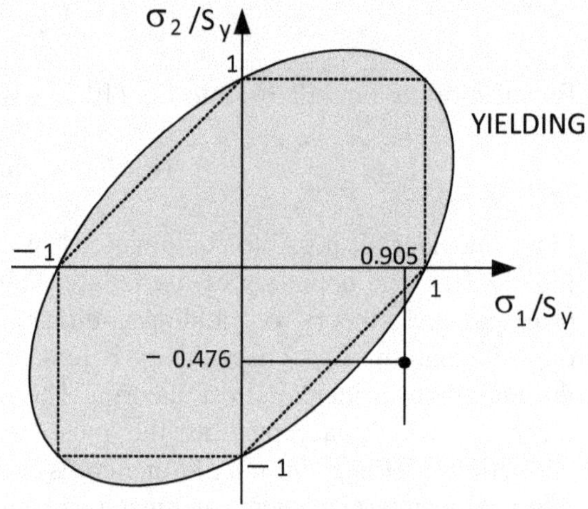

Fig. E12.11

12.6 STRESSES IN THIN WALLED PRESSURE VESSELS

Thin walled[1] pressure vessels (shells) are widely employed as containers for liquids and gasses. They are usually subjected to internal pressure, which generates membrane stresses in the walls of the pressure vessels. Bending stresses do not occur unless the shell is designed with discontinuities such as nozzles or skirts. Because the state of stress is primarily membrane, the pressure vessels are very efficient structures with stresses that are uniformly distributed through the thickness of the wall.

Applications involving pressure vessels vary widely. Pressure vessels are fabricated in various sizes and shapes. They can be small such as a can for soda, or large such as a storage tank for natural gas. While the most common shape is cylindrical, they may be designed with spherical, conical or toroidal shapes.

12.6.1 Spherical Pressure Vessels

Let's consider a portion of a thin spherical shell to illustrate the concept of membrane stresses. The portion of a spherical shell with a radius r and wall thickness t subjected to an internal pressure p is illustrated in Fig. 12.11. Writing the equilibrium equation $\Sigma F_y = 0$ yields:

$$p\pi r_x^2 - (N \sin\phi)(2\pi r_x) = 0 \quad (12.38)$$

where N is the membrane force per unit length and r_x is the radius defined in Fig. 12.11

Solving Eq. (12.38) for N and noting that $r_x/\sin\phi = r$ gives:

[1] A pressure vessel is considered to be thin walled if the ratio of $r/t > 10$. The membrane stresses in thin walled pressure vessels are uniformly distributed through the thickness of the wall.

$$N = pr_x/(2 \sin \phi) = pr/2 \quad (12.39)$$

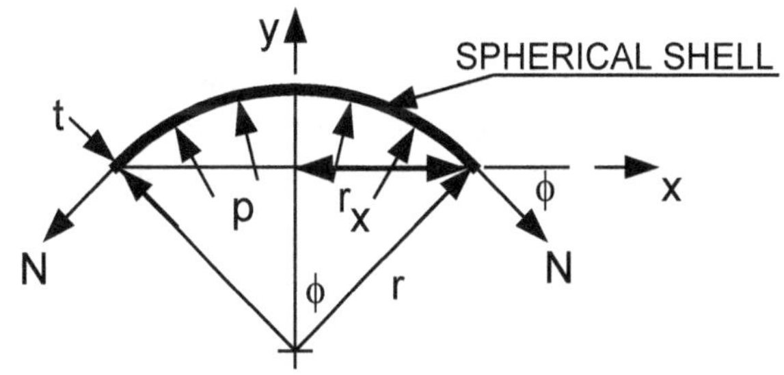

Fig. 12.11 A portion of a spherical shell subjected to an internal pressure p.

It is evident from Eq. (12.39) that the value of N is independent of the angle ϕ; hence the membrane force per unit length is independent of position. The membrane stresses σ_m for a spherical pressure vessel are obtained from Eq. (12.39) as:

$$\sigma_m = N/t = pr/2t \quad (12.40)$$

Because the spherical pressure vessel is symmetric, it is evident that the membrane stresses in the meridian σ_{my} and tangential σ_{mx} directions are the same, and the stress state is biaxial with:

$$\sigma_1 = \sigma_2 = \sigma_{my} = \sigma_{mx} = pr/2t \quad (12.41)$$

These stresses are also principal stresses because axes of symmetry always coincide with principal planes.

EXAMPLE 12.12

A spherical pressure vessel with a radius r = 20 ft and a wall thickness of 3/8 in. is subjected to a pressure of 75 psi. Determine the membrane stresses in the longitudinal and meridian directions. Also determine the maximum shear stress.

Solution:

From Eq. (12.41), we write:

$$\sigma_{my} = \sigma_{mx} = pr/2t = (75)(20)(12)/[(2)(3/8)] = 24{,}000 \text{ psi} \quad (a)$$

Also $\tau_{xy} = 0$ because the meridian and longitudinal membrane stresses are principal stresses.

From Eq. (12.18), the maximum shear stress is given by:

$$\tau_{Max} = \sqrt{\left(\left(\frac{\sigma_{xx} - \sigma_{yy}}{2}\right)^2 + \tau_{xy}^2\right)} = \sqrt{\left(\frac{24000 - 24000}{2}\right)^2 + 0} = 0$$

The maximum shear stress is zero for in-plane stresses. For out of plane stresses τ is given by:

$$\tau_{Max} = (\sigma_1 - \sigma_3)/3 = \sigma_1/2 = pr/(4t) \quad (12.41a)$$

12.6.2 Cylindrical Pressure Vessels

Cylindrical pressure vessels are the most common type employed because the cylindrical shape is easy to fabricate. Thin metal plates are formed into rounds and the longitudinal (axial) seam is welded shut to produce a thin walled tube. End caps are welded on this tube to produce a thin walled vessel capable of maintaining internal pressures.

To determine the membrane stresses σ_h (hoop or circumferential direction) and σ_a (axial or longitudinal direction) in a thin walled cylindrical pressure vessel consider a cylinder with a radius r and a wall thickness t as illustrated in Fig. 12.12. Next remove a semi-circular segment of the cylinder as a FBD. The forces due to pressure p and the membrane forces acting on this FBD are shown in Fig. 12.13.

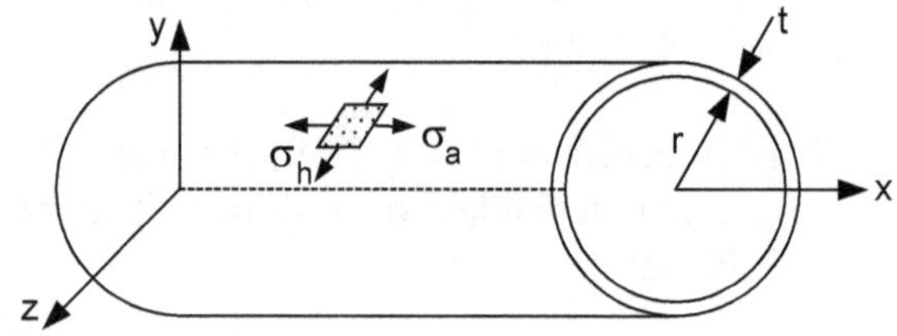

Fig. 12.12 A cylindrical pressure vessel with a radius r and a wall thickness of t.

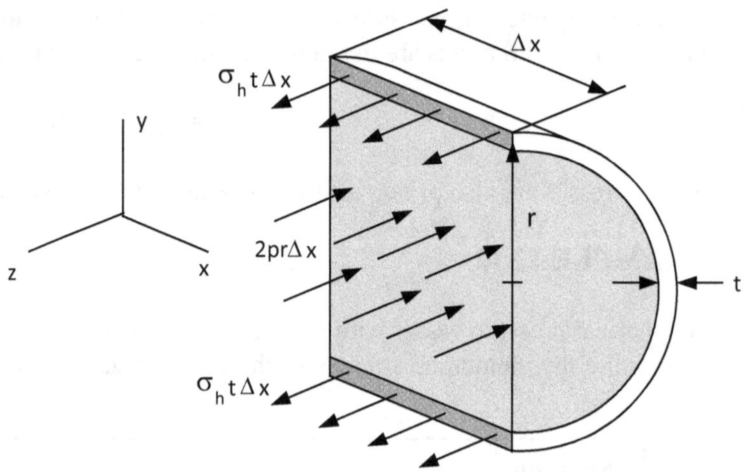

Fig. 12.13 A FBD of a semi-circular segment from the thin walled cylindrical pressure vessel.

To solve for the hoop stress σ_h, write $\Sigma F_z = 0$:

$$2(\sigma_h t\Delta x) - 2pr\Delta x = 0 \qquad \text{(a)}$$

$$\sigma_h = pr/t \qquad (12.42)$$

The axial stresses are determined by considering a segment of the cylinder formed by a section cut perpendicular to the axis of the cylinder as shown in Fig. 12.14. This FBD shows the forces due to pressure and the stresses in the axial direction. Writing $\Sigma F_x = 0$ yields:

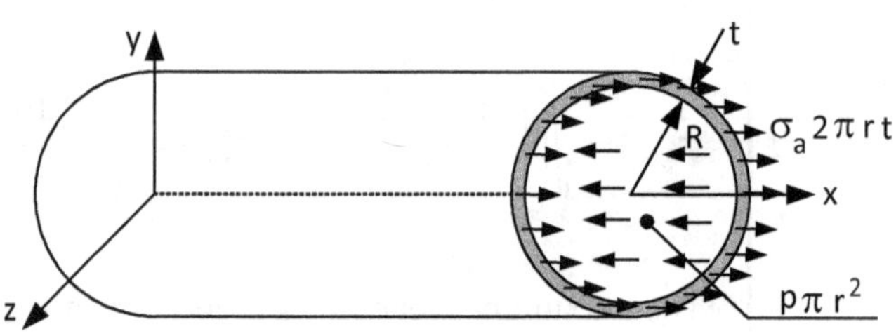

Fig. 12.14 FBD of a cylindrical pressure vessel showing the forces in the axial direction.

$$\sigma_a (2\pi rt) - p(\pi r^2) = 0 \qquad \text{(b)}$$

Solving Eq. (b) for σ_a gives:

$$\sigma_a = pr/2t \qquad (12.43)$$

Comparing Eq. (12.42) and Eq. (12.43) indicates that the hoop stress is twice as large as the axial stress in a thin walled circular pressure vessel. Also, from symmetry it is evident that σ_h and σ_a are principal stresses with:

$$\sigma_1 = \sigma_h \qquad \sigma_2 = \sigma_a \qquad \tau_{ha} = 0 \tag{12.44}$$

The maximum shear stress τ_{Max} is determined from Eq. (12.18) as:

$$\tau_{Max} = \sqrt{\left(\left(\frac{\sigma_{xx} - \sigma_{yy}}{2}\right)^2 + \tau_{xy}^2\right)} = \sqrt{\left(\frac{\sigma_1 - \sigma_2}{2}\right)^2} = \frac{pr}{4t} \tag{12.45}$$

EXAMPLE 12.13

A cylindrical pressure vessel, shown in Fig. E12.13, is employed as a water storage tank for a small municipality. Determine the height of the water h that can be maintained in the tank if the city's engineering director specifies a safety factor of 3.0. The tank is fabricated from 1020 HR steel with a diameter of D = 40 ft., a height H = 50 ft. and a thickness t = 0.375 in. The engineering director also indicates that the safety factor should be based on the maximum shear stress theory of yielding.

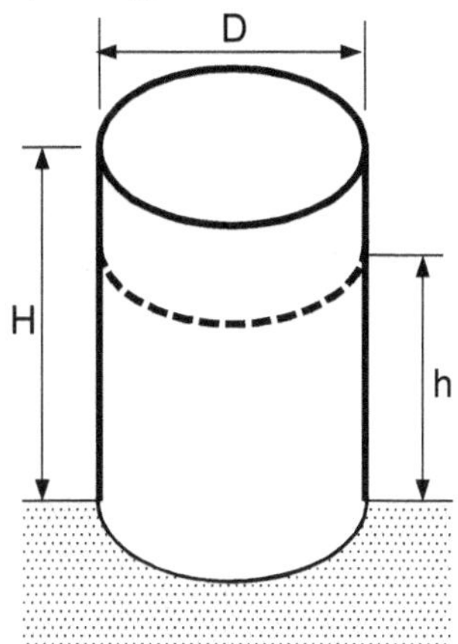

Fig. E12.13

Solution:

The pressure in the water tank varies from zero at the water level to a maximum at the base. The maximum pressure is given by:

$$p_{Max} = \gamma h \tag{12.46}$$

where $\gamma = 62.4$ lb/ft^3 is the density of fresh water.

At the base of the tank, the pressure is:

$$p_{Max} = (62.4)h \quad (\text{lb/ft}^2) \tag{a}$$

The hoop stress is given by Eq. (12.42) as:

$$\sigma_h = \sigma_1 = pR/t = [(62.4)(20)(12)/(0.375)]h = 39{,}940\ h \quad (\text{lb/ft}^2) \tag{b}$$

where h is measured in ft.

From the failure diagram for yielding under maximum shear stress, shown in Fig. E12.13a, we can write:

$$(\sigma_1)_{Design} = S_y/SF \tag{c}$$

Reference to Appendix B-2 for the yield strength of 1020 HR steel gives:

$$(\sigma_1)_{Design} = S_y/SF = 42.0/3.0 = 14.0 \text{ ksi} \tag{d}$$

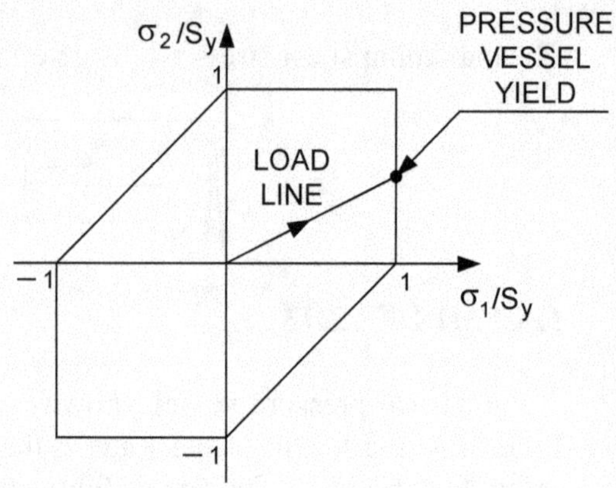

Fig. E12.13a

Converting the units of Eq. (d) from ksi to lb/ft² gives:

$$(\sigma_1)_{Design} = (14 \times 10^3 \text{ lb/in.}^2)(144 \text{ in.}^2/\text{ft}^2) = 2{,}016 \times 10^3 \text{ lb/ft}^2 \quad \text{(e)}$$

From Eq. (b) and Eq. (e), we solve for h as:

$$h = (2{,}016 \times 10^3)/39{,}940 = 50.5 \text{ ft} \quad \text{(f)}$$

Reference to Fig. E12.13 shows that the tank is 50 ft high and if filled to the brim, the stresses would exceed the allowable design limit of 14.0 ksi by about one percent. You may avoid even this slight excess by placing an overflow line at a location 49 ft above the tank's base.

12.7 STRAIN EQUATIONS OF TRANSFORMATION

In Section 12.2, we derived the stress equations of transformation and showed how the stresses $\sigma_{x'x'}$, $\sigma_{y'y'}$ and $\tau_{x'y'}$ varied with the angle of rotation θ of the Ox'y' coordinate system relative to the Oxy coordinate system. A remarkably similar set of transformation equations exist for strain. When deriving the stress equations of transformation, we considered a state of plane stress where $\sigma_{zz} = \tau_{xz} = \tau_{yz} = 0$. This derivation was based solely on the equilibrium equations.

To derive the strain equations of transformation, we will consider a state of plane strain where $\varepsilon_{zz} = \gamma_{xz} = \gamma_{yz} = 0$. This derivation is based only on the geometry of the deformation field. Let's consider the plane body illustrated in Fig 12.15 where the line segment AB is scribed prior to imposing plane strain deformations. After the deformations are imposed, the line segment moves and is designated as A*B*. Point A moves to a new location designated by A*. The displacement in the x direction associated with this movement is u and that in the y direction is v. Point B moves to a new location designated by B* with displacements u + Δu in the x direction and v + Δv in the y direction.

By definition the strain along the line segment AB is given by:

$$\varepsilon_{AB} = (A^*B^* - AB)/AB \quad \text{(a)}$$

From the geometry of Fig. 12.15, it clear that the length of the line segments are given by:

$$(AB)^2 = (\Delta x)^2 + (\Delta y)^2 \quad \text{(b)}$$

$$(A^*B^*)^2 = (\Delta x^*)^2 + (\Delta y^*)^2 \quad \text{(c)}$$

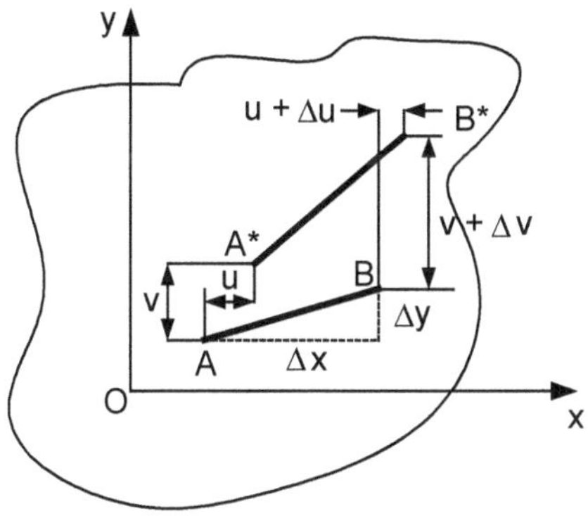

Fig. 12.15 A line segment AB before deformation and A*B* after deformation.

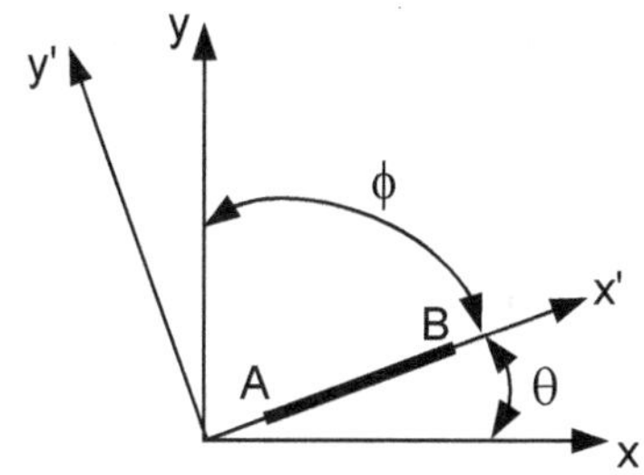

Fig. 12.16 Definition of angles θ and ϕ.

In general, the component Δx* will have a different length than Δx because of the deformation of the body in the x direction. We express this difference in both Δx* and Δy* by writing:

$$\begin{aligned}\Delta x^* &= \left(1+\frac{\partial u}{\partial x}\right)\Delta x + \frac{\partial u}{\partial y}\Delta y \\ \Delta y^* &= \frac{\partial v}{\partial x}\Delta x + \left(1+\frac{\partial v}{\partial y}\right)\Delta y\end{aligned} \qquad \text{(d)}$$

The length of the line segment A*B* can be determined by substituting Eq. (d) into Eq. (c). Because the deformations are extremely small, the products and squares of the derivatives can be neglected. From the substitution and simplification, we obtain:

$$(A*B*) = \left(1+2\frac{\partial u}{\partial x}\right)(\Delta x)^2 + \left(1+2\frac{\partial v}{\partial y}\right)(\Delta y)^2 + 2\left(\frac{\partial u}{\partial y}+\frac{\partial v}{\partial x}\right)(\Delta x)(\Delta y) \qquad \text{(e)}$$

Let's rewrite Eq. (a) as:

$$(\varepsilon_{AB}+1)^2 = \left(\frac{A*B*}{AB}\right)^2 \qquad \text{(f)}$$

Substituting Eq. (b) and Eq. (e) into Eq. (f) and simplifying gives:

$$(\varepsilon_{AB}+1)^2 = \cos^2\theta + \cos^2\phi + 2\frac{\partial u}{\partial x}\cos^2\theta + 2\frac{\partial v}{\partial y}\cos^2\phi + 2\left(\frac{\partial u}{\partial y}+\frac{\partial v}{\partial x}\right)\cos\theta\cos\phi \qquad \text{(g)}$$

where θ and ϕ are defined in Fig. 12.16.

Expand the left side of Eq. (g) and neglect ε_{AB}^2 because it is a second order term and note that

$$\cos^2\theta + \cos^2\phi = 1 \qquad \text{(h)}$$

We can then express the strain along the line segment AB as:

$$\varepsilon_{AB} = \frac{\partial u}{\partial x}\cos^2\theta + \frac{\partial v}{\partial y}\cos^2\phi + \left(\frac{\partial u}{\partial y} + \frac{\partial v}{\partial x}\right)\cos\theta\cos\phi \qquad (12.47)$$

The relationship between strains and displacements can be expressed as:

$$\partial u/\partial x = \varepsilon_{xx} \qquad \partial v/\partial y = \varepsilon_{yy} \qquad (\partial u/\partial y + \partial v/\partial x) = \gamma_{xy} \qquad (12.48)$$

If we substitute the strain displacement relations of Eq. (12.48) into Eq. (12.47), we obtain the relation for ε_{AB} as:

$$\varepsilon_{AB} = \varepsilon_{xx}\cos^2\theta + \varepsilon_{yy}\sin^2\theta + \gamma_{xy}\cos\theta\sin\theta \qquad (12.49)$$

If we select the line segment AB so that it is parallel to the x' axis, then it is evident that Eq. (12.49) may be written as:

$$\varepsilon_{x'x'} = \varepsilon_{xx}\cos^2\theta + \varepsilon_{yy}\sin^2\theta + \gamma_{xy}\cos\theta\sin\theta \qquad (12.50)$$

Selecting the line segment AB parallel to the y' axis gives:

$$\varepsilon_{y'y'} = \varepsilon_{yy}\cos^2\theta + \varepsilon_{xx}\sin^2\theta - \gamma_{xy}\cos\theta\sin\theta \qquad (12.51)$$

A similar but somewhat more involved derivation can be used to establish the strain equations of transformation for $\gamma_{x'y'}$. If we consider the angular change in an arbitrary right angle formed by two line segments AB_1 and AB_2 and follow the same procedure, the shearing strain $\gamma_{x'y'}$ can be written as:

$$\frac{\gamma_{x'y'}}{2} = -\varepsilon_{xx}\cos\theta\sin\theta + \varepsilon_{yy}\cos\theta\sin\theta + \frac{\gamma_{xy}}{2}\left(\cos^2\theta - \sin^2\theta\right) \qquad (12.52)$$

The strain equations of transformation presented in Eqs. (12.50), (12.51) and (12.52) are the same form as the stress equations of transformation given by Eqs. (12.8) and (12.9). A comparison shows that σ_{xx}, σ_{yy}, $\sigma_{x'x'}$ and $\sigma_{y'y'}$ correspond to ε_{xx}, ε_{yy}, $\varepsilon_{x'x'}$ and $\varepsilon_{y'y'}$. The comparison also shows that τ_{xy} and $\tau_{x'y'}$ correspond to $\gamma_{xy}/2$ and $\gamma_{x'y'}/2$.

It is possible to change Eqs. (12.50), (12.51) and (12.52), which show the angle θ into a double angle representation with suitable trigonometric substitutions. The results of these substitutions are:

$$\varepsilon_{x'x'} = \frac{\varepsilon_{xx} + \varepsilon_{yy}}{2} + \frac{\varepsilon_{xx} - \varepsilon_{yy}}{2}\cos 2\theta + \frac{\gamma_{xy}}{2}\sin 2\theta$$

$$\frac{\gamma_{x'y'}}{2} = -\frac{\varepsilon_{xx} - \varepsilon_{yy}}{2}\sin 2\theta + \frac{\gamma_{xy}}{2}\cos 2\theta \qquad (12.53)$$

$$\varepsilon_{y'y'} = \frac{\varepsilon_{xx} + \varepsilon_{yy}}{2} - \frac{\varepsilon_{xx} - \varepsilon_{yy}}{2}\cos 2\theta - \frac{\gamma_{xy}}{2}\sin 2\theta$$

Because of the correspondence of the stress and strain transformation equations, it is apparent that a Mohr's strain circle can be constructed to graphically represent the state of strain at a point. The procedure for constructing Mohr's circle for strain is exactly the same as that followed previously for the stress circle. The only difference is to change the abscissa from σ to ε and the ordinate from τ to γ/2. We show a typical Mohr's circle for strain in Fig. 12.17.

The center of the strain circle is located at position C where:

$$C = (\varepsilon_{xx} + \varepsilon_{yy})/2 \qquad (12.54)$$

The radius R of the circle is given by:

$$R = \sqrt{\left(\frac{\varepsilon_{xx} - \varepsilon_{yy}}{2}\right)^2 + \left(\frac{\gamma_{xy}}{2}\right)^2} \tag{12.55}$$

It is also evident from Fig. 12.17 that the principal strains are given by:

$$\varepsilon_{1,2} = C \pm R = \frac{\varepsilon_{xx} + \varepsilon_{yy}}{2} \pm \sqrt{\left(\frac{\varepsilon_{xx} - \varepsilon_{yy}}{2}\right)^2 + \left(\frac{\gamma_{xy}}{2}\right)^2} \tag{12.56}$$

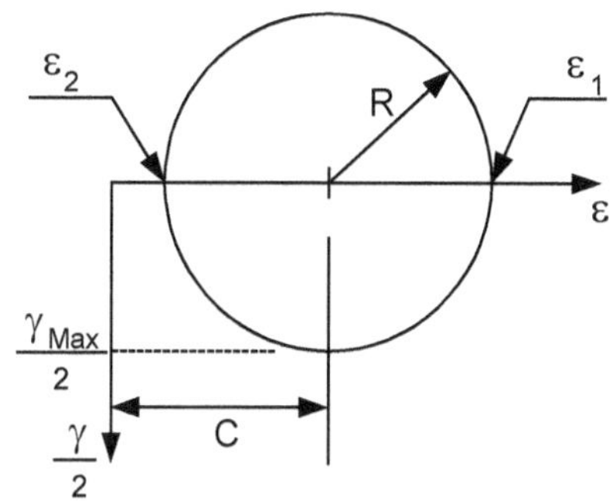

Fig. 12.17 Mohr's circle for strain.

The maximum shearing strain γ_{Max} is given by:

$$\frac{\gamma_{Max}}{2} = R = \sqrt{\left(\frac{\varepsilon_{xx} - \varepsilon_{yy}}{2}\right)^2 + \left(\frac{\gamma_{xy}}{2}\right)^2} \tag{12.57}$$

Because of the correspondence in the stress and strain equations of transformation the principal planes are given by:

$$\tan 2\theta_p = \frac{\gamma_{xy}}{\varepsilon_{xx} - \varepsilon_{yy}} \tag{12.58}$$

The planes upon which the maximum shear strain act are rotated 45° relative to the principal planes.

Let's consider an example, which demonstrates the application of these results for determining the state of strain on arbitrary planes.

EXAMPLE 12.14

A state of plane strain exists at point A in a plane body where the Cartesian components of strain are given by:

$\varepsilon_{xx} = 900 \times 10^{-6}$ $\quad$ $\varepsilon_{yy} = -400 \times 10^{-6}$ $\quad$ $\gamma_{xy} = 300 \times 10^{-6}$

Determine the state of strain on an element rotated counterclockwise 40° relative to the Oxy axes.

Solution:

The strains $\varepsilon_{x'x'}$, $\varepsilon_{y'y'}$ and $\gamma_{x'y'}$ are determined from the strain equations of transformation as:

$$\varepsilon_{x'x'} = \frac{\varepsilon_{xx}+\varepsilon_{yy}}{2} + \frac{\varepsilon_{xx}-\varepsilon_{yy}}{2}\cos 2\theta + \frac{\gamma_{xy}}{2}\sin 2\theta$$

$$\varepsilon_{x'x'} = (\frac{900-400}{2} + \frac{900+400}{2}\cos 80^\circ + \frac{300}{2}\sin 80^\circ)\times 10^{-6} = 510.6\times 10^{-6}$$

$$\frac{\gamma_{x'y'}}{2} = -\frac{\varepsilon_{xx}-\varepsilon_{yy}}{2}\sin 2\theta + \frac{\gamma_{xy}}{2}\cos 2\theta$$

$$\gamma_{x'y'} = [-(900+400)\sin 80^\circ + 300\cos 80^\circ]\times 10^{-6} = -1228\times 10^{-6}$$

$$\varepsilon_{y'y'} = \frac{\varepsilon_{xx}+\varepsilon_{yy}}{2} - \frac{\varepsilon_{xx}-\varepsilon_{yy}}{2}\cos 2\theta - \frac{\gamma_{xy}}{2}\sin 2\theta$$

$$\varepsilon_{y'y'} = (\frac{900-400}{2} - \frac{900+400}{2}\cos 80^\circ - \frac{300}{2}\sin 80^\circ)\times 10^{-6} = -10.6\times 10^{-6}$$

EXAMPLE 12.15

For the state of plane strain described in Example 12.14, determine the principal strains, the maximum shearing strain and the planes upon which they act.

Solution:

The magnitudes of the principal strains are given by Eq. (12.56) as:

$$\varepsilon_{1,2} = C \pm R = \frac{\varepsilon_{xx}+\varepsilon_{yy}}{2} \pm \sqrt{\left(\frac{\varepsilon_{xx}-\varepsilon_{yy}}{2}\right)^2 + \left(\frac{\gamma_{xy}}{2}\right)^2}$$

$$\varepsilon_{1,2} = \left(\frac{900-400}{2} \pm \sqrt{\left(\frac{900+400}{2}\right)^2 + \left(\frac{300}{2}\right)^2}\right)\times 10^{-6}$$

$$\varepsilon_{1,2} = [250 \pm 667.1]\times 10^{-6}$$

$$\varepsilon_1 = 917.1\times 10^{-6} \quad \text{and} \quad \varepsilon_2 = -417.1\times 10^{-6}$$

The principal planes are given by Eq. (12.58) as:

$$\tan 2\theta_p = \frac{\gamma_{xy}}{\varepsilon_{xx}-\varepsilon_{yy}} = \frac{300}{900+400} = 0.2308$$

$$2\theta_p = 13.00^\circ \qquad \theta_p = 6.50^\circ$$

The maximum shear strain is given by Eq. (12.57) as:

$$\frac{\gamma_{Max}}{2} = R = \sqrt{\left(\frac{\varepsilon_{xx} - \varepsilon_{yy}}{2}\right)^2 + \left(\frac{\gamma_{xy}}{2}\right)^2} = \sqrt{\left(\frac{900+400}{2}\right)^2 + \left(\frac{300}{2}\right)^2} \times 10^{-6}$$

$$\frac{\gamma_{Max}}{2} = 667.1 \times 10^{-6} \qquad \gamma_{Max} = 1334.2 \times 10^{-6}$$

The plane upon which the maximum shearing strain acts is given by:

$$\theta_s = \theta_p \pm 45° = 6.50° \pm 45° = 51.50° \text{ or } -38.50°$$

It is usually worth the time and effort to quickly check the results of a calculation like this one by constructing Mohr's circle. We show the Mohr's circle for this state of strain in Fig. E12.15.

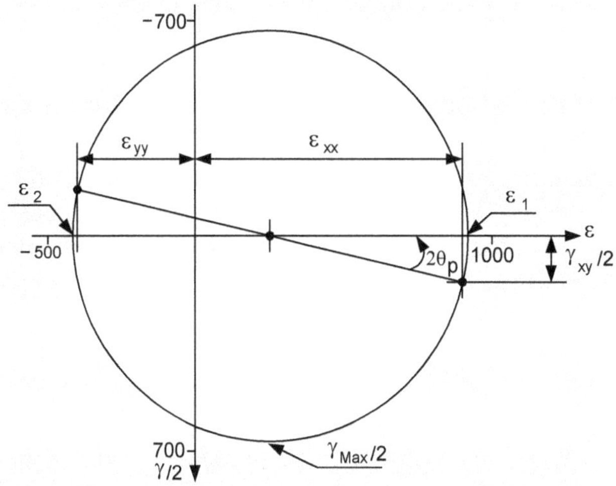

Fig. E12.15 Mohr's strain circle drawn to scale. Multiply strains shown by 10^{-6}.

12.8 DETERMINING PRINCIPAL STRESS WITH STRAIN GAGES

In the laboratory or the field, we usually measure strain with a metal foil resistor called a strain gage. The pattern of the resistor for a typical strain gage is shown in Fig. 12.18. A grid configuration is photoetched from the metallic foil to increase the resistance of the gage and to give the gage a high directional sensitivity. Large tabs at each end of the metal foil conductors are provided to accommodate the lead wire connections. The foil is very thin (about 200 μ in.) and fragile; hence, it requires support to avoid damage in normal handling. This support is provided by a thin plastic carrier (illustrated in Fig. 12.18), which facilitates handling the gage.

Fig. 12.18 A typical strain gage with a grid pattern and thin film plastic carrier.

To employ a strain gage, we bond it to a mechanical member with an adhesive. The adhesive serves a vital function in the strain measuring system because it transmits the strain from the mechanical component to the gage-sensing element (the foil) without loss. The singularly unimpressive feat of adhesively bonding a strain gage to a mechanical component is one of the most critical steps in the entire process of measuring strain with a bonded resistance strain gage.

A strain gage acts as a variable resistor when subjected to a surface strain in the direction of its grid elements. The relation between the change in resistance ΔR with the strain ε is given by:

$$\Delta R/R_0 = (R_f - R_0)/R_0 = (GF)\,\varepsilon_g \quad (12.59)$$

where R_0 is the initial resistance of the gage, usually 120 or 350 Ω (ohms), R_f is the final resistance of the gage after the application of a strain ε, GF is the gage factor or calibration constant for the gage, usually about 2.0 and ε_g is the strain in the direction of the gage axis.

A strain gage is a sensor that converts strain into a change in the electrical resistance ΔR. Because strains, in structural elements fabricated from common engineering materials, are small quantities (of the order of 10^{-3}), a Wheatstone bridge circuit is employed to transform the change in resistance to a voltage that is proportional to the strain.

12.8.1 Determining Principal Stresses from Strain Measurements

In the analysis of certain structural elements, the directions of the principal stresses are known quantities. Consider axially loaded tension members or beams in bending where a uniaxial state of stress exists and $\sigma_{yy} = \tau_{xy} = 0$. In this case, the strain gage is oriented with its axis coincident with the x-axis of the structural element. Then the principal stress σ_1 is given by:

$$\sigma_1 = \sigma_{xx} = E\varepsilon_g \qquad \sigma_2 = \sigma_{yy} = 0 \quad (12.60)$$

where ε_g is the strain measured in the x direction along the axis of the structural member.

Next, consider a spherical pressure vessel where an isotropic state of stress exists where $\sigma_{xx} = \sigma_{yy} = \sigma_1 = \sigma_2$ and $\tau_{xy} = 0$. In this case the strain gage can be mounted with its axis in any direction because all directions are principal. The magnitude of the principal stresses is given by:

$$\sigma_{xx} = \sigma_{yy} = \sigma_1 = \sigma_2 = E\varepsilon_g/(1-\nu) \quad (12.61)$$

where ε_g is the strain measured in any direction in the isotropic stress field.

When less is known beforehand regarding the state of stress in the structural element, it is necessary to employ either two or three strain gages to establish the magnitude and directions of the principal stresses. If the structure has an axis of symmetry, the principal directions are known and the magnitude of the stresses can be determined from the output of two orthogonal strain gages that are mounted with their axes coincident with the axes of symmetry. In this case, the principal stresses are given by:

$$\sigma_1 = \frac{E}{1-\nu^2}(\varepsilon_1 + \nu\varepsilon_2)$$
$$\sigma_2 = \frac{E}{1-\nu^2}(\varepsilon_2 + \nu\varepsilon_1) \quad (12.62)$$

Special strain gage configurations are available for measurements when either two or three strain determinations are required to establish the strain field. When the principal stress directions are known the two-element rectangular rosette shown in Fig. 12 19a is usually employed in measuring the two principal strains ε_1 and ε_2. When the principal stress directions are not known, there are three unknown quantities σ_1, σ_2 and θ_p. In this instance, a three-element strain gage rosette, illustrated in Fig. 12.19b, is employed to completely determine the stress field.

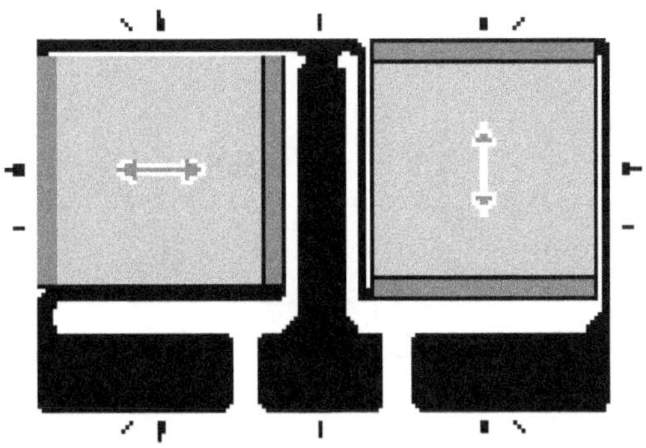

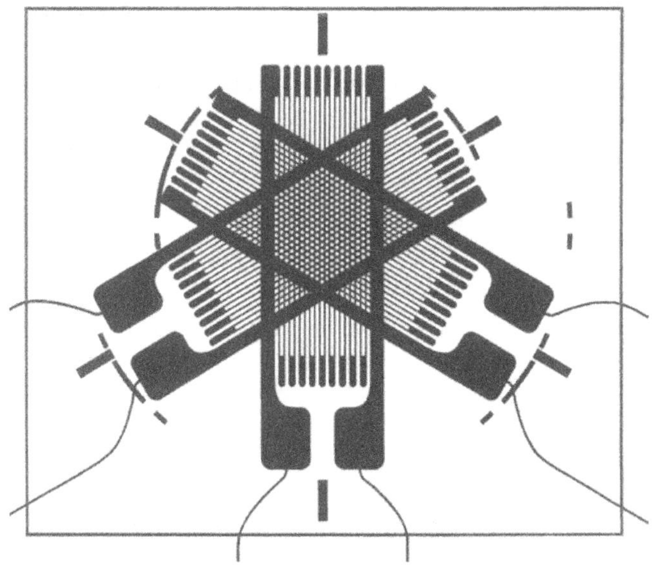

Fig. 12.19 Two and three-element strain gage rosettes. Courtesy of Vishay Intertechnolgy, Inc., Malvern, PA.

12.8.2 Strain Gage Rosettes

In the most general situation, the directions of the principal stresses are not known prior to conducting an experimental analysis. Three element rosettes that provide three independent strain measurements are necessary in determining the three unknown quantities σ_1, σ_2 and θ_p. To show that three strain measurements are sufficient, consider three strain gages aligned along axes A, B and C as shown in Fig. 12.20.

Fig. 12.20 Three strain gages placed at arbitrary angles relative to the x and y-axes.

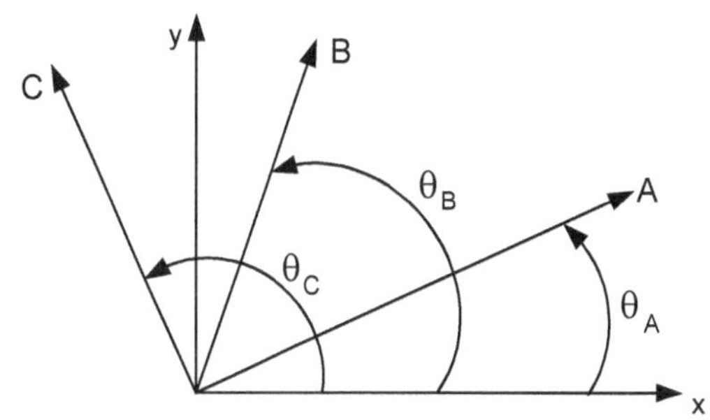

If we recall the strain equation of transformation given by Eq. (12.50), we can write:

$$\varepsilon_A = \varepsilon_{xx}\cos^2\theta_A + \varepsilon_{yy}\sin^2\theta_A + \gamma_{xy}\cos\theta_A\sin\theta_A$$

$$\varepsilon_B = \varepsilon_{xx}\cos^2\theta_B + \varepsilon_{yy}\sin^2\theta_B + \gamma_{xy}\cos\theta_B\sin\theta_B \qquad (12.63)$$

$$\varepsilon_C = \varepsilon_{xx}\cos^2\theta_C + \varepsilon_{yy}\sin^2\theta_C + \gamma_{xy}\cos\theta_C\sin\theta_C$$

In practice, three-element rosettes with θ_A, θ_B and θ_C fixed at specified values are employed to provide sufficient data to establish the stress field. One of the most popular rosettes is the three-element rectangular rosette that is described in the following subsection.

The Three-Element Rectangular Rosette

The three-element rectangular rosette utilizes gages placed at 0°, 45° and 90° positions as indicated in Fig. 12.21. With this selection of θ_A, θ_B and θ_C, Eqs. (12.63) reduce to:

$$\varepsilon_A = \varepsilon_{xx} \qquad \varepsilon_B = ½(\varepsilon_{xx} + \varepsilon_{yy} + \gamma_{xy}) \qquad \varepsilon_C = \varepsilon_{yy} \tag{12.64}$$

From Eq. (12.64), it is clear that:

$$\gamma_{xy} = 2\varepsilon_B - \varepsilon_A - \varepsilon_C \tag{12.65}$$

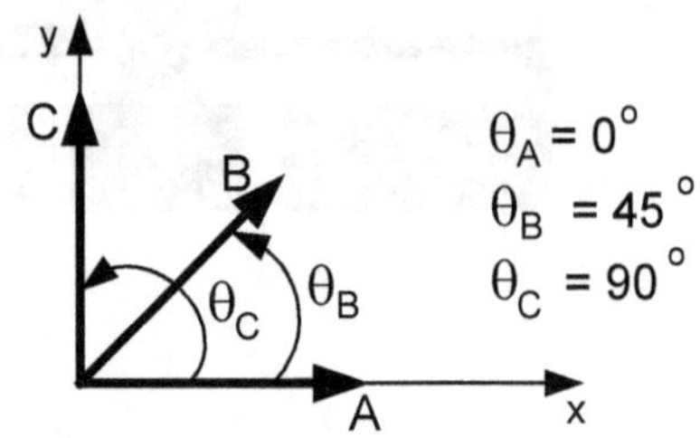

Fig. 12.21 Gage positions in a three-element rectangular rosette.

From Eqs. (12.64) and (12.65), we can determine the Cartesian components of strain from the measurements ε_A, ε_B and ε_C. Next, by employing Eq. (12.56), we can write the expression for the principal strains as:

$$\varepsilon_{1,2} = \frac{\varepsilon_A + \varepsilon_C}{2} \pm \sqrt{\left(\frac{\varepsilon_A - \varepsilon_C}{2}\right)^2 + \left(\frac{2\varepsilon_B - \varepsilon_A - \varepsilon_C}{2}\right)^2} \tag{12.66}$$

The principal angle is established from Eq. (12.58) as:

$$\tan 2\theta_p = \frac{(2\varepsilon_B - \varepsilon_A - \varepsilon_C)}{\varepsilon_A - \varepsilon_C} \tag{12.67}$$

Finally, the expression for the principal stresses is derived by substituting Eq. (12.66) into Eq. (12.6) and simplifying to obtain:

$$\sigma_{1,2} = E\left[\frac{\varepsilon_A + \varepsilon_C}{2(1-\nu)} \pm \frac{1}{2(1+\nu)}\sqrt{(\varepsilon_A - \varepsilon_C)^2 + (2\varepsilon_B - \varepsilon_A - \varepsilon_C)^2}\right] \tag{12.68}$$

By employing Eqs. (12.66) to (12.68) together with the results from the three-element rosette, we can determine the principal strains and stresses and the principal angle.

EXAMPLE 12.16

A two-element rectangular rosette is aligned with the principal stress directions. If the strain measurements are $\varepsilon_1 = 1920 \times 10^{-6}$ and $\varepsilon_2 = -780 \times 10^{-6}$, determine the principal stresses. The structural element upon which the strain gages are mounted is fabricated from 1045 HR steel.

Solution:

The principal stresses can be determined from the principal strains by using Eq. (12.62) as:

$$\sigma_1 = \frac{E}{1-\nu^2}(\varepsilon_1 + \nu\varepsilon_2) = \frac{207\times10^9}{1-(0.3)^2}[1920+(0.3)(-780)]\times10^{-6} = 383.5 \text{ MPa}$$

$$\sigma_2 = \frac{E}{1-\nu^2}(\varepsilon_2 + \nu\varepsilon_1) = \frac{207\times10^9}{1-(0.3)^2}[-780+(0.3)(1920)]\times10^{-6} = -46.40 \text{ MPa}$$

EXAMPLE 12.17

A three-element rectangular rosette is mounted on a steel structure. The strain readings from each gage element are: $\varepsilon_A = 1800 \times 10^{-6}$, $\varepsilon_B = 600 \times 10^{-6}$and $\varepsilon_C = -400 \times 10^{-6}$. Determine the principal strains, the principal stresses and the principal angle.

Solution:

The principal stresses can be determined by substituting the strain gage readings into Eq. (12.66) to obtain:

$$\varepsilon_{1,2} = \frac{\varepsilon_A + \varepsilon_C}{2} \pm \sqrt{\left(\frac{\varepsilon_A - \varepsilon_C}{2}\right)^2 + \left(\frac{2\varepsilon_B - \varepsilon_A - \varepsilon_C}{2}\right)^2}$$

$$\varepsilon_{1,2} = \left(\frac{1800-400}{2} \pm \sqrt{\left(\frac{1800+400}{2}\right)^2 + \left(\frac{2(600)-1800+400}{2}\right)^2}\right)\times10^{-6}$$

$$\varepsilon_{1,2} = [700 \pm 1105]\times10^{-6} \qquad \varepsilon_1 = 1805\times10^{-6} \qquad \varepsilon_2 = -405\times10^{-6}$$

To determine the principal stresses, we will substitute the strain gage readings into Eq. (12.68) to obtain:

$$\sigma_{1,2} = E\left[\frac{\varepsilon_A + \varepsilon_C}{2(1-\nu)} \pm \frac{1}{2(1+\nu)}\sqrt{(\varepsilon_A - \varepsilon_C)^2 + (2\varepsilon_B - \varepsilon_A - \varepsilon_C)^2}\right]$$

$$\sigma_{1,2} = 30\ \times10^6\left[\frac{1800-400}{2(1-0.3)} \pm \frac{1}{2(1+0.3)}\sqrt{(1800+400)^2 + (2(600)-1800+400)^2}\right]\times10^{-6}$$

$$\sigma_{1,2} = 30.00 \pm 25.49 \text{ ksi} \qquad \sigma_1 = 55.49 \text{ ksi} \qquad \sigma_2 = 4.51 \text{ ksi}$$

Finally, the principal angle is determined from Eq. (12.67) as:

$$\tan 2\theta_p = \frac{(2\varepsilon_B - \varepsilon_A - \varepsilon_C)}{\varepsilon_A - \varepsilon_C} = \frac{(2(600)-1800+400)}{1800+400} = -0.0909$$

$$2\theta_p = -5.19^\circ \qquad \theta_p = -2.60^\circ$$

The minus sign associated with θ_p indicates that the principal element is rotated clockwise relative to the Oxy coordinate system.

12.9 COMBINED LOADING

In most structures fabricated from beams, struts and columns, a single type of load such as an axial or transverse force or a bending moment produces the stresses. However, in the design of vehicles or machines, it is common to encounter two or more different types of loads acting on a machine component. For example, shafts that transmit power are subjected to torsion and transverse forces that produce internal bending moments and shear forces. When dealing with components subjected to two or three different types of loading, locating the point of maximum normal or shearing stresses is often difficult. In addition when stresses are due to two or more different types of load, determining the combined stress at a given point must be done with care. A guide to assist you in dealing with both of these difficulties is given below:

1. The bending stress is a maximum at a point located the maximum distance from the neutral axis of the beam.
2. The bending stress is a maximum at a point where the bending moment M is a maximum, which coincides with the location where the shear force V is zero.
3. The transverse and longitudinal shear stresses are equal and a maximum at a point located at the neutral axis of the beam where Q is also a maximum.
4. The transverse and longitudinal shear stresses are a maximum at a point where the shear force V is a maximum, which occurs at a point of application of a transverse load.
5. The shear stress due to torsion is a maximum at the point where the torsion T is a maximum.
6. The maximum shear stress on a shaft occurs on its outside surface.

Even with these guides, the location of the maximum stress is not apparent. In many analyses, it is necessary to evaluate the stresses at several points to discover the most critical location. When two or more normal stresses or shear stresses act on the same plane, they can be added together using the method of superposition providing the combined value does not exceed the yield strength of the material. However, when the stresses act on mutually orthogonal planes, appropriate stress equations of transformation must be used to determine the maximum principal and shear stresses. Let's consider three examples to demonstrate the procedure for performing this type of analysis.

EXAMPLE 12.18

A U shaped machine component, illustrated in Fig. E12.18, is subjected to opposing forces of 20 kN. If the component is fabricated from 1045 HR steel, determine the safety factor based on yield strength using the maximum normal stress theory. Comment on the appropriateness of this theory to predict the onset of yielding for this example.

Fig. E12.18

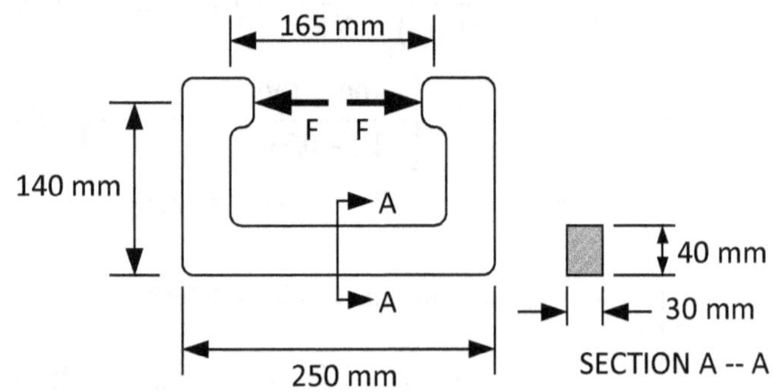

Solution:

Let's begin the analysis by constructing the FBD of the left side of the U shaped member as shown in Fig. E 12.18a.

Fig. E12.18a

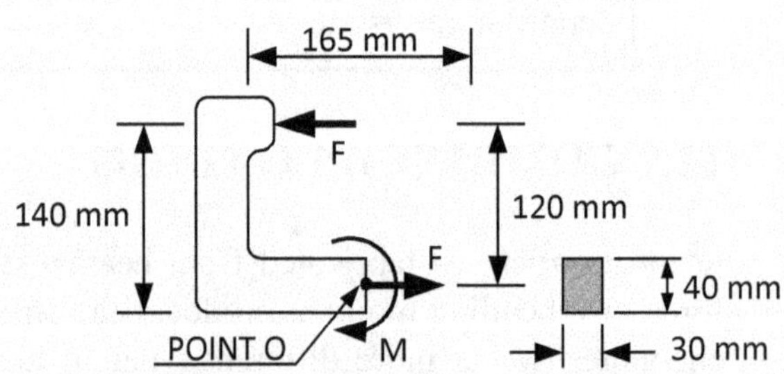

It is evident from the equilibrium relation $\Sigma F_x = 0$, that $P = F = 20$ kN. This internal force produces a uniformly distributed axial stress σ_{xx} that is given by:

$$\sigma_{xx} = P/A = (20 \times 10^3)/1200 = 16.67 \text{ MPa} \quad \text{(a)}$$

The equilibrium relation for the moments about the point O gives:

$$\Sigma M_o = 0 \qquad 120F - M = 0 \qquad M = (120)(20 \times 10^3) = 240 \times 10^4 \text{ N-mm} \quad \text{(b)}$$

The moment M produces a bending stress σ_{xx} that is linearly distributed over the cross section of the component at section A-A. The maximum and minimum bending stress is given by:

$$\sigma_{xx} = \pm Mc/I = \pm 6M/bh^2 \quad \text{(c)}$$

Substituting the results from Eq. (b) into Eq. (c) yields:

$$\sigma_{xx} = \pm\frac{(6)(240 \times 10^4)}{(30)(40)^2} = \pm 300 \text{ MPa} \quad \text{(d)}$$

The stresses due to P and M both act on the same plane and are in the x direction; hence, the results can be superimposed to give:

$$\sigma_{Max} = 16.67 + 300 = 316.7 \text{ MPa} \qquad \sigma_{Min} = 16.67 - 300 = -283.3 \text{ MPa} \quad \text{(e)}$$

The superposition process and the resulting distribution of stresses across the section A-A is presented in Fig. 12.18b.

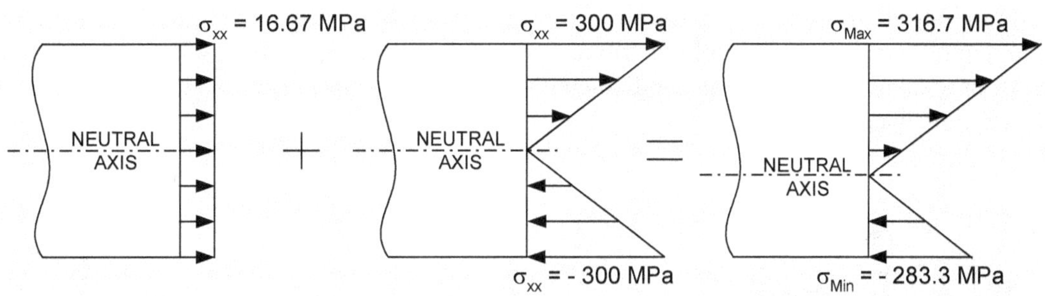

Fig. E12.18b

From Appendix B-2, we determine that S_y = 414 MPa for 1045 HR steel. The maximum normal stress theory for yielding indicates that the onset for yielding occurs when $\sigma_{Max} \geq S_y$. Accordingly the safety factor is determined from:

$$SF = S_y/\sigma_{Max} = 414/316.7 = 1.307$$

The maximum normal stress theory is appropriate in this case because $\sigma_2 = \sigma_3 = 0$. For a uniaxial state of stress, all three of the failure theories described in this chapter predict yielding to occur when $\sigma_{Max} \geq S_y$.

We note that the safety factor is relatively low and warn the design engineer of this fact.

EXAMPLE 12.19

The design engineer responsible for the U shaped machine component, described in Example 12.18, must drill a hole 6 mm in diameter through the member at Section A-A to permit passage of a signal wire. She knows that the hole should pass through the neutral axis to minimize its effects on the maximum stresses and the safety factor. She also is aware that the neutral axis has shifted and requests that you locate it.

Solution:

We locate the position of the neutral axis by using the similar triangles depicted in Fig. E12.19. Let's redraw the two triangles and define the location of the neutral axis relative to the bottom edge of the U shaped member as illustrated in Fig. E12.19.

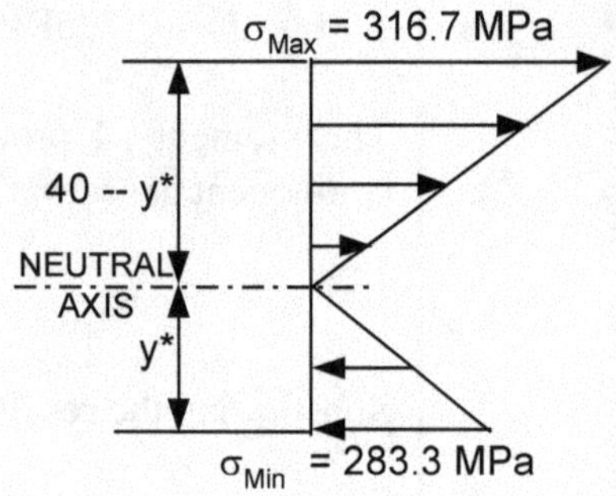

Fig. E12.19

From the proportionality of similar triangles, we write:

$$\frac{y^*}{283.3} = \frac{40 - y^*}{316.7} \qquad \text{(a)}$$

$$y^* = 18.89 \text{ mm}$$

You also indicate that drilling a 6 mm hole at the neutral axis will elevate the stresses slightly and reduce the safety factor.

EXAMPLE 12.20

A traffic sign along a city street is cantilevered from a pole as illustrated in Fig. E12.20. The pressure due to the wind impinging on the sign is uniformly distributed over its area and equal to 12 lb/ft^2 in the x direction. The pole is fabricated from a tube with a 10 in outside diameter and a 0.25 in. wall thickness. Determine the stresses at points A and B located at the base of the pole.

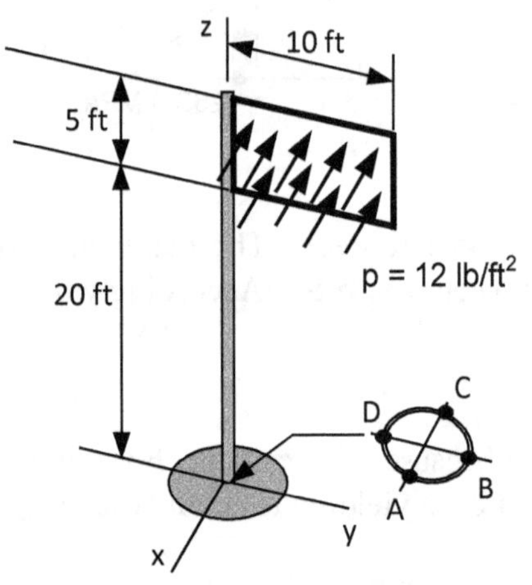

FBD of pole showing all reaction forces and moments.

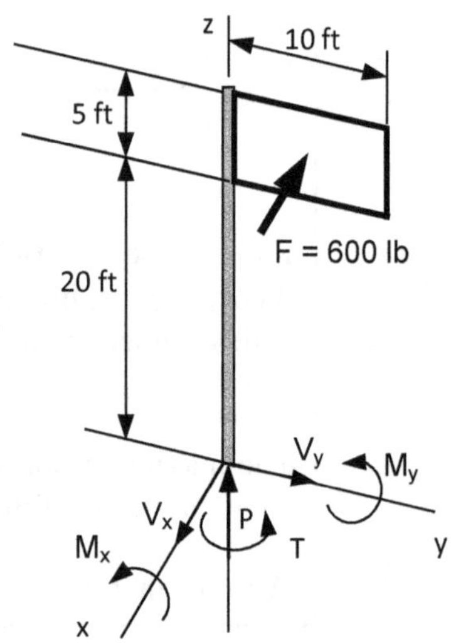

Force and moment directions on the ground.

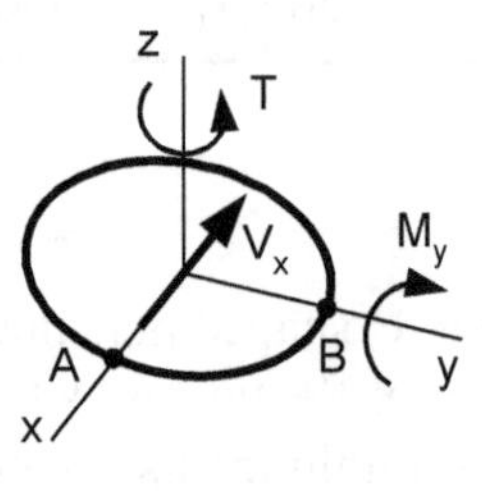

Fig. E12.20a

Solution:

The effective force due to the wind pressure acts at the center of the sign and is given by:

$$F_x = pA = (12)(5)(10) = 600 \text{ lb} \quad \text{(a)}$$

Next let's prepare a FBD showing the forces at the base of the pole as indicated in Fig. E12.20a. Then we write the six equations of equilibrium to obtain:

$$\Sigma F_x = 0 \qquad V_x = 600 \text{ lb} \quad \text{(b)}$$

$$\Sigma F_y = 0 \qquad V_y = 0 \quad \text{(c)}$$

$$\Sigma F_z = 0 \qquad P = 0 \quad \text{(d)}$$

$$\Sigma M_x = 0 \qquad M_x = 0 \quad \text{(e)}$$

$$\Sigma M_y = 0 \qquad M_y = (600)(22.5)= 13{,}500 \text{ ft-lb} \quad \text{(f)}$$

$$\Sigma M_z = 0 \qquad T = -(600)(5)= -3{,}000 \text{ ft-lb} \quad \text{(g)}$$

The geometric parameters that are needed to determine the stresses at the base of the pole are:

$$A = (\pi/4)(d_o^2 - d_i^2) = (\pi/4)[(10)^2 - (9.5)^2] = 7.658 \text{ in}^2. \quad \text{(h)}$$

$$I = I_{xx} = I_{yy} = (\pi/64)(d_o^4 - d_i^4) = (\pi/64)[(10)^4 - (9.5)^4] = 91.05 \text{ in.}^4 \quad \text{(i)}$$

$$J = 2I = 182.1 \text{ in.}^4 \quad \text{(j)}$$

$$r = c = 5.0 \text{ in.} \quad \text{(k)}$$

For circular shafts or tubes it can be shown that Q is given by:

$$Q = (2/3)/(r_o^3 - r_i^3) \quad \text{(m)}$$

$$Q = (2/3)/[(5.0)^3 - (4.75)^3] = 11.885 \text{ in.}^3 \quad \text{(n)}$$

Let's determine the stresses due to the three internal forces M_y, V_x and T and, in the process, use the geometric parameters listed above.

$$\sigma_{zz} = M_y c/I = (13{,}500)(12)(5)/91.05 = 8{,}896 \text{ psi} \quad \text{(o)}$$

$$\tau_T = Tr/J = (3{,}000)(12)(5)/182.1 = 988.4 \text{ psi} \quad \text{(p)}$$

$$\tau_{Vx} = (V_x Q)/(It) = [(600)(11.885)]/[(91.05)(2)(0.25)] = 156.6 \text{ psi} \quad \text{(q)}$$

Consider point A, which is on the tension side of the pole, and show the stresses on an elemental volume removed from the annular ring at the base of the pole.

$$(\sigma_{zz})_A = \sigma_{zz} = 8{,}896 \text{ psi} \quad \text{(r)}$$

$$(\tau_T)_A = \tau_T = 988.4 \text{ psi} \quad \text{(s)}$$

$$(\tau_{Vx})_A = 0 \quad \text{(t)}$$

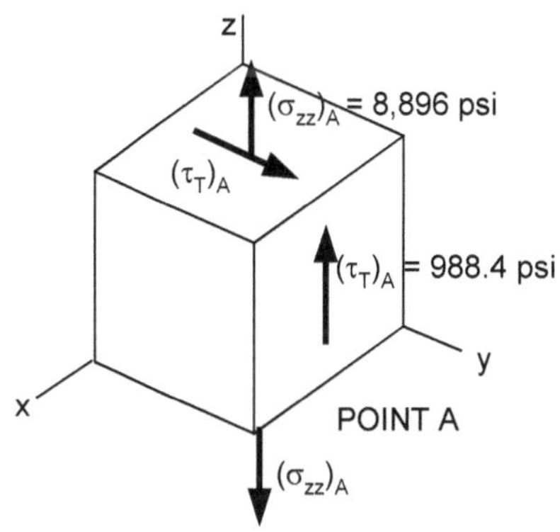

Fig. E12.20b

For point A, only the shear stress (τ_T) due to the torque appears and it acts in the + y direction. The shear stress (τ_{Vx}) due to the shear force V_x does not appear because of the location of point A. Recall from the analysis of beams that the shear stress due to transverse loadings was a maximum at the neutral axis and zero at the top and bottom edges of the beam. For the shear force V_X, point A is at the top edge of the cross section; hence, $\tau_{Vx})_A = 0$.

Finally consider point B, which is on the neutral axis of the pole and show the stresses on an elemental volume removed from the annular ring at the base of the pole.

$$(\sigma_{zz})_B = 0 \quad \text{(s)}$$

$$(\tau)_B = -\tau_T - \tau_{Vx} = -1{,}145 \text{ psi} \quad \text{(q)}$$

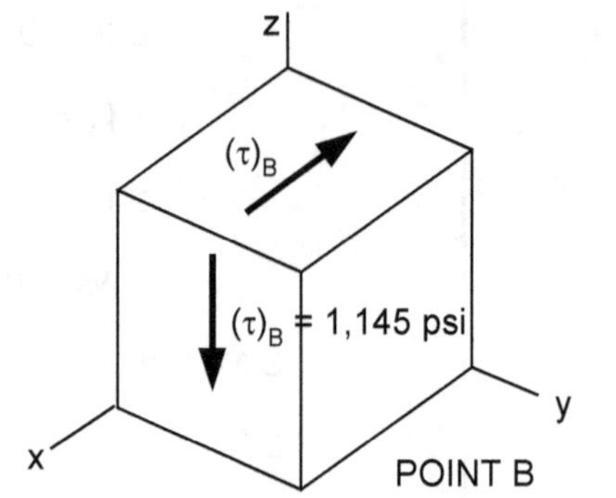

For point B, the normal stress component $(\sigma_{zz})_A$ vanishes because point B is located on the y-axis, which acts as the neutral axis for M_y. Also both shear stress components are present and algebraically sum because they both act in the same direction (– x axis).

EXAMPLE 12.21

A machine component is fabricated from a tube that has been formed into a 90° bend as illustrated in Fig. E12.21. The outside diameter of the tube is 60 mm and its inside diameter is 50 mm. A design engineer wants to place an access hole in the tube at a location 150 mm from its base and is concerned with the possibility of failure by doing so. He requests you to determine the maximum principal stress and the maximum shear stress at point A at this elevation on the tube. An exercise is included in the list of problems to determine the same quantities at point B.

Fig. E12.21

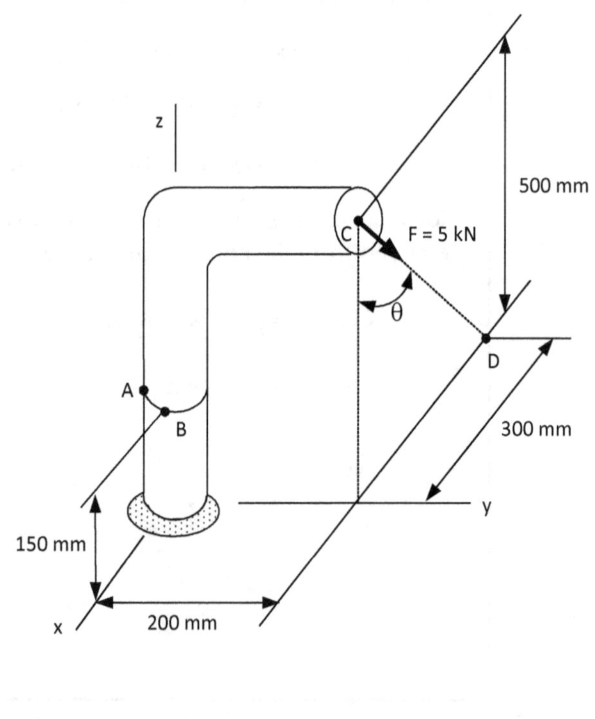

Solution:

Let's first determine the angle θ and resolve the force **F** into its components F_x and F_z.

$$\theta = \tan^{-1}(300/500) = 30.96° \quad \text{(a)}$$

$$F_x = F \sin\theta = 5 \sin(30.96°) = 2.572 \text{ kN} \quad \text{(b)}$$

$$F_z = F \cos\theta = 5 \cos(30.96°) = 4.287 \text{ kN} \quad \text{(c)}$$

Next prepare a FBD of the segment of the tube with a section cut at the 150 mm elevation as shown in Fig. E12.21a.

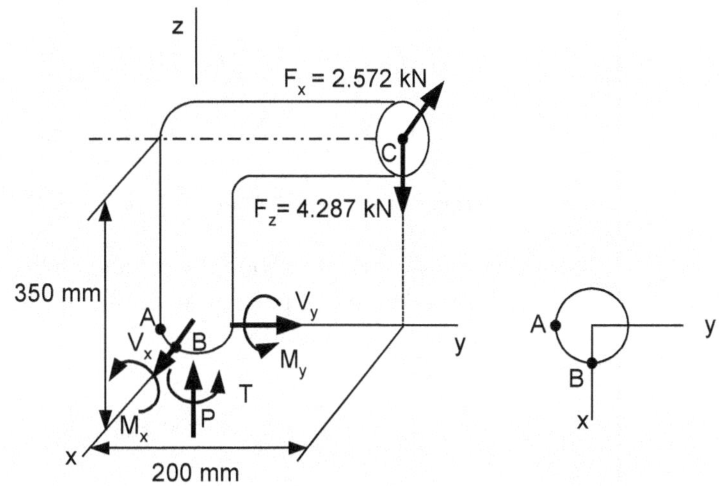

Fig. E12.21a

Let's use this FBD together with the equilibrium relations to determine the internal forces at the section cut.

$$\Sigma F_x = 0 \qquad V_x = 2.572 \text{ kN} \quad \text{(d)}$$

$$\Sigma F_y = 0 \qquad V_y = 0 \quad \text{(e)}$$

$$\Sigma F_z = 0 \qquad P = 4.287 \text{ kN} \quad \text{(f)}$$

$$\Sigma M_x = 0 \qquad M_x = (4.287 \times 10^3)(0.200) = 857.4 \text{ N-m} \quad \text{(g)}$$

$$\Sigma M_y = 0 \qquad M_y = (2.572 \times 10^3)(0.350) = 900.2 \text{ N-m} \quad \text{(h)}$$

$$\Sigma M_z = 0 \qquad T = -(2.572 \times 10^3)(0.200) = -514.4 \text{ N-m} \quad \text{(i)}$$

The geometric parameters that are needed to determine the stresses at points A and B are:

$$A = (\pi/4)(d_o^2 - d_i^2) = (\pi/4)[(60)^2 - (50)^2] = 863.9 \text{ mm}^2. \quad \text{(j)}$$

$$I = I_{xx} = I_{yy} = (\pi/64)(d_o^4 - d_i^4) = (\pi/64)[(60)^4 - (50)^4] = 329.4 \times 10^3 \text{ mm}^4 \quad \text{(k)}$$

$$J = 2I = 658.8 \times 10^3 \text{ mm}^4 \quad \text{(l)}$$

$$r = c = 30 \text{ mm}. \quad \text{(m)}$$

$$Q = (2/3)(r_o^3 - ri^3) = (2/3)[(30)^3 - (25)^3] = 7{,}583 \text{ mm}^3 \quad \text{(n)}$$

Let's determine the stresses at point A due to the five internal forces P, M_x, M_y, V_x and T. In the process, we will use the geometric parameters listed above.

$$\sigma_{zz} = P/A = (4{,}287)/863.9 = 4.962 \text{ MPa} \qquad \text{(compressive)} \tag{o}$$

$$\sigma_{zz} = M_x c/I = (857.4 \times 10^3)(30)/(329.4 \times 10^3) = 78.09 \text{ MPa} \tag{p}$$

$$\sigma_{zz} = M_y y/I = 0 \tag{q}$$

$$\sigma_{xx} = \sigma_{yy} = 0 \tag{r}$$

$$(\sigma_{zz})_{Total} = 78.09 - 4.962 = 73.13 \text{ MPa} \tag{s}$$

$$\tau_T = Tr/J = (514.4 \times 10^3)(30)/(658.8 \times 10^3) = 23.42 \text{ MPa} \tag{t}$$

$$\tau_{Vx} = (V_x Q)/(It) = [(2.572 \times 10^3)(7{,}583)]/[(329.4 \times 10^3)(10)] = 5.921 \text{ MPa} \tag{u}$$

$$(\tau)_{Total} = \tau_T - \tau_{Vx} = 23.42 - 5.921 = 17.50 \text{ MPa} \tag{v}$$

Before proceeding to determine the principal stresses and the maximum shear stress, let's draw an element removed from the tube at point A as shown in Fig. E12.21b. Let's also examine a two dimensional element representing the x-z plane at point A.

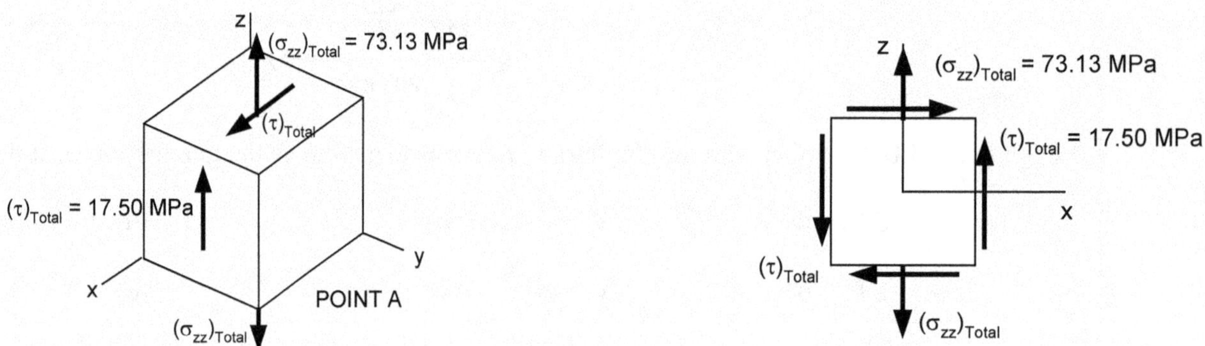

Fig. E12.21b

The principal stresses are given by:

$$\sigma_{Max,\,Min} = \frac{\sigma_{xx} + \sigma_{zz}}{2} \pm \sqrt{\left(\left(\frac{\sigma_{xx} - \sigma_{zz}}{2}\right)^2 + \tau_{xz}^2\right)} \tag{w}$$

$$\sigma_{Max,\,Min} = \frac{0 + 73.13}{2} \pm \sqrt{\left(\left(\frac{0 - 73.13}{2}\right)^2 + (17.50)^2\right)} = 36.56 \pm 40.54 \text{ MPa} \tag{x}$$

$$\sigma_{Max} = 77.10 \text{ MPa} \qquad \sigma_{Min} = -3.980 \text{ MPa} \tag{y}$$

The maximum shear stress is given by:

$$\tau_{Max} = \sqrt{\left(\left(\frac{\sigma_{xx} - \sigma_{zz}}{2}\right)^2 + \tau_{xz}^2\right)} = \sqrt{\left(\left(\frac{0 - 73.13}{2}\right)^2 + (17.50)^2\right)} = 40.54 \text{ MPa} \tag{z}$$

12.10 SUMMARY

In earlier chapters, we considered only uniaxial states of stress that occurred in tension members and beams. In this chapter multi-dimensional states of stress were introduced that give rise to nine Cartesian components of stress. However, we used equilibrium conditions to show a shear stress relation that reduced the number of unique Cartesian components of stress to six.

$$\tau_{xy} = \tau_{yx} \qquad \tau_{yz} = \tau_{zy} \qquad \tau_{xz} = \tau_{zx} \tag{12.1}$$

For the three-dimensional state of stress, Hooke's law must be generalized to give the stress-strain relations presented in Eq. (12.3).

$$\begin{aligned}
\sigma_{xx} &= \frac{E}{(1+\nu)(1-2\nu)}\left[(1-\nu)\varepsilon_{xx} + \nu(\varepsilon_{yy}+\varepsilon_{zz})\right] \\
\sigma_{yy} &= \frac{E}{(1+\nu)(1-2\nu)}\left[(1-\nu)\varepsilon_{yy} + \nu(\varepsilon_{zz}+\varepsilon_{xx})\right] \\
\sigma_{zz} &= \frac{E}{(1+\nu)(1-2\nu)}\left[(1-\nu)\varepsilon_{zz} + \nu(\varepsilon_{xx}+\varepsilon_{yy})\right] \\
\tau_{xy} &= G\gamma_{xy} \qquad \tau_{yz} = G\gamma_{yz} \qquad \tau_{xz} = G\gamma_{xz}
\end{aligned} \tag{12.3}$$

The geometric and loading conditions for plane stress were introduced and the stress strain relations for plane stress were written as:

$$\begin{aligned}
\sigma_{xx} &= \frac{E}{(1-\nu^2)}\left(\varepsilon_{xx} + \nu\varepsilon_{yy}\right) \\
\sigma_{yy} &= \frac{E}{(1-\nu^2)}\left(\varepsilon_{yy} + \nu\varepsilon_{xx}\right) \\
\tau_{xy} &= G\gamma_{xy}
\end{aligned} \tag{12.6}$$

The equations of stress transformation were derived to provide expressions for $\sigma_{x'x'}$, $\sigma_{y'y'}$ and $\tau_{x'y'}$ in terms of σ_{xx}, σ_{yy} and τ_{xy}.

$$\begin{aligned}
\sigma_{x'x'} &= \frac{\sigma_{xx}+\sigma_{yy}}{2} + \frac{\sigma_{xx}-\sigma_{yy}}{2}\cos 2\theta + \tau_{xy}\sin 2\theta \\
\sigma_{y'y'} &= \frac{\sigma_{xx}+\sigma_{yy}}{2} - \frac{\sigma_{xx}-\sigma_{yy}}{2}\cos 2\theta - \tau_{xy}\sin 2\theta \\
\tau_{x'y'} &= -\frac{\sigma_{xx}-\sigma_{yy}}{2}\sin 2\theta + \tau_{xy}\cos 2\theta
\end{aligned} \tag{12.11}$$

After establishing the stress equations of transformation, we showed that the principal stresses were given by:

$$\begin{aligned}
\sigma_{Max,\,Min} &= \frac{\sigma_{xx}+\sigma_{yy}}{2} + \frac{\sigma_{xx}-\sigma_{yy}}{2}\cos 2\theta_p + \tau_{xy}\sin 2\theta_p \\
\sigma_{Max,\,Min} &= \frac{\sigma_{xx}+\sigma_{yy}}{2} \pm \sqrt{\left(\left(\frac{\sigma_{xx}-\sigma_{yy}}{2}\right)^2 + \tau_{xy}^2\right)}
\end{aligned} \tag{12.15}$$

Also, we showed that the principal planes were located at θ_p:

$$\tan 2\theta_p = \frac{2\tau_{xy}}{\sigma_{xx} - \sigma_{yy}} \quad (12.13)$$

In a similar manner, we derived the relation for the maximum shear stress and the location of the plane θ_s upon which this stress acts as:

$$\tau_{Max} = \sqrt{\left(\left(\frac{\sigma_{xx} - \sigma_{yy}}{2}\right)^2 + \tau_{xy}^2\right)} \quad (12.18)$$

$$\tan 2\theta_s = -\frac{\sigma_{xx} - \sigma_{yy}}{2\tau_{xy}} \quad (12.16)$$

A graphical technique, Mohr's circle, was described for visualizing how the Cartesian components of stress vary with the angle of the plane upon which these stresses act. Examples were presented demonstrating the use of Mohr's circle for determining principal stresses, principal planes and maximum shearing stresses.

Three different failure theories were introduced including:

1. The maximum principal stress theory.
2. The maximum shear stress theory.
3. The maximum distortional energy theory.

The maximum principal stress theory indicates that yielding occurs when:

$$|\sigma_1| \geq S_y \qquad |\sigma_2| \geq S_y \quad (12.29)$$

The maximum shear stress theory indicates yielding occurs when:

$$|\sigma_1 - \sigma_2| \geq S_y \quad \text{Same sign stresses} \quad (12.32)$$
$$|\sigma_1| \geq S_y \quad \text{Opposite sign stresses} \quad (12.34)$$

The maximum distortional energy theory indicates yielding occurs when:

$$(\sigma_1/S_y)^2 - (\sigma_1/S_y)(\sigma_2/S_y) + (\sigma_2/S_y)^2 \geq 1 \quad (12.37)$$

The maximum distortional energy theory is widely accepted as the most accurate method for determining the onset of yielding in ductile materials commonly used to fabricate structural element and machine components.

Equations for the stresses in thin-walled pressure vessels were derived. For spherical pressure vessels the stress state is isotropic with $\sigma_1 = \sigma_2$.

$$\sigma_1 = \sigma_2 = \sigma_{my} = \sigma_{mx} = pr/2t \quad (12.41)$$

For cylindrical pressure vessels, we showed that the hoop stress σ_h and axial stress σ_a are given by:

$$\sigma_h = pr/t \quad (12.42)$$

$$\sigma_a = pr/2t \qquad (12.43)$$

The strain equations of transformation were also derived. While the derivation was based only on geometric considerations the results obtained were very similar to the equations of transformation for stresses. You will note the similarity between Eqs. (12.53) and Eqs. (12.10).

$$\begin{aligned}
\varepsilon_{x'x'} &= \frac{\varepsilon_{xx}+\varepsilon_{yy}}{2} + \frac{\varepsilon_{xx}-\varepsilon_{yy}}{2}\cos 2\theta + \frac{\gamma_{xy}}{2}\sin 2\theta \\
\varepsilon_{y'y'} &= \frac{\varepsilon_{xx}+\varepsilon_{yy}}{2} - \frac{\varepsilon_{xx}-\varepsilon_{yy}}{2}\cos 2\theta - \frac{\gamma_{xy}}{2}\sin 2\theta \qquad (12.53) \\
\frac{\gamma_{x'y'}}{2} &= -\frac{\varepsilon_{xx}-\varepsilon_{yy}}{2}\sin 2\theta + \frac{\gamma_{xy}}{2}\cos 2\theta
\end{aligned}$$

Principal strains and the planes upon which principal strains act were also described and equations provided for their determination.

Strain gages were introduced as a method for experimentally determining stresses from gage readings. A strain gage acts as a variable resistor and undergoes a change in resistance when subjected to strain. This change in resistance is related to strain by:

$$\Delta R/R_0 = (R_f - R_0)/R_0 = (GF)\,\varepsilon_g \qquad (12.59)$$

The number of strain gages required for determining the stress at a point depends on the number of unknown quantities. For uniaxial states of stress, a single gage is sufficient. For biaxial states of stress, two gages are sufficient if the principal directions are known. However, for the biaxial state of stress, three gages are required when the directions of the principal stresses are not known. The three strain measurements are made with a three-element strain gage rosette with three strain gages oriented at angles θ_A, θ_B and θ_C. The readings from the gages are related to the Cartesian components of strain by:

$$\begin{aligned}
\varepsilon_A &= \varepsilon_{xx}\cos^2\theta_A + \varepsilon_{yy}\sin^2\theta_A + \gamma_{xy}\cos\theta_A \sin\theta_A \\
\varepsilon_B &= \varepsilon_{xx}\cos^2\theta_B + \varepsilon_{yy}\sin^2\theta_B + \gamma_{xy}\cos\theta_B \sin\theta_B \qquad (12.63) \\
\varepsilon_C &= \varepsilon_{xx}\cos^2\theta_C + \varepsilon_{yy}\sin^2\theta_C + \gamma_{xy}\cos\theta_C \sin\theta_C
\end{aligned}$$

A three-element strain gage rosette is made with gages oriented with $\theta = 0°$, $\theta = 45°$ and $\theta = 90°$. Strain measurements from these three gages give the principal stresses and the principal planes as:

$$\sigma_{1,2} = E\left[\frac{\varepsilon_A+\varepsilon_C}{2(1-\nu)} \pm \frac{1}{2(1+\nu)}\sqrt{(\varepsilon_A-\varepsilon_C)^2 + (2\varepsilon_B-\varepsilon_A-\varepsilon_C)^2}\right] \qquad (12.68)$$

$$\tan 2\theta_p = \frac{(2\varepsilon_B-\varepsilon_A-\varepsilon_C)}{\varepsilon_A-\varepsilon_C} \qquad (12.67)$$

Finally, a section on stresses due to combined loading of machine components was presented. While no new equations were derived in this section, several examples were included to demonstrate the method of superposition of stresses and combining shear and normal stresses.

PROBLEMS

12.1 For the cantilever beams shown in the figure to the right, write the equation for the elastic curves. Also determine the deflection and slope at the free end.

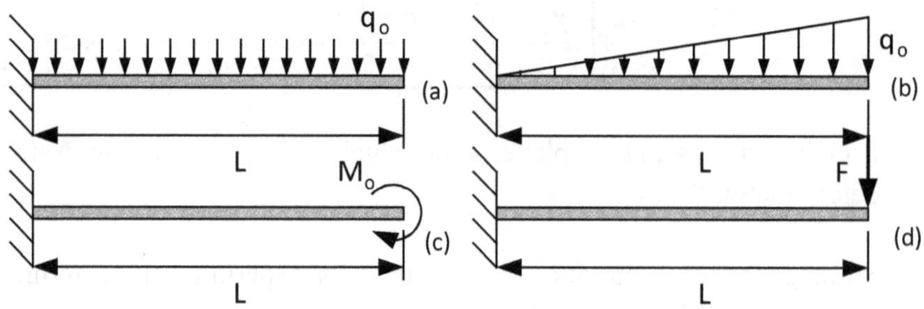

12.2 For the steel cantilever beam, shown in the figure to the left, determine the deflection and slope at the free end if the beam is fabricated from a W305 × 74 wide flange section.

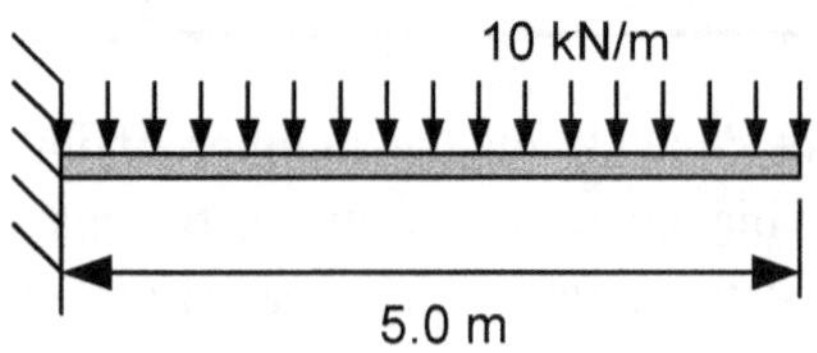

12.3 For the steel cantilever beam, shown in the figure to the right, determine the deflection and slope at the free end if the beam is fabricated from an American Standard section S254 × 38.

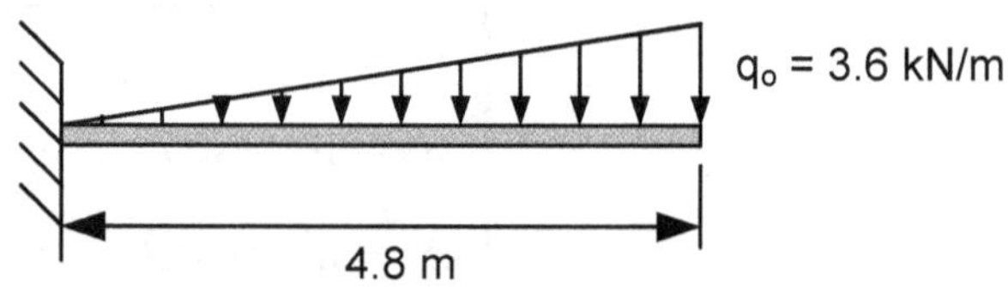

12.4 For the steel cantilever beam, shown in the figure to the right, determine the deflection and slope at the free end if the beam is fabricated from a W18 × 60 wide flange section.

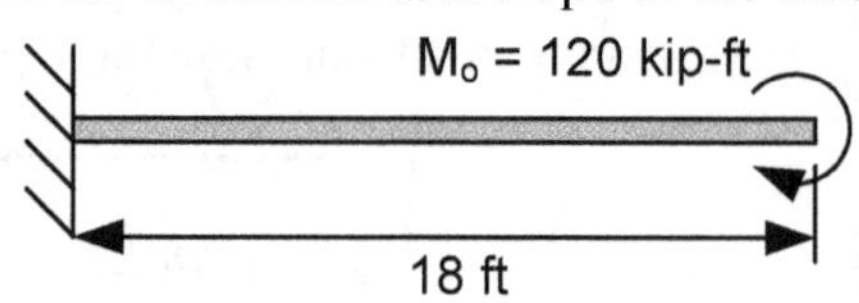

12.5 For the steel cantilever beam, shown in the figure to the left, determine the deflection and slope at the free end if the beam is fabricated from an American Standard section S381 × 64.

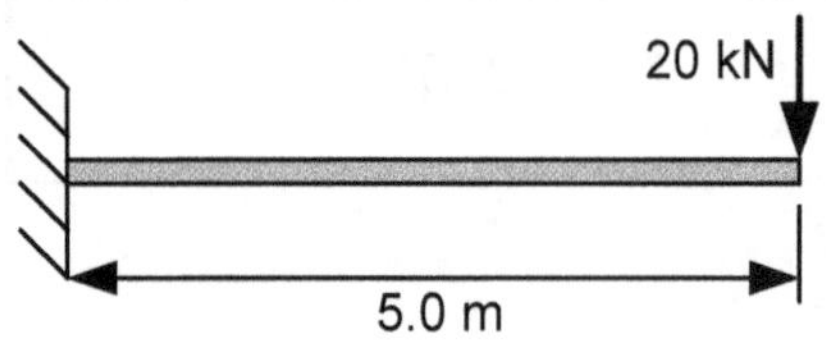

12.6 For the simply supported beams, shown in the figure below, write the equation for the elastic curves. Also determine the deflection at mid-span and the slope at both ends.

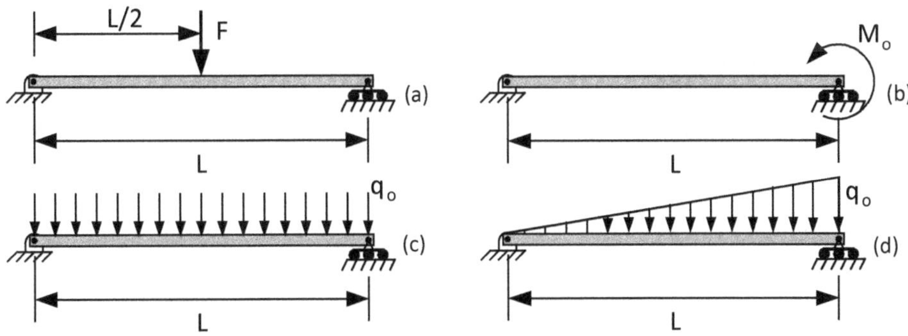

12.7 For the steel simply supported beam shown in the figure to the right, determine the mid-span deflection and slope at both ends if the beam is fabricated from a S203 × 27 American Standard section.

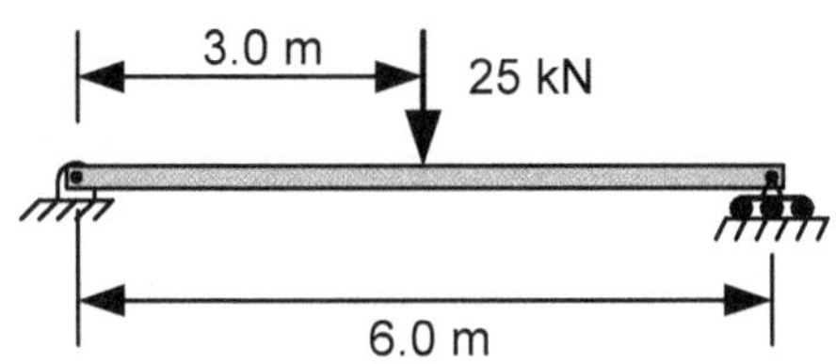

12.8 For the steel simply supported beam, shown in the figure to the left, determine the mid-span deflection and slope at both ends if the beam is fabricated from a W152 × 37 wide flange section.

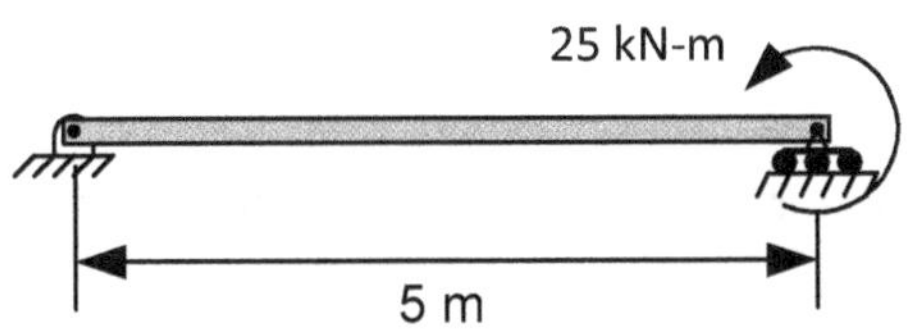

12.9 For the steel simply supported beam, shown in the figure to the right, determine the mid-span deflection and slope at both ends if the beam is fabricated from a W10 × 60 wide flange section.

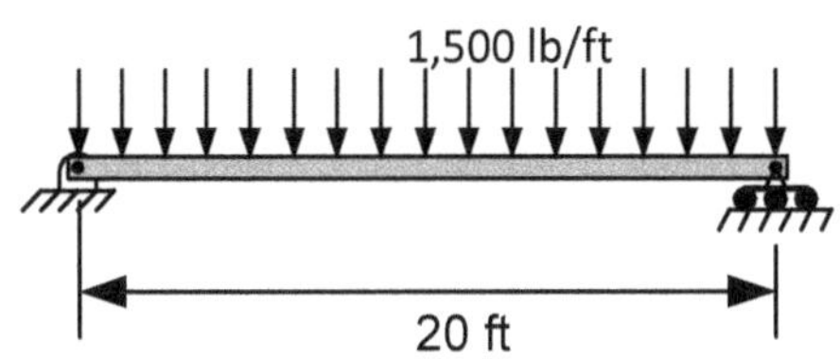

12.10 For the steel simply supported beam, shown in the figure to the left, determine the mid-span deflection and slope at both ends if the beam is fabricated from a S254 × 38 American Standard section.

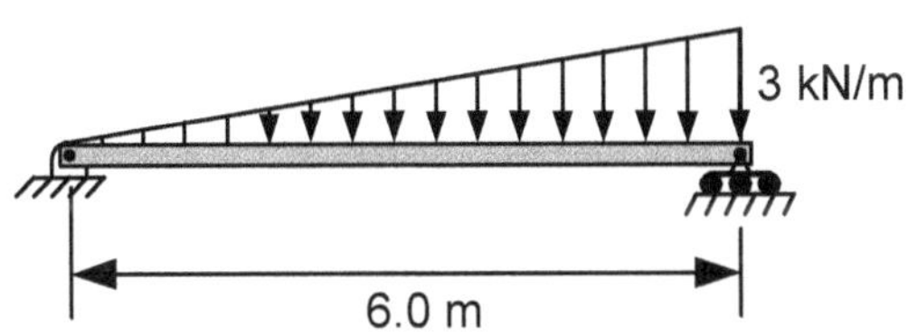

12.11 For the simply supported beam, shown in the figure to the right, determine the equation of the elastic curve and the slope at both ends.

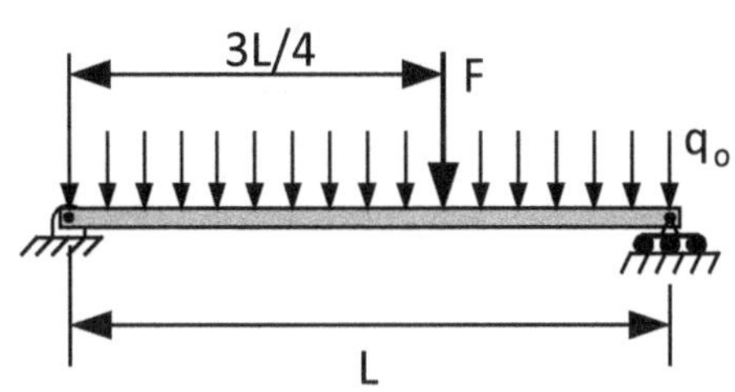

12.12 For the steel simply supported beam, shown in the figure to the left, determine the mid-span deflection and the slope at both ends if the beam is fabricated from a W10 × 60 wide flange section.

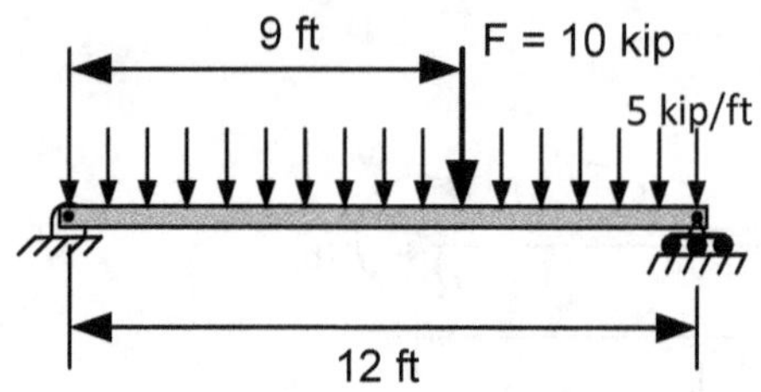

12.13 For the simply supported beam, shown in the figure to the right, determine the equation of the elastic curve and the slope at both ends.

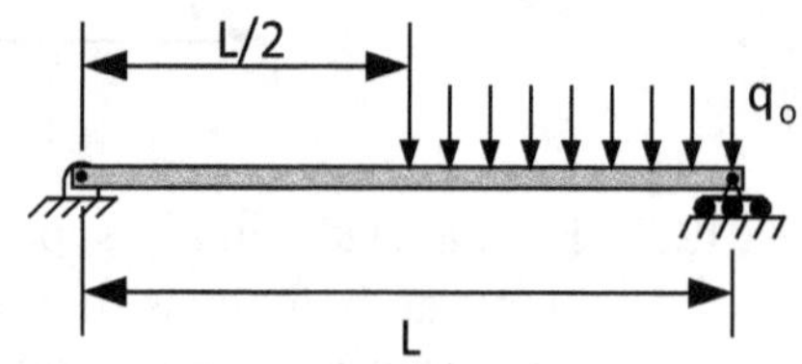

12.14 For the steel simply supported beam, shown in the figure to the left, determine the mid-span deflection and the slope at both ends if the beam is fabricated from a S18 × 70 American Standard section.

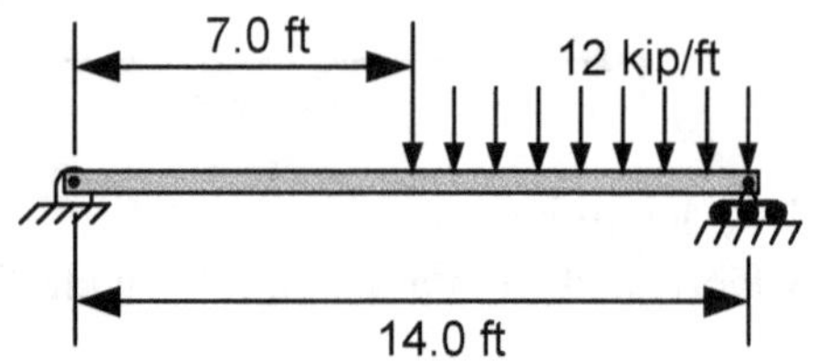

12.15 For the simply supported beam, shown in the figure to the right, determine the equation of the elastic curve and the slope at both ends.

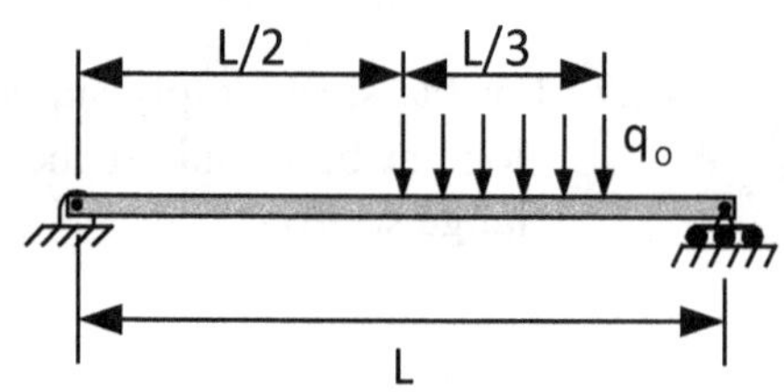

12.16 For the aluminum simply supported beam, shown in the figure to the left, determine the mid-span deflection and the slope at both ends if the beam is fabricated from a W457 × 89 wide flange section.

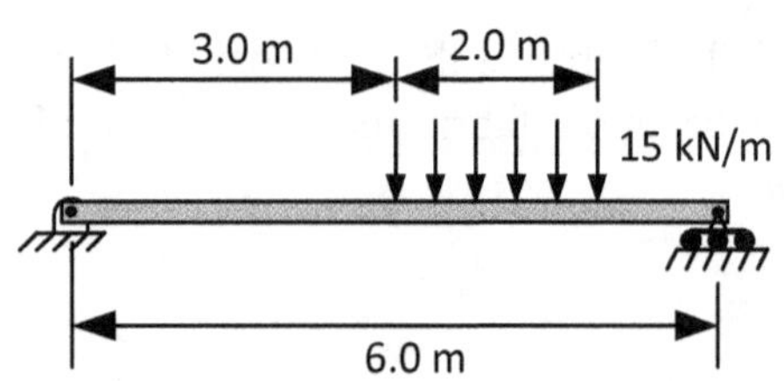

12.17 For the simply supported beam, shown in the figure to the right, determine the equation of the elastic curve and the slope at both ends.

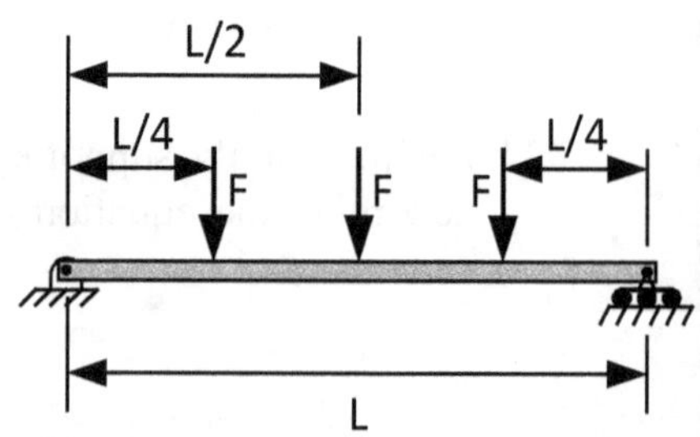

12.18 For the steel simply supported beam, shown in the figure to the left, determine the mid-span deflection and the slope at both ends if the beam is fabricated from a W203 × 46 wide flange section.

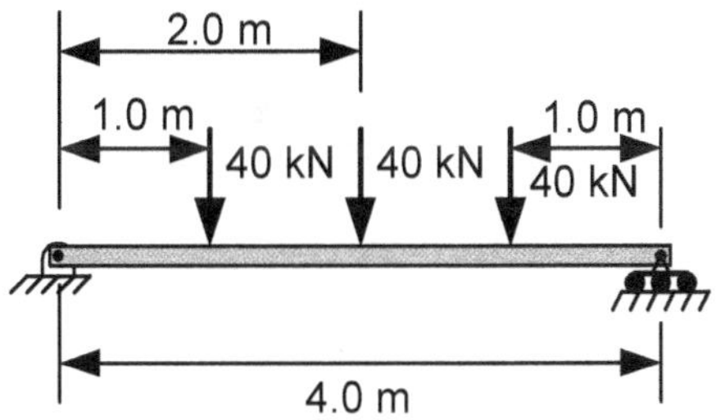

12.19 For the simply supported beam, shown in the figure to the right, determine the equation of the elastic curve and the slope at both ends.

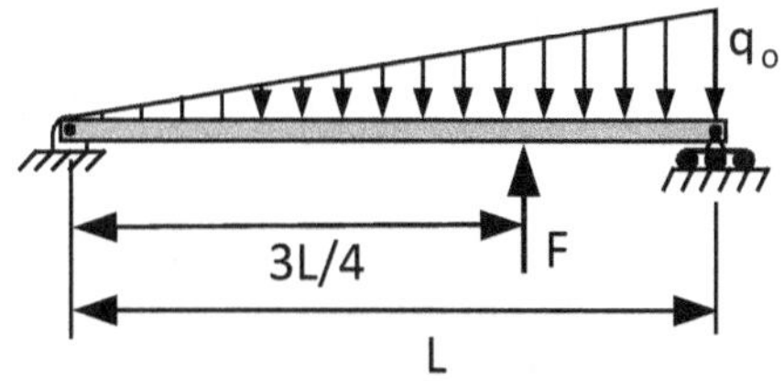

12.20 For the steel simply supported beam, shown in the figure to the left, determine the mid-span deflection and the slope at both ends if the beam is fabricated from a W533 × 92 wide flange section.

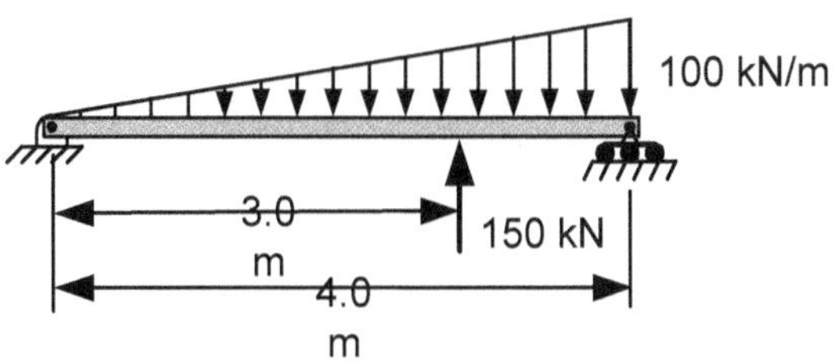

12.21 For the cantilever beam, shown in the figure to the right, write the equation for the elastic curve. Also determine the equations for deflection and slope at the free end.

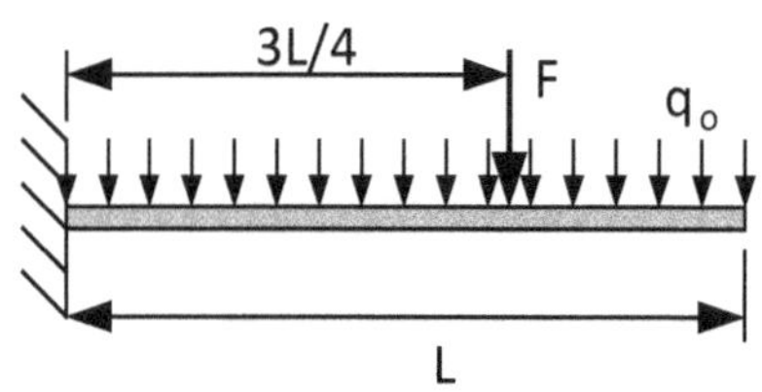

12.22 For the steel cantilever beam, shown in the figure to the right, determine the deflection and the slope at the free end if the beam is fabricated from a S305 × 74 American Standard section.

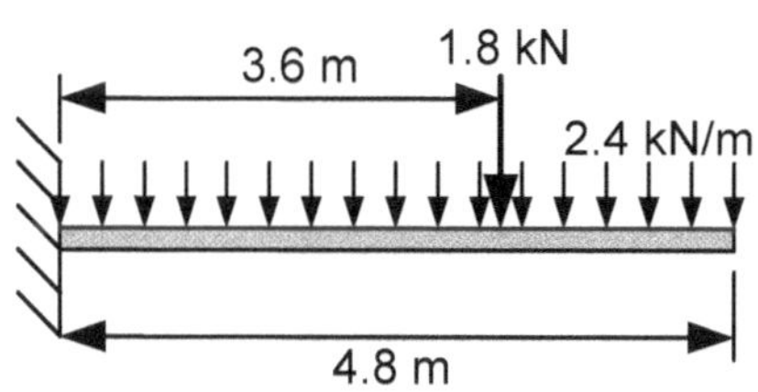

12.23 For the cantilever beam, shown in the figure to the left, write the equation for the elastic curve. Also determine the equations for the deflection and slope at the free end.

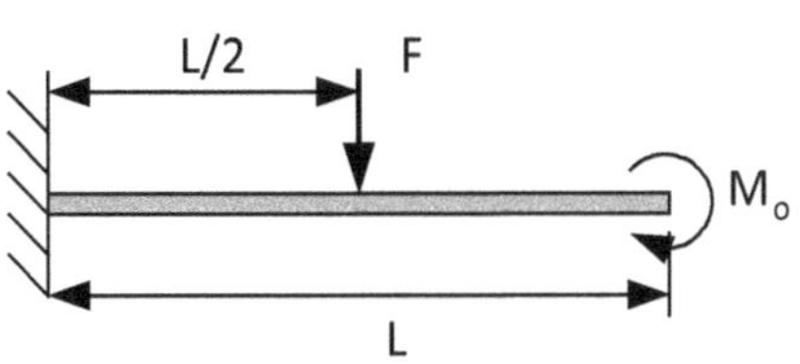

12.24 For the steel cantilever beam, shown in the figure to the right, determine the minimum weight wide flange section if the deflection at the free end is limited to 1.4 in.

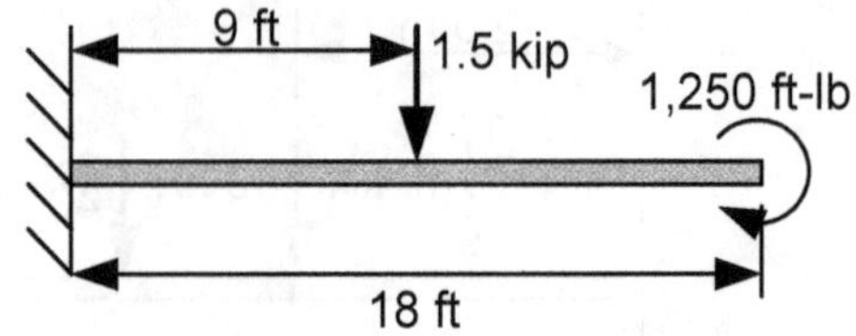

12.25 For the cantilever beam, shown in the figure to the left, write the equation for the elastic curve. Also determine the equations for the deflection and slope at the free end.

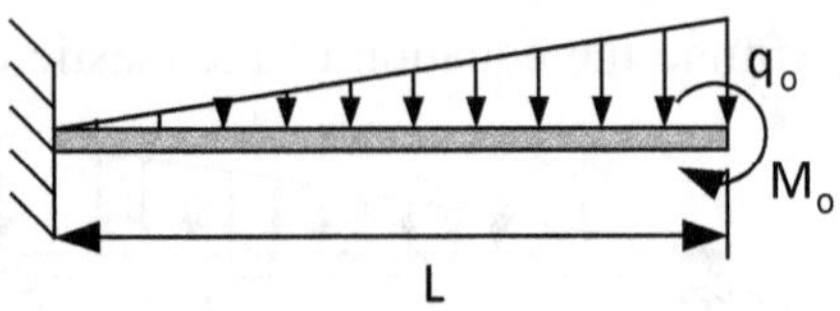

12.26 For the steel cantilever beam, shown in the figure to the right, determine the minimum weight American Standard section if the deflection at the free end is limited to 26 mm.

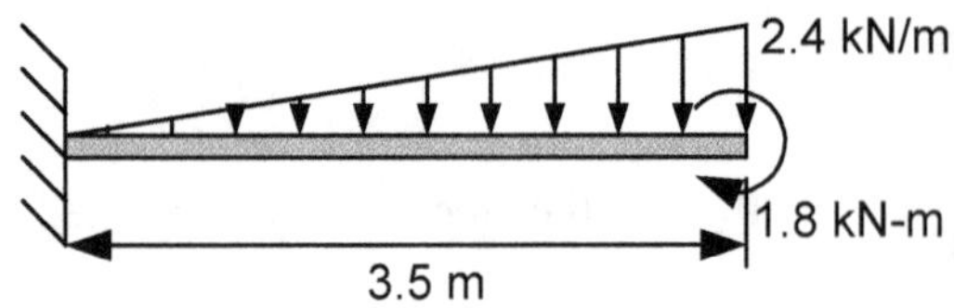

12.27 For the cantilever beam, shown in the figure to the left, write the equation for the elastic curve. Also determine the deflection and slope at the free end.

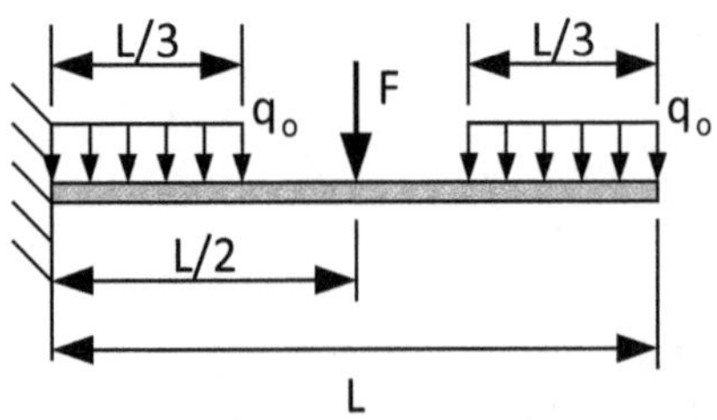

12.28 For the steel cantilever beam, shown in the figure to the right, determine the minimum weight structural tee section if the deflection at the free end is limited to 1.1 in.

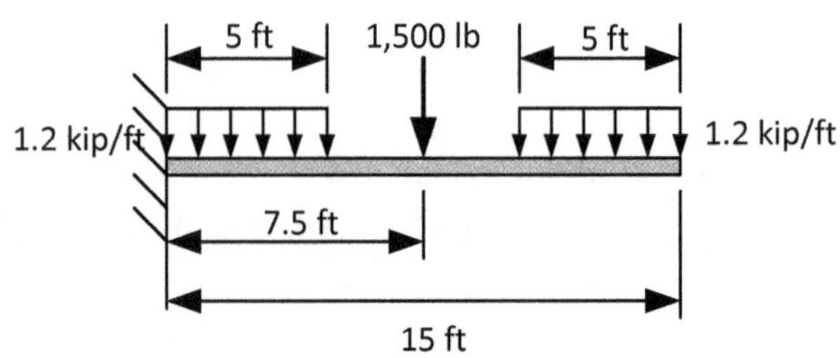

12.29 For the cantilever beam, shown in the figure to the left, write the equation for the elastic curve. Also determine the deflection and slope at the free end.

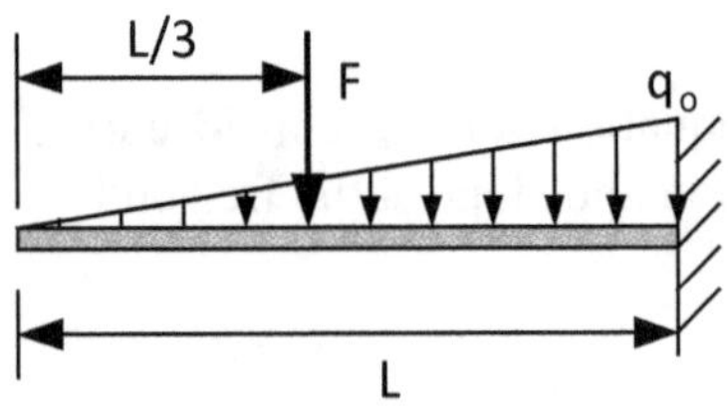

12.30 For the steel cantilever beam, shown in the figure to the right, determine the minimum weight wide flange section if the deflection at the free end is limited to 15 mm.

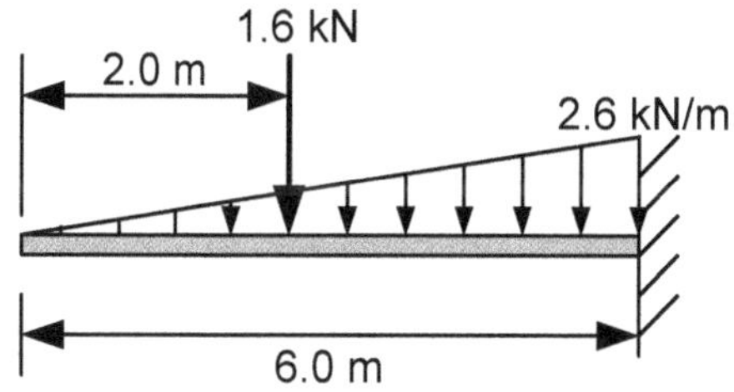

12.31 Using the method of superposition, for the simply supported beam shown in the figure to the left, determine the deflection at mid-span and the slope at the left end of the beam. The beam is steel with a W12 × 65 wide flange section.

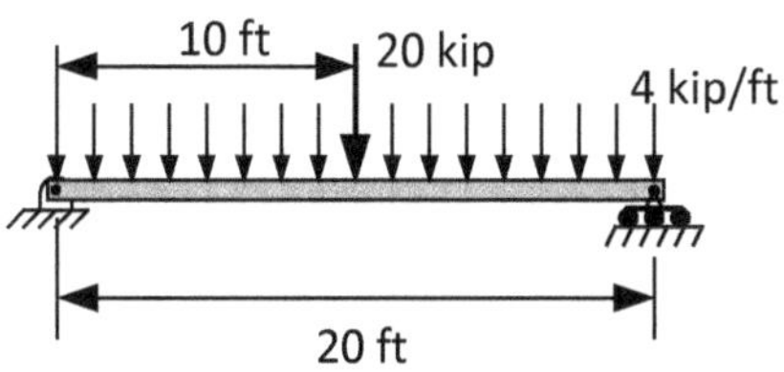

12.32 Using the method of superposition, for the simply supported beam shown in the figure to the right, determine the deflection at mid-span and the slope at the left end of the beam.

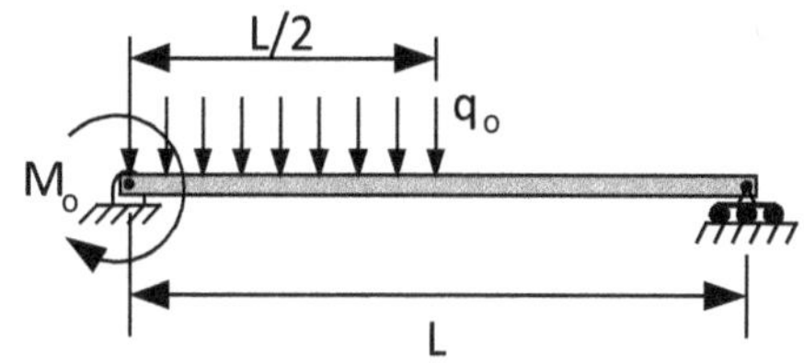

12.33 Using the method of superposition, for the simply supported beam shown in the figure to the left, determine the deflection at mid-span and the slope at the right end of the beam. The beam is steel with a W305 × 97 wide flange section.

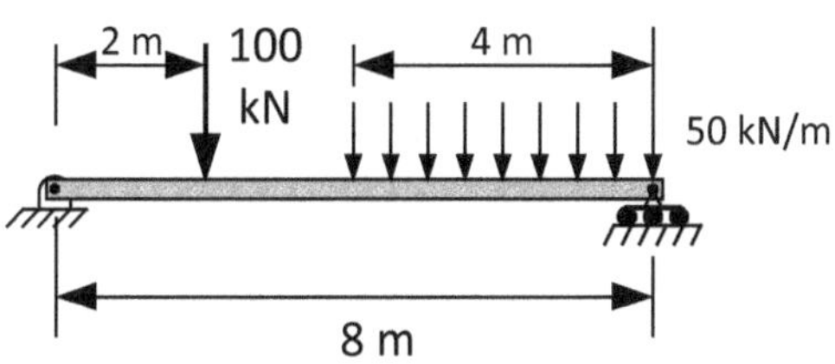

12.34 Using the method of superposition, for the simply supported beam shown in the figure to the right, determine the deflection at mid-span and the slope at the right end of the beam.

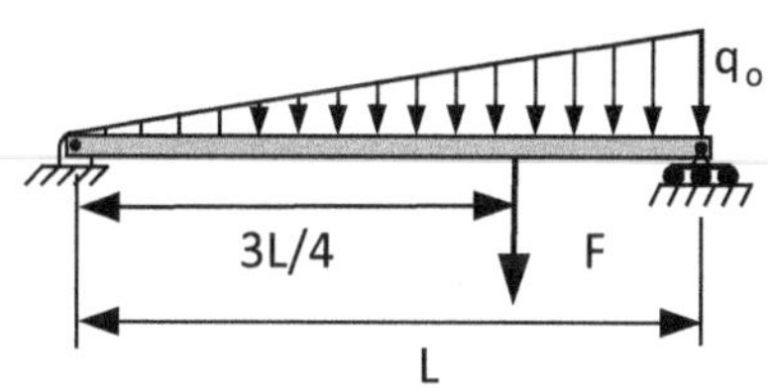

12.35 Using the method of superposition, for the cantilever beam shown in the figure to the left, determine the deflection and the slope at the free end of the beam.

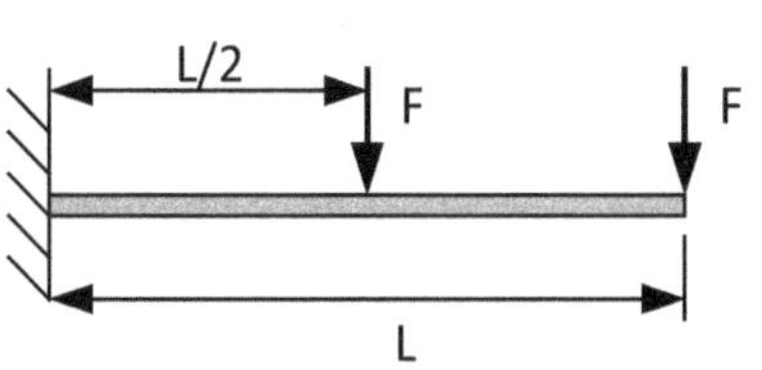

12.36 Using the method of superposition, for the cantilever beam shown in the figure to the right, determine the deflection and the slope at the free end of the beam. The beam is steel with a S457 × 81 American Standard section.

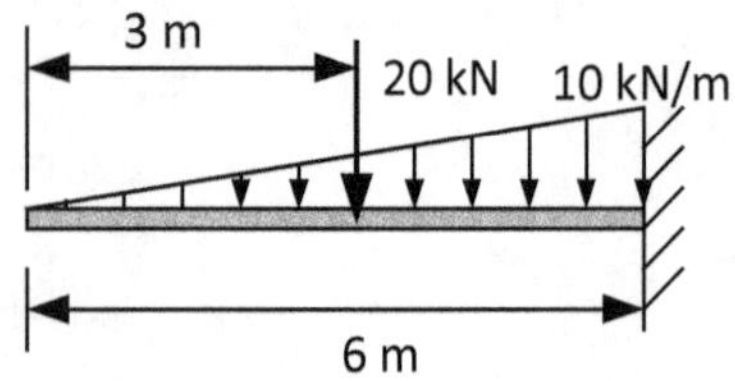

12.37 Using the method of superposition, for the cantilever beam shown in the figure to the left, determine the deflection and the slope at the free end of the beam. The beam is steel with a S20 × 66 American Standard section.

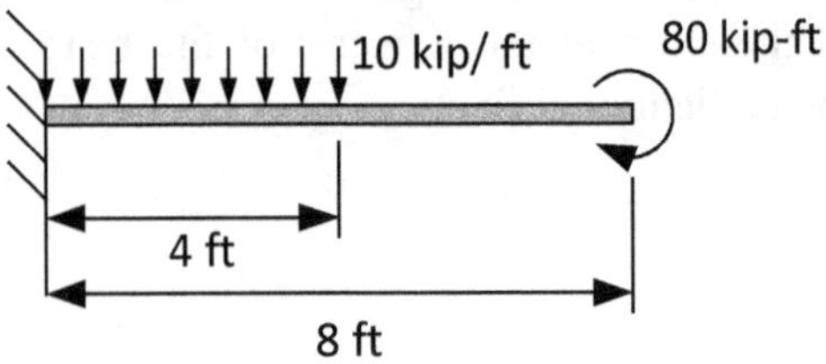

12.38 Using the method of superposition, for the cantilever beam shown in the figure to the right, determine the deflection and the slope at the free end of the beam.

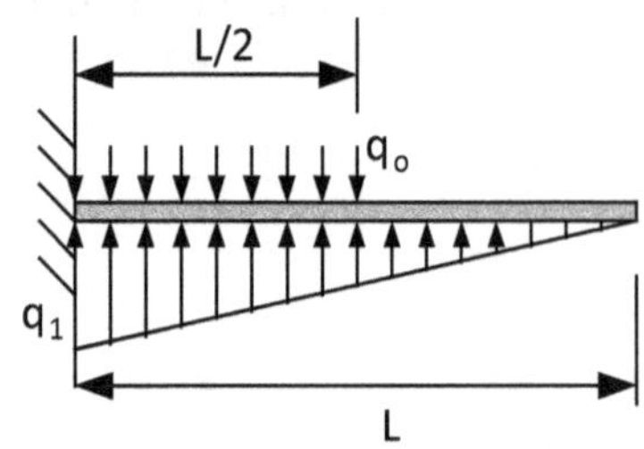

12.39 Determine the reactions at the prop and the wall on the cantilever beam shown in the figure to the left.

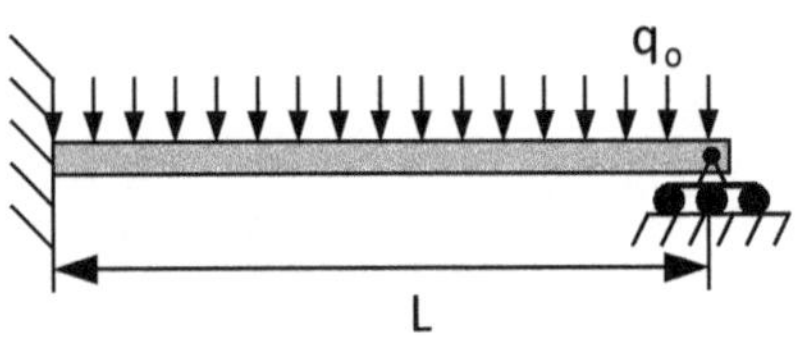

12.40 Determine the reactions at the prop and the wall on the cantilever beam shown in the figure to the right.

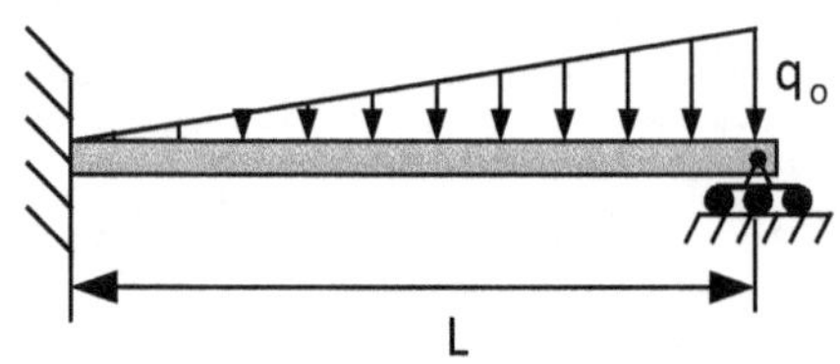

12.41 Determine the reactions at the prop and the wall on the cantilever beam shown in the figure to the left.

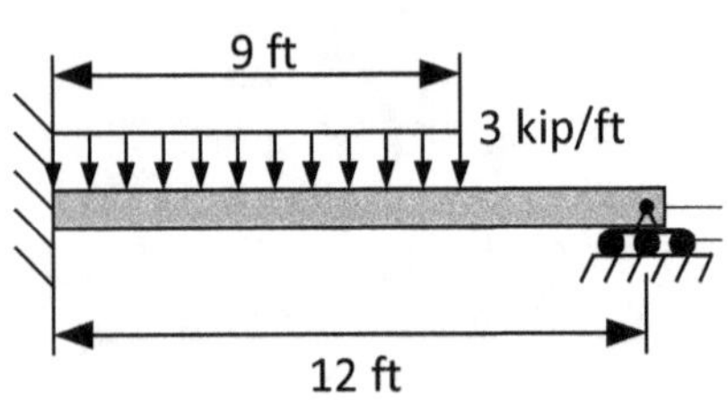

12.42 Determine the reactions at the prop and the wall on the cantilever beam shown in the figure to the right.

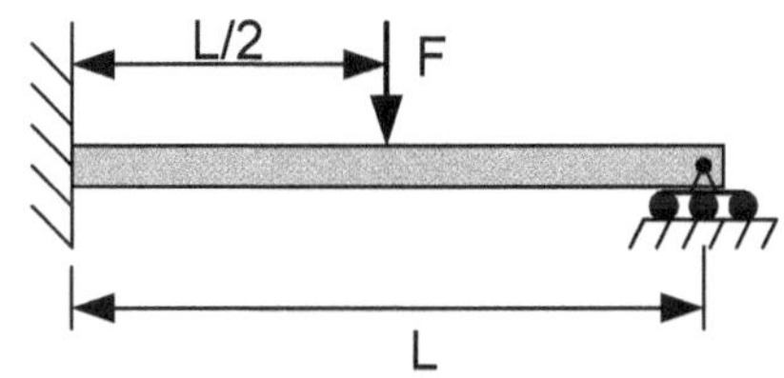

12.43 Determine the reactions at the prop and the wall on the cantilever beam shown in the figure to the left.

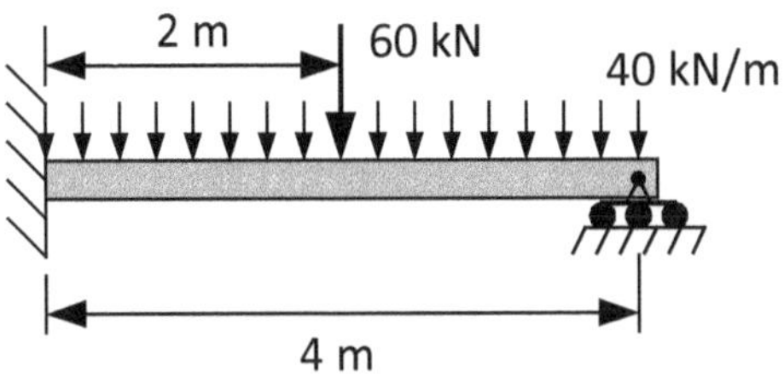

12.44 Determine the reactions at the prop and the wall on the cantilever beam shown in the figure to the right.

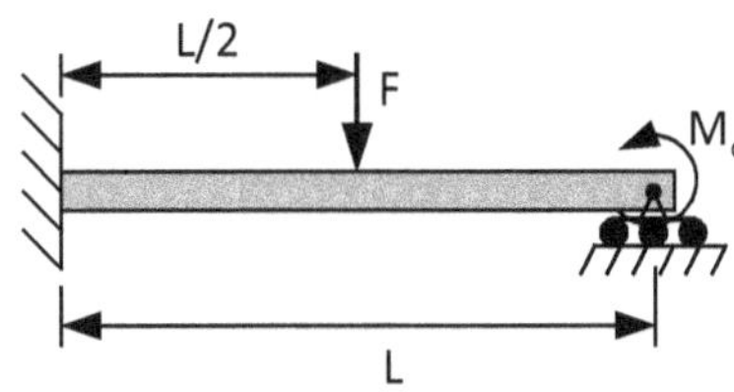

12.45 Determine the reactions at the supports of the continuous beam shown in the figure below.

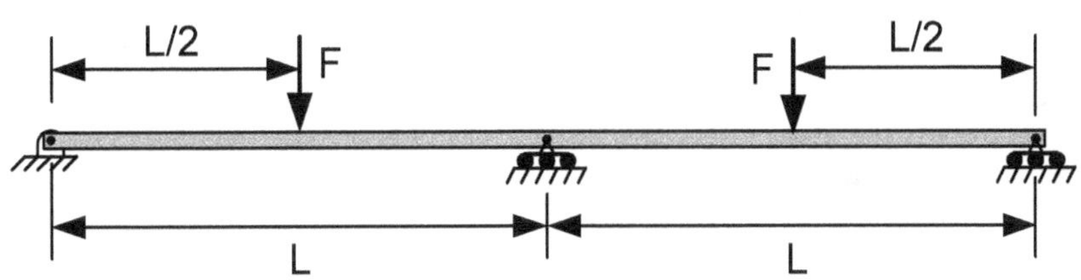

12.46 Determine the reactions at the supports of the continuous beam shown in the figure below.

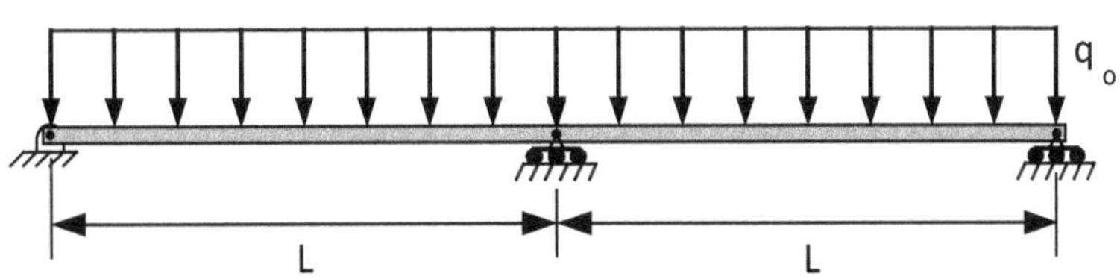

12.47 Determine the reactions at the supports of the continuous beam shown in the figure below.

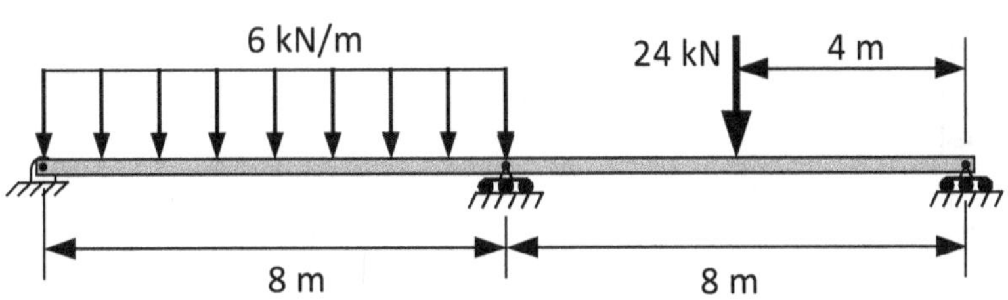

12.48 Determine the reactions at the supports of the continuous beam shown in the figure below.

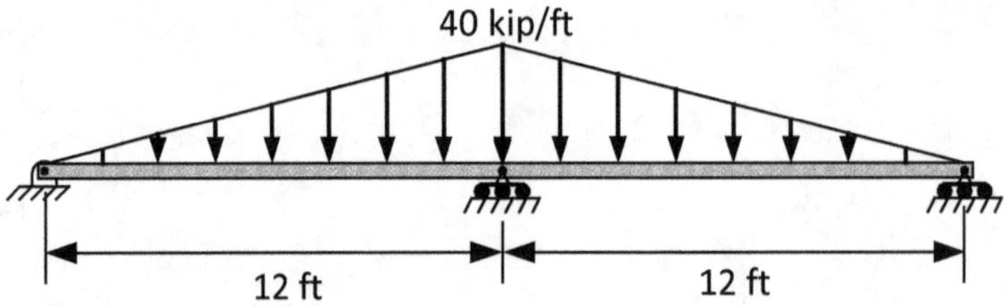

12.49 Determine the reactions at the supports of the continuous beam shown in the figure below.

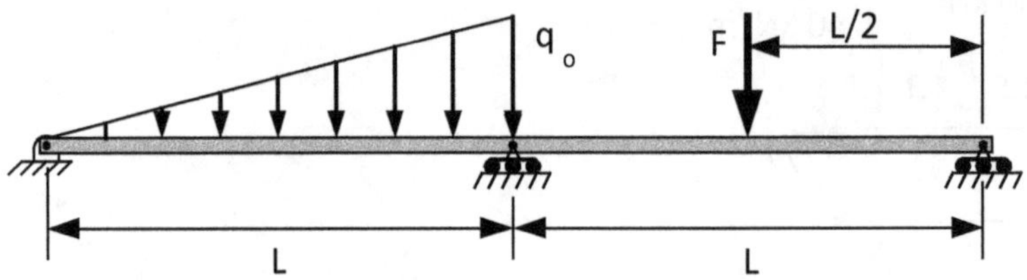

12.50 Determine the reactions at the supports of the built-in beam shown in the figure to the right.

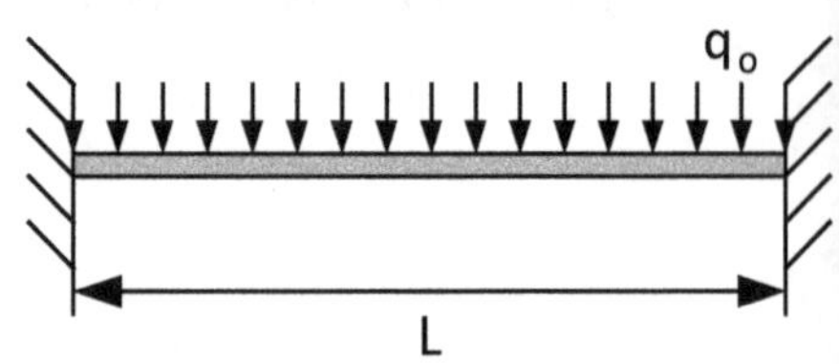

12.51 Determine the reactions at the supports of the built-in beam shown in the figure to the left.

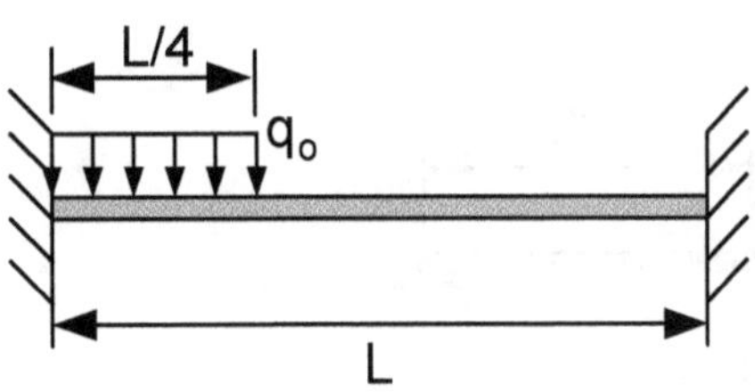

12.52 Determine the reactions at the supports of the built-in beam shown in the figure to the right.

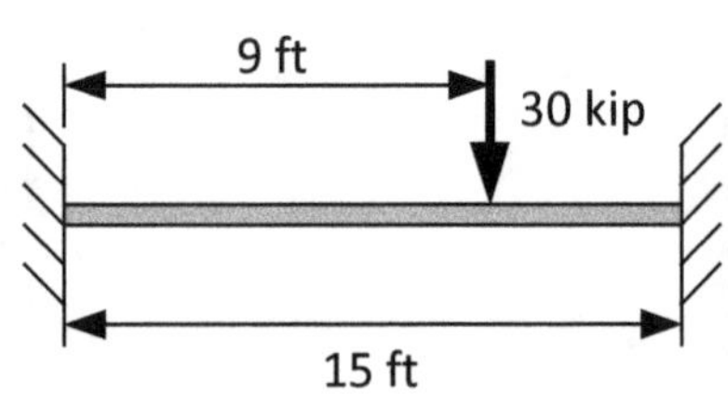

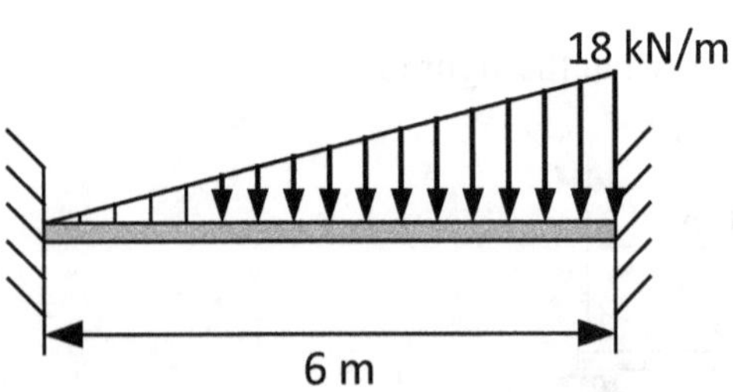

12.53 Determine the reactions at the supports of the built-in beam shown in the figure to the left.

CHAPTER 13

DEFLECTION OF BEAMS

13.1 INTRODUCTION

In Chapter 10, we studied the bending stresses produced in beams subjected to transverse loading. These uniaxial stresses σ_x were generated by the bending moments M(x) produced by the transverse loads. When a beam is subjected to a moment as shown in Fig. 13.1, it bends into a circular arc and deflects. As the beam deforms, the neutral axis becomes a circular arc and plane sections remain plane.

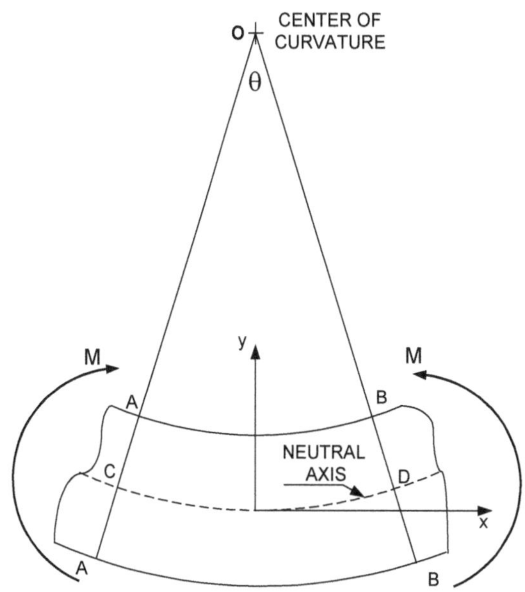

Fig. 13.1 A bending moment M causes the neutral axis of the beam to form a circular arc and deflect.

In Section 10.2, we used the geometry of the beam's deformation and equilibrium relations to position the neutral axis and to establish a relation between the bending moment and the radius of curvature ρ as:

$$M(x) = (EI_z / \rho) \qquad (10.5 \text{ bis})$$

$$\kappa = 1/\rho = M(x)/[EI_z] \qquad (10.6 \text{ bis})$$

where $\kappa = 1/\rho$ is the curvature of the beam and M(x) is the moment that varies with x.

13.2 THE ELASTIC CURVE FOR A BEAM

When a beam deflects under the action of a bending moment, the neutral axis becomes curved. Locally it is a circular arc with a radius of curvature ρ that depends on the bending moment M(x). As the local curvature varies with the bending moment, the neutral axis generates an elastic curve over the length of the beam. If we can write an equation for the elastic curve (the deformed shape of the neutral axis), the deflection of the beam will be apparent.

Let's rewrite Eq. (10.5) as:

$$1/\rho = (M/EI_z) \qquad (13.1)$$

A well-known relation from calculus gives the curvature κ of any curve as a function of the derivatives of y as:

$$\kappa = \frac{1}{\rho} = \frac{\dfrac{d^2y}{dx^2}}{\left[1+\left(\dfrac{dy}{dx}\right)^2\right]^{3/2}} \tag{13.2}$$

Well-designed beams are stiff and consequently the deflections are very small compared to the length of the beam. Because of this fact, the slope of the elastic curve (dy/dx) is small when compared to 1 and $(dy/dx)^2$ is extremely small compared to 1. Hence, Eq. (13.2) can be closely approximated with:

$$\kappa = \frac{1}{\rho} = \frac{d^2y}{dx^2} \tag{13.3}$$

Substituting Eq. (13.1) into (13.3) yields:

$$\frac{d^2y}{dx^2} = \frac{M(x)}{EI_z} \tag{13.4}$$

If we recall Eq. (10.18), which states that shear force V(x) = dM(x)/dx, and substitute this relation into Eq. (13.4), we obtain:

$$\frac{d^3y}{dx^3} = \frac{V(x)}{EI_z} \tag{13.5}$$

Similarly recall Eq. (10.20) which states that the load q(x) = – dV(x)/dx and substitute this relation into Eq. (13.5) to obtain:

$$\frac{d^4y}{dx^4} = \frac{-q(x)}{EI_z} \tag{13.6}$$

Clearly the elastic curve y(x) can be determined by integrating any one of the three equations given above. We usually employ Eq. (13.4) because only two integrations are necessary to obtain the results for the elastic curve y(x) for the beam.

13.3 ESTABLISHING THE ELASTIC CURVE BY INTEGRATION

There are three different approaches for determining the equation describing the elastic curve which include:

1. Integrating the equation for the bending moment M(x) with respect to x
2. Integrating the equation for the shear force V(x) with respect to x
3. Integrating the equation for the transverse loading q(x) with respect to x

We usually integrate Eq. (13.4) which involves M(x) because only two integrations are needed to convert d^2y/dx^2 into y(x). The mathematics give rise to two integration constants that are determined so as to satisfy constraints to the beam by its supports. In most examples, the integration is not difficult because M(x) is usually a polynomial in x. The difficulty arises in the form of the moment equation. In many instances, a single equation cannot be written that expresses M(x) over the entire length L of the beam. Instead, it is often necessary to write two or more different equations for M(x), each valid over a segment of the beam. Each of these moment equations must be integrated twice and the two constants of integration must be determined for each beam segment. The result is a time-consuming and tedious approach in which there are many opportunities for mathematical errors.

It is possible to integrate Eq. (13.5) to obtain the equation for the elastic curve y(x). In this case, we integrate the function for the shear force V(x) with respect to x. Because the unknown y occurs as the third derivative in Eq. (13.5), it is clear that the solution for the differential equation requires three integrations giving rise to three constants of integration. Because all of the mathematical difficulties mentioned in the previous paragraph occur with this approach, there is little merit in pursuing the topic.

It is also possible to integrate Eq. (13.6) to obtain the equation for the elastic curve y(x). In this case, we integrate the function for the load q(x) with respect to x. As the unknown y occurs as the fourth derivative in Eq. (13.6), it is clear that the solution for the differential equation requires four integrations giving rise to four constants of integration. Because all of the mathematical difficulties mentioned in the previous two paragraphs become more severe with this approach, there is very little merit in pursuing the topic.

Regardless of the approach employed to determine the equation of the elastic curve by integration we make several assumptions, which include:

- The deflections are small compared to the length of the beam so that the slope (dy/dx) is small and the square of the slope $(dy/dx)^2$ is negligible when compared to unity.
- The stresses in the beam are lower than the yield strength so that the modulus of elasticity E can be treated as a constant.
- The beam's cross-section does not change along the length of the beam so that the second moment of inertia I_z is a constant with respect to x.
- The distortion due to the shearing stress τ_{xy} is sufficiently small that the out of plane deformations of the beam's section can be ignored. (Plane sections remain plane.)

Let's consider six examples that demonstrate the integration method, describe the effect of beam constraints (boundary conditions), and make use of compatibility and symmetry conditions. We will test the small slope approximation and illustrate some of the difficulties encountered with this approach.

EXAMPLE 13.1

Determine the elastic curve for the simply supported beam subjected to a uniformly distributed load q as shown in Fig. E13.1.

Fig. E13.1

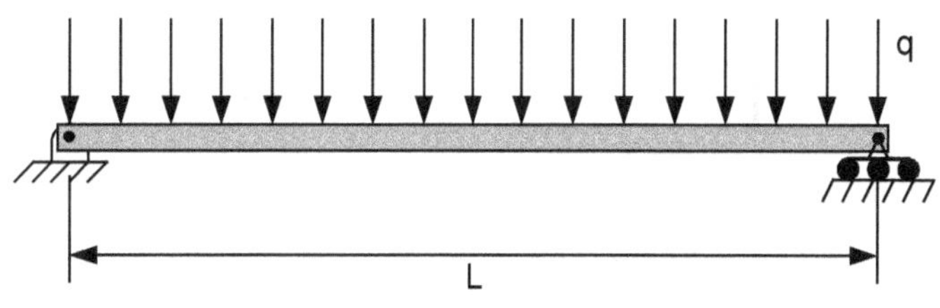

Solution:

Step 1: Prepare a FBD of the entire beam and employ the equilibrium relations to determine the reactions at the beam's simple supports. We have performed this task in Example 10.7 and obtained the results shown below for the reaction forces R_L and R_R as:

$$R_L = R_R = qL/2 \qquad \text{(a)}$$

Step 2: Prepare a FBD of a segment removed from the left side of the beam, and solve for the internal shear force V(x) and the internal moment M(x). See Fig. E10.7a.

$$V(x) = (q/2)(L - 2x) \quad \Rightarrow \quad 0 < x < L \qquad \text{(b)}$$

$$M(x) = (qx/2)(L - x) \quad \Rightarrow \quad 0 < x < L \qquad \text{(c)}$$

Step 3: Write the differential equation for the elastic curve by substituting Eq. (c) into Eq. (13.4) to obtain:

$$EI_Z \frac{d^2y}{dx^2} = \frac{qx}{2}(L - x) \qquad \text{(d)}$$

Step 4: Integrate both sides of Eq. (d) and simplify to obtain:

$$EI_Z \frac{dy}{dx} = \frac{q}{2}\left(\frac{x^2L}{2} - \frac{x^3}{3}\right) + C_1 = \frac{qx^2}{12}(3L - 2x) + C_1 \qquad \text{(e)}$$

$$EI_Z y = \frac{q}{12}\left(x^3L - \frac{x^4}{2}\right) + C_1x + C_2 = \frac{qx^3}{24}(2L - x) + C_1x + C_2 \qquad \text{(f)}$$

Step 5: The integration process has generated two constants of integration—C_1 and C_2. To determine these constants, we examine the constraints placed on the beam because these constraints control the shape of the elastic curve. The beam is constrained at its supports. On the left end, where x = 0, the beam is constrained by a pin and clevis arrangement which provides the following constraint:

$$y(0) = 0 \qquad \text{(g)}$$

At the left end, where x = L, the beam is constrained by a set of rollers, which provides the following constraint:

$$y(L) = 0 \qquad \text{(h)}$$

Step 6: The relations presented in Eqs. (g) and (h) are known as boundary conditions. Let's substitute the boundary requirement of Eq. (g) into Eq. (f) to obtain:

$$y(0) = 0 \qquad \Rightarrow \qquad C_2 = 0$$

Now substitute Eq. (h) into Eq. (f) to obtain:

$$EI_Z y = \frac{qL^3}{24}(2L - L) + C_1 L = 0 \qquad \text{(i)}$$

Solving for C_1 yields:

$$C_1 = -\frac{qL^3}{24} \qquad \text{(j)}$$

Substituting Eq. (j) into Eq. (f) gives the relation for the elastic curve from $0 < x < L$ as:

$$y = \frac{qx}{24EI_Z}\left[x^2(2L - x) - L^3\right] \qquad (13.7)$$

Following a similar process, we can write the expression for the slope (dy/dx) as:

$$\frac{dy}{dx} = \frac{q}{24EI_Z}\left(6x^2L - 4x^3 - L^3\right) \qquad (13.8)$$

We have evaluated Eq. (13.7) to determine $24EI_Zy/q$ as a function of x where x varied from 0 to L with L = 1.0 units. The result is shown in Fig. E13.1a.

Fig. E13.1a

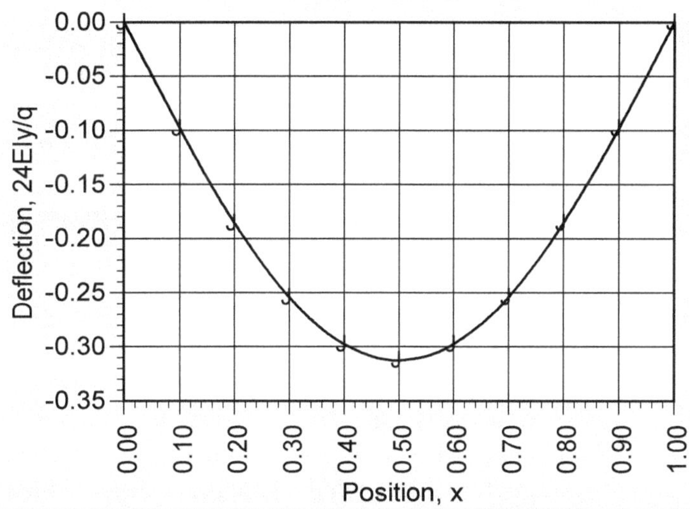

The normalized value of y associated with the elastic curve is a negative quantity because we defined y positive upward previously in Chapter 10. The elastic curve presented in Fig. E13.1a indicates that the deflection y is zero at x = 0 and x = L as required by the constraints of the simply supported beam. The deflection is a maximum at the center of the span of the beam.

EXAMPLE 13.2

The simply supported beam in Example 13.1 is fabricated from a wide flange rolled steel shape. The architect designing the building has indicated a W8 × 40 beam from 1020 HR steel. The beam is 20 ft long and the uniformly distributed load q = 800 lb/ft. Determine the maximum stresses in the beam and the safety factor against failure by yielding, the maximum deflection and the maximum slope. Interpret the results.

Solution:

Step 1: The maximum bending stress σ_{Max} in the beam is calculated from Eq. (10.10) as:

$$\sigma_{Max} = M_{Max}/Z = qL^2/(8Z) \tag{a}$$

where $M_{Max} = qL^2/8$ and Z is the section modulus.

Reference to Appendix D indicates that the section modulus Z and moment of inertia I_z for a W8 × 40 beam are:

$$Z = 35.5 \text{ in.}^3 \qquad I_z = 146 \text{ in.}^4 \tag{b}$$

From Eq. (a) and (b), we write:

$$\sigma_{Max} = qL^2/(8Z) = (800)(20)^2(12)/[(8)(35.5)] = 13.52 \text{ ksi} \tag{c}$$

From Appendix B, we find $S_y = 42$ ksi for 1020 HR steel. Then we can determine the safety factor as:

$$SF = S_y/\sigma_{Max} = 42/13.52 = 3.11 \tag{d}$$

This is a reasonable safety factor considering that the loads in a building are well known and not subject to significant changes over the anticipated service life of the structure.

Step 2: The maximum deflection y is calculated by setting x = L/2 in Eq. (13.7) to obtain:

$$y_{Max} = -\frac{5qL^4}{384EI_Z} \tag{13.9}$$

Substituting numerical parameters into Eq. (13.9) gives:

$$y_{Max} = -\frac{5qL^4}{384EI_Z} = -\frac{(5)(800)(20)^4(12)^4}{(384)(12)(30\times10^6)(146)} = -0.6575 \text{ in.} \tag{e}$$

While this deflection is small compared to the length of the beam it is relatively large compared to the 8.25 in. height of the beam. A suggestion to the architect to consider a W16 × 40 section is in order. This beam has the same weight of 40 lb/ft, but much higher values for Z and I_z. The increase in cost would be modest for significant improvements in both strength and rigidity. The only disadvantage is the increased height of the structure that would be required to accommodate the increased height of the beam.

Step 3: It is evident from an examination of Fig. E13.1a that the maximum slope occurs at the supports for the beam where x = 0 and L. At x = 0, Eq. (13.8) reduces to:

$$\left(\frac{dy}{dx}\right)_{Max} = -\frac{qL^3}{24EI_Z} = -\frac{(800)(20)^3(12)^3}{(24)(12)(30\times10^6)(146)} = -8.767\times10^{-3} \tag{f}$$

The negative sign in Eq. (f) indicates that the deflection y is becoming more negative as x is increasing. Let's square Eq. (f) and compare the result to unity.

$$\left(\frac{dy}{dx}\right)^2_{Max} = \left(-8.767\times10^{-3}\right)^2 = 7.686 \times10^{-5} \tag{g}$$

Clearly, the square of the maximum slope for this practical example is extremely small when compared with unity. This fact verifies the assumption made in reducing the form of Eq. (13.2) to the much simpler version presented as Eq. (13.3).

EXAMPLE 13.3

Derive the equation for the elastic curve for the simply supported beam loaded with a single concentrated force F as shown in Fig. E13.3.

Fig. 13.3

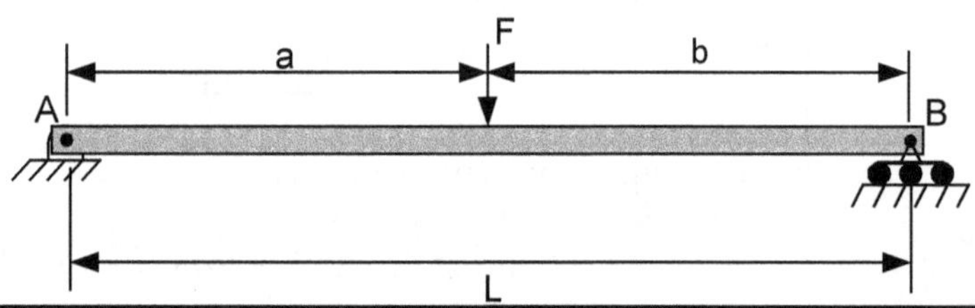

Solution:

Step 1: Determine the reaction forces at the simple supports. Reference to Section 10.4.1 shows the approach and gives the results as:

$$R_L = Fb/L \qquad R_R = Fa/L \tag{a}$$

Step 2: Write equations for the bending moment M as a function of x covering the range in x from zero to L. The procedure followed is illustrated in Section 10.4.1.

$$M(x) = Fbx/L \qquad 0 \le x < a \tag{b}$$

$$M(x) = Fbx/L - F(x - a) \qquad a \le x \le L \qquad (c)$$

These results differ from those presented in Example 13.1 where a single relation was written to describe the distribution of the bending moments over the entire length of the beam. The fact that it is necessary to write two equations to express M(x) complicates the solution.

Step 3: Write the differential equations for the elastic curve by substituting Eqs. (b) and (c) into Eq. (13.4) to obtain:

$$EI_z \frac{d^2y_1}{dx^2} = M(x) = \frac{Fbx}{L} \qquad 0 \le x \le a \qquad (d)$$

$$EI_z \frac{d^2y_2}{dx^2} = M(x) = \frac{Fbx}{L} - F(x-a) \qquad a \le x \le L \qquad (e)$$

Step 4: Integrate these second order differential equations once to obtain the equation for the slopes of the elastic curve as:

$$EI_z \frac{dy_1}{dx} = \frac{Fbx^2}{2L} + C_1 \qquad 0 \le x \le a \qquad (f)$$

$$EI_z \frac{dy_2}{dx} = \frac{Fbx^2}{2L} - \frac{F(x-a)^2}{2} + C_2 \qquad a \le x \le L \qquad (g)$$

Step 5: Before proceeding, let's use one of the compatibility conditions to evaluate the constants of integration in Eqs. (f) and (g). The elastic curve is continuous at the point of load application. Hence, the slope of the elastic curve dy_1/dx is the same as the slope dy_2/dx at x =a where we transition from one solution for the slope to another. Accordingly, we equate Eq. (f) and Eq. (g) and set x = a to obtain:

$$\frac{Fba^2}{2L} + C_1 = \frac{Fba^2}{2L} - 0 + C_2 \qquad \text{at } x = a$$

$$C_1 = C_2 \qquad (h)$$

Step 6: Integrate Eqs. (f) and (g) to obtain the equations for the elastic curve as:

$$EI_z y_1 = \frac{Fbx^3}{6L} + C_1x + C_3 \qquad 0 \le x \le a \qquad (i)$$

$$EI_z y_2 = \frac{Fbx^3}{6L} - \frac{F(x-a)^3}{6} + C_1x + C_4 \qquad a \le x \le L \qquad (j)$$

Step 7: Let's again recognize the compatibility conditions that exist at x = a. Since the elastic curve is continuous, it is clear that $y_1 = y_2$ at x = a. Hence, we can equate Eq. (i) and Eq. (j) and set x = a to obtain:

$$\frac{Fba^3}{6L} + C_1a + C_3 = \frac{Fba^3}{6L} - \frac{F(a-a)^3}{6} + C_1a + C_4 \qquad \text{at } x = a$$

$$C_3 = C_4 \qquad (k)$$

We have used the two compatibility conditions to establish that $C_1 = C_2$ and $C_3 = C_4$, which reduces the four constants of integration to two.

Step 8: Let's consider the constraints placed on the elastic curve by the simple supports at each end of the beam. Clearly, the simple supports constrain the deflection at both ends of the beam to zero. Hence, we can write:

$$y_1(0) = 0 \qquad \text{(l)}$$

$$y_2(L) = 0 \qquad \text{(m)}$$

Substituting Eq. (l) into Eq. (i) gives:

$$C_3 = 0 = C_4 \qquad \text{(n)}$$

Substituting Eq. (m) into Eq. (j) gives:

$$0 = \frac{FbL^3}{6L} - \frac{F(L-a)^3}{6} + C_1 L + 0$$

$$C_1 = -\frac{Fb}{6L}(L^2 - b^2) \qquad \text{(o)}$$

Step 5: With the four constants of integration established, we write the solutions for the elastic curve as:

$$EI_z y_1 = -\frac{Fbx}{6L}(L^2 - x^2 - b^2) \qquad 0 \le x \le a \qquad (13.10)$$

$$EI_z y_2 = -\frac{Fbx}{6L}(L^2 - x^2 - b^2) - \frac{F(x-a)^3}{6} \qquad a \le x \le L \qquad (13.11)$$

We will determine the maximum deflection of a simply supported beam with a concentrated load in the next example. However, you should be aware that the double integration method often gives rise to four or even six constants of integration. As such, it is tedious and time consuming to execute solutions to even simple problems. We will introduce a method involving singularity functions in Section 13.4 that involves the determination of only two constants of integration.

EXAMPLE 13.4

For the simply supported beam described in Example 13.3, determine the maximum deflection y_{Max}. The beam spans 6 m and is made from an American Standard shape with a designation of S203 × 34. The beam is fabricated from 1020 HR steel. A concentrated force of F = 10 kN is applied at a = 4 m.

Solution:

Step 1: Let's begin by deriving the equation for the maximum deflection by considering the slope dy_1/dx. At the position x where the deflection is a maximum[1], $dy_1/dx = 0$. Differentiating Eq. (13.10) and setting the result to zero gives the location of the maximum deflection as:

---

[1] Since a > b the maximum deflection will occur on the left portion of the beam where x<a.

$$EI_z \frac{dy_1}{dx} = -\frac{Fb}{6L}(L^2 - 3x^2 - b^2) = 0 \qquad \text{(a)}$$

$$x = \sqrt{\frac{L^2 - b^2}{3}}$$

Evaluating Eq. (a) gives:

$$x = \sqrt{\frac{L^2 - b^2}{3}} = \sqrt{\frac{36 - 4}{3}} = 3.266 \text{ m} \qquad \text{(b)}$$

Step 2: Determine the bending stresses in the beam to insure that the beam is elastic. The moment is a maximum at x = a = 4.0 m. Hence, we write:

$$M_{Max} = Fba/L = (10 \times 10^3)(2.0)(4.0)/6.0 = 13.33 \text{ kN-m} \qquad \text{(c)}$$

And the bending stress is given by:

$$\sigma_{Max} = M_{Max}/Z = [13.33 \times 10^3 \text{ (N-m)}(10^3 \text{ mm/m})/(265 \times 10^3 \text{ mm}^3) = 50.31 \text{ MPa}$$

where the section modulus $Z = 265 \times 10^3$ mm^3 is listed in Appendix D.

Because σ_{Max} is less than the yield strength of 1020 HR steel (290 MPa), we conclude that the beam under load is elastic and Eq. (13.10) is valid.

Step 3: To determine the expression for the maximum deflection, substitute Eq. (a) into Eq. (13.10) to determine the expression for y_{Max} as:

$$y_{Max} = -\frac{Fb}{9\sqrt{3}LEI_z}\left(L^2 - b^2\right)^{3/2} \qquad (13.12)$$

Evaluating Eq. (13.12) with the numerical parameters listed in the problem statement and in Appendices B and D gives:

$$y_{Max} = -\frac{(10 \times 10^3)(2000)}{9\sqrt{3}(6000)(207 \times 10^3)(27 \times 10^6)}\left[(6000)^2 - (2000)^2\right]^{3/2}$$

$$y_{Max} = -6.926 \text{ mm} \qquad \text{(d)}$$

Again we note that the deflection is small compared to both the length and the height of the beam.

EXAMPLE 13.5

Determine the equation for the elastic curve for the cantilever beam presented in Fig. E13.5. Also write the equation for the maximum deflection and the maximum slope.

Solution:

Step 1: Prepare a FBD of a segment of the beam where $0 \leq x \leq L$ and write an equation for M(x) as:

$$M(x) = -Fx \qquad \text{(a)}$$

Fig. E13.5

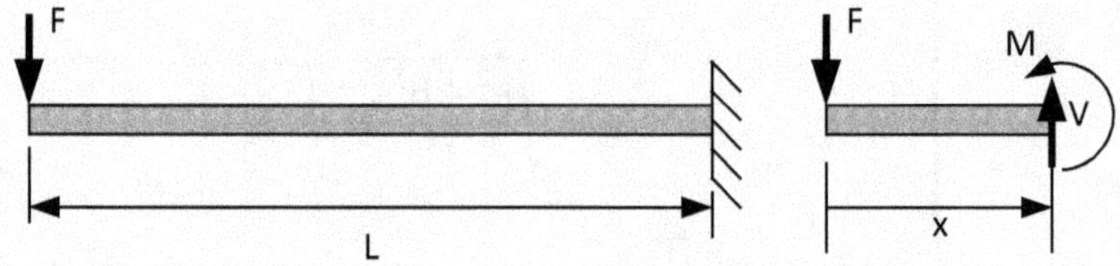

Step 2: Write the differential equation for the elastic curve by substituting Eq. (a) into Eq. (13.4) to obtain:

$$EI_z \frac{d^2y}{dx^2} = M(x) = -Fx \qquad 0 \leq x \leq L \qquad \text{(b)}$$

In this problem a single moment equation is valid over the entire length of the beam indicating that only two constants of integration will be required in the integration process.

Step 3: Integrating Eq. (b), we obtain:

$$EI_z \frac{dy}{dx} = -\frac{Fx^2}{2} + C_1 \qquad 0 \leq x \leq L \qquad \text{(c)}$$

Step 4: The constraint condition (a built-in end) at the right support of the beam, where x = L, forces the slope at the wall to be zero. Hence, Eq. (d) evaluated at x = 0 reduces to:

$$EI_z \frac{dy}{dx} = 0 = -\frac{FL^2}{2} + C_1 \quad \Rightarrow \quad C_1 = FL^2/2 \qquad \text{(d)}$$

Substituting the value for C_1 into Eq. (c) gives the equation for the slope of the elastic curve at any location along its length as:

$$\frac{dy}{dx} = \frac{F(L^2 - x^2)}{2EI_z} \qquad 0 \leq x \leq L \qquad (13.13)$$

Step 5: Integrate Eq. (c) to obtain the expression for the elastic curve as:

$$EI_z y = -\frac{Fx^3}{6} + \frac{FL^2 x}{2} + C_2 \qquad \text{(e)}$$

Step 6: The constraint condition on the right end of the beam due to the built-in support is that the deflection at the wall is zero. Evaluating Eq. (e) with y = 0 at x = L gives:

$$EI_z y = 0 = -\frac{FL^3}{6} + \frac{FL^3}{2} + C_2 \quad \Rightarrow \quad C_2 = -\frac{FL^3}{3} \qquad \text{(f)}$$

Substituting for the constant C_2 into Eq. (e) gives:

$$EI_z y = -\frac{Fx^3}{6} + \frac{FL^2x}{2} - \frac{FL^3}{3} \qquad (13.14)$$

Step 7: Inspection of Eq. (13.14) shows that the deflection is a maximum at the free end where x = 0.

$$y_{Max} = -\frac{FL^3}{3EI_z} \qquad (13.15)$$

The maximum slope also occurs at the free end. Setting x = 0 in Eq. (13.13) gives the relation for the maximum slope as:

$$\left(\frac{dy}{dx}\right)_{Max} = \frac{FL^2}{2EI_z} \qquad (13.16)$$

EXAMPLE 13.6

For the simply supported beam shown in Fig. E13.6, determine the equation for the elastic curve. Also determine the maximum deflection and the maximum slope of the beam.

Fig. E13.6

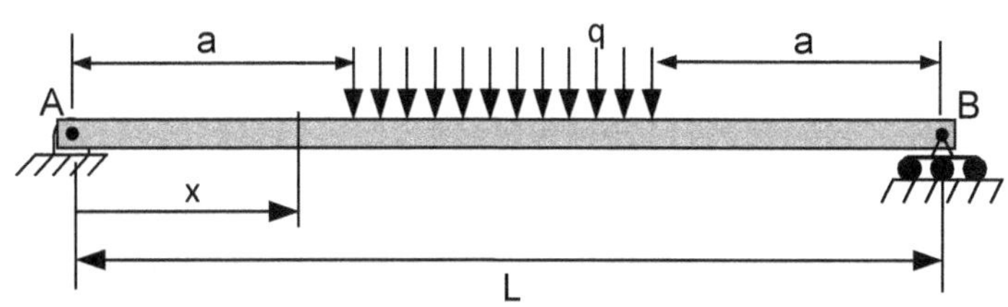

Solution:

Step 1: Write the equations for the reaction forces at the beam's supports and the moments as a function of position x. FBDs of the entire beam and segments of the beam and the application of the equations of equilibrium lead to:

$$R_L = R_R = qb/2 \qquad \text{with } b = (L - 2a) \qquad (a)$$

$$M(x) = qbx/2 \qquad 0 \le x \le a \qquad (b)$$

$$M(x) = qbx/2 - q(x - a)^2/2 \qquad a \le x \le a + b \qquad (c)$$

$$M(x) = qbx/2 - qb(2x - L)/2 \qquad a + b \le x \le L \qquad (d)$$

We note that three different equations are necessary for describing the bending moments over the entire length of this beam. If we double integrate all of these equations, six constants of integration will be generated. Clearly, this approach will be long and tedious and should be avoided if possible. Fortunately, we recognize that the loading and the supports of the beam are symmetric. Symmetry enables us to solve for the equation of the elastic curve for the left half of the beam and to infer the solution for the right half of the beam. With this approach, we will integrate only the first two of the three moment equations.

Step 2: Substituting Eqs (b) and (c) into Eq. (13.4) gives a second order differential equation for the elastic curve as:

$$EI_z \frac{d^2y_1}{dx^2} = M(x) = \frac{qbx}{2} \qquad 0 \le x \le a \qquad \text{(e)}$$

$$EI_z \frac{d^2y_2}{dx^2} = M(x) = \frac{qbx}{2} - \frac{q(x-a)^2}{2} \qquad a \le x \le a+b \qquad \text{(f)}$$

Step 3: Integrating Eq. (e) twice gives:

$$EI_z \frac{dy_1}{dx} = \frac{qbx^2}{4} + C_1 \qquad 0 \le x \le a \qquad \text{(g)}$$

$$EI_z y_1 = \frac{qbx^3}{12} + C_1 x + C_2 \qquad 0 \le x \le a \qquad \text{(h)}$$

To satisfy the boundary condition at the left support where y = 0 at x = 0, it is evident that:

$$C_2 = 0 \qquad \text{(i)}$$

Step 4: Integrating Eq. (f) twice gives:

$$EI_z \frac{dy_2}{dx} = \frac{qbx^2}{4} - \frac{q(x-a)^3}{6} + C_3 \qquad a \le x \le a+b \qquad \text{(j)}$$

$$EI_z y_2 = \frac{qbx^3}{12} - \frac{q(x-a)^4}{24} + C_3 x + C_4 \qquad a \le x \le a+b \qquad \text{(k)}$$

Step 5: Determine C_3 by noting that the beam is symmetric; hence, $(dy_2/dx) = 0$ at $x = L/2$. Substituting this information into Eq. (j) and simplifying the result yields:

$$C_3 = -\frac{qb(3L^2 - b^2)}{48} \qquad \text{(l)}$$

To evaluate the remaining two constants of integration, we use the compatibility conditions which indicate that $(dy_1/dx) = (dy_2/dx)$ and $y_1 = y_2$ at $x = a$. From Eqs. (g) and (j) and the compatibility condition $(dy_1/dx) = (dy_2/dx)$ at $x = a$, it is apparent that:

$$C_1 = C_3 = -\frac{qb(3L^2 - b^2)}{48} \qquad \text{(m)}$$

From Eqs. (h) and (k) and the compatibility condition $y_1 = y_2$ at $x = a$, we find that:

$$C_2 = C_4 = 0 \qquad \text{(n)}$$

Step 6: Substituting the results for the four constants into Eqs. (h) and (k) yields:

$$y_1 = \frac{qbx}{48EI_z}\left[4x^2 - 3L^2 + b^2\right] \qquad 0 \le x \le a \qquad \text{(13.17)}$$

$$y_2 = \frac{q}{48EI_z}\left[4bx^3 - 2(x-a)^4 - bx(3L^2 - b^2)\right] \qquad a \le x \le a+b \qquad (13.18)$$

Step 7: We write the relation for the maximum deflection by observing that $y_{Max} = y_2$ at $x = L/2$. Substituting $x = L/2$ into Eq. (13.18) gives:

$$y_{Max} = -\frac{qb}{384EI_z}\left[8L^3 + b^3 - 4b^2L\right] \qquad (13.19)$$

Substituting the results for the constants $C_1$and C_3 into Eqs. (g) and (j) gives the equations for the slopes of the elastic curve as:

$$\frac{dy_1}{dx} = \frac{qb}{48EI_z}\left[12x^2 - 3L^2 + b^2\right] \qquad 0 \le x \le a \qquad (13.20)$$

$$\frac{dy_2}{dx} = \frac{q}{48EI_z}\left[12bx^2 - 8(x-a)^3 - 3bL^2 + b^3\right] \qquad a \le x \le a+b \qquad (13.21)$$

The maximum slope occurs at the supports. We write the relation for the maximum slope by setting $x = 0$ in Eq. (13.20) to obtain:

$$\left(\frac{dy_1}{dx}\right)_{Max} = -\frac{qb}{48EI_z}\left[3L^2 - b^2\right] \qquad (13.22)$$

These six examples have shown that the double integration of the moment equation enables one to establish the equation of the elastic curve and the equation for the slope over the entire length of the beam. The difficulty that often arises is in determining the constants of integration that are generated during the integration process. When several different moment equations must be written to cover the span of the beam, the number of constants of integration increases and the work involved in the solution becomes excessive. We will introduce an approach in Section 13.4 that avoids this difficulty.

13.3.1 Boundary and Compatibility Conditions

In the solutions to the previous examples, we used information describing the beam's constraints in deriving the equation for the elastic curve of a beam in bending. Beams are supported (constrained) by simple supports or by built-in supports. If simple supports are used, a minimum of two supports is necessary for stability. Usually the supports are placed at the ends of the beam, but this is not required if overhangs provide some design advantage. If the beam has a built-in support at one end, the other end may be free. Fixed and simple supports may both be employed in constraining the deformation of a beam. However, this practice produces a statically indeterminate beam.

Simple supports are designed with several different arrangements that permit free rotation of the beam over the support ($M = 0$) while constraining the deflection so that $y = 0$. Several of the designs for simple supports are illustrated in Fig. 13.2. The roller and the step configurations are the most common because they are less expensive to implement in construction.

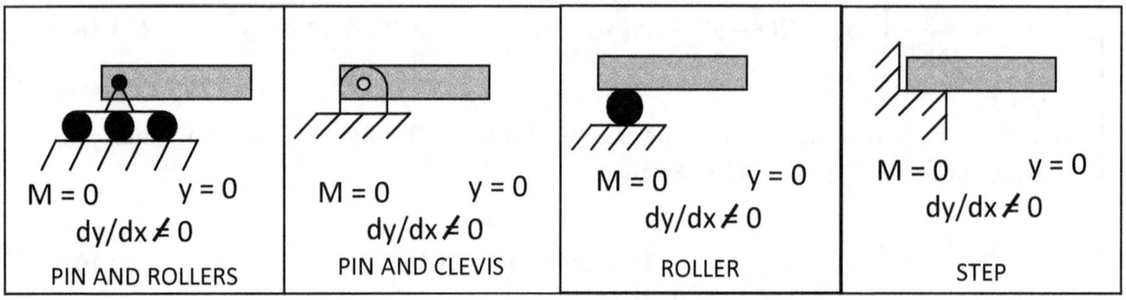

Fig. 13.2 Illustrations of different simple supports for a beam that yield the same boundary conditions.

Fixed or built-in supports are more difficult to achieve in practice. Two methods are illustrated in Fig. 13.3 together with the boundary conditions. Both the slope and the deflection are constrained so that $(dy/dx) = 0$ and $y = 0$. However, the moment at the support is not equal to zero.

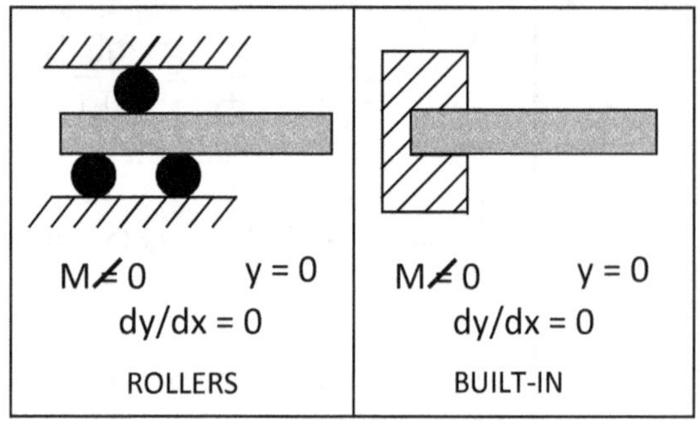

Fig. 13.3 Illustration of fixed or built-in supports for a beam.

With a built-in or fixed support at one end of a beam, the other end may be free as is the case for a cantilever beam. The free end will deflect and rotate so that $(dy/dx) \neq 0$ and $y \neq 0$. However the moment M and the shear force V both are zero at this point.

Compatibility and Symmetry Conditions

When it is necessary to write two or more moment equations to describe M(x) over the entire length of the beam, the integration method produces two or more equations for the elastic curve. Each of these equations is valid over some segment of the beam. For instance in Example 13.3, the expression for y_1 was valid for $0 \leq x \leq a$, and the relation for y_2 was valid for $a \leq x \leq L$. However at $x = a$, the results for both y_1 and y_2 must be identical. In addition the slopes dy_1/dx and dy_2/dx must also be identical at $x = a$. The deflection given by both solutions must be continuous as we transition from one solution to another. We illustrate the concept of continuity of the slope and the deflection by observing the elastic curve in Fig. 13.4

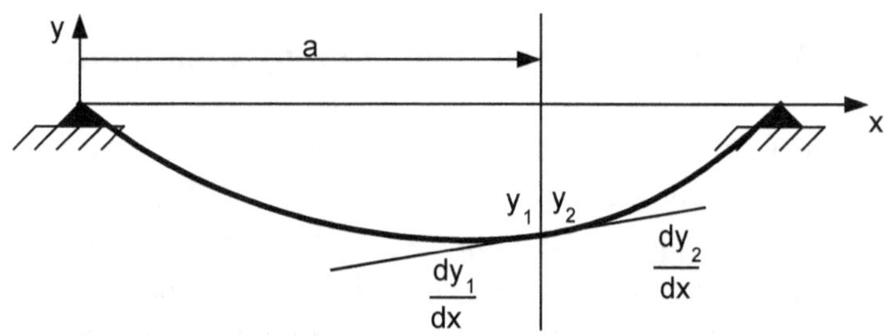

Fig. 13.4 For the elastic curve to be continuous, $y_1 = y_2$ and $(dy_1/dx) = (dy_2/dx)$ at $x = a$.

When the loading on a uniform beam is symmetric, the elastic curve is also symmetric as illustrated in Fig. 13.5. Because of this symmetry, the slope at the center of the beam where $x = L/2$ is zero. We will employ boundary conditions, compatibility conditions, and symmetry to establish constants of integration when establishing the equation for the elastic curve.

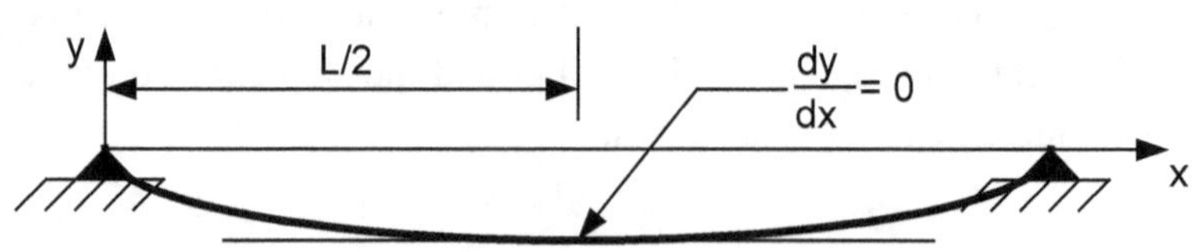

Fig. 13.5 The slope is zero at the mid point of a symmetrically loaded beam.

13.3.3 Integrating the Shear Force Relation and the Load Relation

It is possible to write one or more equations describing the variation of the shear force V as a function of position x over the span of the beam. Substituting V(x) into Eq. (13.5) yields a third order differential equation in terms of the deflection y. By integrating this differential equation three times, we can obtain the equation for the elastic curve. The difficulty with this approach is the number of constants of integration that are generated during the integration process. For a simply supported beam with a single concentrated force, two equations are needed to express the shear force over the length of the beam. Integrating these equations three times produces six constants of integration that can be determined from boundary and compatibility conditions. However, the time and effort required is excessive.

Similarly it is possible to write equations describing the forces q applied to the beam as a function of x. Substituting q(x) into Eq. (13.6) yields a fourth order differential equation in terms of the deflection y. By integrating this differential equation four times, we can obtain the equation for the elastic curve. We encounter the same difficulty with this approach—the number of constants of integration that are generated require an excessive effort to determine. A much better approach for determining the equation for the elastic curve is presented in the next section.

13.4 DEFLECTIONS AND SLOPES WITH SINGULARITY FUNCTIONS

13.4.1 Singularity Functions

In Section 13.3, we found that the integration method was tedious and time consuming when two or more equations were required to express the bending moment over the entire length of the beam. To avoid this difficulty, we will consider a family of singularity functions that will enable us to express the moment over the length of the beam in a single equation regardless of the complexity of the loading. Macauley first introduced singularity functions in 1919 and they were adapted by the mechanics community in the 1960-70 time frame.

To begin, consider a family of singularity functions given by:

$$f_n(x) = \langle x - a \rangle^n \qquad (13.23)$$

When $n \geq 0$ the function f_n is evaluated with the following rule:

$$\begin{aligned} f_n(x) &= 0 && \text{if } x < a \\ f_n(x) &= (x - a)^n && \text{if } x \geq a \end{aligned} \qquad (13.24)$$

Clearly, the pointed brackets are identical to ordinary brackets if the quantity within the pointed brackets is zero or positive. However, if the quantity within the pointed brackets is negative, we set f_n equal to zero. Think of the pointed brackets as a switch that turns off a term when the quantity within the pointed brackets is negative.

Let's determine $f_n(x)$ if n = 0, 1 and 2 as a function of x. The three results are presented as graphs in Fig. 13.6. The function $\langle x - a \rangle^0 = 1$ when $x \geq a$ is called the unit step function, and $\langle x - a \rangle^1$ is called the ramp function, which begins at x = a. We integrate these functions using the following rule:

$$\int_{-\infty}^{x} \langle x - a \rangle^n \, dx = \frac{\langle x - a \rangle^{n+1}}{n+1} \qquad n \geq 0 \qquad (13.25)$$

Different rules apply for $f_n(x)$ when n is either –1 or – 2.

$$\begin{aligned} f_n(x) &= 0 && \text{if } x < a \\ f_n(x) &= 0 && \text{if } x > a \\ f_n(x) &\Rightarrow \infty && \text{if } x = a \end{aligned} \qquad (13.26)$$

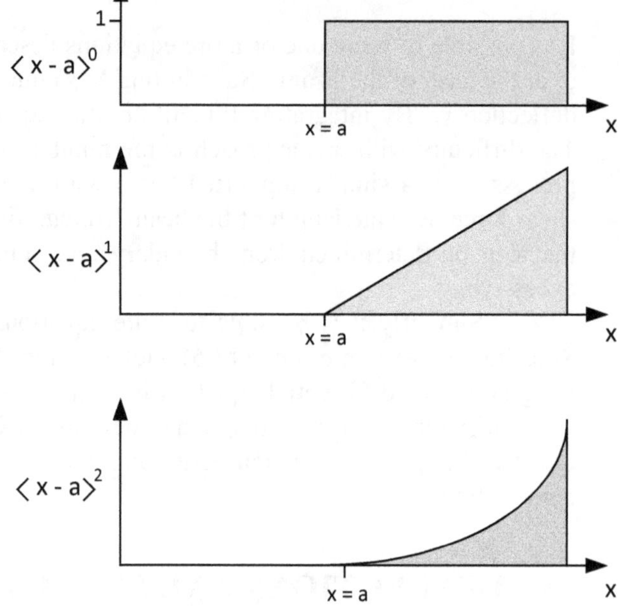

Fig. 13.6 Singularity functions for n = 0, 1 and 2.

The singularity in f_n at x = a does not lead to mathematical difficulty because when $<x - a>^{-1}$ and $<x - a>^{-2}$ are integrated, the result is finite as indicated below:

$$\int_{-\infty}^{x} <x - a>^{-2} \, dx = <x - a>^{-1} \tag{13.27}$$

$$\int_{-\infty}^{x} <x - a>^{-1} \, dx = <x - a>^{0} \tag{13.28}$$

The function $<x - a>^{-1}$ gives rise to a concentrated load and the function $<x - a>^{-2}$ produces a concentrated moment at x = a as shown in Fig. 13.7.

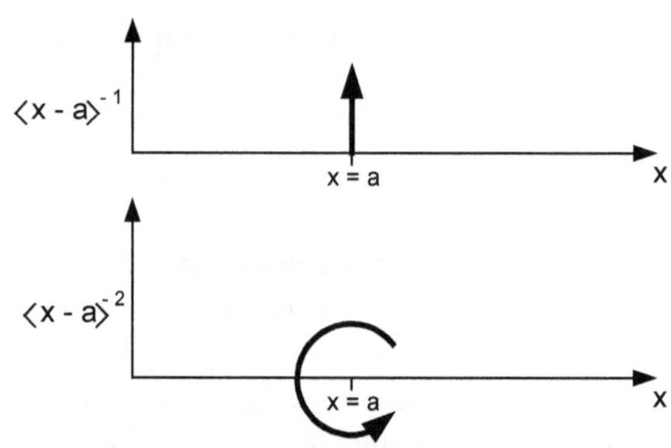

Fig. 13.7 Singularities produced by the function $<x - a>^{-1}$ and $<x - a>^{-2}$ at x = a.

Next, let's show the application of the new singularity function to writing equations for the load, shear force and bending moments as a function of position over the entire length of a beam.

13.4.2 Writing Equations with Singularity Functions

Typical loading q(x) applied to beams includes the four cases listed below. The equation for the load q(x) for each of these different loads can be written using singularity functions as:

1. A concentrated moment at some location x = a.

$$q(x) = M_o <x - a>^{-2} \tag{13.29}$$

2. A concentrated force F at some location x = a.

$$q(x) = F\langle x - a\rangle^{-1} \tag{13.30}$$

3. A uniformly distributed load q_o beginning at x = a.

$$q(x) = q_o\langle x - a\rangle^{0} \tag{13.31}$$

4. A linearly increasing (ramp) load q(x) beginning at x = a.

$$q(x) = (q_o/b)\langle x - a\rangle^{1} \tag{13.32}$$

where q_o/b is the slope of the ramp function.

Equations for the shear force V(x) can be written by substituting Eq. (13.29) into Eq. (10.20) and integrating using Eq. (13.27) to obtain:

$$V(x) = -\int_{-\infty}^{x} q(x)dx = -M_o\int_{-\infty}^{x} \langle x - a\rangle^{-2}\,dx = -M_o\langle x - a\rangle^{-1} \tag{13.33}$$

Similarly, equations for the bending moment can be obtained by substituting V(x) into Eq. (10.18) and integrating using Eq. (13.28) to obtain:

$$M(x) = \int_{-\infty}^{x} V(x)dx = -M_o\int_{-\infty}^{x} \langle x - a\rangle^{-1}\,dx = -M_o\langle x - a\rangle^{0} \tag{13.34}$$

Using this same procedure, the equations for V(x) and M(x) were derived for each of the four types of beam loading functions. The summary of the results obtained is presented in Fig. 13.8.

BEAM LOADING	LOAD EQUATION	SHEAR FORCE EQ.	MOMENT EQUATION
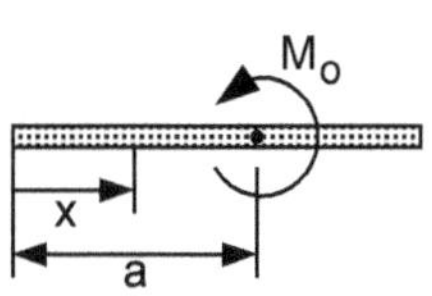	$q(x) = M_o\langle x-a\rangle^{-2}$	$V(x) = -M_o\langle x-a\rangle^{-1}$	$M(x) = -M_o\langle x-a\rangle^{0}$
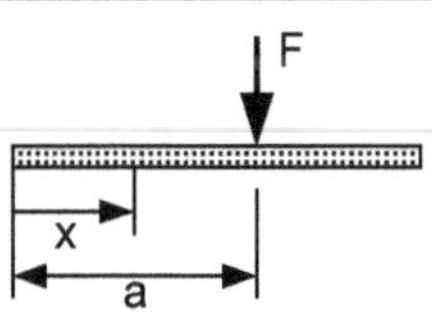	$q(x) = F\langle x-a\rangle^{-1}$	$V(x) = -F\langle x-a\rangle^{0}$	$M(x) = -F\langle x-a\rangle^{1}$
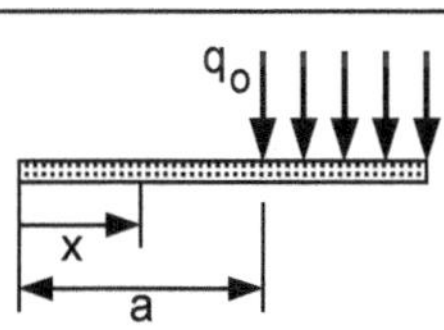	$q(x) = q_o\langle x-a\rangle^{0}$	$V(x) = -q_o\langle x-a\rangle^{1}$	$M(x) = -\frac{q_o}{2}\langle x-a\rangle^{2}$
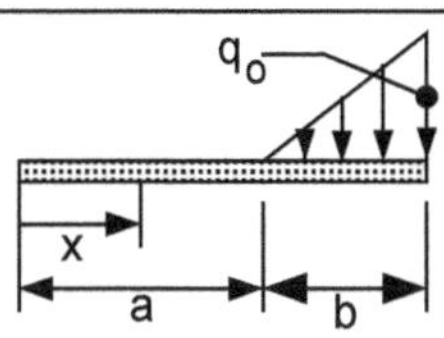	$q(x) = \frac{q_o}{b}\langle x-a\rangle^{1}$	$V(x) = -\frac{q_o}{2b}\langle x-a\rangle^{2}$	$M(x) = -\frac{q_o}{6b}\langle x-a\rangle^{3}$

Fig. 13.8 Equations for q(x), V(x) and M(x) for four different types of beam loading.

13.4.3 Applying Singularity Functions to Beams

Let's consider five examples to demonstrate the economy of effort in solving for the equation of the elastic curve of a beam when using singularity functions. The examples will illustrate several concepts including:

1. Writing shear and bending moment equations for different beam loadings using singularity functions.
2. Writing the equations for the elastic curve and its slope for different beam loadings using singularity functions.
3. Solving for constants of integration that are generated in the integration process.
4. Integrating the pointed brackets that are inherent when using singularity functions.

EXAMPLE 13.7

For the simply supported beam shown in Fig. E13.7, write the equation for the loading function q(x). Then integrate this function to determine V(x) and M(x). Solve for the integration constants that arise in the integration process.

Fig. E13.7

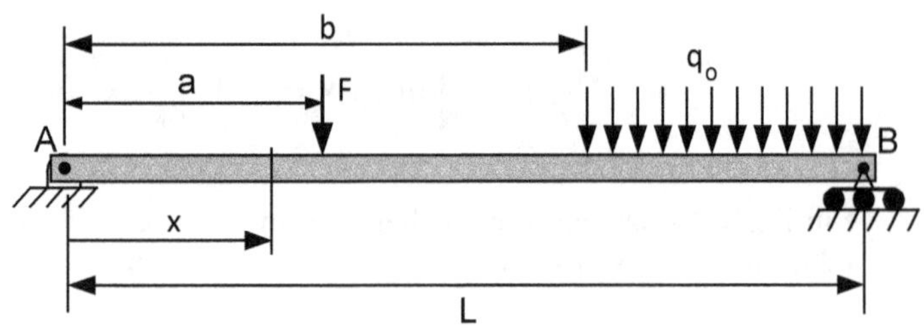

Solution:

Replace the pin and clevis support with a concentrated force R_L directed in the positive y direction. Beginning at x = 0, we must accommodate three loadings in writing q(x)—R_L, F and q_o. Using the results from Fig. 13.8, we write:

$$q(x) = -R_L\langle x-0\rangle^{-1} + F\langle x-a\rangle^{-1} + q_o\langle x-b\rangle^{0} \qquad \text{(a)}$$

where the sign of the term containing R_L is negative because it is in the positive y direction.

Integrate Eq. (a) using Eq. (10.20) and the integration rules for pointed brackets to obtain the shear force V(x) as:

$$V(x) = R_L\langle x-0\rangle^{0} - F\langle x-a\rangle^{0} - q_o\langle x-b\rangle^{1} + C_1 \qquad \text{(b)}$$

Integrate Eq. (b) using Eq. (10.18) and the integration rules to obtain the bending moment M(x) as:

$$M(x) = R_L\langle x-0\rangle^{1} - F\langle x-a\rangle^{1} - (q_o/2)\langle x-b\rangle^{2} + C_1x + C_2 \qquad \text{(c)}$$

The constants of integration C_1 and C_2 both are zero because M(0) = 0 and V(0) = R_L. In addition, because R_L occurs at the left end of the beam where x = 0, the pointed brackets for the first term in Eqs. (b) and (c) can be omitted. Hence, we write:

$$V(x) = R_L - F\langle x-a\rangle^{0} - q_o\langle x-b\rangle^{1} \qquad \text{(d)}$$

$$M(x) = R_Lx - F\langle x-a\rangle^{1} - (q_o/2)\langle x-b\rangle^{2} \qquad \text{(e)}$$

EXAMPLE 13.8

For the simply supported beam shown in Fig. E13.7, write the equation for the elastic curve and the slope of this curve as a function of x. Locate the position of the maximum deflection and determine y_{Max}. Assume EI_z is a constant with respect to x and the dimensions in Fig. E13.7 are a = 2 m, b = 4 m and L = 6m. The force F = 30 kN and the distributed load q_o = 12 kN/m. The beam is fabricated from a steel American Standard S381×64 section.

Solution:

Prepare a FBD of the entire beam and solve for the reaction load R_L at x = 0 by writing a moment equation to obtain:

$$R_L = 2F/3 + 2q_o/6 = (2)(30)/3 + (2)(12)/6 = 24 \text{ kN} \quad \text{(a)}$$

Substitute the result for M(x) from Example 13.7 into Eq. (13.4) and integrate to obtain:

$$EI_z \frac{d^2y}{dx^2} = 24x - 30 < x-2 >^1 - 6 < x-4 >^2 \quad \text{(b)}$$

$$EI_z \frac{dy}{dx} = 12x^2 - 15 < x-2 >^2 - 2 < x-4 >^3 + C_3 \quad \text{(c)}$$

$$EI_z y = 4x^3 - 5 < x-2 >^3 - \frac{1}{2} < x-4 >^4 + C_3 x + C_4 \quad \text{(d)}$$

The constants of integration are determined from the boundary conditions, which are:

$$y(0) = 0 \qquad y(6) = 0 \quad \text{(e)}$$

From the boundary condition y(0) = 0 it is evident that C_4 = 0. From the boundary condition y(6) = 0, we determine C_3 from Eq. (d) as:

$$EI_z y = 4(6)^3 - 5(4)^3 - \frac{1}{2}(2)^4 + C_3(6) = 0 \quad \Rightarrow \quad C_3 = -89.33 \text{ kN - m}^2 \quad \text{(f)}$$

We can then substitute Eq. (f) into Eqs. (c) and (d) and write the equations for the elastic curve and the slope of the elastic curve as:

$$EI_z \frac{dy}{dx} = 12x^2 - 15 < x-2 >^2 - 2 < x-4 >^3 - 89.33 \quad \text{(g)}$$

$$EI_z y = 4x^3 - 5 < x-2 >^3 - \frac{1}{2} < x-4 >^4 - 89.33x \quad \text{(h)}$$

To locate the position of the maximum deflection we use the fact that (dy/dx) = 0 when y is a maximum. Also, the position of the force F and its magnitude compared to q_o leads us to conclude that the maximum deflection will occur in the region 2 m < x < 4 m. Substituting this information into Eq. (g), we obtain:

$$EI_z \frac{dy}{dx} = 12x^2 - 15(x-2)^2 - 89.33 = 0 \quad \text{(i)}$$

Note that the term $<x - 4>^3 = 0$ for $x \leq 4$ m.

Solving the quadratic equation gives the position for the maximum deflection as:

$$x = 2.913 \text{ m} \qquad \text{(j)}$$

This result verifies the conclusion that the maximum location[2] would occur from 2 m < x < 4 m. The other root of Eq. (i)—x = 17.087 m—is neglected because it not realistic as the beam is only 6 m long.

Substituting Eq. (j) into Eq. (h) gives $EI_z\, y_{Max}$ as:

$$EI_z y_{Max} = \frac{4}{6}(2.913)^3 - 5(0.913)^3 - 0 - (89.33)(2.913) = -165.1 \text{ kN-m}^3 \qquad \text{(k)}$$

The negative sign is to be expected because the beam deflects downward while the y-axis is upward. Let E = 207 GPa for steel and reference Appendix D to find $I_z = 186 \times 10^6$ mm^4. Substituting these values into Eq. (k) yields:

$$y_{Max} = \frac{-(165.1 \times 10^3)(10^3)}{(207 \times 10^3)(186 \times 10^6)} = -4.288 \text{ mm} \qquad \text{(l)}$$

This deflection is very small compared to the dimensions of the beam (height h = 381 mm and length L = 6000mm.

EXAMPLE 13.9

A simply supported beam with a constant cross sectional area over its entire length is subjected to loads F and q as shown in Fig. E13.9. Determine the equation for the elastic curve and its slope. Also determine the maximum deflection y_{Max}. The dimension in Fig. E13.9 are a = 6 ft and L = 24 ft. The loads are F = 20 kip and q_o = 6 kip/ft. Assume the beam is fabricated from a steel wide flange W16×67 section.

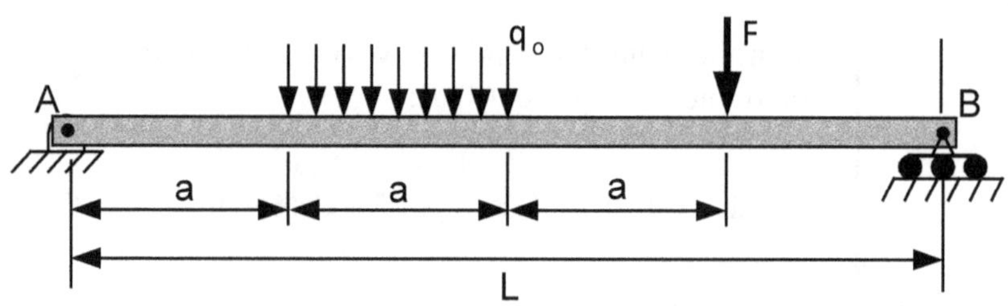

Fig. E13.9

Solution:

As we examine the uniformly distributed load in Fig. E13.9, we are faced with a problem. We can use a singularity function to activate q_o at x = a, but how do we turn it off at x = 2a? The answer is to introduce a new loading equivalent to the loading shown in Fig. E13.9 that is more amendable to using singularity functions. This new loading involves a downward acting q_o from $a \le x \le L$ and an upward acting q_o from $2a \le x \le L$. The two distributed loads cancel each other for $x \ge 2a$ producing the loading shown in Fig. E13.9a.

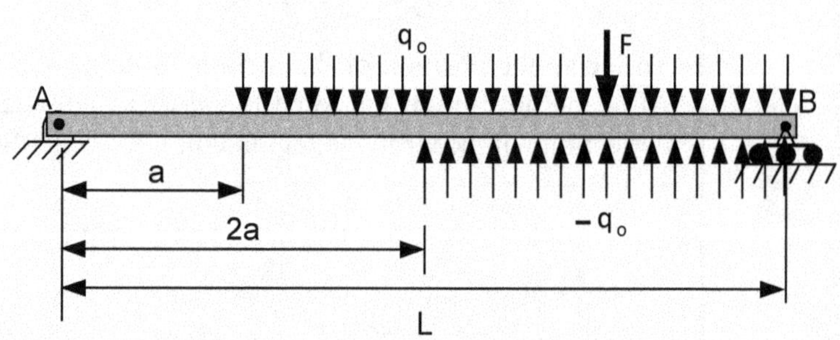

Fig. E13.9a

[2] The maximum deflection is usually near the center of the span of the beam even when the loading is not symmetric.

Let's write the equation for the loading function q(x) by using the results in Fig. 13.8 as:

$$q(x) = -R_L <x-0>^{-1} + q_o <x-a>^0 - q_o <x-2a>^0 + F <x-3a>^{-1} \quad \text{(a)}$$

where the sign of the term containing R_L is negative because it is in the positive y direction.

Integrate Eq. (a) using Eq. (10.20) and the integration rules for pointed brackets to obtain the shear force V(x) as:

$$V(x) = R_L - q_o <x-a>^1 + q_o <x-2a>^1 - F <x-3a>^0 + C_1 \quad \text{(b)}$$

Integrate Eq. (b) using Eq. (10.18) and the integration rules to obtain the bending moment M(x) as:

$$M(x) = R_L x - (q_o/2) <x-a>^2 + (q_o/2) <x-2a>^2 - F <x-3a>^1 + C_1 x + C_2 \quad \text{(c)}$$

The constants of integration C_1 and C_2 both are zero because M(0) = 0 and V(0) = R_L. Hence, we write:

$$V(x) = R_L - 6 <x-6>^1 + 6 <x-12>^1 - 20 <x-18>^0 \quad \text{(d)}$$

$$M(x) = R_L x - 3 <x-6>^2 + 3 <x-12>^2 - 20 <x-18>^1 \quad \text{(e)}$$

Prepare a FBD of the entire beam and solve for the reaction load R_L at x = 0 by writing a moment equation to obtain:

$$R_L = (q_o/4)(15) + F/4 = (6/4)(15) + 20/4 = 27.5 \text{ kip} \quad \text{(f)}$$

Substitute the result for M(x) from Eq. (e) into Eq. (13.4) and integrate to obtain:

$$EI_z \frac{d^2y}{dx^2} = 27.5x - 3 < x-6 >^2 + 3 < x-12 >^2 - 20 < x-18 >^1 \quad \text{(g)}$$

$$EI_z \frac{dy}{dx} = 13.75x^2 - < x-6 >^3 + < x-12 >^3 - 10 < x-18 >^2 + C_3 \quad \text{(h)}$$

$$EI_z y = 4.583x^3 - 0.250 < x-6 >^4 + 0.250 < x-12 >^4 - 3.333 < x-18 >^3 + C_3 x + C_4 \quad \text{(i)}$$

The constants of integration are determined from the boundary conditions, which are:

$$y(0) = 0 \qquad y(24) = 0 \quad \text{(i)}$$

From the boundary condition y(0) = 0 it is evident that $C_4 = 0$. From the boundary condition y(24) = 0, we determine C_3 as:

$$C_3 = -1{,}732 \text{ kip-ft}^2 \quad \text{(k)}$$

We can then substitute Eq. (k) into Eqs. (h) and (i) and write the equations for the elastic curve and the slope of the elastic curve as:

$$EI_z \frac{dy}{dx} = 13.75x^2 - < x-6 >^3 + < x-12 >^3 - 10 < x-18 >^2 - 1732 \quad \text{(l)}$$

$$EI_z y = 4.583x^3 - 0.250 < x-6 >^4 + 0.250 < x-12 >^4 - 3.333 < x-18 >^3 - 1732x \quad \text{(m)}$$

To locate the position of the maximum deflection we use the fact that (dy/dx) = 0 when y is a maximum. Also, the position of the force F and its magnitude compared to q_o leads us to conclude that the maximum deflection will occur in the region 6 ft < x < 12 ft. Substituting this information into Eq. (l), we obtain:

$$EI_z \frac{dy}{dx} = 13.75(x)^2 - (x-6)^3 - 1732 \; = 0 \qquad \text{(n)}$$

Note the terms $\langle x - 12\rangle^3$ and $\langle x - 18\rangle^2$ go to zero for x ≤ 12 ft.

Solving the cubic equation obtained by expanding Eq. (n) gives the position for the maximum deflection as:

$$x = 11.86 \text{ ft} \qquad \text{(o)}$$

The two other roots—x = 25.01 ft and – 5.113 ft)—are impossible because they do not lie on the beam. This result verifies the conclusion that the maximum location would occur slightly to the left of the center of the beam where 6 ft < x < 12 ft.

Substituting Eq. (o) into Eq. (m) gives $EI_z\, y_{Max}$ as:

$$EI_z y_{Max} = 4.583(11.86)^3 - 0.250(11.86-6)^4 - (1732)(11.86) = -13{,}191 \text{ kip-ft}^3 \qquad \text{(p)}$$

The negative sign is to be expected because the beam deflects downward while the y-axis is upward. Let E = 30 × 10^6 psi for steel and reference Appendix D to find I_z = 954 in.^{4} for the wide flange section. Substituting these values into Eq. (p) yields:

$$y_{Max} = \frac{-(13{,}191)(12^3)}{(30\times 10^3)(954)} = -0.7964 \text{ in.} \qquad \text{(q)}$$

Again it is noted that the magnitude of the deflection of the beam is small compared to its height.

EXAMPLE 13.10

A cantilever beam with a constant cross sectional area over its entire length is subjected to a concentrated force F and a ramp load with a maximum distributed force of q_o as shown in Fig. E13.10. Determine the equation for the elastic curve and its slope. Also determine the maximum deflection y_{Max}. The dimensions are a = 4 m and L = 12 m and the loads are F = 24 kN and q_q = 8 kN/m. Assume the beam is aluminum with a W686 × 217 wide flange section.

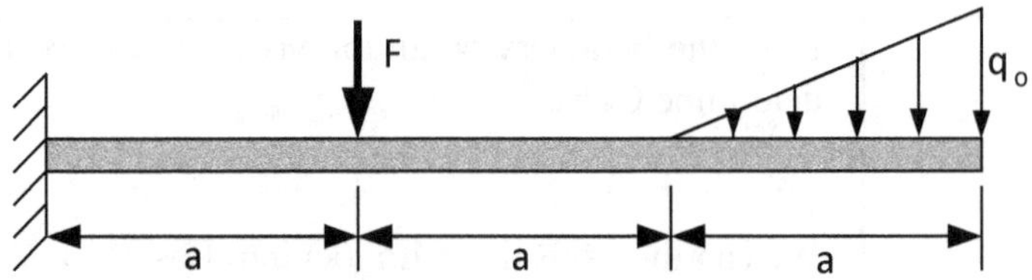

Fig. E13.10

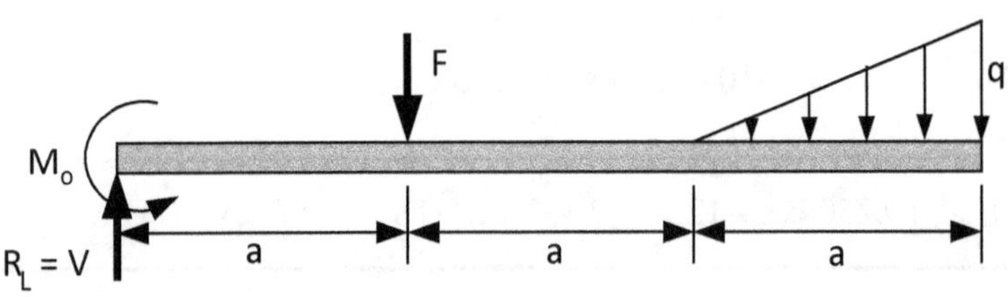

Fig. E13.10a

Solution:

Let's begin by drawing a FBD of the entire beam to show the moment and shear force at the built-in support as shown above in Fig. E13.10a.

At the support on the left side of the beam $\Sigma F = 0$ and $\Sigma M = 0$ yields:

$$R_L = V(0) = F + q_o a/2 \qquad \text{(a)}$$

$$M_o = Fa + (4q_o a^2)/3 \qquad \text{(b)}$$

Let's write the equation for the loading function q(x) by using the results in Fig. 13.4 as:

$$q(x) = M_o \langle x - 0\rangle^{-2} - R_L \langle x - 0\rangle^{-1} + F \langle x - a\rangle^{-1} + (q_o/a) \langle x - 2a\rangle^{1} \qquad \text{(c)}$$

Integrate Eq. (c) using Eq. (10.20) and the integration rules for pointed brackets to obtain the shear force V(x) as:

$$V(x) = - M_o \langle x - 0\rangle^{-1} + R_L \langle x - 0\rangle^{0} - F \langle x - a\rangle^{0} - (q_o/2a) \langle x - 2a\rangle^{2} + C_1 \qquad \text{(d)}$$

Integrate Eq. (b) using Eq. (10.18) and the integration rules to obtain the bending moment M(x) as:

$$M(x) = - M_o \langle x - 0\rangle^{0} + R_L \langle x - 0\rangle^{1} - F \langle x - a\rangle^{1} - (q_o/6a) \langle x - 2a\rangle^{3} + C_1 x + C_2 \qquad \text{(e)}$$

The constants of integration C_1 and C_2 both are zero because $M(0) = - M_o$ and $V(0) = R_L$. Hence, we write:

$$V(x) = - M_o /x + R_L - F \langle x - a\rangle^{0} - (q_o/2a) \langle x - 2a\rangle^{2} \qquad \text{(f)}$$

$$M(x) = - M_o + R_L x - F \langle x - a\rangle^{1} - (q_o/6a) \langle x - 2a\rangle^{3} \qquad \text{(g)}$$

Substitute numerical values for F and q_o in Eqs. (a) and (b) to obtain:

$$R_L = V(0) = F + q_o a/2 = 24 + (8)(4)/2 = 40 \text{ kN} \qquad \text{(h)}$$

$$M_o = Fa + (4q_o a^2)/3 = (24)(4) + (4)(8)(4)^2/3 = 266.7 \text{ kN-m} \qquad \text{(i)}$$

Substitute the result for M(x) from Eq. (g) into Eq. (13.4) and simplify using Eqs. (h) and (i). Then integrate the result to obtain:

$$EI_z \frac{d^2y}{dx^2} = -266.7 + 40x - 24 \langle x - 4\rangle^{1} - \frac{1}{3} \langle x - 8\rangle^{3} \qquad \text{(j)}$$

$$EI_z \frac{dy}{dx} = -266.7x + 20x^2 - 12 \langle x - 4\rangle^{2} - \frac{1}{12} \langle x - 8\rangle^{4} + C_3 \qquad \text{(k)}$$

$$EI_z y = -133.4x^2 + 6.667x^3 - 4 \langle x - 4\rangle^{3} - \frac{1}{60} \langle x - 8\rangle^{5} + C_3 x + C_4 \qquad \text{(l)}$$

The constants of integration are determined from the boundary conditions, which are:

$$y(0) = 0 \qquad (dy/dx)(0) = 0 \qquad \text{(m)}$$

From the boundary condition y(0) = 0 it is evident that $C_4 = 0$, and from the boundary condition (dy/dx)(0) = 0, we determine $C_3 = 0$:

We can set $C_3 = C_4 = 0$ in Eqs. (l) and (k) and write the equations for the elastic curve and the slope of the elastic curve as:

$$EI_z \frac{dy}{dx} = -266.7x + 20x^2 - 12 < x-4 >^2 - \frac{1}{12} < x-8 >^4 \quad \text{(n)}$$

$$EI_z y = -133.4x^2 + 6.667x^3 - 4 < x-4 >^3 - \frac{1}{60} < x-8 >^5 \quad \text{(o)}$$

It is evident from inspection that the maximum deflection of the cantilever beam is at its free end where x = L = 12 m. Accordingly, we write:

$$EI_z y_{Max} = -133.4(12)^2 + 6.667(12)^3 - 4(8)^3 - \frac{1}{60}(4)^5 = -9{,}754 \text{ kN - m}^3 \quad \text{(p)}$$

Let E = 72 GPa for aluminum and reference Appendix D to find $I_z = 2345 \times 10^6$ mm^4 for the wide flange section. Substituting these values into Eq. (p) yields:

$$y_{Max} = \frac{-(9{,}754 \times 10^3)(10^3)^3}{(72 \times 10^3)(2345 \times 10^6)} = -57.77 \text{ mm} \quad \text{(q)}$$

EXAMPLE 13.11

A simply supported beam with a constant cross sectional area over its entire length is designed with an overhang as shown in Fig. E13.11. The beam carries a uniformly distributed load q_o and a concentrated force F. Determine the equation for the elastic curve and its slope. Also determine the maximum deflection y_{Max}. The dimensions are a = 5 ft and L = 20 ft and the loads are F = 15 kip and q_q = 1.2 kip/ft. Assume an aluminum beam fabricated from a 12 × 52 structural tee section.

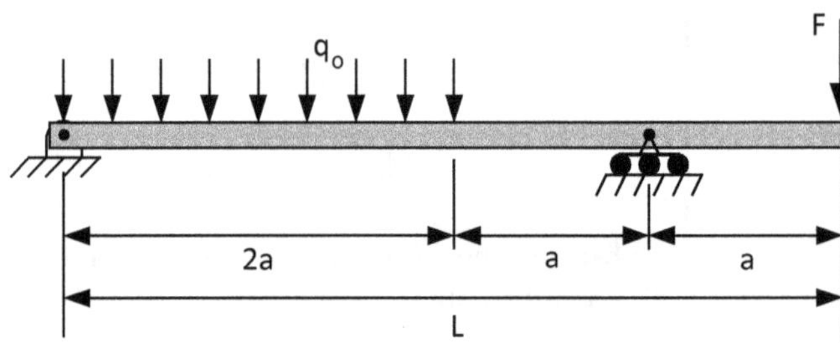

Fig. E13.11

Solution:

Let's begin by determining the reactions R_L and R_R at the two supports. Prepare a FBD and take moments about the right hand support to obtain:

$$R_L = [4aq_o - F]/3 = [(4)(5)(1.2) - 15]/3 = 3 \text{ kip} \quad \text{(a)}$$

$$R_R = 2aq_o + F - R_L = 12 + 15 - 3 = 24 \text{ kip} \quad \text{(b)}$$

When we attempt to write the equation for the loading function q(x) by using the results in Fig. 13.4 we note a difficulty. If we accommodate the uniform load with the term $q_o < x - 0 >^0$ it applies that loading over the entire length of the beam. To remove the uniform loading from the right half of the beam, we must introduce an equal and opposite loading that begins at x = 2a and continues to the right side of the beam as shown in Fig. E13.11a.

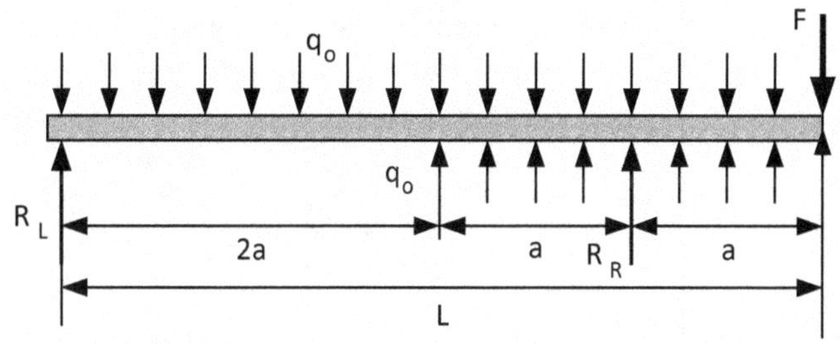

Fig. E13.11a

Writing the relation for the loading shown in Fig. E13.11a yields:

$$q(x) = -R_L \langle x-0\rangle^{-1} + q_o \langle x-0\rangle^{0} - q_o \langle x-2a\rangle^{0} - R_R \langle x-3a\rangle^{-1} \qquad \text{(c)}$$

Integrate Eq. (c) using Eqs. (10.20) and the integration rules for pointed brackets to obtain the shear force V(x). Also substitute numerical values for a, F, q_o, R_L and R_R into Eq. (c).

$$V(x) = +3 \langle x-0\rangle^{0} - 1.2 \langle x-0\rangle^{1} + 1.2 \langle x-10\rangle^{1} + 24 \langle x-15\rangle^{0} + C_1 \qquad \text{(d)}$$

Integrate Eq. (d) using Eqs. (10.18) and the integration rules to obtain the bending moment M(x) as:

$$M(x) = +3 \langle x-0\rangle^{1} - 0.6 \langle x-0\rangle^{2} + 0.6 \langle x-10\rangle^{2} + 24 \langle x-15\rangle^{1} + C_1 x + C_2 \qquad \text{(e)}$$

The constants of integration C_1 and C_2 both are zero because M(0) = 0 and V(0) = R_L. Hence, we write:

$$V(x) = +3 - 1.2x + 1.2 \langle x-10\rangle^{1} + 24 \langle x-15\rangle^{0} \qquad \text{(f)}$$

$$M(x) = +3x - 0.6x^2 + 0.6 \langle x-10\rangle^{2} + 24 \langle x-15\rangle^{1} \qquad \text{(g)}$$

Substitute the result for M(x) from Eq. (g) into Eq. (13.4) and then integrate to obtain:

$$EI_z \frac{d^2y}{dx^2} = 3x - 0.6x^2 + 0.6 \langle x-10\rangle^{2} + 24 \langle x-15\rangle^{1} \qquad \text{(h)}$$

$$EI_z \frac{dy}{dx} = 1.5x^2 - 0.2x^3 + 0.2 \langle x-10\rangle^{3} + 12 \langle x-15\rangle^{2} + C_3 \qquad \text{(i)}$$

$$EI_z y = 0.5x^3 - 0.05x^4 + 0.05 \langle x-10\rangle^{4} + 4 \langle x-15\rangle^{3} + C_3 x + C_4 \qquad \text{(j)}$$

To solve for C_3 and C_4, we use the boundary conditions that y(0) = 0 and y(15) = 0. The first of these boundary conditions yields $C_4 = 0$. The second boundary condition gives:

$$0 = 0.5(15)^3 - 0.05(15)^4 + 0.05(5)^4 + 4(0) + C_3(15) \quad \Rightarrow \quad C_3 = 54.16 \text{ kip-ft}^2 \qquad \text{(k)}$$

We can then substitute Eq. (k) into Eqs. (i) and (j) and write the equations for the elastic curve and the slope of the elastic curve as:

$$EI_z \frac{dy}{dx} = 1.5x^2 - 0.2x^3 + 0.2 \langle x-10\rangle^{3} + 12 \langle x-15\rangle^{2} + 54.16 \qquad \text{(l)}$$

$$EI_z y = 0.5x^3 - 0.05x^4 + 0.05 \langle x-10\rangle^{4} + 4 \langle x-15\rangle^{3} + 54.16x \qquad \text{(m)}$$

There are two possibilities for the location of the maximum deflection of the beam—between the supports or at the free end of the overhanging beam. If y_{Max} occurs between the supports, then $(dy/dx) = 0$ when y is a maximum. Let's first assume that the maximum will be near the center of the beam with $x \leq 15$ so that we can write Eq. (l) as:

$$EI_z \frac{dy}{dx} = 1.5x^2 - 0.2x^3 + 0.2(x-10)^3 + (0) + 54.16 = 0 \qquad \text{(n)}$$

Expanding and solving Eq. (n) gives the location of y_{Max} as:

$$x = 10.14 \text{ ft} \qquad \text{or} \qquad 3.196 \text{ ft} \qquad \text{(o)}$$

Both roots of Eq. (n) must be checked to determine the one which is correct. Substituting Eq. (o) into Eq. (m) gives $EI_z\, y_{Max}$ as:

$$EI_z y_{Max} = 0.5(10.14)^3 - 0.05(10.14)^4 + 0.05(0.14)^4 + 0 + 54.16(10.14) = 541.9 \text{ kip-ft}^3 \qquad \text{(p)}$$

$$EI_z y_{Max} = 0.5(3.196)^3 - 0.05(3.196)^4 + 0 + 0 + 54.16(3.196) = 184.2 \text{ kip-ft}^3 \qquad \text{(q)}$$

The larger of the two values is given by Eq. (p) as 541.9 kip-ft^3.

In this problem the maximum deflection is positive indicating the beam deflects in the positive y direction with a deflection curve of the shape shown in Fig. E13.11a. The result given by Eq. (p) is the maximum deflection between the supports, but it may not be the maximum for the beam. It is necessary to check the deflection at the free end of the beam.

Fig. E13.11b

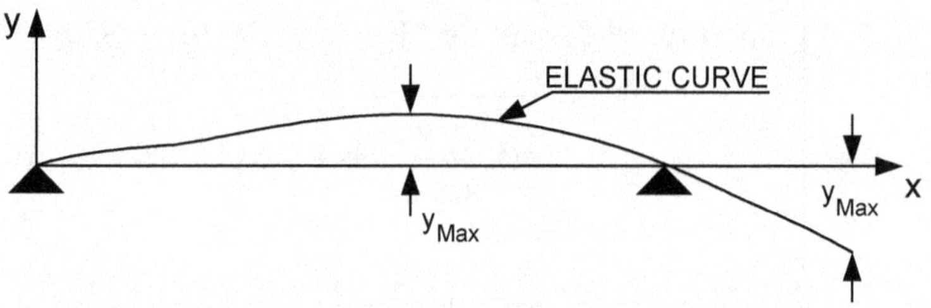

To determine the deflection at the end of the beam where x = 20 ft, we employ Eq. (m), set x = 20 ft and write:

$$EI_z y_{Max} = 0.5(20)^3 - 0.05(20)^4 + 0.05(10)^4 + 4(5)^3 + 54.16(20) = -1917 \text{ kip-ft}^3 \qquad \text{(r)}$$

Comparing the results from Eq. (p) and (r) indicates that the maximum deflection occurs at the free end of the overhanging beam and is given by Eq. (r).
Let $E = 10.4 \times 10^3$ ksi for aluminum and reference Appendix D to find $I_z = 189$ in.^{4} for the structural tee section. Substituting these values into Eq. (r) yields:

$$y_{Max} = \frac{-(1917)(12)^3}{(10.4 \times 10^3)(189)} = -1.685 \text{ in.} \qquad \text{(s)}$$

13.5 SUPERPOSITION CONCEPTS

The principle of superposition is useful in solving for the deflection of beams that are loaded with two or more forcing functions. We have presented the integration method that is effective when the bending moment equation M(x) is valid from $0 \leq x \leq L$. This method becomes cumbersome when two or more bending moment equations are necessary to describe M(x) over the entire length of the beam. We introduced singularity functions that were employed with the integration method to significantly reduce the mathematical complexities involved in solving for constants of integration. Superposition represents a third method for determining the equation of the elastic curve.

Superposition involves combining together the results of the deflection from two or more separate solutions to yield a new solution containing all of the loading functions involved in the separate solutions. This concept of superposition of displacement functions is demonstrated better by the graphic shown in Fig. 13.9. Three concentrated forces—F1, F2 and F3—load the beam on the left in Fig. 13.9. These forces are located a distance a, b and c from the left side of the beam, respectively. If the solution is known for the elastic curve for a beam loaded with a single concentrated force, we can use superposition to add together the solutions for the displacement y(x) associated with the three separate forces.

One of the advantages of the method of superposition is that we make use of the many solutions that already exist. Mechanics of materials is a subject with a long history, and the classic problems of bending of beams were solved many years ago. We have provided a catalog of several of these solutions for your convenience in Appendix

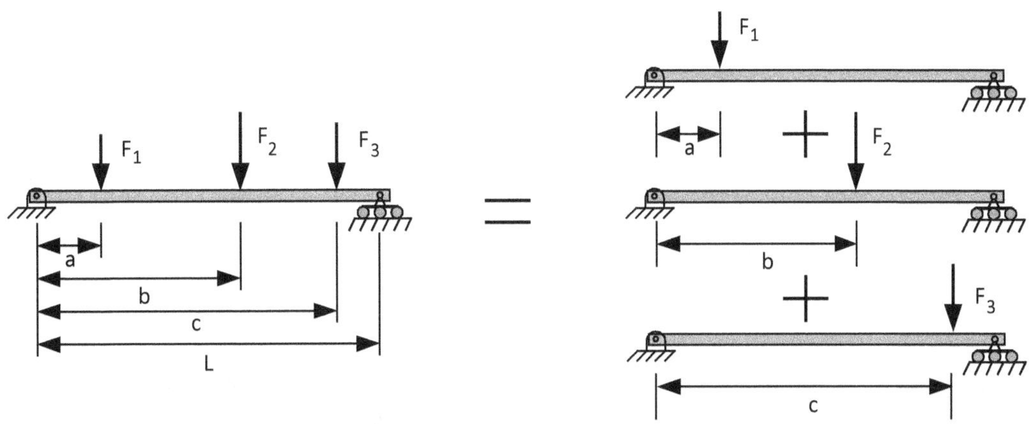

Fig. 13.9 Superposition is adding solutions for y(x) for a single loading function to obtain solutions for two or more loading functions.

There are two requirements that must be satisfied before using the principle of superposition. The first is the linearity requirement. The solution for y(x) must be linear with respect to the loading functions. We have solved several different problems determining y(x) for several different loading functions including concentrated forces, concentrated moments, uniformly distributed loads and ramp functions. In all cases we found that y(x) was linear with respect to the loading function. The second requirement pertains to the integrity of the geometry of the beam during the loading. Does the beam change its dimensions? The beam on the left in Fig. 13.9 must have essentially the same geometry as each of the three beams on the right of Fig. 13.9. If the beam deformations are small, we can conclude that the geometry does not change significantly as the beam is loaded. We have solved for the maximum deflection in several problems and noted that the deflection of the beam was always small when compared to its length. Hence, we conclude that the principle of superposition is valid and can be employed as another method for determining the deflection of a beam in bending.

We will demonstrate the methods of superposition with several examples that show combinations of different loading functions applied to beams supported in different ways.

EXAMPLE 13.12

Consider the simply supported beam subjected to a uniformly distributed load and a concentrated force as illustrated in Fig. E13.12. Determine the deflection at mid span and the slope at x = L. The beam is fabricated from a steel wide flange section W203×60 with a length L = 6 m. The force F = 50 kN is applied at a = 4 m and q_o = 10 kN/m = 10 N/mm.

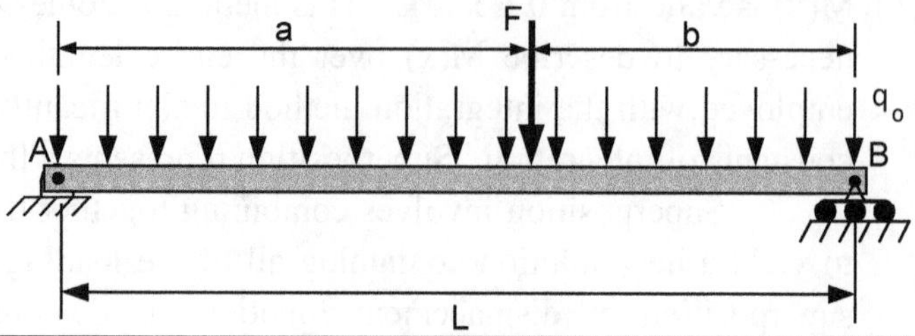

Fig. E13.12

Solution:

Let's employ the principle of superposition to solve this problem. Two identical simply supported beams each with a single load as illustrated in Fig. E13.12a represent the simply supported beam with the two different loadings.

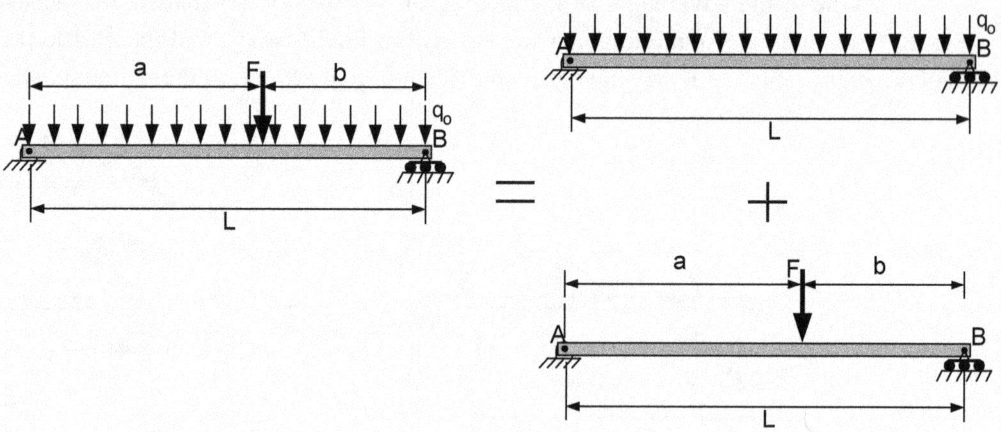

Fig. E13.12a

Consider first the simply supported beam with the uniformly distributed load. Reference to the solutions presented in Appendix E gives the equations for the deflection and slope as:

$$y = -\frac{q_0 x}{24EI_z}\left(x^3 - 2Lx^2 + L^3\right) \tag{a}$$

$$\left(\frac{dy}{dx}\right)_{x=0,L} = \pm\frac{q_0 L^3}{24EI_z} \tag{b}$$

Note that $I_z = 60.8 \times 10^6$ mm^4 is given in Appendix D for the specified wide flange section. Substituting numerical parameters into Eq. (a) and Eq. (b) gives

$$y_{x=L/2} = -\frac{q_0 L}{48EI_z}\left(\frac{L^3}{8} - \frac{L^3}{2} + L^3\right) = -\frac{5q_0 L^4}{384EI_z} \tag{c}$$

$$y_{x=L/2} = -\frac{5q_0 L^4}{384EI_z} = -\frac{(5)(10)(6\times10^3)^4}{(384)(207\times10^3)(60.8\times10^6)} = -13.41 \text{ mm} \tag{d}$$

$$\left(\frac{dy}{dx}\right)_{x=L} = \frac{q_0 L^3}{24EI_z} = \frac{(10)(6\times10^3)^3}{(24)(207\times10^3)(60.8\times10^6)} = 7.150\times10^{-3} \text{ Radians} \tag{e}$$

Next, reference the solution for the simply supported beam subjected to a concentrated load at position x= a in Appendix E and write the relations for the displacement and the slope as:

$$y = -\frac{Fbx}{6EI_zL}\left(L^2 - b^2 - x^2\right) \tag{f}$$

$$\left(\frac{dy}{dx}\right)_{x=L} = +\frac{Fab(L+a)}{6EI_zL} \tag{g}$$

Substituting x = L/2 and numerical parameters into Eq. (f) and solving gives:

$$y_{X=L/2} = -\frac{Fb}{48EI_z}\left(3L^2 - 4b^2\right) = -\frac{(50\times10^3)(2\times10^3)}{(48)(207\times10^3)(60.8\times10^6)}\left[3(36)-4(4)\right]\times10^6 = -15.23 \text{ mm} \tag{h}$$

Substituting numerical parameters Into Eq. (g) gives:

$$\left(\frac{dy}{dx}\right)_{x=L} = \frac{Fab(L+a)}{6EI_zL} = \frac{(50\times10^3)(4\times10^3)(2\times10^3)(10\times10^3)}{6(207\times10^3)(60.8\times10^6)(6\times10^3)} = 8.828\times10^{-3} \text{ Radians} \tag{i}$$

The final step in this process is to superimpose the results from the two solutions. Adding the results from Eqs. (d) and (h) gives the total displacement at mid span as:

$$y_{x=L/2} = -(13.41 + 15.23) = -28.64 \text{ mm} \tag{j}$$

Similarly, the slope at x = L is given by adding the results from Eqs. (e) and (i) to obtain:

$$(dy/dx)_{x=L} = (7.150 + 8.828)\times10^{-3} = 0.01598 \text{ Radians} = 0.9155° \tag{k}$$

The use of the superposition principle enables one to solve for deflections in beams subjected to multiple loads by considering identical beams subjected to a single load function. The solutions for the beams with single loadings are then added to obtain the results for the beam with multiple loads. It is a simple technique that makes use of the existing solutions that are tabulated in Appendix E.

EXAMPLE 13.13

Consider the simply supported beam subjected to a ramp type distributed load and a concentrated moment as illustrated in Fig. E13.13. Determine the deflection at midspan and the slope at both ends of the beam. The steel beam is fabricated from an American Standard flange section S10×35 with a length L = 24 ft. The concentrated moment M_o = 2,400 ft-lb is applied at x = 0. The ramp load is specified with q_o = 1.44 kip/ft = 120 lb/in.

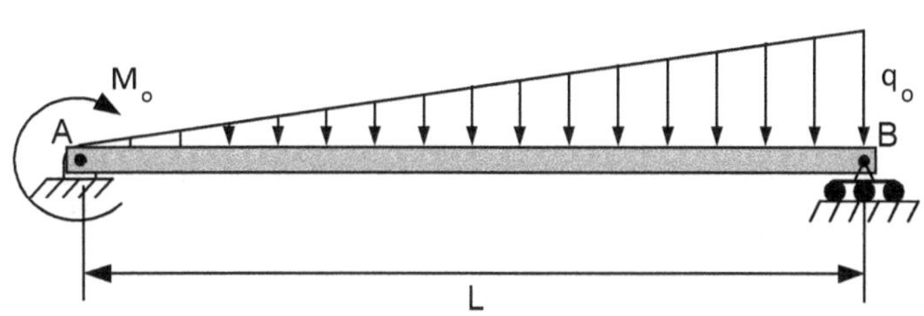

Fig. E13.13

Solution:

We again use the principle of superposition in solving this problem. Two identical simply supported beams each with a single load, as illustrated in Fig. E13.13a, represent the simply supported beam with the ramp distribution and the concentrated moment loads.

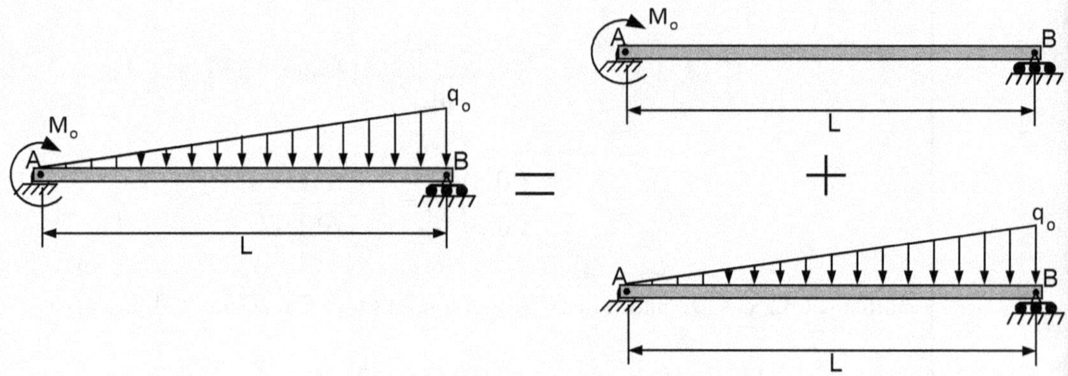

Fig. E13.13

First, consider the solution for the simply supported beam subjected to a concentrated moment at x = 0 that is given in Appendix E. Write the relations for the displacement and the slope as:

$$y = -\frac{M_0 x}{6EI_z L}\left(x^2 - 3Lx + 2L^2\right) \tag{a}$$

$$\left(\frac{dy}{dx}\right)_{x=0} = -\frac{ML}{3EI_z} \tag{b}$$

$$\left(\frac{dy}{dx}\right)_{x=L} = \frac{ML}{6EI_z} \tag{c}$$

At midspan x = L/2 and Eq. (a) reduces to:

$$y = -\frac{3M_0L^2}{48EI_z} \tag{d}$$

Note that $I_z = 147$ in.^{4} is given in Appendix D for the specified American Standard section. Substituting numerical parameters into Eqs. (b), (c) and (d) gives:

$$y = -\frac{3M_0L^2}{48EI_z} = -\frac{3(2400)(12)(24)^2(12)^2}{48(30\times10^6)(147)} = -0.03385 \text{ in.} \tag{e}$$

$$\left(\frac{dy}{dx}\right)_{x=0} = -\frac{ML}{3EI_z} = -\frac{(2400)(12)(24)(12)}{3(30\times10^6)(147)} = -6.269\times10^{-4} \text{ Radians} \tag{f}$$

$$\left(\frac{dy}{dx}\right)_{x=L} = \frac{ML}{6EI_z} = 3.135\ \times10^{-4} \text{ Radians} \tag{g}$$

Consider next the simply supported beam with the ramp type distributed load. Reference to the solutions presented in Appendix E gives the equations for the deflection and slope as:

$$y = -\frac{q_0 x}{360EI_zL}\left(3x^4 - 10L^2x^2 + 7L^4\right) \quad \text{(h)}$$

$$\left(\frac{dy}{dx}\right)_{x=0} = -\frac{7q_0L^3}{360EI_z} \quad \text{(i)}$$

$$\left(\frac{dy}{dx}\right)_{x=L} = \frac{q_0L^3}{45EI_z} \quad \text{(j)}$$

Substitute x = L/2 into Eq. (h) to obtain:

$$y_{x=L/2} = -\frac{q_0}{(2)(360)EI_z}\left(\frac{3L^4}{16} - \frac{10L^4}{4} + 7L^4\right) = -\frac{75q_0L^4}{11{,}520EI_z} \quad \text{(k)}$$

Substituting numerical parameters into Eqs. (i), (j) and (k) gives:

$$y_{x=L/2} = -\frac{75q_0L^4}{11{,}520EI_z} = -\frac{(75)(120)(24)^4(12)^4}{11{,}520(30\times10^6)(147)} = -1.219 \text{ in.} \quad \text{(l)}$$

$$\left(\frac{dy}{dx}\right)_{x=0} = -\frac{7q_0L^3}{360EI_z} = -\frac{7(120)(24)^3(12)^3}{360(30\times10^6)(147)} = -0.01264 \text{ Radians} \quad \text{(m)}$$

$$\left(\frac{dy}{dx}\right)_{x=L} = \frac{q_0L^3}{45EI_z} = \frac{(120)(24)^3(12)^3}{45(30\times10^6)(147)} = 0.01444 \text{ Radians} \quad \text{(n)}$$

The final step in this process is to superimpose the results from the two solutions. Adding the results from Eqs. (e) and (l) gives the total displacement at mid span as:

$$y_{x=L/2} = -(0.03385 + 1.219) = -1.253 \text{ in.} \quad \text{(o)}$$

Similarly the slope at x = 0 and x = L is obtained by adding the results to obtain:

$$(dy/dx)_{x=0} = -(6.269\times10^{-4} + 0.01264) = -0.01327 \text{ Radian} = 0.7601° \quad \text{(p)}$$

$$(dy/dx)_{x=L} = (0.0003135 + 0.01444) = 0.01475 \text{ Radian} = 0.8453° \quad \text{(q)}$$

The use of the superposition principle enables one to solve for deflections in beams subjected to multiple loads by considering identical beams subjected to a single load function. The solutions for the beams with single loadings are then added to obtain the results for the beam with multiple loads.

EXAMPLE 13.14

Consider the cantilever beam subjected to a concentrated force applied at its free end and a uniformly distributed load over the left half of the beam as shown in Fig. E13.14. Determine the deflection and the slope at the free end of the beam. The steel beam is fabricated from a structural tee section WT178 × 61 with a length L = 5.0 m. The concentrated force is F = 2 kN and the uniformly distributed load is specified with q_o = 2.0 kN/m = 2.0 N/mm.

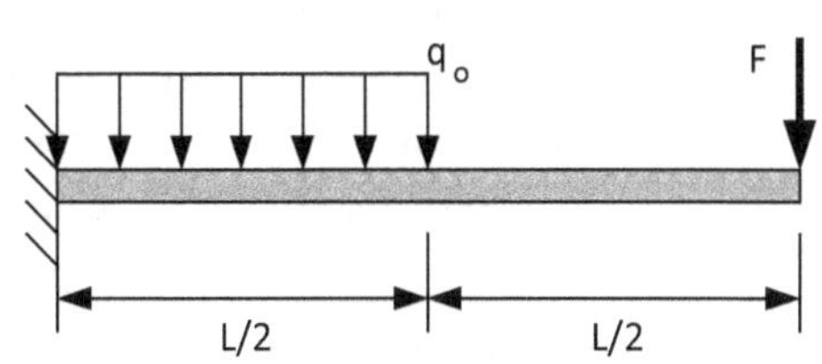

Fig. E13.14

Solution:

Apply the principle of superposition by dividing the loading into two separate parts as shown in Fig. E13.14a.

Fig. E13.14a

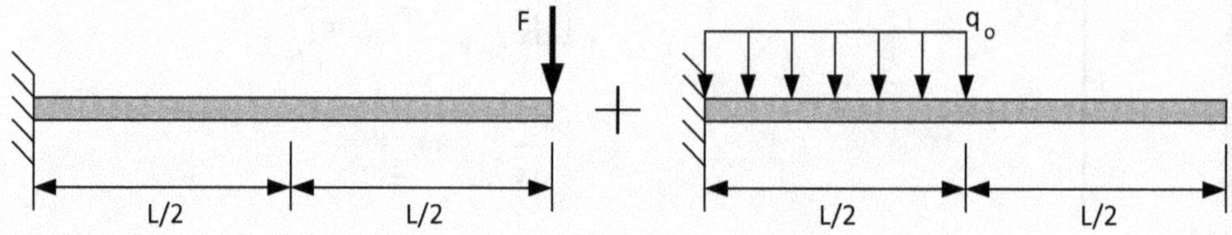

Consider first the solution for the cantilever beam with a concentrated load applied at its free end that is presented in Appendix E. We write the expressions for the deflection and the slope at the free end and substitute numerical parameters to obtain:

$$y_{x=L} = -\frac{FL^3}{3EI_z} = -\frac{2000(5000)^3}{3(207\times10^3)(17.1\times10^6)} = -23.54 \text{ mm} \qquad \text{(a)}$$

$$\left(\frac{dy}{dx}\right)_{x=L} = -\frac{FL^2}{2EI_z} = -\frac{2000(5000)^2}{2(207\times10^3)(17.1\times10^6)} = -7.063\times10^{-3} \text{ Radians} \qquad \text{(b)}$$

The value of $I_z = 17.1 \times 10^6$ mm^4 was determined from Appendix D.

Next consider the solution for the cantilever beam with a uniformly distributed load applied over its left half. Using the solution presented in Appendix E, we write the expressions for the deflection and the slope at the free end and substitute numerical parameters to obtain:

$$y_{x=L} = -\frac{7q_0L^4}{384EI_z} = -\frac{7(2)(5000)^4}{384(207\times10^3)(17.1\times10^6)} = -6.437 \text{ mm} \qquad \text{(c)}$$

$$\left(\frac{dy}{dx}\right)_{x=L} = -\frac{q_0L^3}{48EI_z} = -\frac{2(5000)^3}{48(207\times10^3)(17.1\times10^6)} = -1.471\times10^{-3} \text{ Radians} \qquad \text{(d)}$$

Finally we add the results for the deflections and the slopes to obtain:

$$y_{x=L} = -(23.54 + 6.437) - 29.98 \text{ mm} \qquad \text{(e)}$$

$$\left(\frac{dy}{dx}\right)_{x=L} = -(7.063 + 1.471)\times10^{-3} = 0.008534 \text{ Radian} = 0.4890° \qquad \text{(f)}$$

13.6 STATICALLY INDETERMINATE BEAMS

In most instances beams are statically determinate. The reactions at the supports may be determined from the three applicable equations of equilibrium—$\Sigma F_x = 0$, $\Sigma F_y = 0$ and $\Sigma M_o = 0$. However, in some cases beams are designed with additional supports that produce additional reaction forces making the beam statically indeterminate. Two common examples of statically indeterminate beams are illustrated in Fig. 13.10—the propped cantilever beam and the continuously supported beam.

The propped cantilever has one reaction force R_{By} at the prop and three possible reaction forces R_{Ax}, R_{Ay} and M_A at the built-in end. Three possible equations and four unknowns yield a statically indeterminate beam. For the continuous beam, two reaction forces are possible at the clevis and one at each of the roller supports at locations B and C. Clearly, the continuous beam is statically indeterminate.

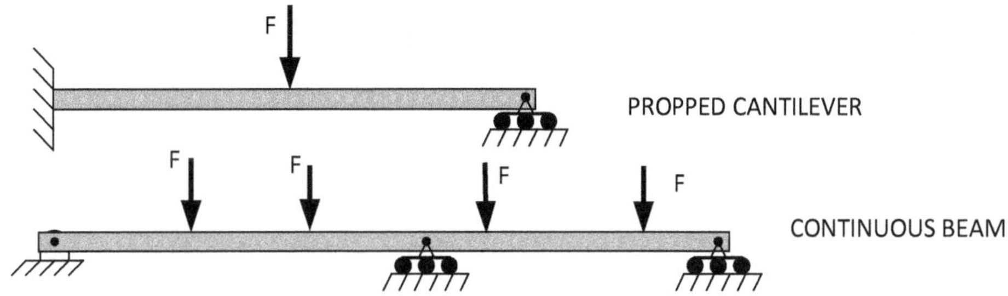

Fig. 13.10 Two examples of statically indeterminate beams.

There are several methods to solve for the additional reaction force; however, we supplement the equilibrium relations with a deformation equation to enable a solution. We will demonstrate two different methods to determine the reaction forces for statically indeterminate beams including the method of superposition and the method of integration with singularity functions.

13.6.1 Superposition Methods

In applying the method of superposition to the solution of statically indeterminate beams, we supplement the equilibrium equations by writing additional equations based on the deformation of the beam. This is accomplished by representing the statically indeterminate beam with two beams, each of which is statically determinate. The deflections of these two beams are added to yield the boundary conditions imposed on the statically indeterminate beam. This process yields the reaction force at one of the supports. Let's consider the propped cantilever beam as an example to demonstrate this approach.

EXAMPLE 13.15

Determine the reaction force at the prop of the cantilever beam shown in Fig. E13.15. The force F = 3,000 lb. The steel beam, 18 ft long, is fabricated from a wide flange section specified as W6 × 25. After solving for the reaction force, solve for the deflection at mid span.

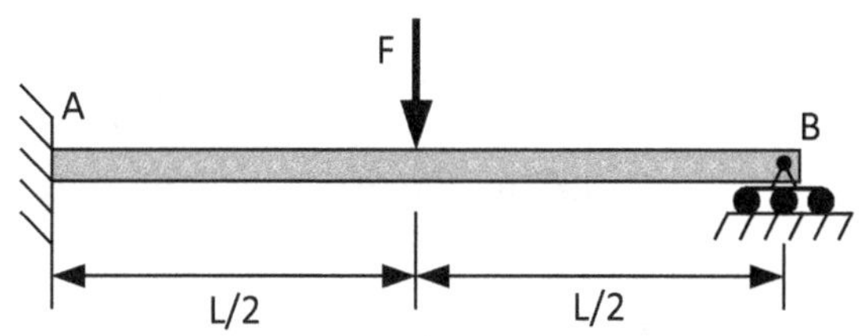

Fig. E13.15

Solution:

The prop at B causes the cantilever beam to be statically indeterminate. It may have been added to reduce the deflection or the stress of the beam; however, the prop complicates the analysis.

We begin the analysis by replacing the prop with a reaction force R and noting the boundary condition at x = L as:

$$y(L) = 0 \tag{a}$$

We then apply the principle of superposition and represent the beam in Fig. E13.15 with the two statically determinate beams shown in Fig. E13.15a.

Fig. E13.15a

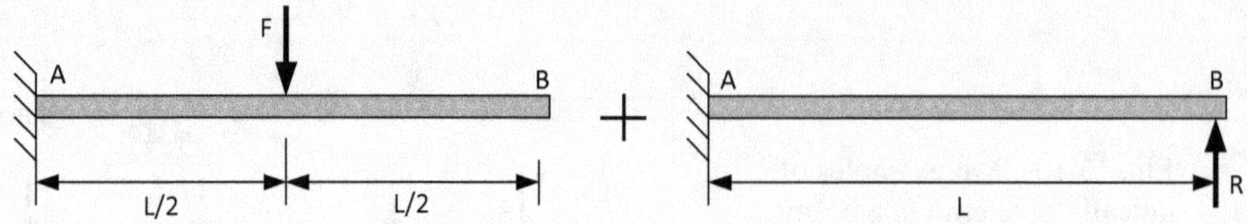

Let's determine the deflection at the free end for each of the two beams using the solutions tabulated in Appendix E. For the cantilever beam with the force F applied at midspan, we write:

$$y_{x=L} = -\frac{5FL^3}{48EI_z} \quad \text{(b)}$$

And for the cantilever beam with the upward directed reaction force, we write:

$$y_{x=L} = \frac{RL^3}{3EI_z} \quad \text{(c)}$$

However, the end of the cantilever beam is propped and y(L) = 0. Hence, it is clear from Eqs. (b) and (c) that:

$$y_{x=L} = \frac{RL^3}{3EI_z} - \frac{5FL^3}{48EI_z} = 0 \quad \text{(d)}$$

Solving Eq. (d) and inserting numerical parameters gives:

$$R = 5F/16 = 5(3000)/16 = 937.5 \text{ lb} \quad \text{(e)}$$

It is interesting that the partitioning of the force F between the built-in support and the prop is independent of the length of the beam and its stiffness (EI_z).

Let's continue the analysis using the result for the reaction force R and the method of superposition to determine the deflection at mid span. From Appendix E, we write the expressions for the deflection at x = L/2 for the two beams illustrated in Fig. 13.15a as:

$$y_{x=L/2} = -\frac{FL^3}{24EI_z} \quad \text{(f)}$$

$$y_{x=L/2} = \frac{5RL^3}{48EI_z} \quad \text{(g)}$$

Substituting Eq. (e) into Eq. (g) yields:

$$y_{x=L/2} = \frac{25FL^3}{768EI_z} \quad \text{(h)}$$

Superimposing the results from Eqs. (f) and (h) and substituting numerical parameter yields:

$$y_{x=L/2} = \frac{FL^3}{EI_z}\left(\frac{25}{768} - \frac{1}{24}\right) = -\frac{7FL^3}{768EI_z} = -\frac{7(3000)(18)^3(12)^3}{768(30\times10^6)(53.4)} = -0.1720 \text{ in.} \quad \text{(i)}$$

In this example, we employed the principle of superposition twice—first to find the reaction force and then to determine the mid span deflection. In both applications, the solutions tabulated in Appendix E greatly reduced the computational effort required.

13.6.2 Integration Method with Singularity Functions

Singularity functions may also be applied to establish the deformation equation needed to solve for the additional reaction force in statically indeterminate beams. With this approach, we remove one of the supports (usually a roller) and replace the support with an unknown reaction force R. This substitution enables us to write expressions for q(x), V(x) and M(x) over the entire length of the beam. We then integrate twice the M(x) relation to obtain y(x). Of course, the equation y(x) contains the unknown force R and two integration constants. These three unknowns are determined from three boundary conditions.

Let's consider an example of a continuous beam to demonstrate the process followed with this approach.

EXAMPLE 13.16

Determine the reaction force at the center support of the continuous beam shown in Fig. E13.16. The forces are both equal to 12 kN. The steel beam is 2L long where L = 6 m and a = b = 3 m. The beam is fabricated from a wide flange section specified as W203 × 36. After solving for the reaction force solve for the deflection at mid span (x = L/2).

Fig. E13.16

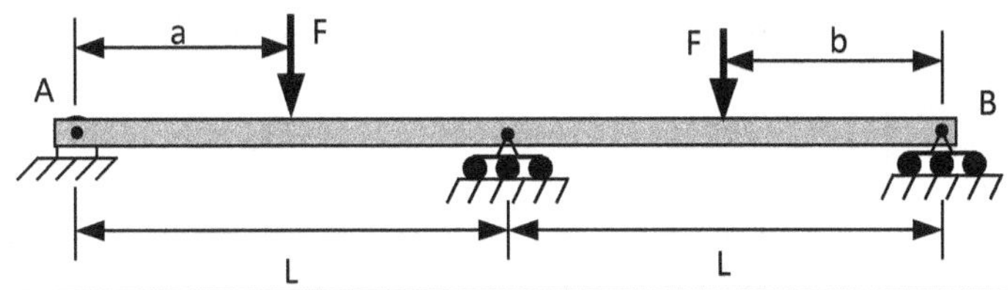

Solution:

Let's remove the center support and replace it with a reaction force R directed upwards as illustrated in Fig. E13.16a.

Fig. E13.16a

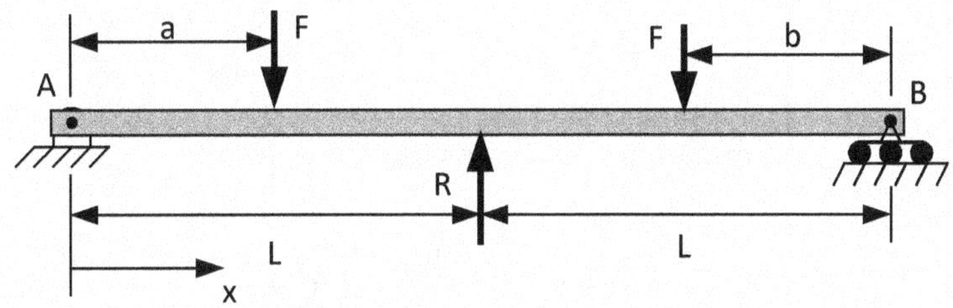

Next write the expression for q(x) with the singularity functions described in Fig. 13.8.

$$q(x) = -R_A <x-0>^{-1} + F<x-a>^{-1} - R<x-L>^{-1} + F<x-3a>^{-1} \quad (a)$$

where R_A is the reaction force at the left side support.

Integrate Eq. (a) to obtain the equation for the shear force V(x) as:

$$V(x) = R_A <x-0>^{0} - F<x-a>^{0} + R<x-L>^{0} - F<x-3a>^{0} + C_1 \quad (b)$$

Integrate Eq. (b) to obtain the equation for the bending moment M(x) as:

$$M(x) = R_A <x-0>^{1} - F<x-a>^{1} + R<x-L>^{1} - F<x-3a>^{1} + C_1x + C_2 \quad (c)$$

The constants of integration C_1 and C_2 both are zero because M(0) = 0 and V(0) = R_A. Hence, we write:

$$M(x) = R_Ax - F<x-a>^{1} + R<x-L>^{1} - F<x-3a>^{1} \quad (d)$$

Substitute Eq. (d) into Eq. (13.4) and integrate to obtain:

$$EI_z \frac{d^2y}{dx^2} = R_A x - F < x - a >^1 + R < x - L >^1 - F < x - 3a >^1 \quad \text{(e)}$$

$$EI_z \frac{dy}{dx} = \frac{R_A}{2} x^2 - \frac{F}{2} < x - a >^2 + \frac{R}{2} < x - L >^2 - \frac{F}{2} < x - 3a >^2 + C_3 \quad \text{(f)}$$

$$EI_z y = \frac{R_A}{6} x^3 - \frac{F}{6} < x - a >^3 + \frac{R}{6} < x - L >^3 - \frac{F}{6} < x - 3a >^3 + C_3 x + C_4 \quad \text{(g)}$$

Examination of Eq. (g) indicates that we have four unknowns—R_A, R, C_3 and C_4. Boundary conditions for the deflection y listed below enable us to write three independent equations.

$$y(0) = 0 \qquad y(L) = 0 \qquad y(2L) = 0 \quad \text{(h)}$$

An equilibrium relation $\Sigma M_B = 0$ gives the fourth equation as:

$$2R_A + R = 2F \quad \text{(i)}$$

From the boundary condition y(0) = 0, it is evident that $C_4 = 0$. From the boundary condition y(L) = 0, we determine C_3 as:

$$EI_z y_{x=L} = \frac{R_A}{6}(L)^3 - \frac{F}{6}\left(L - \frac{L}{2}\right)^3 + 0 - 0 + C_3 L = 0$$

$$C_3 = (L^2/48)[F - 8\, R_A] \quad \text{(j)}$$

From the boundary condition y(2L) = 0, we can determine another independent equation in terms of the remaining unknowns R_A and R as:

$$EI_z y_{x=2L} = \frac{R_A}{6}(2L)^3 - \frac{F}{6}\left(\frac{3L}{2}\right)^3 + \frac{R}{6}(L)^3 - \frac{F}{6}\left(\frac{L}{2}\right)^3 + \frac{2L^3}{48}(F - 8R_A) = 0$$

$$6R_A + R = (13/4)F \quad \text{(k)}$$

From Eqs. (i) and (k), it is clear that:

$$R_A = (5/16)F \qquad \Rightarrow \qquad R = (11/8)F = (11/8)(12)\ 16.5\ \text{kN} \quad \text{(l)}$$

It is interesting that the load partitioning between the supports is independent of the stiffness of the beam and the length of the beam. The partitioning of the loads among the three supports depends only on the placement of the two concentrated forces F as indicated by a = b = L/2.

Let's continue the analysis and determine the deflection at mid span between the first and second supports located by x = a = L/2 = 3 m. Begin by substituting the values for R and R_A from Eq. (l) into Eq. (g) to give the expression for y(x) as:

$$EI_z y = \frac{5F}{96} x^3 - \frac{F}{6} < x - a >^3 + \frac{11F}{48} < x - L >^3 - \frac{F}{6} < x - 3a >^3 - \frac{FL^2}{32} x \quad \text{(m)}$$

Evaluating Eq. (m) at mid span where x = a = L/2 gives:

$$EI_z y_{x=L/2} = -\frac{7FL^3}{768} \qquad \text{(n)}$$

Reference to Appendix D gives $I_z = 34.5 \times 10^6$ mm^4 for the W203 × 36 wide flange section. Then substituting numerical parameters into Eq. (n) gives:

$$y_{x=L/2} = -\frac{7FL^3}{768EI_z} = -\frac{7(12{,}000)(6000)^3}{768(207\times 10^3)(34.5\times 10^6)} = -3.308 \text{ mm} \qquad \text{(o)}$$

13.7 SUMMARY

In this chapter, we returned to study beams subjected to transverse forces with an emphasis on the deflection. The geometry of the deformation of the beam (plane sections remain plane) enabled writing an expression relating curvature κ to the bending moment.

$$\kappa = 1/\rho = M(x)/[EI_z] \qquad \text{(10.6 bis)}$$

We referred to an equation derived in analytical geometry that gives the curvature as a function of the derivatives of the elastic curve and simplified this equation by assuming the square of the slope of the beam was small compared to unity.

$$\kappa = \frac{1}{\rho} = \frac{\dfrac{d^2y}{dx^2}}{\left[1+\left(\dfrac{dy}{dx}\right)^2\right]^{3/2}} \cong \frac{d^2y}{dx^2} \qquad \text{(13.2)}$$

We showed that derivatives of the elastic curve y(x) were related to M(x), V(x) and q(x) as:

$$\frac{d^2y}{dx^2} = \frac{M(x)}{EI_z} \qquad \text{(13.4)}$$

$$\frac{d^3y}{dx^3} = \frac{V(x)}{EI_z} \qquad \text{(13.5)}$$

$$\frac{d^4y}{dx^4} = -\frac{q(x)}{EI_z} \qquad \text{(13.6)}$$

There are three different approaches for determining the equation for the elastic curve y(x) which include:

1. Integrating Eq. (13.4) with the moment M(x) with respect to x
2. Integrating Eq. (13.5) with the shear force V(x) with respect to x
3. Integrating Eq. (13.6) with the transverse loading q(x) with respect to x

Several examples were presented to show the application of Eq. (13.4) to determine the equation of the elastic curve. To determine the constants of integration, we made use of boundary conditions, symmetry and compatibility conditions. It was clear for beams with multiple loadings that the double integration method was time consuming and tedious because of the large number of constants of integration that developed during the integration process. To reduce the number of constants of integration, we introduced the concept of singularity functions $f_n(x)$ given by:

$$f_n(x) = \langle x - a \rangle^n \qquad (13.23)$$

When the exponent $n \geq 0$, the pointed brackets were interpreted either as zero or as regular brackets depending upon the value of the constant a as indicated below:

$$\begin{aligned} f_n(x) &= 0 && \text{if } x < a \\ f_n(x) &= (x - a)^n && \text{if } x \leq a \end{aligned} \qquad (13.24)$$

When integrating pointed brackets with $n \geq 0$, the integration rule is:

$$\int_{-\infty}^{x} \langle x - a \rangle^n \, dx = \frac{\langle x - a \rangle^{n+1}}{n+1} \qquad n \geq 0 \qquad (13.25)$$

However, when n is negative (– 1 or –2) a singularity exists and we define $f_n(x)$ as:

$$\begin{aligned} f_n(x) &= 0 && \text{if } x < a \\ f_n(x) &= 0 && \text{if } x > a \\ f_n(x) &\Rightarrow \propto && \text{if } x = a \end{aligned} \qquad (13.26)$$

The singularity in f_n at $x = a$ does not lead to mathematical difficulty because when $f_n(x) = \langle x - a \rangle^{-1}$ and $f_n(x) = \langle x - a \rangle^{-2}$ are integrated, the result is finite as indicated below:

$$\int_{-\infty}^{x} \langle x - a \rangle^{-2} \, dx = \langle x - a \rangle^{-1} \qquad (13.27)$$

$$\int_{-\infty}^{x} \langle x - a \rangle^{-1} \, dx = \langle x - a \rangle^{0} \qquad (13.28)$$

We then described the use of singularity functions for writing the expression for four common loading functions as:

1. A concentrated moment at $x = a$: $\Rightarrow q(x) = M_o\langle x - a \rangle^{-2}$ (13.29)
2. A concentrated force F at $x = a$: $\Rightarrow q(x) = F\langle x - a \rangle^{-1}$ (13.30)
3. A uniformly distributed load q_o beginning at $x = a$: $\Rightarrow q(x) = q_o\langle x - a \rangle^{0}$ (13.31)
4. A linearly increasing (ramp) load q(x) beginning at $x = a$:
   $q(x) = (q_o/b)\langle x - a \rangle^{1}$ where q_o/b is the slope of the ramp. (13.32)

A graphic in Fig. 13.8 is presented to aid you in writing expressions for loads, shear forces and moments in terms of singularity functions. Several examples are presented to demonstrate the techniques used in applying singularity methods to solve for beam deflections.

We introduced another technique for solving for beam deflections called superposition. Superposition involves combining together the results of the deflection from two or more separate solutions to yield a new solution containing all of the loading functions involved in the separate solutions. One of the advantages of the method of superposition is that we make use of the many solutions that already exist. Mechanics of materials is a subject with a long history, and the classic problems in bending of beams were solved many years ago. We have provided a catalog of several of these solutions for your convenience in Appendix E.

There are two requirements that must be satisfied before using the principle of superposition. First is a linearity requirement. The solution for y(x) must be linear with respect to the loading functions, which is satisfied. The second requirement pertains to the integrity of the geometry of the beam during the loading. The beam does not undergo significant changes in its dimensions, and we conclude that the principle of superposition is valid. The methods of superposition were demonstrated in several examples.

In most instances beams are statically determinate. The reactions at the supports may be determined from the three applicable equations of equilibrium. However, in some cases beams are designed with additional supports that produce additional reaction forces making the beam statically indeterminate. There are several methods to solve for the additional reaction forces; however, we often supplement the equilibrium relations with a deformation equation to enable a solution. Two different methods to determine the reaction forces for statically indeterminate beams were demonstrated—superposition and integration with singularity functions.

PROBLEMS

13.1 For the cantilever beams shown in the figure to the right, write the equation for the elastic curves. Also determine the deflection and slope at the free end.

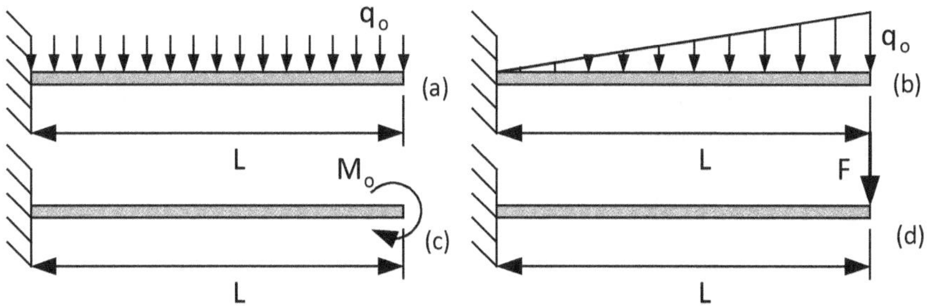

13.2 For the steel cantilever beam, shown in the figure to the right, determine the deflection and slope at the free end if the beam is fabricated from a W305 × 74 wide flange section.

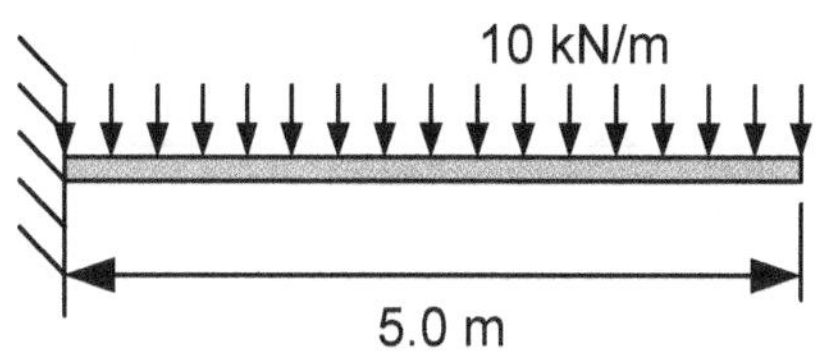

13.3 For the steel cantilever beam, shown in the figure to the left, determine the deflection and slope at the free end if the beam is fabricated from an American Standard section S254 × 38.

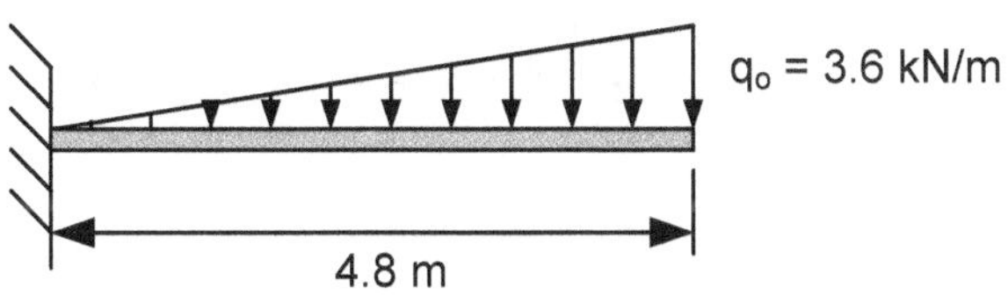

13.4 For the steel cantilever beam, shown in the figure to the right, determine the deflection and slope at the free end if the beam is fabricated from a W18 × 60 wide flange section.

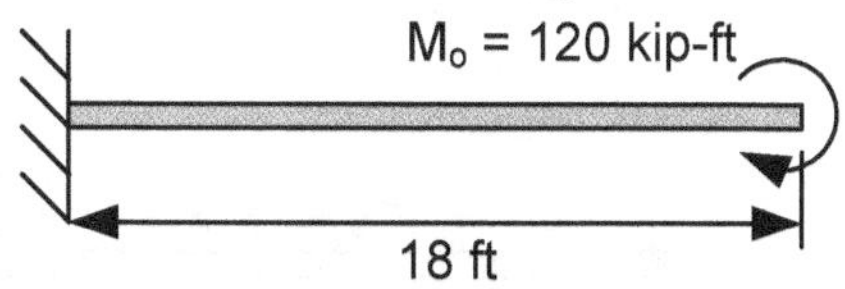

13.5 For the steel cantilever beam, shown in the figure to the left, determine the deflection and slope at the free end if the beam is fabricated from an American Standard section S381 × 64.

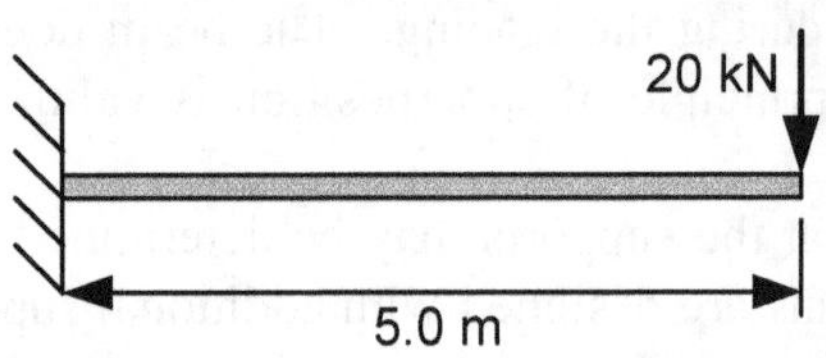

13.6 For the simply supported beams, shown in the figure to the right, write the equation for the elastic curves. Also determine the deflection at mid-span and the slope at both ends.

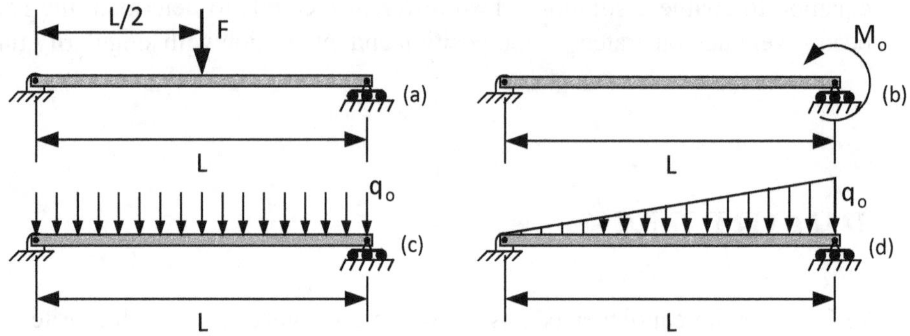

13.7 For the steel simply supported beam shown in the figure to the right, determine the mid-span deflection and slope at both ends if the beam is fabricated from a S203 × 27 American Standard section.

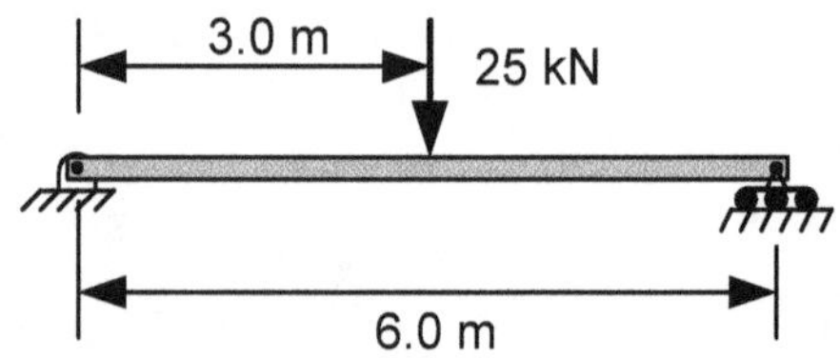

13.8 For the steel simply supported beam, shown in the figure to the left, determine the mid-span deflection and slope at both ends if the beam is fabricated from a W152 × 37 wide flange section.

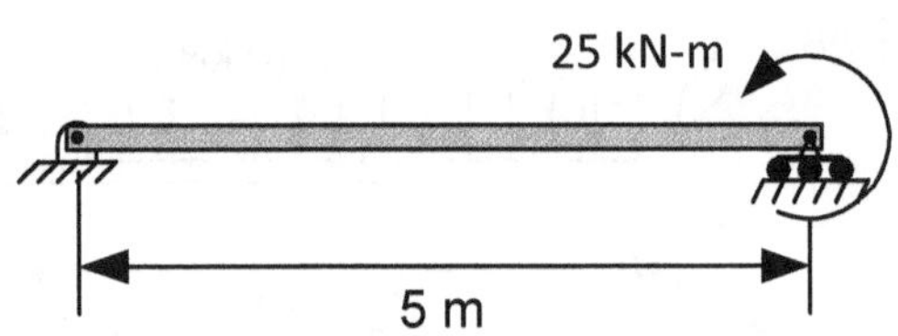

13.9 For the steel simply supported beam, shown in the figure to the right, determine the mid-span deflection and slope at both ends if the beam is fabricated from a W10 × 60 wide flange section.

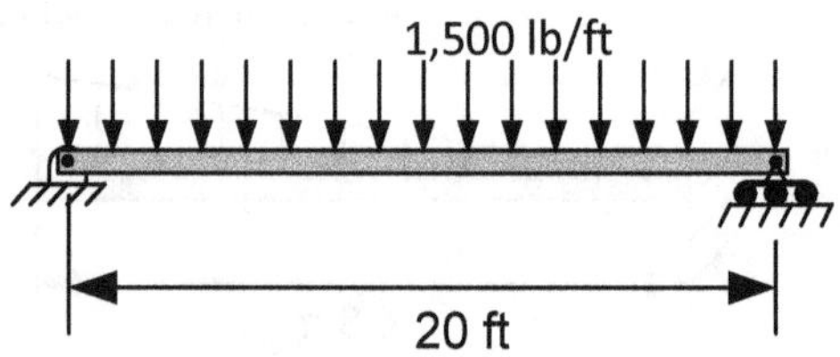

13.10 For the steel simply supported beam, shown in the figure to the left, determine the mid-span deflection and slope at both ends if the beam is fabricated from a S254 × 38 American Standard section.

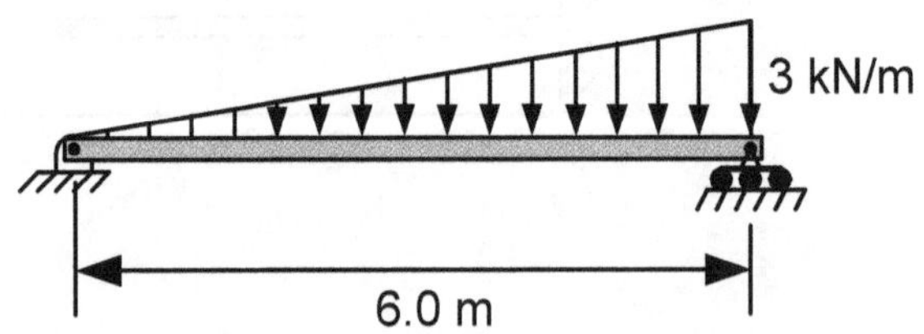

13.11 For the simply supported beam, shown in the figure to the right, determine the equation of the elastic curve and the slope at both ends.

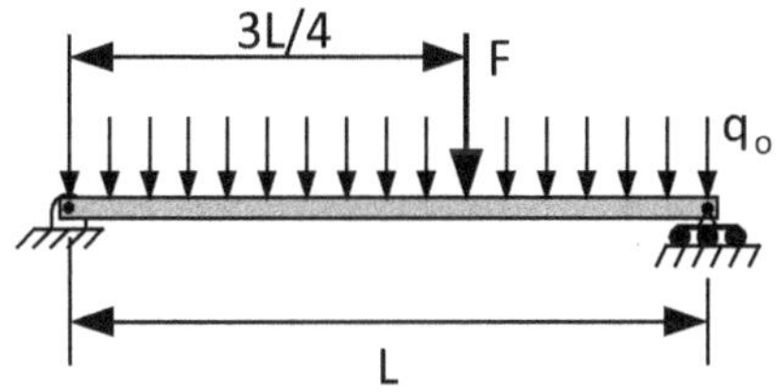

13.12 For the steel simply supported beam, shown in the figure to the left, determine the mid-span deflection and the slope at both ends if the beam is fabricated from a W10 × 60 wide flange section.

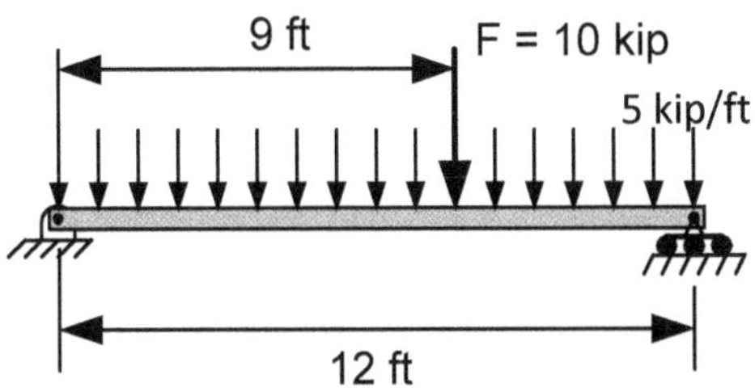

13.13 For the simply supported beam, shown in the figure to the right, determine the equation of the elastic curve and the slope at both ends.

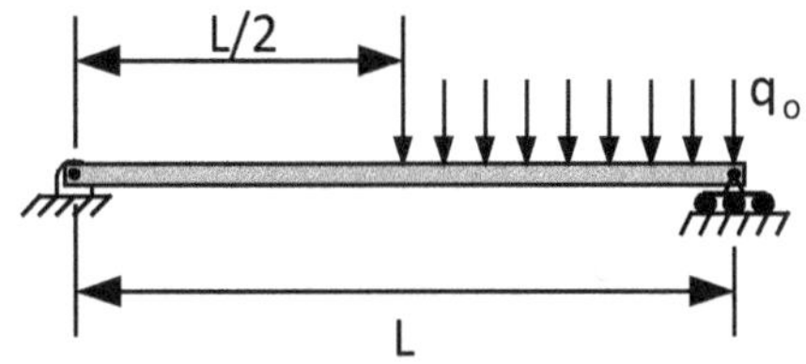

13.14 For the steel simply supported beam, shown in the figure to the left, determine the mid-span deflection and the slope at both ends if the beam is fabricated from a S18 × 70 American Standard section.

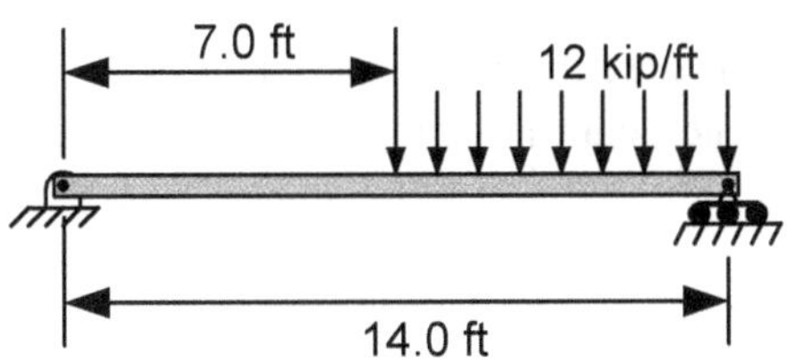

13.15 For the simply supported beam, shown in the figure to the right, determine the equation of the elastic curve and the slope at both ends.

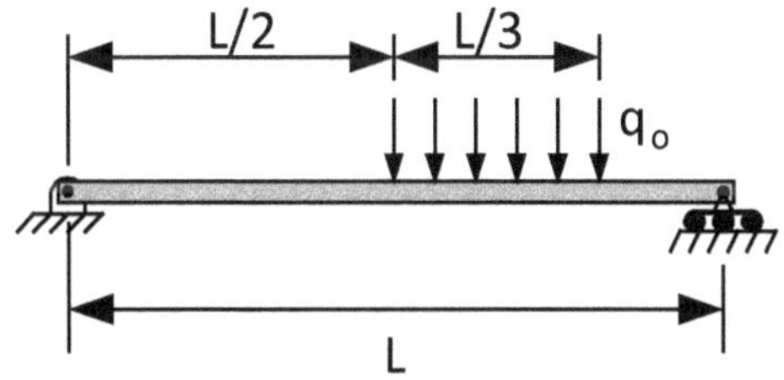

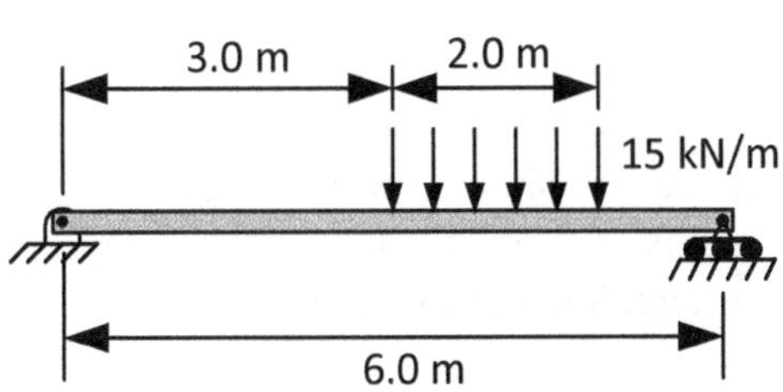

13.16 For the aluminum simply supported beam, shown in the figure to the left, determine the mid-span deflection and the slope at both ends if the beam is fabricated from a W457 × 89 wide flange section.

13.17 For the simply supported beam, shown in the figure to the right, determine the equation of the elastic curve and the slope at both ends.

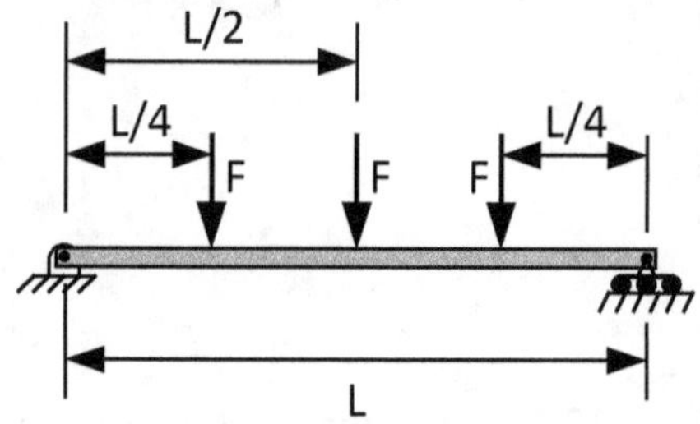

13.18 For the steel simply supported beam, shown in the figure to the left, determine the mid-span deflection and the slope at both ends if the beam is fabricated from a W203 × 46 wide flange section.

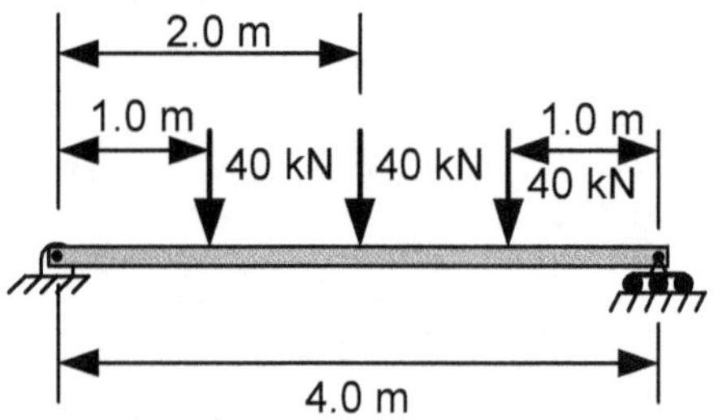

13.19 For the simply supported beam, shown in the figure to the right, determine the equation of the elastic curve and the slope at both ends.

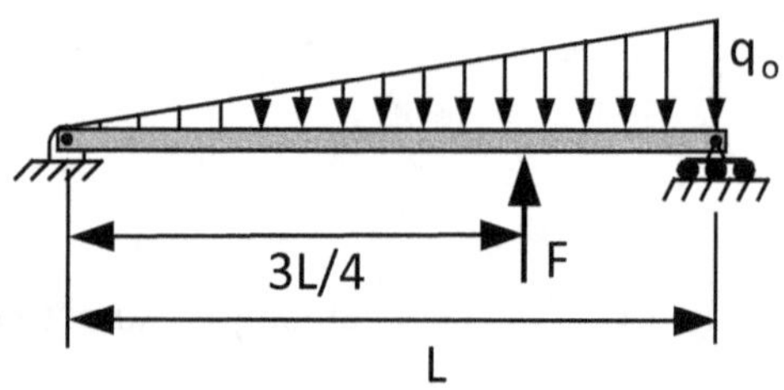

13.20 For the steel simply supported beam, shown in the figure to the left, determine the mid-span deflection and the slope at both ends if the beam is fabricated from a W533 × 92 wide flange section.

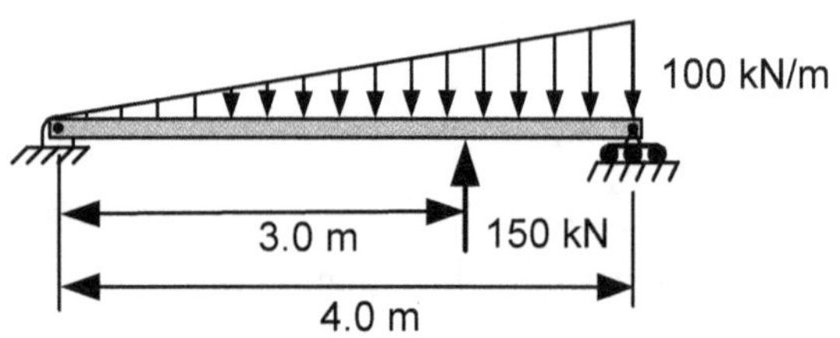

13.21 For the cantilever beam, shown in the figure to the right, write the equation for the elastic curve. Also determine the equations for deflection and slope at the free end.

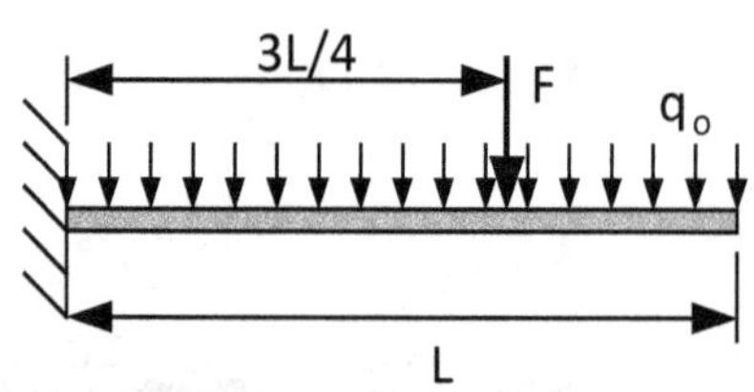

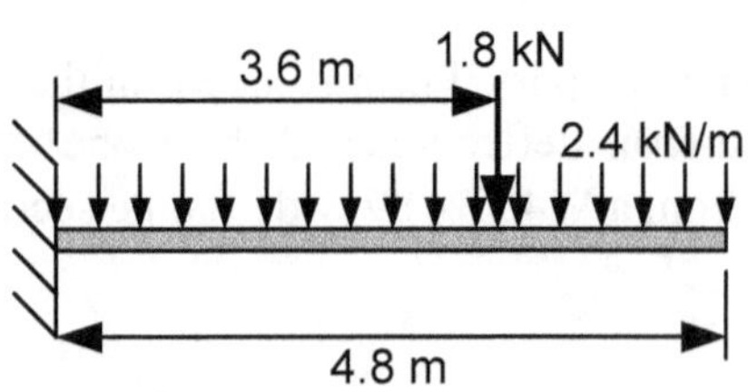

13.22 For the steel cantilever beam, shown in the figure to the left, determine the deflection and the slope at the free end if the beam is fabricated from a S305 × 74 American Standard section.

13.23 For the cantilever beam, shown in the figure to the left, write the equation for the elastic curve. Also determine the equations for the deflection and slope at the free end.

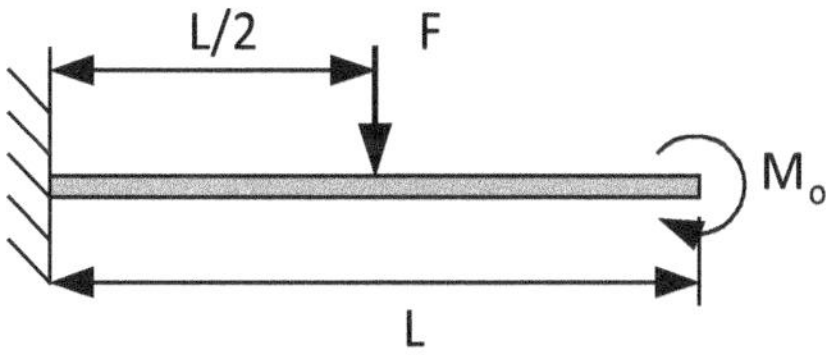

13.24 For the steel cantilever beam, shown in the figure to the right, determine the minimum weight wide flange section if the deflection at the free end is limited to 1.4 in.

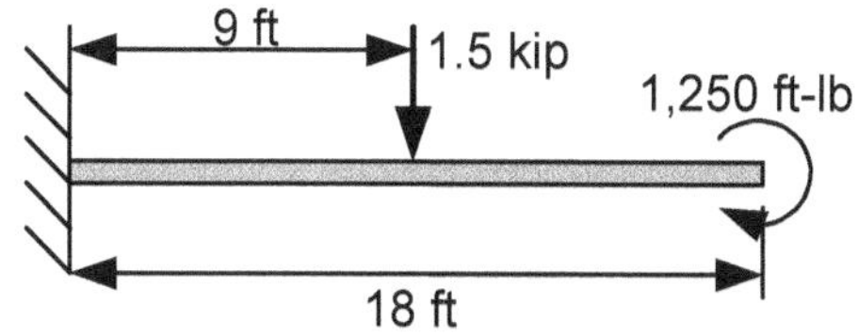

13.25 For the cantilever beam, shown in the figure to the left, write the equation for the elastic curve. Also determine the equations for the deflection and slope at the free end.

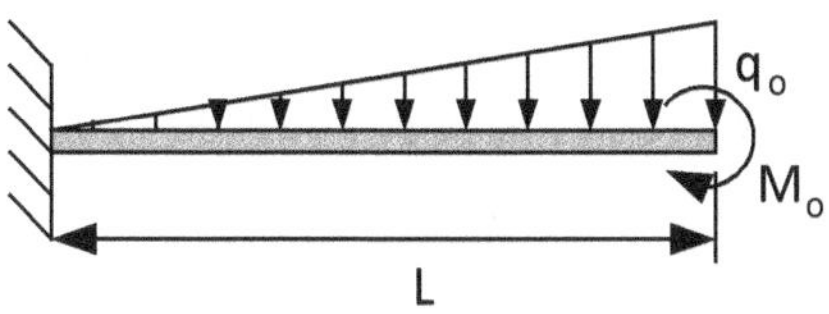

13.26 For the steel cantilever beam, shown in the figure to the right, determine the minimum weight American Standard section if the deflection at the free end is limited to 26 mm.

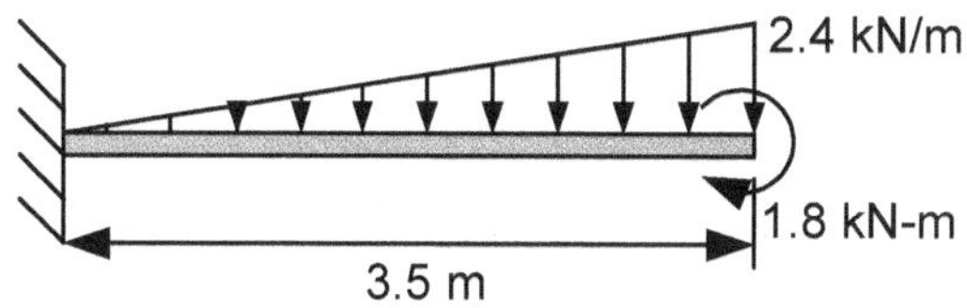

13.27 For the cantilever beam, shown in the figure to the left, write the equation for the elastic curve. Also determine the deflection and slope at the free end.

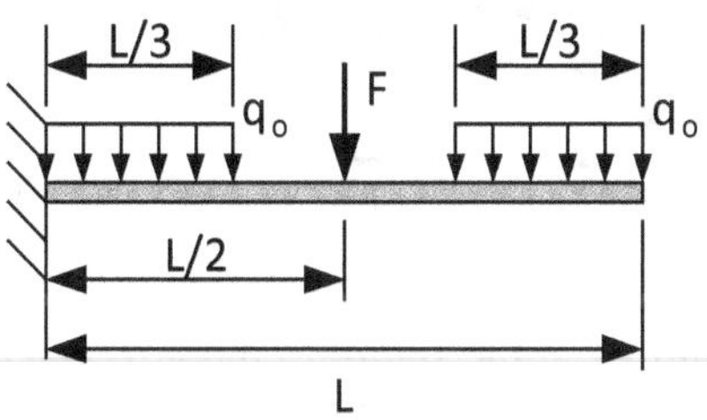

13.28 For the steel cantilever beam, shown in the figure to the right, determine the minimum weight structural tee section if the deflection at the free end is limited to 1.1 in.

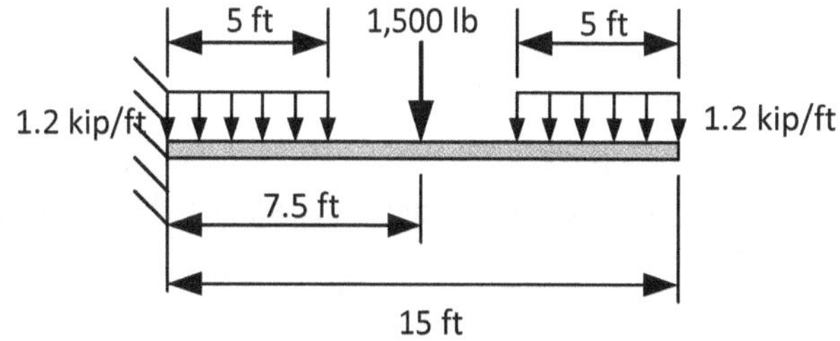

13.29 For the cantilever beam, shown in the figure to the left, write the equation for the elastic curve. Also determine the deflection and slope at the free end.

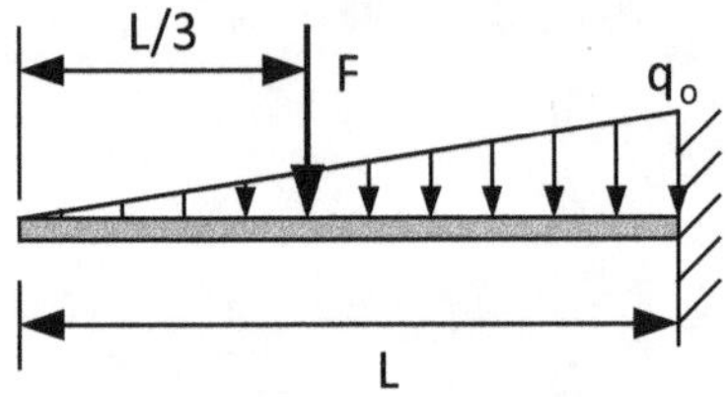

13.30 For the steel cantilever beam, shown in the figure to the right, determine the minimum weight wide flange section if the deflection at the free end is limited to 15 mm.

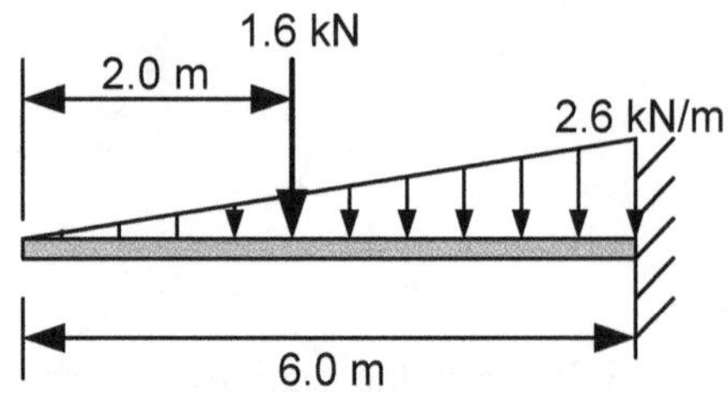

13.31 Using the method of superposition, for the simply supported beam shown in the figure to the left, determine the deflection at mid-span and the slope at the left end of the beam. The beam is steel with a W12 × 65 wide flange section.

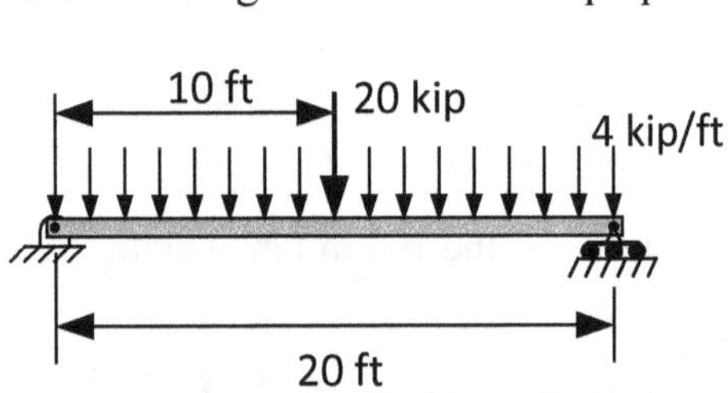

13.32 Using the method of superposition, for the simply supported beam shown in the figure to the right, determine the deflection at mid-span and the slope at the left end of the beam.

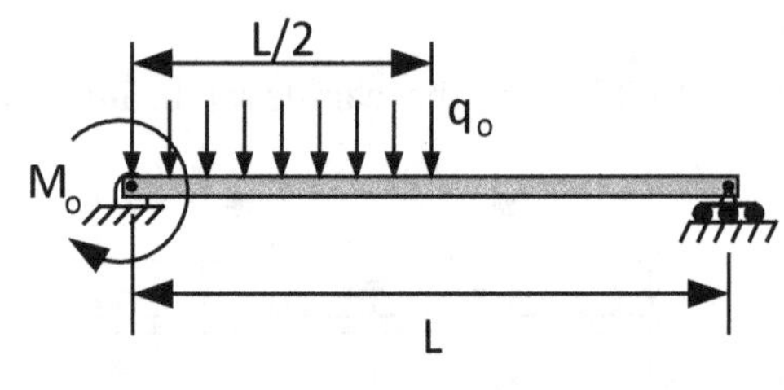

13.33 Using the method of superposition, for the simply supported beam shown in the figure to the left, determine the deflection at mid-span and the slope at the right end of the beam. The beam is steel with a W305 × 97 wide flange section.

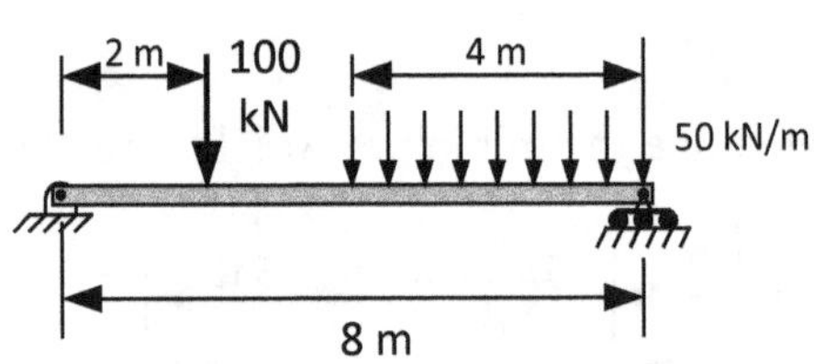

13.34 Using the method of superposition, for the simply supported beam shown in the figure to the right, determine the deflection at mid-span and the slope at the right end of the beam.

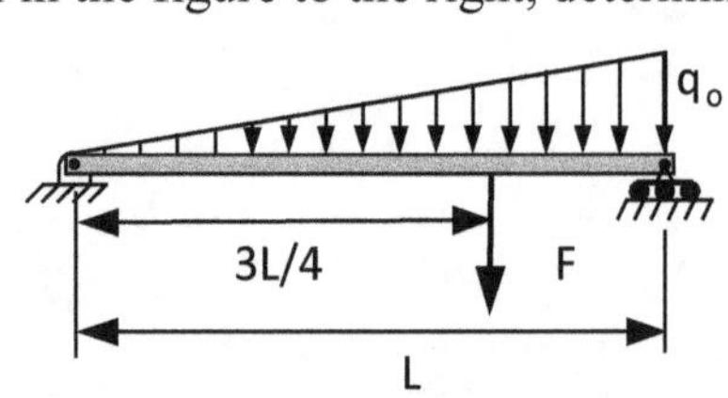

13.35 Using the method of superposition, for the cantilever beam shown in the figure to the left, determine the deflection and the slope at the free end of the beam.

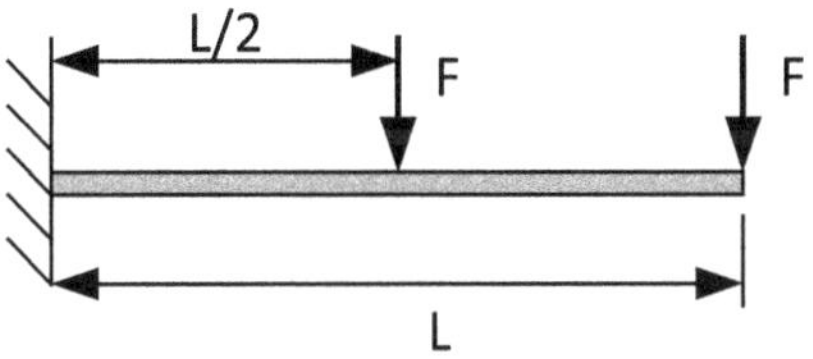

13.36 Using the method of superposition, for the cantilever beam shown in the figure to the right, determine the deflection and the slope at the free end of the beam. The beam is steel with a S457 × 81 American Standard section.

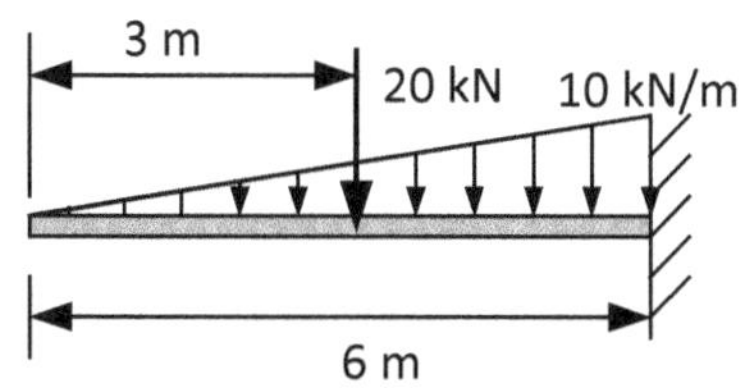

13.37 Using the method of superposition, for the cantilever beam shown in the figure to the left, determine the deflection and the slope at the free end of the beam. The beam is steel with a S20 × 66 American Standard section.

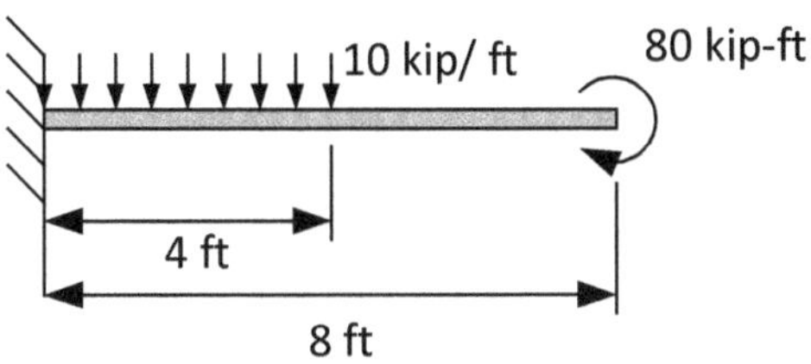

13.38 Using the method of superposition, for the cantilever beam shown in the figure to the right, determine the deflection and the slope at the free end of the beam.

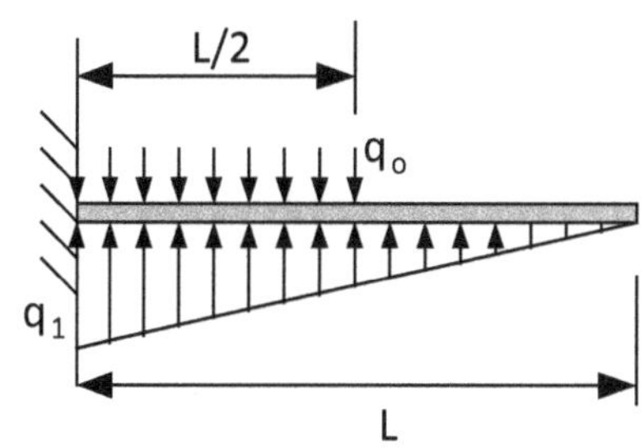

13.39 Determine the reactions at the prop and the wall on the cantilever beam shown in the figure to the left.

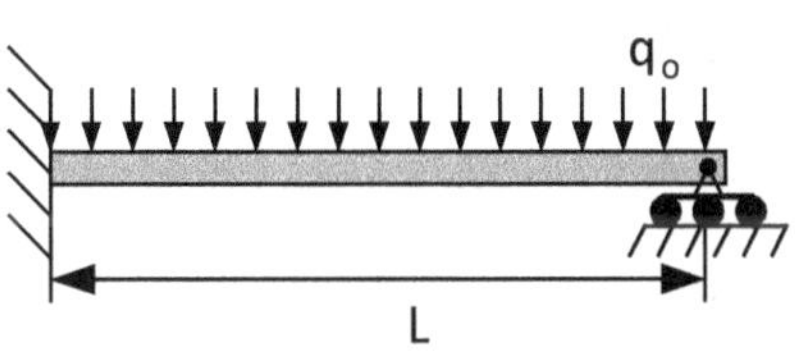

13.40 Determine the reactions at the prop and the wall on the cantilever beam shown in the figure to the right.

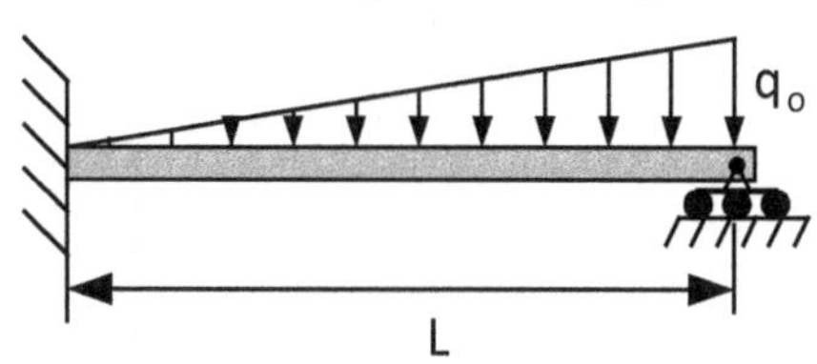

13.41 Determine the reactions at the prop and the wall on the cantilever beam shown in the figure to the left.

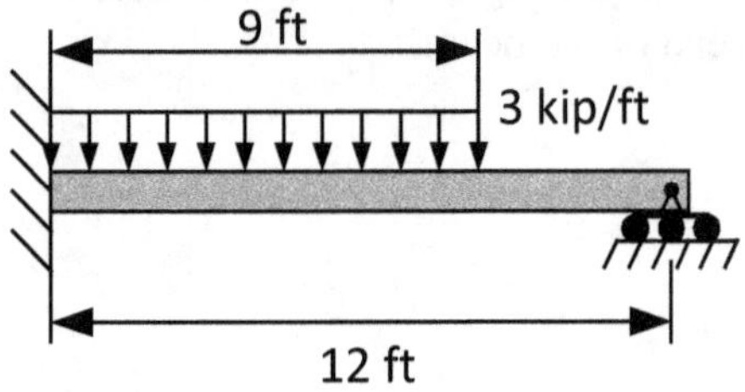

13.42 Determine the reactions at the prop and the wall on the cantilever beam shown in the figure to the right.

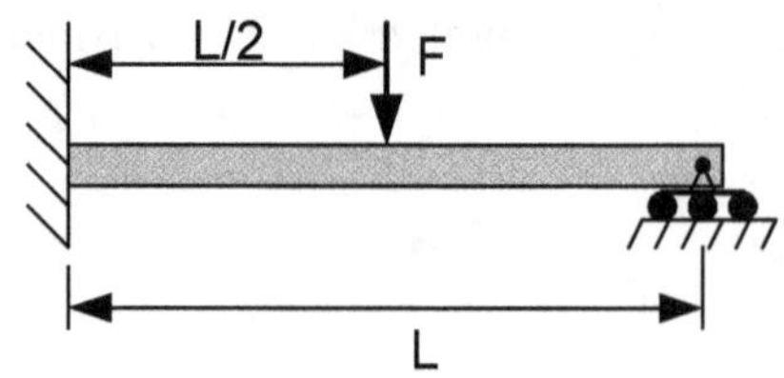

13.43 Determine the reactions at the prop and the wall on the cantilever beam shown in the figure to the left.

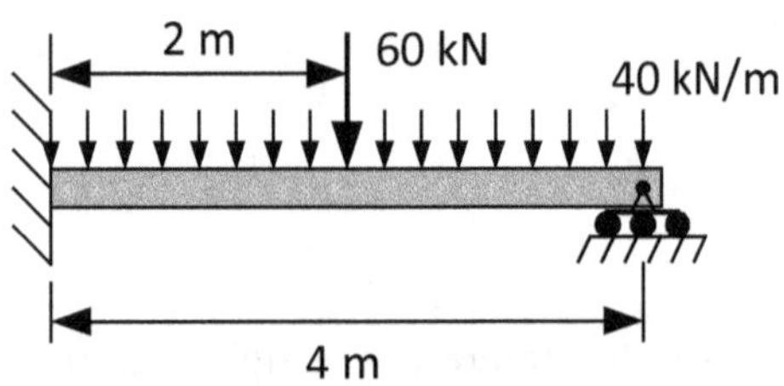

13.44 Determine the reactions at the prop and the wall on the cantilever beam shown in the figure to the right.

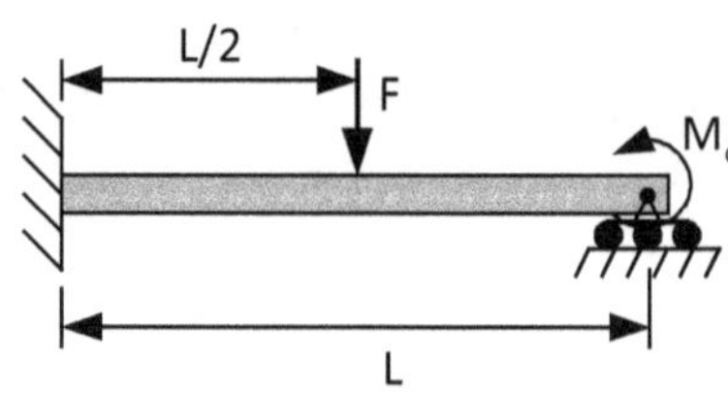

13.45 Determine the reactions at the supports of the continuous beam shown in the figure below.

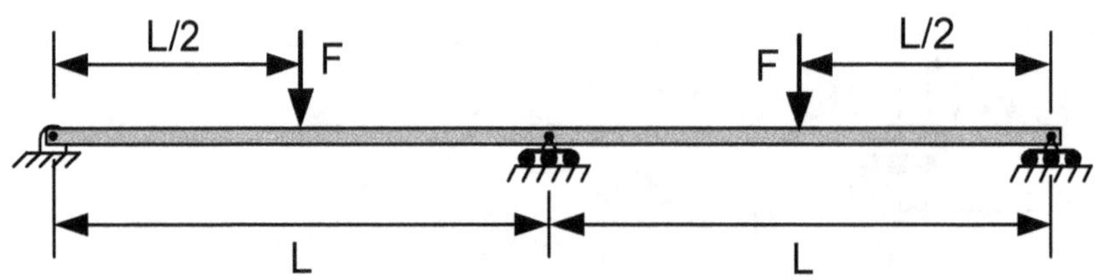

13.46 Determine the reactions at the supports of the continuous beam shown in the figure below.

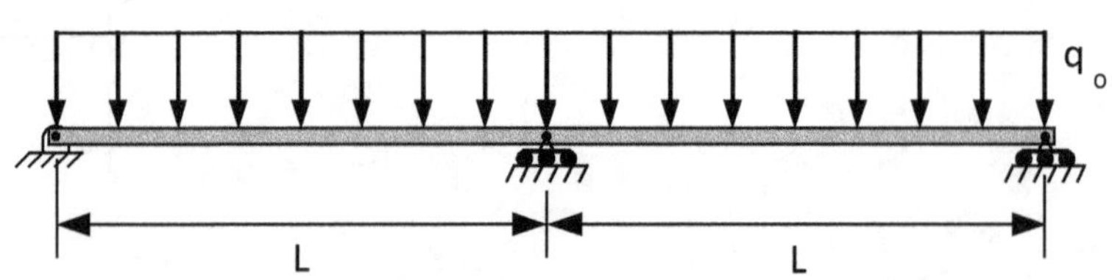

13.47 Determine the reactions at the supports of the continuous beam shown in the figure below.

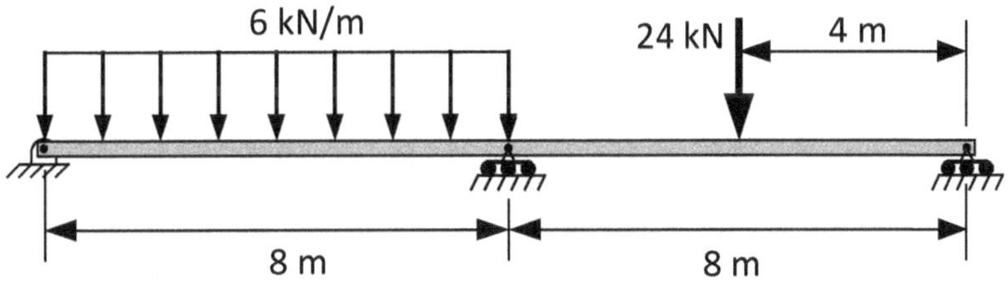

13.48 Determine the reactions at the supports of the continuous beam shown in the figure to the right.

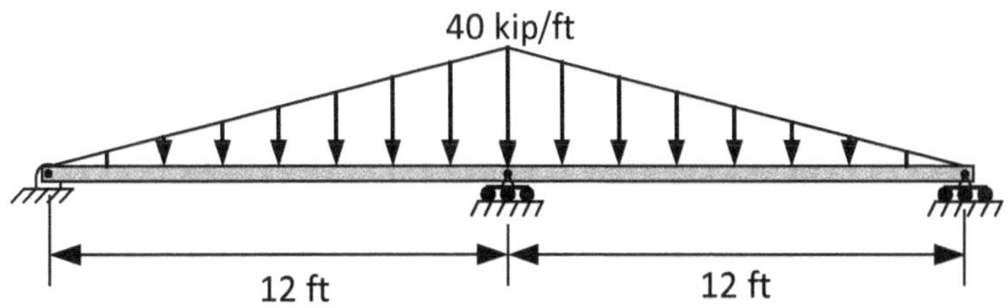

13.49 Determine the reactions at the supports of the continuous beam shown in the figure to the left.

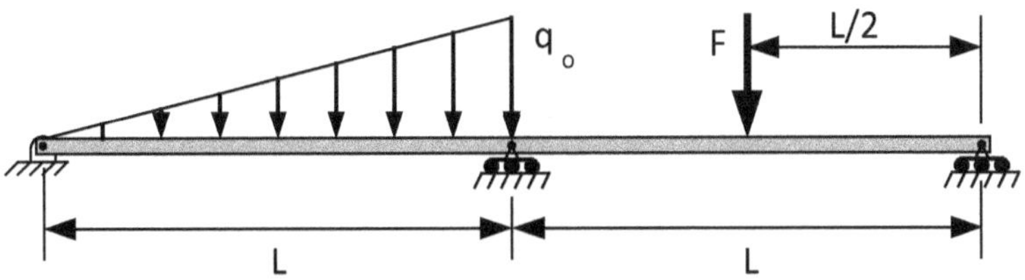

13.50 Determine the reactions at the supports of the built-in beam shown in the figure to the right.

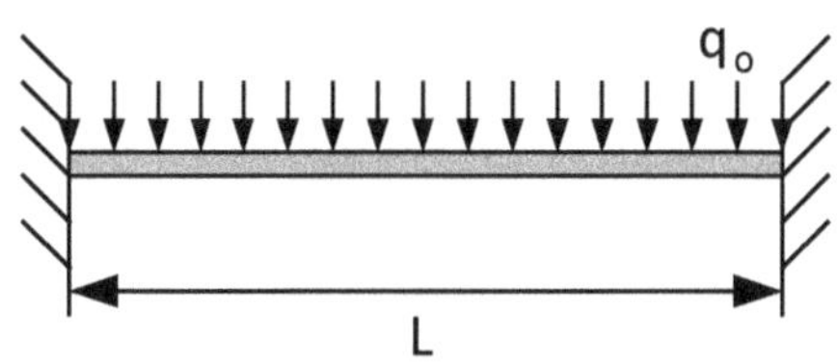

13.51 Determine the reactions at the supports of the built-in beam shown in the figure to the left.

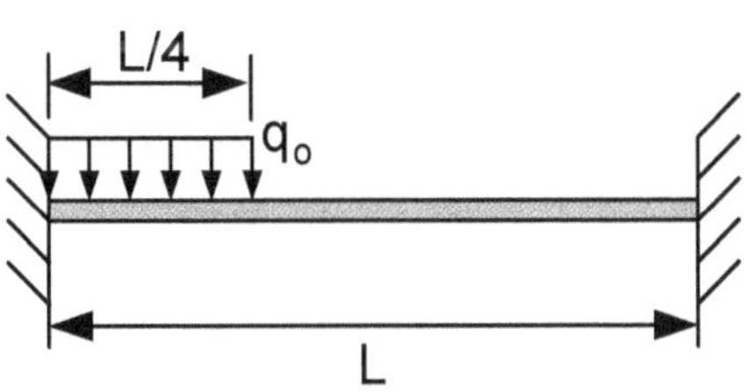

13.52 Determine the reactions at the supports of the built-in beam shown in the figure to the right.

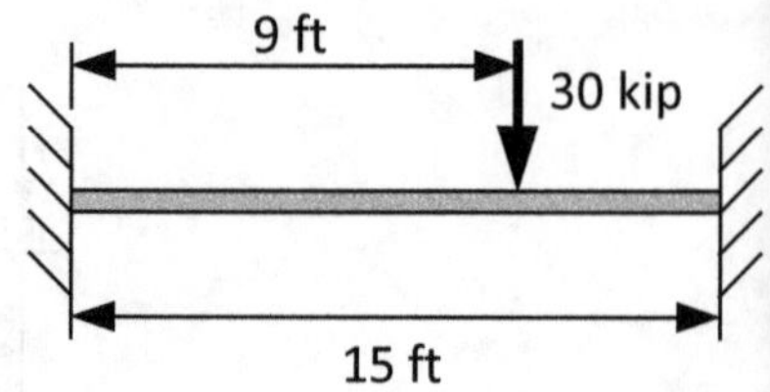

13.53 Determine the reactions at the supports of the built-in beam shown in the figure to the right.

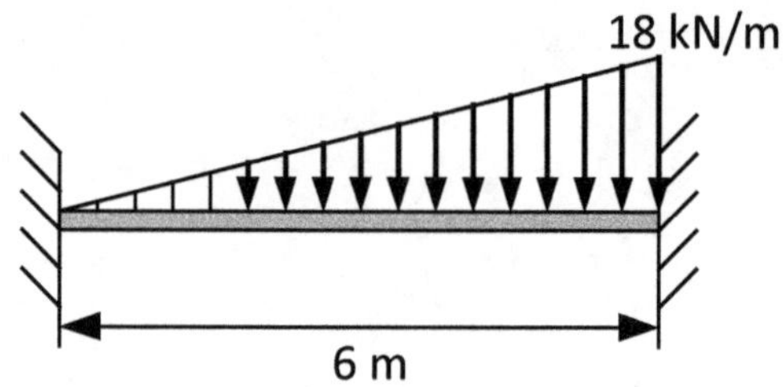

CHAPTER 14

BUCKLING OF COLUMNS

14.1 INTRODUCTION

In this chapter, the concept of instability of columns under the action of compressive force is introduced. This is a new concept that leads to a markedly different mode of failure of many structural elements such as columns, plates and shells. Even the analysis employed differs from our normal approach. In the previous chapters, we were concerned with stresses and deflections. If the stresses were less than the material's strength, the structural element was safe. If the deflections did not exceed some specified limit, the structural element was satisfactory. The concept of stability requires a designer to determine a critical load to insure that the structural element will not become unstable and collapse prior to either yield or fracture.

When the compressive force on a column exceeds some critical value P_{CR}, the column becomes unstable and collapses. This mode of failure is extremely dangerous because the collapse is catastrophic. When a column buckles, without shedding load, it initially deforms in the shape of a half sine wave and then simply folds when a plastic hinge forms at its center as illustrated in Fig. 14.1.

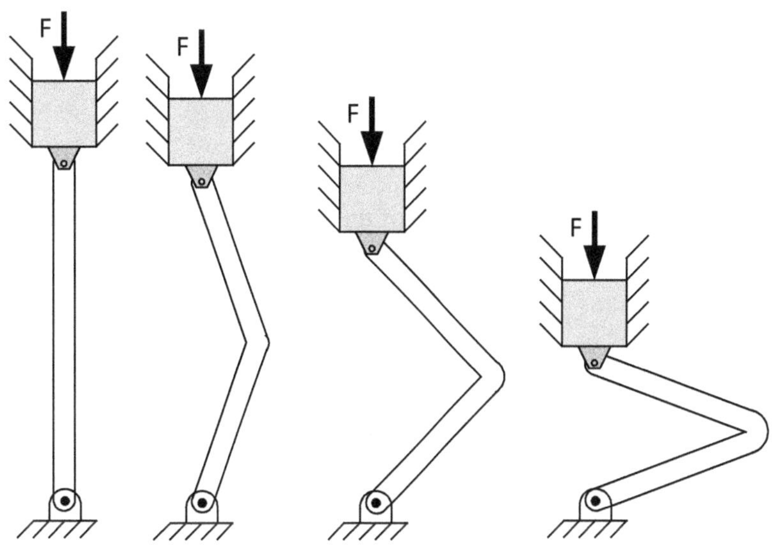

Fig. 14.1 Catastrophic collapse of a column with pinned ends during a buckling event.

In developing the equations for predicting the buckling of a column, we deviate from our previous practice of neglecting the small deformations of a structural element under the action of a load. When computing stress from $\sigma = P/A$, we do not take into account the change in A due to the strains induced by the force P. However, the entire concept of elastic instability is based on load-induced deformations. In analyzing columns, we consider the moment produced by the axial force P times the deflection y. To show this new analytical approach, we consider a column with both ends pinned subjected to a centric force P in the next section.

14.2 BUCKLING OF COLUMNS WITH BOTH ENDS PINNED

Consider a column of length L with both ends pinned subjected to a centric force P as illustrated in Fig. 14.2a. As the magnitude of the centric force increases, the column will bow slightly as indicated in Fig 14.2b. The column is at this stage of loading elastic and stable. To perform the analysis, we model the pinned end supports and the column as shown in Fig. 14.2c. We have selected the x-axis in the vertical direction and the y-axis in the horizontal direction to adapt the column coordinate system to that previously used for a beam. The origin is placed at the top of the column.

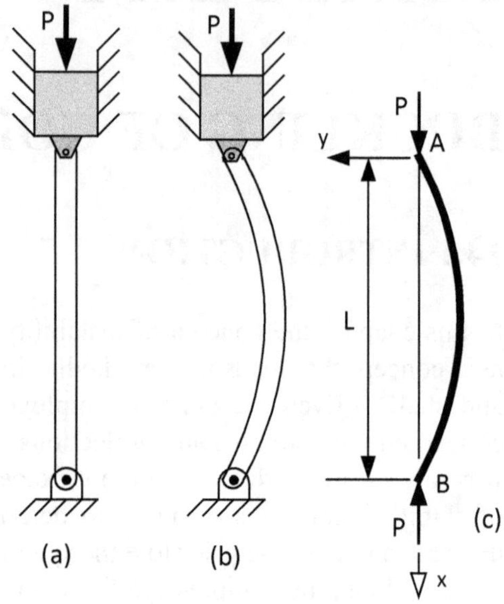

Fig. 14.2 Modeling a column with pinned ends.

Next, let's make a section cut and remove a portion of thc top of the column as a FBD as indicated in Fig. 14.3.

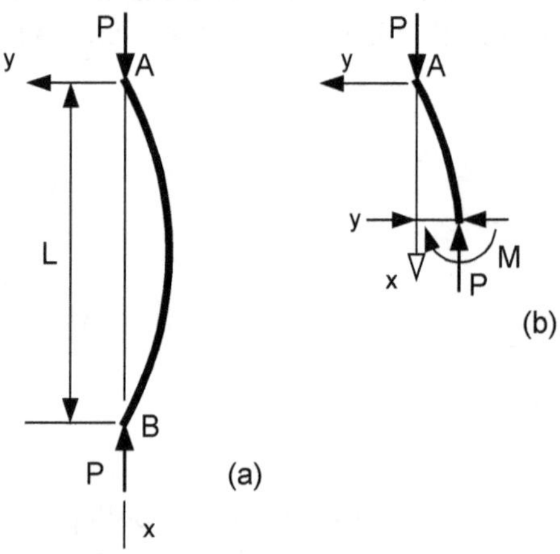

Fig. 14.3 FBD of the top portion of the column.

An examination of the FBD shows a couple, which produces a moment M = – Py. This is a bending moment and Eq. (13.4) is valid. Hence, we write:

$$\frac{d^2y}{dx^2} = \frac{M(x)}{EI} = -\frac{Py}{EI} \tag{14.1}$$

Rearranging Eq. (14.1) to write the differential equation in standard format yields:

$$\frac{d^2y}{dx^2} + p^2y = 0 \tag{14.2}$$

where $p = \sqrt{\frac{P}{EI}}$ (14.3)

Equation (14.2) is a linear, second-order differential equation with constant coefficients, which can be solved for y to obtain:

$$y = C_1 \sin(px) + C_2 \cos(px) \qquad (14.4)^{1}$$

where C_1 and C_2 are constants of integration.

We follow the procedure established in Chapter 13 and solve for the constants of integration using boundary conditions. Since both ends of the column are pinned, we write:

$$y(0) = 0 \qquad y(L) = 0 \qquad (14.5)$$

It is clear from the first of the boundary conditions listed in Eq. (14.5) that $C_2 = 0$. Substituting the second boundary condition ($y = 0$ at $x = L$) into Eq. (14.4) yields:

$$C_1 \sin pL = 0 \qquad (14.6)$$

There are two interpretations for Eq. (14.6). First, the equation is satisfied if $C_1 = 0$, which gives $y = 0$ for all of x. This is not realistic since it implies that the column remains straight and will never buckle. Because buckling has been demonstrated in the laboratory, this solution is discarded. The remaining solution to Eq. (14.6) is obtained by seeking values of pL which permit $\sin pL = 0$.

If $\quad pL = n\pi \quad$ then $\quad \sin pL = 0$

Then from Eq. (14.3), we write:

$$p = \sqrt{\frac{P}{EI}} = \frac{n\pi}{L}$$

$$P = \frac{n^2 EI \pi^2}{L^2} \qquad (14.7)$$

Again, it is necessary to interpret Eq. (14.7) because it contains a counter n, which can be 0, 1, 2, etc. We seek the lowest meaningful value of n. Clearly $n = 0$ does not produce a realistic result because this value gives $P = 0$. We seek a value for $P > 0$. If we select $n = 1$, then:

$$P_{CR} = \frac{\pi^2 EI}{L^2} \qquad (14.8)$$

where P_{CR} is the critical centric compression load that initiates buckling and catastrophic collapse of the column. I is the minimum value of the moment of inertia for the cross sectional area of the column.

Equation (14.8) is known as the Euler relation for elastic buckling. It is valid for long slender columns[2] with both ends pinned. Returning to Eq. (14.4), we may use Eq. (14.7) and express the deflection of the column as:

$$y = C_1 \sin\frac{\pi x}{L} \qquad (14.9)$$

This result indicates that the column deflects as a sinusoid when the load $P = P_{CR}$. The fact that we cannot determine the amplitude of the sin wave (C_1) is not important because the column has become unstable and will quickly collapse.

---

[1] This solution can be verified by differentiating Eq. (14.4) twice with respect to x to obtain d^2y/dx^2. Then substitute d^2y/dx^2 and y into Eq.(14.2).

[2] We will discuss the distinction among long slender columns, intermediate columns and short columns later in this Chapter.

Let's apply the Euler buckling theory to determine the critical load for two different columns both with pinned ends.

EXAMPLE 14.1

A column 18 feet long is fabricated from an aluminum tube with a 6 in. outside diameter and a 5.6 in. inside diameter. Determine the critical buckling force if the tube is pinned at both ends.

Solution:

We employ Eq. (14.8) to write the expression for the critical buckling force as:

$$P_{CR} = \frac{\pi^2 EI}{L^2} \qquad \text{(a)}$$

The modulus of elasticity $E = 10.4 \times 10^6$ psi for an aluminum alloy is from Appendix B-1.

The minimum value of the moment of inertia for a tube is:

$$I = (\pi/64)[d_o^4 - d_i^4] = (\pi/64)[(6)^4 - (5.6)^4] = 15.34 \text{ in.}^4 \qquad \text{(b)}$$

Substituting Eq. (b) into Eq. (a) yields:

$$P_{CR} = \frac{\pi^2 EI}{L^2} = \frac{\pi^2 (10.4 \times 10^6) 15.34}{(18)^2 (12)^2} = 33.75 \text{ kip}$$

EXAMPLE 14.2

A column 10 m long is fabricated from a square tube (1020 HR steel) shown in Fig. E14.2. Determine the critical buckling force if the tube is pinned at both ends. Also determine the compressive stress in the tube and comment on its magnitude.

Solution:

We employ Eq. (14.8) to write the expression for the critical buckling force as:

$$P_{CR} = \frac{\pi^2 EI}{L^2} \qquad \text{(a)}$$

The modulus of elasticity E = 207 GPa for the steel tube is from Appendix B-1.

The minimum value of the moment of inertia for the square tube is:

$$I = (1/12)[h_o^4 - h_i^4] = (1/12)[(60)^4 - (40)^4] = 0.8667 \times 10^6 \text{ mm}^4 \qquad \text{(b)}$$

Substituting Eq. (b) into Eq. (a) yields:

$$P_{CR} = \frac{\pi^2 EI}{L^2} = \frac{\pi^2(207\times10^3)(0.8667\times10^6)}{(10\times10^3)^2} = 17.71 \text{ kN} \qquad \text{(c)}$$

The axial stress produced by the critical force P_{CR} is:

$$\sigma_{CR} = P_{CR}/A = 17.71 \times 10^3/[(60)^2 - (40)^2] = 8.855 \text{ MPa} \qquad \text{(d)}$$

The yield strength of 1020 HR steel is listed in Appendix B-2 as 290 MPa; hence, the stress at the critical load that produces buckling is only 3.0% of the yield strength. This example illustrates the danger of collapse due to buckling when columns are long and slender.

14.3 INFLUENCE OF END CONDITIONS

The critical buckling load is profoundly affected by the end conditions of the column. Increasing the constraint at either end markedly increases the critical load required for the onset of elastic instability. We will consider three additional sets of boundary conditions to show the influence of constraint on buckling load.

1. Top — pinned and bottom — built-in.
2. Top — free and bottom — built-in.
3. Top — built-in and bottom — built-in.
4.

14.3.1 Column Buckling with One Pinned and One Built-in End.

Consider the column with its bottom built-in and its top pinned as shown in Fig. 14.4. It is subjected to an axial load P as shown in Fig. 14.5. To begin the analysis, we draw the deformed shape of the column in the initial stages of instability. In preparing this drawing, we recognize that the slope of the elastic curve at the built-in end and the deflection at both ends must be zero.

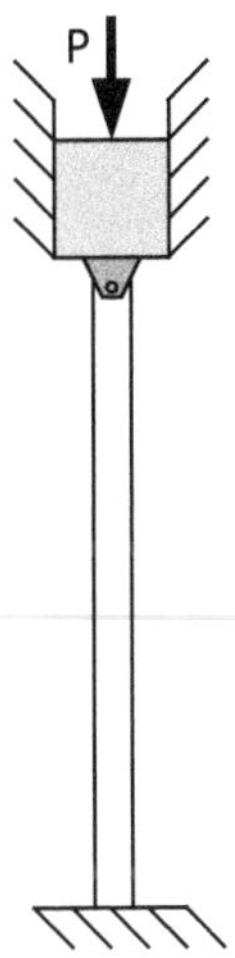

Fig. 14.4 A column with one end pinned and the other built-in.

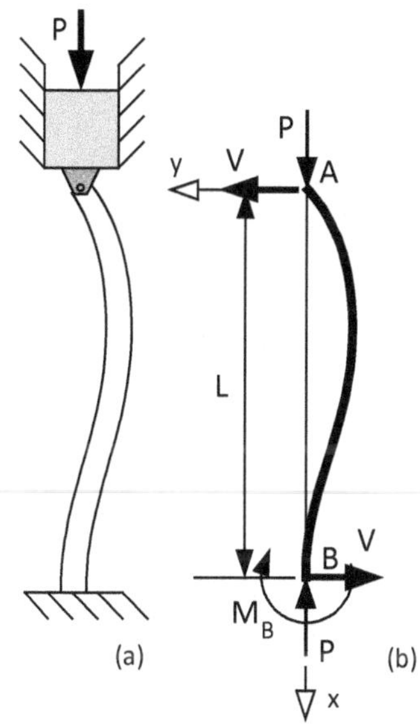

Fig. 14.5 Deformation of the column at initiation of elastic instability.

Next, let's make a section cut and remove a portion of the top of the column as a FBD as indicated in Fig. 14.6.

An examination of the FBD shows two couples produced by the equal and opposite forces V and P. These couples result in bending moment given by:

$$M = -Py - Vx \qquad (14.10)$$

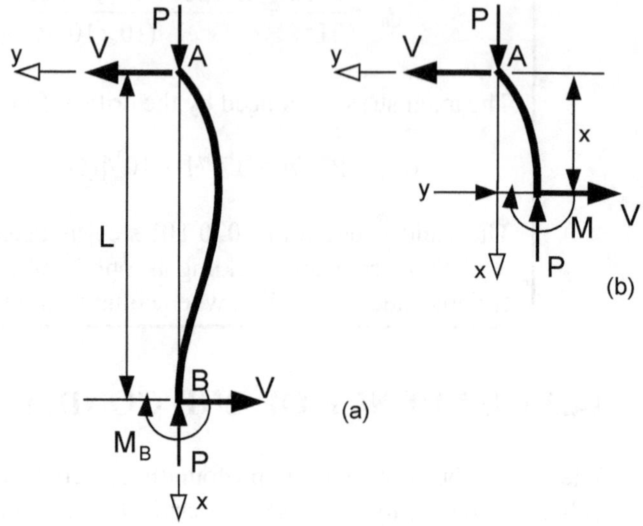

Fig. 14.6 FBD of the top portion of the column.

Substituting Eq. (14.10) into Eq. (13.4) yields:

$$\frac{d^2y}{dx^2} = \frac{M(x)}{EI} = -\frac{1}{EI}(Py + Vx) \tag{14.11}$$

Rearranging Eq. (14.11) to write the differential equation in standard format gives:

$$\frac{d^2y}{dx^2} + p^2 y = -\frac{Vx}{EI} \tag{14.12}$$

where $p = \sqrt{\frac{P}{EI}}$

Equation (14.12) is a linear, second-order differential equation with constant coefficients; however, it is not homogeneous. We obtain the homogeneous solution by setting the right hand side of Eq. (14.12) equal to zero and solving for y_h to obtain:

$$y_h = C_1 \sin(px) + C_2 \cos(px) \tag{14.13}$$

where C_1 and C_2 are constants of integration.

The particular solution is an expression for y_p that satisfies Eq. (14.12). Consider the expression for y_p shown below:

$$y_p = -Vx/P \tag{14.14}$$

The expression in Eq. (14.14) satisfies Eq. (14.12) and hence it represents the particular solution. The general solution is given by:

$$y = y_h + y_p = C_1 \sin(px) + C_2 \cos(px) - Vx/P \tag{14.15}$$

We follow the usual procedure to solve for the constants of integration using boundary conditions. Because the top of the column is pinned, we write:

$$y(0) = 0 \tag{14.16}$$

The boundary conditions for the built-in end are:

$$y(L) = 0 \qquad dy/dx\,(L) = 0 \tag{14.17}$$

It is clear from the boundary conditions listed in Eq. (14.16) that $C_2 = 0$. Substituting the first boundary condition (y = 0 at x = L) from Eq. (14.17) into Eq. (14.15) yields:

$$C_1 \sin pL = VL/P \qquad (14.18)$$

Next differentiate Eq. (14.15) with respect to x to obtain:

$$dy/dx = pC_1 \cos (px) - V/P \qquad (14.19)$$

Substituting the boundary condition (dy/dx = 0 at x = L) from Eq. (14.17) into Eq. (14.19) gives:

$$pC_1 \cos pL = V/P \qquad (14.20)$$

Dividing Eq. (14.18) by Eq. (14.20) leads to the transcendental equation:

$$\tan pL = pL \qquad (14.21)$$

Solving this equation by trial and error leads to the first meaningful solution for pL as:

$$pL = 4.4934 \qquad (14.22)$$

Then substituting Eq. (14.3) into Eq. (14.22) and solving, we obtain:

$$P_{CR} = \frac{20.19EI}{L^2} \approx \frac{\pi^2 EI}{(0.7L)^2} \qquad (14.23)$$

Comparing the result in Eq. (14.23) with that in Eq. (14.8) shows that they differ only by the coefficient multiplying the term (EI/L^2). Recognizing this similarity, let's write a general equation for the buckling of columns as:

$$P_{CR} = \frac{\pi^2 EI}{(kL)^2} \qquad (14.24)$$

For the column with two pinned ends k = 1, and for the column with one end pinned and the other built–in k = 0.7. Clearly, building-in one end of the column increased the critical load required for buckling by a factor of more than two.

14.3.2 Column Buckling with One Free and One Built-in End.

This case deals with a column with a free end, where the load is applied, and a built-in end as depicted in Fig. 14.7

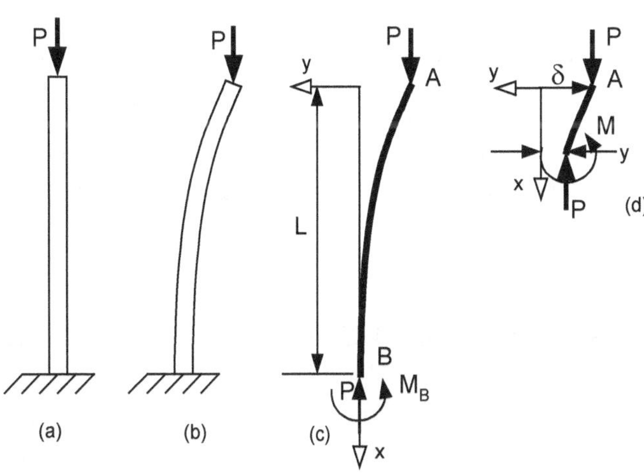

Fig. 14.7 A cantilever type column with one end free and the other built-in.

A FBD of the free end of the cantilever column is shown in Fig. 14.7d. From this FBD we write the moment equation as:

$$M = P(\delta - y) \tag{14.25}$$

Substituting Eq. (14.25) into Eq. (13.4) and rearranging to write the differential equation in standard format gives:

$$\frac{d^2y}{dx^2} + p^2y = p^2\delta \tag{14.26}$$

Using similar procedures to solve Eq. (14.26), we obtain:

$$y = y_h + y_p = C_1 \sin(px) + C_2 \cos(px) + \delta \tag{14.27}$$

We solve for the constants of integration using boundary conditions. Because the top of the column undergoes a deflection δ, we write:

$$y(0) = \delta \tag{14.28}$$

The boundary conditions for the built-in end are:

$$y(L) = 0 \qquad dy/dx\,(L) = 0 \tag{14.29}$$

It is clear from the boundary condition listed in Eq. (14.28) that $C_2 = 0$. Next differentiate Eq. (14.27) with respect to x to obtain:

$$dy/dx = pC_1 \cos(px) \tag{14.30}$$

Substituting the boundary condition ($dy/dx = 0$ at $x = L$) into Eq. (14.30) gives:

$$C_1 \cos pL = 0 \tag{14.31}$$

It is evident that Eq. (14.31) is satisfied when $pL = n\pi/2$. The first bucking mode occurs when $n = 1$; hence, $pL = \pi/2$. Substituting this value into Eq. (14.3) gives the critical buckling load as:

$$P_{CR} = \frac{\pi^2 EI}{4L^2} \tag{14.32}$$

Comparison of Eq. (14.32) with Eq. (14.23) shows a remarkable reduction in the critical buckling load when the pin at the top of the column is removed and the axial compression load is applied to the column's free end. The reduction is a factor of about eight.

14.3.3 Column Buckling with Two Built-in Ends.

The final case considered is the column with both ends built-in as shown in Fig. 14.8. The procedure for deriving the Euler equation for the critical load for buckling is similar to that described previously. We begin with the FBD as shown in Fig. 14.8d and write the moment equation as:

$$M = -Py + M_A \tag{14.33}$$

Substituting Eq. (14.33) into Eq. (13.4) and rearranging gives the differential equation in standard format as:

$$\frac{d^2y}{dx^2} + p^2y = \frac{M_A}{EI} \tag{14.34}$$

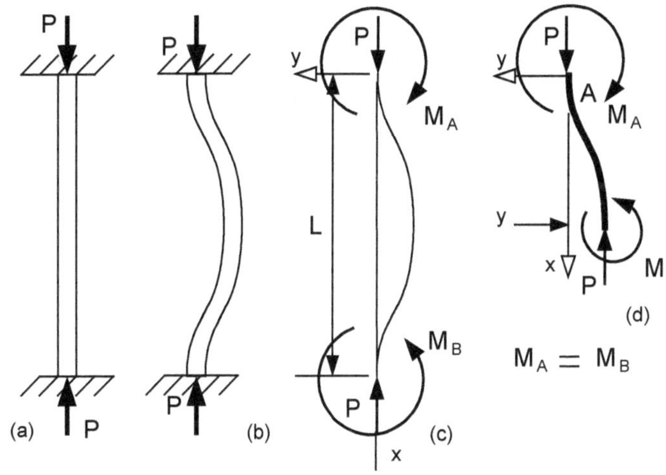

Fig. 14.8 A column with both ends built-in under axial compression loading.

Solving Eq. (14.34) for the homogeneous and particular solutions gives the general solution as:

$$y = y_h + y_p = C_1 \sin (px) + C_2 \cos (px) + M_A/P \tag{14.35}$$

We follow the usual procedure to solve for the constants of integration using boundary conditions. At the top of the column (x = 0), y = 0 and dy/dx = 0. At the middle of the column symmetry exists so that dy/dx = 0 at x L/2. At the bottom of the column (x = L), y = 0 and dy/dx = 0. It is clear from the boundary condition (y = 0 at x = 0) and Eq. (14.35) that:

$$C_2 = -M_A/P \tag{14.36}$$

Differentiating Eq. (14.35) with respect to x and substituting Eq. (14.36) yields:

$$dy/dx = C_1 p \cos (px) + (M_A/P)p \sin (px) \tag{14.37}$$

Substituting the boundary condition (dy/dx = 0 at x = 0) into Eq. (14.37) gives $C_1 = 0$. We can now write Eq. (14.34) as:

$$y = C_2[\cos (px) - 1] \tag{14.38}$$

From the boundary condition (y = 0 at x = L) and Eq. (14.38), we can write:

$$\cos (pL) - 1 = 0 \tag{14.39}$$

Equation (14.39) is satisfied when pL = 2nπ. The first mode of buckling occurs with n = 1, which leads to:

$$pL = 2\pi \tag{14.40}$$

Then substituting Eq. (14.3) into Eq. (14.40), we obtain:

$$P_{CR} = \frac{4\pi^2 EI}{L^2} \tag{14.41}$$

Let's apply these new relations to determine the critical load for two different columns with various end conditions.

EXAMPLE 14.3

A column 15 feet long is fabricated from a steel tube with the rectangular cross section presented in Fig. E14.3. The column is pinned at one end and built-in at the other. Determine the critical buckling force.

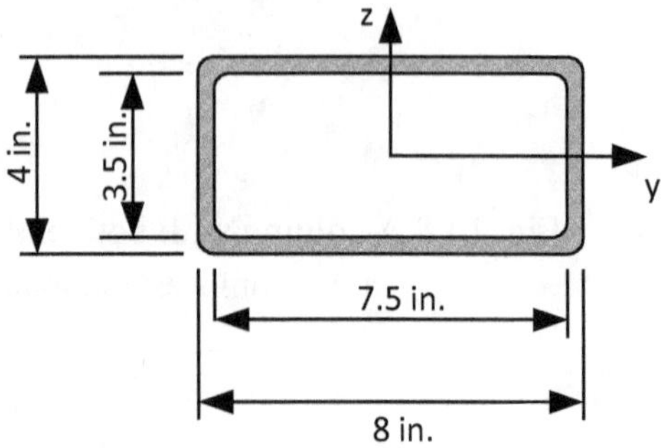

Fig. E14.3

Solution:

We employ Eq. (14.23) to write the expression for the critical buckling force for this column as:

$$P_{CR} = \frac{20.19EI}{L^2} \quad \text{(a)}$$

The modulus of elasticity $E = 30 \times 10^6$ psi for the steel tube is from Appendix B-1.

The cross section is rectangular; hence the moment of inertia is a function of the choice of axis. For this reason, we will determine both I_y and I_z below:

$$I_y = (1/12)[b_o h_o^3 - b_i h_i^3] = (1/12)[(8)(4)^3 - (7.5)(3.5)^3] = 15.87 \text{ in.}^4 \quad \text{(b)}$$

$$I_z = (1/12)[b_o h_o^3 - b_i h_i^3] = (1/12)[(4)(8)^3 - (3.5)(7.5)^3] = 47.62 \text{ in.}^4 \quad \text{(c)}$$

Substituting the lower value of the moment of inertia from Eq. (b) into Eq. (a) yields:

$$(P_{CR})_y = \frac{20.19EI_y}{L^2} = \frac{20.19(30 \times 10^6)15.87}{(15)^2(12)^2} = 296.7 \text{ kip} \quad \text{(d)}$$

Examination of the results indicates the importance of the choice of axis when determining the moment of inertia when the cross section is not circular or square. In this case, the difference in the inertia is about a factor of three.

EXAMPLE 14.4

A column 6.4 m long is fabricated from a steel wide flange section with a designation of W152 × 24. Determine the critical buckling force for the following two cases: (a) the column is pinned at one end and built-in at the other, and (b) the column is built-in at both ends.

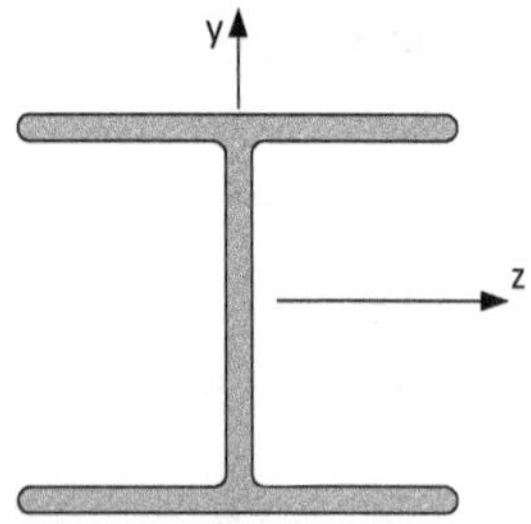

Fig. E14.4

Solution:

Employ Eq. (14.32) to write the expression for the critical buckling force for the first column as:

$$P_{CR} = \frac{\pi^2 EI}{4L^2} \tag{a}$$

The modulus of elasticity (207 GPa) for the steel column is from Appendix B-1. The moment of inertia is a function of the choice of axis. Both I_y and I_z are given in Appendix D as:

$$I_y = 1.84 \times 10^6 \text{ mm}^4 \qquad I_z = 13.4 \times 10^6 \text{ mm}^4 \tag{b}$$

We select the minimum value of the moment of inertia I_y given in Eq. (b) and substitute this value into Eq. (a) to obtain:

$$(P_{CR})_y = \frac{\pi^2 EI_y}{4L^2} = \frac{\pi^2(207\times10^3)(1.84\times10^6)}{(4)(6400)^2} = 22.94 \text{ kN} \tag{c}$$

Examination of the results indicates the importance of the choice of the end conditions for column buckling. In this case, the difference in the critical buckling load is in excess of a factor of sixteen.

Next, employ Eq. (14.41) to obtain the expression for the critical buckling force for the second column:

$$(P_{CR})_y = \frac{4\pi^2 EI_y}{L^2} = \frac{4\pi^2(207\times10^3)(1.84\times10^6)}{(6400)^2} = 367.1 \text{ kN} \tag{d}$$

Again, we have selected the minimum value of the moment of inertia I_y for evaluating the critical buckling force.

14.3.4 Summary of Equations For Critical Column Buckling Loads

We have derived four equations for the buckling of columns under the action of centric compressive loads. These equations all are of the form $P_{CR} = \pi^2 EI/(kL)^2$. We summarize the results for k as a function of the various end constraints imposed on the column in Table 14.1

Table 14.1 Constant k as a function of column end constraints.

Case No.	End Constraints	k
1	Pinned—Pinned	1
2	Pinned—Built-in	0.7
3	Free—Built-in	2
4	Built-in—Built-in	0.5

The determination of the critical buckling load for columns subjected to centric compressive forces involves identifying the end constraints and substituting the correct value of k into Eq. (14.24). Care must also be exercised in selecting the correct axis about which to determine the inertia I as two choices exist in most cases. In most instances the minimum inertia is employed in the determination. Four examples have been described to demonstrate the method for computing the Euler buckling loads.

We have not emphasized the critical buckling stress σ_{CR} in this section, as we were more concerned with determining the critical buckling force. The topic of stresses produced in columns is discussed in much more detail in the next section.

14.4 COLUMN STRESSES AND LIMITATIONS OF EULER'S THEORY

As a centric load is applied to a column, an axial compressive stress develops that increases until the column becomes unstable. At the inception of buckling the stress is called the critical stress and is given by:

$$\sigma_{CR} = P_{CR}/A \tag{14.42}$$

If we consider the column pinned at both ends, then $P_{CR} = \pi^2 EI/L^2$. Substituting this value into Eq. (14.42) and rearranging symbols yields:

$$\sigma_{CR} = \frac{\pi^2 EI}{L^2 A} = \frac{\pi^2 EAr^2}{L^2 A} = \frac{\pi^2 E}{(L/r)^2} \tag{14.43}$$

Note $I = Ar^2$ in the expansion of Eq. (14.43), where A is the cross sectional area of the column and r is the radius of gyration of the area.

The term (L/r) is known as the slenderness ratio of the column. As (L/r) increases the column becomes more flexible and will buckle at lower and lower stress levels. Let's consider an example to illustrate the reduction in the critical stress σ with increasing (L/r).

EXAMPLE 14.5

Consider a steel column with E = 207 GPa and S_y = 300 MPa. Prepare a graph showing σ_{CR} as a function of the slenderness ratio (L/r).

Solution:

The solution involves evaluating Eq. (14.43) as (L/r) is varied from 10 to 200 in steps of 10.

$$\sigma_{CR} = \frac{\pi^2 E}{(L/r)^2} = S_y \tag{a}$$

The results obtained from a spreadsheet calculation using Eq. (a) are presented in Fig. E14.5.

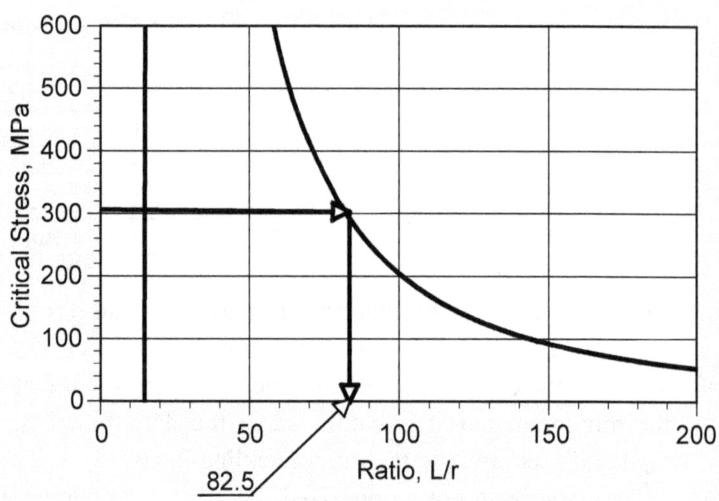

Fig. E14.5 Critical stress σ_{CR} as a function of slenderness ratio (L/r).

Examination of Fig. E14.5 indicates that the critical stress σ_{CR} exceeds the yield strength S_y of the steel specified in the example statement for (L/r) < 82.5. This fact means that the Euler theory for buckling is valid only when $(L/r) \geq 82.5$. For slenderness ratios less than this value, failure of the column will occur by compressive yielding.

The limiting value of (L/r) depends on the column constraint and both the modulus of elasticity and the yield strength of the material from which the column is fabricated. We will explore this dependency in the next example.

EXAMPLE 14.6

Consider a column with both ends pinned. Determine the limiting value of (L/r) as a function of yield strength for columns fabricated from both steel and aluminum. Consider the yield strength varying from 20 to 200 ksi.

Solution:

Set $\sigma_{CR} = S_y$ in Eq. (14.43) and solve for (L/r) to obtain:

$$L/r = \sqrt{\frac{\pi^2 E}{S_y}} \qquad \text{(a)}$$

Let $E = 30 \times 10^6$ pi and 10.4×10^6 psi for steel and aluminum, respectively. Using a spreadsheet to evaluate Eq. (a) and varying S_y from 20 to 200 ksi in steps of 10, we obtain the results presented in Fig. E14.6.

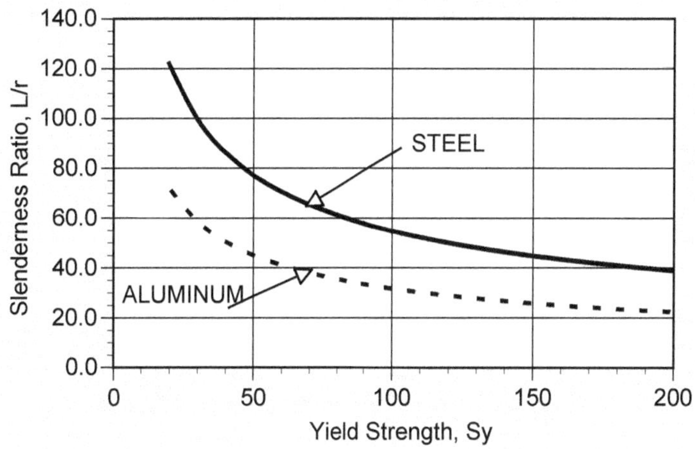

Fig. E14.6 The limiting value of the slenderness ratio as a function of yield strength for a column with both ends pinned.

The marked difference in the limiting value of the slenderness ratio for steel and aluminum alloy is due to the fact that aluminum is less resistant to buckling because of its much lower modulus of elasticity.

These two examples have illustrated that the Euler theory of buckling has limits of applicability. Long slender columns with high values of (L/r) will buckle under the action of centric compressive loads and Euler's equations are valid. However, if (L/r) for the column is less than the limit value, as is the case for shorter columns with larger radii of gyration, then the Euler theory is not valid. In these cases, the column does not buckle. The failure mode is one of yielding due to excessive compressive stress. The different modes of failure are illustrated in Fig. 14.9.

Fig. 14.9 Safety and failure regions for a centric loaded, pin-pin ended column. Yield strength 300 MPa.

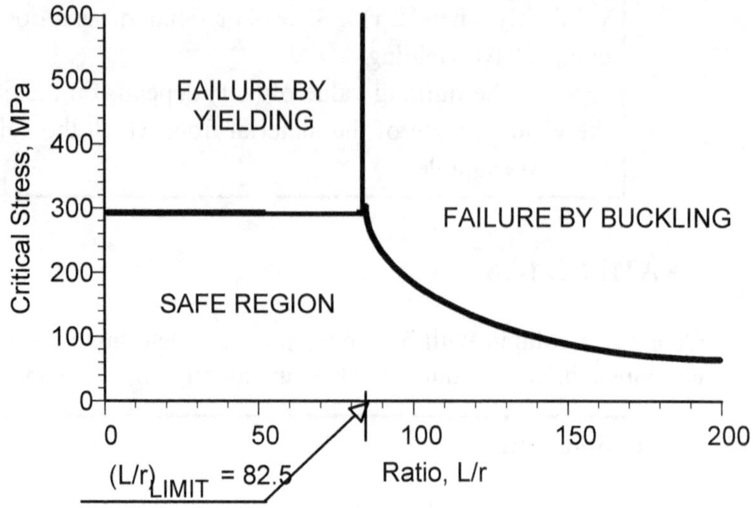

14.5 ECCENTRICALLY LOADED COLUMNS

The developments in the previous sections of this chapter assumed that the axial compressive load was applied to the column through the centroid of its cross section. In practice this centric loading is nearly impossible to achieve. Even if the load is placed at the centroid of the cross section, columns are never perfectly straight. As a consequence the loading always exhibits some small eccentricity e. Let's explore the effect of the eccentricity of the load on the buckling relations by considering the pin-pin ended column shown in Fig. 14.10.

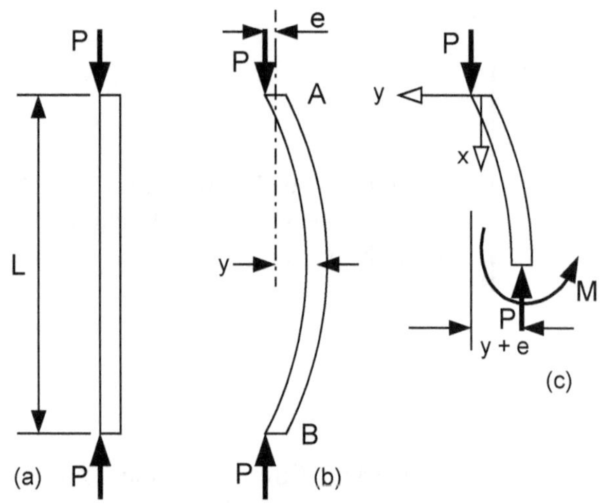

Fig. 14.10 A column with an eccentric axial load P.

The analysis follows the same procedure as described previously. We deform the column as a sine wave because the ends are pinned. The column is cut at a location x from the top end and a FBD of the end is made as shown in Fig. 14.10c. Writing the equation for the moment M gives:

$$M = -P(y + e) \tag{14.44}$$

Substituting Eq. (14.44) into Eq. (13.4) yields:

$$\frac{d^2y}{dx^2} = \frac{M(x)}{EI} = -\frac{1}{EI}(Py + Pe) \tag{14.45}$$

Rearranging Eq. (14.45) to write the differential equation in standard format gives:

$$\frac{d^2y}{dx^2} + p^2 y = -p^2 e \tag{14.46}$$

where p is given in Eq. (14.3)

Solving Eq. (14.46) yields:

$$y = y_h + y_p = C_1 \sin(px) + C_2 \cos(px) - e \tag{14.47}$$

We follow the usual procedure to solve for the constants of integration using boundary conditions. Because both ends of the column are pinned, we write:

$$y(0) = 0 \qquad\qquad y(L) = 0 \tag{14.48}$$

From the first boundary condition ($y = 0$ at $x = 0$), it is clear that $C_2 = e$.

From the second boundary condition ($y = 0$ at $x = L$), we can show that:

$$C_1 = e[1 - \cos(pL)]/\sin(pL) \tag{14.49}$$

Using the following trigometric identities:

$$1 - \cos(pL) = 2\sin^2(pL/2) \qquad\qquad \sin(pL) = 2\sin(pL/2)\cos(pL/2)$$

the expression for C_1 can be rewritten as:

$$C_1 = e\tan(pL/2) \tag{14.50}$$

Substituting the values for C_1 and C_2 into Eq. (14.47) yields:

$$y == e[\tan(pL/2)\sin(px) + \cos(px) - 1] \tag{14.51}$$

This relation shows that the amplitude of the deflection of the column prior to the initiation of buckling depends upon the eccentricity e. We will show a graph illustrating this fact later in this section. The maximum value of y occurs at the mid point of the column where $x = L/2$. At that location, we can write the expression for y_{Max} as:

$$y_{Max} = e[\sec(pL/2) - 1] \tag{14.52}$$

The column will buckle when y_{Max} becomes large. For small eccentricity, y_{Max} becomes large when:

$$\sec(pL/2) \rightarrow \propto \qquad\qquad \Rightarrow \qquad\qquad pL/2 = \pi/2$$

From this result and Eq. (14.3), it is clear that:

$$P_{CR} = \frac{\pi^2 EI}{L^2} \tag{14.53}$$

The buckling load is identical with that derived for centrically loaded columns when $e = 0$. While the end result is the same, the bucking process differs. With centric loading the column remained straight as the load P was increased until P_{CR} was achieved. At that point the column became unstable and suddenly buckled (collapsed). With eccentricity the column deflects laterally with increasing magnitude until the critical load is achieved. We will illustrate this behavior in the next example.

EXAMPLE 14.7

Consider a column with both ends pin loaded with an axial force P. The force P is applied with an eccentricity e. Determine y_{Max} as a function of P as the quantity pL/2 varies from 0 to π/2. Let e = 1 and 2 in. The column is 30 ft long and fabricated from a square steel tube depicted in Fig. E14.7.

Fig. E14.7

Solution:

Before determining y_{Max}, establish the critical buckling load from Eq. (14.53) as:

$$P_{CR} = \frac{\pi^2 EI}{L^2} = \frac{\pi^2(30\times10^6)[(6)^4-(5)^4]}{(12)(30)^2(12)^2} = 127.8 \text{ kip} \quad \text{(a)}$$

This result enables us to bound the calculations needed in preparing a graph of y_{Max} as a function of P. Recall Eq. (14.52) and rewrite it as:

$$y_{Max} = e\left(\sec\left(\frac{pL}{2}\right)-1\right) = e\left(\sec\sqrt{\frac{PL^2}{4EI}}-1\right) \quad \text{(b)}$$

Substituting numerical parameters into Eq. (b) gives:

$$y_{Max} = e\left(\sec\sqrt{\frac{PL^2}{4EI}}-1\right) = e\left(\sec\sqrt{\frac{P(360)^2}{4(30\times10^6)(55.92)}}-1\right) = e\left(\sec\sqrt{19.31\times10^{-6}P}-1\right) \quad \text{(c)}$$

Next, employ a spreadsheet to evaluate Eq. (c) for P varying from 0 to 130 kip to obtain the graph shown in Fig. E14.7a.

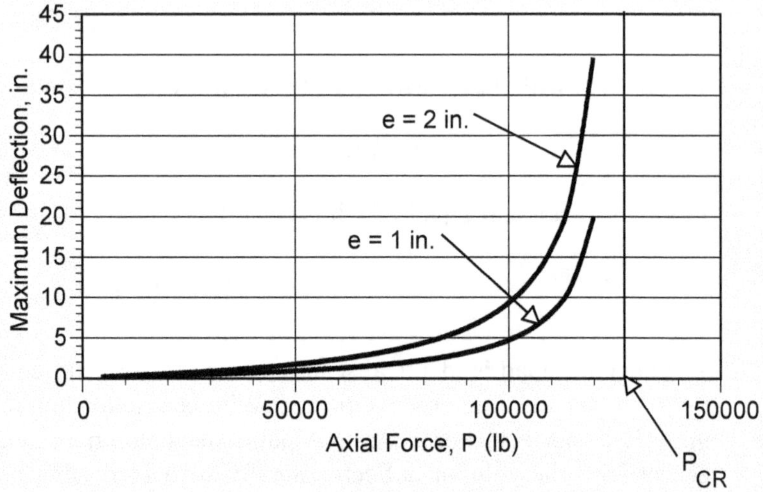

Fig. E14.7a

Examination of the results in Fig. E14.7a shows that the deflection is not a linear function of the axial load P. Instead the deflection increases exponentially as the critical buckling load is approached. This non-linear behavior is due to the moment M = P(e + y). As P increases, the deflection y also increases; hence, the moment increases in a non-linear manner.

14.6 STRESSES IN COLUMNS WITH ECCENTRIC LOADING

In some structural applications, it is necessary to apply loads to columns with significant eccentricity. While this practice does not affect the critical buckling load, as determined by the Euler theory, it may result in failure of the column due to excessive compressive stress at loads less than P_{CR}. Let's consider the free-ended column, shown in Fig. 3.11, with an axial force P that is applied with an eccentricity e.

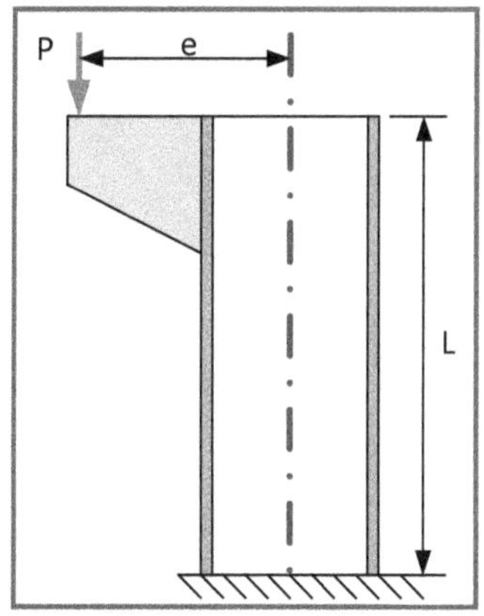

Fig. 4.11 A free-ended column with an eccentric load P.

The load P generates an internal force and a bending moment both of which are constant over the length of the column. A FBD of the top portion of the column, presented in Fig. 14.12, shows this force and moment.

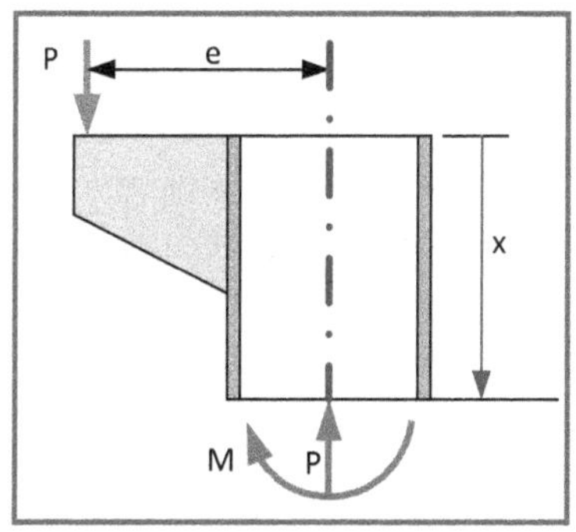

Fig. 14.12 A FBD showing an internal force P and a bending moment M.

Recall that the bending moment M depends on both the eccentricity e and the lateral deflection y. Considering equilibrium permits us to write:

$$M = P\,(y + e) \tag{14.54}$$

The axial stresses produced in the column due to P and M are:

$$\sigma_c = \sigma_a + \sigma_b = -\,P/A \pm Mc/I \tag{14.55}$$

It is evident from Eq. (14.55) that the axial stresses due to the axial force and the bending moment combine. The combined distribution of axial stresses across the column is illustrated in Fig. 14.13.

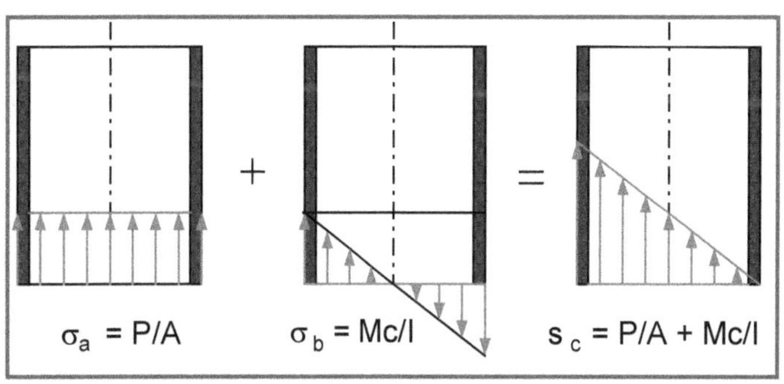

Fig. 14.13 Combined stresses due to eccentric loading of a column.

For an axial compressive force applied to a column, the maximum stress is due to the combined axial and bending stresses given by:

$$\sigma_{Max} = \sigma_a + \sigma_b = -(P/A + M_{Max}c/I) \tag{14.56}$$

Substituting Eq. (14.54) into Eq. (14.56) gives:

$$\sigma_{Max} = -[P/A + P(y_{Max} + e)c/I] \tag{14.57}$$

Using the value for y_{Max} from Eq. (14.52), we can write:

$$\sigma_{Max} = -P\left(\frac{1}{A} + \frac{ce}{I}\sec\left(\frac{pL}{2}\right)\right) \tag{14.58}$$

An alternate formula for σ_{Max} can be obtained by substituting Eqs. (14.3) and (14.53) into Eq. (14.58). After simplification:

$$\sigma_{Max} = -P\left(\frac{1}{A} + \frac{ce}{I}\sec\left(\frac{\pi}{2}\sqrt{\frac{P}{P_{CR}}}\right)\right) \tag{14.59}$$

This equation is the well-known secant formula for eccentrically loaded columns. The equation may be used for columns with any end conditions, providing the appropriate formula for P_{CR} is selected. Also observe that Eq. (14.59) is non-linear in terms of the load P. Thus, it is often necessary to solve the equation by trial and error or by using advanced methods (i.e. programmable calculator, computer software, etc.). For the same reason, the safety factor pertains only to the critical load and not to the stresses.

In the design of an eccentrically loaded column, the procedure is to determine the maximum stress using Eq. (3.59) and the critical buckling stress using the appropriate relation that depends on the end conditions of the column. We then compare these two values to determine the mode of failure for the column. We will illustrate this procedure in the next example.

EXAMPLE 14.8

Consider a column with both ends pin loaded with an axial force P as shown in Fig. E14.8. The force P is applied with an eccentricity e = 350 mm. Determine the maximum load P that can be applied if a safety factor SF = 2.5 is to be maintained against failure by excessive compressive stress or by buckling. The steel column is 8.0 m long and is fabricated from an American Standard section with a designation of S127 × 22. The steel is 1020 HR with S_y = 290 MPa.

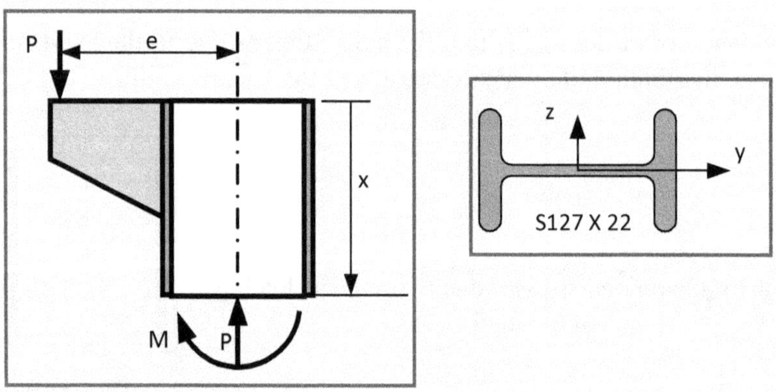

Fig. E14.8

Solution:

We employ Eq. (14.8) to write the expression for the critical buckling force for the pinned ended column as:

$$P_{CR} = \frac{\pi^2 EI}{L^2} \tag{a}$$

The modulus of elasticity E = 207 GPa for the steel column is from Appendix B-1.

The moment of inertia is a function of the orientation of the pin. If the pin is inserted along the y axis (see Fig. E14.8), then the end of the column will rotate about this axis. However, if the pin is inserted in the z direction, the column will rotate about this axis. Both I_y and I_z are given in Appendix C as:

$$I_y = 0.695 \times 10^6 \text{ mm}^4 \tag{b}$$

$$I_z = 6.33 \times 10^6 \text{ mm}^4 \tag{c}$$

Substituting the minimum value I_y into Eq. (a) yields:

$$\left(P_{CR}\right)_y = \frac{\pi^2 EI_y}{L^2} = \frac{\pi^2(207\times10^3)(0.695\times10^6)}{(8{,}000)^2} = 22.19 \text{ kN} \tag{d}$$

Examination of the results indicates the importance of the direction of the pin, when the cross section is not circular or square. In this case, the difference in the moment of inertia I is in excess of a factor of nine. Clearly, for the section shown in Fig. E14.8, the pin should be inserted in the z direction. Many conservative designers will base their recommendations on the minimum critical load, even with the column end free to rotate about the z-axis. Let's follow this conservative approach. Then the allowable load for buckling is given by:

$$(P_{Allowable})_{CR} = P_{CR}/SF = 22.19/2.5 = 8.876 \text{ kN} \tag{e}$$

The maximum stress is given by Eq. (14.59) as:

$$\sigma_{Max} = -S_y = -P_y\left(\frac{1}{A} + \frac{ce}{I_y}\sec\left(\frac{\pi}{2}\sqrt{\frac{P_y}{P_{CR}}}\right)\right) \tag{f}$$

Additional properties for the American Standard section are obtained from Appendix C as:

$$A = 2{,}800 \text{ mm}^2 \tag{g}$$

$$c = w_f/2 = (83.4)/2 = 41.7 \text{ mm} \tag{h}$$

Substituting all numerical parameters for the S127 × 22 section and the yield strength of the 1020 HR steel into Eq. (f) gives:

$$290 = P_y\left(\frac{1}{2{,}800} + \frac{(41.7)(350)}{0.695\times10^6}\sec\left(\frac{\pi}{2}\sqrt{\frac{P_y}{22{,}190}}\right)\right) \tag{i}$$

Solving Eq. (i) by trial and error:

$$P_y = 8{,}018 \text{ N} = 8.018 \text{ kN} \tag{j}$$

Then the allowable load for yielding is:

$$(P_{Allowable})_y = P_y/SF = 8.018/2.5 = 3.207 \text{ kN} \quad \text{(k)}$$

Compare both allowable loads from Eqs. (e) and (k) and choose the smallest one, which gives:

$$P_{Allowable} = 3.207 \text{ kN} \quad \text{(l)}$$

In this case, the column will fail by yielding rather than buckling.

14.7 DESIGN CODES

Design codes have been developed by several industrial associations to facilitate the design of columns fabricated from steel, aluminum and wood. These codes accommodate both long slender columns and intermediate length columns. Local governments controlling building permits have adopted these codes and structural engineers employ the formulas contained in the codes when sizing columns.

14.7.1 Steel Columns

The American Institute for Steel Construction (AISI) has adopted two different equations for the design of steel columns. These equations permit a designer to determine the maximum allowable stress for centric loaded columns for specified slenderness ratios. The first of these two equations for long slender columns is based on Euler's theory, but with a safety factor SF = 1.92.

$$\sigma_{Allowable} = \frac{\pi^2 E}{1.92(kL/r)^2} \quad (14.60)$$

where k depends on the column's end conditions and is given in Table 14.1.

Equation (14.60) is applied for specified values of (kL/r) as indicated below:

$$SR_C = \sqrt{\frac{2\pi^2 E}{S_y}} \le \left(\frac{kL}{r}\right) \le 200 \quad (14.61)$$

The limitations Eq. (14.60) imposed by Eq. (14.61) insure that the column will remain elastic and that $\sigma_{Allowable} \le S_y$ even when the residual stresses in the rolled steel section approach one half of the yield strength of the steel.

For columns with slenderness ratios less than $SR_C = [(2\pi^2E)/S_y]^{1/2}$, a more conservative equation is employed to determine the allowable stress. Because of plastic behavior of the steel with these lower slenderness ratios, a larger safety factor is incorporated in the design equation. The safety factor specified in the design code is a variable depending on (kL/r).

$$SF = 1.667 + 0.375\,[(kL/r)/SR_c] - 0.125\,[(kL/r)^3/SR_C^{\,3})] \quad (14.62)$$

The allowable stress in this range of (kL/r) is given by:

$$\sigma_{Allowable} = \left(1 - \frac{(kL/r)^2}{2(SR_C)^2}\right)\left(\frac{S_y}{SF}\right) \quad (14.63)$$

Equation (14.63) is employed for columns with (kL/r) < SR_c. A graph showing the allowable stress as a function of the slenderness ratio (kL/r) is presented in Fig. 14.14.

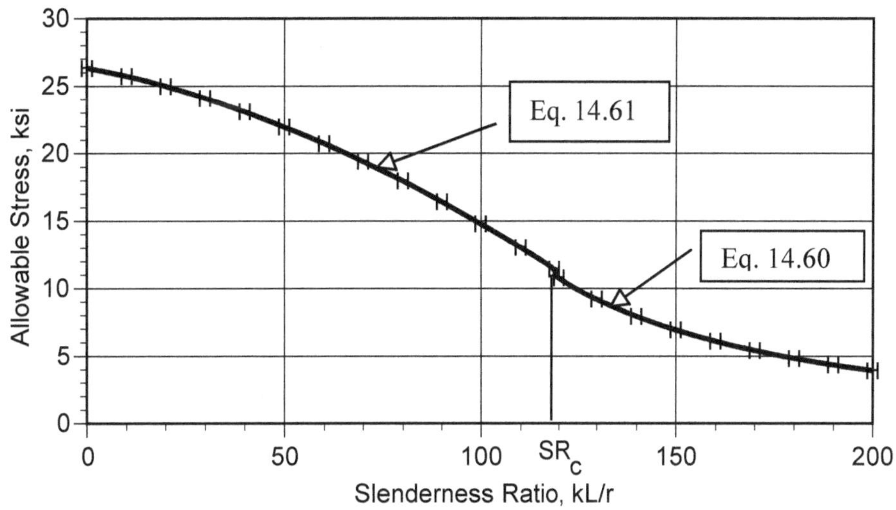

Fig. 14.14 Design curve for pin ended column fabricated from 1020 HR steel with S_y = 42 ksi.

A smooth transition from one equation to the other is clear at the critical slenderness ratio that occurs at SR_c = 118.7 for a steel with $S_y = 30 \times 10^3$ ksi.

EXAMPLE 14.9

A column fabricated from a steel wide flange section W 12 × 50 is pinned at both ends. Use the curve resulting from the code equations that is shown in Fig. 14.14 to determine the allowable centric load applied to the column. The column is 24 ft long.

Solution:

First reference Appendix D and note for the W 12 × 50 section that the minimum value of r = 1.96 in. and A = 14.7 in^2. Also note k = 1 for a column with pinned ends. Determine the slenderness ratio (kL/r):

$$(kL/r) = (1.0)(24)(12)/1.96 = 146.9 \qquad \text{(a)}$$

Employing the graph in Fig. 14.14, we locate the point (kL/r) = 146.9 and draw a line upward to intersect the curve as shown in Fig. E14.9. We then construct a horizontal line from the intersection point to the ordinate to obtain the value of $\sigma_{Allowable}$ = 7.15 ksi.

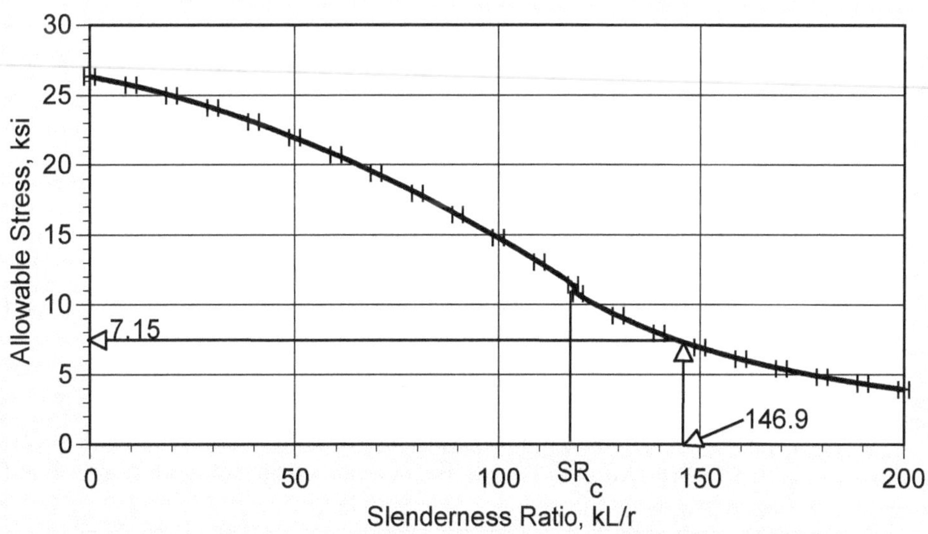

Fig. E14.9

The maximum allowable load on the column is then given by:

$$P_{Allowable} = \sigma_{Allowable} A = (7.15)(14.7) = 105.1 \text{ kip} \quad \text{(b)}$$

Clearly, the use of the design curve and the geometric properties of the wide flange section from Appendix D enable a rapid determination of the maximum allowable centric force applied to the column.

14.7.2 Aluminum Columns

The Aluminum Association has also developed a design code for aluminum columns. It accounts for the elastic and plastic behavior of aluminum by using three different equations to determine the allowable stress. Because the physical properties of aluminum alloys vary considerably from one alloy to another, three equations are provided by the Aluminum Association for each alloy commonly employed in construction. The relations presented below are for the alloy 2014 T-6.

$$\sigma_{Allowable} = 28 \text{ ksi} \qquad 0 \leq (kL/r) \leq 12 \qquad (14.64)$$

$$\sigma_{Allowable} = [30.7 - 0.23(kL/r)] \text{ ksi} \qquad 12 \leq (kL/r) \leq 55 \qquad (14.65)$$

$$\sigma_{Allowable} = [54{,}000/(kL/r)^2] \text{ ksi} \qquad 55 \leq (kL/r) \qquad (14.66)$$

The allowable stress $\sigma_{Allowable}$ as a function of (kL/r) was determined from these three equations to develop the design curve presented in Fig. 14.15. Recall that this curve is valid only for the aluminum alloy 2014 T-6.

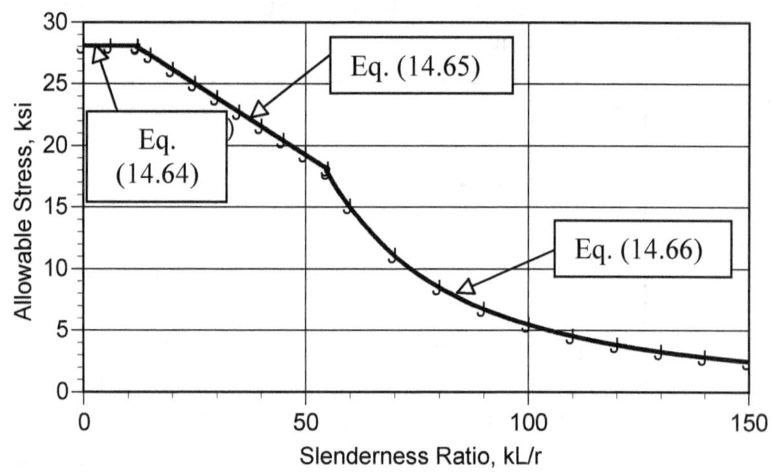

Fig. 14.15 Design curve for a column fabricated from aluminum alloy 2014 T-6 (S_y = 41 ksi).

EXAMPLE 14.10

A column fabricated from an aluminum (2014 T-6) structural T section WT 6 × 48 is pinned at one end and built-in at the other. Use the curve resulting from the code equations that is shown in Fig. 14.15 to determine the allowable centric load applied to the column. The column is 18 ft long.

Solution:

First reference Appendix D and note for the WT 6 × 48 section that the minimum value of r = 1.51 in. Also the area A = 14.1 in^2. Note k = 0.7 for a column pinned at one end and built-in at the other. Determine the slenderness ratio (kL/r):

$$(kL/r) = (0.7)(18)(12)/(1.51) = 100.1 \quad \text{(a)}$$

Employing the graph in Fig. 14.15, we locate the point (kL/r) = 101.1 and draw a line upward to intersect the curve as shown in Fig. E14.10. We then construct a horizontal line from the intersection point to the ordinate to obtain the value of $\sigma_{Allowable}$ = 5.40 ksi.

Fig. E14.10

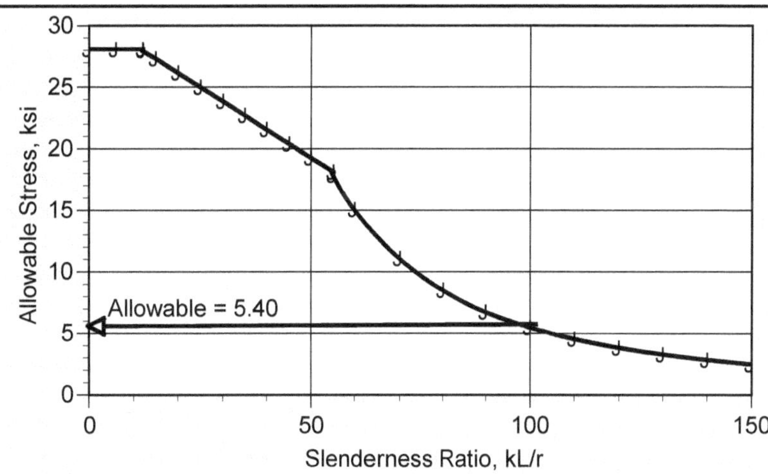

The maximum allowable load on the column is then given by:

$$P_{Allowable} = \sigma_{Allowable}\, A = (5.40)(14.1) = 76.14 \text{ kip} \qquad \text{(b)}$$

Clearly, the use of the design curve and the geometric properties of the structural tee section from Appendix D enable a rapid determination of the maximum allowable centric force applied to the column.

14.7.2 Wood Columns

Design of both columns and beams fabricated from wood is difficult because the physical properties of wood depend on the type of tree used to produce the structural members and even the part of the country where the tree was grown. Even the moisture content of the wood affects its properties. To deal with these many variables, the American Forest and Paper Association has published the National Design Specification for Wood Construction. The Western Woods Products Association has incorporated these design specifications in an excellent reference for designing structural members fabricated from many types of woods in their Western Woods Use Book[3].

A computer program is provided with the Western Woods Use Book that has merged the physical properties of several types of woods with the design relations adopted by the wood products and construction industries. This program enables the user to determine allowable column loading and beam loading while taking into account several factors affecting the physical properties of the wood such as section size, duration of the load, temperature and moisture content.

14.8 SUMMARY

The concept of elastic instability was introduced. Columns can fail by buckling at stress levels lower than the yield strength of the materials from which they are fabricated. Failure by buckling is catastrophic as the column collapses suddenly. The critical load P_{CR} depends on the end conditions of the column with the critical load increasing as the constraint at the ends increase.

Equations for P_{CR} were derived for four different end conditions. The results are given by:

$$P_{CR} = \frac{\pi^2 EI}{(kL)^2} \qquad (14.24)$$

Where k, given in the table below, accounts for the constraint provided by the column supports.

[3] Available from Western Woods Products Association, 522 SW 5th Avenue, Suite 400, Portland, OR 97204-3930

Case No.	End Constraints	k
1	Pinned—Pinned	1
2	Pinned—Built-in	0.7
3	Free—Built-in	2
4	Built-in—Built-in	0.5

The inertia I of the cross section of the column is almost always taken as the minimum value.

The critical stress σ_{CR} in centric loaded columns was defined as:

$$\sigma_{CR} = P_{CR}/A \qquad (14.42)$$

The critical stress may be written as:

$$\sigma_{CR} = \frac{\pi^2 EI}{L^2 A} = \frac{\pi^2 EAr^2}{L^2 A} = \frac{\pi^2 E}{(L/r)^2} \qquad (14.43)$$

where A is the cross sectional area of the column and r is the radius of gyration of the area.

The term (L/r) is known as the slenderness ratio of the column. As (L/r) increases, the column becomes more flexible and will buckle at lower and lower stress levels.

When the slenderness ratio is large, columns buckle in accordance with Eq. (14.24); however, for shorter and stiffer columns $\sigma_{CR} > S_y$. In these cases the column fails by yielding. The transition from the Euler theory of buckling to yielding occurs at a limit value of (L/r), which depends on the modulus of elasticity and yield strength of the column material as indicated in Fig. E14.6. Safety and failure regions for a typical column are illustrated in Fig. 14.9.

When the load applied to the column exhibits some eccentricity, a bending moment is produced which affects the stresses in the column. However, the eccentricity does not affect the critical bucking load. While the result for P_{CR} is the same, the bucking process differs. With centric loading the column remained straight as the load P was increased until P_{CR} was achieved. At that point the column became unstable and suddenly buckled (collapsed). With eccentricity the column deflects laterally with increasing magnitude until the critical load is achieved.

The stress in an eccentrically loaded column is due to a combination of a compressive stress σ_a due to P and a bending stress σ_b due to the moment M = Pe. The two stresses superimpose as indicated by:

$$\sigma_c = \sigma_a + \sigma_b = -(P/A + Mc/I) \qquad (14.56)$$

An alternate formula for σ_{Max} was derived as:

$$\sigma_{Max} = -P\left(\frac{1}{A} + \frac{ce}{I}\sec\left(\frac{\pi}{2}\sqrt{\frac{P}{P_{CR}}}\right)\right) \qquad (14.59)$$

This equation, known as the secant formula, may be used for columns with any end conditions by calculating P_{CR} from Eq. (14.24). Because the stress is not linearly related to load, it is often necessary to solve the equation by trial and error or by using advanced methods, such as a programmable calculator or computer software. Also, any safety factor should only be applied to the loads and not to the stresses. For eccentrically loaded columns, both the maximum stress and the critical buckling stress must be computed and compared to determine the actual mode of failure.

Finally, a sampling of the design codes for columns fabricated from steel and aluminum were described. These codes are extensive and in practice control the design of columns placed in most major structures. Reference was also given for design codes for columns and beams fabricated from timber.

PROBLEMS

14.1 Write an engineering brief explaining why understanding buckling is important in ensuring public safety.

14.2 An 18 ft long column is fabricated from an aluminum tube with an outside diameter of 8.0 in. and an inside diameter of 7.6 in. as shown in the figure to the right Determine the critical buckling force P_{CR} if the tube is pinned at both ends.

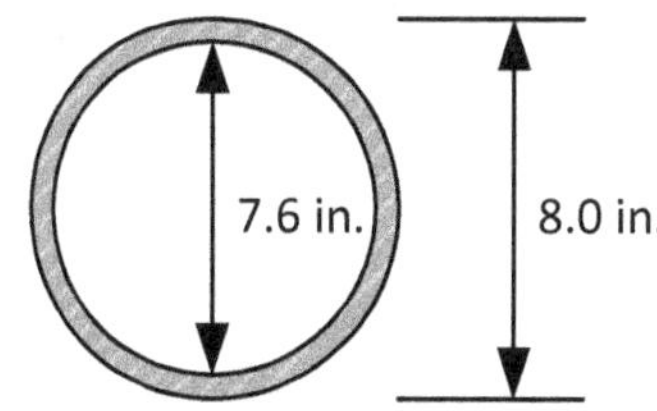

14.3 A 8 m long column is fabricated from a square steel tube as illustrated in the figure to the left. Determine the critical buckling force P_{CR} if the tube is pinned at both ends. Also determine the compressive stress in the tube at the critical load and comment on its magnitude.

14.4 A 5 m long column is fabricated from an aluminum tube with the rectangular cross section defined in the figure to the left. The column is pinned at one end and built-in at the other. Determine the critical buckling force P_{CR}.

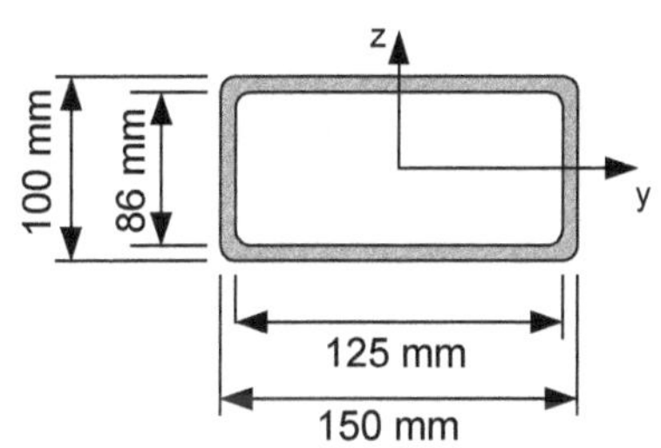

14.5 Design a round tubular column for a construction firm that is to be fabricated from aluminum. The column is 6.0 m long and is expected to exhibit a critical load of 300 kN. The ratio of its inside to outside diameter is 0.85. Both of its ends are supported by pins.

14.6 Design a column fabricated from an aluminum, square-tubular section for a construction firm. The column is 24 ft long and is expected to exhibit a critical load of 100 kip. The ratio of its inside to outside dimensions is 0.90. Both of its ends are supported by pins.

14.7 Design a column that is fabricated from a steel rectangular section. The column is 26 ft long and is expected to exhibit a critical load of 175 kip. The ratio of its inside to outside dimensions is 0.88 ($w_i/w_o = d_i/d_o = 0.88$). Both of its ends are built-in.

14.8 Select the most suitable wide flange section for the design of a steel column 7.1 m long. The column is expected to support a load of 500 kN and is built-in on both ends.

14.9 Select the most suitable American Standard section for the design of an aluminum column 16 ft long. The column is expected to support a load of 100 kip and is built-in on one end and pinned on the other.

14.10 Select the most suitable structural tee section for the design of a steel column 5.3 m long. The column is expected to support a load of 260 kN and is built-in on both ends.

14.11 A control rod in a vehicle is subjected to axial forces of ± 2,800 lb. If it is fabricated from stainless steel determine its diameter if the design incorporates a safety factor of 2.5. Consider both ends pinned.

14.12 A lever is used to provide an axial force to a 30 in. long vertical member as shown in the figure to the right. Determine the maximum force F that may be exerted if the vertical link is fabricated from an aluminum bar with a rectangular cross section. The dimensions of the rectangular bar are 15 mm by 40 mm. Note all of the joints are pinned.

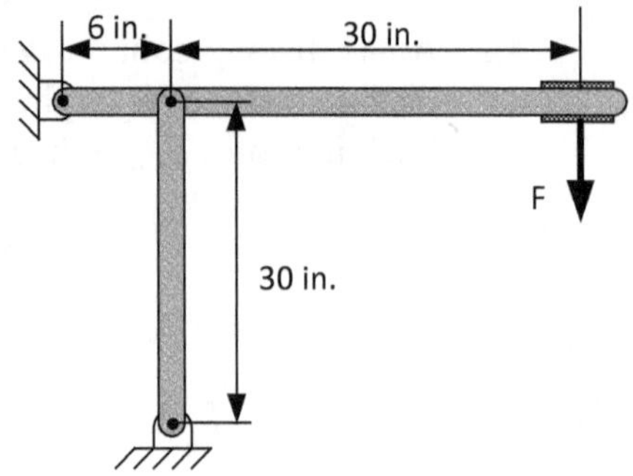

14.13 The truss shown in the figure to the left is fabricated from round, steel tubes. If the tubes all have the same outside diameter of 80 mm, determine the inside diameter of the compression members required to prevent buckling. Employ a safety factor of 4.0.

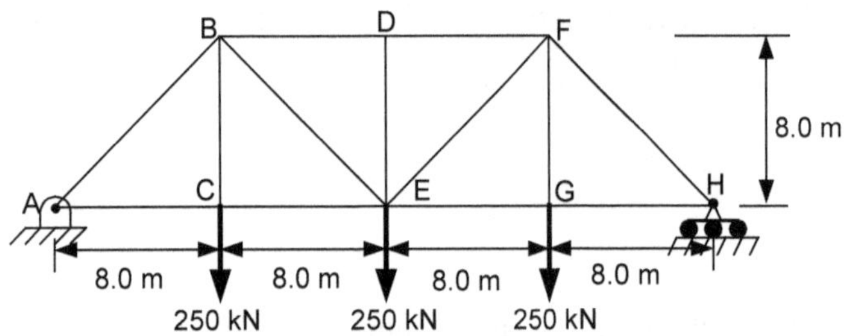

14.14 The truss, shown in the figure to the right, is fabricated from square, steel tubes. If the tubes all have the same outside dimension of s_o = 4.0 in., determine the inside dimension of the compression members required to prevent buckling. Employ a safety factor of 2.5.

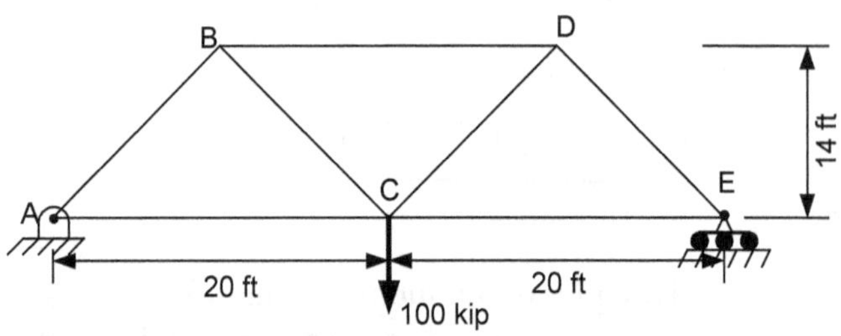

14.15 A force F acts on a simple truss containing a horizontal tie rod and an inclined strut, as shown in the figure to the right. The strut, fabricated from an aluminum alloy, is rectangular with dimensions w and d. Determine those dimensions if w = 2d and a safety factor of 2.5 is required.

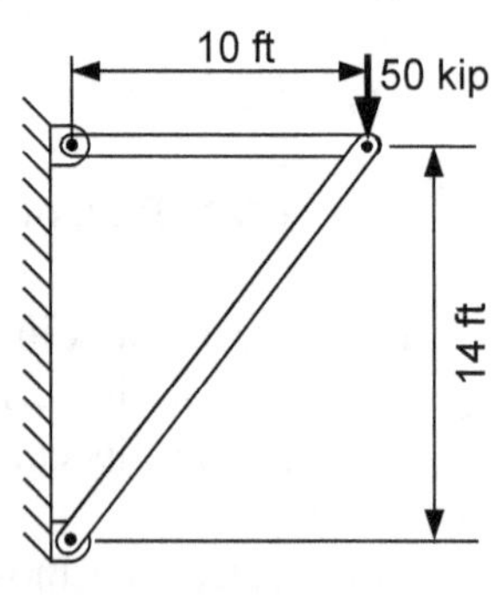

14.16 For the frame, shown in the figure to the left, determine the maximum force that can be supported before the inclined member buckles. The inclined member is fabricated from a rectangular steel bar with dimensions w and d where w/d = 1.5.

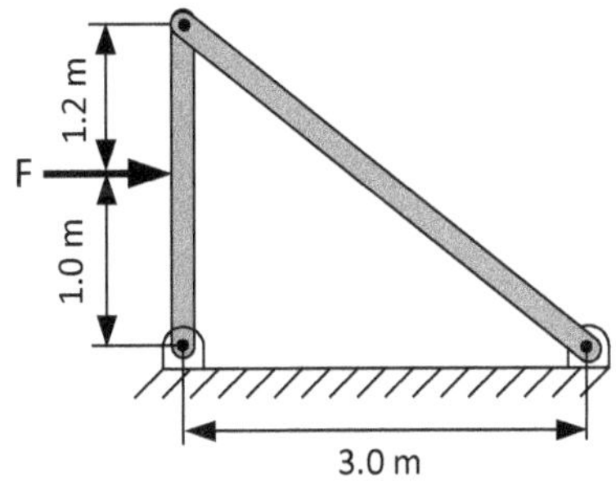

14.17 The truss, shown in the figure to the right, is fabricated from rectangular, steel tubes. If the tubes all have the same outside dimension of w_o = 150 mm in. and d_o = 75 mm determine the wall thickness of the compression members required to prevent buckling. Employ a safety factor of 4.0.

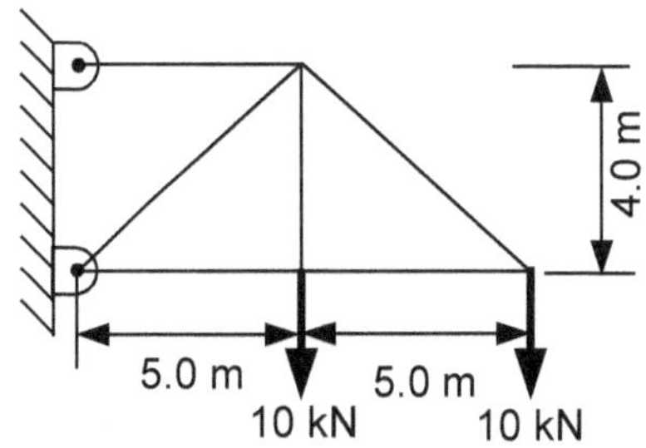

14.18 A wooden column, with a square cross section, is subjected to an eccentric load as shown in the figure to the left. The column is embedded in a concrete footer at one end and free at the other. Determine the load that can be applied to the column without causing it to fail by either buckling or breaking. The column is fabricated from Western white pine.

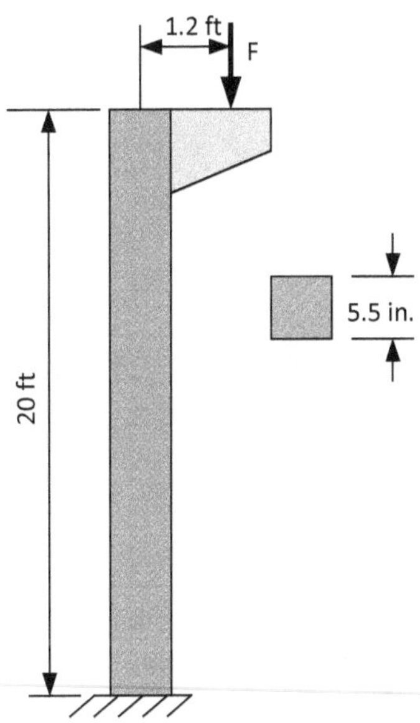

14.19 A column fabricated from a W305 × 74 wide flange section is loaded eccentrically as indicated in the figure to the right. Determine the force F required for the column to fail by either buckling or yielding. The wide flange section is fabricated from 1020 HR steel. The column is braced so that it does not buckle about its minimum inertia axis. Assume the top of the column is free.

14.20 Determine the deflection y_{Max} at the top of the column, shown in the figure to the right, when the applied load F is equal to 80% of the failure load.

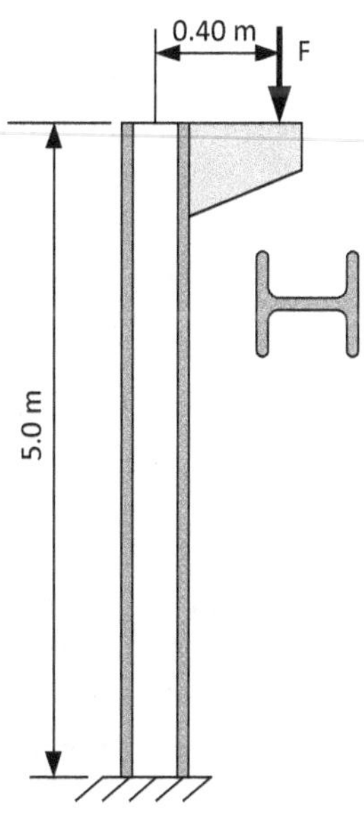

14.21 A column, fabricated from a S 203 × 34 American Standard section, is loaded eccentrically as indicated in the figure to the left. Determine the force F required for the column to fail by either buckling or yielding. The wide flange section is fabricated from 1020 HR steel. Assume both ends of the column are built-in, and that bracing prevents buckling about the minimum inertia axis.

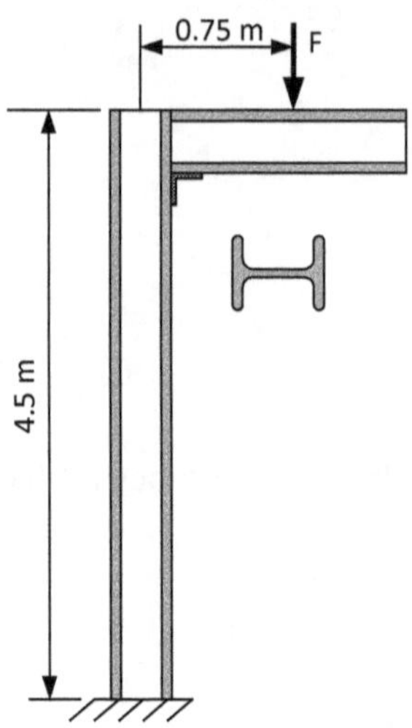

14.22 A column is fabricated from a solid aluminum rod with a diameter of 60 mm as shown in the figure to the right. It is loaded with a force F with an eccentricity e = 30 mm. Determine the maximum force F that can be applied to the column before it fails by either buckling or yielding. The aluminum alloy employed is 2024-T-4.

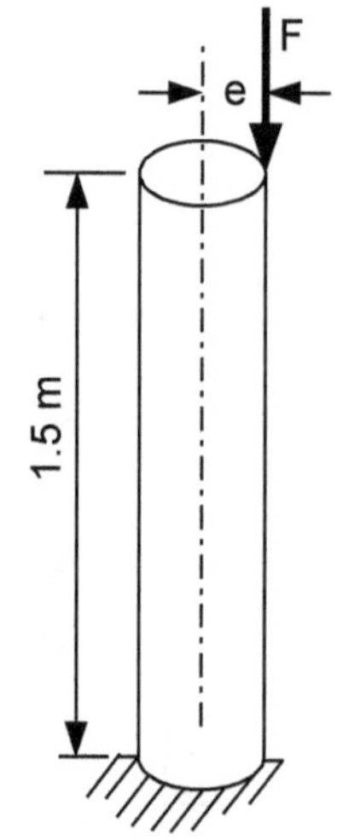

14.23 Develop a design curve similar to the one shown in Fig. 14.14 for steel columns with pinned/pinned ends and with a yield strength of 50 ksi. Also determine the limiting value of the slenderness ratio SR_c for this case.

14.24 Determine the safety factor as a function of (kL/r) that is employed using the design code for steel columns. Consider yield strengths of 35, 40, 45, and 50 ksi.

14.25 A 8.4 m long column is fabricated from a W 254 × 45 steel wide flange section with pinned/pinned ends. Use the curve resulting from the code equations that is shown in Fig. 14.14 to determine the allowable centric load applied to the column.

14.26 A 8.0 m long column is fabricated from a S 178 × 23 steel American Standard section with pinned/pinned ends. Use the curve resulting from the code equations that is shown in Fig. 14.14 to determine the allowable centric load applied to the column.

14.27 A 4.3 m long column is fabricated from aluminum (2014 T-6) in the form of a WT 178 × 36 structural tee section. Use the curve resulting from the code equations that is shown in Fig. 14.15 to determine the allowable centric load applied to the column. The ends are both built-in.

14.28 An aluminum (2014-T6) tube with a square cross section, shown in the figure to the right, is to be employed in a building lobby. The column length is 5.6 m and its ends are both built in. Determine the largest axial (centric) load that it can carry.

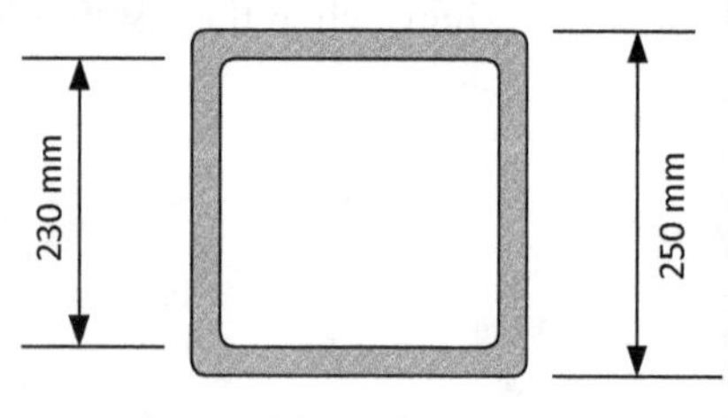

CHAPTER 15

ENERGY METHODS

15.1 INTRODUCTION

To introduce the concept of energy in the analysis of structural elements, let's consider the tie bar subjected to an axial tension force F. The tie bar extends an amount δ as the force is applied. Because we have moved a force F through a distance δ, work $\mathbf{W}$ has been performed. The amount of work is established from the force-displacement graph presented in Fig. 15.1.

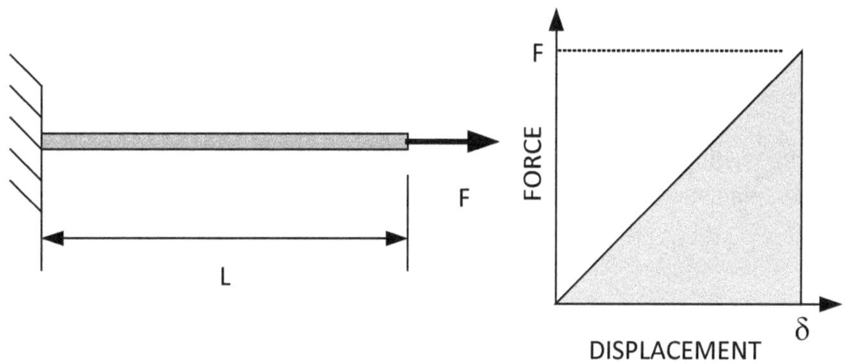

Fig. 15.1 Force-displacement graph for a tie rod with an applied force F.

The work performed on the tie rod is the area under the F-δ curve in Fig. 15.1. Hence:

$$\mathbf{W} = F\delta/2 \qquad (15.1)$$

The tie bar may be considered as a system in which energy is conserved. This fact implies that work performed on the bar is conserved as strain energy $\mathbf{E}_\varepsilon$. Accordingly, we equate the work and strain energy and express the strain energy stored in the tie rod as:

$$\mathbf{E}_\varepsilon = F\delta/2 \qquad (15.2)$$

In later sections of this chapter, we will derive the equations for the strain energy stored in beams and torsion bars. The ability to compute the strain energy stored in structural members is important because three different energy theorems can be employed to extend our methods of analysis including:

1. Theorem of virtual work.
2. Theorem of Castigliano.
3. Theorem of least work (minimum potential energy).

15.2 THE ENERGY THEOREMS

15.2.1 The Theorem of Virtual Work

Consider a structural member subjected to a number of forces that maintain the member in equilibrium as illustrated in Fig. 15.2.

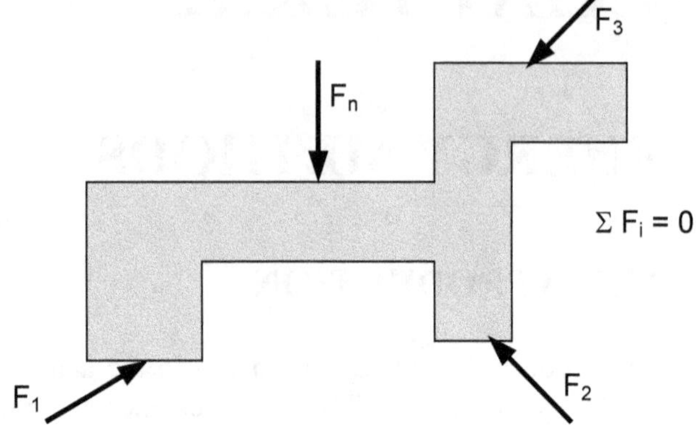

Fig. 15.2 A structural member in equilibrium subjected to forces F_1, F_2, F_3 and F_n.

Suppose we apply a virtual[1] displacement $\Delta\delta_n$ at the location and in the direction of F_n. Then the strain energy is increased by a small amount given by:

$$\Delta \mathbf{E}_\varepsilon = F_n \Delta\delta_n \qquad (15.3)$$

In writing Eq. (15.3), we have assumed that the other external forces F_1, F_2 and F_3 do not move as the virtual displacement is applied. With this assumption, we can rewrite Eq. (15.3) as:

$$\frac{\partial \mathbf{E}_{\mathbf{e}}}{\partial \delta_n} = F_n \qquad (15.4)$$

In applying Eq. (15.4), $\mathbf{E}_\varepsilon$ is expressed in terms of displacements so that the applied loads F_1, F_2 and F_3 do not appear and are not an issue. The application is in the determination of the force F_n when the loading of a structural element is accomplished by applying displacements at specified locations.

15.2.2 The Theorem of Castigliano

Consider again the structural member in Fig. 15.2, which is in equilibrium when subjected to forces F_1, F_2, F_3 and F_n. Let's increase the force F_n by a virtual amount ΔF_n. Accordingly, the increase in the strain energy is given by:

$$\Delta \mathbf{E}_\varepsilon = \delta_n \Delta F_n \qquad (15.5)$$

Again we have assumed that ΔF_n is sufficiently small that the external forces F_1, F_2 and F_3 do not move as ΔF_n is applied. We can then rewrite Eq. (15.5) as:

$$\frac{\partial \mathbf{E}_{\mathbf{e}}}{\partial F_n} = \delta_n \qquad (15.6)$$

In applying Eq. (15.6), we write the expression for $\mathbf{E}_\varepsilon$ so that the displacements δ_1, δ_2, δ_3 and δ_n do not appear. Castigliano's theorem is useful in determining displacements at selected locations on structural members or in solving for reactions in statically indeterminate structures.

---

[1] The word virtual means the displacement $\Delta\delta_n$ is small when compared to δ_n.

15.2.3 The Theorem of Least Work[2]

Consider a number of different stress distributions that are possible for some structural member subjected to external loading. Moreover, assume that all of these different stress distributions yield internal forces that satisfy the equations of equilibrium. Only one of these stress distributions is correct and the others violate the compatibility conditions[3].

The correct (compatible) stress distribution results in the least amount of strain energy stored in the structural member. We can express this fact by writing:

$$\partial \mathbf{E}_s = 0 \qquad (15.7)$$

All three of the theorems are useful in the analysis of structures. Castigliano's theorem is often employed to determine the reaction at the redundant support in a statically indeterminate structure. The theorem of virtual work is helpful in solving for the critical load in the buckling of structural members. The theorem of least work is used with finite elements to determine displacement fields that satisfy the compatibility conditions. Energy methods can also be used to solve for forces generated during impact events. We will demonstrate many of these applications later in this chapter. The theorems of virtual work and least work are deferred until a later course.

15.3 STRAIN ENERGY DENSITY

Consider an arbitrary element removed from the tie rod shown previously in Fig. 15.1. When the tie rod is subjected to an axial force F, a normal stress σ_{xx} results on opposites sides of the element as shown in Fig. 15.3a.

Fig. 15.3 An element removed from the tie rod is subjected to a uniaxial stress σ_{xx}.

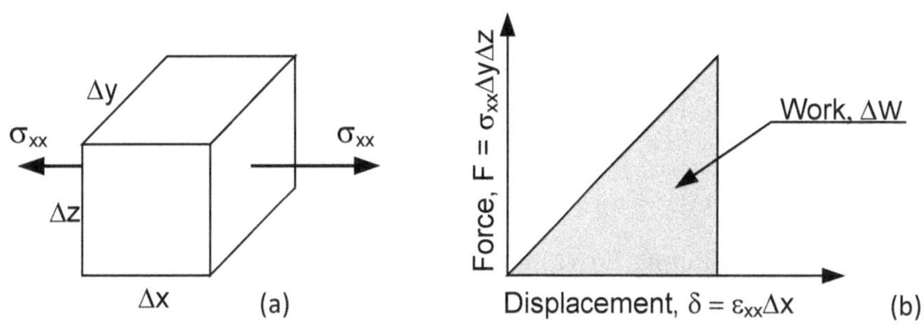

The work performed on the element, illustrated in Fig. 15.3b, is the shaded area under the force-displacement line. We equate this work to strain energy and write:

$$\Delta \mathbf{W} = \Delta \mathbf{E}_\varepsilon = (1/2)(\sigma_{xx}\,\varepsilon_{xx})\Delta x \Delta y \Delta z \qquad (15.8)$$

Let's rewrite Eq. (15.8) as:

$$\mathbf{e}_\varepsilon = \Delta \mathbf{E}_\varepsilon / \mathbf{V} = (1/2)\sigma_{xx}\,\varepsilon_{xx} \qquad (15.9)$$

where $\mathbf{e}_\varepsilon$ is the strain energy density, which is the amount of strain energy stored per unit volume of the member, and $\mathbf{V} = \Delta x \Delta y \Delta z$ is the volume of the element.

Equation (15.9) is valid for a uniaxial state of stress where $\sigma_{yy} = \sigma_{zz} = \tau_{xy} = \tau_{yz} = \tau_{zx} = 0$. However, if the member is subjected to a general three-dimensional state of stress, we write the expression for $\mathbf{e}_\varepsilon$ as:

$$\mathbf{e}_\varepsilon = (1/2)(\sigma_{xx}\,\varepsilon_{xx} + \sigma_{yy}\,\varepsilon_{yy} + \sigma_{zz}\,\varepsilon_{zz} + \tau_{xy}\gamma_{xy} + \tau_{yz}\gamma_{yz} + \tau_{zx}\gamma_{zx}) \qquad (15.10)$$

[2] The theorem of least work is also referred to as the theorem for minimum potential energy.

[3] Many different stress distributions can be devised to satisfy equilibrium; however, only the true (correct) distribution will yield the correct displacement field for the member. This requirement is called the compatibility condition.

If the state of stress in the structural member is a principal state, then x, y and z are principal axes and $\gamma_{xy} = \gamma_{yz} = \gamma_{zx} = 0$. Then Eq. (15.10) reduces to:

$$\mathbf{e}_\varepsilon = (1/2)(\sigma_1\,\varepsilon_1 + \sigma_2\,\varepsilon_2 + \sigma_3\,\varepsilon_3) \tag{15.11}$$

where $\sigma_{xx} = \sigma_1$, $\varepsilon_{xx} = \varepsilon_1$, etc.

Let's consider two examples to illustrate the methods for computing strain energy density.

EXAMPLE 15.1

A point in a machine component is subjected to a three-dimensional state of stress that is described in terms of three principal stresses given by $\sigma_1 = 16{,}200$ psi, $\sigma_2 = 9{,}300$ psi and $\sigma_3 = -\,12{,}500$ psi. Determine the strain energy density for this point if the machine component is fabricated from steel.

Solution:

Let's begin by writing the expression of the strain energy density in terms of the principal stresses and strains, which is given by Eq. (15.11) as:

$$\mathbf{e}_\varepsilon = (1/2)(\sigma_1\,\varepsilon_1 + \sigma_2\,\varepsilon_2 + \sigma_3\,\varepsilon_3) \tag{a}$$

Recall the stress-strain relations as:

$$\varepsilon_1 = (1/E)[\sigma_1 - \nu(\sigma_2 + \sigma_3)]$$

$$\varepsilon_2 = (1/E)[\sigma_2 - \nu(\sigma_1 + \sigma_3)] \tag{b}$$

$$\varepsilon_3 = (1/E)[\sigma_3 - \nu(\sigma_1 + \sigma_2)]$$

Substituting Eq. (b) into Eq. (a) gives:

$$\mathbf{e}_\varepsilon = (1/2E)[\sigma_1^2 + \sigma_2^2 + \sigma_3^2 - 2\nu(\sigma_1\,\sigma_2 + \sigma_2\,\sigma_3 + \sigma_3\,\sigma_1)] \tag{15.12}$$

Substituting numerical parameters from the problem statement into Eq. (15.12) gives:

$$\mathbf{e}_e = \frac{1}{2(30\times10^6)}\left\{(16.2)^2 + (9.3)^2 + (\text{-}12.5)^2 - (2)(0.30)\left[(16.2)(9.3) + (9.3)(\text{-}12.5) + (\text{-}12.5)(16.2)\right]\right\}(10^6)$$

$$\mathbf{e}_e = 10.11\ \text{in.-lb/in}^3$$

We have used the units in.-lb/in.^{3} instead if lb/in^2 for the strain energy density because the former conveys the concept of energy per unit volume. Actually both representations of the units are correct.

EXAMPLE 15.2

A point in a structural member is subjected to a three dimensional state of stress described by the six Cartesian stress components σ_{xx} = 185 MPa, σ_{yy} = – 114 MPa, σ_{zz} = 75.6 MPa, τ_{xy} = 105 MPa, τ_{yz} = 82.6 MPa, τ_{zx} = 67.9 MPa. Determine the strain energy density if the member is fabricated from aluminum alloy 2024 T-4.

Solution:

Let's begin by writing the expression for the strain energy density in terms of the Cartesian components of stress, which is given by Eq. (15.10) as:

$$\mathbf{e}_\varepsilon = (1/2)(\sigma_{xx}\,\varepsilon_{xx} + \sigma_{yy}\,\varepsilon_{yy} + \sigma_{zz}\,\varepsilon_{zz} + \tau_{xy}\gamma_{xy} + \tau_{yz}\gamma_{yz} + \tau_{zx}\gamma_{zx}) \qquad \text{(a)}$$

Recall the stress strain relations as:

$$\varepsilon_{xx} = (1/E)[\sigma_{xx} - \nu(\sigma_{yy} + \sigma_{zz})] \qquad \gamma_{xy} = \tau_{xy}/G$$

$$\varepsilon_{yy} = (1/E)[\sigma_{yy} - \nu(\sigma_{xx} + \sigma_{zz})] \qquad \gamma_{yz} = \tau_{yz}/G \qquad \text{(b)}$$

$$\varepsilon_{zz} = (1/E)[\sigma_{zz} - \nu(\sigma_{xx} + \sigma_{yy})] \qquad \gamma_{zx} = \tau_{zx}/G$$

Substituting Eq. (b) into Eq. (a) gives:

$$\mathbf{e}_\varepsilon = (1/2E)[\sigma_{xx}^2+\sigma_{yy}^2+\sigma_{zz}^2 - 2\nu(\sigma_{xx}\sigma_{yy}+\sigma_{yy}\sigma_{zz}+\sigma_{zz}\sigma_{xx})] + (1/2G)[\tau_{xy}^2+\tau_{yz}^2+\tau_{zx}^2] \qquad (15.13)$$

Substituting numerical parameters from the problem statement into Eq. (15.13) gives:

$$\mathbf{e}_e = \frac{10^{12}}{2(72\times10^9)}\Big[(185)^2+(-114)^2+(75.6)^2-(2)(0.32)\big[(185)(-114)+(-114)(75.6)+(75.6)(185)\big]\Big]$$

$$+\frac{10^{12}}{2(27\times10^9)}\Big[(105)^2+(82.6)^2+(67.9)^2\Big]$$

$$\mathbf{e}_e = 853.4\times10^3\ \text{N-m/m}^3$$

15.4 STRAIN ENERGY IN STRUCTURAL ELEMENTS

This section describes methods employed to determine the strain energy stored in common structural elements such as tie rods, beams in bending and torsion bars. First, consider the tie rod from Fig. 15.1.

15.4.1 Strain Energy in Tie Rods

Previously we described a technique for determining the strain energy stored in an axially loaded rod in Section 15.1. Let's consider a second approach based on strain energy density. Recall Eq. (15.9) and substitute Hooke's law into this relation to obtain:

$$\mathbf{e}_\varepsilon = (1/2)(\sigma_{xx}\,\varepsilon_{xx}) = (1/2)\,E\varepsilon_{xx}^2 = [1/(2E)]\sigma_{xx}^2 \qquad (15.14)$$

Integrate Eq. (15.14) over the volume $\mathbf{V}$ of the tie rod to obtain:

$$\mathbf{E}_e = \frac{1}{2E}\int \sigma_{xx}^2\,dV \qquad (15.15)$$

Because $\sigma_{xx} = F/A$ over each volume element in the tie rod, we can rewrite Eq. (15.15) as:

$$\mathbf{E}_e = \frac{1}{2E}\frac{F^2}{A^2}AL = \frac{F^2L}{2AE} \qquad (15.16)$$

Let's consider two examples to demonstrate this approach for determining the strain energy in a tie bar.

EXAMPLE 15.3

A tie rod, 16 ft long fabricated from steel with a yield strength S_y = 42 ksi is subjected to an axial force of 9,500 lb. If the bar diameter is 0.75 in., determine the strain energy stored in the tie rod and the strain energy density.

Solution:

We employ Eq. (15.16) to determine the strain energy E_ε as:

$$\mathbf{E_e} = \frac{F^2L}{2AE} = \frac{(9500)^2(16)(12)}{2(\pi)(0.375)^2(30\times10^6)} = 653.7 \text{ in.-lb} \qquad \text{(a)}$$

The strain energy density is given by Eq. (15.14) as:

$$\mathbf{e_e} = \frac{\sigma_{xx}^2}{2E} = \left(\frac{9{,}500}{\pi(0.375)^2}\right)^2\left(\frac{1}{(2)(30\times10^6)}\right) = 7.707 \text{ in.-lb/in.}^3 \qquad \text{(b)}$$

It is difficult to physically interpret the significance of these results because we do not have a basis for comparison. What is a large strain energy or strain energy density? The total strain energy increases with the volume of the member. Consequently, the strain energy stored in a member depends upon both the square of the stress and the rod volume. To assess the results of this analysis, it is advantageous to examine the strain energy density, which is independent of the volume. Hence we seek a maximum value of the strain energy density to be used for comparison purposes. From Eq. (15.14) it is evident that:

$$(\mathbf{e}_\varepsilon)_{Max} = [1/(2E)]S_y^2 \qquad (15.17)$$

$$(\mathbf{e}_\varepsilon)_{Max} = (42{,}000)^2/[(2)(30\times10^6)] = 29.40 \text{ in.-lb/in.}^3 \qquad \text{(c)}$$

Comparing the results of Eq. (b) with Eq. (c) indicates that the tie rod in this example was loaded to a strain energy density of (7.707/29.40) = 0.2621, or 26.21% of the maximum value possible was achieved.

EXAMPLE 15.4

A tie rod, fabricated from stainless steel is shown in Fig. E15.4. If the axial loading is F = 450 kN, determine the strain energy stored in the tie rod and the strain energy density.

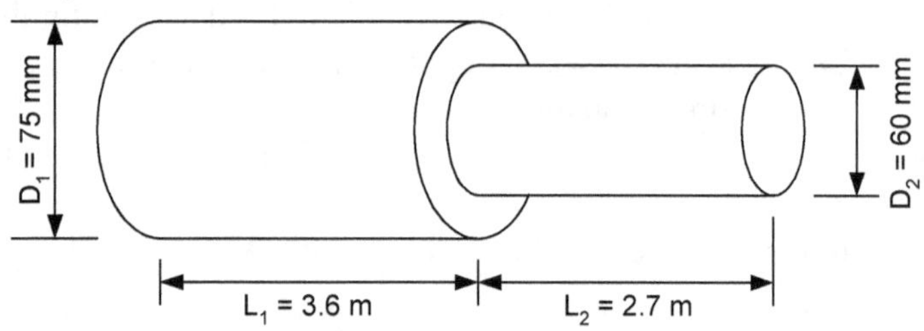

Fig. E15.4

Solution:

Let's begin by computing the stress in each segment of the tie rod as:

$$\sigma_1 = \frac{4F}{\pi D_1^2} = \frac{(4)(450\times10^3)}{\pi(75)^2} = 101.9 \text{ MPa} \qquad \text{(a)}$$

$$\sigma_2 = \frac{4F}{\pi D_2^2} = \frac{(4)(450\times 10^3)}{\pi(60)^2} = 159.2 \text{ MPa}$$

Using Eq. (15.14) and Eq. (a), we determine the strain energy density for each segment of the bar as:

$$(\mathbf{e_e})_1 = \frac{\sigma_1^2}{2E} = \frac{(101.9)^2\times 10^{12}}{2(190\times 10^9)} = 27.33\times 10^3 \text{ N-m/m}^3$$

$$(\mathbf{e_e})_2 = \frac{\sigma_2^2}{2E} = \frac{(159.2)^2\times 10^{12}}{2(190\times 10^9)} = 66.70\times 10^3 \text{ N-m/m}^3 \qquad \text{(b)}$$

The total strain energy stored in the tie rod is determined from Eq. (15.15) as:

$$\mathbf{E}_\varepsilon = (\mathbf{e}_\varepsilon)_1 \mathbf{V}_1 + (\mathbf{e}_\varepsilon)_2 \mathbf{V}_2 = (27.33\times 10^3)(.0375)^2\pi(3.6)$$

$$+ (66.70\times 10^3)(.03)^2\pi(2.7) = 943.9 \text{ N-m}$$

Because the bar was not of constant diameter, it was necessary to divide it into two segments and determine the strain energy density in each segment. The total strain energy is then computed by multiplying each value of e_ε by the appropriate volume and summing the two quantities.

15.4.2 Strain Energy for a Beam in Bending

For a beam subjected to transverse loading, as shown in Fig. 15.4, the flexural stress σ_{xx} can be determined from:

$$\sigma_{xx} = My/I_z \qquad (15.18)$$

where M is the bending moment at any location defined by x.

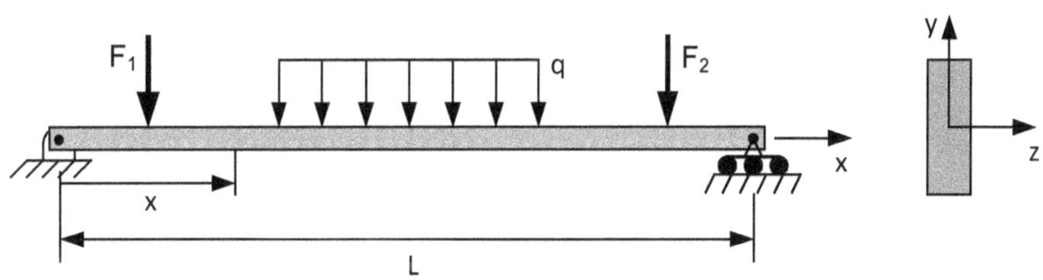

Fig. 15.4 A beam subjected to transverse forces.

To determine the strain energy density, we use Eq. (15.14), which is valid for uniaxial states of stress. Combing Eq. (15.14) and Eq. (15.18), we obtain:

$$\mathbf{e_e} = \left(\frac{1}{2E}\right)\left(\frac{My}{I_z}\right)^2 \qquad (15.19)$$

An examination of Eq. (15.19) shows that the strain energy density varies with the location x over the length of the beam because M is a function of x. The strain energy density also depends on the position y measured from the neutral axis of the beam.

We integrate Eq. (15.19) over the volume to obtain the strain energy stored in the beam as:

$$\mathbf{E_e} = \int \frac{1}{2E}\left(\frac{My}{I_z}\right)^2 dAdx = \int \frac{1}{2E}\left(\frac{M}{I_z}\right)^2 \left[\int y^2 dA\right] dx \qquad (15.20)$$

In Eq. (15.20), we recognize $\int y^2\, dA = I_z$. Hence, we rewrite Eq. (15.20) as:

$$\mathcal{E}_e = \left(\frac{1}{2EI_z}\right)\int_0^L M^2 dx \qquad (15.21)$$

We have assumed that both E and I_z are constant over the volume of the beam in the derivation of Eq. (15.21). Let's consider two examples to demonstrate the use of Eq. (15.21) in determining the strain energy stored in a beam subjected to transverse loading.

EXAMPLE 15.5

For the beam shown in Fig. E15.5, determine the expression for the strain energy due to the normal stress σ_{xx}.

Fig. E15.5

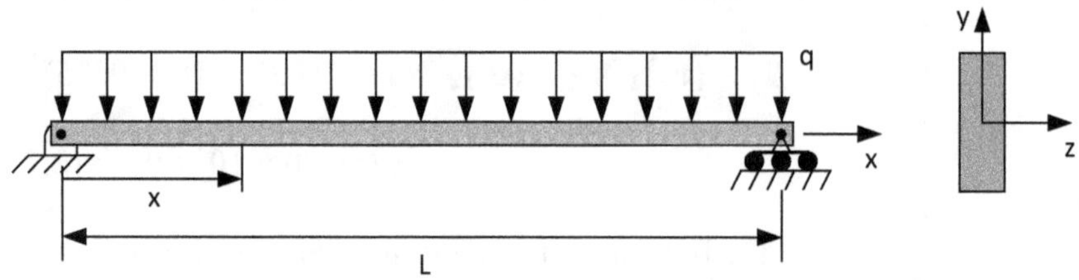

Solution:

The reactions at the simple supports are determined from an appropriate FBD and the application of the equilibrium equations as:

$$\Sigma F_x = 0 \quad \Rightarrow \quad R_{Lx} = 0$$

$$\Sigma M_L = 0 \quad \Rightarrow \quad R_R = qL/2 \qquad (a)$$

$$\Sigma F_y = 0 \quad \Rightarrow \quad R_{Ly} = qL/2$$

Next prepare a FBD of the left-hand portion of the beam as shown in Fig. E15.5a.

Fig. E15.5a

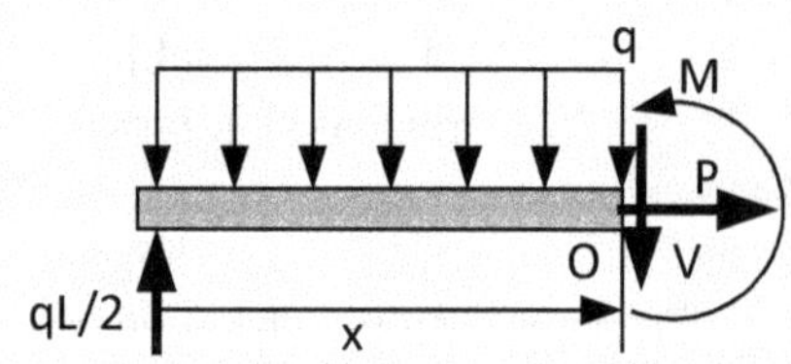

Applying the equilibrium equation yields:

$$\Sigma F_x = 0 \quad \Rightarrow \quad P = 0$$

$$\Sigma F_y = 0 \quad \Rightarrow \quad qL/2 - qx - V = 0 \qquad (b)$$

$$\Sigma M_O = 0 \quad \Rightarrow \quad (qL/2)x - (qx)(x/2) + M = 0$$

Solving Eqs. (b) for V and M yields:

$$V = (q/2)(L - 2x) \qquad (c)$$

$$M = (qx/2)(x - L) \qquad (d)$$

Substituting Eq. (d) into Eq. (15.21) gives:

$$\mathbf{E_e} = \frac{1}{2EI_z}\int_0^L \left[\left(\frac{qx}{2}\right)(x-L)\right]^2 dx \qquad \text{(e)}$$

Integrating Eq. (e) and simplifying gives:

$$\mathbf{E_e} = \frac{q^2L^5}{240EI_z} \qquad (15.22)$$

It is important to recognize that the strain energy represented in Eq. (15.22) is due to the normal stress σ_{xx}, which in turn is produced by the bending moment M. However, a shear stress τ_{xy} due to the shear force V exists. This shear stress τ_{xy} produces additional strain energy so that the total strain energy includes contributions from both σ_{xx} and τ_{xy}. We will describe methods to determine the strain energy due to shearing stresses later in this chapter.

EXAMPLE 15.6

For the beam shown in Fig. E15.6, determine the expression for the strain energy due to the normal stress σ_{xx}.

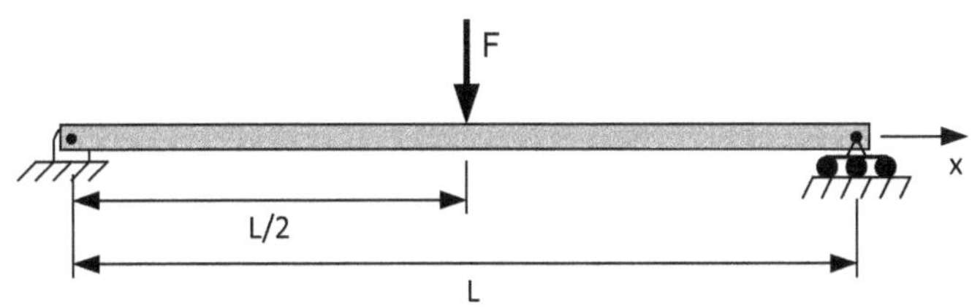

Fig. E15.6

Solution:

The reactions at the simple supports are determined from an appropriate FBD and the application of the equilibrium equations to obtain:

$$R_{Lx} = 0 \qquad R_{Ly} = F/2 \qquad R_R = F/2 \qquad \text{(a)}$$

Next prepare a FBD of the left-hand portion of the beam as shown in Fig. E15.6a.

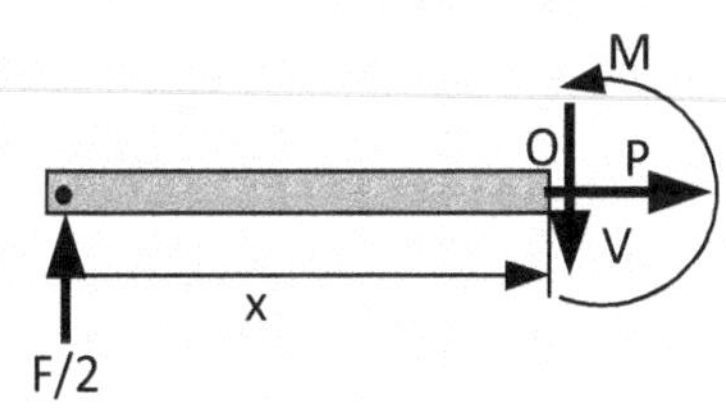

Fig. E15.6a

Applying the equilibrium equations yield:

$$\Sigma F_x = 0 \quad \Rightarrow \quad P = 0 \quad 0 \leq x \leq (L/2)$$

$$\Sigma F_y = 0 \quad \Rightarrow \quad V = F/2 \quad 0 \leq x \leq (L/2) \qquad \text{(b)}$$

$$\Sigma M_O = 0 \quad \Rightarrow \quad M = (F/2)x \quad 0 \leq x \leq (L/2)$$

Substituting the expression for the moment M in Eq. (b) into Eq. (15.21) gives a relation that is valid for the left half of the beam.

$$\mathcal{E}_e = \frac{1}{2EI_z}\int_0^{L/2}\left(\frac{Fx}{2}\right)^2 dx \qquad (c)$$

Integrating Eq. (c) gives:

$$\mathcal{E}_e = \frac{F^2L^3}{192EI_z} \qquad (d)$$

The result in Eq. (d) represents the strain energy due to σ_{xx} in the left half of the beam. Because the loading on the beam is symmetric, we multiply this result by two to obtain the strain energy over the entire volume of the beam as:

$$\mathcal{E}_e = \frac{F^2L^3}{96EI_z} \qquad (15.23)$$

Again, it is important to recognize that the strain energy represented in Eq. (15.23) is due to the normal stress σ_{xx}; however, a shear stress τ_{xy} due to the shear force V exists. This shear stress τ_{xy} produces additional strain energy so that the total strain energy includes contributions from both σ_{xx} and τ_{xy}.

15.4.3 Strain Energy Due to Shearing Stresses

The most general expression for strain energy density is given in Eq. (15.10) where the contributions due to all six Cartesian components of stress are included. To develop a simpler form of this expression, consider the state of stress where $\sigma_{xx} = \sigma_{yy} = \sigma_{zz} = \tau_{yz} = \tau_{zx} = 0$. Then we can rewrite Eq. (15.10) as:

$$\mathbf{e}_\varepsilon = \tau_{xy}\gamma_{xy}/2 = \tau_{xy}^2/(2G) \qquad (15.24)$$

Integrate Eq. (15.24) over the volume V of a structural member to obtain the relation for the strain energy due to τ_{xy} as:

$$\mathcal{E}_e = \frac{1}{2G}\int \tau_{xy}^2 dV \qquad (15.25)$$

Let's consider a circular shaft subjected to a torsional load to demonstrate the application of Eq. (15.25) in the determination of strain energy due to shear stress.

EXAMPLE 15.7

A 55 in. long steel shaft is subjected to a torque T = 20,000 in.-lb. If the shaft is 1.5 in. in diameter, compute the strain energy stored in the shaft.

Solution:

Let's begin by writing the relation for the shear stress τ_{xy} as a function of the radial position ρ.

$$\tau_{xy} = T\rho/J \qquad (a)$$

Substituting Eq. (a) into Eq. (15.25) gives:

$$\mathcal{E}_e = \frac{1}{2G}\int\left(\frac{T\rho}{J}\right)^2 dA dx = \frac{1}{2G}\int\left(\frac{T}{J}\right)^2\left[\int \rho^2 dA\right]dx \qquad (b)$$

Recognizing that $\int \rho^2 dA = J$, we rewrite Eq. (b) as:

$$\mathcal{E}_e = \frac{1}{2GJ}\int_0^L T^2 dx = \frac{T^2 L}{2GJ} \qquad (15.26)$$

Note that T and J are considered constant over the length of the shaft in Eq. (15.26). Substituting numerical parameters into this relation yields:

$$\mathcal{E}_e = \frac{T^2 L}{2GJ} = \frac{(20\times10^3)^2(55)(32)}{(2)(11.5\times10^6)\pi(1.5)^4} = 1925\ \text{in.-lb} \qquad (c)$$

EXAMPLE 15.8

Let's again consider the beam with the uniformly distributed load q that we studied in Example 15.5. In this re-examination, determine the relation for the strain energy stored in the beam due to the shear stress τ_{xy}. The beam has a rectangular cross section as shown in Fig. E15.8.

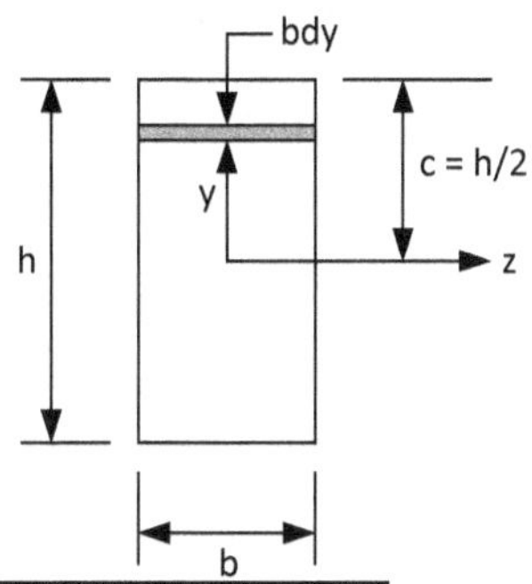

Fig. E15.8

Solution:

For a beam with a rectangular cross section, it can be shown that:

$$\tau_{xy} = \frac{3V}{2A}\left(1 - \frac{y^2}{c^2}\right) \qquad (a)$$

Recall from Example 15.5 that the shear force V was given by:

$$V = qL/2 - qx \qquad (b)$$

Substitute Eq. (b) into Eq. (a) to obtain:

$$\tau_{xy} = \frac{3q}{2A}\left(1 - \frac{y^2}{c^2}\right)\left(\frac{L}{2} - x\right) \qquad (c)$$

Substitute Eq. (c) into Eq. (15.25) and note that dV = bdydx to obtain:

$$(\mathcal{E}_e)_\tau = \frac{1}{2G}\left(\frac{3q}{2A}\right)^2 \int \left(1 - \frac{y^2}{c^2}\right)^2 \left(\frac{L}{2} - x\right)^2 bdydx \qquad (d)$$

Expanding Eq. (d) gives:

$$(\mathbf{E}_e)_t = \frac{1}{8G}\left(\frac{9q^2}{bh^2}\right)\int_{-c}^{+c}\left(1-\frac{2y^2}{c^2}+\frac{y^4}{c^4}\right)dy\int_0^L\left(\frac{L^2}{4}-Lx+x^2\right)dx \qquad \text{(e)}$$

Integrating and substituting the upper and lower limits into the results yields:

$$(\mathbf{E}_e)_t = \frac{q^2L^3}{20Gbh} \qquad (15.27)$$

A comparison of the strain energies stored in the beam due to shear and normal stress is obtained from the ratio of the strain energies given by Eqs. (15.22) and (15.27) as:

$$\frac{(\mathbf{E}_e)_t}{(\mathbf{E}_e)_s} = \frac{240q^2L^3EI_z}{20bhGq^2L^5} \qquad \text{(a)}$$

Noting that $I_z = bh^3/12$ and $G = E/[2(1 + \nu)]$, we can reduce Eq. (a) to:

$$\frac{(\mathbf{E}_e)_t}{(\mathbf{E}_e)_s} = 2(1+\nu)\left(\frac{h}{L}\right)^2 \qquad (15.28)$$

Evaluating Eq. (15.28) for a series of beams with different ratios of h/L gives the strain energy ratio for uniformly distributed loaded beams, shown in Table 15.1.

Table 15.1
Strain Energy Ratio for Steel Beams with Different Ratios of h/L.

h/L	$(E_\varepsilon)_\tau/(E_\varepsilon)_\sigma$	$(E_\varepsilon)_\tau/(E_\varepsilon)_\sigma$	Type of Beam
0.2	0.104	10.4%	Deep
0.1	0.026	2.6%	Long and Slender
0.05	0.0065	0.65%	Long and Slender
0.02	0.0010	0.10%	Long and Slender

Examination of the results of Table 15.1 shows that the fraction of strain energy due to shear stresses is very small if the beam is long and slender. The shear stress becomes an important contributor to strain energy only for short deep beams. For this reason, we neglect the contribution of the shear stresses in beams to the total strain energy stored when h/L < 1/10.

15.5 DYNAMIC (IMPACT) LOADING

All of the methods introduced thus far in this textbook, pertain to structures with loadings, which are gradually applied. Typical force-time and force-displacement curves for gradual application of loads to structures are presented in Fig. 15.5.

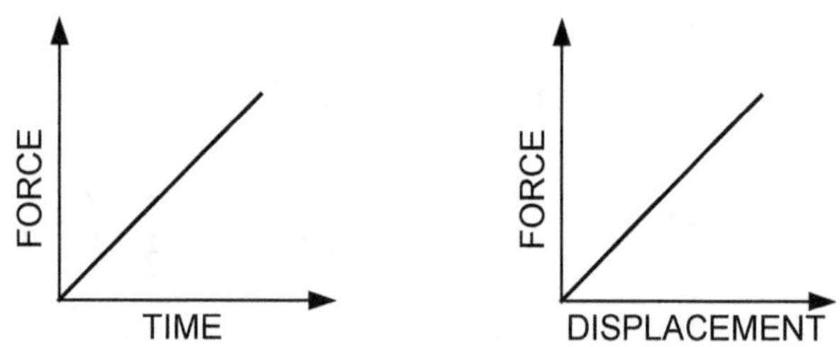

Fig. 15.5 Force-time and force-displacement functions associated with gradual (static) loading.

For dynamically applied forces, sometimes called impact forces, we treat the force as if it were applied instantaneously with the force-time trace as illustrated in Fig. 15.6. The instantaneous application of load on a structural element has a dramatic effect on the magnitude of the stresses generated.

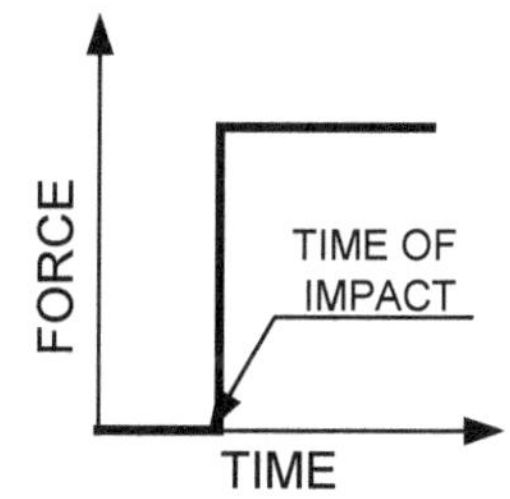

Fig. 15.6 Force-time trace for impact loading.

We are able to use energy methods to solve for the stresses in rods and beams when they are subjected to impact loading. With this approach, two assumptions are made:

1. Energy is conserved, which implies that the impact is elastic and that no work is converted to heat or to produce plastic deformation.
2. The impacting body remains in contact with the structure during the impact period and transfers all of its energy to the structure.

Next, we will describe the application of energy methods for determining the maximum stresses developed due to impact of both rods and beams.

15.5.1 Impact Loading of Rods

Consider the axial rod shown in Fig. 15.7 that is impacted by a ball traveling with a velocity v at the instant it strikes the rod. If the ball remains in contact with the rod during the entire impact period, the kinetic energy in the ball is converted into strain energy stored in the rod. To determine the stresses produced by the impact of the ball, recall the expression for the kinetic energy E_k of the ball is given by:

$$\mathbf{E_K} = mv^2/2 \qquad (15.29)$$

where: m is the mass of the ball and v is the velocity of the ball at the instant of impact.

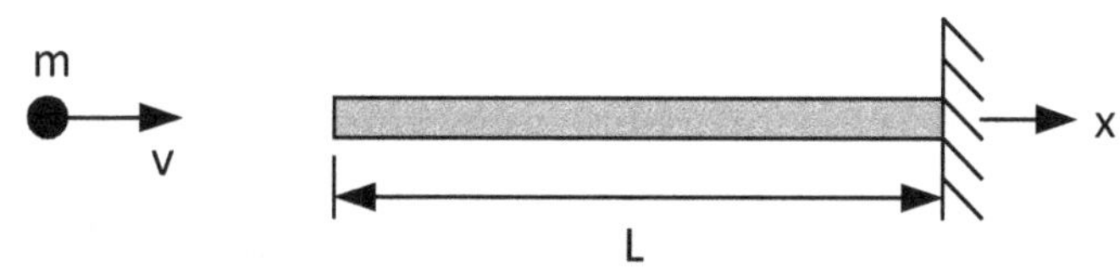

Fig. 15.7 Ball impacting rod with kinetic energy (½)mv².

If energy is conserved, the kinetic energy is converted to strain energy and we can write:

$$\mathbf{E_K} = \mathbf{E}_\varepsilon \qquad (15.30)$$

Substituting Eq. (15.16) into Eq. (15.30) yields:

$$(F^2L)/(2AE) = mv^2/2 \qquad (15.31)$$

Solving Eq. (15.31) for the maximum impact force gives:

$$F_{Max} = \sqrt{\frac{mv^2AE}{L}} \qquad (15.32)$$

The axial stress developed during impact is then given by:

$$\sigma_{Max} = \frac{F_{Max}}{A} = \sqrt{\frac{mv^2E}{AL}} = \sqrt{\frac{mv^2E}{\mathbf{V}}} \qquad (15.33)$$

Let's consider an example to demonstrate the application of this approach to the solution of impact problems involving rods.

EXAMPLE 15.9

A hammer strikes a steel rod with an axial impact. Determine the maximum force and the maximum stress produced if the hammer weighs 1.6 lb. The rod is 32 in. long with a rectangular cross section having dimensions of 1.25 by 1.5 in. The velocity of the head of the hammer at impact was 15 ft/s.

Solution:

Substitute the numerical parameters from the problem statement into Eq. (15.32) to obtain the impact force as:

$$F_{Max} = \sqrt{\frac{mv^2AE}{L}} = \sqrt{\frac{1.6(15\times 12)^2(1.25)(1.5)(30\times 10^6)}{(386)(32)}} = 15{,}360 \text{ lb} \qquad \text{(a)}$$

Note that $g = 386$ in./s^2 must be used in Eq. (a) to convert the 1.6 lb weight into mass.

The axial stress developed during impact is determined from Eq. (15.33) as:

$$\sigma_{Max} = \frac{F_{Max}}{A} = \frac{15{,}360}{(1.25)(1.5)} = 8{,}195 \text{ psi} \qquad \text{(b)}$$

The result of this very simple example clearly demonstrates that very large impact forces can be generated with small masses during an impact event. In this example, a dynamic force in excess of seven tons was produced with a 1.6 lb hammer traveling with a velocity of 10.23 MPH.

15.5.2 Impact Loading of Beams

Consider a simply supported beam that is subjected to impact by a falling weight W as illustrated in Fig. 15.8.

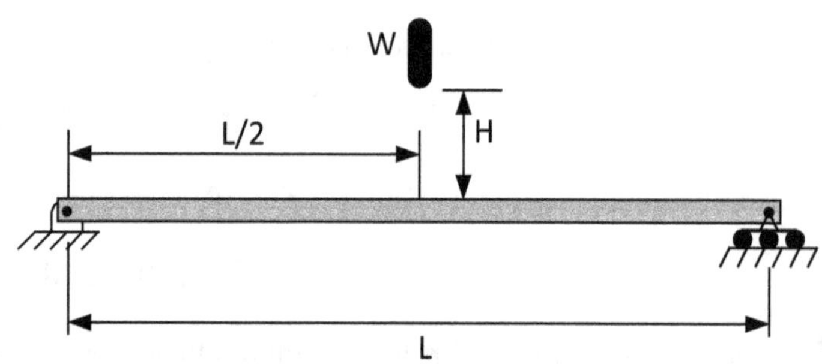

Fig. 15.8 A simply supported beam subjected to impact by a falling weight W.

We approach this problem by recognizing that the potential energy of the weight E_P is converted to kinetic energy prior to impact. Upon impact the kinetic energy E_K is converted to strain energy E_ε stored in the beam. Accordingly, we can write:

$$E_P = E_K = E_\varepsilon = W(H + \delta) \qquad (15.34)$$

where δ is the deflection of the beam at the point of impact.

The strain energy stored in the beam at impact is determined from Eq. (15.21) as:

$$\mathcal{E}_e = \left(\frac{1}{2EI_z}\right)\int_0^L M^2 dx \qquad \text{(a)}$$

For the simply supported beam with a force F applied at the beam's midpoint, we can write the moment equation as:

$$M = Fx/2 \qquad 0 \leq x \leq (L/2) \qquad \text{(b)}$$

This equation is identical to the moment relation written for the beam in Example 15.6. Using the results from Example 15.6, we recall:

$$\mathcal{E}_e = \frac{F^2L^3}{96EI_z} \qquad \text{(c)}$$

Let's first assume that the deflection of the beam is small with respect to the drop height H. Then the potential energy is reduced to E_p = WH. Equating Eq. (c) and Eq. (15.34) gives:

$$WH = \frac{F^2L^3}{96EI_z} \qquad \text{(d)}$$

Solving Eq. (d) for the maximum force gives:

$$F_{Max} = \sqrt{\frac{96EI_zWH}{L^3}} \qquad (15.35)$$

The maximum bending stress occurs at the midpoint of the beam under the impact point. The bending moment at this location is given by:

$$M_{Max} = \frac{F_{Max}L}{4} = \sqrt{\frac{6EI_zWH}{L}} \qquad (15.36)$$

The stress at the midpoint of the beam's span is given by:

$$(\sigma_{xx})_{Max} = M_{Max}c/I_z = M_{Max}/Z \qquad \text{(e)}$$

where $Z = I_z/c$ is the section modulus.

If the beam's cross section is rectangular with $c = h/2$ and $I_z = bh^3/12$, then Eq. (15.36) can be rewritten as:

$$(\sigma_{xx})_{Max} = \sqrt{\frac{18EWH}{Lbh}} = \sqrt{\frac{18EWH}{\mathbf{V}}} \qquad (15.37)$$

where **V** is the volume of the beam.

EXAMPLE 15.10

An experimental impact machine is to be employed in testing the integrity of the weld made in steel beams. A schematic diagram of the impact machine is shown in Fig. E15.10. The cross section of the test beam is b = 40 mm, h = 60 mm and its length L = 400 mm. For a hammer height H set at 1.5 m, determine the effective weight of the hammer necessary to break the weld if its estimated strength is 500 MPa.

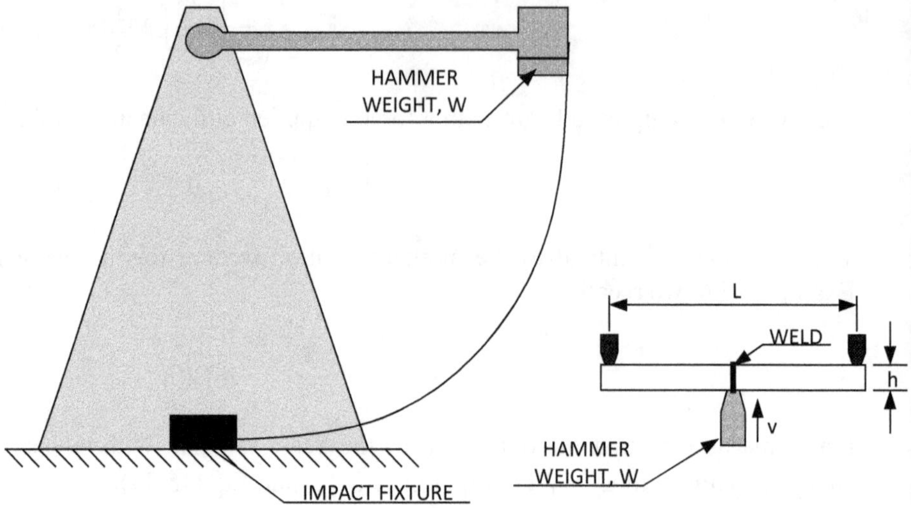

Fig. E15.10

Solution:

While the hammer swings through an arc, the equivalent drop of 1.5 meters must be sufficient to generate a maximum stress that is equal to or greater than the strength of the weld material. Accordingly, we can write:

$$(\sigma_{xx})_{Max} = (S_u)_{Weld} \tag{a}$$

Substituting Eq. (a) into Eq. (15.37) and squaring both sides yields:

$$\frac{18EWH}{Lbh} = S_u^2 \tag{b}$$

Solving Eq. (b) for the hammer weight W using the numerical parameters in the problem statement gives:

$$W = \frac{S_u^2 bhL}{18EH} = \frac{(500\times10^6)^2(0.04)(0.06)(0.4)}{(18)(207\times10^9)(1.5)} = 42.94 \text{ N} \tag{c}$$

This solution shows that an impact machine incorporating a relatively small weight (43 N ≈ 10 lb) dropped through a short distance (1.5 m) can develop sufficiently high stress to produce failure of a relatively large steel test specimen. Of course, we are assuming an elastic impact. If plastic bending occurs prior to the failure of the weld, the amount of energy necessary to deform the beam increases dramatically.

The maximum force developed during impact is given by Eq. (15.35) as:

$$F_{Max} = \sqrt{\frac{96EI_zWH}{L^3}} = \sqrt{\frac{8EWHbh^3}{L^3}} = \sqrt{\frac{8(207\times10^9)(42.94)(1.5)(0.04)(0.06)^3}{(0.4)^3}} = 120.0 \text{ kN}$$

In this example, we observe that the impact force is more than 2500 times larger than the weight attached to the impact machine. Clearly, dynamic loading is a very important consideration in the design of structures and machine components.

EXAMPLE 15.11

A weight W is positioned at the midpoint on a simply supported beam as shown in Fig. E15.11. While the weight is in contact with the beam (H = 0), its gravitational force has not been applied to the beam. The weight is suddenly released to produce an impact event, although the height of the drop is zero. Determine the equations for the maximum stress, the displacement of the load point and the maximum force developed during the impact.

Fig. E15.11

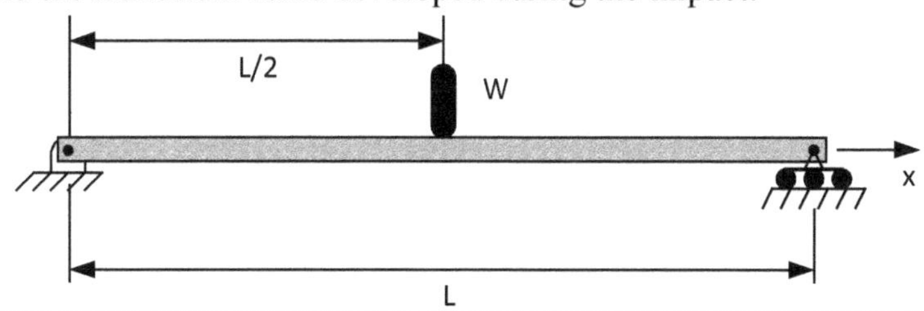

Solution:

The potential energy of the weight in contact with the beam before it is released is given by:

$$\mathbf{E_p} = W\delta \qquad \text{(a)}$$

where δ is the beam's deflection at midpoint due to the sudden application of the weight W.

Let's assume that the elastic curve for the beam during impact is identical to that produced by static loading. Then reference to Appendix E indicates that the force F required to produce the deflection during impact is given by:

$$F = \left(\frac{48EI_z}{L^3}\right)\delta \qquad \text{(b)}$$

The strain energy stored in the beam is equal to the work performed on the beam by this force and is given by:

$$\mathbf{E}_\varepsilon = \mathbf{W} = F\delta/2 \qquad \text{(c)}$$

Substituting Eq. (b) into Eq. (c) yields:

$$\mathbf{E_e} = \left(\frac{24EI_z}{L^3}\right)\delta^2 \qquad \text{(d)}$$

Noting that $\mathbf{E_p} = \mathbf{E}_\varepsilon$, we can write:

$$W\delta = \left(\frac{24EI_z}{L^3}\right)\delta^2 \qquad \text{(e)}$$

Solving Eq. (e) for δ gives:

$$\delta = \frac{WL^3}{24EI_z} = 2\delta_{st} \qquad \text{(f)}$$

where δ_{st} is the static deflection of the beam as given in Appendix E.

The maximum force F_{Max} during this impact event is given by substituting Eq. (f) into Eq. (b) to obtain:

$$F_{Max} = \left(\frac{48EI_z}{L^3}\right)\left(\frac{WL^3}{24EI_z}\right) = 2W \qquad (g)$$

The results from Eqs. (f) and (g) show the effect of suddenly applying a weight W to a simply supported beam doubles both the maximum force and the beam deflection when compared to the case where the weight W is slowly applied.

Now that we have established the importance of the potential energy due to the deflection of the beam during an impact event let's reconsider the assumption made in the derivation of Eq. (15.36). In this derivation, we assumed $\delta \ll H$ and dropped δ in writing the energy balance. Let's repeat this derivation retaining δ in the energy balance. Begin by recalling Eq. (15.34) as:

$$\mathcal{E}_P = \mathcal{E}_K = \mathcal{E}_\varepsilon = W(H + \delta) \qquad (a)$$

For the simply supported beam impacted by a weight W at its midpoint, we recall Eq. (15.23) and write:

$$\mathcal{E}_e = \frac{F^2L^3}{96EI_z} \qquad (b)$$

Again, let's assume that the elastic curve for the beam during impact is identical to that produced by static loading. Then the force F required to produce the deflection during impact is given by:

$$F = \left(\frac{48EI_z}{L^3}\right)\delta \qquad (c)$$

Substituting Eq. (c) into (b) and equating the strain energy with the potential energy gives:

$$\left(\frac{48EI_z}{2L^3}\right)\delta^2 - W\delta - WH = 0 \qquad (d)$$

Recall from Appendix E that the static deflection δ_{st} is given by:

$$\delta_{st} = \frac{WL^3}{48EI_z} \qquad (e)$$

Combining Eqs. (d) and (e) and rearranging yields a quadratic equation in terms of δ:

$$\delta^2 - 2\delta_{st}\delta - 2\delta_{st}H = 0 \qquad (f)$$

Solving Eq. (f) gives:

$$\delta = \delta_{st}\left(1 + \sqrt{1 + 2\left(\frac{H}{\delta_{st}}\right)}\right) \qquad (15.38)$$

$$\delta = \mathbf{M}\delta_{st} \qquad (15.38a)$$

where the term $\mathbf{M} = \left(1 + \sqrt{1 + 2\left(\frac{H}{\delta_{st}}\right)}\right)$ is a multiplier on the static deflection.

We examine the magnitude of this multiplier as a function of the ratio H/δ_{st} in Table 15.2.

Table 15.2
Multiplier for beam deflection for different ratios of H/δ_{st}.

H/δ_{st}	M	H/δ_{st}	M
0	2.00	50	11.05
1	2.73	100	15.17
2	3.24	200	21.02
5	4.32	500	32.64
10	5.58	1000	45.73
20	7.40	2000	64.25

This multiplier is the same for the maximum force developed during impact. Because δ_{st} is usually a small quantity, the ratio of H/δ_{st} becomes large even for relatively small values of the drop height H.

15.6 CASTIGLIANO'S THEOREM

In a previous section, we have developed relations for the strain energy for rods in tension or compression, beams in bending, and shafts in torsion. These expressions are useful because they may be employed in the application of Castigliano's theorem. Castigliano's theorem enables us to determine the expressions for the displacement of a structural member at any point where a force is applied. To illustrate this concept consider the strain energy stored in a tie rod, which is given by Eq. (15.16) as:

$$\mathbf{E_e} = \frac{F^2 L}{2AE} \tag{a}$$

If we differentiate this relation with respect to the force F, in accordance with Castigliano's theorem and Eq. (15.6), we obtain:

$$\frac{\partial \mathbf{E_e}}{\partial F} = \frac{FL}{AE} = \delta \tag{b}$$

Differentiating the expression for the strain energy with respect to a force gives the displacement at the point of application of the force. Moreover, the displacement is in the direction of the applied force.

The general procedure for applying Castigliano's theorem is to write an expression for the strain energy of a structural member in terms of the applied loads F_1, F_2, F_3 and F_n. The expression for the strain energy is then differentiated with respect to one of the forces to obtain the displacement at the point of application of that force. The differentiation is repeated for other forces to obtain additional solutions for displacements.

Several examples will be described to demonstrate the application of Castigliano's theorem.

EXAMPLE 15.12

For the simply supported beam shown in Fig. E15.12, determine the deflection at midspan using Castigliano's theorem.

Fig. E15.12

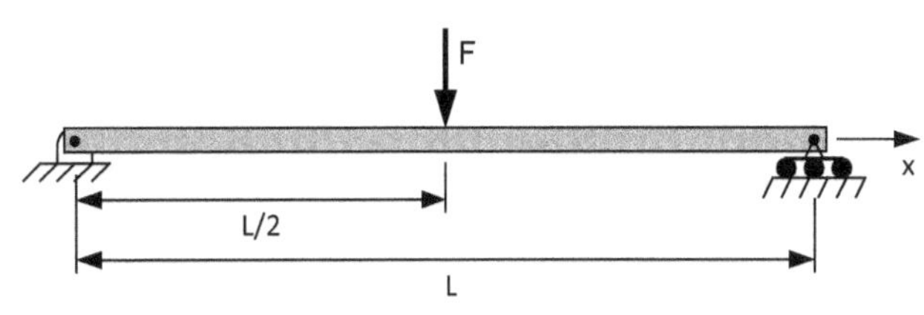

Solution:

We make use of Eq. (15.23) for the strain energy stored in this beam, which was derived in Example 15.6 as:

$$\mathbf{E}_e = \frac{F^2L^3}{96EI_z} \tag{a}$$

Employ Eq. (15.6) with Eq. (a) to obtain the deflection at the midspan of the beam as:

$$\delta = \frac{\partial \mathbf{E}_e}{\partial F} = \frac{FL^3}{48EI_z} \tag{b}$$

This result matches the answer given in Appendix E for the identical case.

It is clear that Castigliano's method provides a useful approach to determine the deflection at a point on a structure where a concentrated load is applied. But what if we find it necessary to determine the displacement at some other point on the structure? The next example describes a clever approach to provide a solution for this class of problems.

EXAMPLE 15.13

For the cantilever beam, shown in Fig. E15.13, determine the deflection at its free end.

Fig. E15.13

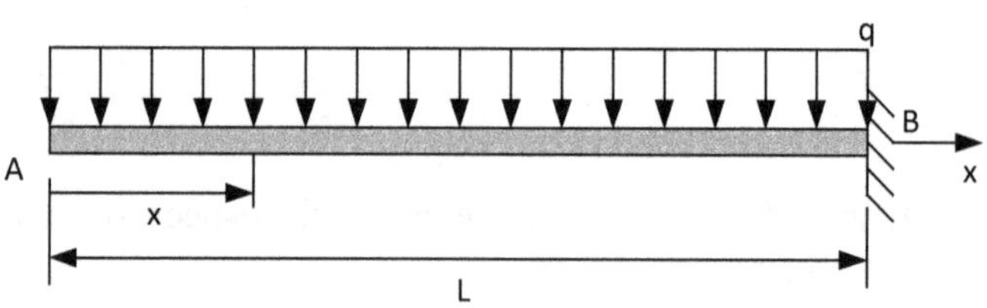

Solution:

The application of Castigliano's theorem to the solution of this problem requires a concentrated force to be located at the point where the deflection is to be determined. However, in this case the cantilever beam is subjected to a uniformly distributed load. To circumvent this difficulty, we apply a fictitious concentrated force f at the free end of the beam as shown in Fig. E15.13a.

Fig. E15.13a

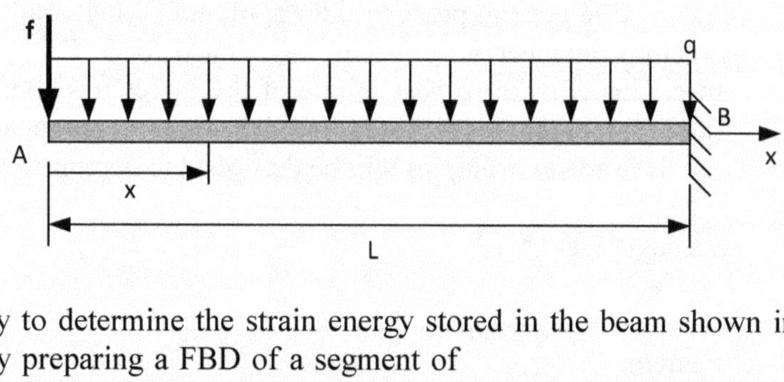

To apply Castigliano's theorem it is necessary to determine the strain energy stored in the beam shown in Fig. E15.13a. We begin this determination by preparing a FBD of a segment of the beam as shown in Fig. E15.13b.

Fig. 15.13b

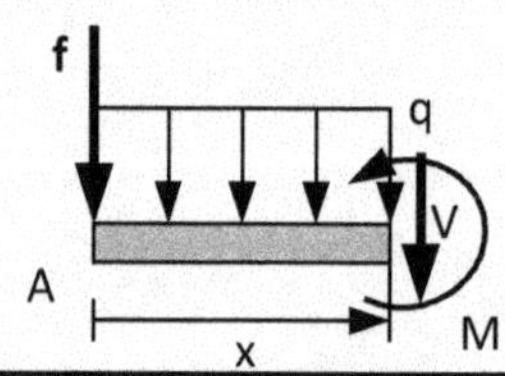

Writing the moment equation for this segment, we obtain:

$$M = -\,fx - qx^2/2 \qquad \text{(a)}$$

Next, use Eq. (15.21) to express the relation for the strain energy stored in the beam as:

$$\mathbf{E}_e = \left(\frac{1}{2EI_z}\right)\int_0^L M^2 dx \qquad \text{(b)}$$

If we differentiate Eq. (b) with respect to f and employ Eq. (15.6), it is possible to write the expression for the deflection at the free end of the beam as:

$$\delta = \frac{\partial \mathbf{E}_e}{\partial f} = \left(\frac{1}{EI_z}\right)\int_0^L M\frac{\P M}{\P f}dx \qquad (15.39)$$

From Eq. (a), it is evident that:

$$\frac{\partial M}{\partial f} = -x \qquad \text{(c)}$$

Substituting Eqs. (a) and (c) into Eq. (15.39) gives:

$$\delta = \left(\frac{1}{EI_z}\right)\int_0^L \left(fx^2 + \frac{qx^3}{2}\right)dx \qquad \text{(d)}$$

Integrating Eq. (d) yields:

$$\delta = \left(\frac{1}{EI_z}\right)\left[\frac{fx^3}{3} + \frac{qx^4}{8}\right]_0^L = \left(\frac{1}{EI_z}\right)\left[\frac{fL^3}{3} + \frac{qL^4}{8}\right] \qquad \text{(e)}$$

Let the fictitious force $f \Rightarrow 0$ and Eq. (e) becomes:

$$\delta = \left(\frac{qL^4}{8EI_z}\right) \qquad (15.40)$$

Again, this result agrees with the answer given in Appendix E for the identical case.

EXAMPLE 15.14

For the simply supported beam shown in Fig. E15.14, use Castigliano's theorem to determine the deflection at the midpoint of the span. Note that $b = L - 2a$.

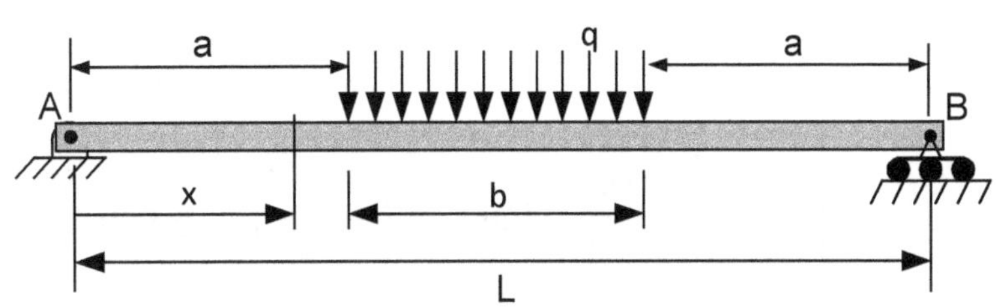

Fig. 15.14

Solution:

We first add the fictitious force f at the midpoint of the span and prepare a FBD of the entire beam as shown in Fig. E15.14a.

Fig. E15.14a

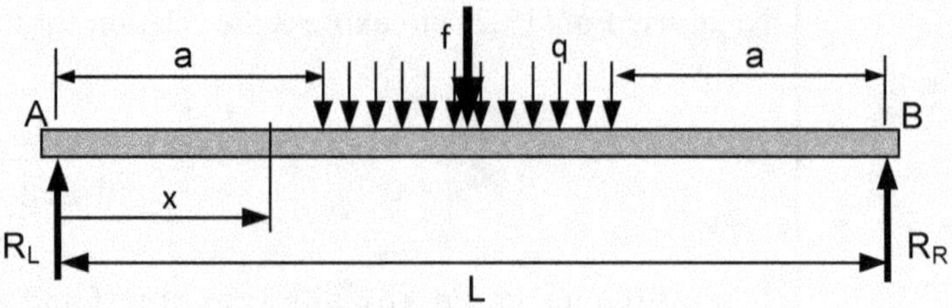

From the equilibrium relations and the statement that b = L – 2a, it is clear that:

$$R = R_L = R_R = (1/2)(qb + f) \qquad \text{(a)}$$

An FBD of the beam segment for x ≤ a is presented in Fig. E15.14b. Writing the equation for the moment M from this FBD yields:

$$M_1 = \frac{x}{2}(qb+f) \qquad \text{for } x \le a \qquad \text{(b)}$$
$$\frac{\partial M_1}{\partial f} = \frac{x}{2}$$

Fig. E15.14b

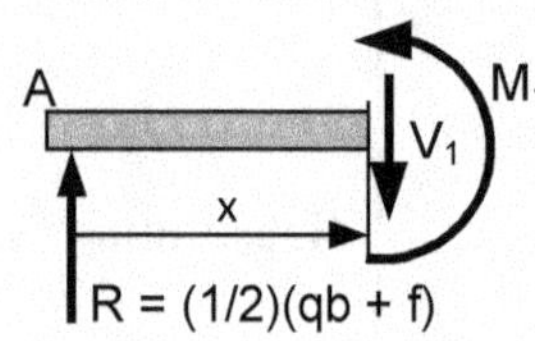

The FBD presented in Fig. E15.14c is helpful in writing the equation for the moment M which is valid from a ≤ x ≤ L/2.

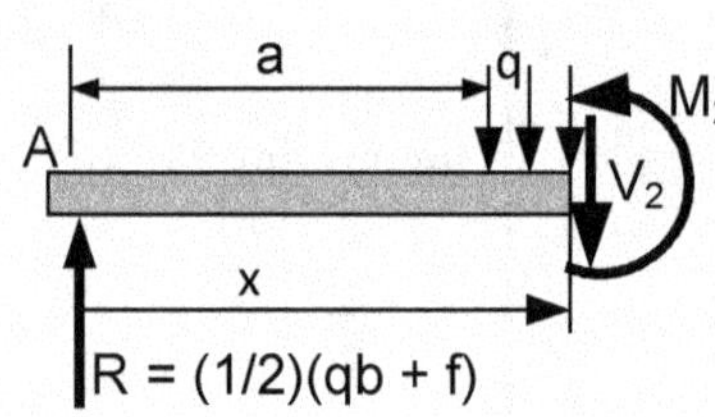

Fig. E15.14c

$$M_2 = \frac{x}{2}(qb+f) - \frac{q}{2}(x-a)^2 \qquad \text{for } a \le x \le L/2 \qquad \text{(c)}$$
$$\frac{\partial M_2}{\partial f} = \frac{x}{2}$$

Note that the loading of the beam is symmetric with respect to x even after the addition of the fictitious force f. We will use this symmetry in simplifying the calculations needed to determine the midspan deflection. Let's write Eq. (15.39) and adapt it to account for the symmetry of loading by setting the limits of integration from zero to L/2 and multiplying the resulting relation by two.

$$\delta = \left(\frac{1}{EI_z}\right)\left\{\int_0^a M_1 \frac{\partial M_1}{\partial f} dx + \int_a^{L/2} M_2 \frac{\partial M_2}{\partial f} dx\right\}2 \qquad \text{(d)}$$

Substituting Eqs. (b) and (c) into Eq. (d) yields:

$$\delta = \left(\frac{2}{EI_z}\right)\left\{\int_0^a \left(\frac{x}{2}\right)(qb+f)\left(\frac{x}{2}\right)dx + \int_a^{L/2}\left(\left(\frac{x}{2}\right)(qb+f) - \frac{q}{2}(x-a)^2\right)\left(\frac{x}{2}\right)dx\right\} \qquad \text{(e)}$$

Let's simplify Eq. (e) by letting $f \Rightarrow 0$ to obtain:

$$\delta = \left(\frac{2}{EI_z}\right)\left\{\int_0^a \left(\frac{qbx^2}{4}\right)dx + \int_a^{L/2}\left(\frac{qbx^2}{4} - \frac{qx(x-a)^2}{4}\right)dx\right\}$$
$$= \left(\frac{q}{2EI_z}\right)\left\{\int_0^a bx^2 dx + \int_a^{L/2}\left[bx^2 - x(x-a)^2\right]dx\right\} \qquad \text{(f)}$$

Integrating Eq. (f) gives:

$$\delta = \left(\frac{q}{2EI_z}\right)\left\{\left[\frac{bx^3}{3}\right]_0^a + \left[\frac{bx^3}{3}\right]_a^{L/2} - \left[\frac{x^4}{4} - \frac{2ax^3}{3} + \frac{a^2x^2}{2}\right]_a^{L/2}\right\} \qquad \text{(g)}$$

Substitute the upper and lower limits in Eq. (g), replace a with (1/2)(L – b), and simplify to obtain:

$$\delta = \left(\frac{qb}{384EI_z}\right)\left(8L^3 + b^3 - 4Lb^2\right) \qquad \text{(h)}$$

The result is the same as the deflection determined by the double integration method in Chapter 13 [see Eq. (13.19)]. In this case, the sign of the deflection is positive because the fictitious force was directed downward. The positive sign indicates that the deflection is in the same direction of the fictitious force.

15.6.1 Castigliano's Method Applied to Statically Indeterminate Beams

In Chapter 13, we described techniques for solving for the reaction forces in statically indeterminate beams by employing either the superposition or the integration method. We can also solve for the reaction forces by using Castigliano's theorem. This approach utilizes Eq. (15.6) applied at one of the structure's supports. We rewrite Eq. (15.6) as:

$$\frac{\partial \mathbf{\mathcal{E}_e}}{\partial F_n} = \frac{\partial \mathbf{\mathcal{E}_e}}{\partial R} = \delta_R = 0 \qquad (15.41)$$

where $\delta_R = 0$ is the displacement of the reactive force R at the support.

In applying Eq. (15.41), we differentiate the expression for the strain energy in the beam with respect to an appropriate reaction force R. The result of the differentiation gives the displacement at the support; however, because the support is fixed this displacement is zero. Thus Eq. (15.41) provides the additional relation that can be used together with the usual equilibrium equations to solve for unknown reaction forces in statically indeterminate beams. Let's consider two examples to demonstrate this method of analysis.

EXAMPLE 15.15

For the uniformly loaded cantilever beam, shown in Fig. E15.15, use Castigliano's theorem to determine the reaction force at the prop.

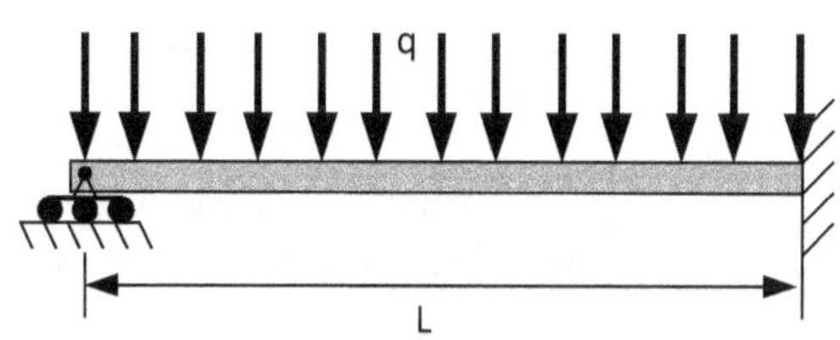

Fig. E15.15

Solution:

The propped cantilever beam is statically indeterminate because we have three unknown reactive forces—the reaction R at the prop, and the moment and the reaction force at the built-in end of the beam. Let's begin the solution by drawing a FBD of a portion of the left end of the beam that includes the reaction force R at the prop as shown in Fig. E15.15a.

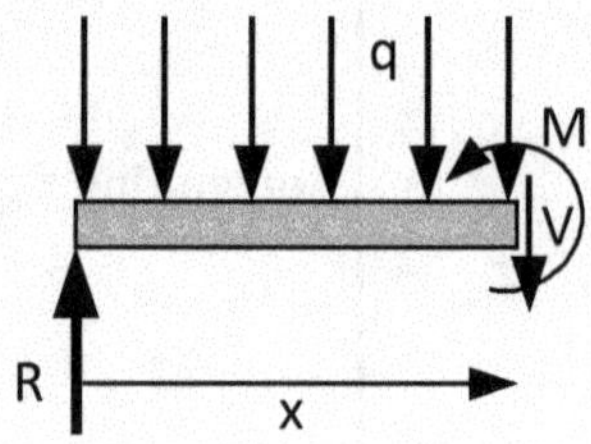

Fig.15.15a

Writing the equation for the moment M from this FBD yields:

$$M = Rx - \frac{qx^2}{2} \qquad \frac{\partial M}{\partial R} = x \tag{a}$$

Let's write Eq. (15.41) and use Eq. (15.39) to obtain:

$$\delta = \frac{\partial E_e}{\partial R} = \left(\frac{1}{EI_z}\right)\int_0^L M\frac{\partial M}{\partial R}dx = 0 \tag{b}$$

Substituting Eq. (a) into Eq. (b) and integrating gives:

$$\int_0^L M\frac{\partial M}{\partial R}dx = \int_0^L \left(Rx - \frac{qx^2}{2}\right)xdx = 0 \qquad \left[\frac{Rx^3}{3} - \frac{qx^4}{8}\right]_0^L = 0 \tag{c}$$

Substituting the limits into Eq. (c) and simplifying yields:

$$R = 3qL/8 \tag{d}$$

This example illustrates another useful approach for determining unknown reactive forces acting on statically indeterminate structures.

EXAMPLE 15.16

For the uniformly loaded continuous beam, shown in Fig. E15.16, use Castigliano's theorem to determine the reaction force at the center support.

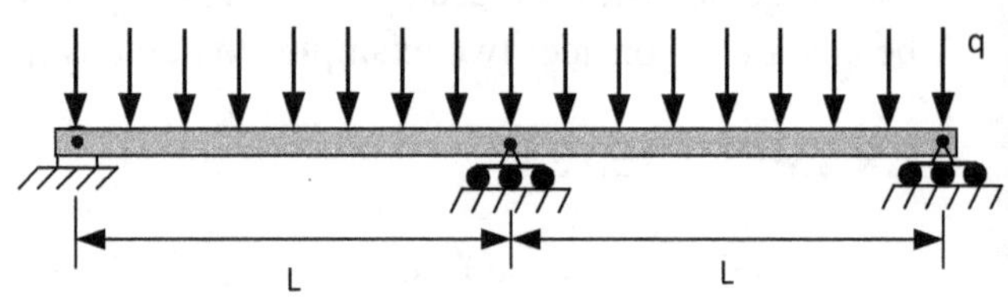

Fig. E15.16

Solution:

The propped continuous beam is statically indeterminate because we have three unknown reactive forces—the three reactions at the simple supports—and only two useful equations of equilibrium. Let's begin the solution by writing the equilibrium relation $\Sigma F_y = 0$ for the entire beam to obtain:

$$R_L + R_C + R_R = 2qL \qquad \text{(a)}$$

Because of symmetry $R_L = R_R$, and we can rewrite Eq. (a) as:

$$R_C = 2qL - 2R_L \qquad \text{(b)}$$

Now draw a FBD of a portion of the left end of the beam that includes the reaction force R_L at the left hand end as shown in Fig. E15.16a.

Fig. E15.16a

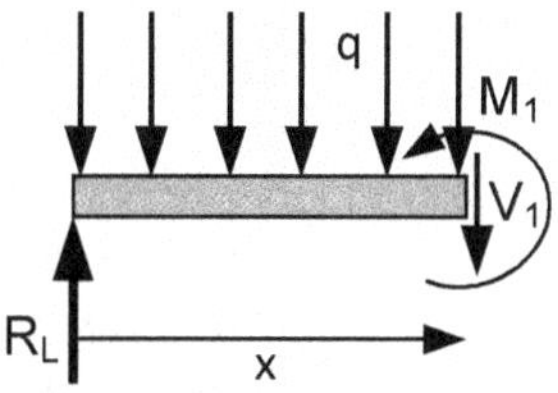

Writing the equation for the moment from this FBD yields:

$$M_1 = R_L x - \frac{qx^2}{2} \qquad \text{for } 0 \le x \le L \qquad \text{(c)}$$
$$\frac{\partial M_1}{\partial R_L} = x$$

Next, we prepare another FBD that includes the reaction R_C at the center support as shown in Fig. E15.16b.

Fig. E15.16b

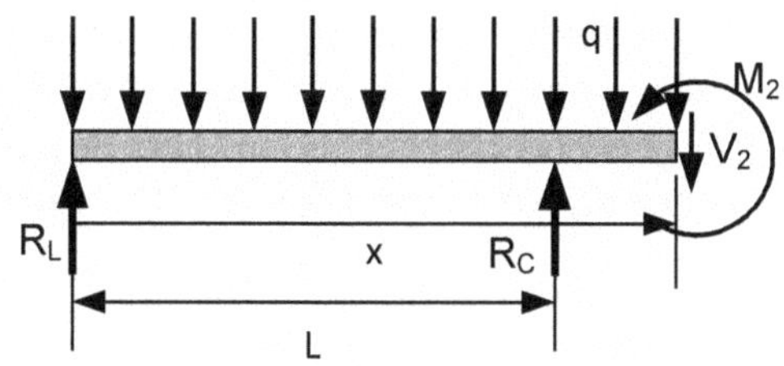

Writing the equation for the moment from this FBD yields:

$$M_2 = R_L x - \frac{qx^2}{2} + R_C(x - L) \qquad \text{for } L \le x \le 2L \qquad \text{(d)}$$

Substituting Eq. (b) into Eq. (d) and simplifying gives:

$$M_2 = R_L(2L - x) - \frac{qx^2}{2} + 2qL(x - L) \qquad \text{for } L \le x \le 2L \qquad \text{(e)}$$
$$\frac{\partial M_2}{\partial R_L} = 2L - x$$

Let's write Eq. (15.41) and adopt the form of the strain energy shown in Eq. (15.39) to obtain:

$$\delta_L = \frac{\partial \mathbf{E_e}}{\partial R_L} = \left(\frac{1}{EI_z}\right)\left\{\int_0^L M_1 \frac{\partial M_1}{\partial R_L}dx + \int_L^{2L} M_2 \frac{\partial M_2}{\partial R_L}dx\right\} = 0 \qquad \text{(f)}$$

Substituting Eqs. (c) and (e) into Eq. (f) and integrating gives:

$$\delta_L = \int_0^L \left(R_L x - \frac{qx^2}{2}\right)(x)dx + \int_L^{2L}\left(R_L(2L - x) - \frac{qx^2}{2} + 2qL(x - L)\right)(2L - x)dx = 0$$
$$\left[\frac{R_L x^3}{3} - \frac{qx^4}{8}\right]_0^L + \left[R_L\left(4L^2x - 2Lx^2 + \frac{x^3}{3}\right) + q\left(\frac{x^4}{8} - Lx^3 + 3L^2x^2 - 4L^3x\right)\right]_L^{2L} = 0 \qquad \text{(g)}$$

Substituting the limits into Eq. (g) and simplifying yields:

$$(2L^3/3)R_L - qL^4/4 = 0 \qquad \text{(h)}$$

Solving for R_L yields:

$$R_L = R_R = 3qL/8 \qquad \text{(i)}$$

Finally from Eq. (i) and Eq. (b), it is clear that:

$$R_C = 5qL/4 \qquad \text{(j)}$$

This example illustrates that the center support in a continuous beam carries a large portion of the loading. In this case, the center support carries over three times as much load as each end support. Note that these same results can be obtained by using the methods (i.e. superposition and singularity functions) presented in Chapter 13.

15.6.2 Castigliano's Method Applied to Determining Rotation

In previous sections, we considered a structural element subjected to forces F_1, F_2, F_3 and F_n to derive the relation:

$$\frac{\partial \mathbf{U}_e}{\partial F_n} = \delta_n \qquad (15.6)$$

In the derivation of Eq. (15.6), we equated the incremental changes in the work $\Delta\mathbf{W} = \delta_n \Delta F_n$ with the incremental changes in the strain energy $\Delta\mathbf{U}_\varepsilon$. However, Castigliano's method is more general and can also accommodate incremental changes in the work due either to the application of a concentrated moment M or a torque T. When a concentrated moment is applied to a structural member, the work is given by:

$$\mathbf{W} = (1/2)M\theta \qquad (15.42)$$

where θ is the angle through which the moment rotates.

Similarly when applying a torque T to a shaft, the resulting work is:

$$\mathbf{W} = (1/2)T\phi \qquad (15.43)$$

where ϕ is the angle of twist of the shaft.

Recognizing these alternative forms for expressing work enables us to write the expressions for the angular displacements (rotations) as:

$$\theta_n = \partial \mathbf{U}_\varepsilon / \partial M_n \qquad (15.44)$$

$$\phi_n = \partial \mathbf{U}_\varepsilon / \partial T_n \qquad (15.45)$$

Let's demonstrate the use of these relations in the solution of two example problems.

EXAMPLE 15.17

A cantilever beam is formed in the shape of a quarter circle as illustrated in Fig. E15.17. If the cantilever beam is loaded with a concentrated moment M_o and a concentrated force F at its free end, determine the angular displacement and the beam's deflection in the direction of the force F.

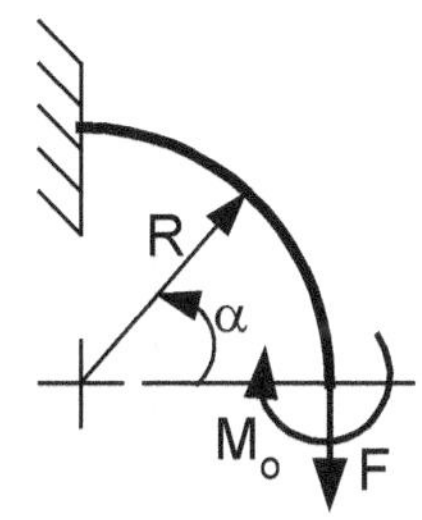

Fig. E15.17

Solution:

Let's begin by preparing a FBD of a segment of the circular arc as shown in Fig. E15.17a.
Writing the equation for the moment M at some arbitrary position defined by the angle α as shown in the FBD gives:

$$M = M_0 + FR(1 - \cos\alpha) \qquad \text{(a)}$$

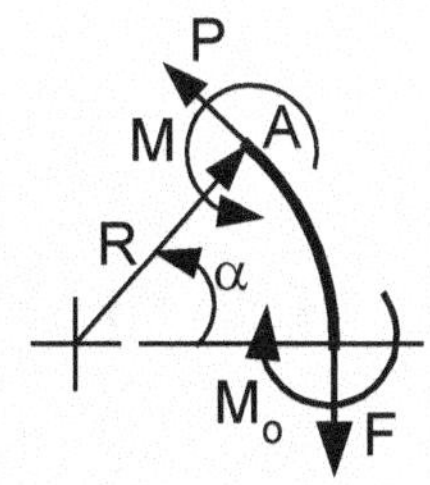

Fig. E15.17a

The cantilever beam in this example is in the form of a circular arc; hence, it is necessary to modify Eq. (15.21) to read as:

$$\mathbf{E}_e = \left(\frac{1}{2EI_z}\right)\int_0^{\pi/2} M^2 R d\alpha \qquad (15.46)$$

where Rdα = dx

In this case, the upper limit on the integral is π/2 because the cantilever is formed as a quarter circle. The upper limit is adjusted to accommodate the shape of the circular arc. Let's determine the deflection downward δ_y in the direction of the force F by using Eq. (15.6) to obtain:

$$\delta_y = \frac{¶\mathbf{E}_e}{¶\mathbf{F}} = \left(\frac{1}{EI_z}\right)\int_0^{\pi/2} M\frac{\partial M}{\partial F} R d\alpha \qquad \text{(b)}$$

Substitute Eq. (a) into Eq. (b) and integrate to obtain:

$$\delta_y = \left(\frac{1}{EI_z}\right)\int_0^{\pi/2}\left[M_0 + FR(1-\cos\alpha)\right]\left[R(1-\cos\alpha)\right]Rd\alpha$$

$$\delta_y = \left(\frac{R^2}{EI_z}\right)\int_0^{\pi/2}\left[M_0(1-\cos\alpha) + FR(1-\cos\alpha)^2\right]d\alpha \qquad \text{(c)}$$

$$\delta_y = \left(\frac{R^2}{EI_z}\right)\left[M_0\alpha - M_0\sin\alpha + FR\left(\alpha - 2\sin\alpha + \frac{\alpha}{2} + \frac{\sin 2\alpha}{4}\right)\right]_0^{\pi/2}$$

Substituting the limits into Eq. (c) and simplifying yields:

$$\delta_y = \frac{R^2}{EI_z}\left\{\frac{M_0}{2}(\pi - 2) + \frac{FR}{4}(3\pi - 8)\right\} \qquad \text{(d)}$$

To determine the rotation at the free end of the cantilever beam, substitute Eq. (a) into Eq. (15.44) to obtain:

$$\theta = \frac{\partial \mathbf{U_e}}{\partial M_0} = \left(\frac{1}{EI_z}\right)\int_0^{\pi/2} M \frac{\partial M}{\partial M_0} R d\alpha \qquad \text{(e)}$$

Integrating Eq. (e) gives:

$$\theta = \frac{\partial \mathbf{U_e}}{\partial M_0} = \left(\frac{1}{EI_z}\right)\int_0^{\pi/2} [M_0 + FR(1-\cos\alpha)][1]R d\alpha$$

$$\theta = \left(\frac{R}{EI_z}\right)[M_0\alpha + FR\alpha - FR\sin\alpha]_0^{\pi/2}$$

$$\theta = \left(\frac{R}{EI_z}\right)\left[M_0\frac{\pi}{2} + FR\left(\frac{\pi}{2} - 1\right)\right] \qquad \text{(f)}$$

Castigliano's theorem has provided an approach for solving for the deflection and rotation at select points on **thin** curved beams. Additional exercises of this type will be given in the problems section of this chapter.

EXAMPLE 15.18

A shaft is fabricated by brazing together two sections of round tubing as shown in Fig. E15.18. The shaft is fixed at one end and a torque T is applied at its free end. Determine the equation for the angle of twist ϕ. Assume the thickness of the braze metal is negligible.

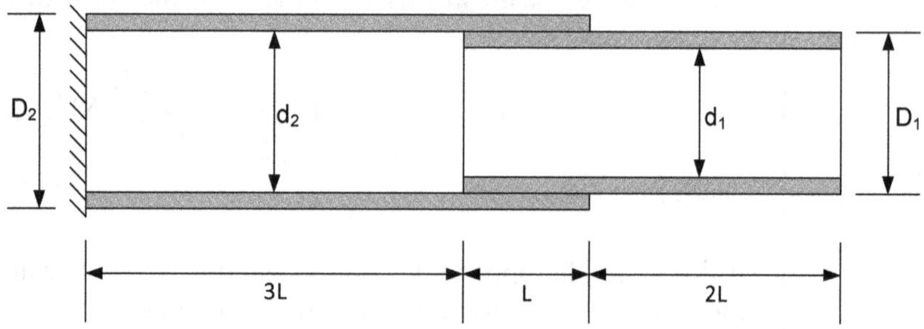

Fig. E15.18

Solution:

The strain energy for a shaft subjected to torsion loading is given by Eq. (15.26) as:

$$\mathbf{U_e} = \frac{1}{2GJ}\int_0^{L} T^2 dx = \frac{T^2 L}{2GJ} \qquad \text{(a)}$$

We can apply this relation if it is adjusted to account for the variation of the polar moment of inertia J along the length of the shaft. Accordingly we divide the shaft into three parts—the left side, the center portion, and the right side and rewrite Eq. (a) as:

$$\mathbf{U_e} = \frac{T^2 L}{2G}\left[\frac{3}{(\pi/32)(D_2^4 - d_2^4)} + \frac{1}{(\pi/32)(D_2^4 - d_1^4)} + \frac{2}{(\pi/32)(D_1^4 - d_1^4)}\right] \qquad \text{(b)}$$

Recalling Eq. (15.45), substitute Eq. (b) into this relation and simplify to obtain:

$$\phi = \frac{\partial \mathbf{U_e}}{\partial T} = \frac{32TL}{\pi G}\left[\frac{3}{(D_2^4 - d_2^4)} + \frac{1}{(D_2^4 - d_1^4)} + \frac{2}{(D_1^4 - d_1^4)}\right] \qquad \text{(c)}$$

Again Castigliano's method provides another independent approach to determining displacements (angular or linear) at one point on the structural member.

15.6.3 Castigliano's Method Applied to Truss Deflections

In our previous discussion of trusses, we were concerned with determining forces, stresses, and safety factors for the individual members. However, in many instances, it is important to determine the displacement of the joints in the truss. There are two approaches for determining the displacement. The first is to calculate the extension or contraction of the individual members, and then to solve a complex geometry problem to establish the movement of each of the joints. The second approach, which is described in this section, is the application of Castigliano's method for this determination.

We have derived the relation for the strain energy stored in a tie rod in Eq. (15.16). This equation is also valid for truss members because the forces are always applied along the axis of the individual members. Let's rewrite this relation as:

$$\mathcal{E}_e = \frac{P_i^2 L_i}{2A_i E} \tag{15.47}$$

where P_i is the internal force and A_i is the cross sectional area in truss member I and L_i is the length of member i.

The total strain energy in the truss system is then obtained by summing the strain energy stored in the individual members. Hence, we can write:

$$\mathcal{E}_e = \sum_{i=1}^{n} \frac{P_i^2 L_i}{2A_i E} \tag{15.48}$$

where n is the number of members in the truss.

Substituting Eq. (15.48) into Eq. (15.6) yields:

$$\delta_k = \frac{\partial \mathcal{E}_e}{\partial F_k} = \frac{\partial}{\partial F_k} \sum_{i=1}^{n} \frac{P_i^2 L_i}{2A_i E} \tag{15.49}$$

Let's differentiate Eq. (15.49) to obtain:

$$\delta_k = \sum_{i=1}^{n} \frac{P_i L_i}{A_i E} \left(\frac{\partial P_i}{\partial F_k} \right) \tag{15.50}$$

Care must be exercised in the application of Eq. (15.50). To determine the $(\partial P_i/\partial F_k)$, it is necessary to write equations for the internal forces P_i in each member of the truss in terms of the external force F_k applied at a specified joint in the truss. Numbers must not be substituted into the equations until after Eq. (15.50) has been executed in symbolic format. Let's consider three examples to demonstrate this approach for determining the displacement of a joint in a truss.

EXAMPLE 15.19

Determine the displacement of joint B in the vertical direction for the two-member truss depicted in Fig. E15.19 using Castigliano's method. Truss members AB and BC are fabricated from the same material, with identical length L and area A.

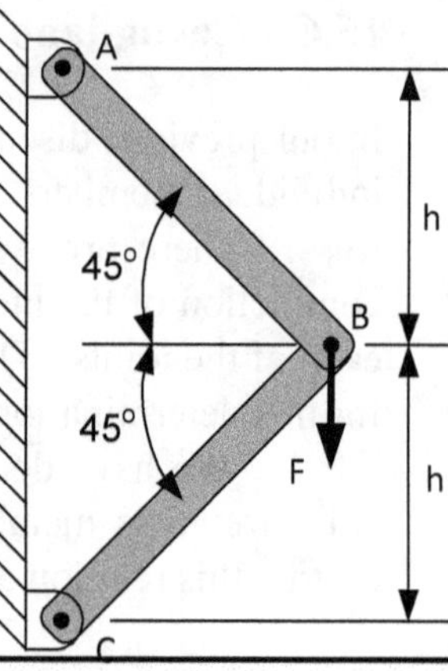

Fig. E15.19

Solution:

We begin by writing the equations for the internal forces P_{AB} and P_{BC} in terms of the force F and the geometric parameters describing the truss. From the FBD of the pin at joint B, we write:

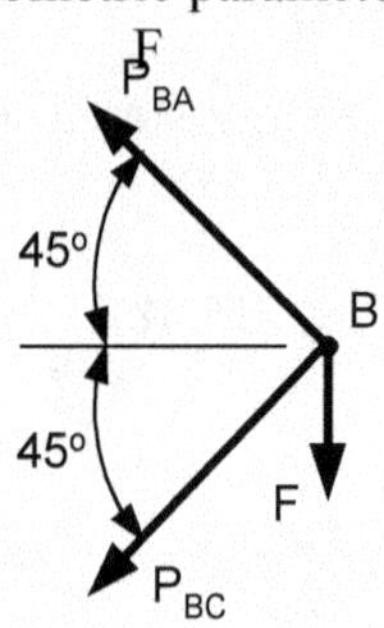

$$\Sigma F_x = -P_{BA} \cos 45° - P_{BC} \cos 45° = 0$$
$$P_{BA} = -P_{BC} \qquad \text{(a)}$$

$$\Sigma F_y = P_{BA} \sin 45° - P_{BC} \sin 45° - F = 0$$
$$P_{BA} = F/\sqrt{2} \qquad P_{BC} = -F/\sqrt{2} \qquad \text{(b)}$$

Next let's take the partial derivatives of the two internal forces to obtain:

$$\partial P_{BA}/\partial F = 1/\sqrt{2} \qquad \partial P_{BC}/\partial F = -1/\sqrt{2} \qquad \text{(c)}$$

Substituting the results of Eqs. (b) and (c) into Eq. (15.50) yields:

$$\delta_B = \frac{L}{AE}\sum_{1}^{2}\left(\frac{\partial P_i}{\partial F_B}\right)P_i = \frac{L}{AE}\left[\left(\frac{F}{\sqrt{2}}\right)\left(\frac{1}{\sqrt{2}}\right)+\left(\frac{-F}{\sqrt{2}}\right)\left(\frac{-1}{\sqrt{2}}\right)\right] \qquad \text{(d)}$$

Reducing Eq. (d) gives:

$$\delta_B = FL/AE \qquad \text{(e)}$$

This result gives the deflection of the pin at joint B in the downward direction. The question of whether the pin at joint B moves in the horizontal direction is given as a problem at the end of the chapter.

EXAMPLE 15.20

Determine the displacement in the vertical direction at joint C for the seven-member truss depicted in Fig. E15.20 using Castigliano's method. All of the truss members are fabricated from the same material. The cross sectional area of members BC, AE, BD and DE is A; the area of AB is 2A, and the area of members BE and CD is $(2/\sqrt{2})A$.

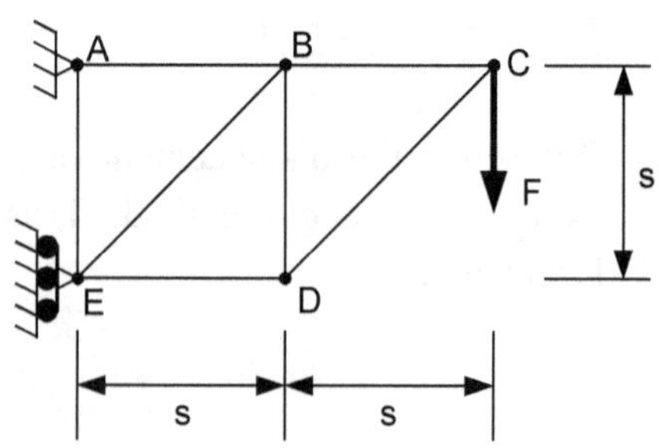

Fig. 15.20

Solution:

Prepare FBDs for the entire truss and for the individual pins A, B, C and E and write the equilibrium relations to determine the reaction forces and the forces in the seven truss members. The FBDs are shown below, and the forces determined by applying the equilibrium relations are listed in Table E15.20.

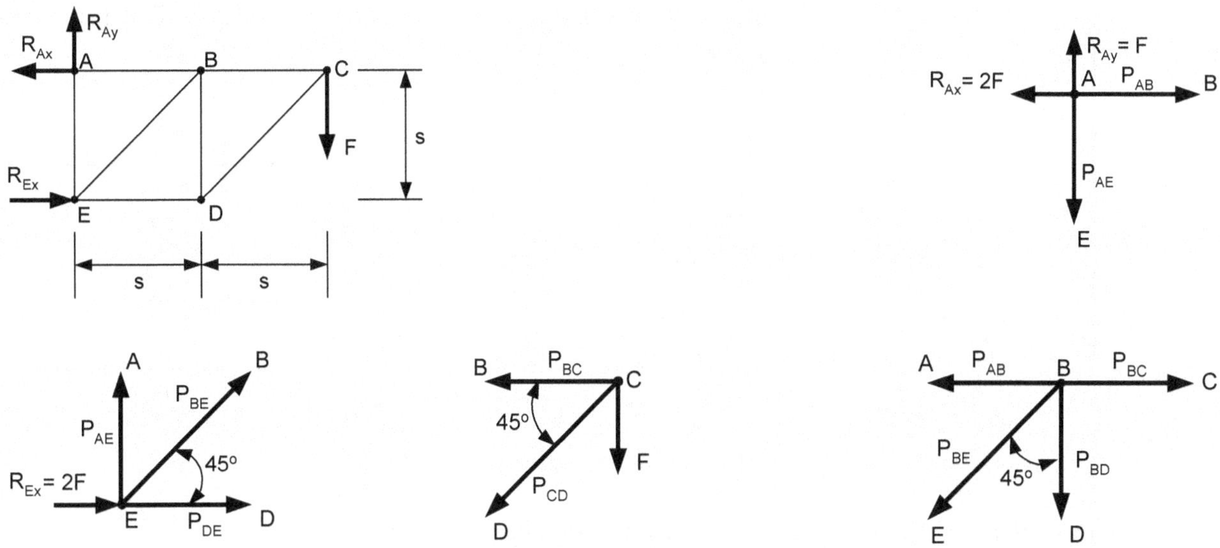

Table E15.20
Arrangement of results for internal forces, length and areas of truss members.

Member	P_i	$\partial P_i/\partial F$	Length. L_i	Area, A_i	$(P_iL_i/A_i)(\partial P_i/\partial F)$
AB	2F	2	s	2A	2Fs/A
BC	F	1	s	A	Fs/A
AE	F	1	s	A	Fs/A
BE	$-2F/\sqrt{2}$	$-2/\sqrt{2}$	$\sqrt{2}$ s	$(2/\sqrt{2})A$	2Fs/A
BD	F	1	s	A	Fs/A
DE	– F	– 1	s	A	Fs/A
CD	$-2F/\sqrt{2}$	$-2/\sqrt{2}$	$\sqrt{2}$ s	$(2/\sqrt{2})A$	2Fs/A
					$\Sigma = 10(Fs/A)$

In the last row in Table E15.20, we have determined the sum of $(P_iL_i/A_i)(\partial P_i/\partial F)$ for the seven different truss members. Then it is evident from Eq. (15.50) that the downward deflection of the joint at C is given by:

$$\delta_y = (10Fs)/(AE)$$

The question remains regarding the horizontal deflection of the joint at C. Let's determine this deflection in the next example.

EXAMPLE 15.21

Determine the displacement in the horizontal direction at joint C for the seven-member truss depicted in Fig. E15.20 using Castigliano's method. All of the truss members are fabricated from the same material. The cross sectional area of members BC, AE, BD and DE is A; the area of AB is 2A, and the area of members BE and CD is $(2/\sqrt{2})A$

Solution:

To determine the horizontal displacement of the pin at joint C, it is necessary to apply a fictitious force f acting in the horizontal direction at that location. The approach is the same as with Example 15.20. We prepare FBDs for the entire truss and for the individual pins A, B, C and E and write the equilibrium relations to determine the reaction forces and the forces in the seven truss members. Because the reactions and the internal forces due to the force F have already been determined, we will consider only the fictitious force f. We will then superimpose the result due to f on the previous results for F in Table E15.21.

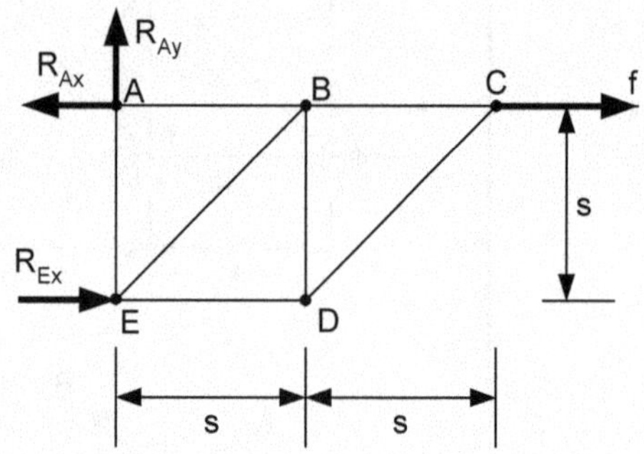

Fig. E15.21

$R_{Ax} = f \quad R_{Ay} = 0 \quad R_{Ex} = 0$

Considering the pin at joint A, we establish that $P_{AB} = f$ and $P_{AE} = 0$.

Considering the pin at joint C, we establish that $P_{BC} = f$ and $P_{CD} = 0$.

Considering the pin at joint B, we establish that $P_{BE} = 0$ and $P_{BD} = 0$.

Finally, considering the pin at joint E, we establish that $P_{DE} = 0$.

We superimpose these results on those obtained in Example 15.20 to give:

Table E15.21
Revision of Table E15.20 showing results for internal forces, length and areas of truss members.

Member	P_i	$\partial P_i/\partial f$	Length. L_i	Area, A_i	$(P_iL_i/A_i)(\partial P_i/\partial f)$
AB	2F + f	1	s	2A	(2F + f)(s/2A)
BC	F + f	1	s	A	(F + f)(s/A)
AE	F	0	s	A	0
BE	– 2F/√2	0	√2 s	(2/√2)A	0
BD	F	0	s	A	0
DE	– F	0	s	A	0
CD	– 2F/√2	0	√2 s	(2/√2)A	0
					Σ = (2Fs/A)+(3fs/2A)

In the last row in Table E15.21, we have determined the sum of $(P_iL_i/A_i)(\partial P_i/\partial f)$ for the seven different truss members. Then it is evident from Eq. (15.50) that the deflection in the positive x direction of the joint at C is given by:

$$\delta_y = (2Fs)/(AE) \qquad \text{(a)}$$

Of course, in writing Eq. (a), we permitted the fictitious force f to go to zero. In this solution, we have only considered the displacement of one joint in the truss. We could determine the displacement of any joint by applying fictitious forces at the joint in question and repeating the analysis described in Examples 15.20 and 15.21.

15.7 SUMMARY

When a force F is applied to a structural member, it undergoes a displacement δ and work is performed. This work is conserved and stored as strain energy $\mathbb{E}_\varepsilon$.

$$\mathbb{E}_\varepsilon = F\delta/2 \qquad (15.2)$$

The ability to compute the strain energy stored in structural members is important because three different energy theorems can be employed to extend our methods of analysis. These include: the theorem of virtual work, the theorem of Castigliano and the theorem of least work (minimum potential energy). We described several applications of Castigliano's theorem in this chapter.

The application of Castigliano's theorem involves the partial differentiation of the strain energy with respect to an applied force F to give the displacement δ as:

$$\frac{\partial \mathbb{E}_e}{\partial F_n} = \delta_n \qquad (15.6)$$

The concept of strain energy density was introduced and methods for computing this quantity were demonstrated based on the following equations:

$$\mathbf{e}_\varepsilon = \Delta\mathbb{E}_\varepsilon / V = (1/2)\sigma_{xx}\,\varepsilon_{xx} \qquad (15.9)$$

$$\mathbf{e}_\varepsilon = (1/2)(\sigma_{xx}\,\varepsilon_{xx} + \sigma_{yy}\,\varepsilon_{yy} + \sigma_{zz}\,\varepsilon_{zz} + \tau_{xy}\gamma_{xy} + \tau_{yz}\gamma_{yz} + \tau_{zx}\gamma_{zx}) \qquad (15.10)$$

$$\mathbf{e}_\varepsilon = (1/2)(\sigma_1\,\varepsilon_1 + \sigma_2\,\varepsilon_2 + \sigma_3\,\varepsilon_3) \qquad (15.11)$$

The strain energy density served a useful function because it could be integrated over the volume of the structural member to establish the strain energy stored in the element.

$$\mathbb{E}_e = \frac{1}{2E}\int \sigma_{xx}^2\,dV \qquad (15.15)$$

We showed that the strain energy stored in a tie rod is given by:

$$\mathbb{E}_e = \frac{1}{2E}\frac{F^2}{A^2}AL = \frac{F^2 L}{2AE} \qquad (15.16)$$

We also showed the strain energy stored in a beam is given by:

$$\mathbb{E}_e = \left(\frac{1}{2EI_z}\right)\int_0^L M^2 dx \qquad (15.21)$$

And the strain energy stored in a shaft subjected to a torque T is given by:

$$\mathbb{E}_e = \frac{1}{2GJ}\int_0^L T^2 dx = \frac{T^2 L}{2GJ} \qquad (15.26)$$

After describing and demonstrating techniques for determining the strain energy in structural members, we used energy concepts to write relations for impact stresses developed in rods and beams. For a rod subjected to axial impact by an object with a mass m traveling with a velocity v, we derived the relation for the stress as:

$$\sigma_{Max} = \sqrt{\frac{mv^2E}{AL}} = \sqrt{\frac{mv^2E}{V}} \qquad (15.33)$$

We also considered impact of a beam by a weight W falling through a distance H. If the deflection of the beam was small relative to H, then the maximum stress is given by:

$$(\sigma_{xx})_{Max} = \sqrt{\frac{18EWH}{Lbh}} = \sqrt{\frac{18EWH}{V}} \qquad (15.37)$$

When the displacement of the beam is not negligible relative to the height H of the falling weight, both the stress and the displacement is increased. The relation for the displacement is given by:

$$\delta = \delta_{st}\left(1+\sqrt{1+2\left(\frac{H}{\delta_{st}}\right)}\right) \qquad (15.38)$$

where $\delta_{st} = \dfrac{WL^3}{48EI_z}$ and $\left(1+\sqrt{1+2\left(\dfrac{H}{\delta_{st}}\right)}\right)$ is a multiplier **M** on the static deflection.

Castigliano's theorem and different variations of Eq. (15.6) were applied to four different types of problems, which included determining the displacement of beams under an applied load, solving for unknown reactions in statically indeterminate beams, determining rotations in curved beams and shafts, and determining displacements of select pins in trusses. These applications were demonstrated with many example problems.

PROBLEMS

15.1 An aluminum rod with a cross sectional area A = 400 mm^2 and length L = 1.0 m is subjected to an axial force of 32.5 kN as shown in the figure to the right. Determine the work **W** performed on the rod and the strain energy stored in the rod.

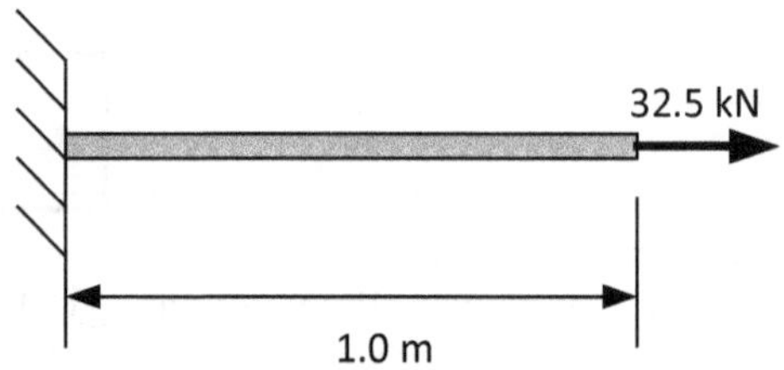

15.2 A point in a machine component is subjected to a three-dimensional state of stress that is described in terms of three principal stresses σ_1 = 120 MPa, σ_2 = 85 MPa and σ_3 = – 99.4 MPa. Determine the strain energy density at this point if the machine component is fabricated from steel.

15.3 A point in a structural member is subjected to a three dimensional state of stress described by the six Cartesian stress components σ_{xx} = 60.2 MPa, σ_{yy} = 51.1 MPa, σ_{zz} = 34.2 MPa, τ_{xy} = – 25.3 MPa, τ_{yz} = 48.7 MPa and τ_{zx} =38.8 MPa. Determine the strain energy density at this point if the structural member is fabricated from brass.

15.4 A tie rod, 25 ft long with a 10 in. diameter is fabricated from steel. If the tie rod is subjected to an axial force of 1570 ton, determine the strain energy stored in the tie rod and the strain energy density.

15.5 For the simply supported beam shown in the figure below, determine the strain energy stored in it due to the normal stresses σ_{xx} produced by the uniformly distributed load q = 500 N/m. The beam is fabricated from steel.

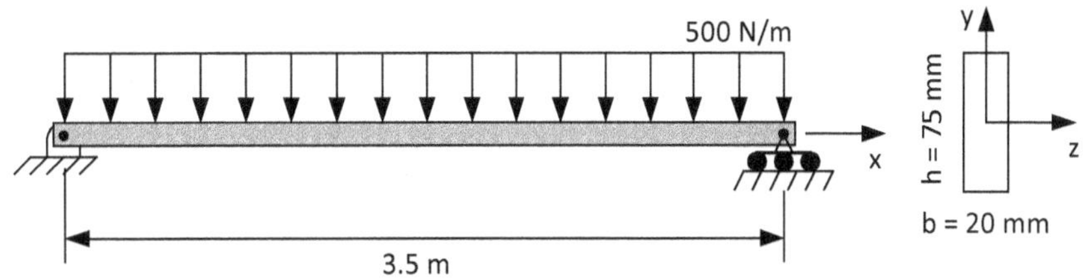

15.6 For the simply supported beam shown in the figure below, determine the expression for the strain energy due to the normal stress σ_{xx}. Assume that EI_z is a constant.

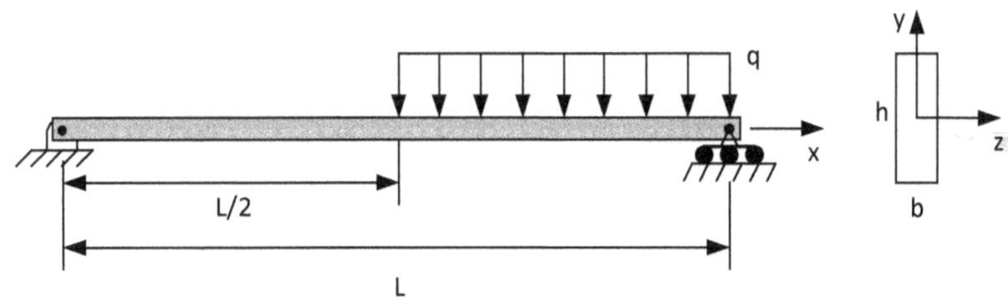

15.7 For the simply supported beam shown in the figure below, determine the expression for the strain energy due to the normal stress σ_{xx}. Assume that EI_z is a constant.

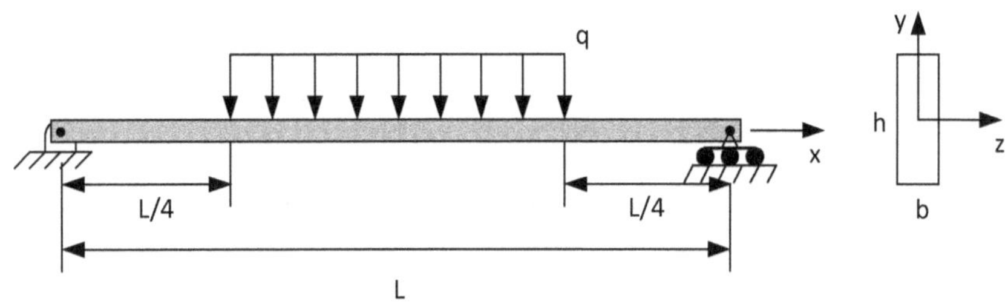

15.8 For the simply supported beam shown in the figure below, determine the expression for the strain energy due to the normal stress σ_{xx}. Assume that EI_z is a constant.

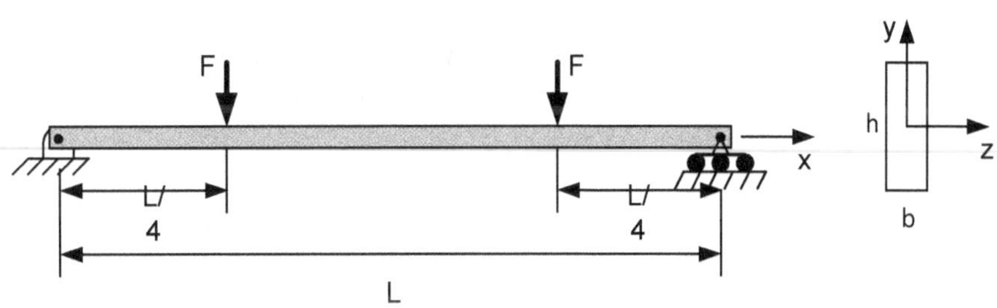

15.9 For the simply supported beam shown in the figure below, determine the expression for the strain energy due to the normal stress σ_{xx}. Assume that EI_z is a constant.

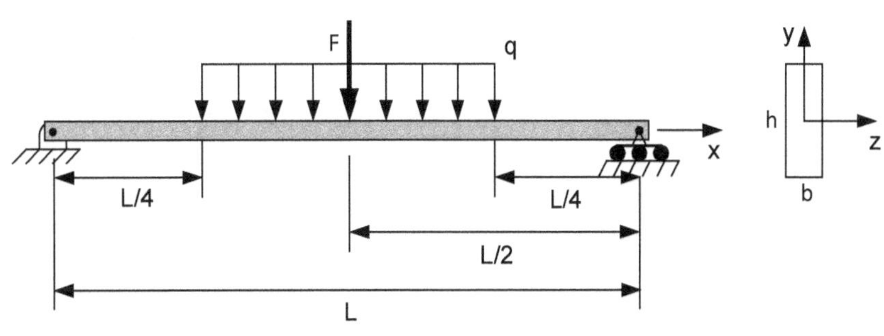

15.10 For the cantilever beam shown in the figure below, determine the expression for the strain energy due to the normal stress σ_{xx}. Assume that EI_z is a constant.

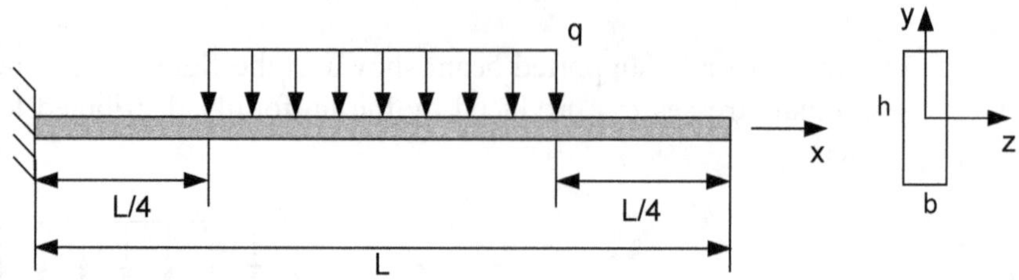

15.11 For the cantilever beam shown in the figure below, determine the expression for the strain energy due to the normal stress σ_{xx}. Assume that EI_z is a constant.

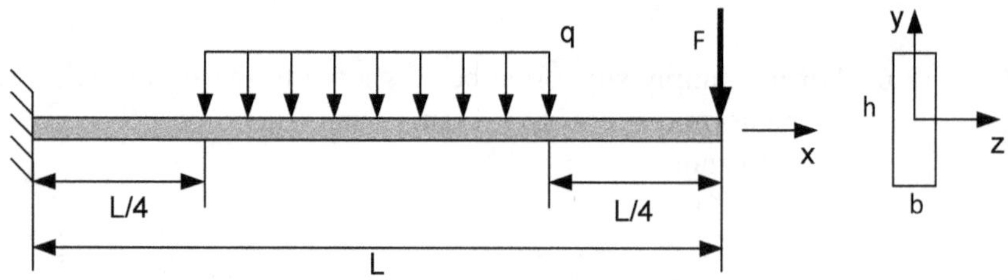

15.12 For the simply supported beam shown in the figure below, determine the relation for the strain energy stored due to the shear stress τ_{xy}. The beam has a rectangular cross section with dimensions h and b.

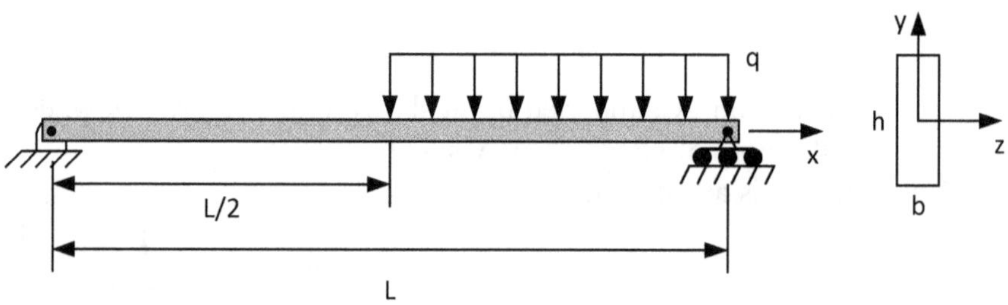

15.13 For the simply supported beam in shown in the figure below, determine the relation for the strain energy stored due to the shear stress τ_{xy}. The beam has a rectangular cross section with dimensions h and b.

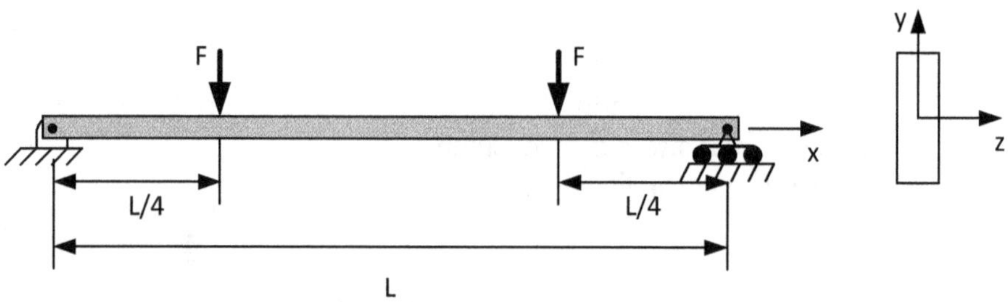

15.14 For the simply supported beam shown in the figure below, determine the relation for the strain energy stored due to the shear stress τ_{xy}. The beam has a rectangular cross section with dimensions h and b.

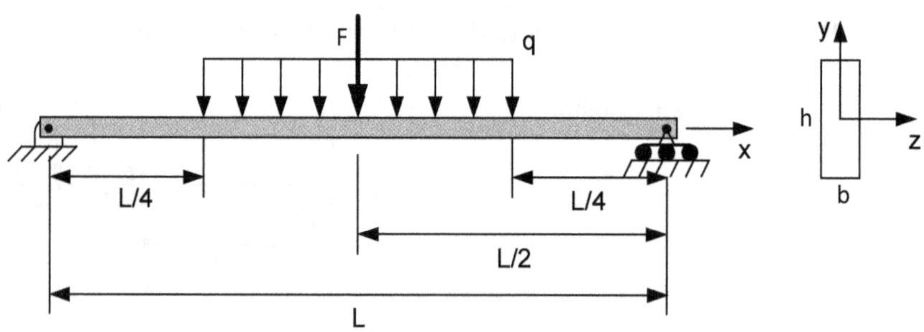

15.15 For the cantilever beam in shown in the figure below, determine the relation for the strain energy stored due to the shear stress τ_{xy}. The beam has a rectangular cross section with dimensions h and b.

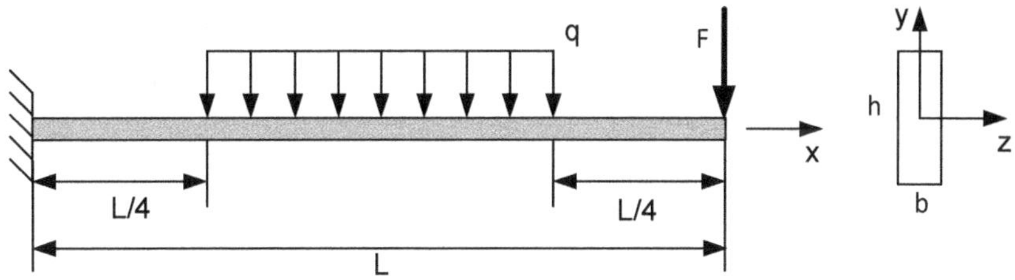

15.16 A 3.0 m long shaft with a 75 mm diameter, shown in the figure below, is subjected to a torque T = 450 N-m. Compute the strain energy stored in the shaft. The shaft is fabricated from brass.

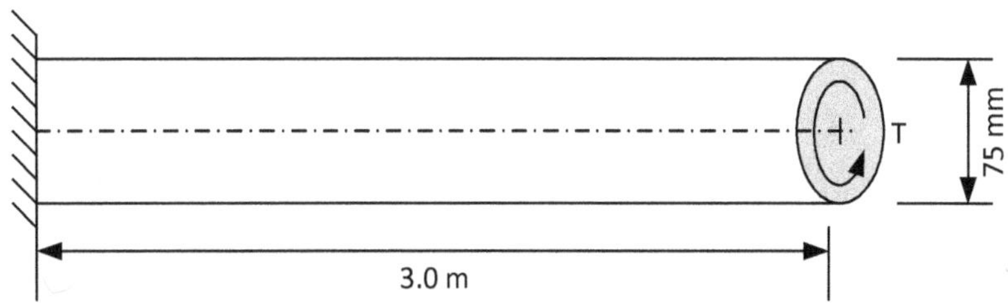

15.17 A stepped shaft, with lengths and diameters defined in the figure below, is subjected to a torque T = 110 ft-lb N-m. Compute the strain energy stored in the shaft. The shaft is fabricated from stainless steel.

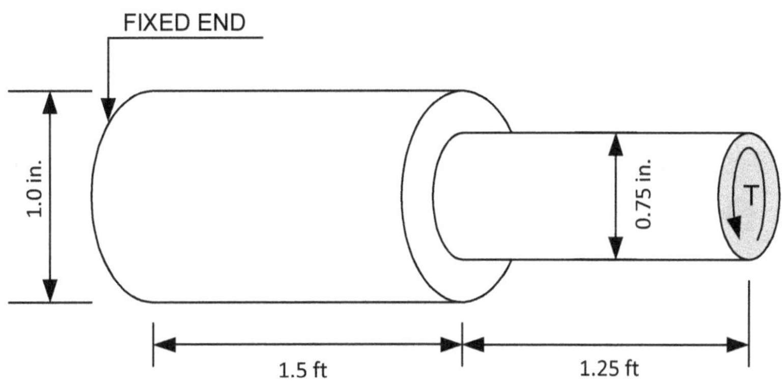

15.18 A sphere fired from an air gun strikes an aluminum rod with an axial impact as illustrated in the figure below. Determine the maximum force and the maximum stress produced if the sphere weighs 150 gm. The velocity of the sphere at the instant of impact is 200 m/s. The rod has a solid, circular cross section.

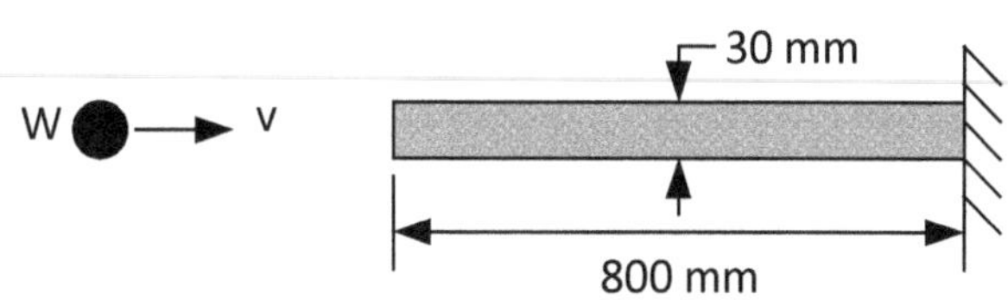

15.19 A short, steel cylinder of diameter d = 0.750 in. and length l = 2.25 in. is fired from an air gun and strikes a long, steel rod with an axial impact as shown in the figure below. Determine the maximum force and the maximum stress produced if the cylinder has a density of γ = 0.284 lb/in.^{3} The velocity of the cylinder at the instant of impact is 500 ft/s. The rod has a solid, circular cross section.

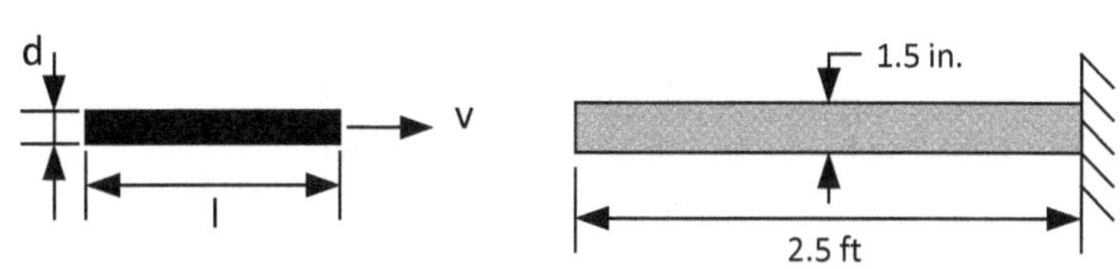

15.20 Consider a simply supported beam that is subjected to impact by a falling weight W as illustrated in the figure to the right. Derive the relation for the maximum force and the maximum bending stress in the beam. Assume that the deflection of the beam is small relative to the drop height H.

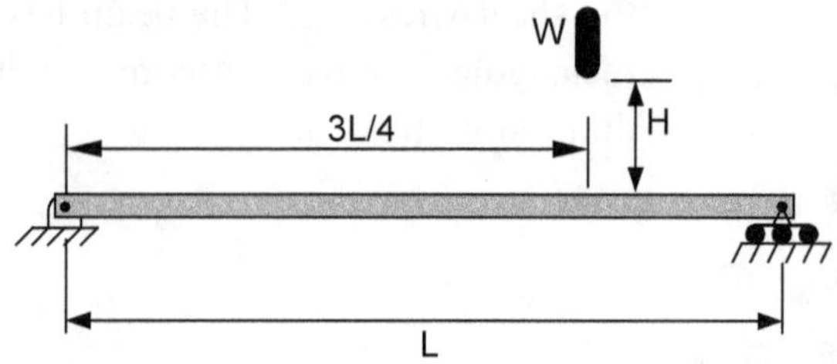

15.21 A weight W is positioned at the free end of a cantilever beam as shown in the figure to the left. While the weight is in contact with the beam (H = 0), its gravitational force has not been released. The weight is suddenly released to produce an impact event, although the height of the drop is zero. Determine the equations for the maximum stress, the displacement of the load point and the maximum force developed during the impact.

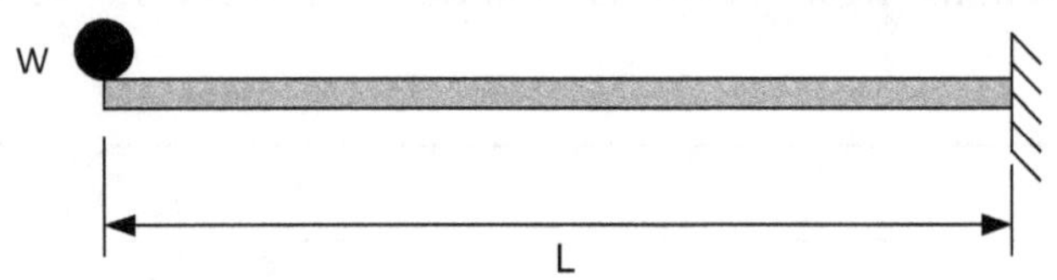

15.22 For the simply supported steel beam shown in the figure to the right, determine the maximum force and the maximum stress when the weight impacts the beam. The beam is fabricated from a W 12 × 50 wide flange section.

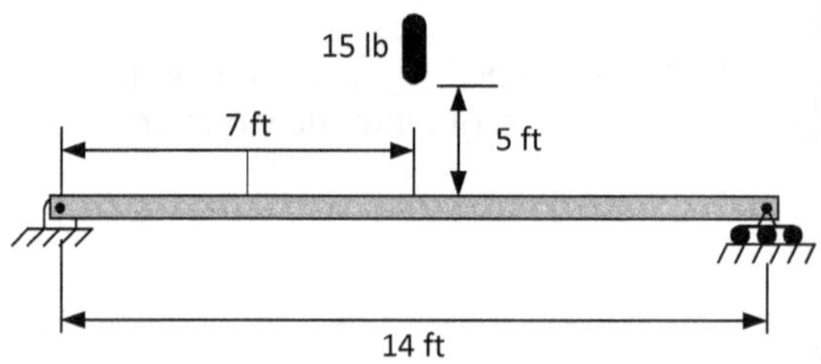

15.23 For the aluminum alloy cantilever beam shown in the figure to the left, determine the maximum force and maximum stress when the mass impacts the beam. The beam is fabricated from a S 127 × 22 American Standard section.

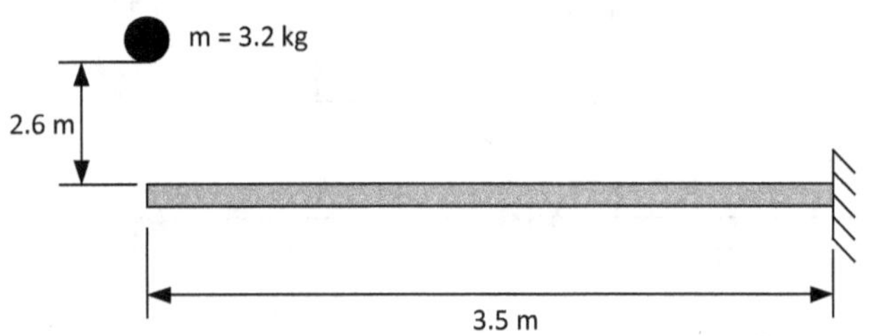

15.24 For the simply supported beam shown in the figure to the right, write the equation for the deflection at the quarter span (x = L/4) using Castigliano's theorem.

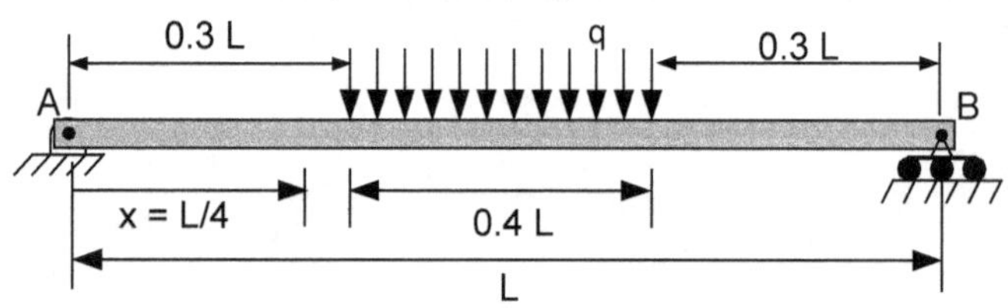

15.25 For the cantilever beam shown in the figure to the left, write the equation for the deflection at its free end using Castigliano's theorem.

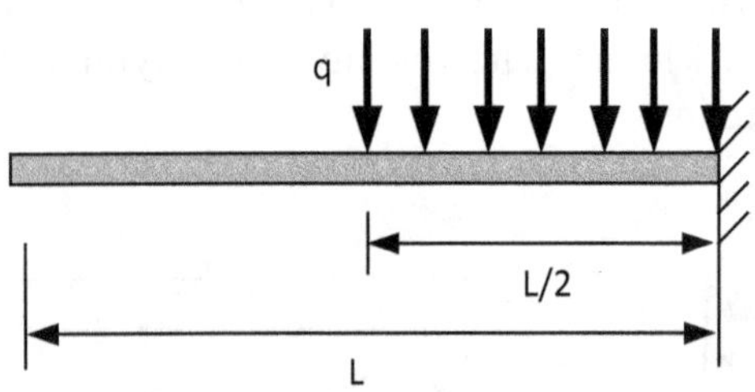

15.26 For the cantilever beam shown in the figure to the right, write the equation for the deflection at midspan and its free end using Castigliano's theorem.

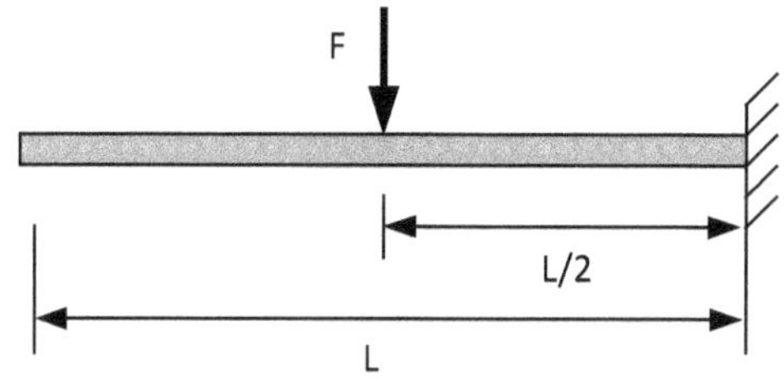

15.27 For the simply supported beam shown in the figure to the left, use Castigliano's theorem to write the equation for the deflection at the midpoint of the span.

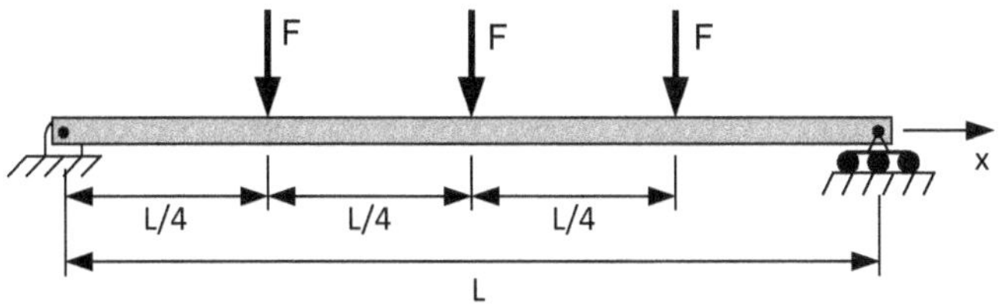

15.28 For the simply supported beam shown in the figure to the right, write the equation for the deflection at midspan and both quarter points using Castigliano's theorem.

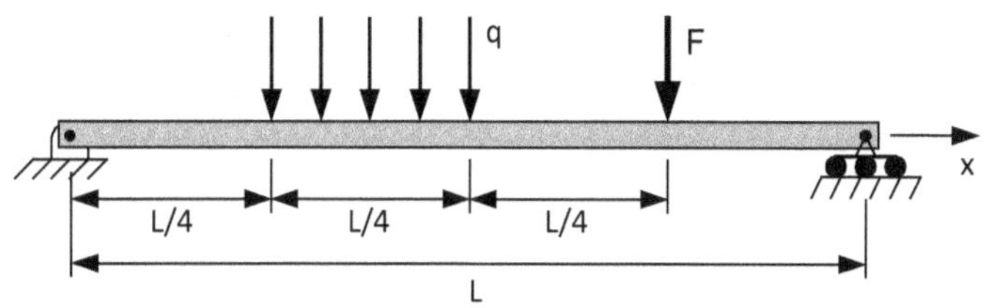

15.29 For the cantilever beam shown in the figure to the left, write the equation for the deflection at midspan and its free end using Castigliano's theorem.

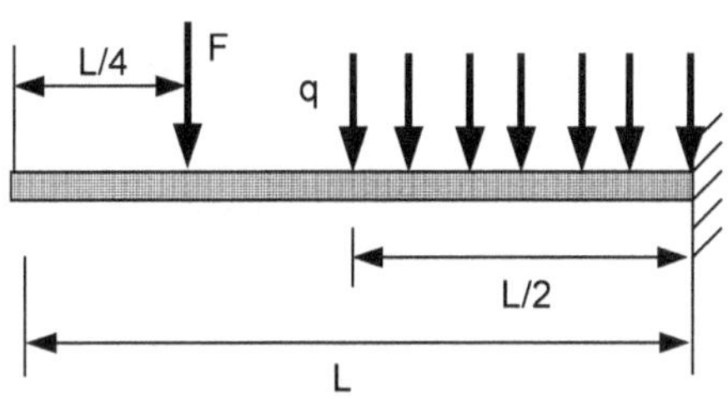

15.30 For the cantilever beam shown in the figure to the right, use Castigliano's theorem to write the equation for the reaction force at the prop.

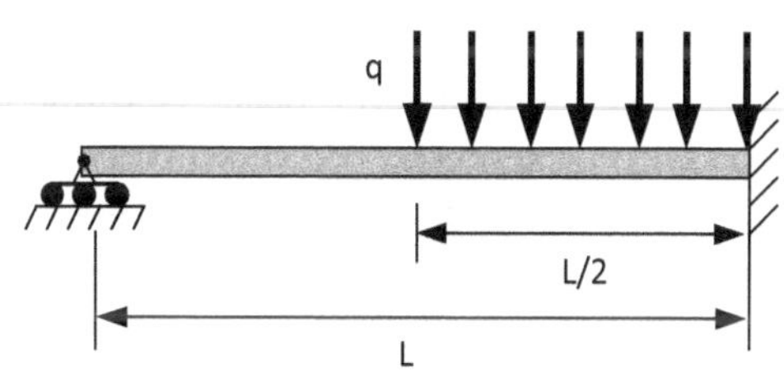

15.31 For the cantilever beam shown in the figure to the left, use Castigliano's theorem to write the equation for the reaction force at the prop.

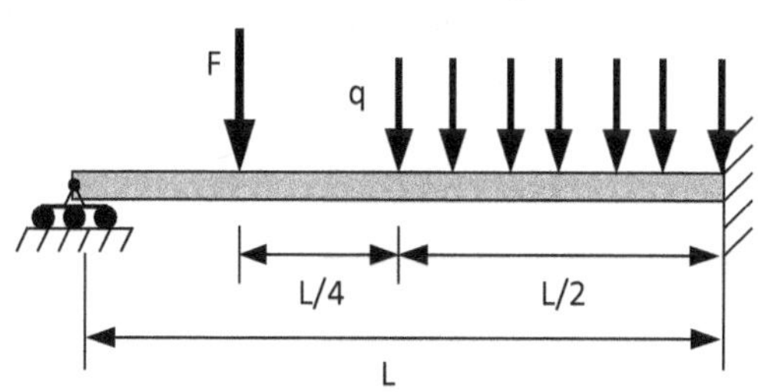

15.32 For the continuous beam shown in the figure to the right, use Castigliano's theorem to write the equation for the reaction force at the center support.

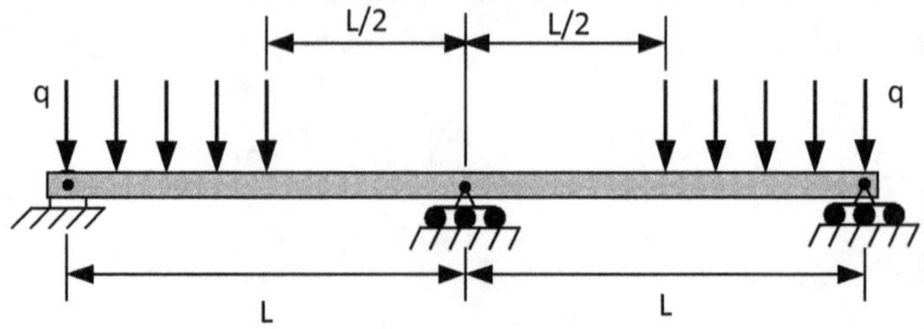

15.33 For the continuous beam, shown in the figure to the left, use Castigliano's theorem to write the equation for the reaction force at the center support.

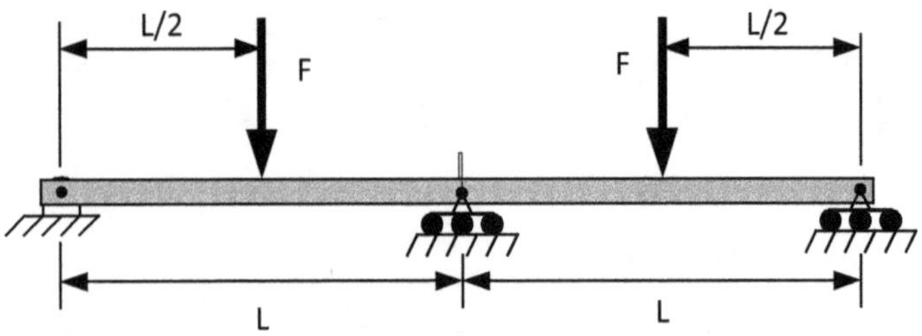

15.34 A cantilever beam is formed in the shape of a quarter circle as illustrated in the figure to the right. If the cantilever beam is loaded with a vertical force F at its free end, write the equation for the beam's deflection in the direction of the applied force. Assume that EI_z is a constant.

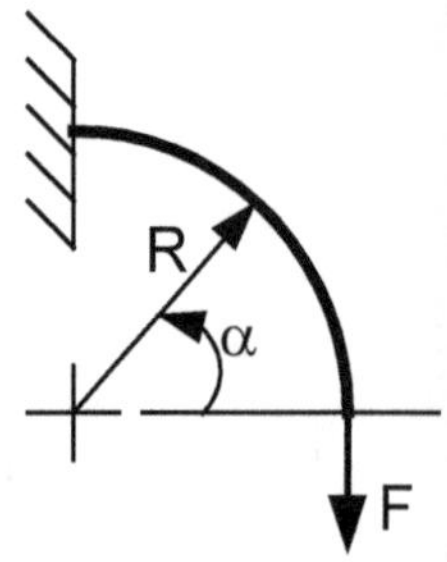

15.35 A cantilever beam is formed in the shape of a quarter circle as illustrated in the figure to the right. If the cantilever beam is loaded with a horizontal force F at its free end, write the equation for the beam's deflection in the direction of the applied force. Assume that EI_z is a constant.

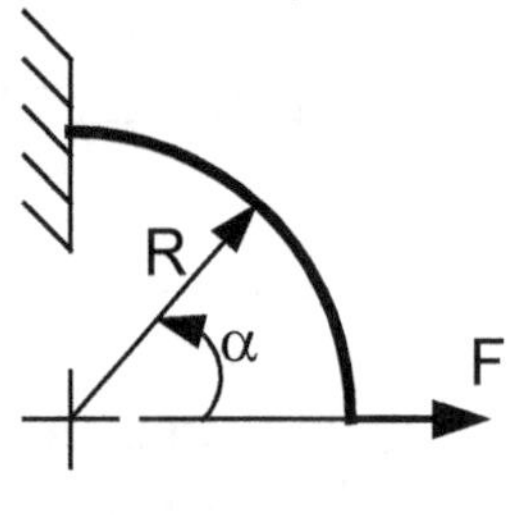

15.36 A cantilever beam is formed in the shape of a semicircle as illustrated in the figure to the right. If the cantilever beam is loaded with a horizontal force F at its free end, write the equation for the beam's deflection in the direction of the applied force. Assume that EI_z is a constant.

15.37 A cantilever beam is formed in the shape of a semicircle as illustrated in the figure to the right. If the cantilever beam is stretched by an amount δ in the horizontal direction, determine the force that must be applied to produce this deflection.

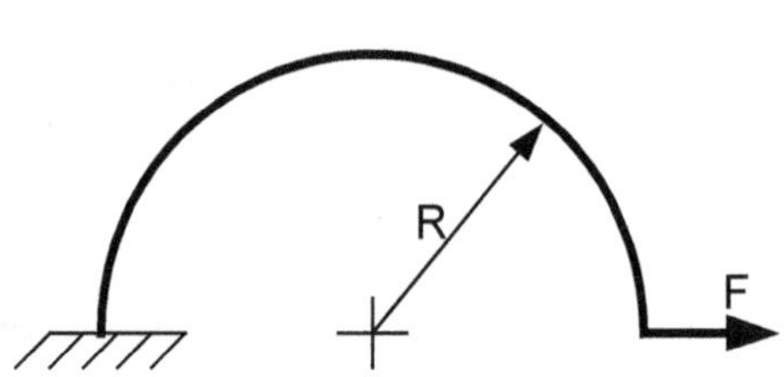

15.38 A shaft is fabricated by brazing together two sections of round tubing as shown in the figure below. The shaft is fixed at one end and a torque T is applied to it at its free end. Use Castigliano's theorem to write the equation for the angle of twist ϕ. Assume the thickness of the braze metal is negligible.

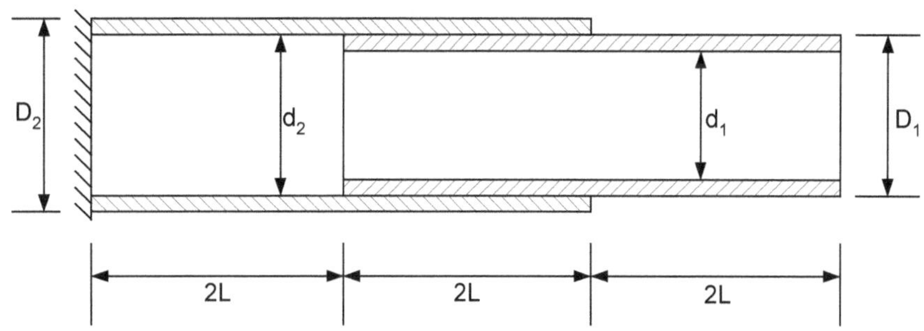

15.39 A shaft is fabricated by brazing together two sections of round tubing as shown in the figure below. The shaft is fixed at one end and a torque T is applied to it at its free end. Use Castigliano's theorem to write the equation for the angle of twist ϕ. Assume the thickness of the braze metal is negligible.

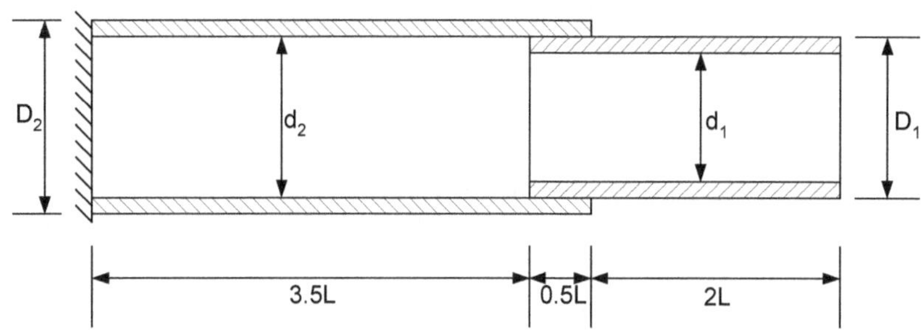

15.40 Write the equation for the displacement in the (a) vertical and (b) horizontal direction at joint B for the two-member truss, shown in the figure to the right, using Castigliano's method. Truss members AB and BC are identical of the same material, length L, and area A. The ratio of h/w is 4/3.

15.41 Determine the displacement in the (a) vertical and (b) horizontal direction at joint B for the two-member truss, shown in the figure to the right, using Castigliano's method. Truss members AB and BC are identical of the same material, length L, and area A. The ratio h/w = 1.00.

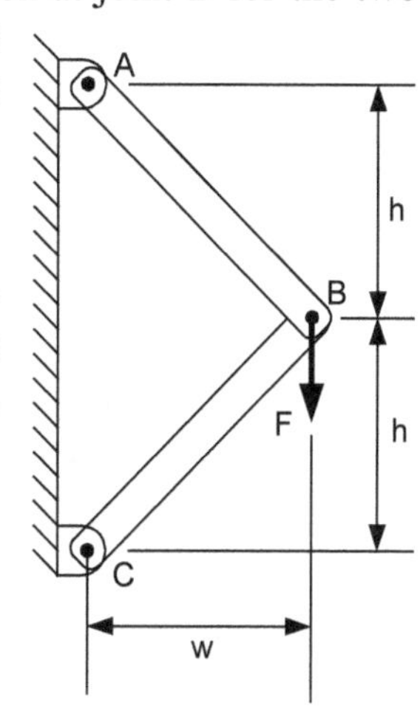

15.42 Write the equation for the displacement in the vertical direction at joint C for the seven-member truss, shown in the figure to the left, using Castigliano's method. All of the truss members are fabricated from the same material. The cross sectional area of members BC, AE, BD and DE is 2A; the area of AB is A; and the area of members BE and CD is 3A.

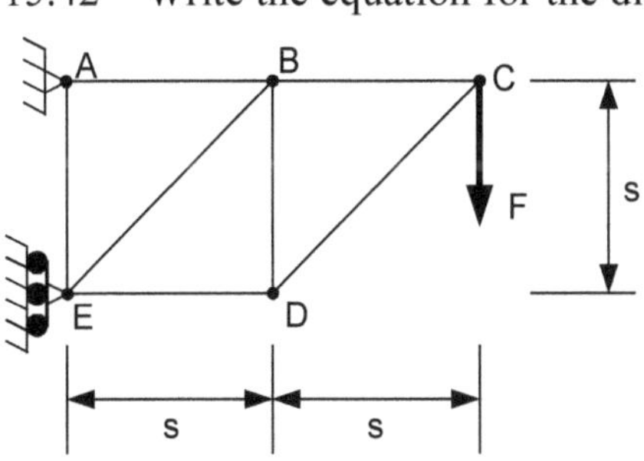

15.43 Write the equation for the displacement in the vertical direction at joint D for the five-member truss, illustrated in the figure to the right, using Castigliano's method. All of the truss members are fabricated from the same material with the same cross sectional area of A. The ratio h/s = 1.5.

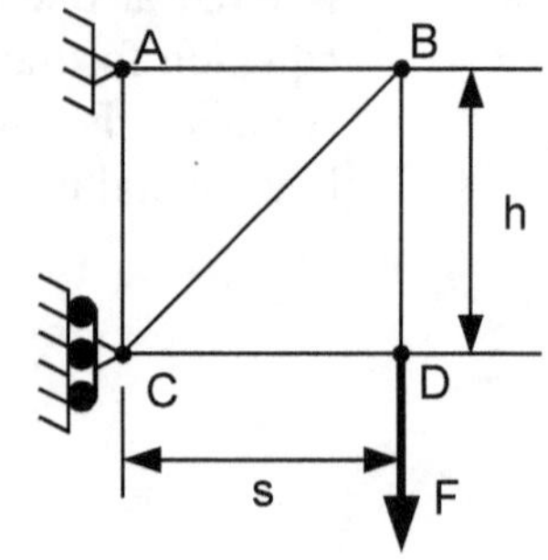

15.44 Write the equation for the displacement in the vertical direction at joint C for the seven-member truss, shown in the figure to the right, using Castigliano's method. All of the truss members are fabricated from the same material with the same cross sectional area of A. The ratio of h/s is 0.85.

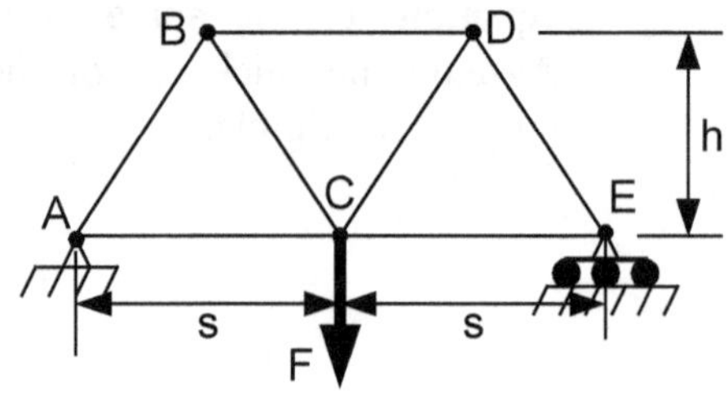

15.45 Write the equation for the displacement in the vertical direction at joint B for the seven-member truss, illustrated in the figure to the right using Castigliano's method. All of the truss members are fabricated from the same material with the same cross sectional area of A. The ratio of h/s is 0.85.

15.46 Write the equation for the displacement in the vertical direction at joint D for the five-member truss, shown in the figure to the right, using Castigliano's method. All of the truss members are fabricated from the same material with the same cross sectional area of A. The ratio of h/s is 1.20.

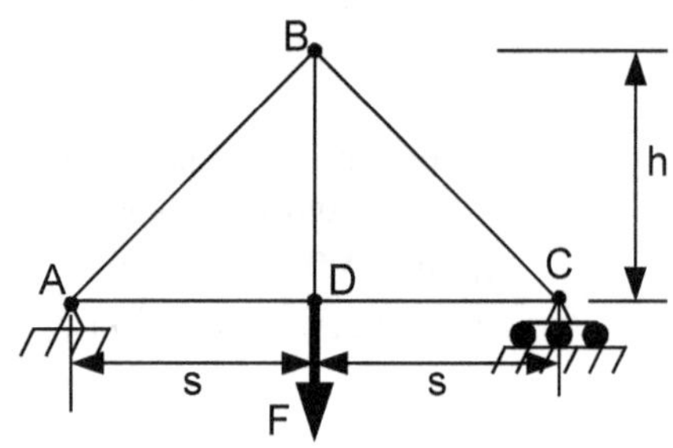

15.47 Write the equation for the displacement in the horizontal direction at joint C for the five-member truss, illustrated in the figure to the right, using Castigliano's method. All of the truss members are fabricated from the same material with the same cross sectional area of A. The ratio of h/s is 1.20.

CHAPTER 16

FRACTURE MECHANICS

16.1 INTRODUCTION

In the all of the previous chapters, theories were described to show procedures for determining stresses and strains in structural members that were free of flaws. However, when flaws such as cracks exist in the structure, these theories are not sufficient to predict the onset of failure. The difficulty is due to the geometry at the crack tip. The crack tip is sharp with a radius of curvature approaching zero and this sharpness produces local stresses σ_{xx}, σ_{yy} and τ_{xy}, which tend to infinity as one approaches the crack tip. Because these stresses approach infinity for any loading of the structure, the theories of failure, such as Tresca or Von Mises, cannot be applied and the load required producing either localized yielding or the onset of crack propagation cannot be predicted.

To analyze structural members containing cracks, it is necessary to introduce a method that deals with the singular state of stress at the crack tip, where the stresses approach infinity for even the very small loads. Fracture mechanics, developed by Irwin [1] from the earlier work of Inglis [2], Griffith [3], and Westergaard [4], treats singular stress fields by introducing a quantity known as a **stress intensity factor** K_I defined as:

$$K_I = \lim_{r \Rightarrow 0}(\sqrt{2\pi r}\,\sigma_{yy}) \quad (16.1)$$

The coordinate system for Eq. (16.1) is defined in Fig. 16.1 and σ_{yy} is evaluated in the limit along the $\theta = 0°$ line.

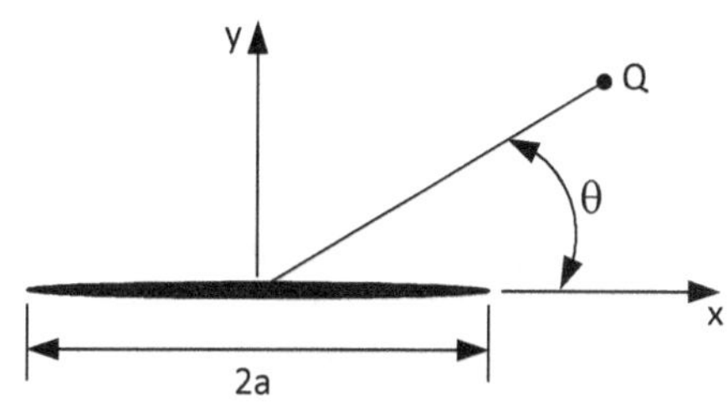

Fig. 16.1 Coordinate system defining a double-ended crack.

The limit process in Eq. (16.1) gives a stress intensity factor K_I that is a linear function of the loads applied to the structure. The stress intensity factor remains finite and provides a basis for determining the critical load. The critical condition in fracture mechanics is the onset of crack initiation when the crack extends suddenly at high velocity in cleavage failure or when the crack extends at a much lower velocity by tearing in a shear failure. With either type of crack extension, the structure has failed. The introduction of the stress intensity factor K_I provides a means, which is dependent upon the applied state of stress (i.e. σ_o) and the crack length, for judging the criticality of a crack.

The second and equally important aspect of fracture mechanics is the characterization of a material property that defines the onset of crack initiation. This property is termed **toughness,** and it is determined using well-defined testing procedures. For example, **the plane strain crack initiation toughness** K_{Ic} defines the resistance of the material to crack initiation.

$$K_I \geq K_{Ic} \quad (16.2)$$

When the applied stress intensity factor K_I equals or exceeds the plane strain crack initiation toughness K_{Ic}, the crack initiates and extends into the structure. Because K_I is a linear function of the applied load, say F, the failure criterion given in Eq. (16.2) provides an approach for predicting the critical load F_{CR} associated with crack initiation. This concept is illustrated in Fig. 16.2.

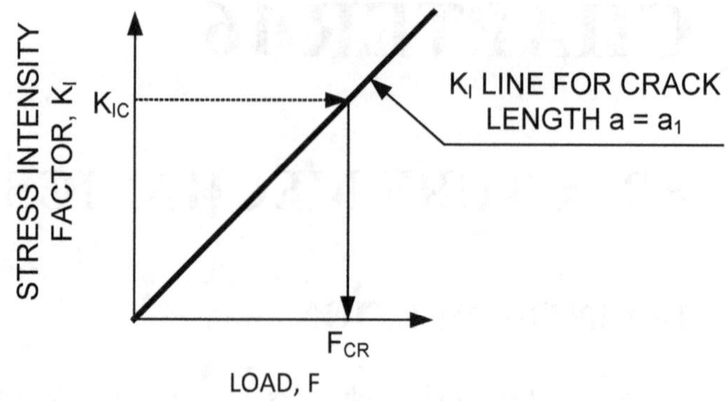

Fig. 16.2 An approach for predicting the critical load F_{CR} associated with crack initiation.

Before beginning a study of fracture mechanics, it is important to understand singular stress fields, where all of the Cartesian components of stress tend to infinity at the crack tip while the loads applied to the structural element remain finite. To better describe this singular stress state, let's consider the solution for a wide plate under uniaxial load with an elliptical hole.

16.2 STRESSES DUE TO AN ELLIPTICAL HOLE IN A UNIFORMLY LOADED PLATE

In previous chapters, we have introduced the concept of stress concentration factors where discontinuities in a structural element, such as holes and notches, elevate the stresses in the vicinity of the discontinuity. We treated the increase in the stresses by introducing a stress concentration factor K that was a multiplier on the nominal stress as indicated in Eq. (16.3).

$$\sigma_{Max} = K\,\sigma_{Nom} \tag{16.3}$$

where σ_{Max} represents the maximum stress adjacent to the discontinuity, K is the stress concentration factor that depends on the geometry of the discontinuity, the structural element, and the type of loading and σ_{Nom} is the nominal stress at some location removed from the discontinuity.

In typical structural applications, where discontinuities are inevitable, stress concentration factors usually range from 2 to 5. The stresses are elevated, but they remain finite. Theories of failure can be applied to predict the onset of local yielding with considerable success.

Let's now consider a discontinuity that produces much higher stress concentrations than the typical values referenced above. Suppose we have a very wide plate with a centrally located elliptical hole subjected to uniaxial loading as shown in Fig. 16.3.

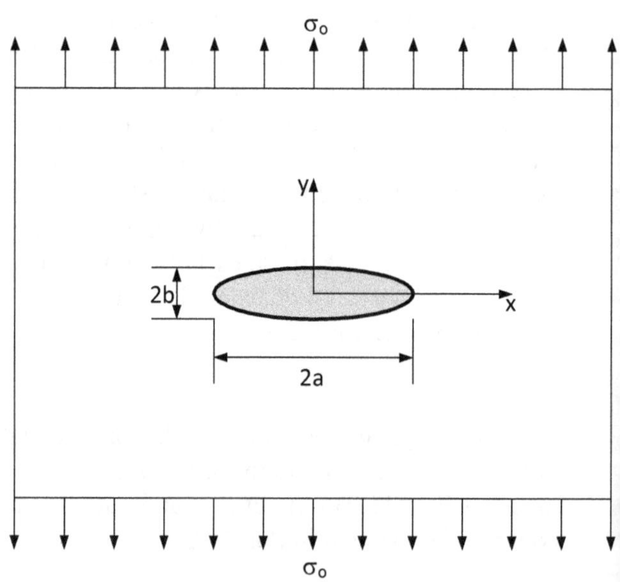

Fig. 16.3 Elliptical hole in an infinite plate with a uniform stress σ_o applied perpendicular to the crack.

Inglis considered an infinite plate with an elliptical hole subjected to a uniform stress σ_o, as shown in Fig. 16.3. He found that σ_{Max} occurs at the ends of the major axis (± a) of the ellipse and is given by:

$$\sigma_{Max} = \sigma_0\left(1+\frac{2a}{b}\right) \tag{16.4}$$

It is instructive to examine the results of Eq. (16.4) for the two limit cases. First, when a = b, the elliptical hole becomes circular and $\sigma_{Max} = 3\sigma_o$. This result confirms the stress concentration of K = 3 for a circular hole in an infinite plate with a uniaxial load as indicated in Chapter 4. Second, when $b \Rightarrow 0$, the elliptical hole approaches a flat opening that represents a crack. In this case, Eq. (16.4) shows that $\sigma_{Max} \Rightarrow \propto$ as $b \Rightarrow 0$. Note that σ_{Max} goes to infinity regardless of the magnitude of the applied stress σ_o. This fact indicates that localized yielding will occur at the crack tips for any nonzero load. The commonly employed failure theories such as Tresca and von Mises predict yielding for any level of load; therefore, they cannot provide useful information regarding the stability of the crack as the applied stress σ_o is increased from zero to some value sufficient for crack initiation.

When the applied stress σ_o is applied in a direction parallel to the major axis of the elliptical hole, the value of σ_{Max} on the boundary of the hole is at the ends of the minor axes (± b) and is given by:

$$\sigma_{Max} = \sigma_0\left(1+\frac{2b}{a}\right) \tag{16.5}$$

In the limit as $b \Rightarrow 0$, the ellipse represents a crack and the stress $\sigma_{Max} = \sigma_o$; thus, concentration of the stress does not occur. At the ends of the major axis of the elliptic hole, Eq. (16.5) does not apply; however, $\sigma = -\sigma_o$ at these two points for all values of b/a.

Equations (16.4) and. (16.5) describe the stresses at the tip of a crack in the limit as $b \Rightarrow 0$. It is evident that the stresses at the crack tip are singular when the crack is perpendicular to the applied stress σ_o. This fact indicates local yielding will occur for any nonzero stress σ_o and that methods for predicting structural stability based on classical failure theories for yielding are inadequate. We must introduce concepts of crack stability based on fracture mechanics to predict if the crack will initiate at a specified value of the applied stress σ_o.

EXAMPLE 16.1

If we define the stress concentration for the elliptical hole in a very wide (infinite) plate as:

$$K = \left(1+\frac{2a}{b}\right) \tag{a}$$

Prepare a graph of the stress concentration K as $b \Rightarrow 0.05$ units.

Solution:

Let's arbitrarily set a = 1 unit and permit b to decrease from 1 to 0.05 in steps of 0.05 units. Evaluating Eq. (a) in a spreadsheet and preparing a graph of the results yields Fig. E16.1.

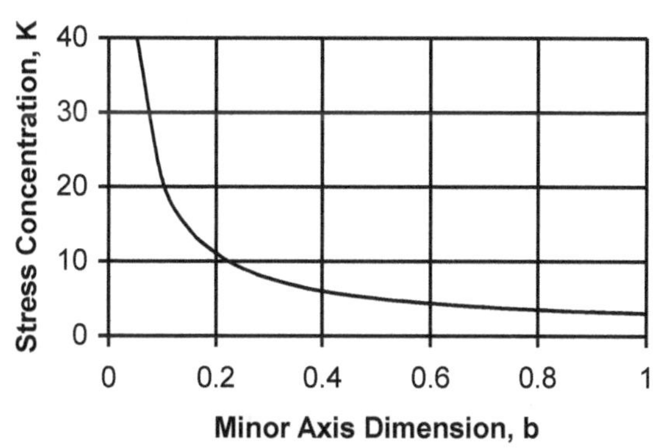

Fig. E16.1

The stress concentration K increases exponentially to a value exceeding 40 as the minor dimension of the ellipse approaches 0.05 unit.

EXAMPLE 16.2

Let's repeat the calculation of the stress concentration factor defined in Eq. (16.4), but in this case explore the value of K as the dimension b decreases from 0.05 to 0.001 unit.

Solution:

Let's arbitrarily set a = 1 unit and permit b to decrease from 0.05 unit to 0.001 unit. Evaluating the stress concentration K and preparing a graph of the results in a spreadsheet yields Fig. E16.2.

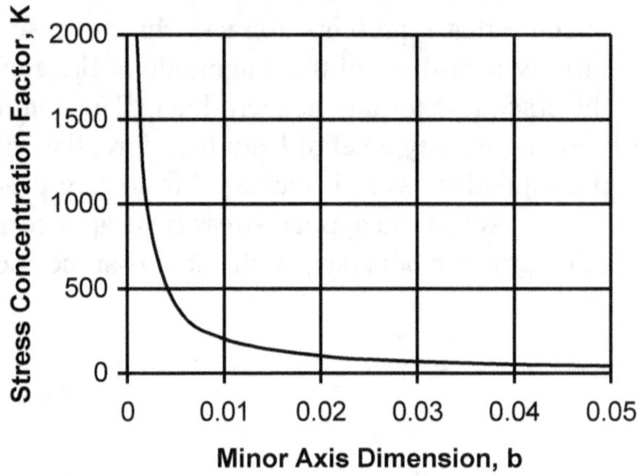

Fig. E16.2

The stress concentration K increases exponentially to a value exceeding 2000 as the minor dimension of the ellipse approaches 0.001 unit.

As the elliptical hole is collapsed into a crack by letting the dimension of the minor axis $b \Rightarrow 0$, it is clear that the stress σ_{Max} becomes infinite. These two examples illustrate that point with numerical values of the stress concentration as the elliptical hole is collapsed. Because of the stress singularity, traditional methods of predicting failure are not applicable. This dilemma baffled engineers responsible for designing safe structures for many years and it was not until the early 1950s that we began to understand a method for dealing with the stress singularities that exist at crack tips.

16.3 STRESSES NEAR THE CRACK TIP

Westergaard [4] employed advanced theories of elasticity and mathematics to develop a solution for the stress near a crack tip. While this development is beyond the scope of this book, we will use his results to develop an expression for the stress intensity factor.

The classic problem in fracture mechanics, shown in Fig 16.4a, is an infinite plate with a central crack of length 2a. The plate is subjected to biaxial tension with σ_o applied in both the x and y directions at the remote boundaries. It is more convenient to relocate the origin of the coordinate system and place it at the tip of the crack as shown in Fig. 16.4b.

Next, consider a small region near the crack tip where $r \ll a$. Then, the Westergaard results enable us to write an equation for σ_{yy} that is valid along the crack line where θ and y are both equal to zero.

$$\sigma_{yy} = \sqrt{\frac{a}{2r}}\,\sigma_0 \qquad (16.6)$$

Equation (16.6) shows that the stress $\sigma_{yy} \Rightarrow \propto$ and is singular with order $1/\sqrt{r}$ as we approach the crack tip along the x-axis. Finally, Eq. (16.6) can be substituted into Eq. (16.1). Taking the limit as $r \Rightarrow 0$ in accordance with Eq. (16.1) gives the stress intensity factor K_I as:

$$K_I = \sqrt{\pi a}\ \sigma_0 \qquad (16.7)$$

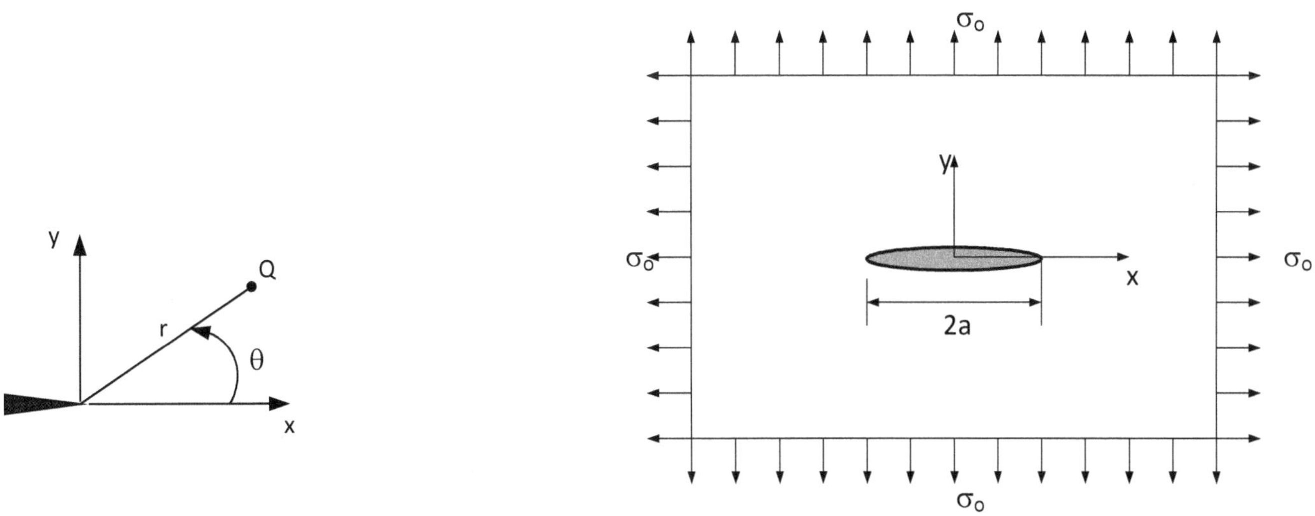

(a) (b)

Fig. 16.4 (a) Coordinates defined with the origin at the crack tip and (b) A crack of length 2a in an infinite plate subjected to biaxial stress σ_o,.

This result shows that the stress intensity factor K_I varies as a linear function of the applied stress σ_o and increases with crack length as a function of $\sqrt{a}$, as illustrated in Fig. 16.5.

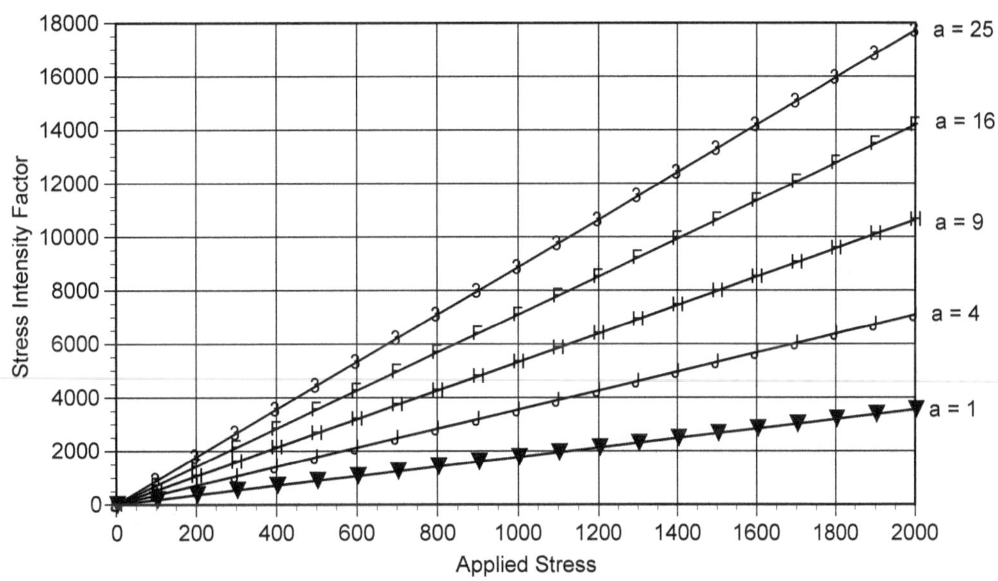

Fig. 16.5 Stress intensity factor K_I as a function of applied stress σ_o.

The stress intensity factor K_I and the stress concentration factor K are two totally different quantities and they should not be confused. The stress concentration factor K is used to predict the increase in stresses due to geometric discontinuities that lead to elevated but finite stresses. The stress intensity factor K_I is used to accommodate the stresses at crack tips, which are singular in nature. The units of the two quantities differ. The stress concentration factor is dimensionless. The stress intensity factor has units of psi $\sqrt{in}$ or MPa $\sqrt{m}$.

EXAMPLE 16.3

The quality control inspectors discover a crack in the weld of a wide steel plate that is to be subjected to biaxial loading with a stress σ_o = 8,500 psi. They estimate that the crack is 5.6 in. long. You are asked to determine the stress intensity factor.

Solution:

We will employ Eq. (16.7) to estimate the stress intensity associated with the weld crack as:

$$K_I = \sqrt{\pi a}\ \sigma_0 = \sqrt{\pi \frac{5.6}{2}}(8{,}500) = 25.21 \text{ ksi}\sqrt{\text{in.}} \qquad \text{(a)}$$

At this stage in the development, we cannot interpret the significance of this result. Clearly, a comparison to some material property is necessary before an assessment of the stability of the weld crack can be made. We will discuss material properties and crack initiation toughness in the next section.

EXAMPLE 16.4

Your supplier of large aluminum castings indicates that cracks up to 40 mm long may be generated as the castings cool to room temperature. The castings are subjected to loads that produce a stress of σ_o = 72 MPa. You are asked to estimate a stress intensity factor for the worst case.

Solution:

In the worst case, the crack will be perpendicular to the applied stress and located away from the boundaries of the casting; hence, difficult for the inspectors to locate. We will assume that the result of Eq. (16.7) represents the worst-case scenario. Then the estimate of the stress intensity factor associated with the cooling crack is given by:

$$K_I = \sqrt{\pi a}\ \sigma_0 = \sqrt{\pi(20)}(72) = 570.7 \text{ MPa}\sqrt{\text{mm}} \qquad \text{(a)}$$

Again, we will postpone interpretation of this result until the information on the aluminum's resistance to crack initiation has been introduced.

16.4 FRACTURE TOUGHNESS

In the previous section, we showed that the stress intensity factor K_I could be determined for a wide plate with a central crack. However, we were not able to predict the safety of the structure (wide plate) because the material property defining resistance to crack initiation was not available. We must have a second quantity—the material's fracture toughness—to predict crack behavior.

Fracture toughness K_c is a measure of a material's ability to resist crack initiation. Fracture toughness is a material property, which depends on the type of material (steel, aluminum, titanium, etc.), the alloy, its heat treatment and operating temperature. Fracture toughness also depends on the thickness of the structural element containing the crack or the thickness of the test specimen used in measuring the toughness.

We noted previously that the stresses were singular at the crack tip. Accordingly any non-zero stress applied to the structure should cause the crack to initiate and propagate. Fortunately, **this statement is not true**. Cracks remain stable until the applied stress σ_o is increased to some critical value. At this critical loading the cracks initiate and extend. We describe the condition for crack initiation with:

$$K_I \geq K_c \qquad (16.8)$$

where K_c is the fracture toughness.

After initiation, the crack often propagates at high speed across the width of the structural element causing catastrophic failure. Cracks are stable when the applied stress σ_o produces a stress intensity factor K_I that is:

$$K_I < K_c \qquad (16.9)$$

The reason cracks are stable, when the stresses at the crack tip are singular, is the presence of local plastic deformation at the crack tip. When the stresses increase at the crack tip, localized yielding occurs in the immediate neighborhood of the crack tip. A plastic zone is formed and the crack tip becomes blunted as illustrated in Fig. 16.6.

The size and shape of the plastic zone depends on the magnitude of the applied stress σ_o and the material thickness. As the material thickness is increased the material in the center of the plate is constrained, the size of the plastic zone diminishes, and the blunting of the crack tip is less pronounced. These changes affect the material toughness K_c as shown in Fig. 16.7. As the specimen thickness increases the fracture toughness K_c decreases until it reaches a limiting value K_{Ic}. This limiting value K_{Ic} is known as the **plane strain fracture toughness**. This material property is often used to predict the safety of structures because it provides the most conservative estimates of the allowable loading.

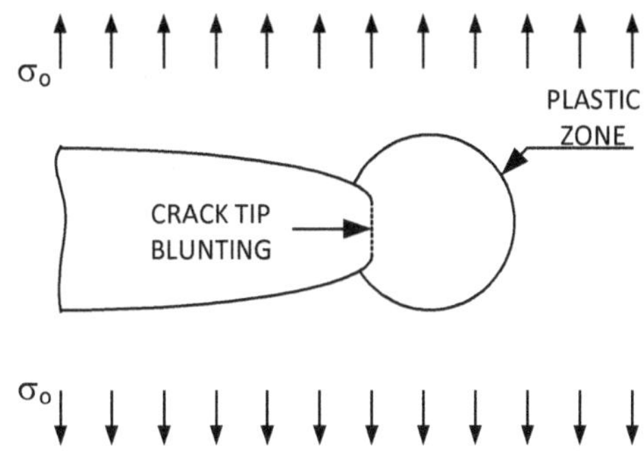

Fig. 16.6 Localized yielding produces a small plastic zone that blunts the crack tip.

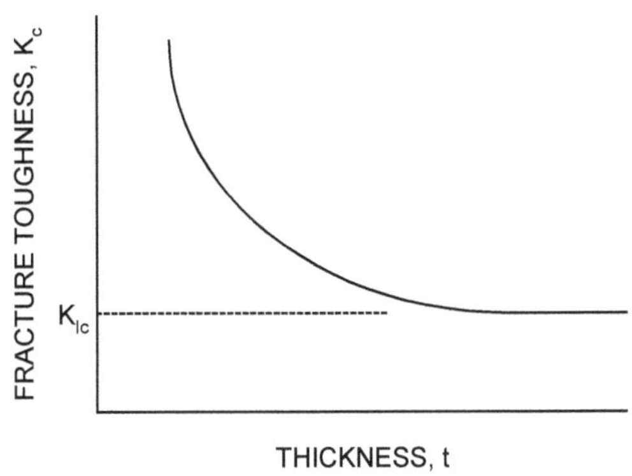

Fig. 16.7 Fracture toughness decreases with increasing specimen thickness.

A table showing the fracture properties of several common engineering materials is presented in Table 16.1. A more complete listing of these properties is presented in a recent book by R. J. Sanford [5].

EXAMPLE 16.5

In Example 16.3, the quality control inspectors discover a 5.6 in. long crack in the weld of a wide steel plate that is subjected to biaxial loading with a stress σ_o = 8,500 psi. We determined the stress intensity factor as $K_I = \sqrt{\pi a}\ \sigma_0 = \sqrt{\pi(2.8)}(8{,}500) = 25.21 \text{ ksi}\sqrt{\text{in.}}$; however, in executing this solution we could not predict the stability of the crack. If the steel used is 1020 HR, predict if the crack is stable or unstable.

Solution:

Let's use the stability condition presented in Eq. (16.9) and write:

$$K_I < K_c \qquad (a)$$

Substituting numerical results from Table 16.1 and Example 16.3 into this equation gives:

$$K_I = 25.21 \text{ ksi}\sqrt{\text{in.}} < K_{Ic} = 70 \text{ ksi}\sqrt{\text{in.}} \qquad \text{(b)}$$

The stress intensity factor is less than the plane strain fracture toughness; hence, the crack is stable.

Table 16.1
Strength and fracture toughness of several engineering materials.

Material	S_y, ksi	S_u, ksi	K_{Ic}, ksi√in.	S_y, MPa	S_u, MPa	K_{Ic}. MPa√mm
Steel—1010 HR	25	45	70	172	310	2432
Steel—1020 HR	30	58	70	207	400	2432
Steel—4340	195	210	80	1344	1448	2780
Steel—18 Ni	240	260	75	1655	1793	2606
Stainless Steel 301 ½ Hard	125	165	100	862	1138	3475
Stainless Steel 440C Tempered	270	290	17	1862	1199	591
Aluminum 2024-T852	57	70	28	393	483	973
Aluminum 2124-T851	53	55	31	365	379	1077
Aluminum 6061-T651	44	47	27	303	324	938
Titanium—70	70	80	50	483	552	1737
Titanium 6Al-6V-2Sn	150	160	50	1034	1103	1737
Al-Bronze—630	82	117	53	565	807	1842
Magnesium—AM503	16	29	12	110	200	417

EXAMPLE 16.6

In Example 16.4 your supplier of large aluminum castings indicated that cracks up to 40 mm long may be generated as the castings cool to room temperature. The casting are subjected to loads that produce a stress of σ_o = 72 MPa. You determined the stress intensity factor for the worst case as:

$$K_I = \sqrt{\pi a}\ \sigma_0 = \sqrt{\pi(20)}\ (72) = 570.7 \text{ MPa}\sqrt{\text{mm}}$$

If the casting is made from an alloy with a fracture toughness of K_c = 15 MPa√m, determine if the crack is stable.

Solution:

Let's use the stability condition presented in Eq. (16.9) and write:

$$K_I < K_c \qquad \text{(a)}$$

Substituting numerical results from the problem statement and Example 16.3 into this equation gives:

$$K_I = 570.7 \text{ MPa}\sqrt{\text{mm}} < K_c = 474.3 \text{ MPa}\sqrt{\text{mm}} \qquad \text{(b)}$$

Note in Eq. (b), we converted the quantity 15 MPa√m to 474.3 MPa√mm. The stress intensity factor is greater than the fracture toughness K_c; hence, the crack is unstable. The structure must be redesigned to lower the applied stress to prevent catastrophic failure.

16.5 STRESS INTENSITY FACTORS FOR SELECT GEOMETRIES

Cracks in plane bodies of finite size are important since the crack poses a threat to the stability and safety of the entire structure even though only one member may be cracked. It is important to determine the stress intensity factor for the specific geometry and loading involved to assess the safety factor for the cracked member. Many solutions for boundary value problems with cracks exist and a few of the more commonly encountered geometries and loadings will be presented in this section. A much more complete listing of solutions is given in reference [6].

First, consider the tension strip with a centrally located crack as shown in Fig. 16.8. The loading at the ends of the strip, well removed from the crack, is a uniaxial stress σ_o. An approximate solution for K_I is:

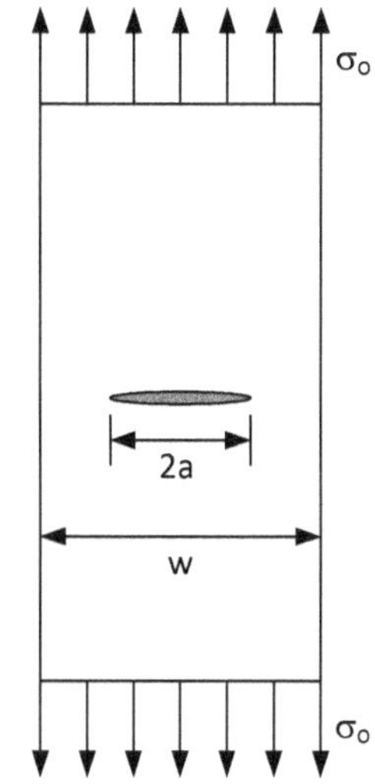

Fig. 16.8 A tension strip with a central crack of length 2a.

$$K_I = \sqrt{\frac{w}{\pi a}\tan\frac{\pi a}{w}}\ \sqrt{\pi a}\ \sigma_0 \qquad (16.10)$$

Note that Eq. (16.10) may be written as:

$$K_I = \alpha\, K_{I\infty} = \alpha\ \sqrt{\pi a}\ \sigma_0 \qquad (16.11)$$

where: $\alpha = \sqrt{\frac{w}{\pi a}\tan\frac{\pi a}{w}}$ is a multiplier[1]. (16.12)

and $K_{I\infty}$ is the stress intensity factor for an infinite plate given by Eq. (16.7)

With K_I expressed in the form shown in Eq. (16.11), the effect of the finite boundaries of the member is given in terms of a multiplying factor α that acts on $K_{I\infty}$. The magnitude of α depends on a/w and ranges from 1.00 for (a/w) $\Rightarrow$ 0 to 4.00 for (a/w) = 0.487.

Next, consider single and double edge cracks in a tension strip as shown in Figs. 16.9a and 16.9b. Series solutions for K_I for these two problems yield the following expressions for the multiplier α in terms of (a/w). For the tension panel with a single edge crack (SEC), the multiplier term (in Eq. (16.11) is given by:

$$\alpha = 1.12 - 0.231\frac{a}{w} + 10.55\left(\frac{a}{w}\right)^2 - 21.71\left(\frac{a}{w}\right)^3 + 30.38\left(\frac{a}{w}\right)^4 \qquad (16.13)$$

For the tension panel with a double edge crack (DEC), the multiplier term (in Eq. (16.11) is given by:

---

[1] A slightly more accurate result for α associated with long cracks can be determined from $\sqrt{\sec\frac{\pi a}{w}}$. However, the two equations for α agree within 5% for (a/w) $\leq$ 0.25.

$$\alpha = 1.12 + 0.429\frac{a}{w} - 4.78\left(\frac{a}{w}\right)^2 + 15.44\left(\frac{a}{w}\right)^3 \tag{16.14}$$

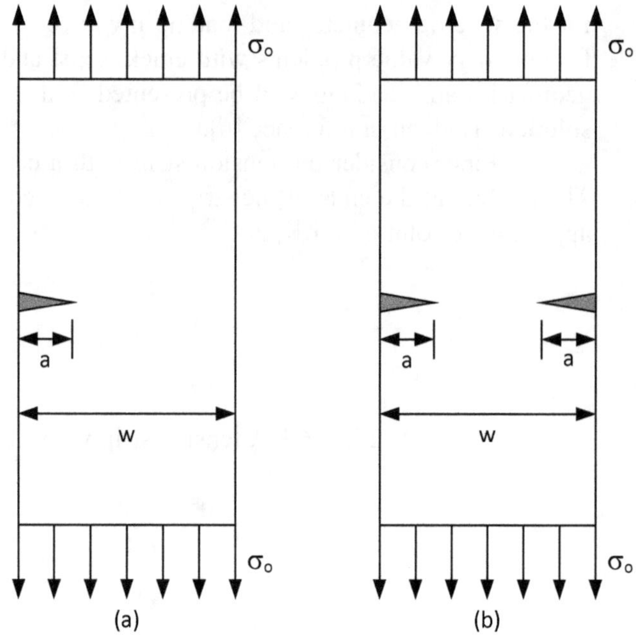

Fig. 16.9 Tension panels with edge cracks: (a) Single-edge crack. (b) Double edge crack.

Results for multipliers of tension panels with cracks are presented in Table 16.2.

Table 16.2
Multiplier (as a function of (a/w) for single (SEC) and double edge cracked (DEC) tension panels.

(a/w)	(SEC)	(DEC)
0	1.12	1.12
0.10	1.18	1.13
0.20	1.37	1.14
0.30	1.66	1.24
0.40	2.10	1.52
0.45	2.42	1.75

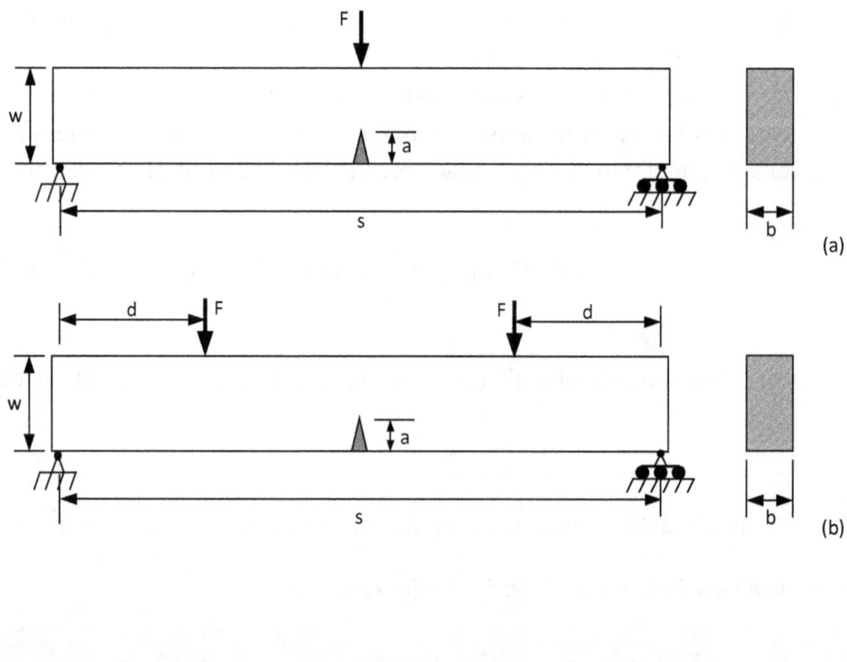

Fig. 16.10 Beams in bending with crack on tension side:
(a) Symmetric three-point bending and
(b) Four-point (pure) bending.

Another common case is a beam with a crack subjected to three-and four-point bending as shown in Figs. 16.10a and 16.10b. With bending, the form of the solution changes and the multiplying factor (is not used. Instead, the stress intensity factor KI is expressed in terms of the beam parameters s (span), w (height), and b (thickness), as defined in Fig. 16.10, and a series in terms of the (a/w) ratio. For the beam subjected to three-point bending, KI is given by:

$$K_I = \frac{Fs}{bw^{3/2}}\left[2.9\left(\frac{a}{w}\right)^{1/2} -4.6\left(\frac{a}{w}\right)^{3/2} +21.8\left(\frac{a}{w}\right)^{5/2} -37.6\left(\frac{a}{w}\right)^{7/2} +38.7\left(\frac{a}{w}\right)^{9/2}\right] \quad (16.15)$$

For the beam subjected to four-point (pure) bending, KI is given by:

$$K_I = \frac{6M}{bw^2}\sqrt{\pi a}\left[1.122 - 1.4\left(\frac{a}{w}\right) + 7.33\left(\frac{a}{w}\right)^2 - 13.08\left(\frac{a}{w}\right)^3 + 14.0\left(\frac{a}{w}\right)^4\right] \quad (16.16)$$

EXAMPLE 16.7

Table 16.2 provides the multiplier α for tension panels containing a single and double edge crack as a function of the (a/w) ratio. Prepare a similar table for the tension panel containing a central crack of length 2a.

Solution:

Calculate the value of the multiplier α from Eq. (16.12) as:

$$\alpha = \sqrt{\frac{w}{\pi a}\tan\frac{\pi a}{w}} \quad (a)$$

Evaluate Eq. (a) as (a/w) ⇒ 0 to 0.45 and tabulate the results as shown below:

(a/w)	Multiplier, α
0	1.000
0.10	1.017
0.20	1.075
0.30	1.208
0.40	1.565
0.45	2.113

You may prepare a graph with these results to obtain a curve showing α as a continuous function of (a/w). Alternatively, you can use a spreadsheet to prepare a more complete table with many more values of (a/w).

EXAMPLE 16.8

A tension panel with a central crack is subjected to a uniaxial loading of $\sigma_o = 6.8$ ksi. Determine the value of K_I if the crack length 2a = 1.8 in. and the panel width w = 6.2 in.

Solution:

To calculate K_I, let's use Eq. (16.11):

$$K_I = \alpha\, K_{I\infty} \quad (a)$$

Next, calculate the value of the multiplier α from Eq. (16.12) as:

$$\alpha = \sqrt{\frac{w}{\pi a}\tan\frac{\pi a}{w}} = \sqrt{\frac{6.2}{\pi(0.9)}\tan\frac{\pi(0.9)}{6.2}} = 1.037 \qquad \text{(b)}$$

Calculate $K_{I\infty}$ as:

$$K_{I\infty} = \sqrt{\pi a}\;\sigma_0 = \sqrt{\pi(0.9)}\;(6.8) = 11.43 \text{ ksi}\sqrt{\text{in.}} \qquad \text{(c)}$$

Substituting Eqs. (b) and (c) into Eq. (a) yields:

$$K_I = \alpha\, K_{I\infty} = (1.037)(11.43) = 11.86 \text{ ksi}\sqrt{\text{in.}} \qquad \text{(d)}$$

EXAMPLE 16.9

A tension strip, 210 mm wide, contains an edge crack that is 70 mm deep. If K_I is limited to 500 MPa$\sqrt{\text{mm}}$, determine the maximum axial stress σ_o that can be applied to the tension strip.

Solution:

For a tension panel with a single edge crack (SEC), the multiplier term α in Eq. (16.11) is given by Eq. (16.13) as:

$$\alpha = 1.12 - 0.231\frac{a}{w} + 10.55\left(\frac{a}{w}\right)^2 - 21.71\left(\frac{a}{w}\right)^3 + 30.38\left(\frac{a}{w}\right)^4 \qquad \text{(a)}$$

Substitute (a/w) = (70/210) = 0.3333 into Eq. (a) to obtain:

$$\alpha = 1.12 - 0.231(0.3333) + 10.55(0.3333)^2 - 21.71(0.3333)^3 + 30.38(0.3333)^4 = 1.785 \qquad \text{(b)}$$

From Eq. (16.11), we can write:

$$K_{I\infty} = K_I/\alpha \qquad \text{(c)}$$

Substituting the results from Eq. (b) and the problem statement into Eq. (c) gives:

$$K_{I\infty} = 500/1.785 = 280.1 \text{ MPa}\sqrt{\text{mm}} \qquad \text{(d)}$$

Recall Eq. (16.7) and write:

$$\sigma_o = K_{I\infty}/(\pi a)^{1/2} = 280.1/\sqrt{[\pi(70)]} = 18.88 \text{ MPa} \qquad \text{(e)}$$

EXAMPLE 16.10

A tension strip, 4.8 in. wide, contains a double edge crack—each 0.60 in. long. If an axial stress σ_o = 9.6 ksi is applied to the tension strip, determine the stress intensity factor. Comment on the stability of these cracks if the tension strip is fabricated from 440C tempered stainless steel.

Solution:

For the tension panel with a double edge crack (DEC), the multiplier term α in Eq. (16.11) is given by Eq. (16.14) as:

$$\alpha = 1.12 + 0.429\frac{a}{w} - 4.78\left(\frac{a}{w}\right)^2 + 15.44\left(\frac{a}{w}\right)^3 \quad \text{(a)}$$

Substitute (a/w) = (0.60/4.8) = 0.125 into Eq. (a) to obtain:

$$\alpha = 1.12 + (0.429)(0.125) - 4.78(0.125)^2 + 15.44(0.125)^3 = 1.129 \quad \text{(b)}$$

Calculate $K_{I\infty}$ as:

$$K_{I\infty} = \sqrt{\pi a}\ \sigma_0 = \sqrt{\pi(0.6)}\ (9.6) = 13.18 \text{ ksi}\sqrt{\text{in.}} \quad \text{(c)}$$

Substituting Eqs. (b) and (c) into Eq. (a) yields:

$$K_I = \alpha\ K_{I\infty} = (1.129)(13.18) = 14.88 \text{ ksi}\sqrt{\text{in.}} \quad \text{(d)}$$

Reference to Table 16.1 shows that K_{Ic} = 17 ksi$\sqrt{\text{in}}$ for 440C tempered stainless steel. Hence, $K_I < K_{Ic}$ and we conclude that the crack is stable. However, if the applied stress σ_o is increased from 9.6 ksi to 10.97 ksi, the crack will initiate.

EXAMPLE 16.11

A simply supported beam with a span of 6.0 m is loaded with a force F applied at its centerline. Determine the maximum length of a stable crack located below the load point. The beam, fabricated from aluminum 6061-T651, has a rectangular cross section with a height of 120 mm and a thickness of 40 mm. The applied force F = 10 kN.

Solution:

For the beam subjected to three-point bending, K_I is given by Eq. (16.15) as:

$$K_I = \frac{Fs}{bw^{3/2}}\left[2.9\left(\frac{a}{w}\right)^{1/2} - 4.6\left(\frac{a}{w}\right)^{3/2} + 21.8\left(\frac{a}{w}\right)^{5/2} - 37.6\left(\frac{a}{w}\right)^{7/2} + 38.7\left(\frac{a}{w}\right)^{9/2}\right] \quad \text{(a)}$$

Let's rewrite Eq. (a) as:

$$K_I = \frac{Fs}{bw^{3/2}}\mathbf{M}_1 \quad \text{(b)}$$

where $\mathbf{M_1}$ is given by:

$$\mathbf{M}_1 = \left[2.9\left(\frac{a}{w}\right)^{1/2} - 4.6\left(\frac{a}{w}\right)^{3/2} + 21.8\left(\frac{a}{w}\right)^{5/2} - 37.6\left(\frac{a}{w}\right)^{7/2} + 38.7\left(\frac{a}{w}\right)^{9/2}\right] \quad \text{(c)}$$

Substituting numerical parameters into the first term in Eq. (b) yields:

$$\frac{Fs}{bw^{3/2}} = \frac{(10\times10^3)(6\times10^3)}{(40)(120)^{3/2}} = 1141\ MPa\sqrt{mm} \qquad (d)$$

We predict crack instability when $K_I \geq K_{Ic}$; hence, we can use Eq. (b) to write:

$$\mathbf{M}_1 = \frac{K_{Ic}}{\dfrac{Fs}{bw^{3/2}}} \qquad (e)$$

From Table 16.1 we note that $K_{Ic} = 938\ MPa\sqrt{mm}$. Then substituting this value and the result from Eq. (d) into Eq. (e) yields:

$$\mathbf{M_1} = 938/1141 = 0.8221 \qquad (f)$$

Finally, we use Eq. (f) in Eq. (c) and solve for (a/w) by trial and error to obtain:

$$(a/w) \approx 0.1 \qquad (g)$$

The maximum crack length that can be tolerated prior to crack initiation is then predicted as:

$$a = (0.1)w = (0.1)(120) = 12\ mm \qquad (h)$$

EXAMPLE 16.12

A simply supported beam with a span of 20 ft is loaded in four-point (pure) bending with the two active loads spaced a distance d = 5 ft from the beam's supports. The beam, fabricated from aluminum 2124-T851, has a rectangular cross section with a height of 4.0 in. and a thickness of 0.75 in. The inspection department assures you that they can find cracks that are at least 1.25 in. long. Determine the maximum force F that can be applied at each of the two load points before the crack initiates.

Solution:

For the beam subjected to four-point (pure) bending, K_I is given by Eq. (16.16) as:

$$K_I = \frac{6M}{bw^2}\sqrt{\pi a}\left[1.122 - 1.4\left(\frac{a}{w}\right) + 7.33\left(\frac{a}{w}\right)^2 - 13.08\left(\frac{a}{w}\right)^3 + 14.0\left(\frac{a}{w}\right)^4\right] \qquad (a)$$

Let's determine the value of the term in the brackets of Eq. (a) for (a/w) = (1.25/4) = 0.3125.

$$\left[1.122 - 1.4(0.3125) + 7.33(0.3125)^2 - 13.08(0.3125)^3 + 14.0(0.3125)^4\right] = 1.135 \qquad (b)$$

We employ the data for the fracture toughness of this alloy listed in Table 16.1 and write:

$$(K_I)_{Max} = K_{Ic} = 31\ ksi\sqrt{in.} \qquad (c)$$

Substituting Eqs (b) and (c) into Eq. (a) yields:

$$(K_I)_{Max} = \frac{6M}{bw^2}\sqrt{\pi a}\,[1.135] = 31\ ksi\sqrt{in.} \qquad (d)$$

Note the bending moment M = Fd = (5)(12)F and then substitute numerical parameters into Eq. (d) and solve for F_{Max} to obtain:

$$F_{Max} = 459.4 \text{ lb}$$

16.5.1 Cracks at Holes in Tensile Panels

In Chapter 4, we introduced the concept of stress concentration factors that arise when discontinuities such as holes occur in bars subjected to axial loading. When structural members containing holes are subjected to cyclic loading, fatigue cracks often initiate and grow slowly with an increase in the number of cycles of applied load. When a crack becomes sufficiently long, $K_I \geq K_{Ic}$ and it becomes unstable and propagates at high speed causing catastrophic failure.

Let's consider a crack emanating from a centrally located circular hole in a tension strip as shown in Fig. 16.11. If the crack is shallow relative to the diameter of the hole, the crack tip is located in a region of high stress concentration where $\sigma = 3\sigma_o$. In this instance, we can write an expression approximating K_I as:

$$K_I = (1.12)(3\sigma_o)\sqrt{(\pi a)} = (3.36)\sigma_o\sqrt{(\pi a)} \quad (16.17)$$

Equation (16.17) is a good approximation for very short cracks; however, when the cracks become longer this relation is prone to error. A more accurate formula due to Bowie [7] incorporates a factor β that accounts for the length of the crack relative to the radius of the hole as indicated below:

$$K_I = \beta\sigma_o\sqrt{(\pi a)} \quad (16.18)$$

where the factor β is a function of a/R given in Table 16.3.

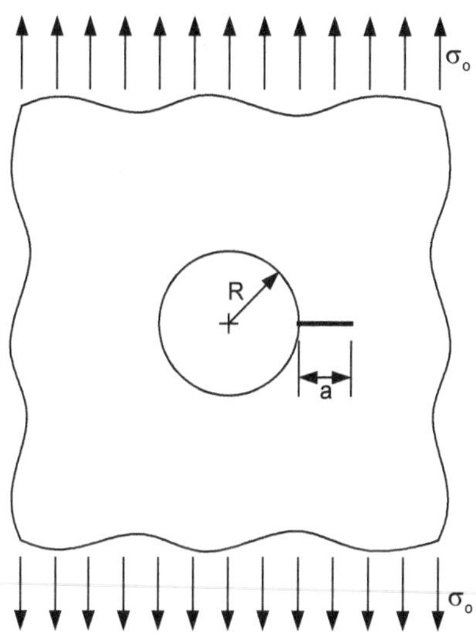

Fig. 16.11 Remotely loaded tension panel with fatigue cracks emanating from a circular hole.

Table 16.3
β Factors for a single crack extending from a circular hole in a tension panel.

a/R	β		a/R	β
0.00	3.39		0.80	1.47
0.10	2.73		1.00	1.37
0.20	2.30		1.50	1.18
0.30	2.04		2.00	1.06
0.40	1.86		3.00	0.94
0.50	1.73		5.00	0.81
0.60	1.64		10.00	0.75

EXAMPLE 16.13

An inspector locates a crack adjacent to a 50 mm diameter circular hole in a tension panel, fabricated from aluminum 2024-T852, that is occasionally loaded with a uniaxial stress σ_o = 90 MPa. She has discovered a 10 mm crack and recommends replacing the panel. You are to review her recommendation.

Solution

Let's first determine the stress intensity factor due to the crack adjacent to a hole by employing Eq. (16.18).

$$K_I = \beta\sigma_o\sqrt{(\pi a)} \qquad \text{(a)}$$

Note that a/R = 10/25 = 0.4; hence from Table 16.3 we find that β = 1.86. Substituting numerical parameters into Eq. (a) yields:

$$K_I = (1.86)(90)\sqrt{(\pi 10)} = 938.3 \text{ MPa}\sqrt{\text{mm}} \qquad \text{(b)}$$

Reference to Table 16.1 reveals that the fracture toughness K_{Ic} = 973 MPa√mm for aluminum 2024-T852. You conclude that the panel is safe, but by a very narrow margin because K_I is almost as large as K_{Ic}. You agree with the inspector's recommendation.

16.5.2 The Penny-Shaped Crack

In the previous sections, we considered surface cracks in two-dimensional bodies and provided solutions for the stress intensity factor K_I. However, cracks occur in three-dimensional bodies and we may be called upon to determine the safety of structures with embedded cracks. In some structural members, the cracks extend only a fraction of the way through the thickness of the structural member creating another three-dimensional crack stability condition.

The penny shaped crack, illustrated in Fig. 16.12, is a flat crack with a circular shape of diameter 2a that is embedded within an infinite body. The applied stress σ_o is perpendicular to the embedded crack surface.

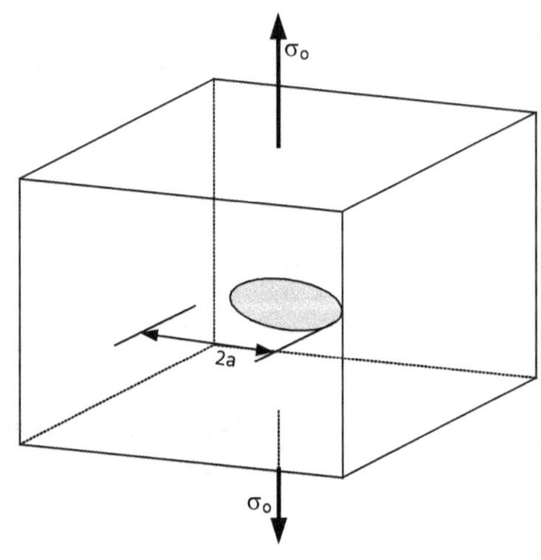

Fig. 16.12 Penny shaped crack embedded in an infinite body.

The stress intensity factor is a constant around the crack edge and is given by:

$$K_I = \frac{2}{\pi}\sigma_0\sqrt{\pi a} \qquad (16.19)$$

As the crack grows under the action of cyclic loading, it advances as a series of expanding concentric circles.

EXAMPLE 16.14

Determine the maximum size of embedded shrinkage cracks that can form in a casting of 18 Ni steel if the casting is to be subjected to an applied stress of 160 ksi.

Solution:

Let's begin by using Eq. (16.19) and write the expression for the stress intensity factor for the embedded crack as:

$$K_I = \frac{2}{\pi}\sigma_0\sqrt{\pi a} \qquad \text{(a)}$$

Next we note the fracture toughness for this steel is K_{Ic} = 75 ksi√in. Setting $K_{Ic} = K_I$ in Eq. (a) and solving for the maximum crack length a gives:

$$a_{Max} = \frac{\pi}{4}\left(\frac{K_{Ic}}{\sigma_0}\right)^2 \qquad \text{(b)}$$

Substituting numerical parameters in Eq. (b) yields:

$$a_{Max} = \frac{\pi}{4}\left(\frac{75}{160}\right)^2 = 0.1726 \text{ in.} \qquad \text{(c)}$$

If the applied stress is perpendicular to the crack face, then the length of the crack that can be tolerated at this relatively high stress level is a very short 0.1726 in.

16.5.3 Semicircular Corner Crack

The stress intensity factor for a semicircular corner crack, illustrated in Fig. 16.13, is approximated by combining the results from the penny shaped crack and the edge cracked panel to give:

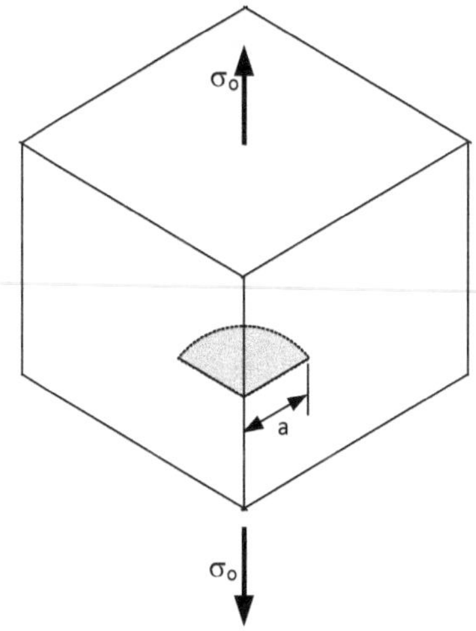

Fig. 16.13 Semicircular corner crack with a radius a.

$$K_I = (1.12)^2\frac{2}{\pi}\sigma_0\sqrt{\pi a} = \frac{2.509}{\pi}\sigma_0\sqrt{\pi a} \qquad (16.20)$$

where $(1.12)^2$ is the correction factor to account for the two surface flaws and ($2/\pi$) is the correction factor to account for the embedded circular crack.

EXAMPLE 16.15

Damage to external corners of very heavy machine components often occurs prior to installation when they are handled in chain slings before and after shipping. If a surface flaw is introduced on one of these corners with a maximum radius of 8 mm, determine the applied stress the component can withstand before the crack (flaw) becomes unstable. The component is fabricated from 4340 steel.

Solution:

The stress intensity factor for a corner crack is given by Eq. (16.20) as:

$$K_I = (1.12)^2 \frac{2}{\pi} \sigma_0 \sqrt{\pi a} = \frac{2.509}{\pi} \sigma_0 \sqrt{\pi a} \tag{a}$$

Next we note the fracture toughness for this steel is K_{Ic} = 2780 MPa√mm. Setting $K_{Ic} = K_I$ in Eq. (a) and solving for the maximum stress σ_o that can be applied to the component gives:

$$(\sigma_0)_{Max} = \left(\frac{K_{Ic}}{2.509}\right)\sqrt{\frac{\pi}{a}} \tag{b}$$

Substituting numerical parameters in Eq. (b) yields:

$$(\sigma_0)_{Max} = \left(\frac{K_{Ic}}{2.509}\right)\sqrt{\frac{\pi}{a}} = \frac{2780}{2.509}\sqrt{\frac{\pi}{8}} = 694.3 \text{ MPa} \tag{c}$$

The stress that may be imposed before the flaw becomes unstable and initiates is 694.3 MPa, which is only about 50% of the yield strength of 1344 MPa for this type of steel. Clearly, even small flaws are detrimental and significantly compromise the strength of structures.

16.5.4 Elliptical Corner Crack at a Circular Hole

The formation of an elliptical crack emanating from a circular hole is depicted in Fig 16.14. This geometry is complex because of the presence of the hole and the shape of the crack; hence the solution for the stress intensity factor K_I incorporates many correction factors as indicated in Eq. (16.21).

Fig. 16.14 An elliptical crack emanating from a circular hole.

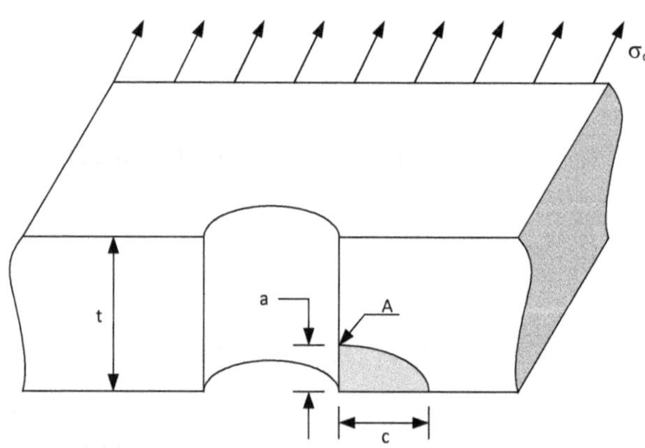

The stress intensity factor K_I varies with position around the elliptical crack front. Its maximum value occurs at point A and is given by:

$$K_I = (1.12)(3\sigma_0)\sqrt{\frac{\pi a}{Q}}\sqrt{\sec\frac{\pi a}{2t}} \qquad (16.21)$$

where a is the depth of the elliptical crack, c is the width of the crack, t is the thickness of the plate, (1.12) is the surface flaw correction factor, ($3\sigma_o$) is the correction factor for the presence of the circular hole, $\sqrt{\sec\frac{\pi a}{2t}}$ is the correction factor for the effect of the finite plate thickness and Q is the elliptical flaw correction factor defined in Table 16.4.

Table 16.4
Flaw shape parameter Q.

Flaw Shape, (a/2c)	Correction Factor, Q		Flaw Shape, (a/2c)	Correction Factor, Q
0	0.91		0.30	1.53
0.05	0.95		0.35	1.62
0.10	1.02		0.40	1.82
0.15	1.11		0.45	2.04
0.20	1.23		0.50	2.23
0.25	1.37			

EXAMPLE 16.16

An inspector has located a shallow surface crack at the edge of hole in a tension panel that is fabricated from aluminum 2024-T852. His measurements indicate a crack depth a = 0.22 in. and a crack width c = 0.56 in. The plate is 1.25 in. thick and the hole diameter is 1.00 in. The tension panel is loaded with an applied stress σ_o = 10 ksi. You are to assess the safety of the tension panel.

Solution:

The stress intensity factor K_I for this example is determined from Eq. (16.21) with appropriate substitutions for the various correction factors.

$$K_I = (1.12)(3\sigma_0)\sqrt{\frac{\pi a}{Q}}\sqrt{\sec\frac{\pi a}{2t}} \qquad (a)$$

The shape factor Q = 1.23 is found in Table 16.4 corresponding to a/2c = 0.1964. Substituting numerical parameters into Eq. (a) yields:

$$K_I = (1.12)(3)(10.0)\sqrt{\frac{\pi(0.22)}{1.23}}\sqrt{\sec\frac{\pi(0.22)}{2(1.25)}} = 25.68\ \text{ksi}\sqrt{\text{in.}} \qquad (b)$$

Reference to Table 16.1 gives K_{Ic} = 28 ksi√in. for the aluminum 2024-T852 used in fabricating the plate. Comparing K_I = 25.68 ksi√in. with K_{Ic} = 28 ksi√in. indicates that the tension panel is safe, but the margin of safety is very small. The equation employed is an approximation containing many correction factors. A more accurate analysis using advanced methods of three-dimensional finite element analysis may be justified if many tension panels are damaged and replacement of all of them is expensive.

16.6 THE THREE MODES OF LOADING

A crack in a structural member can be subjected to three different types of loading. The opening mode, illustrated in Fig. 16.15a, involves loads that produce displacements of the crack surfaces perpendicular to the plane of the crack. In this case, the crack line or the x-axis, as shown in Fig. 16.4b, is a principal axis. The subscript I on K_I indicates that the stress intensity factor is due to opening mode loading.

The shearing mode, illustrated in Fig. 16.15b, is due to in-plane shear loads, which cause the two crack surfaces to slide on one another. The displacement of the crack surfaces is in the plane of the crack and perpendicular to the leading edge of the crack. The subscript II on K_{II} implies that the stress intensity factor is due to shear mode loading. Of course, opening and shear mode loading conditions can occur together. In this case the loading is defined as mixed-mode where both K_I and K_{II} exist in the region near the crack tip.

The tearing mode, shown in Fig. 16.15c, is due to out-of-plane shear loading. The displacements of the crack surfaces are in the plane of the crack and parallel to the leading edge of the crack. The subscript III on K_{III} is used to depict the tearing mode. While the superposition of the three modes of loading gives the most general loading condition, one usually is interested in mode I or mixed-mode loading since they occur more frequently in engineering applications. In this book, we will consider only opening mode loading.

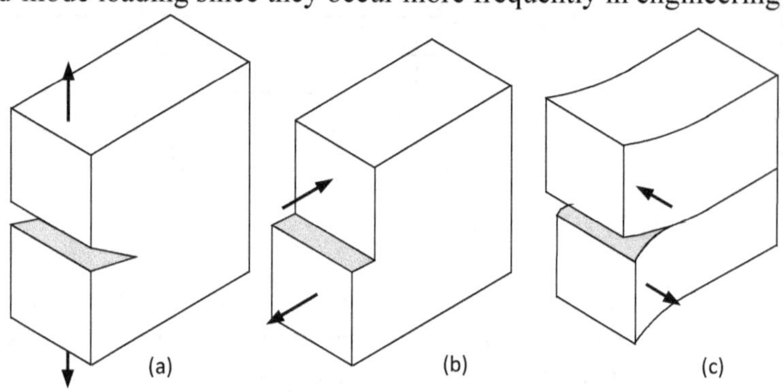

Fig. 16.15 Three modes of crack loading: (a) Opening mode; (b) Shearing mode; (c) Tearing mode.

16.7 CRACK TIP PLASTICITY

The elasticity solutions of structural members with cracks show that a stress singularity exists at the tip of the crack. This fact implies that the material, at the tip of the crack and in a small region near the tip, has exceeded the yield strength of the specimen material. In this region, plastic deformation occurs, and because the stresses are limited by yielding, a stress singularity cannot occur. This region of plastic deformation is called the crack tip plastic zone.

The size of the plastic zone is important because most of the developments in fracture mechanics are valid only if the size of the plastic zone is small relative to the length of the uncracked ligament from the crack tip to the nearest boundary. To study the size of the plastic zone, consider first the magnitude of σ_{yy} given by Eq. (16.6) along the x-axis where $\theta = 0$. Setting $\sigma_{yy} = S_y$, where S_y is the yield strength of the material, gives:

$$\sigma_{yy} = \sqrt{\frac{a}{2r_p^*}}\sigma_0 = S_y \tag{16.22}$$

where r_p^* is the apparent distance along the x-axis to the boundary of the plastic zone.

Solving Eq. (16.22) for r_p^* gives:

$$r_p^* = \frac{a}{2}\left(\frac{\sigma_0}{S_y}\right)^2 \tag{16.23}$$

The apparent distance r_p^* is shown in Fig. 16.16a where the distribution of σ_{yy} along the x-axis is presented. The shaded area shown in Fig. 16.16a indicates load-shedding P_s due to yielding. Thus,

$$P_s = t\int_0^{r_p^*}(\sigma_{yy} - S_y)dx \qquad (16.24)$$

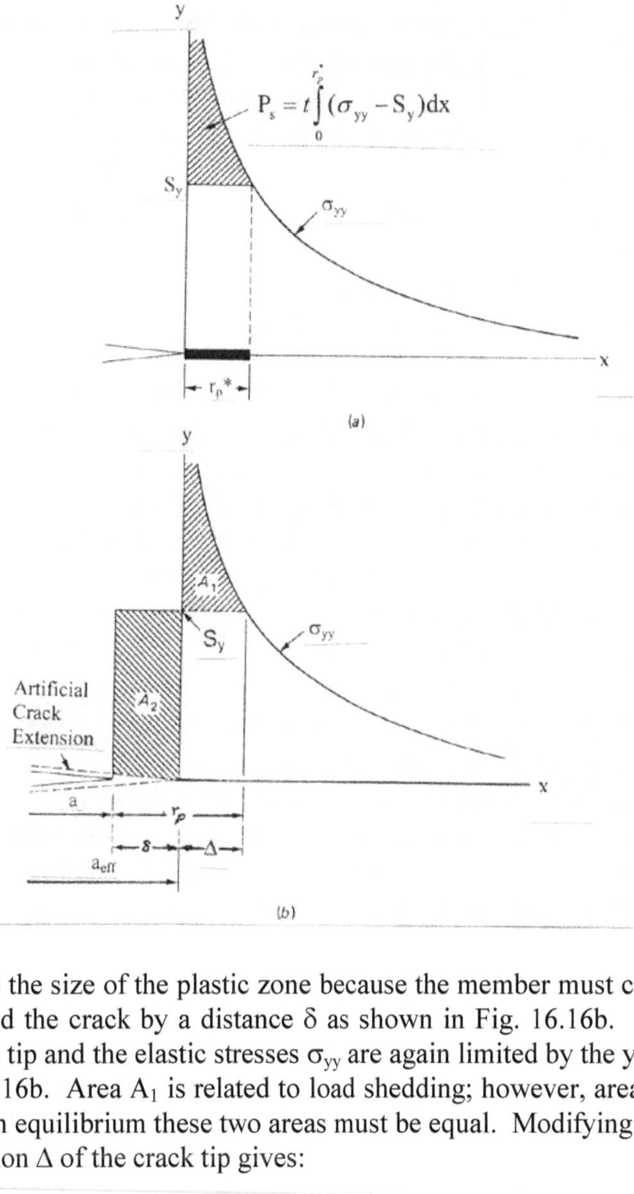

Fig. 16.16 (a) Stress distribution showing yielding in a region at the front of the crack tip. (b) An improved yield model incorporating artificial crack extension.

The result of load loss due to local yielding is to increase the size of the plastic zone because the member must carry P_s. The effect of the plastic zone is to artificially extend the crack by a distance δ as shown in Fig. 16.16b. The coordinate system is translated by δ to the extended crack tip and the elastic stresses σ_{yy} are again limited by the yield strength S_y. Two areas, A_1 and A_2, are shown in Fig. 16.16b. Area A_1 is related to load shedding; however, area A_2 is related to load loss by the crack extension. To maintain equilibrium these two areas must be equal. Modifying Eq. (16.23) to account for changes due to the artificial extension Δ of the crack tip gives:

$$\Delta = \frac{a+\delta}{2}\left(\frac{\sigma_0}{S_y}\right)^2 \qquad (16.25)$$

Because $\delta \ll a$, it is evident by comparing Eqs. (16.23) and (16.25) that $\Delta \approx r_p^*$. The equilibrium requirement that $A_1 = A_2$ together with Eqs. (16.24) and (16.22) can be used to give:

$$\sigma_0\int_0^{r_p^*}\sqrt{\frac{a+\delta}{2x}}dx = S_y(r_p^* + \delta) \qquad \text{(a)}$$

Integrating and simplifying Eq. (a) yields:

$$(r_p^* + \delta)^2 = 2(a + \delta)(r_p^*)\left(\frac{\sigma_0}{S_y}\right)^2 \qquad \text{(b)}$$

Since δ is small when compared to the original crack length a, Eq. (b) and Eq. (16.23) can be used to obtain:

$$r_p^* + \delta = 2r_p^* \qquad \Rightarrow \qquad r_p^* = \delta \qquad \text{(c)}$$

Then the true length r_p of the plastic zone in front of the crack is given by:

$$r_p = \delta + \Delta = 2r_p^* \qquad (16.26)$$

To account for the effect of the plastic zone, an effective crack length a_{eff} is introduced such that:

$$a_{eff} = a + r_p^* \qquad (16.27)$$

The value of a_{eff} is used to determine K_I and relations like Eq. (16.7) are modified to give

$$K_I = \sqrt{\pi(a + r_p^*)}\ \sigma_0 \qquad (16.28)$$

Note that $r_p^* \approx \Delta$ and Eq. (16.25) gives:

$$r_p^* = \frac{a}{2}\left(\frac{\sigma_0}{S_y}\right)^2 \qquad (16.29)$$

Substituting Eq. (16.29) into Eq. (16.28) yields K_I corrected for the plastic deformation at the crack tip as:

$$K_I = \sqrt{\pi a\left[1 + \frac{1}{2}\left(\frac{\sigma_0}{S_y}\right)^2\right]}\ \sigma_0 \qquad (16.30)$$

Similar corrections are made to Eqs. (16.10) thru (16.16) to determine K_I for ductile materials, where localized yielding in the region about the crack tip occurs.

The value of r_p is the measure of the size of the plastic zone along the x-axis. Apart from correcting the determination of K_I, r_p is important in establishing if the body is in a state of small scale yielding or large scale yielding. When r_p is small compared to the length of the uncracked ligament, the structural member is subjected to small scale yielding. In this state it responds in nearly an elastic manner with strains and displacements, outside the small plastic enclave, varying linearly with respect to load. At some critical load, the crack becomes unstable and propagates at a very high velocity (500 to 1000 m/s in steel). The failure is a cleavage type often with little or no evidence of ductility. This behavior is classified as **linear elastic fracture mechanics** (LEFM) and the stability of the crack may be predicted using the methods introduced in previous sections of this chapter.

In large scale yielding, r_p is large compared to the length of the unbroken ligament. The structural member does not respond elastically due to the significant redistribution of stress caused by large scale yielding. With large plastic deformation, the crack tip becomes blunted. In some materials, the crack will become unstable and extend at relatively low velocities by tearing. This behavior is classified as elastic plastic fracture mechanics (EPFM). Failure by crack tearing is a current research topic and is not discussed in this book.

In other cases of large scale yielding, the size of the plastic zone increases until it reaches a boundary. The specimen then fails by plastic instability. For failure by plastic instability, limit load analysis is the preferred approach for predicting the critical load. We will describe this approach in the next section of this chapter.

These three classifications of the effect of localized yielding on crack stability are illustrated in Fig. 16.17 where a graph is presented that shows stress intensity factor as a function of net section stress. When the local yielding is small, LEFM methods are applicable and crack instability occurs when $K_I = K_{Ic}$. When yielding becomes global, the entire net section yields, and failure is by plastic collapse when $\sigma_{yy} = S_y$. For yielding, which is significant but not global, failure occurs by plastic tearing with considerable ductility. In this case, the approach employed for analysis is EPFM. The regions inside the bounding curves are considered safe because the cracks will remain stable for these conditions.

Fig. 16.17 A failure analysis diagram for ductile materials.

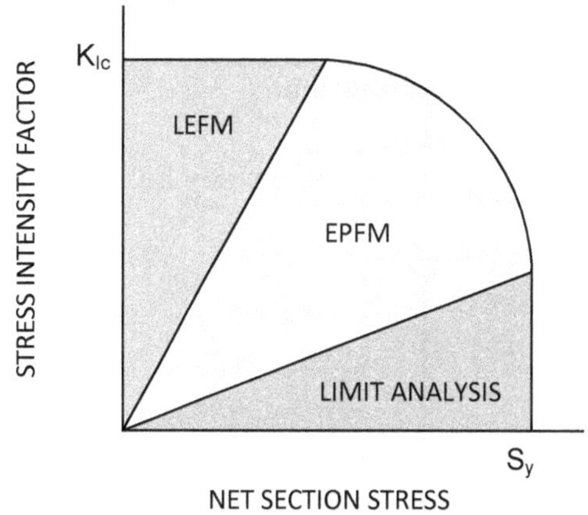

EXAMPLE 16.17

Let's reconsider Example 16.4 taking into account the effect of the plastic zone size on the determination of the stress intensity factor K_I. Recall that your supplier of large aluminum castings (aluminum 2124-T851) indicates that cracks up to 40 mm long may be generated as the castings cool to room temperature. The castings are subjected to loads that produce a stress of $\sigma_o = 72$ MPa. You are asked to estimate a stress intensity factor for the worst case.

Solution:

In the worst case, the crack will be perpendicular to the applied stress and located some distance from the boundaries of the casting; hence, difficult for the inspectors to locate. We will assume that the result of Eq. (16.28) and Eq. (16.30) will represent the worst-case scenario. Then the estimate of the stress intensity associated with the cooling crack is given by:

$$K_I = \sqrt{\pi(a + r_p^*)}\ \sigma_0 \qquad \text{(a)}$$

$$K_I = \sqrt{\pi a\left[1 + \frac{1}{2}\left(\frac{\sigma_0}{S_y}\right)^2\right]}\ \sigma_0 \qquad \text{(b)}$$

Reference to Table 16.1 gives $S_y = 365$ MPa for the aluminum alloy 2124-T851. Substituting the numerical parameters into Eq. (b) yields:

$$K_I = \sqrt{\pi(20)\left[1 + \frac{1}{2}\left(\frac{72}{365}\right)^2\right]}\ (72) = \sqrt{\pi(20)\,[1.0195]}(72) = 576.2 \text{ MPa}\sqrt{\text{mm}} \qquad \text{(c)}$$

The term in the brackets under the square root symbol represents the correction for the effect of the small plastic zone in front of the crack tip. In this example, the correction is small because the bracketed term is only 1.0195—a correction of the crack length of about 2%.

Reference to Table 16.1 indicates that K_{Ic} = 1,077 MPa√mm for this aluminum alloy. Hence the crack should remain stable, but the margin of safety is not large.

EXAMPLE 16.18

Let's reconsider Example 16.3 taking into account the effect of the plastic zone size on the determination of the stress intensity factor K_I. Recall that the quality control inspectors discovered a crack in the weld of a wide steel plate that is to be subjected to biaxial loading with a stress σ_o = 8,500 psi. They estimate that the crack is 5.6 in. long. You are asked to determine the stress intensity factor for 4340 steel.

Solution:

We will employ Eq. (16.30) to estimate the stress intensity associated with the weld crack as:

$$K_I = \sqrt{\pi a \left[1 + \frac{1}{2}\left(\frac{\sigma_0}{S_y}\right)^2\right]}\ \sigma_0 \tag{a}$$

Reference to Table 16.1 gives S_y = 195 ksi for 4340 steel. Substituting the numerical parameters into Eq. (a) yields:

$$K_I = \sqrt{\pi(2.8)\left[1 + \frac{1}{2}\left(\frac{8.5}{195}\right)^2\right]}\ (8.5) = 25.22\ \text{ksi}\sqrt{\text{in.}} \tag{b}$$

In this case, the correction term in the brackets (1.00095) was negligible. The value of K_I determined in Examples 16.3 and 16.18 are essentially the same.

Reference to Table 16.1 indicates that K_{Ic} = 80 ksi$\sqrt{\text{in}}$. for this alloy steel. Hence, the crack should remain stable with a comfortable margin of safety.

16.8 LIMIT ANALYSIS OF PLATES WITH CRACKS

For materials that exhibit sufficient ductility, the plastic zone size that develops in the front of the crack becomes large. Consequently, the crack tip is blunted and the stress singularity is inhibited in its development. In such cases, linear elastic fracture mechanics (LEFM) is not a valid approach. As the load applied to a cracked plate is increased, the plastic zone size increases and if the crack does not initiate by tearing, this plastic zone will extend across the entire net section. We will discuss two approaches to describing the plastic behavior of cracked plates—the Dugdale model [8] and the limit analysis method.

16.8.1 The Dugdale Model

The Dugdale model provides a method for determining the extent of the plastic zone as it grows with increasing load across the uncracked ligament. The method is applied to relatively thin plates where plane stress conditions exist and to materials that yield without significant strain hardening. Dugdale considered the plastic zone to take the form of thin triangular strips as shown in Fig. 16.18.

Dugdale proved that the length of the plastic strips is given by:

$$\frac{r_p}{c} = \sec\left(\frac{\pi}{2}\frac{\sigma_0}{S_y}\right) - 1 \tag{16.31}$$

We will evaluate the ratio r_p/c as a function of σ_o/S_y in the next example to illustrate the extension of the plastic zone across a ligament with increasing load σ_o.

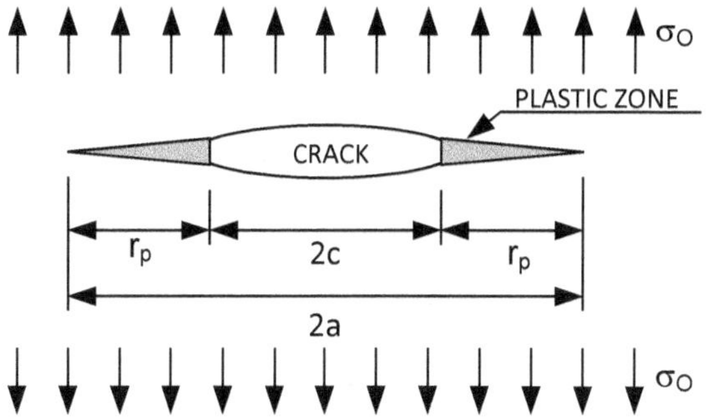

Fig. 16.18 The Dugdale model exhibits plastic zones as thin triangular strips ahead of the crack tips.

EXAMPLE 16.19

Prepare a table showing the extension of the Dugdale plastic strip as a function of the ratio of σ_o/S_y.

Solution:

Use a spreadsheet to evaluate Eq. (16.31) as shown below.

Ratio, σ_0/S_y	r_p/c	Ratio, σ_0/S_y	r_p/c	Ratio, σ_0/S_y	r_p/c
0.00	0.0000	0.35	0.1728	0.70	1.2027
0.05	0.0031	0.40	0.2361	0.75	1.6131
0.10	0.0125	0.45	0.3151	0.80	2.2361
0.15	0.0284	0.50	0.4142	0.85	3.2837
0.20	0.0515	0.55	0.5398	0.90	5.3925
0.25	0.0824	0.60	0.7013	0.95	11.7455
0.30	0.1223	0.65	0.9139	0.99	62.6646

Examination of this table shows that extension of the plastic strip is highly non-linear with increasing load σ_o. For small ratios of σ_o/S_y, the extension of the plastic strip is a small fraction of the crack length. However, when this ratio σ_o/S_y becomes large, the extension of the plastic strip is much more significant with each increment of load.

16.8.2 Limit Load Analysis

Let's consider a center cracked tension panel of width w that is fabricated from a ductile material subjected to an axial stress σ_o. Because of plastic flow, the crack tips have been blunted. Moreover, we will assume that the cracks do not extend by tearing. The maximum applied stress prior to rupture by excessive plastic flow can be determined with limit analysis. In a limit analysis, we assume that the material is perfectly plastic with a yield strength of S_y. The analysis treats the uncracked ligaments of the tension panel as uniformly stressed as indicated in Fig. 16.19.

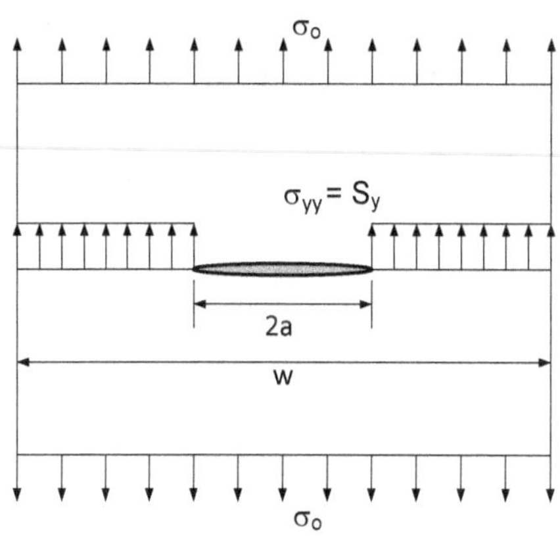

Fig. 16.19 The ligament of the cracked plate is stressed to the yield limit with $\sigma_{yy} = S_y$.

To begin let's cut the tension panel at the crack line, and write the equilibrium equation for forces in the y direction to obtain:

$$(w - 2a)(t)S_y = (wt)\sigma_o \qquad \text{(a)}$$

Solving Eq. (a) for the limit stress $(\sigma_o)_{Limit}$ gives:

$$(\sigma_o)_{Limit} = [1 - (2a/w)]S_y \qquad (16.32)$$

EXAMPLE 16.20

Determine the limit stress that can be achieved prior to plastic rupture of a cracked tension plate. Use a spreadsheet to calculate $(\sigma_o)_{Limit}/S_y$ as a function of (2a/w) as this ratio varies from 0 to 0.99.

Solution:

Evaluating Eq. (16.32) in a spreadsheet gives the results presented in the table below:

(2a/w)	$(\sigma_o)_{Limit}/S_y$	(2a/w)	$(\sigma_o)_{Limit}/S_y$	(2a/w)	$(\sigma_o)_{Limit}/S_y$
0.00	1.00	0.35	0.65	0.70	0.30
0.05	0.95	0.40	0.60	0.75	0.25
0.10	0.90	0.45	0.55	0.80	0.20
0.15	0.85	0.50	0.50	0.85	0.15
0.20	0.80	0.55	0.45	0.90	0.10
0.25	0.75	0.60	0.40	0.95	0.05
0.30	0.70	0.65	0.35	0.99	0.01

These results show the linear decrease in the limit stress for a cracked tension panel fabricated from a perfectly plastic material.

16.9 FACTORS THAT AFFECT FRACTURE TOUGHNESS

In the preceding sections, we have treated fracture toughness as a material property without addressing many of the factors that affect it. These factors include plate thickness, alloy heat treatment (strength), temperature, and rate of loading (strain rate). The influence of the first three of these factors on fracture toughness will be described in some detail in subsequent subsections. The effect of strain rate is to decrease the fracture toughness because high stress levels can be achieved at the tip of a crack before yielding by slip mechanics can occur.

16.9.1 Plate Thickness

We showed in Fig. 16.7 that the fracture toughness K_c decreased with increasing plate thickness. For sufficiently thick plates, the value of K_c reached a minimum called the plane strain fracture toughness K_{Ic}. The reason for the marked difference in fracture toughness with plate thickness is the amount of deformation in the local neighborhood about the crack tip. For thin plates fabricated from ductile material, plastic slip occurs at the crack front and the material shears on 45° planes as shown in Fig. 16.20. The formation of these slip planes enables local necking to occur along the crack line with extensive plastic deformation. This plastic deformation increases the energy required to initiate the crack and markedly increases the fracture toughness.

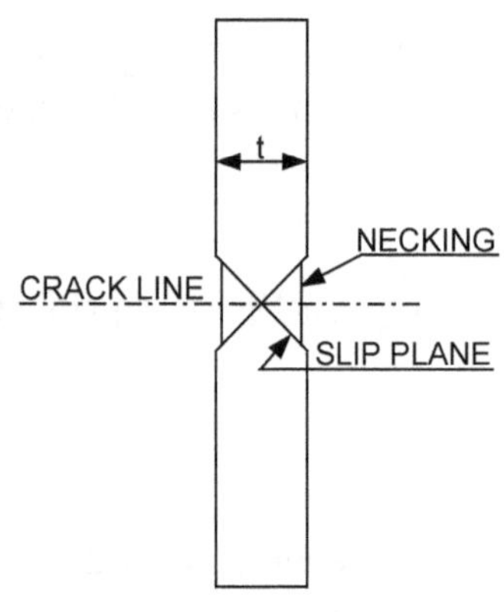

Fig. 16.20 Plastic slip on 45° planes through the plate thickness enables local necking at the crack line.

As the plate thickness increases, plane stress conditions no longer exist and an out-of-plane tensile stress σ_{zz} begins to develop at an interior region of the body near the crack tip. The presence of σ_{zz} constrains yielding at the interior region, and slip cannot occur in the presence of sufficiently large out-of-plane tension stress. As the plate thickness increases, the size of the interior region where yielding is constrained grows and the amount of plastic deformation is suppressed. We show the effect of the constraint, due to the plate thickness, on the appearance of the fracture surface in Fig. 16.21.

The effect of the thickness-induced constraint is to suppress the slip planes in the center region of the specimen. As a consequence, the shear lips become smaller and the plastic zone size is suppressed. The flat surface fails with significant amounts of cleavage. The energy consumed in this fracture process is reduced and the fracture toughness diminishes. For sufficiently thick plates, the fracture toughness reaches a minimum that is referred to as the plane strain fracture toughness K_{Ic}.

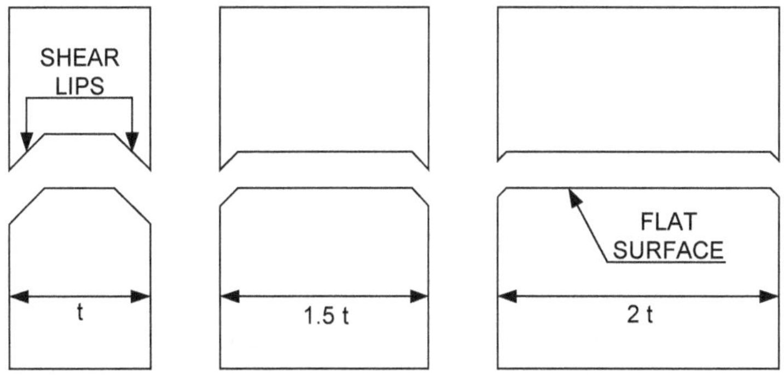

Fig. 16.21 Appearance of fracture surface of a typical steel with increasing plate thickness.

16.9.2 Strength

Designers dealing with structural failures will sometimes redesign the structure by changing only the material selection. Increasing the yield strength should fix the problem. **Perhaps not.** Selecting a higher strength alloy or a heat treatment that will increase strength will often eliminate structural failures if the problem was due to stresses that exceeded the yield strength of the material specified in the original design. However, if failure was due to the initiation of inherent flaws in the structure, then increasing the strength of the material will exacerbate the problem.

A typical curve showing fracture toughness as a function of tensile strength of high quality alloy steel is presented in Fig. 16.22. It is clear that the plane strain fracture toughness decreases markedly as the tensile strength increases.

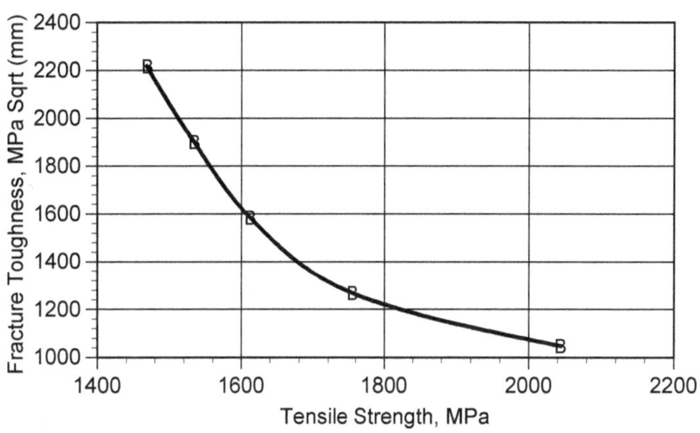

Fig. 16.22 Plane strain fracture toughness with tensile strength for a high quality alloy steel.

16.9.3 Temperature

At low temperatures, ductile steel often behaves in a brittle manner with fractures that exhibit reduced plastic deformation. We conduct Charpy impact tests to determine the temperature for the transition from ductile to brittle behavior. Charpy tests involve striking a small, notched three-point beam specimen with a pendulum-hammer swung from a height sufficient to cause the specimen to fracture. The energy absorbed by the beam in the failure process is measured. At higher temperatures, the failed beams exhibit plastic deformation and the energy required to fracture them is high. However, at lower temperatures the plastic deformation is reduced significantly and the energy required for fracture is greatly diminished. A typical graph presenting Charpy energy with temperature is shown in Fig. 16.23.

Examination of Fig. 16.23 shows that at higher temperatures the energy absorbed is relatively high, with upper shelf absorbed energy levels in the neighborhood of 100 ft-lb. At lower temperatures, the absorbed energy is much lower and often is as low as 5 to 10 ft-lb. There is a range of temperatures between the upper and lower shelves that are termed transition temperatures where the fracture process is mixed. At low temperatures, the fracture process is dominated by cleavage with little plastic flow. At high temperatures, the fracture process is dominated by slip with large plastic deformations. In the transition regions, these two mechanisms of failure compete, and the fracture surfaces show evidence of both processes. In specifying a steel for structural applications, it is essential that the service temperature, to which the structure is exposed, be above the lower shelf temperature. In fact, 15 ft-lb is often the specified energy level because it is usually located above the lower shelf.

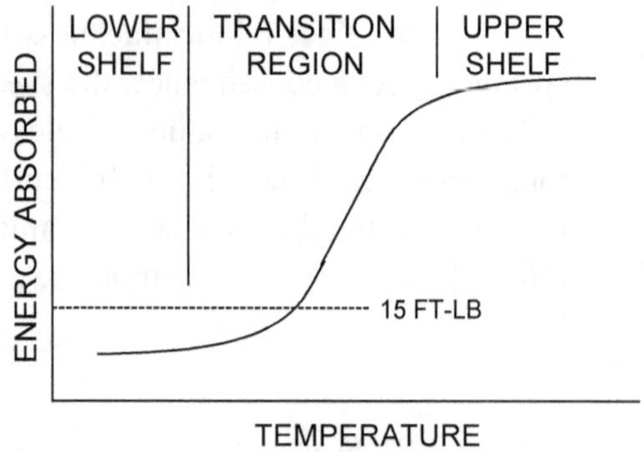

Fig. 16.23 Charpy test results showing energy absorbed at different temperatures for a typical steel.

A second means for ascertaining the effect of temperature on toughness is to measure K_c or K_{Ic} as a function of temperature. The graph showing fracture toughness as a function of temperature is similar to the results from a Charpy test as indicated in Fig. 16.24.

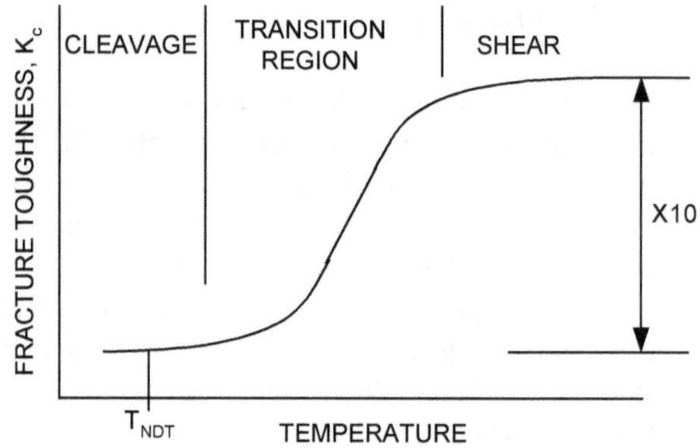

Fig. 16.24 Variations in fracture toughness over a wide range of temperature.

The region with relatively large levels of fracture toughness in Fig. 16.24 is dominated by failure with significant evidence of plastic deformation induced by local shearing. The region with relatively low levels of fracture toughness is dominated by cleavage type fracture with little evidence of plastic deformation. In the transition region, these two failure mechanisms compete and the failure region is considered as mixed. The temperature where the fracture toughness reaches a minimum is called the nil-ductility-temperature and is given the symbol T_{NDT}. Allowable service temperatures are often based on the T_{NDT}. For example, a service temperature of T_{NDT} + 30° F is allowable if the design stress is less than (S_y/2), and T_{NDT} + 60° F if the design stress is less than S_y.

Unfortunately tests to measure fracture toughness are expensive to conduct. To avoid these significant costs, we often convert Charpy data to provide an approximation of the fracture toughness. One of the approximations for the plane strain fracture toughness of bridge steel in the temperature transition region is given by:

$$K_{Ic} = \sqrt{655 E \mathcal{E}_{CVN}} \qquad (16.33)$$

where K_{Ic} is the plane fracture toughness, with units of Pa√m, E is the modulus of elasticity, in units of Pa, and $\mathcal{E}_{CVN}$ is the absorbed energy in a Charpy test, in units of joules J.

EXAMPLE 16.21

Determine an approximate value for the fracture toughness of a bridge steel if the Charpy V-notch test results for the absorbed energy were 30 J. The test temperature and the service temperature are both in the transition range.

Solution:

Let's substitute the numerical parameters into Eq. (16.33) to obtain:

$$K_{Ic} = \sqrt{(655)EE_{CVN}} = \sqrt{(655)(207 \times 10^9)(30)} = 63.78 \text{ MPa}\sqrt{m} \quad (a)$$

$$K_{Ic} = \left(63.78 \text{ MPa}\sqrt{m}\right)\sqrt{1000 \text{ mm/m}} = 2017 \text{ MPa}\sqrt{mm} \quad (b)$$

16.10 CRACK GROWTH UNDER CYCLIC LOADING

Material failure by cyclic loading was introduced in Chapter 5, and the use of the S-N diagram to predict failure or safety of structural components was described. The S-N diagram is employed when the structural component is subjected to large numbers of cycles (many thousand if not millions). It is also employed for components that are initially flaw free. For high cycle fatigue, a significant fraction of the cycles to failure are expended in generating a fatigue crack large enough to find under close inspection. However, in some structural applications, flaws are inherent and fatigue life is measured in hundreds of cycles[2]. In these cases it is important to track the size of the cracks in structures.

Initially, the cracks inherent in structures are small and stable. However, the cracks grow under cyclic loading and become longer. Eventually one of the cracks will grow to critical size, initiate under load, and propagate. In most instances, this crack will propagate across the structure and result in catastrophic failure. Clearly, we seek to avoid these failures by exercising fracture control procedures. Part of the process of fracture control is to predict the growth rates for each of the cracks in the structure.

Cracks grow a small amount each time they undergo a stress cycle. A typical graph showing the growth of a crack as a function of the number of stress cycles is presented in Fig. 16.25. Inspection of Fig. 16.25 indicates two important aspects of crack growth. First, that the rate of crack growth (da/dN), given by the slope of each curve, increases as the cracks grow longer. Second, that cracks exposed to higher stress levels grow faster than cracks exposed to lower level stresses. Attempts have been made to develop empirical relations for (da/dN) as a function of the applied stress and crack length; however, too many variables affected the growth rate, and these empirical relations found limited application.

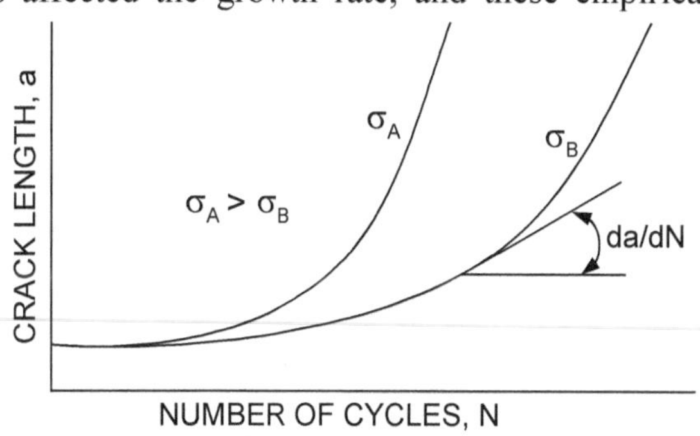

Fig. 16.25 Cyclic crack growth at two different stress levels.

More recently, Paris [9] related crack growth rate to the range in the stress intensity factor that occurs under the action of cyclic loading as:

$$(da/dN) = C(\Delta K)^n \quad (16.34)$$

where (da/dN) is the crack growth rate C and n are material properties ΔK is the range in the stress intensity factor ($K_{Max} - K_{Min}$).

[2] An example is the pressure vessel that houses a nuclear reactor. One cycle of loading is produced by the thermal stress generated each time a unit is shut down and restarted. Over a thirty-year design life with three or four cycles per year, we are concerned with about 100 or so cycles of thermal stress.

Many experiments have been conducted to prove the validity of Eq. (16.34). Typical results of these crack growth experiments reveal three different phases of the crack growth process as indicated in Fig. 16.26.

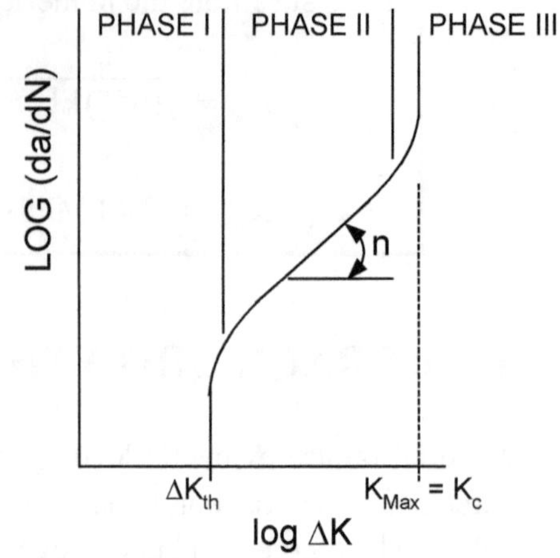

Fig. 16.26 Log-log graph of growth rate with range in stress intensity factor.

The results in Fig. 16.26 indicate that there is a threshold value of ΔK. This fact implies that cracks will remain stable for essentially an infinite number of cycles if $\Delta K < \Delta K_{th}$. During phase II, the cracks grow at a rate predicted by Eq. (16.34). The slope of the log-log curve over this linear region of the log-log relation gives the exponent n in Eq. (16.34). Phase III defines rapid growth behavior where K_{Max} approaches the fracture toughness K_c.

In a fracture control procedure, it is necessary to predict the number of cycles to failure. It is possible to make this prediction by substituting Eq. (16.11) into Eq. (16.34) to obtain:

$$\frac{da}{dN} = C\Delta K^n = C\left(\alpha\sqrt{\pi a}\,\Delta\sigma_0\right)^n \tag{16.35}$$

Integrating Eq. (16.35) yields:

$$\int_{a_i}^{a_c} a^{\frac{-n}{2}}\,da = C\left(\alpha\sqrt{\pi}\Delta\sigma_0\right)^n \int_0^{N_f} dN \tag{a}$$

Completing the integration and solving for the number of cycles to failure N_f gives:

$$N_f = \left(\frac{2}{2-n}\right)\left(a_c^{\left(1-\frac{n}{2}\right)} - a_i^{\left(1-\frac{n}{2}\right)}\right)\left(\frac{1}{C\left(\alpha\sqrt{\pi}\Delta\sigma_0\right)^n}\right) \quad \text{not valid for } n = 2 \tag{16.36}$$

where a_c is the critical crack length, and a_i is the initial crack length.

The constants C and n are determined by conducting crack growth experiments under cyclic loading. Sanford [5] provides an extensive listing of these constants for a number of different materials. For more advanced work, it is recommended that a computer program for predicting life developed by NASA [10] be employed. This program known as NASGRO contains three modules that include:

a. Numerical stress analysis to determine geometric stress intensity factors.
b. Material database.
c. Analysis of critical crack size and determination of safe life.

In the following examples, we will demonstrate the techniques for predicting the remaining life in a structure that is subjected to cyclic loading where crack growth by fatigue processes occurs.

EXAMPLE 16.22

A wide and thick steel plate contains an edge crack 0.20 in. long. The plate is subjected to alternating stresses between 0 and 25 ksi. Experiments conducted with a similar steel provided data for the material constants in Eq. (16.35) as $C = 5 \times 10^{-10}$ and $n = 4$ when ΔK is expressed in units of ksi$\sqrt{\text{in}}$. The plane strain fracture toughness for this steel is $K_{Ic} = 48$ ksi$\sqrt{\text{in}}$. Determine the number of cycles for this crack to grow to a critical size that produces failure.

Solution:

Let's first determine the length of the crack that will initiate under the maximum stress (critical crack length). Using Eqs. (16.11) and (16.13) with (a/w) ⇒ 0 for the edge cracked plate, we write:

$$K_I = K_{Ic} = (1.12)\sqrt{\pi a_c}\,\sigma_0 = 48 \text{ ksi}\sqrt{\text{in}}. \qquad \text{(a)}$$

Solve Eq. (a) for the critical crack length a_c as:

$$a_c = \frac{1}{\pi}\left(\frac{K_{Ic}}{(1.12)\sigma_0}\right)^2 = \frac{1}{\pi}\left(\frac{48}{(1.12)(25)}\right)^2 = 0.9354 \text{ in.} \qquad \text{(b)}$$

Substituting numerical values into Eq. (16.36) gives:

$$N_f = \left(\frac{2}{2-4}\right)\left(a_c^{(1-2)} - a_i^{(1-2)}\right)\left(\frac{1}{5\times 10^{-10}\left((1.12)\sqrt{\pi}(25)\right)^4}\right) \qquad \text{(c)}$$

$$N_f = -\left(\frac{1}{0.9354} - \frac{1}{0.20}\right)(329.7) = 1{,}296 \text{ cycles} \qquad \text{(d)}$$

EXAMPLE 16.23

A finite width plate (w = 200 mm) contains a central crack with 2a = 12 mm. The plate is subjected to alternating stresses between 90 and 180 MPa. Experiments conducted with a similar material provided data for the material constants in Eq. (16.35) as $C = 3 \times 10^{-10}$ and $n = 3$ when ΔK is expressed in units of MPa$\sqrt{\text{mm}}$. The plane strain fracture toughness for this steel is $K_{Ic} = 2{,}000$ MPa$\sqrt{\text{mm}}$. Determine the number of cycles for this crack to grow to a critical size, which produces failure.

Solution:

Let's first determine the length of the crack that will initiate under the maximum stress (critical crack length). Using Eq. (16.11) for the center cracked plate, we write:

$$K_I = K_{Ic} = \alpha\sqrt{\pi a_c}\,\sigma_0 = 2{,}000 \text{ MPa}\sqrt{\text{mm}}. \qquad \text{(a)}$$

Substituting numerical parameters into the multiplier term α given by Eq. (16.12) yields:

$$\alpha = \sqrt{\frac{W}{\pi a}\tan\frac{\pi a}{W}} = \sqrt{\frac{200}{\pi(6)}\tan\frac{\pi(6)}{200}} = 1.0015 \qquad \text{(b)}$$

Solve Eq. (a) for the critical crack length a_c as:

$$a_c = \frac{1}{\pi}\left(\frac{K_{Ic}}{\alpha\sigma_0}\right)^2 = \frac{1}{\pi}\left(\frac{2000}{(1.0015)(180)}\right)^2 = 39.18 \text{ mm} \qquad \text{(c)}$$

where the maximum value for σ_o (180 MPa) is used in Eq. (c).

Substituting numerical values into Eq. (16.36) gives:

$$N_f = \left(\frac{2}{2-3}\right)\left(a_c^{(1-1.5)} - a_i^{(1-1.5)}\right)\left(\frac{1}{3\times10^{-10}\left((1.0015)\sqrt{\pi}(90)\right)^3}\right) \qquad \text{(d)}$$

$$N_f = -2\left(\frac{1}{\sqrt{39.18}} - \frac{1}{\sqrt{6.0}}\right)(817.5) = 406 \text{ cycles} \qquad \text{(e)}$$

Note that the stress range $\Delta\sigma = (180 - 90) = 90$ MPa is required in Eq. (d).

16.11 FRACTURE CONTROL

Large complex structures usually have inherent flaws of one type or another. These flaws are often introduced in fabrication during the welding process where incomplete welds, embedded slag, holes, inadequate fusion bonding, and shrinkage cracks are common. Flaws are sometimes introduced during service. Mechanisms that produce flaws include fatigue, stress-corrosion cracking, and impact damage. Regardless of the source of the flaws, they are common and the history of technology contains many examples of dramatic and catastrophic failures, which caused significant losses in life and property. Indeed as recently as 1972, a 584-ft-long tank barge [11] broke almost completely in half while in port with calm seas. The vessel was only one year old indicating that relatively modern methods of design, material procurement, and welding procedures do not always insure safe structures.

Because flaws in structures and components are common, it is essential in the design, fabrication, and maintenance of a structure to establish a fracture-control procedure. If properly implemented, the fracture-control plan can insure the safety of the structure during its entire service life. Fracture control involves first, the recognition that flaws exist, and second, that these flaws must be maintained in a stable state. To describe the technical aspects of fracture control, consider a simple structure, namely a large plate, with the stress-intensity factor given by:

$$K_I = \sqrt{\pi a}\ \sigma_0 \qquad (16.7)$$

The crack in this structure will remain stable if:

$$K_I < K_{Ic} \qquad (16.2)$$

Combining these two equations gives the stability relation for this simple structure as:

$$\sqrt{\pi a}\ \sigma_0 < K_{Ic} \qquad (16.37)$$

This stability relation illustrates the three elements in all fracture-control procedures; namely,

1. Flaw size, a
2. Load or stress imposed, σ_o
3. Material fracture toughness, K_{Ic}

Clearly, to insure the integrity of a structure requires that the variables in the stability relation be maintained so that equality in Eq. (16.37) never occurs during the life of a structure.

A typical fracture-control plan usually considers material resistance (i.e. the crack-initiation toughness K_c or K_{Ic}) as the first step in maintaining cracks in a stable state. The material selection sets the upper limit for K_I and it is controlled by the designer in the specification of the material for the structure. Typical values of K_{Ic} for common engineering materials are presented in Table 16.1. To ensure that the material meets the specification, the supplier provides a certification indicating that tests have been conducted using representative samples and that the test results exceed the minimum required strength and toughness.

With welded structures, selection and certification of the base plate is only the first step in insuring the specified toughness. Weld materials and weld processes must be selected and then certified. The weld material is often under matched (lower yield strength) relative to the base plate with a higher toughness. The difficulty is in certifying the weld process because this certification requires extensive testing of specimens containing the weld and the heat-affected zone. Testing is expensive but necessary because experience shows that welds often degrade the toughness of the base plate in the heat-affected zone.

In some applications, K_c or K_{Ic} decreases with service life. For instance, in nuclear reactor pressure vessels, radiation damage due to high neutron flux occurs and the toughness degrades with extended service. Also, some alloys that age-harden will exhibit losses in K_c or K_{Ic} during service life, which is often measured in decades.

The second step in a fracture-control plan involves inspection to determine the flaw size a. It is evident from the examples illustrated in the previous section that the critical crack length a_c is given by:

$$a_c = \frac{1}{\pi}\left(\frac{K_{Ic}}{\alpha\sigma_0}\right)^2 \qquad (16.38)$$

Inspection with x-rays, eddy current sensors, and ultrasonic waves are employed to locate cracks. Large cracks, which approach or exceed a_c, are repaired. However, small cracks with a length a_d below the minimum detectable length or with a length a_r below the repairable limit remain in the structure. These three crack lengths are depicted in Fig. 16.27 as a function of the time the structural element is in service. These cracks may grow during service due to fatigue loading or stress-corrosion cracking, and they degrade the strength of the structure as shown in Fig. 16.28. The growth period represents the time interval during which one or more cracks extend from a_r to a_c. This period is usually long and provides the opportunity for periodic inspections during the service life. It is important in the fracture-control plan to specify inspection intervals, which are sufficiently short relative to the growth period to provide two or more opportunities to detect the cracks before they achieve critical length. When longer cracks are detected, they are repaired (in the field) and the growth period is extended.

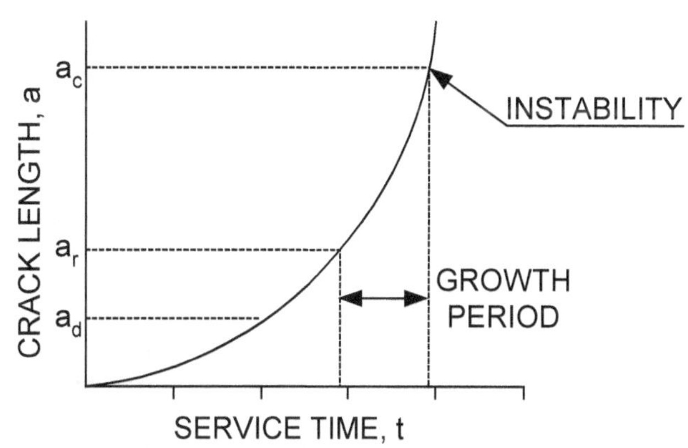

Fig. 16.27 Graph of crack growth with time in service showing three stages of crack development.

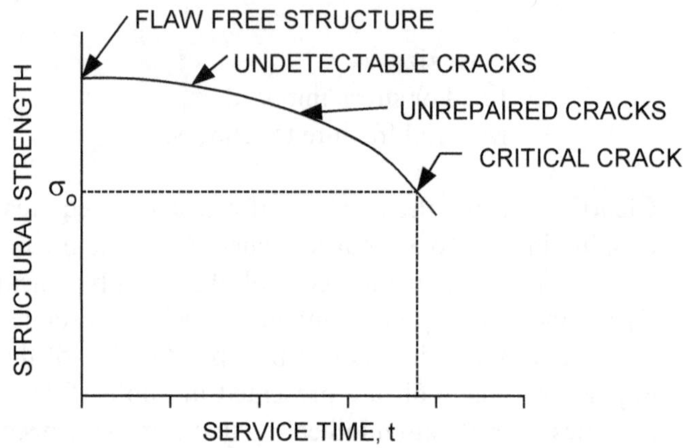

Fig. 16.28 Graph of degradation of structural strength with time in service as the crack grows.

The third step in the fracture-control plan involves the applied loads, which establishes the stress σ_o. From Fig. 16.28, it is evident that the structural strength will exceed σ_o provided $a < a_c$. In many types of structures, periodic inspections and field repair insure that crack growth is controlled and that the cracks do not approach critical length. In these structures, structural stability is achieved simply by maintaining the loads so that $\sigma < \sigma_o$. However, in some structures, accessibility is limited and complete inspection is difficult if not impossible. In these instances, a proof-testing procedure can be applied to the structure to insure safety. In proof testing, loads are applied to the structure, which are larger than the normal loading that produces σ_o. These proof loads produce an elevated stress σ_p and if the structure does not fail, the cracks in the structure are less than the critical length as shown in Fig. 16.29. The single cycle of proof load provides confidence that the cracks in the structure are short and stable. Indeed, depending upon the ratio σ_p/σ_o a growth period can be predicted where the safety of the structure can be ensured. It is important to divide the growth period into two or more parts to determine the proof-test interval. One repeats proof testing at these intervals to continue to extend the growth period. In this sense, repeated proof testing is similar to repeated inspections. However, with an inspection procedure, the cracks which grow with time in service can be repaired; but with a proof test procedure, the cracks continue to grow and when $a_p \geq a_c$, the structure will fail during the proof test.

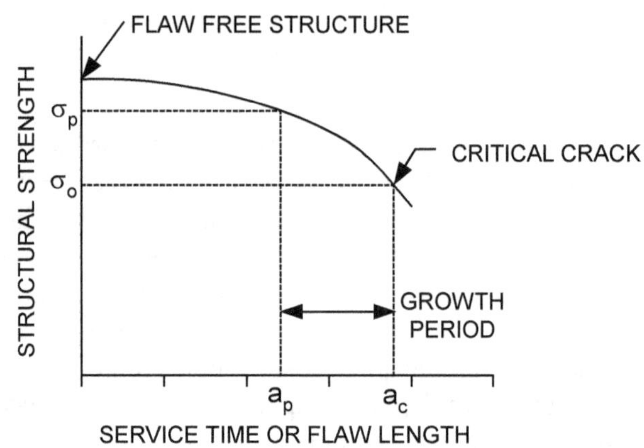

Fig. 16.29 An illustration of proof testing at σ_p to ensure that $a_p < a_c$.

16.12 SUMMARY

The concept of singular stresses and the stress intensity factor K_I was introduced as a means for analyzing structures containing cracks. The stress intensity factor is compared to some fracture toughness K_c or K_{Ic} to determine if the crack is stable or unstable. To introduce the concept of singular stresses at a crack tip, an equation for the stress concentration K at an elliptical hole in a tension panel was given. As the minor axis b approaches zero, the stress concentration K approaches infinity. Clearly, a stress singularity exists at the crack tip.

$$K = \left(1 + \frac{2a}{b}\right) \tag{16.4}$$

The stress σ_{yy} at the crack tip was determined by Westergaard as:

$$\sigma_{yy} = \sqrt{\frac{a}{2r}}\ \sigma_0 \qquad (16.6)$$

Using the definition of the stress intensity factor, we showed that K_I for a central crack of length 2a in a very wide plate is given by:

$$K_I = \sqrt{\pi a}\ \sigma_0 \qquad (16.7)$$

We then presented relations giving the stress intensity factor K_I for a number of different two-dimensional geometries. These relations were similar to Eq. (16.7) except for the multiplying term α in the expression to account for the effects of the finite boundaries of these geometries.

$$K_I = \alpha\, K_{I\infty} = \alpha\ \sqrt{\pi a}\ \sigma_0 \qquad (16.11)$$

The crack initiation criterion was introduced as:

$$K_I \geq K_c \qquad (16.8)$$

Fracture toughness K_c is a material property; however, many factors affect its value. Plate thickness is one of these factors. When the plate becomes sufficiently thick, the toughness decreases to a minimum known as the plane strain fracture toughness K_{Ic}.

A few solutions for the stress intensity factor associated with cracks in three-dimensional structural members containing cracks were given in Eqs. (16.19) to (16.21).

Three modes of loading a structural member containing a crack were introduced. These included the opening, shear and tearing modes. The treatment in this book is limited to the opening mode.

The effect of plastic flow at the crack tip when σ_{yy} exceeds the yield strength S_y was described. The result of load loss due to local yielding is to increase the size of the plastic zone, which in turn artificially extends the crack by a distance δ. To account for the effect of the plastic zone, an effective crack length a_{eff} is introduced such that:

$$a_{eff} = a + r_p^* \qquad (16.27)$$

The value of a_{eff} is used to determine a corrected estimate for K_I as:

$$K_I = \sqrt{\pi (a + r_p^*)}\ \sigma_0 \qquad (16.28)$$

$$r_p^* = \frac{a}{2}\left(\frac{\sigma_0}{S_y}\right)^2 \qquad (16.29)$$

Three different treatments of yielding on crack stability were described. When the local yielding is small, LEFM methods are applicable and crack instability occurs when $K_I \geq K_{Ic}$. When yielding becomes global, the entire net section yields, and failure is by plastic collapse when $\sigma_{yy} \geq S_y$. For yielding, which is significant but not global, failure occurs by plastic tearing with considerable ductility. Two approaches describing the plastic behavior of cracked plates—the Dugdale model and the limit analysis method—were described.

A discussion was provided for the many factors that affect fracture toughness, which included—plate thickness, strength and temperature.

The concept of crack growth under cyclic loading was introduced. The Paris law that estimates the cyclic rate of crack growth was given as:

$$(da/dN) = C(\Delta K)^n \qquad (16.34)$$

An expression predicting the life of a cracked structure was derived from this law as:

$$N_f = \left(\frac{2}{2-n}\right)\left(a_c^{\left(1-\frac{n}{2}\right)} - a_i^{\left(1-\frac{n}{2}\right)}\right)\left(\frac{1}{C\left(\alpha\sqrt{\pi}\Delta\sigma_0\right)^n}\right) \quad \text{not valid for } n = 2 \qquad (16.36)$$

Finally, a method for insuring the safety of cracked structures over an extended service life was discussed. A typical fracture-control plan usually considers the crack-initiation toughness K_c or K_{Ic} as the first step in maintaining cracks in a stable state. The material selection sets the upper limit for K_I and it is controlled by the designer in the specification of the material for the structure. The second step in a fracture-control plan involves inspection to determine the flaw size a. Inspection with x-rays, eddy current sensors, and ultrasonic waves are employed to locate cracks. Large cracks, which approach or exceed the critical crack length a_c, are repaired and opportunity for frequent periodic inspections during the service life is provided. When cracks cannot be detected, proof testing is performed to insure that the cracks will remain stable under design loads.

REFERENCES

1. Irwin, G. R.: Fracture I, Handbuch der Physik VI, Flügge Ed., pp. 558-590, Springer, 1958.
2. Inglis, C. E.: “Stresses in a Plate Due to the Presence of Cracks and Sharp Corners,” Trans. Inst. Naval Architects, 55, pp. 219-241, 1913.
3. Griffith, A. A.: "The Phenomena of Rupture and Flow in Solids,” Phil. Trans. Roy. Soc. of London, A 221, pp 163-197, 1921.
4. Westergaard, H. M.: "Bearing Pressures and Cracks", J. Applied Mechanics, 61, pp A49-A53, 1939.
5. Sanford, R. J. Principles of Fracture Mechanics, Prentice Hall, Upper Saddle River, NJ, 2003.
6. Tada, H., Paris, P. and G. R. Irwin: The Stress Analysis of Cracks Handbook, Del Research, Hellertown, PA, 1973.
7. Bowie, O. L., “Analysis of an Infinite Plate Containing Radial Cracks Originating at the Boundary of an Internal Circular Hole,” J. Mathematics and Physics, Vol. 35, pp. 60-71, 1956.
8. Dugdale, D. S., “Yielding of Steel Sheets Containing Slits,” J. Mechanics and Physics of Solids, Vol. 8, pp. 100-104, 1960.
9. Paris, P., and Erdogan, F., “A Critical Analysis of Crack Propagation Laws,” J. Basic Engineering, Vol. 85, pp. 528-534, 1963.
10. NASA/NASGRO Computer Program, http:/mmptdpublic.jsc.nasa.gov/nasgro/NasgroMain.html.
11. Marine Casualty Report, "Structural Failure of the Tank Barge I.O.S. 3301 Involving the Motor Vessel Martha R. Ingram on 10 January 1972 Without Loss of Life", Report No. SDCG/NTSB, March 1974.

PROBLEMS

16.1 Your manager directs you to perform a stress analysis of a machine component containing a well-defined crack, and to employ the von-Mises theory of yielding to predict when the structure will fail. How do you respond to this directive?

16.2 Determine the stress concentration at an elliptical hole of length 2a when a uniaxial stress σ_o is applied in a direction perpendicular to the length of the hole. Let b = a/50 and b = a/100. Where does σ_{max} occur?

16.3 Determine the stress concentration at an elliptical hole of length 2a when a uniaxial stress σ_o is applied in a direction parallel to the length of the hole. Let b = a/50 and b = a/100. Where does σ_{max} occur?

16.4 A crack 60 mm long develops in a very wide plate, which is to support a uniaxial stress of 70 MPa. Determine the stress intensity factor K_I. What assumption did you make in this determination?

16.5 If the crack in Problem 16.4 grows at a rate of 10 mm/month, determine the service life remaining before crack initiation if K_{Ic} = 1265 MPa$\sqrt{\text{mm}}$.

16.6 For a very wide plate subjected to a uniaxial stress of 100 MPa, prepare a graph showing the stress intensity factor as a function of crack length 2a.

16.7 Prepare a graph showing α as a function of (a/w) for a tension strip with a centrally located crack. Let (a/w) vary from 0.05 to 0.80.

16.8 A steel strap 2.0 mm thick and 20 mm wide with a central crack of 2.0 mm long is loaded to failure. Determine the critical load F_c if K_{Ic} for the strap material is 2780 MPa$\sqrt{\text{mm}}$.

16.9 Determine the failure stress σ_o applied to the strap of Problem 16.8. Compare this stress to the yield and tensile strengths of 4340 alloy steel. How will the strap fail?

16.10 Use a spreadsheet program to compute the multiplier α for SEC and DEC cracked tension strips. Evaluate α for $0 < (a/w) < 0.49$ in increments of 0.01. Display the results in a listing similar to those shown in Table 16.2.

16.11 A steel tension bar, 6.0 mm thick and 20 mm wide with a single edge crack 3.0 mm long, is subjected to a uniaxial stress of 90 MPa. Determine the stress intensity factor K_I.

16.12 If K_{Ic} is 2,000 MPa$\sqrt{\text{mm}}$ for the steel used to fabricate the tension bar in Problem 16.11, will the crack initiate?

16.13 Determine the critical crack length a_c for the tension bar in Problem 16.11 if K_{Ic} = 2,000 MPa$\sqrt{\text{mm}}$.

16.14 Determine the critical load F_c for the tension bar in Problem 16.11 if K_{Ic} = 2,000 MPa$\sqrt{\text{mm}}$.

16.15 A steel tension bar, 1.0 in. thick and 10.0 in. wide with double edge cracks each 0.15 in. long, is subjected to a uniaxial stress of 25 ksi. Determine the stress intensity factor K_I.

16.16 If K_{Ic} is 65 ksi$\sqrt{\text{in}}$. for the steel used to fabricate the tension bar in Problem 16.15, will the crack initiate?

16.17 Determine the critical crack length a_c for the tension bar in Problem 16.15 if K_{Ic} = 65 ksi$\sqrt{\text{in}}$.

16.18 Determine the critical load F_c for the tension bar in Problem 16.15 if K_{Ic} = 65 ksi$\sqrt{\text{in}}$.

16.19 Determine the normalized stress intensity factor $\frac{K_I b w^{3/2}}{Fs}$ as a function of (a/w) for a beam subjected to three-point bending. Let (a/w) vary from 0.05 to 0.50.

16.20 Determine the normalized stress intensity factor $\frac{K_I b w^2}{M\sqrt{\alpha}}$ as a function of (a/w) for a beam subjected to pure bending. Let (a/w) vary from 0.05 to 0.50.

16.21 Find the critical load F_c that can be applied to a beam subjected to three-point bending as illustrated below. Note K_{Ic} = 1,500 MPa$\sqrt{}$mm.

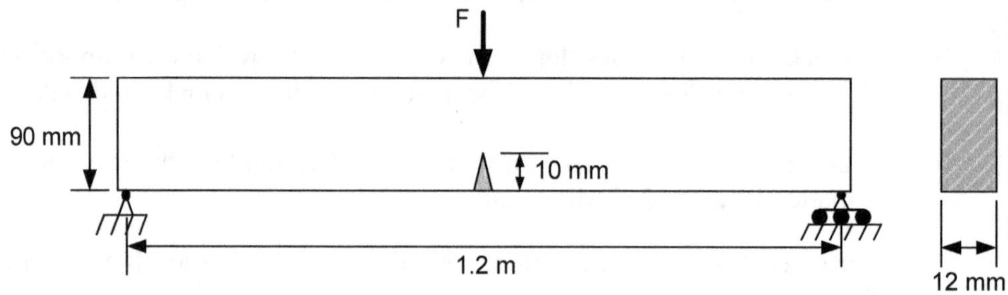

16.22 Find the critical moment $M_c = F_c d$ that can be applied to a beam in pure bending as illustrated in the figure below. Note K_{Ic} = 75 ksi $\sqrt{}$in.

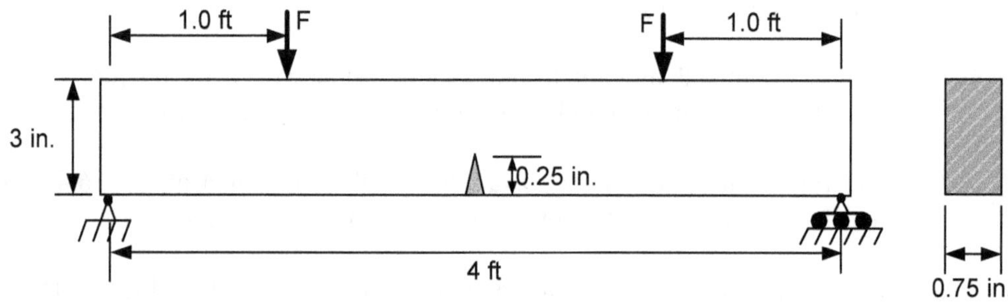

16.23 An inspector locates a 7.0 mm long crack adjacent to a 30 mm diameter circular hole in a tension panel, fabricated from aluminum 2124-T851. The tension panel is occasionally loaded with a uniaxial stress σ_o = 150 MPa. He recommends replacing the panel. You are to review his recommendation. Do you agree with his recommendation? Justify your answer.

16.24 Determine the critical crack length a_c for the geometry shown in the figure to the right. Note that K_{Ic} = 55ksi$\sqrt{}$in.

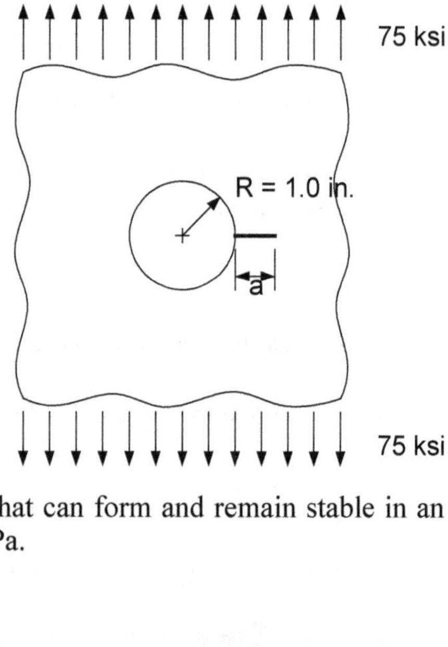

16.25 Determine the maximum size of an embedded shrinkage crack a_{Max} that can form and remain stable in an 18 Ni-steel casting if the crack is subjected to an applied stress of 800 MPa.

16.26 A surface flaw is introduced on an external corner of a 630-aluminum bronze component as shown in the figure to the right. Determine the applied stress σ_o the component can withstand before the crack (flaw) becomes unstable.

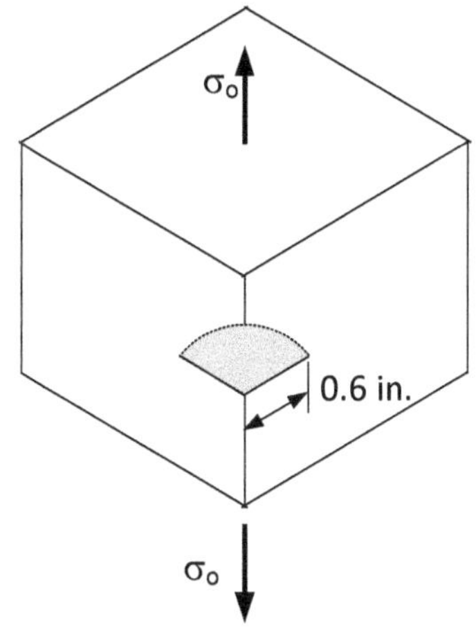

16.27 An inspector has located a shallow surface crack at the edge of hole in a tension panel as illustrated in the figure to the left. The crack dimensions are given in the figure. You are to establish the stress σ_o that will initiate the crack. Note the panel material is 70-titanium.

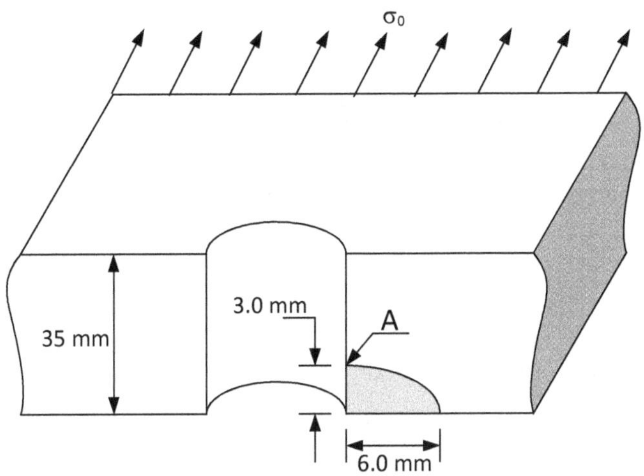

16.28 Give an example of:

a. Opening mode loading.
b. In-plane shear mode loading.
c. Out-of-plane shear mode loading.
d. Mixed mode (I and II) loading.

16.29 Determine r_p at the critical load for the beam shown in the figure below if S_y = 700 MPa. Compare r_p to the crack length and the ligament length. Comment on this comparison.

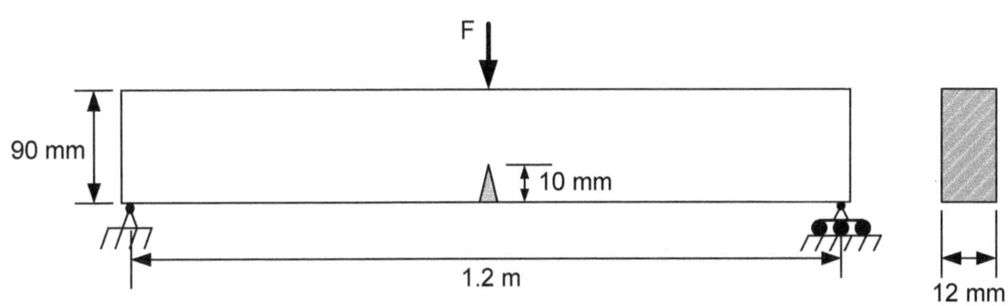

16.30 For the beam in shown below, which is the same as the one defined Problems 16.21 and 16.29, adjust the crack length to accommodate the effect of yielding if S_y = 700 MPa and then determine the new critical load F_c. Compare this result to that found in the original solution for Problem 16.21. Note that K_{Ic} = 1,500 MPa$\sqrt{\text{mm}}$.

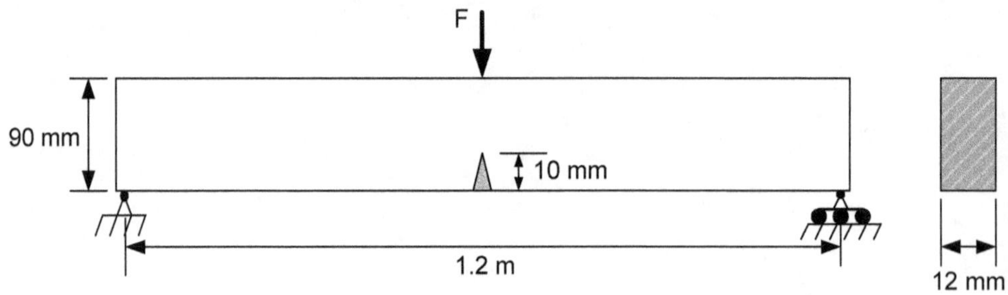

16.31 Consider a thin tension panel shown in the figure below. The panel is fabricated from a ductile material that is perfectly plastic when it yields at a stress of 120 ksi. Use the Dugdale model to determine the length of the plastic zone r_p.

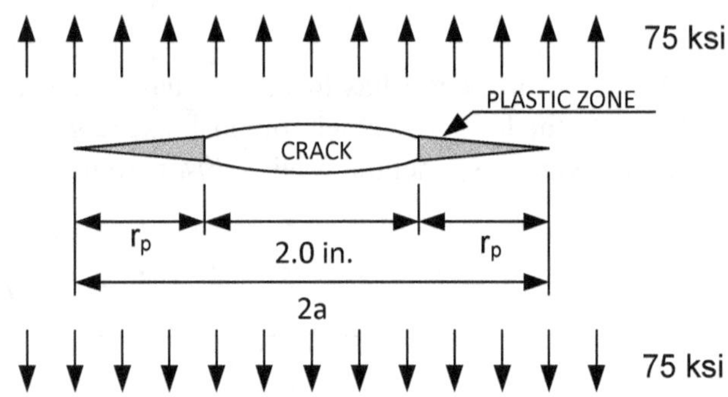

16.32 Determine the limit moment M_{Limit} that can be achieved prior to plastic rupture of a cracked beam subjected to pure bending. Use a spreadsheet to calculate M_{Limit}/S_y as a function of (a/w) as this ratio varies from 0 to 0.75.

16.33 Describe the effect of plate thickness on the fracture toughness of ductile metals.

16.34 Describe the effect of increasing the tensile strength of a metallic alloy on its fracture toughness.

16.35 Describe the effect of decreasing temperature on fracture toughness of typical structural steels.

16.36 Determine an approximate value for the fracture toughness of a structural steel if the Charpy V-notch test result for the absorbed energy is 25.0 J. The test temperature and the service temperature are both in the transition range.

16.37 A wide and thick steel plate contains an edge crack 0.65 in. long. The plate is subjected to alternating stresses between σ_{Min} = 0 ksi and σ_{Max} = 22 ksi. Experiments conducted with similar steel provided data for the material constants in Eq. (16.35) as $C = 5.8 \times 10^{-10}$ and n = 4.0 when ΔK is expressed in units of ksi$\sqrt{\text{in}}$. The plane strain fracture toughness for this steel is K_{Ic} = 60 ksi$\sqrt{\text{in}}$. Determine the number of cycles N_f for this crack to grow to a critical size that produces failure.

16.38 A wide and thick steel plate contains an edge crack 21 mm long. The plate is subjected to alternating stresses between σ_{Min} = 0 MPa and σ_{Max} = 210 MPa. Experiments conducted with similar steel provided data for the material constants in Eq. (16.35) as $C = 9.8 \times 10^{-15}$ and n = 4.0 when ΔK is expressed in units of MPa$\sqrt{\text{mm}}$. The plane strain fracture toughness for this steel is K_{Ic} = 3,210 MPa$\sqrt{\text{mm}}$. Determine the number of N_f cycles for this crack to grow to a critical size that produces failure.

16.39 Describe the three steps involved in fracture control plans.

16.40 Construct a graph showing the critical crack length a_c as a function of K_{Ic}/σ_o over the range from 0.01 to 0.15 $\sqrt{m}$. Assume $\alpha = 1$ for simplicity.

16.41 Write a fracture control plan for a large natural gas storage tank that is to be located adjacent to a densely populated metropolitan area.

16.42 Answer why proof testing is used to insure structural integrity?

Notes:

APPENDIX A

WIRE AND SHEET METAL GAGES
DIMENSIONS ARE SHOWN IN INCHES

Gage No. of Wire	B & S[1] Non Ferrous Metals	American S. & W Steel Wire	American S. & W Music Wire	Steel Manufactures' Sheet	Gage No. of Wire
7-0s	0.651354	0.4900	......	......	7-0s
6-0s	0.580049	0.4615	0.004	......	6-0s
5-0s	0.516549	0.4305	0.005	......	5-0s
4-0s	0.460	0.3938	0.006	......	4-0s
000s	0.40964	0.3625	0.007	......	000s
00	0.3648	0.3310	0.008	......	00
0	0.32486	0.3065	0.009	......	0
1	0.2893	0.2830	0.010	......	1
2	0.25763	0.2625	0.011	......	2
3	0.22942	0.2437	0.012	0.2391	3
4	0.20431	0.2253	0.013	0.2242	4
5	0.18194	0.2070	0.014	0.2092	5
6	0.16202	0.1920	0.016	0.1943	6
7	0.14428	0.1770	0.018	0.1793	7
8	0.12849	0.1620	0.020	0.1644	8
9	0.11443	0.1483	0.022	0.1495	9
10	0.10189	0.1350	0.024	0.1345	10
11	0.090742	0.1205	0.026	0.1196	11
12	0.080808	0.1055	0.029	0.1046	12
13	0.071961	0.0915	0.031	0.0897	13
14	0.064084	0.0800	0.033	0.0747	14
15	0.057068	0.0720	0.035	0.0763	15
16	0.05082	0.0625	0.037	0.0598	16
17	0.045257	0.0540	0.039	0.0538	17
18	0.040303	0.0475	0.041	0.0478	18
19	0.03589	0.0410	0.043	0.0418	19
20	0.031961	0.0348	0.045	0.0359	20
21	0.028462	0.0317	0.047	0.0329	21
22	0.025347	0.0286	0.049	0.0299	22
23	0.022571	0.0258	0.051	0.0269	23
24	0.0201	0.0230	0.055	0.0239	24
25	0.0179	0.0204	0.059	0.0209	25
26	0.01594	0.0181	0.063	0.0179	26
27	0.014195	0.0173	0.067	0.0164	27
28	0.012641	0.0162	0.071	0.0149	28
29	0.011257	0.0150	0.075	0.0135	29
30	0.010025	0.0140	0.080	0.0120	30
31	0.008928	0.0132	0.085	0.0105	31
32	0.00795	0.0128	0.090	0.0097	32
33	0.00708	0.0118	0.095	0.0090	33
34	0.006304	0.0104	......	0.0082	34
35	0.005614	0.0095	......	0.0075	35
36	0.005	0.0090	......	0.0067	36
37	0.004453	0.0085	......	0.0064	37
38	0.003965	0.0080	......	0.0060	38
39	0.003531	0.0075	......	......	39
40	0.003144	0.0070	......	......	40

[1] Courtesy of Brown and Sharpe Manufacturing Company.

APPENDIX B-1

PHYSICAL PROPERTIES OF COMMON STRUCTURAL MATERIALS

MATERIAL	ELASTIC MODULUS, E		SHEAR MODULUS, G		POISSON'S RATIO, ν	THERMAL EXPANSION COEFFICIENT, α	
	Mpsi	GPa	Mpsi	GPa	—	$\times 10^{-6}/°F$	$\times 10^{-6}/°C$
METAL							
Aluminum Alloy	10.4	72	3.9	27	0.32	12.9	23.2
Brass, Bronze	16	110	6.0	41	0.33	11.1	20.0
Copper	17.5	121	6.6	46	0.33	9.4	16.9
Cast Iron - Gray	15	103	6.0	41	0.26	6.7	12.1
Cast Iron - Malleable	25	170	9.9	68	0.26	6.7	12.1
Magnesium Alloy	6.5	45	2.4	17	0.35	14.4	25.9
Nickel Alloy	30	207	11.5	79	0.30	7.8	14.0
Steel	30	207	11.5	79	0.30	6.3	11.3
Stainless Steel	27.5	190	10.6	73	0.30	9.6	17.3
Titanium Alloy	16.5	114	6.2	43	0.33	4.9	8.8
WOOD[1]							
Douglas Fir	1.9	13	0.1	0.7	—	2.1	3.8
Sitka Spruce	1.5	10	0.07	0.5	—	2.1	3.8
Western White Pine	1.5	10	—	—	—	2.1	3.8
White Oak	1.8	12	—	—	—	2.1	3.8
Red Oak	1.8	12	—	—	—	2.1	3.8
Redwood	1.3	9	—	—	—	2.1	3.8
CONCRETE							
Medium Strength	3.6	25	—	—	—	5.5	9.9
High Strength	4.5	31	—	—	—	5.5	9.9
PLASTIC							
Nylon Type 6/6	0.4	2.8	—	—	—	80	144
Polycarbonate	0.35	2.4	—	—	—	68	122
Polyester, PBT	0.35	2.4	—	—	—	75	135
Polystyrene	0.45	3.1	—	—	—	70	125
Vinyl, Rigid PVC	0.45	3.1	—	—	—	75	135
STONE							
Granite	10	70	4	28	0.25	4	7.2
Marble	8	55	3	21	0.33	6	10.8
Sandstone	6	40	2	14	0.50	5	9.0
GLASS	9.6	65	4.1	28	0.17	44	80
RUBBER	0.22[2]	0.0015	0.073[2]	0.0005	0.50	125	225

The values for the properties given above are representative. Because processing methods and exact composition of the material influence the properties to some degree, the exact values may differ from those presented here.

1. Wood is an orthotropic material with different properties in different directions. The values given here are parallel to the grain.
2. The modulus for rubber is given in ksi.

APPENDIX B-2

TENSILE PROPERTIES OF COMMON STRUCTURAL MATERIALS

MATERIAL	ULTIMATE TENSILE STRENGTH, S_u		YIELD STRENGTH, S_y		DENSITY, ρ	
	ksi	MPa	ksi	MPa	lb/in.^{3}	Mg/m^3
CARBON & ALLOY STEELS						
1010 A	44	303	29	200	0.284	7.87
1018 A	49.5	341	32	221	0.284	7.87
1020 HR	66	455	42	290	0.284	7.87
1045 HR	92.5	638	60	414	0.284	7.87
1212 HR	61.5	424	28	193	0.284	7.87
4340 HR	151	1041	132	910	0.283	7.84
52100 A	167	1151	131	903	0.284	7.87
STAINLESS STEELS						
302 A	92	634	34	234	0.286	7.92
303 A	87	600	35	241	0.286	7.92
304 A	83	572	40	276	0.286	7.92
440C A	117	807	67	462	0.286	7.92
CAST IRON						
Gray	25	170	—	—	0.260	7.20
Malleable	50	340	32	220	0.266	7.37
ALUMINUM ALLOYS						
1100-0	12	83	4.5	31	0.098	2.71
2024-T4	65	448	43	296	0.100	2.77
6061-T6	38	260	35	240	0.098	2.71
7075-0	34	234	14.3	99	0.100	2.77
7075 T6	86	593	78	538	0.100	2.77
MAGNESIUM ALLOYS						
HK31XA-0	25.5	176	19	131	0.066	1.83
HK31XA-H24	36.2	250	31	214	0.066	1.83
NICKEL ALLOYS						
Monel 400 A	80	550	32	220	0.319	8.83
Cupronickel A	53	365	16	110	0.323	8.94
COPPER ALLOYS						
Oxygen-free (99.9%) A	32	220	10	70	0.322	8.91
90-10 Brass A	36.4	251	8.4	58	0.316	8.75
80-20 Brass A	35.8	247	7.2	50	0.316	8.75
70-30 Brass A	44.0	303	10.5	72	0.316	8.75
Naval Brass	54.5	376	17	117	0.316	8.75
Tin Bronze	45	310	21	145	0.318	8.80
Aluminum Bronze	90	620	40	275	0.301	8.33
TITANIUM ALLOY						
Annealed	155	1070	135	930	0.167	4.63

A = Annealed and HR = Hot Rolled

APPENDIX B-3

TENSILE PROPERTIES OF NON METALLIC MATERIALS

MATERIAL	ULTIMATE TENSILE STRENGTH, S_u		YIELD STRENGTH, S_y		DENSITY, ρ	
	ksi	MPa	ksi	MPa	lb/in.^{3}	Mg/m^3
WOOD						
Douglas Fir	15	100	—	—	0.017	0.470
Sitka Spruce	8.6	60	—	—	0.015	0.415
Western White Pine	5.0	34	—	—	0.014	0.390
White Oak	7.4	51	—	—	0.025	0.690
Red Oak	6.8	47	—	—	0.024	0.660
Redwood	9.4	65	—	—	0.015	0.415
CONCRETE[1]						
Medium Strength	4.0	28	—	—	0.084	2.32
High Strength	6.0	40	—	—	0.084	2.32
PLASTIC						
Nylon Type 6/6	11	75	6.5	45	0.0412	1.14
Polycarbonate	9.5	65	9	62	0.0433	1.20
Polyester, PBT	8	55	8	55	0.0484	1.34
Polystyrene	8	55	8	55	0.0374	1.03
Vinyl, Rigid PVC	6.5	45	6	40	0.0520	1.44
STONE[1]						
Granite	35	240	—	—	0.100	2.77
Marble	18	125	—	—	0.100	2.77
Sandstone	12	85	—	—	0.083	2.30
GLASS[1]						
98% Silica	7	50	—	—	0.079	2.19
RUBBER						
Natural, Vulcanized	4	28	—	—	0.034	0.95

1. The tensile strength of concrete, stone and bulk glass is negligible. The compressive strength for these materials is reported in this table.

APPENDIX C

PROPERTIES OF AREAS

C.1 AREA

The cross sectional area, A of structural members plays an extremely important role in the efficiency and the adequacy of any member in the design of the structure. For example, the stress in a uniaxial tension or compression member is given by:

$$\sigma = P/A$$

The force P is in the numerator and the area A of the cross section is in the denominator of the relation for the stresses. To lower the stresses we have only two options—decrease the force P or increase the area A. The area A of the cross section depends on its shape as indicated in Fig. C.1:

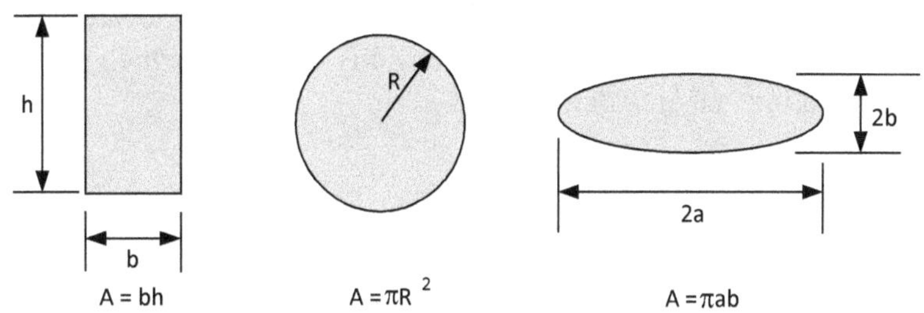

Fig. C.1 Dimensions and areas for common cross-sections.

For areas of arbitrary shape, we determine their area by integration as indicated in Fig. C.2:

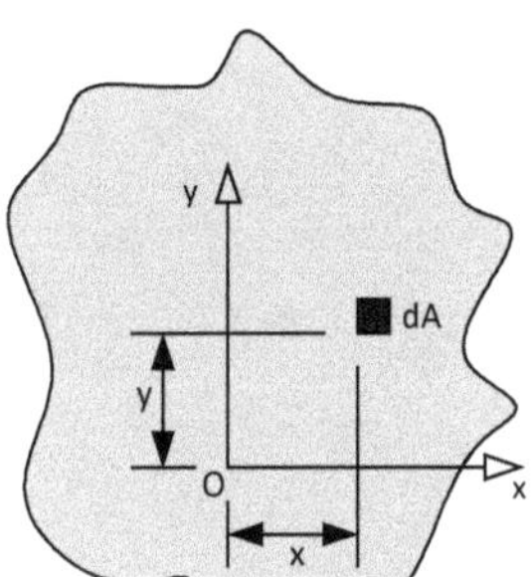

Fig. C.2 An arbitrary area A.

The area, A is determined by summing the incremental area dA in an integration process:

$$A = \int_A dA \tag{C.1}$$

If the boundaries of the area are not known in terms of well-defined mathematical expressions, it is not possible to integrate to determine the area A. However, it is always possible to divide the area into many small squares or rectangles each with an area ΔA. These squares or rectangles follow the boundary of the shape in question and completely fill the interior region. The area is then given by:

$$A = \sum_{i=1}^{N} \Delta A_i = N\Delta A \tag{C.2}$$

where N is the number of small squares or rectangles.

C.2 FIRST MOMENT OF AN AREA

The first moment of an area is important because it is useful in locating the position of the centroid of a given cross sectional area. Let's consider an arbitrary area with the coordinate system Oxy as shown in Fig. C.2. The first moment of the area A about the x-axis is defined as:

$$Q_x = \int_A y\,dA \tag{C.3}$$

Also, the first moment of the area A with respect to the y-axis is defined as:

$$Q_y = \int_A x\,dA \tag{C.4}$$

Depending on the location of the coordinate system relative to the area, the numerical values obtained for the first moment Q_x and Q_y may be either positive or negative. The units for Q_x and Q_y are mm^3 in the SI system or in^3 in the U. S. Customary system.

C.3 CENTROID OF THE AREA A

The centroid of an area A is defined by point C located relative to an arbitrary coordinate system Oxy in Fig. C.3. Indeed, the definition of a centroid is the point, which is considered the center of gravity of a line, an area or a volume.

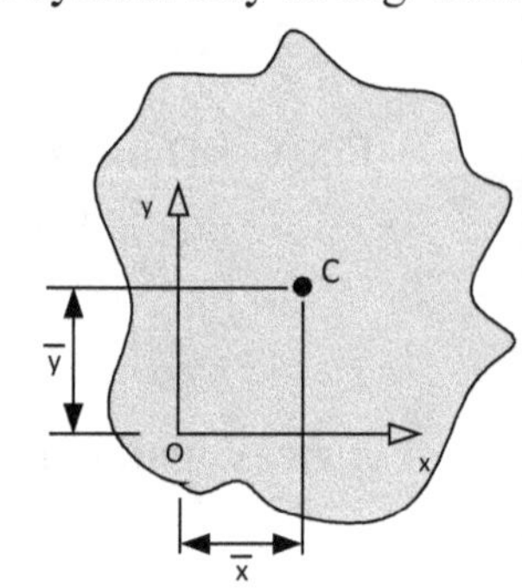

Fig. C.3 The centroid C of an area A is located with coordinates x and y.

The coordinates $\overline{x}$ and $\overline{y}$ of a centroid of an area are determined from the first moments as:

$$Q_y = \int_A x\,dA = A\overline{x}$$
$$Q_x = \int_A y\,dA = A\overline{y} \tag{C.5}$$

where $\overline{x}$ and $\overline{y}$ are dimensions locating the centroid as shown in Fig. C.3.

Let's illustrate the method for determining the first moment of the area and the location of a centroid by considering a few simple shapes the examples presented below.

EXAMPLE C.1

Consider the rectangular area illustrated in Fig C.4, with the origin of the Oxy coordinate system positioned at its corner. Determine the first moments of the rectangular area and the location of its centroid relative to this coordinate system.

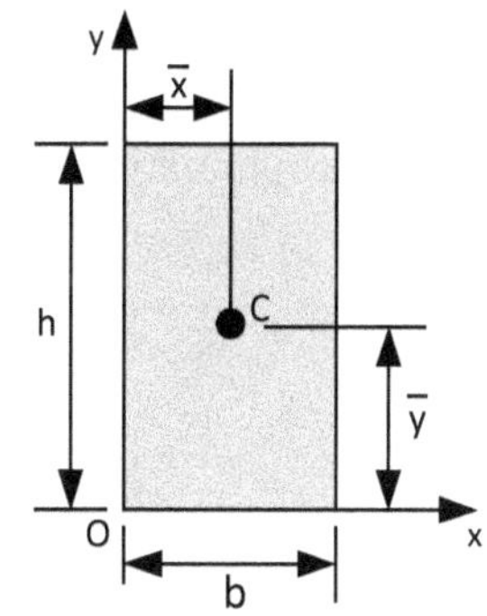

Fig C.4 A rectangle with the coordinate system located along its edges.

Solution:

For the rectangle area presented in Fig. C.4, the first moments of the area about are given by Eq. (C.5) as:

$$Q_x = A\bar{y} = (bh)\frac{h}{2} = \frac{bh^2}{2}$$
$$Q_y = A\bar{x} = (bh)\frac{b}{2} = \frac{b^2h}{2} \qquad \text{(a)}$$

It was possible to quickly solve for the first moments, Q_x and Q_y, because we recognized the location of the centroid for the rectangular area. When an axis of symmetry exists for a given area, the centroid is located somewhere on this axis of symmetry. With the rectangle, two axes of symmetry exist, and the location of the centroid is obviously at the intersection of the axes.

For the circular cross section shown in Fig. C.5, the center of the circle clearly locates the centroid. The center also serves as the origin for the centroidal axes x_c and y_c.

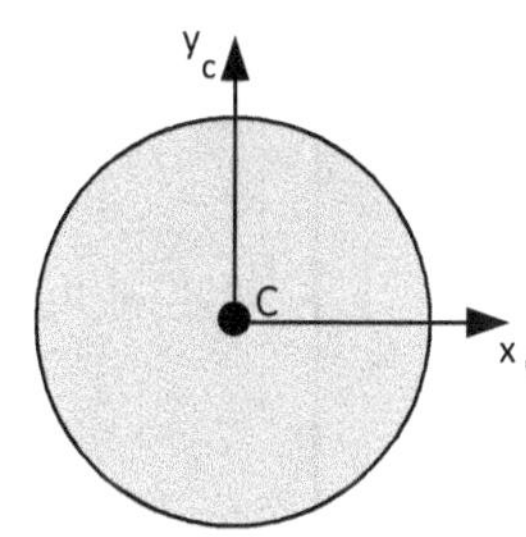

Fig. C.5 The centroid serves as the origin for the centroidal axes Cx_cy_c.

For cross sectional shapes such as ellipses, circles, squares and rectangles, the center may be located by inspection because these geometries have two axes of symmetry. However, for non-symmetric figures, such as triangles, portions of circles, parabolic areas, etc., locating the center of the area is not obvious. We will show methods for determining the location of the centroid of areas that do not exhibit two axes of symmetry.

With respect to a centroidal coordinate system, the first moment of the area must vanish for both axes. Therefore:

$$Q_{\bar{x}} = \int_A y\,dA = 0 \qquad Q_{\bar{y}} = \int_A x\,dA = 0 \qquad \text{(C.6)}$$

These relations are employed to locate the centroid of an area of any shape providing its boundary is defined with some mathematical function.

EXAMPLE C.2

For a right triangle, determine the first moment of the area about its base and vertical side and the position of its centroid relative to its two sides. The right triangle with a base b and a height h is illustrated in Fig. C.6:

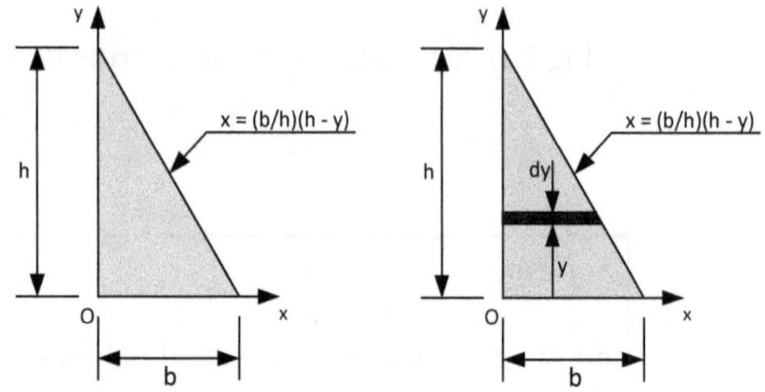

Fig C.6 A right triangle with a coordinate system coincident with its base and vertical side.

Solution:

To begin, let's determine the first moment of the area, A of the right triangle relative to the x-axis (its base side). Writing Eq. C.3 gives:

$$Q_x = \int_A y\,dA = \int_y xy\,dy \qquad \text{(a)}$$

Note, the equation for the diagonal boundary of the triangle is given by:

$$x = (b/h)(h - y) \qquad \text{(b)}$$

The limits on the integral go from 0 to h to encompass the area of the triangle. We substitute Eq. (b) and the limits on the integral into Eq. (a), and write:

$$Q_x = \frac{b}{h}\int_0^h (h - y)y\,dy \qquad \text{(c)}$$

Integrating Eq. (c) gives:

$$Q_x = \frac{b}{h}\left[\frac{hy^2}{2} - \frac{y^3}{3}\right]_0^h \qquad \text{(d)}$$

Completing the integration gives:

$$Q_x = bh^2/6 \qquad \text{(C.7)}$$

By using Eq. C.4 and following the same procedure, we find the first moment about the side of the triangle is given by:

$$Q_y = b^2 h/6 \qquad \text{(C.8)}$$

Equation C.7 gives the first moment of the area of a right triangle about its base. This is an interesting exercise in calculus, but what does it have to do with determining the location of the centroid of the right triangle? The result presented in Eq. (C.7) is an intermediate step. We continue the solution by combining the results of Eqs. (C.7) and (C.8) with Eqs. (C.5) to obtain:

$$Q_x = bh^2/6 = A\,\bar{y} = (bh/2)\,\bar{y} \qquad \text{(e)}$$

$$Q_y = b^2h/6 = A\bar{x} = (bh/2)\bar{x} \qquad \text{(f)}$$

where $\bar{x}$ and $\bar{y}$ locate the Cx_cy_c coordinates relative to the Oxy coordinates (see Fig. C.7).

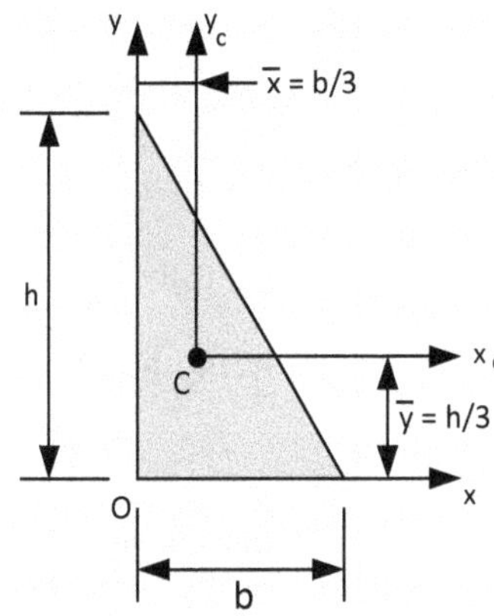

Fig. C.7 A right triangular area with a base axes Oxy and centroidal axes Cx_cy_c.

To determine the position of the centroid, let's solve Eqs. (e) and (f) for x and y to obtain:

$$\bar{y} = h/3 \qquad \bar{x} = b/3 \qquad \text{(C.9)}$$

We employ Eqs. (C.3) and (C.4) to determine the first moment of the area Q relative to either the x or y axes. The location of the centroid is then established from Eq. (C.5). The location of the centroid for common shapes has been determined, and the results are presented together with the definition of the shape of the area in Fig. C.8. This illustration is also shown the back cover.

C.4 LOCATING THE CENTROID OF A COMPOSITE AREA

In many cases, the shape of a cross section is unusual and differs from the common geometries described in Fig. C.8. To analytically locate the centroid of areas with uncommon shapes, we divide its area into several different common shapes for which we have solutions for the location of the centroid. Then we combine the product of these individual areas and their centroid locations to give the location of the centroid of the composite. Let's consider the uncommon shape defined in Fig. C.9, and locate the position of its centroid by employing the composite area technique.

Fig. C.9 The unusual shaped area on the left is divided into common shaped areas on the right.

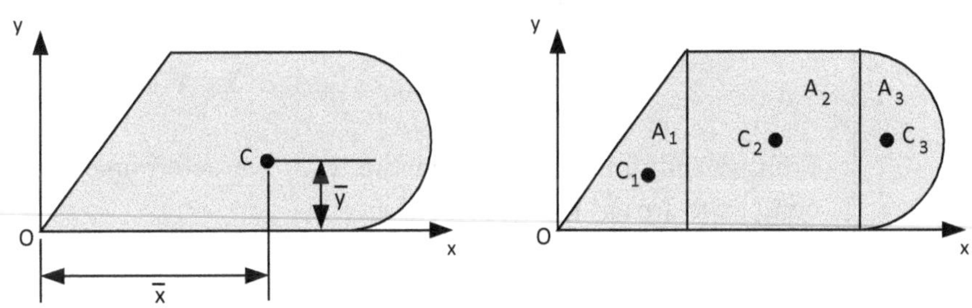

EXAMPLE C.3

Determine the location of the centroid of the odd shape defined in Fig. C.9.

Solution:

To begin the solution, we divide the area into three sub-regions—A_1, A_2 and A_3 as shown in Fig. C.9. Note the sub-regions are a right triangle, rectangle and semicircle. The dimensions of the three different shapes are:

- The triangle—base b = 4 units and height h = 6 units.
- The rectangle—width w = 8 units and height h = 6 units.
- The semicircle—radius R = 3 units.

We apply Eq. (C.5) to the composite area, and write:

$$\Sigma Q_x \quad \Rightarrow \quad \bar{Y}\, A_t = \Sigma\, \bar{y}_n\, A_n \qquad \text{(C.10)}$$

$$\Sigma Q_y \quad \Rightarrow \quad \bar{X}\, A_t = \Sigma\, \bar{x}_n\, A_n \qquad \text{(C.11)}$$

where A_t is the total area of the composite shape.

Let's first determine $\bar{Y}$ from Eq. (C.10):

$$\bar{Y}(A_1 + A_2 + A_3) = \bar{y}_1 A_1 + \bar{y}_2 A_2 + \bar{y}_3 A_3 \qquad \text{(a)}$$

Solving for $\bar{Y}$ yields:

$$\bar{Y} = (\bar{y}_1 A_1 + \bar{y}_2 A_2 + \bar{y}_3 A_3)/(A_1 + A_2 + A_3) \qquad \text{(b)}$$

Substituting results from Fig. C.8 into Eq. (b) gives:

$$\bar{Y} = \frac{\left(\frac{h}{3}\frac{bh}{2} + \frac{h}{2}wh + R\frac{\pi R^2}{2}\right)}{\left(\frac{bh}{2} + wh + \frac{\pi R^2}{2}\right)} \qquad \text{(c)}$$

Substituting b = 4, h = 6, w = 8 and R = 3 units into Eq. (c) yields:

$$\bar{Y} = 2.838 \text{ units}$$

This result is slightly less than h/2 as we anticipated. The presence of the triangle shifts the location of the centroid downward from the centerline of the rectangle and semi-circle.

Next, let's determine the position of the centroid in the direction of the x-axis. We begin by using Eq. (C.11), and write:

$$\bar{X}(A_1 + A_2 + A_3) = \bar{x}_1 A_1 + \bar{x}_2 A_2 + \bar{x}_3 A_3 \qquad \text{(d)}$$

Solving for $\bar{X}$ yields:

$$\bar{X} = (\bar{x}_1 A_1 + \bar{x}_2 A_2 + \bar{x}_3 A_3)/(A_1 + A_2 + A_3) \qquad \text{(e)}$$

From the information listed in Fig. C.8, we determine the centroid location for each of the shapes in the composite area as indicated below:

$$\bar{X} = \frac{\frac{2b}{3}\frac{bh}{2} + \left(b + \frac{w}{2}\right)wh + \left(b + w + \frac{4R}{3\pi}\right)\frac{\pi R^2}{2}}{\frac{bh}{2} + wh + \frac{\pi R^2}{2}} \qquad \text{(f)}$$

Substituting b = 4, h = 6, w = 8 and R = 3 units into Eq. (f) yields:

$$\bar{X} = 8.142 \text{ units}$$

We note that the location of the centroid of the unusual area is slightly to the right of the center of the rectangular area. This position is to be expected because the orientation of the right triangle with its area concentrated toward the right side tends to shift the centroid to the right.

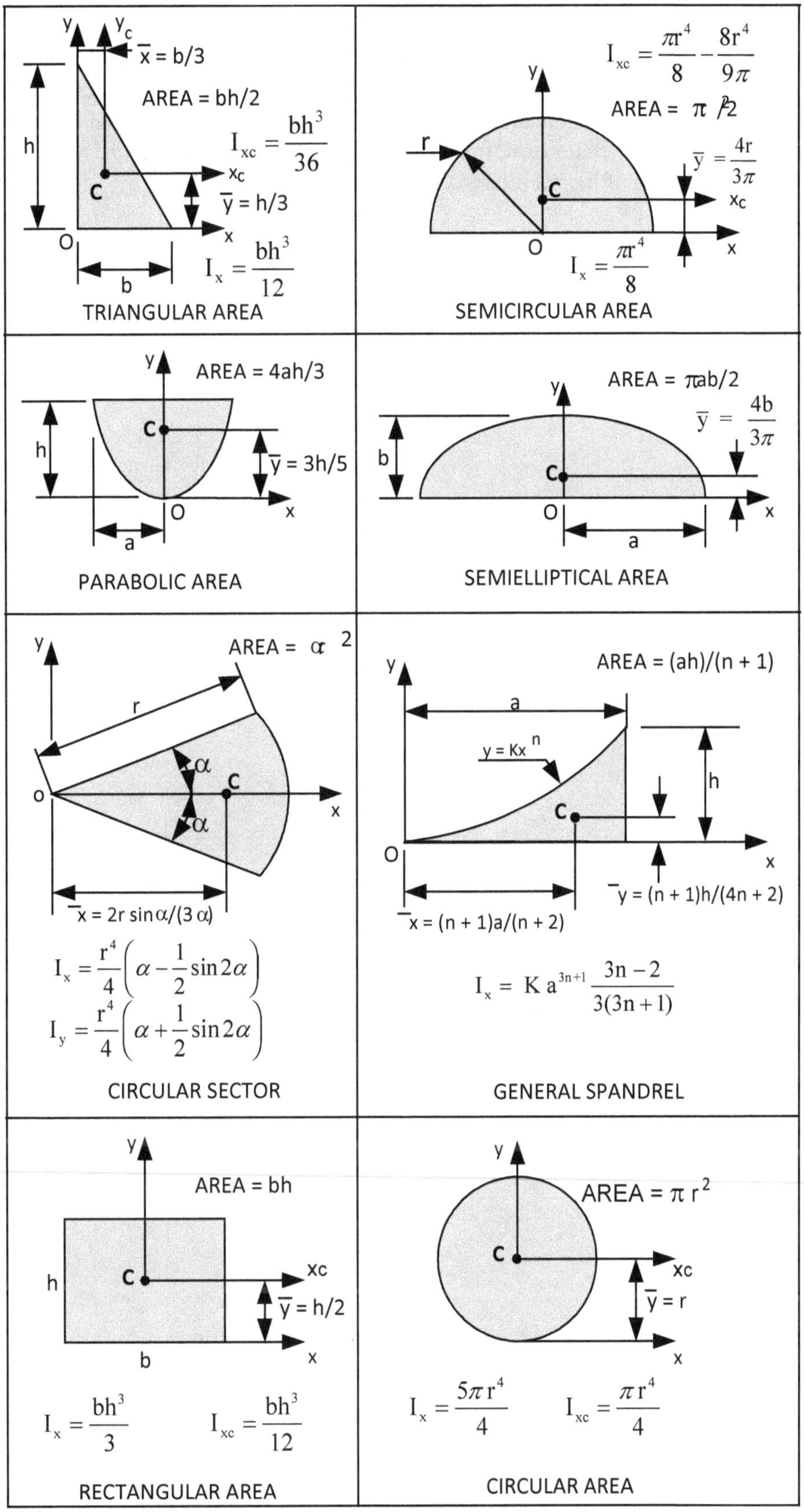

Fig. C.8 Area properties of some common shapes.

C.5 SECOND MOMENT OF THE AREA

The second moment of the area is also know as the moment of inertia. In our discussion of beam theory in Chapter 8, we indicated the importance of the moment of inertia I_z in determining the stresses produced by internal moments. Several different second moments of the area A, illustrated in Fig. C.10, are defined in this Section.

The second moment of the area, often called the moment of inertia, is referenced to one or both of the coordinate axes. The moment of inertia relative to the y and z axes is defined by:

$$I_z = \int_A y^2 dA \qquad I_y = \int_A z^2 dA \tag{C.12}$$

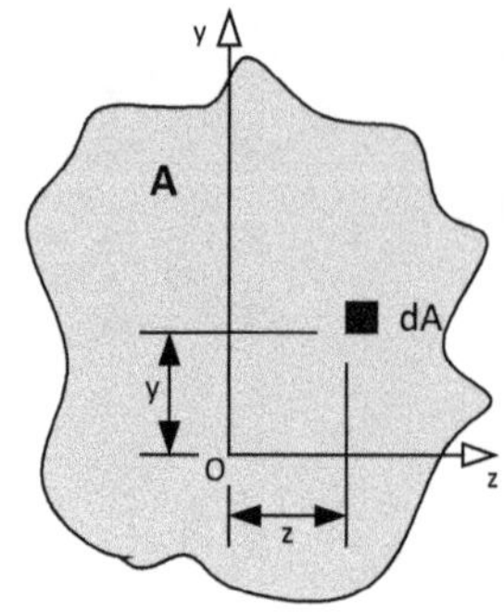

Fig. C.10 An elemental area dA is used when integrating to determine the second moment of the area A.

We also define a polar moment of inertia of the area A relative to the origin O of the coordinate system as indicated in Fig. C.11 as:

$$J_O = \int_A r^2 dA = \int_A (z^2 + y^2) dA = \int_A z^2 dA + \int_A y^2 dA$$

$$J_O = I_y + I_z \tag{C.13}$$

The units for the second moments of the area are in.^{4} in the U. S. Customary system and m^4 or mm^4 in the SI system.

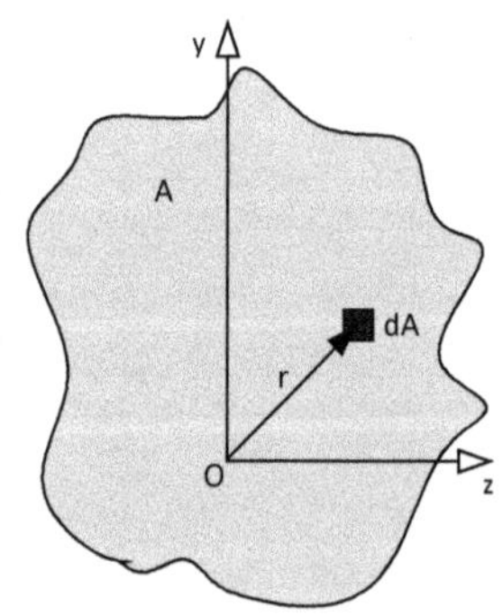

Fig. C.11 Coordinate system for the polar moment of inertia J_O.

Finally, the radius of gyration of an area A with respect to its axes is defined by:

$$I_z = r_z^2 A \qquad I_y = r_y^2 A \qquad J_O = r_O^2 A \tag{C.14}$$

The symbols r_z and r_y reference the radii of gyration relative to the z and y-axes, respectively. The radius of gyration for the polar moment of inertia is r_O.

Let's consider a few examples to demonstrate the integration of Eq. (C.12) to determine the second moment of the area.

EXAMPLE C.4

For the rectangular cross sectional area and coordinate system as defined in Fig C.12, determine the equations for I_z and I_y.

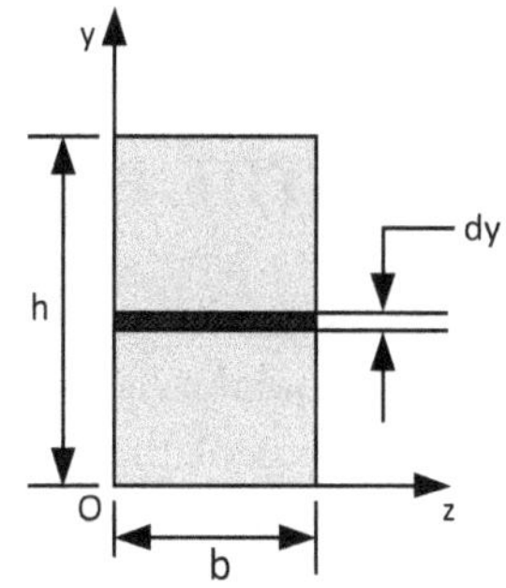

Fig. C.12 A rectangular area with a coordinate system along its edges.

Solution:

To determine I_z, we write Eq. (C.12) and observe that dA = b dy. Then:

$$I_z = \int_A y^2 dA = b\int_0^h y^2 dy = b\left[\frac{y^3}{3}\right]_0^h = \frac{bh^3}{3} \tag{C.15}$$

Similarly for I_y, we write Eq. C.12 and observe that dA = h dz. Then:

$$I_y = \int_A z^2 dA = h\int_0^b z^2 dz = h\left[\frac{z^3}{3}\right]_0^b = \frac{hb^3}{3} \tag{C.16}$$

To show the importance of the location of the coordinate system in determining the second moment of the area, let's shift the origin of the coordinate system to the center of the rectangle as shown in Fig C.13.

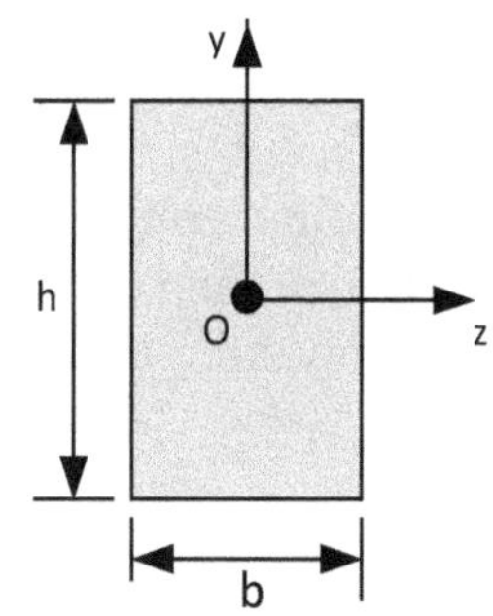

Fig. C.13 A rectangular area with a centrally located coordinate system.

EXAMPLE C.5

Determine the moment of inertia of a rectangular area relative to its centroidal axes.

Solution:

To determine the moment of inertia I_z, we write Eq. (C.12) and note that dA = b dy.

$$I_z = \int_A y^2 dA = b\int_{-h/2}^{h/2} y^2 dy = b\left[\frac{y^3}{3}\right]_{-h/2}^{h/2} = \frac{bh^3}{12} \tag{C.17}$$

To determine the moment of inertia I_y, we write Eq. (C.12) and note that dA = h dz.

$$I_y = \int_A z^2 dA = h\int_{-b/2}^{b/2} z^2 dz = h\left[\frac{z^3}{3}\right]_{-b/2}^{b/2} = \frac{hb^3}{12} \tag{C.18}$$

A comparison of the results for the moments of inertia for the rectangle clearly indicates the importance of the location of the coordinate system relative to the area in question. As we may be required to determine the moment of inertia about axes with arbitrary locations, a method for accounting for shifting the position of coordinates is useful.

C.6 THE PARALLEL AXIS THEOREM

Let's again consider an arbitrary area A positioned some distance from the z-axis as shown in Fig C.14. The centroidal axis of the area A is known and is identified with the z' axis. Assume that we have determined the second moment of the area with respect to the centroidal axis z'. Suppose we are required to compute the inertia I_z with respect to an axis z that is parallel to the centroidal axis. We begin again with Eq. C.12 and write:

$$I_z = \int_A y^2 dA = \int_A (y_1 + d)^2 dA$$

$$I_z = \int_A y_1^2 dA + 2d \int_A y_1 dA + d^2 \int_A dA \qquad \text{(C.19)}$$

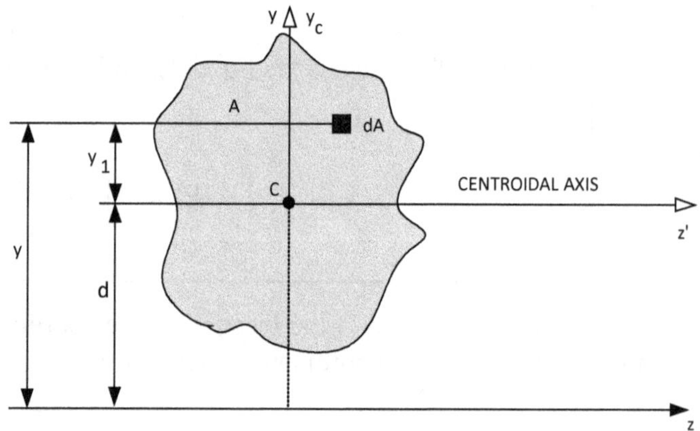

Fig. C.14 An arbitrary area A position a distance d from the z-axis.

The first term in Eq. (C.19) is the moment inertia $I_{z'}$ of the area about the centroidal axis. The second term $\int y_1 \, dA$ is the first moment of the area about the centroidal axis, which is zero by definition of the centroid. The final term is simply Ad^2. Hence, the parallel axis theorem is given by:

$$I_z = I_{z'} + Ad^2 \qquad \text{(C.20)}$$

where $I_{z'}$ is the second moment of the area about the centroidal axis.

EXAMPLE C.6

To demonstrate the use of Eq. (C.20), let's determine the moment of inertia I_z of the rectangle shown in Fig C.15.

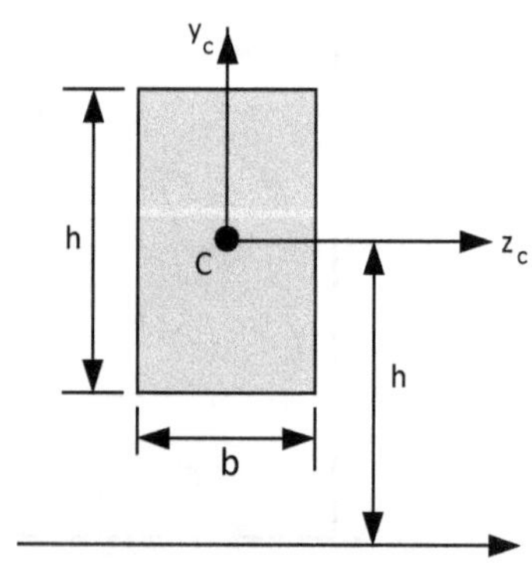

Fig. C.15 A rectangular area shifted by an amount d = h relative to the z-axis.

Solution:

To determine the moment of inertia I_z, we recall Eq. (C.20) and write:

$$I_z = I_{z'} + Ad^2$$

From Eq. (C.17) it is clear that $I_{z'} = bh^3/12$. Then we substitute into Eq. (C.20) to obtain:

$$I_z = bh^3/12 + (bh)(h^2) = (13/12)\, bh^3$$

This example illustrates two points. First, the parallel axis theorem is very helpful in determining the increase in the moment of inertia when the reference axis is some distance removed from the centroidal axis. Second, the moment of inertia is very sensitive to the movement of the reference axis relative to the centroidal axis. In this example, we moved the reference axis by an amount equal to the height h of the section and increased the inertia by a factor of 12.

C.7 MOMENTS OF INERTIA OF COMPOSITE AREAS

To determine the moment of inertia of areas with complex shapes, we divide the area into subsections to form a composite area. Each subsection is a simple shape, such as a square, rectangle, circle or semi-circle with known properties. Let's consider the structural tee with a web and a flange as illustrated in Fig. C.16 and demonstrate the procedure for determining the moment of inertia.

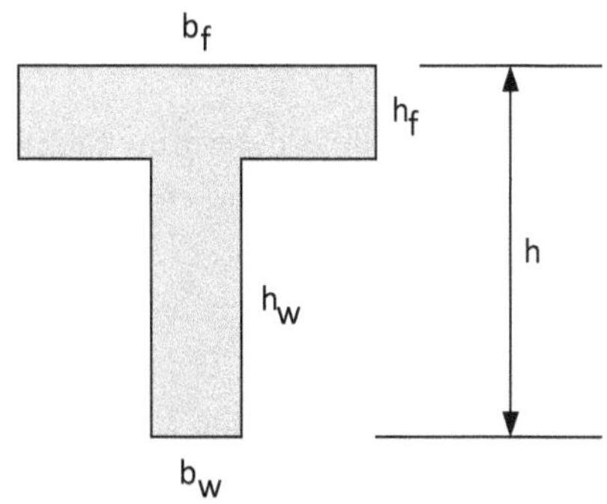

Fig. C.16 A structural tee is divided into two areas A_{web} and A_{flange}.

The procedure for determining the properties of the composite area representing the structural tee involves the three steps listed below:

- Determine the location of the centroidal axis of the composite area.
- Determine the moment of inertia of each area of the composite section about its centroidal axis.
- Employ the parallel axis theorem to determine the moment of inertia of the total section relative to the centroidal axis.

EXAMPLE C.6

Determine the moment of inertia I_z of the structural tee shown in Fig. C.16 relative to its centroidal axis. The dimensions of the structural tee are given by:

$h_f = 0.1$ h, $h_w = 0.9$ h, $b_w = 0.1$ b, and b = 150 mm and h = 200 mm.

Solution:

To determine the location of the centroidal axis of the composite area, subdivide the structural tee into two areas, namely A_1 and A_2 as shown in Fig. C.17. Then apply Eq. (C.10) to determine the location of the centroid relative to the z (reference) axis.

Fig. C.17 Subdivide the structural tee to form two rectangular areas and establish a convenient reference axis.

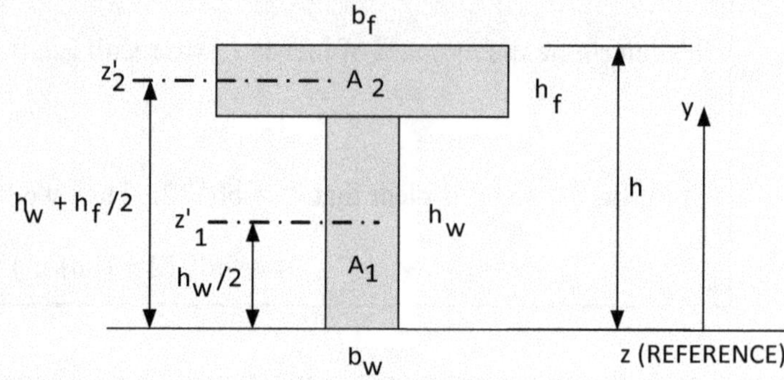

From Eq. (C.10), we write:

$$\Sigma Q_{z(REFERENCE)} = \overline{Y} A_t = \Sigma\, \overline{y}_n A_n = \overline{y}_1 A_1 + \overline{y}_2 A_2 \qquad \text{(a)}$$

Substituting the dimensional quantities from Fig. C.17 into Eq. (a) yields:

$$\overline{Y} = \frac{A_1 \dfrac{h_w}{2} + A_2 (h_w + \dfrac{h_f}{2})}{A_1 + A_2} = \frac{\dfrac{b_w h_w^2}{2} + b_f h_f h_w + \dfrac{b_f h_f^2}{2}}{b_w h_w + b_f h_f} \qquad \text{(b)}$$

Recall $h_f = 0.1$ h, $h_w = 0.9$ h, $b_w = 0.1$ b, $b_f =$ b, b = 150 mm, and h = 200 mm. Substitute these quantities into Eq. (b) to obtain:

$$\overline{Y} = 0.713 \text{ h} = 0.713 \times 200 = 142.6 \text{ mm} \qquad \text{(c)}$$

Next determine the moment for inertia of each area of the composite section about its centroidal axis. For A_1 (the web) with a width $b_w = 0.1$, b = 15 mm, and a height $h_w = 0.9$ h = 180 mm, we use Eq. (C.17) and write:

$$I_{z'1} = b_w h_w^3/12 = (15)(180)^3/12 = 7.29 \times 10^6 \text{ mm}^4 \qquad \text{(d)}$$

For A_2 (the flange) with a width $b_f = 150$ mm, and a height $h_f = 0.1$ h = 20 mm, we write:

$$I_{z'2} = b_f h_f^3/12 = (150)(20)^3/12 = 0.100 \times 10^6 \text{ mm}^4 \qquad \text{(e)}$$

Finally, employ the parallel axis theorem to determine the moment of inertia of the total area relative to the centroidal axis. Note for the composite area, we express the moment of inertia due to the two areas as:

$$I_z = I_{z1} + I_{z2} \qquad \text{(f)}$$

We use Eq. (C.20) to expand Eq. (f) as:

$$I_z = I_{z1'} + A_1 d_1^2 + I_{z2'} + A_2 d_2^2 \qquad \text{(g)}$$

The dimensions d_1 and d_2 are given by:

$$d_1 = \overline{Y} - h_w/2 = 142.6 - 90 = 52.6 \text{ mm} \qquad \text{(h)}$$

$$d_2 = h_w + (h_f/2) - \overline{Y} = 180 + 10 - 142.6 = 47.4 \text{ mm} \qquad \text{(i)}$$

Substituting numerical values for the terms in Eq. (g) yields the final result:

$$I_z = (7.290 \times 10^6) + (15)(180)(52.6)^2 + (0.1 \times 10^6) + (150)(20)(47.4)^2 = 21.60 \times 10^6 \text{ mm}^4$$

The determination of the moment of inertia for complex shapes is simple but tedious. Care must be exercised to avoid numerical errors in computing each of the quantities shown in Eq. (g). The moment of inertia is important in determining the bending stresses in beams because I_z occurs in the denominator of the well know flexural formula $\sigma = -My/I_z$. The procedure for establishing the position of the centroidal axis is also important because it locates the neutral axis about which bending occurs. When the position of the neutral axis is known, we can establish y_{max}, I_z and the maximum bending stress σ.

EXAMPLE C.7

Determine the following quantities for the unsymmetrical wide flanged section shown in Fig. C.18.

1. The location of the centroid relative to a defined reference axis.
2. The moment of inertia of the web, top flange, and bottom flange relative to the centroidal axis.
3. The moment of inertia of the section relative to the centroidal axis.

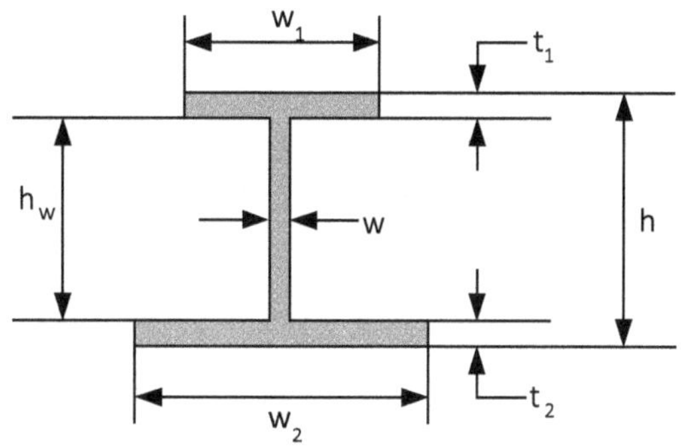

Fig. C.18 Dimensions of the unsymmetrical wide flanged section.

Solution:

To determine the location of the centroidal axis of the composite area, divide the unsymmetrical wide flanged section into three different areas consisting of the top flange, web and bottom flange as shown in Fig. C.19. Also establish a reference axis, z that is used as the datum for dimensioning the location of the centroid.

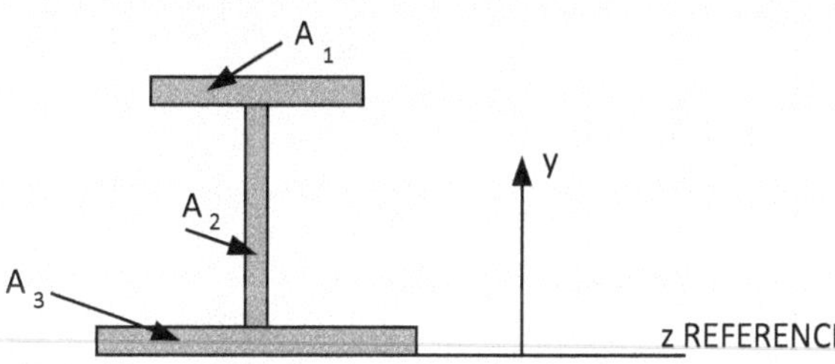

Fig. C.19 Dividing the unsymmetrical wide flanged section into three parts.

Let's apply Eq. (C.10) to determine the location of the centroid of this section relative to the z (reference) axis.

$$\Sigma Q_{z(REFERENCE)} = \overline{Y} A_t = \Sigma\, \overline{y}_n A_n = \overline{y}_1 A_1 + \overline{y}_2 A_2 + \overline{y}_3 A_3 \qquad \text{(a)}$$

Substituting the symbols for the dimensions from Fig. C.18 into Eq. (a) yields:

$$\overline{Y} = \frac{A_1(h_w + t_2 + t_1/2) + A_2(t_2 + h_w/2) + A_3(t_2/2)}{A_1 + A_2 + A_3}$$

$$\overline{Y} = \frac{1}{2}\left[\frac{t_1 w_1(2h_w + 2t_2 + t_1) + h_w w(2t_2 + h_w) + w_2 t_2^2}{w_1 t_1 + h_w w + w_2 t_2}\right] \qquad \text{(b)}$$

Suppose we assume the proportions of the cross section as:

$$w_1 = 10\ w;\ w_2 = 16\ w;\ t_1 = 0.15\ h;\ t_2 = 0.10\ h;\ \text{and}\ h_w = 0.75\ h \qquad \text{(c)}$$

Then Eq. (b) reduces to:

$$\overline{Y} = 0.4736\ h \qquad \text{(d)}$$

Finally if the height h = 300-mm and web thickness w = 15 mm, the location of the centroidal axis is given by:

$$\overline{Y} = 142.1\ \text{mm} \qquad \text{(e)}$$

To determine the moment of inertia of each area of the composite section about its centroidal axis, apply Eq. (C.17) to each of the three rectangular areas to obtain:

$$I_{z'1} = (1/12)w_1t_1^3 = (1/12)(150)(45)^3 = 1.139 \times 10^6\ \text{mm}^4 \qquad \text{(f)}$$

$$I_{z'2} = (1/12)wh_w^3 = (1/12)(15)(225)^3 = 14.24 \times 10^6\ \text{mm}^4 \qquad \text{(g)}$$

$$I_{z'3} = (1/12)\ w_2t_2^3 = (1/12)(240)(30)^3 = 0.540 \times 10^6\ \text{mm}^4 \qquad \text{(h)}$$

Next use the parallel axis theorem [Eq. (C.20)] to determine the moment of inertia of each of the three areas relative to the centroidal axis. For the top flange (Area 1):

$$I_{z1} = I_{z'1} + A_1d_1^2 = I_{z'1} + (w_1\ t_1/4)(2h - t_1 - 2\overline{Y})^2$$

$$I_{z1} = 1.139 \times 10^6 + (150)(11.25)(600 - 45 - 284.2)^2$$

$$I_{z1} = 1.139 \times 10^6 + 123.7 \times 10^6 = 124.8 \times 10^6\ \text{mm}^4 \qquad \text{(i)}$$

For the web (Area 2):

$$I_{z2} = I_{z'2} + A_2d_2^2 = I_{z'2} + (wh_w/4)(2t_2 + h_w - 2\overline{Y})^2$$

$$I_{z2} = 14.24 \times 10^6 + (15)(56.25)(60 + 225 - 284.2)^2$$

$$I_{z2} = 14.24 \times 10^6 + 0.0005 \times 10^6 = 14.24 \times 10^6\ \text{mm}^4 \qquad \text{(j)}$$

For the bottom flange (Area 3):

$$I_{z3} = I_{z'3} + A_3d_3^2 = I_{z'3} + (w_2\ t_2/4)(2\ y - t_2)^2$$

$$I_{z3} = 0.540 \times 10^6 + (240)(7.5)(284.2 - 30)^2$$

$$I_{z3} = 0.540 \times 10^6 + 116.3 \times 10^6 = 116.8 \times 10^6\ \text{mm}^4 \qquad \text{(k)}$$

Before we complete the computation, examine the results listed in Eqs. (i – k). For both of the flanges, we note that the contribution of the $I_{z'}$ term was negligible compared to the Ad^2 term. For the web the $I_{z'}$ term was dominant and the $A_2d_2^2$ term was negligible. To complete the solution for the moment of inertia of this composite area, we sum the moments of inertia due to the three areas as:

$$I_z = I_{z1} + I_{z2} + I_{z3} \qquad \text{(l)}$$

Substituting the results for Eqs. (i – k) into Eq. (l) yields:

$$I_z = (124.8 + 14.24 + 116.8) \times 10^6 = 255.8 \times 10^6\ \text{mm}^4 \qquad \text{(m)}$$

The contribution of the web to the moment of inertia about the centroidal axis of this unsymmetrical wide flange section was a small part of the total (less than 6%). This example shows that the main purpose of the web is to move the flanges outward from the neutral (centroidal) axis.

PROBLEMS

C.1 Determine the area of a right triangle if its base is 30 mm and its height is 70 mm.

C.2 Determine the area of an isosceles triangle if its base is 30 mm and its height is 70 mm.

C.3 Determine the area of a right triangle if its base is 4in. and its height is 7 in.

C.4 Determine the area of an isosceles triangle if its base is 5 in. and its height is 9.5 in.

C.5 Determine the area of a circle with a diameter of 14 mm.

C.6 Determine the area of a circle with a diameter of 3.2 in.

C.7 Determine the area of a circle with a diameter of 88 mm.

C.8 Determine the area of a circle with a diameter of 3.6 in.

C.9 Determine the area of an ellipse if the major and minor axes are defined by a = 2b = 16 mm.

C.10 Determine the area of an ellipse if the major and minor axes are defined by a = 3b = 6 in.

C.11 Determine the area of an ellipse if the major and minor axes are defined by a = 4b = 20 mm.

C.12 Determine the area of an ellipse if the major and minor axes are defined by a = (3.5)b = 7.0 in.

C.13 Draw a right triangle with a base of 30 mm and height of 70 mm and then locate and dimension its centroid.

C.14 Draw an isosceles triangle with a base of 30 mm and height of 70 mm and then locate and dimension its centroid.

C.15 Draw a right triangle with a base of 4in. and height of 7 in. and then locate and dimension its centroid.

C.16 Draw an isosceles triangle with a base of 5 in. and height of 9.5 in. and then locate and dimension its centroid.

C.17 Verify the location of the centroid for the semicircular area defined in the figure to the right.

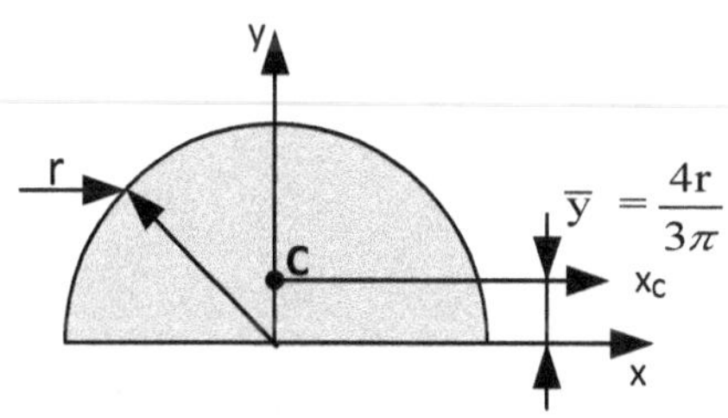

C.18 Verify the location of the centroid for the semi-elliptical area defined in the figure to the right.

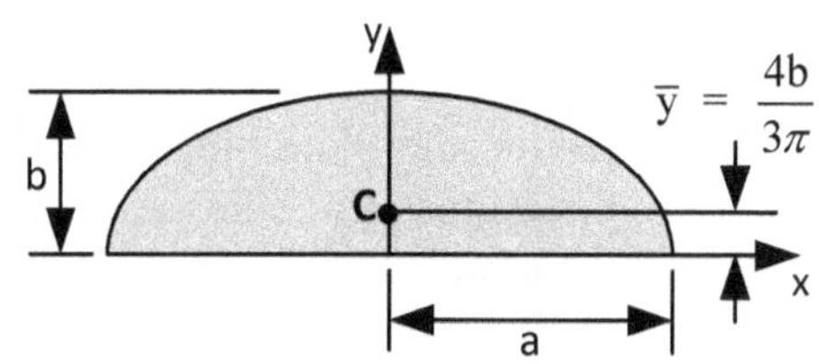

C.19 Verify the location of the centroid for the parabolic area defined in the figure to the left.

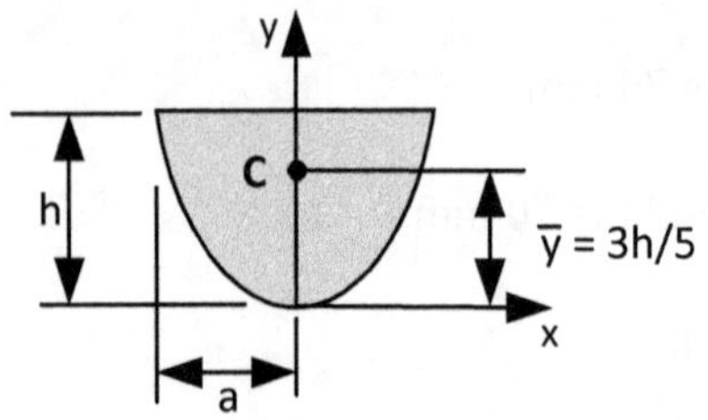

C.20 Verify the location of the centroid for the circular sector defined in the figure to the right.

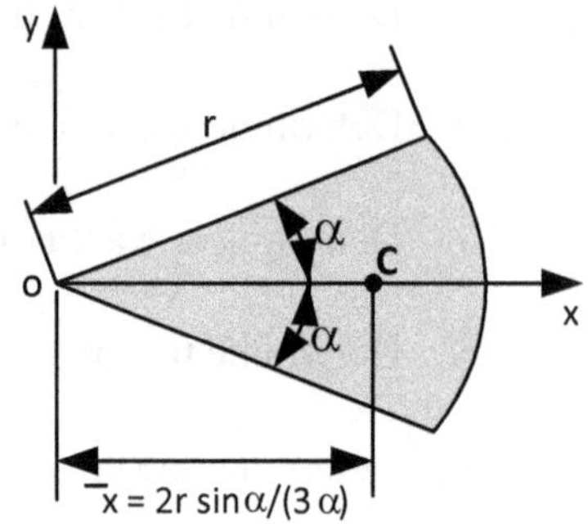

C.21 Verify the location of the centroid for the general spandrel defined in the figure to the left.

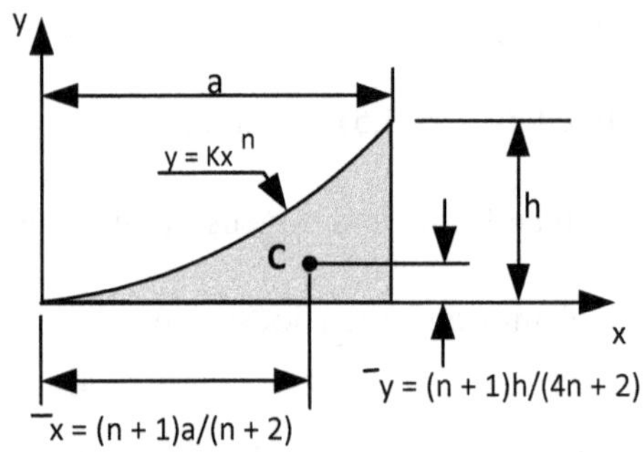

C.22 For the unusual plate shown in the figure to the right, derive an equation locating the centroid relative to point A. The elliptical hole is centered in the rectangular portion of the plate.

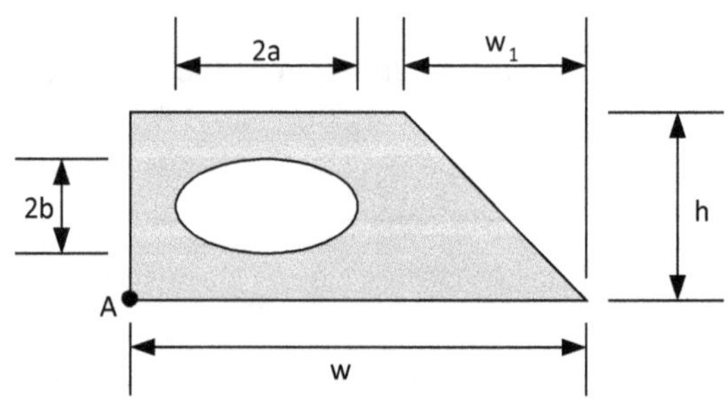

C.23 Determine the location of the centroid relative to point A for the plate defined in the figure to the right. All dimensions are given in mm, and the elliptical hole is centered in the rectangular portion of the plate.

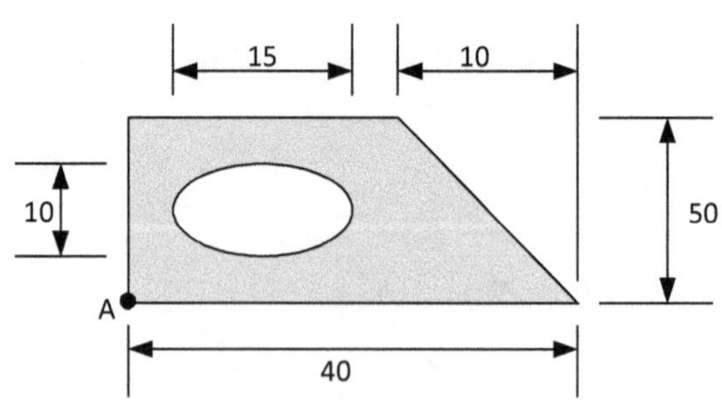

C.24 Divide the plate, illustrated in the figure to the right, into three areas (an ellipse, rectangle and triangle). Determine the moment of inertia for each area about its centroidal axis.

C.25 Verify that the moment of inertia of an ellipse about its centroidal axis is given by:

$$I_x = \pi a b^3 /4 \qquad \text{and} \qquad I_y = \pi a^3 b/4$$

C.26 Determine the moment of inertia for the total area of the plate, shown in the figure to the left, about its centroidal axis.

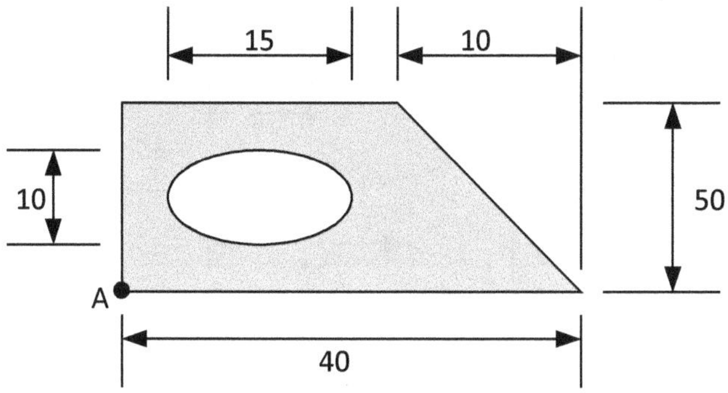

C.27 For the U shaped section, shown in the figure to the right, determine the location of the centroid $\overline{y}$ relative to an axis along its base. All dimensions are in inches.

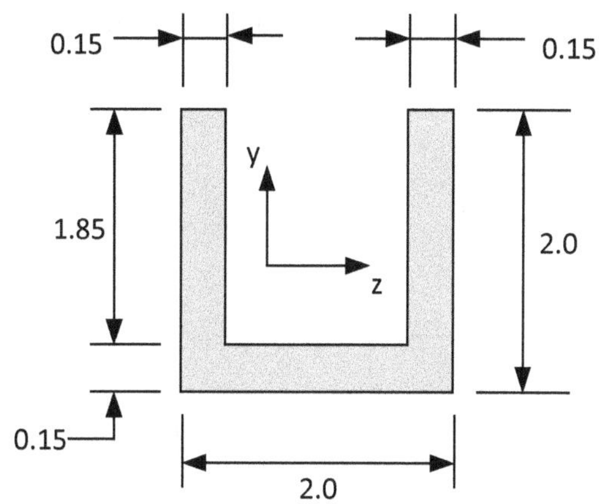

C.28 Determine the location of the centroid $\overline{y}$, relative to an axis along its base, for the U shaped section shown in the figure to the left. All dimensions are in inches.

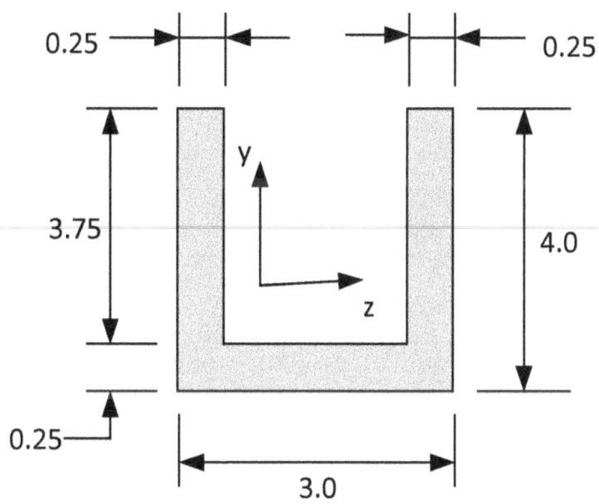

C.29 Determine the moment of inertia I_z about the centroid of the U shaped member defined in (a) Problem C.27 and (b) Problem C.28.

C.30 Verify that the polar moment of inertia of a quarter circle area with a radius R is $J_o = \pi R^4/8$.

C.31 For the uncommon shape presented in the figure to the right, determine the centroid and the moment of inertia relative to the centroidal axis. The rectangular hole is centered in the elliptical area.

C.32 For the uncommon shape presented in the figure to the right, determine the moment of inertia relative to the z-axis.

C.33 For the uncommon shape presented in the figure to the right, determine the moment of inertia relative to the y-axis.

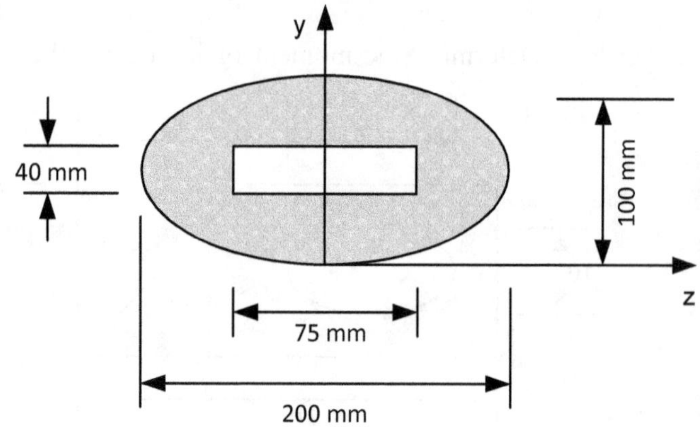

C.34 For the irregular cross sectional area, described in the figure to the left, locate the centroid relative to both the x and y axes. The dimensions are given in mm.

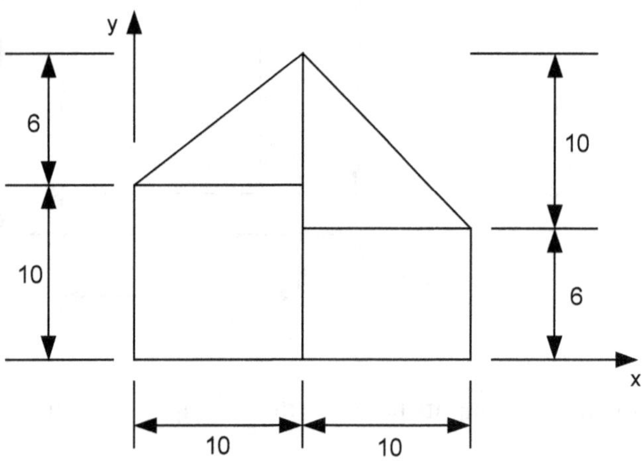

APPENDIX D

GEOMETRIC PROPERTIES OF ROLLED STEEL SHAPES

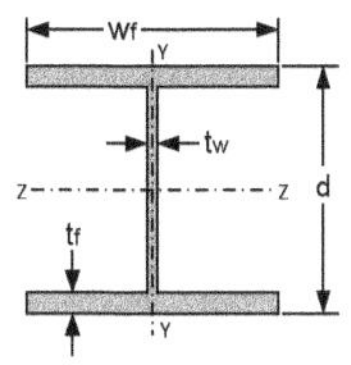

Wide-Flange Beams (U.S. Customary Units)

			Flange		Web	AXIS Z-Z			AXIS Y-Y		
Designation*	Area	Depth	Width	Thickness	Thickness	I	Z	r	I	Z	r
	(in.^{2})	(in.)	(in.)	(in.)	(in.)	(in.^{4})	(in.^{3})	(in.)	(in.^{4})	(in.^{3})	(in.)
W36 × 230	67.6	35.90	16.470	1.260	0.760	15000	837	14.9	940	114	3.73
× 160	47.0	36.01	12.000	1.020	0.650	9750	542	14.4	295	49.1	2.50
W33 × 201	59.1	33.68	15.745	1.150	0.715	11500	684	14.0	749	95.2	3.56
× 152	44.7	33.49	11.565	1.055	0.635	8160	487	13.5	273	47.2	2.47
× 130	38.3	33.09	11.510	0.855	0.580	6710	406	13.2	218	37.9	2.39
W30 × 132	38.9	30.31	10.545	1.000	0.615	5770	380	12.2	196	37.2	2.25
× 108	31.7	29.83	10.475	0.760	0.545	4470	299	11.9	146	27.9	2.15
W27 × 146	42.9	27.38	13.965	0.975	0.605	5630	411	11.4	443	63.5	3.21
× 94	27.7	26.92	9.990	0.745	0.490	3270	243	10.9	124	24.8	2.12
W24 × 104	30.6	24.06	12.750	0.750	0.500	3100	258	10.1	259	40.7	2.91
× 84	24.7	24.10	9.020	0.770	0.470	2370	196	9.79	94.4	20.9	1.95
× 62	18.2	23.74	7.040	0.590	0.430	1550	131	9.23	34.5	9.80	1.38
W21 × 101	29.8	21.36	12.290	0.800	0.500	2420	227	9.02	248	40.3	2.89
× 83	24.3	21.43	8.355	0.835	0.515	1830	171	8.67	81.4	19.5	1.83
× 62	18.3	20.99	8.240	0.615	0.400	1330	127	8.54	57.5	13.9	1.77
W18 × 97	28.5	18.59	11.145	0.870	0.535	1750	188	7.82	201	36.1	2.65
× 76	22.3	18.21	11.035	0.680	0.425	1330	146	7.73	152	27.6	2.61
× 60	17.6	18.24	7.555	0.695	0.415	984	108	7.47	50.1	13.3	1.69
W16 × 100	29.4	16.97	10.425	0.985	0.585	1490	175	7.10	186	35.7	2.52
× 67	19.7	16.33	10.235	0.665	0.395	954	117	6.96	119	23.2	2.46
× 40	11.8	16.01	6.995	0.505	0.305	518	64.7	6.63	28.9	8.25	1.57
× 26	7.68	15.69	5.500	0.345	0.250	301	38.4	6.26	9.59	3.49	1.12
W14 × 120	35.3	14.48	14.670	0.940	0.590	1380	190	6.24	495	67.5	3.74
× 82	24.1	14.31	10.130	0.855	0.510	882	123	6.05	148	29.3	2.48
× 43	12.6	13.66	7.995	0.530	0.305	428	62.7	5.82	45.2	11.3	1.89
× 30	8.85	13.84	6.730	0.385	0.270	291	42.0	5.73	19.6	5.82	1.49
W12 × 96	28.2	12.71	12.160	0.900	0.550	833	131	5.44	270	44.4	3.09
× 65	19.1	12.12	12.000	0.605	0.390	533	87.9	5.28	174	29.1	3.02
× 50	14.7	12.19	8.080	0.640	0.370	394	64.7	5.18	56.3	13.9	1.96
× 30	8.79	12.34	6.520	0.440	0.260	238	38.6	5.21	20.3	6.24	1.52
W10 × 60	17.6	10.22	10.080	0.680	0.420	341	66.7	4.39	116	23.0	2.57
× 45	13.3	10.10	8.020	0.620	0.350	248	49.1	4.33	53.4	13.3	2.01
× 30	8.84	10.47	5.810	0.510	0.300	170	32.4	4.38	16.7	5.75	1.37
× 22	6.49	10.17	5.750	0.360	0.240	118	23.2	4.27	11.4	3.97	1.33
W8 × 40	11.7	8.25	8.070	0.560	0.360	146	35.5	3.53	49.1	12.2	2.04
× 31	9.13	8.00	7.995	0.435	0.285	110	27.5	3.47	37.1	9.27	2.02
× 24	7.08	7.93	6.495	0.400	0.245	82.8	20.9	3.42	18.3	5.63	1.61
× 15	4.44	8.11	4.015	0.315	0.245	48.0	11.8	3.29	3.41	1.70	0.876
W6 × 25	7.34	6.38	6.080	0.455	0.320	53.4	16.7	2.70	17.1	5.61	1.52
× 16	4.74	6.28	4.030	0.405	0.260	32.1	10.2	2.60	4.43	2.20	0.967
W5 × 16	4.68	5.01	5.000	0.360	0.240	21.3	8.51	2.13	7.51	3.00	1.27
W4 × 13	3.83	4.16	4.060	0.345	0.280	11.3	5.46	1.72	3.86	1.90	1.00

Courtesy of the American Institute of Steel Construction.
*W is the symbol for a wide-flange beam, followed by the nominal depth in inches, then the weight in pounds per foot of length.

GEOMETRIC PROPERTIES OF ROLLED STEEL SHAPES

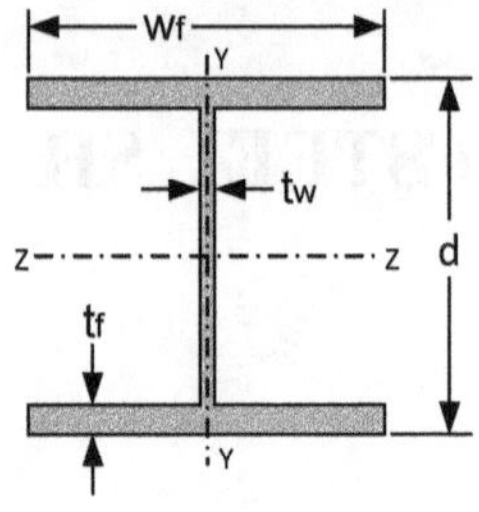

Wide-Flange Beams (SI Units)

Designation*	Area (mm^2)	Depth (mm)	Width (mm)	Flange Thickness (mm)	Web Thickness (mm)	AXIS Z-Z I $(10^6\ mm^4)$	AXIS Z-Z Z $(10^3\ mm^3)$	AXIS Z-Z r (mm)	AXIS Y-Y I $(10^6\ mm^4)$	AXIS Y-Y Z $(10^3\ mm^3)$	AXIS Y-Y r (mm)
W914 × 342	43610	912	418	32.0	19.3	6245	13715	378	391	1870	94.7
× 238	30325	915	305	25.9	16.5	4060	8880	366	123	805	63.5
W838 × 299	38130	855	400	29.2	18.2	4785	11210	356	312	1560	90.4
× 226	28850	851	294	26.8	16.1	3395	7980	343	114	775	62.7
× 193	24710	840	292	21.7	14.7	2795	6655	335	90.7	620	60.7
W762 × 196	25100	770	268	25.4	15.6	2400	6225	310	81.6	610	57.2
× 161	20450	758	266	19.3	13.8	1860	4900	302	60.8	457	54.6
W686 × 217	27675	695	355	24.8	15.4	2345	6735	290	184	1040	81.5
× 140	17870	684	254	18.9	12.4	1360	3980	277	51.6	406	53.8
W610 × 155	19740	611	324	19.1	12.7	1290	4230	257	108	667	73.9
× 125	15935	612	229	19.6	11.9	985	3210	249	39.3	342	49.5
× 92	11750	603	179	15.0	10.9	645	2145	234	14.4	161	35.1
W533 × 150	19225	543	312	20.3	12.7	1005	3720	229	103	660	73.4
× 124	15675	544	212	21.2	13.1	762	2800	220	33.9	320	46.5
× 92	11805	533	209	15.6	10.2	554	2080	217	23.9	228	45.0
W457 × 144	18365	472	283	22.1	13.6	728	3080	199	83.7	592	67.3
× 113	14385	463	280	17.3	10.8	554	2395	196	63.3	452	66.3
× 89	11355	463	192	17.7	10.5	410	1770	190	20.9	218	42.9
W406 × 149	18970	431	265	25.0	14.9	620	2870	180	77.4	585	64.0
× 100	12710	415	260	16.9	10.0	397	1915	177	49.5	380	62.5
× 60	7615	407	178	12.8	7.7	216	1060	168	12.0	135	39.9
× 39	4950	399	140	8.8	6.4	125	629	159	3.99	57.2	28.4
W356 × 179	22775	368	373	23.9	15.0	574	3115	158	206	1105	95.0
× 122	15550	363	257	21.7	13.0	367	2015	154	61.6	480	63.0
× 64	8130	347	203	13.5	7.7	178	1025	148	18.8	185	48.0
× 45	5710	352	171	9.8	6.9	121	688	146	8.16	95.4	37.8
W305 × 143	18195	323	309	22.9	14.0	347	2145	138	112	728	78.5
× 97	12325	308	305	15.4	9.9	222	1440	134	72.4	477	76.7
× 74	9485	310	205	16.3	9.4	164	1060	132	23.4	228	49.8
× 45	5670	313	166	11.2	6.6	99.1	633	132	8.45	102	38.6
W254 × 89	11355	260	256	17.3	10.7	142	1095	112	48.3	377	65.3
× 67	8580	257	204	15.7	8.9	103	805	110	22.2	218	51.1
× 45	5705	266	148	13.0	7.6	70.8	531	111	6.95	94.2	34.8
× 33	4185	258	146	9.1	6.1	49.1	380	108	4.75	65.1	33.8
W203 × 60	7550	210	205	14.2	9.1	60.8	582	89.7	20.4	200	51.8
× 46	5890	203	203	11.0	7.2	45.8	451	88.1	15.4	152	51.3
× 36	4570	201	165	10.2	6.2	34.5	342	86.7	7.61	92.3	40.9
× 22	2865	206	102	8.0	6.2	20.0	193	83.6	1.42	27.9	22.3
W152 × 37	4735	162	154	11.6	8.1	22.2	274	68.6	7.12	91.9	38.6
× 24	3060	160	102	10.3	6.6	13.4	167	66.0	1.84	36.1	24.6
W127 × 24	3020	127	127	9.1	6.1	8.87	139	54.1	3.13	49.2	32.3
W102 × 19	2470	106	103	8.8	7.1	4.70	89.5	43.7	1.61	31.1	25.4

Courtesy of The American Institute of Steel Construction.

*W is the symbol for a wide-flange beam, followed by the nominal depth in mm, then the mass in kg per meter of length.

GEOMETRIC PROPERTIES OF ROLLED STEEL SHAPES

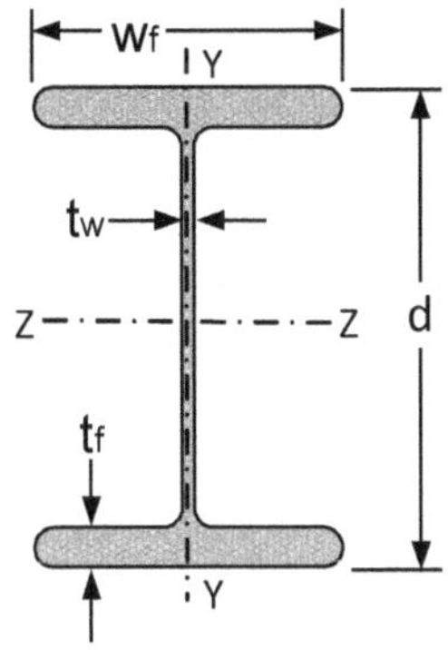

American Standard Beams (U.S. Customary Units)

Designation*	Area (in^2)	Depth (in.)	Flange Width (in.)	Flange Thickness (in.)	Web Thickness (in.)	AXIS Z-Z I ($in.^4$)	AXIS Z-Z Z ($in.^3$)	AXIS Z-Z r (in.)	AXIS Y-Y I ($in.^4$)	AXIS Y-Y Z ($in.^3$)	AXIS Y-Y r (in.)
S24 × 121	35.6	24.50	8.050	1.090	0.800	3160	258	9.43	83.3	20.7	1.53
× 106	31.2	24.50	7.870	1.090	0.620	2940	240	9.71	77.1	19.6	1.57
× 100	29.3	24.00	7.245	0.870	0.745	2390	199	9.02	47.7	13.2	1.27
× 90	26.5	24.00	7.125	0.870	0.625	2250	187	9.21	44.9	12.6	1.30
× 80	23.5	24.00	7.000	0.870	0.500	2100	175	9.47	42.2	12.1	1.34
S20 × 96	28.2	20.30	7.200	0.920	0.800	1670	165	7.71	50.2	13.9	1.33
× 86	25.3	20.30	7.060	0.920	0.660	1580	155	7.89	46.8	13.3	1.36
× 75	22.0	20.00	6.385	0.795	0.635	1280	128	7.62	29.8	9.32	1.16
× 66	19.4	20.00	6.255	0.795	0.505	1190	119	7.83	27.7	8.85	1.19
S18 × 70	20.6	18.00	6.251	0.691	0.711	926	103	6.71	24.1	7.72	1.08
× 54.7	16.1	18.00	6.001	0.691	0.461	804	89.4	7.07	20.8	6.94	1.14
S15 × 50	14.7	15.00	5.640	0.622	0.550	486	64.8	5.75	15.7	5.57	1.03
× 42.9	12.6	15.00	5.501	0.622	0.411	447	59.6	5.95	14.4	5.23	1.07
S12 × 50	14.7	12.00	5.477	0.659	0.687	305	50.8	4.55	15.7	5.74	1.03
× 40.8	12.0	12.00	5.252	0.659	0.462	272	45.4	4.77	13.6	5.16	1.06
× 35	10.3	12.00	5.078	0.544	0.428	229	38.2	4.72	9.87	3.89	0.980
× 31.8	9.35	12.00	5.000	0.544	0.350	218	36.4	4.83	9.36	3.74	1.000
S10 × 35	10.3	10.00	4.944	0.491	0.594	147	29.4	3.78	8.36	3.38	0.901
× 25.4	7.46	10.00	4.661	0.491	0.311	124	24.7	4.07	6.79	2.91	0.954
S8 × 23	6.77	8.00	4.171	0.426	0.441	64.9	16.2	3.10	4.31	2.07	0.798
× 18.4	5.41	8.00	4.001	0.426	0.271	57.6	14.4	3.26	3.73	1.86	0.831
S7 × 20	5.88	7.00	3.860	0.392	0.450	42.4	12.1	2.69	3.17	1.64	0.734
× 15.3	4.50	7.00	3.662	0.392	0.252	36.7	10.5	2.86	2.64	1.44	0.766
S6 × 17.25	5.07	6.00	3.565	0.359	0.465	26.3	8.77	2.28	2.31	1.30	0.675
× 12.5	3.67	6.00	3.332	0.359	0.232	22.1	7.37	2.45	1.82	1.09	0.705
S5 × 14.75	4.34	5.00	3.284	0.326	0.494	15.2	6.09	1.87	1.67	1.01	0.620
× 10	2.94	5.00	3.004	0.326	0.214	12.3	4.92	2.05	1.22	0.809	0.643
S4 × 9.5	2.79	4.00	2.796	0.293	0.326	6.79	3.39	1.56	0.903	0.646	0.569
× 7.7	2.26	4.00	2.663	0.293	0.193	6.08	3.04	1.64	0.764	0.574	0.581
S3 × 7.5	2.21	3.00	2.509	0.260	0.349	2.93	1.95	1.15	0.586	0.468	0.516
× 5.7	1.67	3.00	2.330	0.260	0.170	2.52	1.68	1.23	0.455	0.390	0.522

Courtesy of The American Institute of Steel Construction.

*S is the symbol for a standard beam, followed by the nominal depth in inches, then the weight in pounds per foot of length.

GEOMETRIC PROPERTIES OF ROLLED STEEL SHAPES

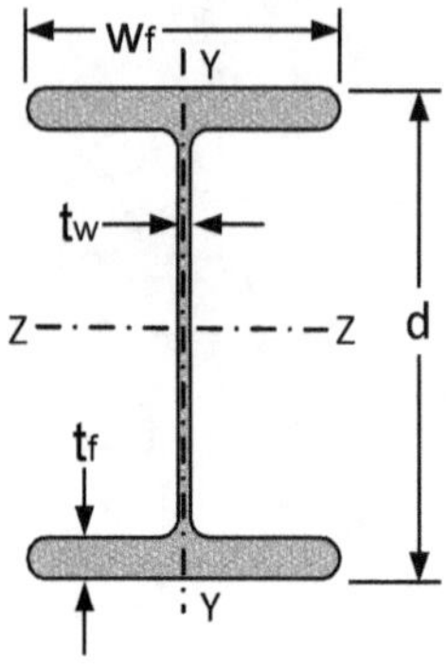

American Standard Beams (SI Units)

Designation*	Area (mm²)	Depth (mm)	Flange Width (mm)	Flange Thickness (mm)	Web Thickness (mm)	AXIS Z-Z I (10^6 mm⁴)	AXIS Z-Z Z (10^3 mm³)	AXIS Z-Z r (mm)	AXIS Y-Y I (10^6 mm⁴)	AXIS Y-Y Z (10^3 mm³)	AXIS Y-Y r (mm)
S610 × 180	22970	622.3	204.5	27.7	20.3	1315	4225	240	34.7	339	38.9
× 158	20130	622.3	199.9	27.7	15.7	1225	3935	247	32.1	321	39.9
× 149	18900	609.6	184.0	22.1	18.9	995	3260	229	19.9	216	32.3
× 134	17100	609.6	181.0	22.1	15.9	937	3065	234	18.7	206	33.0
× 119	15160	609.6	177.8	22.1	12.7	874	2870	241	17.6	198	34.0
S508 × 143	18190	515.6	182.9	23.4	20.3	695	2705	196	20.9	228	33.8
× 128	16320	515.6	179.3	23.4	16.8	658	2540	200	19.5	218	34.5
× 112	14190	508.0	162.2	20.2	16.1	533	2100	194	12.4	153	29.5
× 98	12520	508.0	158.9	20.2	12.8	495	1950	199	11.5	145	30.2
S457 × 104	13290	457.2	158.8	17.6	18.1	358	1690	170	10.0	127	27.4
× 81	10390	457.2	152.4	17.6	11.7	335	1465	180	8.66	114	29.0
S381 × 74	9485	381.0	143.3	15.8	14.0	202	1060	146	6.53	91.3	26.2
× 64	8130	381.0	139.7	15.8	10.4	186	977	151	5.99	85.7	27.2
S305 × 74	9485	304.8	139.1	16.7	17.4	127	832	116	6.53	94.1	26.2
× 61	7740	304.8	133.4	16.7	11.7	113	744	121	5.66	84.6	26.9
× 52	6645	304.8	129.0	13.8	10.9	95.3	626	120	4.11	63.7	24.1
× 47	6030	304.8	127.0	13.8	8.9	90.7	596	123	3.90	61.3	25.4
S254 × 52	6645	254.0	125.6	12.5	15.1	61.2	482	96.0	3.48	55.4	22.9
× 38	4815	254.0	118.4	12.5	7.9	51.6	408	103	2.83	47.7	24.2
S203 × 34	4370	203.2	105.9	10.8	11.2	27.0	265	78.7	1.79	33.9	20.3
× 27	3490	203.2	101.6	10.8	6.9	24.0	236	82.8	1.55	30.5	21.1
S178 × 30	3795	177.8	98.0	10.0	11.4	17.6	198	68.3	1.32	26.9	18.6
× 23	2905	177.8	93.0	10.0	6.4	15.3	172	72.6	1.10	23.6	19.5
S152 × 26	3270	152.4	90.6	9.1	11.8	10.9	144	57.9	0.961	21.3	17.1
× 19	2370	152.4	84.6	9.1	5.9	9.20	121	62.2	0.758	17.9	17.9
S127 × 22	2800	127.0	83.4	8.3	12.5	6.33	99.8	47.5	0.695	16.6	15.7
× 15	1895	127.0	76.3	8.3	5.4	5.12	80.6	52.1	0.508	13.3	16.3
S102 × 14	1800	101.6	71.0	7.4	8.3	2.83	55.6	39.6	0.376	10.6	14.5
× 11	1460	101.6	67.6	7.4	4.9	2.53	49.8	41.7	0.318	9.41	14.8
S76 × 11	1425	76.2	63.7	6.6	8.9	1.22	32.0	29.2	0.244	7.67	13.1
× 8.5	1075	76.2	59.2	6.6	4.3	1.05	27.5	31.2	0.189	6.39	13.3

Courtesy of The American Institute of Steel Construction.

*S is the symbol for a standard beam, followed by the nominal depth in mm, then the mass in kg per meter of length.

GEOMETRIC PROPERTIES OF ROLLED STEEL SHAPES

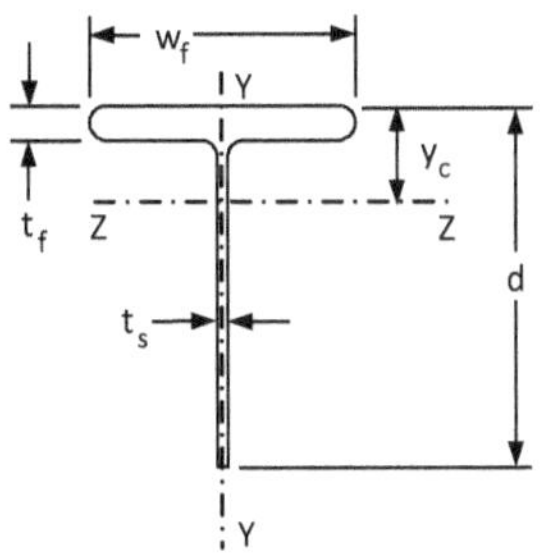

Structural Tees (U.S. Customary Units)

Designation*	Area	Depth of Tee	Flange		Stem Thickness	AXIS Z-Z			y_c	AXIS Y-Y		
	(in.^{2})	(in.)	Width (in.)	Thickness (in.)	(in.)	I (in.^{4})	Z (in.^{3})	r (in.)	(in.)	I (in.^{4})	Z (in.^{3})	r (in.)
WT18 × 115	33.8	17.950	16.470	1.260	0.760	934	67.0	5.25	4.01	470	57.1	3.73
× 80	23.5	18.005	12.000	1.020	0.650	740	55.8	5.61	4.74	147	24.6	2.50
WT15 × 66	19.4	15.155	10.545	1.000	0.615	421	37.4	4.66	3.90	98.0	18.6	2.25
× 54	15.9	14.915	10.475	0.760	0.545	349	32.0	4.69	4.01	73.0	13.9	2.15
WT12 × 52	15.3	12.030	12.750	0.750	0.500	189	20.0	3.51	2.59	130	20.3	2.91
× 47	13.8	12.155	9.065	0.875	0.515	186	20.3	3.67	2.99	54.5	12.0	1.98
× 42	12.4	12.050	9.020	0.770	0.470	166	18.3	3.67	2.97	47.2	10.5	1.95
× 31	9.11	11.870	7.040	0.590	0.430	131	15.6	3.79	3.46	17.2	4.90	1.38
WT9 × 38	11.2	9.105	11.035	0.680	0.425	71.8	9.83	2.54	1.80	76.2	13.8	2.61
× 30	8.82	9.120	7.555	0.695	0.415	64.7	9.29	2.71	2.16	25.0	6.63	1.69
× 25	7.33	8.995	7.495	0.570	0.355	53.5	7.79	2.70	2.12	20.0	5.35	1.65
× 20	5.88	8.950	6.015	0.525	0.315	44.8	6.73	2.76	2.29	9.55	3.17	1.27
WT8 × 50	14.7	8.485	10.425	0.985	0.585	76.8	11.4	2.28	1.76	93.1	17.9	2.51
× 25	7.37	8.130	7.070	0.630	0.380	42.3	6.78	2.40	1.89	18.6	5.26	1.59
× 20	5.89	8.005	6.995	0.505	0.305	33.1	5.35	2.37	1.81	14.4	4.12	1.57
× 13	3.84	7.845	5.500	0.345	0.250	23.5	4.09	2.47	2.09	4.80	1.74	1.12
WT7 × 60	17.7	7.240	14.670	0.940	0.590	51.7	8.61	1.71	1.24	247	33.7	3.74
× 41	12.0	7.155	10.130	0.855	0.510	41.2	7.14	1.85	1.39	74.2	14.6	2.48
× 34	9.99	7.020	10.035	0.720	0.415	32.6	5.69	1.81	1.29	60.7	12.1	2.46
× 24	7.07	6.985	8.030	0.595	0.340	24.9	4.48	1.87	1.35	25.7	6.40	1.91
× 15	4.42	6.920	6.730	0.385	0.270	19.0	3.55	2.07	1.58	9.79	2.91	1.49
× 11	3.25	6.870	5.000	0.335	0.230	14.8	2.91	2.14	1.76	3.50	1.40	1.04
WT6 × 60	17.6	6.560	12.320	1.105	0.710	43.4	8.22	1.57	1.28	172	28.0	3.13
× 48	14.1	6.355	12.160	0.900	0.550	32.0	6.12	1.51	1.13	135	22.2	3.09
× 36	10.6	6.125	12.040	0.670	0.430	23.2	4.54	1.48	1.02	97.5	16.2	3.04
× 25	7.34	6.095	8.080	0.640	0.370	18.7	3.79	1.60	1.17	28.2	6.97	1.96
× 15	4.40	6.170	6.520	0.440	0.260	13.5	2.75	1.75	1.27	10.2	3.12	1.52
× 8	2.36	5.995	3.990	0.265	0.220	8.70	2.04	1.92	1.74	1.41	0.706	0.773
WT5 × 56	16.5	5.680	10.415	1.250	0.755	28.6	6.40	1.32	1.21	118	22.6	2.68
× 44	12.9	5.420	10.265	0.990	0.605	20.8	4.77	1.27	1.06	89.3	17.4	2.63
× 30	8.82	5.110	10.080	0.680	0.420	12.9	3.04	1.21	0.884	58.1	11.5	2.57
× 15	4.42	5.235	5.810	0.510	0.300	9.28	2.24	1.45	1.10	8.35	2.87	1.37
× 6	1.77	4.935	3.960	0.210	0.190	4.35	1.22	1.57	1.36	1.09	0.551	0.785
WT4 × 29	8.55	4.375	8.220	0.810	0.510	9.12	2.61	1.03	0.874	37.5	9.13	2.10
× 20	5.87	4.125	8.070	0.560	0.360	5.73	1.69	0.988	0.735	24.5	6.08	2.04
× 12	3.54	3.965	6.495	0.400	0.245	3.53	1.08	0.999	0.695	9.14	2.81	1.61
× 9	2.63	4.070	5.250	0.330	0.230	3.41	1.05	1.14	0.834	3.98	1.52	1.23
× 5	1.48	3.945	3.940	0.205	0.170	2.15	0.717	1.20	0.953	1.05	0.532	0.841
WT3 × 10	2.94	3.100	6.020	0.365	0.260	1.76	0.693	0.774	0.560	6.64	2.21	1.50
× 6	1.78	3.015	4.000	0.280	0.230	1.32	0.564	0.861	0.677	1.50	0.748	0.918
WT2 × 6.5	1.91	2.080	4.060	0.345	0.280	0.526	0.321	0.524	0.440	1.93	0.950	1.00

Courtesy of The American Institute of Steel Construction.

*WT is the symbol for a structural T-section (cut from a W-section), followed by the nominal depth in inches, then the weight in pounds per foot of length.

GEOMETRIC PROPERTIES OF ROLLED STEEL SHAPES

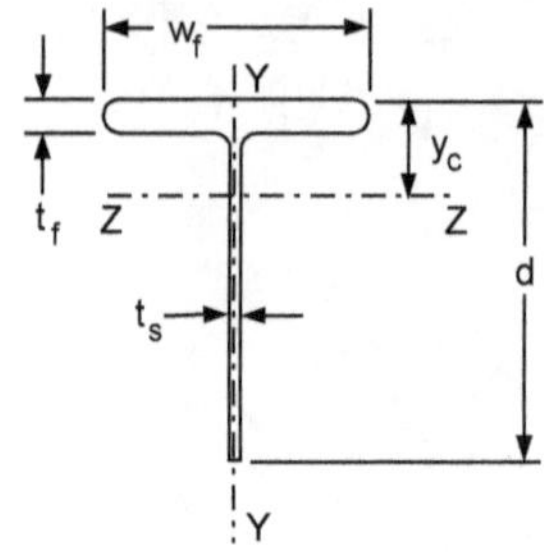

Structural Tees (SI Units)

Designation*	Area (mm^2)	Depth of Tee (mm)	Flange Width (mm)	Flange Thickness (mm)	Stem Thickness (mm)	AXIS Z-Z I (10^6 mm^4)	AXIS Z-Z Z (10^3 mm^3)	AXIS Z-Z r (mm)	y_c (mm)	AXIS Y-Y I (10^6 mm^4)	AXIS Y-Y Z (10^3 mm^3)	AXIS Y-Y r (mm)
WT457 × 171	21805	455.9	418.3	32.0	19.3	389	1098	133	102	196	936	94.7
× 119	15160	457.3	304.8	25.9	16.5	308	914	142	120	61.2	403	63.5
WT381 × 98	12515	384.9	267.8	25.4	15.6	175	613	118	99.1	40.8	305	57.2
× 80	10260	378.8	266.1	19.3	13.8	145	524	119	102	30.4	228	54.6
WT305 × 77	9870	305.6	323.9	19.1	12.7	78.7	328	89.2	65.8	54.1	333	73.9
× 70	8905	308.7	230.3	22.2	13.1	77.4	333	93.2	75.9	22.7	197	50.3
× 63	8000	306.1	229.1	19.6	11.9	69.1	300	93.2	75.4	19.6	172	49.5
× 46	5875	301.5	178.8	15.0	10.9	54.5	256	96.3	87.9	7.16	80.3	35.1
WT229 × 57	7225	231.3	280.3	17.3	10.8	29.9	161	64.5	45.7	31.7	226	66.3
× 45	5690	231.6	191.9	17.7	10.5	26.9	152	68.8	54.9	10.4	109	42.9
× 37	4730	228.5	190.4	14.5	9.0	22.3	128	68.6	53.8	8.32	87.7	41.9
× 30	3795	227.3	152.8	13.3	8.0	18.6	110	70.1	58.2	3.98	51.9	32.3
WT203 × 74	9485	215.5	264.8	25.0	14.9	32.0	187	57.9	44.7	38.8	293	63.8
× 37	4755	206.5	179.6	16.0	9.7	17.6	111	61.0	48.0	7.74	86.2	40.4
× 30	3800	203.3	177.7	12.8	7.7	13.8	87.7	60.2	46.0	5.99	67.5	39.9
× 19	2475	199.3	139.7	8.8	6.4	9.78	67.0	62.7	53.1	2.00	28.5	28.4
WT178 × 89	11420	183.9	372.6	23.9	15.0	21.5	141	43.4	31.5	103	552	95.0
× 61	7740	181.7	257.3	21.7	13.0	17.1	117	47.0	35.3	30.9	239	63.0
× 51	6445	178.3	254.9	18.3	10.5	13.6	93.2	46.0	32.8	25.3	198	62.5
× 36	4560	177.4	204.0	15.1	8.6	10.4	73.4	47.5	34.3	10.7	105	48.5
× 22	2850	175.8	170.9	9.8	6.9	7.91	58.2	52.6	40.1	4.07	47.7	37.8
× 16	2095	174.5	127.0	8.5	5.8	6.16	47.7	54.4	44.7	1.46	22.9	26.4
WT152 × 89	11355	166.6	312.9	28.1	18.0	18.1	135	39.9	32.5	71.6	459	79.5
× 71	9095	161.4	308.9	22.9	14.0	13.3	100	38.4	28.7	56.2	364	78.5
× 54	6840	155.6	305.8	17.0	10.9	9.66	74.4	37.6	25.9	40.6	265	77.2
× 37	4735	154.8	205.2	16.2	9.4	7.78	62.1	40.6	29.7	11.7	114	49.8
× 22	2840	156.7	165.6	11.2	6.6	5.62	45.1	44.5	32.3	4.25	51.1	38.6
× 12	1525	152.3	101.3	6.7	5.6	3.62	33.4	48.8	44.2	0.587	11.6	19.6
WT127 × 83	10645	144.3	264.5	31.8	19.2	11.9	105	33.5	30.7	49.1	370	68.1
× 65	8325	137.7	260.7	25.1	15.4	8.66	78.2	32.3	26.9	37.2	285	66.8
× 45	5690	129.8	256.0	17.3	10.7	5.37	49.8	30.7	22.5	24.2	188	65.3
× 22	2850	133.0	147.6	13.0	7.6	3.86	36.7	36.8	27.9	3.48	47.0	34.8
× 9	1140	125.3	100.6	5.3	4.8	1.81	20.0	39.9	34.5	0.454	9.03	19.9
WT102 × 43	5515	111.1	208.8	20.6	13.0	3.80	42.8	26.2	22.2	15.6	150	53.3
× 30	3785	104.8	205.0	14.2	9.1	2.39	27.7	25.1	18.7	10.2	99.6	51.8
× 18	2285	100.7	165.0	10.2	6.2	1.47	17.7	25.4	17.7	3.80	46.0	40.9
× 13	1695	103.4	133.4	8.4	5.8	1.42	17.2	29.0	21.2	1.66	24.9	31.2
× 7	955	100.2	100.1	5.2	4.3	0.895	11.7	30.5	24.2	0.437	8.72	21.4
WT76 × 15	1895	78.7	152.9	9.3	6.6	0.733	11.4	19.7	14.2	2.76	36.2	38.1
× 9	1150	76.6	101.6	7.1	5.8	0.549	9.24	21.9	17.2	0.624	12.3	23.3
WT51 × 10	1230	52.8	103.1	8.8	7.1	0.219	5.26	13.3	11.2	0.803	15.6	25.4

Courtesy of The American Institute of Steel Construction.
*WT is the symbol for a structural T-section (cut from a W-section), followed by the nominal depth in mm, then the mass in kg per meter of length.

APPENDIX E

EQUATIONS FOR DEFLECTIONS OF SIMPLY SUPPORTED BEAMS

SIMPLY SUPPORTED BEAMS LENGTH, L	SLOPE	DEFLECTION EQUATIONS
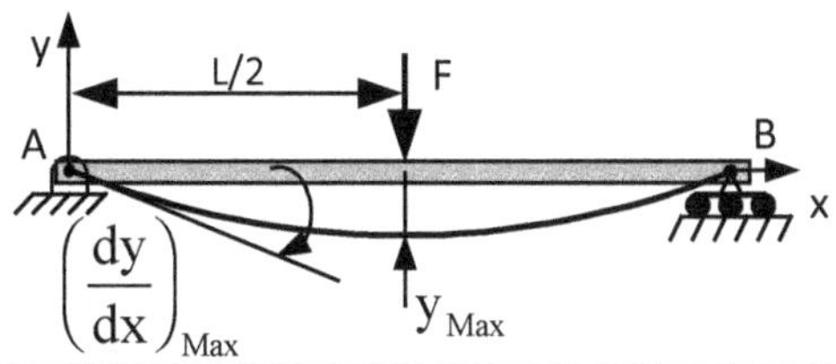	$\left(\frac{dy}{dx}\right)_{Max} = \frac{-FL^2}{16EI_z}$	$y = \frac{-Fx}{48EI_z}(3L^2 - 4x^2)$ $0 \le x \le L/2$ $y_{Max} = \frac{-FL^3}{48EI_z}$
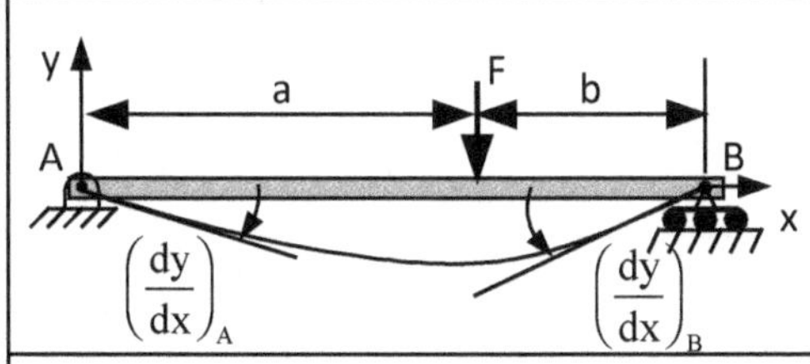	$\left(\frac{dy}{dx}\right)_A = \frac{-Fab(L+b)}{6EI_zL}$ $\left(\frac{dy}{dx}\right)_B = \frac{Fab(L+a)}{6EI_zL}$	$y = \frac{-Fbx}{6EI_zL}(L^2 - b^2 - x^2)$ $0 \le x \le a$ $y_{x=a} = \frac{-Fba}{6EI_zL}(L^2 - b^2 - a^2)$
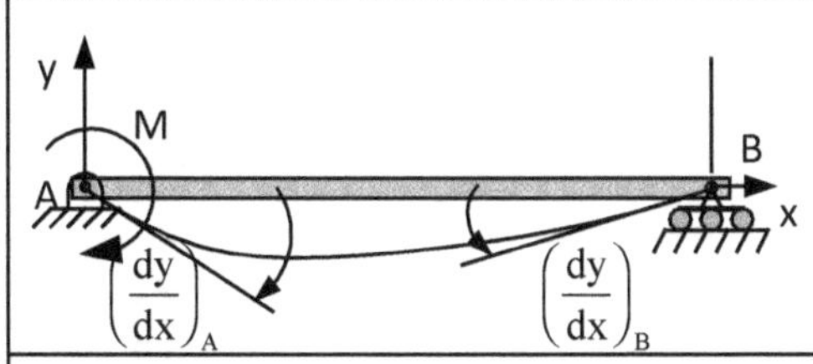	$\left(\frac{dy}{dx}\right)_A = -\frac{ML}{3EI_z}$ $\left(\frac{dy}{dx}\right)_B = \frac{ML}{6EI_z}$	$y = \frac{-Mx}{6EI_zL}(x^2 - 3Lx + 2L^2)$ $y_{Max} = \frac{-ML^2}{\sqrt{243}EI_z}$
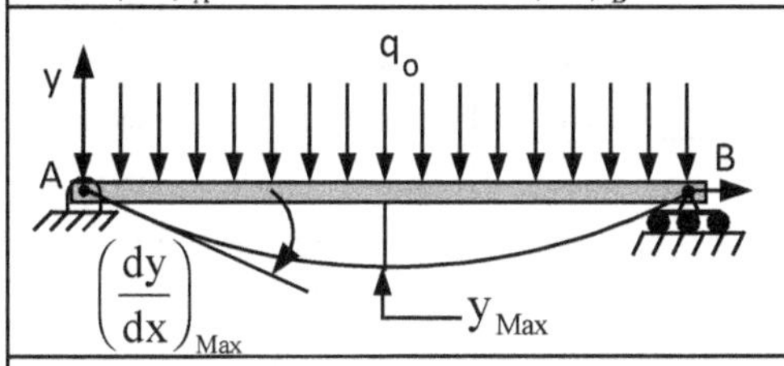	$\left(\frac{dy}{dx}\right)_{Max} = -\frac{q_0L^3}{24EI_z}$	$y = \frac{-q_0x}{24EI_z}(x^3 - 2Lx^2 + L^3)$ $y_{Max} = \frac{-5q_0L^4}{384EI_z}$
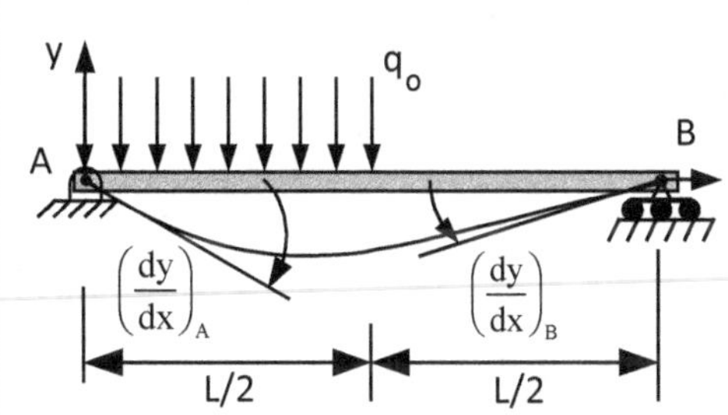	$\left(\frac{dy}{dx}\right)_A = -\frac{3q_0L^3}{128EI_z}$ $\left(\frac{dy}{dx}\right)_B = \frac{7q_0L^3}{384EI_z}$	$y = \frac{-q_0x}{384EI_z}(16x^3 - 24Lx^2 + 9L^3)$ $0 \le x \le L/2$ $y = \frac{-q_0L}{384EI_z}(8x^3 - 24Lx^2 + 17L^2x - L^3)$ $L/2 \le x \le L$ $y_{Max} = \frac{-6.563 \times 10^{-3} q_0L^4}{EI_z}$ at $x = 0.4598L$
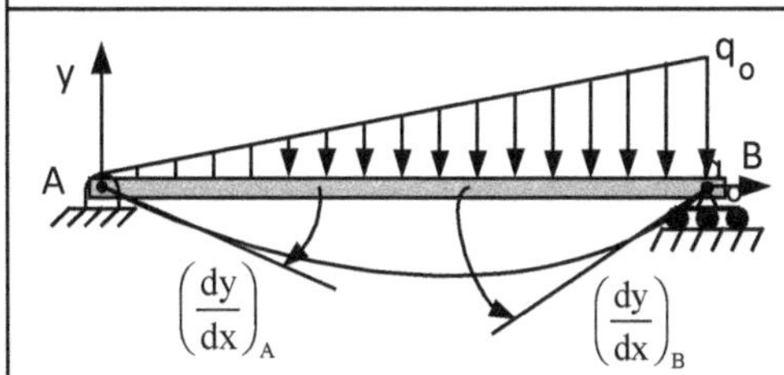	$\left(\frac{dy}{dx}\right)_A = \frac{-7q_0L^3}{360EI_z}$ $\left(\frac{dy}{dx}\right)_B = \frac{q_0L^3}{45EI_z}$	$y = \frac{-q_0x}{360EI_zL}(3x^4 - 10L^2x^2 + 7L^4)$ $y_{Max} = \frac{-6.52 \times 10^{-3} q_0L^4}{EI_z}$ at $x = 0.5193$

EQUATIONS FOR DEFLECTIONS OF CANTILEVER BEAMS

CANTILEVER BEAMS LENGTH, L	SLOPE	DEFLECTION EQUATIONS
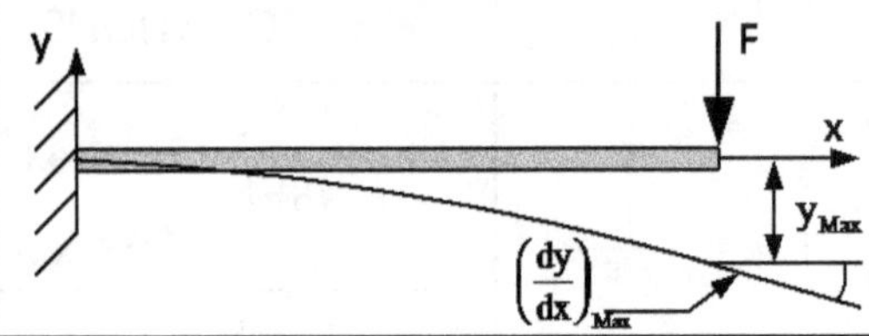	$\left(\frac{dy}{dx}\right)_{Max} = \frac{-FL^2}{2EI_z}$	$y = \frac{-Fx^2}{6EI_z}(3L - x)$ $y_{Max} = \frac{-FL^3}{3EI_z}$
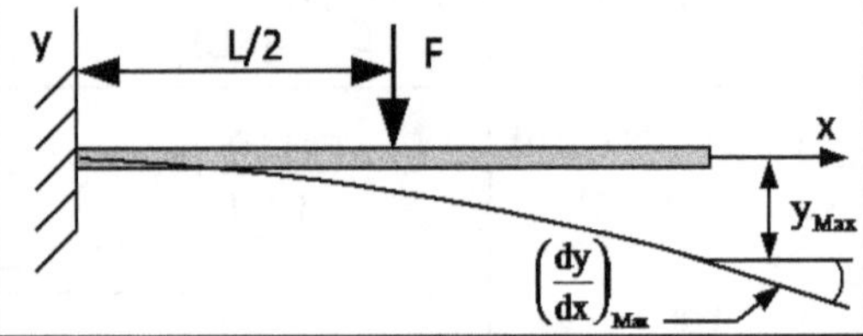	$\left(\frac{dy}{dx}\right)_{Max} = \frac{-FL^2}{8EI_z}$	$y = \frac{-Fx^2}{6EI_z}\left(\frac{3}{2}L - x\right)$ $\quad 0 \le x \le L/2$ $y = \frac{-FL^2}{24EI_z}\left(3x - \frac{1}{2}L\right)$ $\quad L/2 \le x \le L$
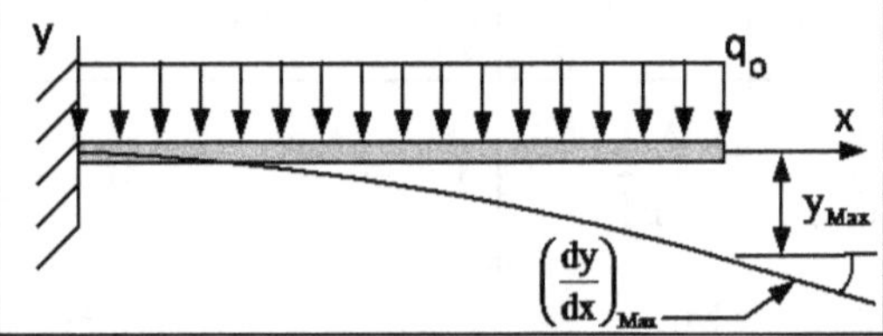	$\left(\frac{dy}{dx}\right)_{Max} = \frac{-q_0L^3}{6EI_z}$	$y = \frac{-q_0x^2}{24EI_z}\left(x^2 - 4Lx + 6L^2\right)$ $y_{Max} = \frac{-q_0L^4}{8EI_z}$
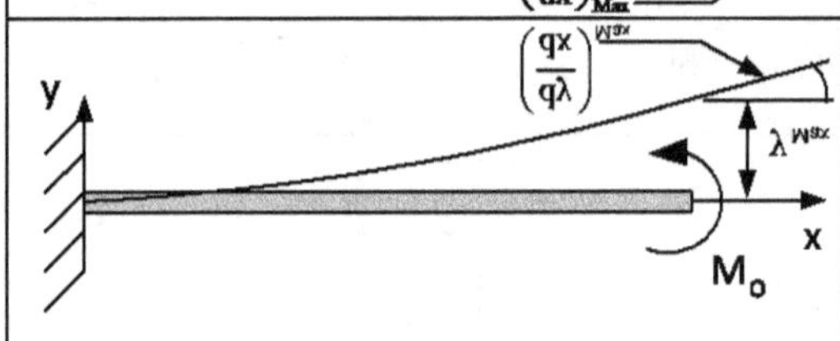	$\left(\frac{dy}{dx}\right)_{Max} = \frac{ML}{EI_z}$	$y = \frac{Mx^2}{2EI_z}$ $y_{Max} = \frac{ML^2}{2EI_z}$
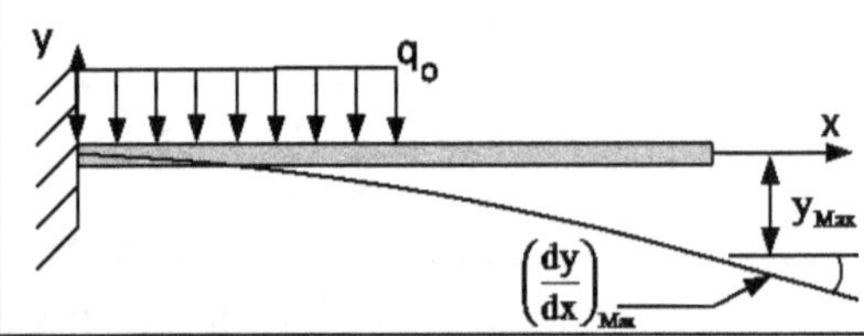	$\left(\frac{dy}{dx}\right)_{Max} = \frac{-q_0L^3}{48EI_z}$	$y = \frac{-q_0x^2}{24EI_z}\left(x^2 - 2Lx + \frac{3}{2}L^2\right)$ $\quad 0 \le x \le L/2$ $y = \frac{-q_0L^3}{192EI_z}\left(4x - \frac{L}{2}\right)$ $\quad L/2 \le x \le L$
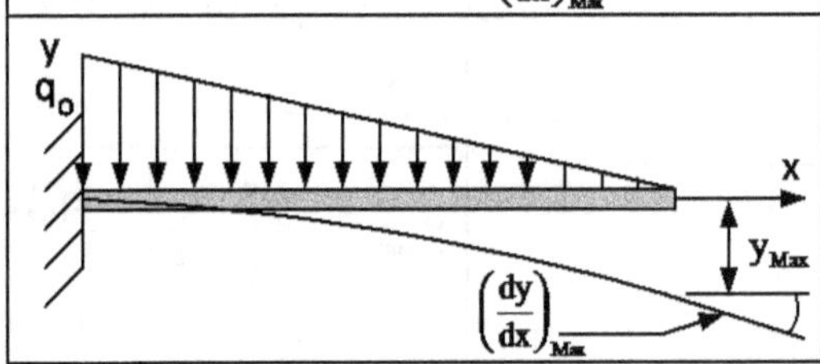	$\left(\frac{dy}{dx}\right)_{Max} = \frac{-q_0L^3}{24EI_z}$	$y = \frac{-q_0x^2}{120EI_zL}\left(10L^3 - 10L^2x + 5Lx^2 - x^3\right)$ $y_{Max} = \frac{-q_0L^4}{30EI_z}$

ANSWERS

Chapter 1

1.2 103.0 N on Earth
39.02 N on Mars

1.3 $W_s = 11.43$ N

1.14 p = 152.8 psi

1.16 d = 2.985 in.

1.21(a) F= 10,680 N
1.21(f) F = 422.6 kN

1.22(a) F = 502.2 lb
1.22(c) F = 2.968 kip

1.27(a) $\sigma = 82.74$ MPa
1.27(d) $\sigma = 913.6$ MPa

1.28(a) $\sigma = 145.0$ ksi
1.28(d) $\sigma = 203.0$ psi

Chapter 2

2.16(a) $S_v = 8{,}660$ N, $\theta_{Sv} = 90°$
2.16(b) $D_v = 13{,}230$ N, $\theta_{Dv} = -40.9°$
2.16(c) $D_v = 13{,}230$ N, $\theta_{Dv} = 139.1°$

2.17(a) $S_v = 913.9$ N, $\theta_{Sv} = 70.04°$
2.17(b) $D_v = 3{,}093$ N, $\theta_{Dv} = -46.93°$
2.17(c) $D_v = 3{,}093$ N, $\theta_{Dv} = 133.1°$

2.18 $F_x = 15{,}105$ lb
$F_{y'} = -3{,}202$ lb

2.19 $F_{AB} = 1{,}949$ N; $F_{CB} = 1{,}647$ N

2.21 $S_v = 15.61$ kN; $\theta = 209.7°$

2.22 F = 8.640 kN; $\theta = 46.67°$

2.25 $F_1 = 7{,}072$ lb; $\theta = 49.0°$

2.26 $F_1 = 10{,}120$ lb; $\theta = 55.87°$

2.33 $M_O = 438.8$ N-m (CCW)

2.34 $M_O = 86.6$ N-m (CCW)

2.37(a) $\bar{r} = (7\bar{i} + 4\bar{j} + 6\bar{k})$ in.
r = 10.05 in.
$\alpha = 45.85°$, $\beta = 66.55°$, $\gamma = 53.34°$

2.37(b) $\bar{r} = (110\bar{i} + 65\bar{j} + 70\bar{k})$ mm
r = 145.7 mm
$\alpha = 40.98°$, $\beta = 63.50°$, $\gamma = 61.29°$

2.39(b) $F_x = 388.2$ lb, $F_y = 964.2$ lb
$F_z = 1082$ lb
$\bar{F} = (388.2\bar{i} + 964.2\bar{j} + 1082\bar{k})$ lb

2.39(d) $F_x = 16.07$ kip, $F_y = -19.15$ kip
$F_z = 0$
$\bar{F} = (16.07\bar{i} - 19.15\bar{j})$ kip

2.40 $\bar{F} = \bar{F}_1 + \bar{F}_2 = (26.25\bar{i} + 37.23\bar{j} - 15.9\bar{k})$ kN

2.41(a) $\theta = 87.35°$
2.41(b) $\theta = 18.28°$

2.43(a) $\overline{M}_O = (-6.4\bar{i} + 14.95\bar{j} - 7.85\bar{k})$ kN - m
2.43(b) $\overline{M}_O = (-14.5\bar{i} - 13.28\bar{j} + 3.5\bar{k})$ ft - kip

2.45 $x_B = -58.62$ ft; $y_B = -3.307$ ft

2.47 $\bar{F} = (5.057\bar{i} + 2.528\bar{j} - 20.23\bar{k})$ kN
See equation above for F_x, F_y, F_z.
$\overline{M}_O = (-80.91\bar{i} + 20.23\bar{j} - 17.70\bar{k})$ kN - m
See equation above for M_x M_y, M_z.
$M_O = 85.26$ kN-m; $\theta = 28.15°$
$\bar{u}_{M_0} = (-0.949\bar{i} + 0.237\bar{j} - 0.208\bar{k})$
$F_{OQ} = 18.52$ kN(Compression)

2.48 $\bar{F} = (741.8\bar{i} + 3{,}338\bar{j} - 6{,}677\bar{k})$ lb
See equation above for F_x, F_y, F_z.

2.48 Continued:

$\overline{M}_O = (-73.44\bar{i} + 33.38\bar{j} + 8.530\bar{k})$ ft - kip

See equation above for M_x M_y, M_z.

$M_O = 81.12$ ft-kip

$\bar{u}_{Mo} = (-0.9053\bar{i} + 0.4115\bar{j} + 0.1052\bar{k})$

$\theta = 36.09°$;

$F_{OQ} = 6.061$ kip (Compression)

Chapter 3

3.6 $F_{CB} = 220.9$ lb; $F_{CA} = 585.3$ lb

3.7 $F_{AC} = 11.97$ kN; $F_{BC} = 9.807$ kN

3.9 $T_{AC} = 1{,}189$ N; $T_{BC} = 1{,}158$ N

3.10 $F = 1.879$ ton
Pulley system is not effective.

3.11 $F = 4.358$ kN
Pulley system is effective.

3.15 $F = W(1 - (r_2/r_1))/2$

3.16 $F_{AB} = 421.7$ N; $F_{BD} = 730.4$ N;
$F_{CG} = 1461$ N; $F_{BC} = F_{CE} = 843.4$ N

3.18 $T_{AB} = 186.0$ N; $T_{BC} = 173.9$ N
$T_{CD} = 212.0$ N; $m_C = 7.413$ kg

3.19 $W_{max} = 6{,}200$ lb > 140 lb
Stuntman will remain dry.

3.21 $\delta_T = 3.851$ in.; $k_{eff} = 6.076$ lb/in.

3.22 $\delta_T = 48.57$ mm; $k_{eff} = 1.441$ N/mm

3.23 $\delta_T = 24.36$ mm; $k_{eff} = 47.00$ N/mm

3.26 $A_x = 0$; $A_y = 10$ kN; $B_y = 30$ kN

3.27 $A_x = 0$; $A_y = 1{,}406$ lb; $B_y = 2{,}344$ lb

3.28 $A_x = 0$; $A_y = 26.16$ kN; $B_y = 53.16$ kN

3.29 $A_x = 0$; $A_y = 11$ kip;
$M_{RA} = 121$ kip-ft (CCW)

3.30 $A_x = 0$; $A_y = 37.33$ kN;
$M_{RA} = 161.8$ kN-m (CCW)

3.31 $A_x = 625.8$ lb; $A_y = 491.8$ lb
$B_y = 559.0$ lb

3.33 $A_x = 0$; $A_y = B_y = 40$ kN

3.36 $W_{max} = 1.892$ ton

3.37 $R_x = 0$; $R_y = 38.25$ kN
$M_z = 860.6$ kN-m

3.38 $A_x = -13.5$ kip; $B_x = 7.5$ kip;
$A_y = B_y = 4.5$ kip

3.39 $P_A = 36$ kN; $P_B = 18$ kN

3.40 $V = F/4$; $M = FL/8$

3.42 $V = q_0L/6$; $M = q_0L^2/9$

3.44 $V = -F$; $M = -(3/4)$ FL

3.46 $V = -4$ kip; $M = 76.8$ kip-ft

3.48 $P = 250$ lb, $V = 0$; $M_z = -375$ in-lb

Chapter 4

4.1 $\sigma = 19.10$ ksi; $\varepsilon = 0.0006367$
$L_f = 22.014$ ft

4.4 $\sigma_{max} = 1.104$ MPa, $\sigma_{12} = 764.8$ kPa
$\sigma_{23} = 320.7$ kPa

4.7 $P_{design} = 11.77$ kN

4.9 $\sigma = 1744$ psi

4.13 4340 Hr — Gage # 4-Os
52100 A — Gage # 4-Os

4.18 $\sigma = 22$ MPa; $\delta = 0.1913$ mm

4.19 $\sigma_{design} = 16.88$ ksi; $A = 1.185$ in.^{2}

4.22 $\tau = 21.33$ ksi

4.25 $P_{max} = 1.651$ kip

4.27 $\sigma_\phi = (6.365)\sin^2 \phi$ MPa
$\tau_\phi = (6.365)\sin \phi \cos \phi$ MPa

4.28 $\sigma = 90.0$ MPa; $\theta = 161.6°$
F = 15.75 kN

4.32 $\delta = 2.235$ mm

4.35 $\sigma_1 = 30.30$ ksi; $\sigma_2 = -1.538$ ksi
$\delta = 0.01841$ in.

4.36 $\sigma_{nom} = 20$ ksi; $\sigma_{max} = 46.6$ ksi

4.37 $\sigma_{nom} = 25$ MPa; $\sigma_{max} = 43.75$ MPa

4.41 $P_{max} = 343.5$ kN

4.42 $t_{min} = 0.320$ in.

4.43 $\sigma_a = -174$ MPa; $\sigma_f = -20$ kPa;
$\delta = 4.833$ mm (shorter)

4.44 $\sigma_s = 33.03$ ksi; $\sigma_b = -8.256$ ksi
$\delta = 0.006192$ in. (shorter)

4.45 $\sigma_s = -7.761$ ksi; $\sigma_c = -1.164$ ksi;
$\delta = 0.02483$ in (shorter)

4.46 $\sigma_a = 19.04$ MPa; $\sigma_b = 761.7$ kPa;
$\delta_a = 0.1983$ mm (longer);
$\delta_b = 0.001731$ mm (longer)

4.47 $\sigma_c = -20.49$ MPa; $\sigma_s = -512.1$ MPa
$\delta_c = 0.4647$ mm (longer)
$\delta_s = 0.4647$ mm (shorter)

4.48 $\sigma_s = 1.784$ ksi; $\sigma_a = -467$ psi

Chapter 5

5.5 $S_y = 46.1$ ksi; $S_u = 62.2$ ksi
S_y (0.2%) = 56.7 ksi

5.6 $S_y = 36$ ksi; $S_u = 44$ ksi
S_y (0.2%) = 38.5 ksi

5.7 %e = 32.8%; %A = 31.38%

5.9 d = 0.4997 in.

5.12 $V = V_0 [1 + (1 - 2\nu)\varepsilon_a] = V_0$
when $\nu = ½$ and higher order strain terms are neglected.

5.13 $\varepsilon_y = 0.001267$

5.14 $\varepsilon_a = 0.003153$; $\varepsilon_t = -0.001009$

5.15 $\varepsilon_x = \varepsilon_y = 0.0004807$

5.18 $\sigma_a = 52.8$ MPa; $\sigma_h = 105.6$ MPa

5.19 $\sigma = 57.55$ ksi; $\sigma_T = 107.4$ ksi

5.23 $N = \infty$ infinite life in fatigue

5.26 $\sigma_{max} = 255$ MPa

5.33 E = 30,000 ksi; G = 11,490 ksi
$\nu = 0.3054$

5.34 $E = 9.888 \times 10^6$ psi, $\sigma_u = 59.08$ ksi
$\sigma_f = 48.89$ ksi; $\varepsilon_f = 0.2100$
σ_y (0.2%) = 40.21 ksi
%e = 21.00%; %A = 31.77%

5.35 E = 199.6 GPa, $\sigma_u = 300.0$ MPa
$\sigma_f = 150.0$ MPa; $\varepsilon_f = 0.1864$
σ_y (0.2%) = 199.6 MPa
%e = 18.64%; %A = 49.78%

Chapter 6

6.8 $P_{AC} = 26.25$ kip (T); $P_{BC} = 52.5$ kip (T)
$P_{CD} = 36.34$ kip (C); $P_{CE} = 42.5$ kip (T)

6.10 $P_{AC} = 67.48$ kip (T); $P_{BC} = 135.0$ kip (T)
$P_{CD} = 67.08$ kip (C); $P_{CE} = 97.48$ kip (T)

6.12 $A_{AB} = 6.261$ in.^{2}; $A_{AC} = 2.8$ in.^{2}
$A_{BD} = 2.8$ in.^{2}; $A_{BC} = 5.6$ in.^{2}

6.13 $A_{AB} = 1,933$ mm^2; $A_{AF} = 1,694$ mm^2;
$A_{BC} = 1,607$ mm^2; $A_{BF} = 285.7$ mm^2

6.17 $P_{DE} = 0$; $P_{DF} = P_{DB} = 30$ kip (C)

6.20 $P_{DC} = 100$ kN (T); $P_{DE} = 131.7$ kN (T)
$P_{DF} = 257.1$ kN (C)

6.23 $P_{AC} = 200(s/h)$;
$P_{AD} = -200[(s/h)^2 + 1]^{1/2}$

6.25 $P_{AC} = 3.907$ kN (T); $P_{CD} = 0.2504$ kN (C)
$P_{CE} = 4.206$ kN (T); $P_{DF} = 4.206$ kN (C)

6.26 $P_{CD} = 41.08$ kN (C); $P_{CJ} = 11.27$ kN (T)
$P_{KJ} = 34.53$ kN (T)

6.30 $\sigma_{BC} = 10.4$ ksi (T), $SF_{BC} = 4.038$

6.35 $A_{AB} = 4{,}340$ mm^2; $A_{BD} = 4{,}340$ mm^2
$A_{DE} = 5{,}482$ mm^2; $A_{EF} = 2{,}856$ mm^2
$A_{CF} = 2{,}284$ mm^2

Chapter 7

7.4 $P_{EB} = -8.944$ kips; $d_{EB} = 1.033$ in.
$P_{EC} = -4.472$ kips; $d_{EC} = 0.7305$ in.
$P_{ED} = 15.56$ kips; $d_{ED} = 1.363$ in.

7.13 $MOS_{AB} = 165.1\%$
$MOS_{AP} = MOS_{AQ} = 181.1\%$

7.14 $d_{AB} = 8.148$ mm; $d_{AC} = 7.374$ mm
$d_{AD} = 13.98$ mm

7.17 Gage # 2; steel wire

7.18 $P_{AD} = 165.9$ lb; $P_{BD} = 437.6$ lb
$P_{CD} = 441.1$ lb

7.19 Gage # 000s steel wire

7.20 W = 1007 lb; $A_x = -419.6$ lb
$A_y = 2{,}518$ lb; $A_z = -881.2$ lb
$P_{BC} = 1{,}730$ lb; $P_{DE} = 1{,}841$ lb

7.21 F= 104.2 kN; A_x =41.7 kN
$A_y = 0$; $A_z = 250$ kN
$B_x = -145.9$ kN; $B_z = 250$ kN

Chapter 8

8.4 F = 36.84 N

8.5 $G_y = 2{,}591$ N; MA = 8.637

8.12 $P_C = 30$ ton; MA 6

8.13 $F_C = 165$ lb; W = 8.4 in-lb
$\Delta d_C = 0.0509$ in.

8.14 $A_x = -4.5$ kip; $A_y = 7.3$ kip
$M_{RA} = 64.35$ kip-ft; $C_y = -1.443$ kip
$C_x = -12.75$ kip; $P_{BD} = -15.46$ kip

8.15 $B_y = 3.84$ kN; $C_x = 0$; $C_y = -9.6$ kN;
$E_x = 0$; $E_y = 0.96$ kN; $P_{AF} = 0$

8.17 $P_{AC} = 1{,}178$ N; $P_{BC} = -834.4$ N
$P_{CE} = 627.9$ N; D = 699.3 N

8.21 $C_x = 2.5$ kip; $C_y = 14.59$ kip; $A_x = 0$;
$A_y = 34$ kip; $M_{RA} = 50$ kip-ft (CCW);
T = 13.35 kip; F = 15.61 kip

Chapter 9

9.5(a) $\mu_{Ave} = 0.4156$; $\mu_{Range} = 0.0515$

9.5(b) $\mu_{Ave} = 0.4530$; $\mu_{Range} = 0.0656$

9.7 F = 1,845 N

9.9 $(F_{fW})_{max} = 57.95$ lb $< F_{fW} = 64.48$ lb
Worker slips and cannot move crate.

9.11 $(F_1)_{max} = 1094$ lb

9.12 $F_{fA} = 26.17$ lb < 57.93 lb $= (F_{fA})_{max}$
The ladder is stable.

9.13 F = 240 N; The crate begins to tip.

9.14 $(F_{fA})_{max} = 156.4$ N < 210 N $= F_{fA}$
Cylinder slips and remains in corner.

9.30 $F_{fB} = 21.25$ lb < 50 lb $= (F_{fB})_{max}$
$x_B = 0.4808$ ft < 1.5 ft
$F_{fA} = 21.25$ lb < 38.25 lb $= (F_{fA})_{max}$
Both blocks are in equilibrium.

9.31 $(W_1)_{max} = 1.667$ kN; $(W_1)_{min} = 0.3$ kN

Chapter 10

10.3 $\varepsilon_{max} = 0.002660$; $\sigma_{max} = 191.5$ MPa

10.4 $M_{max} = 480$ kN-m

10.5 $\sigma_A = -2.17$ ksi; $\sigma_B = -0.7232$ ksi
$\sigma_C = 1.446$ ksi; $\sigma_D = 1.808$ ksi

10.6 $M_{max} = 117.5$ kip-in (+ moment)
$M_{max} = 249.6$ kip-in (– moment)

10.7 $I_z = 394.6 \times 10^6$ mm^4
$Z = 1{,}902 \times 10^3$ mm^3

10.9 $I_z = 2{,}113$ in.^{4}; $I_y = 115.2$ in.^{4}

10.10 $I_z = 3{,}013$ in.^{4}; $I_y = 176.0$ in.^{4}

10.11 $M_{Modified}/M_{Original} = 2.106$

10.14 $V_1(x) = 6.444$ kip; $M_1(x) = 6.444x$ kip-ft
$V_2(x) = 1.444$ kip;
$M_2(x) = [1.444x + 30]$ kip-ft
$V_3(x) = -6.556$ kip;
$M_3(x) = [-6.556x + 118]$ kip-ft

10.15 $V(x) = -10$ kN; $M(x) = -10x$ kN-m

10.16 $V(x) = [3.15 - x]$ kN
$M(x) = [3.15x - x^2/2]$ kN-m

10.17 $V_1(x) = [5.192 - 0.07x]$ kip
$M_1(x) = [5.192x - 0.035x^2]$ kip-ft
$V_2(x) = [-1.808 - 0.07x]$ kip
$M_2(x) = [35 - 1.808x - 0.035x^2]$ kip-ft

10.18 $V_1(x) = 1.6$ kN
$M_1(x) = [1.2 + 1.6x]$ kN-m
$V_2(x) = [3.4 - 1.2x]$ kN
$M_2(x) = [-0.15 + 3.4x - 0.6x^2]$ kN-m
$V_3(x) = -2$ kN
$M_3(x) = [12 - 2x]$ kN-m

10.19 $Z_{min} = 984.1 \times 10^3$ mm^3
W 406 × 60 for minimum weight beam

10.20 $Z_{min} = 750 \times 10^3$ mm^3
S381 × 64 for minimum weight beam

10.21 $Z_{min} = 52.22 \times 10^3$ mm^3
WT 203 × 19 for minimum weight beam

10.22 $\tau_{max} = 1.5$ ksi

10.23 $\tau_{max} = 16.98$ MPa

10.24 $\tau_a = 1.2$ MPa; $\tau_{max} = 1.4$ MPa

10.26 s = 1.767 in.

10.28 $q_y = 5{,}734$ N/m

10.30 $h_f = 245.6$ mm; $t_m = 0.5080$ mm
for h/b ratio = 3/1

10.31 $(M_z)_{max} = 692.7$ kip-in.

10.32 $(M_y)_{max} = 120$ kip-in.

10.35 MOS = 4.167%

10.36 $\sigma_c = 20.82$ MPa (C); $\sigma_s = 156.2$ MPa (T)

10.44 $\sigma_{nom} = 85.5$ ksi; $\sigma_{max} = 179.6$ ksi

10.46 $\sigma_{nom} = 71.29$ MPa; $\sigma_{max} = 114.1$ MPa

Chapter 11

11.2 $\gamma_{max} = 312.2\ \mu\varepsilon$

11.4 $\tau_{max} = 17.39$ ksi

11.6 $d_0 = 14.26$ mm

11.10 $\tau_{nt} = 4.676$ ksi; $\sigma_n = 4.515$ ksi

11.13 $\phi = 3.464°$

11.15 $\tau_{max} = 847.5$ MPa in DE
$\phi_{AE} = 32.83°$; $\phi_{AC} = -3.871°$

11.17 $\tau_{nom} = 15.65$ ksi; $\tau_{max} = 18.47$ ksi

11.18 $T_{max} = 215.5$ N-m

11.19 d = 20 mm; D = 27 mm
r = 3 mm for 4340 HR steel

11.20 $d = 0.875$ in.; $D = 1.75$ in.
$r = 0.175$ in.

11.21 $\tau_{max} = 9.978$ ksi; $\phi = 8.036°$

11.24 $\phi = 3.213°$; $\tau_{max} = 60.07$ MPa

Chapter 12

12.1(a) $\varepsilon_x = 427.5\ \mu\varepsilon$; $\gamma_{xy} = 1{,}565\ \mu\varepsilon$
$\varepsilon_y = 687.5\ \mu\varepsilon$; $\gamma_{yz} = 1{,}096\ \mu\varepsilon$
$\varepsilon_z = -818.3\ \mu\varepsilon$; $\gamma_{zx} = -934.8\ \mu\varepsilon$

12.2(a) $\sigma_x = 7.212$ ksi; $\tau_{xy} = 10.35$ ksi
$\sigma_y = -15.87$ ksi; $\tau_{yz} = -7.475$ ksi
$\sigma_z = -10.10$ ksi; $\tau_{zx} = -5.175$ ksi

12.4(a) $\sigma_{x'} = 2.283$ ksi; $\sigma_{y'} = 33.72$ ksi;
$\tau_{x'y'} = -9.272$ ksi

12.5 $\sigma_1 = 35.02$ ksi; $\sigma_2 = -29.02$ ksi
$\theta_1 = -19.33°$; $\theta_2 = 70.67°$
$\tau_{max} = 32.02$ ksi; $\sigma_n = 3$ ksi
$\theta_s = 25.67°$

12.6 $\sigma_1 = 112.1$ MPa; $\sigma_2 = -67.12$ MPa;
$\theta_1 = 15.07°$; $\theta_2 = -74.93°$;
$\tau_{max} = 89.62$ MPa; $\sigma_n = 22.5$ MPa;
$\theta_s = -29.93°$

12.10 $p_{max} = 750$ kPa

12.12 $h = 26.92$ ft

12.13 $\sigma_w = 131.3$ MPa; $\tau_w = -43.75$ MPa

12.14 $t_{min} = 0.8037$ in.

12.17 $\varepsilon_{x'} = 456.5\ \mu\varepsilon$; $\varepsilon_{y'} = 593.4\ \mu\varepsilon$
$\gamma_{x'y'} = -832.7\ \mu\varepsilon$

12.20 $\varepsilon_1 = 960.8\ \mu\varepsilon$; $\varepsilon_2 = -495.8\ \mu\varepsilon$;
$\theta_1 = 4.941°$; $\theta_2 = -85.06°$;
$\gamma_{max} = 1{,}457\ \mu\varepsilon$; $\varepsilon_n = 232.5\ \mu\varepsilon$;
$\theta_s = 49.94°$

12.23 $\sigma_1 = 21.03$ ksi; $\sigma_2 = -3.890$ ksi

12.25 $\varepsilon_1 = 1{,}805\ \mu\varepsilon$; $\varepsilon_2 = -404.5\ \mu\varepsilon$;
$\sigma_1 = 19.41$ ksi; $\sigma_2 = 2.004$ ksi;
$\theta_1 = -2.597°$

12.29 $\sigma_A = 41.80$ ksi; $\tau_A = 5.045$ ksi
$\sigma_B = 0$; $\tau_B = -5.502$ ksi

12.30 $\sigma_C = -41.80$ ksi; $\tau_C = -5.045$ ksi
$\sigma_D = 0$; $\tau_D = 4.588$ ksi

12.31 At point A:
$\sigma_1 = 43.24$ MPa; $\sigma_2 = -1.085$ MPa
$\tau_{max} = 22.16$ MPa
At point B:
$\sigma_1 = 41.06$ MPa; $\sigma_2 = -2.737$ MPa
$\tau_{max} = 21.90$ MPa

12.32 $SF_y = 1.222$

12.33 $\sigma_{max} = 14.73$ ksi

12.34 At point A:
$\sigma_1 = 1.741$ MPa; $\sigma_2 = -0.1628$ MPa
$\tau_{max} = 0.9519$ MPa
At point B:
$\sigma_1 = 1.838$ MPa; $\sigma_2 = -0.2618$ MPa
$\tau_{max} = 1.050$ MPa
At point C:
$\sigma_1 = 1.737$ MPa; $\sigma_2 = -0.1631$ MPa
$\tau_{max} = 0.9501$ MPa

12.35 $\sigma_1 = 32.29$ MPa; $\sigma_2 = -2.944$ MPa
$\tau_{max} = 17.62$ MPa

12.36 At point A:
$\sigma_1 = 0$; $\sigma_2 = -11.24$ ksi
At point B:
$\sigma_1 = 0.9493$ ksi; $\sigma_2 = -1.053$ ksi
At point C:
$\sigma_1 = 11.04$ ksi; $\sigma_2 = 0$

Chapter 13

13.1(b) $y(x) = \frac{q_0}{120LEI}\left(-20L^3x^2 + 10L^2x^3 - x^5\right)$

$$y_B = -\frac{11q_0L^4}{120EI}$$

$$y'_B = -\frac{q_0L^3}{8EI}$$

13.5 $y_B = -21.64$ mm, $y'_B = -0.3720°$

13.6 $y(x) = \frac{M_0}{6LEI}\left(x^3 - L^2x\right)$

$$y_{mid} = -\frac{M_0L^2}{16EI} \qquad y'_A = -\frac{M_0L}{6EI}$$

$$y'_B = \frac{M_0L}{3EI}$$

13.14 $y_C = -0.1867$ in.; $y'_A = -0.1782°$
$y'_B = -0.2292°$

13.15

$$y(x) = \frac{q_0}{1944EI}\left(36Lx^3 - 81\left\langle x - \frac{L}{2}\right\rangle^4 + 81\left\langle x - \frac{5L}{6}\right\rangle^4 - 31L^3x\right)$$

$$y'_A = -\frac{31q_0L^3}{1944EI} \qquad y'_B = \frac{19q_0L^3}{972EI}$$

13.20 $y_C = -0.2543$ mm; $y'_A = -0.01534°$
$y'_B = 0.005482°$

13.23

$$y(x) = \frac{1}{EI}\left(\frac{F}{6}x^3 - \left(\frac{FL}{4} + \frac{M_0}{2}\right)x^2 - \frac{F}{6}\left\langle x - \frac{L}{2}\right\rangle^3\right)$$

$$y_B = -\frac{L^2}{48EI}\left(5FL + 24M_0\right)$$

$$y'_B = -\frac{L}{8EI}\left(FL + 8M_0\right)$$

13.28 $I_{min} = 312.4$ in.^{4}; WT 15×54
for minimum weight beam

13.32

$$y_C = -\frac{L^2}{768EI}\left(5q_0L^2 + 48M_0\right)$$

$$y'_A = -\frac{L}{384EI}\left(9q_0L^2 + 128M_0\right)$$

13.34

$$y_C = -\frac{L^3}{768EI}\left(5q_0L + 11F\right)$$

$$y'_B = \frac{L^2}{5760EI}\left(128q_0L + 315F\right)$$

13.35

$$y_B = -\frac{7FL^3}{16EI} \qquad y'_B = -\frac{5FL^2}{8EI}$$

13.38

$$y_B = \frac{L^4}{3840EI}\left(128q_1 - 70q_0\right)$$

$$y'_B = \frac{L^3}{48EI}\left(2q_1 - q_0\right)$$

13.39 $A_y = (5/8)q_0L$; $B_y = (3/8)q_0L$
$M_{RA} = -(1/8)q_0L^2$

13.44 $A_y = (11/16)F + (3/2)(M_0/L)$
$B_y = (5/16)F - (3/2)(M_0/L)$
$M_{RA} = -(3/16)FL - (1/2)M_0$

13.49 $A_y = (2q_0L/15) - (3/32)F$
$B_y = (2q_0L/5) + (11/16)F$
$C_y = -(q_0L/30) + (13/32)F$

13.51 $A_y = (121q_0L/512)$
$B_y = (7q_0L/512)$
$M_{RA} = -(67q_0L^2/3072)$
$M_{RB} = -(13q_0L^2/3072)$

Chapter 14

14.2 $P_{CR} = 82.06$ kip

14.4 $P_{CR} = 340.7$ kN

14.8 $(I_y)_{min} = 3.084 \times 10^6$ mm^4; W127 × 24
for minimum weight column

14.12 $P_{max} = 513.8$ lb

14.15 $d = 4.422$in.; $w = 8.844$ in.

14.17 $I_{Required} = 2.570 \times 10^6\ mm^4$
$t_{min} = 6.558$ mm

14.18 $F_{max} = 4.900$ kip (Buckling)

14.21 $F_{max} = 95.05$ kN (Yielding)

Chapter 16

16.4 $K_I = 679.6$ MPa-mm$^{1/2}$

16.9 $\sigma_0 = 1{,}562$ MPa (Yielding)

16.17 $a_c = 1.684$ in.

16.22 $M_c = 90.73$ kip-in.

16.23 $K = 1{,}247$ MPa-mm$^{1/2}$
Replace panel as $K_I > K_{Ic}$

16.27 $\sigma_0 = 196.2$ MPa

16.37 $N_f = 478$ cycles

16.38 $N_f = 104$ cycles

Appendix C

C.17 $\bar{x} = 0$ $\bar{y} = \dfrac{4r}{3\pi}$

C.18 $\bar{x} = 0$ $\bar{y} = \dfrac{4b}{3\pi}$

C.23 $\bar{x} = 17.81$ mm $\bar{y} = 23.72$ mm

C.25 $I_x = (\pi/4)ab^3$; $I_y = (\pi/4)ba^3$

C.27 $\bar{y} = 0.7241$ in

C.28 $\bar{y} = 1.554$ in.

C.29(a) $I_z = 0.3535\ in^4$

C.29(b) $I_z = 4.343\ in.^4$

C.34 $\bar{x} = 9.445$ mm $\bar{y} = 6.278$ mm

INDEX

Unit Conversion Factors

Quantity	U. S. Customary	SI Equivalent
Acceleration	ft/s^2 in/s^2	$0.3048\ m/s^2$ $0.0254\ m/s^2$
Area	ft^2 in^2	$0.0929\ m^2$ $645.2\ mm^2$
Distributed Load	lb/ft lb/in.	14.59 N/m 0.1751 N/mm
Energy	ft-lb	1.356J
Force	kip = 1000 lb lb	4.448 kN 4.448N
Impulse	lb-s	4.448 N-s
Length	ft in mi	0.3048 m 25.40 mm 1.609 km
Mass	lb mass slug ton mass	0.4536 kg 14.59 kg 907.2 kg
Moments or Torque	ft-lb in-lb	1.356 N-m 0.1130 N-m
Area Moment of Inertia	in^4	$0.4162 \times 10^6\ mm^4$
Power	ft-lb/s hp	1.356 W 745.7 W
Stress and Pressure	lb/ft^2 lb/in^2 (psi) ksi = 1000 psi	47.88 Pa 6.895 kPa 6.895 MPa
Velocity	ft/s in/s mi/h (mph)	0.3048 m/s 0.0254 m/s 0.4470 m/s
Volume	ft^3 in^3 gal	$0.02832\ m^3$ $16.39\ cm^3$ 3.785 L
Work or Energy	ft-lb	1.356 J

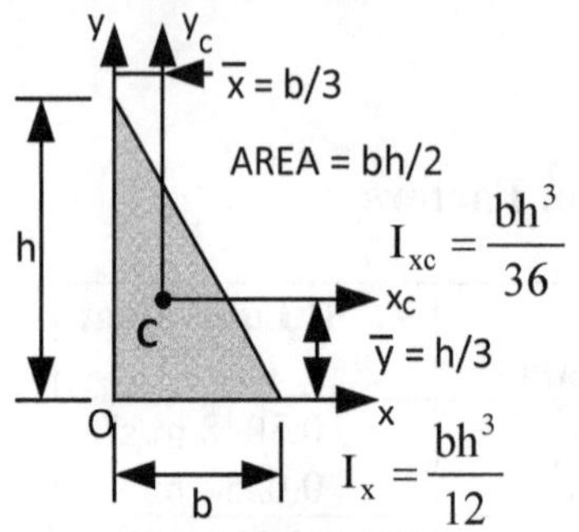

TRIANGULAR AREA

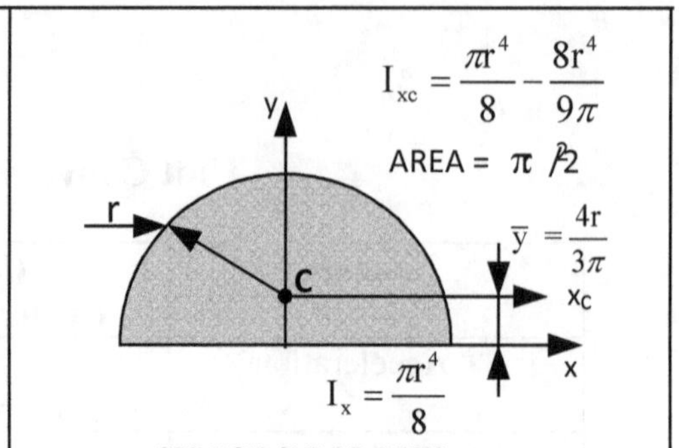

SEMICIRCULAR AREA

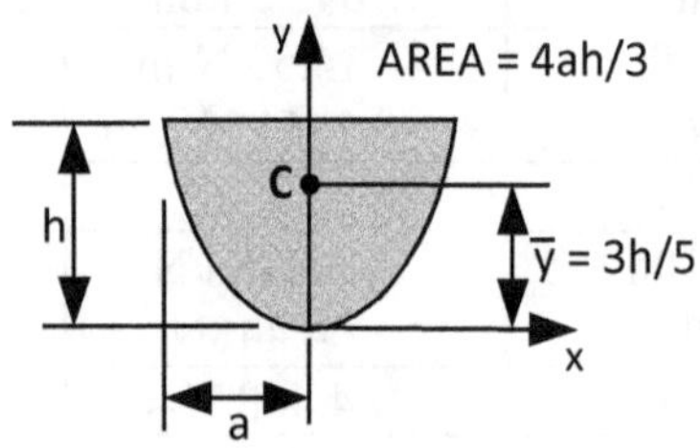

PARABOLIC AREA

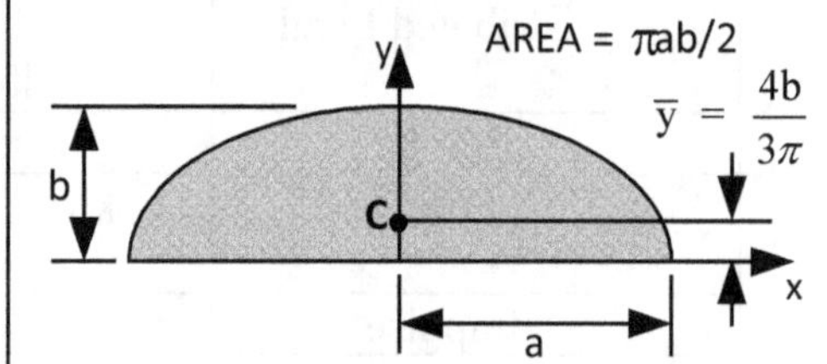

SEMIELLIPTICAL AREA

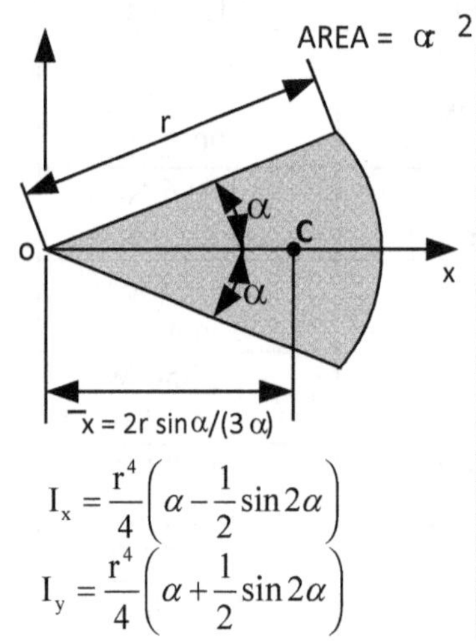

CIRCULAR SECTOR

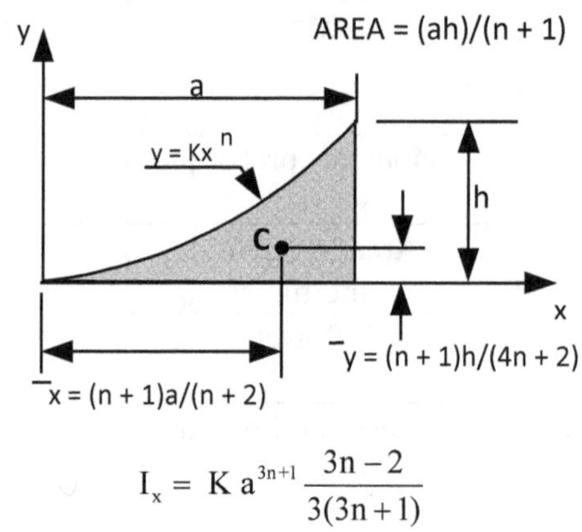

GENERAL SPANDREL

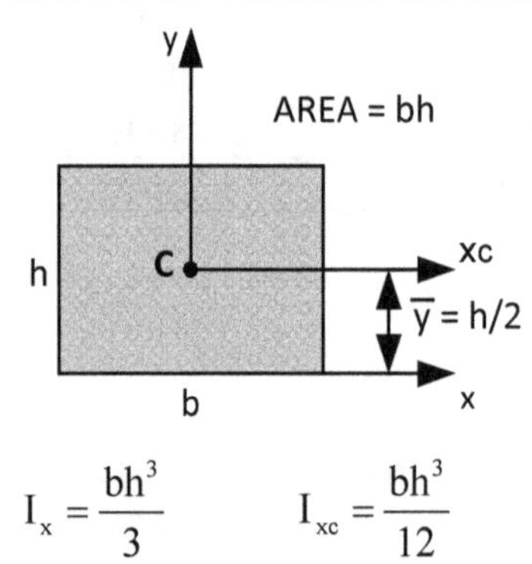

RECTANGULAR AREA

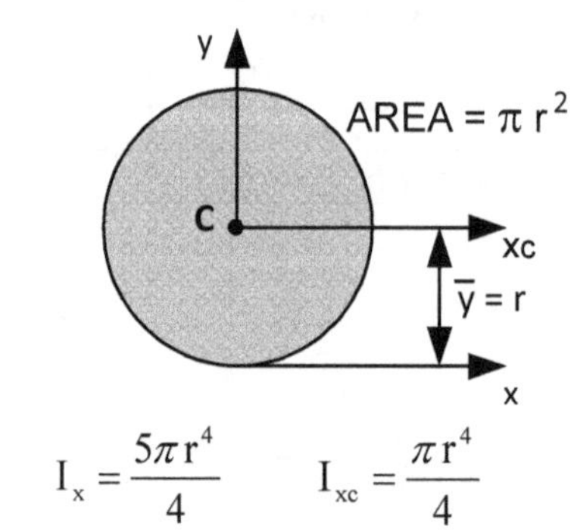

CIRCULAR AREA

Basic Quantities and Units

System of Units	Length	Time	Mass	Force	g
International System of Units (SI)	meter (m)	second (s)	kilogram (kg)	newton (N) $(kg\text{-}m)/s^2$	9.807 m/s^2
U. S. Customary (FPS)	foot (ft)	second (s)	slug $(lb\text{-}s^2)/ft$	pound (lb)	32.17 ft/s^2

Units of Other Frequently Used Quantities

System of Units	Moment M	Stress σ	Strain ε
SI	newton-meter (N-m)	pascal (Pa) = (N/m^2) mega pascal MPa = N/mm^2	dimensionless
U. S. Customary (FPS)	foot-pound (ft-lb)	pound/square foot (lb/ft^2) pound/square in. $(lb/in.^2)$ = psi	dimensionless

SI Prefixes

Multiplication Factor	Prefix Name	Prefix Symbol
10^{18}	exa	E
10^{15}	peta	P
10^{12}	tera	T
10^{9}	giga	G
10^{6}	mega	M
10^{3}	kilo	k
10^{2}	hecto*	h
10^{1}	deka*	da
10^{-1}	deci*	d
10^{-2}	centi*	c
10^{-3}	milli	m
10^{-6}	micro	μ
10^{-9}	nano	n
10^{-12}	pico	p
10^{-15}	femto	f
10^{-18}	atto	a

*To be avoided when possible.

www.ingramcontent.com/pod-product-compliance
Lightning Source LLC
LaVergne TN
LVHW082010200926
848481LV00010B/2827